Fishes of the Western North Atlantic

MEMOIR

SEARS FOUNDATION FOR MARINE RESEARCH

Number I

Fishes of the Western North Atlantic

PART NINE

Volume One: *Orders* Anguilliformes and Saccopharyngiformes

Volume Two: Leptocephali

VOLUME ONE

Order Anguilliformes

ANGUILLIDAE, HETERENCHELYIDAE, MORINGUIDAE, CHLOPSIDAE, MYROCONGRIDAE, MURAENIDAE, SYNAPHOBRANCHIDAE, OPHICHTHIDAE, COLOCONGRIDAE, DERICHTHYIDAE, MURAENESOCIDAE, NEMICHTHYIDAE, CONGRIDAE, NETTASTOMATIDAE, SERRIVOMERIDAE

Order Saccopharyngiformes

CYEMATIDAE, SACCOPHARYNGIDAE, EURYPHARYNGIDAE, MONOGNATHIDAE

NEW HAVEN, 1989

SEARS FOUNDATION FOR MARINE RESEARCH, YALE UNIVERSITY

Library of Congress Cataloging-in-Publication Data

Fishes of the western North Atlantic
Volume One. Orders Anguilliformes and Saccopharyngiformes
Volume Two. Leptocephali

Edited by Eugenia B. Böhlke

1. Order Anguilliformes
2. Order Saccopharyngiformes
3. Leptocephali

Includes Bibliography and Index

QH91.Y3

ISBN 0-935868-45-3

Printed in the United States of America by
Allen Press, Inc., Lawrence, Kansas

Fishes of the Western North Atlantic

Editor

EUGENIA B. BÖHLKE

Authors

EUGENIA B. BÖHLKE
Academy of Natural Sciences of Philadelphia
Philadelphia, Pennsylvania

JAMES E. BÖHLKE†
Academy of Natural Sciences of Philadelphia
Philadelphia, Pennsylvania

MARK M. LEIBY
Florida Department of Natural Resources
St. Petersburg, Florida

JOHN E. McCOSKER
California Academy of Sciences
San Francisco, California

E. BERTELSEN
Zoological Museum, University of Copenhagen
Copenhagen, Denmark

CATHERINE H. ROBINS
Rosenstiel School of Marine and
Atmospheric Sciences, Miami, Florida

C. RICHARD ROBINS
Rosenstiel School of Marine and
Atmospheric Sciences, Miami, Florida

DAVID G. SMITH
Museum of Comparative Zoology, Harvard
University, Cambridge, Massachusetts

KENNETH A. TIGHE
National Museum of Natural History
Washington, D.C.

JØRGEN G. NIELSEN
Zoological Museum, University of Copenhagen
Copenhagen, Denmark

WILLIAM H. HULET
Rosenstiel School of Marine and Atmospheric Sciences
Miami, Florida

Illustrator

MARY H. FUGES
Academy of Natural Sciences of Philadelphia
Philadelphia, Pennsylvania

NEW HAVEN, 1989

SEARS FOUNDATION FOR MARINE RESEARCH, YALE UNIVERSITY

Table of Contents - Volume One

Dedication

JAMES E. BÖHLKE (1930–1982) came from Stanford University to the Academy of Natural Sciences of Philadelphia in 1954. Already a much published scientist, he was to spend his entire professional career at Philadelphia during which time he developed major research programs on tropical western Atlantic marine fishes and South American freshwater fishes, and he pursued world revisions of various fish families. During this time he became Curator of Fishes, held the Chaplin Chair of Ichthyology, and chaired the Department of Ichthyology.

He brought many interests with him to Philadelphia, all of which he pursued vigorously. One of his major fascinations, certainly among marine fishes, was the eels. He published regularly and often on this group—in all, 30 papers between 1949 and 1984 (posthumous). Also, several contributions to this volume include him as a coauthor because of extensive research which he had completed at the time of his death. His book with Charles C. G. Chaplin, on Bahamian fishes included a substantial section on eels. All of Jim's publications on eels are cited in this volume and these citations need not be repeated here. They cover virtually all eel families.

Completion of the volume on eels for "Fishes of the Western North Atlantic" was one of Jim's chief goals and, in furtherance of it, he successfully sought funding from the National Science Foundation, rounded up research material, and at one time or another provided support, encouragement, and advice to all the authors of this work as well as to many other ichthyologists who confronted eel problems.

In his role as Adjunct Professor in the Division of Biology and Living Resources at the University of Miami, he served on doctoral committees of students whose research was on eels: David M. Dean, William H. Hulet, and David G. Smith. He provided the stimulus for many meetings, essentially informal seminars, among those ichthyologists interested in eels. These discussions were held in Philadelphia, Miami, or at the annual meetings of the American Society of Ichthyologists and Herpetologists. It is regretted that these free-wheeling discussions were not recorded. Many of the novel ideas incorporated in this volume were first tested under the peer pressure of these discussion groups.

In appreciation of his spirit, objectivity, and genuine enthusiasm for the eels, we dedicate this volume with much affection to James Erwin Böhlke.

C. Richard Robins
28 July 1987

Introduction

This ninth part of FISHES OF THE WESTERN NORTH ATLANTIC treats the eels and the gulper eels of the orders Anguilliformes and Saccopharyngiformes, and the leptocephali (larvae) of eels and other elopomorph fishes. This study was originally undertaken by James E. Böhlke, the accounts of several sections to be authored by or co-authored with colleagues who previously had worked on those groups. Involved in this initial group were Eugenia B. Böhlke (families Muraenidae and Ophichthidae), Mark M. Leiby (ophichthid leptocephali), Catherine H. Robins (Derichthyidae and Synaphobranchidae), C. Richard Robins (Synaphobranchidae and general classification), David G. Smith (Anguillidae, Congridae, Nettastomatidae, and leptocephali), and Kenneth A. Tighe (Serrivomeridae). After the death in 1982 of J. E. Böhlke, these authors decided to continue with their original commitments and to work toward completion of the volume. John E. McCosker joined as coauthor of the families Muraenidae and Ophichthidae; D. G. Smith agreed to complete work on the remaining families originally to be studied by J. E. Böhlke, and to expand the section on leptocephali; E. B. Böhlke would coordinate the project, supervise preparation of illustrations by Mary H. Fuges, and edit the volume. Several additional colleagues (William H. Hulet, E. Bertelsen and Jørgen G. Nielsen) later joined in coauthoring specific sections.

The resulting study, published in two volumes, treats a total of 180 species in 85 genera of eels and gulper eels, and 171 species of leptocephali. The first volume includes accounts of 174 species in 81 genera in the 15 currently recognized families of the order Anguilliformes, as well as six species in four genera in the four families of the order Saccopharyngiformes; in the second volume are descriptions of 157 eel leptocephali as follows:

- Anguillidae (one genus, one species, two leptocephali)
- Heterenchelyidae (one genus, one species, an eastern Atlantic leptocephalus)
- Moringuidae (two genera, two species, three leptocephali)
- Chlopsidae (six genera, eight species, eight leptocephali)
- Myrocongridae (one genus, one species, no leptocephalus known)
- Muraenidae (eight genera, 24 species, 14 leptocephali)
- Synaphobranchidae (eight genera, 14 species [including ten extralimital species], 17 leptocephali)
- Ophichthidae (25 genera, 54 species, 53 leptocephali)
- Colocongridae (one genus, one species, no leptocephalus known)
- Derichthyidae (two genera, three species, three leptocephali)
- Muraenesocidae (one genus, one species, no leptocephalus known)
- Nemichthyidae (three genera, four species, four leptocephali)
- Congridae (14 genera, 33 species, 32 leptocephali)
- Nettastomatidae (six genera, 14 species, 18 leptocephali)

Serrivomeridae (two genera, three species, three leptocephali)
Cyematidae (one genus, one species, two leptocephali)
Saccopharyngidae (one genus, three species, one leptocephalus)
Eurypharyngidae (one genus, one species, one leptocephalus)
Monognathidae (one genus, one species, one leptocephalus).

In addition, three leptocephali of the order Notacanthiformes, six of the order Elopiformes, and five leptocephali whose adults remain unidentified are also described in Volume Two. The accounts were prepared by the above listed authors; illustrations were prepared by Mary H. Fuges (most adult species) and by the authors (osteology, maps and leptocephali), plus some reproduced from previous publications. All original illustrations have been deposited in the Manuscript Collection of the Library of the Academy of Natural Sciences of Philadelphia.

Acknowledgments. Throughout the years various sources have provided funding. The study was supported in its initial stage by National Science Foundation Grants GB 17736 and DEB 76-20325 awarded to James E. Böhlke. Continuation and completion of the project by the collaborating authors was partially supported by NSF Grant BSR 83-00279 to Eugenia B. Böhlke, and by additional funding provided to specific authors, which is acknowledged in individual chapters. Copy-editing was partially supported by the National Marine Fisheries Service. Publication of the volume was made possible by the Sears Foundation for Marine Research.

This volume is the result of many years of study and interaction between the authors. Each author is greatly indebted to the others for the free exchange of ideas, detailed information, and assistance in locating specimens and literature. The editors and all authors are indebted to the many colleagues and institutions who have extended professional courtesies during the course of this study. For permission to examine specimens, for prompt response to inquiries and loan requests, for help during visits to museums, and for other assistances, we thank the following (institutional abbreviations follow): M. N. Feinberg, C. J. Ferraris, G. Nelson, D. E. Rosen†, C. L. Smith (AMNH); T. C. Hopkins, D. G. Clarke (AMRL); E. B. Böhlke, J. E. Böhlke†, B. Chernoff*, W. G. Saul, W. F. Smith-Vaniz (ANSP); D. F. Markle*, L. van Guelpen (ARC); P. H. Greenwood, G. Howes, A. Wheeler, P. J. P. Whitehead (BMNH); J. E. Randall, A. Susimoto (BPBM); M. E. Anderson, D. Catania, L. J. Dempster, W. N. Eschmeyer, W. I. Follett, T. Iwamoto, J. E. McCosker, P. Sonoda (CAS); M. M. Leiby (FDNR); R. K. Johnson* (FMNH); W. Forman (FSC); R. W. Yerger (FSU); C. E. Dawson*, S. G. Poss (GCRL); W. D. Anderson, Jr., W. A. Roumillat (GMBL); D. Guitart (IOH); R. G. Gilmore, P. A. Hastings* (IRCZM); R. J. Lavenberg (LACM); F. Mago-Leccia (MBUCV); M. M. Dick*, W. L. Fink*, K. E. Hartel, K. F. Liem, D. G. Smith (MCZ); V. Mahnert (MHNG); M. Kottelat (MHNN); J. Blache, M. L. Bauchot, C. Karrer* (MNHN); G. Nunan, H. Travassos† (MNRJ); M. L. Azzaroli (MZUF); H. A. Britski, N. A. Menezes (MZUSP); V. D. Vladykov† (NMC); T. Clarke (NMFSH); P. H. J. Castle (NMNZ); H. Ahnelt, K. Hacker†, P. Kahsbauer†, H. J. Paepke (NMW); M. Boeseman (RMNH); R. Winterbottom (ROM); S. W. Ross*, B. W. Stender (SCMRI); R. H. Rosenblatt, H. J. Walker (SIO); W. Klausewitz (SMF); J. McEachran (TCWC); R. D. Suttkus (TU); W. A. Bussing (UCR); G. H. Burgess, C. R. Gilbert (UF); C. H. Robins, C. R. Robins, K. Sulak* (UMML); R. M. Bailey, R. R. Miller (UMMZ); P. L. Colin*, D. A. Hensley (UPRM); R. L.

† Deceased.
* At institution at time of assistance.

Shipp (USA); D. M. Cohen*, B. B. Collette, M. P. Fahay, T. C. Potthof, W. J. Richards (NMFS); W. H. Krueger (URIZ); R. H. Gibbs†, J. R. Gomon, S. Jewett, E. A. Lachner, E. O. Murdy, V. G. Springer, W. A. Starnes, W. R. Taylor, K. A. Tighe, R. P. Vari, S. H. Weitzman, J. T. Williams (USNM); J. A. Musick, J. E. Olney, C. Wenner* (VIMS); D. Merriman, K. Thomson* (YPM); I. Isbrucker, H. Nijssen (ZMA); K. Decker, H. J. Paepke (ZMB); A. Post (ZMH); E. Bertelsen, J. G. Nielsen (ZMUC). Additional specific acknowledgements are given in individual chapters.

Institutional abbreviations and Standard Symbolic Codes follow those published by Leviton et al., 1985, and are used throughout this volume as follows:

AMNH	American Museum of Natural History, New York, New York.
AMRL	Alabama Marine Resources Laboratory, Dauphin Island, Alabama.
ANSP	Academy of Natural Sciences of Philadelphia, Pennsylvania.
ARC	Atlantic Reference Centre, St. Andrews, New Brunswick, Canada.
BMNH	British Museum (Natural History), London, England.
BPBM	Bernice P. Bishop Museum, Honolulu, Hawaii.
CAS	California Academy of Sciences, San Francisco, California.
CAS-IU	California Academy of Sciences-Indiana University Collection; San Francisco, California.
CAS-SU	California Academy of Sciences-Stanford University Collection; San Francisco, California.
FDNR	Florida Department of Natural Resources, St. Petersburg, Florida.
FDNR-SML	Florida Department of Natural Resources, Gulf States Marine Fisheries Commission, Seamap Reference Collection, St. Petersburg, Florida.
FMNH	Field Museum of Natural History, Chicago, Illinois.
FSC	Freeport Sulphur Company, Belle Chasse, Louisiana.
FSU	Florida State University, Tallahassee, Florida.
GCRL	Gulf Coast Research Laboratory, Ocean Springs, Mississippi.
GMBL	Grice Marine Biological Laboratory, Charleston, South Carolina.
IBH	Instituto de Biologia, Universidad Nacional Autonoma de Mexico, Mexico City, Mexico.
IFAN	Institut Fondamental d'Afrique Noire, Dakar, Senegal.
IOH	Instituto de Oceanologia, Academia de Ciencias, Habana, Cuba.
IRCZM	Harbor Branch Foundation, Fort Pierce, Florida.
IZAC	Instituto de Zoologia, Academia de Ciencias, Habana, Cuba.
LACM	Los Angeles County Museum of Natural History, Los Angeles, California.
MB	Museu Bocage, Universidade de Lisboa, Lisboa, Portugal.
MBI	Marine Biomedical Institute, University of Texas, Galveston, Texas.
MBUCV	Museo de Biologia, Universidad Central de Venezuela, Caracas, Venezuela.
MCZ	Museum of Comparative Zoology, Harvard University, Cambridge, Massachusetts.
MHNG	Museum d'Histoire Naturelle, Geneve, Switzerland.
MHNI	Museo de Historia Natural, Seccion Ictiologia y Herpetologia, Universidad Federal de Minas Gerais, Belo Horizonte, Brazil.
MHNN	Musee d'Histoire Naturelle de Neuchatel, Neuchatel, Switzerland.
MMF	Museu Municipal do Funchal, Funchal, Madeira.
MNHN	Museum National d'Histoire Naturelle, Paris, France.

MNHNM	Museo Nacional de Historia Natural, Mexico, D.F., Mexico.
MNRJ	Museu Nacional, Universidade Federal do Rio de Janeiro, Rio de Janeiro, Brazil.
MOM	Musee Oceanographique Monaco, Monaco.
MZUF	Museo Zoologico de la Specola, Universita di Firenze, Florence, Italy.
MZUSP	Museu de Zoologia, Universidade de Sao Paulo, Sao Paulo, Brazil.
NMC	National Museums of Canada, Ottawa, Canada.
NMFSH	National Marine Fisheries Service, Honolulu, Hawaii.
NMFSM	National Marine Fisheries Service, Miami, Florida.
NMFSPC	National Marine Fisheries Service, Panama City, Florida.
NMNZ	National Museum of New Zealand, Wellington, New Zealand.
NMW	Naturhistorisches Museum, Wien, Austria.
RMNH	Rijksmuseum van Natuurlijke Historie, Leiden, Holland.
ROM	Royal Ontario Museum, Toronto, Ontario.
SCMRI	South Carolina Marine Resources Research Institute, Charleston, South Carolina.
SIO	Scripps Institution of Oceanography, La Jolla, California.
SMF	Natur-Museum und Forschungs-Institut Senckenberg, Frankfurt-am-Main, Germany.
TCWC	Texas Cooperative Wildlife Collection, Texas A&M University, College Station, Texas.
TFMC	Museo de Ciencias Naturales de Santa Cruz de Tenerife, Canary Islands.
TNHM	Texas Natural History Museum, University of Texas, Austin, Texas.
TU	Tulane University, New Orleans, Louisiana.
UANL	Universidad Autonoma de Nuevo Leon, Nuevo Leon, Mexico.
UCR	Universidad de Costa Rica, San Jose, Costa Rica.
UF	University of Florida, Gainesville, Florida.
UF-FSU	University of Florida-Florida State University Collection, Gainesville, Florida.
UMML	Rosenstiel School of Marine and Atmospheric Sciences, University of Miami, Miami, Florida.
UMMZ	University of Michigan Museum of Zoology, Ann Arbor, Michigan.
UPRM	University of Puerto Rico at Mayaguez, Mayaguez, Puerto Rico.
URIZ	University of Rhode Island, Department of Zoology, Kingston, Rhode Island.
USA	University of South Alabama, Mobile, Alabama.
USF	University of South Florida, St. Petersburg, Florida.
USNM	National Museum of Natural History, Washington, D.C.
VIMS	Virginia Institute of Marine Science, Gloucester Point, Virginia.
VUW	Victoria University, Wellington, New Zealand.
YPM	Yale University, Peabody Museum, New Haven, Connecticut.
ZMA	Zoologisch Museum, Universiteit van Amsterdam, Amsterdam, Holland.
ZMB	Universitat Humboldt Museum fur Naturkunde, Berlin, Germany.
ZMH	Zoologisches Institut und Museum, Universitat Hamburg, Germany.
ZMUB	Zoologisk Museum, Universitetet i Bergen, Bergen, Norway.
ZMUC	Zoologisk Museum, Københavns Universitet, Kobenhavn, Denmark.
ZMUU	Zoologiska Museet, Uppsala Universitet, Uppsala, Sweden.

The editors are grateful to all authors for their contributions, and acknowledge the inter-institutional cooperation of curators of collections in museums, both in the United States and abroad, which has made this comprehensive study possible. We acknowledge the Academy of Natural Sciences for long-term support of this study. We thank Dr. Alfred E. Schuyler of the Academy for preparing the manuscript for the printer, and Patricia R. Schuyler for preparing the index.

Eugenia B. Böhlke
Academy of Natural Sciences
of Philadelphia
Philadelphia, Pennsylvania

Bruce B. Collette
Systematics Laboratory
National Marine Fisheries Service
Washington, D.C.

Methods and Terminology

EUGENIA B. BÖHLKE

Families of the orders Anguilliformes and Saccopharyngiformes (sensu Robins, pp. 9–15) are treated in the first section of this volume; genera and species found in the western Atlantic are described and illustrated, with extralimital taxa included in some sections. Larval forms (leptocephali) of Anguilliformes and other orders of elopomorph fishes found in the western Atlantic are described in the second section.

Keys. Keys are based on easily observable external characters when possible, and kept as brief as is practical. They are artificial and are not meant to imply phylogenetic significance. Keys to genera are all-inclusive (except in the Congridae which is for the western Atlantic only); keys to species may include worldwide, New World, Atlantic, or only western Atlantic species. Characters given in the keys and descriptions of higher categories (family and generic accounts) are not always repeated in descriptions of individual species.

Synonymies. Only references to new names, new name combinations, major revisional studies, and papers containing important new information are included in the synonymies. Minor spelling changes are mentioned but are not detailed. Primary synonymies listing all names, including those based on larvae, are in species accounts of adults; only larval names are listed under leptocephali.

Diagnoses and Descriptions. Standard morphometric characters used in the study of fishes are included in the following accounts. However, it should be noted that these are highly variable in eels and frequently of limited use in distinguishing species. Total length is the reference measurement used throughout this volume (except for some leptocephali), and is basic for comparative purposes. However, total length is often difficult to obtain with accuracy, for a variety of reasons. Specimens may be tightly curled for storage in jars and become difficult to straighten for measuring; muscles of some species (especially of large muraenids, congrids and ophichthids) are strongly contracted when preserved—two or more persons are required to unroll and hold large specimens for straight measurements, which even then are not accurate; sometimes the measurement can only be estimated by use of a string which must be snaked (eeled?) along tightly curled individuals. Conversely, in poorly preserved specimens, particularly those obtained before the use of formalin as a fixative, muscles have decomposed and a few centimeters may be added to the total length if the specimen is tightly stretched. Sometimes it is impossible to obtain a total length measurement: the posterior part of the tail is frequently damaged or missing (especially in those species with long, attenuate tails), and sometimes has been regenerated; dredged specimens are often severed into two or more pieces. For these reasons, other standards of comparison and additional characters are often utilized. It is standard practice to compare head characters to the head length rather than the total length (the elongate trunk and tail render comparison with total length insignificant); preanal distance is sometimes used for comparing proportional differences for those species with easily damaged tails.

Counts and measurements utilized in this study are defined for eels as listed below and as illustrated in Figure 1. Measurements are made from point-to-point. (Additional terms specific

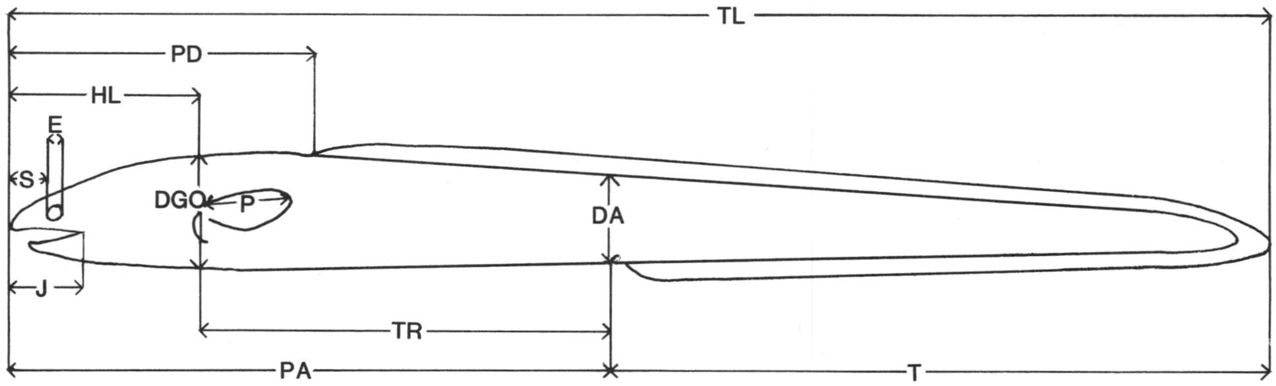

FIGURE 1. Schematic diagram showing typical measurements used for eels. Abbreviations: DA, depth at anus; DGO, depth at gill opening; E, eye diameter; HL, head length; J, jaw length (snout to rictus); P, length pectoral fin; PA, preanal length; PD, predorsal length; S, snout length; T, tail; TL, total length; TR, trunk.

for various families are defined in those accounts.)

total length (TL)—from tip of snout to tip of tail, including caudal fin.
head length (HL)—from tip of snout to upper edge of pectoral base, or to posterodorsal point of gill opening, depending upon family.
trunk—from posterodorsal point of gill opening to anus.
preanal—from tip of snout to anus.
tail—anus to tail tip.
predorsal—tip of snout to base of first dorsal ray.
depth—greatest depth of body, excluding median fins (taken just behind gill opening, at or behind anus, or elsewhere at greatest depth—see *Methods* in individual chapters).
snout—tip of snout to anterior margin of eye.
eye—horizontal diameter of eye.
snout to rictus (upper jaw)—tip of snout to angle of mouth.
lower jaw—tip of lower jaw to angle of mouth.
interorbital width—dorsal distance between clear margins of eyes.
pectoral-fin length—base of first (dorsal-most) pectoral ray to tip of longest ray.
gill opening—distance between upper and lower ends of gill opening.
isthmus (interbranchial distance)—ventral distance between lower ends of gill opening.

Vertebral counts have been found to be particularly useful in identifying certain eels, especially the vertebral formula (see Böhlke, 1982). This formula (expressed as the range of counts of predorsal, preanal, and total vertebrae), or the mean vertebral formula (MVF, the means of those counts) is characteristic of species within genera, and is often broadly characteristic of genera within families. Counts are defined as follows and are obtained from radiographs (Fig. 2):

total (T)—total number of vertebrae including the first reduced vertebra and the hypural.
predorsal (PD)—from the first vertebra to and including the vertebra intersected by a line drawn perpendicularly from the base of the first dorsal-fin ray.
preanal (PA)—from the first vertebra to and including the vertebra intersected by a line drawn perpendicularly from the base of the first anal-fin ray.
precaudal (PC)—from the first vertebra to and including the anteriormost vertebra with a closed haemal arch.

The last count is not always visible on radiographs, but can be seen on cleared and stained specimens. An alternative method which can be used on intact specimens, is to dissect the abdomen posterior to the anus to find the terminus of the kidney, and mark that point with a pin before x-raying. This gives a reliable precaudal count which has been used to compare with the prenephric count of leptocephali to match adults and leptocephali.

Pores in the lateralis system, especially the arrangement and number of pores on the head,

FIGURE 2. Radiograph of a moray showing vertebral counts. Arrows indicate vertebrae counted as first (1), predorsal (PD) and preanal (PA); last vertebra, the hypural, not visible on print, but discernable microscopically on radiograph.

are often used as a diagnostic character for eels (although within certain families there may be some intraspecific variation). Pores are defined and counted as follows (Fig. 3):

adnasal (AD)—pore at the end of the short, ascending branch of the infraorbital canal, which lies just behind the anterior nostril; not always present.

branchial (B)—pores in branchial region of lateral-line canal.

ethmoid (EM)—pore in the ethmoid canal, included in the count for supraorbital pores.

frontal (F)—a single median pore found in the frontal canal, present in some families.

infraorbital (IO)—pores in infraorbital canal, including pores in lateral canal along upper jaw below eye, and postorbital pores (sometimes listed separately) in ascending canal behind eye.

interorbital—pores in the supraorbital canal which are located between the eyes.

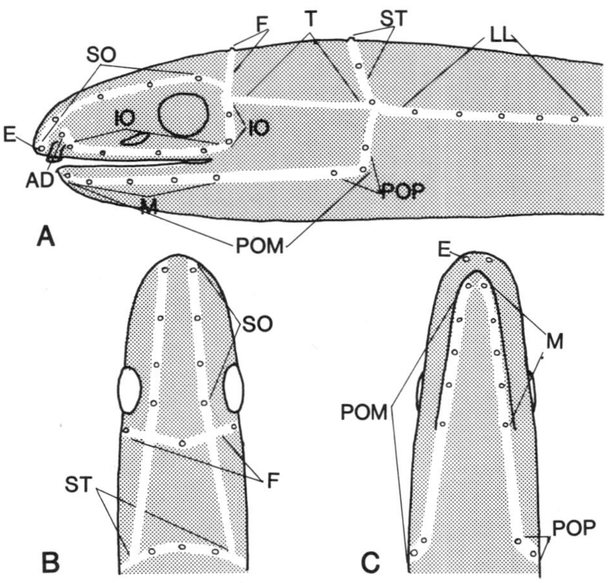

FIGURE 3. Schematic diagrams showing lateralis system of canals and pores for eels. A. Lateral view. B. Dorsal view. C. Ventral view. Abbreviations: AD, adnasal; E, ethmoid; F, frontal; IO, infraorbital; LL, lateral-line; M, mandibular; POM, preoperculomandibular; POP, preopercular; SO, supraorbital; ST, supratemporal, T, temporal.

lateral line (LL)—pores in lateral-line canal (sometimes counted to anus, sometimes total count given).

mandibular (M)—pores along lower jaw, part of preoperculomandibular canal series.

postorbital (PO)—pores in ascending canal behind eye, part of infraorbital canal series.

preoperculomandibular (POM)—pores in preoperculomandibular canal, including mandibular pores along lower jaw and those in preopercular canal (both sometimes listed separately).

supraorbital (SO)—pores on dorsal surface of snout and head, including those in ethmoid canal at tip of snout (sometimes listed separately) and those on dorsal surface of snout and head; the posteriormost pores situated between the eyes are sometimes referred to as interorbital pores.

supratemporal (ST)—total number of pores in supratemporal canal (usually including one median pore).

temporal (T)—temporal canal, which connects the lateral line with the postorbital portion of the infraorbital canal; rarely contains pores.

The number of pores in the lateral-line canal has been used as a diagnostic character in the past; both the total count and the number of pores to anus have been used. However, the lateral-line pores of many eels are difficult to observe and to count; they are affected by the method of preservation, the age and condition of the specimen, and the presence of a waxy exudate which often clogs and obscures them. Also, pores do not always extend to the tail tip, or are not visible, and a complete count cannot be obtained. Lateral-line pores correspond to the number of vertebrae in most eels (one pore and one vertebra developing from each myomere of the leptocephalus, a few exceptions). Thus vertebral counts, which can be obtained more easily and more accurately from radiographs, are utilized throughout this study.

Other characters used to diagnose and identify species of teleosts are frequently subject to variation in eels, and thus are of limited usefulness. Dentition (size, shape, and placement of teeth) varies with growth, and teeth often are missing in large or mangled specimens. Coloration, described for preserved specimens (life colors included when known), is variable, and often dependent on substrate and conditions at time of capture, and on methods of preservation and storage (frequently faded and indiscernable on old specimens).

Sex and sexual condition was observed by gross (macroscopic) examination. The sex of females can be determined at an earlier age than that of males because the ovaries are more easily identified as such. Specimens were termed "ripe" when their body cavities were filled by gonads; egg size given is the greatest diameter of the largest eggs observed, and appears not to correlate with the size of the species; eggs of a small species, such as *Pseudomyrophis fugesae* (a ripe female measuring 95 mm TL) may be as large as those of the largest ripe *Ophichthus rex* (1687 mm TL).

Osteological characters are described and illustrated (in family or generic accounts), based on cleared and stained specimens (using the trypsin-preparation method of Taylor (1967) and subsequent modifications) and on radiographs (exposed as described in Böhlke, 1982). Bone terminology is based on that of Asano (1962) and subsequent modifications as noted in each chapter; gill arch terminology is based on that of Nelson (1966a). Abbreviations used are as follows, or are given under individual illustrations.

AC	Actinost
AN	Anterior nostril
AR	Anal ray
ART	Articular
B	Basibranchial
BA	Basal element of pterygiophore
BO	Basioccipital
BR	Branchiostegal rays
BS	Basisphenoid
C	Ceratobranchial
CE	Centrum
CH	Ceratohyal
CL	Cleithrum
CM	Coronomeckelian
CO	Coracoid
CR	Caudal ray
CTP	Caudal transverse process
CX	Cartilaginous extension of terminal vertebra
D	Dentary
DR	Dorsal ray
E	Epibranchial
EH	Epihyal
EM	Ethmoid section of EPM
EN	Epineural
EP	Epipleural
EPM	Ethmoid-premaxillary complex
EPO	Epiotic
EV	Ethmo-vomer complex
EX	Exoccipital
F	Frontal
FC	Frontal commissure
GH	Glossohyal
H	Hypobranchial
HA	H(a)emal arch
HH	Hypohyal
HM	Hyomandibula(r)
HY	Hypural
I	Infrapharyngobranchial
IM	Intramuscular bone
IOP	Interopercle
LEP	Lateral ethmoid process
LP	Lower pharyngeal tooth plate
M	Mandible
MX	Maxilla
MY	Myorhabdoi
N	Nasal
NA	Neural arch
NS	Neural spine
OP	Opercle
OR	Orbit
P	Pharyngeal
PA	Parietal
PAL	Palatine
PB	Pharyngobranchial
PL	Pleural rib
PMX	Premaxilla
PN	Posterior nostril
PO	Postorbital
POP	Preopercle
PP	Parapophysis
PR	Pectoral ray
PRE	Preorbital
PRO	Prootic
PS	Parasphenoid
PT	Pterygiophore
PTG	Pterygoid
PTO	Pterotic
PTS	Pterosphenoid
Q	Quadrate
R	Rib
RA	Radial element of pterygiophore
RT	Replacement teeth
SA	Sagittal otolith
SBO	Suborbital
SBP	Subopercle
SC	Scapula
SCL	Supracleithrum
SO	Supraorbital
SOC	Supraoccipital
SP	Sphenotic
UH	Urohyal

UP Upper pharyngeal tooth plate
V Vomer

Study Material. Specimens examined in the course of these studies are listed at the end of each species account. Complete data are included for rare or poorly-known species; only catalog number, number and lengths of specimens, and an abbreviated locality are listed for the majority of commonly collected species.

Illustrations. Three illustrations are included for each of most of the species treated: a wash drawing of the whole fish illustrating general appearance and color pattern; a line drawing of the head and anterior trunk region showing the head, nostrils, and head pores in greater detail (median supratemporal and frontal pores sometimes indicated by a slanted line); and line drawings of the upper and lower jaws showing the dentition. These illustrations may depict features which are not described in detail in the text, and are to be used in conjunction with the written descriptions. The majority of the species illustrations were prepared by Mary H. Fuges; credit is given separately for those prepared by others.

Osteological illustrations were prepared by individual authors, some of them taken from published papers; abbreviations used are as listed on pp. 5–6 (or are given in the figure captions for those previously published). Distribution maps were prepared by the authors, usually based on the study material. Illustrations of leptocephali were prepared by the authors, some taken from previous publications.

References. A combined bibliography for all chapters is presented at the end of this volume. It should be noted that the bibliography is part of each chapter, and it should be appended to each if individual family accounts are to be cited.

Summary of New Taxa. During the course of this study, which formally began in 1969, a great amount of material was examined, much of it assembled at the Academy of Natural Sciences of Philadelphia. Following studies resulted in the discovery of new genera and species of eels. Many of these formed the basis of papers published during the past 15 years, and some are first published in this volume. Names are listed below (those marked * published for the first time in this volume).

Genera:

Acromycter Smith and Kanazawa, 1977
Asarcenchelys McCosker, 1985
Atractodenchelys Robins and Robins, 1970
Haptenchelys Robins and Martin, 1976
Hyphalophis McCosker and Böhlke, 1982
Kertomichthys McCosker and Böhlke, 1982
Lethogoleos McCosker and Böhlke, 1982
Linkenchelys Smith*
Monopenchelys Böhlke and McCosker, 1982
Mixomyrophis McCosker, 1985
Stictorhinus Böhlke and McCosker, 1976

Species:

Acromycter atlanticus Smith*
Ariosoma coquettei Smith and Kanazawa, 1977
Asarcenchelys longimanus McCosker, 1985
Atractodenchelys phrix Robins and Robins, 1970
Bascanichthys inopinatus McCosker, Böhlke, and Böhlke*
Dysomma goslinei Robins and Robins, 1976
Dysomma tridens Robins, Böhlke, and Robins*
Enchelycore carychroa Böhlke and Böhlke, 1976
Ethadophis akkistikos McCosker and Böhlke, 1984
Gnathophis bathytopos Smith and Kanazawa, 1977
Gnathophis bracheatopos Smith and Kanazawa, 1977
Gnathophis tritos Smith and Kanazawa, 1977
Gordiichthys ergodes McCosker, Böhlke, and Böhlke*
Gordiichthys leibyi McCosker and Böhlke, 1984
Gordiichthys randalli McCosker and Böhlke, 1984
Gymnothorax bacalladoi Böhlke and Brito, 1987
Gymnothorax hubbsi Böhlke and Böhlke, 1977
Gymnothorax kolpos Böhlke and Böhlke, 1980
Haptenchelys texis Robins and Martin, 1976
Heteroconger luteolus Smith*
Hoplunnis megista Smith*
Hoplunnis similis Smith*
Hyphalophis devius McCosker and Böhlke, 1982
Ilyophis arx Robins, 1976
Japonoconger caribbeus Smith and Kanazawa, 1977
Lethogoleos andersoni McCosker and Böhlke, 1982
Linkenchelys multipora Smith*

Myrophis anterodorsalis McCosker, Böhlke, and Böhlke*
Mixomyrophis pusillipinnis McCosker, 1985
Nettastoma syntresis Smith and Böhlke, 1981
Nettenchelys exoria Böhlke and Smith, 1981
Nettenchelys inion Smith and Böhlke, 1981
Nettenchelys pygmaea Smith and Böhlke, 1981
Ophichthus hyposagmatus McCosker and Böhlke, 1984
Ophichthus menezesi McCosker and Böhlke, 1984
Ophichthus omorgmus McCosker and Böhlke, 1984
Ophichthus rex Böhlke and Caruso, 1980
Parabathymyrus oregoni Smith and Kanazawa, 1977
Pseudomyrophis fugesae McCosker, Böhlke, and Böhlke*
Rhechias bullisi Smith and Kanazawa, 1977
Rhechias polypora Smith and Kanazawa, 1977
Saurenchelys cognita Smith*
Stictorhinus potamius Böhlke and McCosker, 1976
Xenomystax austrinus Smith and Kanazawa*
Xenomystax congroides Smith and Kanazawa*

In addition, various name changes have resulted from this study; family and generic concepts have been revised, species have been synonymized, and some species previously synonymized have been found to be valid. An effort has been made to use and correct all names to conform with the rules and recommendations in the 1985 edition of the International Code of Zoological Nomenclature (ICZN).

The Phylogenetic Relationships of the Anguilliform Fishes

C. RICHARD ROBINS

Acknowledgments. My concept of the eels developed over many years of study and I have profited greatly from thoughtful discussions with those who were my students and are now my cherished colleagues: David M. Dean, William H. Hulet, Catherine H. Robins (nee Hale), and David G. Smith. The late James E. Böhlke was a frequent participant in these discussions and he and I shared many thoughts about eels. Peter H. J. Castle, William A. Gosline, Mark M. Leiby, John E. McCosker, Gareth Nelson, Solomon Raju, the late Donn Eric Rosen, Richard H. Rosenblatt, the late J. L. B. Smith, and Kenneth A. Tighe all have contributed important information and allowed me to test ideas with them. I especially thank Catherine H. Robins and Eugenia B. Böhlke for their help in assembling this information and for providing critical comment.

Introduction. The idea that certain elongate fishes, lacking the pelvic fins and girdle, and generally called eels, are related is pre-Linnaean. In early classifications, however, starting with Linnaeus (1758), emphasis was placed on the absence of the pelvic girdle rather than shape of body, and such diverse fish genera as *Xiphias, Anarhichas, Stromateus, Trichiurus,* and *Gymnotus,* were placed together in *Muraena,* in a group called the Apodes. As more structures were studied, these genera were placed elsewhere, leaving the elongate, tubular fishes without pelvic fins as the Apodes, which, nevertheless, remained a catch-all. This is not surprising because eels are so easily recognized by their long bodies and mode of swimming that the word eel-like is used for all animals of like form. Even Kaup (1856a,b) who prepared the first major treatise on apodal fishes, included such disparate groups as the cusk-eels (Ophidiidae), the pearlfishes (Carapidae), the swamp-eels (Synbranchidae), the spinybacks (Mastacembelidae), and others. In the latter part of the 19th century, many ichthyologists, but especially Gill, Jordan, and Regan, began to characterize the various families and refine the concept of the Apodes. Their ideas were embodied in the works by Jordan and Davis (1891) and, most notably, by Regan (1912b), who eliminated most of the extraneous groups from his order Apodes and subdivided the group into two suborders, on the basis of fused vs paired frontal bones. Some families continued to be included improperly in the Apodes because of lack of information. The Aoteidae turned out to be true eels, but not a distinctive family, for Castle (1967: 2) and Whitley (1968: 31) showed that the monotypic *Aotea* was nothing more than a dried and badly shriveled ophichthid. Springer and Fraser (1976), identified the Alabetidae as elongate, specialized clingfishes (Gobiesocidae), a family remote from the eels.

The failure of Regan and successors to properly define the various families of eels was due in large part to the scarcity of material of many families and a reluctance to dissect, properly, rare specimens. Thus, for example, errors were made in assigning families to the group with paired frontals when they, in fact, had united frontals. Subsequent authors stated that genera or families had united or separated frontals, because Regan or others had said so, without noting that they had not themselves examined the structure. Thus, erroneous information was not only carried along, but given the credibility of having been "independently" supported. It is not the purpose of this paper to review the history of changes in the composition of the various eel families or to trace the considerable wanderings of certain genera from family to family as first one, and then another, author ejected them from families which were the subjects of their reports. These matters are discussed in varying amount of detail by the other authors of this volume (see especially the account of the Ophichthidae by McCosker et al., this volume).

In the light of this history, it is reasonable to ask, once more, if the true eels, Anguilliformes, are a natural or monophyletic group of fishes. I believe that it will be demonstrated below that the Anguilliformes are monophyletic, although, as here defined, the order is further restricted. There are three fundamental questions which must be addressed: (1) What are the Anguilliformes and how are they characterized? (2) How do the Anguilliformes, as here characterized, differ from other orders of the Elopomorpha, and can any group be identified as the sister group of the Anguilliformes? (3) To what extent can we define lineages within the order based on shared derived characters, and thus provide a sound phylogenetic infrastructure? These will be discussed in sequence.

Definition and Characterization of the Anguilliformes

Numbers are assigned to each characteristic in the order discussed, without regard to any perceived importance. These numbers are used in the subsequent discussion. All families of the order Anguilliformes share the features listed. That certain genera should have lost or modified some of them is not surprising in a group as large and as old as the Anguilliformes, and this should not detract from the utility of these characters in addressing the three questions posed above.

General Features of Adult Eels. 1. The vertical fins are confluent. Other elopomorph groups do not share this characteristic. Although most ophichthids have a hardened, rayless tip to the tail, primitive ophichthids have a caudal fin which is confluent with the dorsal and anal fins. Gosline (1951a) showed that the basic caudal structure, though considerably modified, is detectable, upon dissection, inside the tail tip of the juveniles of several ophichthid genera. The Nemichthyidae are secondarily very elongate and the tip of the caudal region is usually lost in *Nemichthys*. Other non-related groups of fishes have the vertical fins confluent (e.g., Ophidiidae) and are convergent in this character.

2. The pelvic fins and pelvic girdle are absent.

3. The pectoral fins, when present, are high on the side, at least midlateral in position, rather than low on the body. According to McDowell (1973: 23), this placement is above the *rectus cervicis* and their musculature arises from the endochondral girdle.

4. The gill membranes are united to the isthmus, and the opening is restricted to a slit or circular opening of varying size. The location of the gill opening varies between and within families. The united midventral opening in most species of *Synaphobranchus* is clearly a specialization within that genus.

5. The nostrils are widely separate except in the highly derived Serrivomeridae, in which they are close together. The anterior nostril is near the tip of the snout. The posterior nostril is in front of the eye in most groups, but is displaced to the upper lip in most ophichthids and one xenocongrid (*Chilorhinus*), and high on the head (above the eye) in some muraenids and nettastomatids. In some nettastomatids the posterior nostril opens at the end of a long tube which may extend to the occipital line or even to a point just behind the dorsal-fin origin (*Nettenchelys taylori*), a situation unique in fishes. The usual placement of the nostrils in the Anguilliformes is a primitive state shared with other elopomorph fishes, and many other fish groups.

6. Scales are absent, or, if present, non-imbricate, embedded, and mostly arranged in basket-weave fashion. (They may be rounded or polygonal in certain species of *Synaphobranchus*.) This scale pattern is shared only with certain ophidiid fishes, a clear case of convergence.

7. The body is very elongate, variously round-

ed or compressed in cross-section. *Coloconger* (family Colocongridae) is the least elongate of all the eels.

8. There are no pyloric caeca. This feature is seldom mentioned in eel descriptions. I have examined many genera in this regard, including at least one genus in every family and most subfamilies.

9. The head including the gill region is elongate, the gills extending well behind the neurocranium.

Osteological Features. 10. There is a 1:1 relationship of the fin rays to the interspinal elements and a 2 or 3 (rarely 4):1 ratio of these to the vertebrae.

11. The pectoral girdle is displaced posteriorly and there is no bony connection between it and the skull.

12. The posttemporal bone is absent. This bone provides, in most fishes, the bony connection between the shoulder girdle and the skull. The loss of this bone is therefore correlated with no. 11 (above).

13. The mesocoracoid bone is absent.

14. There is no gular plate.

15. The preopercle, opercle, subopercle, and interopercle are present as separate elements though they are variously reduced in size and concealed by the skin.

16. There is no ossified symplectic. A cartilaginous symplectic is present in the Synaphobranchidae. (What apparently is an unossified symplectic in the leptocephalus becomes fused with the quadrate in the juvenile or it is lost in all other eels.)

17. There is no orbitosphenoid. Attribution of an orbitosphenoid to eels is erroneous and is based on misidentification of the basisphenoid. The teleost basisphenoid is not homologous with that of higher vertebrates, but I reject use of the name "porpitual" for it (see discussion by McCosker, 1977: 22).

18. There is no opisthotic.

19. The anterior end of the ceratohyal is elongated in response to the general elongation of the region.

20. The branchiostegal rays vary in number, but are usually fairly numerous, with at least the *uppermost rays curving behind and above the opercle.* In some genera the majority of the branchiostegals do this. Branchiostegals are added anteriorly to the elongating ceratohyal in some families, but in others, this fails to happen with the result that a lengthy part of the anterior shaft of the ceratohyal is without branchiostegals. In the Ophichthidae many branchiostegals are added anteriorly, but their bases are remote from the ceratohyal. These overlapping, non-attached branchiostegals were called "jugostegalia" by Parr (1930a), and they represent a derived or synapomorphic feature of the Ophichthidae. Although added, they are true branchiostegals and there is little reason to perpetuate a special term like "jugostegalia" for them. In all eels, the attached branchiostegals are attached both to the epihyal and, at least, the rear part of the ceratohyal.

21. There are 4 gill arches.

22. The slit behind the last gill is reduced.

23. There are no medial processes on the pharyngobranchials.

24. The first infrapharyngobranchial is absent. This bone connects the gill arches to the cranium in most fishes. With posterior displacement of the gill arches, this connection is lost and with it the first infrapharyngobranchial.

25. There is no pseudobranch in the metamorphosed eel. For a description and discussion of this structure in the leptocephalus, see Hulet and Robins, this volume. The fate of the pseudobranch has not been studied in eels. Parry and Holliday (1960: 344) state that it is absent in eels. This is untrue for the larvae, and the basis of their statement with regard to adults is not given. Dissection of a large specimen of *Synaphobranchus brevidorsalis* revealed no trace of any vascularized gland or any other structure that could be construed as a remnant or modification of the pseudobranch. Presumably, the pseudobranch is lost at metamorphosis but its fate has not been studied. The chorioid gland, which Walls (1963: 581) stated to be rigidly dependent upon the presence of a pseudobranch, and which also is said to be lacking in metamorphosed eels, presumably also disappears during metamorphosis

(e.g., Wittenberg and Wittenberg, 1962; Barnett, 1951: 117). Ali and Anctil (1976: 51–57) provided a summary of references to the structure of the retinas of eels.

26. There are no gill rakers.

27. The maxilla is a simple, tooth-bearing bone, which forms most of the rim of the mouth above.

28. There is no supramaxilla.

29. The premaxillae are fused to each other and to the ethmovomer complex or they are absent (Ophichthoidea). It is generally stated that eels lack premaxillae. However, in some eels (e.g., the Synaphobranchoidea) there is a narrow bony isthmus connecting the head of the vomer to an expanded dentigerous area in front. The relationships of the pterygoid arcade (in those eels having this structure complete) and the maxilla to these areas are precisely what one would expect if the anterior section were the premaxilla and the posterior section the head of the vomer. The absence of that part of the lateral canal associated with the premaxilla is a separate matter and has not been properly investigated. The absence of premaxillae may be a reduction characteristic only of the Ophichthoidea.

30. The articular and angular bones are fused. These bones are called the angular and retroarticular bones by some authors (e.g., Smith, 1984a; Harrington, 1955), or the articular and retroarticular (e.g., Nelson, 1973b). The pattern of fusion of the bony elements in the posterior part of the lower jaw is similar in the Albuliformes and Elopiformes (Nelson, 1973b) and may be shared by the Anguilliformes but detailed studies are lacking.

31. There is no free palatine except in the Serrivomeridae. This bone is either absent in adults or appears as a process fused to the side of the vomer. The serrivomerid condition may be neotenous and an adaptation for a very expandible gape. (In *Hoplunnis,* family Nettastomatidae, the palatine may be separated with difficulty. The suturing is much stronger at both ends than in *Serrivomer,* K. Tighe, pers. comm.) Too little is known about the ontogeny of the bones of this region to comment further.

32. A complete pterygoid strut extending from the quadrate to the palatine process fused to the vomer (see 29 above) is probably the primitive condition in eels (see Robins and Robins, 1971: 145–147). Most living genera have a reduced pterygoid, with no bony connection anteriorly, or a slender splint with no connection fore or aft, or they have lost it completely. A few primitive genera (representing various superfamilies) retain the complete bony strut, and in some it is toothed (a secondary specialization).

33. The parietals (no extrascapular bones) are large and meet on the midline of the skull behind the frontal(s).

34. The hyomandibular in primitive eels is oriented such that a line from the joint at the quadrate up the main strut of the hyomandibular is angled behind a vertical. That is, the suspension is slightly toward the rear. In various eels, the suspension shifts toward a vertical and in the Congroidea it is shifted forward, dramatically so in the most derived families (the Nettastomatidae and Serrivomeridae).

Soft Anatomical Structures. 35. There is no oviduct in the adult eel (a condition termed gymnovarian). The eggs reach the outside through a single pore (two slitlike pores have been attributed to the Muraenidae by Bond, 1979: 58).

36. A gasbladder, either physostomus or secondarily physoclistous, is present.

37. The sexes are separate in adult eels. This basic condition is modified in *Rhinomuraena,* a highly evolved muraenid, which was shown by Shen et al. (1979) to be a synchronous but not simultaneous protandrous hermaphrodite. Adult males and females of serrivomerids scarcely overlap in size, the female being larger (Tighe, this volume), but sex reversal has not been demonstrated.

38. Allis's loop, the connection between the anterior end of the supraorbital and infraorbital canals, is present. A rostral commissure may connect the two Allis's loops (see McDowell, 1973: 9, for discussion).

Characters of the Larvae. 39. All eels have a leptocephalus larva with a simple rounded caudal fin (not forked) and no pelvic fins, but with a pectoral fin (even in eels which lack this fin in the juvenile and adult) (see Hulet and Robins, this volume, for discussion of the structure of a leptocephalus).

40. The dorsal and anal fin folds, as they appear, are continuous with the caudal fin.

41. The myomeres are W-shaped, not V-shaped.

42. A rostral commissure is present. This persists in the adults of many (perhaps all?) eels.

Discussion. Although all of the above characters are true of all Anguilliformes, most are not specializations (synapomorphies) of the order, but are variously shared with other primitive fishes. In addition, the general literature on anatomy and physiology of chordates attributes many conditions to "eels" (e.g., the absence of a retractor lentis muscle), but these prove to be based on one or a few studies of *Anguilla,* and simply have not been investigated in other eels or in other primitive teleosts apart from the salmon. These characters deserve further study but I cannot attribute them generally to the order Anguilliformes and therefore do not discuss them further. The numbered characters listed earlier are discussed below relative to other living orders of the Elopomorpha. There is disagreement in the literature concerning the number and composition of these orders, but for purposes of this discussion, I recognize the Elopiformes, Albuliformes (including the Notacanthiformes), and Saccopharyngiformes, the last comprising the Saccopharyngidae, Eurypharyngidae, Cyematidae, and Monognathidae. Some comment on each of these orders is essential to the discussion that follows.

Comments on the Saccopharyngiformes

The history of this order, exclusive of the Cyematidae, was ably discussed by Böhlke (1966b) and it is unnecessary to review any of the matters relative to the fact that gulpers bear no relation to higher groups of bony fishes. Böhlke (1966b) concluded that the gulpers and the Anguilliformes shared a common ancestry, although he retained the order Lyomeri (= Saccopharyngiformes). Recent authors (e.g., Greenwood, 1977; Nelson, 1984) have placed the gulpers as a suborder, Saccopharyngoidei, of the Anguilliformes. The basis of this treatment rests largely on the fact that gulpers have a leptocephalus stage, that they are elongate fishes, and that, supposedly (fide Böhlke, 1966b), "there is one or another eel that shows a tendency toward the gulper character state in having a large mouth, a backward-directed jaw suspension, black velvety skin, tubular projections from the lateral line," etc. The referral by Raju (1974a) of the Cyematidae to the Saccopharyngoidei, polarized the matter, for many of the similarities to the Anguilliformes were based on *Cyema.* Saccopharyngiformes share certain characteristics with the Anguilliformes (see discussion below) and others with all elopomorph fishes. The specializations of the group (here based, except as noted, on my study of the Saccopharyngidae, Eurypharyngidae, and Cyematidae; the Monognathidae are commented on at the end) are given below. For a more detailed description of the order, and a fuller list of other structures absent in these fishes, but not discussed here, see Böhlke (1966b).

a) A strongly posteriorly projecting and very elongate suspensorium.

b) A very deep-bodied leptocephalus, with V-shaped myomeres.

c) Absence of all branchiostegal elements.

d) No preopercle, subopercle, interopercle, and opercle. Trewavas (1933) reported that *Cyema* had a reduced opercle ("a ventral nodule") which was fused to the hyomandibular and she described an "interoperculum." However, Tchernavin (1947a: 345) noted that the latter is the ossified tendon of the *m. abductor mandibulae* and not an interopercle and that the bony lamella was more likely a part of the hyomandibular than a remnant of the opercle. This structure is lacking in *Neocyema* (see Castle, 1977: 75). The same tendon is ossified in my cleared and stained material of *Saccopharynx,* and was illustrated by Tchernavin (1947a: fig. 13) for that genus. There is no reason to identify it as an opercular element.

e) There is no maxilla or premaxilla. Central to this argument is the interpretation of the upper jaw of *Cyema.* The tooth-bearing bone of the "upper jaw" in *Cyema* is closely articulated with the quadrate. In this region it consists of two parts, both entirely paved with teeth. The outer part ends in a curious, tooth-covered, ball-like structure, below an expanded, smooth articulating surface. The inner part, though fused with the outer for most of its length, is separate posteriorly as a broad lobe. The two sections form a long, broad, toothed shelf that is oriented horizontally, the teeth facing downward on the roof of the mouth, the upper surface facing the underside of the neurocranium. Anteriorly, this bone articulates with, but is not attached to, a lateral expansion of the vomer. The vomer continues far forward to form the slender beak of *Cyema*. The vomer is bordered above by the forward continuation of the ethmoid. Trewavas (1933) interpreted the large, tooth-bearing bone of the upper jaw as the maxilla and the lateral expansions and all of the toothed underside of the beak forward of their rear margin as the premaxillae. I believe that the former is the pterygoid and that the lateral expansion of the vomer is either merely that or the remnant of a palatine. Both the expansion and the beak are absent in *Neocyema*, and there is no question that the central tooth-bearing bone of *Neocyema* is a vomer, as stated by Castle (1977b). Castle identified the medial anterior teeth with the vomer and continued to term the lateral bone as a maxilla. If my interpretation is correct, then the Cyematidae, like other Saccopharyngiform eels, lack a secondary upper jaw, namely the maxilla and premaxilla.

f) The hyomandibular articulates with the neurocranium by a single condyle.

g) The lateral-line system on the body has projecting sensory papillae.

h) The gills are small, the arches rounded, and the filaments plumelike (see Tchernavin, 1947b: pl. 8) and set off from the pharynx, into which they open through distinct clefts. A fifth gill is present in the Eurypharyngidae.

i) The efferent branchial arteries are united by commissural vessels above and below the gill clefts, thus forming loops, and they are connected to each other by additional loops (see Tchernavin (1947b: figs. 3–5).

j) The median fin rays are ossified and bilaterally paired, but not segmented or branched.

k) The frontals are paired (Eurypharyngidae, Saccopharyngidae), partly fused (*Neocyema*), or fused (*Cyema*). In *Cyema*, the frontal bone articulates with the ethmoid through a single tubular process (K. Tighe, pers. comm.). The nature of this connection differs fundamentally from that in the elongate anguilliform *Nemichthys*, even though that genus also has anterior process single.

l) Light organs are present toward the tail end of *Saccopharynx* and *Eurypharyx* (see Beebe, 1932: 65; Böhlke, 1966b: fig 220; Owre and Bayer, 1970: fig. 3). Also the tissue associated with the lateral line pores is said to be luminous in these genera.

The fourth saccopharyngiform family, the Monognathidae, is the most evolved and degenerate of the order. Too little is currently known of these odd fishes to comment extensively about them. Raju (1974a) added considerable information and noted that monognathids had 4 gills (previous statements concerning reduction in gill number being erroneous), and that the gills were small and similar in arrangement and in the nature of the filaments to the other saccopharyngiform eels. He considered them to be most closely related to the Saccopharyngidae. The nostrils of monognathids are closer together and higher on the head than those of *Cyema*. The prominent ethmoid tooth and associated gland are unique to the family.

Relationships of the Anguilliformes and Saccopharyngiformes. Many of the supposed similarities between the Anguilliformes and Saccopharyngiformes disappear once the Cyematidae are included within the Saccopharyngiformes. The long association of the Cyematidae with the anguilliform family Nemichthyidae, which is noted even in the common name of *Cyema*, the bobtail snipe-eel, is strictly superficial, as any review of the osteology will reveal.

The relationship between the orders Anguilliformes and Saccopharyngiformes is unclear. The

Saccopharyngiformes are highly modified bathypelagic fishes, yet with a primitive stamp (e.g., the very posteriorly directed hyomandibular with its single dorsal condylar connection to the neurocranium, the V-shaped myomeres of the leptocephalus, and the unsegmented fin rays). The connected efferent branchial arteries are unique to this order among bony fishes. The pouched and plumelike gills are also a shared derived feature of the order, and there are many shared losses such as the opercular and hyal bones, the branchiostegals, and the secondary upper jaw. The principal similarity between the two orders is the leptocephalus larva, a stage common to all elopomorph fishes, but regarded as a shared primitive state by Hulet and Robins, this volume. Also the leptocephalus differs in the shape of the myomeres, and its great depth of body. At present, I regard the Saccopharyngiformes as an order of highly specialized fishes derived from some early group of Elopomorph fishes. There is no basis, in shared derived features (other than the losses of structures), to consider them as a sister group of any of the other living orders of the Elopomorpha. To continue to regard them as degenerate eels requires an evolutionary return to a more primitive state in a suite of characters, as well as the evolvement of complicated structures of which there is no hint in the Anguilliformes. At the same time, it is increasingly clear that there are valid reasons to view the gulpers as a monophyletic order. That is, the order is not a catchall of odd bathypelagic eel-like fishes. Certainly, there is no reason to continue to search for relationships among higher groups of fishes such as the Cetomimiformes. Such considerations have been laid to rest.

The following characters, among elopomorph fishes, are unique to the Anguilliformes: 1, 6 (in part, the non-imbricate basketweave arrangement of scales; absence of scales is a shared loss with the Saccopharyngiformes), 9 (although the gills and the pectoral fin and girdle are far back in the Saccopharyngiformes, the head region is not elongate; also, the associated structures such as the hyal and branchial elements, rather than becoming elongate as in eels, were lost; the cartilaginous remnants of branchial elements in saccopharyngiforms show no elongation), 19, 20, 23, 29.

If one accepts the definitions of the Saccopharyngiformes above, and eliminates it from further consideration, then the long series of losses of structures (2, 8, 11, 12, 13, 14, 15, 16, 17, 18, 24, 25, 26, 28, 30, 31) stands in sharp contrast to the Elopiformes, Albuliformes, and even the Notacanthiformes. While reductions and losses are easily convergent, in sum, this suite of characters is arguably very helpful in defining the Anguilliformes and they add strong support for its monophyly.

Comments on the Anguillavidae and Other Supposed Fossil Eels

Hay (1903) described 2 new families of eels, based on fossils from the Upper Cretaceous of Syria (the localities, two villages about 12 miles northeast of the present town of Jubayl, in Lebanon). The Anguillavidae comprised 2 species: *Anguillavus bathshebae* Hay and *A. quadripinnis* Hay. In describing *Anguillavus,* Hay (1903: 437) noted: "The characters of this genus are included in those of the family. The type of the genus is *Anguillavus quadripinnis.*" Regan (1912b: 380) acted improperly when he designated *Anguillavus bathshebae* Hay as the type of the genus. The Encheliidae were based on *Enchelion montium* Hay. In addition, Hay described *Urenchelys germanus* which he placed in the Anguillidae. Hay did not elaborate on these relationships, being content to describe the fossils in some detail. These descriptions appeared at a time when Darwinian theory was in first bloom and scientists eagerly sought for ancestors and missing links. Inevitably, these elongate fishes with small pelvic fins were identified as early eels and they figured prominently in discussions of eels and eel relationships until recently. These taxa are therefore discussed below in some detail.

Anguillavus bathshebae Hay: The comments

which follow are based on examination of all of the material listed below although most are identified with individual specimens. The fossils are variously incomplete and their state of preservation varies. Except for AMNH 3868 (see comments below) all material seems to represent one species. AMNH 3704: The vertical fins are clearly separated. The caudal fin, while small in size, and rounded in shape (fanlike), retains a full complement of principal rays and hypural elements.There are 19 principal caudal rays and additional procurrent elements (1 above, 3 below). There are 8 pelvic-fin rays. The dorsal-fin origin is above the 13th vertebra, the pelvic-fin origin is below the 40th vertebra, and the anal-fin origin is below the 48th vertebra. The last dorsal ray is above vertebra 101, the last anal ray below vertebra 102. There are 107 vertebrae including that bearing the hypurals. The nature of the urostylar part cannot be determined. There are 16 branchiostegal rays. These curve slightly as in most fishes, but are not modified as in eels. The hyomandibular is vertical, the cleithrum is curved and strongly developed but the area of the pectoral fin is missing. In AMNH 6019, the pectoral girdle is below the 6th vertebra. There does not appear to be a posttemporal but a bone which may be the supracleithrum is visible above the vertebral column and extends nearly to the skull. There are about 18 pectoral rays and 20 branchiostegals. There is no elongation or posterior displacement of the branchial region. The palatopterygoid is complete and there appears to be a separate premaxilla although this area is distorted. AMNH 6018 is not especially helpful though the ratio of 2–3:1 (divided nearly evenly) between vertebrae and interneurals and interhemals is clear. In AMNH 3685 there are 26 total caudal-fin rays including 3 procurrent elements above and 4 below and 19 principal rays. The caudal fin is entirely separate from the dorsal and anal fins. The pelvic fin, with 7–8 rays, starts under the 66th vertebra from the rear. Seventy eight vertebrae are present but the entire front end of the fish is missing. AMNH uncat. is an excellent fossil showing the underside of the head region of *Anguillavus.* The separate maxilla and premaxilla, free from the vomer, the pterygoid arch, and unmodified branchial region, are clearly not eel-like. Comparison of this specimen with the others indicates that it is correctly identified with *Anguillavus.*

Anguillavus quadripinnis Hay: AMNH 3796 shows the lower side of the head and front end of the body. The snout tip is pointed with an expanded vomer and separate maxillae and premaxillae. There are 16 pectoral rays. The fin is about opposite the 6th vertebra but the cleithral elements reach nearly to the cranium. There is no eel-like elongation or posterior displacement of the branchial region and the simple branchiostegals show no anguilliform characters.

Hay (1903: 436) refers to "a row of enlarged plates on each side, probably on the lateral line." These are evident as a row of impressions in some specimens of *A. bathshebae* but no structure is present. If they are plates, then they are reminiscent of the enlarged lateral-line scales of the Halosauridae (Notacanthiformes). This simply cannot be confirmed from this material, but, whatever the nature of the structure, it is present in both species of *Anguillavus.* Regan's decision to ignore Hay's designation of *A. bathshebae* as type species of *Anguillavus* was based largely on his erroneous conclusion, upon studying only the published figures, that only *A. quadripinnis* had this structure.

Woodward (1900: 322) assigned his genus *Urenchelys,* type species *Anguilla hakelensis* Davis, to the Muraenidae (then used to include most eels), an action he continued to follow in 1901. Hay (1903: 434) placed *Urenchelys* in the Anguillidae, also broad in concept. Regan (1912b: 380) erected the Urenchelidae (= Urenchelyidae) to include both *Urenchelys* and *Anguillavus bathshebae.*

Urenchelys hakelensis (Davis): I have examined a series of specimens from the American Museum, all from near the village of Hakel, Lebanon. Not all features are evident on any one specimen. The caudal fin is small and rounded but entirely separate from the dorsal and anal fins. There are 21 caudal rays, the branchiostegals are not eel-like and there is no elongation or displacement of the branchial region although the cleithrum appears to be well behind the skull on one specimen. A premaxilla and maxilla are evident in one specimen. The type specimen of *hakelensis* was illustrated and described by Woodward. He also noted that the caudal fin is clearly seen in a specimen on display in Beirut. Again a separate premaxilla is evident, a supercleithrum is present, and the suspensorium is sharply inclined forward!

Urenchelys germanus Hay: The anal fin begins under the 37th (holotype), 40th, or 43rd vertebra. (The type of *U. avus* Woodward, from Hakel, has the anal-fin origin at about the 37th vertebra, not the 35th as stated, so that this supposed difference does not hold true.) The type of *germanus* shows the pectoral girdle to be closely connected to the skull through what certainly appears to be a posttemporal bone. The premaxilla and maxilla are both present. The dorsal fin begins above the 12th vertebra (about over the 7th or 8th in *avus*). There are about 10 branchiostegals and these start at the forward end of the ceratohyal. The head is heavy-boned and the jaws very broad. In AMNH 3695 the dorsal fin begins above the 10th vertebra and the cleithrum is about opposite the 6th vertebra. It was originally identified with *U. avus* but seems to be the same as the other specimens of *germanus* and was so identified by Hay.

Hay established the family Encheliidae for *Enchelion montium* Hay, based on 2 incomplete specimens from Hakel, Lebanon. This extremely elongate fish has no fins or evident fin supports, and no pectoral girdle. The tiny head is not elongate. The centra are very short and high and very numerous. The branchiostegals are not eel-like. Hay noted that there are no opercular bones but this area is not well preserved. The family is not mentioned by Regan (1912b), but Berg (1940: 451) referred it to the Congridae without explanation.

All of the fossils discussed above are very elongate fishes. *Anguillavus* and *Urenchelys* share a reduced, rounded, but separate caudal fin, and all appear to have a complete secondary upper jaw with both a maxilla and premaxilla. None of the taxa discussed shows any displacement of the branchial region or eel-like modification of the branchiostegals. Where they can be seen (except in *Enchelion*), the opercular elements are broad and not oriented as in eels. *Anguillavus* differs from *Urenchelys* in having pelvic fins, seemingly in having the pectoral girdle less closely associated with the skull, and in having a vertical rather than a forward sloping suspensorium. *Anguillavus* and *U. germanus* have separate premaxillae. This may not be true of *U. hakelensis*, and, if so, could cast doubt on the generic affinities of *germanus*. *Anguillavus* also has a row of "plate-like" impressions along the side of the body. The Anguillavidae and Urenchelyidae could represent elongate members of various elopomorph orders other than eels or they could represent a separate unsuccessful lineage. If they have any relationship to eels they are likely to be survivors of ancestral groups rather than direct ancestors for they are contemporary with or, more likely, post date the origins of the various living Elopomorph orders. Goody (1969: 235) suggested, for example, that "the elopiform lineage was in existence throughout the early Jurassic." There is no basis to align them with the Anguilliformes with which they share no derived characters apart from a general elongation of the body. Bertin and Arambourg (1958: 2314) placed both in the family Urenchelyidae. *Enchelion* (and the Encheliidae) is simply unidentifiable. There is no basis for Berg's assignment of it to the Congridae. It may not be a teleost.

Many other fossils, attributed to the Anguilliformes, have been described, but none has figured so prominently in the discussion of the evolutionary history of eels as have *Anguillavus* and *Urenchelys*. Basically, these fossils fall into two groups: eels that belong to extant groups (e.g., *Voltaconger*) but lacking sufficient detail to be a real help in analyzing relationships, and other elongate fishes of no demonstrable relationship to eels. Much of the literature was reviewed by Blot (1978: 3–6).

Fossil material examined (all from Upper Cretaceous deposits). Anguillavus bathshebae: AMNH 3704 (red label 4513), holotype, and AMNH 3685 (red label 4513), paratype; Hakel, Lebanon. AMNH 3868 and AMNH 6018; Hajula. AMNH 6019; Mt. Lebanon, Lebanon.

Anguillavus quadripinnis: AMNH 3796 (red label 4512), holotype; Mt. Lebanon, Lebanon.

Anguillavus sp.: AMNH uncat., Mt. Lebanon, Lebanon.

Urenchelys germanus (all from Hajula, Lebanon): AMNH 3654 (red label 4515), holotype; AMNH 3860 (red label 4515e); AMNH 3762 (red label 4515b); AMNH 3695 (red label 4515c); AMNH 3790 (red label 4515); AMNH 3861 (red label 4515).

Urenchelys hakelensis: AMNH 3791 (red label 4514); Hakel, Lebanon. AMNH 3826; Hakel, Lebanon. AMNH 6009; Sahil Alma, Lebanon. AMNH 6010; Hakel, Lebanon; AMNH 6012; Hakel, Lebanon. AMNH 8432; Hakel, Lebanon).

Urenchelys: AMNH uncat.; Hakel, Lebanon (2 other fish, *Nematonotus bottae* are more prominent in this section). AMNH 5820; Hajula, Lebanon.

Enchelion montium: AMNH 3765 and AMNH 3766 (red label 4514); cotypes; Hakel, Lebanon.

Comments on the Albuliformes

Several recent treatises have dealt with this order in varying detail. These are by Goody (1969), Forey (1973a,b), Greenwood (1977), Patterson and Rosen (1977) and Smith (1984b). For purposes of this discussion, I adopt the classification put forth by Greenwood (1977: fig. 23). The principal difference between the classifications of Forey (1973a,b) and Greenwood (1977) is that Forey considered the Pterothrissidae to be close to the Notacanthidae whereas Greenwood regarded the Pterothrissinae to be a subfamily of the Albulidae. The comments below apply only to the Albuliformes. I have given reasons above for disagreeing with the close association of the Saccopharyngiformes and Anguilliformes.

Forey (1973b: 151–152) gave abundant evidence for synonymizing *Istieus* and *Pterothrissus* but did not do so because *Istieus* is based on fossils. He noted that "the minor differences between *Istieus* and *Pterothrissus* are far outweighed by the many important features of similarity in neurocranial anatomy, the hyopalatine series, the infraorbital series, the dermal jaws, the form of the teeth, the cephalic sensory canal system and the fins. . . ." To this may be added the observation that the two fossil and two Recent species show as great a range of differences within each pair as there is between the two pairs. I regard *Istieus* Agassiz, 1844, as a senior synonym of *Pterothrissus* Hilgendorf, 1877. The Pterothrissinae (-idae) is the earliest family group name, however. *Pterothrissus* is not an important fish to fisheries scientists or biologists generally and there are few publications concerning it so that no case can be made for conserving the name *Pterothrissus.*

Careful study of the papers cited above reveals that no system of classification works well where there is fundamental disagreement with regard to the nature of the characters or where the classifications are based on largely different suites of characters. With regard to the Albuliformes per se, Greenwood (1977) presented the most persuasive arguments. There is no shared derived character unique to his Albuliformes and the Anguilliformes as defined above. Although the two orders represent about the same level of organization the same can be said for the Anguilliformes and Elopiformes (including the Megalopiformes of Patterson and Rosen, 1977) (see below). As noted above, the Anguillavidae may be related to the Albuliformes. The persuasiveness of the various authors, cited above in arguing for one or another classification, is supported more by similarity in losses in structures in characters obviously undergoing reduction (e.g., the number of uroneural and hypural elements) rather than by newly acquired derived characters. The facts given by Patterson and Rosen (1977) may be variously interpreted (witness their disagreement with Nelson, 1973—see next paragraph) and are not compelling despite the eloquent support given them by the authors.

Comments on the Elopiformes

Again, Goody (1965), Forey (1973a,b), Nelson (1973), Greenwood (1977), Patterson and Rosen (1977), and Smith (1984b) have discussed the composition and relationships of the Elopiformes, the principal differences being that Patterson and Rosen (1977) separated the Megalopiformes and Elopiformes in recognition of what they evidently regarded as an early dichotomy (the new ordinal name Megalopiformes is introduced, without explanation, in their summary table), and in the relative placement of the various component orders of the Elopomorpha. Nelson (1973) argued for an ichthyodectiform–elopomorph relationship, a view dismissed by Patterson and Rosen (1977: 101) as being based on a misinterpretation of features of the lower jaw. Nelson's views deserve careful consideration.

These problems emphasize that we need a better understanding of the anatomy of living and fossil forms and that as long as reasonable scientists differ in their interpretation of character states, it is premature to expect agreement on a classification, no matter on what basis, or with what procedure, it is derived. The integrity of the Elopomorpha continues to rest on the significance of the leptocephalus stage, a feature regarded as derived by Patterson and Rosen, as primitive by Hulet and Robins (this volume), and of questionable significance by Forey (1973a,b).

Accepting the elopomorph concept for the present, we regard the Elopiformes (including the Megalopiformes), Albuliformes (including the Notacanthiformes), and the Anguilliformes as an unresolved trichotomy. The Saccopharyngiformes are remote from these three orders.

Infraordinal Relationships of the Anguilliformes

Regan (1912b) first emphasized the character of fused versus divided frontal bones to subdivide the Anguilliformes although he erred with regard to the nature of this bone in his characterization of some families. The paired condition is clearly the primitive condition. Although it may be argued that the condition of having the frontal bones fused could have occurred more than once, the general relationships of eels as reported elsewhere in this volume suggest that the fusion occurred only once, that it is a derived condition, and that the three superfamilies of eels having this condition (Synaphobranchoidea, Congroidea, Ophichthoidea) represent a monophyletic lineage. This lineage may be recognized as the suborder Congroidei. The two superfamilies (Anguilloidea, Muraenoidea) sharing the primitive condition of divided frontals could be united, for convenience, as the suborder Anguilloidei, although this unit shares no known derived characters and would be paraphyletic at best. Recognizing that our knowledge of the details of eel structure is still fragmentary, I choose to recognize three suborders whose relationships are unresolved: Congroidei, Muraenoidei, and Anguilloidei. In choosing names for the superfamilies, I have chosen the oldest of the family group names accepted by the authors of the accounts of the constituent families of this volume. Priority does not extend to subordinal levels. In selecting Congroidei over Ophichthoidei, I have followed convention.

Suborder Congroidei. The derived feature uniting this suborder is the fusion of the frontal bones. Apart from this, the primitive families in each superfamily share many primitive features particularly in the nature of the pharyngeal bones and the nature of the suspensorium when compared to the Anguilloidea. The three constituent superfamilies are readily identified on the basis of shared specializations, except, perhaps, for the Congroidea. It is not possible, at present, to relate the three superfamilies of the Congroidei inter alia on the basis of specialized characters shared by any two of them.

Superfamily Synaphobranchoidea: This group contains the single family Synaphobranchidae (see Robins and Robins, this volume, for an analysis of characters and infrafamilial trends). The Synaphobranchoidea are defined most clearly by two derived features, the diagonally elongate eye of their leptocephali and the forward directed orientation of the third hypobranchials. Because the Synaphobranchoidea retain more of the primitive character states (Robins and Robins, this volume), we agree with Nelson (1966a: 398) in regarding this superfamily as most primitive in the suborder. To derive them from either of the other superfamilies would require a series of character reversals.

Superfamily Ophichthoidea: This group also

consists of a single family, the Ophichthidae, which has been characterized by McCosker et al., this volume, and Leiby, this volume. In this family, there has been a posterior shift in the branchial region with an anterior elongation of the ceratohyals and, most importantly, the addition of several to many unattached anterior branchiostegals. The branchiostegals of the two sides broadly overlap and provide support for the muscles which operate the branchial pump. Although a few short interspersed bony splints have been seen in the branchiostegal membranes of the chlopsid, *Chilorhinus suensonii* (pers. obs. based on ANSP 116139), no other eel has either the unattached anterior branchiostegals or the overlapping branchiostegals that mark the ophichthoids.

Superfamily Congroidea: To this group I assign the following families: Colocongridae, Congridae, Derichthyidae, Muraenesocidae, Nemichthyidae, Nettastomatidae, and Serrivomeridae. All of these families share a forward inclination of the hyomandibular. The history of the separation of the Colocongridae from the Congridae is discussed fully by Smith, this volume. *Coloconger*, sole genus in the Colocongridae, has the supracleithrum very close to the rear of the skull (the separation equal to about the diameter of the pupil). Also, only two branchiostegals circle the opercle and then barely extend above the upper rear margin, the palatopterygoid arcade is complete, the premaxillae, although fused to each other, are nearly separated from the vomer, and the hyomandibular is the most nearly vertical of any congroid. The Colocongridae may be the sister group of the rest of the congroids although the data base is too insecure to merit cladistic analysis. Some additional comments are in order. The Serrivomeridae have a high central ridge on the frontal bone, an extreme forward orientation of the hyomandibular, a complete palatopterygoid arcade, an unfused palatine, and branchiostegals whose bases overlay and extend forward of the hyal bones (in a manner curiously analogous to that of the Acanthodii). Several of these features (the ridged frontal, the complete palatopterygoid arcade, and orientation of the hyomandibular) are very similar to those of the nettastomatid *Saurenchelys*. As it presently stands, the Nettastomatidae are paraphyletic with this one lineage being closer to the serrivomerids than to some of the other nettastomatid genera. Also family lines between the Congridae, Muraenesocidae, and Nettastomatidae have been shifting and are not yet firmly established. In an early discussion between authors of this volume, the condition of frontal bone in the Nemichthyidae came into question. Dissection of cleared and stained material by Kenneth A. Tighe and the author showed that there is a single fully united frontal in this family.

Suborder Muraenoidei. This suborder comprises the families Myrocongridae, Chlopsidae (= Xenocongridae), and Muraenidae. The characters of these three families were analyzed in detail by Smith (1984b) based on his osteological studies of *Myroconger* and analysis of the published literature on various chlopsids and muraenids. His cladogram (Smith, 1984b: fig. 15) recognizes the monotypic Myrocongridae as the sister group of the Chlopsidae and Muraenidae. Certainly this is supported by the analysis of characters presented by Smith. The only disturbing feature is that a posteriorly sloping suspensorium is a primitive feature of bony fishes and this condition is extreme in the Muraenidae among the three constituent families of the suborder. However, this feature seems evolutionarily labile within the Anguilliformes and it may be regarded as a convergent specialization of the muraenids, perhaps associated with the development of the very strong pharyngeal jaws which also characterize the family. Cope (1871: 452) placed the muraenids (except for *Gymnothorax* which he placed in the Gymnothoracidae in his Enchelycephali) in their own order, the Colocephali, although the reasons for his having done so were erroneous in part (e.g., absence of pterygoid) or based on features which proved to be common to other eels.

Suborder Anguilloidei. This suborder comprises the families Anguillidae, Moringuidae, and Heterenchelyidae. The characters and relationships of these three families were discussed in detail by Smith and Castle (1972) and elsewhere in this volume by Smith. That the three are closely related seems beyond question but no reasonable cladogram is yet possible because of the mix of primitive features and losses associated with the strongly fossorial habits of the Moringuidae and Heterenchelyidae. The upper arm of the preopercle is most reduced and the hyomandibular is narrowed and strongly angled forward in *Anguilla.* The catadromous habit is unique to *Anguilla* but its retention of scales and a palatopterygoid arcade are primitive. Arguably, the Moringuidae represents the sister group of the other two, despite its many losses of structure.

Except for the fact that the Muraenoidei and Anguilloidei share the primitive character of having the frontals divided, there is no basis, at present, for resolving their relationships or that of either of them to the Congroidei.

Ichthyologists studying the various eels have been so involved with the clarification of species and genera, that there has been insufficient time to do the detailed anatomical studies necessary to postulate and test familial and higher relationships. Nor would it have been timely to have attempted to study this problem for the compositions of virtually all eel families, which resulted from studies of species and genera, and which are exemplified by the various papers in this volume, are at present dramatically different from our concepts of the 1950's and 1960's, let alone those of Regan and other earlier authors. For example, the Congridae of Ginsburg (1951) contained *Neoconger* (now in the Moringuidae), *Hoplunnis* (Nettastomatidae), *Dysommina* and *Dysomma* (Synaphobranchidae). At present, our data base is too weak to provide proper support for cladistic analysis. As Jordan (1923: 80) observed, "... we must not expect a degree of accuracy which the subject in question does not permit." We hope the present volume will provide a more reasonable basis for new, detailed, anatomical studies which will answer many questions left unresolved here.

All 15 families of eels of the order Anguilliformes currently recognized occur in the Atlantic (the Myrocongridae found in the eastern and mid-Atlantic). Most of the 600 or so species of eels belong to the three families Ophichthidae, Congridae, and Muraenidae. Recent experiences, both in attempting to associate types of leptocephali with adults and in collecting along steep rocky slopes with gear operated from submarines, suggest that many species remain to be described. Some of the definitions given here and by other authors in this volume are already being compromised by new discoveries.

Anguillids, congrids, and muraenesocids are important commercially and the anguillids are being cultured. Muraenids are also harvested, especially in southern Europe, but in the tropics they have been known to cause serious, even fatal, cases of ciguatoxic disease. Eels are important elements in certain ecosystems. Synaphobranchids (unpublished data of Robins and Robins) comprise more than 40 percent of the fish fauna of the floor of the Tongue of the Ocean (Bahamas) and other deep benthic communities from a standpoint of numbers of individuals, and 55 percent on the basis of biomass. Data given by Brock et al. (1979) show that muraenids comprise about 50 percent of the biomass of piscivores in Hawaii. Certain large eels, especially muraenids, ophichthids, muraenesocids, and congrids, have powerful jaws and strong teeth and can inflict serious wounds if mishandled or if attacked. Various eels are also commonly displayed in public aquaria, and a few have become popular in the marine aquarium trade.

Key to the Families of Anguilliformes[1]

1a. Jaws extremely elongate, slender, maxillary teeth tiny and sandpapery 2
1b. Jaws not elongate and slender, teeth variable 4
2a. Jaws extremely slender, fragile; pectoral fin present 3
2b. Jaws strong; pectoral fins absent, or if present, large teeth on vomer . . Nettastomatidae*, p. 568
3a. Body very long, ribbonlike, generally brown to tan; jaws bend away from each other; anus far forward, at throat; gill openings separate; nostrils separate, the anterior one tubular . . . . . . Nemichthyidae (in part)*, p. 441
3b. Body not ribbonlike, blackish, with silvery sides; jaws together throughout their length; anus far removed from throat; gill openings connected ventrally; nostrils close together, non-tubular . . . . . . Serrivomeridae*, p. 613
4a. Body very long, ribbonlike; front part of head truncate or with short beak; anus at throat . . . . . . Nemichthyidae (in part)*, p. 441
4b. Body not ribbonlike, or, if so, anus far behind throat 5
5a. Caudal end of body ending in hardened tip, no caudal-fin rays visible externally; dorsal and anal rays not confluent; posterior nostril always at rim of mouth, covered by flap . . . . . . Ophichthidae (in part)*, p. 254
5b. Caudal end of body with dorsal, caudal, and anal fins confluent; caudal fin obvious, even if small; posterior nostril variously located 6
6a. Anterior branchiostegals greatly overlapping those of opposite side, not attached basally to hyal bones, the branchial region often appearing enlarged or inflated; posterior nostril at rim of mouth in most species . . . . . . Ophichthidae (in part)*, p. 254
6b. Anterior branchiostegals never overlapping those of opposite side, all at least loosely articulated basally to hyal arch 7
7a. Scales present over most of body 8
7b. Head and body always entirely naked (dermal ridges, or impressions made by nets, may give the appearance of scales) 9
8a. Gill opening low on body, sometimes united ventrally; body flabby, usually dark brown with blackish belly; peritoneum dark . . . . . . Synaphobranchidae (in part)*, p. 207
8b. Gill opening lateral, extending up in front of lower half of pectoral fin; body firm, usually with pale belly and underside of head . . . . . . Anguillidae, p. 25
9a. Eye tiny, often difficult to see, or, if well-developed, dorsal and anal fins with high anterior lobe, the margin falcate 10
9b. Eyes normally developed, dorsal and anal fin edges never falcate 12
10a. Vomer and front part of dentary with large compound teeth . . . . . . Synaphobranchidae (in part)*, p. 207
10b. No large compound teeth on vomer or dentary 11
11a. Pectoral fins absent, anus well in front of midpoint of body . . . . . . Heterenchelyidae, p. 48
11b. Pectoral fins present, anus well behind midpoint of body or, if farther forward, teeth uniserial . . . . . . Moringuidae, p. 55

[1] Based on adults only. Those with an * have the frontal bones fused; all others have the frontal divided by a suture—this feature is not used as a primary key character as it is not easily determined. In so far as possible, this key is based on easily observed features, all of which are illustrated elsewhere in this volume. Technical diagnoses are provided for each family in the various other papers in this volume.

12a. Pectoral fins absent; no pores in lateral line behind the 1–4 (usually 1–2) above the branchial region 13
12b. Pectoral fins present; pores variable 14
13a. Posterior nostril high on head, before uppermost part of eye . . . . Muraenidae, p. 104
13b. Posterior nostril low on head, near lip or inside mouth . . . . Chlopsidae (part), p. 72
14a. Lateral line incomplete, ca 1–7 pores in anteriormost part of canal 15
14b. Lateral line with pores for most of its length 16
15a. Posterior nostril high on head, before uppermost part of eye . . . . Myrocongridae, p. 98
15b. Posterior nostril low on head, before lower part of eye or near lip . . . . Chlopside (part), p. 72
16a. Body short and stout (depth between 10–25% TL); anus well behind midpoint; most lateral-line pores in short tubes . . . . Colocongridae*, p. 413
16b. Body elongate (depth usually less than 10% TL); anus usually near or before midpoint; lateral-line pores not tubular 17
17a. Body narrows just behind head, becoming deeper and more compressed under dorsal-fin origin; adults pelagic . . . . Derichthyidae*, p. 420
17b. Body not distinctly narrower just behind head; adults benthic . . . . Muraenesocidae and Congridae*, pp. 432, 460 (the former, restricted by Smith, 1984a, to *Muraenesox, Congresox, Cynoponticus,* and *Sauromuraenesox,* is not separable from the Congridae on the basis of characters that are useful in a key of this sort)

Order Anguilliformes

Family Anguillidae

FRESHWATER EELS

DAVID G. SMITH

Acknowledgments. The studies I present in this volume began some two decades ago with a masters thesis and then a doctoral dissertation at the University of Miami. James E. Böhlke and C. Richard Robins, co-chairmen of my dissertation committee, provided much encouragement and support throughout the course of my graduate studies, and in the years that followed I continued to enjoy a close personal and professional association with both of them. It is also appropriate at this point to acknowledge my debt to the late Edward C. Raney, through whom I acquired my appreciation for the fascinating group of animals we call fishes; my interest in eels began with a term paper I wrote for his ichthyology course at Cornell. It is probably impossible to mention by name all the people who helped me in one form or another in the studies leading to the present volume. The curators and others associated with major museums and fish collections are cited elsewhere; I thank them all. Robert H. Kanazawa of the USNM (now retired) generously shared with me the information he had accumulated on congrids and nettastomatids. Much of this was published in an earlier paper (Smith and Kanazawa, 1977), but some was withheld pending further work. I thank Drs. F.-W. Tesch and James D. McCleave for inviting me to participate in cruises under their direction to the Sargasso Sea. Eugenia B. Böhlke deserves a special acknowledgment from everyone associated with this volume. Without her dedication, organization, and leadership, the project could never have been completed. Bruce B. Collette reviewed the mountain of manuscripts I sent him with his usual care and attention to detail. The National Science Foundation has played a major role in supporting these studies. Among the grants that benefitted me directly or indirectly are GB-7015 (C. R. Robins, principal investigator), GB-17736 (J. E. Böhlke, principal investigator), BMS 75-08675 (W. H. Hulet and D. G. Smith, principal investigators), INT 77-12568 (D. G. Smith, principal investigator), BSR 83-00279 (E. B. Böhlke, principal investigator), and BSR 8501268 (K. F. Liem, principal investigator). Much of the work was done while I was at the Marine Biomedical Institute, University of Texas Medical Branch at Galveston. I thank the directors of that institute, Drs. Stewart G. Wolf and William D. Willis, Jr., and in particular Dr. William H. Hulet, chief of the Marine Medicine Division, for supporting my work during that time.

Introduction. This is the most important of the eel families, at least in terms of its relationship to man. It is also by far the best known; the literature dealing with anguillids is enormous and exceeds that of all other eels put together. Anguillids are exceptional among eels because they spend most of their adult lives in fresh water, returning to the sea only to breed. They are a small, homogeneous group, relatively unspecialized morphologically, and highly adaptable. The fifteen recognized species belong to a single genus.

The family name dates from the Anguillidi of Rafinesque (1810b: 37).

Characters. Body moderately elongate, anus slightly before midbody, tail flexible and broadly rounded. Dorsal and anal fins well developed, confluent with caudal fin; dorsal fin begins between anus and gill opening. Pectoral fin well developed, broadly rounded. Gill opening crescentic, on side of head just below lateral midline. Small, embedded scales present. Lateral line complete on head and body; no frontal commissure; adnasal pore absent.

Snout relatively short, moderately acute to

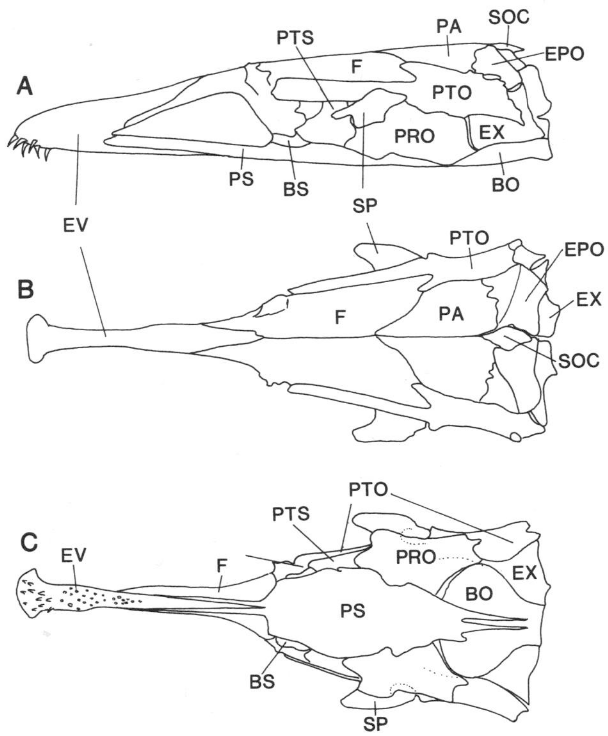

FIGURE 4. Neurocranium of *Anguilla rostrata*. A. Left lateral view. B. Dorsal view. C. Ventral view. Abbreviations given on pp. 5–6.

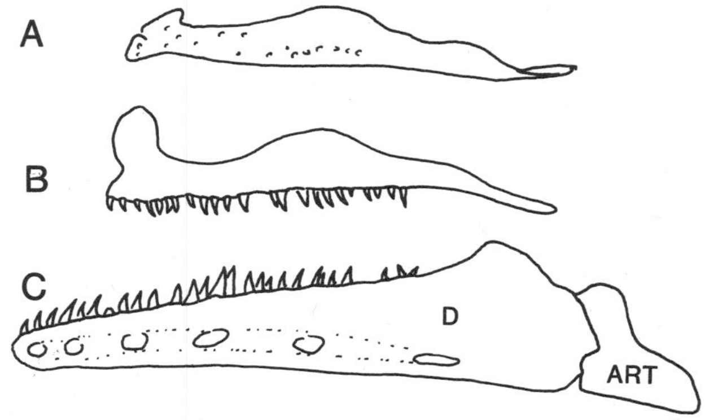

FIGURE 5. Jaws of *Anguilla rostrata*. A. Maxilla in dorsal view. B. Maxilla in lateral view. C. Mandible in lateral view.

rounded in lateral profile, lower jaw projects beyond upper; well developed flange on upper and lower lips. Eye large. Anterior nostril tubular, near tip of snout, directed anterolaterally; posterior nostril round, in front of mideye. Teeth small, granular, in bands; vomerine tooth patch long.

Osteology (based on *Anguilla rostrata*). Neurocranium (Fig. 4) moderately elongate and wedge-shaped in dorsal profile; orbit relatively large, preorbital section not greatly prolonged; small otic bullae present. Ethmovomer forms front of neurocranium and overlaps frontals dorsally and parasphenoid ventrally. Roof of skull formed by large paired frontals and parietals and small, median supraoccipital; supraoccipital bears a crest. Epiotics, exoccipitals, and basioccipital enclose neurocranium posteriorly. Elongate pterotic lies laterally and bears temporal canal. Immediately below are exoccipital, prootic, sphenotic, and pterosphenoid. Parasphenoid lies along ventral axis.

Maxilla bears a well developed pedicel at its anterior end and a broad, secondary hump at about midlength (Fig. 5A,B). Mandible without notable modifications (Fig. 5C).

Suspensorium slender and inclined sharply forward (Fig. 6). Pterygoid well developed and closely associated with quadrate and hyomandibula.

Hyoid elements consist of a median glossohyal and urohyal, paired ceratohyals and epihyals, and about 11 branchiostegal rays (Fig. 7). No hypohyals. Glossohyal elongate. Urohyal very short, consisting of a broad head and a small posterior projection. All branchiostegal rays except first associated with epihyal, posterior ones curve up and around opercle posteriorly.

Gill arches well developed and nearly complete (Fig. 8); only third basibranchial absent. First four ceratobranchials large and stout, fifth reduced and slender. Lower pharyngeals undivided, upper pharyngeals divided.

All four opercular elements present (Fig. 9). Opercle broadly oval and smooth-margined. Subopercle sickle-shaped and lies immediately below opercle. Preopercle nearly as large as oper-

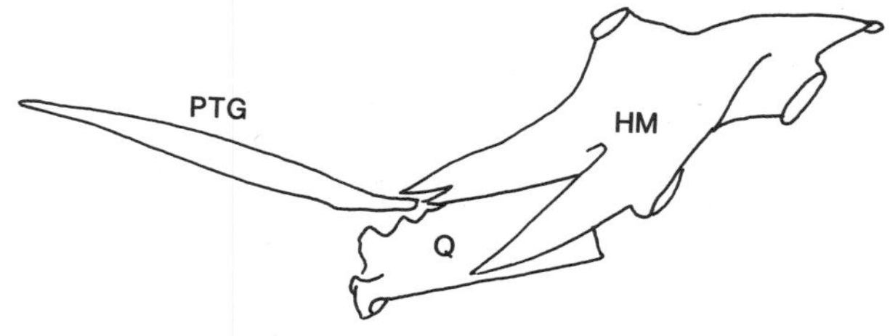

FIGURE 6. Suspensorium of *Anguilla rostrata*.

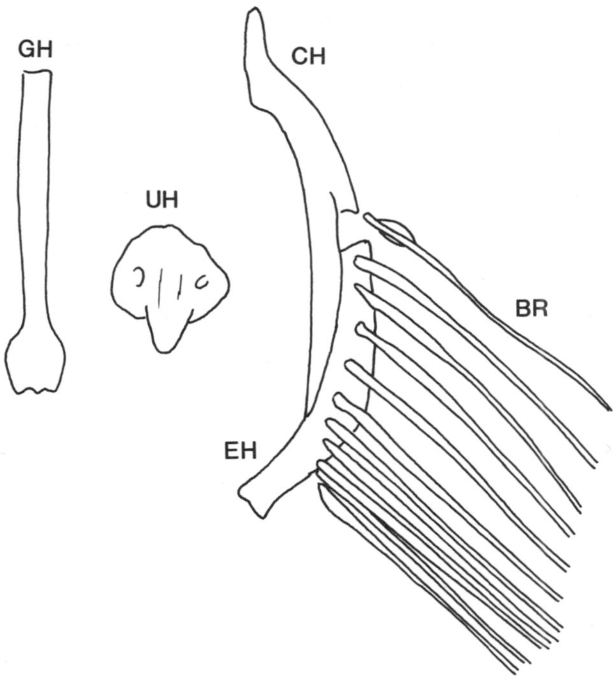

FIGURE 7. Hyoid arch of *Anguilla rostrata.*

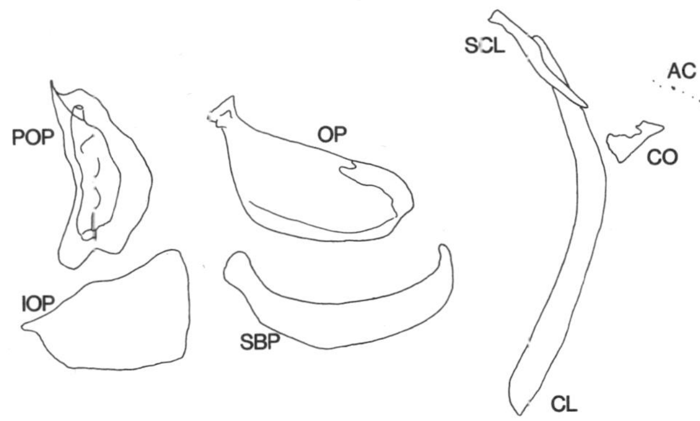

FIGURE 9. Opercular apparatus and pectoral girdle of *Anguilla rostrata.*

cle and somewhat elongate dorso-ventrally; it bears a portion of preopercular canal. Interopercle trapezoidal in shape and almost as large as opercle.

Pectoral girdle consists of a large cleithrum, a smaller supracleithrum, and a coracoid (Fig. 9); no clearly developed scapula present in specimen examined. Seven small actinosts visible in same specimen.

Circumorbital bones not well ossified (Fig. 10). Elongate nasal and preorbital present, and a small supraorbital located between their anterior ends. Four suborbitals behind eye.

Epineurals of first five vertebrae fused to neural arch, forming strong, posterolaterally directed spines (Fig. 11). Second through eighth vertebrae with complex neural spines (Fig. 11). Precaudal vertebrae (Fig. 12) compose about 42% of total (45 of 108 in specimen examined). Ribs present from about 7th–38th vertebra, epineurals from 1st–87th, and epipleurals from 34th–87th. No epicentral processes on caudal vertebrae (Figs. 12, 13).

Discussion. Anguillids are the most familiar of all eels and have long been associated with man. Indeed, to most people "the eel" means a member of the Anguillidae. This is fortunate, because few other eels typify the order so well. Aside from certain congrids, anguillids are the most

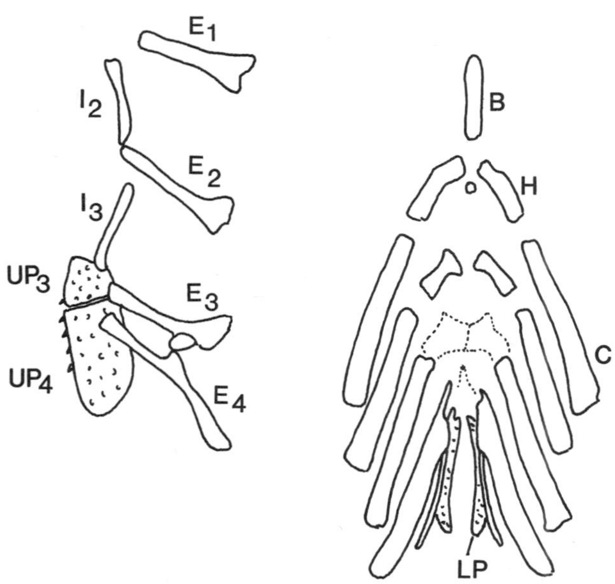

FIGURE 8. Gill arches of *Anguilla rostrata,* ventral elements in ventral view, dorsal elements in dorsal view.

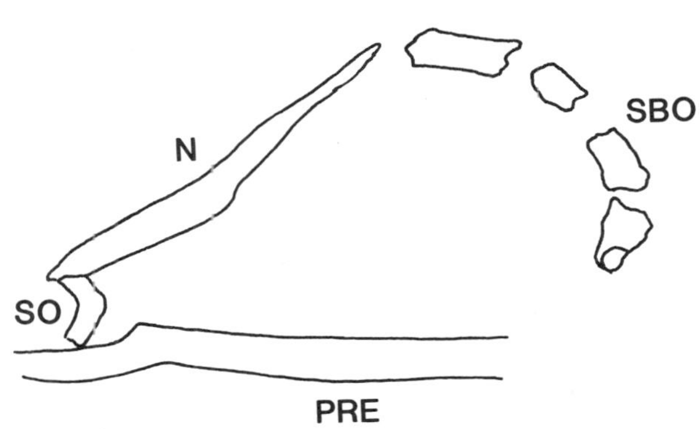

FIGURE 10. Circumorbital bones of *Anguilla rostrata.*

FIGURE 11. Anterior few vertebrae of *Anguilla rostrata.*

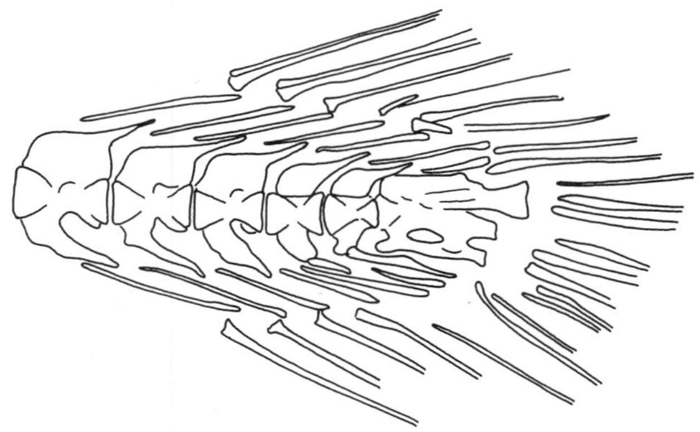

FIGURE 13. Caudal skeleton of *Anguilla rostrata.*

generalized of eels morphologically. They are neither excessively elongate nor short and stubby. The vertical fins are well developed and continuous around the tail. The pectoral fins are well developed. The eye is large and the mouth is moderate. Anguillids retain one especially primitive feature, scales, although these are small and inconspicuous.

The most remarkable and characteristic feature of the Anguillidae is its life history. Anguillids are catadromous, and they have carried this lifestyle to an extreme unmatched among fishes. The adults spend their lives in fresh water or along the immediate coast in estuaries and tidal marshes. They can penetrate far inland, reaching headwaters of river systems hundreds of kilometers from the ocean. There they pass several years, feeding and growing. At maturity they stop feeding, move downstream, and strike out to sea. Somehow they find their way to a certain area in the middle of the ocean and spawn. The leptocephali make their way back to the coast on the currents, metamorphose, and enter fresh water as elvers, thus starting the cycle all over again.

As spectacular as this migration is, it seems clearly derived from the simple inshore-offshore

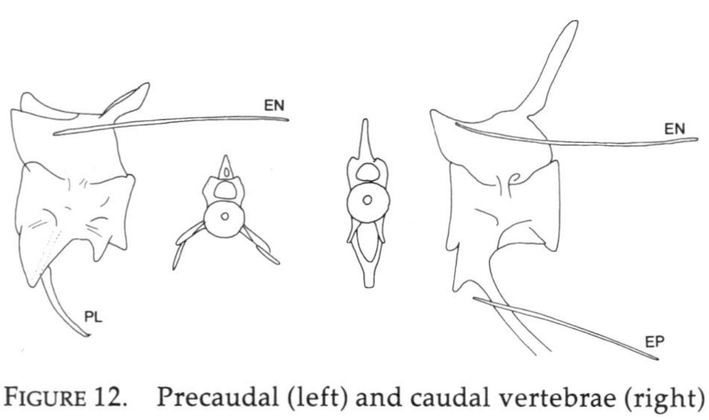

FIGURE 12. Precaudal (left) and caudal vertebrae (right) of *Anguilla rostrata;* lateral and frontal view.

migratory pattern seen in some other eels and in more primitive elopomorph fishes such as *Elops.* Adults of *Elops saurus,* for example, live along the coast and in bays and estuaries. They travel offshore to spawn, and the leptocephali work their way inshore as they grow. The metamorphic leptocephali are found in tidal streams and brackish water some distance from the coast. *Myrophis punctatus* (Ophichthidae) shows a similar pattern. The anguillids have simply extended the pattern at both ends. When they come inshore they do not stop at the limit of tidal influence but continue far upstream into purely fresh water. And when they return to the ocean, they don't stop at the edge of the continental shelf; they continue on into the open sea.

Anguillids are by far the most important of the eels commercially. The principal markets are in Europe and Japan, where eels have long been a delicacy. According to FAO statistics (FAO, 1984: 125), the total yearly catch of eels in Europe from 1979 to 1982 ranged from 12,899 to 16,877 metric tons. In the Far East the demand is even greater; catches in the same period ranged from 29,439 to 35,847 metric tons. Smaller fisheries exist in North America (1081–2430 metric tons) and Australia and New Zealand (869–1685 metric tons), primarily for export to Europe and Japan. In Japan, the wild catch has been supplemented in recent years by eels raised on "farms." Elvers are collected during their upstream migrations and placed in holding ponds, where they are fed until they reach marketable size. Smaller, more experimental culture activities are taking place in Europe, the United States, Taiwan, and New Zealand as well. In addition to their use as food, eels are also valued for their skin, from which a fine, soft leather is made. Eelskin shoes, wallets, etc. command high prices.

The association of eels and humans goes back to ancient times. The eel was a familiar inhabitant of European waters, and by Roman times it was already established as a favorite item of food. Its peculiar shape made it stand out from other fishes, but perhaps what most fascinated people was the problem of its generation. Unlike other fishes, eels never contained eggs or milt and were never seen to spawn or give birth. The ancients, not yet constrained by the scientific method, devised a number of ingenious explanations for its origin. These usually involved some form of spontaneous generation: horsehairs, rotting vegetation, water insects, or bits of skin rubbed from the eel's body.

The idea of spontaneous generation was abandoned in the 17th and 18th centuries, but the mystery of the eel's reproduction remained as stubborn as ever. Some investigators, including Leeuwenhoek and Linnaeus, found parasitic worms in the eel's body cavity and thought that these were unborn eels. A major breakthrough came in 1777 when Mondini found and described the ovary of the eel (Lane, 1978: 2). Another hundred years passed before the testis was discovered, but it was now clear that eels reproduced like other fishes, by laying eggs. But the question remained: where? Some time earlier Redi had observed that eels leave fresh water in the fall and go to the sea (Lane, 1978: 2). In the spring young eels return to the streams. Therefore, the eel must lay its eggs in the sea, and the elvers seen in the spring must be the offspring of the adults that went to sea the previous autumn.

Nevertheless, when the young eels return to fresh water they are already some 6 to 7 cm in length and can hardly be considered newly hatched. Precisely where were the eels hatched, where did they spend the intervening months, and what did the very early stages look like? The first piece of the puzzle was put in place in 1856, although no one realized it at the time. Kaup (1856b), in his catalog of apodal fishes, described several new species of the peculiar genus *Leptocephalus.* Among them was one he named *Leptocephalus brevirostris.* In 1892 Grassi and Calandruccio announced that *Leptocephalus brevirostris* was in fact the larval stage of the common eel. They had observed the entire process of metamorphosis in an aquarium (Grassi, 1896).

The story then shifted to Denmark and a young fisheries worker named Johannes Schmidt. Schmidt had already completed a series of studies on the life histories of gadid fishes in the North Atlantic. This work had impressed him with the importance of hydrographic conditions on the distribution of fish species, especially in regard to spawning areas and larval dispersal. When the Danish Commission for Fishery Investigations began a program to study the life history of the eel in 1905, Schmidt was put in charge.

The investigations began in two parts: collections of leptocephali at sea, and a study of morphological variation of adult eels from various parts of Europe and America. From the latter study, Schmidt showed to his satisfaction that only two species of anguillid eels occur in the Atlantic basin: one in North America, including Greenland and Bermuda, and one in Europe and North Africa, including Iceland and the Azores. Furthermore, he found no evidence of any geographic variation or local races within the European species. To Schmidt this meant that all the eels in Europe belonged to a single spawning population that was completely mixed every year. The implication was that all the eels traveled to a common spawning ground. But where? Grassi believed that the eel spawned in the depths of the Mediterranean, a logical view considering that leptocephali were then known only from that sea. Schmidt was at first willing to accept that, but the discovery of a leptocephalus west of the Faroes in 1904 showed at the very least that the Mediterranean was not the only spawning area. Schmidt soon became convinced that the eel did not spawn in the Mediterranean at all. He led two cruises to the Mediterranean on the Danish fisheries research vessel THOR in the winter of 1908–1909 and the summer of 1910.

Although leptocephali were common throughout the western Mediterranean, they were all full size, some 6 to 8 cm in length. Further cruises in the Atlantic west of Europe revealed that large numbers of fully grown leptocephali were found there, too. Significantly, Schmidt found no leptocephali over the continental shelf. He also discovered a pronounced seasonal variation in the state of maturity of eel larvae. In the spring and early summer, none of the leptocephali collected off Europe had begun metamorphosis; in August and September, by contrast, most of them were undergoing that process (Schmidt, 1922: 184). He also noticed that the metamorphic individuals were always collected closer to the coastal banks than were premetamorphic ones. From this, Schmidt concluded that the spawning ground of the eel was located out in the Atlantic far from the coast.

Up to this point, Schmidt had still not found any leptocephali smaller than about 6 cm. He had no idea how old they were and consequently how far they may have traveled since hatching. In this sense he was no closer to finding the spawning ground than he was when he started. The first distinctly younger leptocephali were collected by the Norwegian vessel MICHAEL SARS southwest of the Azores in 1910. They measured from 4 to 6 cm. In addition, Schmidt found a 41-mm leptocephalus from off Madeira among some old collections in the Zoological Museum in Copenhagen.

To carry the studies further, it would be necessary to sample as widely as possible in the North Atlantic, but the THOR's cruising range was too short to be of much use. Improvising, Schmidt persuaded various Danish commercial and naval vessels to make occasional plankton collections during their Atlantic crossings. Despite the obvious limitations of this method, significant results were obtained. Among the catch was the smallest leptocephalus yet known, 34 mm, and it was collected considerably south and west (25°N, 55°W) of all previous records.

In 1913 Schmidt finally got the use of a vessel capable of working in the open Atlantic. It was the MARGRETHE, a schooner owned by the Vendyssel Packing Co. of Copenhagen. This cruise laid out the framework for the solution to the age-old puzzle. Schmidt found that the number of leptocephali increased from east to west, with the greatest quantities occurring west of 50°W. The average size decreased from east to west and from north to south. No leptocephali larger than 50 mm were collected west of a line from Newfoundland to Cape Verde. The spawning ground of the European eel clearly lay in the western Atlantic, but its precise location was still uncertain. The MARGRETHE's collections were made from October to December and included no specimens smaller than 17 mm; indeed, most of them were 35–45 mm. This indicated that the eel did not spawn during the fall. If the leptocephali were even a few months old, they may have been carried some distance from where they were hatched.

Schmidt had planned to spend the winter in the Virgin Islands (still a Danish possession) and continue the sampling the following spring on the way home. Fate intervened, however, and the MARGRETHE ran aground in the West Indies and was wrecked. Fortunately, the collections were saved, but Schmidt was without a ship again. Undaunted, he continued to work with the fortuitous collections made by trading vessels, but by 1915 the First World War brought a halt even to these. Even so, the large-scale picture was already becoming apparent. The eel spawned in the spring and summer in the general area of the Sargasso Sea. The presence of leptocephali as small as 9 mm in the collections indicated that the steamship routes from the Virgin Islands to the English Channel passed very close to the spawning area.

The opportunity to fill in the details came in 1920. Schmidt turned his persuasive powers on the East Asiatic Company of Copenhagen and obtained the use of a four-masted motor schooner, the DANA. He was now free to explore the potential spawning area at will, following any course that was indicated by examination of the catches on board. The first cruise of the DANA was from April to August, 1920. Beginning at Gibraltar, the ship sampled along a course to St.

Thomas and then spent the bulk of its time criss-crossing the Bermuda-Sargasso Sea area collecting eel larvae and other pelagic fauna. The following year, Schmidt returned to the Sargasso Sea with the DANA and sampled the area from February to May. In 1921 the Danish government purchased a minesweeper from the British navy and converted it for fisheries work. Named DANA II, the ship was given to Schmidt for a third cruise from September, 1921 to June, 1922. This was a more comprehensive cruise, both in the area covered and the variety of work done. It was as much a hydrographic survey of the Atlantic as it was a pelagic faunal survey.

Schmidt's search for the breeding ground of the eel is one of the great scientific detective stories of all time. One of the oldest mysteries of nature was solved, and the solution was one that could hardly have been imagined beforehand. Schmidt showed that the common European eel swims more than halfway across the Altlantic to a spawning ground in the warm waters southeast of Bermuda. The leptocephali that hatch there then embark on a 3-year journey back to Europe, riding the North Atlantic Current. A later expedition around the world in 1928–1930 showed that all anguillid species have a similar catadromous life history, though they do not all travel as far as the European eel.

The DANA expeditions resulted in one of the most important and comprehensive collections of pelagic fauna in existence. The Dana Reports and other publications that have flowed from these collections have added greatly to our knowledge of these organisms. Thus, in addition to its importance as a food source, the eel has contributed indirectly to the advance of marine science.

Distribution. The distribution of anguillid eels is related largely to the accessibility of spawning areas. The North Atlantic eels are found predominantly on the coasts most directly exposed to the Gulf Stream: the east coast of the United States and Canada, and western Europe. On the American side, the abundance of eels falls off markedly south of the Greater Antilles, and they seem to be absent from South America. In Europe the eel occurs commonly in the Atlantic drainage from Norway to Morocco and through most of the Mediterranean basin into the Black Sea.

In the Indian Ocean, eels occur from Cape Agulhas in South Africa to Somalia, and in Madagascar, the Mascarenes, Comoros, and Seychelles, places that are all accessible to warm-water spawning areas. The currents sweeping the southeastern coast of Africa extend to Cape Agulhas, where they diverge seaward and are met by colder currents from the South Atlantic. This marks the farthest point that leptocephali can regularly be deposited on the coast and coincides with the limits of distribution of adult eels. Eels appear to be absent from the Red Sea and the Persian Gulf. This is probably due in part to the arid conditions, but both bodies of water are nearly cut off from the Indian Ocean, and it would be difficult for leptocephali to enter either.

Anguillid eels are found across India and throughout the Indonesian region, but their distribution is directly related to the nearness of deep water. They occur in northern Burma and the Andaman Islands but not along the eastern shore of the Andaman Sea. Eels are present on the southern coasts of Sumatra and Java, which face the open Indian Ocean, but not the northern coasts, which face shallow, continental seas. Eels occur throughout Celebes, which is surrounded by deep water, and eastern Borneo, which fronts the deep Makassar Strait. They are absent from the south and west coasts of Borneo, however, and from the entire area bordering the Gulf of Thailand, including Vietnam, Cambodia, Thailand, and Malaysia. There are eels in northwestern and northeastern Australia, but not in the areas bordering the Timor and Arafura Seas or the Gulf of Carpentaria. All of these eels live only in places where they have direct access to deep-water spawning areas. Unlike the temperate species, these tropical ones do not travel great distances to spawn and apparently cannot cross even modest stretches of shallow water.

In the North Pacific, eels are found in the Philippines, China, Korea, and Japan. In the South Pacific, eels occur along the northern coast of

New Guinea, the entire east coast of Australia, and in New Zealand. Eels inhabit many of the islands of the western and central Pacific as far as Pitcairn and the Marquesas. They are absent from Hawaii and from the west coast of the Americas.

Anguillid eels seem to be basically tropical animals. Several species have managed to colonize temperate regions, but they all need warm salt water to reproduce. For tropical species this presents no problem, but the temperate species must return to a suitable area to breed. The key factors seem to be a combination of temperature and depth along with a current system capable of returning the leptocephali to the coast. The known breeding areas of anguillid eels are characterized by the presence of warm water at relatively great depths. In the Sargasso Sea, temperatures at the 400-m level are above 15°C (Schmidt, 1929: 67–86), a condition not met with elsewhere in the Atlantic, even the equatorial regions. The Sargasso Sea is also the center of the North Atlantic gyre, and the Gulf Stream provides a transport system.

An analogous situation exists in the western North Pacific south of Japan, and here the Japanese eel spawns. The Kuroshio serves the same transport function for leptocephali of this species as the Gulf Stream does for the American and European species. Another area of high temperature at intermediate depth lies to the east and north of Australia and New Zealand. The currents here flow south and west, providing a means of transporting leptocephali back to the coasts. The distribution of eels in Australia is exactly what would be expected if they are using this spawning area: they are common along the east coast but essentially absent west of Bass Strait.

The lack of suitable spawning areas in the South Atlantic and the eastern Pacific undoubtedly explains the absence of eels there.

Genera. A single genus, *Anguilla* Schrank, 1798.

Genus *Anguilla* Schrank, 1798

Anguilla Schrank, 1798: 304 (type species *Muraena anguilla* Linnaeus, 1758, by original designation).

Terpolepis McClelland, 1844: 225 (type species *Anguilla brevirostris* McClelland, 1844, by subsequent designation of Blache et al., 1973).

Tribranchus Müller, 1844: 193 (type species *Tribranchus anguillaris* Müller, 1844, by monotypy).

Discussion of Synonymy. McClelland proposed *Terpolepis* provisionally as a subgenus "should it be thought necessary to separate the Bengal species of the genus from *Anguilla*, Cuv. on account of their naked tessellated scales. . . ." He was obviously using "species" in a plural sense and did not designate a type from among the several species he described. Blache et al. (1973: 220) stated "type: *Anguilla brevirostris* McClelland, 1844, by orig. design." Although the attribution to McClelland was wrong, the statement itself constitutes a type designation.

Characters. Those of the family.

Size. Some species may reach a maximum size of 2 m. Females grow much larger than males.

Distribution. That of the family.

Etymology. Latin *anguilla* (eel). Feminine.

Species. The extreme variability of *Anguilla* over its life history has resulted in a proliferation of nominal species. Thirty-six names have been applied to the European eel alone. In the first comprehensive survey of the genus, Kaup (1856a: 54; 1856b: 32) recognized 45 species; these were reduced to 23 by Günther (1870). Ege (1939), in his definitive revision, recognized 16 species. One of these has since been synonymized (Castle and Williamson, 1974), and the count now stands at 15. Of these, three species are divided into two subspecies each.

Ege divided the species of *Anguilla* into four groups, based on color, dentition, and fin position.

The first group has a variegated or mottled color pattern and broad, undivided maxillary and mandibular tooth bands. It contains three species. *Anguilla celebesensis* Kaup, 1856 occurs in Sumatra, Java, Timor, the Philippines, Celebes, some of the smaller islands of eastern Indonesia, and western New Guinea. *Anguilla interioris* Whitley, 1938 is known only from western New Guinea, and *A. megastoma* Kaup, 1856 occurs from the Solomon Islands and New Caledonia eastward to Pitcairn. The ranges of these three species are nearly if not entirely disjunct, and they appear to be derivatives of a single ancestral species. *Anguilla megastoma* shows a slight differentiation in total vertebrae between populations in Tahiti and Fiji.

The second group is also variegated, but the maxillary teeth are generally bi- or triserial with the inner row of teeth small and separated from the outer by a narrow groove. This group includes three species, one of which is composed of two subspecies. *Anguilla nebulosa labiata* Peters, 1852 is found in eastern Africa from Kenya to South Africa, while *A. nebulosa nebulosa* McClelland, 1844 occurs in India and Ceylon, Burma, the Andaman Islands, and northern Sumatra. *Anguilla marmorata* Quoy and Gaimard, 1824 has the widest distribution of any species of the genus. It occurs in Madagascar and neighboring islands, Indonesia, the Philippines, Taiwan, and eastward through the islands of the western Pacific to the Marquesas. In addition there is a single record from South Africa, the only one from the African mainland. As might be expected, some differentiation has occurred over this vast range. Ege discerned three primary races: one in the Mariana and Caroline Islands, one in the South Pacific from New Caledonia to the Marquesas, and one from Madagascar to New Guinea. Within the latter two areas, further geographic differentiation has occurred. *Anguilla reinhardtii* Steindachner, 1867 inhabits all of eastern Australia from Cape York to Victoria, as well as New Caledonia and Lord Howe Island; it apparently does not occur in Tasmania. Ege described another species in this group, *A. ancestralis,* based on a single collection of elvers from Celebes. Although he placed it in a group defined by color and dentitional characters, he gave no information on the color and dentition of *A. ancestralis.* Castle and Williamson (1974) concluded that *A. ancestralis* is based on juveniles of *A. celebesensis.*

The third group is characterized by a plain color pattern and a dorsal fin that begins at a point about one-half to two-thirds the distance from the gill opening to the anus. This group includes the two Atlantic species as well as four from the Indo-Pacific. *Anguilla rostrata* (Lesueur, 1817) occurs in eastern North America and adjacent regions from Greenland to the West Indies. *Anguilla anguilla* (Linnaeus, 1758) is found in Europe and North Africa from Norway to Morocco and the Canary Islands, and through the Mediterranean basin to the Black Sea. *Anguilla mossambica* Peters, 1852 occurs throughout eastern Africa from Kenya to South Africa and in Madagascar and the Mascarenes. *Anguilla borneensis* Popta, 1924 is known only from eastern Borneo. *Anguilla japonica* Temminck and Schlegel, 1846 is found from northern Vietnam and the northern Philippines to Korea and Japan. *Anguilla dieffenbachii* Gray, 1842 occurs only in New Zealand and the neighboring Auckland and Chatham Islands.

The fourth group is also plain colored, but the dorsal fin begins over or very slightly before the anus. It contains three species, two of which are divided into subspecies. *Anguilla bicolor bicolor* McClelland, 1844 occurs in eastern Africa, Madagascar and adjacent islands, India, Burma, Sumatra, Java, Timor, and northwestern Australia. *Anguilla bicolor pacifica* Schmidt, 1928 occurs in eastern Indonesia and New Guinea. *Anguilla obscura* Günther, 1871 is found from New Guinea to the Society Islands, and there is some differentiation between eastern and western populations. *Anguilla australis australis* Richardson, 1841 is found in eastern Australia from southern Queensland to Victoria, Tasmania, and Lord Howe Island. *Anguilla australis schmidtii* Phillipps, 1925 occurs in New Zealand, the Auckland and Chatham Islands, New Caledonia, and Norfolk Island. In addition there are single records of *A. australis schmidtii* from Fiji and Tahiti, but Ege felt that these were questionable.

The species of *Anguilla* can also be divided ecologically into tropical and temperate species. The temperate species include the northern hemisphere *A. anguilla, A. rostrata,* and *A. japonica.* Two southern hemisphere species, *A. dieffenbachii* and *A. australis* are also clearly temperate. *Anguilla reinhardtii* should probably also be considered a temperate species, although it ranges well into the tropics in northern Australia. *An-*

guilla mossambica may fall into this category as well, as it is common in South Africa.

The most significant difference between tropical and temperate eels is the extent of their migrations and their choice of spawning areas. Tropical species tend to spawn relatively close to home. Often there is little more than an offshore-inshore movement as in *Elops*. In the Indo-Malayan region, for instance, a spawning area is located over the deep water just off the west coast of Sumatra (Jespersen, 1942: 101). The eels travel no farther than this and in fact are unable to colonize the opposite side of Sumatra. There are probably numerous spawning areas throughout the tropical Indo-West Pacific, each used by eels inhabiting nearby land areas. Because of this, several of the widely distributed species have differentiated into geographic races. Although data are still not widely available, tropical eels seem to have extended spawning periods, possibly year-round.

Temperate eels, on the other hand, are the true wanderers. Unable to spawn locally, they must travel in search of suitable breeding grounds. These breeding grounds, as we have seen, are limited by factors of depth, temperature, and current pattern to certain areas in the western, subtropical portions of the major ocean basins. As a result, temperate species have developed much greater powers of migration than their tropical cousins. This is evident at all stages of the life cycle. Not only do the adults journey hundreds of kilometers to reach the spawning area, but the elvers can cross large areas of shallow coastal waters to colonize areas that may be far from the deep ocean. Because individuals from different parts of the range mix more or less completely in the spawning area, temperate species do not show geographic differentiation; each species constitutes a single breeding population. Finally, the temperate eels are distinctly seasonal in their spawning habits. Adults go to sea in the fall, and the elvers appear in the spring.

Whether the temperate-vs-tropical division has any phylogenetic significance is uncertain, although four of the five clearly temperate species belong to Ege's plain-colored, long-finned group.

Although leptocephali of both *Anguilla rostrata* and *A. anguilla* occur in the western North Atlantic, *A. rostrata* is the only one that metamorphoses and lives here as an adult.

Anguilla rostrata (Lesueur, 1817)

AMERICAN EEL

Figures 14–18

Muraena rostrata Lesueur, 1817: 81 (original description, Cayuga Lake, New York, holotype unknown).

Muraena bostoniensis Lesueur, 1817: 81 (original description, Boston, holotype unknown).

Muraena serpentina Lesueur, 1817: 81 (original description, Newport, Rhode Island, holotype unknown).

Muraena argentea Lesueur, 1817: 82 (original description, Boston, holotype unknown).

Muraena macrocephala Lesueur, 1817: 82 (original description, Saratoga, New York, holotype unknown).

Anguilla chrysypa Rafinesque, 1817: 120 (original description, Hudson River, Lake George, and Lake Champlain, New York, holotype unknown). Originally spelled *chrisypa,* emended by Jordan and Davis, 1891: 668.

Anguilla blephura Rafinesque, 1817: 120 (original description, Long Island, New York, holotype unknown).

Anguilla laticauda Rafinesque, 1818: 447 (original description, Ohio River, holotype unknown). Originally spelled *laticanda,* see discussion of synonymy below.

Anguilla aterrima Rafinesque, 1820: 78 (original description, Tennessee and Cumberland Rivers, holotype unknown).

Anguilla xanthomelas Rafinesque, 1820: 78 (original description, Ohio River, holotype unknown).

Anguilla lutea Rafinesque, 1820: 78 (original description, Ohio River, holotype unknown).

Anguilla tenuirostris De Kay, 1842: 310, pl. 53, fig. 173 (original description, New York, holotype unknown).

Anguilla novaeorleanensis Kaup, 1856a: 55 (original description, New Orleans, holotype unknown); 1856b: 43, fig. 33.

Anguilla punctatissima Kaup, 1856a: 55 (original description, Niagara River, holotype unknown); 1856b: 44.

Anguilla cubana Kaup, 1856a: 55 (original description, Cuba, holotype unknown); 1856b: 44.

Anguilla novaeterrae Kaup, 1856a: 55 (original description, Newfoundland, holotype unknown); 1856b: 45, fig. 35.

Anguilla texana Kaup, 1856a: 56 (original description, Texas, holotype unknown); 1856b: 45, fig. 36.

Anguilla wabashensis Kaup, 1856a: 56 (original description, Wabash River, holotype unknown); 1856b: 46.

Anguilla tyrannus Girard, 1859: 75 (original description, Rio Grande, holotype unknown).

Anguilla vulgaris rostrata. Meek, 1883: 430.
Leptocephalus grassii Eigenmann and Kennedy, 1902: 84, fig. 1 (original description, 38°25′N, 72°40′W, two syntypes USNM 49751 and 49752). Leptocephalus.

Discussion of Synonymy. The name used for this species has shifted back and forth over the years. Lesueur and Rafinesque between them proposed seven names in 1817. The first authors who selected one of these names over the others, and who treated the American eel as a single species, apparently were Jordan and Gilbert (1882a: 361). They used *rostrata,* implicitly recognizing Lesueur's priority and probably using line priority within the latter's publication. Jordan and Davis (1891: 668), however, gave Rafinesque priority and recognized his name *chrysypa,* again presumably based on line priority. Bean (1909: 871) and Jordan (1917a: 86) later decided that Lesueur had priority and recognized *rostrata,* again based on line priority.

The criterion accepted by the ICZN, however, is not page and line priority but rather the decision of the first reviser. Jordan (1929: 56) recognized this and, citing Günther (1870: 31) as the first reviser, accepted *bostoniensis.* Günther had chosen *bostoniensis* and clearly treated *rostrata* as a synonym. American workers largely followed Jordan in using *bostoniensis,* but others, including Schmidt and his Danish colleagues, continued using *rostrata.* Confusion reigned once again.

Günther, however, did not include all of Lesueur's names in his (Günther's) synonymy of *bostoniensis.* He specifically excluded *serpentina* and listed it separately (1870: 23) among several others as a valid or possibly valid species. By this action he eliminated himself from consideration as a reviser. Günther's treatment of *Anguilla* was mainly a compilation of previous work, and he clearly did not intend it to be a definitive revision. He realized how poorly defined most of the nominal species were and simply did the best he could with the information available. He recorded *bostoniensis* from Asia as well as from North America, confusing it with *A. japonica.* He recognized two other American species as valid (*serpentina* and *texana*) and recorded *A. vulgaris* (= *A. anguilla*) from North America.

Ginsburg (1951: 436) pointed out the weaknesses of Günther's treatment and felt that Ege (1939), who had produced by far the most comprehensive treatment of the genus *Anguilla,* should be considered the first reviser. Ege had used *rostrata.* Whether Ege was the first reviser is open to question. The ICZN simply requires that the reviser be "the first author to have subsequently cited together such names . . . published on the same date . . . and to have chosen one of them to have precedence over the other(s). . ." (Article 24 [b]). Jordan and Gilbert (1882a) did not actually cite all the names, but they clearly were aware of them and chose *rostrata* over the others. By this measure, both Bean (1909) and Jordan (1917a) could also be considered revisers. At any rate, *rostrata* has become universally accepted and can be justified regardless of who is considered the first reviser.

Rafinesque's *Anguilla laticanda* is almost certainly a printer's error for what was intended to be *laticauda.* In his *Ichthyologia Ohiensis* Rafinesque (1820: 77) spelled it *laticauda* and gave it the common name "broadtail eel," which is the correct translation of *laticauda* (*latus* = broad, *cauda* = tail). *Laticanda* has no meaning.

Distinctive Characters. Anguilla rostrata belongs to the group of species with plain color and a long dorsal fin. Within this group it is distinguished from *A. borneensis, A. mossambica,* and *A. dieffenbachii* by its smaller gape (ca 25–27% HL vs ca 32–36). It has fewer vertebrae (103–111) than *A. anguilla* (110–119) and *A. japonica* (112–119). It is the only species of *Anguilla* found in the western Atlantic and is best recognized there by the characters of the genus.

Description (data from Ege, 1939; Boëtius, 1980; plus 17 additional specimens). Total vertebrae 103–112 (n = 1609, $\bar{x}$ = 107.9), precaudal vertebrae 41–45 (99, 42.85). Branchiostegal rays 9–13 (846). Pectoral-fin rays 14–20 (291). Pores: LL to anus 31–33 (5), POM 7–9 (12), IO 4 + 0 + 0 (12), E 1 (12), SO 3 (12), STC 0 (12).

Proportions as % of TL: preanal 39–47 (117), predorsal 32–36 (17), dorso-anal distance (distance between verticals through anus and dorsal-fin origin) 6–12 (100), predorsal without head 18–

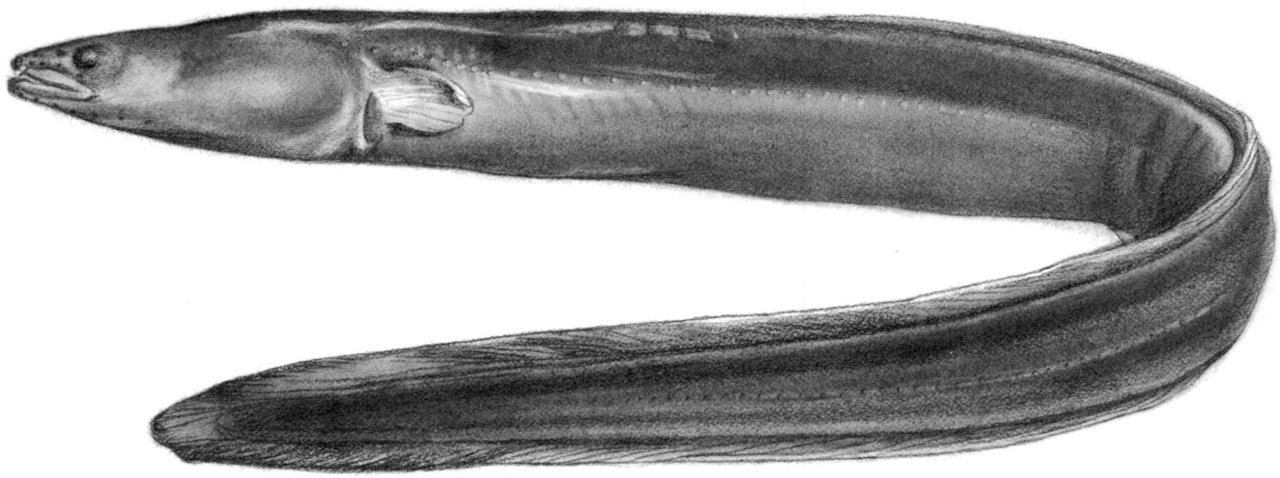

FIGURE 14. *Anguilla rostrata*: ANSP 139680, 278 mm TL (immature).

25 (100), head 11–16 (115), depth at anus 5–7 (13), caudal fin 1–2 (11). Of HL: snout 14–20 (17), eye 9–23 (17), interorbital 13–21 (15), snout-rictus 19–27 (17), gape (tip of lower jaw to rictus) 22–31 (98), gill opening 9–14 (17), interbranchial 15–27 (17), pectoral fin 20–43 (16). Length of intermaxillary-vomerine tooth band as % of intermaxillary-maxillary tooth band: specimens 300–399 mm TL 68.94 (59), specimens 400–499 mm 69.67 (38).

Four supraorbital pores: first (ethmoidal) pore near edge of upper lip before anterior nostril; second just anterodorsal to base of anterior nostril; third just posterodorsal to anterior nostril; fourth posterodorsal to preceding one, slightly closer to anterior nostril than to posterior nostril. First infraorbital pore along upper lip just behind anterior nostril; second between anterior and posterior nostril; third between anterior nostril and eye; fourth beneath eye; no pores behind rictus or behind eye. Six or seven pores on lower jaw before rictus, two pores following these behind rictus. No pores in supratemporal commissure.

Maxillary teeth in narrow bands about 2–5 teeth across, medial teeth somewhat larger than inner and outer ones. Intermaxillary and vomerine tooth patches continuous with each other, forming an elongate band ending slightly but distinctly before posterior end of maxillary teeth. Intermaxillary teeth indistinctly separated from maxillary teeth. Mandibular teeth similar to maxillary.

Color. Immatures brownish or yellowish green, lighter ventrally. Mature individuals gray to black above, white below, sometimes with bronze reflections on flank. Intensity of color can change from dark to pale depending on light conditions and background

Size. Anguilla rostrata can grow to a meter or more in length; Bigelow and Schroeder (1953: 151) stated that it reaches 4 feet (1.2 m) and 16.5 pounds (7.5 kg). Individuals of this size are always females. Males are much smaller, usually 300–350 mm TL. Any eel greater than 400 mm is almost certainly a female.

Development and Growth. Anguilla rostrata spends about a year in the leptocephalus stage and metamorphoses near the edge of the continental shelf. The initial postmetamorphic stage is the so-called glass eel and is passed while the eel is still at sea. At this point, the young eel is without pigment and has a length of about 40–50 mm. Soon afterward, the adult pigmentation appears, and the eel, now called an elver, begins its ascent into fresh water, where it will spend several years feeding and growing. By its second summer in fresh water, the eel has doubled its length to about 100 mm. In this stage it is called a yellow eel, although the color is really more yellowish-green or yellowish-brown. The yellow phase can last anywhere from 4–20 years, during

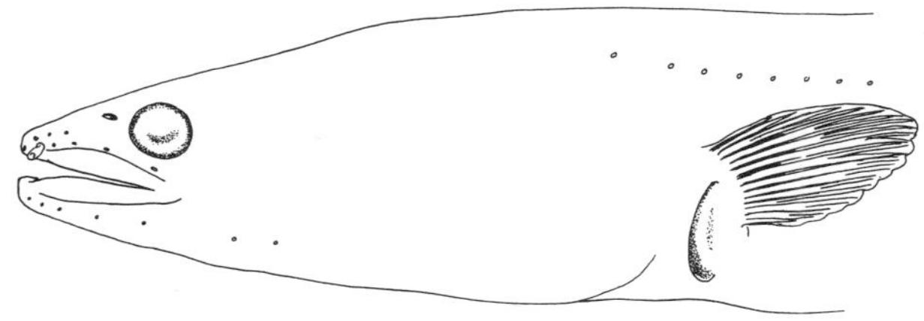

FIGURE 15. *Anguilla rostrata*: ANSP 139680, 278 mm TL (immature); head.

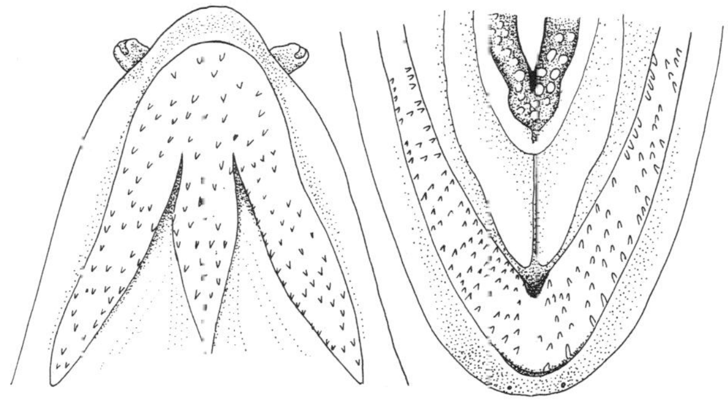

FIGURE 16. *Anguilla rostrata*: ANSP 139681, 290 mm TL (immature); dentition.

which the eel grows to about 300–400 mm in males, to 1 m or more in females.

At the end of this period, the eel stops feeding and begins the irreversible process that will lead to sexual maturation. The changes are so profound that they can be thought of as a second metamorphosis. The most obvious change is in the pigmentation. The yellowish-green is replaced by a sharply bicolored pattern, gray to black above and pure white below. Often a silvery or bronze sheen appears on the flank. The eyes enlarge and the pectoral fin becomes larger and darker. The eel is now called a silver eel, and it begins its migration back to the Sargasso Sea. The external changes are paralleled by equally profound internal and physiological changes as the eel transforms from a freshwater to an oceanic fish.

The modification of the eye involves not only enlargement but an alteration of the visual pigments. The retina of the yellow eel contains two pigments: a porphyropsin with a peak spectral absorption of 523 nm and a rhodopsin with a peak spectral absorption of 501 nm. As the eel transforms to the silver phase and begins its migration to the sea, the porphyropsin is lost and a new rhodopsin appears with a peak spectral absorption of 482 nm (Beatty, 1975). The result is to shift the spectral sensitivity of the eye toward shorter wavelengths. Porphyropsins, with their greater red-sensitivity, are commonly found in freshwater and euryhaline vertebrates. Rhodopsins are most common in marine and terrestrial vertebrates (Wald, 1958). The loss of a porphyropsin and its replacement with a rhodopsin in the eel can be viewed as an adaptation to the oceanic habitat, where blue light is prevalent.

The gas bladder also undergoes modification. First there is an increase in the guanine content of the gas-bladder wall. The guanine layer forms the primary barrier to loss of gas by passive diffusion. The guanine content in the gas bladder of the silver eel is some 47–63% greater than that of the yellow eel, and the gas conductance is correspondingly reduced by 17–27% (Kleckner, 1980b). At the same time, the retial capillaries increase both in length and diameter (Kleckner and Krueger, 1981). This increases the countercurrent exchange efficiency by about 340% over that of yellow eels (Kleckner, 1980b). The combination of reduced permeability and enhanced countercurrent exchange means that the eel can maintain the volume of its gas bladder at higher external pressures and hence greater depths. Kleckner (1980a) found that a 545-mm silver eel could maintain its gas-bladder volume at a depth of 149 m compared to 58 m for a yellow eel of the same size.

The final phases of maturation take place at sea and have not been observed in nature. By treating silver eels with certain gonadotropic hormones, however, it is possible to induce maturation in the laboratory. In this way we can get a glimpse of what must be happening in the Sargasso Sea. In the males (Fig. 17), the most prominent change is the great increase in the size of the eye, which can reach 20–25% HL and exceed the snout length. The snout becomes shorter and blunter and almost squared off in dorsal profile, and the flange on the upper lip becomes less prominent. The pectoral fin in-

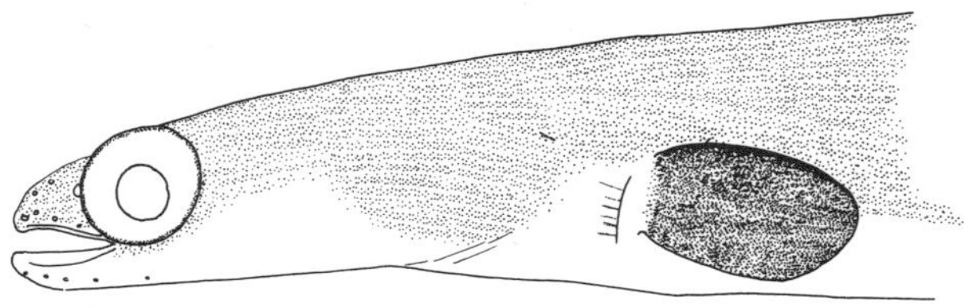

FIGURE 17. *Anguilla rostrata*: head of mature male (hormone treated).

creases to 35–40% HL or more and becomes very dark. The teeth are reduced and eventually lost. The males at this stage are attractive little fishes, almost black above and pure white below with metallic bronze reflections on the flank. They remain alert and active in the aquarium.

The females show similar changes but less extreme. The eye diameter is about 15% HL and remains less than the snout length. The pectoral fin is about the same relative size as the male's and is also dark. As in the male the flange on the upper lip is reduced and the teeth are eventually lost. The most striking changes in the female occur as she approaches spawning condition. As the eggs mature and absorb water, the abdomen swells up greatly (Fig. 18). At the same time, the muscles lose their firmness, the bones soften, and the alimentary canal is reduced to an inconspicuous strand. At the end, the ovaries occupy almost the entire body cavity and can constitute more than half the total weight of the fish. The female is then in a greatly debilitated condition, having converted almost all her resources into the production of eggs. Indeed, she is scarcely more than a sack of skin and bones filled with eggs. The mature eggs are about a millimeter in diameter, somewhat smaller than the eggs of most anguilliforms, and contain one to several oil droplets. A large female can produce several million eggs.

Habits and Behavior. Yellow eels live in a wide variety of habitats from salt marshes and estuaries to rivers and lakes. One of the remarkable qualities of the species is its great adaptability. Eels inhabit large rivers and small streams, large lakes and small ponds, warm water and cold, still or running, clear or muddy. They can tolerate almost any salinity from fresh water to salt and climates from semitropical to subarctic. They are both predators and scavengers and will eat almost anything they can swallow. Small fishes, insects, crustaceans, worms, molluscs, and carrion are all part of the eel's diet. The eels in turn serve as food for a variety of larger predators, although a large eel probably has few enemies.

Eels tend to be nocturnal in habit, spending the daylight hours buried in the bottom sediment and coming out at night to forage. They are strongly thigmotactic and like to rest with the body in contact with the substratum or the walls of crevices and holes. In temperate regions, eels become inactive during the winter and may spend the cold months in a state of virtual hibernation. It is often said that eels can travel overland. This may be an exaggeration, but they can survive long periods out of water—24 hours or more if kept cool and moist.

The beginning of the eel's life is shrouded in mystery. Neither spawning adults nor eggs have ever been collected at sea. Based on the distribution of the smallest leptocephali, we know that the American eel spawns in an area south and southwest of Bermuda bounded approximately by 20°N and 30°N and 60°W and 75°W. The eastern part of this area overlaps the spawning ground of European eel, and there is a substantial area in which leptocephali of both species are found. A line drawn from northwest to southeast passing through 25°N and 63°W will bisect the common spawning area. Northeast of this line European leptocephali predominate, southwest of it American leptocephali predominate.

The depth at which spawning takes place is still unknown. Schmidt (1922: 206) said that the smallest leptocephali (7–15 mm, presumably *A. anguilla*) "float in water-layers about 200–300 meters from the surface. . . ." On the ANTON DOHRN in 1979, the majority of 0-class leptocephali caught in closing nets came from the 50–100-m level at night, somewhat deeper during the day (Schoth and Tesch, 1982: 311). None of these leptocephali were newly hatched, however. The American eel begins spawning in Feb-

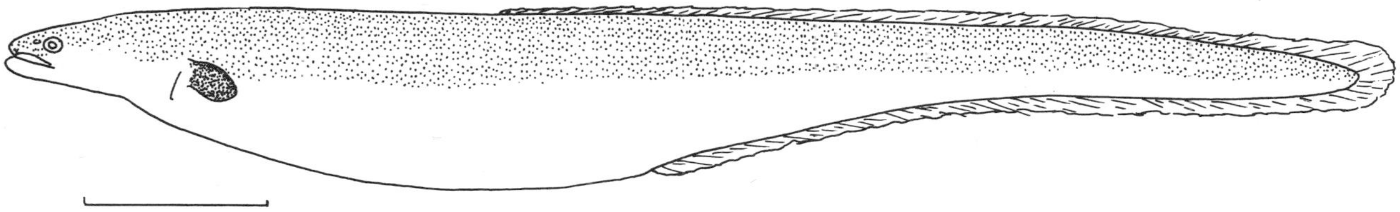

FIGURE 18. *Anguilla rostrata*: mature female (hormone treated). Scale = 100 mm.

ruary (Schmidt, 1922: 203; Kleckner and McCleave, 1985: 72, 81; Wippelhauser et al., 1985: 97) and continues with decreasing frequency through the spring. The spawning season of the European eel is a little later, from March into the summer, with a peak probably in April (Schmidt, 1922: 199, 206; Boëtius and Harding, 1985: 149).

The smallest leptocephali yet known are about 5 mm for *A. anguilla* (Schmidt, 1935: 4) and 4 mm for *A. rostrata* (Kleckner et al., 1983: 290). About the only way to estimate their age is to compare them to the leptocephali of *Anguilla japonica* which have been reared in the laboratory (Yamamoto and Yamauchi, 1974; Yamamoto et al., 1976). The youngest leptocephalus of *A. anguilla* illustrated by Schmidt (1935: fig 6) seems comparable in development to Japanese leptocephali 6 days after hatching: approximately 6 mm in length, short head, downwardly directed mouth, tooth anlagen present, remains of the yolk sac, and no retinal pigment. By 14 days, the Japanese leptocephali measured 7 mm, had lost the yolk sac, and had black pigment in the eye; the mouth was horizontal, and four pairs of teeth were present in each jaw. At this point, they resembled the 7-mm leptocephalus illustrated by Schmidt. Admittedly, this comparison is based on a different species raised under artificial conditions, but the three north-temperate *Anguilla* species are so similar that they probably do not differ greatly in their development. By this criterion, then, the youngest leptocephali collected in the Sargasso Sea are about 1 to 2 weeks old.

Kleckner et al. (1983) presented evidence that the spawning area of *Anguilla rostrata* is bounded on the north by a thermal front. On the west, spawning may occur up to the Bahamas, but the southern boundary is uncertain. The presence of leptocephali as small as 10 mm near the island of Cozumel off the Yucatan Peninsula raises the possibility that some spawning may take place in the Caribbean (see p. 898). It is also possible, however, that these leptocephali may have been carried there from a spawning area just north of the Antilles.

As the leptocephali grow, they begin their dispersal toward the coast. The direction of transport is predominantly to the northwest, and the bulk of the leptocephali seem to enter the Gulf Stream north of the Bahamas (Kleckner and McCleave, 1982; Power and McCleave, 1983). The presence of some leptocephali in the Caribbean, however, as well as the existence of adults in the West Indies, indicates that some transport to the south must occur. Growth continues until September or October when the leptocephali average about 40–60 mm TL. Although unmetamorphosed leptocephali are occasionally collected through the winter and into the following spring, the average size does not increase, suggesting that growth largely ceases after October. At about 55–65 mm TL, the leptocephali are ready to metamorphose (Kleckner and McCleave, 1985: 81–82).

The onset of metamorphosis is controlled by factors that are not fully understood. It seems to be triggered by proximity to land, but this occurs only after a certain stage of growth is reached. The swarms of European leptocephali that pass Bermuda every year never settle out there, presumably because they are still too young. Only the American leptocephali remain that far west until they are ready to metamorphose. It seems equally true that metamorphosis can be delayed if proper conditions are not encountered. The majority of American eel leptocephali begin metamorphosing off the North American continent in the fall of the year in which they were spawned. Kleckner and McCleave (1985: 72, 82) found metamorphosing leptocephali from October to March and glass eels from January to June.

At this point the young eels cease to be waifs drifting passively with the currents (if, indeed, they are ever totally passive) and take active control of their destiny. They make their way shoreward, probably by a combination of active swimming and selective use of tidal currents, and the following spring begin their ascent into fresh water. Elver runs begin as early as January in Florida, as late as June and July in Canada. The average size of elvers increases from south to north, reflecting the difference in age (Vladykov, 1966: fig. 2; Smith, 1968: 290).

Glass eels enter the estuaries initially by rising into the water column during flood tide and retreating to the bottom during the ebb. This process takes about 2 weeks and brings the eels to the upper limits of tidal influence. During this period they become fully pigmented and probably complete their acclimation to fresh water (Sheldon, 1974: 11). They are now elvers, and they exchange a passive for an active mode of travel. Swimming directly into the current, they head upstream, generally keeping close to the shoreline. The strength and tenacity of migrating elvers are legendary. They "... clamber over falls, dams, and other obstructions, even working their way up over damp rocks ..." (Bigelow and Schroeder, 1953: 152), often in huge numbers.

As powerful as the upstream urge may be, it does not seen to be universal. A certain percentage of elvers will remain behind in the estuaries and coastal waters. The eels thus colonize nearly the entire range of waters available to them. How far the elvers travel before they stop and settle down probably varies greatly. Eels are found hundreds of kilometers upstream, but it is not known how long it takes them to reach there. It has been suggested that the extent of inland migration depends on sex, with females being found mainly in fresh water and males mainly in brackish (Schmidt, 1922: 207; Bigelow and Schroeder, 1953: 152; Bertin, 1956; Vladykov, 1966: 1012). So many exceptions to this rule are found, however, that its validity must be questioned. Vladykov (1966: 1013) further suggested that females are more common in northern latitudes and males more common in southern. Again, observation does not bear this out.

Whether they stay near the coast or wander far upstream, eels spend several years in the yellow stage feeding and growing. Relatively little information is available on the movement of yellow eels. Gunning and Shoop (1962) found that marked and recaptured eels in two Louisiana localities remained within some 200–450 linear feet of stream. On the other hand, the presence of eels far up the St. Lawrence, Mississippi, and other rivers indicates that some of them continue to travel great distances during the yellow stage.

After several years in the yellow stage, the eel undergoes a second metamorphosis to the silver stage. The factors that determine when a particular eel metamorphoses are uncertain. It cannot be simply age or size, because silver eels come in a wide variety of both. I have seen females as small as 472 mm and as large as 916 mm. Gray and Andrews (1970: 485) reported ages of 9–17 years, based on otoliths, in females from Newfoundland. Silver males are generally between 300 and 350 mm. Whether males mature earlier than females or whether they simply grow more slowly is uncertain. In any event, the metamorphosis occurs in the late summer or fall.

Perhaps the most significant change that occurs at the yellow-silver transition involves behavior. Yellow eels are more or less neutral in response to currents, but silver eels develop a strong negative rheotaxis in fresh water (Hain, 1975: 215). Their migratory instinct is once more aroused, but now their travels take them downstream. At the same time, they show a positive rheotaxis in salt water. This combination guides their movements unerringly though the estuary and toward the sea. The migration of silver eels takes place at night and primarily during the dark of the moon. Hain (1975: 222) found that activity was greatest within the first 2 hours of darkness and completed before midnight. Rainfall also seems to stimulate movement of silver eels. Silver-eel runs may begin as early as August in northern latitudes and as late as December farther south. The peak period seems to be September–November.

Once the silver eels enter the sea, they essentially disappear. A few individuals turn up in trawl catches over the continental shelf (Wenner, 1973), but after they reach the open sea they are not seen again. We know that they arrive in the spawning area by February, but the route they take and how they navigate remain mysteries. Based on their external appearance and the condition of their eyes and gas bladder, they would seem to be adapted to traveling at moderate depths, perhaps 100–200 m. They do not look like abyssal fishes, but one was recently photographed on the bottom of the Tongue of the Ocean in the Bahamas at a depth of 2000 m (Robins et al., 1979).

The final phases of maturation probably take place after the eels have reached the spawning ground. The reserves of fat and muscle are no longer needed for migration and in the female are converted rather rapidly to egg production. Based on observations of hormone-treated eels in the laboratory, males remain more active than females and probably take the initiative in locating mates. A simple form of courtship behavior has been observed in such laboratory eels (Boëtius and Boëtius, 1980 for *Anguilla anguilla*, and my own observations of *A. rostrata*), but the spawning act was not completed. It is assumed that eels die after spawning, although in the laboratory they may survive for weeks or months after maturity has been induced. Males have even been brought to maturity more than once, with feeding resumed between maturations (Dollerup and Graver, 1985). No return migration of silver eels from the sea has ever been seen, however, and the debilitated condition of ripened females argues against survival under natural conditions.

From the point of view of the species, it is immaterial whether the spawned-out adults live or not. They have produced countless eggs and larvae, and these will eventually return and repopulate the continents.

Relationship to Man. The American eel has never achieved the popularity of its European and Japanese cousins as a food item, but a modest fishery has existed since colonial times, centered mainly in the northeastern states. The total catch in 1975 was 1285 tons, of which 900 tons came from the United States and 384 tons from Canada (Wenner, 1977). From 1979 to 1982 the total catch varied from 1081 to 2430 metric tons (FAO, 1984: 125). Yellow eels are taken by baited traps and trotlines. Silver eels are caught during their seaward migration by stationary nets in estuarine areas. A significant portion of the catch is exported to Europe and the Far East. In recent years the fishery has been extended farther south. Silver eels have been caught in Florida and Louisiana, primarily for export to Europe just before Christmas.

For a time in the 1970's there was a growing market for elvers to be shipped to Japan, where they were used in eel culture. Demand has now fallen off, however, because the American elvers apparently do not do as well in the Japanese culture systems as the native species.

The eel has been widely used as an experimental animal. Some of the first studies on the mechanisms of salt and water metabolism in marine fishes were done on *Anguilla rostrata* (H. W. Smith, 1930).

Variation. Like all anguillid eels, *Anguilla rostrata* shows an enormous amount of sexual and ontogenetic variation. Size, color, eye diameter, pectoral-fin length, snout length, and head shape are among the characters that vary over the life cycle of a single individual.

Williams, Koehn, and Mitton (1973) found geographic variation in several enzyme loci, but they interpreted it as due to selection within a single generation. They were unable to demonstrate variation due to reproductive isolation of local populations and concluded that *Anguilla rostrata* consists of a single panmictic population.

The Systematic Status of Anguilla rostrata. The relationship between the American and European eel has been a point of contention for many years. Up to the time of Kaup (1856a,b) numerous species were recognized on both sides of the Atlantic. Günther (1870) reduced the number considerably, but his treatment was too confused to be of much use. Meek (1883) considered the two forms merely varieties of a single species. Jordan and Davis (1891: 668) considered them separate species, without further comment. Jordan and Evermann (1896: 347) also treated them as separate species, based on a few proportional measurements, but added in a footnote that "probably our eel should be regarded as a subspecies . . . of *A. anguilla.*" Schmidt (1913: 13, 27) demonstrated a nearly complete separation in the number of vertebrae between American and European eels (103–111 and 110–119, respectively) and concluded that they represented distinct species. Ege (1939: 100, 122, 130, 137) added minor differences in precaudal vertebrae (41–45 vs 44–47 in the American and European eels, respectively), pectoral-fin rays (14–20, mean 16.73 vs 15–21, mean 17.37), dorso-anal distance (mean 9.1 vs 11.2% TL), and the ratio between the length of the intermaxillary-vomerine tooth band and the intermaxillary-maxillary tooth band (mean 68.94 vs 76.16% in females 300–399 mm TL).

The two-species model was accepted without serious argument until 1959, when Denys Tucker of the British Museum proposed a radical new hypothesis (Tucker, 1959a). Puzzled over the problems faced by eels migrating from Europe to the Sargasso Sea, Tucker suggested that the American and European eels in fact represented a single species. He proposed that the difference in vertebral numbers was an eco-phenotypic effect caused by the temperature in different parts of the spawning area. Further, he maintained that all the eels, American and European, were offspring of American parents; the European eels never succeeded in returning to the spawning area at all and represented a dead-end population.

He built his case on a series of observations and assumptions. First, he noted that the European eel upon its descent to the sea seemed more mature than the American eel at the same point. He contrasted the color of the European silver eel—black and white with metallic reflections—with that of the American form, which was ". . . little more than a darkening of the normal colour. . . ." The eye and the pectoral fin of the European eel were enlarged, whereas those of the American eel were not. The European silver eel had a more degenerate gut; indeed, in one specimen from the North Sea the anus was occluded. To Tucker, this was a paradox: the European eel was more advanced than the American eel at the start of its journey, yet it had much farther to go. It was, said Tucker, ". . . a fish degenerate and unable to feed, having a fat reserve which must provide for the maturation and vast enlargement of the gonads as well as for the propulsion of the enfeebled body on a presumed 3500 mile journey."

The rest of his argument was based on oceanography. The current systems in the eastern North Atlantic are very complicated and lead to a great variety of conditions to be overcome by the migrating eel. If one presumes that the eel reacts to gradients in temperature, salinity, light, and current flow, the variation in these conditions over its range would require different reaction patterns from different starting points. An eel departing from the Mediterranean, for instance, would face vastly different conditions from one leaving the Baltic. Since all the European eels come from the same spawning area, there could be no opportunity to evolve local races adapted to navigating a particular set of conditions.

Furthermore, he noted, no adult eel had ever been caught in the open seas around Europe despite the intensive trawl fisheries there. In particular, he cited the Strait of Gibraltar; all the eels from the Mediterranean must pass through this bottleneck on their way to the Sargasso Sea, yet none had ever been caught. American eels had never been caught at sea either, but Tucker thought this was simply because their migration routes did not coincide with major trawl fisheries.

When he looked at the American eel, Tucker saw a model of simplicity. It was less advanced at the start of its migration, and hence had plenty of reserves to complete the trip. Besides, it had less distance to travel. The currents off the east coast of North America were much more uniform than those off Europe and the temperature and salinity gradients were more consistent. He also mentioned the recent discovery of a deep countercurrent under the Gulf Stream.

How, then, could one explain the consistent difference in vertebral counts between the American and European eels? Tucker said it was a temperature effect and cited experiments showing that the number of vertebrae in young trout were affected by the temperature at which they were raised. He examined hydrographic charts of the Atlantic and took an oblique section through the spawning area running from about 32°N to 20°N and between 65°W and 70°W. The temperatures at the 400–700 m level, where spawning was postulated to take place, were fairly constant over the entire transect. Above that depth, however, the northern and southern ends of the section differed markedly. In the southern part, where the American eel originates, the temperature rose steeply from about 12–14°C at depth to nearly 25°C at the surface. There was a particularly sharp increase of 4° between 200 and 100 m. In the northern part, however, where European leptocephali are found, the temperature rose more gradually from about 14° to 20°. Tucker proposed that the eels spawned at about 400–700 m over the entire area and that the eggs slowly rose to the surface. Those in the southern part would be subjected to a sharp rise in temperature as they ascended, which would end the segmentation of somites at the 103–110 interval. Eggs in the northern part would experience only a gradual rise in temperature, and consequently segmentation would continue to what may be a genetically determined limit of 111–119. The currents are such that southern leptocephali would be carried to North America and northern ones to Europe.

Tucker's hypothesis was original and ingenious, and it stimulated a great deal of interest, but few of his arguments hold up under close examination. Objections were raised immediately and the idea was debated in a series of letters in *Nature* (D'Ancona, 1959; Jones, 1959; Tucker, 1959b,c, 1960; Deelder, 1960) and in a more complete critique by Bruun (1963).

It is now clear that one of Tucker's first assumptions was wrong. The European eel is no more advanced at the outset of its migration than is the American eel, nor is it as degenerate as Tucker believed. Silver eels in captivity are healthy and active and can remain so for months without feeding (Deelder, 1960; Boëtius and Boëtius, 1985). Boëtius and Boëtius (1980) determined that the energy reserves available to a silver eel in the protein and lipid content of its body are quite sufficient to carry it back to the Sargasso Sea and to achieve full sexual development as well.

Tucker's argument about the difficulty of European eels finding their way back to the spawning ground is based largely on lack of evidence. We still do not know how eels navigate in the open sea or to what physico-chemical factors they are responding. In light of such ignorance, it is premature to dismiss entirely the possibility of the eel's overcoming complicated hydrographic conditions. At the same time, Tucker seemed to oversimplify the migratory task of the American eel by proposing that it simply rode the southward flowing countercurrent under the Gulf Stream. Bruun (1963: 143) pointed out that the countercurrent is a cold, abyssal current; to use it, the eel would have to reverse its temperature responses, heading first for cold water and then for warmer water.

Tucker made a major point of the fact that eels have not been caught in the open Atlantic and the Strait of Gibraltar. He asked: "Are agility and this or that depth of swimming adequate explanations?" The answer to this question is almost certainly yes, plus the limitations of the fishing gear used. Deelder (1960) cited his own experience in the IJsselmeer, a large embayment on the Dutch coast. Although many eels were caught there in stationary fyke nets, a newly designed trawl failed to catch any. If silver eels cannot be captured with advanced gear in such a favorable location, said Deelder, what chance is there to take them in the open ocean? As for Gibraltar, Bruun (1963: 146) pointed out that the Danish collections there had been made with plankton nets designed to catch leptocephali. No one on board ever expected to take adult eels with such equipment. On the ANTON DOHRN in 1979 a large midwater trawl was used in an attempt to catch adult eels in the spawning area. The catches were characterized not only by the absence of *Anguilla* but by the conspicuous absence of any agile, fast-swimming fishes. Even a net of this size captured only weak-bodied, feeble-swimming animals and those whose teeth or spines entangled in the mesh. Once again, Tucker seems to have underestimated the eel.

If the European eels do not return to breed, they represent an unprecedented waste of gametes on the part of the American stock. Jones (1959) brought up the obvious point that such wastefulness should be strongly selected against. Tucker answered this argument by saying that conditions at depth in the spawning area are so uniform that the eels probably could not distinguish one part from another. He went on to cite examples in other species of expatriate populations maintained by recruitment from outside. Expatriate populations certainly exist, but none on such a scale as the eel. The European eel is an enormously abundant animal. Countless millions of them exist all over the continent. It strains credulity to suppose that they represent an expendable surplus. Again, Tucker placed an arbitrary constraint on the eel by assuming that it cannot differentiate regions within the spawning area.

The heart of Tucker's case, however, involved the differences between the American and European eels and how these differences were determined. Tucker concerned himself with only one character, the number of vertebrae, and pointed to evidence that such meristic features can be affected by temperature during early development. He referred to experiments by Tåning (1952), who exposed two groups of young *Salmo trutta* from the same parents to different temperatures and obtained mean vertebral numbers of 56.86 and 60.06. This was proportional to the difference between the two eels. Tucker did not explain fully, however, how those results were obtained. Tåning had reared the low-count group at 2.4°C until the sensitive period, subjected it to a temperature shock of 16°C for 1–2 days, and then returned it to 2.4°C. The high-count group was reared at 12°C, subjected to 2.7°C for 7–8 days, and returned to 12°C. These conditions obviously bear no resemblance to anything encountered in the Sargasso Sea. Furthermore, even if a temperature effect were present, it is difficult to see how the eels could be divided so neatly into two groups. Isotherms slope gradually over the spawning area, and many intermediate forms should be present.

Although Tucker did not acknowledge it, the vertebral number is not the only difference between the two eels. There are also the slight differences in proportional measurements mentioned above. No evidence has ever been presented to show that these characters are affected by temperature. Furthermore, it is now known that certain biochemical and chromosomal differences exist as well (Sick et al., 1967; De Ligny and Pantelouris, 1973; Jamieson and Turner, 1980; Passakas, 1981; Williams and Koehn, 1984). One could make a final appeal on behalf of Tucker that such genetic differences are due to selection and differential survival (Williams, Koehn, and Mitton, 1973). Comparini and Rodino (1980) and Comparini and Schoth (1982), however, showed that the characteristic electrophoretic patterns in three enzymes—MDH, LDH, and PGI—are present even in small leptocephali collected in the spawning area.

The evidence, then, is overwhelmingly against Tucker's single-species hypothesis. The known facts can be explained more parsimoniously by assuming that two closely related but distinct species are present, each with its own spawning area and migration routes.

Williams and Koehn (1984; Williams, Koehn, and Thorsteinsson, 1984) have recently proposed a new version of the single-species hypothesis. Unlike Tucker, they conceded that the American and European eels represent two largely distinct populations, each breeding panmictically. They differed from the standard approach in maintaining that the two populations are not full species but merely geographic varieties of a single species. They presented two sources of evidence to support their view: the close genetic similarity of the two forms, and an apparent intergradation in northern Europe and Iceland.

The genetic difference between the eels on the two sides of the Atlantic is very slight. The only clear biochemical difference is in the frequency of two alternate alleles at one locus of the enzyme malate dehydrogenase (MDH-2). Though significant, this is a quantitative difference only; qualitative differences have not been demonstrated conclusively. In most animals, the genetic similarity among local populations of a species is between 0.89 and 1.00. Distinct species, on the other hand, have values around 0.47–0.78. In the case of the American and European eels, the genetic similarity is 0.896, a level of differentiation usually found among local populations. Williams and Koehn concluded that the MDH-2 locus in *Anguilla* is evidence of partial but not complete isolation.

Boëtius (1980) observed that eels from northern Europe had slightly fewer vertebrae than those from central and southern Europe; the mean values were 114.438 (n = 6059), 114.504 (n = 3283), and 114.744 (n = 2515), respectively. He further noted that discrete samples existed with unusually low counts—below 111—and that these samples occurred almost excusively in northern Europe and Iceland. Boëtius explained these data by postulating an influx of American leptocephali and the occasional presence of hybrids. Williams and Koehn (1984) suggested a more conventional explanation, namely that "morphologically intermediate specimens from regions geographically intermediate between closely similar organisms . . . should be regarded as intergrades and as evidence that the forms thus connected are members of the same species." They pointed to their own work in Iceland (Williams, Koehn, and Thorsteinsson, 1984), where they had found 108–118 vertebrae in a sample of more than 550 specimens. Sixteen specimens—2.9% of the total—had counts of 111 or fewer. This is a higher percentage than would be expected by chance; of the 11,857 specimens examined by Boëtius (1980) from all over Europe, only 1.4% had fewer than 111 vertebrae. They tested their hypothesis by examining the frequency of MDH-2 alleles. If the low-count specimens were intergrades, they should show a shift in this frequency toward the American mean. The frequency of this allele is 0.96 in eels from North America and 0.10 in eels from Europe. The low-count (108–111) Icelandic eels had a frequency of 0.344. Even the eels with 112 vertebrae showed a frequency of 0.250. This confirmed the presence in Iceland of eels that are intermediate in two characters between *Anguilla rostrata* and *A. anguilla.*

Boëtius and Harding (1985: 147) raised doubt about another presumed difference between the American and European eels. They showed that Schmidt's (1922) account of a difference in growth rate between the leptocephali of the two forms does not hold up to examination. Indeed, their graphs plotting size against date of capture (Boëtius and Harding, 1985: figs. 10–12) show essentially identical slopes for the two species.

Williams and Koehn's hypothesis does not explain several other observations, however. They believed that the spawning areas of the two eels were largely separate and that this fact accounted for the genetically based differences between them. Their unstated assumption was that when the spatial separation of spawning areas broke down, breeding in the zone of overlap would be indiscriminate. This seems at odds with the observations of Comparini and Rodino (1980) and Comparini and Schoth (1982) that the MDH-2 frequencies of 0-group leptocephali (young of the year) in the Sargasso Sea are of either pure American or pure European type. Schoth and Tesch (1982: table 1) recorded specimens as small as 5 mm of both species from the same haul. This indicates that the two forms are spawning in the same area and not interbreeding.

Another observation involves the dispersal of leptocephali from the Sargasso Sea to the continents. Although both forms occur together in the Sargasso, they sort themselves out almost completely between North America and Europe. Kleckner and McCleave (1985: 88) pointed out that this process is difficult to explain by passive transport alone. They noted that leptocephali of both types occur in the Gulf Stream, but those found in North American Slope Water are almost exclusively American. This suggests a behavioral difference between the two forms, a fact incompatible with a single-species hypothesis.

Finally, Avise et al. (1986) discovered significant differences in mitochondrial DNA between American and European eels.

Many questions clearly remain to be answered about the relationship between *Anguilla rostrata* and *A. anguilla*. Until more answers are available, I prefer to take the traditional course and treat them as separate species.

Distribution. Anguilla rostrata occurs all the way from Greenland to the West Indies, but it is most abundant along the Atlantic drainage of the United States and southern Canada, the area most directly accessible to leptocephali.

The presence of the species in Greenland (Jensen, 1937; Boëtius, 1985) is difficult to understand. The currents in the Labrador Sea are predominantly cold and southward flowing and would seem to pose a barrier to direct transport of leptocephali from the south. The only obvious route is by way of the Irminger Current, which makes a counterclockwise loop south of Iceland and then flows around the southern tip of Greenland. If this is the case, however, some of the leptocephali should land at Iceland, which is about as close. In the same way, some of the leptocephali of *A. anguilla*, which populate Iceland, should be carried on to Greenland. Neither of these events seems to take place. Boëtius (1980) reported a few *A. rostrata* from Iceland, but Williams and Koehn (1984) have interpreted these as intergrades. *Anguilla anguilla* is unknown in Greenland. Active migration of adult *A. rostrata* across the Labrador Sea seems unlikely. The only other explanation that suggests itself is an active component to the migration of the leptocephali, with American larvae swimming west and European larvae swimming east. This was suggested by Williams and Koehn (1984: 534) and Kleckner and McCleave (1985: 88).

Eels are present in nearly every river and stream, as well as in coastal lagoons and tidal marshes, from Newfoundland to Florida. They go up the St. Lawrence and are common in Lake Ontario and its tributaries. They are much less common in the upper Great Lakes and were probably not present at all before the opening of various canals. Eels occur around the Gulf of Mexico and throughout the Mississippi drainage as far as Minnesota. Although common, there is some doubt whether their numbers here match those along the east coast. Schmidt (1909b: 8) analyzed fishery statistics and found that almost 98% of the catch came from the eastern states. This is consistent with the pattern of dispersal of leptocephali from the spawning area. Leptocephali can reach the Gulf of Mexico only by being carried south into the Caribbean and then into the Gulf on the Loop Current. Such leptocephali could represent only a small percentage of the total number produced each year. Within the Gulf, eels seem to be more common east of the Mississippi than west of it, although figures are hard to come by. Small fisheries exist or have existed in western Florida and Louisiana but not in Texas or Mexico. The Loop Current is confined mainly to the eastern Gulf, and it would be expected to deposit leptocephali mainly there.

The southern boundary of the eel's range is somewhat uncertain. In the Gulf of Mexico, it occurs in the Rio Grande and in northern Mexico at least to Nautla, Veracruz (Castro-Aguirre, 1978: 53). The presence of leptocephali near the island of Cozumel (see p. 898) suggests that adults should be present on the Yucatan Peninsula, but none have been recorded. The species is absent or nearly so from Central America, although Seale (1917) and Breder (1925) recorded specimens from Panama. It is present, however, in the Bahamas and most of the West Indies, including Cuba, Jamaica, and Puerto Rico. Schmidt (1909b: 11) quoted Regan as saying the eel occurs all the way to Grenada, and Vladykov (1966) recorded specimens from Trinidad. It appears to be absent, however, from the mainland of South America. Schmidt (1909b: 13) gave second-hand accounts of eels from Guyana and Suriname, but these have not been confirmed. According to Dr. Francisco Mago L. (pers. comm.) the species does not occur in Venezuela.

Vladykov (1964) suggested that the spawning area of the eel must be farther south than Schmidt proposed in order to explain the presence of eels in Trinidad and the Guianas. It seems clear, however, that eels are not common south of the Antilles, and Vladykov's proposal seems unnecessary.

Finally, eels occur in Bermuda, and they are all *A. rostrata* (Schmidt, 1909b: 18; Boëtius and Boëtius, 1967). This represents another paradox, because the leptocephali around Bermuda are predominantly of the European species (Schmidt, 1922: 204). The most reasonable explanation for this situation is that the European leptocephali in the Bermuda area are too young to metamorphose. By the time they are ready to transform, they have been carried far to the east. Only American leptocephali are found so far to the west in such late stages (Schmidt, 1922: 204).

Etymology. Latin *rostratus* (beaked, curved). The significance is not apparent. Lesueur described five species of *Anguilla* from North America. The name presumably was based on some distinguishing feature. An adjective.

Study Material. 33 specimens, 111–581 mm TL.

MASSACHUSETTS: uncatalogued (7 males, 302–328, hormone treated, specimens lost). RHODE ISLAND: ANSP 156851 (2 males, 299–331, hormone treated). PENNSYLVANIA: ANSP 139680 (1, 278), 139681 (1, 290). NEW JERSEY: ANSP 125919 (2, 128–156). MARYLAND: ANSP 118404 (1, 219), 129834 (1, 218). SOUTH CAROLINA: ANSP 95158 (2, 197–234). LOUISIANA: ANSP 156849 (2 females, 566–571, hormone treated). TEXAS: TNHC 590 (2, 340–362), 597 (1, 137), 618 (1, 205), 1642 (1, 441), 3140 (1, 111), 4500 (1, 476), 4546 (1, 368), 7337 (1, 581). LOCALITY UNCERTAIN: ANSP 156850 (3 females, 455–507), 156852 (2 males, 296–298).

Family Heterenchelyidae

DAVID G. SMITH

Acknowledgments. See Family Anguillidae.

Introduction. The Heterenchelyidae is a small family of highly modified burrowing eels that may be related to the Moringuidae and Anguillidae (see below). Because of their retiring habits, heterenchelyids were practically unknown before 1912, even though one species occurs in the Mediterranean.

The family name Heterenchelyidae was coined by Regan (1912a) to house his new genus *Heterenchelys.* Böhlke (1966a) later showed that *Heterenchelys* was a junior synonym of *Pythonichthys* Poey, 1868, which had always been placed in the Muraenidae. Article 40 (a) of the ICZN requires that the family name be retained, even though the genus on which it is based is synonymized (if such synonymization occurred after 1960). Hence the name Heterenchelyidae must still be used although *Heterenchelys* is no longer recognized.

Characters. Body moderately elongate to elongate, cylindrical anteriorly, compressed posteriorly; tail distinctly longer than head and trunk; tip of tail soft, blunt. Scales absent. Dorsal and anal fins low, confluent with well developed caudal fin. Dorsal fin begins near level of gill opening; pectoral fin absent. Gill opening vertical or slightly oblique, low on side. Lateral line obsolete, no pores on head or body.

Head blunt in profile, jaws equal or lower jaw protrudes slightly; rictus well behind eye. Eye reduced, covered by semi-transparent skin. Anterior nostril near tip of snout, in a low tube; posterior nostril in front of eye, with raised rim. Upper and lower lips with a groove separating them from remainder of cheek and lower jaw. Tongue free at tip.

Teeth conical to molariform. Maxillary teeth bi- or triserial, continuous with those of intermaxillary patch. Outer intermaxillary teeth continuous with external row of maxillary teeth; a variable number of teeth medial to these. Vomerine teeth in one to several rows; length of patch variable, continuous or not with intermaxillary teeth. Mandibular teeth in two to four rows.

Osteology (based on *Pythonichthys asodes,* largely from Smith and Castle, 1972). Neurocranium (Figs. 19, 20) relatively stout; orbit small, slightly elongate in profile; large otic bullae present. Frontals divided, each with two large arches which enclose a portion of cephalic sensory canal. Supraoccipital present, with median crest.

Maxilla stout, with broad pedicel anteriorly, posterior end somewhat expanded and compressed (Fig. 21A,B). Mandible stout, posterior end markedly deeper than anterior end (Fig. 21C).

Suspensorium (Fig. 22) deeper than long, vertical or inclined slightly forward. Pterygoid broad, well developed, closely appressed to quadrate and hyomandibula.

Hyoid arch (Fig. 23) complete except for hypohyals, which are absent. Glossohyal moderate in length (Fig. 23A). Urohyal with expanded central portion and short posterior extension (Fig. 23B). Approximately 10–12 branchiostegal rays, first two associated with ceratohyal, remainder with epihyal (Fig. 23D). Branchiostegals curve up and around posterior end of opercle and extend forward above it (Fig. 19).

Gill arches complete, well developed (Fig. 24). First two basibranchials ossified, third and fourth cartilaginous. First four ceratobranchials stout, fifth small and slender. Upper pharyngeals fused into a single large element.

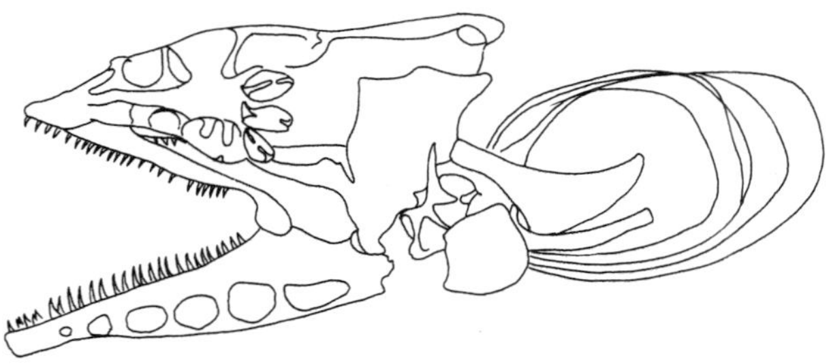

FIGURE 19. Head skeleton of *Pythonichthys asodes,* in situ.

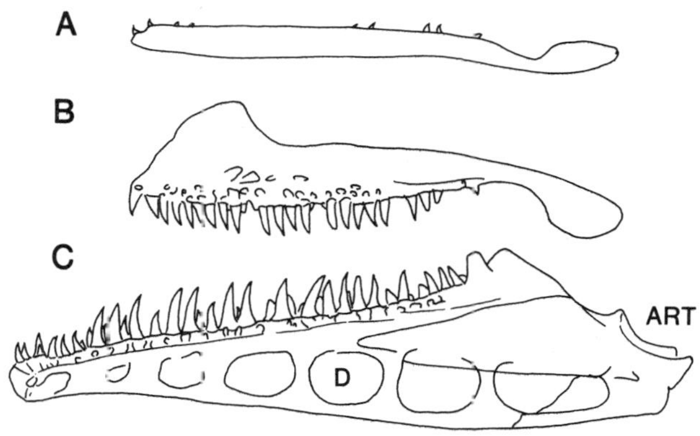

FIGURE 21. Jaws of *Pythonichthys ascdes.* A. Maxilla in dorsal view. B. Maxilla in lateral view. C. Mandible.

Opercular apparatus complete (Fig. 25). Opercle relatively narrow, sickle-shaped. Subopercle only slightly narrower than opercle, little curved. Preopercle a tubular structure with a small dorsal projection. Interopercle roughly square, relatively large.

Circumorbital bones large, cavernous (Fig. 26). Nasal relatively short and broad. Preorbital more slender. Four suborbitals, beginning below eye and extending up behind eye. Supraorbital not evident.

Pectoral girdle consists of a stout cleithrum and supracleithrum (Fig. 27). Scapula, coracoid, and actinosts absent.

First vertebra short and high, second vertebra also distinctly shorter than those following it (Fig. 28). Anteriormost epineurals not fused to body of vertebra. Vertebrae of branchial region with broad, prominent epicentral processes (Fig. 29A). Neural spines relatively short on both precaudal and caudal vertebrae, arise from posterior end of neural arch (Fig. 29B,C). Transition between precaudal and caudal vertebrae not sharp; in specimen examined first completely closed hemal arch on 104th vertebra out of 123 total. Ribs present on 2nd–46th vertebrae, epineurals on 1st–113th, epipleurals on 26th–111th. Except for anterior and posteriormost few, epineurals and epipleurals forked.

Discussion. Heterenchelyids are wholly fossorial eels well adapted to their burrowing habits.

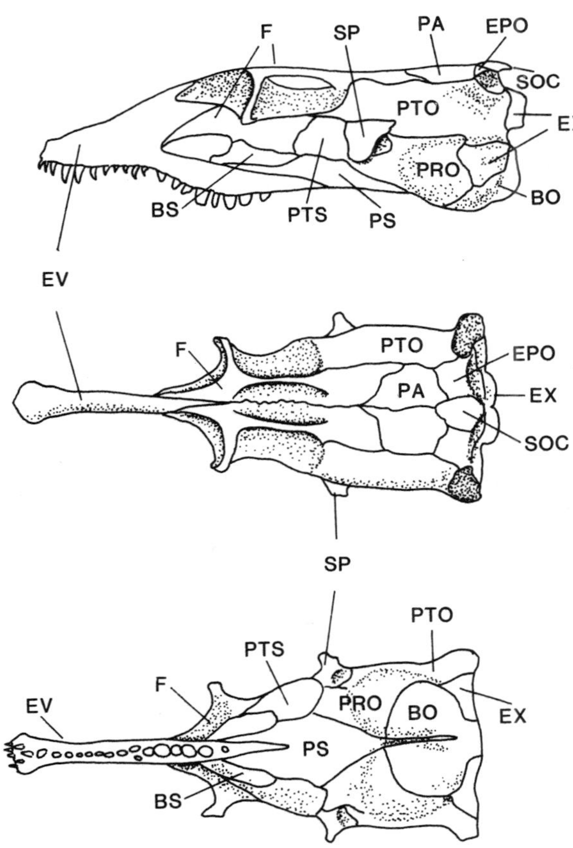

FIGURE 20. Neurocranium of *Pythonichthys asodes.* Abbreviations on pp. 5–6.

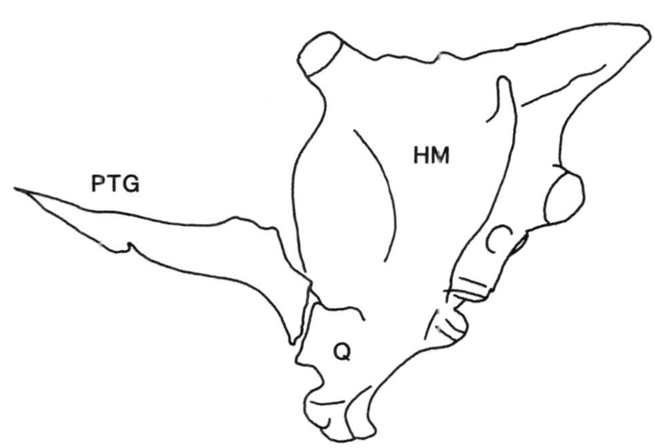

FIGURE 22. Suspensorium of *Pythonichthys asodes.*

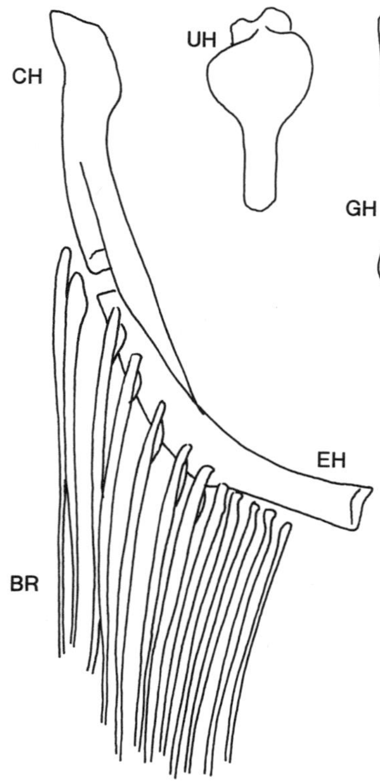

FIGURE 23. Hyoid arch of *Pythonichthys asodes*.

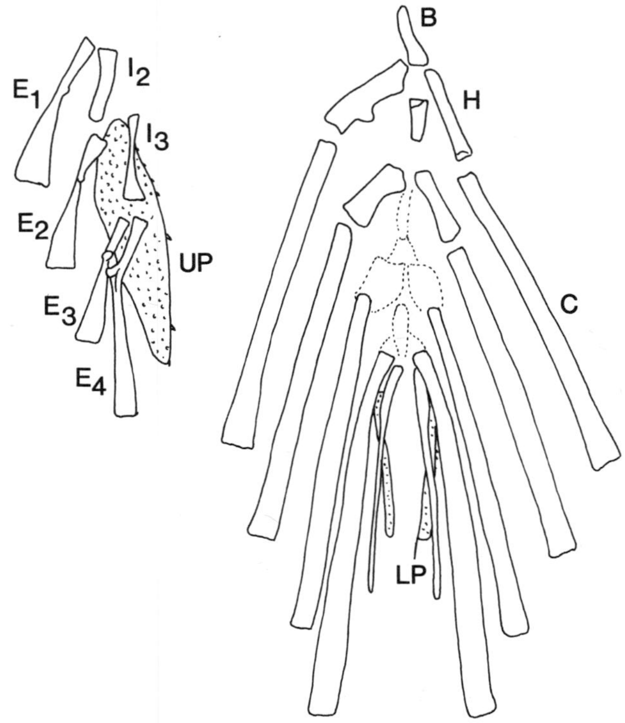

FIGURE 24. Gill arches of *Pythonichthys asodes*, ventral elements in ventral view and dorsal elements in dorsal view.

Superficially they resemble moringuids but can be distinguished by several characters. In heterenchelyids the pectoral fin is totally absent; in moringuids it is always present, though sometimes greatly reduced. The dorsal fin in heterenchelyids begins approximately over the gill opening; in moringuids it begins far behind the gill opening much closer to the anus. In heterenchelyids the anus is well ahead of midlength; in moringuids it is at or behind midlength. In addition there are a number of osteological differences, and the heterenchelyid leptocephalus is quite distinct from that of the moringuids.

Virtually nothing is known about the habits and behavior of heterenchelyids. Blache (1968c) described them as living on sandy or silty bottoms and feeding on small worms, crustaceans and molluscs. There is no evidence of sexual dimorphism in heterenchelyids or that they become pelagic at any stage of their adult lives. They appear to remain buried in the substratum except, perhaps, for brief excursions in search of food.

How closely the Heterenchelyidae is related to the Moringuidae is a matter of conjecture. The resemblances between them, though considerable, are primarily in primitive characters or in characters related to their fossorial habits. It may be informative to compare them with certain ophichthids, which are the only other strongly modified fossorial eels. Characters shared by all three groups can be considered general adapta-

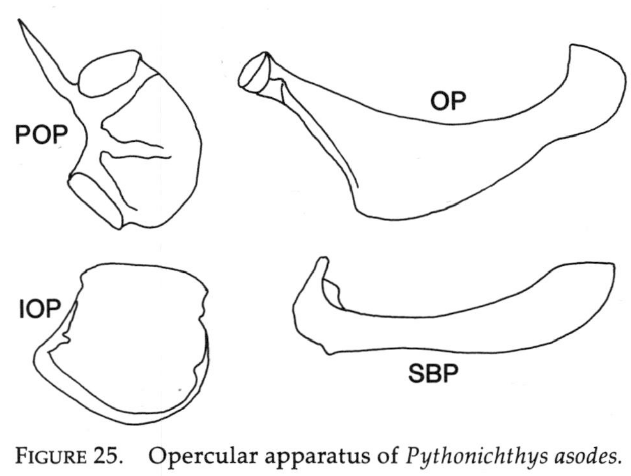

FIGURE 25. Opercular apparatus of *Pythonichthys asodes*.

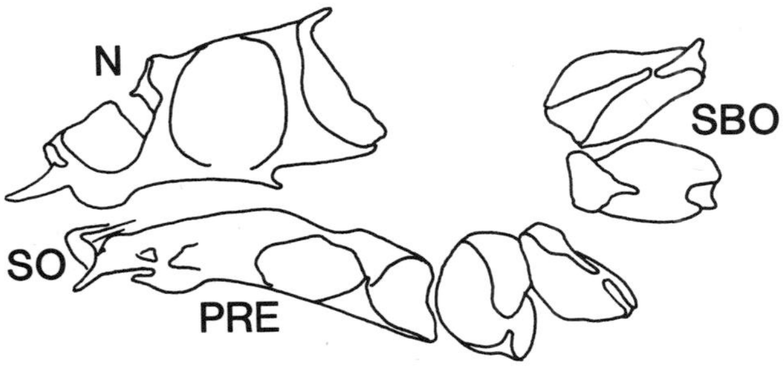

FIGURE 26. Circumorbital bones of *Pythonichthys asodes.*

tions for burrowing. Such shared characters include a reduced eye, absence of pectoral fins, and a cylindrical body. Ophichthids, however, have developed a tail-first burrowing habit. Consequently, the tip of the tail is stiffened, and the caudal fin is absent. In fact, ophichthids can probably travel forward or backward through the sediment with equal facility. Heterenchelyids and moringuids, on the other hand, have a soft, pliable tail and a caudal fin and probably are limited to head-first burrowing. This, however, seems to be simply a less specialized condition. There are other eels—myrophine ophichthids and *Anguilla,* for example—that burrow head-first. Ophichthids show a distinct reduction in the gill arches; heterenchelyids and moringuids retain more complete gill arches—another plesiomorphic condition. On the other hand, heterenchelyids and moringuids show a greater reduction in the lateral line than do ophichthids. Although moringuids have at least some pores on the body, both they and the heterenchelyids lack pores in the supraorbital and infraorbital canals. Ophichthids, by contrast, retain pores in

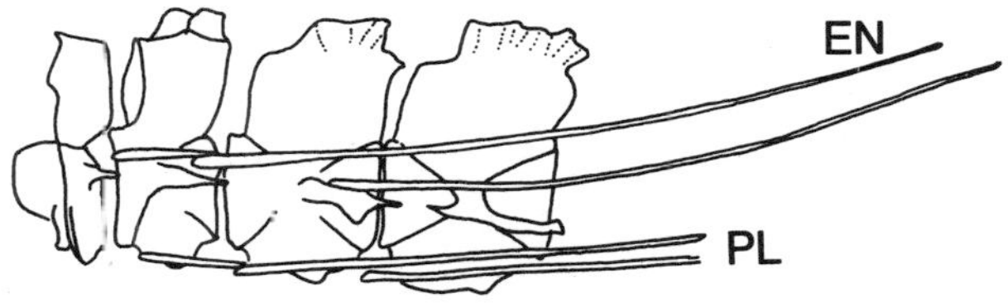

FIGURE 28. Anteriormost vertebrae of *Pythonichthys asodes.*

all canals. The absence of hypohyals is another derived feature shared by the Heterenchelyidae and Moringuidae.

There are also, however, substantial differences between the Heterenchelyidae and Moringuidae. The absence of the pectoral fin in heterenchelyids and the difference in the dorsal-fin origin between the two families has already been mentioned. The large frontal arches in the heterenchelyid skull are without parallel in the Moringuidae. The maxilla in heterenchelyids has a pedicel at the anterior end, that of moringuids does not. The pterygoid is much larger in heterenchelyids than it is in moringuids. The branchiostegal rays curve up and over the opercle in heterenchelyids but not in moringuids. The upper pharyngeals are fused in the Heterenchelyidae, separate in the Moringuidae. The leptocephali of the two families are also quite different. In some of these characters, the heterenchelyids have a more derived condition, for example, the loss of the pectoral fin and the obsolete lateral line. In other characters, the moringuids are more derived, for example, the reduced pterygoid.

Although a close relationship between the

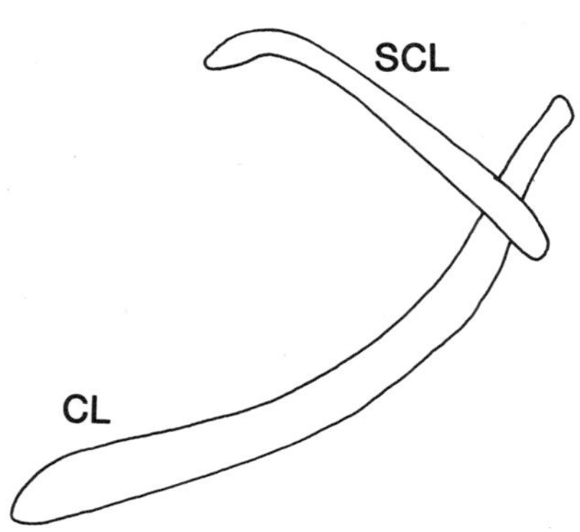

FIGURE 27. Pectoral girdle of *Pythonichthys asodes.*

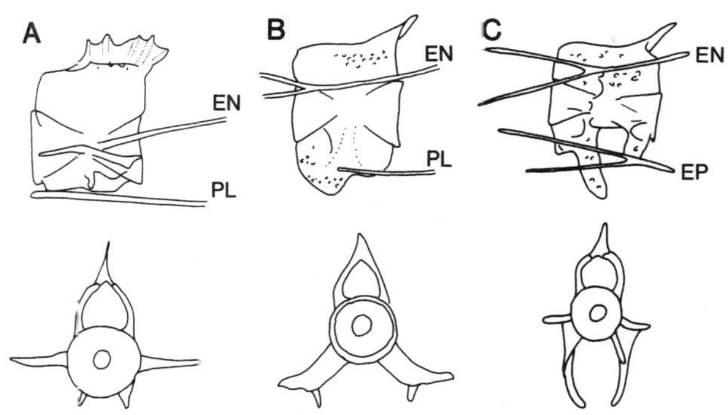

FIGURE 29. Vertebrae of *Pythonichthys asodes,* in lateral and frontal view. A. From branchial region. B. From abdominal region. C. Anterior caudal vertebra.

Heterenchelyidae and Moringuidae is plausible and may well be true, it is difficult to find synapomorphies that are not based on reductions or losses of characters. Modifications of this kind are strongly susceptible to convergence.

Distribution. The center of distribution of the Heterenchelyidae is the eastern Atlantic. At least five species occur off the west coast of Africa, and a sixth is found in the eastern Mediterranean. By contrast the western Atlantic and eastern Pacific contain a single species each. The family appears to be absent from the Indo-West Pacific, although otoliths purportedly from heterenchelyids have been found in Tertiary deposits of Australia and New Zealand (Stinton, 1957, 1958).

Genera. The Heterenchelyidae contains two genera, *Pythonichthys* Poey, 1868 and *Panturichthys* Pellegrin, 1913. Blache (1968c: 1542) listed six differences between them. (1) *Panturichthys* has a longitudinal dermal crest on the top of the head, which *Pythonichthys* lacks. (2) In *Pythonichthys* the inner row of maxillary teeth is about the same length as the outer row, whereas in *Panturichthys* the inner row is shorter. (3) In *Pythonichthys* the vomerine teeth are separated from the intermaxillary teeth; in *Panturichthys* they are continuous. (4) *Panturichthys* has prominent fleshy folds in the roof and floor of the mouth, *Pythonichthys* does not. (5) In *Pythonichthys* the dorsal fin originates over or behind the posterior extremity of the arc formed by the pectoral girdle; in *Panturichthys* the dorsal fin begins distinctly before this point (this character can be seen only in radiographs). (6) The preanal length is 22–37% TL in *Pythonichthys,* 12–20% in *Panturichthys.*

Pythonichthys contains four species. *Pythonichthys microphthalmus* (Regan, 1912) and *P. macrurus* (Regan, 1912) occur off west Africa from Morocco to Angola, *P. sanguineus* Poey, 1868 occurs in the western Atlantic, and *P. asodes* Rosenblatt and Rubinoff, 1972 occurs in the eastern Pacific. *Panturichthys* also contains four species: *P. fowleri* (Ben-Tuvia, 1953) from the eastern Mediterranean, *P. mauritanicus* Pellegrin, 1913 from off Mauritania and Senegal, *P. longus* (Ehrenbaum, 1915) from the eastern Gulf of Guinea, and *P. isognathus* Poll, 1953 from off the Congo and Angola (Blache, 1968c, 1975).

Genus *Pythonichthys* Poey, 1868

Pythonichthys Poey, 1868a: 265 (type species *Pythonichthys sanguineus* Poey, 1868, by monotypy).
Heterenchelys Regan, 1912a: 323 (type species *Heterenchelys microphthalmus* Regan, 1912, by subsequent designation of Jordan, 1920: 545).

Characters. The characters of *Pythonichthys* are largely those of the family, already given above. The distinctions between *Pythonichthys* and *Panturichthys* have also been given. *Pythonichthys* is generally less elongate than *Panturichthys* and has a more posterior anus.

Size. Less than 1 m TL. Blache (1968c: 1551) recorded a specimen of *P. macrurus* of 800 mm TL.

Distribution. Tropical eastern Atlantic, Mediterranean, western North Atlantic, and eastern Pacific.

Etymology. From the Greek *python* (a kind of snake) and *ichthys* (fish). Masculine.

Species. The four species of *Pythonichthys* have been given above. They are distinguished by the number of vertebrae, the form of the dentition, and certain proportional measurements. *Pythonichthys sanguineus* Poey is the only one found in the western North Atlantic.

Pythonichthys sanguineus Poey, 1868
Figures 30–33

Pythonichthys sanguineus Poey, 1868a: 265, pl. 2, fig. 7 (original description, Matanzas, Cuba, holotype lost). Böhlke, 1966a (synonymy, illustration).
Muraena sanguinea. Günther, 1870: 126.
Heterenchelys biaggii Böhlke, 1956b: 8, fig. 2A (original description, Mayagüez, Puerto Rico, holotype FMNH 61774).

Distinctive Characters. The only western Atlantic eel likely to be confused with *Pythonichthys sanguineus* is *Neoconger mucronatus,* but the two can quickly be separated by the following characters. *Neoconger* has a small but well formed

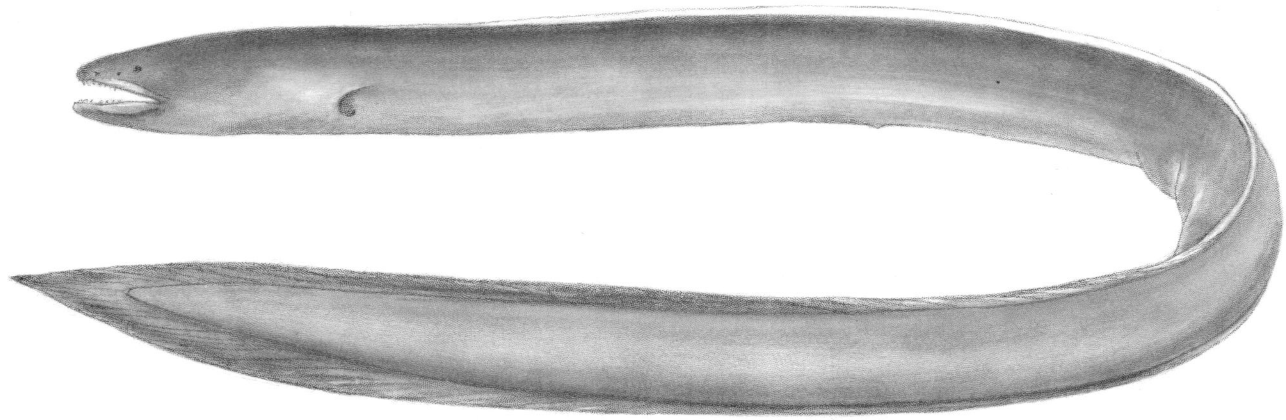

FIGURE 30. *Pythonichthys sanguineus*: ANSP 73872, 346 mm TL.

pectoral fin, which *Pythonichthys* lacks. In *Pythonichthys* the jaws are equal or the lower jaw projects; in *Neoconger* the upper jaw is longer. In *Pythonichthys* the dorsal fin begins near the level of the gill opening; in *Neoconger* the dorsal fin begins far behind the gill opening, slightly before the anus. In *Pythonichthys* the preanal length is about one-third the TL; in *Neoconger* it is about half the TL. *Moringua* also goes through a fossorial stage, but *Moringua* is much more slender than *Pythonichthys* and its anus is located well behind midlength.

Pythonichthys sanguineus is close to the eastern Atlantic *P. microphthalmus* in vertebral counts and most proportional measurements. It differs primarily in dentition. In *P. microphthalmus* the vomerine teeth are in one series instead of 2–3, and the maxillary and mandibular teeth are in two series each instead of three and 3–5, respectively. The other two species, *P. macrurus* and *P. asodes*, have distinctly different vertebral counts and proportional measurements.

Description (including data from Böhlke, 1956b). Total vertebrae 110–111 (n = 3), predorsal vertebrae 8–9 (4), preanal vertebrae 36–38 (4).

Proportions as % of TL: preanal 33–37 (9), predorsal 10–15 (7), head 11–13 (9), depth at anus 4–6 (8), caudal fin 3 (2). Of HL: snout 15–19 (10), eye 2–5 (10), snout-rictus 27–36 (10), interorbital 12–15 (8), gill opening 9–14 (6), interbranchial 15–25 (2).

Maxillary teeth in three series, outer one slightly separated from inner; teeth of outer row pointed, those of inner row pointed anteriorly and becoming blunter posteriorly. Intermaxillary teeth with one or two enlarged conical teeth medially and a semicircle of smaller teeth in front of them, the latter continuous with maxillary teeth. Vomerine teeth in two rows anteriorly,

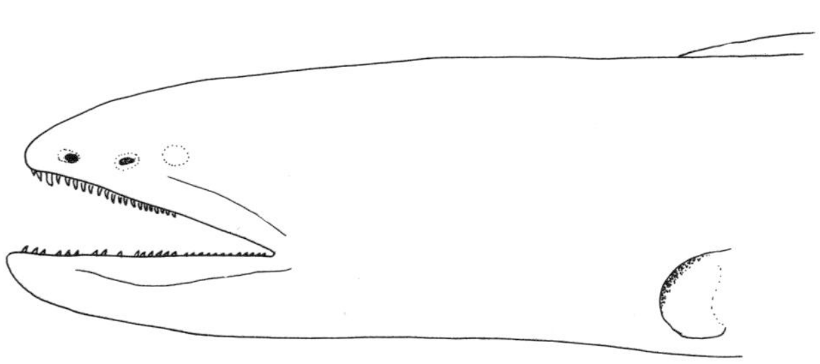

FIGURE 31. *Pythonichthys sanguineus*: ANSP 73872, 346 mm TL; head.

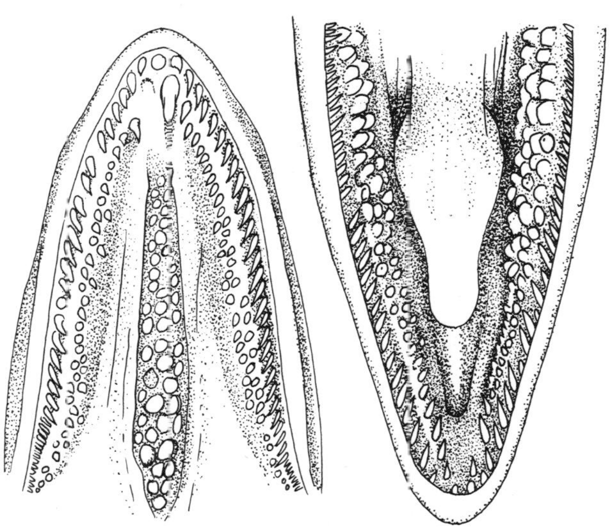

FIGURE 32. *Pythonichthys sanguineus*: ANSP 73872, 346 mm TL; dentition.

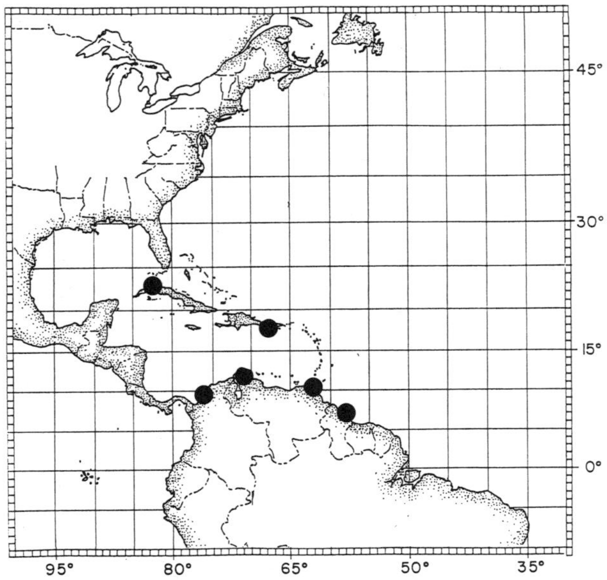

FIGURE 33. Distribution of *Pythonichthys sanguineus.*

increasing to 3–4 rows near midlength of patch, mostly molariform, but a few anteriormost teeth conical; vomerine tooth patch extends posteriorly slightly beyond end of maxillary teeth. Mandible with an outer series of sharp teeth separated from an inner patch of teeth which starts as two rows of relatively sharp teeth anteriorly and increases to approximately four rows of progressively more molariform teeth posteriorly.

Stomach long, ends slightly behind anus. Gas bladder apparently absent.

Color. In life "pinkish cream, more pink on head" (Böhlke, 1966a: 4). In preservative a pale gray, some brown along dorsal margin and around tip of tail to extreme posterior end of anal fin. The reddish color in fresh specimens probably results from blood in the superficial capillaries showing through the unpigmented skin.

Size. The largest specimen examined was 620 mm TL (MCZ 32805). Poey gave a TL of 800 mm for the holotype.

Development and Growth. Although the leptocephali of several heterenchelyid species from the eastern Atlantic have been identified and described (Blache, 1977), that of *Pythonichthys sanguineus* is unknown.

Distribution. Pythonichthys sanguineus has been recorded from scattered localities in the West Indies and off northern South America from Cuba to Guyana in depths of 12–55 m. It is uncommon.

Etymology. Latin *sanguineus* (bloody, blood-red), referring to the color of fresh specimens. An adjective.

Study Material. Eleven specimens, 271–620 mm TL. The MBUCV, Fernandez Yepez, and BMNH specimens are included here based on notes taken by J. E. Böhlke, who examined them and recorded counts and measurements.

CUBA: MCZ 32805 (1, 620), Havana, 1931. PUERTO RICO: ANSP 73872 (1, 346; paratype of *Heterenchelys biaggii* Böhlke), FMNH 61774 (1, 370; holotype of *Heterenchelys biaggii* Böhlke), and FMNH 61775 (1, 271; paratype of *Heterenchelys biaggii* Böhlke), Mayagüez pier, 1934. VENEZUELA: MBUCV 293 (4, 291–501), Puerto Hierro, Sucre, 12 m, 16 Nov. 1969. Fernandez Yepez Collection (1, 538+), Gulf of Venezuela, Modano Blanco, 10 July 1968. COLOMBIA: UMML 22150 (1, 269), 8°20.1′N, 76°53.6′W, 51–55 m, 11 July 1966. GUYANA: BMNH 1961.8.31.99 (1, ca 405), 8°20′N, 58°55′W, 35 m, 21 Jan. 1959.

Family Moringuidae

SPAGHETTI EELS

DAVID G. SMITH

Acknowledgments. James E. Böhlke had a long-standing interest in moringuid eels and over the years compiled an impressive amount of research on the complex nomenclatural history of this family. Sometime in the late 1950's he put much of this information into a manuscript entitled "Notes on the ratabourid eels of the Great Bahama Bank." Events soon rendered the manuscript obsolete, and it was never published, but it and the notes accompanying it provided a good starting point for my own study. For general acknowledgments see family Anguillidae.

Introduction. The Moringuidae is a small family of burrowing eels that appears to be related to the Anguillidae and the Heterenchelyidae. Along with the heterenchelyids and some ophichthids, moringuids are the only eels that show pronounced morphological adaptations to fossorial life. These adaptations include a reduced eye, lack of color, low vertical fins, and reduced head pores. Like heterenchelyids but unlike ophichthids, moringuids burrow head first.

The earliest family-group name applied to these eels was the Ptyobranchidae of McClelland (1844: 199), based on his genus *Ptyobranchus. Ptyobranchus* was soon recognized as a synonym of *Moringua* Gray, 1831 (Cantor, 1853: 228), and the family name based on it seems never to have been used again except by Günther (1870) in the form Ptyobranchina. Cope (1872: 335) created a family name Rataburidae, giving only a brief diagnosis and without mentioning any genera (although it was obviously based on *Rataboura* Gray, a synonym of *Moringua*). The name Moringuidae apparently was first used by Gill (1885: 107). It seems to have been common practice at the time to base family names on the earliest recognized genus, in this case *Moringua.* Moringuidae has been used almost universally since Gill's time, except for names based on genera thought to be distinct (Stilbiscidae, Anguillichthyidae, etc.). Article 40 (b) of the ICZN states: "If a family-group name has been replaced before 1961 because of such synonymy [i.e., of the type genus], and the replacement name has won general acceptance, it is to be maintained." Since Moringuidae has enjoyed general acceptance for many years, I follow Article 40 (b) and recognize it as the valid name of the family.

Cope, incidentally, published two nearly identical versions of his 1872 paper. The one in the Transactions of the American Philosophical Society (14: 445–483) does not mention Rataburidae.

Characters. Body moderately elongate to elongate, cylindrical except near tip of tail; scales absent; anus near or behind midlength; tip of tail soft, blunt. All fins present though sometimes reduced; dorsal and anal fins confluent with caudal; dorsal fin begins far behind head, slightly before or well behind midlength; anal fin begins at or distinctly behind anus.

Eye reduced except in mature *Moringua.* Anterior nostril with a low tube or without a tube; posterior nostril in front of eye. Upper lip without a flange; lower lip with or without a groove separating it from remainder of lower jaw.

Teeth conical, small to moderately enlarged; in one or two series on maxilla, mandible, and vomer; intermaxillary teeth generally the largest, arranged in two longitudinal rows or in a semicircle.

Lateral line on body complete or incomplete, pores present at least preanally; on head, pores present only on lower jaw.

FIGURE 34. *Neoconger mucronatus*, anterior skeletal elements in situ (from Smith and Castle, 1972: fig. 3).

Osteology (largely from Smith and Castle, 1972). Neurocranium (Figs. 34–36): *Neoconger* (Fig. 35): moderately stout, wedge-shaped; ethmovomer narrow at anterior tip, not markedly expanded where intermaxillary teeth are attached; orbit small, elongate in lateral profile; well developed otic bullae; frontals divided; supraoccipital present, bearing a long, slender crest. *Moringua* (Fig. 36): relatively short, ethmovomer not greatly prolonged anteriorly; orbit small, rounded in lateral profile; small otic bullae present; frontals divided; supraoccipital present; basisphenoid absent (although Trewavas, 1932 described and figured a basisphenoid).

Jaws (Figs. 37, 38): *Neoconger*: maxilla (Fig. 37A) moderately elongate, without a pedicel anteriorly; mandible (Fig. 38A) with dentary, articular, and coronomeckelian; ventral margin of mandible uniformly convex in lateral view. *Moringua*: maxilla (Fig. 37B) short, without a pedicel anteriorly; mandible (Fig. 38B) with dentary, articular, coronomeckelian; openings in mandibular sensory canal small; ventral margin of mandible convex in lateral view.

Suspensorium (Fig. 39): *Neoconger* (Fig. 39A): slightly deeper than long, nearly vertical; pterygoid present, somewhat reduced. *Moringua* (Fig. 39B): slightly longer than deep, nearly vertical; pterygoid reduced to a splint.

Hyoid arch (Figs. 40, 41): Median glossohyal and urohyal, paired ceratohyals and epihyals, no hypohyals. Urohyal with a moderately short posterior extension, somewhat longer in *Neoconger* than in *Moringua*. Approximately 18 branchiostegal rays in *Neoconger*, 11 in *Moringua*, the last

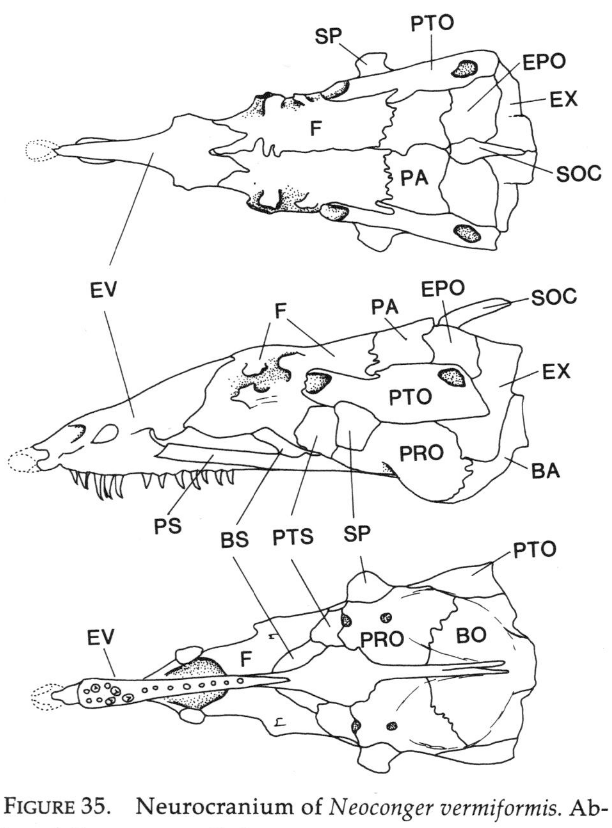

FIGURE 35. Neurocranium of *Neoconger vermiformis*. Abbreviations on pp. 5–6.

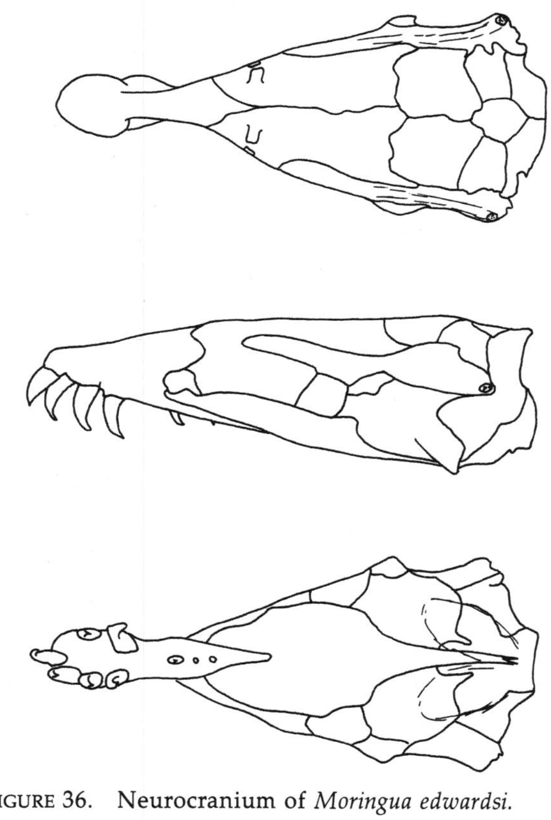

FIGURE 36. Neurocranium of *Moringua edwardsi*.

FIGURE 37. Maxilla in dorsal view (top) and lateral view (bottom). A. *Neoconger vermiformis*. B. *Moringua edwardsi*.

one flattened and expanded near tip; branchiostegals do not curve up behind and over opercle.

Gill arches (Fig. 42A,B): *Neoconger* (Fig. 42A): relatively complete, but second through fourth basibranchials reduced and cartilaginous; lower pharyngeals united; upper pharyngeals separate, the anterior one with an anteriorly directed process. *Moringua* (Fig. 42B): slightly reduced, second through fourth basibranchials absent; ceratobranchials short and broad; lower pharyngeals united; upper pharyngeals separate, the anterior one with an anteriorly directed process.

Opercular apparatus (Fig. 43): Complete in both genera. Opercle broad and triangular, slightly more elongate in *Moringua*. Subopercle broad and crescent-shaped, lying under posterior margin of opercle. Preopercle irregularly shaped, with a dorsal projection; bears portion of preopercular canal. Interopercle triangular, slightly smaller than opercle.

Pectoral girdle (Fig. 44): *Neoconger* (Fig. 44A): complete, with cleithrum, supracleithrum, scapula, and coracoid; cleithrum forked at ventral end; two actinosts in specimen examined. Immature *Moringua* (Fig. 44B): somewhat reduced, with slender cleithrum and supracleithrum, each

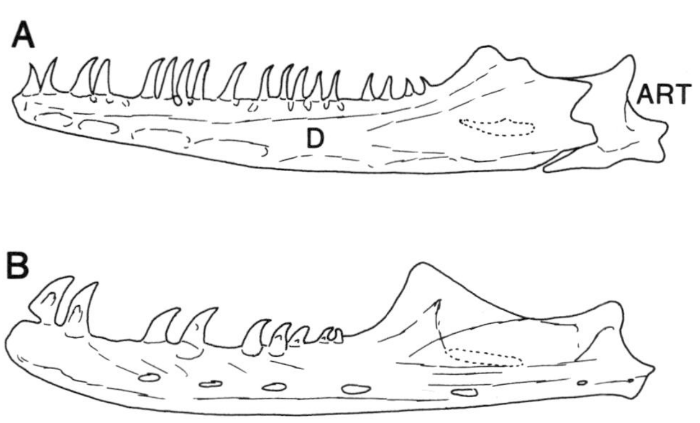

FIGURE 38. Mandible. A. *Neoconger vermiformis*. B. *Moringua edwardsi*.

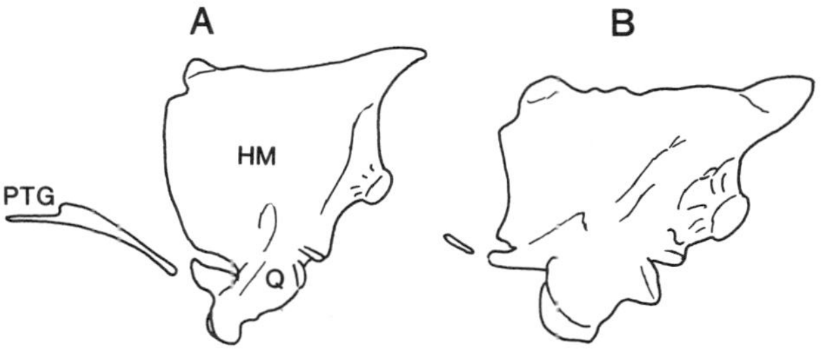

FIGURE 39. Suspensorium. A. *Neoconger vermiformis*. B. *Moringua edwardsi*.

bent slightly in middle; small scapula and coracoid; no actinosts. Mature male *Moringua* (Fig. 44C): well developed, with large cleithrum and supracleithrum, cleithrum bent sharply in middle; massive scapula and coracoid; one large actinost.

Circumorbital bones (Fig. 45): Well developed but incompletely ossified; large nasal, small supraorbital, long preorbital, three suborbitals; somewhat stouter and more substantial in *Neoconger* than *Moringua*.

Vertebrae (Figs. 46, 47): *Neoconger*: first vertebra consists of a centrum and a neural arch, articulates with skull; next few vertebrae with complex neural spines; specimen examined has 51 precaudal and 45 caudal vertebrae, ribs on 4th–49th vertebrae, epineurals on 2nd–78th vertebrae, epipleurals on 39th–79th vertebrae; vertebrae approximately as high as long; three hypurals, six caudal rays. *Moringua*: anteriormost vertebrae similar to those of *Neoconger*, but neural spines in form of a long, scalloped ridge; specimen examined has 82 precaudal and 39 caudal vertebrae, ribs on 3rd–80th vertebrae, epineurals on 2nd–107th vertebrae, epipleurals on 75th–105th vertebrae; except for anteriormost few, vertebrae distinctly longer than high; three hypurals, five caudal rays.

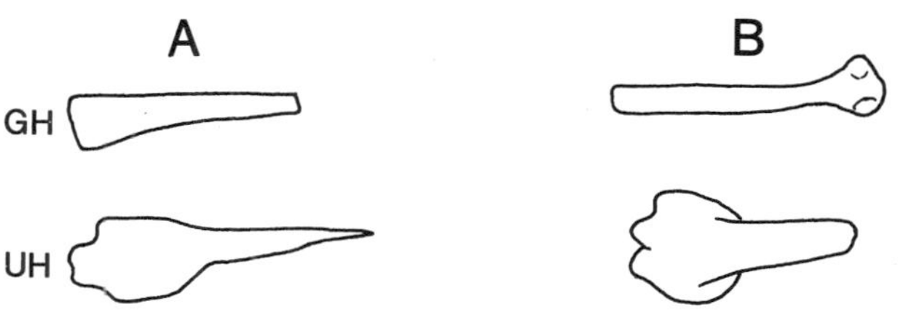

FIGURE 40. Glossohyal (top) and urohyal (bottom). A. *Neoconger vermiformis*. B. *Moringua edwardsi*.

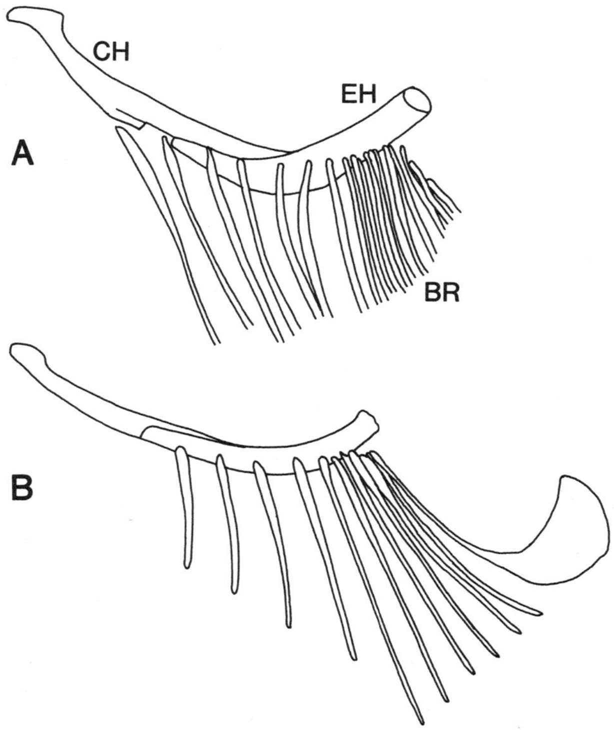

FIGURE 41. Ceratohyal, epihyal, and branchiostegal rays. A. *Neoconger vermiformis.* B. *Moringua edwardsi.*

Discussion. The Moringuidae contains two rather different genera, one of which shows great ontogenetic variation, and is thus rather difficult to define at the family level. The common characters are given above, but *Neoconger* and *Moringua* are best recognized individually. Among the specialized characters they share are the absence of supraorbital and infraorbital pores, the lack of a pedicel on the maxilla, the anteriorly directed process on the upper pharyngeal, and the flattened and expanded last branchiostegal ray. The most striking similarity between them, however, is the form of the leptocephalus, which led to the initial recognition of their relationship (Smith and Castle, 1972).

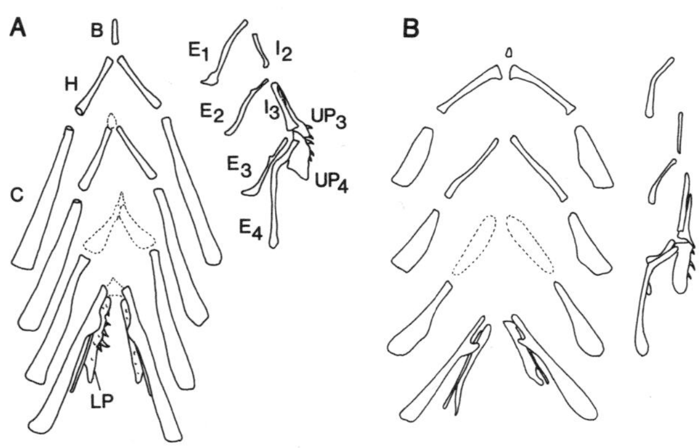

FIGURE 42. Gill arches, ventral elements in ventral view, dorsal elements in dorsal view. A. *Neoconger vermiformis.* B. *Moringua edwardsi.*

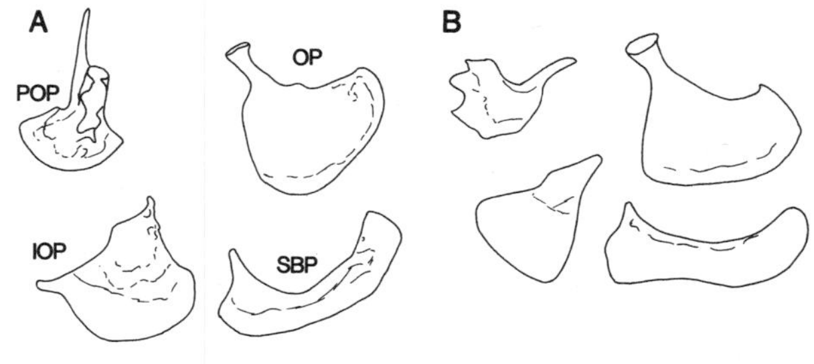

FIGURE 43. Opercular apparatus. A. *Neoconger vermiformis.* B. *Moringua edwardsi.*

In almost all ways, *Neoconger* is the more primitive genus. It does not show great reduction in osteological elements, and its heart and gill arches are in the normal (for eels) position. Aside from a slight enlargement of the eye and a few minor proportional differences, *Neoconger* shows no sexual dimorphism at maturity, and adults do not become pelagic. *Moringua,* on the other hand, shows a number of reductions in its skel-

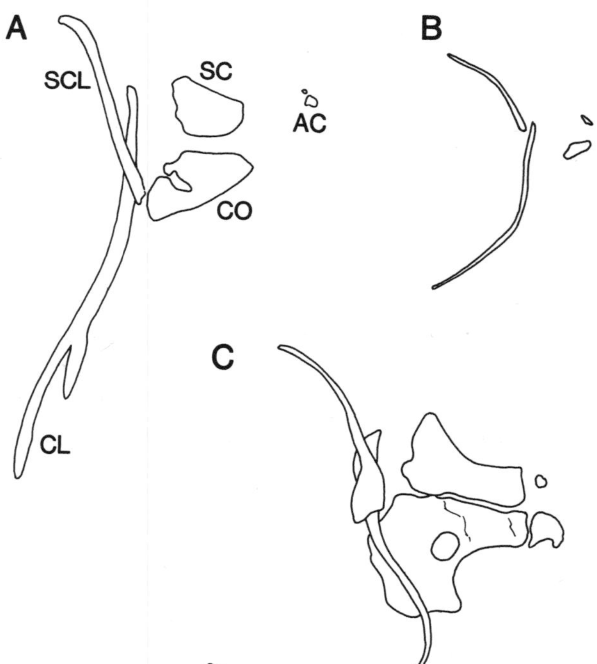

FIGURE 44. Pectoral girdle. A. *Neoconger vermiformis.* B. *Moringua edwardsi,* immature. C. *Moringua edwardsi,* mature male.

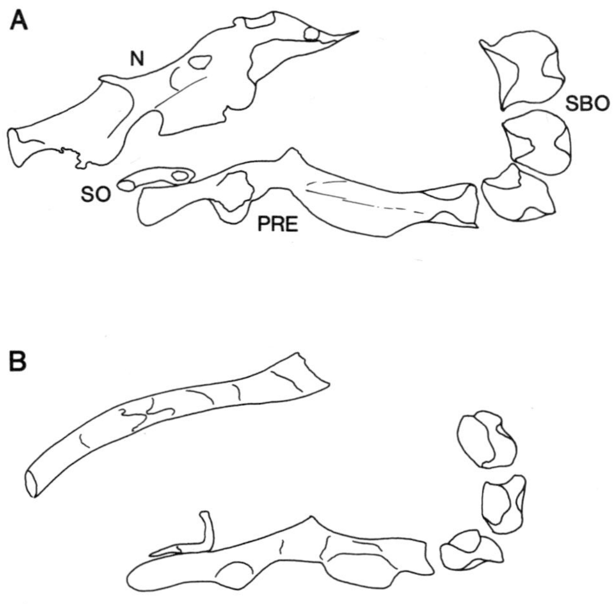

FIGURE 45. Circumorbital bones. A. *Neoconger vermiformis.* B. *Moringua edwardsi.*

etal elements (Smith and Castle, 1972), and the heart and gill arches are displaced posteriorly. *Moringua* shows the most extreme ontogenetic polymorphism of any known eel, so much so that different stages have been described as distinct genera. It is the only eel known that shows sexual dimorphism in the number of vertebrae. Mature males and females leave the bottom at night and swim near the surface. These are all

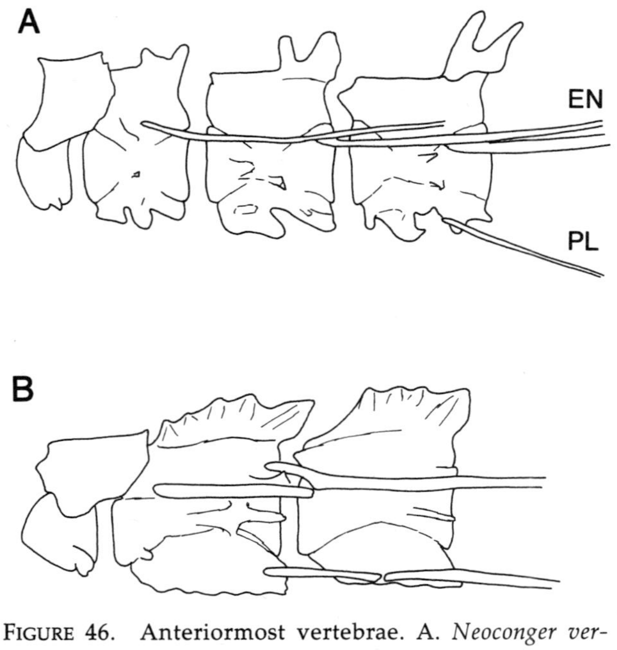

FIGURE 46. Anteriormost vertebrae. A. *Neoconger vermiformis.* B. *Moringua edwardsi.*

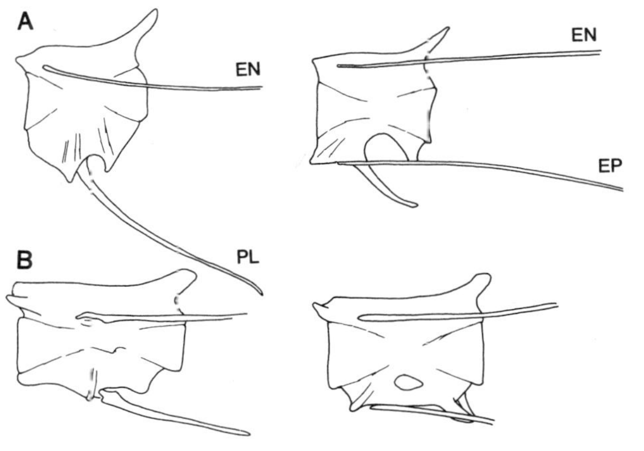

FIGURE 47. Precaudal (left) and caudal (right) vertebrae. A. *Neoconger vermiformis.* B. *Moringua edwardsi.*

specializations that could have been derived from a *Neoconger*-like ancestor.

Distribution. All tropical seas except the eastern Atlantic. The two genera differ in their habitat, *Neoconger* preferring mud bottoms and areas with heavy runoff, and *Moringua* areas with clear water and sand.

Genera. Moringua Gray, 1831, and *Neoconger* Girard, 1858.

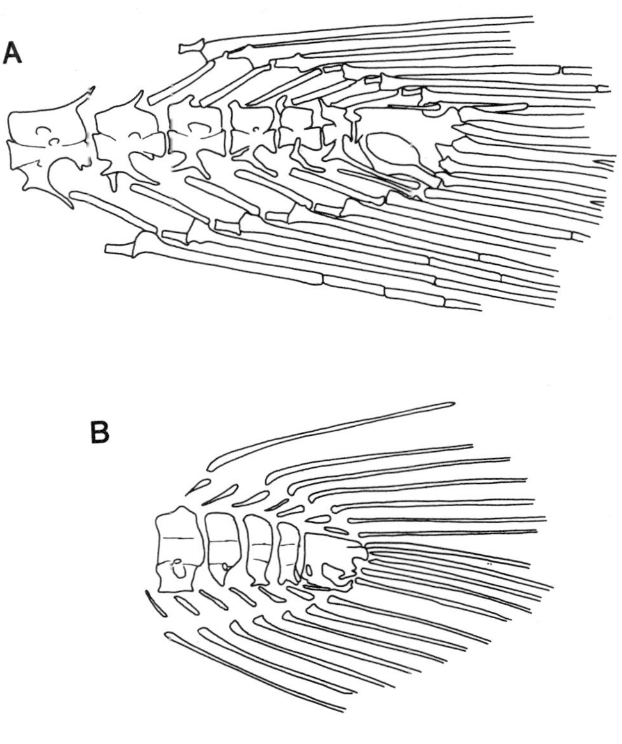

FIGURE 48. Caudal skeleton. A. *Neoconger vermiformis.* B. *Moringua edwardsi.*

Key to the Genera of Moringuidae

1a. Upper jaw projects beyond lower; dorsal fin begins slightly before level of anus; lateral line ends near level of anus .... *Neoconger*, p. 60
1b. Lower jaw projects beyond upper; dorsal fin begins behind level of anus; lateral line complete .... *Moringua*, p. 65

Genus *Neoconger* Girard, 1858

Neoconger Girard, 1858: 171 (type species *Neoconger mucronatus* Girard, 1858, by monotypy).
Chrinorhinus Howell Rivero, 1932: 8 (type species *Chrinorhinus torrei* Howell Rivero, 1932, by monotypy).

Characters. Body moderately elongate, cylindrical anteriorly, strongly compressed near end of tail; anus near midlength; tip of tail soft, blunt, caudal fin well developed and pliable. Dorsal and anal fins low, fleshy, confluent with caudal fin; dorsal fin begins slightly anterior to level of anus; pectoral fin small but well developed. Gill opening crescentic, transverse, on lower half of side, upper corner touching middle to upper edge of pectoral-fin base. Lateral line incomplete, ending near level of anus; head pores present only on lower jaw.

Head conical, smoothly tapering anteriorly, snout projecting beyond lower jaw, rictus behind level of eye. Eye small, covered by transparent skin. Anterior nostril usually not tubular, on side of snout somewhat behind tip; posterior nostril larger, immediately before eye, elongate in an anteroventral to posterodorsal direction, with a raised rim highest on anterior side and lowest at posterodorsal corner. Upper lip continuous with skin of cheek; lower lip separated from skin of lower jaw by a slight groove. Tongue adnate.

Maxilla with one or two series of small, pointed, recurved teeth. Intermaxillary teeth somewhat larger, pointed, in two irregular longitudinal rows, not distinctly separated from maxillary and vomerine teeth. Vomerine teeth in a single row reaching not quite to posterior end of maxillary row. Mandibular teeth pointed, recurved, in one or two series.

Stomach moderate in length, ends some distance before anus. Gas bladder very small, thin walled, located near anus.

Size. These are small eels, seldom exceeding 300 mm TL.

Distribution. Western tropical Atlantic, eastern tropical Pacific, and southwestern Pacific.

Etymology. From the Greek *neo* (new) and *conger*, the conger eel. Masculine.

Species. Three species are recognized: *Neoconger mucronatus* Girard, 1858 from the western Atlantic, *N. vermiformis* Gilbert, 1890 from the eastern Pacific, and *N. tuberculatus* (Castle, 1965) from the southwestern Pacific. *Neoconger mucronatus* and *N. vermiformis* differ only in a few meristic and morphometric characters (Smith and Castle, 1972). *Neoconger tuberculatus* is known from two leptocephali found near Sydney, Australia (Castle, 1965d: 131) and an adult from Fiji (J. E. McCosker, pers. comm.). Evidence from leptocephali suggests that a second species is present in the western Atlantic, but adults have not yet been found (see pp. 63 and 702).

Neoconger mucronatus Girard, 1858
RIDGED EEL
Figures 49–51

Neoconger mucronatus Girard, 1858: 171 (original description, St. Joseph's Island, Texas, lectotype USNM 861). Smith and Castle, 1972: 200 (redescription, synonymy, osteology, life history).
Chrinorhinus torrei Howell Rivero, 1932: 8 (original description, Havana, Cuba, holotype MCZ 32786).
Neoconger torrei. Ginsburg, 1951: 446.
Leptocephalus anaelisae Tommasi, 1960: 93 (original description, 2°27.5′N, 44°02.5′W, holotype unknown). Leptocephalus.

Misidentifications:

Not *Neoconger mucronatus* Girard. Goode and Bean, 1879b: 155 (misidentified *Ahlia egmontis*). Woods, 1942: 191 (probably a large ophichthid).

Distinctive Characters. Those of the genus.

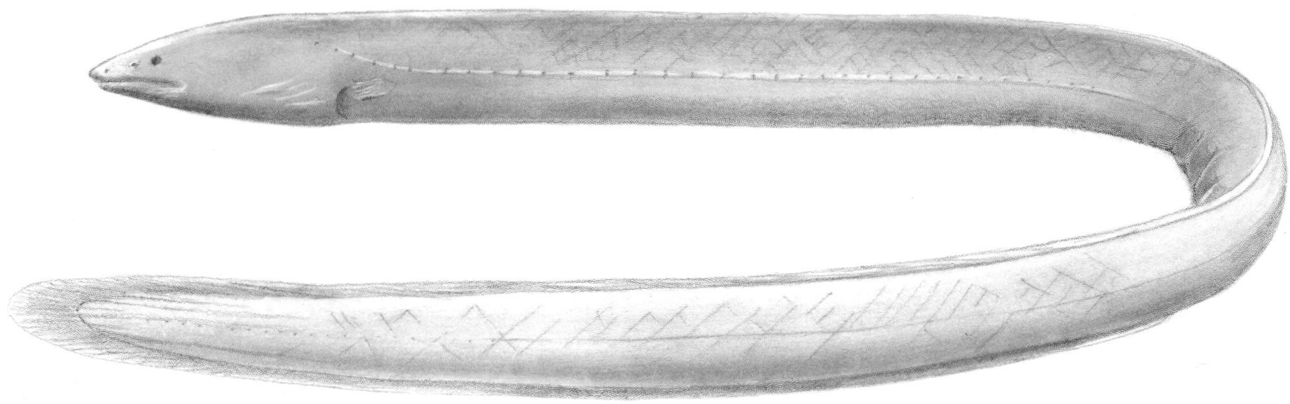

FIGURE 49. *Neoconger mucronatus*: ANSP 94213, 256 mm TL.

Description. Total vertebrae 96–107 (n = 13), preanal vertebrae 43–55 (8), predorsal vertebrae 32–48 (9), precaudal vertebrae ca 49–58 (9). Total LL pores 21–41 (14). Branchiostegal rays 18 (1). Dorsal-fin rays 151–153 (2), anal-fin rays 133 (1), pectoral-fin rays 9–13 (12).

Proportions as % of TL: preanal 48–53 (20), predorsal 37–54 (20), head 9–11 (20), depth at anus 3–5 (19). Of HL: snout 18–22 (21), eye 3–8 (21), interorbital 11–17 (11), snout-rictus 26–35 (20), gill opening 10–19 (21), interbranchial 9–19 (21), pectoral fin 15–34 (21).

Color. In preservative gray or brown, sometimes with dark reticulations. A freshly caught specimen (TCWC 6097.1) was similar, but with some red on posterior part of vertical fins, anterior part of head, and pectoral fin; red appeared caused by blood in superficial vessels and tissue. Pores on lower jaw sometimes surrounded by whitish area. Stomach and intestine pale.

Size. The largest specimen examined was the lectotype, 302 mm TL.

Development and Growth. Two of the specimens examined were males, both from the Gulf of Mexico (FDNR 9078 and TCWC 3981.1). They differ from the females collected in the same area as follows (vertebral counts available only for FDNR 9078): larger eye (8 vs 3–7% HL), longer pectoral fin (33–34 vs 16–25% HL), lesser preanal length (48–50 vs 51–53% TL), lesser predorsal length (45–46 vs 49–54% TL), fewer preanal vertebrae (43 vs 46–47), and fewer predorsal vertebrae (38 vs 41–44). They are also relatively small. The larger specimen (FDNR 9078) is 159 mm TL and has enlarged testes, although it is not fully ripe. The smaller specimen (TCWC 3981.1) is 107 mm and is considerably less advanced. The larger specimen has enlarged nostrils and is the only specimen in which the anterior nostril is tubular.

A similar pattern is found in *Moringua*, where the male is also smaller than the female and has

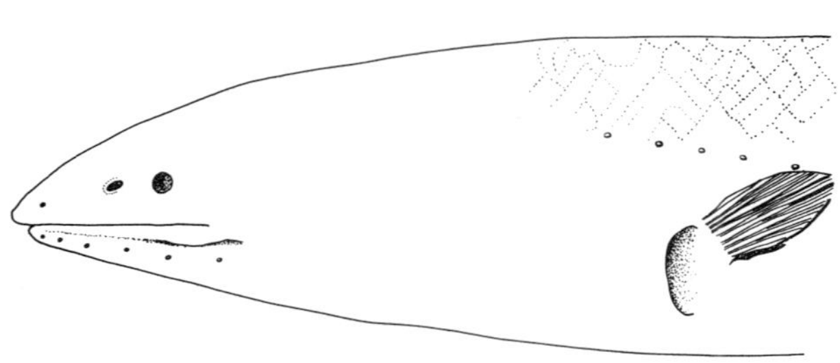

FIGURE 50. *Neoconger mucronatus*: ANSP 94213, 256 mm TL; head.

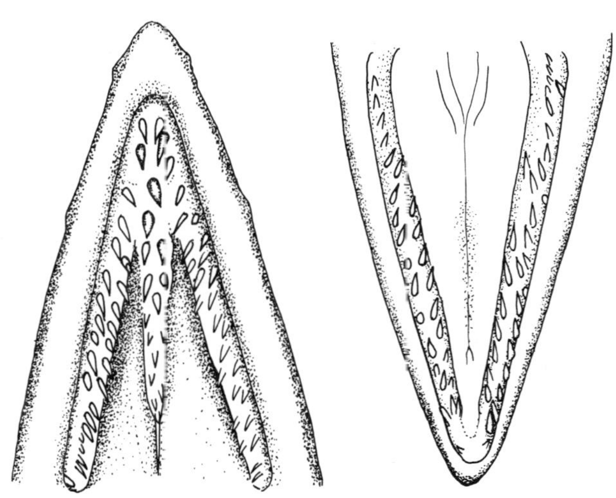

FIGURE 51. *Neoconger mucronatus*: ANSP 94213, 256 mm TL; dentition.

TABLE 1. Numbers of vertebrae in *Neoconger mucronatus* from various locations in the western Atlantic and *N. vermiformis* from the eastern Pacific.

	Total																
	93	94	95	96	97	98	99	100	101	102	103	104	105	106	107	n	x̄
N. mucronatus																	
Gulf of Mexico				2	1	2	3									8	97.8
Caribbean												1	-	-	1	2	105.5
Amazon												1				1	104.0
South Atlantic															1	1	107.0
N. vermiformis	1	-	2	6	5	6	2	11	4	1						38	98.3

	Preanal																	
	40	41	42	43	44	45	46	47	48	49	50	51	52	53	54	55	n	x̄
N. mucronatus																		
Gulf of Mexico				1	-	-	2	1									4	45.5
Caribbean									1	1							2	48.5
Amazon					1												1	44.0
South Atlantic																1	1	55.0
N. vermiformis	1	3	12	12	2												30	42.4

	Predorsal																		
	32	33	34	35	36	37	38	39	40	41	42	43	44	45	46	47	48	n	x̄
N. mucronatus																			
Gulf of Mexico							1	-	-	1	1	1	1					5	41.6
Caribbean											1	-	-	1				2	43.5
Amazon	1																	1	32.0
South Atlantic																	1	1	48.0
N. vermiformis		1	3	12	12	1												29	36.3

a larger eye and pectoral fin. In *Moringua* the changes are more extreme, however, and the mature female is modified as well. Mature females of *Neoconger* are unchanged externally from the immature condition. In addition, mature individuals of *Neoconger* do not seem to become pelagic as those of *Moringua* do.

Both males were collected during the spawning season of this species in the Gulf of Mexico (September–December). Smith and Castle (1972: 199) reported finding no males of either species of *Neoconger* studied. This is similar to the situation found in the Nemichthyidae, where most specimens are either females or indeterminate (Nielsen and Smith, 1978: 12). Nielsen and Smith (1978: 13) suggested as an explanation that maturation in males is more rapid than in females and takes place shortly before spawning. If spawning of *Neoconger mucronatus* is a highly seasonal affair, as it is in the Gulf of Mexico, then perhaps collections must be made at that time if any recognizable males are to be found.

The number of lateral-line pores increases with growth up to a point. Four specimens of 107–159 mm TL had 22–35 pores; eight specimens of 187–302 mm TL had 35–42. Within the latter group there was no clear correlation of pore number with TL.

The leptocephalus has been identified (see p. 701).

Habits and Behavior. From all indications, *Neoconger mucronatus* is a burrowing eel that spends much of its life within the bottom sediment. The reduced, almost vestigial eyes, low dorsal and anal fins, heavy conical head, pointed snout, and uniform gray color all seem adaptations for a fossorial existence. Specimens have generally been collected on soft mud bottoms.

TABLE 2. Predorsal length in *Neoconger mucronatus* from various locations in the western Atlantic and *N. vermiformis* from the eastern Pacific, expressed in % TL.

	Predorsal																			
	37	38	39	40	41	42	43	44	45	46	47	48	49	50	51	52	53	54	n	x̄
N. mucronatus																				
Gulf of Mexico									1	1	–	–	1	4	6	1	–	1	15	50.1
Caribbean												1	1	1					3	49.0
Amazon	1																		1	37.0
South Atlantic																1			1	52.0
N. vermiformis					2	2	12	17	10	5	1								49	44.0

Mature males differ from females in having a larger eye and enlarged nostrils, with the anterior nostril bearing a well developed tube. This suggests that the male actively searches for the female at breeding time, using both visual and olfactory clues.

In the western Gulf of Mexico the spawning season lasts from September to December, with September probably being the peak month (see p. 702).

Variation. Only 24 adult and juvenile *Neoconger mucronatus* were available for study, and 18 of these came from the Gulf of Mexico. Of the remainder, four came from the Caribbean and one each from the Amazon region of Brazil and the Brazilian South Atlantic (see Study Material). Only two of the Caribbean specimens yielded complete vertebral counts (ANSP 139683, UMML 22149). As small as it is, the sample still reveals four distinct populations. Specimens from the Gulf of Mexico have distinctly fewer total vertebrae than any of the others (Table 1). In preanal vertebrae (Table 1), the Gulf of Mexico and Amazonian specimens are about the same, the Caribbean specimens are slightly higher, and the South Atlantic specimen is sharply higher. Predorsal vertebrae show a still different pattern (Table 1), with the Gulf of Mexico and Caribbean specimens broadly overlapping, the South Atlantic specimen distinctly higher, and the Amazonian specimen much lower. In predorsal length (Table 2), the Amazonian specimen falls well below the other three populations.

Leptocephali are much more commonly col-

TABLE 3. Numbers of myomeres in leptocephali of *Neoconger mucronatus* from different areas.

	Total myomeres																			
Area	93	94	95	96	97	98	99	100	101	102	103	104	105	106	107	108	109	110	n	x̄
Gulf of Mexico	1	2	2	3	5	2	3	3	1	–	1								23	97.4
Caribbean					2	–	2	5	9	5	7	10	13	4	1	1			59	103.0
Guianas-Brazil			1	–	–	1	2	2	5	5	4	1	3	–	–	–	1		25	101.8

	Preanal myomeres													
Area	49	50	51	52	53	54	55	56	57	58	59	60	61	n	x̄
Gulf of Mexico				2	6	4	4	4	3					23	54.5
Caribbean				1	5	7	6	9	9	9	7	3		56	56.3
Guianas-Brazil	2	1	2	3	3	2	5	3	1	1	1	–	1	25	54.0

	Predorsal myomeres																			
Area	39	40	41	42	43	44	45	46	47	48	49	50	51	52	53	54	55	56	n	x̄
Gulf of Mexico						1	3	–	3	3	7	2	1						20	47.9
Caribbean			1	1	1	1	–	1	3	3	9	6	5	5	4	3	1	1	45	49.8
Guianas-Brazil	1	6	3	4	3	–	1	–	–	–	–	2	–	1	1				22	43.1

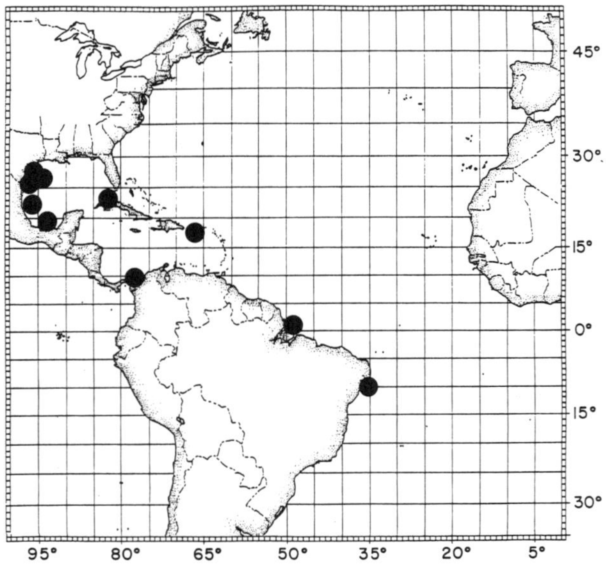

FIGURE 52. Distribution of *Neoconger mucronatus*.

lected than adults, and it may be informative at this point to examine the distribution of myomere counts and see how they compare with the picture just drawn from the limited adult material (Table 3; some of these data contributed by M. M. Leiby). Three populations are evident: one in the Gulf of Mexico, the second in the Caribbean, Straits of Florida, and the Gulf Stream, and the third from the Lesser Antilles to Brazil (no specimens were available from south of the hump of Brazil).

Table 3 is arranged by geographic location for convenience, but it is clear that two distinct forms occur together in the extreme eastern Caribbean and off the Guianas. These are most clearly revealed by the predorsal myomeres, which show little or no overlap. Within the Caribbean, the low-count form has been recorded only at the eastern extremity, in the Lesser Antilles. The high-count form extends southeastward to about the northern border of Brazil, but beyond that point only the low-count form has been collected. The low-count form differs further in that the intestinal loop is much flatter than in either the high-count or the Gulf of Mexico leptocephali. Preanal and total myomeres also differ, but to a lesser degree. The single adult available from the Amazon region of Brazil (USNM 214062) has a much more anterior dorsal-fin origin than any of the specimens from the Gulf of Mexico or the Caribbean (Table 1), paralleling the condition of the leptocephali.

The sympatry of the two forms of leptocephali in the Antilles-Guianas area suggests that they are specifically distinct, but until more adult material from this area is available it may be best to avoid formally recognizing them.

The single specimen from the Brazilian South Atlantic (around 10°S, ZMUC P312001) differs markedly from the Amazonian specimen in preanal vertebrae (55 vs 44), predorsal vertebrae (48 vs 32), and predorsal length (52 vs 37% TL) (Table 1). Evidently there is a sharp discontinuity between the populations of the Brazilian North Atlantic and South Atlantic. In fact, the entire range of variation among populations of *Neoconger* within the western Atlantic is about as great as the variation across the Isthmus of Panama.

Distribution. Gulf of Mexico, Caribbean, and south to Brazil. Depth 3–183 m, mean 61 m.

This species is probably more common than would be inferred from the scarcity of adult records. The leptocephali are plentiful at certain times and places, and Smith and Castle (1972: 23) reported 739 specimens from a single haul. The fossorial habits of the adults probably keep them out of most trawl samples.

Etymology. Latin *mucronatus* (sharp, pointed), perhaps in reference to the conical snout. An adjective.

Study Material. 24 specimens, 98–302 mm TL.

Lectotype: USNM 861 (female: 302), St. Joseph's Island, Texas. Selected by Smith and Castle, 1972: 200.

Paralectotypes: USNM 204928 (3: 263–267), same data as lectotype.

Other Material: TEXAS: ANSP 94213 (1, 256), 110071 (1, 265+), 136001 (1, 176). FDNR 9078 (1, 159). USNM 154997 (1, 98), 154998 (1, 209). MEXICO GULF: TCWC 3981.1 (1, 107), 6097.1 (6, 187–265). UMML Oregon 4810 (1, cleared and stained). CUBA: MCZ 32786 (1, 168; holotype of *Chrinorhinus torrei* Howell Rivero). PUERTO RICO: ANSP 139683 (1, 143). COLOMBIA: UMML 22149 (1, 193), 23294 (1, 188+). BRAZIL: USNM 214062 (1, 156). ZMUC P312001 (1, 261).

Moringua Gray, 1831

Moringua Gray, 1831a: pl. 95, fig. 3 (type species *Moringua linearis* Gray, 1831, by monotypy); 1831b: 9.

Rataboura Gray, 1831a: pl. 95, fig. 4 (type species *Rataboura hardwickii* Gray, 1831, by monotypy); 1831b: 9.

Pterurus Swainson, 1839: 196, 334 (type species *Pterurus maculatus* Swainson, 1839, by subsequent designation of Swain, 1883: 283).

Pachyurus Swainson, 1839: 196, 335 (substitute name for *Moringua* Gray, to take the same type species).

Ptyobranchus McClelland, 1844: 200 (type species *Ptyobranchus arundinaceus* McClelland, 1844, by subsequent designation of Jordan, 1919: 220).

Aphthalmichthys Kaup, 1856a: 68 (type species *Aphthalmichthys javanicus* Kaup, 1856, by monotypy); 1856b: 105.

Pseudomoringua Bleeker, 1864a: 114 (type species *Moringua lumbricoidea* Richardson, 1845, by original designation).

Stilbiscus Jordan and Bollman, 1889: 549 (type species *Stilbiscus edwardsi* Jordan and Bollman, 1889, by original designation).

Mayerina Silvester, 1916: 214 (type species *Mayerina mayeri* Silvester, 1916, by original designation).

Anguillichthys Mowbray in Breder, 1927: 10 (type species *Anguillichthys bahamensis* Mowbray, 1927, by monotypy).

Merinthichthys Howell Rivero, 1935: 343 (type species *Merinthichthys sanchezi* Howell Rivero, 1935, by original designation).

Discussion of Synonymy. *Moringua* and *Rataboura* were described twice by Gray in publications that may have appeared as little as a month apart. *Moringua linearis* Gray and *Rataboura hardwickii* Gray appeared on plates but without text in his "Illustrations of Indian Zoology . . ." (Gray, 1831a), which was published no later than 25 January 1831 (Sawyer, 1953: 51). Full descriptions appeared in "Zoological Miscellany" (Gray, 1831b), published in February, 1831 (Neave, 1940a: 218). Cantor (1853: 228, 234) synonymized all the various names of Gray, Swainson, and McClelland and recognized *Moringua* as the valid genus. Cantor is clearly the first reviser under ICZN rules (Article 24) and must be considered to have established the validity of *Moringua*. Subsequent speculations on line priority (Jordan, 1917: 138; Castle, 1968a: 13) are irrelevant. Later workers almost universally accepted *Moringua*, the major exception being Fowler, who used *Rataboura*.

Characters. Body elongate to very elongate, cylindrical, slightly compressed near tip of tail; anus well behind midlength, preanal length about ⅔ to ¾ TL; tip of tail soft, blunt. Anal fin begins a distinct distance behind anus, dorsal fin begins at about same level; dorsal and anal fins low in immature individuals, confluent with low caudal fin; in mature males and females, both dorsal and anal fins become elevated anteriorly, remain low at midlength, and become elevated again near tip of tail, where they join caudal fin to form a truncate or emarginate tail fin. Pectoral fin minute in immatures, well developed in mature males and females. Lateral line complete on body, generally one pore per vertebra except near tip of tail; head pores apparently absent.

Head relatively short but slender, branchial region somewhat expanded. Mouth moderate, rictus below or very slightly behind eye; lower jaw projects beyond upper. Eye minute and buried beneath semi-opaque skin in immatures, well developed and exposed in mature males and females. Anterior nostril a low tube near tip of snout; posterior nostril a simple opening in front of eye. Anterior border of gill opening extended posteriorly into a small flap that covers gill aperture when closed.

Teeth moderately enlarged, few in number, conical with recurved tips. Intermaxillary teeth relatively large, arranged in a semicircle of approximately six teeth. Vomerine teeth distinctly smaller than intermaxillary teeth, arranged in a single series behind them. Maxillary and mandibular teeth uniserial. Teeth at least partially lost at maturity.

Stomach and intestine begin far behind gill opening (43% of distance from gill opening to anus in specimen examined, ANSP 148162). Stomach large, thin walled, ends before anus. Gas bladder apparently absent.

Discussion. The eels of the genus *Moringua* show an extraordinary sexual and ontogenetic variability. So different in appearance are the immatures, the mature females, and the mature males that they have been placed in separate genera and even separate families.

The immatures of both sexes are yellowish, worm-like, burrowing animals that appear to spend all their time buried in the sand. The eyes are reduced to small dark spots buried beneath the skin and can probably function only to distinguish light from dark. The pectoral fin is rudimentary and often concealed by the flap on the gill opening. The dorsal and anal fins are low and hardly visible, and the caudal fin is low and rounded.

Mature females, by contrast, are strongly countershaded, dark gray or black above and white below, and have a prominent pectoral fin. The eyes are relatively large and appear fully functional, lying beneath a transparent "spectacle" of skin. The dorsal and anal fins are distinctly elevated anteriorly, the elevated parts being directly above and below each other. Posterior to this point, both fins become low again, but just before meeting the caudal fin their rays lengthen to form a truncate or emarginate tail fin. The form of the vertical fins gives the fish an arrow-like appearance and approaches the condition found in more typical, non-anguilliform fishes. The maxillary and vomerine teeth are lost.

Mature males resemble females but are much smaller and the form of the dorsal and anal fins is even more exaggerated. The eye and pectoral fin are larger than those of the mature female, and the upper lip projects ventrally, flap-like, to cover part of the lower jaw. Mature males also tend to lose the maxillary and vomerine teeth. Both mature males and mature females are free swimming and are usually caught at the surface around night lights.

In addition to the superficial differences, there are major internal differences. In immatures and mature females, the heart is displaced posteriorly behind the pectoral fin. In mature males, however, the heart is usually located forward in the normal position. Finally, there is a difference in vertebral number between males and females, a feature that seems to be unique among eels.

Parr (1930) apparently was the first to suspect the nature of the variation found in *Moringua* when he suggested that *Anguillichthys bahamensis* Mowbray and *Stilbiscus edwardsi* Jordan and Bollman were male and female of the same species. He pointed out that specimens of the former were invariably small (140 mm or less) and were all males, whereas his only specimen of the latter was larger and a female. He also noted that the differences were of size and proportion only, not of kind, and could be sexual dimorphism. Gordon (1954) examined three males and four females collected at Bimini, Bahamas and confirmed Parr's suspicions. Neither Parr nor Gordon thought to connect *edwardsi-bahamensis* with the worm-like immature form, which was passing under the generic name *Aphthalmichthys.* That remained for Gosline and Strasburg (1956), who assembled all the relevant information on the Hawaiian *Moringua* and concluded, though somewhat tentatively, that there was but a single species that went through three distinct morphological phases. They also discovered sexual dimorphism in the number of vertebrae. Böhlke and Chaplin (1968: 67) suggested that the same situation prevailed in the western Atlantic and that a single species, *Moringua edwardsi,* was present. Castle and Böhlke (1976) confirmed this and also showed that the sexual dimorphism in vertebral number is reflected in the myomere counts of leptocephali.

Not only do the mature males and females differ greatly from each other in size, but the females themselves show an unusual variation in the size at which they reach maturity. In the specimens of *M. edwardsi* examined for this study, mature females ranged from 193 to 512 mm TL, a difference of more than two-and-a-half times. This could explain the observation of Gosline and Strasburg (1956: 17) and Böhlke and Chaplin (1968: 68) that some immatures are larger than the largest sexually mature females. Mature females (and males as well) are less common in collections than the immatures. This may be because maturity is a terminal phase through which the eel passes rather briefly before spawning and dying. If females mature over a wide range of sizes, and if at all times mature individuals are outnumbered by immatures, then it would be statistically unlikely that the extreme sizes would be represented by the less numerous mature eels. At any rate, after examining more than 600 specimens of *Moringua edwardsi*, I found that the size difference between mature and immature females was not very great. In fact, the largest specimen, 512 mm, was a semi-ripe female.

Gosline and Strasburg raised another question that can probably be answered, namely, where are the immature males? They assumed that all their immature specimens were females and felt it unlikely that the males ever went through such a fossorial stage. Their reasoning was based on such characters as the greater head length and eye size in males, and the unlikeliness that the heart should move backward (where it is in immatures) and then forward again (where it is in mature males). These assumptions seem unwarranted. Castle and Böhlke (1976) showed that the sex, as determined by myomere counts, is established as early as the leptocephalus stage, and that male and female leptocephali occur in roughly equal numbers. It is reasonable to assume that males go through the same fossorial phase as females but simply mature at a smaller size. This hypothesis could easily be tested by x-raying a large number of small, immature *Moringua* to see how the vertebral counts are distributed.

If males are so much smaller than females, are they younger as well? Do they simply mature earlier, short-circuiting much of the juvenile, fossorial phase, or do they live as long as females but without growing? Direct evidence is not available. At least some mature males retain traces of larval pigmentation, especially the melanophores located on the intestinal loop (see page 000 below). It is uncertain, however, whether the retention of larval pigment in eels is a function of age or merely of size.

Finally, there is the chicken-or-egg question of which represents the primitive form, the fossorial juvenile or the free-swimming adult. Gosline and Strasburg (1956: 16) felt that the mature males were the least specialized and hence the most primitive. This is a reasonable conclusion; certainly the well developed eye and fins and the anterior position of the heart are more general—more "normal"—conditions than the reduced eye and fins and the posteriorly displaced heart. Nevertheless, each of these "primitive" characters is derived ontogenetically from the more highly modified state found in the juveniles. Furthermore, the other moringuid genus, *Neoconger*, which is more primitive overall than *Moringua*, is entirely fossorial; it has no pelagic stage. Another observation: although the eye in *Moringua* enlarges late in life, the orbit remains very small. A small orbit is characteristic of small-eyed fossorial eels such as *Neoconger*, *Pythonichthys*, and some ophichthids. Large-eyed eels such as *Anguilla* have a much larger orbit. Thus, one could argue that the free-swimming mature form of *Moringua* is a secondary development from a wholly fossorial ancestor.

Size. Maximum size slightly greater than 1 m TL.

Distribution. Tropical western Atlantic and Indo-West Pacific.

Etymology. Evidently a vernacular (perhaps Portuguese) corruption of *Muraena*. Feminine.

Species. The number of valid species of *Moringua* is uncertain, but it is probably small. A single species, *Moringua edwardsi* (Jordon and Bollman, 1899), is now recognized in the western Atlantic. In the Indo-Pacific the situation is less clear and needs further study. Castle (1968a) suggested that three species are present in the western Indian Ocean.

Moringua edwardsi (Jordon and Bollman, 1889)

SPAGHETTI EEL

Figures 53–55

Stilbiscus edwardsi Jordan and Bollman, 1889: 549 (original description, Green Turtle Cay, Bahamas, holotype USNM 41735).

Aphthalmichthys caribbeus Gill and Smith, 1900: 974 (original description, Puerto Rico, holotype USNM 49720).

Moringua edwardsi. Gill and Smith, 1900: 974.

Leptocephalus diptychus Eigenmann and Kennedy, 1901: 831, fig. (original description, New Providence, Bahamas, holotype USNM 49753). Leptocephalus.

Mayerina mayeri Silvester, 1916: 214 (original description, Puerto Rico, holotype PU 3037, lost).

Aphthalmichthys mayeri. Silvester, 1918: 19, pl. 1.
Moringua boeki Metzelaar, 1919: 18, fig. 8 (original description, Curaçao, holotype ZMA 112.739).
Anguillichthys bahamensis Mowbray in Breder, 1927: 10, fig. 5 (original description, Green Cay, Bahamas, holotype YPM 34).
Stilbiscus bahamensis. Parr, 1930.
Merinthichthys sanchezi Howell Rivero, 1934: 343, fig. 3 (original description, Cuba, holotype MCZ 33453).

Discussion of Synonymy. In their description of *Leptocephalus diptychus,* Eigenmann and Kennedy (1901: 831) chose a 51-mm metamorphic specimen from New Providence (USNM 49753, although the authors did not give catalog numbers) as the holotype ("type"). The other specimen on which the description was based, a 38-mm premetamorphic individual from the Gulf Stream southeast of Nantucket (USNM 49754), was not given any type designation. In a later paper (1902: 84) they followed this scheme in the body of the text, but in the list of specimens at the beginning of the account they switched the two, calling the 38-mm specimen the "type" and the 51-mm specimen the "cotype." Castle (1979: 3) added to the confusion by reversing the total lengths of the two specimens and citing the wrong catalog number as the type. Eigenmann and Kennedy (1902) made another error. They cited "Science, XII, p. 401, 1900" as the original description. That reference, however, is a paper by Eigenmann on putative eggs and larvae of the conger eel and contains no mention whatever of *Leptocephalus diptychus.* Castle (1979: 3) repeated this error.

Metzelaar (1918: 18) intended *Moringua boeki* as a provisional name "if any adult specimen might be found identifiable with the juvenile in hand." The specimen is a mature male, however, with a large eye, high fins, and a strongly countershaded color pattern.

For a complete list of references to this species, see the synonymy in Castle, 1979: 3.

Distinctive Characters. Those of the genus.

Description (including data from Castle and Böhlke, 1976). Total vertebrae 109–123 (n = 175), predorsal vertebrae 74–84 (71), preanal vertebrae 74–84 (71). Branchiostegal rays 11 (1). LL pores to anus 68–75 (26).

Proportions as % TL: preanal 66–71 (38), anus to anal-fin origin 6–10 (38), predorsal 74–78 (38), head 7–12 (38), depth at anus 1–3 (36). Of HL: snout 13–16 (38), eye 2–9 (38), snout-rictus 14–22 (38), gill opening 4–11 (37), interbranchial 10–18 (37), pectoral-fin length 1–22 (34).

Color. In life, immatures nearly uniform yellow or orange above, yellow or white below. Silvester (1918: 19) described the ventral pigmentation as "slate-blue, merging into whitish in anterior part of tail region." In preservative, a featureless light gray to tan. Mature males and females countershaded, dark gray to black above, pale below, the demarcation line approximately at middepth from shortly behind head to origin of anal fin, then descending and becoming obscured about halfway to tip of tail. Vertical fins dark basally, light distally. Two specimens of 69 and 96 mm TL (ANSP 105483) retain some larval pigment.

Size. Females mature over a wide range of sizes. The largest specimen examined was a semi-ripe female from Bermuda, 512 mm TL (ANSP 152000). This is probably close to the maximum size for the species; Böhlke and Chaplin (1968: 69) stated that it grows to about 20 inches (508 mm). The smallest mature female was 193 mm (ANSP 101106) and was actually further along in development than the 512-mm specimen, with large eggs filling the body cavity.

Mature males ranged from 114–149 mm TL. Thus it appears that any *M. edwardsi* much larger than 150 mm is likely to be a female.

Development and Growth. Leptocephali of *Moringua edwardsi* (see p. 699 for description) grow to about 50 mm before metamorphosing. They shrink somewhat during metamorphosis; Castle (1979: 4) illustrated a metamorphic specimen of 44.2 mm, although the one described by Eigenmann and Kennedy (1902: 84, fig. 2c–e) was 51 mm. The newly transformed juvenile is thus between 40 and 50 mm TL. Growth commences again, and the eel remains in the fossorial "aphthalmichthys" stage until it reaches about 110–150 mm in males, about 200–500 mm in females. The duration of this stage is unknown. Eventually the reproductive phase begins, and

FIGURE 53. *Moringua edwardsi*: ANSP 101540, immature female, 347 mm TL; inset: mature male, 117 mm TL. Illustration by S. P. Gigliotti.

the eel undergoes a second metamorphosis, becoming the large-eyed, large-finned, countershaded adult form. It is not known whether further growth occurs at this point or whether the fish simply spawns and dies.

Habits and Behavior. The immature form of *Moringua edwardsi* appears to spend all its time buried in the sand. There is no evidence that it regularly emerges from the bottom and swims freely. Indeed, its structure would seem to argue against any such behavior. Observations of living specimens indicate that they burrow headfirst. The morphology of the head and body are well adapted for such behavior; the stout, conical snout and the heavy, protruding lower jaw make an excellent plow, and the cylindrical body provides power. Böhlke and Chaplin (1968: 69) reported that *M. edwardsi* "... is capable of making fairly rapid progress through the sand, just beneath the surface." It undoubtedly hunts by smell and touch, feeding on whatever benthic or burrowing organisms it comes across.

Mature males and females, by contrast, are commonly taken by dip nets on the surface at night. They are also taken sometimes by rotenone along with the immatures. Böhlke and Chaplin suggested that they burrow during the day and come out at night. Many eels have similar habits. It is not known whether free-swimming adults feed during their excursions or whether this behavior is related entirely to reproduction.

There is no direct information on spawning itself. Although mature males and females are frequently seen and collected at the surface, no one has reported an actual spawning act. Nor is it known precisely where spawning may occur. Castle (1979: 8) found the smallest leptocephali (less than 20 mm) in an area centered on the

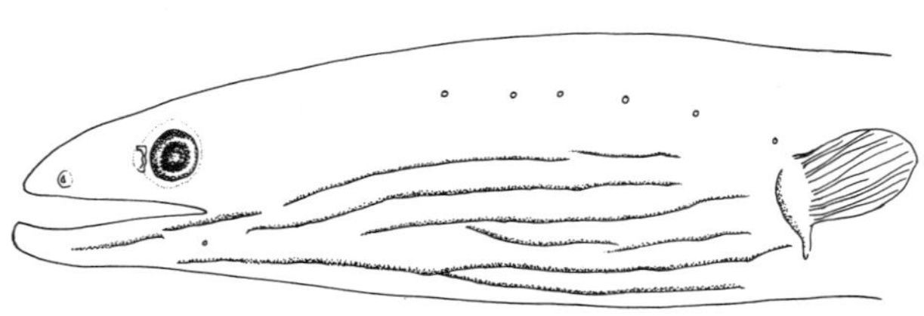

FIGURE 54. *Moringua edwardsi*: ANSP 103421, mature female, 380 mm TL; head.

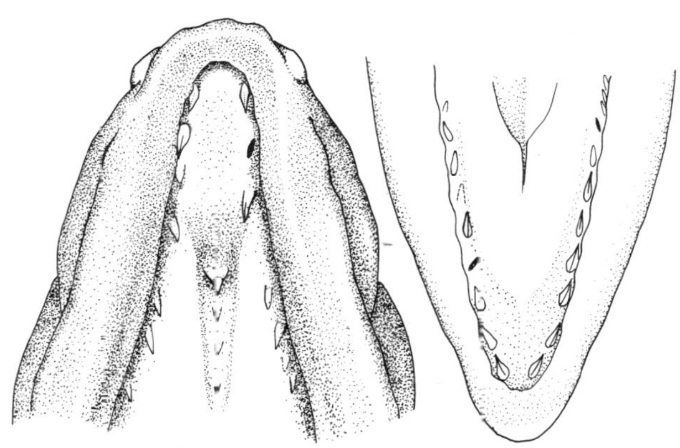

FIGURE 55. *Moringua edwardsi*: ANSP 158293, female, 391 mm TL; dentition.

TABLE 4. Differences in meristic and morphometric characters between immature females, mature females, and mature males of *Moringua edwardsi.* HL and predorsal length in % TL, snout, eye, and pectoral fin in % HL.

	HL	Predorsal length	Snout length	Eye	Pectoral fin	LL pores	Total vertebrae	Predorsal vertebrae	Preanal vertebrae
Immature ♀	7	75–78	13–16	2–3	1–5	70–75	—	—	—
Mature ♀	6–8	74–77	13–16	3–9	8–17	70–75	—	—	—
Combined ♀	—	—	—	—	—	—	116–123	80–84	79–84
Mature ♂	9–12	74–76	13–14	5–9	12–22	68–71	109–117	74–79	74–80

Bahamas and the islands of the West Indies. This suggests that the adults do not travel great distances to spawn. Perhaps a simple movement offshore is sufficient. There is no evidence of a distinct seasonal pattern of spawning.

The fate of adults after spawning is also unknown. The extent of the modifications suggests that spawning is a terminal event and that the individuals die soon after. The males, at least, lose some of their teeth, as do males of *Anguilla* and the nemichthyids, which are one-time spawners. A study of stomach contents of mature *Moringua* could establish whether or not they feed. If they do not, then they almost certainly die after spawning.

Variation. The differences between immatures, mature females, and mature males of *Moringua* have been discussed above in the account of the genus. Table 4 shows the numerical values that can be placed on the variation in *Moringua edwardsi.* Immature females are defined as those that have not developed the countershaded color pattern.

Geographic variation in *Moringua edwardsi* has not been demonstrated, but specimens from the Brazilian South Atlantic have not been examined.

Remarks. Moringua edwardsi is similar in several characters to the species from the western Indian Ocean treated by Castle (1968a) as *M. ferruginea* Bliss. The latter has 115–125 vertebrae, 73–75 preanal lateral-line pores, and a head length comparable to that of *M. edwardsi.* Until the species structure of Indo-Pacific *Moringua* is better understood, the precise relationship of *M. edwardsi* cannot be determined.

Distribution. Known from Bermuda, the Bahamas, and Florida south through the Caribbean to Venezuela; also from the state of Bahia in Brazil (recorded at MZUSP by J. E. Böhlke, unpublished). Although Bright and Cashman (1974: 346) reported a specimen from the West Flower Garden Bank, off Texas, the species is not common in the western Gulf of Mexico. Depth from the surface to some 22 m (Böhlke and Chaplin, 1968). Common in areas of clear water and sandy bottom.

Etymology. Named for Dr. Charles Lincoln Edwards of the University of Cincinnati, who collected the type specimen (Jordan and Evermann, 1896: 363).

Study Material. 729 specimens, 51–512 mm TL.

BERMUDA: ANSP 133540 (2, 158–198), 133542 (1, 152), 133543 (1, 222), 133544 (2, 70–80), 133545 (1, 125), 133546 (1, 104), 133547 (12, 100–215), 133548 (5, 79–237), 148161 (2, 300–457), 148162 (4, 244–425), 148167 (1, 387), 152000 (2, 324–512), 158293 (1, 391). BAHAMAS: AMNH 22857 (3, 126–268), 23652 (4, 119–345), 23979 (2, 127–136), 27746 (1, 264), 27773 (1, 130). ANSP 101067 (8, 54–180), 101068 (4, 173–308), 101070 (1, 232), 101079 (5, 140–224), 101082 (3, 156–207), 101085 (6, 111–258), 101087 (15, 81–305), 101090 (26, 68–161), 101091 (2, 135–214), 101096 (3, 145–295), 101098 (33, ca 69–270), 101099 (4, 63–265), 101101 (3, 125–232), 101106 (1, 193), 101107 (23, 92–226), 101108 (5, 68–245), 101112 (4, 102–173), 101115 (70, ca 65–280), 101118 (9, 116–268), 101122 (1, 268), 101126 (4, 63–288), 101127 (1, 161), 101128 (15, 51–262), 101132 (7, 137–245), 101135 (3, 86–200), 101136 (10, 64–226), 101137 (11, 111–256), 101139 (3, 59–73), 101145 (17, 68–228), 101149 (29, ca 122–327), 101151 (5), 101154 (9), 101540 (1, 347), 101541 (101, ca 90–240), 115097 (22, 85–236), 115126 (1, 102), 116304 (3, 152–188), 127078 (11, 76–278), 128796 (7, 78–235), 130553 (3, –92), 137926 (3, 114–221), 140464 (1, 215), 143169 (1, 152), 147265 (2, 296–324). YPM 34 (1, 115;

holotype of *Anguillichthys bahamensis* Mowbray), 35 (1, 115; paratype of *Anguillichthys bahamensis*). JAMAICA: ANSP 114993 (1, 304). HISPANIOLA: ANSP 119044 (3, 135–280), 119051 (8, 69–180), 127266 (2, 88–132), 127296 (20, 65–213), 127979 (1, 286). PUERTO RICO: ANSP 103421 (1, 380), 103422 (1, 145), 115641 (2, 244–295), 118627 (6), 118702 (1, 262), 126343 (20, 65–213), 129901 (8, 116–285), 139837 (5, 109–253), 139844 (6, 150–250). VIRGIN ISLANDS: ANSP 124486 (1, 251). LESSER ANTILLES: ANSP 105483 (2, 263–315), 105484 (1, 319), 105499 (5, 139–141), 105764 (4, 70–202), 101789 (2, 280–292). CURAÇAO: ZMA 112.739 (1, 135; holotype of *Moringua boeki* Metzelaar). VENEZUELA: ANSP 110407 (1, 355). COLOMBIA: ANSP 117557 (1, 258), 117558 (7, 237–331), 117559 (7, 280–345), 117560 (14). PANAMA: 113749 (1, 143), 113758 (1, 285), 113773 (4), 113774 (21, 77–245). MEXICO QUINTANA ROO: ANSP 117325 (1, 66). PROVIDENCIA: ANSP 128804 (1, 143). CAYMAN ISLANDS: ANSP 102267 (1, 122), 104903 (6, 121–231), 104930 (4), 104967 (9, 92–210), 105017 (5, 112–175).

Family Chlopsidae

FALSE MORAYS

DAVID G. SMITH

Acknowledgments. See family Anguillidae.

Introduction. The Chlopsidae is a family of small, moray-like eels found in warm seas worldwide. Chlopsids differ from muraenids primarily in lacking some of the specializations of the latter. These characters include the presence of the fifth ceratobranchial and second infrapharyngobranchial (absent in muraenids), two upper pharyngeals (one in muraenids), a vertical suspensorium (inclined posteriorly in muraenids), small pharyngeal jaws (greatly enlarged in muraenids), and pectoral fin sometimes present (always absent in muraenids). In addition, the posterior nostril is displaced ventrally in chlopsids, whereas in muraenids it is displaced dorsally. The Chlopsidae in turn is somewhat more specialized than the Myrocongridae, which is the primitive sister group of both the Chlopsidae and the Muraenidae (Smith, 1984a). The three families form a fairly cohesive and well defined group within the Anguilliformes.

Böhlke (1956a) accepted the name Xenocongridae Regan, 1912 because it had priority over Chlopsidae Grassi, 1913. Rafinesque (1810b: 42), however, had created a name Chlopsidini, which contained his species *Chlopsis bicolor* and *Nettastoma melanura.* Although Rafinesque ranked the Chlopsidini as an order, it should be considered a family-group name, because it was placed one step above the generic level and was based on an included genus. Rafinesque's classification was not comparable to one that would be used today. He had subclasses, divisions, and sections, and it is not clear that his orders were truly orders in the modern sense. Grassi (1913) was aware of Rafinesque's treatment and even reproduced part of the latter's classification. It is unclear whether Grassi intended his Chlopsidae as an original name or simply as Rafinesque's name put into proper grammatical form. At any rate, Rafinesque's name clearly pre-dates all the others that have been used for this family and is accepted here as the Chlopsidae.

Characters. Body moderately elongate, compressed, anus before midbody. Dorsal and anal fins well developed, confluent with caudal fin; dorsal fin begins over or slightly behind a vertical through gill opening. Pectoral fin present or absent. Gill opening reduced to a small, round, pore-like opening. Scales absent.

Head moderate, eye well developed, snout moderate or short, projecting slightly beyond lower jaw. Upper lip without an upturned flange, lower lip with or without a downturned flange. Anterior nostril tubular, near tip of snout, directed anterolaterally. Posterior nostril variable; it may be a simple half-moon shaped opening on side of snout opposite lower half of eye; it may open on edge of upper lip and be covered dorsally by a flap; or it may open inside upper lip. Sensory papillae on snout.

Lateral line obsolete except for a few pores at anteriormost end of canal in branchial region. Head pores reduced in a consistent pattern as follows. Preoperculomandibular canal with five to seven pores, all on lower jaw, the posteriormost one sometimes behind level of rictus. Infraorbital canal always with four pores (possibly a fifth pore in *Linkenchelys*), all on upper jaw; two of these between anterior and posterior nostril, and two between posterior nostril and rictus

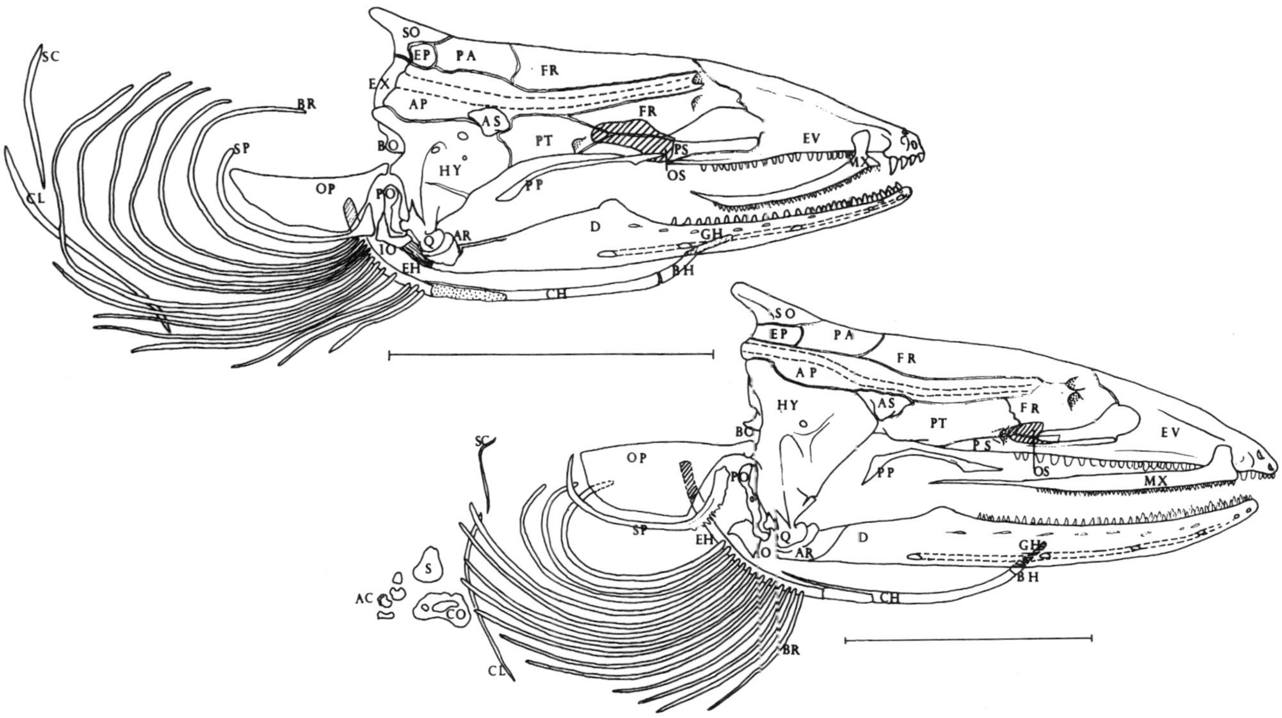

FIGURES 56–57. Figure 56. (top) Anterior skeletal elements of *Chlopsis bicolor,* in situ (from Robins and Robins, 1967: fig. 3). AP = pterotic, AR = articular, AS = sphenotic, BH = hypohyal, EP = epiotic, FR = frontal, HY = hyomandibula, IO = interopercle, OS = basisphenoid, PO = preopercle, PT = pterosphenoid, SC = supracleithrum, SO = supraoccipital, SP = subopercle, other abbreviations as on pp. 5–6. Figure 57. (bottom) Anterior skeletal elements of *Kaupichthys hyoproroides,* in situ (from Robins and Robins, 1967: fig. 5). S = scapula, other abbreviations as in Figure 56. Scale = 5 mm in both figures.

below eye. No adnasal pore (except possibly in *Linkenchelys*). One supraorbital pore (the ethmoidal pore) at anteroventral tip of snout, before anterior nostril; one pore above ethmoidal pore anteromedial to base of anterior nostril, and one behind that one, above and behind anterior nostril; additional supraorbital pores absent except in *Linkenchelys.* No pores behind eye, no pores in supratemporal commissure, and no pores in preopercular section of preoperculomandibular canal. No frontal commissure.

Teeth small and conical to long and needle-like, sometimes recurved, large fangs never present. Maxillary and mandibular teeth in two to several series, the inner teeth generally the largest. Intermaxillary teeth in a round patch or in several transverse rows, usually distinctly separated from maxillary and vomerine teeth. Vomerine teeth in one or two longitudinal series extending to or beyond posterior end of maxillary tooth patches; when two series present, each may be composed of a single row of teeth or of two closely set rows, in which case, the inner teeth are larger.

Osteology. One specimen of *Chilorhinus suensonii* examined (ANSP 116139). Information on *Chlopsis bicolor* and *Kaupichthys hyoproroides* from Robins and Robins, 1967. Additional information from Gosline, 1950 and 1951, and Nelson, 1966a.

Neurocranium (Figs. 56–60) moderately elongate, low and smooth, orbit relatively small. Frontals divided. Parietals divided in *Chlopsis bicolor, Kaupichthys hyoproroides,* and *Chilorhinus platyrhynchus,* although they appeared to be fused in specimen of *Chilorhinus suensonii* examined here. Supraoccipital present, sometimes forming a crest.

Jaws (Fig. 61): Maxilla slender, with a well developed pedicel anteriorly, articulated with ethmovomer just behind its anterior expansion. Mandible nearly as long as neurocranium (somewhat shorter in *Chilorhinus*); the part of dentary

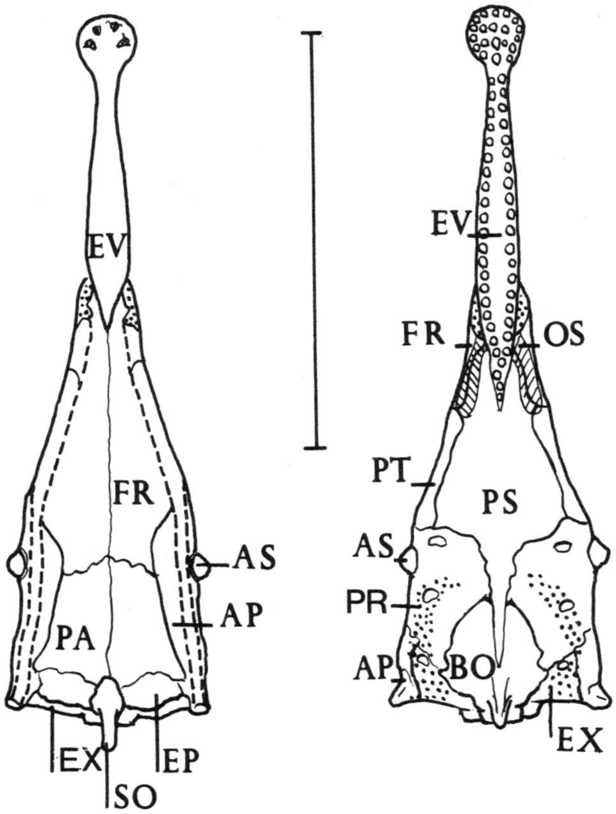

FIGURE 58. Neurocranium of *Chlopsis bicolor* in dorsal (left) and ventral (right) view (from Robins and Robins, 1967: fig. 4). Abbreviations as in Figure 56.

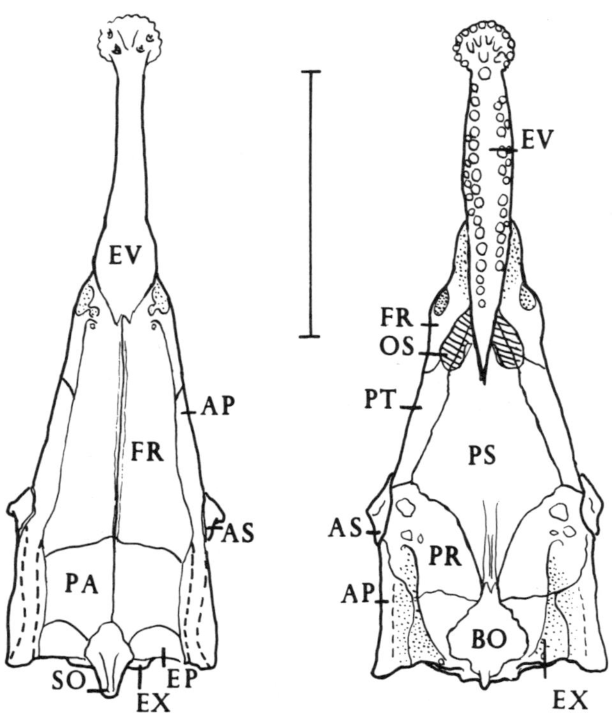

FIGURE 59. Neurocranium of *Kaupichthys hyoproroides* in dorsal (left) and ventral (right) view (from Robins and Robins, 1967: fig. 6). Abbreviations as in Figure 56.

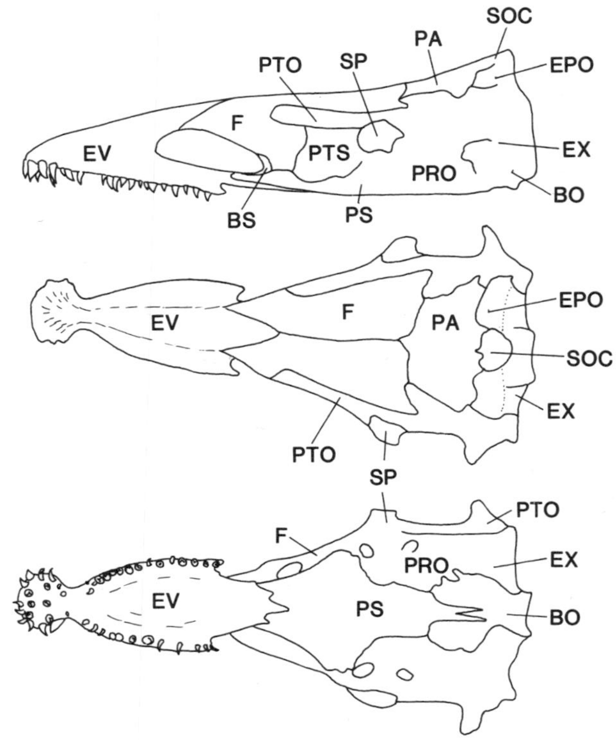

FIGURE 60. Neurocranium of *Chilorhinus suensonii* (note: some of the sutures could not be located). Abbreviations on pp. 5–6.

posterior to dorsal process just behind teeth is particularly elongate; articular smail, restricted to posterior extremity of mandible.

Suspensorium (Figs. 56, 58, 62) nearly vertical in *Chlopsis* and *Kaupichthys*, inclined slightly anteriorly in *Chilorhinus*. Pterygoid somewhat reduced in *Chlopsis* and *Kaupichthys*, stout in *Chilorhinus*, not strongly bound to vomer.

Hyoid arch (Figs. 56, 58, 63): Glossohyal moderately short, somewhat triangular in dorsal view, base of triangle forward. Urohyal in *Chilorhinus* very short, with virtually no posterior extension; presence and condition in other genera unstated. Hypohyals (= basihyals) present in *Chlopsis* and

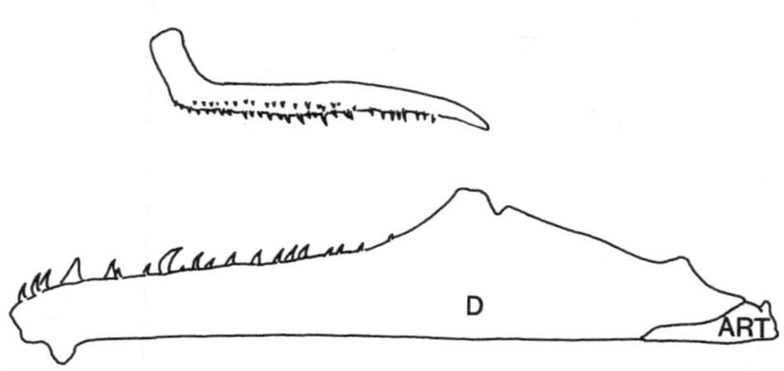

FIGURE 61. Maxilla and mandible of *Chilorhinus suensonii*, drawn to same scale.

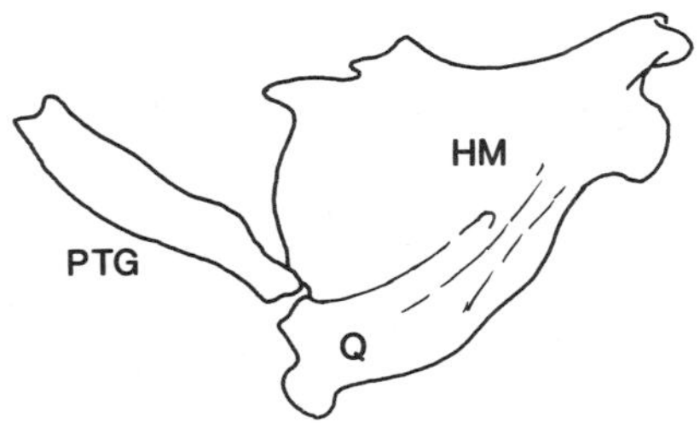

FIGURE 62. Suspensorium of *Chilorhinus suensonii.*

Kaupichthys but absent in *Chilorhinus.* Branchiostegal rays arise from epihyal, posterior ones curve up behind opercle; anterior five or six branchiostegals more widely spaced than remainder.

Gill arches reduced (Fig. 64). Basibranchials absent. First two hypobranchials present, third absent. Five ceratobranchials present, first four stout, the fifth small and slender. Upper pharyngeals divided; lower pharyngeals undivided, tear- or pear-shaped, the narrow end forward. Both infrapharyngobranchials present, although the anterior one (I-2) sometimes reduced. Pharyngeal jaws not greatly enlarged.

Opercular apparatus complete (Figs. 56, 58, 65). Opercle broad and roughly rectangular in *Chilorhinus,* more flattened dorsally in *Chlopsis* and *Kaupichthys.* Subopercle blade-like and curved up around ventral and posterior edge of opercle. Interopercle rather small, irregularly shaped. Preopercle small, housing a portion of preoperculomandibular canal.

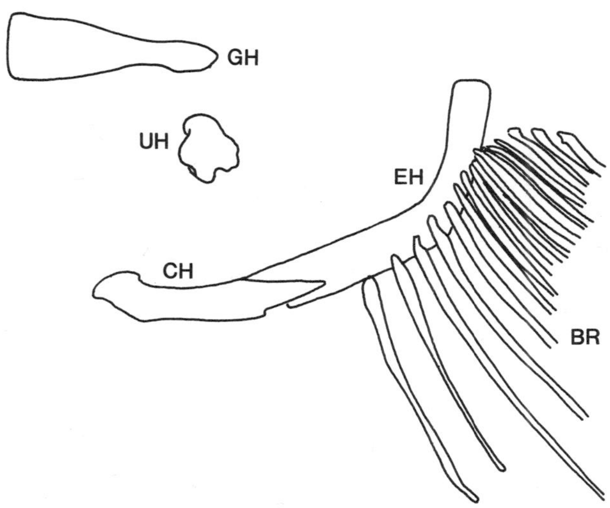

FIGURE 63. Hyoid arch of *Chilorhinus suensonii.*

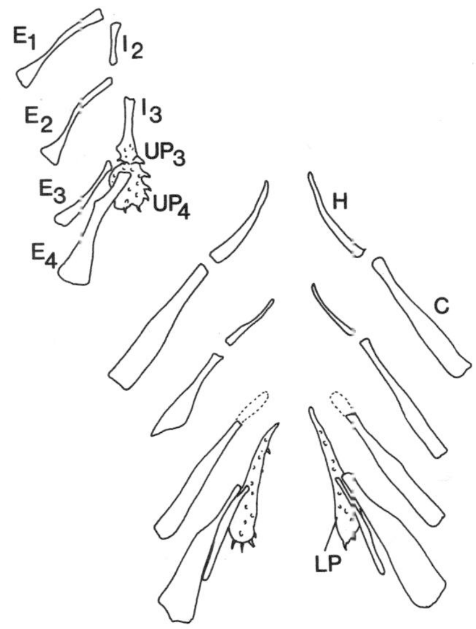

FIGURE 64. Gill arches of *Chilorhinus suensonii,* ventral elements in ventral view, dorsal elements in dorsal view.

Pectoral girdle variable (Figs. 56, 58, 65). Cleithrum and supracleithrum always present; scapula and coracoid present only in those species that have a pectoral fin. Supracleithrum fairly well developed in *Chlopsis,* greatly reduced in *Chilorhinus.*

Circumorbital bones not well ossified (Fig. 66). In *Chilorhinus,* nasal a narrow, tubular structure. Preorbital tubular, with prominent, short tubes leading to pores. Suborbitals present behind eye, but number of elements uncertain. Supraorbital absent. Prominent, finger-like cartilages supporting nasal chamber. Lateral-line ossicles absent.

Vertebrae (Figs. 67–69): Precaudal vertebrae with prominent transverse processes. Caudal vertebrae without epicentral processes. Ribs absent.

Discussion. The common name "false moray" describes the Chlopsidae well. Phylogenetically

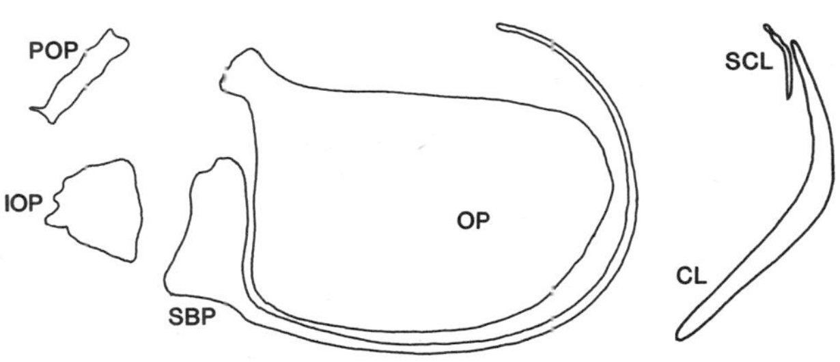

FIGURE 65. Opercular apparatus and pectoral girdle of *Chilorhinus suensonii.*

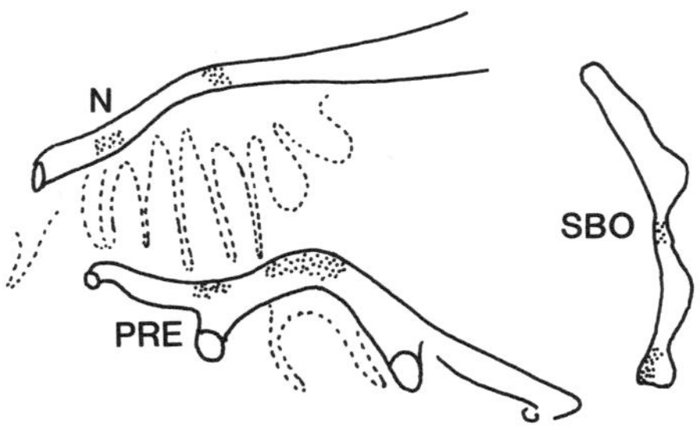

FIGURE 66. Circumorbital bones of *Chilorhinus suensonii* (shaded areas are ossified).

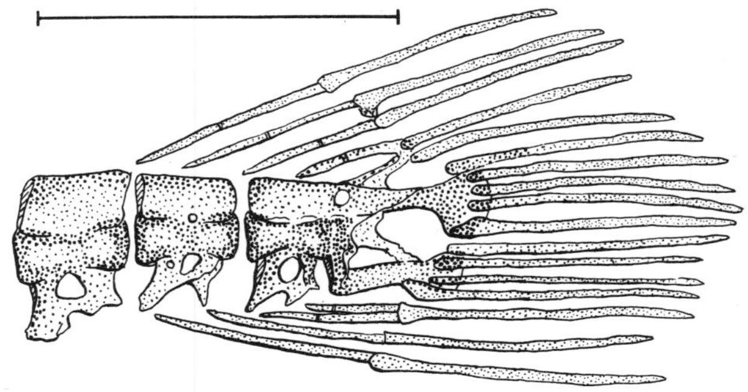

FIGURE 68. Caudal skeleton of *Chlopsis bicolor* (from Robins and Robins, 1967: fig. 7).

they are the closest relatives of the Muraenidae, and in their appearance, habits, and behavior chlopsids greatly resemble small morays. Aside from *Chilorhinus,* which prefers grass beds, chlopsids seem to be cryptic fishes that hide in holes and crevices on coral reefs or areas of rock and rubble. Like many of the smaller morays, they are seldom seen alive, appearing only after rotenone or other ichthyocides have driven them from their refuges. Detailed stomach-content analyses have not been made, but chlopsids are probably active predators, eating whatever happens to be available. The radiograph of one specimen of *Chlopsis dentatus* (ANSP 110002) shows the skeletons of two large fishes, each about one-sixth the total length of the eel, which had been swallowed head first.

Little is known about reproduction in chlopsid eels. An examination of gonad condition in *Kaupichthys hyoproroides* from the Bahamas revealed no traces of seasonal periodicity. The most intriguing hint of reproductive behavior is J. L. B. Smith's (1966) third-hand account of the collection of *Powellichthys ventriosus.* Quoting the Fisheries Officer at Cook Island (Raratonga?), Smith related the following story.

> My assistant was fishing on a dark night offshore in about 500 fathoms. He was using a lantern and flies made from white nylon for *Selar crumenophthalmus.* As he started to fish strange fingerling fish appeared round him in thousands. He dimmed his light and moved off about half a mile away. As soon as he turned up the light these fish appeared again in their thousands. The school lasted about half an hour before it vanished.
>
> This small fish now appears rather eel-like. However, when the school was on the surface these fish had peculiarly distorted abdomens that appeared blown out the full length of the abdominal cavity. Some indication of the extension can be given from the fact that this specimen was retained in the mesh of a dip net of 1½ in. stretched mesh, i.e. ¾ in. square.

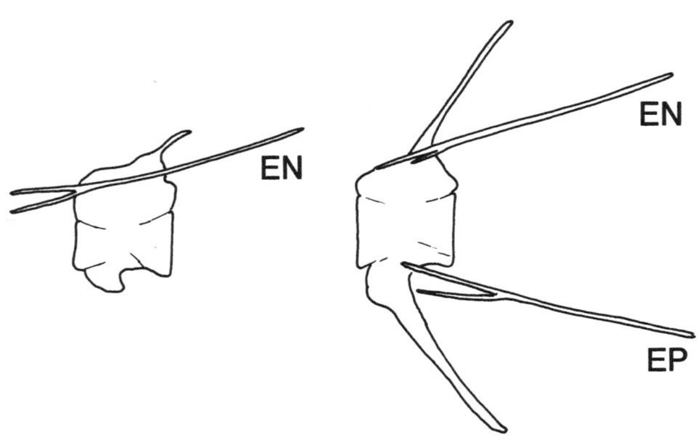

FIGURE 67. Precaudal (left) and caudal (right) vertebra of *Chilorhinus suensonii.*

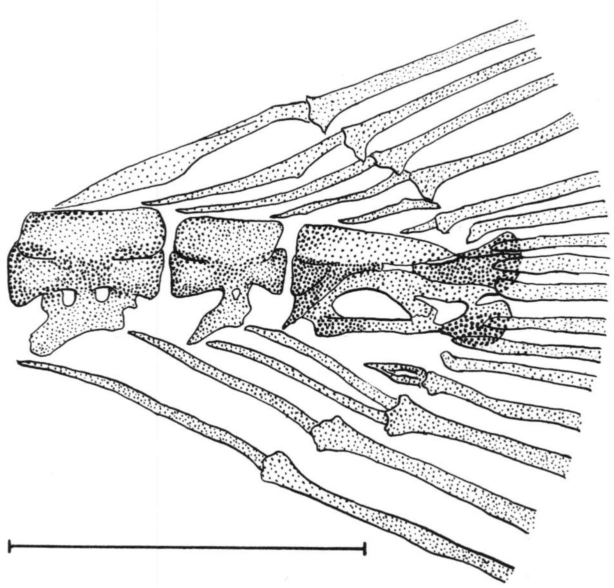

FIGURE 69. Caudal skeleton of *Kaupichthys hyoproroides* (from Robins and Robins, 1967: fig. 8).

The indication is plainly of a spawning aggregation, especially since the eels had distended abdomens and the type specimen was a ripe female. Nothing similar has been reported elsewhere, and it is unknown whether this habit is peculiar to *Powellichthys* or is found in other species. If such a spawning school lasts for no more than half an hour, it could easily escape notice.

One of the peculiarities of the Chlopsidae is the rarity of many of the species. *Powellichthys ventriosus* J. L. B. Smith, and *Catesbya pseudomuraena* Böhlke and Smith are known from single specimens; *Xenoconger fryeri* Regan and *Linkenchelys multipora* D. G. Smith are known from two. Each represents a monotypic genus. In addition, several others are represented by fewer than ten specimens each. Some of this apparent rarity may be because their small size and cryptic habits make them difficult to collect. For example, leptocephali of *Chlopsis olokun* (Robins and Robins) are common in the Gulf of Guinea (Blache, 1977), although only four adults are known. In other cases, though, the scarcity seems real. The holotype of *Catesbya pseudomuraena* remains the only specimen known despite all the collecting that has been done on western Atlantic coral reefs. In the western Atlantic, *Kaupichthys hyoproroides* and *Chilorhinus suensonii* are common, *Kaupichthys nuchalis* is less common, and the others are rare.

Distribution. Throughout the tropical and subtropical world ocean, from the shoreline to about the edge of the continental shelf.

Genera. There are eight nominal genera of Chlopsidae: *Chlopsis* Rafinesque, 1810, *Chilorhinus* Lütken, 1852, *Xenoconger* Regan, 1912, *Kaupichthys* Schultz, 1943, *Powellichthys* J. L. B. Smith, 1965, *Robinsia* Böhlke and Smith, 1967, *Catesbya* Böhlke and Smith, 1968, and *Linkenchelys* D. G. Smith, new genus. All but *Xenoconger* and *Powellichthys* are found in the western Atlantic. Interrelationships among chlopsids are not well understood, and the variety of leptocephali indicates that additional genera may exist.

TABLE 5. Mean vertebral formulae (MVF) and ranges of vertebral counts for western Atlantic species of Chlopsidae.

		Predorsal		Preanal		Total	
Species	MVF	Range	n	Range	n	Range	n
Catesbya pseudomuraena[1]	11-33-141	11	1	33	1	141	1
Chilorhinus suensonii	17-39-100	16–17	14	38–41	13	98–106	63
Chlopsis bicolor	12-39-132	12	1	39	1	127–134	7
Chlopsis dentatus	8-32-122	7–9	2	31–34	2	116–125	8
Kaupichthys hyoproroides[2]	9-27-118	8–9	13	26–28	15	110–120	62
Kaupichthys nuchalis[1]	8-35-122	7–8	12	33-37	12	119–125	17
Linkenchelys multipora[1]	17-22-120	16–18	2	21–22	2	119–122	2
Robinsia catherinae[1]	8-36-132	8	1	36	1	130–ca 137	5

[1] Includes holotype.
[2] Western Atlantic specimens only.

KEY TO THE GENERA OF CHLOPSIDAE IN THE WESTERN NORTH ATLANTIC

1a. Lateral-line pores 4–5; origin of dorsal fin closer to anus than to gill opening .... *Linkenchelys*, p. 78
1b. Lateral-line pores 1–2; origin of dorsal fin closer to gill opening than to anus 2
2a. Vomerine teeth in a single median series .... *Robinsia*, p. 80
2b. Vomerine teeth in two series, one on each side of midline 3

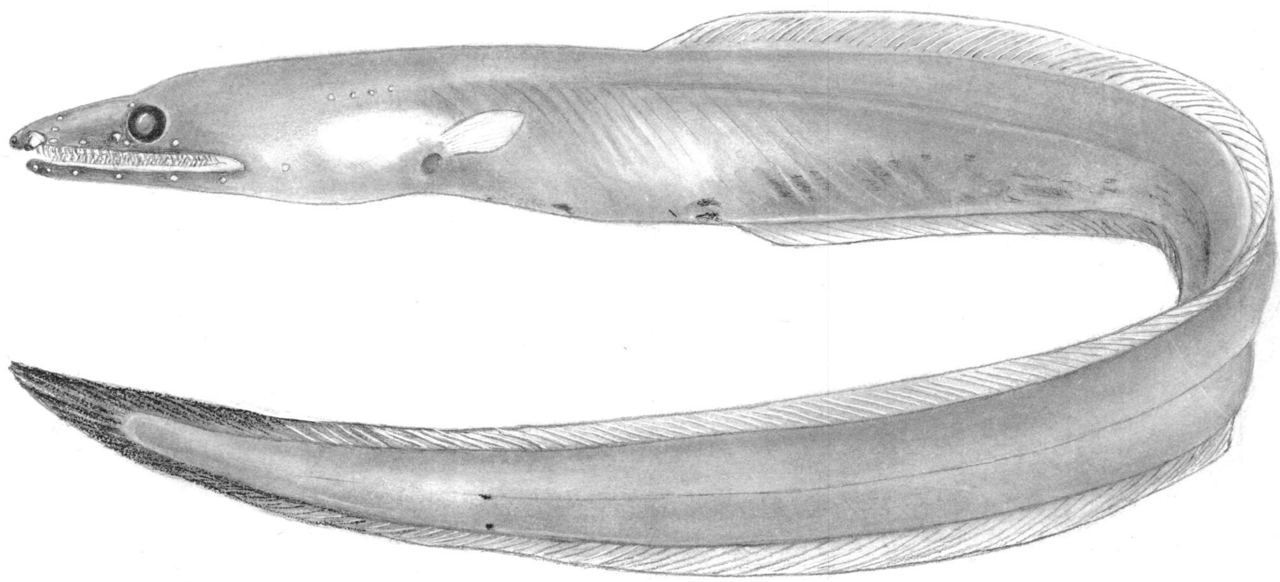

FIGURE 70. *Linkenchelys multipora*: ANSP 156814, holotype, 82 mm TL.

3a. Two rows of teeth in each vomerine series . . . . . *Catesbya*, p. 82
3b. One row of teeth in each vomerine series 4
4a. Lower lip with a downturned flange . . . . . *Chilorhinus*, p. 84
4b. Lower lip without a downturned flange 5
5a. Pectoral fin present . . . . . *Kaupichthys*, p. 86
5b. Pectoral fin absent . . . . . *Chlopsis*, p. 92

Linkenchelys new genus

Type species: *Linkenchelys multipora* n. sp.

Characters. Linkenchelys is distinguished from all other chlopsid genera by having four or five lateral-line pores in the branchial region. Its dorsal fin begins closer to the anus than to the gill opening, another unique feature. It has a single series of vomerine teeth, like *Robinsia,* but unlike *Robinsia* the teeth are slender and needle-like. The pectoral fin is well developed. The posterior nostril is low on the side of the head but above the lip. For other characters see the description of the only known species, *L. multipora.*

Size. The two specimens are both 82 mm TL.

Distribution. Known only from the Bahamas.

Etymology. Named for the Johnson-Sea Link submersible, which collected the only known specimens. Feminine.

Species. A single species, *Linkenchelys multipora* n. sp.

Linkenchelys multipora new species
Figures 70–73, Table 1

Distinctive Characters. Those of the genus.

Description. Total vertebrae ca 119–122 (n = 2), predorsal vertebrae 16–18 (2), preanal vertebrae 21–22 (2). Pectoral-fin rays ca 12 (2). Pores: LL 4–5 (2), POM 7 (2), IO 4 or 5 (2), SO 1 + 4 or 5 (2).

Proportions as % of TL: preanal 24–27 (2), predorsal 22–24 (2), head 15–16 (2), depth at anus 4–6 (2), pectoral fin 3–4 (2). Of HL: snout 27–29 (2), eye 6–10 (2), snout-rictus 55–57 (10), pectoral fin 18–22 (2).

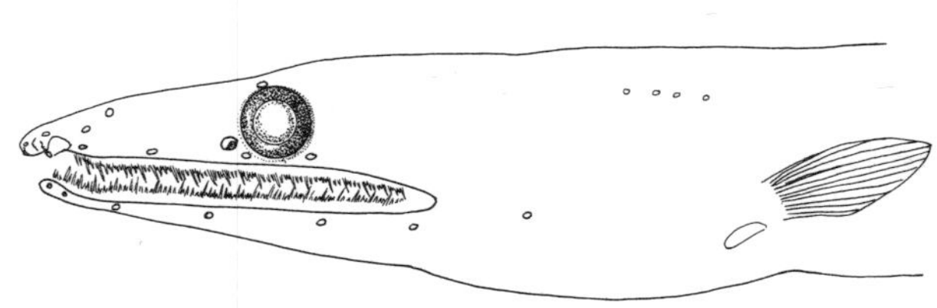

FIGURE 71. *Linkenchelys multipora*: ANSP 156814, holotype, 82 mm TL; head.

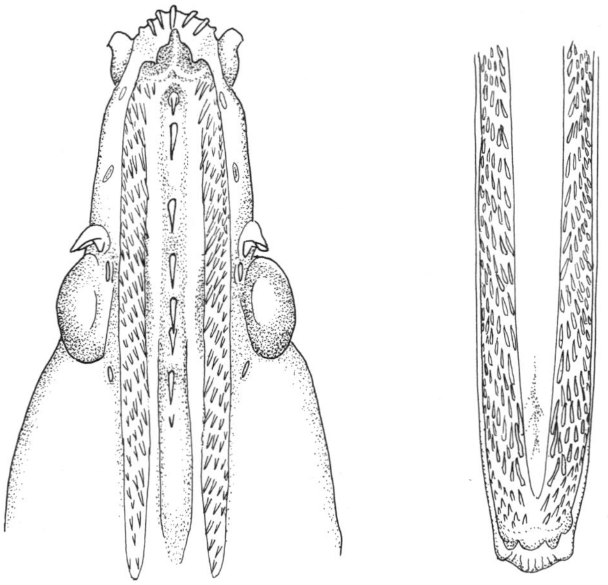

FIGURE 72. *Linkenchelys multipora*: ANSP 156814, holotype, 82 mm TL; dentition.

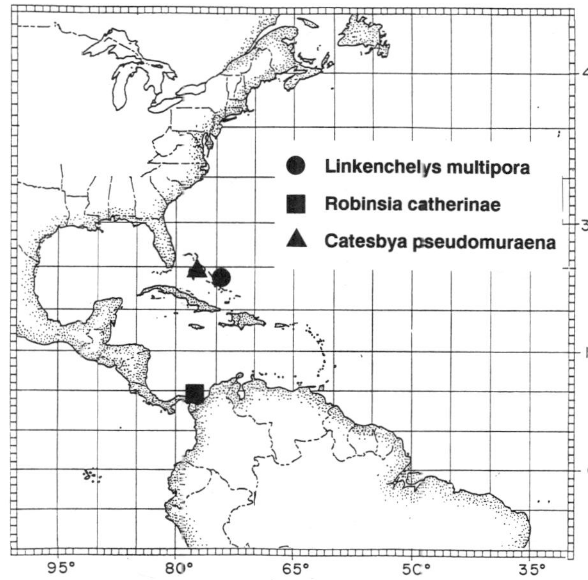

FIGURE 73. Distribution of *Linkenchelys multipora, Robinsia catherinae,* and *Catesbya pseudomuraena.*

Body moderately elongate, strongly compressed. Anus relatively far forward. Dorsal and anal fins well developed, confluent with caudal fin. Dorsal fin begins well behind gill opening, shortly before anus. Pectoral fin well developed.

Head moderate, somewhat narrow, acute in lateral view; gape large, rictus well behind eye. Upper jaw projects beyond lower, fleshy snout projects somewhat beyond bony tip of ethmoid. Tip of snout with longitudinal folds or plicae. Anterior nostril tubular, slightly behind tip of snout, directed anterolaterally. Posterior nostril elliptical, with flap on anterodorsal side; nostril low on head but above lip, just above and before third infraorbital pore.

Lateral line on body obsolete except for 4–5 pores in anteriormost part of canal in branchial region. Supraorbital pores five or six, depending on interpretation of one of them: first (ethmoidal) pore at anteroventral tip of snout; second pore dorsal to this at tip of snout within plicae; third pore just medial to anterior nostril; a pore just behind base of anterior nostril tube, somewhat more lateral in position than the other supraorbital pores, which could be either a supraorbital or an adnasal (infraorbital) pore; a pore behind and medial to the previous one, in a more or less straight line behind third pore; last pore over mid-dorsal margin of eye. Infraorbital canal with four or five pores, depending on interpretation of pore mentioned above: first just behind anterior nostril; second midway between anterior and posterior nostril; third just behind posterior nostril, level with anterior margin of eye; fourth level with posterior margin of eye. Preoperculomandibular canal with six pores before rictus and one behind.

Teeth long, slender, needle-like. Maxillary teeth in narrow bands, arranged in oblique rows trending posterolateral to anteromedial, increasing in size from outside to inside; longest teeth at about first quarter of length of maxilla. Mandibular teeth like those of maxilla, perhaps a little longer. Intermaxillary tooth patch not evident. A single series of eight long, straight, slender teeth widely spaced along midline of vomer, first slightly behind anterior end of maxillary tooth patch, last just about even with posterior margin of eye; anterior and posteriormost teeth smallest, second tooth somewhat larger, third through seventh teeth largest.

Color. In preservative uniform light olive-brown, caudal fin and posteriormost extremity of dorsal and anal fins black. Remnants of larval pigment as follows: several large, irregularly shaped melanophores on ventral midline from

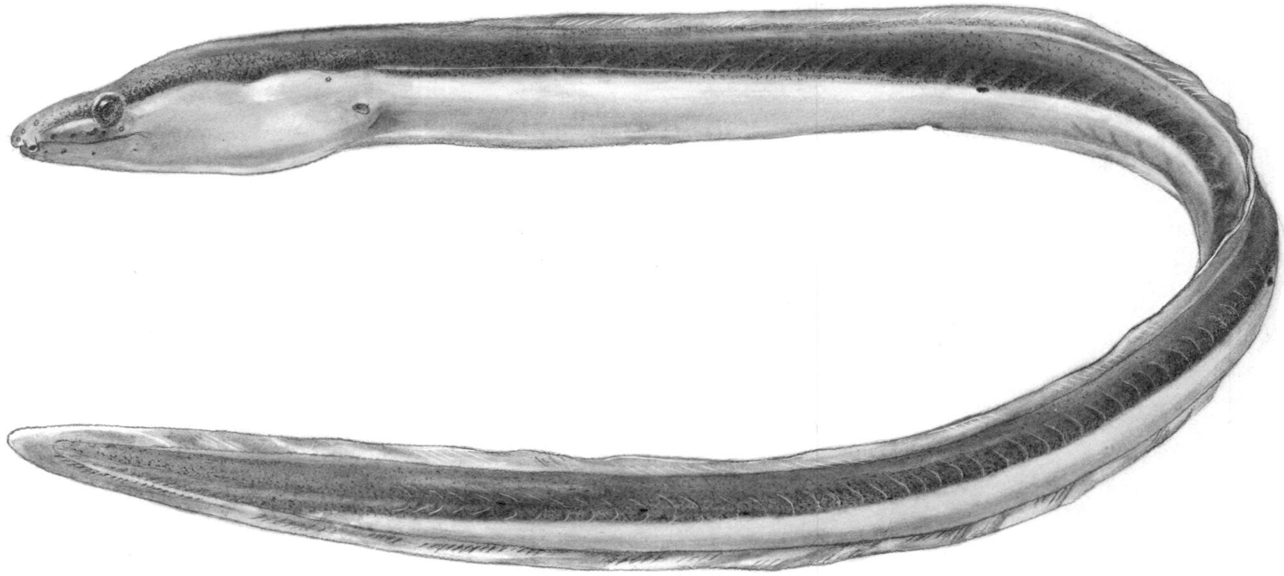

FIGURE 74. *Robinsia catherinae*: ANSP 110001, paratype, 81 mm TL (from Böhlke and Smith, 1967: fig. 1).

throat to anus; a few more large melanophores on ventral midline behind anus; two dark, deep-lying spots midlaterally, one shortly behind tip of appressed pectoral fin and one over anus.

Size. The two specimens are both 82 mm TL.

Development and Growth. The leptocephalus is unknown.

Distribution. Known only from off the island of San Salvador in the Bahamas, at ca 238 m.

Etymology. From the Latin *multi* (many) and *porus* (pore), in reference to the relatively numerous lateral-line pores. To be treated as an adjective.

Study Material. Two specimens, 82 mm TL.

Holotype: ANSP 156814 (82 mm TL), Bahamas, San Salvador, off Cockburn Town, 238 m, coll. R. G. Gilmore, Johnson-Sea Link submersible, dive no. 1289, 17 Oct. 1982.

Paratype: USNM 280144 (1, 82), Bahamas, San Salvador, close to station where holotype was taken, 1 Nov. 1982.

Genus *Robinsia* Böhlke and Smith, 1967

Robinsia Böhlke and Smith, 1967: 2 (type species *Robinsia catherinae* Böhlke and Smith, 1967, by original designation).

Characters. Robinsia catherinae, the only species in the genus, resembles *Chlopsis* in overall appearance and in such characters as the labial posterior nostril and the lack of a pectoral fin. Even its color pattern is remarkably similar to that of the bicolored species of *Chlopsis.* It differs mainly in its uniserial vomerine dentition. On this basis the species was made the type of a new genus, *Robinsia.* For other characters, see the description of *Robinsia catherinae.*

Size. To about 200 mm TL.

Distribution. Western Atlantic and western Indian Oceans.

Etymology. Named for Dr. C. Richard Robins. Feminine.

Species. A single species, *Robinsia catherinae* Böhlke and Smith, 1967.

Robinsia catherinae Böhlke and Smith, 1967
Figures 73–76, Table 5

Robinsia catherinae Böhlke and Smith, 1967: 3, fig. 1 (original description, western Atlantic and western Indian Oceans, holotype ANSP 110000).

Distinctive Characters. Pectoral fin absent; posterior nostril on upper lip; vomerine teeth in a single median series; bicolored, brown above and white below.

Description. Total vertebrae 130–ca 137 (n = 5), predorsal vertebrae ca 8 (1), preanal vertebrae 36 (1). Pores: LL 1 (4), POM 5 (4), IO 4 (4), SO 1 + 2 (4), STC 0 (4).

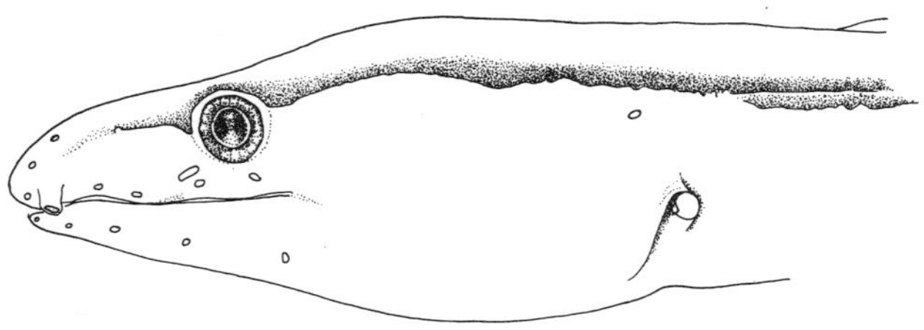

FIGURE 75. *Robinsia catherinae*: ANSP 110001, paratype, 81 mm TL; head.

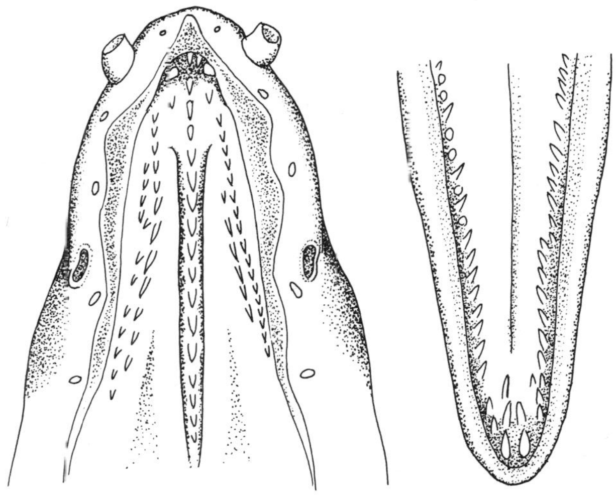

FIGURE 76. *Robinsia catherinae*: ANSP 110001, paratype, 81 mm TL; dentition.

Proportions as % of TL: preanal 32–34 (5), predorsal 13–15 (5), head 11–13 (5), depth at anus 4–5 (4). Of HL: snout 23–27 (5), eye 10–11 (5), interorbital 13–19 (5), snout-rictus 33–38 (3).

Body moderately elongate. Gill opening round or elliptical, placed below midside. Pectoral fin absent, although a vestige of one may be present in form of a fleshy tag on dorsal margin of gill opening. Dorsal fin begins slightly but distinctly behind gill opening.

Head and snout moderately elongate, rictus near posterior margin of eye. Posterior nostril elliptical, on outer edge of upper lip, sometimes covered by a weakly developed flap from its dorsal margin.

A single lateral-line pore in branchial region. Five pores along lower jaw, all before rictus. Small sensory papillae on snout and chin.

Maxillary teeth narrow and conical, biserial, the inner teeth larger. Intermaxillary teeth moderate in size, in a round patch. Vomerine teeth closely set and laterally compressed, with a convex anterior profile and a straight posterior profile, in a single median row extending posteriorly well past end of maxillary tooth patch. Mandibular teeth larger than those of maxilla, compressed and with posteriorly pointed tips, uniserial for most of length, but with two long, slender teeth forming an inner row anteriorly; the anteriormost tooth on lower jaw an enlarged canine.

Color. Sharply bicolored, brown above and white below, dark color most intense along boundary and near tail tip. Demarcation line begins on snout and runs through middle to lower part of eye, thence posteriorly above gill opening and along midlateral line, beginning to descend about midtail, terminating on ventral margin a short distance before tip of tail. This pigmentation is similar to that of *Chlopsis bicolor*, but in *Robinsia catherinae* the white area is more extensive, and the dividing line lies higher on the body. The western Atlantic paratype retains some elements of larval pigmentation: eight spots on ventral midline between gill opening and anus; several spots on midlateral line, eight on left and nine on right.

Size. The largest specimen known is the 198-mm paratype from the Indian Ocean. The holotype, a 139-mm female, contained large eggs.

Development and Growth. The leptocephalus has been identified (see p. 937).

Variation. The western Atlantic specimen has fewer vertebrae (130) than those from the Indian Ocean (134–ca 137). In the western Atlantic specimen, the branchial pore is located in the white area, whereas in the Indian Ocean specimens it is in the brown area. The posterior nostril in the Indian Ocean specimens bears a median projection or flap from the dorsal margin that extends down over the opening; the western Atlantic specimen lacks this projection, but it is much smaller than the Indian Ocean specimens. Böhlke and Smith (1967: 4) felt that these differences were insufficient to warrant a specific distinction between the populations of the two oceans. No additional specimens have been found since then, and there seems no reason to change that conclusion.

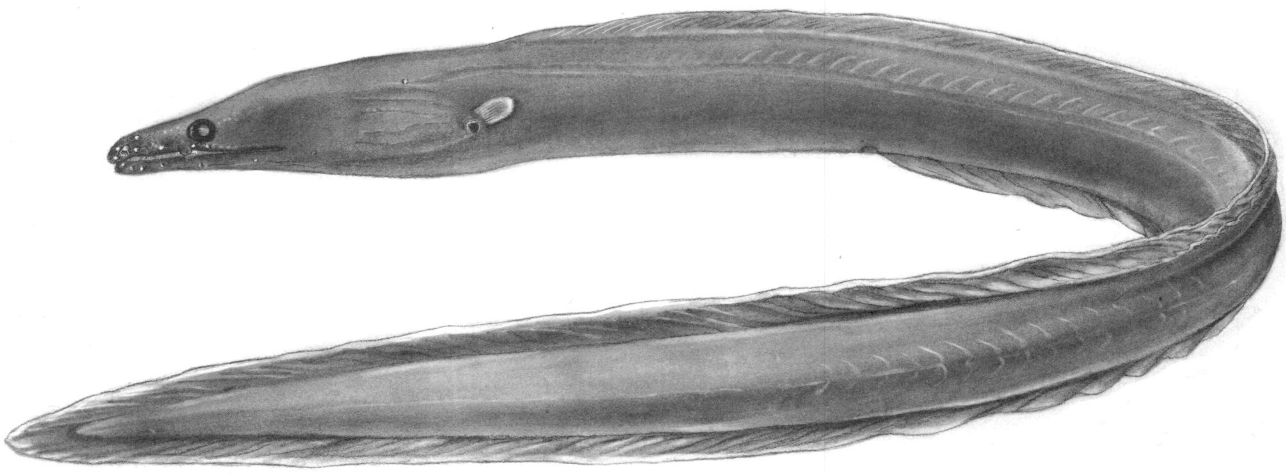

FIGURE 77. *Catesbya pseudomuraena*: ANSP 100634, holotype, 139.5 mm TL (from Böhlke and Smith, 1968: fig. 3).

The same authors noted a slight difference in the mandibular dentition between the two males and the single female.

Distribution. Known from the western Atlantic off Panama and the western Indian Ocean off Kenya. Depth 52–243 m. Rare.

Etymology. Named for Dr. Catherine Hale Robins. A noun in the genitive case.

Study Material. Five specimens, 81–198 mm TL.

Holotype: ANSP 110000 (female, 139), Anton Bruun 420A, Indian Ocean, 2°42′S, 40°53′E, 140 m, 6 Nov. 1964.

Paratypes: PANAMA: ANSP 110001 (1, 81), 9°22′N, 78°08.8′W, 52 m, 19 July 1966. INDIAN OCEAN: Biologische Anstalt Helgoland-Hamburg (3, 133–198), 1°18′S, 41°56′E, 177–243 m, 15 Jan. 1965.

Genus *Catesbya* Böhlke and Smith, 1968

Catesbya Böhlke and Smith, 1968: 35 (type species *Catesbya pseudomuraena* Böhlke and Smith, 1968, by original designation).

Characters. In possessing a pectoral fin and a labial, concealed posterior nostril, *Catesbya* resembles certain species of *Kaupichthys.* It differs from these and all other chlopsids primarily in the nature of its dentition. The species of *Kaupichthys, Chlopsis* and *Chilorhinus* have their vomerine teeth in two single rows, one on each side of the midline, and these rows terminate behind the posterior end of the maxillary tooth patches. In *Catesbya,* each of the vomerine tooth patches consists of two rows of teeth, the inner ones much larger than the outer, and the vomerine tooth patches terminate before the posterior end of the maxillary patches. Furthermore, the teeth in *Catesbya* are slender and needle-like, quite unlike the short, conical teeth of *Kaupichthys, Chlopsis,* and *Chilorhinus.* The maxillary and mandibular teeth of *Catesbya* are arranged in diagonal rows, whereas in the other genera (except *Linkenchelys*) they are set irregularly. Two rare Indo-Pacific chlopsids, *Xenoconger fryeri* Regan and *Powellichthys ventriosus* J. L. B. Smith, also have the vomerine teeth in two double rows, but in neither case are the teeth needle-like. *Xenoconger fryeri* differs further in lacking a pectoral fin and having the posterior nostril above the lip and uncovered. *Powellichthys ventriosus* does have a pectoral fin, but the teeth are not needle-like and the vomerine tooth patches extend beyond the end of the maxillary teeth. The elongate head and snout and the large gape are characters unique to *Catesbya.*

A full description of *Catesbya* is given in the account of its only known species, *C. pseudomuraena.*

Size. The only known specimen is 140 mm TL.

Distribution. Western tropical Atlantic.

Etymology. Named for Mark Catesby, early naturalist who explored the southeastern United States and the Bahamas. Feminine.

Species. A single species, *Catesbya pseudomuraena* Böhlke and Smith, 1968.

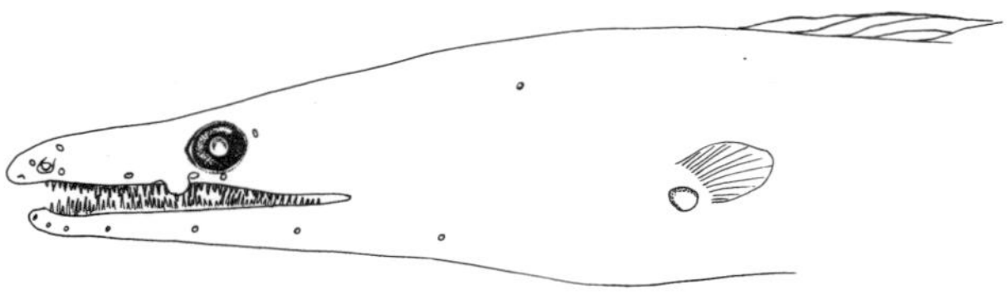

FIGURE 78. *Catesbya pseudomuraena*: ANSP 100634, holotype, 139.5 mm TL; head.

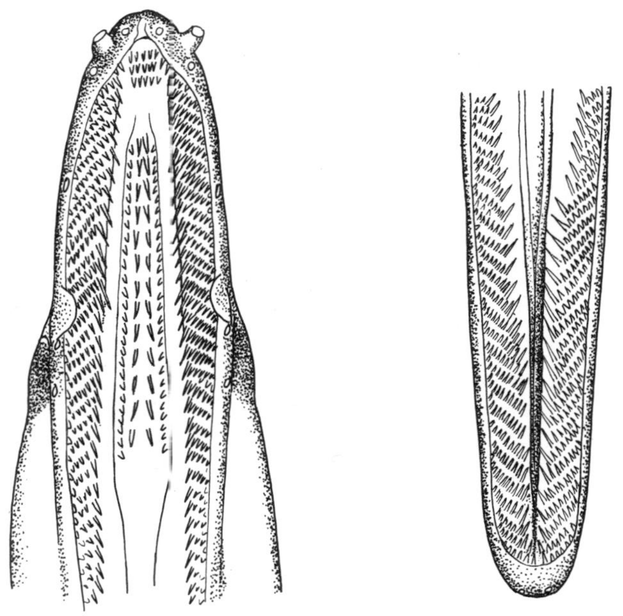

FIGURE 79. *Catesbya pseudomuraena*: ANSP 100634, holotype, 139.5 mm TL; dentition.

Catesbya pseudomuraena
Böhlke and Smith, 1968
Figures 73, 77–79, Table 5

Catesbya pseudomuraena Böhlke and Smith, 1968: 36, fig. 3 (original description, Bahamas, holotype ANSP 110634).

Distinctive Characters. Teeth slender and needle-like; vomerine teeth in two double rows, the rows not diverging, and ending before posterior end of maxillary tooth patches; pectoral fin present; one pore in branchial region; seven pores on lower jaw, one behind rictus; head and snout long and slender, rictus well behind posterior margin of eye; total vertebrae 141.

Description (largely from Böhlke and Smith, 1968). Total vertebrae 141 (n = 1), predorsal vertebrae 11 (1), preanal vertebrae 33 (1). Pores: LL 1 (1), POM 7 (1), IO 4 (1), SO 1 + 2 (1), STC 0 (1).

Proportions as % of TL: preanal 29 (1), predorsal 15 (1), head 13 (1), depth at anus 3 (1). Of HL: snout 26 (1), eye 9 (1), interorbital 14 (1), snout-rictus 49 (1), pectoral fin 12 (1).

Body slender, rather elongate. Dorsal fin begins slightly behind gill opening; pectoral fin well developed, rounded, somewhat longer than eye diameter.

Head and snout slender; gape wide, rictus well behind posterior margin of eye. Posterior nostril horizontally elongate, opening on outer edge of upper lip, covered dorsally by a well developed flap that reaches lower jaw when mouth closed. Small sensory papillae on snout and lower jaw. One lateral-line pore in branchial region. Seven pores on lower jaw, the posteriormost behind rictus.

Teeth all slender and needle-like. Maxillary and mandibular teeth arranged in parallel diagonal rows oriented from anteromedial to posterolateral, teeth in each row increasing in size from outside to inside; about eight teeth in each diagonal row near midlength of tooth patch. Maxillary tooth patches extend forward outside intermaxillary patch to level of anterior margin of intermaxillary patch, the groove between maxillary and intermaxillary patches on each side longitudinal rather than diagonal. Intermaxillary teeth in three transverse rows, increasing in size from front to rear; the median tooth in posterior row largest. Vomerine teeth in two rows on each side, the inner teeth much larger than outer; the patches on each side do not diverge from each other, and they terminate before posterior end of maxillary tooth patches.

Color. Uniform brown, vertical fins pale distally. Near tip of tail basal part of dorsal and anal fins becomes darker while distal part remains pale. Pectoral fin clear. A dark crescentic marking parallels anterior margin of eye.

Size. The only known adult specimen is 140 mm TL. The gonads are undeveloped and the sex is uncertain.

Development and Growth. The leptocephalus has been identified (see p. 935).

Distribution. The only known adult specimen was collected near New Providence Island in the Bahamas. A few leptocephali are known from

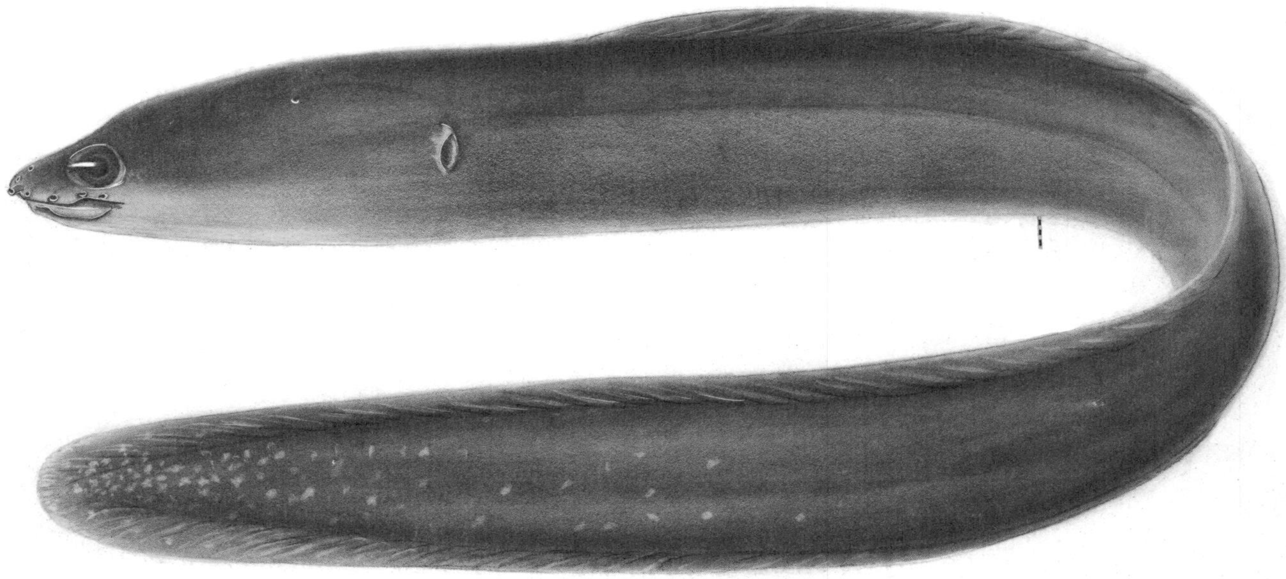

FIGURE 80. *Chilorhinus suensonii*: ANSP 157601, 158 mm TL. Illustration by S. P. Gigliotti.

widely scattered locations in the western Atlantic, including the Straits of Florida, the coast of Colombia, and the Sargasso Sea (Smith, 1969; also see this volume, p. 935). The holotype was collected on a small patch reef surrounded by sand at a depth of about 8 m. Several specimens of *Kaupichthys hyoproroides* were taken along with it.

Etymology. From the Greek *pseudo* (false) and *Muraena* (a genus of moray eels), in reference to the overall moray-like appearance of the species. A noun in apposition.

Study Material. One specimen, 140 mm TL.

Holotype: ANSP 110634 (140, sex indeterminate), Bahamas, New Providence Island, western tip of Lyford Cay, depth to ca 8 m, poison station, coll. C. C. G. Chaplin, S. A. Waterman, C. Limbaugh, and J. E. Böhlke, 8 May 1956.

Genus *Chilorhinus* Lütken, 1852

Chilorhinus Lütken, 1852: 16 (type species *Chilorhinus suensonii* Lütken, 1852, by monotypy).

Brachyconger Norman, 1922: 217 (type species *Brachyconger platyrhynchus* Norman, 1922, by monotypy). Preoccupied by *Brachyconger* Bleeker, 1865, a muraenesocid.

Endeconger Jordan, 1923: 131 (substitute name for *Brachyconger* Norman, 1922).

Characters. Body relatively short and stout, compressed posteriorly, anus slightly before midbody. Dorsal fin begins distinctly behind gill opening; as tail narrows toward tip, dorsal and anal fin increase in height, maintaining relatively straight profile before curving around end of tail to form a broad tail fin. Pectoral fin vestigial or absent.

Posterior half of head about same depth as trunk, anterior half of head narrowing rather abruptly toward tip of snout. Snout short, broadly rounded, somewhat depressed. Mouth small, rictus before posterior margin of eye. Lower lip with a free, downturned flange; upper lip without a flange, but lip itself projects ventrally, flaplike, to cover part of lower jaw when mouth closed. Posterior nostril opens inside mouth on inner edge of upper lip. One lateral-line pore in branchial region; six pores on lower jaw, posteriormost one behind rictus.

Maxillary teeth slender, in two rows, the inner teeth larger. Intermaxillary teeth in a round patch. Vomerine teeth in two single rows, diverging widely and not reconverging posteriorly, extending to about posterior end of maxillary tooth patches. Mandibular teeth in several irregular series anteriorly, uniserial posteriorly.

Discussion. Chilorhinus is the most distinctive of the chlopsid eels, with its stout body, short snout, divergent vomerine tooth rows, flange on the lower lip, and posterior nostril opening in-

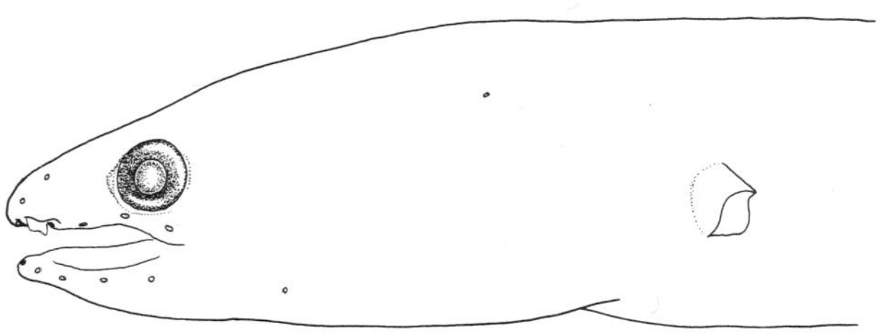

FIGURE 81. *Chilorhinus suensonii*: ANSP 100400, 158 mm TL; head.

side the mouth. J. L. B. Smith (1965, 1966) felt that *Chilorhinus* deserved to be placed in a separate family, or at least a subfamily, but the differences seem greatly outweighed by the similarities. The leptocephalus of *Chilorhinus*, too, is typical of chlopsids, and closely resembles that of *Chlopsis bicolor*. *Chilorhinus* does not merit separation from the other chlopsid genera at either a family or subfamily level.

Size. Maximum size slightly more than 200 mm TL.

Etymology. From the Greek *cheilos* (lip) and *rhinos* (snout, nose), perhaps referring to the ventrally projecting upper lip. Masculine.

Species. *Chilorhinus* contains two nearly identical species: *C. suensonii* Lütken, 1852 in the western Atlantic and *C. platyrhynchus* (Norman, 1922) in the Indo-West Pacific. The only difference between them seems to be the number of vertebrae, 98–106 in *C. suensonii* and 107–112 (based on two specimens) in *C. platyrhynchus* (Smith, 1969: 392).

Chilorhinus suensonii Lütken, 1852

SEAGRASS EEL

Figures 80–82, Table 5

Chilorhinus suensonii Lütken, 1852: 16, pl. 1, fig. 3, 3b, 3c (original description, St. Croix, Virgin Islands, syntypes ZMUC 100, ZMUC 320). Böhlke, 1956a: 83 (the most complete treatment of the species, including diagnosis, illustration, and synonymy). Böhlke and Chaplin, 1968: 74.

Rabula megalops Starks, 1913: 12, pl. 2 (original description, Natal, Brazil, holotype CAS-SU 22217). Böhlke, 1956a: 86 (synonymized with *Chilorhinus suensonii*).

Leptocephalus feathersi Fowler, 1944: 126, fig. 45 (original description, Old Providence Island, western Caribbean, holotype ANSP 70132). Leptocephalus.

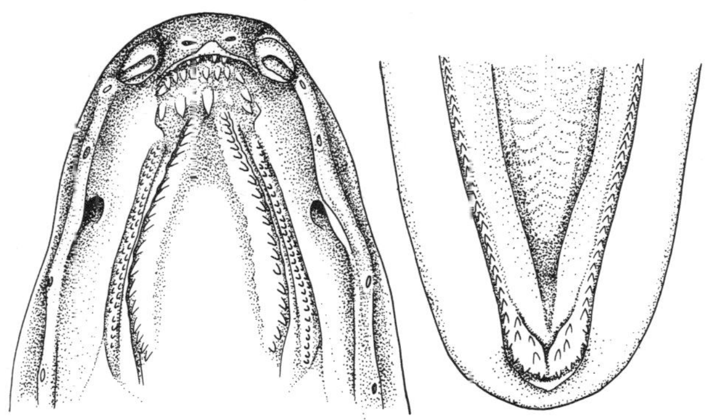

FIGURE 82. *Chilorhinus suensonii*: ANSP 100400, 158 mm TL; dentition.

Distinctive Characters. Those of the genus. Total vertebrae 98–106.

Description (partly from Böhlke, 1956a). Total vertebrae 98–106 (n = 63), predorsal vertebrae 16–17 (14), preanal vertebrae 38–41 (13). Branchiostegal rays 15–21 (51). Pectoral-fin rays 1–3 (51). Pores: LL 1 (12), POM 6 (15), IO 4 (15), SO 1 + 2 (15), STC 0 (15).

Proportions as % of TL: preanal 43–49 (28), predorsal 22–27 (27) plus one specimen at 34, head 18–20 (28), depth at anus 6–9 (28). Of HL: snout 15–24 (28), eye 8–12 (28), interorbital 15–24 (27), snout-rictus 22–32 (28).

Stomach ends about a quarter of its own length before anus. Gas bladder apparently absent.

Color. In preservative brown, paler ventrally, especially under head and breast. In small specimens, the pale ventral part of head is often sharply set off from darker dorsal part by a line that runs through lower part of eye to anterior nostril. Larger specimens often sprinkled with indistinct pale spots, giving a mottled appearance. Pale ring around gill opening. Color not much different in life; Böhlke and Chaplin (1968: 74) described it as purplish brown with the pale belly sometimes silvery. Stomach and intestine pale.

Size. Maximum size slightly greater than 200 mm TL. Females seem to grow larger than males. The largest male examined was 131 mm TL (ANSP 100389). Böhlke and Smith (1968: 34) recorded seven mature females 173–210 mm and one mature male 113 mm.

Development and Growth. The leptocephalus has been identified (see p. 939).

Habits and Behavior. Chilorhinus suensonii is a characteristic inhabitant of grassbeds. Böhlke and Chaplin (1968: 74) reported that in the Bahamas the species is particularly abundant in semi-protected bays and tidal creeks, areas of shallow water and fine-sand bottoms overgrown with turtle grass (*Thalassia testudinium*). Böhlke and Smith (1968: 34) reported that eight specimens trawled from Arrowsmith Bank, off the northeastern tip of the Yucatan Peninsula, came up in a net that contained green and brown algae.

Böhlke (1956a: 91) speculated that *Chilorhinus suensonii* was probably a bottom scavenger, because specimens collected by rotenone had fresh *Jenkinsia lamprotaenia* (Clupeidae) in their stomachs. The eels presumably picked these up off the bottom before they themselves were affected by the poison. *Chilorhinus suensonii* probably eats any small fishes or invertebrates that come along. The radiograph of CAS-SU 22217 shows part of the vertebral column of a fish in the stomach; one specimen from ANSP 100389 contained the remains of amphipods.

Variation. All but one of the specimens examined came from the Bahamas and the Caribbean. The exception is CAS-SU 22217, the holotype of *Rabula megalops* Starks, which was collected at Natal, Brazil. This specimen has slightly more total vertebrae (106) and preanal vertebrae (43) than any of the other 62 specimens whose vertebrae were counted (98–105 and 38–41, respectively). More specimens are needed to establish the extent of the difference between the populations of the two areas.

Böhlke (1956a: 90) reported that a specimen dredged off Great Abaco in the Bahamas (MCZ uncatalogued; it cannot be located today) showed several differences from the other specimens, most notably a more posterior dorsal-fin origin, a broader interorbital space, a longer snout, and a longer upper jaw. He suggested that it might represent an undescribed species, but this seems unlikely. No other comparable specimens have been found, and there is no evidence from leptocephali that more than one species of *Chilorhinus* is present in the Bahamas-Caribbean area (Smith, 1969). Unless contrary evidence appears, the specimen should be treated simply as an anomaly.

One specimen (ANSP 100389) out of 15 examined had six POM pores on one side and five on the other. The remaining specimens had six pores on both sides.

Distribution. Western Atlantic from Bermuda and the Bahamas through the Caribbean to Brazil. To my knowledge, there are no records of adults from Florida or the Gulf of Mexico, although one leptocephalus has been found in Campeche Bay. Shoreline to about 275 m, mainly shallow water. Common.

Etymology. Named for a 19th-century Danish naval officer who collected marine animals during his voyages. A noun in the genitive case.

Study Material. 801 specimens, 37–210 mm TL.
BAHAMAS: ANSP 73703 (1, 54), 73704 (126, to 161 mm), 73705 (29, 52–153), 73706 (52, 42–129), 73707 (10, 42–146), 73708 (4, 42–98), 73709 (3, 47–88), 73710 (142, 37–161), 73711 (10, 40–94), 73712 (2, 56–66), 73713 (98, 39–123), 73714 (4, 39–78), 73715 (1, 49), 100389 (8, 86–158), 100390 (2, 84–112), 100391 (9, 40–74), 100392 (1, 101), 100393 (111, 43–114), 100394 (74, 37–95), 100395 (2, 49–61), 100396 (6, 80–88), 100397 (2, 156–191), 100398 (2, 39–43), 100399 (1, 89), 100400 (1, 158), 100401 (1, 62), 100402 (1, 191), 100403 (21, 47–111), 101284 (13, 101–147), 109010 (1, 120), 115030 (2, 68–75), 115083 (1, 60), 116139 (6, 68–145; cleared and stained), 116314 (1, 85), 116344 (1, 51; metamorphic leptocephalus). CUBA: USNM 37479 (1, 176), 192047 (1, 49). PUERTO RICO: ANSP 115642 (4, 55–108), 117173 (1, 124). USNM 50203 (1, 109), 63077 (1, 43), 126173 (1, 109). VIRGIN ISLANDS: ANSP 1045–1046 (2, 73–79), 110412 (2, 62–64), 117037 (8, 47–80). LESSER ANTILLES: ANSP 105723 (5, 50–101), 105757 (2, 56–83), 105428 (5, 44–90), 105783 (2, 95–111). USNM 170254 (1, 98), 190500 (1, 138). HONDURAS: ANSP 70148–70152 (5, 43–47). MEXICO QUINTANA ROO: ANSP 110989 (4, 112–206). UMML 23485 (4, 173–199). BRAZIL: CAS-SU 22217 (1, 127; holotype of *Rabula megalops* Starks).

Genus *Kaupichthys* Schultz, 1943

Kaupichthys Schultz, 1943: 50 (type species *Kaupichthys diodontus* Schultz, 1943, by original designation).

Characters. Body moderately elongate. Pectoral fin present, but may be reduced. Dorsal fin orig-

TABLE 6. Character states in four species of *Kaupichthys*.

Species	Posterior nostril	Length of pectoral fin	LL pores	Color
hyoproroides	labial	> eye diameter	2	uniform brown
nuchalis	lateral	> eye diameter	2	pale nuchal band
atronasus	labial	> eye diameter	1	mottled, base of anterior nostril black
brachychirus	labial	< ½ eye diameter	1	uniform brown

inates above or slightly behind gill opening. Head and snout moderate, neither greatly elongate nor shortened. Rictus near posterior margin of eye. Lips without a flange. Anterior nostril tubular, near tip of snout; posterior nostril either a half-moon shaped opening on side of head above lip, or a slit opening into edge of lip and covered by a flap dorsally. One or two lateral-line pores in branchial region; six or seven preoperculomandibular pores.

Maxillary teeth small, in several rows. Intermaxillary teeth in a round patch, the inner and posterior ones somewhat enlarged. Vomerine teeth in two single rows, diverging slightly at midlength, converging at posterior end. Mandibular teeth in two to four series, widest anteriorly, the inner teeth largest.

Discussion. Kaupichthys includes some of the most common chlopsid eels, although they remained unknown until the widespread introduction of ichthyocides. Like all chlopsids, they are secretive and easily escape notice, even when present in large numbers. The species of *Kaupichthys* are found mainly in and around coral reefs. There is no indication that they burrow in the bottom sediment; rather, like morays, they probably hide in holes and crevices in the reef structure.

There is still substantial uncertainty about the interrelationships of the species currently assigned to *Kaupichthys*. The genus has been defined largely on the basis of dentition and the presence of pectoral fins. This involves the assumption that dentition accurately reflects relationships. *Kaupichthys nuchalis* resembles *K. hyoproroides* in its dentition and well developed pectoral fin, but its posterior nostril is located above rather than on the lip. The situation has been further complicated by the recent discovery of two undescribed forms in the Indo-Pacific that resemble *K. nuchalis* in most characters but show two radically different dentition patterns. Along with Böhlke's (1956a) observation that *K. hyoproroides* shows sexual dimorphism in its dentition (see below), this suggests that we should be cautious in assessing the systematic importance of dentition. *Kaupichthys brachychirus* resembles *K. hyoproroides* in its dentition and its nostril, but it has a greatly reduced pectoral fin. The leptocephali of *K. nuchalis* and *K. brachychirus* also differ from that of *K. hyoproroides* in pigmentation, a character that often reflects substantial taxonomic differences. Clearly, much work remains to be done before the genus *Kaupichthys* can be properly defined.

Size. These are small eels, probably not exceeding 250 mm TL.

Distribution. Western tropical Atlantic from Bermuda to Venezuela. Tropical Indo-West Pacific from east Africa to Hawaii.

Etymology. Named for J. J. Kaup, 19th-century ichthyologist who wrote extensively on eels. Masculine.

Species. Four species of *Kaupichthys* are currently recognized: *K. hyoproroides* (Strömman, 1896) from the western Atlantic and Indo-West Pacific, *K. nuchalis* Böhlke, 1967 from the western Atlantic, *K. atronasus* Schultz, 1943 and *K. brachychirus* Schultz, 1953, both from the Indo-West Pacific. They differ in the location of the posterior nostril, the length of the pectoral fin, the number of lateral-line pores, and the color (Table 6). In addition, there seems to be at least one undescribed species in the Indo-Pacific.

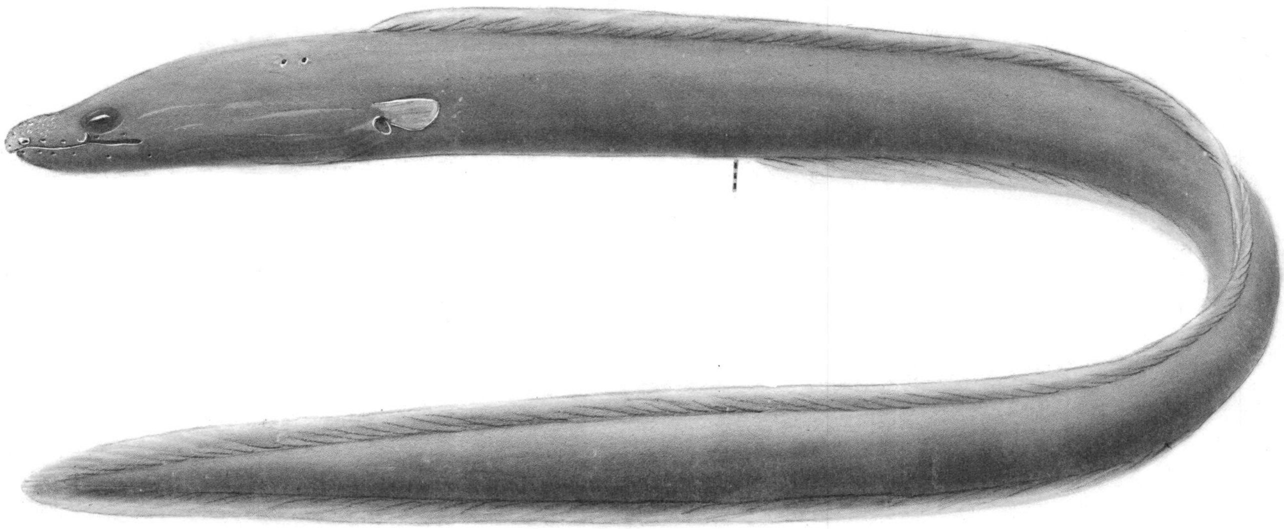

FIGURE 83. *Kaupichthys hyoproroides*: ANSP 111553, 232 mm TL. Illustration by S. P. Gigliotti.

KEY TO THE SPECIES OF *KAUPICHTHYS* IN THE WESTERN NORTH ATLANTIC

1a. Posterior nostril on upper lip, concealed by a flap dorsally; no nuchal band .... *hyoproroides*, p. 88

1b. Posterior nostril a simple opening on side of head opposite lower part of eye; pale nuchal band .... *nuchalis*, p. 91

Kaupichthys hyoproroides (Strömman, 1896)

FALSE MORAY or REEF EEL

Figures 83–85, Tables 5–7

Leptocephalus hyoproroides Strömman, 1896: 39, pl. 4, figs. 5–6 (original description, 27°N, 73°W, holotype ZMUU). Leptocephalus.

Kaupichthys diodontus Schultz, 1943: 50, pl. 6 (original description, Samoa, holotype USNM 115980). Gosline, 1950 (osteology and relationships).

Kaupichthys atlanticus Böhlke, 1956a: 66, figs. 1–2, pl. 7 (original description, Bahamas, holotype ANSP 73687).

Kaupichthys diodontus japonicus Matsubara and Asano, 1959: 293, figs. 1–2, A–B (original description, Japan, holotype unknown).

Kaupichthys hyoproroides. Böhlke and Smith, 1968: 28 (synonymized *Kaupichthys diodontus*, *K. atlanticus*, and *K. diodontus japonicus* with *Leptocephalus hyoproroides*). Smith, 1969: 396 (evidence for identification of *Leptocephalus hyoproroides* with *Kaupichthys*).

Distinctive Characters. Posterior nostril on outer edge of lip, covered by a flap dorsally; pectoral fin well developed; color uniform brown; vomerine teeth in two single rows; two lateral-line pores in branchial region; vertebrae 109–123.

Description (including data from Böhlke, 1956a; Böhlke and Smith, 1968). Total vertebrae 109–123 (n = 101), predorsal vertebrae 8–9 (13), preanal vertebrae 26–28 (15). Branchiostegal rays 14 (2). Pectoral-fin rays 14–15 (1). Dorsal-fin rays ca 350 (1). Pores: LL 2 (22), POM 6 (22), IO 4 (22), SO 1 + 2 (22), STC 0 (22).

TABLE 7. Total vertebrae in *Kaupichthys hyoproroides* (including data from Böhlke and Smith, 1968).

	Vertebrae																
Area	109	110	111	112	113	114	115	116	117	118	119	120	121	122	123	n	x̄
Western Atlantic					1	—	3	8	12	17	4	6				62	117.6
Indian Ocean		1	2	1	2	4	1	1	1	—	2	1	2	—	1	19	115.6
Western Pacific	3	—	1	2	5	4	3	1	—	—	—	—	—	1		20	113.4

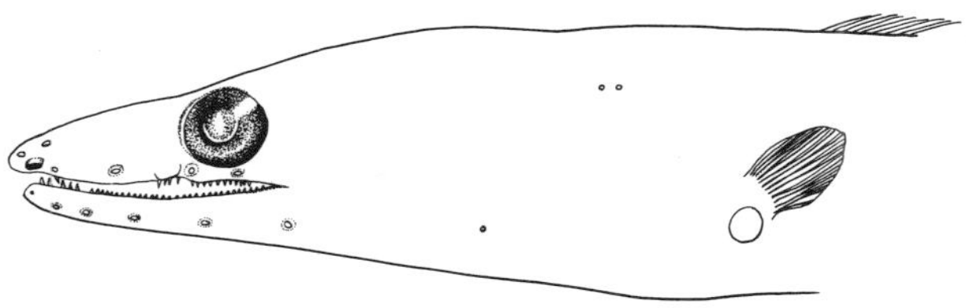

FIGURE 84. *Kaupichthys hyyproroides*: ANSP 117946, 123 mm TL; head.

Proportions as % of TL: preanal 28–32 (24), predorsal 14–16 (23), head 13–15 (23), depth at anus 3–5 (23). Of HL: snout 18–25 (23), eye 7–14 (23), snout-rictus 30–39 (22), pectoral fin 10–19 (23).

Dorsal fin begins over or very slightly behind gill opening. Pectoral fin broadly rounded, length slightly greater than eye diameter.

Posterior nostril elliptical, opening on exterior edge of upper lip and directed posteroventrally, covered dorsally by a flap which touches lower jaw when mouth closed. Two lateral-line pores at anterior end of canal in branchial region; usually six preoperculomandibular pores, all anterior to rictus, although occasionally a small seventh pore is present behind rictus.

Maxillary teeth small, conical, in several rows. Intermaxillary teeth in a round patch, the inner and posterior teeth somewhat enlarged. Vomerine teeth modestly enlarged, bluntly conical, arranged in two single rows that diverge slightly at midlength and converge at posterior end, terminating posterior to end of maxillary tooth patches. Mandibular teeth in two to four series, widest anteriorly, the inner teeth largest.

Stomach relatively long, extends well past anus. Gas bladder short, thin-walled, begins near gill opening, ends at about first quarter of stomach length.

Color. Uniform light brown with a slash of dark pigment at corner of mouth. Vertical fins dusky basally, paler distally. Ventral surface of lower jaw and throat pale, but not sharply set off from surrounding brown. Color differs little between fresh and preserved specimens. In very young specimens, brown most prominent on sides, lighter along dorsal and ventral midlines. A specimen of 48 mm TL (ANSP 140580) has a

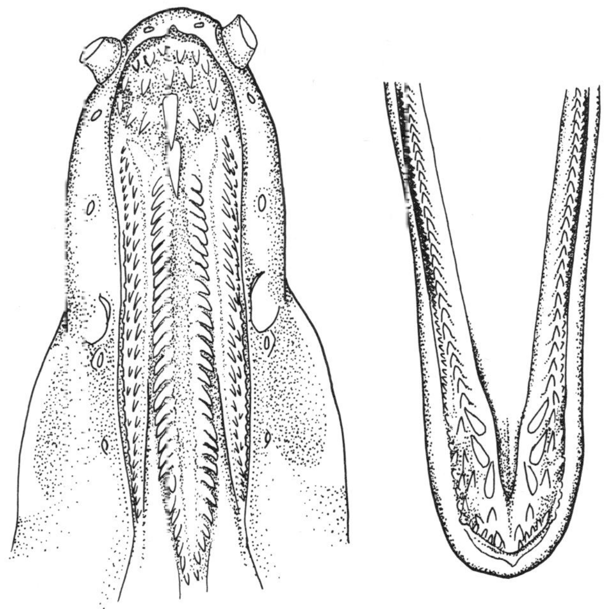

FIGURE 85. *Kaupichthys hyoproroides*: ANSP 117946, 123 mm TL; dentition.

few larval melanophores under gut. Stomach and intestine pale.

Size. The largest specimen examined was 238 mm TL (ANSP 113756). Sexual maturity is reached in females by at least 150 mm.

Development and Growth. The leptocephalus has been identified (see p. 934).

Habits and Behavior. Kaupichthys hyoproroides is found primarily in and around coral reefs. Like all chlopsids, it is cryptic and is never seen except when poisoned. Its habits can only be inferred, but it probably behaves much like a small moray, hiding in holes and crevices and feeding on small invertebrates.

Variation. Specimens from the western Atlantic have significantly more vertebrae ($\bar{x}$ = 117.6) than those from the western Pacific (113.4), although each group has at least one count that stands well apart from the rest (Table 7). Specimens from the western Indian Ocean are intermediate ($\bar{x}$ = 115.6), however, and span nearly the entire range.

Gosline (1950: 312) recorded a vertebral count of "approximately 98" for a cleared and stained specimen of *Kaupichthys hyoproroides* from Bikini Atoll. He later (1951: 200) gave 97 as the vertebral count of *Kaupichthys*, presumably based on the same specimen. This count is well below the

FIGURE 86. *Kaupichthys nuchalis*: ANSP 106376, holotype, 123 mm TL. Illustration by S. P. Gigliotti.

range of all other specimens examined, and until the specimen has been re-examined, it should be accepted with caution. Gosline (1950: 312) seemed somewhat uncertain about the accuracy of the preanal count as well (". . . about 20 [vertebrae] lie ahead of the anus").

Four leptocephali from off Oahu have 116–125 myomeres (ANSP 156847, 156848). This is somewhat higher than the range of vertebral counts for *K. hyoproroides* in the western Pacific. Adult specimens from Hawaii are not available for comparison.

Böhlke (1956a: 67, 71) reported several characters that apparently differed between sexes, most notably dentition. In males the mandibular teeth were in three or four series anteriorly and biserial posteriorly; the inner teeth were much larger than the outer ones, compressed and hooked backward at their tips. In females, the mandibular teeth were in four irregular series, even as far as midway back; the inner teeth were larger than the outer, but the difference was not as great as in the males, and the teeth were broadly conical and not recurved. The maxillary teeth were in three rows in males, in three to five rows in females. The vomerine teeth were more pointed in males than in females, and the rows were more widely separated in males.

It now appears that these differences are not sexual. In later studies (unpublished), Böhlke found that the "male"-type dentition and the "female"-type dentition occurred in both sexes. My own attempts to confirm these observations have been inconclusive. Some specimens clearly have the "male"-type and some clearly the "female"-type dentition, but others are ambiguous. At any rate there is a substantial variation in dentitional pattern that is difficult to explain.

Böhlke (1956a: 71) reported several other sexually dimorphic characters, including the length of the pectoral fin, the interorbital width, and the length of the snout. Because the sex of the individuals in these cases was determined by their dentition rather than by gonad type, the differences may not be strictly sexual.

Distribution. Western Atlantic from Bermuda, southern Florida and the Bahamas through the Caribbean to Venezuela. Indo-West Pacific from Hawaii to east Africa, including Japan and Australia. Common.

Etymology. From *Hyoprorus*, a name applied to the leptocephalus of *Nettastoma melanura*, and *-oides*, like; in reference to a supposed resemblance. An adjective.

Study Material. 246 specimens, 43–238 mm TL.

BERMUDA: ANSP 126290 (1, 160), 133623 (2, 91–113), 133624 (1, 60), 133625 (1, 115), 133626 (2, 115–205), 147864 (1, 185), 147874 (1, 150). FLORIDA KEYS: UMML 8686 (2, 128–207), 21747 (1, 125). BAHAMAS: ANSP 73687 (1, 148; holotype of *Kaupichthys atlanticus* Böhlke), 73692 (4,

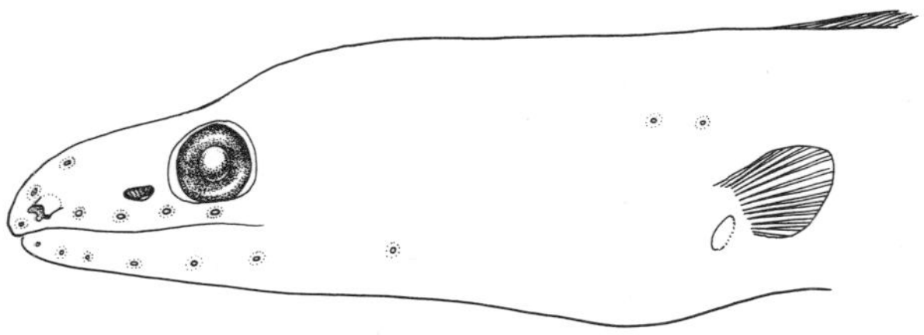

FIGURE 87. *Kaupichthys nuchalis*: ANSP 127984, 131 mm TL; head.

155–225), 73695 (5, 73–201), 73697 (1, 136), 100623 (2, 86–107), 100632 (1, 107), 101176 (2, 120–138), 111553 (1, 227), 111205 (2, 48–65), 114166 (1, 46), 114167 (1, 47), 114168 (1, 45), 114169 (2, 53–54), 114170 (1, 52), 114171 (1, 64), 114172 (1, 48), 114173 (1, 43), 116475 (1, 94), 127077 (3, 43–85), 128797 (2, 53–98), 128814 (1, 97), 130551 (3, 45–120), 128799 (2, 112–139), 140495 (2, 129–183). UMML 1746 (2, 128–141), 7859 (1, 128), 12639 (3, 61–143), 12776 (1, 151), 12957 (9, 45–129). HISPANIOLA: ANSP 111372 (2, 80–141), 117942 (3, 54–143), 117946 (1, 123), 119050 (5, 65–147), 119150 (2, 92–94), 127263 (2, 104–150). UMML 7541 (4, 134–167). PUERTO RICO: ANSP 124386 (2, 120–157), 139833 (4, 116–235), 139842 (2, 61–150), 147108 (1, 96). VIRGIN ISLANDS: UMML 3745 (2, 116–152). LESSER ANTILLES: ANSP 105775 (5, 67–135), 110421 (1, 74), 117979 (1, 74), 140580 (1, 48). VENEZUELA: UMML 14172 (1, 125), 15342 (3, 85–171). COLOMBIA: ANSP 117579 (2, 109–120). PANAMA: ANSP 113755 (8, 84–164), 113756 (15, 63–238), 113759 (29, 62–198). NICARAGUA: ANSP 114919 (1, 108). BELIZE: UMML 9952 (1, 107), 9846 (1, 76). MEXICO QUINTANA ROO: ANSP 114831 (5, 64–101), 123731 (1, 77). PROVIDENCIA: ANSP 128803 (3, 50–82), 128981 (2, 90–168). CAYMAN ISLANDS: ANSP 126416 (1, 88). INDIAN OCEAN: 44 specimens, 53–139 mm TL, see Böhlke and Smith (1968: 30). WESTERN PACIFIC: 21 specimens, 115–171 mm TL, see Böhlke and Smith (1968: 29).

Kaupichthys nuchalis Böhlke, 1967

COLLARED EEL

Figures 86–89, Tables 5, 6

Kaupichthys nuchalis Böhlke, 1967: 95, fig. 2 (original description, Bahamas, holotype ANSP 106376). Böhlke and Smith, 1968: 30 (additional vertebral counts). Bright and Cashman, 1974: 346 (record from northern Gulf of Mexico).

Distinctive Characters. Posterior nostril above lip, not covered by flap; pectoral fin well developed; two lateral-line pores in branchial region; pale nuchal band; vomerine teeth in two single rows.

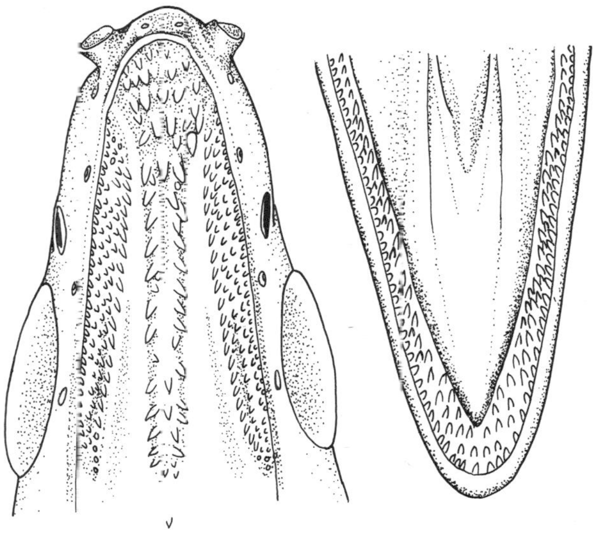

FIGURE 88. *Kaupichthys nuchalis*: ANSP 127984, 131 mm TL; dentition.

Description (including data from Böhlke, 1967; Böhlke and Smith, 1968). Total vertebrae 119–125 (n = 17), predorsal vertebrae 7–8 (12), preanal vertebrae 33–37 (12). Branchiostegal rays 8–9 (1). Pectoral-fin rays 11–12 (1), dorsal-fin rays 376 (1), anal-fin rays 297 (1), caudal-fin rays 7 (1). Pores: LL 2 (11), POM 7 (13), IO 4 (13), SO 1 + 2 (13), STC 0 (13).

Proportions as % of TL: preanal 32–36 (24), predorsal 11–14 (23), head 10–12 (23), depth at anus 3–5 (22). Of HL: snout 20–24 (13), eye 8–14 (13), snout-rictus 33–40 (8), pectoral fin 10–21 (12).

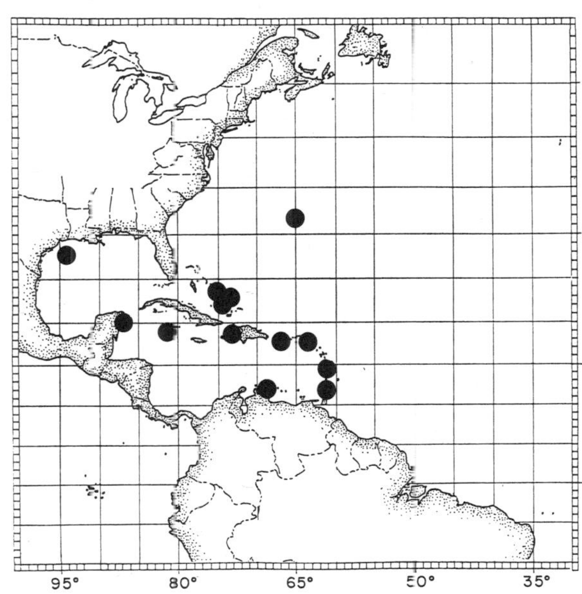

FIGURE 89. Distribution of *Kaupichthys nuchalis*.

Dorsal fin originates over or slightly behind gill opening. Pectoral fin broadly rounded.

Posterior nostril a simple opening on side of head in front of lower part of eye; ventral margin of nostril rounded, dorsal margin flattened. Two lateral-line pores in branchial region; seven pores on lower jaw, six before rictus and one behind.

Maxillary teeth slender, in three or four rows anteriorly, narrower posteriorly, inner teeth larger than outer. Intermaxillary teeth in a round patch, the median and posterior teeth larger. Vomerine teeth in two single rows that diverge slightly at midlength and converge posteriorly, terminating near posterior end of maxillary teeth. Mandibular teeth in three rows anteriorly, two rows posteriorly, inner teeth much larger than outer.

Color. Body brown, underside of head and throat pale. Pale area extends dorsally in a broad band around head behind eye, lightest on sides, slightly obscured on top. Pores surrounded by pale field; the pale fields around lateral-line pores coalesce to form one pale area, connected with pale area on throat. Small specimens have some dark brown melanophores on bottom of head and throat, perhaps remnants of larval pigment.

Size. This is a small species. The largest specimen examined was 131 mm TL (ANSP 127984), a female with large eggs. Bright and Cashman (1974: 346) reported a specimen 163 mm TL.

Development and Growth. The leptocephalus has been identified (see p. 936).

Habits and Behavior. Kaupichthys nuchalis occurs mainly in and around coral reefs. Böhlke (1967: 98) noted that specimens had been collected at Curaçao from cylindrical sponges and suggested that sponges may be the usual habitat of the species.

Remarks. Böhlke (1967) assigned this species to *Kaupichthys* with some hesitation. Although it resembles *K. hyoproroides* in dentition, pore pattern, and pectoral-fin development, it differs markedly in the nature of the posterior nostril. It also has distinctly fewer branchiostegal and pectoral-fin rays. The leptocephalus, too, differs markedly from that of *K. hyoproroides* (see pp. 934, 936). Böhlke chose the conservative course and placed the species in *Kaupichthys* pending further study. That practice is followed here, although future studies may necessitate changes.

Distribution. Bahamas and the islands of the Caribbean. Also recorded from the West Flower Garden Bank in the northern Gulf of Mexico (Bright and Cashman, 1974: 346). Depth from shoreline to 77 m (Böhlke, 1967); one specimen (ANSP 140578) is stated to have come from 604–732 m. Relatively common, but less so than *Kaupichthys hyoproroides* and *Chilorhinus suensonii.*

Etymology. From the Latin *nucha* (nape) and *-alis* (having the nature of), in reference to the nuchal band. An adjective.

Study Material. 29 specimens, 42–131 mm TL.

Holotype: ANSP 106376 (123 mm TL), south shore of Grand Bahama Island, 5 miles (8 km) east of West End, depth 40 feet (12 m), coll. R. E. Schroeder and S. Gross, 1 July 1959.

Paratypes: VIRGIN ISLANDS: ANSP 106380 (1, 101). LESSER ANTILLES: ANSP 106384 (3, 91–100). VENEZUELA: ANSP 106385 (1, 117), 106386 (1, 125).

Other material: BERMUDA: ANSP 133627 (1, 90). FLORIDA STRAITS: USNM 179229 (1, 102). BAHAMAS: ANSP 128795 (2, 52–90), 147553 (1, 53). HISPANIOLA: ANSP 111382 (2, 42–111), 117947 (1, 109), 127984 (1, 131). UMML 21640 (3, 82–105). VIRGIN ISLANDS: ANSP 140578 (1, 85). UMML 8309 (1, 100). MEXICO QUINTANA ROO: ANSP 114831 (5, 64–101), 117330 (1, 94), 117379 (1, 106). UMML 23483 (1, 53).

Genus *Chlopsis* Rafinesque, 1810

Chlopsis Rafinesque, 1810b: 58 (type species *Chlopsis bicolor* Rafinesque, 1810, by monotypy).

Garmanichthys Seale, 1917: 80 (type species *Garmanichthys dentatus* Seale, 1917, by monotypy).

Arenichthys Beebe and Tee Van, 1938: 301 (type species *Arenichthys apterus* Beebe and Tee Van, 1938, by original designation).

Characters. Body moderately elongate. Pectoral fin absent. Dorsal fin originates over or slightly behind gill opening. Head and snout moderate, neither greatly elongate nor shortened. Rictus near posterior margin of eye. Lips without a flange. Anterior nostril tubular, near tip of snout; posterior nostril elliptical, opening on outside of upper lip below and slightly before anterior margin of eye, concealed by a flap from dorsal margin. One lateral-line pore in branchial region, five or six pores in preoperculomandibular canal.

Intermaxillary teeth in a round patch. Vomerine teeth in two single rows, diverging at midlength and converging posteriorly, terminating behind posterior end of maxillary teeth. Maxillary and mandibular teeth mostly biserial, those of the inner series larger; maxillary tooth patch begins near anterior end of vomerine tooth patch.

Discussion. A note of confusion was injected into the literature by Jordan and Davis (1891: 650) when they synonymized *Saurenchelys* Peters, a nettastomatid, with *Chlopsis.* This strange conclusion must have been based on an incomplete reading of two primary sources. Rafinesque (1810b), in his description of *Chlopsis,* compared it only to *Nettastoma,* which he had described in a paper published earlier in the same year. Peters (1864) described a new genus *Saurenchelys,* which he distinguished from *Nettastoma* by the position of its posterior nostril. Apparently assuming that Rafinesque and Peters were referring to the same fish, but without checking the descriptions further—and certainly without examining specimens—Jordan and Davis placed *Saurenchelys* in the synonymy of *Chlopsis.* Workers who had actually studied *Chlopsis* (Grassi and Calandruccio, 1894, 1896; Supino, 1905a; Grassi, 1913) realized that the two genera were quite separate entities, and Grassi (1913) placed *Chlopsis* in its own family. American ichthyologists, however, continued to follow Jordan and Davis in treating *Chlopsis* as a nettastomatid. Böhlke (1956a) finally cleared up the confusion, but for more than half a century the name *Chlopsis* was used for two completely different assemblages of eels.

Size. The largest specimen of *Chlopsis* known, the holotype of *C. olokun,* is 305 mm TL.

Distribution. Eastern and western tropical Atlantic, including the Mediterranean, eastern tropical Pacific, western Indian Ocean, and western Pacific.

Etymology. From the Greek *chlon* (twig) and *-opsis* (in appearance). Masculine.

Species. Six species of *Chlopsis* are recognized: *C. bicolor* Rafinesque, 1810 from the eastern and western Atlantic, *C. dentatus* (Seale, 1917) from the western Atlantic and Indian Oceans, *C. olokun* (Robins and Robins, 1966) from the eastern tropical Atlantic, *C. apterus* (Beebe and Tee Van, 1938) from the eastern tropical Pacific, *C. bicollaris* (Myers and Wade, 1941) from the Galapagos, and *C. kazuko* Lavenberg, 1988 from the eastern Pacific. *Chlopsis dentatus* is easily distinguished by its banded or mottled color pattern; the other species are all bicolored to some degree. *Chlopsis bicollaris* has two pale bands dorsally across the head and neck between the eye and the dorsal fin; the remaining species are without bands of any kind. In *C. olokun* the dorsal fin begins above the gill opening and the bicolored effect is only weakly developed. *Chlopsis kazuko* also has the dorsal fin beginning above the gill opening, but it is strongly bicolored and has fewer vertebrae (123–126 vs 125–139). *Chlopsis bicolor* and *C. apterus* are both sharply bicolored, brown above and white below, and the dorsal fin begins behind the gill opening. *Chlopsis bicolor* has fewer vertebrae (127–134) than *C. apterus* (136), and the pattern of pigmentation is somewhat different in the two species.

The following species described under *Chlopsis* are nettastomatids. *Chlopsis equatorialis* Gilbert, 1891 and *C. gilberti* Garman, 1899 belong to *Facciolella* Whitley. *Chlopsis fierasfer* Jordan and Snyder, 1901 and *C. finitimus* Whitley, 1935 belong to *Saurenchelys* Peters. *Chlopsis taiwanensis* Chen and Weng, 1968 is a synonym of *Gavialiceps taeniola* Alcock, which is here considered to be a congrid close to *Xenomystax* Gilbert.

Key to the Species of *Chlopsis* in the Western North Atlantic

1a. Bicolored, brown above and white below . . . . *bicolor,* p. 94
1b. Banded or mottled . . . . *dentatus,* p. 95

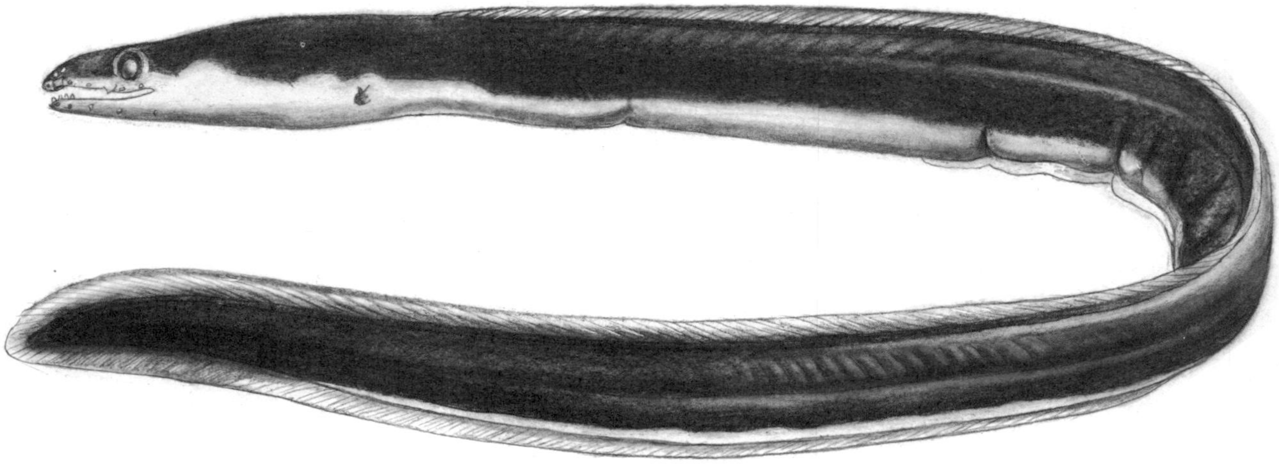

FIGURE 90. *Chlopsis bicolor*: ANSP 128386, 193 mm TL.

Chlopsis bicolor Rafinesque, 1810

BICOLOR EEL

Figures 90–93, Table 5

Chlopsis bicolor Rafinesque, 1810b: 59, pl. 2, fig. 1 (original description, Sicily, holotype unknown). Supino, 1905a: 25 (anatomy). Böhlke, 1956a: 77 (description and synonymy). Robins and Robins, 1967: 232 (occurrence in western North Atlantic). Blache et al., 1973: 233 (synonymy). Maul, 1976: 16 (record from Meteor Bank, eastern North Atlantic). Menezes and Benvegnu, 1976: 270 (record from western South Atlantic).

Misidentifications:

Not *Chlopsis bicolor* Rafinesque. Castle, 1966: 31; 1970a: 14 (*Chlopsis olokun*).

Distinctive Characters. Posterior nostril on lip, covered by a flap; pectoral fin absent; vomerine teeth in two single rows; sharply bicolored, brown above, pale below; dorsal fin begins behind gill opening.

Description (including data from Robins and Robins, 1967). Total vertebrae 127–134 (n = 7), predorsal vertebrae 12 (1), preanal vertebrae 39 (1). Branchiostegal rays 13 (1). Pores: LL 1 (4), POM 4–5 (5), IO 4 (4), SO 1 + 2 (4), STC 0 (4).

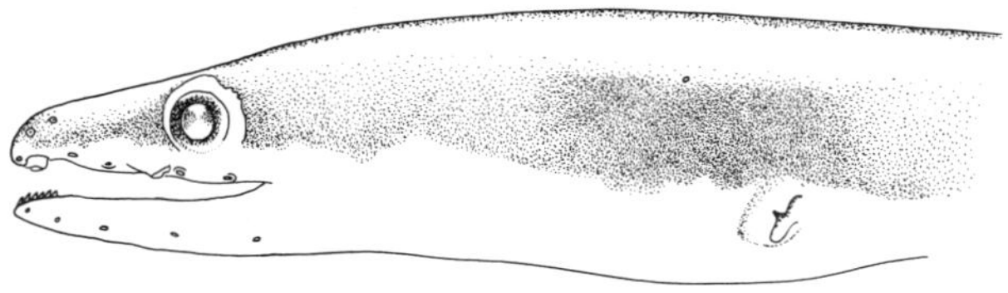

FIGURE 91. *Chlopsis bicolor*: ANSP 128386, 193 mm TL; head.

Proportions as % of TL: preanal 32–36 (8), predorsal 13–16 (8), head 9–13 (8), depth at anus 2–4 (8). Of HL: snout 21–27 (8), eye 9–12 (4), interorbital 12–18 (4), snout-rictus 34–52 (7).

Dorsal fin begins distinctly behind gill opening. A single lateral-line pore in branchial region. Four or five pores on lower jaw, all before rictus.

Maxillary teeth slender, conical, in a narrow band, inner teeth largest. Intermaxillary teeth conical, relatively heavy, in a round patch, located entirely anterior to maxillary teeth. Vomerine teeth relatively broad and heavy, in two series, reaching well behind posterior end of maxillary tooth patches, the two rows diverging

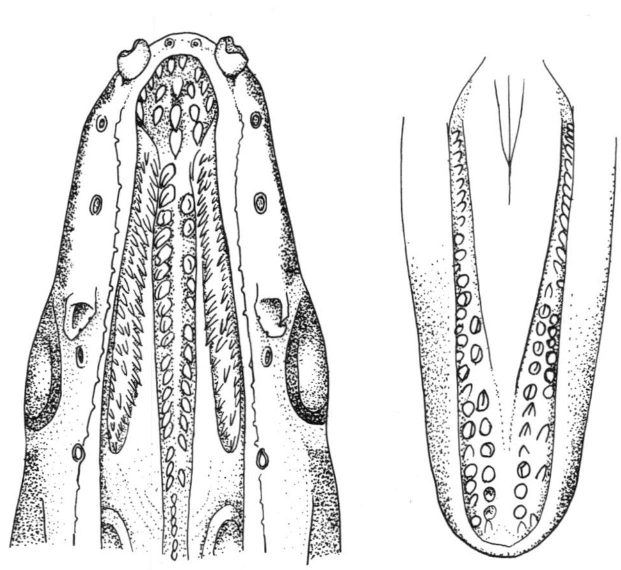

FIGURE 92. *Chlopsis bicolor*: ANSP 128386, 193 mm TL; dentition.

slightly toward midlength, then converging at posterior end. Mandibular teeth broadly conical, biserial anteriorly, uniserial posteriorly.

Color. In life and in preservative, dark brown above and white below, the distinction very sharp. On head, demarcation line begins at base of anterior nostril and passes posteriorly near ventral margin of eye to about level of gill opening. At about midtrunk, demarcation line begins to descend, and posterior to anus the white area restricted to a narrow band along ventral margin, finally ending shortly before end of tail. Posterior extremity of anal-fin base black. Robins and Robins (1967: 233) reported that some specimens have a blackish streak along demarcation line on head.

Size. The largest specimen examined was 235 mm TL (ANSP 150529).

Development and Growth. The leptocephalus has been identified (see p. 940).

Habits and Behavior. Although this species is said to be common in the Mediterranean (Schmidt, 1912b: 55), little is known of its habits and behavior. Like all chlopsids, it is cryptic and seldom seen alive. I had the rare opportunity to observe a living *Chlopsis bicolor* in 1968, when one was dredged up in the Straits of Florida north of Miami by the R/V GERDA. The specimen was placed in a plastic sorting tray, where it swam around vigorously. Despite its small size, it displayed an aggressiveness worthy of its muraenid relatives and at one point bit my finger and held on firmly. It was transferred to an aquarium where it was photographed and lived for several days before dying through an accident.

Variation. In the specimens from the western North Atlantic, the white ventral area extends to a line above the gill opening and cuts across the lower part of the eye. In a Mediterranean specimen (ANSP 67695), the demarcation line passes at about the ventral margin of the eye and runs below the gill opening, although the gill opening itself is pale.

Four specimens had five POM pores, one specimen (ANSP 128386) had four.

Distribution. Known from widely scattered localities on both sides of the Atlantic. In the eastern North Atlantic from the Mediterranean and from Meteor Bank south of the Azores. In the western Atlantic from the Carolinas, southern Florida, and Arrowsmith Bank, off the northeastern Yucatan Peninsula. In the western South Atlantic from off southern Brazil. Depth 32–351 m.

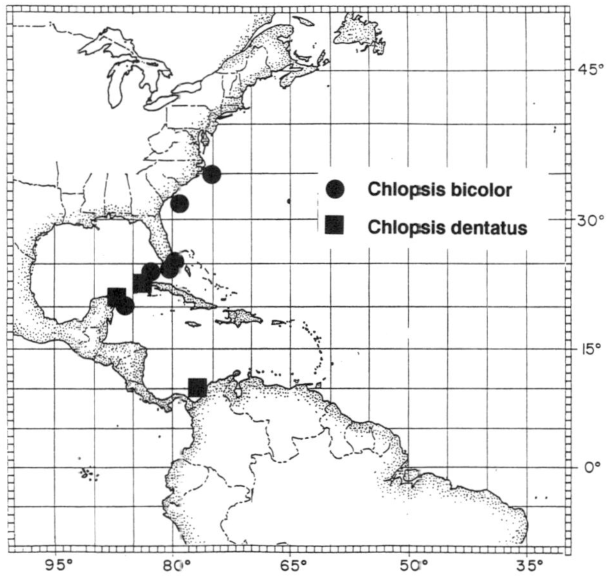

FIGURE 93. Distribution of *Chlopsis bicolor* and *Chlopsis dentatus*.

Etymology. Latin *bicolor* (of two colors). The form is invariable.

Study Material. Eight specimens, 109–235 mm TL.

NORTH CAROLINA: ANSP 150529 (1, 235), 35°06′N, 75°18′W, 32–38 m, 17 Sept. 1982. SOUTH CAROLINA: ANSP 128386 (1, 193), 32°05′N, 79°14′W, surface, 14 July 1973. FLORIDA ATLANTIC: UMML 17377 (1, 159), 25°14′N, 80°09′W, 92 m, 15 Apr. 1965. FLORIDA STRAITS: UMML 16634 (1, 174), 24°20′N, 82°52′W, 342–351 m, 24 Jan. 1965. UMML 16635 (1, 109), 24°29′N, 80°54′W, 201–210 m, 26 Jan. 1965. MEXICO QUINTANA ROO: ANSP 111544 (1, 174), NE edge of Arrowsmith Bank, 275–329 m, 28 Jan. 1968. MEDITERRANEAN: ANSP 67695 (1, 191), Strait of Messina. USNM 92240 (1, 208), Strait of Messina.

Chlopsis dentatus (Seale, 1917)
Figures 93–96, Table 5

Garmanichthys dentatus Seale, 1917: 80 (original description, Barbados, holotype MCZ 28095).

Chlopsis dentatus. Böhlke, 1956a: 76 (synonymized *Garmanichthys* with *Chlopsis*). Böhlke and Smith, 1968: 31 (records from Indian Ocean, illustrations, additional counts and measurements).

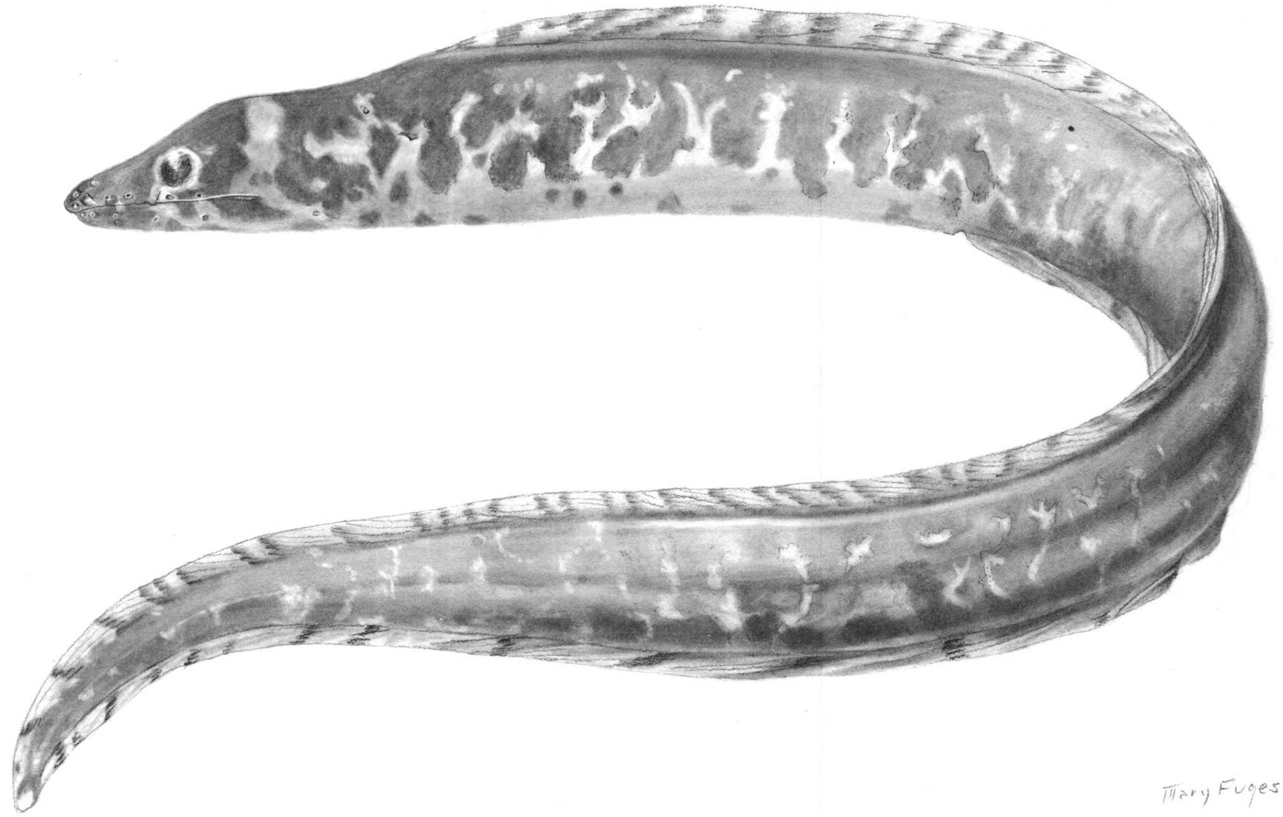

FIGURE 94. *Chlopsis dentatus*: ANSP 110015, 157 mm TL (from Böhlke and Smith, 1968: fig. 1).

Distinctive Characters. Posterior nostril on lip, covered by a flap; pectoral fin absent; vomerine teeth in two single rows; irregularly banded or mottled color pattern.

Description (largely from Böhlke, 1956a; Böhlke and Smith, 1968). Total vertebrae 116–124 (n = 8), predorsal vertebrae 7–9 (2), preanal vertebrae 31–34 (2). Pores: LL 1 (8), POM 6 (8), IO 4 (8), SO 1 + 2 (8), STC 0 (8).

Proportions as % of TL: preanal 32–35 (7), predorsal 13–16 (7), head 12–14 (7), depth at anus 4–7 (7). Of HL: snout 12–14 (7), eye 9–16 (7), interorbital 18–23 (7), snout-rictus 38–42 (7).

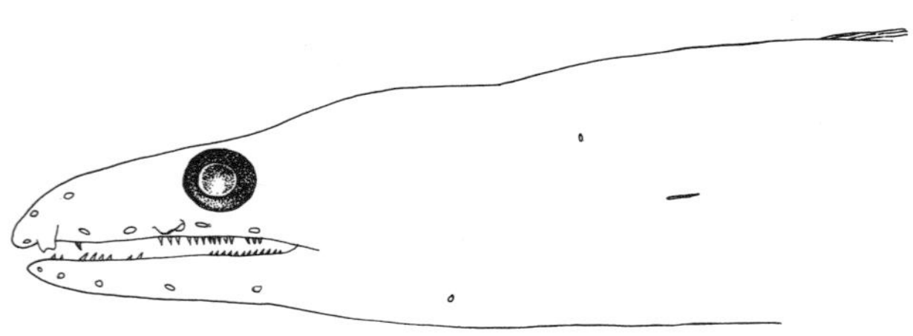

FIGURE 95. *Chlopsis dentatus*: ANSP 110014, 110 mm TL; head.

Dorsal fin begins slightly behind gill opening. A single lateral-line pore in branchial region. Five pores on lower jaw before rictus, a sixth pore in line with these behind rictus. Dentition essentially as in *Chlopsis bicolor*.

Color. The blotched or mottled color pattern of *Chlopsis dentatus* is unique among known chlopsids. The following description is from Böhlke and Smith (1968: 31).

> A transparency received from the Smithsonian Oceanographic Sorting Center, of the larger specimen in ANSP 110002 from off Kenya, shows the dark markings on the body as brown, those on the dorsal fin as dark brown or black. The anterior half of the head appears black. The pale areas on the body behind the anus and on the upper one-fourth of the trunk are pallid or whitish below and laterally to faintly tan above, while those on most of the trunk (the region containing the viscera) and on the posterior half of the head are pale greenish. The anal fin is pale greenish.

Size. The largest specimen known is 199 mm TL (ANSP 110002).

Development and Growth. The leptocephalus has been tentatively identified (see p. 938).

Habits and Behavior. Chlopsis dentatus seems to occupy habitats similar to those of *Chlopsis bicolor,* i.e., moderately deep water amid rock and rubble. Beyond that, nothing is known about its habits and behavior. One specimen (ANSP 110002) had two relatively large fishes in its stomach.

Variation. The specimens from the Indian Ocean have distinctly more vertebrae than those from the western Atlantic (121–124 vs 116–117). The former also have more dark pigment and more extensive bars and blotches than the latter. Böhlke and Smith (1968) chose not to recognize these differences at the specific level, and that decision is followed here.

Distribution. Known in the western Atlantic from off Cuba, Barbados, Colombia, and the Caribbean coast of Mexico; also the western Indian Ocean off Kenya and Mozambique. Blache (1972) recorded leptocephali identified as this species from the Gulf of Guinea. In addition, a small specimen that appears to be *Chlopsis dentatus* has been collected off the Solomon Islands in the western Pacific (AMS I.17284-001). Depth 64–366 m.

Etymology. Latin *dentatus* (toothed). An adjective.

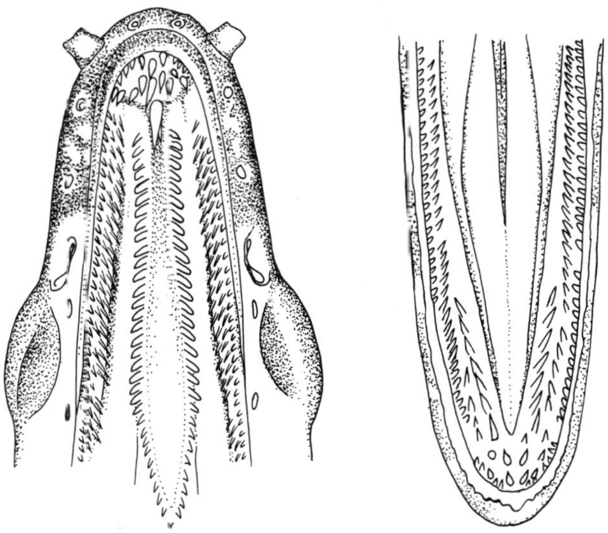

FIGURE 96. *Chlopsis dentatus*: ANSP 110014, 110 mm TL; dentition.

Study Material. Nine specimens, 56–199 mm TL.

Holotype: MCZ 28095 (103), off Barbados, 126 m.

Other material: CUBA: USNM 111595 (1, 57), 23°11′45″N, 82°18′54″W, 355 m, 17 Jan. 1885. COLOMBIA: ANSP 110014 (1, 110), 10°04′N, 76°06′W, 183–220 m, 28 May 1964. MEXICO QUINTANA ROO: UMML 23484 (1, 89), 21°12′N, 86°20′W, 64–73 m, 9 Sept. 1967. INDIAN OCEAN: ANSP 110002 (2, 102–199) and ANSP 110015 (1, 157), 2°42′S, 40°53′E, 140 m, 6 Nov. 1964. ANSP 110003 (1, 96), 26°00′S, 33°05′E, 135 m, 22 Aug. 1964. WESTERN PACIFIC: AMS I.17284-001 (1), 8°24′S, 160°35′E, ca 366 m, 27 Aug. 1973.

Family Myrocongridae

DAVID G. SMITH

Acknowledgments. See Smith (1984a: 593). I also thank Dr. Alasdair Edwards, University of Newcastle-upon-Tyne, for sharing the information he obtained at St. Helena.

Introduction. The Myrocongridae is the smallest family of the order Anguilliformes, containing a single species, *Myroconger compressus* Günther. It is most closely related to the Chlopsidae and Muraenidae and is considered the primitive sister group of those two (Smith, 1984a: 591). For a complete treatment of the family see Smith (1984a).

The family Myrocongridae was established as a subfamily by Gill (1890b: 169).

Characters. See *Myroconger compressus.*

Distribution. Eastern tropical Atlantic and St. Helena.

Genera. A single genus, *Myroconger* Günther, 1870.

Genus *Myroconger* Günther, 1870

Myroconger Günther, 1869: 238 (nomen nudum). Günther, 1870: 93 (type species *Myroconger compressus* Günther, 1870, by monotypy).

Characters. Those of the species *Myroconger compressus.*

Discussion of Synonymy. Günther (1869) listed "*Myroconger compressus,* g. et sp. n." among a collection of fishes sent from St. Helena by J. C. Melliss, but he did not provide a description. He noted that the eels from the collection would be described in his *Catalogue of Fishes* but added that "the new genus *Myroconger* is of great interest, being a *Muraena* with pectoral fins." The full description appeared as promised the following year (Günther, 1870: 93).

Size. That of the species *Myroconger compressus.*

Distribution. That of the species *Myroconger compressus.*

Etymology. From the Greek *myros* (a kind of eel) and *Conger* (the conger eel). Masculine. Since Günther recognized the similarity of this eel to *Muraena,* it is difficult to understand why he chose a name that indicates a relationship to a completely different eel.

Species. A single species, *Myroconger compressus* Günther, 1870.

Myroconger compressus Günther, 1870
Figures 97–110

Myroconger compressus Günther, 1869: 239 (nomen nudum). Günther, 1870: 93 (original description, St. Helena, holotype BMNH 1867.10.8.49). Melliss, 1875: 111 (brief account of occurrence at St. Helena). Osorio, 1893: 180 (record from the island of São Thomé in the Gulf of Guinea). Robins and Robins, 1966: 122 (redescription of holotype and speculation on relationships). Nelson, 1967: 562 (brief notes on osteology of holotype and speculation on relationships). Smith, 1984a (redescription, osteology, analysis of relationships, information on distribution).

Distinctive Characters. Body compressed, moray-like; eye well developed; pectoral fin present; lateral line restricted to a few pores at anteriormost end; posterior nostril level with upper margin of eye.

Description (from Smith, 1984a: 585). Total vertebrae 131 (n = 1), predorsal vertebrae 4–5 (2), preanal vertebrae 47 (1), precaudal vertebrae 52 (1). Branchiostegal rays 11 (1). Pectoral-fin rays 16 (1). Pores: LL 5–7 (2), POM 6 (1), IO 4 (2), SO 1 + 2 (2), STC 0 (2).

Proportions as % of TL: preanal 45 (1), predorsal 12 (1), head 14 (1), depth at anus 8 (1), pectoral fin 3 (1), caudal fin 2 (1). Of HL: snout

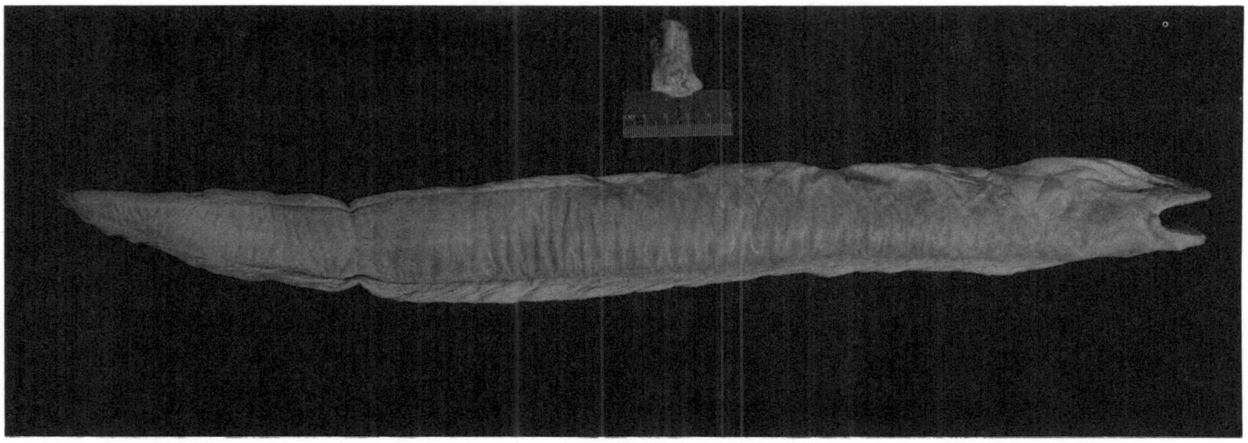

FIGURE 97. *Myroconger compressus*: BMNH 1867.10.8.49, holotype, 538 mm TL.

24 (2), eye 12–14 (2), snout-rictus 48 (1), gill opening 10–12 (2), interbranchial 32 (1), pectoral fin 19–22 (2).

Body moderately elongate, strongly compressed; anus slightly anterior to midbody. Dorsal and anal fins well developed, confluent around tail, dorsal fin originating before pectoral-fin base. Tail blunt but soft. Pectoral fin well developed, broadly rounded. All fin rays unsegmented. Gill opening oblique, small but not greatly restricted. Scales absent.

Head stout, well muscled, slightly depressed; many small hair-like papillae on snout and lower jaw; upper jaw projecting little if at all beyond lower jaw; no flange on either upper or lower lip. Mouth relatively large, rictus behind posterior margin of eye. Eye well developed. Anterior nostril tubular, just behind tip of snout; posterior nostril oval with a low raised rim, high on head near upper margin of eye.

Lateral line incomplete, 5–7 pores in branchial region and above pectoral fin. Four pores in infraorbital canal along upper jaw: two between level of anterior and posterior nostril, one between posterior nostril and eye, and one below eye. Three supraorbital pores: one (the ethmoid pore) at tip of snout anteroventral to anterior nostril, one near tip of snout anterodorsal to anterior nostril, and one above and behind anterior nostril. Six pores in preoperculomandibular canal, all before rictus.

Teeth numerous, moderate in size, sharp. Maxillary teeth in three irregular rows, confluent with intermaxillary teeth, the inner teeth the largest. Mandibular teeth in a band of three or four very irregular series, widest at anterior end, median and inner teeth larger than outer ones. Intermaxillary teeth in a rounded patch of approximately six transverse rows, confluent with maxillary teeth but separated by a slight gap from vomerine teeth. Vomerine teeth in a long, narrow band of two irregular rows anteriorly and one row posteriorly, reaching almost to level of posterior end of maxillary tooth bands.

Color. Melliss (1875: 111) referred to this species as the "Red Eel," with obvious implications. No one has seen the species alive since then, so the live color cannot be confirmed. The two known specimens are uniform light brown or tan in preservative.

Size. The holotype is 538 mm TL. The other specimen (ANSP 140308) is incomplete, having been cut off shortly behind the head. If the head and predorsal length of this specimen are in the

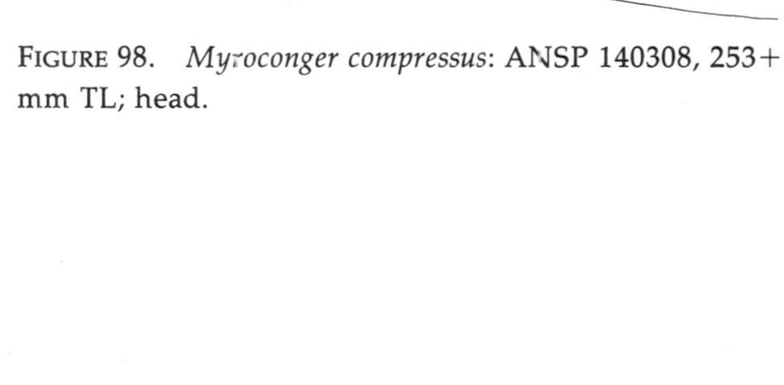

FIGURE 98. *Myroconger compressus*: ANSP 140308, 253+ mm TL; head.

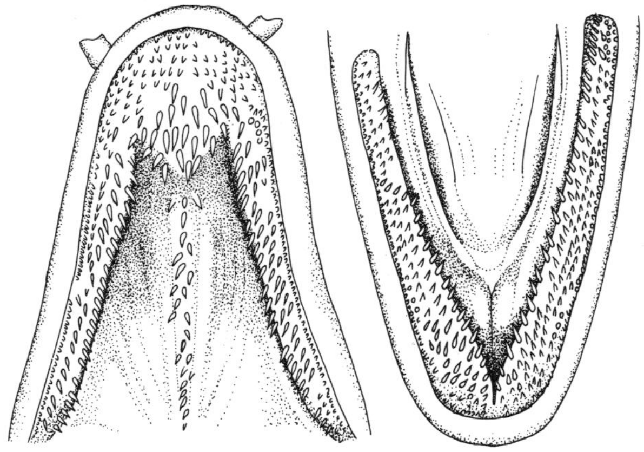

FIGURE 99. *Myroconger compressus*: data as for Figure 98; dentition.

same proportion to TL as those of the holotype, it would have measured about 685–690 mm TL.

Osteology (from Smith, 1984a: 586). Neurocranium (Fig. 100) stout, antorbital part heavy, with broad intermaxillary expansion; posterodorsal margin raised and overhangs posterior wall. Orbit relatively small. Frontals divided. Supraoccipital present, with a low crest. Otic bullae absent. Stout cartilaginous strut loosely attached to each side of ethmovomer immediately before orbit.

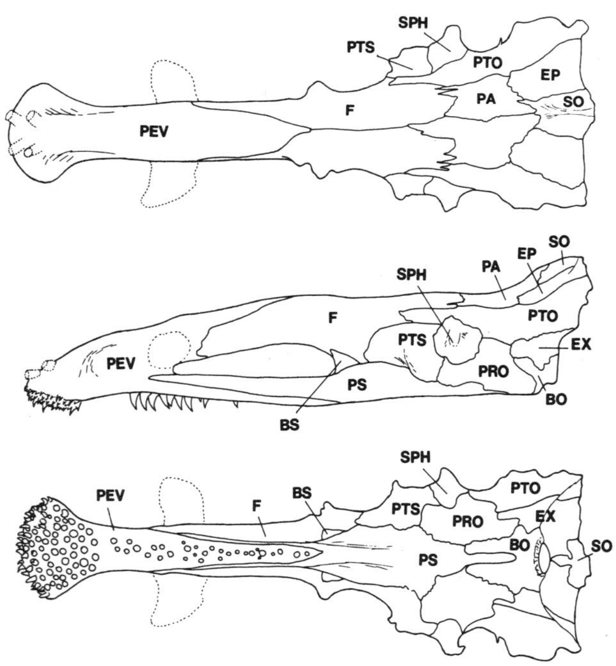

FIGURE 100. Neurocranium of *Myroconger compressus* (from Smith, 1984a: fig. 4). EP = epiotic; PEV = ethmovomer; SO = supraoccipital; SPH = sphenotic; other abbreviations as on pp. 5–6.

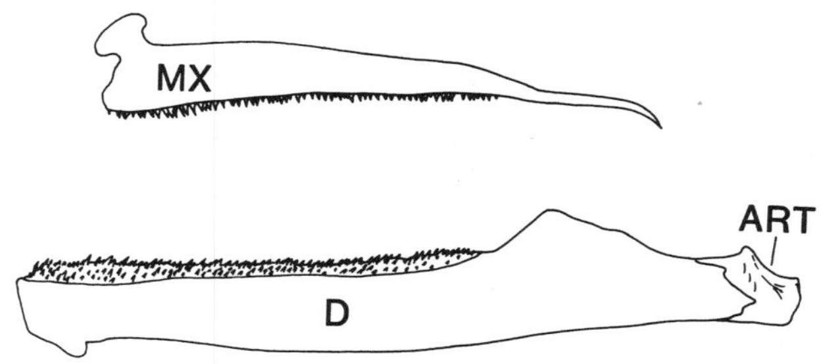

FIGURE 101. Jaws of *Myroconger compressus* (from Smith, 1984a: fig. 5). Upper, maxilla; lower, mandible. AR = articular; DE = dentary.

Maxilla (Fig. 101, top) with club-shaped pedicel on anterodorsal extremity, articulates with ethmovomer just behind intermaxillary expansion, tapers to slender tip posteriorly. Mandible (Fig. 5, bottom) consists of dentary and articular; in lateral view articular occupies only posterior extemity; coronomeckelian apparently absent.

Suspensorium vertical (Fig. 102). Slender pterygoid closely approaches quadrate posteriorly; anteriorly lies alongside parasphenoid but does not come close to it.

Hyoid arch consists of median glossohyal and urohyal, paired hypohyals, ceratohyals, and epihyals (Fig. 103). Glossohyal very short; urohyal with expanded central body and short posterior extension. Branchiostegal rays all arise from epihyal, curve up and around opercle posteriorly, nearly closing a circle.

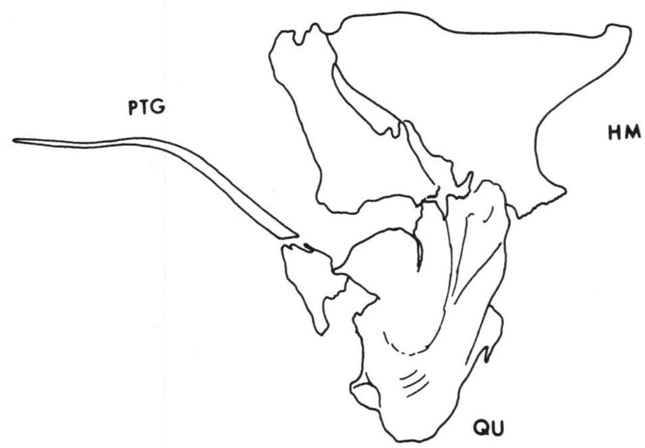

FIGURE 102. Suspensorium of *Myroconger compressus* (from Smith, 1984a: fig. 6). QU = quadrate.

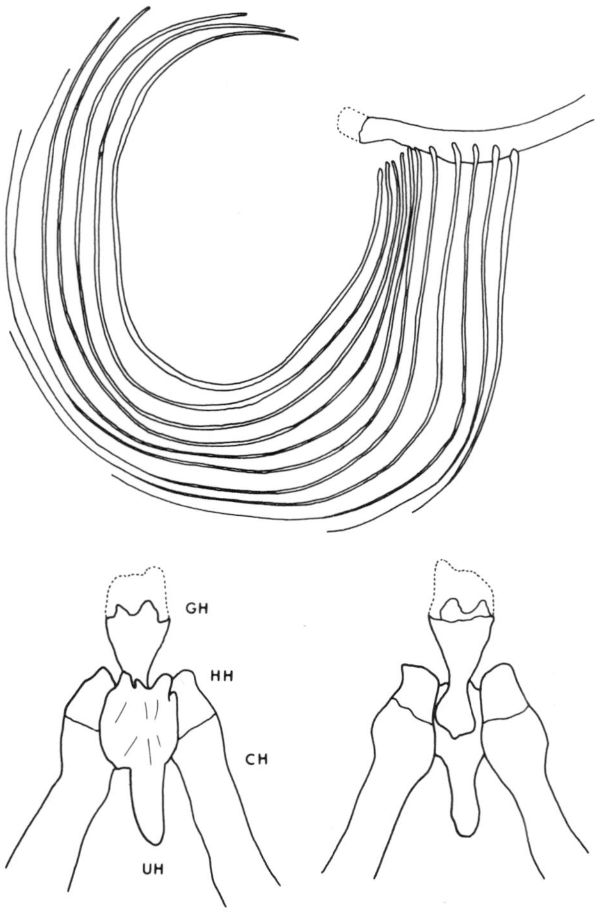

FIGURE 103. Hyoid arch of *Myroconger compressus* (from Smith, 1984a: fig. 7). Upper, posterior end of epihyal and branchiostegal rays; lower, anterior tip of hyoid arch in ventral (left) and dorsal (right) view.

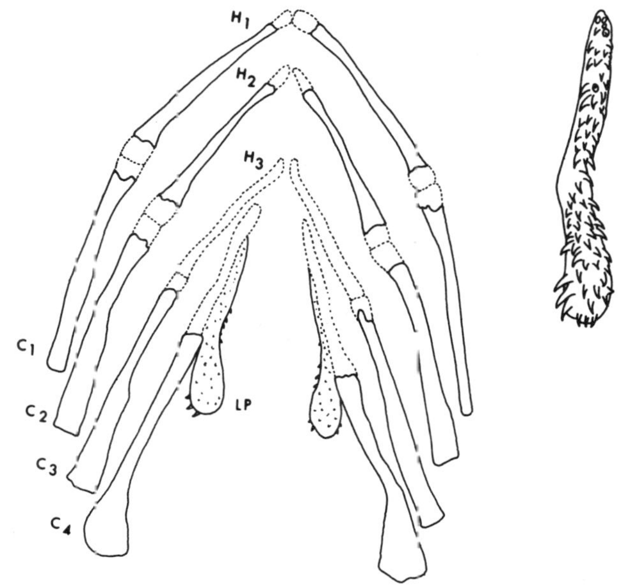

FIGURE 104. Ventral gill-arch elements of *Myroconger compressus* (from Smith, 1984a: fig. 8), ventral view; inset shows left lower pharyngeal in dorsal view.

Gill arches reduced (Figs. 104, 105). Basibranchials absent. First four ceratobranchials present and more or less equal in size, fifth absent. Three hypobranchials, first two ossified, third cartilaginous. Lower pharyngeal elongate, undivided. Dorsal elements well developed but anterior infrapharyngobranchial absent. Upper pharyngeals divided, anterior element much smaller than posterior.

Opercular apparatus complete (Fig. 106). Opercle much larger than other elements. Subopercle does not extend around posterior margin of opercle. Preopercle somewhat elongate. Interopercle irregularly trapezoidal in shape, only partly ossified.

Pectoral girdle well developed (Fig. 107). Supracleithrum relatively long but slender. Scapula and coracoid embedded in a cartilaginous plate. Three actinosts.

Circumorbital bones weakly developed (Fig. 108). A slender nasal and preorbital, each with additional elements immediately behind them. Small supraorbital may be present just below anterior end of nasal. Several small suborbitals behind eye.

Lateral-line ossicles in form of narrow, inverted V- or U-shaped tubes, consisting mainly

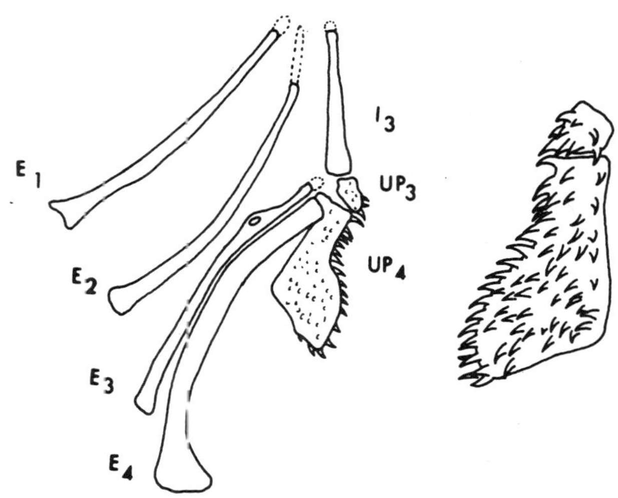

FIGURE 105. Dorsal gill-arch elements of *Myroconger compressus* (from Smith, 1984a: fig. 9), dorsal view; inset shows left upper pharyngeals in ventral view.

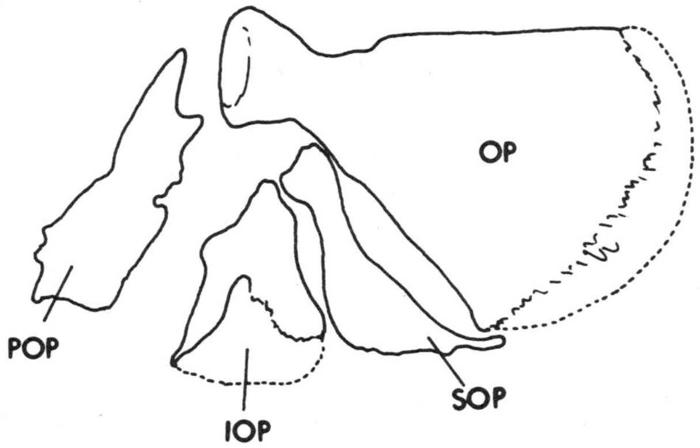

FIGURE 106. Opercular apparatus of *Myroconger compressus* (from Smith, 1984a: fig. 10). SOP = subopercle.

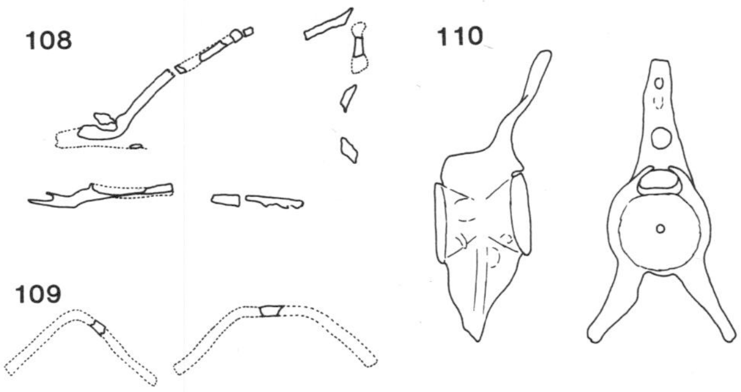

FIGURES 108–110 (from Smith, 1984a: figs. 12, 13 and 14, respectively). *Myroconger compressus.* Figure 108. Circumorbital bones. Figure 109. Lateral-line ossicles. Figure 110. Precaudal vertebra.

of cartilage with a small ossification near middle (Fig. 109).

Precaudal vertebrae moderately short, with broad parapophyses and a long, slender neural spine arising from posterior part of neural arch (Fig. 110). Epineurals present, ribs absent.

Development and Growth. The leptocephalus is unknown.

Remarks. I previously discussed the peculiar history of this species (Smith, 1984a: 591). Although Melliss (1875) gave a common name for it and stated that it was "commonly taken in about eight or nine fathoms of water and used as food," no one has ever reported it from St. Helena again. Citing my correspondence with Arthur Loveridge, I suggested that the specimen may not have come from St. Helena at all and that Melliss may have confused *Myroconger* with *Gymnothorax unicolor,* known locally as the "griggs."

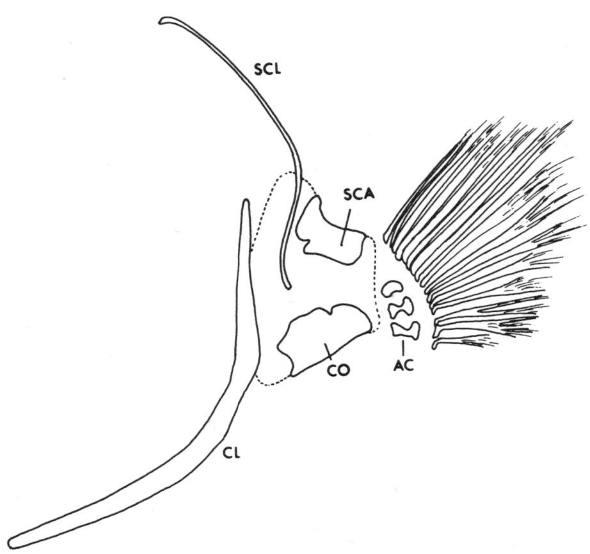

FIGURE 107. Pectoral girdle of *Myroconger compressus* (from Smith, 1984a: fig. 11). SCA = scapula.

I have since corresponded with Dr. Alasdair Edwards of the University of Newcastle-upon-Tyne (U.K.), who has recently collected fishes at St. Helena. Edwards was also unable to find the species or to find anyone who knew of it. He argued convincingly, however, that Günther's specimen did indeed come from St. Helena. According to Edwards (pers. comm., 24 August 1984), "Melliss was meticulous and every sample within his collection of specimens was sent with a perforated metal tag numbered in Roman numerals with his own collection reference number." In the same letter, Edwards further noted that "Melliss recognised 8 different eels at St. Helena and clearly considered the Red Eel as quite distinct from 'the small Brown Eels, called Griggs' which he records from shallower water than the Red Eel where they are indeed most common. . . ."

This, of course, only deepens the mystery. How could something that was common enough to be used as food in 1875 have completely disappeared, leaving not even a memory? Moreover, this disappearance dates at least from Cunningham (1910), who omitted *Myroconger* from the list of fishes he found at the island. Neither Edwards nor I have an explanation, but we feel that some sort of misunderstanding must be involved. Perhaps the "red eel" is a rarely caught straggler from deep water and Melliss confused it with the griggs (which Edwards says is now known as the "greeks"). Edwards has left a "Wanted" notice for the species at the fish cold store in St. Helena, but so far nothing has come in.

Distribution. If the above argument is accepted, the holotype came from the island of St. Helena in the South Atlantic. The only other specimen known (ANSP 140308) came from Dakar, Senegal, on the west coast of Africa. Osorio (1893: 180) recorded a specimen from the island of São Thomé in the Gulf of Guinea, but the specimen has been lost and the record cannot be confirmed (Smith, 1984a: 593). The species thus must have a fairly wide distribution, but so far only those two specimens are known.

Etymology. Latin *compressus* (compressed), referring to the strongly compressed body. An adjective.

Study Material. Two specimens, 538 and 253+ mm TL.

Holotype: BMNH 1867.10.8.49 (538 mm TL), St. Helena, South Atlantic, date of collection unknown but sent to British Museum in 1868 (Günther, 1869: 238).

Other material: ANSP 140308 (1, 253+, cleared and stained), fish market in Dakar, Senegal, West Africa, April, 1972.

Family Muraenidae

MORAYS

EUGENIA B. BÖHLKE, JOHN E. McCOSKER, *and* JAMES E. BÖHLKE

Acknowledgments. The authors wish to acknowledge the interest, assistance, and long-term support by the curatorial staff, technicians and clerical workers in the Ichthyology Departments of their respective institutions (ANSP and CAS). In addition to those colleagues and institutions listed in the introduction to this volume, we would like to thank the following for various and sundry assistances specifically for our studies: Amy Pertschuk (CAS) for illustrations; W. D. L. Ride (Canberra College of Advanced Education) for nomenclatorial assistance; Neal Talbot and Constantine Gillespie for their computer program "Eel Stretcher" ©. McCosker wishes to thank Ira Rubinoff and the staff of the Smithsonian Tropical Research Institute for assistance during his projects in Caribbean Panama, and the Kuna Indians of the Comarca de San Blas for their advice and assistance.

Introduction

Eels of the family Muraenidae, well-known as the morays, are found in all tropical and subtropical oceans and seas, and a few species occur in temperate waters. Some enter freshwater as juveniles and adults. They are perhaps the easiest of eels to recognize by the uninitiated through their general facies and fin reduction to but a confluent dorsal, caudal and anal fin. In like manner, their lack of specific characters has made them the most perplexing to classify; the state of muraenid taxonomy, particularly at the generic level, remains incomplete. We herein recognize, with certain reservations, 23 species within the western Atlantic distributed among eight genera. Worldwide, there are perhaps 200 species within more than 15 genera. A systematic review of muraenid genera and species has not been attempted by modern ichthyologists.

Typically, morays are shallow-water reef and crevice-dwelling eels, but some inhabit deeper waters. Although some morays are known to be nocturnally active (Winn and Bardach, 1959; Starck and Davis, 1966), more recent studies (Hobson, 1968; Chave and Randall, 1971) indicate that, in general, many are not. All morays are benthic as adults and several lineages have apparently secondarily invaded deeper water and live over sand and mud substrates; none are known to occupy midwater. All are carnivores, and their abundance and biomass within shallow reef habitats is generally much greater than casually perceived. With the use of ichthyocides, Brock, Lewis and Wass (1979) determined that morays comprised as much as 46.6% of the carnivorous fish biomass of a patch reef in Kaneohe Bay, Oahu. Large morays are much feared by divers and fishermen because of their malevolent appearance and occasional habit of provoked or unprovoked attacks upon hands thrust in their proximity. This is more a reflection of muraenid myopia than temperament (Randall, 1969). Morays are consumed at many locales but are considered a nuisance by most trap and line fishermen. Large morays are occasionally implicated in tropical fish poisoning and have been the source of ciguatoxic material for studies of ichthyosarcotoxins (Yasumoto and Scheuer, 1969; Hashimoto et al., 1969a,b). Perhaps the most vivid description of muraenid ciguatera poisoning was that of Khlentzos (1950) who described a banquet wherein 57 Filipinos consumed a single freshly-cooked *Gymnothorax flavimarginatus*. All diners exhibited some symptoms of ciguatera poisoning, many had convulsions, eleven became comatose and two died. The bite of certain morays is purported to be venomous; however no conclusive evidence supports the contention that the reaction is anything more than the result of secondary infection. The presence of a crinotoxin and a hemagglutinin in the skin of *Gymnothorax nudivomer* has been established (Randall et al., 1981), however its function has not been demonstrated.

Günther's (1870) treatment of muraenid species was the latest attempt to treat all described species. The plethora of subsequent records and descriptions have not been summarized, and one is thus forced to rely on a few regional treatments, including: northern Gulf of Mexico (Ginsburg, 1951); Bahamas (Böhlke and Chaplin, 1968); eastern Atlantic and Mediterranean (Blache, 1967a–g); Galapagos Islands (McCosker and Rosenblatt, 1975); Easter Island (Randall and McCosker, 1975); Hawaii (Gosline and Brock, 1960); southeast Africa (Castle and McCosker, 1986); and Indian Ocean and Red Sea (Smith, 1962a).

We are satisfied that our treatment of western Atlantic species is nearly complete, and it is unlikely that further collections will significantly alter our conclusions. We suggest, however, that further collecting efforts, particularly in deeper water and in the southwestern Atlantic, may result in discovery of additional species. And finally, we conclude that a worldwide revision of muraenid genera is in order.

History of Muraenid Classification

Morays have not commanded, in post-Linnaean times, the respect they did in the second century BC. Due to their abundance in shallow Mediterranean waters, edibility, and perceived ferocity, Romans had a high regard for *Muraena helena.* Pliny the Elder recounted how morays were reared in vivaria and kept by wealthy Romans. Live eels were decorated with jewels and said to be obedient to their master's beck and call. Caesar once demonstrated his largess after victory by giving 6000 live specimens to his friends. Vaedius Pollio, we are told, cast his recalcitrant or transgressing slaves into eel ponds to serve as eel food or for the entertainment of his guests. Respect for and attention to eels has plummeted since the end of the Roman Empire. (Some argue that all or most references to *Muraena* by Pliny et al. were actually based upon *Anguilla anguilla.* With that caveat however, and in order to save this apodal lore from obscurity, we here include it.) The common and subsequently generic name *Muraena* commemorates Licinius Muraena, a wealthy Roman and eel fancier; the family name Muraenidae is based on the moray, *Muraena helena.*

The general similarity of the muraenid facies allowed eighteenth and nineteenth century ichthyologists to develop a rather coherent early familial classification. Morays were readily assembled into a cohesive group, based particularly on the absence of pectoral fins and the restricted gill openings. With few exceptions, muraenids were rarely improperly included within other apodal families, and a very few non-muraenids were improperly assigned to the family.

Linnaeus (1758) is credited with the first binomial appelation for a muraenid, *Muraena helena,* the first apodal fish in his catalogue. His work was based largely upon Artedi's (1738) *Muraena pinnis pectoralibus carens.* Artedi's first eel listed within *Muraena* was *Muraena unicolor,* which was the common *Anguilla* of Europe. Subsequent naturalists (e.g., Risso, 1826; Yarrell, 1836; Swainson, 1838) variously adopted the "family" Muraenidae for true eels which included *Anguilla,* a category which contemporary systematists would consider to be the Order Anguilliformes. Bonaparte (1831) and Swainson modified Cuvier's (1817) classification and created subdivisions of the Muraenidae, particularly separating the European Ophichthidae and Muraenidae. Swainson's extensive analysis recognized a more limited order of apodal fishes, comprised of the Muraenidae and the Sphagebranchidae (see History of Ophichthid Classification, in this volume), separated on the basis of differences in their gill openings. The Muraenidae were further subdivided, including the "Subfamily Muraeninae" which included both morays and ophichthines that lacked pectoral fins.

McClelland (1844) should be credited with the recognition of the family Muraenidae as it is now understood. The manner in which he treated certain muraenid genera created a host of taxonomic problems, particularly as concerns his synonymy of *Gymnothorax* within *Muraena* and the description of *Lycodontis* (and its subsequent suppression in the same paper). Richardson (1848a) treated all morays in the genus *Muraena*, except for *Gymnomuraena zebra* and *Nettastoma vittata* (= *Channomuraena vittata*), and, in that same work, described the axial skeleton of *Muraena helena.* Kaup (1856a,b) later classified true eels according to their nostril position. Within his Section II: Phaneromycteres, were four families, including the Muraenidae with the subfamilies Muraeninae and the Thyrsoideinae.

Bleeker's (1864a) *Systema Muraenorum Revisum* was the most significant review of that period. In it he designated numerous generic types, synonymized many genera, and created six "families," the last of which, the Gymnothoracoidei, corresponds to the Muraenidae as presently recognized. Members of the family were characterized by their nostril position, lack of paired fins, and branchial openings. It was subdivided into two phalanxes, the Muraenophides (including *Echidna* and *Gymnothorax* and its synonyms, *Priodonophis, Strophidon,* and *Thyrsoidea*) and the Ichthyophides (including *Gymnomuraena, Ichthyophis* = *Uropterygius,* and *Channomuraena*). Günther (1870) based his classification largely on Müller's (1846) concept of Muraenoidei and considered all true eels to occupy the "family Muraenidae" with two subfamilies, the Muraenidae Platyschistae (nine groups) and the Muraenidae Engyschistae (containing only the Muraenidae as now understood). The grouping was based primarily upon the gill opening condition, and included *Muraena* (with several subgenera), *Gymnomuraena, Enchelycore* and *Myroconger.*

Cope (1871) divided true eels into two orders based primarily upon their gill arch conditions, the Colocephali (the morays) and the Enchelycephali (all others). He considered the morays to be a specialized offshoot of a more generalized, *Anguilla*-like, stock (1884). Cope's two groups were considered by subsequent authors as suborders (Gill, 1890c, and Jordan and various coauthors) or as orders (Herre, 1953).

Poey (1868a, 1876b) was inconsistent in his familial groupings, generally adopting the "family Muraenidae" for *Muraena* and "family Gymnothoracidi" for other genera, erroneously including *Pythonichthys* (later shown to be a heterenchelyid) in the Muraenidae. His *Revisium Piscium Cubensium* (1880) included the morays within the Ophichthyidi (previously, his Ophisuridi), and generally added numerous names to an already growing muraenid synonymy.

Perhaps the single most important nineteenth century familial work was that of Gill (1890c). His osteological diagnosis and description of the family greatly expanded Cope's work. The results of his analysis are consistent with most of our findings. Gill erred only in his inclusion of *Pythonichthys* and his doubtful recognition of the poorly known *Myroconger* within the Muraenidae. He included, as well, a nearly complete synonymy of the family name.

Other American ichthyologists, primarily Jordan and his co-authors (Jordan and Davis, 1891; Jordan and Evermann, 1896a; Jordan and Snyder, 1901), followed Cope in recognizing muraenids to be within the series or suborder Colocephali. They recognized *Gymnothorax* (with *reticularis* as its type species) as a valid genus and were the first to exclude *Myroconger* from the family but included *Pythonichthys,* largely on the basis of Poey's inadequate description.

Regan's (1912b) analysis of eel osteology was accurate in its treatment of the morays and has not been significantly modified except by addition. Regan removed *Heterenchelys* and *Myroconger* from the family and recognized most of the genera (or their current synonyms) that are valid today. He placed his emphasis within the family upon the paired frontals, hyomandibular breadth, reduction of the palatopterygoid, and the condition of the neural spines and the caudal transverse processes.

The twentieth century has experienced a vacillation in the usage of familial and generic names within the family, certainly at the expense of nomenclatural stability. Fowler was largely responsible for this moray morass and was rebuked by Myers and Wade (1941) for his actions. Fowler (1912a) used Muraenidae as a familial name until 1925, when he adopted Echidnidae (having replaced Anguillidae with Muraenidae), and created the new subfamilies Echidninae and Uropterygiinae. Other authors (e.g., Schultz, 1943; Smith, 1949) briefly adopted the use of Echidnidae but subsequently returned to the use of Muraenidae. Nelson (1966a), obviously unaware of Fowler's (1925) work, also described as new the subfamily Uropterygiinae. To further complicate this, Fowler (1958) had created the replacement name Uropterygina for the extinct paleoniscid suborder Platysomida, preoccupied, and Uropterygidae for Platysomidae, based upon *Platysomus* Agassiz, preoccupied. This homonymy in family group names is based on different generic names and would benefit from a decision by the International Commission on Zoological Nomenclature (Art. 55a).

Perhaps Fowler's ultimate folly was his (1936) attempt, adopting "first type designations," to change *Anguilla* auctorum to *Muraena*, *Muraena* auctorum to *Muraenophis*, and *Gymnothorax* auctorum to *Lycodontis*. Myers and Wade (1941: 87) fiendishly flailed Fowler, fortunately fending off fatuity, with their assertion that:

> The utter confusion which these shifts would introduce into the nomenclature of several of the commonest genera and best-known families of eels is patent, and anything that can be done to save the old and universally accepted usage is to be commended. However, no especial action seems needed at the present time, owing to several things entirely overlooked by Mr. Fowler.

Myers and Wade continued by pointing out that Opinion 93 of the ICZN (1926), under suspension of the Rules, placed *Gymnothorax* in the International Official List of Generic Names, with *G. reticularis* Bloch as its type species. Furthermore, Opinion 77 of the Commission (1922) had placed *Muraena*, with *M. helena* as its type species, on the Official List, thereby saving *Anguilla* as it had been universally accepted.

The last four decades have not seen any comprehensive attempt to clarify muraenid nomenclature or intrafamilial relationships. Although Jacoby (1867), Gosline (1951b, 1952, 1958), Burton (1956), Nelson (1966a, 1967), and Böhlke and McCosker (1982) have discussed various aspects of moray osteology and myology, the description of the osteology of *Muraena helena* given below is the first comprehensive osteological description of a muraenid eel.

Methods

General characteristics of the family Muraenidae are presented in the following pages. The key to genera is based primarily on our study of Atlantic species, and is subject to revision and expansion when a world-wide revision of the family is undertaken. All valid Atlantic species are included in keys to species, and all western Atlantic species are described and illustrated.

Counts, measurements, terms, and methods of study are defined in the general introduction (pp. 1–6). For morays, preanal length is measured to mid-anus, depth of body is measured at gill opening and at anus, and head length is measured to the upper end of the gill opening.

In addition to the cleared and stained specimens listed in the species accounts, the following material was examined for osteological comparisons: *Anarchias cantonensis*: CAS 57411 (1, 145); *Echidna nebulosa*: CAS 29111 (1, 175); *Enchelycore bayeri*: CAS 28675 (1, 310), 28684 (1, 245); *Gymnomuraena zebra*: CAS 29134 (1, 227); *Gymnothorax reticularis*: CAS 15921 (2, 340–385); *Muraena helena*: ANSP 128117 (1, 460); *Muraena lentiginosa*: CAS 56017 (2, 175–290).

Synopsis of Characters

General Characters. Within the Anguilliformes, the family Muraenidae exists as a discrete and apparently monophyletic assemblage. Several

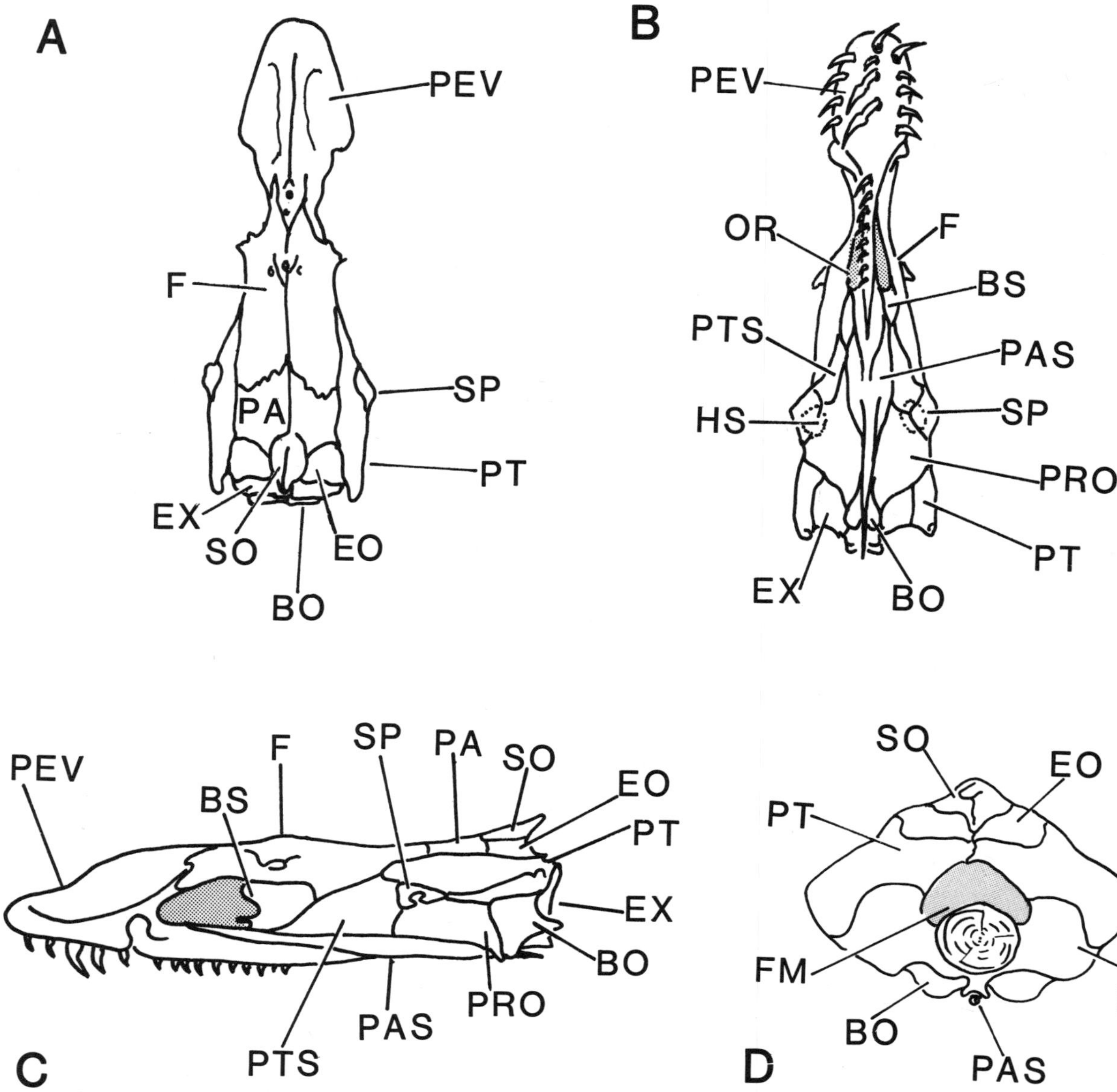

FIGURE 111. Neurocranium of *Muraena helena*, a generalized moray; ANSP 128117, 460 mm TL, Spain; neurocranial length 31.5 mm. A. Dorsal view. B. Ventral view. C. Left lateral view. D. Posterior view. Abbreviations: BO, basioccipital; BS, basisphenoid; EO, epiotic; EX, exoccipital; F, frontal; FM, foramen magnum; HS, socket of hyomandibular attachment to sphenotic, prootic and pterosphenoid; OR, orbit; PA parietal; PAS, parasphenoid; PEV, ethmovomer; PRO, prootic; PT, pterotic; PTS, pterosphenoid; SO, supraoccipital; SP, sphenotic.

character states, some unique to morays, may be used to define the family. They are: (a) gill arches extremely reduced, the ossified members limited to ceratobranchials 1–4, the third infrapharyngobranchial, the tooth plates and, in the subfamily Uropterygiinae, hypobranchials 1–2 (but absent in the Muraeninae); (b) pectoral fins absent; (c) gill openings constricted, nearly mid-lateral; (d) median fins present and confluent with caudal; (e) body scaleless; (f) frontal commissure absent; (g) temporal pore canal absent; (h) palatines absent; (i) pterygoid reduced to a thin sliver; (j) frontal sutured along its midline (this condition not obvious in older specimens); (k) ossified elements of hyoid reduced to slender epihyal and ceratohyal; and (l) branchiostegal rays few (8–10 pairs), slender and non-overlapping.

Osteological Characters. It is the purpose of this section to describe the osteology of *Muraena helena*. Limitations did not allow us to make an extensive comparative study of the various species which represent the 18–20 lineages which we identify within the family, so this analysis must not be considered to represent the range of osteological variation within the Muraenidae. *Muraena helena* was chosen in that it is the type species of the type genus of the family, and it represents, as well, a generalized shallow water muraenid. The work is based primarily upon cleared and stained osteological preparations and dissections of *Muraena helena* and *M. lentiginosa,* a related eastern Pacific species. Cleared and stained specimens of all other recognized muraenid genera were examined (see METHODS) but not critically dissected in most cases.

The neurocranium of *M. helena,* like that of all muraenids, is stout, heavily ossified, and well-fused along the cranial sutures (Fig. 111). It is fairly elongate, and truncate posteriorly. Muraenids range in skull proportions from the deep *Channomuraena* and *Monopenchelys* to the elongate species of *Gymnothorax* and *Enchelycore.* The neurocranial length is short relative to the total length of the fish. That of *M. helena* is about 6.8%. The neurocranium of more elongate species, such as *Thyrsoidea macrura,* is but 2.5% TL. Various aspects of the neurocranium and related bone complexes of *M. helena* are illustrated in Figure 111A–D.

Like many anguilliforms, *Muraena* lacks free premaxillae. Its rostrum is formed from the fusion of the ethmoid and vomer, as well as the region between the insertion of the maxillae (variously called the "intermaxillary" or "premaxillary"), and is here designated the ethmovomer (bearing "intermaxillary" teeth anteriorly and vomerine teeth on the posterior shaft). No sutures within the ethmovomer can be seen. The nasals are not ossified. A large nasal capsule exists in all morays examined; it is weakly cartilaginous and surrounds the nasal rosette. Ventrally, its general appearance is shovel-like, with an inverted Y-shaped rachis forward of the vomer, to which the maxillary articulates at its posterior base. The dentition of the vomer is discontinuous with that of the intermaxillary region and stops well short of the parasphenoid fusion. The ethmovomer articulates posterodorsally with the frontal and posteroventrally with the parasphenoid, forming the anterior half (or more) of the orbit. The parasphenoid is very elongate, toothless, and forms the floor of the orbit. It is spread laterally in its midsection and narrows posteriorly to a narrow point.

The paired prootics are large and fenestrated with numerous nerve openings. Their major axis is about 60° from horizontal. Together with the basioccipital they form the otic bulla. The median basioccipital rests upon the shaft of the parasphenoid and beneath the prootic. It does not form a swollen bulla (as in other anguilliforms), rather it is small and nipple-shaped. The otoliths lie within the bulla, but are barely visible externally due to the thickness of the underlying bone. Muraenid otoliths have not received systematic treatment in the literature. We are advised by Dirk Nolf of the Rijksuniversitet-Gent (in litt., 1978) that muraenid otoliths display intraspecific variability and are not useful for generic distinction. The typical *Muraena* sagitta (the largest of the three pairs) is small, ovate and biconvex. Its length is about 7–8% of the total neurocranial length. A shallow sulcus is grooved into the medial surface. Like the sagittae of ophichthids, those of *Muraena* and *Gymnothorax* are distinctive in having a shallow ostial canal which opens anteriorly rather than turning dorsally and opening from the sulcus, as is typical of congrid otoliths. Minor variation in shape, from slightly rounded to elongate, and lateral edge sculpturing does exist in some *Gymnothorax* species.

The paired pterotics form the dorsolateral edge of the posterior cranium. The cephalic lateralis tract passes through the pterotics, having entered through the frontal. The pterotics of muraenids appear to be proportionately larger than the adjoining bones, as compared to those of other eels. The paired exoccipitals dorsolaterally surround the foramen magnum. They are overlain by the supraoccipital and rest upon the basioccipital, which is fenestrated by numerous nerve tracts.

The single supraoccipital is small, subrectangular in shape, and lies along the posterodorsal midline of the skull. The crest is variously developed in muraenids; in *Muraena helena* it is weakly developed as a small protruding teat. The paired epiotics form the eave of the back of the skull. They are thin and subrectangular in shape.

The paired parietals are bordered posteriorly by the epiotics and the intervening supraoccipital, and anteriorly by the frontal. The middle fusion of the parietals is regular and nearly complete. The articulation to the frontal, however, is jagged and quite irregular. A large nerve opening lies just anterior to the center of the supraoccipital/parietal suture. The frontal is a single dorsal element which, together with the epiotics and parietals, forms the roof of the cranium. The frontal is flat in some species but ridged in most. The elevation is considerable in the elongate species of *Gymnothorax* and *Enchelycore*, providing them the typical narrow-faced appearance of many morays. The frontal of *Muraena helena* is moderately ridged. Anteriorly, the frontal is forked at the insertion of the ethmovomer, and broadly expanded to form the bracing and attachment of the postorbital nerve tract. The frontal of adult morays is fused to a degree that no suture is obvious. Smaller, younger eels show a suture between left and right frontals. However, even after soaking the skull in strong potassium hydroxide (5–8%), the paired bones will not separate. This fusion was considered to be significant by Regan (1912b) in his classification of the eels; we find it to be of limited value in relating eel families. The interorbital pore canal and numerous paired nerve tracts penetrate the frontal. The dorsalmost bone of the supraorbital pore series is fused to the frontal; the next two vertical elements are not fused in *Muraena helena*; however in other morays and certain ophichthids they appear to be, and thereby provide a strut which braces the maxillae to the skull.

The unpaired basisphenoid (= orbitosphenoid of others) is a median bone which forms the posteroventral margin of the orbit. A curious tongue-shaped process protrudes anteriorly from its center into the orbit. This process is flanked by three pairs of large nerve tracts. The paired pterosphenoids (= alisphenoid of others) lie behind the basisphenoid and narrow and turn inward anteriorly. They are stout in muraenids and often possess a sharp dimple, as well as a large neural fossa, just ahead of the sphenotic. The sphenotics are small and triangular, and form the dorsal edge of the anterior socket of the hyomandibular.

The form and appearance of the suspensorium (Fig. 112) of most muraenids is a function of their typically piscivorous diet. The large jaw musculature, responsible for the cheeky appearance and bulging head of the elongate-jawed species of *Enchelycore*, inserts broadly along the enlarged dentary and hyomandibular. The hyomandibular is stout and typically shaped like an inverted triangle. The posterior axis of the hyomandibular of most morays is vertical or nearly so. Jaw articulation with the quadrate is typically beneath the first vertebra, and the lower jaw is about equal in length to the neurocranium. (*Channomuraena vittata* is remarkable in having a mandible nearly twice the length of the skull, articulating with the quadrate beneath the 6th or 7th vertebra.) A strong ridge parallels the leading hyomandibular edge and serves for the attachment of the adductor mandibularis and to encase a nerve tract. A condyle inserts into the sphenotic/pterotic suture and the dorsal margin is ankylosed to the skull. The quadrate fits tightly into the hyomandibular and provides the process that the articular of the mandible rotates against. The articular fits tightly within a sleeve along the inner face of the dentary. There is a remnant of the coronomeckalian along its inner face. The dentary is toothed for most of its length and is joined to its opposite member at the symphysis by a cartilaginous connection. The pterygoid is reduced to a small sliver. The maxilla is toothed and variously developed among muraenids. The maxilla of *Muraena helena* articulates by a large condyle within a socket of the ethomovomer just in advance of the orbit. Posteriorly it tapers to a thin point. Teeth are typically borne on the outer lateral edge and, in species with biserial dentition such as *M. helena*, a secondary inner but shorter row exists. Variation in the location of maxilla/

ethmovomer articulation within the Muraenidae appears to be much more limited than in the ophichthids.

Dentition within the muraenids is quite variable and appears to be useful as an indication of generic and subgeneric relationships. Extraordinary emphasis on dentition, however, has erroneously resulted in polyphyletic assemblages, such as the previous inclusion of *Gymnomuraena zebra* within *Echidna.* Dentition within the family ranges from elongate fangs (*Enchelycore* species) to cobble-like molariform teeth (*Echidna* species and *Gymnomuraena*) to numerous fine needles (*Uropterygius* species) to wedge-shaped, serrated teeth (certain species of *Gymnothorax*). Teeth are borne on the ethmovomer, maxilla, dentary, and upper and lower pharyngeal tooth plates. It appears that certain species of *Gymnothorax, Muraena* and *Enchelycore* have the ability to replace the large intermaxillary teeth. Slender canines can be seen lying horizontally beneath the skin, and occasionally in various stages of protrusion. Recent evidence suggests that the presence and condition of muraenid teeth is variable according to age and sex. Hatooka (1986) has demonstrated that male *Gymnothorax richardsoni* have fewer teeth than females and that males of two Indo-Pacific species of *Echidna* have sharper teeth than those of the females. He noted, as well, that large male *Echidna nebulosa* have serrate teeth, thereby casting doubt upon that character as being unique to *Gymnothorax* (sensu stricto).

The opercular series of muraenids, as in most eels, has been significantly reduced (Fig. 112). This is related to the modified respiratory movement, gill opening restriction, and the difference in anguilliform opercular function, as well as the increased role of the branchiostegal rays in supporting the branchial basket (particularly in the ophichthids). In *Muraena helena,* the opercle is the largest element and is deeply incised and emarginate along its posterior edge. The rectangular subopercle, interopercle, and preopercle are tucked beneath and before the opercle, each smaller and more reduced to cartilage than the last.

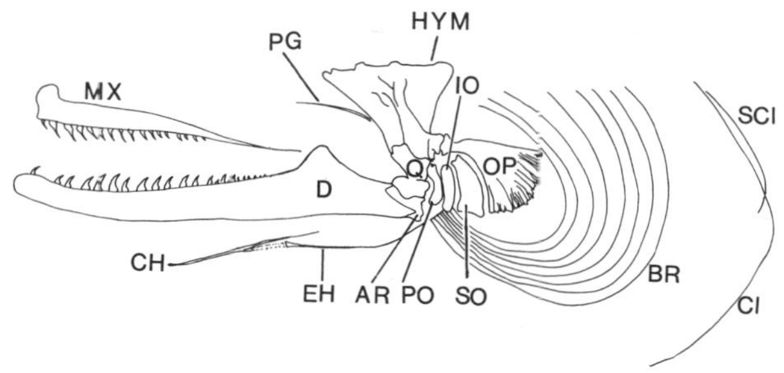

FIGURE 112. Suspensorium, jaws, hyoid apparatus, opercular apparatus and pectoral girdle of *Muraena helena.* Abbreviations: AR, articular; BR, branchiostegal rays; CH, ceratohyal; Cl, cleithrum; D, dentary; EH, epihyal; HYM, hyomandibular; IO, interopercle; MX, maxillary; OP, opercle; PG, pterygoid; PO, preopercle; Q, quadrate; SCl, supracleithrum; SO, subopercle.

The branchial basket of morays is reduced (Fig. 112). We have not systematically examined this condition and base our discussion largely on that of *Muraena helena.* The elongate epihyal is connected by cartilage to the inner face of the quadrate and bears 8–9 slender branchiostegal rays. The rays are non-overlapping and loop behind and above the opercle. They are closely followed by the slender cleithrum and supracleithrum which are the only ossified remains of the pectoral girdle. Those slender elements function, if at all, like additional branchiostegal rays. The epihyal is connected to the ceratohyal by means of a cartilaginous strut. As in ophichthines (but not in myrophines), bracing is provided by a slender posterior elongation of the ceratohyal. The ceratohyals are joined anteriorly, however there is no evidence of an ossified glossohyal, basihyal, hypohyal, or urohyal. This is an extreme reduction among apodal fishes.

The gill arch system of muraenids (Fig. 113) has proven useful to a variety of earlier authors (Cope, 1871, 1884; Popta, 1904). Nelson (1966a) provided the most comprehensive treatment and his findings remain largely unchanged. He included the muraenids within an "anguilloid lineage" and found them more specialized than, but most closely related to chlopsids and *Dysommina.* He noted that:

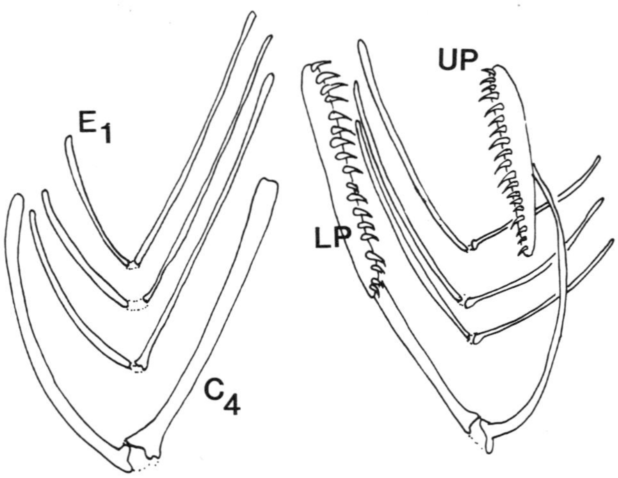

FIGURE 113. Gill arch skeleton of *Muraena helena*. Dorsal view. Gill arches have been cut along the dorsal midline and spread laterally; left upper and lower pharyngeal tooth plates are removed to show underlying bones; stippling indicates cartilage. Abbreviations: C_4, fourth ceratobranchial; E_1, first epibranchial; LP, lower pharyngeal tooth plate; UP, fused upper pharyngeal tooth plates.

> The outstanding feature of the gill arch skeleton of the muraenids is the enlargement of the fourth arch and with the loss of C5 the tooth-bearing bones it comes to support. Teeth on these bones are enlarged, recurved, and generally in two rows (Popta, 1904). These elements form the so-called pharyngeal jaws.

Within the Muraenidae, two structural types exist. One, for which Nelson (1966a) described the subfamily Uropterygiinae (but see previous discussion concerning familial synonymy), is notable in having independent hypobranchials in the first and second arches and the lower pharyngeal tooth plates are ungrooved and lie generally on the dorsal surface of the proximal end of the fourth ceratobranchial. This is seen in species of *Uropterygius, Anarchias, Channomuraena,* and *Scuticaria.* The other condition, shared by the remaining muraenids in the subfamily Muraeninae, is the loss of all hypobranchials. This is apparent in *Muraena helena* (Fig. 113), in which the lower tooth plates lie on the medial surface of the proximal end of the fourth ceratobranchial, which fits into a groove within the plate. All morays examined possess an ossified third infrapharyngobranchial, and the upper tooth plate consists of a fusion of the third and fourth plates.

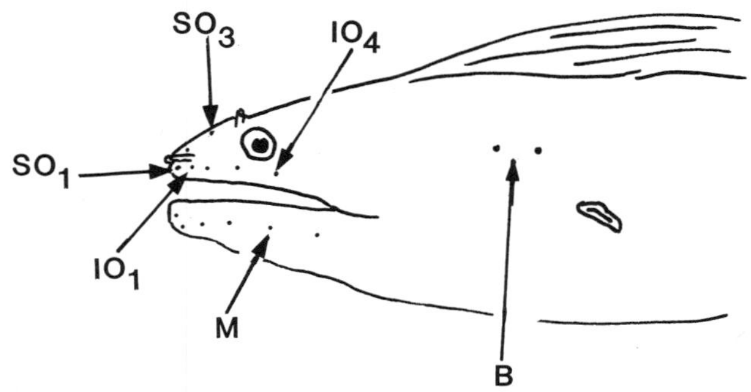

FIGURE 114. Diagrammatic representation of muraenid head pores, adapted from Blache (1967: fig. 4). Abbreviations: B, branchial pores; IO_1, IO_4, first and fourth infraorbital pores; M, mandibular pores; SO_1, SO_3, first (= ethmoidal) and third supraorbital pores.

The lateralis system of morays is reduced as compared to that of other eels. The presence or absence of certain pores appears to be significant as an indicator of relationships. The cephalic lateralis system of Atlantic muraenids is divisible into four major groups (Fig. 114) with the following range of pore numbers: supraorbital (3–4); infraorbital (3–4); mandibular (6–7); and branchial (0–4), the last the remnant of the reduced lateral-line system.

The anterior axial skeleton of *Muraena helena* is illustrated in Figure 115. Vertebral numbers (predorsal, preanal and total) have proven to be useful at the specific level (Böhlke, 1982) and have been utilized throughout our treatment; counts for all western Atlantic species are summarized in Table 9. The caudal skeleton (Fig. 116) is relatively invariable throughout the family. Muraenids do not employ their tail in digging and have therefore not undergone the reduction and modification that ophichthids, heterocongrines, moringuids and other eels have. The transverse processes of the caudal vertebrae, epineurals, and the intramuscular bones are variably developed within the family. Those of *Muraena helena* are moderately developed. Morays appear to lack pleural ribs.

Median fin development and points of origin are variable within the family but follow generic patterns. Species in the subfamily Uropterygiinae are apomorphic in this condition, the median fins being reduced to near the tail tip and

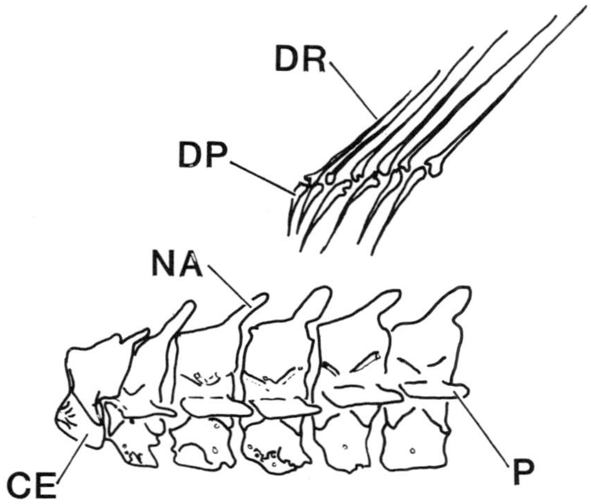

FIGURE 115. Anteriormost six vertebrae of *Muraena helena*, left lateral view. Abbreviations: CE, centrum; DP, dorsal pterygiophore; DR, first dorsal fin ray; NA, neural arch; P, parapophysis.

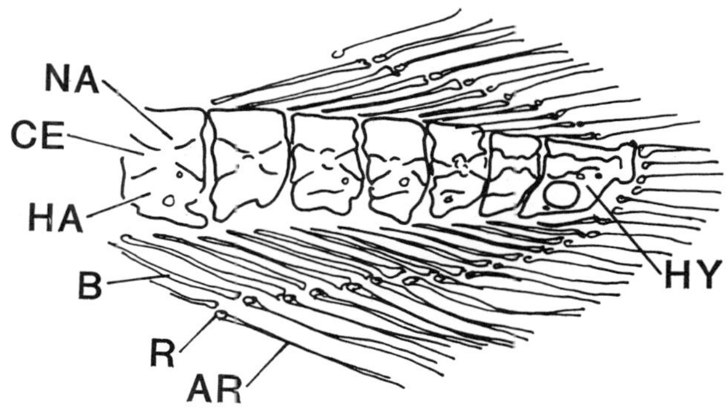

FIGURE 116. Caudal skeleton of *Muraena helena*, left lateral view. Abbreviations: AR, anal-fin ray; B, basal element of pterygiophore; CE, centrum; HA, haemal arch; HY, hypural; NA, neural arch; R, radial element of pterygiophore.

confluent with the caudal. Within the Muraeninae, the condition of species of *Muraena* and *Gymnothorax* appears plesiomorphic (as compared to other anguilloids) with little of no specialization. The mid-body doral-fin origin of *Monopenchelys acuta* and the Indo-Pacific species *"Rabula" fuscomaculata* and *"R." marshallensis* appears to be an apomorphy within the family.

DISTRIBUTION

Too little is presently known about the relationships and worldwide distribution of muraenids to make an adequate appraisal of their evolutionary history. However, the eastern Pacific morays are fairly well known and show numerous similarities to the Atlantic fauna. If one presumes that the present and historic distribution of muraenids is limited to tropical and warm temperate provinces, then it is possible to construct some historical hypotheses.

We broadly recognize 15 genera within the family (realizing that as many as four to six species groups may exist within *Gymnothorax*, sensu lato). Seven genera are circumtropical. Within the very large composite of *Gymnothorax*, there are species groups of varying zoogeographic interest: species of the subgenera *Gymnothorax*, *Lycodontis* and *Taeniophis* (as defined in this volume, based on study of Atlantic material) are circumtropical and appear to share common ancestors with both eastern Atlantic and Indo-Pacific species; species of the subgenus *Neomuraena* are found only in the New World, with closely related species pairs present in the western Atlantic and eastern Pacific. Four genera (*Anarchias, Echidna, Enchelycore* and *Uropterygius*) are relatively speciose, each with 10–19 species, and do not appear to possess New World species pairs, but are present in the eastern Atlantic or have closely related species there. Two monotypic genera, *Channomuraena* and *Monopenchelys*, are curiously known from the Atlantic, western Pacific and Indian oceans. *Muraena*, as here recognized, appears limited to the New World, eastern Atlantic and Mediterranean.

Seven genera are not represented in the Atlantic: *Scuticaria* and *Siderea*, each with two or more Indo-Pacific species; and five monotypic genera, *Enchelynassa, Gymnomuraena, Rhinomuraena, Strophidon* and *Thyrsoidea*. Of these, only three (*Enchelynassa, Gymnomuraena* and *Scuticaria*) are present in the eastern Pacific, limited primarily to the offshore islands (Rosenblatt et al., 1972). Table 8 summarizes the distribution of Atlantic genera and species of the Atlantic and eastern Pacific.

The trans-Pacific and trans-Atlantic distribution of many species probably relates to the muraenid leptocephalus, although, as Rosenblatt et al. (1972) have indicated, transgression of the

TABLE 8. Geographic distribution of species of muraenid genera found in the Atlantic (not including genera and subgenera found only in the Indo-Pacific).

Genus	E. Atlantic	Mid-Atlantic	W. Atlantic	E. Pacific	Indo-Pacific
Anarchias	*longicauda*			*galapagensis*	7 species
	similis	*similis*	*similis*		
Channomuraena	*vittata*	*vittata*	*vittata*	———	*vittata*
Echidna	*catenata*	*catenata*	*catenata*	*nocturna*	9 species
	peli	*peli*			
Enchelycore	*anatina*	*anatina*	*anatina*	*octavianus*	6 species
	carychroa?	*carychroa*	*carychroa*		
	nigricans	*nigricans*	*nigricans*		
Gymnothorax					
(*Gymnothorax*)			*hubbsi*	*panamensis?*	? species
	maderensis	*maderensis*	*maderensis*		
(*Lycodontis*)	*mareei*	*mareei*		———	? species
	miliaris	*miliaris*	*miliaris*		
	unicolor	*unicolor*			
(*Neomuraena*)	———	———	*conspersus*	*angusticeps*	———
			kolpos	*serratidens*	
			nigromarginatus	*equatorialis*	
			ocellatus	? species	
			saxicola		
(*Taeniophis*)	*afer*		*funebris*	*castaneus*	? species
	moringa	*moringa*	*moringa*	*mordax*	
	polygonius	*polygonius*	*polygonius*	*dovii*	
	vicinus	*vicinus*	*vicinus*		
———	*bacalladoi*	———	———	? species	? species
Monopenchelys	———	*acuta*	*acuta*	———	*acuta*
Muraena	*augusti*			*argus*	———
	helena		*retifera*	*clepsydra*	
	melanotis	*pavonina*	*pavonina*	*lentiginosus*	
	robusta		*robusta*		
Uropterygius	*wheeleri*	*macularius*	*macularius*	*polystictus*	ca 14 species
				macrocephalus	

broad Pacific water gap by eels has been limited to morays. Although the Atlantic is not as difficult a barrier for other apodal species to cross, a similar pattern exists.

RELATIONSHIPS WITHIN THE MURAENIDAE

The family Muraenidae has been divided into two subfamilies, the Muraeninae and the Uropterygiinae. Characteristics which divide the two, other than gill arches and median fin development, have not been systematically analyzed. We continue to recognize the two subfamilies based on these characters, pending a complete familial revision.

Four genera are placed in the Uropterygiinae, three (*) present in the Atlantic: *Anarchias**, *Channomuraena**, *Scuticaria*, and *Uropterygius**. Eleven genera are included in the Muraeninae, five (*) of which are found in the Atlantic: *Echidna**, *Enchelycore**, *Enchelynassa*, *Gymnomuraena*, *Gymnothorax*, sensu lato*, *Monopenchelys**, *Muraena**, *Rhinomuraena*, *Siderea*, *Strophidon*, and *Thyrsoidea*.

TABLE 9. Mean vertebral formulae (MVF) and ranges of vertebral counts in western Atlantic species of Muraenidae.

		Predorsal		Preanal		Total	
	MVF	Range	n	Range	n	Range	n
Anarchias similis	97-102-107	94–104	46	95–106	45	104–113	63
Channomuraena vittata	122-123-150	120–126	12	118–127	13	145–157	21
Echidna catenata	6-58-117	5–7	36	56–61	39	114–122	37
*Enchelycore anatina**	7-56-154	6–8	6	52–59	7	151–155	6
*carychroa**	9-48-135	7–11	297	45–51	321	128–140	322
nigricans	8-60-144	6–10	75	58–63	76	141–148	74
*Gymnothorax bacalladoi**	5-55-130	5–6	3	54–56	3	130–131	3
conspersus	6-66-168	5–8	46	62–70	47	163–174	42
funebris	6-60-140	4–7	41	57–61	41	137–142	41
*hubbsi**	7-58-135	6–8	11	56–60	11	133–137	11
*kolpos**	6-63-164	5–7	31	60–65	31	159–167	28
*maderensis**	6-68-153	5–7	9	66–70	9	149–156	9
*miliaris**	5-49-120	3–7	44	47–51	51	117–126	54
moringa	6-54-138	4–7	76	50–57	79	134–142	75
ocellatus	6-50-145	4–8	167	46–54	175	136–151	165
*nigromarginatus**	6-49-142	5–8	49	45–52	54	137–147	48
*saxicola**	6-53-140	4–7	235	47–59	237	134–147	245
*polygonius**	5-53-138	4–6	16	51–55	17	135–142	17
*vicinus**	5-51-133	3–6	52	48–53	60	128–140	52
*Monopenchelys acuta**	73-53-130	67–78	38	50–56	38	124–135	40
*Muraena pavonina**	4-50-127	3–5	20	49–52	26	122–129	25
*retifera**	4-56-131	4–6	41	52–58	41	128–135	39
robusta	4-64-153	3–4	5	62–65	6	149–156	5
Uropterygius macularius	106-110-121	100–110	38	105–115	38	117–126	48

* Includes holotype or lectotype.

KEY TO THE SUBFAMILIES AND GENERA OF MURAENIDAE

(Genera entirely extralimital to the western Atlantic are included in the key but not in the generic diagnoses.)

1a. Dorsal and anal fins restricted to tail tip; hypobranchials 1 and 2 present . . . . Subfamily Uropterygiinae, 2

1b. Dorsal-fin origin near or before anus; anal-fin origin just behind anus; hypobranchials absent . . . . Subfamily Muraeninae, 5

2a. Jaws elongate, the lower protruding; snout short, eye above anterior third of jaw; adults large and robust . . . . *Channomuraena*, p. 122

2b. Jaws not elongate, subequal; snout not reduced, eye about above mid-jaw; adults small to moderate and elongate 3

3a. Posterior nostril contiguous with an enlarged supraorbital pore, appearing as double pore; adults small, not exceeding 200 mm (Atlantic species) . . . . *Anarchias*, p. 117

3b. Posterior nostril not contiguous with a supraorbital pore 4

4a. Preanal length about equal to tail; posterior nostril above or behind mid-eye; adults small to medium-length, not exceeding 300–400 mm . . . . *Uropterygius*, p. 126

4b. Preanal length much greater than tail; posterior nostril before mid-eye; adults large, often exceeding 500 mm . . . . *Scuticaria* Jordan and Snyder, 1901 (two Indo-Pacific species)

5a. Dorsal-fin origin shortly behind anus; a single branchial pore . . . . *Monopenchelys*, p. 191

5b. Dorsal-fin origin before anus; 1–4 branchial pores 6

6a. Some teeth molariform or granular, at least on vomer; no canine teeth 7

6b. Teeth not noticeably molariform; usually some canine teeth present 8

7a. Anus about at mid-body, tail about 50% TL . . . . *Echidna*, p. 130

7b. Anus posterior to mid-body, tail about 30% TL . . . . *Gymnomuraena* Lacepède, 1803 (one Indo-Pacific species)

8a. Anterior nostrils not simple tubes, modified with flaps or leaflike enlargements 9

8b. Anterior nostrils simple tubes 10

9a. Anterior nostrils with broad, foliose enlargements; lower jaw tip with projecting barbels; body slender and elongate, tail about twice preanal length; enlarged canines absent . . . . *Rhinomuraena* Garman, 1888 (one Indo-Pacific species)

9b. Anterior nostrils with bi-lobed flaps on posterior margins; no barbels on lower jaw tip; body stout, tail about equal to preanal length; enlarged canines in upper jaw . . . . *Enchelynassa* Kaup, 1855 (one Indo-Pacific species)

10a. Body elongate, its depth 40 times or more in TL 11

10b. Body stouter, its depth 30 times or less in TL 12

11a. Tail about equal to preanal length; dorsal fin elevated, well developed . . . . *Strophidon* McClelland, 1844 (one Indo-Pacific species)

11b. Tail longer than preanal length, about 60–70% of TL; dorsal fin not well developed . . . . *Thyrsoidea* Bleeker, 1854 (one Indo-Pacific species)

12a. Jaws not curved and elongate, closing completely; inner row of jaw teeth, if present, not greatly enlarged 13

12b. Jaws elongate and arched, meeting only at their tips; inner row of jaw teeth elongate fangs . . . . *Enchelycore*, p. 134

13a. Posterior nostril tubular, above and at anterior margin of eye; teeth of adults uniserial, slender, without serrations; gill opening typically within black spot . . . . *Muraena*, p. 194

13b. Posterior nostril flush or nearly so with head profile; teeth uniserial or multiserial, with or without serrations; gill opening with body coloration 14

14a. Some teeth caniniform; usually one or more depressible fangs in middle of front of upper jaw; teeth on vomer usually uniserial . . . . *Gymnothorax*, p. 145

14b. Teeth short; no median depressible fangs in upper jaw; two rows of low, conical teeth on vomer . . . . *Siderea* Kaup, 1856 (several Indo-Pacific species)

SUBFAMILY UROPTERYGIINAE

Morays of the subfamily Uropterygiinae are readily recognized by the very short dorsal and anal fins which are restricted to the end of the tail and often difficult to detect. Other characters include: head and jaws moderate to elongate; anterior nostril tubular, posterior nostril with raised rim; teeth uniserial to multiserial, sharp canines present; gill opening small and round, low on side; coloration uniform or patterned with spots, blotches, or bands. Neurocranium short; gill arches enlarged, modified to form pharyngeal jaws; hypobranchials 1–2 ossified.

The subfamily Uropterygiinae includes four genera, three of which (*) are present in the Atlantic: *Anarchias**, *Channomuraena**, *Scuticaria*, and *Uropterygius**.

Genus *Anarchias* Jordan and Starks, 1906

Anarchias Jordan and Starks in Jordan and Seale 1906: 204 (type species *Anarchias allardicei* Jordan and Starks, 1906, by original designation).

Characters. Body elongate, slender, laterally compressed posteriorly; head and trunk shorter than tail (42–45% TL in Atlantic species). Fins restricted to tail tip. Snout short, subconical and broad; jaws moderate, subequal and closing completely; eye moderate, its center slightly before midpoint of upper jaw; anterior nostril tubular, anteriorly directed; posterior nostril a rimmed pore above center of eye, contiguous with enlarged posteriormost supraorbital pore, appearing as a double pore. Gill opening small, slit-like, about at midbody. Head pores developed; four supraorbital pores, four infraorbital pores, six mandibular pores, two branchial pores. Teeth numerous, conical, slender, many caniniform; jaw teeth biserial anteriorly, uniserial posteriorly, vomerine teeth uniserial.

Gill arches as described under family account; hypobranchials 1–2 ossified, lower pharyngeal tooth plates on dorsal surface of fourth ceratobranchial. Precaudal vertebrae less numerous than caudal. Peritoneum pale. Coloration uniform or patterned.

Size. Small to large eels; the Atlantic species are small, growing to just over 200 mm TL.

Distribution. Circumtropical, found on both sides of the Atlantic and in the Indo-Pacific.

Etymology. From the Greek *an-* (without) and *archos* (anus); in reference to its apparent lack of an anal fin. Masculine.

Remarks. Species of *Anarchias* are easily recognized by the restriction of the median fins to the end of the tail and the presence of the pore immediately adjacent to the posterior nostril. The genus clearly appears to represent a monophyletic assemblage.

Species. There are about ten currently recognized species distributed in all tropical oceans: two Atlantic species, *similis* (Lea, 1913) (*yoshiae* Kanazawa, 1952 a synonym) primarily in the western Atlantic but also found in the eastern Atlantic, and *longicauda* (Peters, 1876) (*grassi* (Roule, 1916) a synonym) in the eastern Atlantic; in the Indo-Pacific, *allardicei* Jordan and Starks, 1906 (*fuscus* Smith, 1962 a synonym), *cantonensis* (Schultz, 1943), *galapagensis* (Seale, 1940), *leucurus* (Snyder, 1904), *seychellensis* Smith, 1962, and *maldiviensis* Klausewitz, 1964. The genus is currently under revision by E. Holm, who has informed us (in litt., 1986) that there is an undescribed species from the Indo-Pacific, and that *Anarchias vermiformis* Smith, 1962 has been referred to *Uropterygius concolor* Rüppell, 1835.

Key to Atlantic Species of *Anarchias*

1a. Total vertebrae 104–113, predorsal vertebrae 94–105, preanal vertebrae 95–106; 2–5 inner maxillary teeth . . . . *similis*, p. 118

1b. Total vertebrae 112–117, predorsal vertebrae 113–115, preanal vertebrae 112–115; 5–8 inner maxillary teeth . . . . *longicaudis* Peters, 1876)
(an eastern Atlantic species)

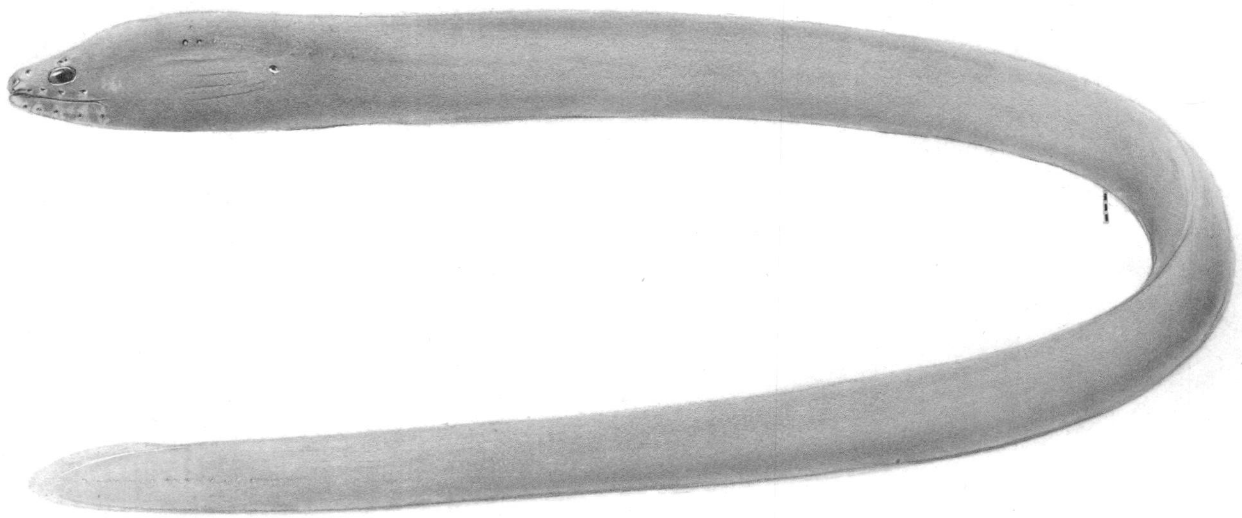

FIGURE 117. *Anarchias similis*: ANSP 111588, 79 mm TL, Bahamas. Illustration by S. P. Gigliotti.

Anarchias similis (Lea, 1913)

PYGMY MORAY

Figures 117–120, Tables 9, 10

Leptocephalus similis Lea, 1913: 26, figs. 23–24, pl. 4 (5–6) (original description, 34°44'N, 47°52'W, syntypes BNM). Leptocephalus.

Anarchias yoshiae Kanazawa, 1952: 75, fig. 10 (original description, Bermuda, holotype FMNH 48729). Blache, 1975a: 708 (Baie de Biafra, Annobon, redescription, illustrations).

Anarchias grassi (not of Roule, 1916). Blache, 1967f: 1707, fig. 2 (in part, Annobon).

Discussion of Synonymy. The single western Atlantic species of *Anarchias* has long been known as *A. yoshiae* Kanazawa. However, *Leptocephalus similis,* described by Lea in 1913, has been found to be the larval form of this species (see p. 000), and the name *similis* must be used for the adults as well.

Distinctive Characters. The smallest western Atlantic moray (known only to 206 mm); coloration dark brown background with irregular pale pattern always present on head and chin, extending variously onto body, head pores in white spots; median fins restricted to tail tip; supraorbital pore adjacent to posterior nostril, appearing as a double pore; MVF 97-102-107.

Description. Total vertebrae 104–113, x̄ = 106.9 (n = 63); predorsal vertebrae 94–104, x̄ = 97.4 (46); preanal vertebrae 95–106, x̄ = 101.5 (45); kidney terminus at 51 (2): dorsal-fin rays 23–34, x̄ = 29 (7); anal-fin rays 11–22, x̄ = 15 (6). Pores: SO 1 + 3; IO 4; POM 6; B 2.

Proportions as % of TL (n = 20, 54–206 mm): preanal 42–45, x̄ = 43; head 11–13, x̄ = 12; depth at gill opening 3.7–7.8, x̄ = 5.6; depth at anus 2.7–5.2, x̄ = 4.3; width at gill opening 2.2–4.3, x̄ = 3.0; width at anus 1.6–4.1, x̄ = 3.0. Of HL: snout 13–17, x̄ = 15; eye 7.3–10, x̄ = 8.8; interorbit 7.6–13, x̄ = 10; upper jaw 31–42, x̄ = 36; lower jaw 29–42, x̄ = 34.

Head moderate, its profile abruptly elevated and muscular behind eye. Snout and jaws short; eye moderate, above mid-jaw. Anterior nostril tubular, near tip of snout, projecting anterolaterally. Posterior nostril a rimmed pore above mid-eye, immediately adjacent to the posteriormost supraorbital pore which is posterodorsad to the nostril and separated by a thin septum in young and by a slight fleshy bridge in adults. Remainder of head pores as listed and illustrated; little variation, one 78-mm female with a pair of pores adjacent to each posterior nostril.

Teeth numerous and sharp. On each side of upper jaw, an outer row of 7–10 small intermaxillary teeth enclosing three rows of enlarged depressible canines; 2–3 fangs increasing in size posteriorly on midline, the last the largest tooth in the mouth, flanked by 3–4 large canines on either side of the median row. A single row of

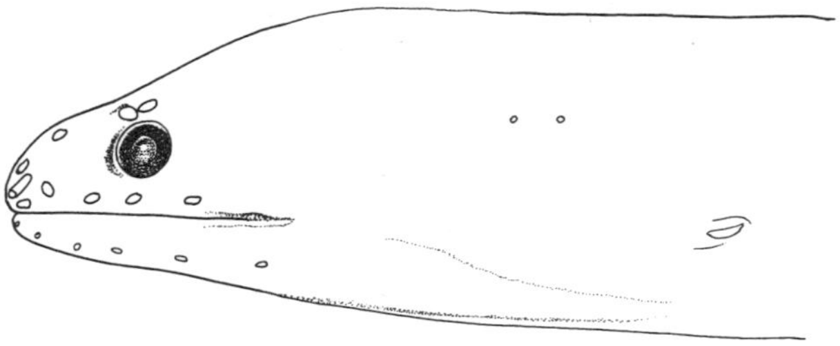

FIGURE 118. *Anarchias similis*: ANSP 136057, 122 mm TL, Bahamas; head.

3–11 small sharp teeth on vomer, well separated anteriorly from intermaxillary teeth. Maxillary teeth in two rows; an outer row of 12–20 small stout teeth which is continuous with outer intermaxillary teeth, and an inner row of 2–5 long, slender, widely-spaced canines. Lower jaw teeth biserial anteriorly and uniserial posteriorly; outer row of 21–31 closely-set, stout teeth with recurved tips, enclosing inner row of 4–7 large, spaced teeth, the first three increasing in size posteriorly.

Median fins beginning shortly before tail tip, not visible externally, the rays apparent on dissection and indistinctly on radiographs; fin-ray counts obtained are approximations based partly on pterygiophores seen on radiographs. Anus shortly before mid-body.

Color. Highly variable coloration, apparently not correlated with size, sex or state of maturity. Head (at least lower jaw) always patterned with pale marks, and head pores in distinct white spots. Body color varies from uniformly dark brown, to dark with scattered distinct pale spots, to dark with small distinct pale spots and diffuse larger pale areas, to a bold overall pattern of irregular pale markings, as in Figure 117.

Size. The largest specimen we have seen is a ripe male, 206 mm, from off the Atlantic coast of Florida at 70–71 m. Of 79 specimens examined for sex, 32 (53–151 mm) were females, 15 (76–136) ripe with 0.6–1.0-mm eggs; of 20 males (62–206), three (99–206) were ripe. In ripe females the eggs fill the abdominal cavity and, because of the small size of the species, appear very large; an 85-mm female had the largest eggs (1.0 mm) in masses of three eggs across in each ovary at its widest; a 95-mm female had 0.6-mm eggs in

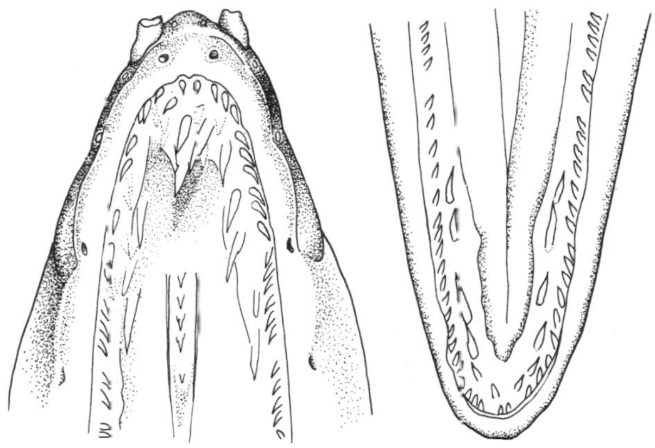

FIGURE 119. *Anarchias similis*: ANSP 100484, 206 mm TL, Florida Atlantic; dentition.

one row in each ovary. Ripe females were taken in July through November, ripe males in February, September and October. Both sexes exhibited all phases of coloration.

Development. The leptocephalus was described by Lea (1913: 26) before the adult was known (see p. 901).

Behavior. Little is known about behavior except what can be inferred from collection data. *Anarchias similis* apparently inhabits moderately deep waters, living on or near the bottom among algae-covered boulders and eel grass. Most of the study material (22 lots, many with more than one specimen) was obtained by dredge at depths

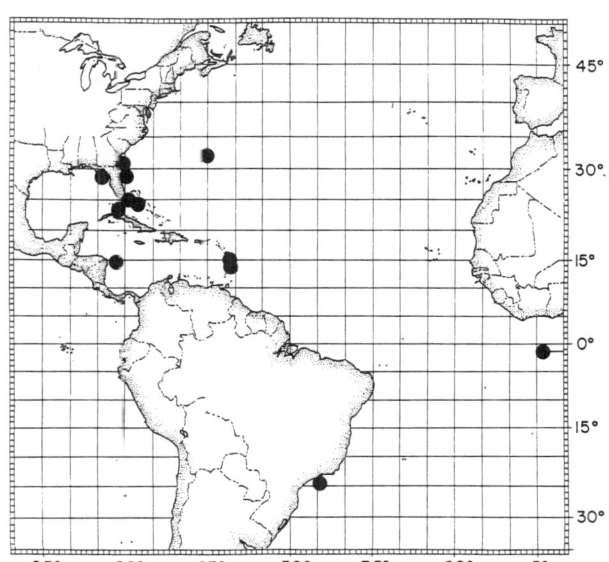

FIGURE 120. Distribution of *Anarchias similis* based on material examined.

between 26–97 m; only five lots, consisting of single specimens, were taken by poison over shallow waters to 5.5 m. Both poison station and dredge records stated that bottom included algae-covered boulders and eel grass.

Distribution (Fig. 120). Found off Bermuda, the east coast of the United States from Georgia to Florida, off Nicaragua, from the West Indies, and off Brazil. In the eastern Atlantic, known from one collection from the island of Annobon in the Gulf of Guinea.

This species was reported only from the western Atlantic until Blache reidentified specimens from Annobon (as *Anarchias yoshiae* in Blache, 1975a: 708). We have examined 20 of his specimens (MNHN 1974-190) and agree that they are conspecific, although some differences were noted (see under *Remarks*).

Etymology. From the Latin *similis* (like), probably referring to its similarity to *Leptocephalus* (= *Anarchias*) *euryurus*, the eastern Atlantic larva described by Lea in the same paper. An adjective.

Remarks. Blache (1975a: 714) separated the eastern Atlantic *Anarchias longicaudus* (as *A. grassi*, see below) and western Atlantic *A. similis* (as *A. yoshiae*) by differences in preanal distance, position of branchial pores, and vertebral counts. Our data show no significant difference in preanal distance (40–45% TL for 25 specimens of *similis*, 39–45% TL for 9 specimens of *longicaudus*) or in the placement of the branchial pores. The latter were incorrectly described by Blache, no doubt due to a prominent artifact apparent in the four larger specimens of MNHN 84-421 that we also examined; a large hole in the branchial region directly below the posteriormost branchial pore which opened straight through the head from side to side (likely resulting from a tag tied onto the fish in the past), was illustrated as the gill opening. The true branchial apertures are posterior to these holes and to the branchial pores, and are situated in pale areas as are the gill openings in other specimens. The two species may be separated by vertebral numbers (Table 10), with some differences in the number and placement of fin rays: 23–34 dorsal-fin rays and 11–22 anal-fin rays flanking the last 8–11 vertebrae in western Atlantic *similis*, 0–6 rudimentary rays on the last 3–4 vertebrae in *longicaudus*. It must be noted that eastern Atlantic specimens of *similis* examined also had fewer fin-rays (8–10 dorsal-fin rays and 4–7 anal-fin rays along the last 3–5 vertebrae), a condition attributed to their small size. In addition, larvae of the two species are distinct.

The eastern Atlantic species of *Anarchias*, described by Roule (1916: 6) as *Uropterygius grassi*, was shown by Blache (1977: 26) to be the adult of *Leptocephalus euryurus* Lea, 1913. Our studies show this species to be conspecific with *Muraena longicauda* described by Peters (1876: 850), and long known only from the holotype. Its status has been uncertain, partly because, despite the fact that the coordinates of capture of the holotype given in the original description clearly placed it in the Cape Verde Islands, Jordan and Davis (1891: 599) wrote: "... taken in the open Atlantic, between the West Indies and Europe," and various subsequent publications on fishes of the Americas included it as from the tropical Atlantic off the West Indies. We have examined and X-rayed the holotype of *Muraena longicauda*, ZMB 9776, and, although it is small, hard and shriveled, it is clearly conspecific with specimens known as *Anarchias grassi*; the mottled color pattern is apparent, the vertebral count is similar, pores are readily observable, including the pore adjacent to the posterior nostril, and, although many are missing, the tooth pattern is similar. The species name *longicauda* is available and the masculine form should be used for the eastern Atlantic species.

Study Material. A total of 99 specimens, 33–206 mm TL, including the holotype and paratypes of *Anarchias yoshiae* Kanazawa.

BERMUDA: ANSP 133632 (1, 54). FMNH 48729 (1, 136); holotype of *Anarchias yoshiae*. USNM 157364 (1, 116); paratype of *A. yoshiae* [formerly CNHM (= FMNH) 48955]. COASTAL UNITED STATES: *Georgia:* GMBL 78-110 (2, 117–120). GMBL 78-113 (1, 115). *Florida Atlantic:* ANSP 94210 (11, 64–110), 100484 (1, 206), 101242 (1, 53), 101313 (9, 85–129), 104681 (9, 60–138), 110424 (2, 76–107), 115527 (1, 128, cleared and stained), 138394 (1, 120), 138409 (4, 68–103), 138412 (6, 58–98), 138417 (1, 33), 158167 (1, 136). USNM 153163 (1, 45); paratype of *A.*

TABLE 10. Frequency distributions of vertebral counts in *Anarchias similis* from the eastern and western Atlantic and in *A. longicaudus*.

	Predorsal																					
	94	95	96	97	98	99	100	101	102	103	104	105	106	107	108	109	110	111	112	113	114	115
similis																						
Western Atlantic	3	3	9	12	5	7	6	-	1†													
Eastern Atlantic									2	1	-	1										
longicaudus																				1	-	1

	Preanal																					
	94	95	96	97	98	99	100	101	102	103	104	105	106	107	108	109	110	111	112	113	114	115
similis																						
Western Atlantic		1	1	-	2	5	3	12	4	9†	2	5	1									
Eastern Atlantic									1	2	-	-	1									
longicaudus																			1	1	-	1

	Total														
	104	105	106	107	108	109	110	111	112	113	114	115	116	117	118
similis															
Western Atlantic	2	1†	8	7	14	11	7	8	2	3					
Eastern Atlantic		2	4	4	3	2	1								
longicaudus									2*	2	1	2	3	1	

* Includes holotype.

† Includes holotype of *Anarchias yoshiae*.

yoshiae. Florida Keys: ANSP 101241 (2, 94–112). USNM 77747 (1, 105) and USNM 77748 (1, ca 72); paratypes of *A. yoshiae.* USNM 116773 (1, 108). *Florida Gulf:* USA 05507 (1, 151). BAHAMAS: ANSP 111588 (1, 79), 130550 (1, 64), 136057 (1, 122), 140720 (1, 61). LESSER ANTILLES: *Martinique:* ANSP 110427 (3, 76–90). *St. Lucia:* ANSP 105258 (1, 53), 105741 (1, 121), 110429 (3, 49–96), 140582 (3, 62–82), 140583 (1, 108). NICARAGUA: ANSP 110419 (1, 97), 133009 (1, 92). BRAZIL: UF 44616 (1, 102).

EASTERN ATLANTIC: *Annobon:* MNHN 1974-190 (20, 56–139).

Comparative material: Anarchias longicaudus. CAPE VERDE ISLANDS: ZMB 9776 (1, ca 107), holotype of *Muraena longicauda* Peters; Atlantischen Ocean, 15°40′1″N, 23°5′8″W, 38 fm; SMZ GAZELLE; 29 June 1874. BMNH 1934.12.19.20 (1, 190.5); 16°4′N, 24°44′30″W, 52 m; 26 July 1901. MNHN 84-419 (3, 93–149.5); Talisman; 1883. MNHN 84-421 (5, 105–ca 150); Talisman; 1883. MNHN B 2683 (2, 120–145); Talisman. AZORES: USNM uncat. (1, 129). MEDITERRANEAN: MZUF 4434 (1, 252); Nice; April 1886.

Genus *Channomuraena* Richardson, 1848

Channo-muraena Richardson, 1848a: 96 (type species *Ichthyophis vittatus* Richardson, 1844, by original designation).

Characters. Body robust, muscular and cylindrical, evenly tapering posteriorly, tail tip rounded; head and trunk longer than tail (63–70% TL). Fins restricted to tail tip. Snout short, not reaching tip of lower jaw; jaws extremely elongate, closing completely; eye minute, in anterior third of jaw; anterior nostril short, tubular; posterior nostril nontubular, with raised rim, above and before eye. Gill opening small, slit-like, at midbody. Head pores developed, small and inconspicuous; typically three supraorbital, four infraorbital, seven mandibular, and two branchial pores. Teeth small, pointed and cardiform, in 3–4 subequal rows on jaws and vomer.

Neurocranium short, stout, flattened, sloping posteriorly. Maxillae elongate, tapering posteriorly; dentary slender and elongate, hyomandibular posteriorly inclined. Gill arches like those of *Anarchias;* hypobranchials 1–2 ossified; tooth plate short relative to elongate branchial-bearing elements; lower pharyngeal tooth plates on dorsal surface of 4th ceratobranchial. Intramuscular bones and caudal transverse processes developed. Precaudal vertebrae more numerous than caudal vertebrae. Coloration strongly banded.

Size. Large robust eels, growing to 1.5 m.

Distribution. Known from the Atlantic and Pacific oceans.

Etymology. From the Greek *channos* (to yawn) and *Muraena* (the moray), in reference to its huge gaping jaws. Feminine.

Remarks. Channomuraena is easily recognized by its rounded and robust body, tiny eye located far forward above huge gaping jaws, and median fins restricted to the end of the tail. It differs so greatly in appearance and in some osteological characters from other muraenid genera that it has been suggested that it perhaps should be removed from the family. One of the reasons given has been the reduction of pharyngeal teeth; however our radiographs show strong pharyngeal dentition; this and other characters support its retention in the family.

Species. A single species, *Channomuraena vittata* (Richardson, 1844).

Channomuraena vittata (Richardson, 1844)

BROADBANDED MORAY

Figures 121–124, Tables 9, 11

Raro. Parra, 1787: 66, pl. 30, fig. 3.

Ichthyophis vittatus Richardson, 1844: 114, pl. 53, figs. 7–9 (original description, "brought from China," holotype BMNH 1855.10.16.280).

Nettastoma vittata. Richardson, 1848a: 96.

Channo-muraena vittata. Richardson, 1848a: 96 ("Hab. Gulf of Mexico? Sea of China"). Kaup, 1856b: 97.

Channomuraena vittata. Kaup, 1856a: 66. Blache, 1967f: 1723, figs. 10–14 (redescription, illustrations).

Channomuraena cubensis Poey, 1868a: 266, pl. 3, fig. 6 (original description, Cuba, holotype MCZ 9191).

Gymnomuraena vittata. Günther 1870: 134.

Distinctive Characters. A large thick muscular moray with 13 to 16 irregular broad dark bands encircling body behind eye; tail short, 30–37% TL; vertical fins restricted to tail tip; gape huge, eye far forward near tip of snout; teeth numerous and multiserial; MVF 122-123-150.

Description. Total vertebrae 145–157, $\bar{x}$ = 150 (n = 21); predorsal vertebrae 120–126, $\bar{x}$ = 122

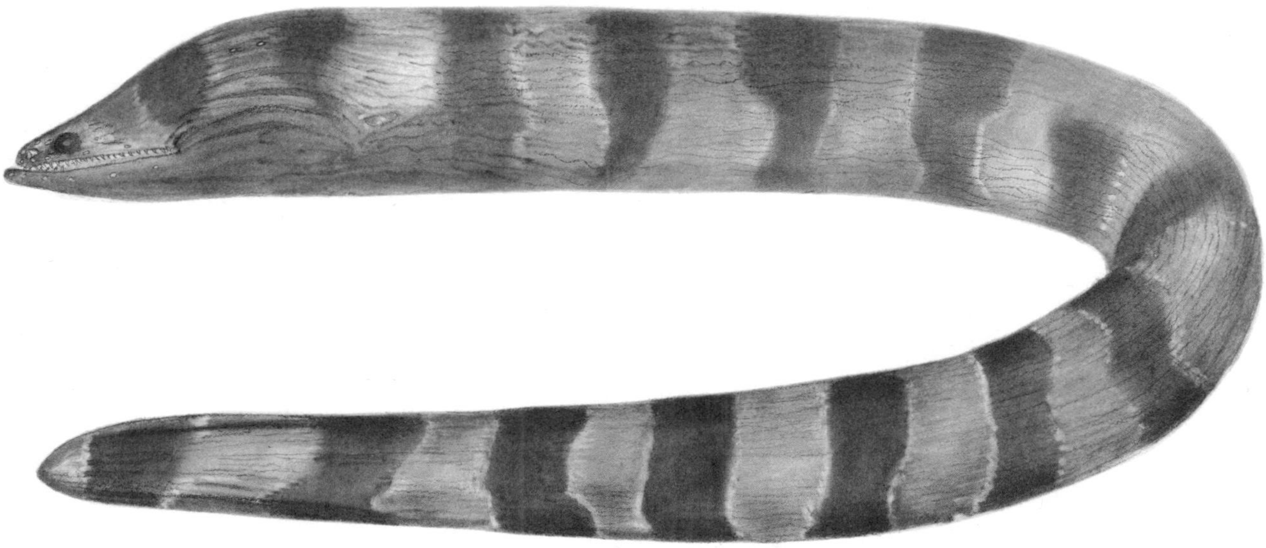

FIGURE 121. *Channomuraena vittata*: USNM 24962, 800 mm TL, Cuba.

(12); preanal vertebrae 118–127, x̄ = 123 (13); dorsal-fin rays 70–86, x̄ = 76 (4); anal-fin rays 60–93, x̄ = 72 (4). Pores: SO 1 + 2; IO 4; POM 7; B 2.

Proportions as % TL (n = 15, 500–1165 mm TL): preanal 63–70, x̄ = 66; head 12–15, x̄ = 14; depth at gill opening 6.1–10, x̄ = 7.3; depth at anus 4.5–6.4, x̄ = 5.3; width at gill opening 3.2–5.6, x̄ = 4.6; width at anus 2.6–4.3, x̄ = 3.4. Of HL: snout 7.4–12, x̄ = 9.4; upper jaw 34–50, x̄ = 43; lower jaw 35–54, x̄ = 46; eye 4.3–7.6, x̄ = 5.3; interorbit 8.6–12, x̄ = 10.

Head large, abruptly enlarging behind short snout; eye small, far forward near tip of snout. Jaws elongate and widely separated posteriorly. Anterior nostril tubular, near tip of snout. Posterior nostril in short tube about one-third height of anterior, situated above front margin of eye. Head pores small and inconspicuous; two branchial pores above and just posterior to rictus, much closer to rictus than to gill opening; one specimen with one additional supraorbital pore, one with five extra infraorbital pores on one side, and one with three branchial pores on one side, the middle of them a double pore. Lips and dorsal surface of snout papillose in large specimens.

Teeth relatively small and numerous for a moray. Intermaxillary teeth in a triangular patch of uniformly-sized, sometimes widely-spaced, sharp teeth. Vomerine teeth continuous with inter-

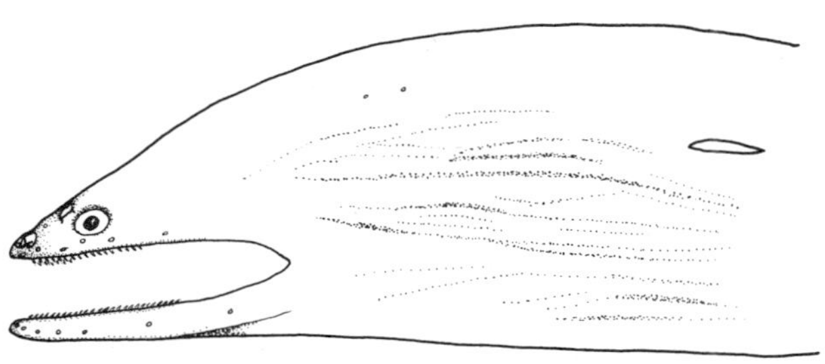

FIGURE 122. *Channomuraena vittata*: ANSP 128125, ca 1165 mm TL, Bermuda; head.

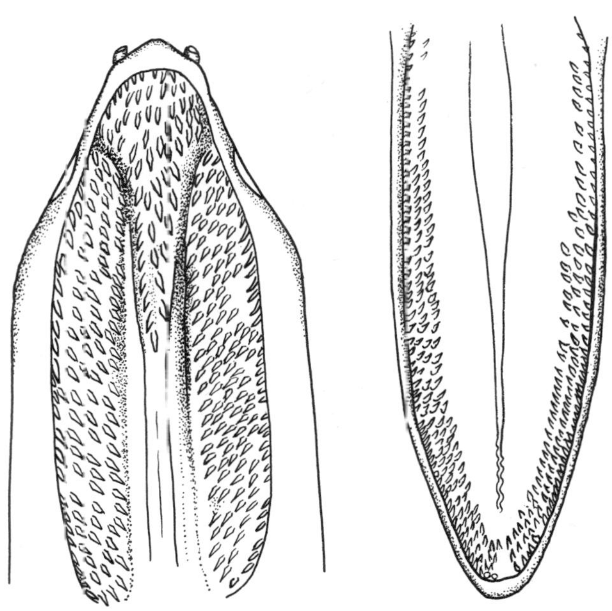

FIGURE 123. *Channomuraena vittata*: data as for Figure 122; dentition.

maxillary teeth, multiserial anteriorly to uniserial posteriorly, terminating shortly behind posterior margin of eye. Maxillary teeth multiserial anteriorly, tapering to fewer rows posteriorly, teeth similar in size, small and sharp. Dentary teeth multiserial, numerous, subequal; a conspicuous separation between tooth patches anteriorly. Largest specimens with about five rows of teeth anteriorly, smallest specimens with three rows decreasing to one row posteriorly. Tips of jaw teeth arrowhead-like in shape and canted slightly back, set on tall bases which are sometimes also broad.

Body round in cross-section, muscular, strongly contracted in preservative; skin described as loose and flabby in life. Median fins present at end of tail, not visible externally but fin-rays countable in stained specimens and on radiographs.

Color. Head and background color pale to medium dark, variously described as grey, tan or olive; thirteen to sixteen broad, irregular and/or broken darker bands encircling head and body behind eye. Color of preserved specimens frequently olivaceous, the bands appearing dark olive, the background and head lighter; in large individuals, the large bands bordered by narrow, almost white, bands. Life colors variable; a live specimen in the Bermuda Aquarium had dark brownish bands on a tan background. Kodachromes taken (by B. Rose) just after capture of ANSP 144118 in the Bahamas show reddish-brown bands and light brown interspaces; field notes for another state "Color: brown (medium) with mottled grey spots."

Size. A large species, growing to 1500 mm. Our study material included two very small specimens, ca 91 and 122 mm TL, from the Atlantic and Pacific oceans, respectively, and three slightly larger (146–253 mm) from the Pacific; all others from the Atlantic were large (330 to ca 1500). Of 23 specimens examined, four (800–1163) were males and six (500–1500) were females, one (1500) filled with small eggs.

Development. Leptocephalus described by Castle (1970a: 15); see p. 904.

Behavior. Channomuraena vittata was relatively unknown and little was known about its behavior until SCUBA gear enabled divers to explore areas where it lives. Large specimens were observed lurking on the bottom around and under rocks and ledges where they presented a very threatening appearance. Because of its unique snake-like form and habits, we received repeated inquiries and kodachromes from divers and from boatmen in the Bahamas and Ascension Island, who described its unusual habit of puffing up its head, much like a cobra does, when approached or threatened.

Mr. R. A. Kuhlman described a specimen from Ascension Island: "skin very loose . . . like a thin cloth bag . . . wrinkled during movement in water. Extremely strong but acted very sluggish even after being speared." Mr. Benyan Rose of Grand Bahama Island wrote: "The specimen that I sent you [ANSP 144118] was found . . . on the outside of the reef in 40 ft of water. The bottom was a flat, rock bar with potholes and ledges on it . . . between two reefs and with a very strong tidal flow. It was flat, rolling, hard bottom with an algae mat and gorgonians. There were shallow ledges and potholes. All three of the specimens that I have seen in the water were found in this type of hole with spiny lobster . . . . Whenever I approached either with the [lobster] tail or my hand he puffed his head. I caught it with a gaff and he put up quite a fight. It stayed alive for 2–3 days in a cooler. Each time I lifted the lid to change the water he puffed his head."

Distribution (Fig. 124). In the western Atlantic, *Channomuraena vittata* has been found off Bermuda, the Bahamas, Cuba, the Antillean chain, the Yucatan Peninsula, and Ascension Island; in the eastern Atlantic reported from the Cape Verde Archipelago (Blache, 1967f: 1724) and the islands of Annobon and São Tome in the Gulf of Guinea. In the Pacific, specimens recorded from Palmyra, Christmas Island, and the Carolines. The exact locality of the holotype of *Channomuraena vittata* remains in some doubt; Richardson wrote "brought from China" (1844: 114), but the Banks and Solander manuscript lists "Edutha" (Oceanic Pacific) as the origin. However, there is no longer reason to doubt that it came from the Pacific (see

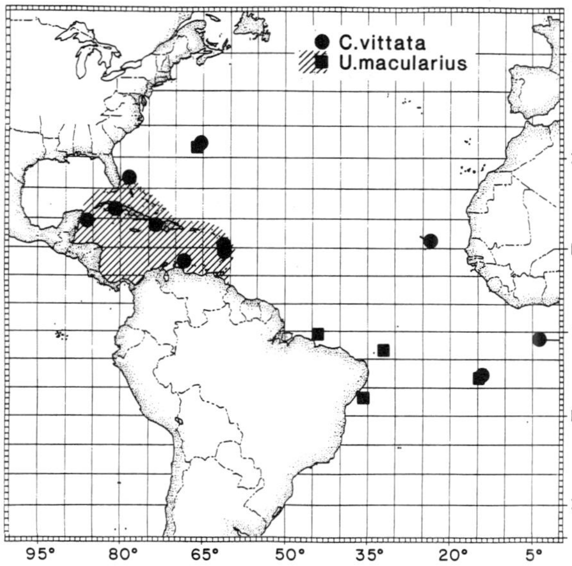

FIGURE 124. Distribution of *Channomuraena vittata* and *Uropterygius macularius* based on material examined and records in Blache, 1967f.

Remarks below). The species has been taken by hook and line, trap and spear from recorded depths of 7–37 m.

Etymology. From the Latin *vittatus* (banded), in reference to the color pattern. An adjective.

Remarks. The holotype was well described and illustrated in 1844 by Richardson, based on a specimen "brought from China." The species was redescribed by him in 1848 with notes on a second specimen labeled as "brought from the West Indies," and, beginning with Günther (1870: 135), the locality of the holotype was regarded as questionable. However, in 1959 a small (122 mm) eel that appeared extremely similar to *C. vittata* was taken at Palmyra Island, and three additional specimens (139–253 mm) were collected in the western Pacific. Vertebral counts were similar (Table 11) and no morphological or meristic differences could be found between specimens from the two oceans (minor differences in some head characters is attributed to the small size of the Pacific specimens); the doubt concerning the type locality can be eliminated. If the specimens were not conspecific, the Pacific form would be *Channomuraena vittata* and the Atlantic one would take Poey's name of *C. cubensis*. However, our data indicate them to represent one species.

Study Material. A total of 32 specimens, 91–ca 1500 mm TL, including the holotype.

Holotype: BMNH 1855.10.16.280 (48½ inches (1232 mm) TL, stuffed specimen); "brought from China"; Haslar coll. 3.

Other material: ATLANTIC: BERMUDA: ANSP 128125 (1, ca 1165). UMML 30539 (1, ca 1110). MEXICO, YUCATAN: ANSP 148100 (1, ca 91). BAHAMAS: ANSP 144118 (1, 1163). CUBA: MCZ 9191 (1, 857), holotype of *Channomuraena cubensis* Poey. MCZ 9191a (1, 790); paratype of *Channomuraena cubensis* Poey. MCZ 9190 (1, 792). USNM 24962 (1, 800). PUERTO RICO: ANSP 134515 (1, ca 500). WEST INDIES: BMNH 1856.2.24.1 (1, stuffed). ZMC J-279 (1, 809). LESSER ANTILLES: *Dominica:* USNM 247319 (1, head only). USNM 263576 (1, ca 920). *Martinique:* MNHN B.2135 (2, 649–815). MNHN B.1883-1131 (1, 716), *Curaçao:* UMML 32503 (1, ca 860). ZMA 113.017 (1, ca 588). ZMA 113.018 (2, ca 718–728). ASCENSION ISLAND: ANSP 102567 (2, ca 1000–ca 1500), 102741 (1, ca 1000), 158083 (1, ca 330), 158902 (1, 315+). BMNH

TABLE 11. Frequency distributions of vertebral counts in *Channomuraena vittata*.

	Predorsal							Preanal									
	120	121	122	123	124	125	126	118	119	120	121	122	123	124	125	126	127
C. vittata																	
Atlantic	3	3	1	1	1	1	1	1	-	1	1	4	1	-	2	2	
Pacific			1														1

	Total												
	145	146	147	148	149	150	151	152	153	154	155	156	157
C. vittata													
Atlantic	1	1	2	4	1	2	-	2	-	1	1	1	1
Pacific					1	1	-	-	1	1			

1979.1.5.12 (1, 366, color slide). EASTERN ATLANTIC: *Annobon:* MNHN 1974-100 (1).

PACIFIC: *Palmyra Island:* UH 1715 (1, ca 122); 30 Dec. 1959. *Christmas Island:* WAM P.26110-001 (1, 253); Smith Point, 7–10 m; 4 June 1978. *Caroline Islands:* CAS 13974 (1, 146) and CAS 13975 (1, 139); Kapingamarangi Atoll.

Genus *Uropterygius* Rüppell, 1838

Uropterygius Rüppell, 1838: 83 (type species *Uropterygius concolor* Rüppell, 1838, by monotypy).

Scutica Jordan and Evermann, 1896a: 403 (type species *Gymnomuraena nectura* Jordan and Gilbert, 1882 = *Gymnomuraena macrocephalus* Bleeker, 1865, by monotypy). Described as a subgenus of *Uropterygius* Rüppell, 1837.

Characters. Body slender, moderately elongate, laterally compressed posteriorly; head and trunk slightly shorter than tail (45–49% TL in Atlantic species). Median fins restricted to tail tip. Snout short, subconical and broad; jaws short, subequal and closing completely. Eye moderate, its center above mid-jaw. Anterior nostril tubular, near tip of snout; posterior nostril a small round opening above anterior margin of eye, in a short tube in some species. Gill opening small and slit-like, at or slightly below mid-side. Head pores developed; typically three supraorbital, four infraorbital, six mandibular, and one or two branchial pores. Teeth conical, slender and sharp, depressible, often multiserial on jaws.

Gill arches like those of *Anarchias* and *Channomuraena*; hypobranchials 1–2 and ceratobranchials 1–4 ossified, lower pharyngeal tooth plates with a lateral groove. Gasbladder very reduced. Peritoneum pale. Coloration usually dark, either uniform or patterned.

Size. Small to large species. The Atlantic species are small, growing to about 300 mm TL.

Distribution. Circumtropical. Found on both sides of the Atlantic and in the mid-Atlantic islands.

Etymology. From the Greek *oura* (tail) and *pterygion* (a little fin), in reference to the restricted fins. Masculine.

Remarks. *Uropterygius* is a large cosmopolitan genus that has typically included the large Indo-Pacific species *Scuticaria tigrinus* (Lesson) and *Scuticaria bennettii* (Günther). McCosker et al. (1984) examined the species of the subgenus *Scuticaria* and concluded that "... after a careful examination of all species within the genus *Uropterygius, (Scuticaria)* may deserve generic status." We concur and herein exclude those species from *Uropterygius.*

Species. *Uropterygius,* sensu stricto, contains about 19 valid species (including two or three undescribed eastern and western Pacific species) (McCosker and Rosenblatt, in prep.). Two species are found in the Atlantic: *Uropterygius macularius* (Lesueur, 1825) in the western and mid-Atlantic, and *U. wheeleri* Blache, 1967 in the eastern ocean. They are extremely similar in appearance and are separated by vertebral counts.

Key to the Atlantic Species of *Uropterygius*

1a. Total vertebrae 117–126, predorsal vertebrae 100–110, preanal vertebrae 105–115 .... *macularius,* p. 126

1b. Total vertebrae 127–135, predorsal vertebrae ca 117, preanal vertebrae ca 121 .... *wheeleri,* Blache, 1967 (an eastern Atlantic species)

Uropterygius macularius (Lesueur, 1825)

MARBLED MORAY

Figures 124–127, Tables 9, 12

Muraenophis macularia Lesueur, 1825: 107, pl. 4, fig. 1 (original description, Barbadoes, neotype ANSP 106286).

Gymnomuraena nectura (not of Jordan and Gilbert). Metzelaar, 1919a: 20 (Bonaire).

Leptocephalus juliae Tommasi, 1960: 91 (original description, off Recife, Brazil, specimen presumably in the Institute Oceanografico de la Universidad de São Paulo). Leptocephalus.

"*Uropterygius* Bohlke (n. sp. in manuscript)." Rubinoff and Rubinoff, 1962: 3 (Galeta Pt., Panama).

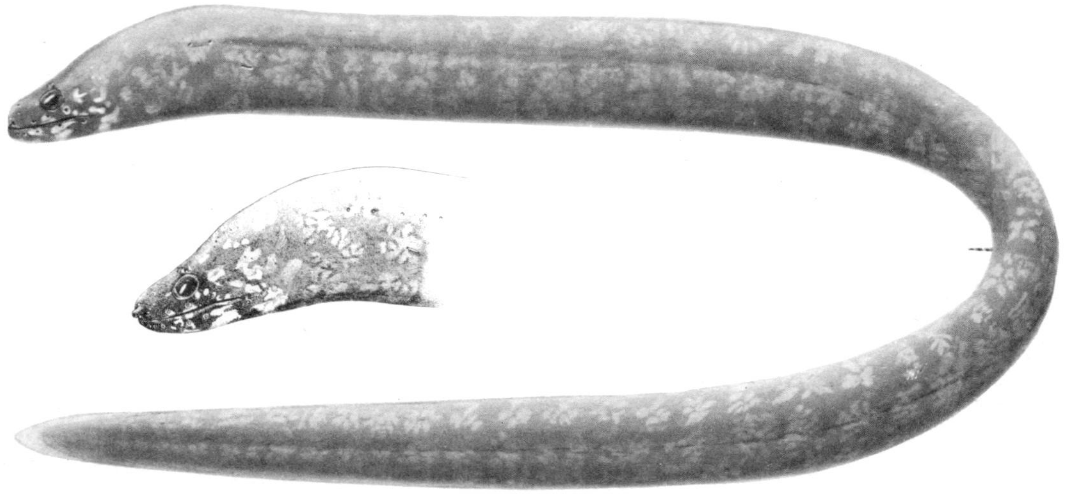

FIGURE 125. *Uropterygius macularius*: ANSP 106286, 224 mm TL, Bahamas, holotype of *Uropterygius diopus*. Illustration by S. P. Gigliotti.

Uropterygius diopus Böhlke, 1967: 91, fig. 1 (original description, Bahama Islands, holotype ANSP 106286).
Uropterygius juliae. Eldred, 1968c: 1 (matched larva and adult).

Discussion of Synonymy. The name *Uropterygius diopus* has commonly been used for this species. Eldred (1968c: 1) matched the larval *Leptocephalus juliae* Tommasi, 1960, with the adult *U. diopus* Böhlke, 1967. We now believe that the species described as *Muraenophis macularia* by Lesueur in 1825 is conspecific, and we resurrect his name as the oldest available for the species (see *Remarks* below).

Distinctive Characters. A small brown moray with pale mottling on head and chin, variously extending onto body; no pore adjacent to posterior nostril; tail 51–55% TL; jaw teeth biserial, vomerine teeth uniserial; MVF 106-110-121.

Description. Total vertebrae 117–126, x̄ = 121 (n = 48); predorsal vertebrae 100–110, x̄ = 106 (38); preanal vertebrae 105–115, x̄ = 110 (38); dorsal-fin rays 32–37, x̄ = 34 (5); anal-fin rays 20–27, x̄ = 24 (5). Pores: SO 1 + 2; IO 4; M 6, B 2.

Proportions as % (n = 23, 101–232 mm): Of TL: preanal 45–49, x̄ = 47; head 10–12, x̄ = 11; depth at gill opening 3.6–6.3, x̄ = 4.7; depth at anus 3.1–5.6, x̄ = 4.1. Of HL: snout 15–21, x̄ = 17; upper jaw 32–46, x̄ = 38; eye 6.2–10, x̄ = 8.3; interorbit 11–16, x̄ = 12.

Snout short; jaws short and subequal; eye above mid-jaw. Anterior nostril tubular; posterior nostril in a short tube located between and above front half of eye. Gill opening small, linear or oblique, its anterior end somewhat the higher,

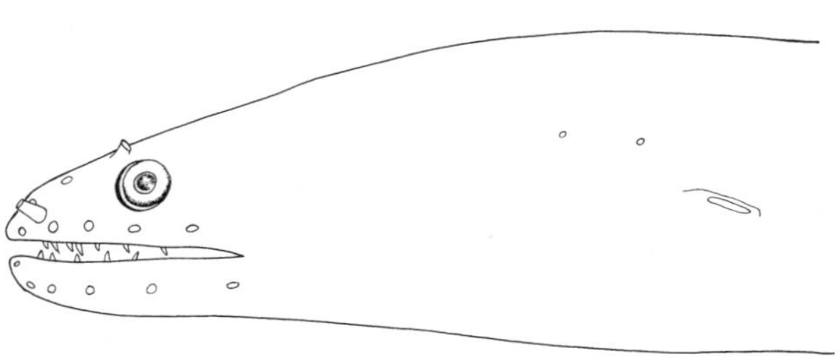

FIGURE 126. *Uropterygius macularius*: ANSP 128806, 217 mm TL, Providencia; head.

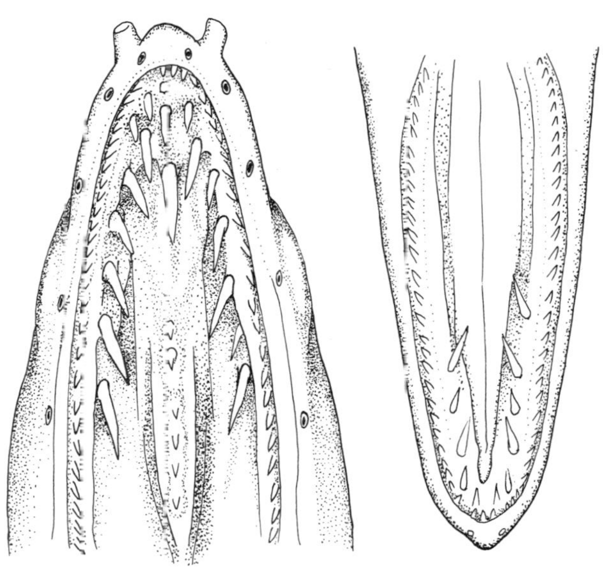

FIGURE 127. *Uropterygius macularius*: data as for Figure 126; dentition.

at or just below mid-side. Head pores as listed; branchial pores closer to gill opening than to rictus; the fourth infraorbital pore is paired on one side of one specimen.

Teeth numerous and sharp. On each side of upper jaw an outer row of 10–13 intermaxillary teeth closely followed by two rows of maxillary teeth; outer row of 14–20 small, close-set teeth, inner row of 2–7 enlarged, depressible, widely-spaced canines that increase in length from front to rear. Between outer intermaxillary teeth, three irregular rows of long depressible canines, 3–4 teeth in each series, the posteriormost median tooth the longest in the mouth. Series of 3–11 small, short teeth on vomer, the row well separated from intermaxillary teeth. Lower jaw teeth similar to those on maxillary; 25–35 short teeth in outer row, 4–8 canines in inner series.

Color. Color variable, from almost uniformly brown to highly patterned as illustrated in Figure 125; pale marks always present on head; head pores (except branchial pores) with pale edges. Fins pale.

Size. The largest specimen examined was 290 mm TL. Of 13 females (188–288 mm), eight (198–248) were ripe with eggs 0.9–1.5 mm diameter; most of these were taken during March through May, one in November. Of seven males (150–261), one 150-mm specimen taken in March was ripe.

Development. The leptocephalus was described by Tommasi in 1960 (see p. 903).

Behavior. Little is known about the habits of *Uropterygius macularius.* It has been collected in coral or rocky areas at moderate depths, usually by fish toxicants. We know of no sight records by snorkelers or skin divers and infer that it lives within the substrate. Nearly all specimens have been taken in poison collections, and few were recorded before such collection techniques were common (the holotype recorded by Lesueur, and ZMA 109.181 collected by J. Boeke in Bonaire and reported as *Gymnomuraena nectura* by Metzelaar in 1919).

Distribution (Fig. 124). Known from Bermuda, the Bahamas, the Florida Keys, throughout the Caribbean and off South America to Brazil. Collections were from near shoreline to 137 m, most often collected with rotenone-based toxicants at depths of 15–30 m; at the two deepest stations (67 and 137 m) specimens were captured with a tumbler dredge.

Etymology. From the Latin *macula* (spotted) and the suffix *aria* (connected), in reference to the coloration. An adjective.

Remarks. Muraenophis macularia was described and illustrated by Lesueur in a paper largely overlooked by subsequent authors (see additional comments under *Echidna catenata* and *Ophichthus puncticeps,* pp. 131 and 402). Four eels were described, three of them illustrated. The illustration of *macularia* depicts an elongate, slender moray (about 155 mm TL) with marbled coloration which greatly resembles the above species. The description includes applicable characters "head narrow, snout pretty long,: mouth deeply cleft, armed with long and fine teeth," although it also states that the dorsal fin commences "on the nuchae" and the spots "form a kind of chain," characters which might suggest *Echidna catenata.* However, one of the other eels described in the same paper is clearly *E. catenata,* and it is obvious from the illustrations that the species were different.

No type specimens exist for Lesueur's eels (only two of his type specimens are known according to Fowler, 1925a: 416). Therefore, following ICZN Article 75E, in order to establish nomenclatorial stability we herein designate ANSP 106286, the holotype of *Uropterygius diopus* Böhlke, as neotype of *Muraenophis macularia* Lesueur, 1825.

Uropterygius macularius is very similar in appearance and size to *Anarchias similis,* and they apparently occupy similar habitats. Both have the vertical fins restricted to the tail tip, and the coloration is almost identical. They are easily separated by the generic character of the pore adjacent to the posterior nostril which is present in *Anarchias* and absent in *Uropterygius.* There are additional differences in meristics and vertebral counts.

The closely related eastern Atlantic species, *Uropterygius wheeleri,* is distinguished from *U. macularius* solely on the basis of vertebral counts

TABLE 12. Frequency distributions of vertebral counts in Atlantic species of *Uropterygius*.

	Predorsal																	
	100	101	102	103	104	105	106	107	108	109	110	111	112	113	114	115	116	117
macularius	1	-	-	1	8	5	4	6*	5	2	6							
wheeleri																		1†

	Preanal																
	105	106	107	108	109	110	111	112	113	114	115	116	117	118	119	120	121
macularius	1	1	2	3	8	5	8*	2	5	1	2						
wheeleri																	1†

	Total																		
	117	118	119	120	121	122	123	124	125	126	127	128	129	130	131	132	133	134	135
macularius	2	2	7	9	8	8*	6	2	3	1									
wheeleri‡											1	-	-	-	3†	-	1	-	1

* Neotype.
† Paratype.
‡ Data from Blache, 1967: 1722 (except paratype).

(Table 8 and Blache, 1975a: 715). The existence of clearly separable larval forms confirms the presence of two species in the Atlantic.

Study Material. A total of 159 specimens, 51–290 mm TL, including the neotype of *Muraenophis macularia* Lesueur, and the holotype and 74 paratypes of *Uropterygius diopus* Böhlke.

Neotype: ANSP 106286 (224 mm TL); Bahamas, Great Bahama Bank, Nassau vicinity, N side of the easternmost of a series of rocks continued E of Sandy Cay, surface to 6 m; J. E. Böhlke et al., sta. 303; 16 May 1956; holotype of *Uropterygius diopus*.

Other material: The 74 paratypes of *Uropterygius diopus* as listed in Böhlke (1967: 103–104) from the Florida Keys, Bahamas, Jamaica, Haiti, Puerto Rico, Virgin Islands, Lesser Antilles, Providencia, Serrana Bank, Grand Cayman, Mexico, Panama, and Brazil.

BERMUDA: ANSP 133628 (1, 213), 148151 (1, 147). BAHAMAS: ANSP 115073 (1, 111), 144113 (1, 73). HAITI: ANSP 127298 (5, 138–230), 127992 (1, 103). PUERTO RICO: FMNH 74709 (1). BRITISH VIRGIN ISLANDS: BMNH 1976.7.14.14.16 (3, 122–202). BMNH 1976.14.17. 22 (6, 174–250). USNM 274344 (2, 165–189). DOMINICA: USNM 274345 (1, 181). LOS ROQUES: MBUCV 7545 (1). BONAIRE: ZMA 109.181 (1). COLOMBIA: ANSP 117580 (2, 204–214), 117587 (1, 283), 127464 (1, 290, cleared and stained). PANAMA: ANSP 113746 (3, 157–192). CAS 31218 (2, 101–138), 31673 (1, 220). PROVIDENCIA: ANSP 128806 (1, 217). UF 21333 (2, 110–236). SERRANA BANK: ANSP 144112 (1, 87). FMNH 90716 (1, 157), 90717 (2, 116–120). USNM 200788 (1, 80). HONDURAS: ANSP 137637 (1, 248). BELIZE: USNM 278165 (1, 107), 278169 (2, 66–159), 278170 (1, 106). MEXICO: ANSP 126038 (1, 197). BRAZIL: MZUSP 11922 (1, 215). USNM 214064 (1, 69). *Atol das Rocas:* ANSP 123293 (1, 261). UF uncat. (3, 183–227). ASCENSION ISLAND: GMBL 80-26 (1, 175), 80-28 (2, 133–219), 80-50 (13, 43–231), 80-54 (4, 46–94), 80-63 (8, 40–170), 80-71 (1, 52).

Comparative material: Paratype of *Uropterygius wheeleri* Blache: MNHN 56-88 (1, 320); island of Annobon (Bay of Biafra); 13 June 1950; CALYPSO.

Subfamily MURAENINAE

Included in the subfamily Muraeninae are most of the often-seen and well-known morays which possess a long dorsal fin beginning before or shortly behind the level of the anus which is confluent with the caudal and anal fins. Other characters include: head and jaws moderate to elongate; anterior nostril tubular, posterior nostril tubular or a simple pore; teeth uniserial to multiserial, sharp canines frequently present; gill opening small and round, at or below mid-side; coloration uniform or with bold patterns. Neurocranium short to elongate; gill arches enlarged and modified to form pharyngeal jaws; hypobranchials 1–2 absent.

Included in the Muraeninae are 11 genera, five (*) of which are in the Atlantic: *Echidna**, *Enchelycore**, *Enchelynassa*, *Gymnomuraena*, *Gymnothorax**, *Monopenchelys**, *Muraena**, *Rhinomuraena*, *Siderea*, *Strophidon*, and *Thyrsoidea*.

Genus *Echidna* Forster, 1777

Echidna Forster, 1777: 181 (type species *Echidna variegata* Forster in Lichtenstein, 1844 = *Muraena echidna* Gmelin = *Muraena nebulosa* Ahl, 1789, by virtual tautonomy).

Megaderus Rafinesque, 1815: 93 (substitute name for *Echidna* Forster, 1777).

Molarii Richardson, 1848a: 79 (type species *Muraena ophis* Rüppell, 1828 = *Muraena nebulosa* Ahl, 1789, by Jordan and Evermann, 1896, as first revisors).

Poecilophis Kaup, 1856a: 66 (type species *Gymnothorax catenatus* Bloch, 1795, by subsequent designation of Jordan and Evermann, 1906, as first revisers).

Leihala Jordan, 1925: 5 (type species *Poecilophis tritor* Vaillant and Sauvage, 1875 = *Muraena polyzona* Richardson, 1844, by original designation).

Characters. Body moderately elongate, laterally compressed posteriorly; head and trunk equal to or shorter than tail (43–47% TL in Atlantic species); dorsal-fin origin in anterior trunk region, in vicinity of gill opening. Snout short, subconical, broad; jaws moderate to short, stout, subequal and closing completely. Eye moderate, its center at midpoint of upper jaw. Anterior nostril tubular, near tip of snout and anteriorly directed; posterior nostril a small round opening above anterior margin of eye, often with a low raised rim. Gill opening mid-lateral, small and slit-like. Head pores developed, small; typically three supraorbital, four infraorbital, six mandibular and two branchial pores. Teeth molariform or conical, always stout and fixed, none caniniform; jaw and vomerine teeth biserial to multiserial, maxillary tooth rows short.

Gill arches like those of *Muraena* (see p. 112); hypobranchials 1–2 absent; tooth plates stout. Peritoneum pale. Coloration sometimes uniform, often with conspicuous, contrasting patterns of bands and spots.

Size. Small to moderate eels; Atlantic species recorded to 684 mm TL.

Distribution. Circumtropical, shallow-water species, found in all oceans.

Etymology. From the Greek *Echidna* (a viper). Feminine.

Remarks. Species of *Echidna* are extremely similar to species of *Gymnothorax*, *Muraena*, and *Siderea* in their general appearance (body shape, coloration and dorsal-fin origin). They differ mainly in dentition; teeth of *Echidna* are multiserial, short and blunt, with some rounded molariform teeth and never any caniniform teeth. The dorsal-fin origin of Atlantic species is on the head, well before the gill openings; it varies considerably in Indo-Pacific species from before to well behind the gill openings.

Species. Currently known from 12 species: in the mid- and western Atlantic, *Echidna catenata* (Bloch, 1795); in the eastern Atlantic, *E. peli* (Kaup, 1856); in the eastern Pacific, *E. nocturna* (Cope, 1872); in the Indo-Pacific, *E. amblyodon* (Bleeker, 1856), *E. delicatula* (Kaup, 1856), *E. fascigula* (Peters, 1855), *E. leucotaenia* Schultz, 1943, *E. polyzona* (Richardson, 1844), *E. rhodochilus* Bleeker, 1863, *E. unicolor* Schultz, 1953, *E. xanthospilus* (Bleeker, 1859); and pan-Indo-Pacific, *E. nebulosa* (Ahl, 1789). We regard the widespread Indo-Pacific species *Gymnomuraena zebra* (Shaw, 1797) as generically distinct from *Echidna*.

Key to the Atlantic Species of *Echidna*

1a. Color dark brown, with faint reticulations or widely-spaced small pale spots .... *peli* (Kaup, 1856) (an eastern Atlantic species)

1b. Bold coloration of contrasting reticulations forming a chain-like pattern .... *catenata*, p. 131

Echidna catenata (Bloch, 1795)

CHAIN MORAY

Figures 128–131, Table 9

Muraena seu conger brasiliensis Seba, 1738: 72, pl. 69, figs. 4, 5 (Brazil).
Gymnothorax catenatus Bloch, 1795: 84, pl. 415, fig. 1 (type locality Suriname, holotype ZMB 4012).
Muraenophis catenula Lacepède, 1803: 628, 641 (based on *catenatus* of Bloch).
Muraena sordida Cuvier, 1817: 233 (based on *Muraena seu conger brasiliensis* of Seba).
Muraenophis zebra (not of Shaw) Lesueur, 1825: 107, pl. 4, fig. 2 (original description, Barbados, holotype unknown).
Muraena catenata. Richardson, 1848a: 95. Günther, 1870: 130.
Muraena alusis Bleeker, 1856: 67 (replacement name for *catenatus* of the Atlantic, *catenatus* thought to be Indo-Pacific).
Poecilophis catenatus. Kaup, 1856a: 66; Kaup, 1856b: 100.
Echidna fuscomaculata Poey, 1867b: 263 (original description, Cuba, holotype unknown).
Echidna flavofasciata Poey, 1867b: 263 (original description, Cuba, holotype unknown).
Echidna catenata. Jordan and Davis, 1891: 612.
Echidna flavoscripta. Jordan and Davis, 1891: 612 (lapsus for *Echidna flavofasciata*).
Muraena punctatofasciata (not of Bleeker, 1863). Murray, 1902: 434 (misidentification).
Echidna catenata flavofasciata. Howell Rivero, 1932: 21 (*flavofasciata* a subspecies).

Discussion of Synonymy. Günther (1870: 131) correctly placed two of Poey's species, *Echidna fuscomaculata* and *E. flavofasciata* in the synonymy of *E. catenata*; he was followed in this by Poey (1876a: 202). Jordan and Davis (1891: 612) apparently inadvertently changed the name *E. flavofasciata* to *E. flavoscripta*, and the latter name was used by Jordan and Evermann (1896a: 403), Jordan, Evermann, and Clark (1930: 94) and Torre y Huerta (undated, unpaginated). (Poey did, in fact, name a moray species *Gymnothorax flavoscriptus* (1876a: 200), that species a synonym of *Gymnothorax moringa*.) In addition, in the collection at USNM there are two specimens from Cuba (received from Poey) with two more (apparently unpublished) names: USNM 9802, *Echidna flavomaculata* (Poey's original number 71, cataloged in 1873) and USNM 13061, *Echidna maculata* (original number 161, cataloged 7 Dec. 1874). Both specimens are *E. catenata*.

Muraenophis zebra Lesueur (not of Shaw), here added to the synonymy of *Echidna catenata*, was described and illustrated by Lesueur (1825: 107) in a paper apparently overlooked by all subsequent authors except Richardson, who incorrectly applied one of the four species names to an ophichthid (see pp. 128, 402). Lesueur's description and illustration of *Muraenophis zebra* from the Barbados accurately depicts *Echidna catenata*.

Distinctive Characters. A medium-size dark moray with conspicuous yellowish chain-like pattern on head, body and fins; snout short and blunt; teeth conical, blunt or molariform, no canines; MVF 6-58-117.

Description. Total vertebrae 114–122, $\bar{x}$ = 117.4 (n = 37); predorsal vertebrae 5–7, $\bar{x}$ = 6.1 (36); preanal vertebrae 56–61, $\bar{x}$ = 58.5 (39); kidney terminus at 65–66 (2); dorsal-fin rays 292–302, $\bar{x}$ = 296 (3); anal-fin rays 159–165, $\bar{x}$ = 162 (3). Pores: SO 1 + 2; IO 4; M 6; B 2.

Proportions expressed as % (n = 20, 284–680 mm): Of TL: predorsal 8.9–13, $\bar{x}$ = 12; preanal 53–57, $\bar{x}$ = 55; head 12–14, $\bar{x}$ = 14; depth at gill opening 5.5–7.9, $\bar{x}$ = 6.7; width at gill opening 2.9–6.0, $\bar{x}$ = 4.4; depth at anus 5.0–7.3, $\bar{x}$ = 6.2; width at anus 3.2–5.1, $\bar{x}$ = 4.0. Of HL: snout 15–20, $\bar{x}$ = 17; upper jaw 28–37, $\bar{x}$ = 32; lower jaw 25–35, $\bar{x}$ = 30; eye 7.0–10, $\bar{x}$ = 8.2; interorbit 9.3–14, $\bar{x}$ = 12.

Head short; snout short and blunt; eye moderate, over or slightly behind mid-jaw. Anterior nostril tubular, near tip of snout, the tube short, not reaching edge of upper jaw; posterior nostril with raised, crenulate margin, situated above anterior margin of eye. Head pores small, round, all but branchial and anteriormost lower jaw pore in indented "dimple." Branchial pores very small and close together, frequently difficult to discern, the anteriormost below dorsal-fin origin. Little variation in head pores; one specimen with four

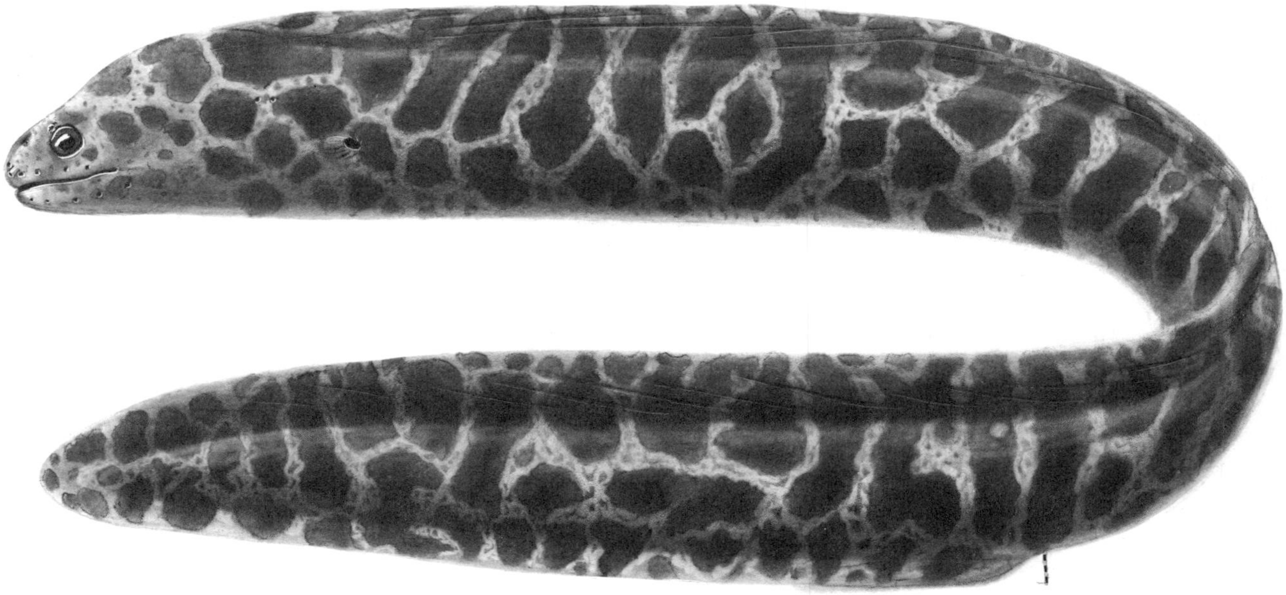

FIGURE 128. *Echidna catenata*: ANSP 100525, 330 mm TL, Bahamas. Illustration by S. P. Gigliotti.

supraorbital pores, one with five infraorbital pores on one side only, and one with three branchial pores on both sides.

In young, all teeth except those of inner row on maxillary are broad-based, some are rounded and some have narrow pointed tips. With growth, the teeth become increasingly blunt, many with flattened tips, but those of the inner maxillary row remain somewhat narrow and pointed. On upper jaw an anterior median intermaxillary tooth; 6–11 teeth on each side, additional tiny teeth sometimes present, enclosing 1–3 longitudinal rows of larger teeth; a median row of three teeth in small individuals, the posteriormost tooth the largest (sometimes four teeth in this row in large fish); 3–4 teeth on either side of median row in large specimens. Two short rows of teeth on maxillary, continuous with outer intermaxillary teeth; outer row of 9–12 blunt teeth, inner row of 4–8 taller pointed teeth. Vomerine teeth mostly biserial; 4–11 pairs followed by 1–4 uniserial teeth. Dentary teeth partly to completely biserial, plus 2–3 teeth forming an additional third row anteriorly in large specimens. In young, the anterior outer row is short, of 3–5 teeth, and the

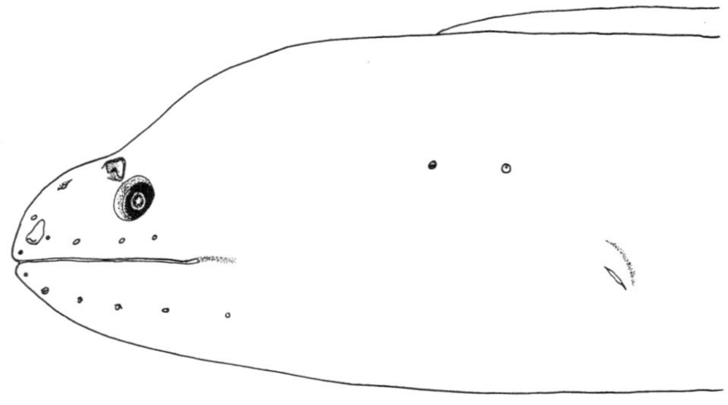

FIGURE 129. *Echidna catenata*: ANSP 130816, 242 mm TL, Grand Cayman; head.

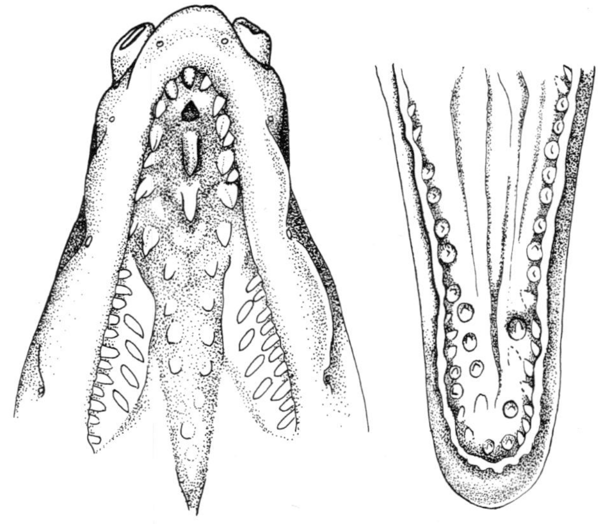

FIGURE 130. *Echidna catenata*: ANSP 104905, 180 mm TL, Grand Cayman; dentition of young specimen.

long inner row of 14–17 teeth continues all the way back; in adults, the outer series is also complete; 17–26 smaller teeth in outer series, 16–18 larger ones in inner series.

Color. Illustrated fish with typical coloration. The pattern is highly variable, especially the widths of pale areas (the "chains") and degree of dark spotting on them. Pale areas usually are narrower and more linear than dark ones, but sometimes they are about equal. The largest specimen examined (MNRJ 2328, 684 mm TL, from Brazil) has a reversed color pattern in which the broad areas are pale and narrow dark areas form a dark chain on a light background. Coloration is of same intensity on all of fish, including head and fins, except that individuals smaller than about 100 mm may have pale and patternless fins. Anterior nostril with body color pattern; posterior nostril with dark rim. Peritoneum freckled and spotted. In life, brown or black with a yellow chain pattern.

Size. The largest specimen examined was 684 mm TL; Randall (1968: 38) stated the species grows to 2 ft 4 inches (ca 710 mm). Of 29 females (216–680 mm), five (475–680 mm) were ripe with 0.9–1.3-mm eggs, the latter taken in April, July and August. Of 10 males (243–610 mm), two (438–618 mm) captured in June and October were ripe.

Development. Although *Echidna catenata* is one of the most common morays found on coral reefs of the western Atlantic, and *E. peli* is common in the eastern Atlantic, to date, no leptocephalus referable to the genus has been found in the Atlantic. In the western Atlantic, *E. catenata* has a very distinctive vertebral count; its larvae could hardly be confused with those of any other species.

Behavior. This species abounds in rock and sand habitats, often in surprisingly shallow water where it is seen by snorkelers and skin divers. Almost all of our study material came from such shallow areas. Most of the over 400 study specimens were examined for sexual condition, but only seven ripe individuals were found; where mature *Echidna catenata* go to spawn and where the leptocephali develop remains a mystery to date.

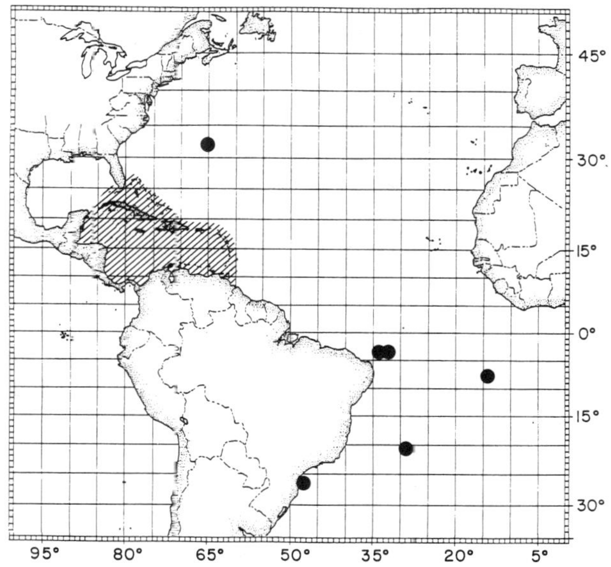

FIGURE 131. Distribution of *Echidna catenata* based on material examined.

The chain moray is reported to feed on crustaceans and small fishes, but its primary food source is small crabs (Randall, 1968: 38). The stout molariform teeth of the species are well adapted for this diet. When a tidepool on Ascension Island was baited with freshly-killed crabs, 57 *Echidna catenata* (most of them juveniles four to eight inches long) quickly congregated to feast on the spoil (W. V. Vickrey, in litt., 1964).

Distribution (Fig. 131). *Echidna catenata* is abundant in shallow waters throughout the islands of the western and mid-Atlantic, as listed in the *Study Material.* It was recorded from "West Africa" by Fowler (1919: 275), but Blache (1967c: 709) found no examples of the species in the eastern Atlantic and considered this record erroneous. We note that the three specimens on which the record is based (USNM 42387) were collected by the Eclipse Expedition, which had stopped at Ascension Island on its way to Africa. The specimens are good examples of *Echidna catenata,* and it is most likely that they were actually collected at Ascension Island.

Most of the study material was taken with rotenone-based fish toxicants, frequently in tide pools at depths less than 2 m. One specimen was taken at 20 m among rocks at Ascension Island (Lubbock, 1980: 284), and a few have been taken by trawl.

Etymology. From the Latin *catenatus* (chained), in reference to the chain-like color pattern. An adjective.

Remarks. The teeth of *Echidna catenata* have been described as blunt and molariform, distinguishing it from all other morays in the area which have sharp teeth. Actually, the only molariform teeth are those on the vomer; the others are less sharp than those in other muraenids, and become rounded to quite blunt in large adults, but they do not approach the pavement-like teeth of some of the Indo-Pacific species placed in *Echidna.*

Study Material. A total of 408 specimens, 35–684 mm TL, including the holotype.

Holotype: ZMB 4012 (259); Suriname.

Other material: BERMUDA: ANSP 133868 (8, 86–218), 148150 (1, 221). USNM 164747 (1, 252). FLORIDA: MCZ uncat. (1, 315). UMML 4644 (1, 232), 5701 (1, 260). BAHAMAS: ANSP 72303 (1, 298), 72304 (1, 255), 72305 (1, 395), 100525 (1, 330), 100526 (1, 194), 100527 (2, 238–281), 100528 (1, 555), 100530 (10, 88–342). MCZ 33918 (1, 225). USNM 38474 (3, 185–188). CUBA: ANSP 92922 (1, ca 610). CAS-SU 54213 (3, 322–470). MCZ 27220 (1, 615). USNM 9802 (1, 428), 13061 (1), 33091 (1, 370), 33093 (1, 291), 37443 (1, 110), 132941 (1, 155). HAITI: ANSP 124691 (2, 61–142). FMNH 74757 (1, 170). MCZ 9127 (1, 342), 9173 (4, 218–650). DOMINICAN REPUBLIC: MCZ 34693 (2, 199–395). PUERTO RICO: ANSP 87892 (1, 475), 115547 (6, 103–184), 115643 (7, 48–290), 118629 (9, 53–221), 118703 (1, 118), 122922 (1, 346), 129899 (4, 146–228), 129927 (5, 76–216), 139743 (18, 33–56), 139744 (1, 55). FMNH 74715 (1, 197), 74716 (1, 188). USNM 117442 (1, 164), 126093 (1, 165), 147643 (1, 183), 274352 (5, 182–480), 274353 (1, 242), 274357 (1, 166), 274360 (1, 335), 274362 (1, 140). VIRGIN ISLANDS: ANSP 100531 (1, 208), 110413 (1, 158). MCZ 27210 (1, 230).

LESSER ANTILLES: *St. Martins:* ANSP 16036–16037 (2, 304–455), 38159 (1, 157). *St. Barthelemy:* ANSP 105496 (3, 114–147). *Dominica:* ANSP 105766 (1, 143). USNM 274350 (1, 580), 274351 (3, 102–254), 274355 (4, 120–230), 274356 (6, 90–365), 274358 (1, 241), 274359 (3, 152–290), 274361 (4, 103–136), 274363 (3, 57–110). *St. Lucia:* ANSP 105488 (1, 237), 105493 (1, 310), 105725 (1, 152), 105747 (1, 123), 158796 (44, 35–89). CAS-IUM 8777 (1, 164), 8846 (1, 192). CAS-SU 1716 (3, 150–247). USNM 41305 (3, 185–282). *Grenadines:* ANSP 105491 (2, 125–247). *Barbados:* ANSP 88738 (3, 204–337), 126408 (19, 72–241). CAS-SU 37266 (1, 105). USNM 50593 (1, 218). *Grenada:* ANSP 106181 (1, 343). *Los Roques:* ANSP 75192 (1, 322). MBUCV 313 (1), 884 (1), 1498 (1), 2115 (1). *Bonaire:* ANSP 113695 (2, 277–ca 407). *Curaçao:* CAS 59206 (4, 362–495). USNM 274364 (1, 310). *Tobago:* ANSP 97936 (1, 265). BMNH 1920.12.22.21 (1, 402). USNM 274349 (1, 323). *Trinidad:* ANSP 76386 (1, 438). CARIBBEAN SEA: *Serrana Bank:* ANSP 103035 (16, 82–318), 115512 (4, 116–300, cleared and stained). FMNH 74735 (1, ca 60). UMML uncat. (6, 145–300). USNM 187769 (9, 111–183), 193609 (26, 115–430), 198759 (20, 96–300). *Swan Island:* MCZ 31430 (1, 325). *Grand Cayman:* ANSP 104905 (1, 180), 130816 (1, 242). *Providencia:* USNM 38644 (1, ca 530). PANAMA: ANSP 113766 (1, 450). CAS-SU 18509 (1, 207). FMNH 8208 (1). MCZ 43892 (1, 65), 43893 (1, 410), 43894 (1, ca 500), 43895 (1, 136). USNM 79462 (1, 510), 79463 (1, 420). HONDURAS: NMW 60106 (1). MEXICO: ANSP 114911 (1, 145). COLOMBIA: ANSP 117556 (8, 107–ca 555), 117569 (1, 290), 117571 (3, 112–313), 117578 (1, 64). VENEZUELA: MBUCV 1 (1), 95 (1), 245 (1), 366 (1), 1187 (1), 2512 (1), 2523 (1). MNHN 1956-587 (1). BRAZIL: MNRJ 2328 (3, 563–684). MZUSP uncat. (1, 420). *Atol das Rocas:* MZUSP 11819 (6). USNM 194019 (1, 168). *Fernando Noronha:* USNM 274348 (5, 187–410). *Ilha da Trinidade:* MNRJ 9362 (1, 354). ASCENSION ISLAND: ANSP 103034 (3, 370–630), 103107 (1, ca 680). GCRL 15499 (1). GMBL 80-41 (1, 195). GMBL 80-371 (1, 285). USNM 274354 (2, 43–164).

LOCALITY UNKNOWN: MNHN 2144 (1). MNHN B.2367 (1). MNHN B.2368 (1). LOCALITY QUESTIONED (Africa): USNM 42387 (3, 157–270).

Genus *Enchelycore* Kaup, 1856

Enchelycore Kaup, 1856a: 60 (type species *Enchelycore euryrhina* Kaup, 1856 = *Muraena nigricans* Bonnaterre, 1788, by monotypy); 1856b: 71.

Eurymyctera Kaup, 1856a: 59 (type species *Eurymyctera crudelis* Kaup, 1856 = *Muraena schismatorhynchus* Bleeker, 1853 by monotypy); 1856b: 72.

Enchelycotte Dumeril, 1856: 200 (lapsus for *Enchelycore* Kaup, 1856).

Aemasia Jordan and Snyder, 1901: 883 (type species *Aemasia lichenosa* Jordan and Snyder, 1901, by original designation).

Rhinechidna Barbour, 1908:41 (type species *Rhinomuraena eritima* Jordan and Seale, 1906 = *Muraena schismatorhynchus* Bleeker, 1853, by monotypy).

Fimbrinares Whitley, 1948: 72 (type species *Fimbrinares mosaica* Whitley, 1948 = *Gymnothorax ramosus* Griffin, 1926, by original designation).

Characters. Body elongate, laterally compressed throughout trunk and tail, and tapering posteriorly; head and trunk equal to or shorter than tail (38–50% TL in western Atlantic species). Dorsal-fin origin above or behind gill opening. Head muscular; snout long and narrow; jaws slender and elongate, strongly arched and not closing completely, the teeth exposed in lateral aspect when mouth is closed; eye moderate to large, about at midpoint of jaw. Anterior nostril tubular, anteriorly directed; posterior nostril a round, oval, or elongate opening usually above and before eye, slightly raised and crenulate in some species. Gill opening about midbody, small and slit-like. Head pores developed; typically three supraorbital, four infraorbital, six mandibular, and two branchial pores. Teeth conical, sharp and prominent, long fangs alternating with shorter teeth along length of jaws; jaw teeth at least partly biserial.

Neurocranium elongate, similar to that of *Muraena*. Gill arches like those of *Muraena*, hypobranchials 1–2 absent. Gasbladder small, about 5% of TL. Peritoneum pale. Coloration dark, with or without pattern or spots.

Size. Species varying from small to large. In the Atlantic, the smallest species grows to about 350 mm TL, the largest to over 1100 mm TL.

Distribution. Throughout all tropical and subtropical oceans; most species inhabit shallow coral reefs and rocky areas, a few are found in deeper waters.

Etymology. From the Greek *enchelys* (eel) and *kore* (girl), the application not evident. Feminine.

Remarks. Species of the genus *Enchelycore* are distinguished from other morays of the subfamily Muraeninae by their arched jaws and exposed fang-like teeth, and the dorsal-fin origin over or behind the gill opening (in contrast to the more anterior placement of the fin in similar-appearing species of *Gymnothorax*, and *Muraena*). (The posterior nostril condition, once used as a diagnostic character, was reevaluated by Böhlke and Böhlke (1976); it varies from a simple round pore near the upper margin of the eye to a large oval opening well in front of the eye, with various intermediate conditions, and is no longer useful to characterize the genus.) In addition, one species, *Enchelycore schismatorhynchus* (Bleeker, 1853), differs from all others in having the dorsal-fin origin before the gill opening and the posterior nostril in a low tube.

Species. The species of *Enchelycore* have recently been treated by Randall and McCosker (1975), McCosker and Rosenblatt (1975), and Böhlke and Böhlke (1976, 1980). Ten species are currently recognized: three in the Atlantic, *anatina* (Lowe, 1838), *carychroa* Böhlke and Böhlke, 1976, and *nigricans* (Bonnaterre, 1788); one in the eastern Pacific, *octaviana* (Myers and Wade, 1941); and six in the Indo-Pacific, *bayeri* (Schultz, 1953), *bikiniensis* (Schultz, 1953), *kamara* Böhlke and Böhlke, 1980, *lichenosa* (Jordan and Snyder, 1901), *ramosa* (Griffin, 1926), and *schismatorhynchus* (Bleeker, 1853). In addition, we provisionally include the widespread Indo-Pacific species *Muraena pardalis* Schlegel, 1846 within this genus, pending further osteological study.

Key to the Atlantic Species of *Enchelycore*

1a. Posterior nostril large and elongate, well before eye; 2–5 inner dentary teeth; color variably patterned or uniformly brown, jaw pores not set in white spots; total vertebrae 141–148 . . . . *nigricans*, p. 136

1b. Posterior nostril not elongated, situated above anterior margin of eye; 6–14 inner dentary teeth; color patterned or uniformly brown; total vertebrae 128–140 or 145–155 2

2a. Color uniform brown, jaw pores set in white spots; 6–13 inner dentary teeth; total vertebrae 128–140 . . . . *carychroa*, p. 140

2b. Color pattern of light spots and blotches on brown body; 10–14 inner dentary teeth; total vertebrae 145–154 . . . . *anatina*, p. 142

TABLE 13. Frequency distribution of vertebral counts in Atlantic species of *Enchelycore*.

	Predorsal					
	6	7	8	9	10	11
anatina	1	3[1]	2[2]			
carychroa		7	130	134[3]	22	3
nigricans	2	15	30	12	1	

	Preanal																		
	45	46	47	48	49	50	51	52	53	54	55	56	57	58	59	60	61	62	63
anatina								1[1]	–	1	–	2	1	–	2[2]				
carychroa	2	30[3]	96	131	62	8	1												
nigricans														2	19	20	17	3	1

	Total																											
	128	129	130	131	132	133	134	135	136	137	138	139	140	141	142	143	144	145	146	147	148	149	150	151	152	153	154	155
anatina																			1[1]	–	–	–	–	2	–	–	1[2]	3
carychroa	2	–	3	8	27	51	68[3]	62	59	25	13	3	1															
nigricans														2	8	12	14	16	6	3	2							

[1] Holotype of *Muraena sanctaehelenae*.
[2] Neotype of *Muraena anatina*.
[3] Holotype of *Enchelycore carychroa*.

Enchelycore anatina (Lowe, 1838)

FANGTOOTH MORAY

Figures 132–136, Tables 9, 13

Muraena anatina Lowe, 1838: 192 (original description, Madeira, neotype BMNH 1869.3.7.1).

Muraena sanctae helenae Günther, 1869: 239 (listed only); 1870: 115 (original description, St. Helena, holotype BMNH 1868.3.11.3).

Muraena sanctae-helenae. Goode, 1876: 76 (Bermudas, first North American record).

Sidera sanctae-helenae. Jordan, 1887b: 566.

Gymnothorax anatinus. Jordan and Davis, 1891: 602.

Gymnothorax sanctae-helenae. Jordan and Davis, 1891: 602.

Lycodontis anatinus. Fowler, 1936a: 314; 1936b: 1213. Blache, 1967d: 1151–1158, figs. 13–16 (Canary Islands and Madeira; redescribed, illustrated).

Enchelycore anatina. Böhlke and Böhlke, 1976: 141 (referred to *Enchelycore*).

Misidentifications:

Not *Muraena sanctae-helenae.* Bean, 1880: 113 (= *Gymnothorax moringa* (Cuvier)).

Not *Lycodontis sanctae-helenae.* Barbour, 1905: 111 (= *Gymnothorax moringa* (Cuvier)). Bean, 1906: 32 (= *G. moringa*).

Discussion of Synonymy. BMNH 1869.3.7.1 was listed as the holotype by Blache (1967d: 1151), Blache et al. (1973: 225) and by Böhlke (1982: 39). However, Ben-Tuvia and Golani (1984: 542) pointed out that this specimen, cataloged 30 years after the species was described, was most likely not from Lowe's collection and therefore could not be the holotype. Since no type specimen for it could be found in the BMNH collection, they designated the "holotype" to be the lectotype, an invalid designation because the specimen is not considered part of Lowe's collection. The species *anatina* is well known primarily on the basis of the supposed holotype; Lowe's original description, five lines of Latin with no illustration, could conceivably apply to a species other than the currently known *anatina* (although the description of the triserial dentition would exclude most Atlantic morays). In order to establish nomenclatorial stability, we herein designate BMNH 1869.3.7.1 as neotype of *Muraena anatina* Lowe.

Muraena sanctae helenae Günther was considered to be a junior synonym of *Enchelycore an-*

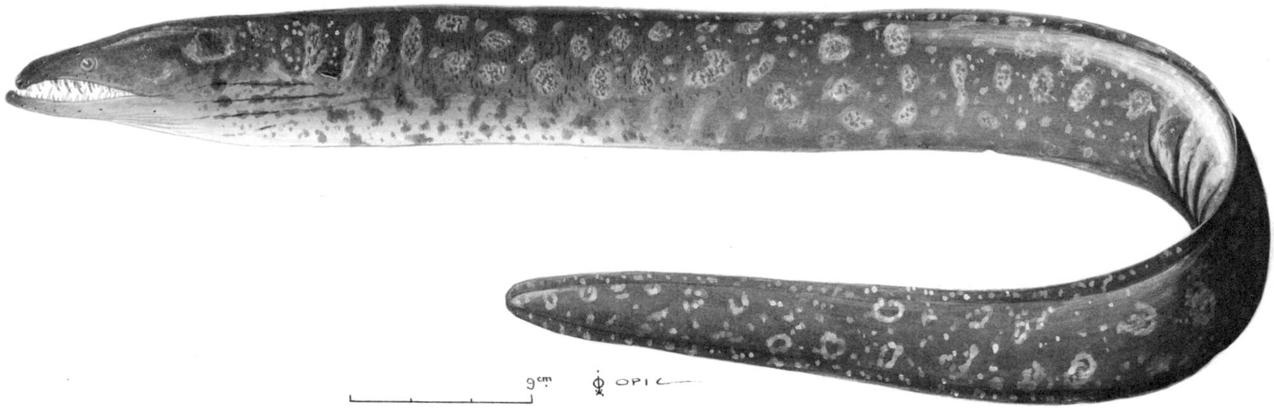

FIGURE 132. *Enchelycore anatina*: MNHN 1883-142, 1120 mm TL, Canary Islands. Illustration by M. Opic, courtesy of M. Opic and J. Blache.

atina by Fowler (1936a: 314) and by Blache (1967d: 1151), based on eastern and mid-Atlantic specimens. Ben-Tuvia and Golani (1984) resurrected the species *sanctaehelenae* on the basis of differences in body depth, coloration, and vertebral count. However, the body depth (6.7% TL given by Ben-Tuvia and Golani) of the holotype of *Muraena sanctaehelenae* lies well within the range for our study specimens of *Enchelycore anatina*; coloration of this species (as for many morays) is extremely variable (see Figs. 134, 135), and the number of vertebrae in the holotype of *sanctaehelenae* (Ben-Tuvia and Golani reported 145; we counted 146 in a radiograph taken by EBB in 1973), while low, lies within the 15% range found in many species of eels. We consider *sanctaehelenae* to be a junior synonym of *Enchelycore anatina*.

Distinctive Characters. A large moray with distinctive color pattern of light spots and mottling on a brown background; tail 55–60% TL; posterior nostril not elongating forward with age; lower jaw teeth biserial, inner series complete; MVF 7-56-154.

Description. Total vertebrae 146–155, $\bar{x}$ = 152.4 (n = 7); predorsal vertebrae 6–8, $\bar{x}$ = 7.2 (6); preanal vertebrae 52–59, $\bar{x}$ = 56.1 (7); dorsal-fin rays 363–418, $\bar{x}$ = 390 (2); anal-fin rays 303–327, $\bar{x}$ = 315 (2). Pores: SO 1 + 2; IO 4; M 6; B 2.

Proportions as % (the 75-mm specimen first, in parentheses, followed by 6, 650–963). Of TL: predorsal (12), 10–13, $\bar{x}$ = 12; preanal (40), 42–45, $\bar{x}$ = 44; head (14), 12–14, $\bar{x}$ = 13; depth at gill opening (6), 5–7, $\bar{x}$ = 6; depth at anus (4), 5–6, $\bar{x}$ = 6; width at gill opening (3), 3–4, $\bar{x}$ = 3; width at anus (3), 3–4, $\bar{x}$ = 3. Of HL: snout (17), 20–25, $\bar{x}$ = 22; upper jaw (44), 46–54, $\bar{x}$ = 49; eye (13), 6–9, $\bar{x}$ = 8; interorbit (14), 8–12, $\bar{x}$ = 10.

Head long, profile uniform to eye-level; snout long and flat; jaws elongate, subequal, strongly arched, teeth exposed in lateral aspect when mouth is closed. Eye moderate, above mid-jaw. Head pores typical for the genus, as illustrated, with little variation; one specimen had an added infraorbital pore behind eye on one side, one

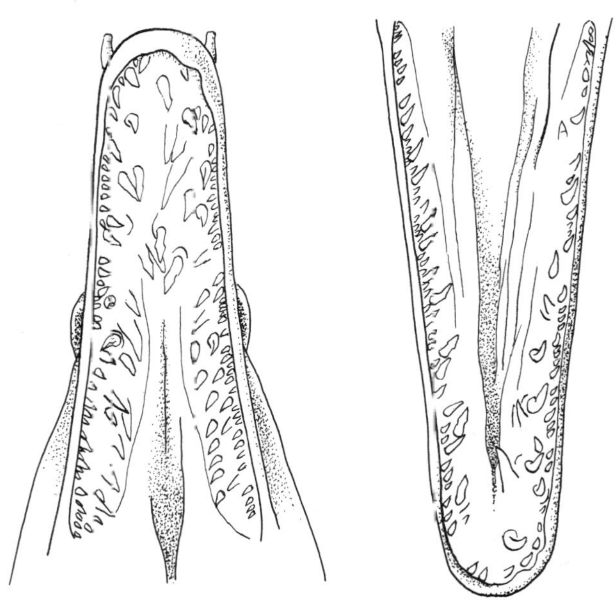

FIGURE 133. *Enchelycore anatina*: data as for Figure 132; dentition (after Blache, 1967d: fig. 15).

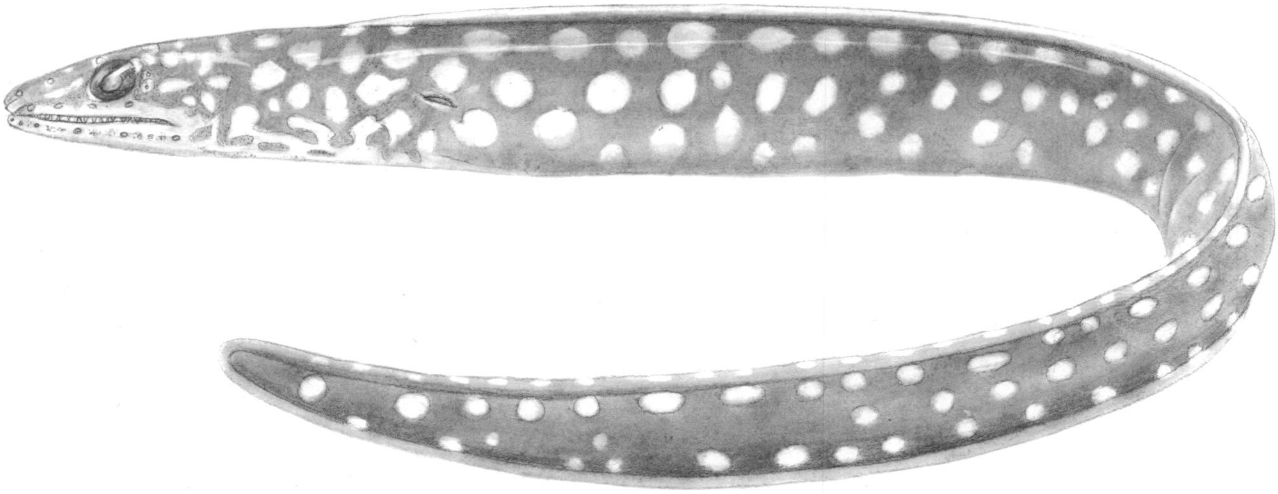

FIGURE 134. *Enchelycore anatina*: ANSP 102472, 75 mm TL, Florida Atlantic; juvenile coloration.

had four supraorbital pores on both sides, and the illustrated specimen (Blache, 1967d: figs. 12, 13, our Fig. 132) had four branchial pores on one side only.

Teeth prominent and menacing, many fang-like teeth exposed. A row of 19–25 intermaxillary teeth on each side of upper jaw, including some long fangs, enclosing three extremely long, slender, depressible median fangs; irregularly 1–2 rows of 3–5 depressible teeth, intermediate in size, between the outer row and median series, appearing as one irregular row or as two short rows. Vomerine teeth very small and difficult to see, sometimes hidden by folds of flesh; a single row, well separated from intermaxillary teeth, of 8–15 teeth counted on three specimens (the illustrated specimen, drawn from Blache, 1967d: fig. 15, shows only three). Maxillary dentition triserial, innermost teeth the longest; outer row of 31–40 small teeth, a few of the anterior ones longer than others; median series of 6–12 slender teeth; and innermost series of 5–13 elongate, depressible teeth. Dentary teeth biserial; outer series of 48–68 small teeth, the anterior ones larger with 3–4 small teeth between them; inner series of 10–14 long, slender, depressible teeth.

Body elongate, tail tapering. Median fins continuous with caudal fin; dorsal-fin origin above gill opening.

Coloration. Brown, with contrasting pale markings on head, body and fins. In adults, a complex pattern of pale areas colored with small dark spots, interspersed with smaller simple spots without dark centers, the spots rounded or with irregular outlines (Fig. 135). Lower jaws and sides of head with fewer spots or sometimes unmarked. Vertical fins with series of spots on distal edge. The very small specimen was brown with large pale spots (Fig. 134).

Life colors of two specimens were observed in the aquarium of the Amsterdam Zoological Society, and from kodachromes of those specimens (courtesy of F. deGraef and I. Isbrucker). The ground color of the body was medium brown, the spots and blotches with considerable yellow; smallest spots were completely yellow, larger

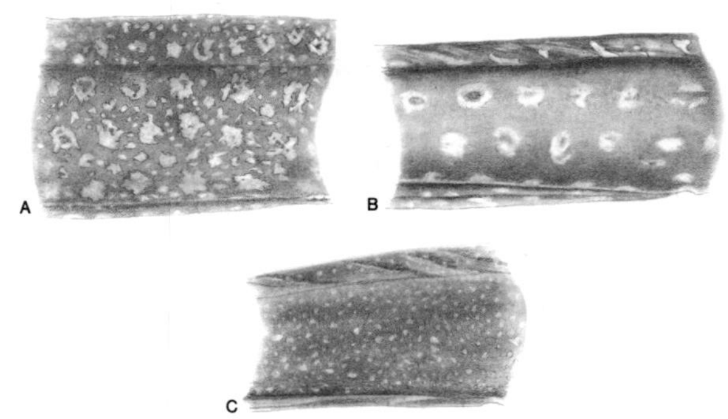

FIGURE 135. *Enchelycore anatina*: variation in color pattern. A. BMNH 1872.8.28.12, 963 mm TL, Bermuda. B. BMNH 1899.1.16.33, 650 mm, Madeira. C. BMNH 1910.9.9.2, 888 mm, St. Helena.

blotches part yellow and part white. The small pale spots on the head were yellow, with pale blue spots also sprinkled over the upper part of the head. Iris yellowish, lower lip pale.

Size. Grows to over 1100 mm TL. The largest of our study specimens was 963 mm; the illustrated specimen (not seen by us) was 1120 mm long (Blache, 1967d: 1152). Of four specimens examined for sex, two were females (815 and 888 mm), the larger one from St. Helena ripe with 1.1-mm eggs, and two were males (650 and 963 mm), neither ripe.

Development. No leptocephalus in the western Atlantic referable to the genus *Enchelycore* has been identified to date. Blache (1977: 53) identified one 30-mm larva from the Gulf of Guinea as this species.

Behavior. Virtually nothing is known of the life history or behavior of *Enchelycore anatina.* It is an inhabitant of deep waters according to available records, not found in shallow-water or reef habitats. Melliss reported it was taken for food at St. Helena (1875: 111), but he said the same of other seldom-captured species.

Distribution (Fig. 136). An uncommon but apparently widely-distributed species. Known from single specimens from Bermuda, Florida, Rio de Janeiro and St. Helena in the western Atlantic, from Madeira and the islands of the Azores, the Canaries and Cape Verde in the eastern Atlantic, plus one specimen recently captured in the eastern Mediterranean (Ben-Tuvia and Golani, 1984). The few records of capture depth vary: the juvenile from off Florida is from 369–380 m; the Mediterranean adult was captured at 50 m; the Amsterdam aquarium specimens from the Canary Islands were obtained in "quite deep water"; Blache (1967a: 1158, quoting from Maul, in litt.) gave a depth of about 10 m for Madeira specimens.

Etymology. From the Latin *anatinus* (pertaining to a duck), called "The Ducks'bill *Muraena*" by the author, probably in reference to its jaw shape. An adjective.

Remarks. Until recently, the species *anatina* was known from very few specimens from the eastern Atlantic. *Gymnothorax sanctaehelenae* was considered to be a synonym of *anatina* or a valid mid- and western Atlantic species. Three of our western Atlantic study specimens (from Bermuda, Rio de Janeiro, and St. Helena), originally identified as *sanctaehelenae,* had the characters and vertebral counts of *anatina,* and a small juvenile was captured off the Florida coast, thus extending the range of *anatina* to include both eastern and western Atlantic. Some specimens identified as *sanctaehelenae* (those listed under *Misidentifications,* plus others in collections not published upon) were misidentified large specimens of *Gymnothorax moringa.* The identification of a moray taken in the eastern Mediterranean as *E. anatina* considerably extends the range for this species. We note that the Pacific species *Enchelycore lichenosa* is very similar to *E. anatina;* based on study of limited material of both species, the two are separable only by differences in the number of vertebrae (136–149 for *lichenosa,* vs 151–155 for *anatina*). *Enchelycore lichenosa* is similarly known from few specimens from widely separated areas in the Pacific (Japan and the Galapagos Islands, cf. McCosker and Rosenblatt, 1975).

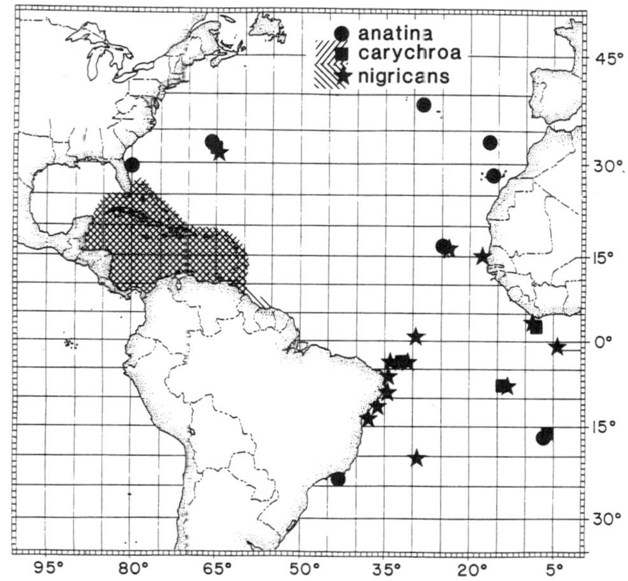

FIGURE 136. Distribution of Atlantic species of *Enchelycore* based on material examined and records in Blache, 1967a and d, and in Bauchot, 1986.

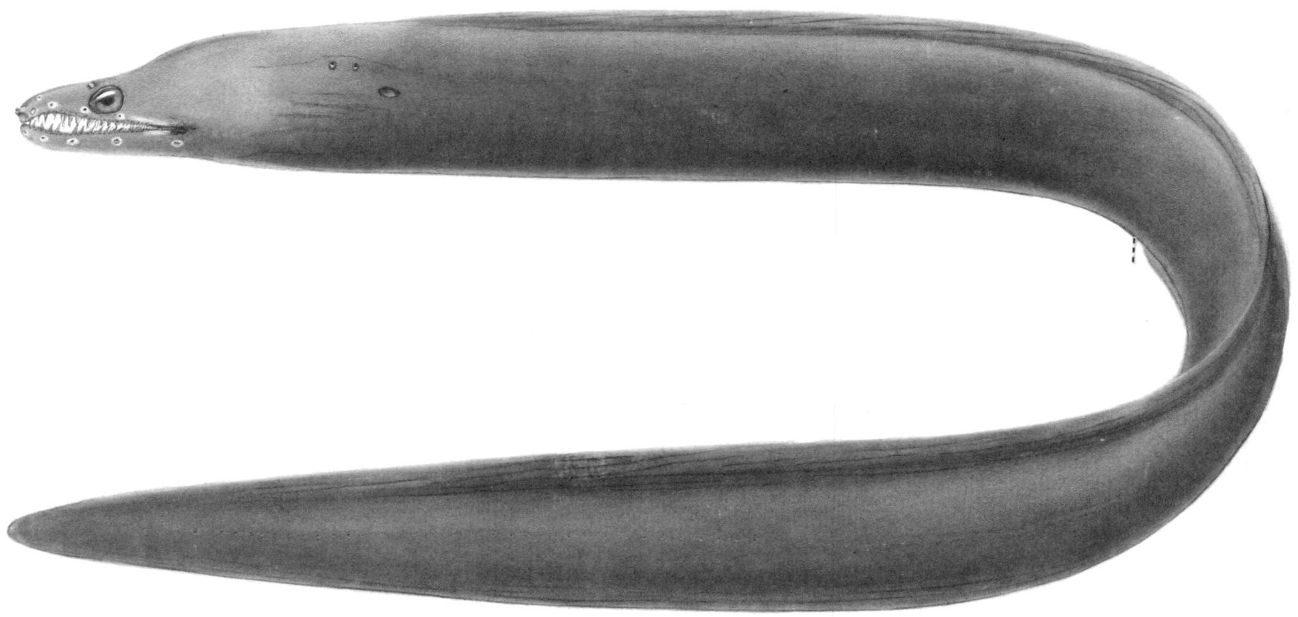

FIGURE 137. *Enchelycore carychroa*: ANSP 111570, paratype, 206 mm TL, Bahamas. Illustration by S. P. Gigliotti.

Study Material. A total of nine specimens, 75–963 mm TL, including the neotype and the holotype of *Muraena sanctaehelenae* Günther.

Neotype: BMNH 1869.3.7.1 (718); Madeira; Moore.

Other material: BERMUDA: BMNH 1872.8.28.12 (1, 963): J. H. Jones. FLORIDA ATLANTIC: ANSP 102472 (1, 75); 29°23′N, 80°05′W; 369–380 m; 1 May 1960. BRAZIL: Rio de Janeiro: MCZ 9068 (1, 815): Thayer Exped., L. Agassiz; 1865. ST. HELENA: BMNH 1868.3.11.2 (1, 785); holotype of *Muraena sanctae helenae*; J. Mellis. BMNH 1910.9.9.2 (1, 888); J. J. Cunningham. MADEIRA: BMNH 1899.1.16.33 (1, 650); A. J. Arendrup. CANARY ISLANDS: NMW 4913 (1); Las Palmas; Simony; 1890. NMW uncat. (1); Teneriffe; Steindachner; 1865.

Enchelycore carychroa Böhlke and Böhlke, 1976

CHESTNUT MORAY

Figures 136–139, Tables 9, 13

Rabula longicauda (not of Peters, 1876). Herre, 1942: 290 (English Harbor, Antigua).

Enchelycore sp. Böhlke and Chaplin, 1968: 81 ("... from the Bahamas south through the Lesser Antilles, and from Grand Cayman Island"). Starck, 1968: 17 (Alligator Reef, Florida). Birdsong and Emery, 1968: 192 (Albuquerque and Courtown Cays off Nicaragua). Barriga-Bonilla et al., 1969: 80 (San Andres). Palacio, 1974: 16 (Colombia). Bright and Cashman, 1974: 346, fig. 6 (West Flower Garden Bank, Gulf of Mexico). Ogden, Yntema, and Clavijo, 1975: 4 (St. Croix, U.S. Virgin Islands).

Enchelycore carychroa Böhlke and Böhlke, 1976: 138 (original description, Bahama Islands, holotype ANSP 100000).

Distinctive Characters. A small dark brown moray with head pores set in conspicuous white spots and with dark grooves in branchial region; tail 56–62% TL; posterior nostril round or oval, above front margin of eye; lower jaw teeth biserial, inner series complete; MVF 9-48-135.

Description. Total vertebrae 128–140, $\bar{x}$ = 134.6 (n = 322); predorsal vertebrae 7–11, $\bar{x}$ = 8.6 (297); preanal vertebrae 45–51, $\bar{x}$ = 47.7 (321); kidney terminus at 57–58 (2); dorsal-fin rays 434–444, $\bar{x}$ = 437.6 (2); anal-fin rays 305–347, $\bar{x}$ = 322.2 (3). Pores: SO 1 + 2; IO 4; M 6; B 2.

Proportions as % (n = 27, 57–315 mm TL) of TL: predorsal 11–15, $\bar{x}$ = 13; preanal 38–44, $\bar{x}$ = 41; head 11–14, $\bar{x}$ = 13; depth at gill opening 4.4–6.5, $\bar{x}$ = 5.2; depth at anus 3.6–6.8, $\bar{x}$ = 4.8. Of HL; snout 16–21, $\bar{x}$ = 18; upper jaw 37–50, $\bar{x}$ = 45; lower jaw 38–48, $\bar{x}$ = 43; eye 8.4–13, $\bar{x}$ = 10; interorbit 9.5–14, $\bar{x}$ = 11; posterior nostril length 1.6–2.7, $\bar{x}$ = 2.0.

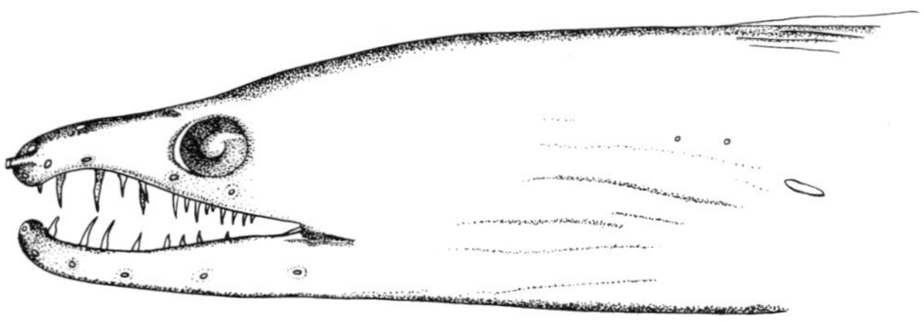

FIGURE 138. *Enchelycore carychroa*: ANSP 128774, paratype, 259 mm TL, Bahamas; head.

General characters similar to those of *Enchelycore anatina.* Posterior nostril round to oval, with slightly raised margin, situated above anterior margin of eye, not elongating or moving anteriorly with age. Head pores as listed; variation in 25 specimens included one specimen with one additional infraorbital pore behind eye on one side, one specimen with five mandibular pores on both sides (the illustrated specimen, Fig. 138), and one with three branchial pores on one side and two on the other. Several rows of tiny papillae present along edges of gape.

An outer irregular ring of 11–17 intermaxillary teeth enclosing 3–5 very long canines along midline. Vomerine tooth patch short, beginning below middle to posterior portion of eye and ending before posterior ends of maxillary tooth patches, the teeth small and irregularly biserial. Maxillary teeth biserial, rows about equal in length; outer row of 25–38 small, close-set, laterally-compressed teeth, an occasional one large and fang-like, especially noticeable in large individuals; the inner row of 4–7 long, widely-spaced, depressible canines. Dentary teeth in two rows; outer row of 40–58 small close-set teeth, 1–2 anteriormost and about every fourth tooth larger than others, except posteriorly where they are more uniform; 6–13 long, widely-spaced, depressible canines in inner row.

Coloration. Head and body dark brown, darkening posteriorly; corner of mouth, grooves in branchial region, margin of eye and longitudinal lines on dorsal fin blackish. Head pores set in prominent white spots. Posterior nostrils with white edge on small individuals, with black edge on large specimens. Pale areas on lip edges, sometimes extending to include several pores on upper and lower lips. The single eastern Atlantic specimen from West Africa does not have the head pores set in white spots, but those from Ascension Island do.

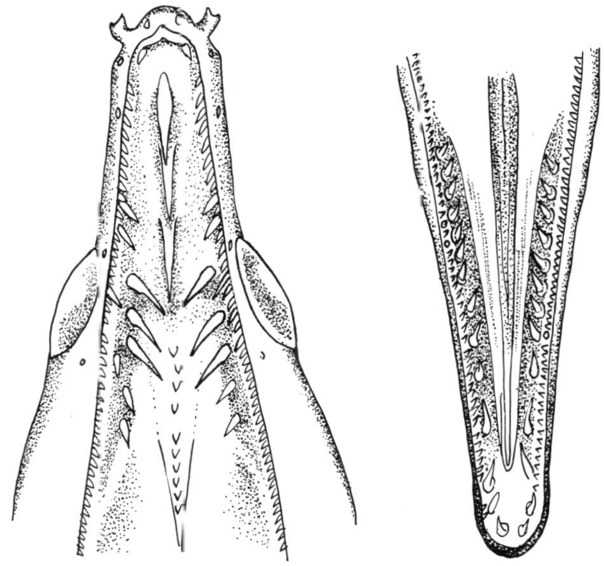

FIGURE 139. *Enchelycore carychroa*: data as for Figure 138; dentition.

Size. The smallest species of *Enchelycore,* known only to 335 mm TL. This species was well-known and abundant in collections for many years before it was described, because all specimens were small and thought to be juveniles. When mature individuals were found of both sexes (23 females, 169–296 mm, filled with eggs to 0.8 mm, and eight ripe males, 195–325 mm), it became apparent that it was simply a small species.

Development. Leptocephalus not known for this or for any *Enchelycore* species in the western Atlantic.

Behavior. Enchelycore carychroa is a common inhabitant of coral reefs and rocks, and is frequently taken in poison stations throughout the tropical western and mid-Atlantic in shallow to moderate depths.

Distribution (Fig. 136). A common insular species. Known from Bermuda, the Bahamas, Florida Keys, and the Flower Garden Bank in the northern Gulf of Mexico; throughout the islands of the West Indies and off northern South America south to Brazil; from Ascension Island, St. Helena, and possibly the coast of West Africa (a

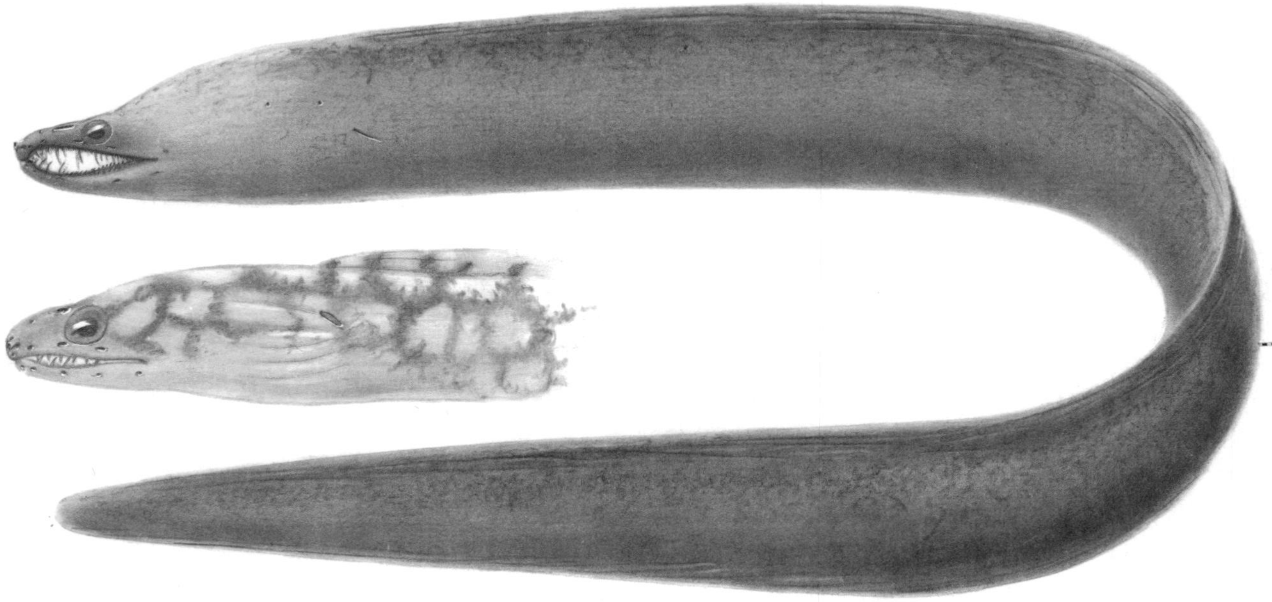

FIGURE 140. *Enchelycore nigricans*: ANSP 100802, 762 mm TL, Bahamas. Insert: head of juvenile. Illustrations by S. P. Gigliotti.

single mutilated specimen). Of 110 collections of *carychroa* for which depth has been recorded, most have been taken between 1.5 and 15 m. It has been taken with rotenone-based poisons from shoreline to 54 m, and twice with a tumbler dredge, at 47 and 63 m.

Etymology. From the Greek *caryon* (nut) and *chroa* (color), referring to the coloration. An adjective.

Study Material. A total of 719 specimens, 40–335 mm TL, including the holotype and 590 paratypes, plus those listed below.

Holotype: ANSP 100000 (213); Bahamas, Great Bahama Bank, New Providence Island, patch reef offshore and just W of Delaporte Point, depth to 12 ft; C. C. G. Chaplin, J. E. Böhlke, sta. 212; 11 Apr. 1955.

Paratypes: 590 specimens, 40–335 mm TL, from Bermuda, the Bahamas, Florida Keys, and throughout the Caribbean islands and the islands off Central and South America, as listed in Böhlke and Böhlke, 1976: 142–143.

Other material: BERMUDA: ANSP 148106 (1, 225). BAHAMAS: ANSP 156823 (1, 76). FLORIDA KEYS: UMML 4245 (1, 267), 719 (1, 235). PUERTO RICO: ANSP 144114 (1, 98). MNHN 1974-130 (1, 136). UPR 2327 (1, 135), 2401 (3, 177–260). ANGUILLA: USNM 183563 (1, 195). BARBADOS: UPR 1840 (5, 66–275). VENEZUELA: MBUCV 1194 (1), 1334 (1). MBUCV 2197 (1). MNHN 1974-106 (4, 48–178). USNM 194039 (2, 135–327). CURAÇAO: MNHN 1974-129 (1, 149). UPR 1577 (8, 105–292), 2576 (16, 46–272). PANAMA: ANSP 113762 (25, 54–316). CAS 13490 (4, 135–281). SERRANA BANK: FMNH 90714 (1, 122). BELIZE: FMNH 77600 (1, 273), 77871 (1, 193). USNM 278167 (4, 57–168), 278168 (1, 104), 278177 (3, 88–194), 178178 (9, 64–157). FERNANDO NORONHA: USNM 274259 (1, 146). ASCENSION ISLAND: GMBL 78-18 (1, 76), 80-20 (3, 145–247), 80-26 (5, 99–ca 325), 80-28 (7, 112–261), 80-50 (2, 169–170), 80-54 (1, 122), 80-57 (6, 70–191), 80-68 (1, 91), 80-71 (1, 96). ST. HELENA: USNM 267882 (1, 287). GULF OF GUINEA, FERNANDO POO: UMML 32872 (1, 211).

Enchelycore nigricans (Bonnaterre, 1788)

VIPER MORAY

Figures 136, 140–142, Tables 9, 13

Muraena unicolor: maxillis elongatis teretiusculis, inferiore longiore: pinnis pectoralibus carens Gronovius, 1763: 39 (original description, "Americam Meridionalem," holotype unknown).

Muraena nigricans Bonnaterre, 1788: 34 (description after Gronow, 1763: 39 [species no. 163]).

Muraenophis nigricans. Lacepède, 1803: 628, 641.

Muraena anguina Gronow in Gray, 1854: 18 (after Gronow, 1763).

Enchelycore euryrhina Kaup, 1856a: 60 (original description, locality unknown, holotype MNHN 2362). Kaup, 1856b: 73.

Enchelycore nigricans. Günther, 1870: 135. Blache, 1967a: 163–177 (eastern Atlantic; redescribed and illustrated).

Gymnothorax nigrocastaneus Cope, 1871: 483 (original de-

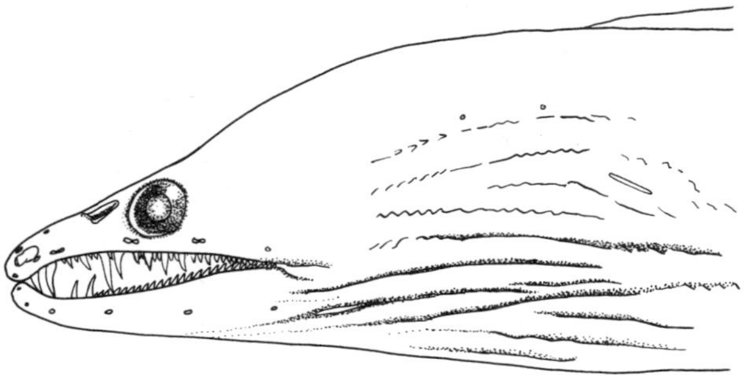

FIGURE 141. *Enchelycore nigricans*: ANSP 108739, ca 560 mm TL, Grenadines; head.

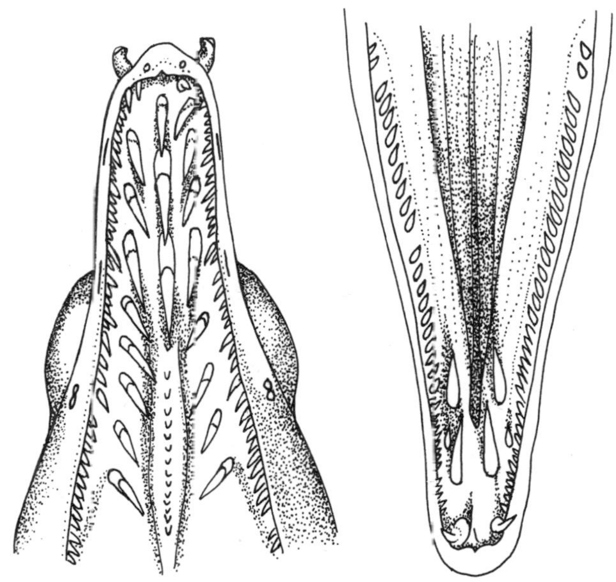

FIGURE 142. *Enchelycore nigricans*: ANSP 100801, 412 mm TL, Bahamas; dentition.

scription, West Indies, St. Martin's, holotype ANSP 16032).

Gymnothorax umbrosus Poey, 1876b: 67, pl. 9, figs. 1–2 (original description, Cuba, holotype USNM 13062).

Sidera nigrocastanea. Jordan, 1887b: 565.

Sidera umbrosa. Jordan, 1887b: 565.

Gymnothorax brunneus Nichols, 1920: 59 (original description, Bermuda, holotype AMNH 7309).

Enchelycore nigrocastaneus. Jordan, Evermann, and Clark, 1930: 90.

Enchelycore brunneus. Beebe and Tee-Van, 1933a: 137. Beebe and Tee-Van, 1933b: 47, 278.

Distinctive Characters. A large moray with variable coloration of pale reticulations in young to uniform dark brown or slightly mottled pattern in adults; jaw pores not set in white spots; tail 51–55% TL; posterior nostril of adults elongated well in front of eye; lower jaw teeth biserial anteriorly only, 3–5 inner teeth; MVF 8-60-144.

Description. Total vertebrae 141–148, $\bar{x}$ = 144 (n = 74); predorsal vertebrae 6–10, $\bar{x}$ = 7.9 (75); preanal vertebrae 58–63, $\bar{x}$ = 60 (76); prenephric vertebrae 68–70 (3); dorsal-fin rays 417–465, $\bar{x}$ = 441 (3); anal-fin rays 267–293, $\bar{x}$ = 277 (3). Pores: SO 1 + 2; IO 4; M 6; B 2.

Proportions expressed as % (n = 19, 132–870 mm). Of TL: predorsal 10–13, $\bar{x}$ = 11; preanal 45–50, $\bar{x}$ = 48; head 11–14, $\bar{x}$ = 12; depth at gill opening 3.8–6.3, $\bar{x}$ = 5.2; depth at anus 3.6–5.8, $\bar{x}$ = 4.7. Of HL: snout 15–21, $\bar{x}$ = 18; upper jaw 37–52, $\bar{x}$ = 46; lower jaw 35–51, $\bar{x}$ = 45; eye 7.7–12, $\bar{x}$ = 9.6; interorbit 8.0–13, $\bar{x}$ = 10; length posterior nostril 4.5–7.8, $\bar{x}$ = 5.6.

General characters similar to those of *Enchelycore anatina.* Posterior nostril a small, round or oval opening above eye in young, becoming greatly elongate and displaced forward in front of eye in adults; with raised and convoluted margins in largest specimens. Head pores as listed; infraorbital pores and, to a lesser extent, mandibular pores become elongate and have convoluted margins in large adults. Considerable variation in infraorbital pores, large specimens with added pore along upper lip behind eye; in 110 specimens less than 230-mm TL, seven had the additional pore; of 65 specimens 230–735 mm, 53 have the added pore, and 13 have an additional postorbital pore behind mid-eye. In the latter sample, five fish have seven mandibular pores (three on one side only) and three have three branchial pores. Very large specimens develop still more pores: A 715-mm individual has seven infraorbital pores and eight mandibular pores on the left side only; an 820-mm individual has four supraorbital pores, 11 infraorbital pores and seven mandibular pores.

Dentition strong, menacing in appearance. On each side of upper jaw, an outer ring of 13–19 intermaxillary teeth, some of the teeth enlarged, with smaller teeth between them. Immediately inside the outer ring, a row of long depressible canines, increasing in length from front to back (in young, part of the outer row rather than distinct from it). Three inner rows of 3–4 long depressible fangs, increasing in size posteriorly, the median row with the longest teeth. Vomerine teeth small; a single row of 8–16 teeth beginning below mid-eye, the teeth hidden in folds of flesh in large specimens. Maxillary teeth in two series; an outer row of 18–25 small recurved teeth, and an inner row of 2–5 long, slender depressible fangs. Dentary teeth in two series; an outer row of 17–40 small teeth, and an anterior inner series of 3–5 large canines.

Color. Young individuals always with contrasting pattern of pale background with brownish bands and streaks demarcating polygonal-shaped blocks (sometimes the pale areas appearing as rosettes, the pattern similar to that of *Uropterygius macularius*); dark areas between eyes and a dark nuchal band; lower jaw pale, a dark slash at mouth angle; fins pale (Fig. 140, inset). Fish darkens with growth, large specimens becoming uniformly dark brown (occasional individuals slaty) overall, or dark but retaining, often faintly, the still darker marblings of young; fins darken with the body color and no longer have pale edges. Adult coloration apparently not correlated with sex or state of maturity.

Size. Largest specimen examined was a male, ca 870 mm SL. Blache (1967a: 173–176) recorded a male 930 mm in length from Cape Verde Islands in the eastern Atlantic. Of 110 individuals longer than 160 mm, 39 (177–780) were females, only three (548–707 mm) with 1.0-mm eggs. All three were taken in June, one from Grenada and two from the Florida Keys; the 707-mm fish contained a volumetrically-estimated 50,000 eggs. None of the 26 males (250–870 mm) were ripe.

Development. Leptocephalus not identified in the western Atlantic, although ripe females have been taken (see above). Blache (1977: 58) listed two leptocephali, 17–53 mm TL, from the Gulf of Guinea, the identification considered questionable by Smith (see p. 900).

Behavior. Enchelycore nigricans is one of the most common inhabitants of coral reefs and rocky shorelines, and numerous specimens of all sizes, from juveniles to adults, are taken at poison stations. It is a secretive species, seldom seen by divers.

Distribution (Fig. 136). A wide-ranging species present throughout both sides of the tropical and subtropical Atlantic. In the western Atlantic, from Bermuda, the Bahamas, the Florida Keys, the off-shore reefs of western and southern Gulf of Mexico, throughout the West Indies, from coral reef areas and off-shore islands of Central and South America, and taken at the mid-Atlantic islands. In the eastern Atlantic, reported from Senegal to Gabon and from the islands of Cape Verde and the Bay of Biafra (Blache, 1967a). Depth of capture recorded from shoreline to 24 m.

Etymology. From the Latin *nigricans* (swarthy or blackish), referring to the dark color phase. An adjective.

Study Material. A total of 450 specimens, 50–870 mm TL, including the holotypes of *Enchelycore euryrhina* Kaup, *Gymnothorax nigrocastaneus* Cope, *Gymnothorax umbrosus* Poey, and *Gymnothorax brunneus* Nichols.

BERMUDA: AMNH 7309 (1, 343), holotype of *Gymnothorax brunneus.* ANSP 126264 (1, 675), 133814 (2, 274–526), 133833 (1, 206), 148152 (1, 440), 148155 (1, 156). FMNH 48149 (2, 213–225), 48150 (1, 344). FLORIDA KEYS: ANSP 89350 (1, 352). UMML 3268 (4, 451–707), 3634 (1, 660), 4366 (1, 145), 4752 (1), 18037 (1, 730), 19217 (2, 186–783). BAHAMAS: ANSP 72296 (3, 55–276), 72297 (2, 76–100), 72298 (1, 93), 100781 (2, 168–232), 100782 (3, 110–355), 100783 (1, 236), 100784 (3, 69–93), 100785 (1, 371), 100786 (1, 68), 100787 (1, 164), 100789 (2, 124–156), 100790 (1, 265), 100791 (2, 51–54), 100792 (1, 172), 100793 (1, 195), 100795 (3, 85–370), 100796 (3, 156–365), 100797 (1, 72), 100798 (6, 78–191), 100800 (2, 107–182), 100801 (1, 412), 100802 (1, 762), 100803 (1, 549), 100804 (1, 727), 100805 (3, 458–607), 100806 (5, 103–746), 100807 (2, 337–601), 100809 (1, 212), 106147 (3, 102–114), 111569 (1, 206), 115107 (1, 104), 115501 (1, 119, cleared and stained), 140463 (2, 228–404), 148991 (1, 670). UF 8942 (1). UMML 798 (1), 28399 (1, ca 870), 31976 (4, 193–616).

CUBA: ANSP 91893 (1, ca 580). CAS-SU 54192 (1, 150). CAS-SU 54209 (13, 141–305). MCZ 32793 (1, 521). USNM 13062 (1, 670), holotype of *Gymnothorax umbrosus.* JAMAICA: FMNH 73890 (3, 101–218). UMML 13231 (3, 170–342). HAITI: ANSP 117928 (1, 568), 117929 (1, 665), 117938 (1, 265), 127292 (1, 521). FMNH 74710 (1, 180), 74720 (1, 214). PUERTO RICO: ANSP 115548 (3, 175–290), 116142 (2, 210–225, cleared and stained), 122691 (4, 297–ca 645), 129935 (2, 199–573), 139745 (19, 35–249), 139829 (1, 280), 139830 (12, 53–384), 139832 (3, 485–575), 139841 (5, 64–164), 143069 (1, 192). FMNH 74707 (1, 112), 74713 (2, 207–217). UMML 1333 (12, 123–580). USNM 274261 (2, 182–196). VIRGIN ISLANDS: ANSP 89355 (1, 83). UMML 3737 (1, 244). USNM 120481 (1, 92).

LESSER ANTILLES: *Anguilla Island:* ANSP 105774 (2, 183–214). *St. Martins:* ANSP 16032 (1, 615), holotype of

Gymnothorax nigrocastaneus. UMML 5592 (1, 161). *St. Barthelemy:* ANSP 106251 (1, 430). *Antigua:* ANSP 105780 (1, 170). *Dominica:* ANSP 106252 (9, 65–443). USNM 274260 (1, 154), 274262 (4, 94–345). *Martinique:* ANSP 105490 (1, 263). *St. Lucia:* ANSP 105726 (1, 165), 105736 (5, 124–318), 136520 (5, 55–102). CAS-IUM 8847 (1, 232). *St. Vincent:* ANSP 105733 (1, 148), 106057 (1, 259). *Grenadines:* ANSP 105479 (1, 237), 105739 (1, 158), 105785 (1, 318), 105786 (1, 226), 108738 (1, 675), 108739 (1, ca 560), 108740 (4, 250–548). UMML uncat. (8, 260–629). *Grenada:* ANSP 105487 (2, 224–279), 105502 (5, 88–277), 108736 (1, 548). *Barbados:* ANSP 89357 (8, 107–463), 122095 (1, 309). *Tobago:* ANSP 97948 (1, 152), 97981 (1, 264). *Curaçao:* CAS uncat (2, 693–755).

SERRANA BANK: ANSP 105587 (12, 97–715), 115505 (1, 410, cleared and stained), 130215 (6, 131–195). FMNH 74729 (10, 139–301), 74730 (26, 70–317). SWAN ISLAND: UMML uncat. (3, 82–232). GRAND CAYMAN ISLAND: ANSP 104919 (1, 461), 104943 (1, 470), 104968 (6, 50–175), 105001 (1, 139), 105007 (1, 71), 105818 (1, 590), 112603 (1, 86, cleared and stained). BELIZE: USNM 278166 (1, 270). NICARAGUA: UMML 24088 (2, 101–128). PANAMA: ANSP 113761 (10, 136–476), 113771 (3, 114–371). FMNH 90718 (1, 130). COLOMBIA: ANSP 117566 (6, 160–302), 117567 (1, 308), 143079 (3, 169–252). UMML uncat. (9, 90–410). VENEZUELA: ANSP 78736 (1, 412). MNHN 1965-583 (1). MBUCV 169 (1), 349 (1), 1066 (1), 1257 (1), 3150 (1), 3206 (1). UMML 14070 (1, 481). SURINAME: ZMB 5513 (1, 332). BRAZIL: CAS 59198 (1, ca 612). CAS 59199 (1, ca 626). MHNG 665-41 (1). MZUSP 11820 (1), 11821 (1), 11822 (2), 11823 (1), 11824 (1). UF 19933 (1).

MID-ATLANTIC ISLANDS: *St. Paul's Rock:* BMNH 1879.5.14.429 (1, ca 825). UF 44623 (11, 169–635), 44626 (7, 56–550). *Atol das Rocas:* ANSP 123288 (1, 304). UMML uncat. (1, ca 715). UF 44620 (1, 65). *Ilha da Trindade:* MNRJ 2148 (2). MNRJ 2321 (1). *Fernando de Noronha:* USNM 274279 (1, 82), 278175 (8, 147–323). *Ascension Island:* GCRL 15496 (1), 15498 (5), 15500 (1). GMBL 78-17 (1, ca 650), 80-26 (1, ca 550), 80-28 (1, ca 550), 80-32 (1, 185), 80-43 (1, 230), 80-49 (1, 107). USNM 274264 (1, 127). USNM 274265 (1, 78).

EASTERN ATLANTIC: *Gulf of Guinea:* ANSP 158109 (4, 133–320). MNHN 1965-585 (1).

LOCALITY UNKNOWN: MNHN B.2362 (1, 604), holotype of *Enchelycore euryrhina.* MNHN B.2363 (1, 582), paratype of *Enchelycore euryrhina.* MNHN B.2501 (1, 320), paratype? of *E. euryrhina* [Kaup mentioned only two examples].

Genus *Gymnothorax* Bloch, 1795

Gymnothorax Bloch, 1795: 83 (type species *Gymnothorax reticularis* Bloch, 1795, under suspension of the Rules by ICZN Opinion 93).

Lycodontis McClelland, 1844: 154, 173 (type species *Lycodontis literata* McClelland, 1844 = *Muraenophis tile* Hamilton-Buchanan, 1822, by Jordan and Evermann, 1896, as first revisers).

Thaerodontis McClelland, 1844: 154, 174 (type species *Thaerodontis reticulata* McClelland, 1844 = *Muraena tesselata* Richardson, 1845, by monotypy).

Polyuranodon Kaup, 1856a: 65 (type species *Polyuranodon kuhlii* Kaup, 1856 = *Muraena polyuranodon* Bleeker, 1853, by monotypy). Kaup, 1856b: 96.

Neomuraena Girard, 1859a: 76 (type species *Neomuraena nigromarginata* Girard, 1859, by monotypy).

Priodonophis Kaup, 1860: 22 (type species *Gymnothorax ocellatus* Agassiz, 1828, by original designation and by monotypy). Also spelled *Priodontophis* by other authors.

Taeniophis Kaup, 1860, Supplement: 1 (type species *Taeniophis westphali* Kaup, 1860 = *Gymnothorax funebris* Ranzani, 1839, by Jordan and Davis, 1891, as first revisors).

Pseudomuraena Johnson, 1862: 167 (type species *Pseudomuraena maderensis* Johnson, 1862, by monotypy).

Rabula Jordan and Davis, 1891: 590 (type species *Muraena aquae-dulcis* Cope, 1872 = *Muraena mordax* Ayres, 1859, by original designation). Described as a subgenus of *Gymnothorax* Bloch, 1795.

Ahynnodontophis Fowler, 1912a: 25 (type species *Gymnothorax stigmanotus* Fowler, 1912 = *Siderea castanea* Jordan and Gilbert, 1882, by original designation). Described as a subgenus of *Gymnothorax* Bloch, 1795.

Verdithorax Whitley, 1931: 311 (type species *Muraena prasina* Richardson, 1848, by original designation).

Notorabula Whitley, 1934: 154 (type species *Muraena callorhyncha* Günther, 1870, by original designation).

Serranguilla Whitley and Phillips, 1939: 228 (type species *Gymnothorax prionodon* Ogilvy, 1895, by monotypy).

Chasmenchelys Fowler, 1944: 270 (type species *Muraena panamensis* Steindachner, 1876, by monotypy).

Characters of Gymnothorax, *sensu lato.* Body elongate, laterally compressed posteriorly, tail tapering; head and trunk shorter than or subequal to tail (39–54% TL in Atlantic species). Dorsal-fin origin on head, in advance of gill opening. Snout and jaws developed, subequal and closing completely; snout subconical. Eye moderate, its center about above midpoint of upper jaw. Anterior nostril tubular; posterior nostril nontubular or in a short tube, usually above anterior margin of eye. Gill opening round or slit-like, at midflank. Head pores developed; three supraorbital, three or four infraorbital, six mandibular and usually two (a few species with one or 3–4) branchial pores. Teeth strong, conical, pointed and smooth, or wedge-shaped and serrated along edge; jaw teeth of adults uniserial to multiserial, variably developed within subgenera.

Gill arches reduced, like those of *Muraena*, limited to ceratobranchials 1–4 and the upper and lower pharyngeal tooth plates. Peritoneum pale. Coloration variably uniform, mottled, spotted and/or banded.

Size. Moderate to large eels; in the Atlantic, recorded to about two m (although one species, *Gymnothorax hubbsi*, attains a maximum size of only 336 mm TL).

Distribution. Circumtropical, common in all oceans, entering temperate waters. Most of the common species are found in shallow waters where they are inhabitants of coral reefs and rocky areas; others live offshore along continental slopes at greater depths, and still others are known from one or a few specimens obtained at depths to 600 m.

Etymology. From the Greek *gymnos* (naked) and *thorax* (breast) in reference to the absence of pectoral fins. Masculine.

Remarks. We provisionally recognize *Gymnothorax* in its broadest sense, until a worldwide analysis is completed. The large assemblage of species in *Gymnothorax*, sensu lato, have been assigned to various generic groups, based in large part on differences in dentition, such as the presence or absence of serrations on the teeth. Some of these differences in dentition may well indicate generic distinction, and some species possess other differences probably deserving of generic rank. On the other hand, some genera which we continue to recognize as valid are based on minor character distinctions (*Muraena* with tubular posterior nostrils, *Siderea* lacking strong caniniform teeth) and may well also belong within *Gymnothorax*, sensu lato. In order not to further complicate the nomenclatorial history of these closely-related genera, we inconsistently, but intentionally, retain those genera which possess distinctive characters and which have long been recognized, and lump together in *Gymnothorax* those with inconsistent and controversial histories until a world-wide study can be accomplished.

Although we recognize *Gymnothorax* in its broadest sense, we can provisionally assign the Atlantic species to subgenera, and we have done so in the key to species. For many years authors have vacillated in placing species in one inclusive genus (*Gymnothorax*), or in separating those with smooth-edged teeth (*Lycodontis*) from those with serrate teeth (*Gymnothorax* or its synonym *Priodonophis*). Both names have had a long and confusing history. *Gymnothorax* was first used by Bloch for his species *reticularis* and *afer*; it was later applied to various generic and family groups, until 1926 when the ICZN placed it on the Official List of Generic Names (Opinion 93, under the Plenary Power for Suspension of the Rules). The opinion (apparently written by D. S. Jordan) is quoted herewith in its entirety:

> As originally given, *Gymnothorax* was simply a substitute name for *Muraena* L. Later, in dividing this extensive genus, Bleeker and after him Günther used the name *Gymnothorax* for one of its great divisions, and this arrangement has been largely followed. The first fixation of type may be held to separate *Gymnothorax* from *Muraena*, and I think that the use of the former name, should be preferred to the later *Lycodontis* McClelland based on one of the species of *Gymnothorax*. The case for the use of *Gymnothorax* is stated in Jordan, Genera of Fishes p. 168, that for its suppression on p. 53.

Recent authors have referred numerous Atlantic species with smooth-edged teeth to *Lycodontis*; however, its generic history requires explanation. McClelland (1844) first described *Lycodontis* on pages 154, 158, 173 and 185–187. Later, within the body of the same paper (p. 202), he rejected the name in an "errata" which stated "For *Lycodontis*, pp. 154 . . . and where ever it occurs in the preceding pages, read *Strophidon*." Technically, therefore, *Lycodontis* was both created and synonymized within the same paper. In the general synopsis that followed (p. 214), he included the latter name within the five genera he listed in the Muraenidae: *Dalophis, Muraena, Strophidon, Thaerodontis* and *Gymnomuraena*.

Most authors apparently either ignored or were unaware of McClelland's rejection of *Lycodontis*. Bleeker (1864a), as first reviser, made *brummeri* the type species of *Strophidon,* and referred other species, including most of McClelland's *Strophidon,* to *Gymnothorax,* selecting *reticularis* as its type. Later authors used *Lycodontis* for species with smooth-edged teeth, based on its type species (*Lycodontis literata* = *Muraenophis tile*), and *Gymnothorax* for those with serrate teeth. Whether there are two genera, separated only by the tooth condition, has long been a subject of controversy and inconsistency. Like many authors before and since, Blache (1967d: 1122) avoided a definitive analysis of the *Gymnothorax*/*Lycodontis* quagmire in stating that "Nous ne prétendons pas . . . donner une révision du genre *Lycodontis,* révision qui s'avère pourtant indispensable, mais qui ne peut être effectuée, avec fruit, qu'à l'échelle mondiale," an opinion which we share.

Species. It is difficult to even estimate the total number of valid species within *Gymnothorax,* sensu lato. Well over 200 nominal species have been described, the majority of them present in the Indo-Pacific; to date, no comprehensive study and comparison with type specimens has been undertaken, except the present study which has been largely limited to the species in the Atlantic. We assign western Atlantic species to four subgenera, with one species (known only from three specimens), which shows affinities mainly with Indo-Pacific species, not assigned. They are separable by differences in morphology, osteological condition, dentition, pores, habits and habitat (Table 14). The subgeneric names used here are tentative and subject to change, pending study and comparison with Indo-Pacific species.

Included in the subgenera *Taeniophis* and *Lycodontis* are many of the well-known and often-observed reef species present throughout the Atlantic and Indo-Pacific. They live in shallow-water coral reef or rock habitats, and are usually collected by fish toxicants. They have smooth teeth, and have been placed in the genus *Lycodontis* by many authors. Species in *Taeniophis* are elongate eels with elevated, muscular heads, long snouts, tapering tails, uniserial jaw teeth (although sometimes a short inner anterior row in young), including some long sharp "fangs." The species are very similar to those currently placed in the genus *Muraena,* which differ in having tubular posterior nostrils and a conspicuous dark blotch around the gill openings. Atlantic species of *Taeniophis* are *afer* (Kaup, 1856); *funebris* Ranzani, 1839; *moringa* (Cuvier, 1829); *polygonius* Poey, 1876; and *vicinus* (Castelnau, 1855) (*afer* known only from the eastern Atlantic). They are separable by coloration and by vertebral counts.

Species of *Lycodontis* are less elongate and have shorter snouts; the teeth are small, sharp, and basically biserial. They are related to species in the Indo-Pacific genus *Siderea* which have similar dentition, but a more posterior dorsal-fin origin. In the Atlantic, three species are present: *mareei* Poll, 1953; *miliaris* (Kaup, 1864); and *unicolor* (Delaroche, 1809); *mareei* and *unicolor* have not been taken in the western Atlantic. They are separable by coloration and vertebral counts.

Species of the subgenera *Gymnothorax* and *Neomuraena* are moderately elongate eels with muscular heads and long tapering tails; the teeth are large, serrate, wedge-shaped and uniserial. They inhabit offshore coastal slopes and deeper waters, and are most often captured by trawl. There are two species of *Gymnothorax* in the Atlantic: *hubbsi* Böhlke and Böhlke, 1976 (found only in the western Atlantic); and *maderensis* Johnson, 1862, separable by coloration and vertebral counts.

Species of *Neomuraena* differ from those of *Gymnothorax* (and all other Atlantic muraenid species) in having three infraorbital pores rather than the usual four. They appear to be limited to the New World. Atlantic species include: *conspersus* Poey, 1876; *kolpos* Böhlke and Böhlke, 1980; *nigromarginatus* (Girard, 1859); *ocellatus* Agassiz, 1831; and *saxicola* Jordan and Davis, 1891.

Gymnothorax bacalladoi Böhlke and Brito, 1987 differs considerably from all other Atlantic species in possessing biserial serrate jaw teeth; it shows closer affinities with several Indo-Pacific species. It also may be generically distinct in the reduction of the lateral-line pore system to a single branchial pore (previously noted only in the monotypic genus *Monopenchelys*).

A similarity to eastern Pacific species exists within certain subgenera and probably reflects

TABLE 14. Summary of characters for Atlantic subgenera and species of *Gymnothorax*.

	Coloration					Pores			
	Body	Head	Dorsal fin	Anal fin	Jaw teeth	IO	B	Habitat	Collection method
(*Gymnothorax*)									
hubbsi	pale reticulations	body color	pale edge	pale edge	serrated, uniserial	4	2	moderately deep	trawl
maderensis	dark reticulations	dusky	pale edge	pale edge				deep	trap, hook & line
(*Neomuraena*)									
conspersus	small white spots, dark tail	body color	white saddles	dark	serrated, uniserial	3	2	deep	trawl, trap, hook & line
kolpos	small white spots, dark tail	body color	dark edge	dark				deep	trawl
saxicola	honeycomb spots	reticulations	patterned	dark edge				moderately deep	trawl
nigromarginatus	small white spots	body color	dark edge	dark edge				moderately deep	trawl
ocellatus	white spots	body color	patterned	patterned				moderately deep	trawl
(*Lycodontis*)									
mareei	uniform dark	white pores	dark	dark	smooth, biserial	4	2	shallow reef	poison
miliaris	variable	body color	body color	body color				shallow reef	poison
unicolor	uniform dark	slightly pale	dark	dark				shallow reef	poison
(*Taeniophis*)									
afer	dark, pale blotches	body color	dark	dark	smooth, uniserial	4	3–4	shallow reef	poison
funebris	uniform dark	body color	dark	dark		4	2	shallow reef	poison
moringa	dark spots	body color	body color, dark edge	body color, dark edge				shallow reef	poison
polygonius	polygons	body color	body color	body color				deep	trap, hook & line
vicinus	diffuse dark spots	body color	dark edge	body color, dark edge				shallow reef	poison
bacalladoi	dark, pale blotches	body color	body color	dark	serrated, biserial	4	1	deep	trap

monophyly within those New World species groups. In particular, the subgenus *Taeniophis* includes the eastern Pacific *dovii* (Günther) and *castaneus* Jordan and Gilbert. The subgenus *Neomuraena* contains *angusticeps* (Hildebrand and Barton), *equatorialis* (Hildebrand), *serratidens* (Hildebrand and Barton), and one or more additional species (at least one valid species, perhaps one of those described by deBuen, holotypes unknown) in the eastern Pacific.

Gymnothorax stigmanotis Fowler, 1912, has been included in faunal lists for the western Atlantic since its description by Fowler (1912a: 26). The holotype (ANSP 16705), in the personal collection of E. D. Cope, was received by ANSP after his death. Fowler wrote "No data—this specimen was in a jar labelled 'Texas' and may have been secured somewhere in the West Indies." It is not identifiable with any western Atlantic species, but agrees in all characters with the eastern Pacific species *Gymnothorax castaneus* Jordan and Gilbert. We consider Fowler's assumed locality to be in error, and herein place the name in the synonymy of *G. castaneus.*

Key to the Atlantic Species of *Gymnothorax*

1a. Snout elongate and acute, narrowing sharply from the bulbous nape; jaws elongate; center of eye slightly before mid-jaw; teeth smooth, anterior teeth caniniform .... 2 (subgenus *Taeniophis*)

1b. Snout short and blunt, not narrowing sharply from nape; jaws short; center of eye above or slightly behind mid-jaw; teeth smooth or serrate, anterior teeth short and pointed or wedge-shaped 6

2a. Color dark brown to black, either uniform or with faint pale areas 3

2b. Bold color pattern of reticulations, spots or blotches 4

3a. Color uniform dark brown to black (green in life); vomerine teeth biserial .... *funebris,* p. 176

3b. Color dark, with faint irregular pale blotches; vomerine teeth uniserial .... *afer* (Kaup, 1856) (an eastern Atlantic species)

4a. Color pattern of large spotted polygonal areas separated by dark reticulations .... *polygonius,* p. 184

4b. Coloration spotted, mottled, or blotched 5

5a. Contrasting pattern of dark overlapping spots on pale background (gold in life); pale edges on fins of young, dark edges on large specimens .... *moringa,* p. 179

5b. Pattern of diffuse dark spots and blotches (purplish in life); pale edges on fins .. *vicinus,* p. 186

6a. Jaw teeth biserial, smooth, recurved and small; median intermaxillary teeth numerous .... 7 (subgenus *Lycodontis*)

6b. Teeth serrate and wedge-shaped; single row of 2–3 median intermaxillary teeth 9

7a. Highly variable color pattern of contrasting spots, blotches or reticulations .... *miliaris,* p. 155

7b. Body and fins uniform dark brown 8

8a. Body, fins and head uniformly dark brown; head pores in bright white spots .... *mareei* Poll, 1953 (an eastern Atlantic species)

8b. Body and fins dark brown, head darker; head pores not in white spots .... *unicolor* (Delaroche, 1809) (an eastern and mid-Atlantic species)

9a. Maxillary teeth biserial, the rows short .... *bacalladoi,* p. 189

9b. Jaw teeth uniserial 10

10a. Four infraorbital pores; eye small, 1½ or more in snout; 1–2 median intermaxillary teeth present (absent in largest specimens) .... 11 (subgenus *Gymnothorax*)

10b. Three infraorbital pores; eye large, 1¼ or less in snout; median intermaxillary teeth absent .... 12 (subgenus *Neomuraena*)

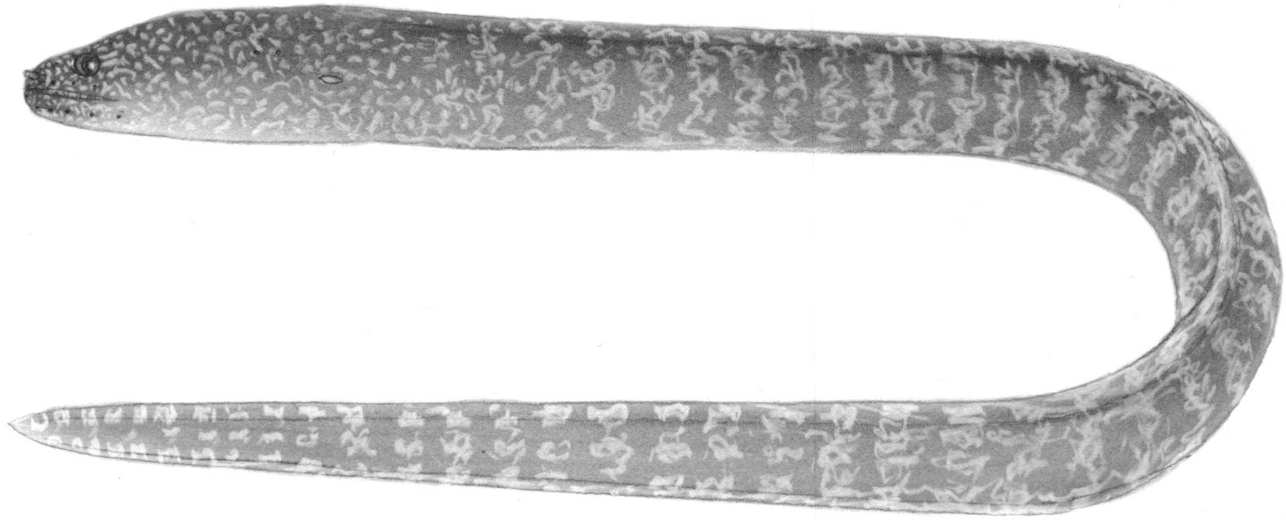

FIGURE 143. *Gymnothorax hubbsi*: USNM 214833, holotype, 273 mm TL, Florida Atlantic.

11a. Overall color pattern of pale dendritic marks on brown background on head, body, and fins; head colored like body . . . . . *hubbsi*, p. 150
11b. Color pattern of dark marks or reticulations on body and fins; head noticeably dusky . . . . . *maderensis*, p. 152
12a. Head and body spotted dorsally and ventrally; tail dark, with 3–4 large white spots 13
12b. Head and body spotted dorsally, pale ventrally; tail not noticeably dark 14
13a. Dorsal fin with body coloration basally, margin dark . . . . . *kolpos*, p. 164
13b. Dorsal fin with bold pattern of white saddle marks on black fin . . . . . *conspersus*, p. 161
14a. Dorsal and anal fins with dark edges, without prominent pattern; spots on head and body small and separate . . . . . *nigromarginatus*, p. 172
14b. Dorsal fin with black and white pattern 15
15a. Anal fin without pattern, with dark edge; body pattern of large polygonal spots separated by dark reticulations . . . . . *saxicola*, p. 167
15b. Anal fin patterned with alternating dark and pale areas; body spots separate, roughly eye-sized, but unequal . . . . . *ocellatus*, p. 174

Subgenus *Gymnothorax*

Gymnothorax hubbsi Böhlke and Böhlke, 1977

LICHEN MORAY

Figures 143–146, Tables 9, 14, 15

Gymnothorax hubbsi Böhlke and Böhlke, 1977: 238 (original description, Florida Atlantic, 38 fm, holotype USNM 214833).

Distinctive Characters. A small moray with overall color pattern of pale hieroglyphic-like marks on brown background; four infraorbital pores; teeth serrate, jaw teeth biserial to uniserial, usually one median intermaxillary fang; MVF 7-58-135.

Description. Total vertebrae 133–137, $\bar{x}$ = 135 (n = 10); predorsal vertebrae 6–8, $\bar{x}$ = 6.8 (10); preanal vertebrae 56–60, $\bar{x}$ = 58 (10); dorsal-fin rays 405–416 (2); anal-fin rays 251–260 (2); kidney terminus at 67–69 (2). Head pores: SO 1 + 2; IO 4; M 6; B 2.

Proportions as % (n = 10, 182–306 mm): Of TL: predorsal 10–13, $\bar{x}$ = 11; preanal 45–51, $\bar{x}$ = 48; head 11–12, $\bar{x}$ = 12; depth at gill opening 5.4–7.3, $\bar{x}$ = 6.6; depth at anus 4.0–5.8, $\bar{x}$ = 4.8. Of HL: snout 17–20, $\bar{x}$ = 19; upper jaw 42–52, $\bar{x}$ = 46; eye 7.9–11, $\bar{x}$ = 10; interorbit 12–16, $\bar{x}$ = 14.

Snout and jaws moderate; eye moderate, above mid-jaw. Anterior nostril tubular, near tip of snout; posterior nostril a simple round opening above anterior margin of eye. Head pores as list-

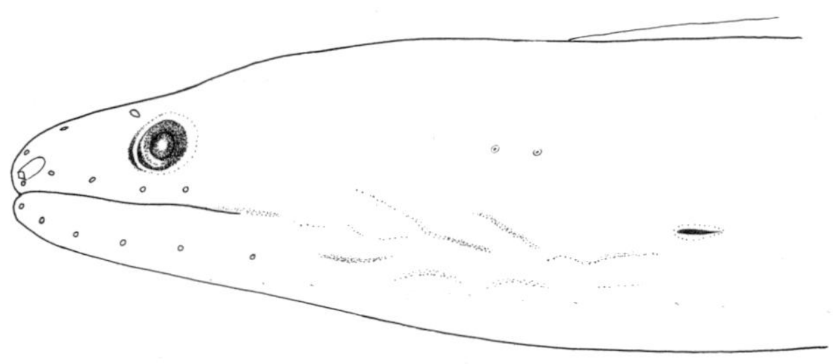

FIGURE 144. *Gymnothorax hubbsi*: ANSP 115293, paratype, 220 mm TL, Florida Atlantic; head.

ed; branchial pores below dorsal-fin origin, closer to gill opening than to rictus; one specimen with one branchial pore on one side. Teeth large, triangular, laterally compressed, posterior margins always and anterior margins sometimes serrate; distal anterior margins canted posteriorly. On each side of upper jaw, 5–6 large intermaxillary teeth with 0–3 tiny teeth between them, followed by main outer series of 8–15 maxillary teeth and, sometimes, 2–3 long slender depressible inner teeth anteriorly. On midline of upper jaw, 0–2 (usually one) large fangs anteriorly; vomerine teeth 8–19, beginning below mid-eye or farther back and extending about to end of maxillary tooth row, sometimes anteriormost teeth biserial or staggered, most of row uniserial. On lower jaw, 13–18 large teeth in main series, plus 1–9 tiny teeth in anterior outer series.

Color. Medium brown background with pattern of pale dendritic marks (Fig. 143), sometimes the marks fewer and more widely-spaced. Head, body and fins uniformly patterned; distal edge of fins white posteriorly. Life colors not known, presumably similar to those of preserved specimens.

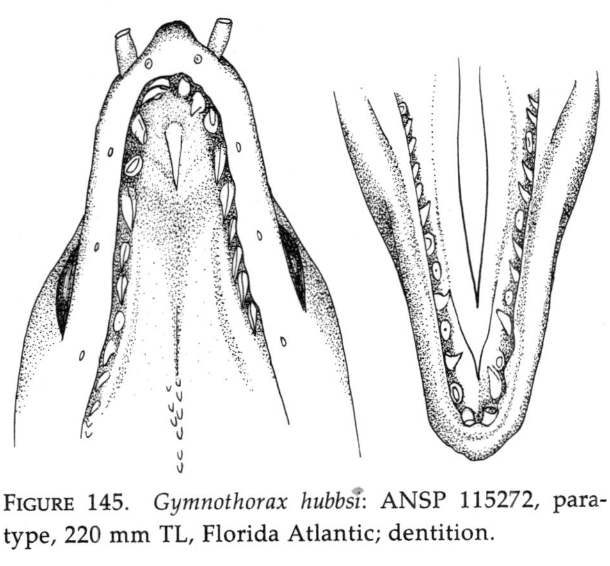

FIGURE 145. *Gymnothorax hubbsi*: ANSP 115272, paratype, 220 mm TL, Florida Atlantic; dentition.

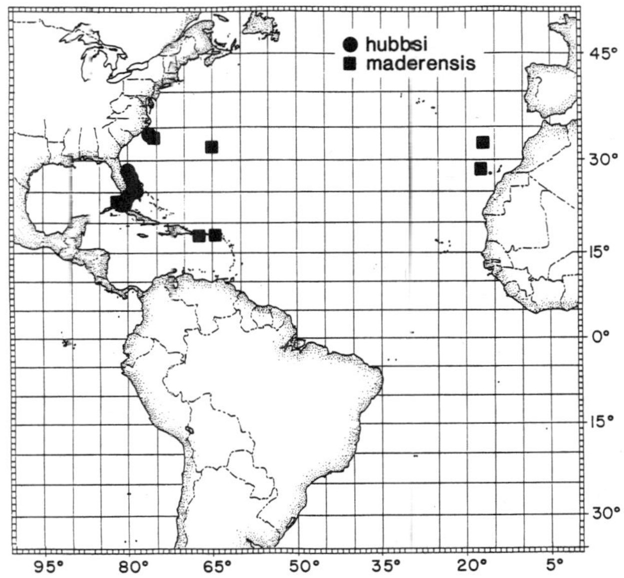

FIGURE 146. Distribution of *Gymnothorax hubbsi* and *G. maderensis* (subgenus *Gymnothorax*) based on material examined.

Size. A small species, the largest individual 336 mm TL. Of the 11 specimens examined, two (220–275 mm) were females, the largest ripe with 0.6-mm eggs (captured in February off Florida); eight specimens (182–336 mm) were males.

Development. Leptocephalus not known.

Behavior. An uncommon moray, with little known of its habits and behavior. Apparently inhabits moderate depths; usually taken by trawl or dredges, once by hook and line; not found in coral or rocky areas and not collected by poison to date. Its small size probably accounts for the few specimens taken, as it could easily escape through the mesh of most commercial trawls.

Distribution (Fig. 146). Infrequently captured, only 11 known specimens. Taken off the Atlantic coast of Florida, with single specimens from North Carolina, off the western edge of the Great Bahama Bank, and from the vicinity of Havana, Cuba. Taken by 6-ft tumbler or 8-ft scallop dredges, and by 50–70-ft trawls at depths of 64–91 m (once recorded as more than 183 m); all but once collected singly. The largest and most recently acquired specimen was taken by hook and line at 150 m off North Carolina.

TABLE 15. Frequency distributions of vertebral counts in western Atlantic species of *Gymnothorax*, subgenus *Gymnothorax*.

	Predorsal				Preanal																
	5	6	7	8	54	55	56	57	58	59	60	61	62	63	64	65	66	67	68	69	70
hubbsi	1	4*	3	3	1	-	-	1	3	4*	2										
maderensis	4*	5	3	-													1	2	5*	3	2

	Total														
	132	133	134	135	136	137	...	149	150	151	152	153	154	155	156
hubbsi	1	3	1	2	3*	1									
maderensis								1	1*	-	-	1	6	2	1

* Holotype.

Etymology. Named for Professor Carl L. Hubbs, "with great respect and affection" (Böhlke and Böhlke, 1977: 238). A noun in the genitive case.

Study Material. A total of 11 specimens, 182–336 mm TL, including the holotype and eight paratypes.

Holotype: USNM 214833 (273); Florida Atlantic, off Cape Canaveral, 28°33′N, 80°03′W, 38 fm (70 m); SILVER BAY sta. 2727; 1 Feb. 1961.

Paratypes: FLORIDA ATLANTIC: ANSP 115256 (1, 282); off Vero Beach 27°40′N, 79°58′W, 50 fm; SILVER BAY 2721; 1 Feb. 1961. ANSP 115270 (1, 247, cleared and stained); off Daytona Beach, 29°02′N, 80°07.5′W; SILVER BAY 5579; 11 Mar. 1964. ANSP 115293 (1, 220); taken with holotype. ANSP 130200 (1, 265); SE of Cape Canaveral, 28°13′N, 80°02′W, 32–40 fm; SILVER BAY 2008; 25 April 1960. FLORIDA KEYS: ANSP 115272 (1, 220); off Key Largo, 25°01′N, 80°19′W, 45–46 fm; SILVER BAY 2369; 25 Oct. 1960. ANSP 133321 (1, 275); off Key Largo, 24°56′N, 80°22′W, 46 fm; SILVER BAY 2362; 25 Oct. 1960. BAHAMAS: ANSP 115269 (1, 182); western edge Great Bahama Bank, outside 100 fm; SILVER BAY sta. no. lost; Nov. 1960. CUBA: ANSP 88787 (1, 306); Havana; L. Howell Rivero; 1932.

Other material: NORTH CAROLINA: ANSP 152248 (1, 336); off Cape Lookout, 34°02.5′N, 76°10.5′W, 150 m; T. Credle; 1 May 1983. FLORIDA ATLANTIC: ANSP 137971 (1, 232); off Boca Raton, 26°21.4′N, 80°02.6′W, 71–75 m; BOWERS Cruise 133, sta. 4; 17 Sept. 1975.

Gymnothorax maderensis (Johnson, 1862)

SHARKTOOTH MORAY

Figures 146–149, Tables 9, 14, 15

Pseudomuraena maderensis Johnson, 1862: 167, fig. (original description, Madeira, holotype BMNH 1862.6.14.10).

Muraena maderensis. Günther, 1870: 125 (redescription).

Gymnothorax madeirensis. Jordan and Davis, 1891: 606.

Ahynnodontophis rosaliae Howell Rivero, 1932: 19 (original description, Cuba "off Cojimar in deep water," holotype MCZ 32785).

Lycodontis unicolor (not of Delaroche, 1809). Fowler 1936a: 309 (*maderensis* synonymized with *unicolor*).

Gymnothorax maderensis. Blache, 1967e: 1697 (redescribed, illustrated; Madeira, Dahomy).

Distinctive Characters. A large moray with color pattern of dark reticulations on pale background, head noticeably dusky, and prominent pale margin on fins; teeth serrate and uniserial in adults, vomerine teeth frequently missing; MVF 6-68-153.

Description. Total vertebrae 149–156, $\bar{x}$ = 153.2 (n = 9); predorsal vertebrae 5–7, $\bar{x}$ = 6.1 (9); preanal vertebrae 66–70, $\bar{x}$ = 68.4 (9); dorsal-fin rays 415–454, $\bar{x}$ = 432 (4); anal-fin rays 240–259, $\bar{x}$ = 250 (4); kidney terminus at 76–79, $\bar{x}$ = 78 (3). Head pores: SO 1 + 2; IO 4; M 6; B 2.

Proportions as % (n = 7, 499–930 mm): Of TL: predorsal 10–11, $\bar{x}$ = 10; preanal 50–54, $\bar{x}$ = 53; head 10–12, $\bar{x}$ = 12; depth at gill opening 6.2–8.7, $\bar{x}$ = 7.2; depth at anus 4.2–5.2, $\bar{x}$ = 4.6; width at gill opening 3.1–4.9, $\bar{x}$ = 4.0; width at anus 2.8–3.8, $\bar{x}$ = 3.1. Of HL: snout 18–22, $\bar{x}$ = 20; upper jaw 37–45, $\bar{x}$ = 42; lower jaw 39–44, $\bar{x}$ = 40; eye 6.4–10, $\bar{x}$ = 8.6; interorbit 13–18, $\bar{x}$ = 15.

Head, snout and jaws moderate; eye moderate, above mid-jaw. Anterior nostril tubular; posterior nostril with slightly raised and crenulate edge, situated above or above and slightly behind

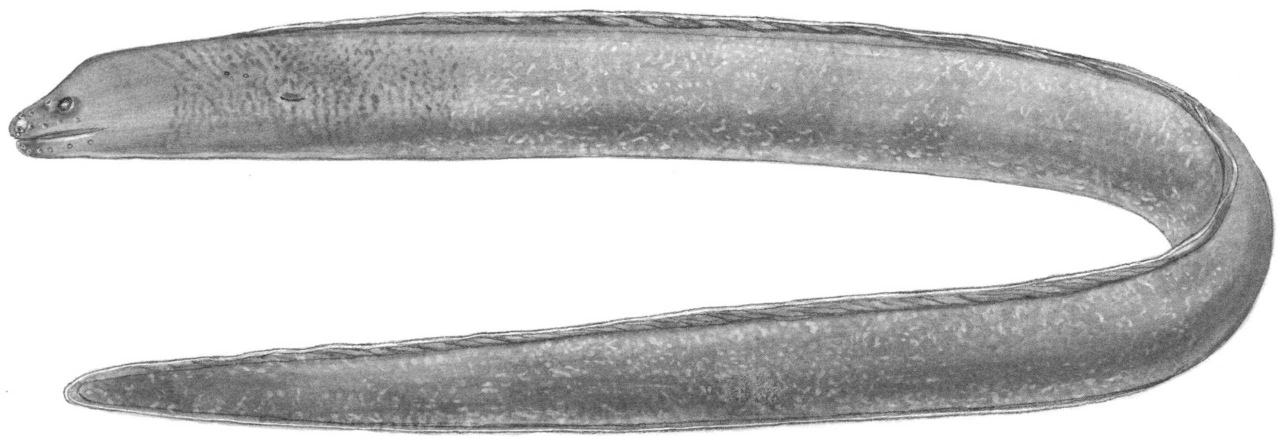

FIGURE 147. *Gymnothorax maderensis*: MCZ 32785, 930 mm TL, Cuba, holotype of *Ahynnodontophis rosaliae*.

anterior margin of eye. Teeth large, triangular, serrate on both sides, the posterior edge with a basal knob (the "tubercle" described on the holotype). On each side of upper jaw, 4–6 large intermaxillary teeth, the anterior pair the smallest, closely followed by 5–10 maxillary teeth which are directed back, decreasing in size posteriorly, the series extending to just behind level of posterior margin of eye. No anterior median (intermaxillary) fangs; vomerine teeth frequently not visible, 2–7 small conical teeth, sometimes biserial anteriorly, beginning just behind level of mid-eye, deeply embedded and covered by muscle folds; one specimen (ANSP 150527) had a longer row of nine teeth plus three additional teeth forming a biserial row anteriorly. On lower jaw, 10–18 stout triangular teeth, strongly serrate and directed back, decreasing in size posteriorly. The absence of anterior inner jaw teeth possibly attributable to the large size of the specimens examined.

Color. Overall pattern of dark brown reticulations on a pale background, pattern extending onto belly and fins; numerous very fine reticulations on head giving a characteristic dusky appearance; distal margins of fins pale. Old alcoholic specimens, notably the holotypes of *Pseudomuraena maderensis* and *Ahynnodontophis rosaliae* (the latter the illustrated specimen, Fig. 147), appear dark overall, the dusky head and dark reticulations hardly visible. Fresh specimens yellowish, the color fading to pale cream in alcohol; reported covered with lime green slime when captured (S. W. Ross, 1982, in litt.).

Size. The largest specimen studied was 1042

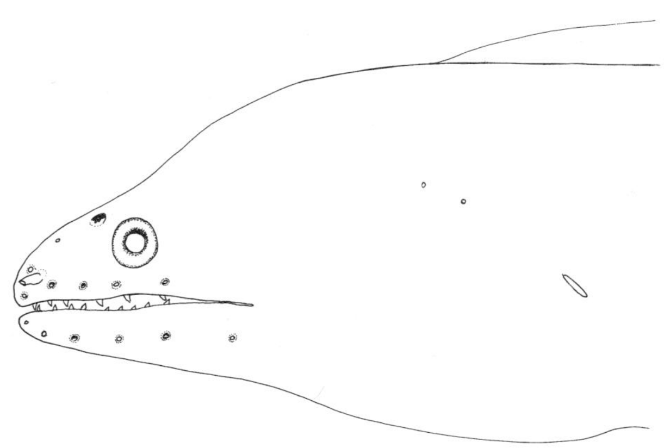

FIGURE 148. *Gymnothorax maderensis*: ANSP 136038, 710 mm TL, Bermuda; head.

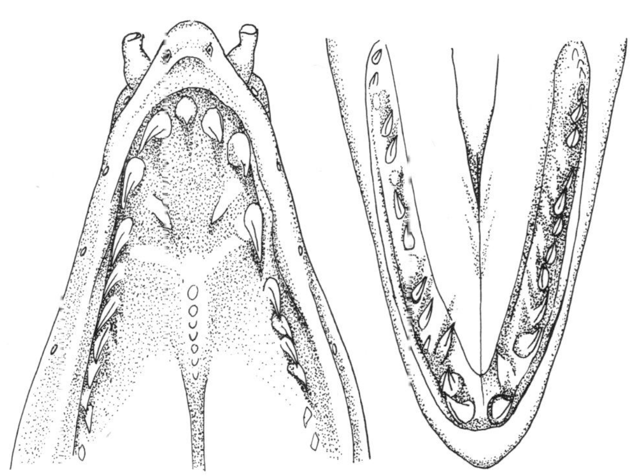

FIGURE 149. *Gymnothorax maderensis*: data as for Figure 148; dentition.

mm TL; an additional larger specimen (not seen and not listed) recently taken off Puerto Rico was reported (D. A. Hensley, in litt., 1986) to measure 1000 mm with end of tail missing. We have seen no specimens smaller than 499 mm. Of those examined, four (720–930) were females, the 820-mm specimen taken off North Carolina in May ripe with 1.5-mm eggs; four were males (620–905 mm), none of them ripe.

Development. Leptocephalus not found in the western Atlantic. Blache (1977: 37) described four larvae, 24–107 mm, from the Gulf of Guinea.

Behavior. Development and early life history remain completely unknown. All specimens we have seen are large adults, and have usually been captured by hook and line, occasionally by trap, always from "deep water." The specimens from North Carolina were captured over rocky reef areas.

Distribution (Fig. 146). In the western Atlantic, found in widely scattered locations off Bermuda, North Carolina, Cuba, Puerto Rico, and the Virgin Islands. In the eastern Atlantic from Madeira, the Canary Islands, and recorded from Dahomey (Blache, 1967e). Rarely captured; recorded depths in the western Atlantic 148–274 m; in the eastern Atlantic 85–200 m (Bauchot, 1986: 540).

Etymology. Named for the island of Madeira where the type specimen was taken. An adjective.

Remarks. The reticulate color pattern of *Gymnothorax maderensis* resembles that of several other morays found in the western Atlantic. Its pale coloration is similar to the pale phase of *Gymnothorax miliaris,* which differs in having biserial dentition, smooth teeth, and head coloration the same as that on the body. In its darker phase, *G. maderensis* resembles *G. hubbsi* (a small species with reticulations appearing uniformly on the head, body and fins) and *Anarchias similis* and *Uropterygius macularius* (both uniformly patterned and with median fins restricted to tip of tail). These species are also separable by vertebral formulae.

Until the last decade, *Gymnothorax maderensis* was known only from the eastern Atlantic. The holotype and one additional specimen from Madeira were described in 1862 by Johnson and later by Günther (1870: 125); both descriptions stated that the vomer was toothless and that the species was closely allied to *Thyrsoidea unicolor* Kaup, from which it differed in possessing uniserial dentition. Fowler synonymized *maderensis* with *unicolor* (1936a: 311), attributing the difference in dentition to variation with size and age. *Gymnothorax unicolor* differs greatly; it has biserial, smooth teeth, a uniformly dark color with no pattern except a darker head, and a different vertebral formula (4–6 predorsal, 50–57 preanal, and 136–144 total vertebrae). Blache (1967e: 1697–1705) recognized *maderensis* as a valid species and redescribed and illustrated it, listing two specimens from Madeira and Dahomey.

In 1932 Howell Rivero described *Ahynnodontophis rosaliae* from Cuba. His description was similar to that for *G. maderensis;* the absence of median intermaxillary and vomerine teeth was noted, but no mention was made of serrations on the teeth. The holotype of *A. rosalie* has serrate teeth and vertebral counts similar to those of *Gymnothorax maderensis;* the two were considered conspecific and the holotype of *A. rosaliae* was illustrated as the only western Atlantic specimen known. Subsequently, nine additional specimens have been taken in the western Atlantic, the occurrence of the species in western waters now well established.

Study Material. A total of 14 specimens, 499–1042 mm TL, including the holotype and the holotype of *Ahynnodontophis rosaliae* Howell Rivero.

Holotype: BMNH 1862.6.14.10 (1042 mm TL); Madeira; J. Y. Johnson.

Other material: WESTERN ATLANTIC: BERMUDA: ANSP 134105 (1, 810); 1975. ANSP 136038 (1, 710); hook and line; 13 Jan. 1975. ANSP 137864 (2, 620–728); hook and line; 1976. ANSP 156875 (1, 560); SW of Bermuda Platform, fish trap (questionably said to be at about 900 m); Sept. 1984. NORTH CAROLINA: ANSP 150527 (1, 905); 33°53′N, 76°22′W, hook and line, 148 m; Oct. 1982. ANSP 157318 (1, 843); 33°49.5′N, 76°30′W, 146 m; 22 May 1985. UF 38898 (1, 850); 33°50′N, 76°29.5′W, hook and line, 146 m; 20 May 1982. CUBA: MCZ 32785 (1, 930); Habana, off Cojimar, deep water; holotype of *Ahynnodontophis rosaliae.* PUERTO RICO: ANSP 130825 (1, 499); 6 mi south of La Parguera, trap at 274 m; 1974. VIRGIN ISLANDS: SIO 76-205 (1, ca 775); off St. Croix, Lang Bank;

1974. UF 34953 (1, 770); 8 mi S of St. Thomas, ca 18°12'N, 65°02'W, 223 m; Nov.–Dec. 1981.

EASTERN ATLANTIC: CANARY ISLANDS: Teneriffe: NMW uncat. (1, examined through bottle only); Steindachner; March, 1865.

Subgenus *Lycodontis*

Gymnothorax miliaris (Kaup, 1856)

GOLDENTAIL MORAY

Figures 150–156, Tables 9, 14, 16, 17

Thyrsoidea miliaris Kaup, 1856a: 63 (original description, Martinique, holotype MNHN B.2456). Kaup, 1856b: 90.
Thyrsoidea flavopicta Kaup, 1856a: 64 (original description, Brazil, holotype unknown). Kaup, 1856b: 90.
Thyrsoidea irregularis Kaup, 1856a: 65 (original description, Brazil, holotype MNHN B.3120). Kaup, 1856b: 95.
Muraena elaborata Poey, 1860: 323 (original description, Cuba, holotype USNM 24961).
Muraena multiocellata Poey, 1860: 324 (original description, Cuba, holotype unknown).
Gymnothorax multiocellata. Poey, 1867b: 260.
Gymnothorax scriptus Poey, 1867b: 261 (original description, Cuba, holotype unknown).
Gymnothorax elaboratus. Poey, 1867b: 262.
Muraena flavopicta. Günther, 1869: 239.
Muraena miliaris. Günther, 1870: 100.
Muraena irregularis. Günther, 1870: 115.
Gymnothorax miliaris. Poey, 1876a: 201.
Gymnothorax multiocellatus. Poey, 1876a: 201.
Sidera miliaris. Jordan, 1887b: 566.
Sidera elaborata. Jordan, 1887b: 566.
Gymnothorax flavopictus. Jordan and Davis, 1891: 605.
Gymnothorax irregularis. Jordan and Davis, 1891: 605.
Lycodontis miliaris. Jordan and Evermann, 1896a: 397.
Lycodontis elaboratus. Jordan and Evermann, 1896a: 398.
Leptocephalus fuliginosus Strömman, 1896: 34 (original description). Leptocephalus.
Muraena grandimaculis Regan, 1903b: 418, pl. 14, fig. 3 (original description, Rio de Janeiro, holotype MHNG 665.39).
Muraena myrialeucostictus Fowler, 1912a: 18, fig. 4 (original description, St. Martins, West Indies; holotype ANSP 16031).
Muraena trinitatis A. Miranda Ribeiro, 1918: 175, fig. 3 (original description, Ilha da Trindade, holotype MNRJ 1922).
Muraena myrialeucosticta. Jordan, Evermann, and Clark, 1930: 90.
Muraena aurea Mowbray, 1931: 2 [unnumbered] (original description, S shore of Bermuda, holotype unknown).
Lycodontis flavopictus. Fowler, 1936a: 315.
Murenophis miliaris. Fowler, 1936b: 1215.
Lycodontis irregularis. Fowler, 1941: 138.

Comments on Synonymy. *Gymnothorax miliaris* has extremely variable coloration, which has resulted in the description of the many nominal species listed in the synonymy. There are three basic color types, all described by Kaup (1856a,b). The common dark phase is "*miliaris*" (synonyms *elaborata, multiocellata, scriptus, myrialeucostictus* and *trinitatis* [two of the three types]) which possesses a dark background covered all over with very small dots (Fig. 150). The other two are pale phases: "*flavopicta*" has a pale cream background with dark hieroglyphic-like reticulations (Fig. 151), and "*irregularis*" (synonyms *grandimacula, trinitatis* [one specimen] and *aurea*) is a cream-colored eel with a few widely-separated large irregular dark blotches (Fig. 152). All gradations and combinations of these patterns exist (Fig. 153A–H); since we find no other observable differences between them, we consider all to be conspecific.

Distinctive Characters. A moderate-size moray with highly variable coloration ranging from many small dots on a dark background to a cream-colored eel with very few widely scattered brown blotches; tip of tail always with pale pattern; posterior nostril in short tube; teeth moderate, smooth-edged, no long canines, jaw teeth biserial; MVF 5-49-120.

Description. Counts and measurements are given separately for the dark spotted and pale patterned phases of *Gymnothorax miliaris* (Tables 16, 17). Kidney terminus 57–59 (n = 3). Head pores: SO 1 + 2; IO 4; M 6; B 2.

Snout and jaws short; eye moderate, above mid-jaw. Anterior nostril tubular; posterior nostril in a short tube, above anterior margin of eye. Head pores small and inconspicuous, as listed above. Teeth numerous, conical and sharp, neither caniniform nor molariform. On anterior upper jaw, many intermaxillary teeth on ethmovomer, the teeth similar in size and shape; about 4–6 outer teeth on each side (flanked by 5–8 tiny teeth), an intermediate row of 2–5 teeth, and a median row of 2–5 teeth along midline. Maxillary

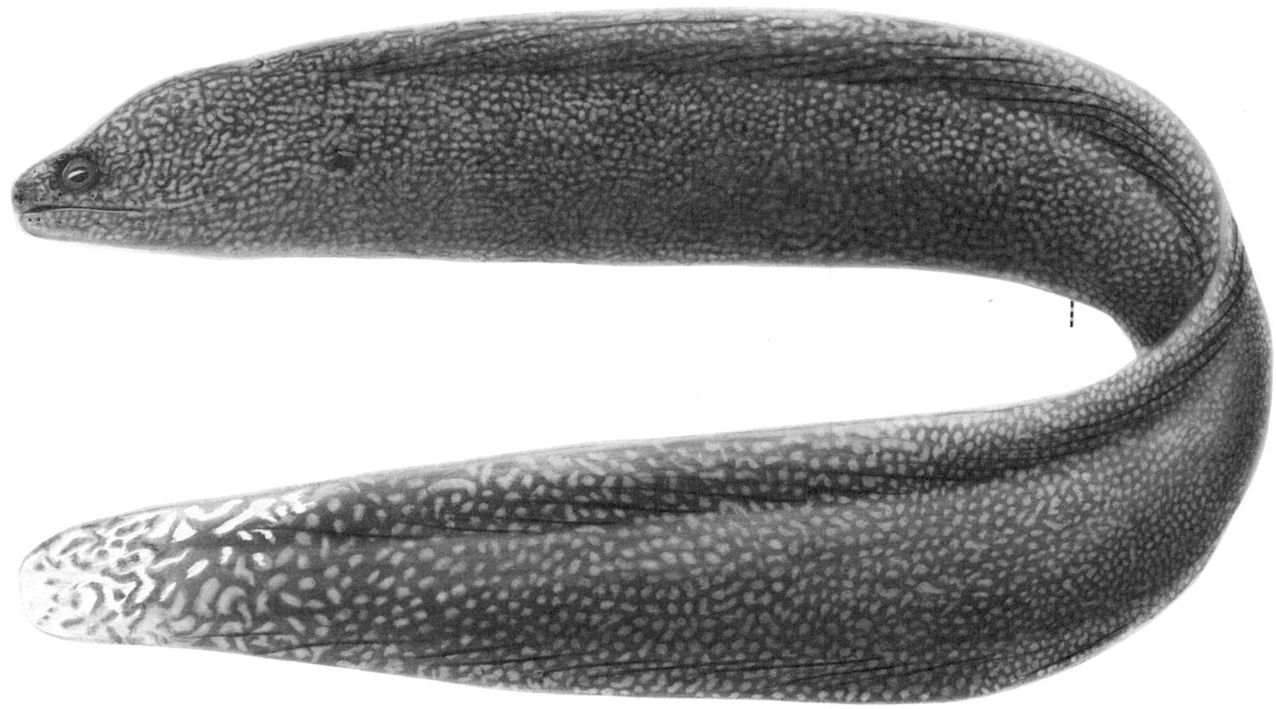

FIGURE 150. *Gymnothorax miliaris*: ANSP 111562, 437 mm TL, Bahamas; typical dark coloration. Illustration by S. P. Gigliotti.

teeth biserial, 15–24 teeth in outer row, flanked anteriorly by 0–9 tiny teeth, and an inner row of 6–12 teeth which are slightly larger than those in outer row. Uniserial row of 14–16 stout teeth on vomerine shaft. On lower jaw, an outer row of 18–30 teeth, flanked by 0–10 tiny teeth anteriorly, and an anterior inner row of 2–5 teeth. Small specimens with more teeth, the flanking tiny teeth few or absent in largest specimens.

Body shape relatively short and stubby; tail

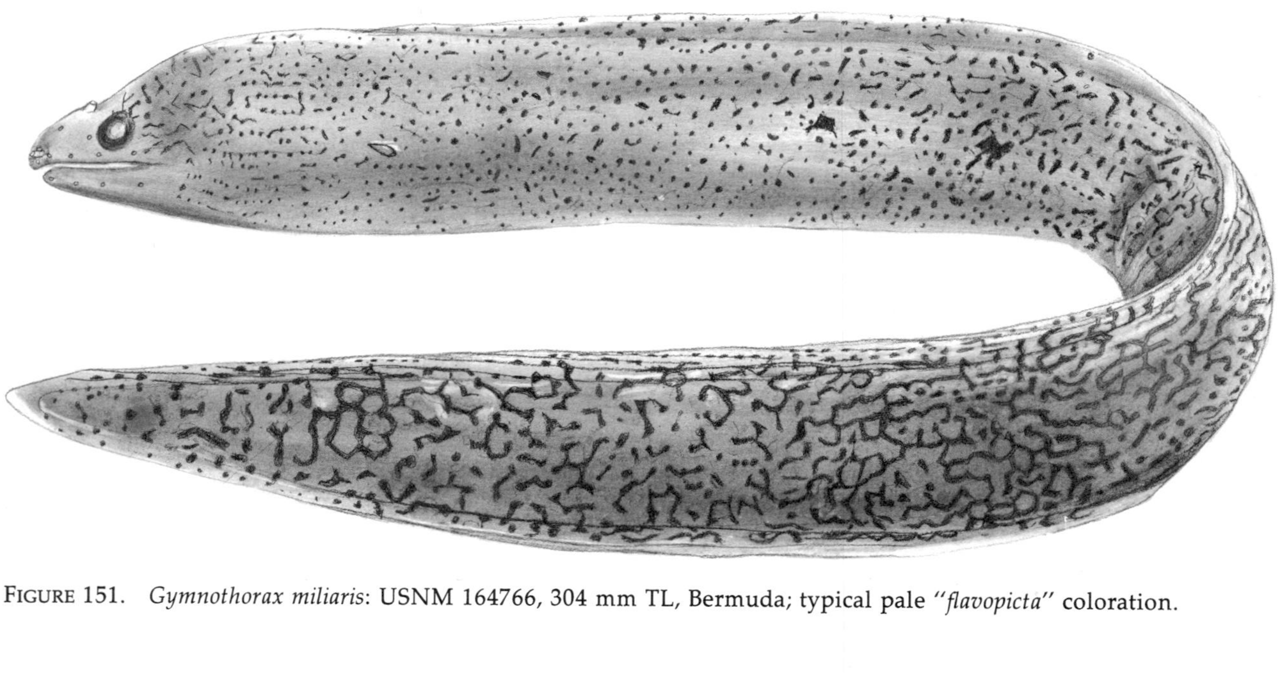

FIGURE 151. *Gymnothorax miliaris*: USNM 164766, 304 mm TL, Bermuda; typical pale "*flavopicta*" coloration.

FIGURE 152. *Gymnothorax miliaris*: MHNG 665.39, 440 mm TL, Brazil, holotype of *Muraena grandimaculis*; typical pale "*irregularis*" coloration.

not tapering, as in most species of *Gymnothorax*, but rounded at tip; body depth about equal at gill openings and at anus.

Color. In the common, dark color phase of *Gymnothorax miliaris*, the dark brown to black background is sprinkled with tiny white dots, the pattern of uniform color and intensity except at the tip of the tail, which has characteristic coalescing pale areas, the rounded tip always pale as illustrated in Figure 150. The number and placement of dots on the body is highly variable, from widely scattered to so close together that the dark background appears as a reticulum (Fig. 153A–C). Spots always pale, small and uniform in size, coloration similar on head, body and fins. All small specimens with dark coloration. In life, the tip of the tail and the spots are yellowish, the background color brown to black.

Pale phases are extremely variable. Closest in transition to the dark phase is the bold contrasting pattern of "*flavopicta*" with broad dark reticulations on a pale cream background (Fig. 153D–H). The names *Muraena elaborata* and *Thyrsoidea irregularis* were based on individuals with less of the dark reticulum showing (Fig. 153D–F), and in *M. grandimaculis* (Fig. 152) and *M. aurea* only a few widely-scattered isolated black blotches

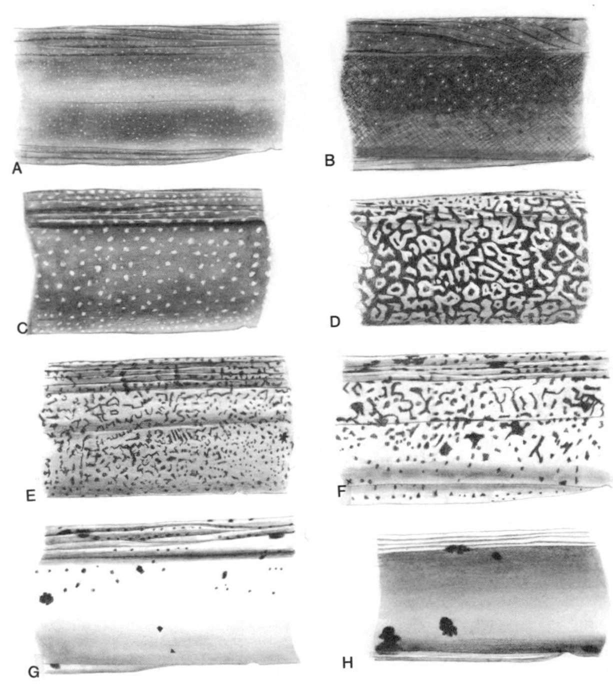

FIGURE 153. *Gymnothorax miliaris*: variation in coloration. Dark phase: A. ANSP 126419, 307 mm TL, Serrana Bank. B. ANSP 123292, 205 mm TL, Atol das Rocas. C. ANSP 75183, 257 mm TL, Bahamas. Pale "*flavopicta*" phase: D. ZMC P311702, 517 mm TL, Venezuela. E. AMNH 19591, 398 mm TL, Bahamas. F. UPR 1219, 355 mm TL, Curaçao. Pale "*irregularis*" phase: G. ANSP 158818, 415 mm TL, Bermuda. H. ANSP 158826, 452 mm TL, Brazil.

TABLE 16. Summary of counts, proportions (in percent), and dentition of dark and pale specimens of *Gymnothorax miliaris.*

	Dark			Pale			Combined		
	Range	x̄	n	Range	x̄	n	Range	x̄	n
Predorsal vertebrae	4–6	4.6	26	3–7	4.6	18	3–7	5.0	44
Preanal vertebrae	47–51	49.0	33	47–51	48.7	18	47–51	48.9	51
Total vertebrae	118–126	120.4	37	117–126	120.8	17	117–126	120.5	54
Dorsal-fin rays	304–326	315	2	293–321	300	4	293–326	305	6
Anal-fin rays	214–227	220	2	195–231	210	4	195–231	213	6
Proportion of TL									
(TL in mm)	95–492	324	20	304–517	401	15	95–517	362	35
Predorsal	8.1–14	10		8.2–13	10		8–14	10	
Preanal	43–48	46		43–50	46		43–50	46	
Head	10–15	13		12–15	13		10–15	13	
Depth at gill opening	5.0–10	7.0		6.4–10	6.9		5.0–10	7.0	
Depth at anus	4.7–8.7	6.8		5.3–9.8	7.2		4.7–9.8	7.0	
Width at gill opening	3.4–7.2	4.6		3.3–7.8	4.9		3.3–7.8	4.8	
Width at anus	3.2–6.3	4.6		3.6–6.2	4.8		3.2–6.3	4.7	
Proportion of HL									
Snout	16–19	18		16–20	18		16–20	18	
Upper jaw	32–43	40		35–42	40		32–43	40	
Lower jaw	32–44	40		34–45	39		32–45	39	
Eye	8.4–14	11		7.1–12	10		7.1–13	12	
Interobit	8.4–18	14		13–18	15		8.4–18	15	
Dentition									
Intermaxillary									
Outer	8–16 + 0–8			5–10 + 3–5					
Median	3 rows 2–5			3 rows 2–3					
Vomerine	9–18			8–16					
Maxillary									
Outer	15–23 + 10			15–24					
Inner	8–11			6–12					
Dentary									
Outer	22–30 + 10			18–29 + 8–15					
Inner	2–5			1–3					

were present (Fig. 153G,H). Pale phase coloration seen only in large specimens (over 300 mm TL).

Size. A moderate-size moray, the largest specimen seen 550 mm TL. Of dark specimens, 11 were females (179–492 mm), seven (270–424) ripe with eggs to 2.3-mm diameter; three males were 358–524 mm TL. Of six pale females (304–415), three (355–408) were ripe; three males were 353–466 mm TL. Ripe females were obtained in the Virgin Islands, Venezuela and the Serrana Bank.

Development. Leptocephalus tentatively identified (p. 906).

Behavior. Gymnothorax miliaris is a fairly common inhabitant of coral reef and rock areas. It has most often been taken by poison on patch reefs near sandy areas; large specimens have been

TABLE 17. Frequency distribution of vertebral counts in *Gymnothorax miliaris*, subgenus *Lycodontis.*

	Predorsal					Preanal				
	3	4	5	6	7	47	48	49	50	51
Dark		11	13*	2		2	9	13	5*	4
Pale	2	7	7	-	2	2	7	5	3	1

	Total									
	117	118	119	120	121	122	123	124	125	126
Dark		4	8	10	8	4*	1	1	-	1
Pale	1	1	3	3	2	4	2	-	-	1

* Holotype.

taken by trawl, spear, hook and line, and "net." Although Meliss (1875: 111) reported *Muraena "flavopicta"* to be abundant at St. Helena, taken in considerable quantity in deep water, and that "Salted and grilled, it forms one of the chief articles of food to the poor," it has not otherwise been recorded as a food source.

Distribution (Fig. 156). An insular species, widely distributed throughout the western and mid-Atlantic. Found in Bermuda, the Bahamas, Florida Keys, throughout the West Indies and Central America, and from islands and reefs off northern South America to Brazil. In the mid-Atlantic, taken at Ascension Island and St. Helena, and reported from St. Paul's Rocks by Lubbock and Edwards (1981: 138). In the eastern Atlantic, recorded from the Cape Verde Islands (Blache, 1967d: 1178). Most often collected at poison stations at depths to 60 m; larger specimens taken by hook and line at unrecorded depths.

Etymology. From the Latin *mille* (thousand), referring to the number of spots. A noun in apposition.

Remarks. The species *miliaris* has often been placed in the genus *Muraena* because the posterior nostril is in a short tube. However, it differs considerably (in general appearance, body shape, dentition, and osteology) from *M. helena,* the type species, and other species referred to *Muraena. Gymnothorax miliaris* possesses a unique combination of characters which differ considerably from those of all other species of *Gymnothorax* in the Atlantic. It is closest to *mareei* and *unicolor* in features of skull osteology and dentition, but those species have a typical tapering body shape, whereas *miliaris* is short and stubby. The dentition is similar that of species placed in the Indo-Pacific genus *Siderea.* We place *miliaris* in the subgenus *Lycodontis* with some reservation, pending additional studies and osteological comparison with the type species and other Indo-Pacific species.

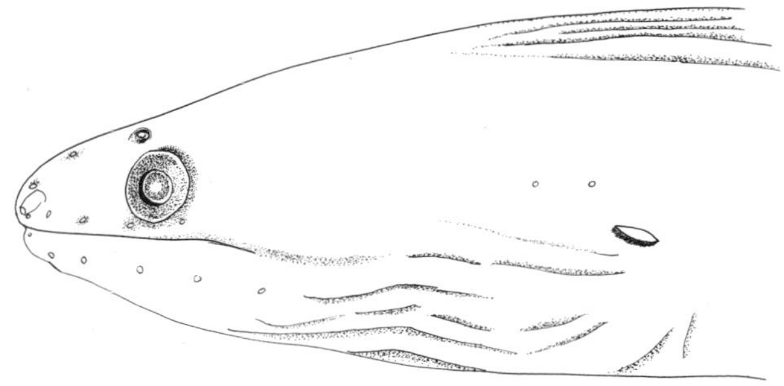

FIGURE 154. *Gymnothorax miliaris*: ANSP 126797, 237 mm TL, Bahamas; head.

Study Material. DARK PHASE: A total of 230 specimens, 53–550 mm TL, including the holotypes of *Thyrsoidea miliaris* Kaup, *Muraena elaborata* Poey, *Muraena myrialeucostictus* Fowler, and two syntypes of *Muraena trinitatis* Miranda Ribeiro.

Holotype: MNHN B.2456 (470); Martinique; M. Achard.

Other material: BERMUDA: ANSP 126291 (1, 100),

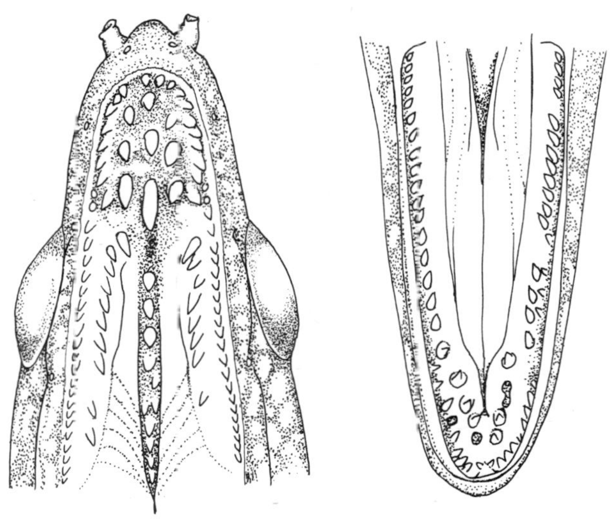

FIGURE 155. *Gymnothorax miliaris*: ANSP 75183, 257 mm TL, Bahamas; dentition.

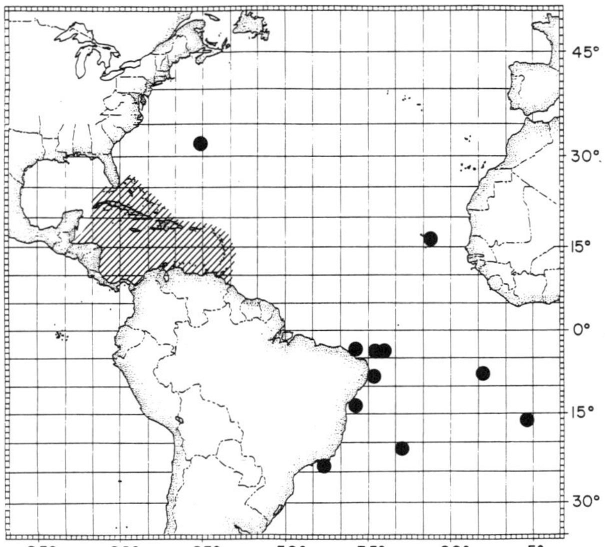

FIGURE 156. Distribution of *Gymnothorax miliaris* (subgenus *Lycodontis*) based on material examined.

133759 (1, 335), 133760 (4, 82–336), 133826 (1, 116), 133829 (1, 107), 148156 (2, 188–382), 148157 (1, 340), 148158 (1, 270), 148159 (1, 360), 148160 (2, 199–240). BMNH uncat. (1); J. M. Jones. BMNH uncat. (1); J. M. Jones, #22. FMNH 48563 (1, 483). USNM 164775 (1, 490), 164780 (1, 497), 175763 (3, 257–465), 175767 (1, 340). FLORIDA KEYS: MCZ 9124 (1, ca 365). SIO 67-86-26 (7 [of 8], 210–390). UMML 2973 (1, 490), 3538 (1, 360), 3731 (1, 440), 4367 (1, 395), 4716 (2, 422–480), 7247 (1, 327), 11654 (1, 142), 18008 (1, 230), 18042 (2, 238–434), 18067 (1, 210), 18943 (1, 345), 19535 (2, 143–270), 19877 (1, 297), 20131 (1), 20266 (5, 180–382), 30580 (1, 515). UMML uncat. (2, 378–400).

BAHAMAS: ANSP 72295 (1, 274), 75183 (2, 181–257), 100779 (1, 55), 100780 (1, 52), 102488 (1, 135), 111562 (1, 437), 113453 (3, 148–195), 113454 (1, 143), 113455 (4, 167–495), 113456 (9, 134–309), 113457 (1, 312), 113458 (5, 114–310), 113459 (1, 129), 113460 (3, 166–308), 113461 (2, 180–357), 113462 (2, 127–196), 113463 (3, 60–87), 113464 (1, 414), 115088 (1, 121), 115139 (1, 335), 116143 (2, 165–186, cleared and stained), 117821 (1, 95), 126775 (1, 165), 126797 (1, 237), 126927 (1, 420), 127083 (1, 53). UMML 5220 (2, 117–397), 6444 (1, 130), 6527 (1, 242), 12612 (1, 303), 12706 (2, 166–235), 22096 (1, 145), 31718 (1), 32566 (1, 98).

CUBA: ANSP 91895 (1, 345). CAS-SU 54205 (3, 190–250). CAS-SU 54206 (2, 257–327). MCZ 27209 (1, 316), 32799 (1, ca 520). USNM 24961 (1, 424); holotype of *Muraena elaborata*. JAMAICA: UMML 13227 (1, 295), 30370 (1, 163). HAITI: ANSP 77220 (1, 432), 111362 (1, 270), 119059 (1, 281), 119163 (1, 90), 126671 (1, 322), 127275 (1, 241), 127297 (9, 104–370), 127990 (1, 114). MCZ 27213 (1, 385). UMML 7459 (1, 240). PUERTO RICO: ANSP 139838 (1, 254). VIRGIN ISLANDS: UMML 3738 (3, 220–435), 4599 (1, ca 320), 5112 (1, 355). USNM 274266 (1, 234).

LESSER ANTILLES: *St. Martins:* ANSP 16031 (1, 407); holotype of *Muraena myrialeucostictus. Antigua:* ANSP 105485 (5, 77–174). *Dominica:* ANSP 105773 (1, 217). USNM 274268 (2, 121–292). *St. Lucia:* ANSP 110095 (1, 390, cleared and stained), 119000 (1, 87). *Curaçao:* CAS 59203 (1, 368). PANAMA: ANSP 113765 (3, 324–362), 113770 (1, 298). CAS 13489 (1, 112). TABL 107197 (3, 172–405). HONDURAS: TABL 107210 (1, 145). USNM 278176 (1, 188). PROVIDENCIA: ANSP 128805 (1, 113). UMML uncat. (1, 330). SERRANA BANK: ANSP 102456 (3, 139–228), 126419 (1, 307). FMNH 74727 (6, 173–302), 74731 (7, 140–236). USNM 178986 (1, 332), 187768 (1, 201). SWAN ISLAND: TABL BCF 2106 (2, 168–228). GRAND CAYMAN: ANSP 104915 (2, 72–215), 105089 (1, 153).

COLOMBIA: ANSP 117570 (2, 270–282), 117577 (1, 71). TABL 107198 (3, 216–357). CAS uncat. (1, 550). UMML 30059 (1, ca 220). VENEZUELA: MBUCV 1333 (1), 2645 (1), 3277 (1). UMML 14936 (1, 523), 22053 (1, 333). BRAZIL: CAS 59202 (1, 402). MZUSP 11921 (1). ZMB 3995 (1). *Atol das Rocas:* ANSP 123291 (1, 205), 123292 (2, 300–301). UF 44619 (1, 265), 44621 (1, 200). *Fernando de Noronha:* USNM 274267 (1, 185). *Ilha de Trindade:* MNRJ 1922 (2, 386–393); two of three syntypes of *Muraena trinitatis* Miranda Ribeiro. ASCENSION ISLAND: ANSP 117433 (1, ca 500).

LOCALITY UNKNOWN: MNHN B.2457 (1). BMNH uncat. (1); Zoological Society Collection.

PALE PHASE: A total of 18 specimens, 304–517 mm TL, including the holotypes of *Thyrsoidea irregularis* Kaup, *Muraena grandimaculis* Regan, and one syntype of *Muraena trinitatis* Miranda Ribeiro.

BERMUDA: ANSP 158818 (1, 415); Castle Harbor; L. S. Mowbray; 28 July 1967. USNM 164766 (2, 304–315); Nonsuch; W. Beebe; 1 July 1930; NYZS 9312. FLORIDA KEYS: ANSP 136554 (1, 466); Big Pine Key, 6–8 m; G. Clark; 6 Jan. 1977. SIO 67-86-26 (1 [of 8]); Looie Key; R. Rosenblatt; 1 Apr. 1969. BAHAMAS: AMNH 19591 (1, 398); Bimini, off S shore S. Bimini; Lerner Marine Laboratory; 27 Jan. 1952. VIRGIN ISLANDS: UMML 5377 (1, 353); St. John's; C. R. Robins et al.; 28 April 1959. LESSER ANTILLES: UMML 31345 (1, 356); Bird Island, 10 m; GILLIS sta. 53; 12 Aug. 1972. CURAÇAO: UPR 1219 (1, 355); 40 m; G. Klay; 1959. VENEZUELA: ZMC P311702 (1, 517); Puerto de la Cruz; Jan. 1968. BRAZIL: ANSP 158826 (1, 452); Recife; Brazilian Trawling Survey; 1944–1945. MHNG 665.37 (1, 365); Bahia. MHNG 665.39 (1, 440); Rio Janeiro; Regan; holotype of *Muraena grandimaculis*. MNHN B.3120 (1, 480); Bahia; exchange with Musee Geneve; holotype of *Thyrsoidea irregularis*. UMIM 4287 (1). UMML 13277 (1). *Isla de Trindade:* MNRJ 1922 (1, 390); one of three syntypes of *Muraena trinitatis*. ST. HELENA: BMNH 1863.3.11.4 (1); J. Meliss.

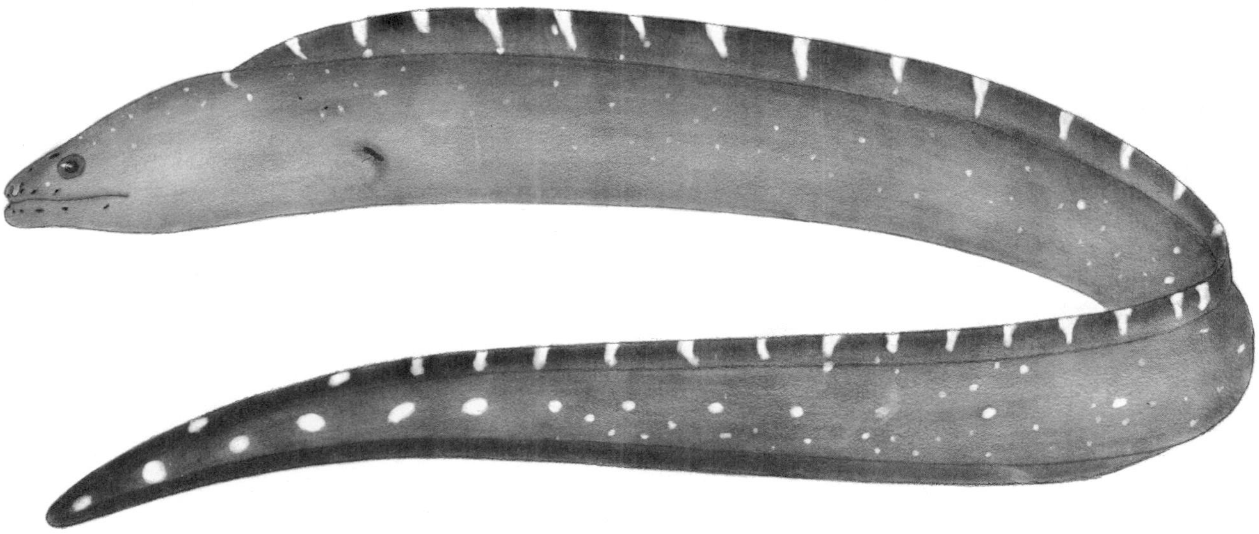

FIGURE 157. *Gymnothorax conspersus*: ANSP 117428, 534 mm TL, Suriname. Illustration by G. H. Jensen.

Subgenus *Neomuraena*

Gymnothorax conspersus Poey, 1867

SADDLED MORAY

Figures 157–160, Tables 9, 14, 18

Gymnothorax conspersus Poey, 1867b: 259 (original description, Cuba, holotype unknown). Böhlke and Böhlke, 1980: 218 (redescribed and illustrated).
Muraena conspersa. Günther, 1870: 102.
Sidera conspersa. Jordan, 1887b: 566.
Lycodontis conspersus. Jordan and Evermann, 1896a: 397.
Muraena ocellata (not of Agassiz, 1829). Regan, 1914: 22 (one of six specimens).
Gymnothorax ocellatus (not of Agassiz, 1829). Howell Rivero, 1932: 18.
?*Gymnothorax nigromarginatus* (not of Girard, 1859). Roux, 1973: 38 (the question in Roux's paper).

Misidentifications:

Not *Gymnothorax conspersus*. Jordan and Davis, 1891: 594, 604 (based on *Muraenophis punctata* Castelnau, 1855 = *Muraena retifera* Goode and Bean, 1882).
Not *Lycodontis conspersus*. Jordan and Evermann, 1896a: 397 (based on *Muraenophis punctata* Castelnau, as above).

Discussion of Synonymy. This species was described in 1867 by Poey, based on one specimen from Matanzas, Cuba, taken in July 1863. Günther (1870: 102) diagnosed the species in an English summary of the characters listed by Poey, mentioning small blue dots on the body but not the markings on the dorsal fin. Jordan and Davis (1892: 594) noted the dorsal-fin markings but (p. 604) concluded that *Muraenophis punctata* of Castelnau (name preoccupied) was the same species because it also was reported to have small blue dots on the body. Castelnau's *M. punctata* = *Muraena retifera* Goode and Bean, a much different species (see p. 200). *Gymnothorax conspersus* is a valid species; Poey's manuscript illustration was published, and the species was redescribed and illustrated by Böhlke and Böhlke in 1980.

Distinctive Characters. A large dark moray with small white spots on head and trunk, dark tail with a single series of large white spots posteriorly, and a series of white slash marks on distal half of dorsal fin; lower jaw projecting beyond upper jaw; three infraorbital pores; teeth serrated and uniserial in adults; MVF 6-66-168.

Description. Total vertebrae 163–174, $\bar{x}$ = 168.3 (n = 42); predorsal vertebrae 5–8, $\bar{x}$ = 6.5 (46); preanal vertebrae 62–70, $\bar{x}$ = 66.2 (47); dorsal-fin rays 351–387, $\bar{x}$ = 368 (3); anal-fin rays 250–267, $\bar{x}$ = 260 (3); kidney terminus at 79–82, $\bar{x}$ = 81 (3). Head pores: SO 1 + 2; IO 3; M 6; B 2.

Proportions as % (n = 20, 428–1054 mm). Of TL: predorsal 9.2–12, $\bar{x}$ = 10; preanal 44–49, $\bar{x}$ = 46; head 12–15, $\bar{x}$ = 13; depth at gill opening 5.0–7.4, $\bar{x}$ = 6.3; depth at anus 3.5–6.4, $\bar{x}$ = 4.8; width at gill opening 2.7–5.9, $\bar{x}$ = 4.3; width at anus 2.9–4.6, $\bar{x}$ = 3.6. Of HL: snout 15–20, $\bar{x}$ = 19; upper jaw 33–47, $\bar{x}$ = 40; lower jaw 33–46, $\bar{x}$ = 40; eye 6.0–9.3, $\bar{x}$ = 7.6; interorbit 12–17, $\bar{x}$ = 14.

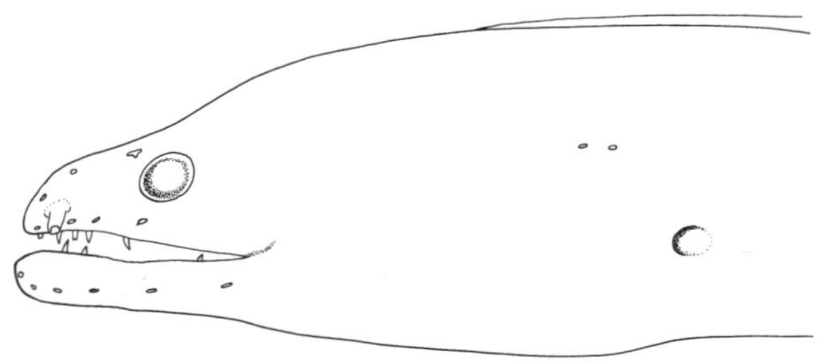

FIGURE 158. *Gymnothorax conspersus*: data as for Figure 157; head.

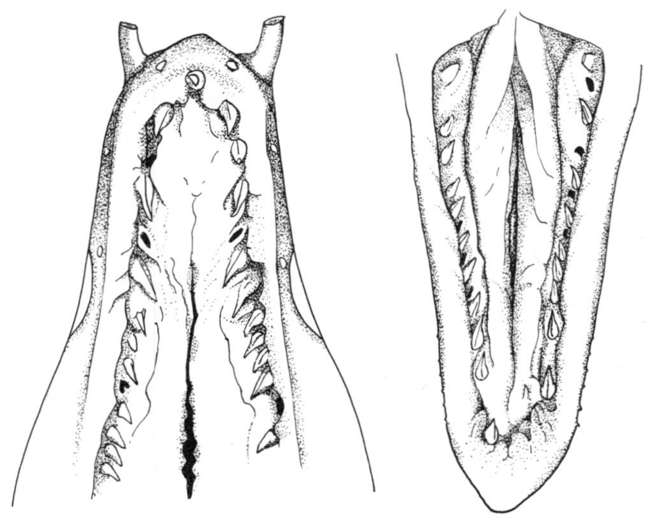

FIGURE 159. *Gymnothorax conspersus*: data as for Figure 157; dentition.

Head and snout moderate; jaws moderate, lower jaw prominent, usually extending forward beyond tip of snout in lateral aspect. Eye moderate, above mid-jaw. Lips papillose. Anterior nostril tubular, reaching edge of snout when depressed. Posterior nostril above anterior margin of eye, slightly raised from profile of head, variously flared, sometimes with small flaps, frills or slits, but not tubular. Head pores as listed above and illustrated in Figure 158; all pores variously irregular in shape, some slightly raised, some with convoluted rims. Variation in pores: one specimen with a fourth infraorbital pore just below posterior margin of eye (this the pore present in other muraenid species); two specimens with an additional mandibular pore on one side. Gill opening a small oval opening slightly below midline of fish.

Teeth large, triangular, laterally compressed and "shark-like," serrate on both edges. On upper jaw, an anterior median tooth, followed on each side by four long intermaxillary teeth, closely followed by 8–11 maxillary teeth, the anteriormost maxillary tooth smaller than the last intermaxillary tooth, the teeth evenly spaced, canted back and decreasing in size posteriorly. On lower jaw, a single row of 13–18 teeth, the anterior three large, similar in size and shape to intermaxillary teeth and fitting evenly between them when mouth is closed, the remainder abruptly smaller, decreasing in size and increasingly angled posteriorly. In small specimens, serrations visible only on posterior basal edge of intermaxillary and largest anterior dentary teeth; serrations increasing in size and extent, becoming conspicuous deep indentations on both anterior and posterior edges of all teeth (always more developed on posterior edge than on anterior). Most specimens with no median (intermaxillary) teeth in upper jaw. Vomerine teeth none or few, hidden by folds of muscle and difficult to see; 1–3 small teeth in three specimens.

Body elongate, with tapering tail; anus shortly before mid-body. Dorsal-fin origin between gill opening and rictus.

Color. Medium to dark brownish black, pale on head, shading to dark on tail; dorsal and ventral coloration similar at any given point. Distinct spots present on head and trunk, small and inconspicuous (particularly on head) on some, numerous but widely spaced on others. Spots increasing in size and decreasing in number posteriorly, ending in one row of large round white spots on tail. A single row of larger spots visible along mid-body of some specimens. Tip of tail dark. Eye and head pores ringed with dark pigment, more noticeable in larger specimens. Dorsal fin dark, with regular white slashes or saddle-shaped marks on edge, extending onto basal half or more of fin. Anal fin dark and unmarked.

Poey's notes described life colors (translated by Drs. Luis Alvarez-Lajonchere and Dario Guitart) "The body green, partly greyish and light, turning blackish toward the tail tip and along the dorsum, except on the head. Spots sky blue, as is the iris. Anterodorsal part of head without spots. Ground color of fins dark."

Size. The 1095-mm specimen examined, a ripe male, is the largest *Gymnothorax conspersus* known to date. Of 20 specimens checked, 13 (480–975 mm) were females and seven (428–832) were males; eight of the females were ripe with eggs to 1.6-mm, these taken throughout the year.

Development. Leptocephalus identified which could be this species or that of *Gymnothorax kolpos* (see p. 909).

Behavior. Little is known of the habits and behavior of this moray beyond what can be inferred from the few specimens collected. We have seen no specimen smaller than 428 mm, and there is a high incidence of ripe females (32%) in the study material, these all caught in deep waters. This suggests that the young live at greater depths than mature individuals, which perhaps enter shallow water prior to spawning.

Like most morays, this species is probably piscivorous, although guts of most study specimens were largely empty; gut contents observed from radiographs showed an entire fish in one specimen, small lengths of vertebrae in another, and a fish hook in a third.

Distribution (Fig. 160). *Gymnothorax conspersus* has been collected once off North Carolina, but is primarily known from the tropical western Atlantic from the Florida Straits through the Caribbean islands, and off the north and east coasts of South America as far south as Rio Grande do Sul, Brazil. Most collections were made by trawl at depths below 200 m; only five specimens (all collected south of 23°S) were in waters 50–200 m. Usually single specimens collected; on four occasions taken with *G. ocellatus.* A special effort to collect deep-water eels by trap off Puerto Rico (by P. L. Colin and D. Arneson) yielded nine specimens of *G. conspersus* obtained on five different occasions at depths of 164 to 310 meters.

Etymology. From the Latin *conspersus* (spotted), in reference to the color pattern. An adjective.

Remarks. Prior to 1980, all references to *Gymnothorax conspersus* were based upon Poey's type specimen which apparently no longer exists. Two

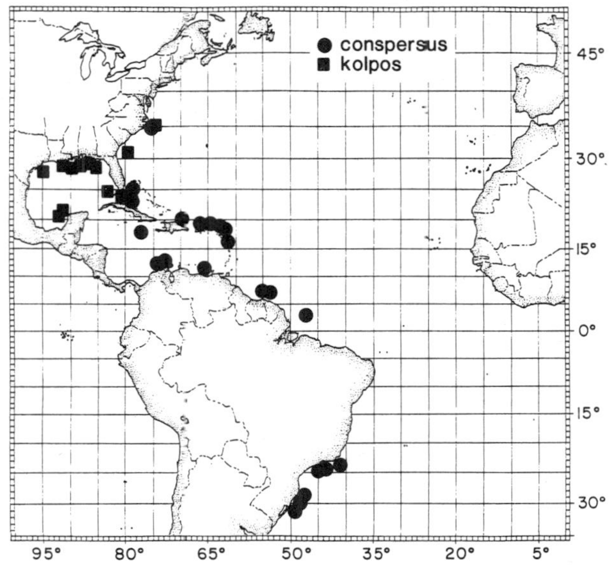

FIGURE 160. Distribution of *Gymnothorax conspersus* and *G. kolpos* (subgenus *Neomuraena*) based on material examined.

specimens listed by Howell Rivero (1932: 18) as *G. ocellatus* "taken off Cojimar in very deep water" are also *G. conspersus* (based on a hand-colored photograph provided by Howell Rivero, the location of the specimens unknown).

Poey's notes and the manuscript illustration of *Gymnothorax conspersus* leave no doubt that this is his species, despite his indecision regarding the vomerine teeth. In manuscript notes, he wrote that it remained a question whether the vomerine teeth were uniserial or biserial, but in the description of the species in the *Repertorio* (1867b: 259) he wrote that they were small and biserial. As noted in the following accounts of morays with serrate teeth, vomerine teeth are frequently not visible or difficult to find in large specimens. Most of our study specimens (all large) had no vomerine teeth; the stained fish had three very small uniserial teeth.

The five species here included in the subgenus *Neomuraena* have a similar coloration of white spots on a dark background. *Gymnothorax conspersus* and *G. kolpos* (and *G. equatorialis* in the Pacific) differ from the other species in possessing a very dark tail marked posteriorly with a single series of 3–4 large white spots. *Gymnothorax conspersus* may be separated from all other

TABLE 18. Frequency distributions of vertebral counts in western Atlantic species of *Gymnothorax*, subgenus *Neomuraena*.

	Predorsal					Preanal								
	4	5	6	7	8	43	44	45	46	47	48	49	50	51
conspersus		3	18	21	4									
kolpos		2	17*	6										
saxicola	4	60	144*	27						1	1	3	5	24
nigromarginatus		2	33	11*	3			1	3	5	15*	17	9	3
ocellatus	1	17	90	53	6	1	–	–	1	5	19	32	44	37

	Total														
	134	135	136	137	138	139	140	141	142	143	144	145	146	147	148
conspersus															
kolpos															
saxicola	1	1	8	22	26	31	50	33	40	14*	10	7	1	1	
nigromarginatus				2	2	4	4	11	12*	7	1	3	1	1	
ocellatus			2	1	5	–	5	2	9	14	41	25	23	23	9

* Includes primary type.

spotted species by the regular, uniformly-spaced, white slash marks on the dark dorsal fin.

Study Material. A total of 61 specimens, 428–1095 mm TL.

NORTH CAROLINA: ANSP 157316 (1, 990). BAHAMAS: ANSP 107096 (1, 790). CUBA: MCZ 38819 (1, 730). JAMAICA: MNHN 1974-199 (1, 815). DOMINICAN REPUBLIC: UMML 32252 (1, 496). PUERTO RICO: ANSP 115242 (1, 658+), 134030 (1, 740), 136686 (1, 774), 138142 (2, 965+), 138368 (4, 738–910), 139684 (1, 948), 139685 (1, 813). FMNH 69199 (1, 532), 69201 (1, 500+). VIRGIN ISLANDS: ANSP 107094 (1, 860), 136093 (1, 1095), 137565 (1, 725). ZMUC CN6 (1, 938).

LESSER ANTILLES: *Anguilla Island:* ANSP 115234 (1, 776). *Barbuda:* ANSP 117440 (4, 567–715). *Martinique:* MNHN 4625 (1, 498). COLOMBIA: ANSP 107091 (2, 808–832), 107095 (2, 573–670), 116364 (1, ca 780), 116365 (1, ca 915). UCR 530 (1, 680). VENEZUELA: FMNH 69198 (1, 708). MBUCV 5675 (1, 1054). USNM 221241 (1, 820). SURINAME: ANSP 106568 (1, 625), 115288 (1, 437, cleared and stained), 116639 (1, 538, cleared and stained), 117428 (1, 534). FMNH 69191 (1, 467), 69197 (1, 455). BRAZIL: BMNH 1913.12.4.208-13 (1 [of 6 specimens], 504). GCRL 16931 (1, 730). MNHN C.478 (1, 503), uncat. (1, 533). MNRJ uncat. (2, 547–738). MZUSP 9853 (1, 561), 9854 (1, 664), 9855 (1, 428), 9856 (1, 653), 9857 (1, 550), 9858 (1, 525), 9859 (1, 565), 9860 (1, 677), 9861 (1, 730). USNM 185965 (1, 480).

EXACT LOCALITY UNKNOWN: ANSP 100493 (1, ca 800).

Gymnothorax kolpos Böhlke and Böhlke, 1980

BLACKTAIL MORAY

Figures 160–163, Tables 9, 14, 18

Gymnothorax nigromarginatus (not of Girard, 1859). Ginsburg, 1951: 462 (in part: two specimens, MCZ 35110 and USNM 154982).

Gymnothorax kolpos Böhlke and Böhlke, 1980: 223 (original description, off Alabama, holotype ANSP 107093).

Distinctive Characters. A large dark moray with small ocellated white spots on head, body and trunk, tail dark with single row of large white spots posteriorly, and dorsal fin with black edge; three infraorbital pores; teeth serrated and uniserial in adults; MVF 6-63-164.

Description. Total vertebrae 159–167, $\bar{x}$ = 163.5 (n = 28); predorsal vertebrae 5–7, $\bar{x}$ = 6.1 (31); preanal vertebrae 60–65, $\bar{x}$ = 63.0 (31); dorsal-fin rays 326–397, $\bar{x}$ = 363 (3); anal-fin rays 228–278, $\bar{x}$ = 254 (3); kidney terminus at 77–82 (2). Head pores: SO 1 + 2; IO 3; M 6; B 2.

Proportions as % (n = 20, 310–790 mm): Of TL: predorsal 7.2–11, $\bar{x}$ = 9.8; preanal 41–47, $\bar{x}$ = 44; head 11–13, $\bar{x}$ = 12; depth at gill opening 2.8–5.5, $\bar{x}$ = 4.2; depth at anus 4.1–7.8, $\bar{x}$ = 5.1; width at gill opening 2.8–5.5, $\bar{x}$ = 4.1; width at anus 2.7–5.9, $\bar{x}$ = 3.8. Of HL: snout 16–22, $\bar{x}$ = 18; upper

TABLE 18. Continued.

Preanal																		
52	53	54	55	56	57	58	59	60	61	62	63	64	65	66	67	68	69	70
										1	-	4	11	12	6	9	2	1
								1	2	9*	2	6	5					
38	48	61*	42	9	2	1	2											
1																		
30	4	2																

Total																			
149	150	151	152	···	160	161	162	163	164	165	166	167	168	169	170	171	172	173	174
								2	-	3	5	6	4	7	7	6	-	-	1
					1	1	2	6	3*	5	3	1							
3	2	1																	

jaw 31–48, x̄ = 40; lower jaw 32–47, x̄ = 40; eye 6.4–9.7, x̄ = 8.6; interorbit 9.3–15, x̄ = 13.

General characters similar to those of *Gymnothorax conspersus.* Variation in head pores: five specimens with a fourth infraorbital pore on one side only; one specimen with five mandibular pores, two with seven; one specimen with a single branchial pore on one side, two with three on one side. Dentition: on upper jaw, anterior median tooth followed by four outer intermaxillary and 8–14 maxillary teeth on each side; 13–19 teeth on each side of lower jaw. No median intermaxillary teeth; vomerine teeth 0–7 (seven on the three smallest specimens examined).

Color. Head and trunk pale to medium greenish brown, becoming darker posteriorly to dark brownish-black on posterior third of tail; one individual (from Campeche Bank) pale overall, light brown rather than dark brownish-black posteriorly. Numerous distinct small pale ocellated spots anteriorly, becoming fewer and larger posteriorly; only three or four large white spots on posterior third of tail. Dorsal and ventral coloration uniform on body at any given point. Small specimens paler in color, spots not obviously ocellated (possibly due in part to fading of the few small specimens). Dorsal fin with body color pattern anteriorly, entire fin brownish-black posteriorly. Anal fin brown anteriorly to black posteriorly, without spots. Margins of pores, nostrils and eye ringed with dark pigment, not noticeable in small specimens, marked with increasing size.

Size. The 925-mm specimen examined, a ripe male from Campeche Bank at 139–143 m, is the largest specimen known to date. Of 33 specimens, 20 (290–890 mm) are females, six (310–780) ripe with 1.0–1.7-mm eggs; 11 (400–925) are males, the 925-mm one ripe.

Development. Leptocephali that are this species or *Gymnothorax conspersus* have been identified (p. 909).

Behavior. Adult *Gymnothorax kolpos* has most often been collected between 75–100 m; like *Gymnothorax conspersus,* no small specimens have yet been taken (the smallest a 290-mm female), and the habitat of young remains unknown. Gut contents on radiographs show a 1-inch section of vertebrae in one eel and an entire fish in another, otherwise nothing identifiable.

Distribution (Fig. 160). When described, *Gymnothorax kolpos* had been collected only on the eastern and northern continental shelf of the Gulf of Mexico (from the Straits of Florida to Louisi-

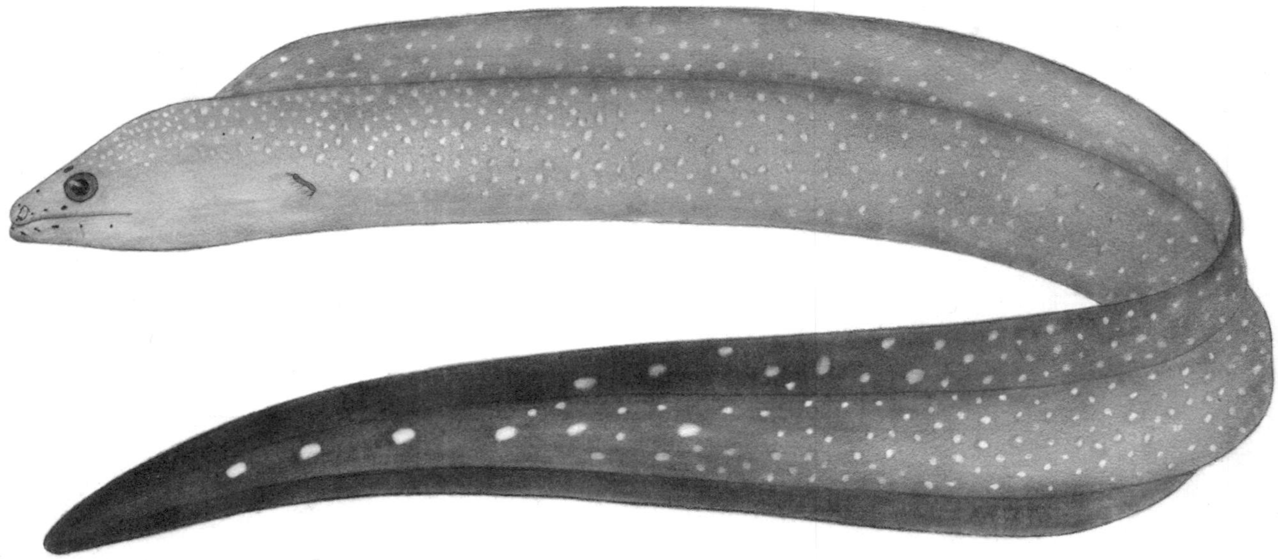

FIGURE 161. *Gymnothorax kolpos*: ANSP 107093, holotype, 645 mm TL, Gulf of Mexico off Alabama. Illustration by G. H. Jensen.

ana), and from the Bay of Campeche, Mexico. Recently, two specimens (one not seen) were taken off North Carolina, and one off Georgia, indicating that it is not restricted to the Gulf of Mexico. Most specimens were taken by trawl at depths of 48–230 m (usually 75–100 m); the North Carolina specimens were obtained by bottom long line at 190–198 m. Usually single specimens taken, only on four occasions were two collected together. One collection had one *G. kolpos* with two *G. nigromarginatus*; it has otherwise not been taken with other moray species.

Etymology. From the Greek *kolpos* (gulf), referring to its Gulf of Mexico habitat. A noun in apposition.

Remarks. The uniformly dark dorsal fin and the very dark tail with three or four sharply contrasting large white spots are diagnostic of this species at all sizes and are reliable characters to separate *Gymnothorax kolpos* from the other spotted species of *Gymnothorax* in the Atlantic. The Pacific species, *Gymnothorax equatorialis*, has a dark tail and sometimes appears similar, but the anterior head and trunk spots are not uniform in size and are not ocellated.

Study Material. A total of 33 specimens, 290–925 mm TL, including the holotype and 24 paratypes.

Holotype: ANSP 107093 (645 mm TL); Alabama, 29°19′N, 88°04′W, 120 m; OREGON sta. 3201; 2 Feb. 1961.

Paratypes: STRAITS OF FLORIDA: ANSP 134031 (1,

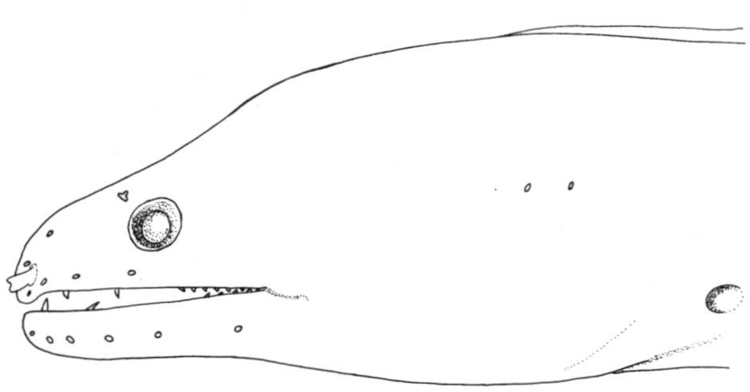

FIGURE 162. *Gymnothorax kolpos*: ANSP 134031, paratype, 463 mm TL, Florida Keys; head.

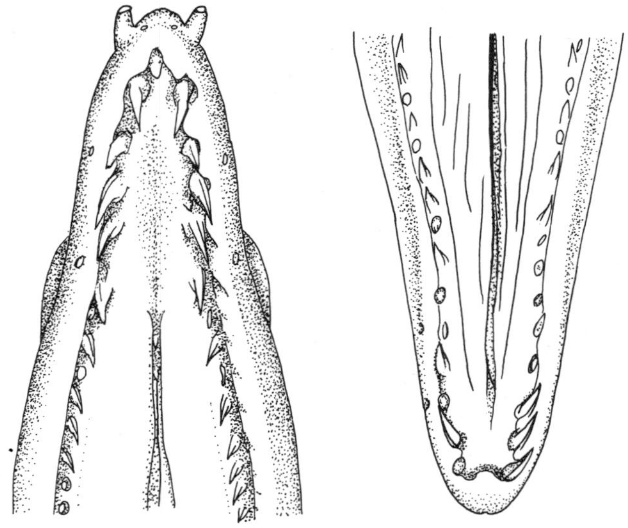

FIGURE 163. *Gymnothorax kolpos*: ANSP 107092, paratype, 607 mm TL, Gulf of Mexico off Louisiana; dentition.

463). FLORIDA KEYS: UMML 17424 (1, 290), 20524 (1, 321). GULF OF MEXICO: *Florida:* MCZ 35110 (1, 607). MNHN 1974-105 (1, 384). UF-FSU 20683 (1, 407), 20721 (1, 310), 22266 (1, 501), 22292 (1, 380). *Alabama:* ANSP 127501 (1, 344, cleared and stained). GCRL 1066 (1, ca 590). USAIC 03127 (2, 319–407). *Louisiana:* ANSP 107092 (1, 607), 116135 (1, 790). GCRL 2790 (1, ca 780). TU 82455 (2, 518–525). USNM 154982 (1, 400), 158739 (1, 567+), 158741 (1, ca 630), 215279 (1, 652). *Texas:* ANSP 107098 (1, 573, cleared and stained). *Mexico:* Yucatan: USNM 158729 (1, 556).

Other material: NORTH CAROLINA: ANSP 157317 (1, 840). GEORGIA: GMBL 84-111 (1, 883). GULF OF MEXICO: *Campeche Bank:* UF 28450 (1, 554), 28453 (2, 765–815), 28455 (1, 925), 28462 (1, 790), 28465 (1, 552).

Gymnothorax saxicola Jordan and Davis, 1891

HONEYCOMB MORAY

Figures 164–167, Tables 9, 14, 18–20

Muraena ocellata (not of Agassiz). Richardson, 1848a: 82 (in part, one specimen from the Gulf of Mexico).
Priodonophis meleagris (not of Shaw). Poey, 1867b: 262.
Gymnothorax ocellatus (not of Agassiz). Goode and Bean, 1879b: 154; 1880: 344.
Muraena ocellata (not of Agassiz). Jordan and Gilbert, 1882a: 260.
Sidera ocellata (not of Agassiz). Jordan, 1884: 42.
Gymnothorax ocellatus saxicola Jordan and Davis, 1891: 606 (replacement name for *meleagris* in the Atlantic, holotype CAS-SU 1658).
Lycodontis ocellatus saxicola. Jordan and Evermann, 1896a: 399.
Gymnothorax saxicola. Ginsburg, 1951: 461 (neotype, USNM 34280, designated).

Discussion of Synonymy. The western Atlantic morays with serrate teeth and spotted coloration, commonly taken by trawl in coastal waters, have variously been considered to be one species (*Gymnothorax ocellatus*), three species (*Gymnothorax saxicola, G. nigromarginatus,* and *G. ocellatus*), or three subspecies of *G. ocellatus.* (See *Remarks* for chronological history of names.) We treat them herein as three species. Synonymies include names and name combinations published in taxonomic works which were based on specimens examined. We do not cite names published in the many check lists and faunal studies based on material not seen by the authors.

Jordan and Davis did not designate a type for their *Gymnothorax ocellatus saxicola* var. nov., but gave "Snapper Banks at Pensacola" as the locality. Ginsburg later (1951: 461) designated USNM 34280 from Cedar Keys, Florida as neotype, stating "As no type appears to have been set aside by Jordan and Davis, the above specimen is hereby designated as the neotype. The authors mention Cedar Keys in their account and the specimen designated is possibly one of those examined by them; but there is no way now of definitely identifying their specimens." However, a Stanford specimen (SU 1658), now in the CAS collection, was listed as the holotype by Böhlke (1953: 47). The catalog entry, confirmed to be in Jordan's handwriting by M. E. Anderson (1987, in litt.), lists "*Lycodontis ocellatus saxicola* [the original entry of *Gymnothorax ocellatus* crossed out]; Pensacola; D. S. Jordan; 1883; Type." There is no reason to doubt the validity of the catalog information for this specimen, and we consider it to be the holotype. According to ICZN Article 75h, "If, after the designation of a neotype, the holotype . . . is found still to exist, the case is to be referred to the Commission to rule whether the neotype is or is not to be retained as the name-bearing type," an action which we will pursue, recommending that the neotype designation be invalidated.

Distinctive Characters. A moderate-size moray with a "honeycomb" color pattern of pale polygonal spots enclosed by brown reticulations, dorsal fin patterned with a bold undulating black and white margin, and anal fin with uniform dark edge; three infraorbital pores; teeth serrate and uniserial; MVF 6-53-140.

Description. Counts and proportions are presented in Table 19. Head pores: SO 1 + 2; IO 3; M 6; B 2.

Head moderately elongate, muscular, its profile from snout tip to shortly behind rictus uniform or slightly convex. Snout blunt and rounded; jaws moderate and subequal, the lower jaw slightly shorter. Eye large, above mid-jaw. Anterior nostril tubular; posterior nostril a rimmed pore above anterior half of eye. Gill opening small, rounded or elliptical, slightly below midside. Head pores rimmed, positioned as illustrated; posteriormost pores on snout and jaws in dimpled depressions; branchial pores very small

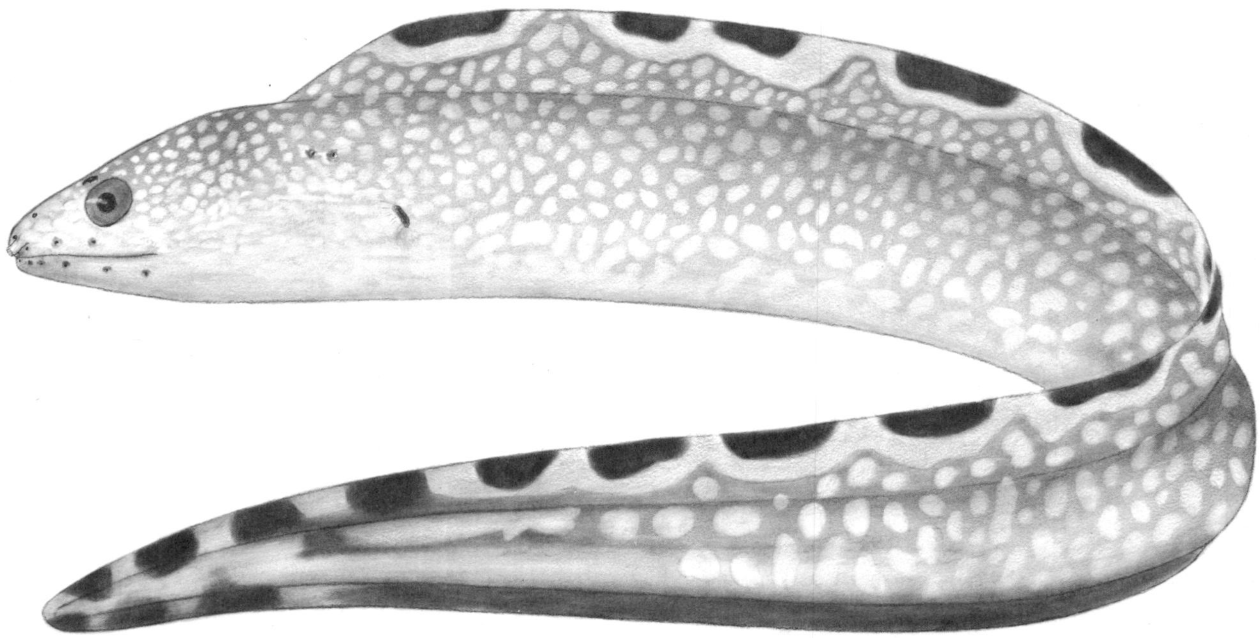

FIGURE 164. *Gymnothorax saxicola*: ANSP 117429, 378 mm TL, Gulf of Mexico off Florida. Illustration by G. H. Jensen.

and difficult to see; no variation noted in numbers of pores.

Dentition uniserial; teeth large, triangular and compressed, shark-like, with serrate edges. On upper jaw, one median tooth at tip of snout; on each side, four increasingly larger stout outer intermaxillary teeth, closely followed by 8–12 increasingly smaller teeth on maxillary. No median intermaxillary teeth; vomerine teeth conical and very small, the series beginning below mid-eye, curving upward in roof of mouth and disappearing in folds of muscle, often not visible ex-

TABLE 19. Summary of counts and proportions (in percent) of *Gymnothorax saxicola, G. nigromarginatus,* and *G. ocellatus.*

	saxicola			*nigromarginatus*			*ocellatus*		
	Range	x̄	n	Range	x̄	n	Range	x̄	n
Predorsal vertebrae	4–7	5.8	235	5–8	6.3	49	4–8	6.3	167
Preanal vertebrae	47–59	53.1	237	45–52	48.6	54	46–54	50.1	175
Total vertebrae	134–147	140.2	245	137–147	141.5	48	136–151	144.6	165
Proportion of TL									
(TL in mm)	212–496	382	30	218–527	392	30	212–557	387	30
Predorsal	10–16	13	30	10–13	12	30	9.0–14	12	30
Preanal	41–50	46	30	39–45	42	30	41–48	44	30
Head	13–16	14	30	11–15	14	30	12–17	14	30
Depth at gill opening	4.6–9.0	7.4	30	5.4–7.3	6.4	30	4.9–9.4	6.9	30
Depth at anus	4.1–7.2	5.5	30	4.0–6.5	5.4	30	3.7–8.0	5.5	30
Proportion of HL									
Snout	16–22	20	30	14–22	18	30	16–21	19	30
Upper jaw	38–51	44	30	34–54	41	30	34–50	43	30
Lower jaw	36–48	41	30	33–51	41	30	32–46	41	30
Eye	7.1–15	11	30	7.4–14	9.8	30	8.1–13	10	30

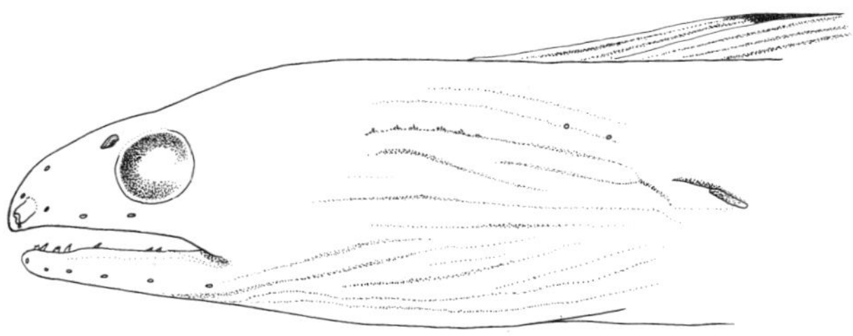

FIGURE 165. *Gymnothorax saxicola*: ANSP 115263, 296 mm TL, Florida Atlantic; head.

cept with dissection. On lower jaw a single row of 9–16 teeth, the first 3–4 increasingly large, the remaining series abruptly smaller.

Body elongate, tail tapering. Dorsal fin high and fleshy, its origin about midway between rictus and gill opening; anal fin well developed.

Color. Typically, brown or yellowish-brown background color forming reticulated pattern around pale or yellowish irregular polygonal spots; honeycomb pattern prominent on back and sides, extending faintly onto belly on highly colored specimens, belly pale on others. On some large specimens a series of uniform round spots extending along mid-body. In some specimens, the honeycomb pattern is less prominent, body coloration appears more as pale spots on a brown background, and the reticulated pattern is seen only on the head (see additional comments under *Remarks*). Large pale areas on tail; tip of tail pale. Reticulated pattern dorsally on head, with two slightly darker areas, one between eyes and one just before dorsal-fin origin; underside of head pale, narrow dark lines visible on large specimens; conspicuous dark ring around eye.

Distal edge of dorsal fin with striking pattern of saddle-shaped black areas outlined by undulating bright white stripe along middle and edge of fin; body color pattern extending onto base of fin. Anal-fin margin dark, without black and white pattern; body color extending onto fin basally, grading to dark edge, occasionally a few pale areas near end of tail.

Size. The largest specimen examined was 620 mm TL. Most of the study material ranged from 300–450 mm, and many specimens had developing or ripe gonads; of 46 females (254–620 mm), 36 (295–620) had 0.5–1.3-mm eggs, and one 523-

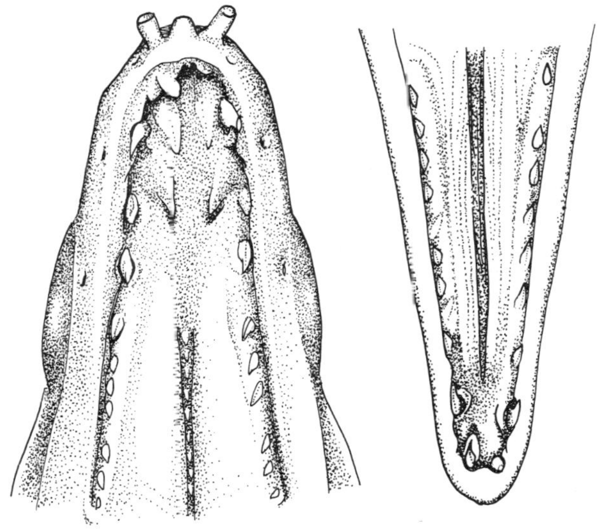

FIGURE 166. *Gymnothorax saxicola*: data as for Figure 165; dentition.

mm specimen was distended with 2.6-mm opaque eggs. Of 38 males (305–555), four (406–555) were ripe. Females were gravid at relatively small sizes, a 275-mm specimen was filled with 1.3-mm eggs.

Development. Leptocephali of the *Gymnothorax saxicola–nigromarginatus–ocellatus* complex have been identified (p. 907). At least two of the species are separable by myomere count, with variation and intergradation similar to that found in vertebral counts of adults.

Behavior. The species in this complex, like other serrate-toothed forms, live at moderate depths along the continental shelf. Almost all specimens were collected by trawl, and most were adults or sub-adults; nothing is known of the habitat or habits of young and juveniles.

Distribution (Fig. 167). *Gymnothorax saxicola* is the common moray obtained by trawls off continental United States along the Atlantic coast, from North Carolina to Florida (the specimen recorded from New Jersey, received long ago from a sea captain, likely a waif or the locality imprecise), and in the eastern Gulf of Mexico off Florida and Alabama to the Mobile Bay drainage. West of that point, it is replaced by *G. nigromarginatus.* Almost all specimens were taken by trawl at depths commonly ranging from 2–86 m (four collections recorded from depths of 128–213 m);

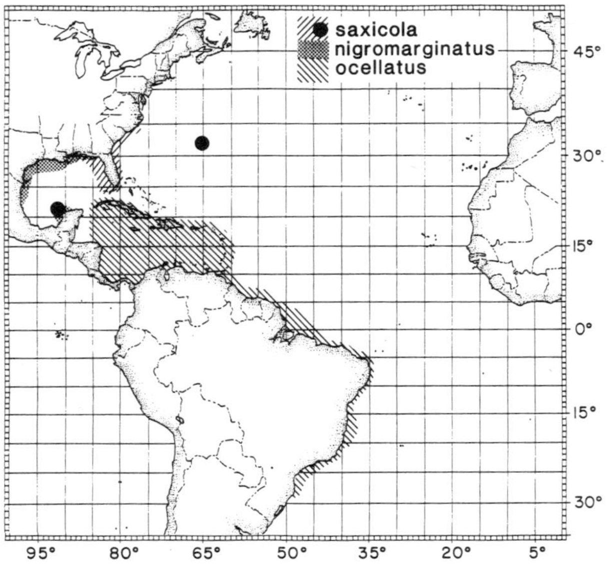

FIGURE 167. Distribution of *Gymnothorax saxicola, G. nigromarginatus,* and *G. ocellatus* (subgenus *Neomuraena*) based on study material.

occasionally taken by hook and line, but no records of poison collections.

Variation. A few specimens from the Mobile Bay boundary area exhibit variation in some characters and do not fit well in our concept of any of the species in this complex (three lots, UF-FSU 15713, UF-FSU 18482, and UF 19606). They had the high preanal vertebral count and median-fin coloration of *saxicola,* the small body spots characteristic of *G. nigromarginatus,* and a slightly higher total vertebral count.

Etymology. From the Latin *saxum* (a rock) and *cola* (inhabitant), in reference to the habitat of the species. A noun in apposition.

Remarks. The earliest record of a species in this complex was that of Quoy and Gaimard (1824), who described and illustrated a spotted moray from Rio de Janeiro under the name of an Indo-Pacific species, *Muraena meleagris* Shaw. In 1829, Agassiz named *Gymnothorax ocellatus* from the same locality. Variation in coloration was noted as early as 1848 when Richardson (1848a: 82) described the aberrant color pattern of one specimen from the Gulf of Mexico (this specimen, examined by us, is a typical example of *G. saxicola*). In 1859 Girard described a third color form from the Texas coast as *Neomuraena nigromarginata.* In 1867 Poey referred two color forms taken off Cuba to *G. ocellatus* and *G. meleagris* (= *saxicola*). In 1891 Jordan and Davis published the name *Gymnothorax ocellatus saxicola* (as a "variety" or subspecies) for Quoy and Gaimard's Atlantic *meleagris,* and described the honeycomb pattern of this species.

Ginsburg (1951) was the first to have adequate samples at hand to compare the three forms, a total of 62 specimens. He presented data showing variation in seven characters (all with considerable overlap and intergradation) in addition to the differences in coloration. He compared three size groups of the three species and demonstrated differences in tail length (*saxicola* with a considerably shorter tail), head length (*ocellatus* with a shorter head), and body depth (*saxicola* with a deeper body). (Values given for his single large 611-mm specimen of "*G. nigromarginatus,*" MCZ 35110, must be discounted; it is a *Gymnothorax kolpos.*) Ginsburg stated that "In general, the differences between them are somewhat like the differences between two subspecies . . . ," but he concluded "the degree of divergence is perhaps greater than what is usual between two coordinate subspecies," and decided to treat them as independent species.

We have studied over 1000 specimens collected throughout the range of the three forms, and have compared vertebral counts as well as the conventional characters studied by Ginsburg. In unpublished studies in 1974 by E. B. and J. E. Böhlke and W. Moss (then at ANSP), 17 character states were evaluated for 117 specimens representative of the three forms, and several computer analyses were carried out. Based on these studies and on all data now available, we conservatively recognize the three species, but note that, while most specimens conform with the descriptions and illustrations here presented, there are also some that present intermediate and confusing combinations of characters.

Table 20 lists and summarizes some of the characters studied and used to separate the species. They are most readily identifiable by color pattern. *Gymnothorax saxicola* is distinguished by the large pale honeycomb-like spots separated by narrow brown reticulations on the

TABLE 20. Summary of distinguishing characters for *Gymnothorax saxicola, G. nigromarginatus,* and *G. ocellatus.*

	saxicola	*nigromarginatus*	*ocellatus*
General characters			
Head	intermediate	shortest	longest
Tail	shortest	longest	intermediate
Body depth	greatest	least	intermediate
Total vertebrae	fewest	intermediate	most
Preanal vertebrae	most	fewest	intermediate
Coloration			
Head			
Spot size (mm)[1]	29	19	24
Spot number[2]	26	17	10
Reticulations	present	absent	absent
Body			
Spot size[3]	52	21	35
Spot number[4]	17	17	10
Between-spot size[5]	13	20	32
Spot shape	elongated	round	round or irregular
Dorsal fin	patterned	dark margin	patterned
Anal fin	dark margin or dark	dark margin or dark	patterned

[1] Diameter of largest spots, in mm × 10.
[2] Number of spots.
[3] Diameter of largest spots, midway between gill opening and anus, in mm × 10.
[4] Number of spots in area midway between gill opening and anus, one eye-diameter wide from midside to dorsal-fin base.
[5] Average distance between spots in area at midside, midway between gill opening and anus, in mm × 10.

head and body, the boldly patterned dorsal fin, and a uniformly dark anal fin. In the eastern Gulf of Mexico where it coexists with *G. nigromarginatus,* the head and body coloration may appear more as large pale spots on a brown background, but the fin coloration is distinctive.

Morphometric and meristic differences are less distinctive. *Gymnothorax saxicola* has the greatest preanal distance, with the corresponding shortest tail and highest preanal vertebral count; it has the greatest body depth and the fewest total vertebrae. *Gymnothorax nigromarginatus* has the shortest preanal distance and preanal vertebrae count, and the longest tail, but it does not differ greatly from *G. ocellatus;* it is the most elongate of the three, and has the least body depth. *Gymnothorax ocellatus* has the longest head, the most total vertebrae, and is intermediate between the other two in other characters. *Gymnothorax saxicola* is the most distinctive and identifiable of the three species, the most definitive character the preanal vertebral count and corresponding preanal distance.

The three species are separated geographically (Fig. 167). They exist sympatrically at the boundaries, where there also are some non-conforming individuals (noted under *Variation* in the species accounts).

No species of the subgenus *Neomuraena* has been found in the eastern Atlantic. However, similar spotted species occur in the eastern Pacific. *Gymnothorax equatorialis* (Hildebrand, 1946) is very similar in appearance to the *saxicola-nigromarginatus-ocellatus* complex; it is separable by subtle differences in coloration. The body spots are most like those of *ocellatus,* but the tail is noticeably dark with only a few large pale spots, much like the tails of *conspersus* and *kolpos;* both dorsal and anal fins have uniform dark edges; and the MVF is 7-53-146. Other eastern Pacific species, *Gymnothorax angusticeps* (Hildebrand and Barton, 1949), *G. serratidens* (Hildebrand and Barton, 1949), and a third unidentified species, have noticeably different color patterns and vertebral counts.

Study Material. A total of 426 specimens, 167–620 mm TL, including the holotype.

Holotype: CAS-SU 1658 (289); Florida, Snapper Banks at Pensacola; D. S. Jordan; 1883.

Other material: BERMUDA: ANSP 126241 (1, 447), 140435 (1, 535). FMNH 48439 (1, 419), 49049 (1, 466). COASTAL UNITED STATES: *New Jersey?:* ANSP 995 (1, 361). *North Carolina:* ANSP 115232 (2, 400–426). GMBL 73-196 (1, 366), 73-197 (1, 408). USNM 35142 (1, 442), 70301 (1, 381), 154990 (1). *South Carolina:* ANSP 86212 (1, 540), 115237 (1, 497), 127790 (2, 380–415), 128385 (1, 412), 131488 (3, 508–609), 131489 (3, 463–532), 131491 (5, 416–518), 131492 (4, 404–483), 131497 (1, 400), 131498 (1, 503), 131500 (1, 450), 131505 (2, 396–442), 131506 (1, 456), 150526 (1, 474). GMBL 56-14 (1, 430), 71-241 (2, 278–279), 73-33 (1, 470+), 73-297 (2, 344–372), 73-305 (4, 392–478). MNHN B.2443 (1, 429). USNM 25383 (1, 500), 25384 (1, 498), 25389 (1, 510), 25390 (1, 483), 25393 (1, 578), 154989 (1). *Georgia:* ANSP 115240 (2, 404–482), 131504 (2, 420–543). USNM 154986 (1), 154987 (1), 154988 (1), 156846 (1, 460), 158721 (1, 424), 190391 (1, 465), 190392 (1, 388). *Florida Atlantic:* ANSP 115507 (1, 395, cleared and stained), 115226 (1, 264), 115233 (1, 430, cleared and stained), 115243 (2, 385–397), 115244 (1, 395), 115252 (1, 412, cleared and stained), 115253 (1, 395+), 115262 (1, 310), 115263 (1, 296), 115265 (1, 212), 115273 (1, 202), 115277 (5, 380–448), 116121 (2, 342–370), 131490 (3, 452–504), 131493 (3, 365–431), 131494 (1, 344), 131495 (1, 167), 131499 (1, 475), 131501 (2, 493–408), 131502 (1, 444), 131503 (1, 372), 138395 (1, 303), 138396 (1, 128), 138397 (1, 334), 138398 (2, 141–249), 138413 (1, 141), 138414 (1, 340), 138415 (1, 285), 138416 (2, 280), 158134 (1, 425), 158135 (1, 368), 158157 (1, 387), 158158 (1, 523). GMBL 73-268 (1, 364), 73-276 (1, 297), 73-278 (1, 370+), 73-281 (2, 345–396), 74-224 (2, 220–240). UF 21337 (1, 187). USNM 156845 (1), 158722 (1, 420), 158723 (2, 355–357), 158724 (1, 368), 159418 (1, 430). *Florida Straits:* FMNH 45345 (1), 69195 (1). MCZ 9012 (1, 430). *Florida Keys:* ANSP 115258 (1, 385), 138399 (1, 415), 158136 (5, 295–395), 158137 (1, 254), 158138 (1, 340), 158139 (1, 370). MCZ 46607 (1, 385), 46659 (1, 375). UMML 537 (1, ca 340). UMML 1177 (2, 299–370), 2353 (1, 355), 4784 (5, 275–350). USNM 6029 (1), 134158 (1, 290), 158699 (1, 330), 158716 (2, 352), 158719 (1, 317).

GULF OF MEXICO: BMNH uncat. (1); J. Richardson. *Florida:* ANSP 115218 (2, 374–422), 115246 (1, 408), 115248 (1, 440), 115267 (1, 236, cleared and stained), 116638 (1, 397, cleared and stained), 117429 (1, 378), 158141 (2, 374–415), 158142 (1, 358), 158143 (1, 450), 158144 (1, 375), 158145 (3, 344–407), 158146 (1, 437), 158147 (1, 407), 158148 (2, 373–395), 158149 (1, 376), 158150 (1, 412), 158151 (2, 400–414), 158153 (1, 402), 158154 (1, 420), 158155 (2, 375–436), 158156 (1, 361), 158159 (1, 360). CAS-IUM 1458 (1, 346), 3549 (1, 370). CAS-SU 7475 (1, 422). MCZ 2461 (1, 442), 9161 (2, 297–353), 45945 (6, 335–459), 45946 (6, 320–422), 45947 (2, 365–396), 45949 (4, 327–377), 45950 (4, 305–401), 45951 (4, 313–420). MNHN 1974-201 (5). NMW 61522 (1), 61649 (1). UF-FSU 3355 (4, 132–184), 5262 (2, 258–324), 10948 (3, 331–348), 13343 (1, 318), 13687 (1, 379), 15713 (1, 350), 15716 (1, 265), 15717 (1, 261), 17003 (1, 238), 17990 (3, 225–327), 18482 (5, 253–423), 18880 (1, 395), 18889 (4, 135–184), 19086 (1, 314), 19606 (1, 198). UMMZ 153418 (1, 186), 154119 (1, 225), 154313 (1, 230), 154316 (1, 300), 154830 (3, 362–445), 154944 (1, 376), 154952 (2, 353–410), 155061 (1, 340), 155079 (1, 419). USNM 5160 (20, 135–417), 5997 (3, 290–444), 23634 (1, 334), 30779 (1, 200+), 32802 (1, 383), 34280 (1, 425, Ginsburg's designated neotype), 39881 (1, 367), 84549 (1, 266), 133800 (4, 248–405), 152241 (3, 410–503), 152255 (2, 309–432), 154991 (1), 157576 (1, 358), 158559 (1), 158701 (9), 158702 (7, 377–464), 158703 (9, 260–405), 158704 (13, 275–376), 158706 (1, 450), 158707 (1, 400), 158708 (3, 317–352), 158709 (1, 440), 158710 (1, 320), 158711 (1, 452), 158712 (2, 366–372), 158713 (10, 353–416), 158714 (3, 381–435), 158715 (1, 296), 158717 (1, 435), 158731 (5), 202588 (1, 384), 278179 (9, 239–383). *Alabama:* GCRL 1067 (1). USNM 101510 (1, 420), 158700 (1, 382), 158705 (2, 320–442), 158718 (1, 410), 158720 (1, 418), 158721 (1, 424). *Mississippi:* ANSP 115220 (1, 496), 115268 (1, 234). GCRL 1068 (1). *Louisiana:* GCRL 1176 (2). *Yucatan, Gulf of Campeche:* FMNH 45345 (1, 382), 69195 (1, 380). UMMZ 102286 (1, 408), 143091 (1, 423), 174108 (1, 400).

EXACT LOCALITY UNKNOWN: ANSP 115261 (2, 319–348), 115264 (1, 287). USNM 158736 (15, 319–415), 158737 (1, 350).

Gymnothorax nigromarginatus (Girard, 1859)

BLACKEDGE MORAY

Figures 167, 168, Tables 9, 14, 18–20

Neomuraena nigromarginata Girard, 1859a: 76, pl. 61 (original description, Texas, holotype USNM 7004).

Gymnothorax ocellatus (not of Agassiz). Goode and Bean, 1879b: 155 (in part).

Sidera nigromarginata. Jordan and Evermann, 1887: 473.

Gymnothorax ocellatus nigromarginatus. Jordan and Davis, 1891: 606.

Lycodontis ocellatus nigromarginatus. Jordan and Evermann, 1896a: 399.

Gymnothorax nigromarginatus. Ginsburg, 1951: 461 ("neotype" designated).

FIGURE 168. *Gymnothorax nigromarginatus*: ANSP 158161, 444 mm TL, Gulf of Mexico off Louisiana. Illustration by G. H. Jensen.

Discussion of Synonymy. Ginsburg (1951: 462) designated USNM 7004 as neotype, noting that "it is very likely that it [USNM 7004] is Girard's original type." There is little doubt that it is indeed the type specimen, which was inadvertently cataloged twice, and we consider it to be the holotype.

See also *Discussion of Synonymy* under *Gymnothorax saxicola.*

Distinctive Characters. A moderate-size moray with uniformly small, well-separated white spots on brown background, the dorsal and anal fins with black margins; three infraorbital pores; teeth serrate and uniserial; MVF 6-49-142.

Description. Counts and proportions are presented in Tables 18 and 19. General characters and dentition similar to those of *Gymnothorax saxicola.*

Color. Body color medium brown with uniform-size, well-separated, small white or yellowish spots, the spots fading on belly and diffuse at end of tail; belly pale. Head brown, with small white spots; dark areas on top of head between eyes and before dorsal-fin origin; dark ring around eye. Dorsal fin with dark margin, sometimes with slight undulations but without prominent black and white pattern, anal fin with uniform dark edge; body color pattern extending onto fins to above mid-fin. Tip of tail pale.

Size. The largest specimen studied was 527 mm TL. Of 13 females (328–620 mm), 10 (328–620) were ripe with 0.6–1.3-mm eggs; of 27 males (292–555), three (404–555) were ripe.

Development. Leptocephali of the *Gymnothorax saxicola–nigromarginatus–ocellatus* complex have been identified (see p. 907).

Distribution (Fig. 167). *Gymnothorax nigromarginatus* occurs in the Gulf of Mexico from the Mobile Bay area west to Texas, off Yucatan and Honduras. It has been taken by trawl at depths of 10–91 m.

Variation. Two specimens (ANSP 1056 and ANSP 115266) from the northern Gulf of Mexico and one (ANSP 115271) from off Yucatan had characteristics and coloration of the species, but had a patterned dorsal fin.

Etymology. From the Latin *niger* (black) and *marginis* (border), with the suffix *atus* (provided with), in reference to the fin color. An adjective.

Remarks. See discussion under *Remarks* for *Gymnothorax saxicola.*

Study Material. A total of 74 specimens, 121–527 mm TL, including the holotype.

Holotype: USNM 7004 (391); St. Joseph Island, Texas; G. Wurdeman.

GULF OF MEXICO: *Florida:* ANSP 1056 (1, 340), 158110 (1, 402), 158111 (1, 379). CAS-SU 7476 (1, 302). UF-FSU 4624 (1, 235), 10947 (1, 450), 16997 (1, 382). UMML 871

FIGURE 169. *Gymnothorax ocellatus*: ANSP 115254, 390 mm TL, Colombia. Illustration by G. H. Jensen.

(2, 283–308), 2689 (1, 420). USNM 44867 (1, 356), 152580 (1, 218), 158702 (1, 400), 158735 (1, 328). *Alabama:* ANSP 115266 (1, 382). USNM 158733 (1, 476). *Mississippi:* GCRL 1069 (1, 404), 1070 (1, 388), 1177 (2, 380–527), 1539 (1, 409). USNM 158728 (1, 540), 158731 (5, 430–500). *Louisiana:* ANSP 158161 (1, 444), 158820 (1, 394). FMNH 78577 (1, 398). MNHN 1974-104 (1, 402). TU 57976 (8, 294–440). USNM 131153 (2, 370–426), 154980 (1, 426), 154983 (1, 336), 158727 (3, 440–502), 158732 (1, 432), 158734 (1, 405), 158740 (1, 436), 162582 (1, 395). *Texas:* ANSP 127459 (1, 361, cleared and stained). FMNH 40299 (1, 295), 45062 (1, 391), 45063 (3, 351–418), 45064 (1, 443). USNM 154981 (1, 348), 154984 (1, 330), 154985 (1, 291). *"Western":* ANSP 158797 (1, 383). MEXICO, YUCATAN: ANSP 115271 (1, 121). FMNH 45347 (1, 477), 45348 (4, 388–396), 45953 (1, 423). HONDURAS: FMNH 69188 (2, 217–325).

EXACT LOCALITY UNKNOWN: USNM 158738 (3, 313–445).

Gymnothorax ocellatus Agassiz, 1831
OCELLATED MORAY
Figures 167, 169, Tables 9, 14, 18–20

Murene pintade = *Murena meleagris* (not of Shaw). Quoy and Gaimard, 1824: 245, pl. 52, fig. 2.

Gymnothorax ocellatus Agassiz in Spix and Agassiz, 1831: 91, pl. 50b (original description, Brazil, syntype MHNN 1193).

Muraena ocellata. Jenyns, 1842: 145.

Muraena tricolor Banks and Solander in Richardson, 1848a: 82 (manuscript name listed in synonymy).

Murenophis ocellata. Castelnau, 1855: 82.

Murenophis variegata Castelnau, 1855: 83, pl. 43, fig. 2 (original description, Rio de Janeiro, holotype MNHN B.2440).

Priodonophis ocellatus. Kaup, 1860: 22.

Sidera ocellata. Jordan, 1886: 34.

Sidera variegata. Jordan, 1887: 566.

Gymnothorax ocellatus ocellatus. Jordan and Davis, 1891: 606.

Gymnothorax ocellatus januarius. Jordan and Davis, 1891: 607 (replacement name for *Murenophis variegata* Castelnau, preoccupied).

Lycodontis ocellatus. Jordan and Evermann, 1896a: 399.

Lycodontis jordani Evermann and Marsh, 1902: 78, pl. 2 (original description, Puerto Rico, holotype USNM 79458).

Gymnothorax jordani. Meek and Hildebrand, 1923: 167.

Discussion of Synonymy. See *Remarks* for *Gymnothorax saxicola.*

Distinctive Characters. A moderate-size moray with color pattern of scattered white spots on a brown background, dorsal and anal fins patterned with alternating black and white marks; three infraorbital pores; teeth serrate and uniserial; MVF 6-50-145.

Description. Counts and proportions are given in Tables 18 and 19. General characters and dentition similar to those of *Gymnothorax saxicola.*

Color. Body color medium brown, darker dorsally shading to pale ventrally, with distinct, well-separated, white spots of varying sizes and shapes, the largest about eye-size, becoming diffuse on tail; a row of uniform round spots along midside.

Head brown, with small white spots; slightly darker areas dorsally between eyes and before dorsal-fin origin; dark ring around eye. Dorsal fin with distinctive bold pattern of black and white marks on distal half of fin, body pattern extending onto base of fin. Anal fin similarly but less distinctly patterned. Tip of tail pale.

Size. The largest specimen examined was 580 mm TL. Of 42 females (278–487 mm), 31 (278–487) contained eggs 0.5–1.4-mm diameter, and three specimens (385–430) were greatly distended with 1.8–2.1-mm opaque eggs. Of 62 males (334–520), two (415–428) were ripe.

Development. Leptocephali of the *Gymnothorax saxicola–nigromarginatus–ocellatus* complex have been identified (see p. 907).

Distribution (Fig. 167). *Gymnothorax ocellatus* is the most common moray taken by trawl in the Caribbean and off northern South America. It is found throughout the Caribbean from the northern islands of the Greater Antilles to the Yucatan peninsula, and off the coasts of Central and South America to southern Brazil. Almost all specimens were taken by trawl at depths commonly ranging from 15–90 m; two collections were recorded at 140 and 160 m, and a few specimens were taken by hook and line. Not recorded from poison stations.

Etymology. From the Latin *ocellus* (a little eye) and the suffix *atus* (with), in reference to the coloration. An adjective.

Variation. Several specimens showed some variation in coloration. Two specimens from Nicaragua (ANSP 158160 and UCR 387-7) and one lot of three specimens from Honduras (UCR 351-50) had unusually small spots which resembled those of *G. nigromarginatus.* Two lots from Colombia (UMML 27253 and UMML 30176) and one lot from Venezuela (ANSP 115219), a total of five specimens, had numerous extremely small spots.

Remarks. See *Remarks* for *Gymnothorax saxicola.*

Study Material. A total of 412 specimens, 230–580 mm TL, including the holotypes of *Murenophis variegata* Castelnau and *Lycodontis jordani* Evermann and Marsh. (32 specimens from Suriname at RMNH not listed.)

CUBA: CAS-IUM 2918 (1, 362). MCZ 9106 (1, 364), 32806 (1, 370). USNM 33120 (1). JAMAICA: CAS-SU 4740 (4, 395–480). USNM 193606 (1, 278). HAITI: ANSP 83883 (1, 379). BMNH 1906.14.4 (1). USNM 133664 (1, 320). DOMINICAN REPUBLIC: ANSP 14539 (1, 415). PUERTO RICO: ANSP 134327 (1, 428), 158112 (1, 520). USNM 49358 (1, 400), holotype of *Lycodontis jordani.* LESSER ANTILLES: *Martinique:* USNM 178635 (1, 396). *Grenada:* ANSP 52407 (1, 290). *Curaçao:* ANSP 158133 (1, 465). *Tobago:* ANSP 115225 (1, 415), 115239 (2, 458–473), 115241 (3, 388–459), 115260 (1, 418), 131507 (2, 410–411). UMML 28838 (1, 411). *Trinidad:* ANSP 76383 (1, 367). PANAMA: ANSP 49836 (4, 342–384), 115222 (1, 395). CAS 30623 (1, 408). SIO 71-264 (6). UMML 26733 (1). UMML 26963 (1). USNM 79458 (1, 314). COSTA RICA: UCR 237-31 (1, 370). NICARAGUA: ANSP 158160 (1, 297). UCR 205-11 (1, 287), 387-7 (1, 290). HONDURAS: ANSP 158127 (2, 360–404), 158128 (1, 455), 158129 (1, 380), 158130 (1, 345). UCR 351-50 (3, 272–290). MEXICO: MCZ 34268 (1, 264).

COLOMBIA: ANSP 14773 (1, 387), 115223 (2, 350–412), 115247 (3, 345–394), 115249 (3, 387–447), 115251 (1, 430), 115254 (1, 390), 115257 (1, 412), 115259 (2, 350–399), 115278 (6, 330–379), 115285 (6, ca 385–455), 115287 (1, 450), 115289 (1, 395), 115509 (1, 370, cleared and stained), 127460 (1, 370, cleared and stained), 131508 (1, 317), 131509 (1, 357+), 131992 (1, 445), 158132 (1, 312). UMML 27253 (1, 310+), 28725 (1, 395), 30176 (3, 308–442). USNM 94760 (1, 315). VENEZUELA: ANSP 103994 (9, 440–580), 115219 (1, 410), 115235 (1, 504), 115236 (2, 519–530), 115245 (1, 396), 115255 (1, 377), 115281 (1, 444), 115282 (3, 404–557), 115283 (1, 492), 158113 (1, 438), 158114 (4, 396–540), 158115 (1, 310), 158126 (2, 318–345), 158131 (1, 455). MCZ 41013 (1, 442). MCZ 45934 (1, 362). UMML 28726 (4, 306–379). BRITISH GUIANA: ANSP 131994 (1, 380), 131998 (1, 417), 158117 (3, 225–274), 158118 (1, 257), 158119 (1, 328). USNM 158947 (1, 360), 158951 (1, 343), 158952 (1, 374), 158954 (1, 358), 158958 (1, 375). SURINAME: ANSP 101759 (18, 289–428). 115284 (1, 458), 115286 (1, 432), 115292 (1, 421), 131995 (1, 461), 138403 (1, 433), 158120 (2, 369–370), 158121 (1, 457), 158122 (4, 365–445). MCZ 45948 (4, 370–405). USNM 158946 (1, 411), 158948 (4, 360–428), 158949 (1, 342), 158950 (1, 437). FRENCH GUIANA: ANSP 100491 (7, 325–399), 115221 (1, 415), 115224 (1, 345), 131990 (1, 314), 131996 (1, 346), 131997 (1, 356), 131999 (2, 305–390), 136635 (2, ca 470–487), 136639 (2, 363–383), 136640 (2, 403–453), 136655 (13, 344–450), 138400 (2, 480–520), 138401 (4, 334–415), 138404 (5, 392–482), 158123 (6, 370–483), 158124 (2, 390–410), 158125 (5, 300–395), 158126 (1, 231). MCZ 45952 (1, 383). MNHN 1900-138 (1). USNM 158953 (1, 406), 158955 (1, 422), 158956 (1, 385), 158957 (2, 363–366).

BRAZIL: ANSP 136636 (4, 342–441), 136641 (2, 302–459), 136643 (1, 424), 136648 (2, 417–440), 136651 (1, 374), 136653 (5, 394–439), 136654 (4, 424–525), 138402 (1, 350),

138405 (18, 337–443), 138406 (4, 374–453). BMNH 1844.5.14.74 (1). BMNH 1847.4.21.19, 24, 25, 26 (4); Brandt. BMNH 1855.9.19.1288 (1); Haslar collection. BMNH 1852.9.13.87 (1); Zoological Society. BMNH 1913.12.4.208–13 (5 of 6). BMNH 1923.7.30.60–61 (1). BMNH uncat. (1); M. Stevens. BMNH uncat. (1); S. Stuart. CAS 11741 (1, 453), 54312 (4, 212–326), 59195 (2, 360–362), 59196 (1, 420), 59197 (1, 430), 59200 (1, 450), 59201 (1, 320). FMNH 69196 (2, ca 398–408). MBUCV 257 (1), 2818 (1, 542). MCZ 4716 (2, 350–379), 9067 (8), 9086 (2, 238–348). MHNG 146-79 (1, 425). MNHN B.2440 (1, 370); holotype of *Murenophis variegata*. MNHN B.2442 (1); Gay. MNHN B.2444 (3); Musee Geneve. MNHN B.2445 (4); Delalande. MNHN B.2446 (2); Castelnau. MNHN B.2447 (1, 338). MNRJ 1660 (1, 466), 1921 (1, 485), 2209 (1, 452), 4130 (1, 356), 9724 (1, 468). MZUSP 17 (1, 382), 26 (1, 324), 1349 (1, 388), 10069 (2, 368–378), 11866 (2, 369–374), 11867 (2, 390–406), 11868 (1, 437), 11869 (1, 410), 11871 (2, 473–488), 11873 (1, 444), 11874 (3, 466–490), 11875 (1, 464), 11876 (1, 448), 11877 (1, 557), 11878 (1, 400), 11879 (4, 355–450), 11880 (1, 471), 11881 (1, 481), 11882 (3, 423–436), 11886 (1, 519), 11887 (1, 282), 11888 (1, 274), 11889 (1, 472), 11890 (1, 373), 11891 (1, 404), 11892 (1, 457). NMW 61632 (3), 61633 (1), 61634 (2, 462–526), 61635 (1), 61637 (1), 61638 (2), 67653 (3). USNM 83385 (1, 394). ZMB 3990 (1); Sello.

EXACT LOCALITY UNKNOWN: ANSP 131509 (1, 355). MNHN B.2441 (1). MNHN B.2448 (1); Eydoux. MNHN B.2449 (2). NMW 61703 (1). USNM 6938 (1, 395).

Subgenus *Taeniophis*

Gymnothorax funebris Ranzani, 1839

GREEN MORAY

Figures 170–173, Tables 9, 14, 21

Muraena maculata nigra et viridis Catesby, 1738: 20, pl. 20.
Gymnothorax funebris Ranzani in Alessandrini, 1838: 433 (*nomen nudum*, listed only).
Gymnothorax funebris Ranzani, 1839: 14, pl. 4, fig. 2a–b (original description, Brazil, holotype at Bologne), a preprint of Ranzani, 1840. Ranzani, 1840: 77, pl. 11, fig. 2a–b.
Muraena lineo-pinnis Richardson, 1848a: 89 (original description, Puerto Cabello, holotype BMNH 1847.5.14.43).
Thyrsoidea aterrima Kaup, 1860: 22, pl. 3, fig. 1 (original description, Puerto Cabello, holotype ZMH 389).
Taeniophis Westphali Kaup, 1860 Supplement: 1 (original description, Caribbean Sea, holotype unknown).
Taeniophis lineopinnis. Kaup, 1860 Supplement: 3.
Taeniophis aterrima. Kaup, 1860 Supplement: 3.
Muraena infernalis Poey, 1861: 347, 354 (original description, Cuba, holotype unknown).
Gymnothorax infernalis. Poey, 1867b: 258.
Muraena afra (in part). Günther, 1870: 123 (West Atlantic specimens).
Muraena aterrima. Günther, 1870: 124.
Sidera funebris. Bean and Dresel, 1884b: 169.
Sidera aterrima. Jordan, 1887b: 565.
Lycodontis funebris. Jordan and Evermann, 1896a: 396.

Discussion of Synonymy. Jordan and Davis (1891: 603) considered *Thyrsoidea concolor* Abbott, 1861 to be a synonym of *Gymnothorax funebris*. We have examined and X-rayed the holotype of *concolor*, and find it to be conspecific with *G. moringa*.

Distinctive Characters. A large, uniformly dark green to brown moray; teeth smooth; jaw teeth uniserial, vomerine teeth biserial in half-grown and adults; MVF 6-60-140.

Description. Total vertebrae 137–142, $\bar{x}$ = 140.3 (n = 41); predorsal vertebrae 4–7, $\bar{x}$ = 5.6 (41); preanal vertebrae 57–61, $\bar{x}$ = 59.6 (41); dorsal-fin rays 340–361, $\bar{x}$ = 351 (5); anal-fin rays 217–237, $\bar{x}$ = 228 (5); kidney terminus at 69–71, $\bar{x}$ = 70 (4). Head pores: SO 1 + 2; IO 4; M 6; B 2.

Proportions as % (n = 33, 124–1250 mm): Of TL: predorsal 10–12, $\bar{x}$ = 11; preanal 43–49, $\bar{x}$ = 46; head 13–16, x = 14; depth at gill opening 4.4–10, $\bar{x}$ = 6.7; depth at anus 4.5–8.0, $\bar{x}$ = 6.0; width at gill opening 2.8–7.0, $\bar{x}$ = 4.2; width at anus 2.7–5.5, $\bar{x}$ = 3.8. Of HL: snout 14–20, $\bar{x}$ = 17; upper jaw 33–44, $\bar{x}$ = 38; lower jaw 30–44, $\bar{x}$ = 38; eye 6–11, $\bar{x}$ = 8.8; interorbit 11–14, $\bar{x}$ = 12.

Snout moderate; jaws moderate; eye above midjaw. Anterior nostril tubular, near tip of snout; posterior nostril pore-like, raised in adults and teardrop-shaped in largest specimens, situated above anterior margin of eye. Variation in head pores of 22 specimens examined; four with seven mandibular pores on one side, one with seven pores on both sides; nine specimens with three branchial pores on one side, one with three pores on both sides.

On each side of upper jaw, a row of 5–6 large stout intermaxillary teeth (flanked by 0–6 tiny teeth outside or between the main series), continuous with maxillary teeth, which consist of a main row of 10–18 triangular, posteriorly-directed teeth (a few tiny teeth sometimes present between or outside them), and a short anterior

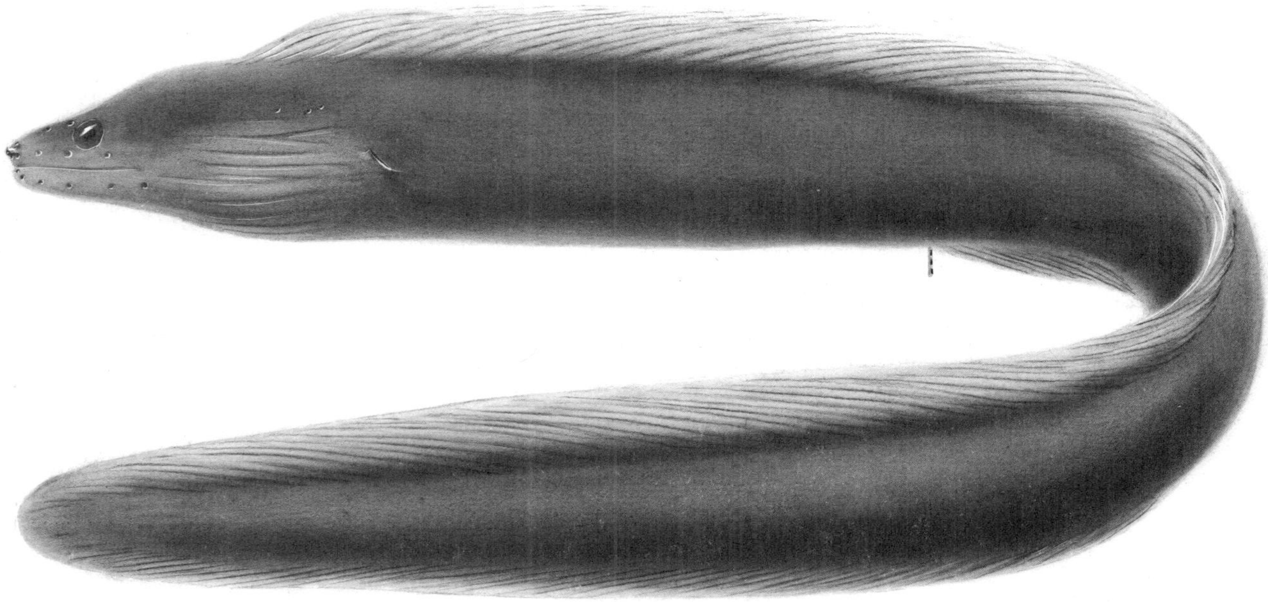

FIGURE 170. *Gymnothorax funebris*: ANSP 111550, 683 mm TL, Bahamas. Illustration by S. P. Gigliotti.

inner series of 1–3 long, slender teeth, the inner series frequently missing on large individuals. Along midline, anteriorly three large intermaxillary fangs, increasing in size posteriorly, the last the largest tooth in the mouth; these well separated from vomerine series which is biserial for most of its extent, tapering to a single row posteriorly (except in very small specimens where teeth are mostly uniserial, with a few pairs or staggered anteriorly); a total of 7–33 small conical vomerine teeth, the number increasing with growth. An 890-mm specimen had two well-separated biserial rows anteriorly, uniting and tapering to one uniserial row posteriorly. Dentary teeth uniserial; anteriorly 1–4 tall, slender teeth which increase in size posteriorly, with 0–7 tiny teeth clustered outside the last two, followed by a series of 11–20 triangular, posteriorly-directed teeth, the first few abruptly smaller than the large anterior teeth, then increasing in size posteriorly; 1–2 tiny teeth outside this row in a few specimens.

Body moderately elongate. Dorsal fin high and fleshy, with oblique folds, its origin nearer gill opening than mouth angle; anal fin fleshy, with longitudinal folds; prominent gular folds. Skin of small to medium specimens smooth, old large individuals with grooves and ridges.

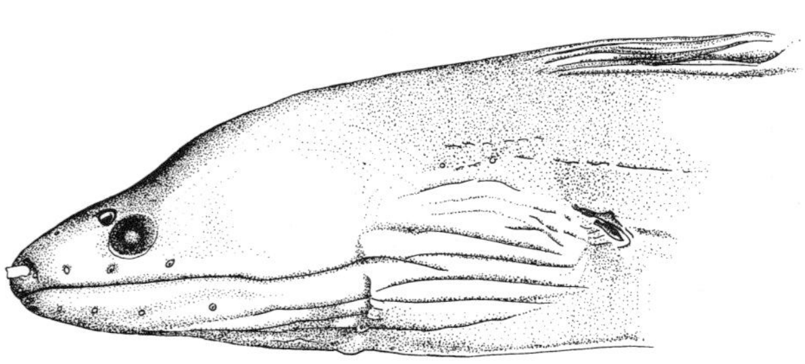

FIGURE 171. *Gymnothorax funebris*: ANSP 106173, 360 mm TL, Bahamas; head.

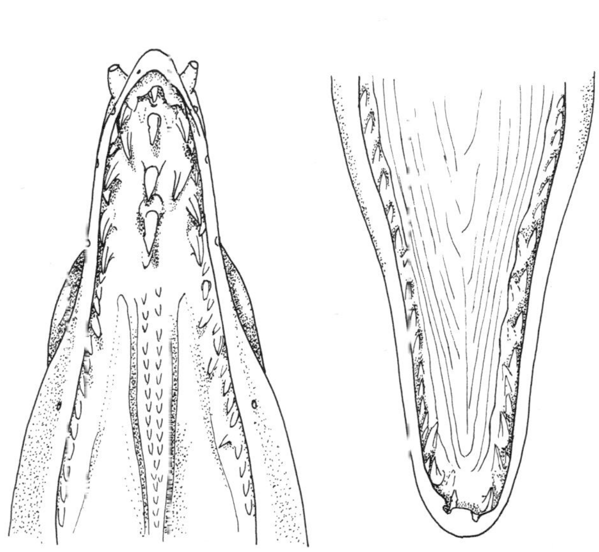

FIGURE 172. *Gymnothorax funebris*: ANSP 117585, 595 mm TL, Colombia; dentition.

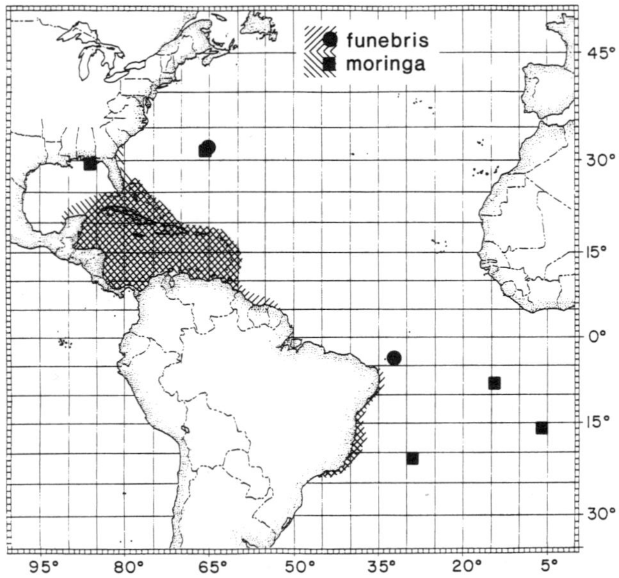

FIGURE 173. Distribution of *Gymnothorax funebris* and *G. moringa* (subgenus *Taeniophis*) based on material examined.

Color. A uniformly dark green to brown moray; head pores, gill opening and anus sometimes darker than body; fins with body color, without pale edge; inside of mouth pigmented. Juveniles uniformly dark, without white lower jaw (unlike most other small *Gymnothorax* and *Muraena* species). Life colors best described as uniform dirty pea-green, the color said to be due to mucous secretion over the dark brown body.

Size. Recorded to 6 feet 2.5 inches [1890 mm] and a weight of 27 pounds, and said to grow larger (Randall, 1968: 35). Of 110 specimens examined, 17 (196–1250 mm) were males and 4 (360–1115 mm) females. Only a single 1115-mm female was ripe, the body cavity filled with 1.0-mm eggs; no date of capture was given.

Development. Leptocephalus not identified.

Behavior. Gymnothorax funebris is a fairly common and well-known species; it has been observed by snorkelers and skin divers and is often exhibited in public aquaria. An inhabitant of rocky tidepools and shallow coral reef areas, it has also been found in brackish-water tidal creeks and mangrove areas. A "large" individual was observed swimming free at about 33 m at midday off Bonaire (James Q. Wilson, pers. comm.). Unprovoked agressiveness toward a diver has been reported (Böhlke and Chaplin, 1968: 75). Said to be used for food in some areas (Cervigón, 1966: 183), the green moray has also been reported to be ciguatoxic (Halstead, 1967: 70). McCosker observed that the Kuna Indians of San Blas, Panama, commonly ate this species and were unaware of reports of its toxicity.

Distribution (Fig. 173). Known from Bermuda, the Bahamas, the Florida Keys, throughout the Caribbean, and from reefs off Central and South America to southern Brazil. Reported from New Jersey (Fowler, 1952: 110), the specimen not found and the locality considered erroneous or the specimen a waif brought north by the Gulf Stream. Most specimens were taken with rotenone-based poisons, at recorded depths of less than 5 m; occasionally taken by hook and line or spear; no records of capture by trawl.

Etymology. From the Latin *funebris* (funereal), alluding to its dark color. A noun in apposition.

Remarks. Gymnothorax funebris exhibits subtle differences in its general appearance and body shape from the other three western Atlantic species of the subgenus *Taeniophis;* the snout and jaws are not as elongate, the head and body appear somewhat deeper, the tail is not as long, and the fins are higher and more fleshy. It differs also in having biserial vomerine dentition in adults. Its affinities appear closer to *Gymnothorax castaneus* in the eastern Pacific, with a similar appearance and vertebral count but which has tiny white to yellow spots superimposed on the dark coloration of the dorsal fin and dorsal part of the head and body.

Study Material. A total of 127 specimens, 55–1250 mm TL. (An additional 44 specimens from Brazil [MZUSP] not listed.)

BERMUDA: ANSP 103936 (1, ca 1195), 126225 (1, 235). FMNH 5166 (1), 5170 (1). FLORIDA KEYS: FMNH 51104 (1). UMML 525 (1, 672), 3086 (1, 505), 7267 (2, 730–855), 13946 (1, 483). USNM 116779 (1, 95). BAHAMAS: ANSP 102541 (1, ca 1250), 106173 (1, 360), 106174 (1, 237), 106309 (2, 124–486), 106310 (1, ca 278, cleared and stained), 111550 (1, 683). UMML 31269 (12, 303–534). USNM 38473 (1, 574).

WEST INDIES: USNM 190521 (1, 650), 192050 (1, 63). CUBA: ANSP 90530 (1, ca 1115). CAS-SU 54204 (2, 170–347). MCZ 32789 (1, 266). JAMAICA: FMNH 2801 (2),

TABLE 21. Frequency distributions of vertebral counts in western Atlantic species of *Gymnothorax* subgenus *Taeniophis*.

	Predorsal					Preanal													
	3	4	5	6	7	48	49	50	51	52	53	54	55	56	57	58	59	60	61
afer		1	2*	2							1	-	1	3*					
funebris		1	17	20	3										1	4	11	17	8
moringa		3	30	36	6			1	1	3	18	26	20	9	1				
polygonius		3*	11	2					1	4*	3	7	3						
vicinus	1	10	32	8*		1	7	16*	25	9	1								

	Total																
	128	129	130	131	132	133	134	135	136	137	138	139	140	141	142	143	144
afer													1		1		1*
funebris										2	1	6	16	7	9		
moringa							2	7	1	11	16	12	16	7	3		
polygonius								2*	2	4	3	4	1	-	1		
vicinus	1	2	2	2	10	14	8*	7	4	1							

* Holotype.

74732 (1), 73892 (2, 190–264). LACM 5111 (1, 325). SANTO DOMINGO: ANSP 1038 (1, 890). PUERTO RICO: ANSP 115549 (2, 203–242), 115611 (5, 180–202), 118630 (3, 133–470), 129902 (1, 445), 129913 (1, 507). FMNH 74708 (1, 137). FMNH 90713 (1, 142). USNM 50152 (1, 105), 147637 (1, 304), 219138 (1, 108). LESSER ANTILLES: *Dominica:* BMNH 1866.1.22.63 (1). *Martinique:* UMML 32809 (1, 116). *St. Lucia:* USNM 41294 (1, 200). *Barbuda:* USNM 274278 (1, 105). *Grenadines:* ANSP 126530 (1, 490).

CENTRAL AMERICA: MEXICO: *Quintana Roo:* ANSP 114910 (1, 161). UANL 986 (2, 230–355). HONDURAS: UMML (TABL 107196) (3, 109–323). COSTA RICA: LACM 6905-18 (1, ca 530). PANAMA: ANSP 49941 (1, head). ANSP 113769 (1, 402). LACM 24346 (1, 157). MCZ 44278 (2, 110–440), 44279 (2, 55–92), 44280 (1, 140), 44281 (1, 78), 44282 (1, 86), 44283 (1, 210), 44284 (1, 306), 44286 (1, 420), 44287 (1, 198), 44290 (1, 520), 44291 (3, 130–155). UMML 22436 (5, 142–335), 26763 (1, 352). USNM 5979 (1, 676). GRAND CAYMAN: ANSP 141335 (1, 45). UF 21334 (1, 86).

SOUTH AMERICA: COLOMBIA: ANSP 117564 (4, 212–380), 117568 (2, 196–275), 117585 (1, 595), 117595 (1, 274), 117596 (2, 300–308). VENEZUELA: BMNH 1847.5.14.43 (1, 647); holotype of *Muraena lineopinnis* Richardson. MBUCV 2451 (2). FRENCH GUIANA: MNHN A.4908 (1), 1904-421 (1). BRAZIL: CAS 59182 (1, 500). CAS-IUM 8540 (1, 361). CAS-SU 22138 (1, 551). FMNH 7228 (1, 456). MHNG 665-38 (1, 582). NMW 61645 (1, 782), 61655 (1). UF 19932 (1, 186). USNM 104317 (1, 790). ZMB 6109 (1, 795). *Fernando Noronha:* USNM 274263 (1, 71).

EXACT LOCALITY UNKNOWN: ANSP 16704 (1, 1050).

Gymnothorax moringa (Cuvier, 1829)

SPOTTED MORAY

Figures 173–176, Tables 9, 14, 21

Muraena maculata nigra Catesby, 1738: 21, pl. 21.

Muraena moringa Cuvier, 1829: 352 (original description, after Catesby, 1738).

Gymnothorax rostratus Agassiz in Spix and Agassiz, 1831: 91, pl. 50a (original description, Brazil, holotype unknown).

Muraena moringua. Richardson, 1848a: 89 (lapsus for *moringa*).

Muraena punctata Gronow in Gray, 1854: 18 (based on Catesby, 1738, name preoccupied).

Murenophis rostrata. Castelnau, 1855: 80, pl. 42, fig. 1 (Rio de Janeiro, Bahia).

Murenophis curvilineata Castelnau, 1855: 81, pl. 42, fig. 2 (original description, Bahia, Brazil, holotype MNHN B.2436).

Murenophis caramuru Castelnau, 1855: 82, pl. 43, fig. 1 (original description, Bahia, Brazil, holotype MNHN B.2437).

Thyrsoidea moringua. Kaup, 1856a: 60 (lapsus). Kaup, 1856b: 79.

Thyrsoidea concolor Abbott, 1861: 470 (original description, Vera Cruz, Mexico, holotype ANSP 970).

Gymnothorax moringa. Goode 1876: 72

Gymnothorax flavoscriptus Poey, 1876a: 200, pl. 9, fig. 2 (original description, Cuba, holotype unknown).

Muraena sanctae-helenae (not of Günther, 1870). Bean, 1880: 113 (misidentification).

Gymnothorax picturatus Poey, 1880: 257, pl. 10, figs. 1-3 (original description, Cuba, holotype unknown).
Sidera moringa. Bean and Dresel, 1884b: 169.
Sidera flavoscripta. Jordan, 1887b: 565.
Lycodontis moringa. Jordan and Evermann, 1896a: 395.
Lycodontis albimentis Evermann and Marsh, 1902: 78, fig. 9 (original description, Porto Rico, holotype USNM 49527).
Leptocephalus strommani Eigenmann and Kennedy, 1902: 89, fig. 8 (original description, holotype USNM 49762). Leptocephalus.
Lycodontis sanctae-helenae (not of Günther, 1870). Barbour, 1905: 111 (misidentification). Bean, 1906: 32 (misidentification).
Gymnothorax moringua. Fowler, 1919: 275 (misspelling).
Lycodontis moringua. Fowler, 1936a: 312 (misspelling).

Misidentifications:

Not *Muraena acutirostris* Abbott, 1861: 476 (original description, "Sandwich [Hawaiian] Islands," holotype ANSP 998). See discussion below.

Discussion of Synonymy. Muraena acutirostris Abbott, 1861 was put in synonymy with *Gymnothorax moringa* by Sutcliffe (1970) who considered the type locality of "Sandwich Islands" to be erroneous. However, there is no reason to question this locality and, on the basis of color, dentition, and vertebral counts, the species could equally well be a synonym of *Gymnothorax flavimarginatus* (Rüppell), a common Indo-Pacific muraenid species.

Thyrsoidea concolor Abbott, 1861 is here added to the synonymy of *Gymnothorax moringa.* This species was listed as a synonym of *G. funebris* by Jordan and Davis (1892: 603). However it lacks the biserial vomerine dentition of *funebris,* and the vertebral formula fits that of *moringa* rather than *funebris.*

Lycodontis albimentis Evermann and Marsh, 1902, could be the young of either *Gymnothorax moringa* or *G. vicinus,* based on the pale lower jaw found on the young of both species. However, its vertebral formula is most similar to that of *G. moringa.*

Distinctive Characters. A large elongate moray with coloration of overlapping contrasting dark spots on a pale background (yellow in life), head pores in pale spots; teeth smooth, jaw and vomerine teeth uniserial in adults; MVF 6-54-138.

Description. Total vertebrae 134–142, $\bar{x}$ = 138.5 (n = 75); predorsal vertebrae 4–7; $\bar{x}$ = 5.5 (76); preanal vertebrae 50–57, $\bar{x}$ = 54.1 (79); dorsal-fin rays 333–356, $\bar{x}$ = 341 (3); anal-fin rays 244–266, $\bar{x}$ = 252 (3); kidney terminus at 63–64, $\bar{x}$ = 63.8 (4). Head pores: SO 1 + 2; IO 4; M 6; B 2.

Proportions as % (n = 25, 126–1025 mm): Of TL: predorsal 11–13, $\bar{x}$ = 12; preanal 42–47, $\bar{x}$ = 44; head 13–15, $\bar{x}$ = 14; depth at gill opening 4.9–9.1, $\bar{x}$ = 6.6; depth at anus 4.9–8.8, $\bar{x}$ = 6.4; width at gill opening 3.3–6.8, $\bar{x}$ = 4.6; width at anus 2.7–5.5, $\bar{x}$ = 4.4. Of HL: snout 17–23, $\bar{x}$ = 20; upper jaw 37–47, $\bar{x}$ = 41; lower jaw 36–47, $\bar{x}$ = 40; eye 8.0–12, $\bar{x}$ = 10; interorbit 8.7–12, $\bar{x}$ = 10.

Snout and jaws moderately elongate; eye moderate, over mid-jaw. Anterior nostril tubular, near tip of snout; posterior nostril round to teardrop-shaped, slightly raised, above and before anterior margin of eye. Little variation in head pores; one specimen with seven mandibular pores on one side only.

Dentition similar to that of *Gymnothorax funebris,* except that vomerine teeth are uniserial. On each side of upper jaw, 5–6 large outer intermaxillary teeth (flanked by 0–7 tiny teeth) continuous with maxillary teeth; on maxillary, an outer row of 14–18, the first four increasing in size, and 1–2 inner teeth (missing in large specimens). Along midline, three long, slender fangs, the last the largest in the mouth, well separated from 5-11 uniserial teeth on vomer. On lower jaw, 15–24 uniserial teeth, the anterior four increasingly larger, plus tiny teeth variously present anteriorly.

Color. Highly variable coloration; pattern typically, but not always, as shown in Figure 174. The fish may be much paler, with pale markings predominant, or the dark areas may be much more extensive with pale areas reduced to disconnected streaks. Many specimens, particularly larger ones, have a row of larger round dark spots along body above mid-side, and some also have another row below mid-side, which are darker than the other dark markings (see photograph opposite p. 50, Beebe and Tee-Van, 1933a); others have these larger, darker spots over the body in no pattern. Fins with body color; margins pale

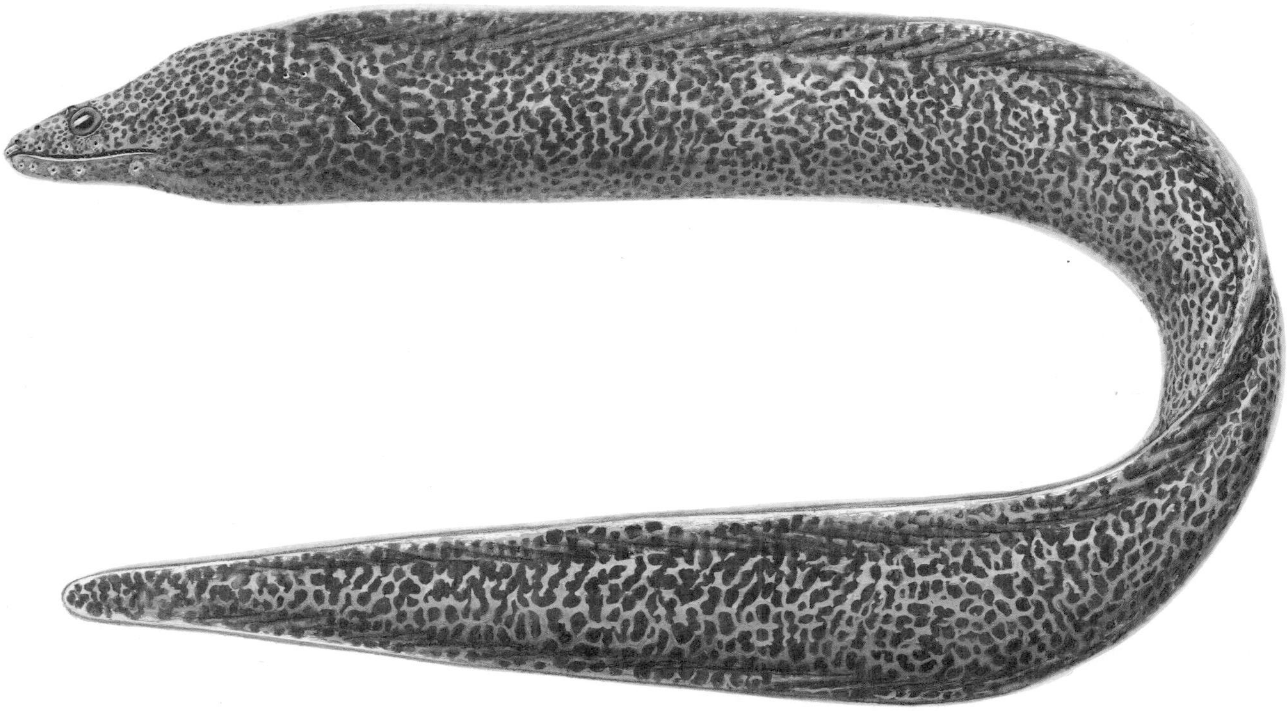

FIGURE 174. *Gymnothorax moringa*: ANSP 111552, 692 mm TL, Bahamas. Illustration by S. P. Gigliotti.

on small and pale specimens, black on dark and on large specimens. Anterior nostrils pale; lower jaw pores typically set in white spots. Large dark specimens with the mid-lateral row of spots and dark fins have often been misidentified; many have been listed as *Gymnothorax sanctaehelenae* (a synonym of *Enchelycore anatina*), or as *Gymnothorax* species (Cervigón, 1966: 180, a 920-mm ripe female with dark fins). Recently metamorphosed specimens are uniform brown with contrasting pale lower jaw; the pattern first appears as pale spots in the head region, spreading back with growth, the demarcation between the colors sharp.

In life, the dark areas in juveniles and adults are brownish to purplish black, and the pale areas are yellowish, the yellow most noticeable dorsally, particularly on dorsal fin.

Size. Reported to grow to an estimated 2–3 m at Ascension Island (Lubbock, 1980: 285); the largest specimen we have seen is 1025 mm. Of 153 specimens whose sex was determined, 97 (165–1025 mm) were females, 8 of them (595–

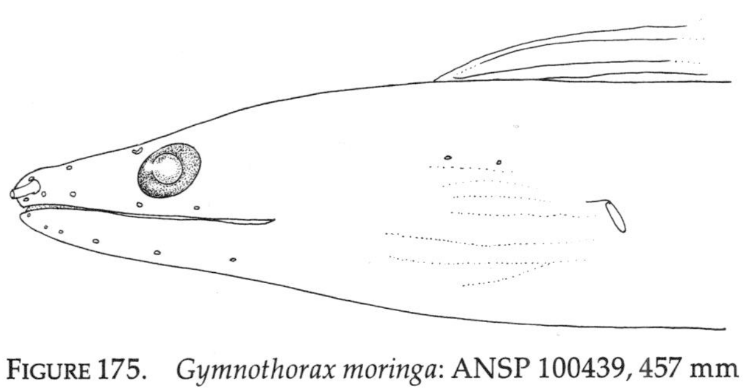

FIGURE 175. *Gymnothorax moringa*: ANSP 100439, 457 mm TL, Bahamas; head.

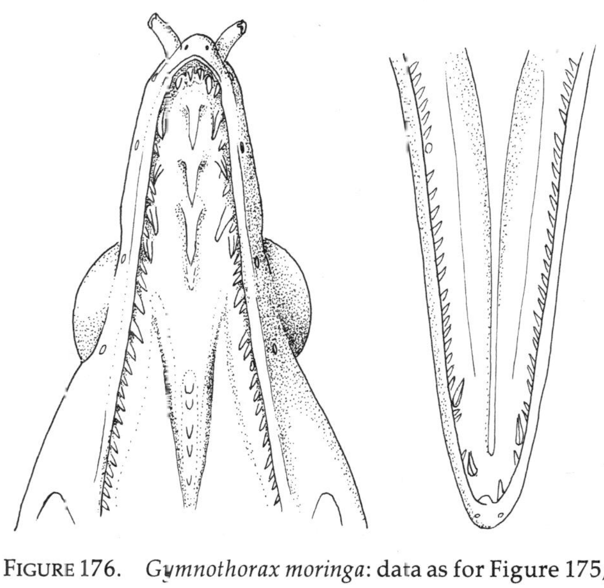

FIGURE 176. *Gymnothorax moringa*: data as for Figure 175; dentition.

1025) ripe with eggs 0.7–1.0 mm, and 19 (252–600) with developing eggs; 56 (252–1000) were males, 6 of them (575–940) ripe. Ripe females were taken in May, July and December off South Carolina and Georgia, and ripe males were taken in May, June, September and October in the Antilles.

Development. Leptocephalus described as *Leptocephalus strömmani* by Eigenmann and Kennedy (1902: 89) (see p. 910).

Behavior. Gymnothorax moringa is the most common moray in the western Atlantic, found in shallow reef, rock and seagrass habitats. It is frequently observed by divers and has been kept in salt-water aquaria. It feeds on small fish; a radiograph of a 235-mm specimen shows a 35.5-mm fish in its gut. An individual was observed and photographed feeding on a smaller, live green moray (*G. funebris*) at about 18 m at midday off Bonaire (James Q. Wilson, pers. comm.). The species has been seen feeding on young turtles during the hatching season on Ascension Island (Lubbock, 1980: 285), where it was reported to be unusually aggressive to man; in two cases, attacked divers were transporting dying fishes and spiny lobsters, but at other times, divers were attacked for no apparent reason. It is reportedly used for food in some areas, but the species has been reported as ciguatoxic (Halstead, 1967: 71).

Distribution (Fig. 173). An insular species, common in the western Atlantic from South Carolina and Bermuda to Brazil, including the islands throughout the West Indies, the Gulf of Mexico, the Caribbean and bordering coastlines, and in the mid-Atlantic islands of St. Helena and Ascension. The coastal West African record of Fowler (1919: 275), was questioned by Blache (1967d: 1150), and was probably based on a specimen from Ascension Island (see *Remarks* under *Echidna catenata*). Most specimens were taken with rotenone-based poisons in shallow waters; occasionally taken by hand-lines and spears, and by trawls at depths ranging to 90 m. A large ripe female was obtained west of Puerto Rico from a fish trap at 200 m.

Etymology. From the Portuguese *moringa*, a corruption of *Muraena.* A noun in the nominative case.

Remarks. The variability of coloration in adults, combined with ontogenetic changes in dentition, has resulted in the descriptions of numerous nominal species which are synonyms of *Gymnothorax moringa.* Very young, recently transformed specimens (50–100 mm) appear uniform medium brown with pale fins and white lower jaws; juveniles are boldly spotted, the fins with pale edges, and have partially biserial jaw teeth. Large individuals have basically uniserial dentition, variable spotted or mottled coloration (sometimes with one or two series of large dark spots along mid-body), and the fins may have dark edges.

Gymnothorax moringa is very similar to *G. vicinus;* the species are separable by subtle differences in vertebral counts (preanal vertebrae 50–57, $\bar{x}$ = 54 in *moringa* versus 48–53, $\bar{x}$ = 51 in *vicinus;* total vertebrae 134–142, $\bar{x}$ = 138 versus 128–140, $\bar{x}$ = 133, respectively), and in coloration. Juveniles may be separated, with difficulty, by differences in the head spots, which are contrasting and with sharp borders in *moringa,* but diffuse and with indistinct demarcations in *vicinus* when viewed under the microscope. Adults of *moringa* are distinguished by the highly contrasting pattern of dark spots on a yellow background.

Study Material. A total of 864 specimens, 39–1025 mm TL, including the holotypes of *Murenophis caramuru* Castelnau, *Murenophis curvilineata* Castelnau, *Thyrsoidea concolor* Abbott, and *Lycodontis albimentis* Evermann and Marsh. (Ten specimens in nine lots from Brazil (MZUSP) not listed.)

BERMUDA: ANSP 1040 (1, 740), 1041 (1, 963), 52238 (1, 480), 126289 (6, 80–241), 126645 (5, 70–165), 133753 (1, 80), 133754 (2, 159–168), 133755 (2, 61–159), 133756 (1, 212), 133757 (1, 97), 133842 (2, 71–775), 133859 (3, 87–126), 133860 (3, 129–750), 133873 (1, 440), 136042 (3, 50–56), 136043 (1, 52), 148148 (2, 229–251), 148149 (3, 166–239), 148153 (2, 204–234), 148154 (1, 120). FMNH 5172 (1, 770), 5174 (1, 415), 5175 (1, ca 750), 5176 (1, ca 645), 48151 (5, 83–250), 48256 (2, 157–185), 48461 (1, 195), 48658 (1, 148). MNHN 1974-114 (1, 429). USNM 20183 (2, 370–775), 21931 (1, 308), 23810 (1, 816), 154903 (1, 765), 164728 (1, 185), 164753 (4, 158–252), 177929 (1, 92), 274286 (1, 152).

BAHAMAS: ANSP 997 (1, ca 670), 72291 (1, 208), 72292 (1, 188), 72293 (2, 158–521), 72294 (1, 292), 72299 (1, 362), 100420 (1, 214), 100421 (2, 132–315), 100422 (2, 162–247), 100423 (4, 95–189), 100424 (1, 98), 100425 (8, 100–152), 100426 (1, 259), 100427 (1, 296), 100428 (2, 138–181), 100429 (2, 107–116), 100430 (1, 269), 100431 (1, 181), 100432 (1, 155), 100433 (1, 224), 100434 (1, 410), 100435 (1, 436), 100436 (1, 403), 100437 (2, 280–443), 100438 (2, 311–412), 100439 (1, 457), 100440 (8, 96–425), 100441 (4, 175–356), 100442 (5, 269–531), 100443 (1, 438), 100444 (2, 74–205), 100445 (1, 196), 100446 (1, 422), 100447 (1, 605), 100448 (1, 810), 100449 (17, 65–475), 100450 (3, 52–513), 100451 (2, 272–381), 100452 (1, 78), 100453 (4, 84–119), 100454 (6, 77–309), 100455 (1, 376), 100456 (9, 121–380), 100457 (1, 373), 100458 (2, 202–396), 100459 (1, 676), 100460 (16, 73–605), 100461 (4, 546–800), 100611 (34, 57–298), 100612 (3, 127–267), 100613 (6, 82–387), 100614 (6, 54–452), 100615 (2, 52–58), 100616 (9, 50–126), 100617 (3, 69–85), 100618 (1, 57), 111552 (1, 692), 113335 (1, ca 685), 113412 (2, 51–81), 115504 (1, 491, cleared and stained), 116145 (8, 107–225, cleared and stained), 120710 (2, 119–208, cleared and stained), 126802 (1, 430), 126806 (1, 532), 127080 (2, 79–98), 128801 (7, 52–135), 140514 (2, 104–109). CAS-IUM 8811 (1, 429). MCZ 30997 (1, 635), 31000 (1, 323), 35791 (1, 550). USNM 185964 (1, 118).

UNITED STATES COAST: *South Carolina:* GMBL 70-35 (1, ca 1040). GMBL 72-124 (1, 1025). GMBL 72-152 (1, 904). MCZ 63258 (1, 525+). *Georgia:* ANSP 102466 (1, 595), 110081 (3, 153–600). GMBL 78-114 (1, 113). *Florida Keys:* ANSP 90312 (10, 188–417), 90317 (2, 372–395), 103205 (1, 652). CAS 19172 (1, 565). CAS uncat. (1, 457). TU 17006 (1, 870). USNM 6794 (3, 680–900), 157937 (1, 262), 167678 (4, 380–ca 850). GULF OF MEXICO: *Florida:* ANSP 157986 (1, 882), 158108 (2, 753–820), 158178 (1, 565). CAS 19169 (1, 828).

CUBA: ANSP 92925 (1, ca 770). MHNG uncat. (1). USNM 12542 (1). JAMAICA: CAS-SU 4911 (1, 540). FMNH 73880 (2). MCZ 31445 (1, ca 250). HAITI: ANSP 90653 (1, 589), 103783 (1, 940), 111365 (2, 203–206), 114418 (1, 218), 126668 (4, 272–350), 126677 (7, 106–263). DOMINICAN REPUBLIC: ANSP 110082 (1, 741). PUERTO RICO: ANSP 122056 (19, 142–653), 136516 (1, 735), 139741 (30, 58–170), 139834 (3, 77–151). CAS-SU 8277 (1, ca 190). FMNH 90715 (1, 177). USNM 49527 (1, 50); holotype of *Lycodontis albimentis.* USNM 147642 (1, 177). USNM 274341 (1, 505). VIRGIN ISLANDS: ANSP 23357 (1, ca 590), 80632 (1, 119), 106567 (1, 614), 110410 (4, 119–304), 110411 (1, 108), 124575 (1, 54). USNM 131970 (1, 156), 147642 (1, 177), 170182 (1, 115), 183442 (3, 212–550), 183541 (1, 275).

LESSER ANTILLES: *Anguilla Island:* ANSP 105738 (5, 70–216), 105755 (1, 62). USNM 183476 (2, 116–127). *St. Martin's:* ANSP 16033 (1, ca 570), 16034 (1, 755), 16035 (1, 172), 105497 (1, 302). *St. Barthelemy:* ANSP 105696 (1, 343), 105698 (15, 66–276), 105748 (1, 94), 105758 (1, 82), 110097 (1, 545). *St. Kitts:* ANSP 992 (1, ca 450). *Nevis:* ANSP 122789 (1, 668). *Dominica:* ANSP 105445 (2, 72–117). USNM 274280 (1, 176), 274281 (6, 65–375), 274282 (2, 282–582), 274283 (1, 560), 274284 (2, 309–325), 274285 (3, 60–153), 274287 (1, 500), 274338 (1, 552), 274342 (4, 137–265), 278173 (1, 686). *Martinique:* ANSP 105732 (3, 72–113). MNHN A.6397 (1, 1083). NMW 61646 (1). *St. Lucia:* ANSP 105730 (2, 90–179), 105777 (2, 80–290), 110106 (1, 320), 110428 (1, 131). *Barbados:* ANSP 88739 (6, 100–ca 350), 88740 (3, 70–127). *St. Vincent:* ANSP 106179 (4, 70–235), 109809 (1, 585). *Grenadines:* ANSP 105724 (1, 144), 110085 (17, 67–ca 725), 110087 (5, 140–565), 110091 (6, 246–573), 110098 (7, 223–565), 110099 (24, 109–480). USNM 170152 (1, ca 800). *Grenada:* ANSP 105750 (1, 39). *Bonaire:* ZMA 844 (1, 70). *Curaçao:* CAS 59183 (15, 430–692). *Tobago:* ANSP 97938 (1, 318), 97982 (1, 375). USNM 274337 (1, 139). *Trinidad:* ANSP 76219 (1, 440).

PANAMA: ANSP 113753 (1, 68). CAS 13488 (1, 135). CAS 31615 (8, 113–286). CAS 31697 (2, 262–326). FMNH 74758 (1, 206). COSTA RICA: LACM 36220-18 (2, 227–425). BELIZE: USNM 278164 (1, 67), 278172 (1, 79). MEXICO: ANSP 970 (1, 237), holotype of *Thyrsoidea concolor.* ANSP 94129 (1, ca 865), 126669 (1, 215). USNM 192238 (1, 285), 274340 (4, 192–480). CAMPECHE BANK: FMNH 45370 (10). COURTOWN CAYS: ANSP 89033 (1, 825). SAN ANDRES: ANSP 86222 (1, ca 480). SERRANA BANK: ANSP 102463 (3, 116–425). FMNH 74723 (12). USNM 187767 (1, 74), 193617 (4, 213–500). GRAND CAYMAN: ANSP 104908 (2, 171–295), 104942 (2, 44–373), 104950 (2, 176–220), 104959 (1, 291), 105578 (3, 58–ca 630), 141337 (9, 42–63).

COLOMBIA: ANSP 117572 (4, 219–413), 117584 (4, 239–420), 117589 (1, 287), 117592 (7, 148–215). VENEZUELA: ANSP 75193 (1, 248), 105825 (1, 662). MBUCV 545 (1), 784 (1), 867 (27), 872 (2), 886 (2), 1067 (1), 1164 (17), 1183 (1), 1193 (1), 1256 (1), 1294 (1), 1364 (1), 1429 (1), 1632 (1), 2418 (2), 2452 (1), 2514 (1), 2524 (1), 2544 (1). NMW 61648 (1). USNM 194011 (1, 83), 194012 (2, 95–193), 194116 (13, 109–ca 580), 194121 (24, 103–790). BRAZIL: CAS 54309 (1, 665), 59310 (1, ca 440), 59311 (1, 632), 59184 (2, 790–ca 1045), 59186 (1, 550), 59187 (1, 490), 59188 (1, 570), 59189 (1, 567), 59190 (1, 558), 59191 (1, 681). CAS-SU 51715 (1, 415). FMNH 92294 (1, ca 560). MCZ 893 (1, 775). MHNG 331-42 (1, 581). MHNG 809-52 (1, 796). MNHN B.2436 (1, 548); holotype of *Murenophis curvilineata.* MNHN B.2437 (1, 503); holotype of *Murenophis caramuru.* MNHN uncat. (2, 512–931). MNRJ 1666 (2, 448–607), 3042 (1). MNRJ uncat. (1, 250). NMW 61647 (1). *Ilha da Trindade:* MNRJ 2322 (1). MNRJ uncat. (1, 720).

ASCENSION ISLAND: ANSP 102740 (1, ca 1000), 102799 (1, ca 800). GCRL 15497 (3), 15501 (1). GMBL 78-17 (3, 48–169), 80-20 (1, 63), 80-36 (1, 95), 80-41 (1, ca 550), 80-43 (11, 99–ca 460), 80-49 (1, 56), 80-54 (2, 55–67). MNHN B.2435 (1). USNM 274339 (2, 102–145), 274343 (1), 274347 (2, 237–295). ST. HELENA: USNM 267879 (1, 325), 267894 (1, 606). ZMC uncat. (1).

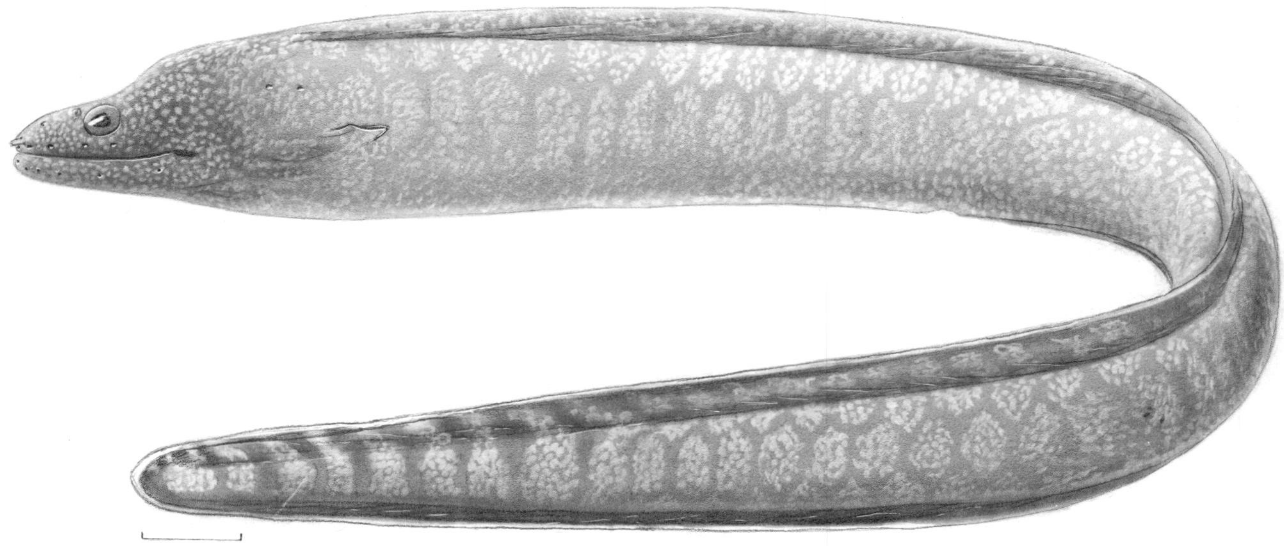

FIGURE 177. *Gymnothorax polygonius*: ANSP 115275, 625 mm TL, Bahamas.

EXACT LOCALITY UNKNOWN: MNHN B.2439 (1). USNM 8131 (1). "West Africa": USNM 42386 (1, 200).

Gymnothorax polygonius Poey, 1876
POLYGON MORAY
Figures 177–180, Tables 9, 14, 21

Gymnothorax polygonius Poey, 1876b: 68, pl. 10 (original description, Cuba, holotype USNM 9825).
Sidera polygonia. Jordan, 1887b: 565.
Lycodontis polygonius. Jordan and Evermann, 1896a: 394.

Distinctive Characters. A large moray with color pattern of spotted polygons separated by dark reticulations; teeth smooth, uniserial on vomer and jaws; MVF 5-53-138.

Description. Total vertebrae 135–142, $\bar{x}$ = 137.8 (n = 17); predorsal vertebrae 4–6, $\bar{x}$ = 4.9 (16); preanal vertebrae 51–55, $\bar{x}$ = 53.4 (17); dorsal-fin rays 336–370, $\bar{x}$ = 350.0 (4); anal-fin rays 253–274, $\bar{x}$ = 262.2 (4); kidney terminus at 59–62, $\bar{x}$ = 60.7 (3). Head pores: SO 1 + 2; IO 4; M 6; B 2.

Proportions as % (n = 16, 575–835 mm): Of TL: predorsal 10–13, $\bar{x}$ = 12; preanal 44–49, $\bar{x}$ = 47; head 13–16, $\bar{x}$ = 14; depth at gill opening 5.8–8.9, $\bar{x}$ = 7.7; depth at anus 5.0–7.0, $\bar{x}$ = 6.2; width at gill opening 3.3–5.6, $\bar{x}$ = 4.7; width at anus 3.6–5.1, $\bar{x}$ = 4.1. Of HL: snout 19–22, $\bar{x}$ = 20; upper jaw 38–54, $\bar{x}$ = 44; lower jaw 38–53, $\bar{x}$ = 44; eye 7.0–11, $\bar{x}$ = 9.0; interorbit 10–14, $\bar{x}$ = 12.

General characters similar to those of *Gymnothorax moringa*. Posterior nostril round or tear-drop-shaped, slightly raised on large specimens, situated above anterior margin of eye. Variation in head pores of 16 specimens: two with an additional infraorbital pore on one side, two with three branchial pores on one side, and one with only one on one side.

Dentition similar to that of *Gymnothorax moringa*, the vomerine teeth uniserial. On each side of upper jaw, 5–6 intermaxillary teeth, flanked by a few tiny teeth, closely followed by 8–12 maxillary teeth; along the midline, 0–3 anterior median intermaxillary fangs (missing in illustrated specimen), and 5–12 vomerine teeth (missing in some specimens, the anteriormost staggered in some). On lower jaw, 16–25 teeth. The

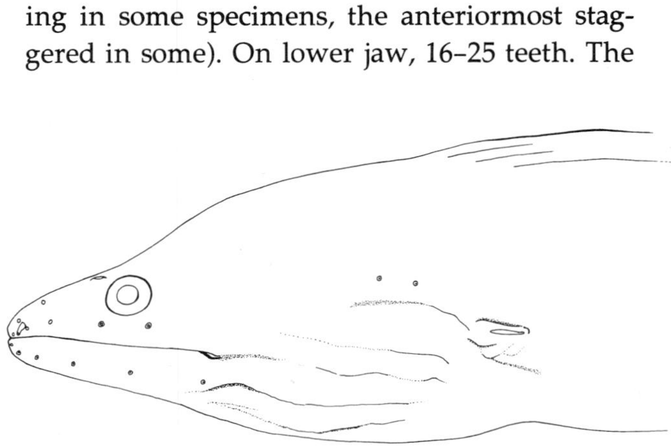

FIGURE 178. *Gymnothorax polygonius*: ANSP 136687, 734 mm TL, Puerto Rico; head.

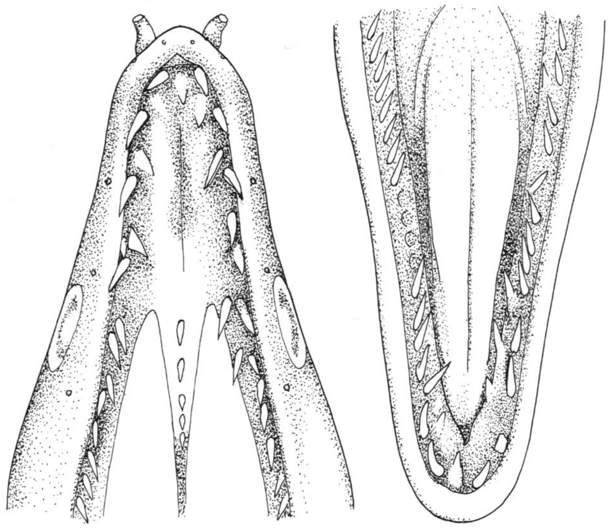

FIGURE 179. *Gymnothorax polygonius*: ANSP 136685, 835 mm TL, Puerto Rico; dentition.

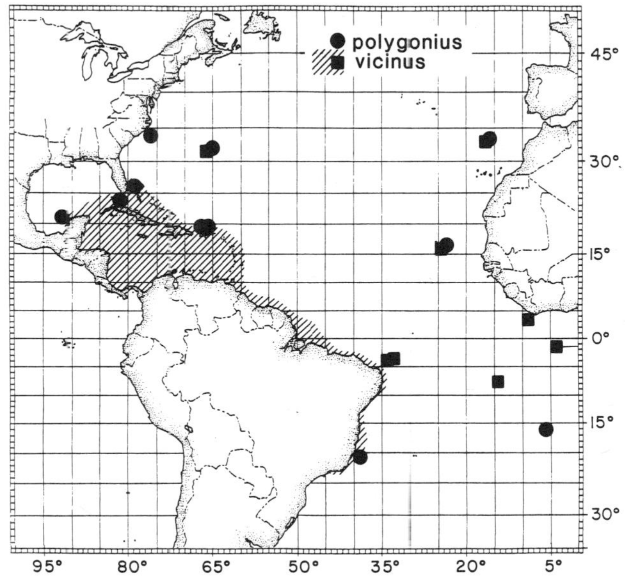

FIGURE 180. Distribution of *Gymnothorax polygonius* and *G. vicinus* (subgenus *Taeniophis*) based on material examined and on records in Blache, 1967d, and in Bauchot, 1986.

absence of inner maxillary and dentary teeth is possibly due to the large size of the study specimens.

Color. Narrow, dark, lilac-brown lines delimiting pale areas of polygonal, squarish, or indeterminate shapes. Within the polygons, numerous small pale spots separated by the same dark color. Posteriorly on tail, dark bars and paler interspaces continued obliquely on to dorsal fin. Head pattern of pale spots separated by dark reticulations; both rictus and gill opening sometimes darker; inside of mouth with reticulated pattern. Dorsal and anal fins with pale margins, that of the anal fin prominent, of the dorsal fin less noticeable. In life, the dark reticulations are brownish-purple, and the inside of the mouth was observed to be bright purple (S. W. Ross, in litt., 1982); margin of anal fin bright yellow, margin of dorsal fin less colorful. A kodachrome of a specimen in a fish market in Salvador, Brazil (courtesy of C. E. Dawson, 1979) showed the dark color to be brown, the pale areas and margins of fins cream-colored or tan.

Size. The largest study specimen, 835 mm, is the largest recorded in collections. Of 18 western Atlantic specimens, four (575–722 mm) were females, three of them (575–722) ripe with 0.9–1.0-mm eggs; 11 (638–832) were males, one 642-mm was ripe. The ripe females were from Bermuda, Cuba, and the Campeche Bank, the ripe male from Puerto Rico; they were taken during August to October.

Development. Leptocephalus not identified.

Behavior. Little is known of the life history of *Gymnothorax polygonius.* Its habitat differs from that of other Atlantic species placed in the subgenus *Taeniophis,* in that all specimens have been from deep water; it has not been taken at poison stations nor seen in reef or rock habitats. We know of no young or juveniles to date; the smallest specimen reported to date is 445 mm TL (Cervigón, 1986: 6).

Distribution (Fig. 180). A rather uncommon moray, infrequently taken from widely separated localities throughout the Atlantic. In the western Atlantic, single specimens taken off Bermuda, North Carolina, the Bahamas, Cuba, Puerto Rico, Campeche Bank, and Brazil, and one specimen from the mid-Atlantic island of St. Helena. Cervigón (1986: 6) reported two taken off Venezuela at Archipelago de los Roques. In the eastern Atlantic, one specimen reported from Ilha de Santo Antao, Cape Verde Archipelago (Blache, 1967g: 339), and one from Madeira reported by Bauchot (1986: 541). One specimen was taken between 90–95 m, the others with data recorded from 165–256 m; they were caught with hook and line, fish traps and a 40-ft. balloon trawl.

Etymology. From the Latin *polygonius* (with polygons), in reference to the color pattern. A noun in apposition.

Remarks. There has been some question as to whether USNM 9825 is indeed the holotype of *Gymnothorax polygonius.* Poey wrote (1876b: 68): "Type #602 (722 mm) at Smithsonian." Jordan and Davis (1891: 600) stated: "The specimen (9825 U.S.N.M., 28 inches long) was sent from Cuba by Poey . . . it may be Poey's original type." We have examined this specimen, which now measures 693 mm, and accept it as the type specimen.

Study Material. A total of 23 specimens, 575–832 mm TL, including the holotype.

Holotype: USNM 9825 (693); Cuba; F. Poey.

Other material: BERMUDA: AMNH 37218 (1, 640); W. Beebe. ANSP 137566 (1, 616); 200 m; 25 Oct. 1976. NORTH CAROLINA: ANSP 150525 (1, 745) and UF 38897 (2, 781–794); 86–97 m; 10 July 1982. ANSP 157319 (1, 680); 146 m; 22 May 1985. BAHAMAS: ANSP 115275 (1, 625); west of Bimini, 229 m; 16 June 1957. CUBA: AMNH 12307 (1, 575); Havana, in deep water; Aug. 1932. ANSP 92921 (1, 660); Havana, in deep water; June 1932. CAS 19174 (1, ca 635); Havana. PUERTO RICO: ANSP 106566 (1, 665); trawled at 183 m; 6 Oct. 1959. ANSP 136514 (1, 692+); fish trap at 200 m; 7 Dec. 1977. ANSP 136515 (3, 630–700); fish trap at 200 m; 22 Nov. 1977. ANSP 136684 (1, 750); fish trap at 256 m; 8 Feb. 1978. ANSP 136685 (1, 835); fish trap at 229 m; 23 Feb. 1978. ANSP 136686 (1, 642); fish trap at 165 m; 25 Aug. 1978. ANSP 136687 (1, 734); fish trap at 220 m; 1 Sept. 1978. CAMPECHE BANK: UF 28461 (1, 722); 24 Aug. 1980. BRAZIL: CAS 59192 (1, 642); Brazilian Trawling Survey; 9 Dec. 1944. ST. HELENA: USNM 267909 (1, 690); 110 m; 19 July 1983.

Gymnothorax vicinus (Castelnau, 1855)

PURPLEMOUTH MORAY

Figures 180–183, Tables 9, 14, 21

Murenophis vicina Castelnau, 1855: 81, pl. 42, fig. 4 (original description, Bahia, Brazil, holotype MNHN B.2423).

Thyrsoidea cormura Kaup, 1860: 23, pl. 3, fig. 2 (original description, Puerto Cabello, holotype unknown).

Taeniophis cormura. Kaup, 1860, Supplement: 4.

Muraena erebus Poey, 1861: 416 (original description, Cuba, holotype unknown).

Gymnothorax erebus. Poey, 1867b: 258.

Muraena vicina. Günther, 1870: 121.

Gymnothorax obscuratus Poey, 1870: 320 (original description, Havana, Cuba, holotype MCZ 27217).

Gymnothorax virescens Poey, 1876a: 198, pl. 9, fig. 1 (original description, Cuba, holotype unknown).

Gymnothorax versipunctatus Poey, 1876a: 199, pl. 10 (original description, Cuba, holotype unknown).

Sidera vicina. Jordan, 1886: 34.

Sidera virescens. Jordan, 1887b: 565.

Sidera obscurata. Jordan, 1887b: 565.

Gymnothorax vicinus. Jordan, 1891: 315.

Leptocephalus forstromi Strömman, 1896: 24, pl. 2, figs. 1–3 (original description, 03°–04°S, 30°–34°W; syntypes ZMUU). Leptocephalus.

Lycodontis vicinus. Jordan and Evermann, 1896a: 394.

Lycodontis obscuratus. Jordan and Evermann, 1896a: 398.

Gymnothorax carcinognathus Fowler, 1912a: 22, fig. 5 (original description, St. Martin's, West Indies, holotype ANSP 38163).

Gymnothorax cyanopunctatus Howell Rivero, 1932: 17 (original description, Matanzas, Cuba, holotype MCZ 32787).

Lycodontis guarapariensis Pinto, 1975: 401, figs. 1–2 (original description, Guarapari, Espirito Santo, Brazil, holotype MHNI 208).

Distinctive Characters. A large moray with mottled coloration varying from diffuse overlapping spots to almost uniformly dark, median fins with pale edges posteriorly, and a dark slash at angle of gape; teeth smooth, uniserial on vomer and jaws; MVF 5-51-133.

Description. Total vertebrae 128–140, $\bar{x}$ = 133.2 (n = 52); predorsal vertebrae, 3–6, $\bar{x}$ = 4.9 (52); preanal vertebrae 48–53, $\bar{x}$ = 50.6 (60); dorsal-fin rays 326–359, $\bar{x}$ = 346 (4); anal-fin rays 254–281, $\bar{x}$ = 269 (4); kidney terminus at 56–60, $\bar{x}$ = 57.3 (3). Head pores: SO 1 + 2; IO 4; M 6; B 2.

Proportions as % (n = 24, 154–860 mm): Of TL: predorsal 10–13, $\bar{x}$ = 11; preanal 41–47, $\bar{x}$ = 44; head 13–16, $\bar{x}$ = 14; depth at gill opening 4.6–8.0, $\bar{x}$ = 6.4; depth at anus 5.2–7.1, $\bar{x}$ = 6.2; width at gill opening 3.2–8.3, $\bar{x}$ = 4.4; width at anus 3.3–6.9, $\bar{x}$ = 4.3. Of HL: snout 17–22, $\bar{x}$ = 20; upper jaw 33–48, $\bar{x}$ = 42; lower jaw 31–46, $\bar{x}$ = 40; eye 8.1–12, $\bar{x}$ = 10; interorbit 7.5–11, $\bar{x}$ = 9.4.

Head and snout elongate, jaws long and slightly curved in large specimens. Anterior nostril tubular, near tip of snout (one fish has two anterior nostrils on the left side); posterior nostril

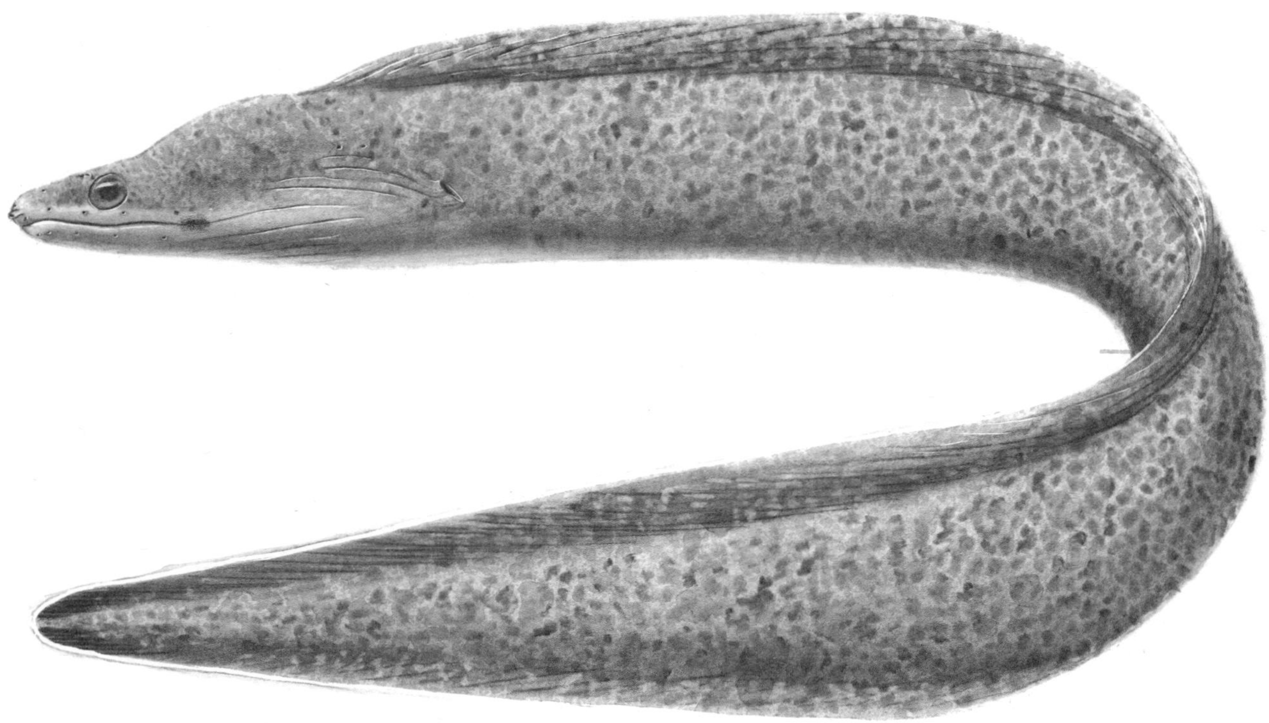

FIGURE 181. *Gymnothorax vicinus*: ANSP 101874, 460 mm TL, Bahamas. Illustration by S. P. Gigliotti.

non-tubular, somewhat raised in adults, situated above or slightly behind anterior margin of eye. Some variation in head pores: in 36 specimens, one had one additional supraorbital pore on both sides, one of them paired; one had three infraorbital pores on both sides, and one had three and one had five infraorbital pores on one side only; eight specimens had three branchial pores on both sides and two had one on one side only.

Dentition similar to that of *Gymnothorax moringa*. On each side of upper jaw, 5–6 intermaxillary teeth, flanked by 3–9 tiny teeth; 15–21 outer maxillary teeth, 1–3 in anterior inner series; three large median intermaxillary fangs, the third the longest tooth in the mouth, separated from a row of 6–14 vomerine teeth, the anteriormost sometimes staggered. On dentary 11–23 teeth, the anterior four enlarged and flanked by 2–8 tiny teeth.

Color. Highly variable, from an extensively mottled pattern of overlapping dark spots (Fig. 181) to an almost uniformly brown color with faint freckles of darker color. When mottled, the body color is continued onto vertical fins; when

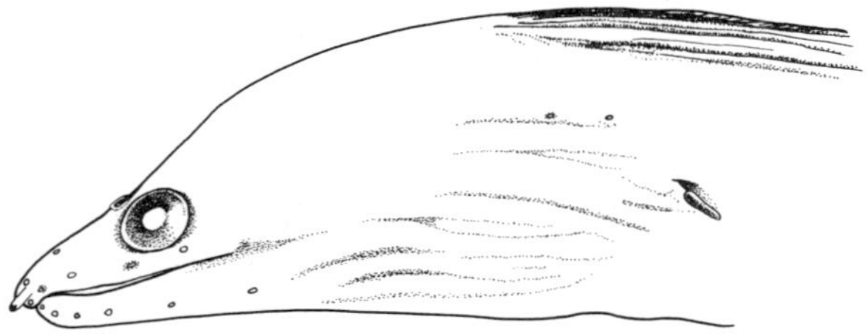

FIGURE 182. *Gymnothorax vicinus*: ANSP 111551, 385 mm TL, Bahamas; head.

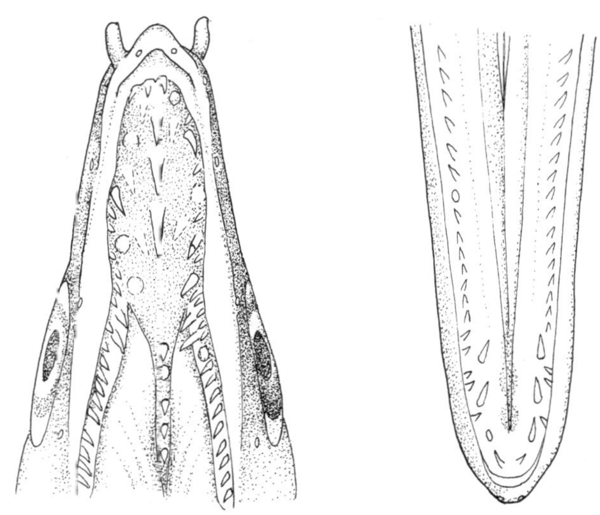

FIGURE 183. *Gymnothorax vicinus*: ANSP 104977, 185 mm TL, Grand Cayman; juvenile teeth.

uniform, fins may be uniformly dark except for their margins. Entire length of anal fin and posterior one-third to entire length of dorsal fin pale-edged. Characteristic of the species is the dark horizontal slash at angle of gape and the pigmented interior of the mouth; gill opening sometimes darker than surrounding area. Young appear uniformly dark, except for white lower jaw and fin edges. In life, dark color is purplish brown, pale areas white, not yellow.

Size. The largest specimen examined was 900 mm; reported to reach about 1.2 m (Randall, 1968: 35). Of those sexed, 84 (136–900 mm) were females, of which nine (380–785) were ripe with eggs 0.5–0.9 mm; 51 (220–860) were males. Ripe females were taken during June through December.

Development. Leptocephalus identified (see p. 912).

Behavior. Gymnothorax vicinus is a common inhabitant of shallow coral reefs and rocky areas, and of seagrass beds where only occasional boulders are present.

Distribution (Fig. 180). A tropical species found on islands and reefs off Bermuda, Florida, and the Bahamas, throughout the Caribbean, off Central and South America, and at Ascension Island; reported from 7½-Fathom Reef in the northern Gulf of Mexico off Texas (Hoese and Moore, 1977: 128). In the eastern Atlantic taken at the offshore islands of Cape Verde and Fernando Poo (Blache, 1967d: 1125) and off Madeira (Bauchot, 1986: 543). Usually captured by fish toxicants at depths to 9 m; large specimens have been speared, and occasional specimens were taken by trawls at depths of 22–73 m.

Etymology. From the Latin *vicinus* (near), in reference to its similarity to *Gymnothorax moringa* (cf. Jordan and Evermann, 1896a: 394). An adjective.

Study Material. A total of 593 specimens, 39–900 mm TL, including the holotype, and the holotypes of *Gymnothorax obscuratus* Poey, *Gymnothorax carcinognathus* Fowler, and *Gymnothorax cyanopunctatus* Howell Rivero. (An additional 31 lots (168 specimens) from Brazil at MZUSP not listed.)

Holotype: MNHN B.2423 (1, 699 mm TL); Brasil; Castelnau.

BERMUDA: ANSP 103959 (5, 71–822), 126240 (1, 297), 126641 (1, ca 750), 126646 (1, ca 550), 133758 (9, 84–ca 700), 148145 (3, 310–422), 148146 (2, 410–ca 560), 148147 (1, 355). BMNH 1872.8.28.11 (1). BMNH 1874.10.31.30 (1). FMNH 48080 (1, 592), 48119 (2, 650–744), 48200 (5, 286–ca 730), 48264 (3, 465–ca 900), 48265 (4, 265–727), 48704 (1, 85), 48760 (1, 299). TU 63153 (7, 199–488). USNM 274275 (4, 190–465).

BAHAMAS: ANSP 72290 (1, 263), 100404 (1, 51), 100405 (1, 49), 100406 (1, 48), 100407 (2, 44–120), 100408 (5, 109–282), 100409 (6, 84–273), 100410 (6, 42–316), 100411 (1, 84), 100412 (1, 668), 100413 (2, 542–617), 100414 (1, 555), 100415 (2, 175–482), 100725 (30, 41–272), 100726 (10, 39–172), 101874 (1, 460), 103358 (4, 139–285), 111551 (1, 385), 113427 (6, 93–ca 640), 116144 (4, 90–185, cleared and stained), 116322 (1, 124). MNHN 1974-104 (1). MNHN 1974-108 (1). USNM 274270 (1, 260).

FLORIDA KEYS: ANSP 89349 (1, 386), 90313 (1, 224), 102447 (1, ca 600), 110148 (1, 300). CUBA: ANSP 92923 (1, ca 785), 92924 (1, 860). CAS-SU 54211 (2, 292–449). MCZ 27217 (1, 565); holotype of *Gymnothorax obscuratus.* MCZ 27219 (1, 750). MCZ 32787 (1, ca 630); holotype of *Gymnothorax cyanopunctatus.* MCZ 32787a (1, 518); paratype of *G. cyanopunctatus.* JAMAICA: FMNH 73882 (2). LACM 5116 (1, 624). HAITI: ANSP 119107 (1, 298), 127274 (1, 221). FMNH 74714 (1, 279). DOMINICAN REPUBLIC: FMNH 74759 (1). PUERTO RICO: ANSP 110083 (1, 567), 115550 (7, 129–429), 115610 (2, 275–325), 115644 (5, 110–317), 118543 (1, 200), 118631 (6, 174–577), 118704 (11, 149–610), 129740 (48, 43–300), 129906 (7, 200–480), 129928 (3, 169–392), 144115 (1, 102). FMNH 74718 (1, 369), 74724 (2, 211–293), 74725 (2, 215–225), 74726 (1, 270), 74728 (9, 76–286), 90712 (3, 62–109). USNM 147629 (1, 114). VIRGIN ISLANDS: ANSP 89353 (1, ca 560), 89354 (1, 169), 89356 (1, 159), 110373 (2, 207–230), 124575 (1, 55).

LESSER ANTILLES: *Anguilla:* ANSP 105746 (2, 43–48), 110094 (1, 556). USNM 183466 (1, ca 800). *St. Martins:* ANSP 38163 (1, 565), holotype of *Gymnothorax carcinognathus. Barbuda:* MNHN 1974-202 (1). *Antigua:* ANSP 105762 (2, ca 47), 126484 (1, ca 830). *Dominica:* USNM 274276 (8, 76–770), 274277 (6, 194–570). *Martinique:* ANSP 105309 (3, 154–292), 110086 (1, ca 730). *St. Lucia:* ANSP 106184 (2, 200–293), 110084 (2, 475–576). *Barbados:* ANSP 100936 (1, 547), 122780 (6, 295–466). *Grenadines:* ANSP 110089 (3, 550–693). *Curaçao:* CAS 59193 (5, 350–ca 510). ZMA uncat. (1, 50).

PANAMA: FMNH 74734 (1, 222). LACM 24203 (7, 168–262). SIO 67-45-26 (4, 80–320). HONDURAS: FMNH 69190 (1, ca 490). MEXICO: BMNH uncat. (1); W. Seale. FMNH 45369 (4, 133–471), 45954 (1), 46308 (1). USNM 192298 (1, 127), 274274 (1, 580). SERRANA BANK: ANSP 117781 (1, 258). FMNH 47412 (1, 290). GRAND CAYMAN: ANSP 104977 (1, 185), 133250 (4, 41–49), 141336 (1, 45). COLOMBIA: ANSP 117588 (1, 234), 117591 (35, 91–405), 117594 (1, 238), 117597 (2, 327–380). VENEZUELA:

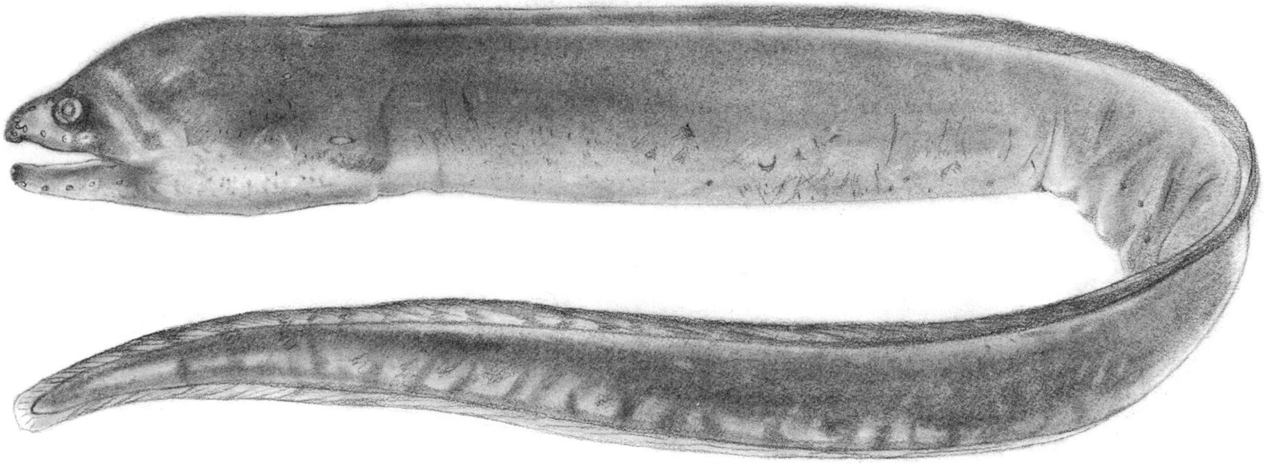

FIGURE 184. *Gymnothorax bacalladoi*: ANSP 159041, paratype, 347 mm TL, Canaries.

MBUCV 333 (1), 1186 (1), 1332 (1), 1430 (1), 1629 (1), 2516 (1), 2450 (1), 2575 (1). MNHN 1965-635 (1). USNM 194048 (2, 320–500). FRENCH GUIANA: MNHN 1882-410 (1). MNHN 1904-421 (1). BRAZIL: ANSP 101248 (1, 122), 102470 (1, 87), 102476 (2, 268–580), 105536 (1, 413), 106517 (1, 293), 136652 (1, 569). CAS 54303 (1, 400), 54304 (1, 552), 54305 (1, 527), 54306 (1, 425), 54307 (1, 415), 54308 (1, 381), 59194 (3, 586–626). CAS-SU 22099 (7, 124–326), 51713 (4, 339–553), 51715 (1, 408), 51716 (1, 410), 51718 (1, 212). FMNH 7202 (1, 216), 7234 (1, 120). MCZ 892 (2, 567–583), 9039 (1, 390), 9040 (1, 590), 34358 (1, 559). MNHN 1974-118 (1, 484). MHNG 809-53 (1, 701). MNRJ 9217 (7, 146–373), MNRJ uncat. (4, 141–524). NMW 61534 (1), 61539 (1), 61540 (3, 516–621). NMW uncat. (1). USNM 274273 (1, ca 680). ZMB 6108 (1, 510). *Atol das Rocas:* ANSP 123290 (2, 408–465). *Fernando Noronha:* BMNH 1888.1.12.156–157 (2). USNM 274272 (6, 122–540). ASCENSION ISLAND: USNM 274271 (1, 45).

EASTERN ATLANTIC: *Cape Verde Islands:* BMNH uncat. (1); Rev. R. Solomon. BMNH uncat. (1); Rev. Lowe. *Fernando Poo:* ANSP 158177 (1, 610). *Annobon:* ANSP 158801 (7, 47–487), 158825 (2, 106–240). *Accra:* BMNH 1930.3.24.6 (1). BMNH 1939.7.12.9 (1).

EXACT LOCALITY UNKNOWN: USNM 6737 (1, 690).

Subgenus undetermined

Gymnothorax bacalladoi Böhlke and Brito, 1987

CANARY MORAY

Figures 184–186, Tables 9, 14

Gymnothorax bacalladoi Böhlke and Brito, 1987: 459, figs. 1-3 (original description, Canary Islands, holotype TFMC (ZM-2)).

Distinctive Characters. A moderately elongate moray with mottled brown coloration and posterior jaw pores in white spots; dorsal-fin origin before gill openings; four infraorbital pores, one branchial pore; teeth serrate, vomerine and maxillary teeth biserial; MVF 5-55-130.

Description. Total vertebrae 130–131, $\bar{x}$ = 130 (3); predorsal vertebrae 5–6, $\bar{x}$ = 5.3 (3); preanal vertebrae 54–56, $\bar{x}$ = 55. Head pores: SO 1 + 2; IO 4; M 6; B 1.

Proportions, expressed as % (n = 3, 260–347 mm). Of TL: predorsal 7.3–10, $\bar{x}$ = 8.6; preanal 48–50, $\bar{x}$ = 49; head 11–12, $\bar{x}$ = 12; depth at gill opening 5.6–7.1, $\bar{x}$ = 63; depth at anus 4.0–4.5, $\bar{x}$ = 4.3. Of HL: snout 15–17, $\bar{x}$ = 16; upper jaw 36–37, $\bar{x}$ = 36; lower jaw 35–37, $\bar{x}$ = 36; eye 9.2–10, $\bar{x}$ = 9.7; interorbit 10–13, $\bar{x}$ = 12.

Head elevated and muscular; snout and jaws short, subequal; eye moderate, above mid-jaw. Anterior nostril tubular; posterior nostril a raised rimmed pore above anterior half of eye. Head pores developed, with irregular or elliptical rims, as listed above; usually one branchial pore, one specimen (ANSP 159041) with 2 branchial pores on right side only. Gill opening small, oval, slightly below mid-side.

Teeth strong and serrate. On upper jaw, an anterior median tooth followed on each side by five intermaxillary teeth; the teeth increasingly larger posteriorly, triangular, stout, and canted back, the posterior margin strongly serrated and with basal knob. On midline of upper jaw, one or two long, slender, smooth fangs, these the

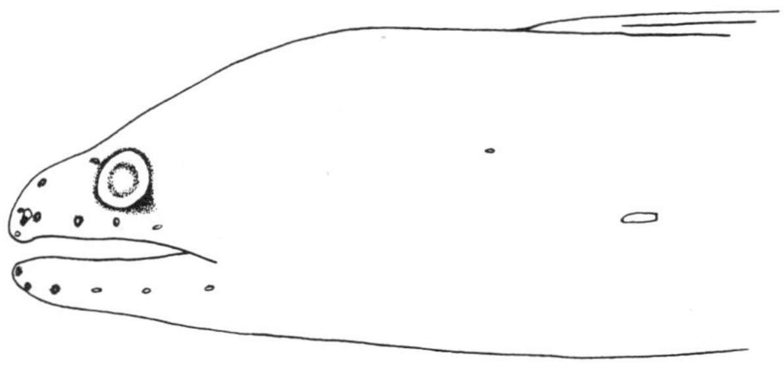

FIGURE 185. *Gymnothorax bacalladoi*: data as for Figure 184; head.

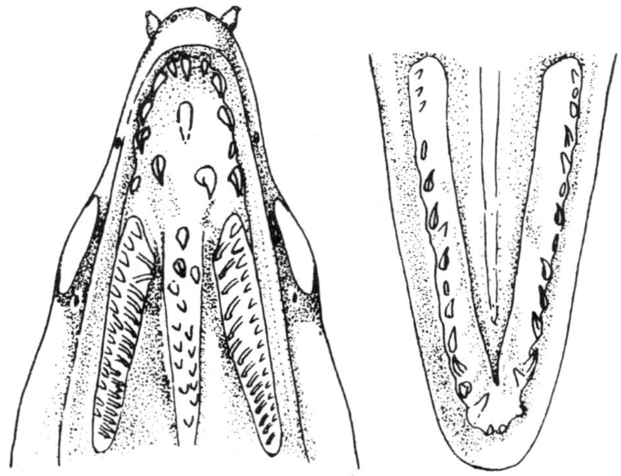

FIGURE 186. *Gymnothorax bacalladoi*: data as for Figure 184; dentition.

longest teeth in the mouth; 9–14 low, conical, stout, almost molariform teeth on vomer, the row biserial for most of its extent or staggered anteriorly, and extending back to just beyond posterior end of maxillary tooth row. Maxillary teeth biserial, the rows short; an outer row of 14–27 close-set, short, triangular, serrate teeth of almost uniform size; an inner row of 14–18 well-spaced, tall, narrow, smooth teeth. Dentary teeth in one row, 10–16 strong, triangular teeth, largest anteriorly; teeth serrate, the serrations stronger posteriorly, posterior basal knob present on anterior teeth.

Body elongate, compressed laterally; tail tapering. Dorsal-fin origin on head, about mid-way between rictus and gill opening; anal fin beginning just behind anus, which is slightly before mid-body.

Color. A mottled brown eel, belly somewhat pale. Body and fins brown with pale and dark mottled areas; margins of fins dark, blackish posteriorly. Head with similar coloration; posteriormost two upper and lower jaw pores in white spots; dark ring around eye, and dark pigmentation on anterior nostril, at mouth angle, around gill openings, anterior head pores and anus. The holotype with a pale nuchal band on dorsal surface of head just behind rictus, and pale blotches alternating with dark longitudinal folds in branchial area.

Size. The three known specimens are 260–347 mm TL. The gonads in all are undeveloped, and the species likely grows to a larger size.

Development. No leptocephalus has been identified.

Distribution. Known only from the Canary Islands. The holotype was captured in a bow net on the rocky bottom at 604 m; the paratypes were captured by hook at night at 17–20 m, also over rocky bottom.

Etymology. Named in honor of Dr. J. J. Bacallado for his contributions to the study of the marine fauna of the Canary Islands. A noun in the genitive case.

Remarks. The general appearance and body shape of this species is typical of other species of *Gymnothorax* with serrate teeth. However, in some characters, notably dentition and cranial osteology, it differs from all other species from the Atlantic and Mediterranean known to date, and it cannot be placed in a subgenus at this time. It should be compared to similar Indo-Pacific species, although we know of none with which it could be conspecific.

The most distinguishing feature is the single branchial pore. Most muraenids have two branchial pores (the monotypic genus *Monopenchelys* was so-named because it was the only known moray species possessing a single branchial pore). Nearly all species of *Gymnothorax* possess two branchial pores (*G. afer* differs from other Atlantic species in having 3 or 4 branchial pores). In addition, the biserial serrate dentition of *G. bacalladoi* is unique in Atlantic species; it is most similar to that of the eastern Pacific *G. panamensis*, which has the dorsal-fin origin well behind the gill openings (above vertebrae 8–12) whereas this species has the "typical" *Gymnothorax* position above vertebrae 5–6. The coloration of *G. bacalladoi* is similar to that of *G. afer*, which possesses uniserial, smooth-edged teeth and 3–4 branchial pores.

Gymnothorax bacalladoi has not been taken in the western or mid-Atlantic to date, and is properly extralimital to this study. However, during the past decade a number of moray species previously considered to be restricted to one side of the Atlantic have been found on both sides, particularly those inhabiting deep waters, and it is quite possible that *G. bacalladoi* may later be taken in the western Atlantic. The holotype came to the attention of the senior author in late 1986, and the two paratypes were obtained in 1987 as this manuscript was ready to go to the printer.

Study Material. The holotype and two paratypes, 260–347 mm TL.

Holotype: TFMC (ZM-2) (260 mm TL, immature); Canary Islands, Tenerife Island, Barranco Hondo, captured with the crustaceans *Plesionika edwardsi* (Brandt), *Paromola cuvieri* (Risso), and *Cancer bellianus* Johnson, by means of a bow net placed on rocky bottom at 604 m; G. Dionis; 10 March 1981.

Paratypes: ANSP 159041 (1, 347) and CAS 60405 (1, 287); Canary Islands, Tenerife Island, Puerto de Santa Cruz, captured with hook in the night over rocky bottom, depth 17–20 m; F. de Armas; January 1987.

Genus *Monopenchelys* Böhlke and McCosker, 1982

Monopenchelys Böhlke and McCosker, 1982: 127 (type species *Uropterygius acutus* Parr, 1930, by original designation).

Characters. Body slender, elongate, laterally compressed; head and trunk shorter than tail, 42–47% TL. Dorsal-fin origin behind level of anus. Snout elongate, subconical, broad; jaws moderate, subequal and closing completely. Eye moderate, above midpoint of upper jaw. Anterior nostril tubular, anteriorly directed; posterior nostril a small round opening above anterior margin of eye. Gill opening small and slit-like, below mid-side. Head pores developed; three supraorbital, four infraorbital, six mandibular and one branchial pore. Teeth conical, slender and depressible; biserial on jaws and uniserial on vomer.

Gill arches like those of *Muraena*; hypobranchials 1–2 absent. Coloration uniform.

Size. Small eels, the largest specimen known just over 200 mm TL.

Distribution. Found in the Atlantic, Indian, and Pacific oceans.

Etymology. From the Greek *mono* (one), *ope* (opening), and *enchelys* (eel), in reference to the diagnostic importance of the single branchial pore. Feminine (ICZN Opinion 915, 1970).

Remarks. The single species was originally described in the genus *Uropterygius*, which has the median fins confined to the tip of the tail. The subgenus *Rabula* Jordan and Davis, 1891, was later created for species with the dorsal-fin origin nearer the vent than the gill opening, its type species *Muraena aquaedulcis* Cope. It was later elevated to generic rank, and two Indo-Pacific species, *Rabula fuscomaculata* Schultz, 1953, and *R. marshallensis* Schultz, 1953, were placed in the genus; Parr's species *acutus* was added by Böhlke and Chaplin (1968: 79), based on the position of the dorsal-fin origin. McCosker and Rosenblatt (1975: 422) demonstrated that the type specimen for *Rabula* was a specimen of *Gymnothorax mordax* with an abnormal dorsal fin, and *Rabula* was placed in the synonymy of *Gymnothorax*. Böhlke and McCosker (1982) described the new genus *Monopenchelys* for *acuta*, based on a suite of characters (the single branchial pore, strong dentition, and other osteological characters), which *fuscomaculata* and *marshallensis* do not share. The affinities of the genus lie within the subfamily Muraeninae, not the Uropterygiinae, as evidenced by the condition of its gill arches.

Species. A single widespread species, *Monopenchelys acuta* (Parr, 1930).

Monopenchelys acuta (Parr, 1930)

REDFACE EEL

Figures 187–190, Tables 9, 22

Uropterygius acutus Parr, 1930b: 16, fig. 2 (original description, West Caicos Island, holotype YPM 2604).

Rabula acuta. Böhlke and Chaplin, 1968: 79.

"Rabula" acuta. Böhlke, 1977: 4.

FIGURE 187. *Monopenchelys acuta*: YPM 2604, holotype, 140 mm TL, Caicos Islands. Illustration by S. P. Gigliotti.

Monopenchelys acuta. Böhlke and McCosker, 1982: 283 (placed in new genus).

Distinctive Characters. A small, slender, brown moray with pale head (red-orange in life) and tail tip; tail 53–58% TL; dorsal-fin origin shortly behind anus; one branchial pore; MVF 73-53-130.

Description. Total vertebrae 124–135, x̄ = 130.3 (n = 40); predorsal vertebrae 67–78, x̄ = 72.7 (38); preanal vertebrae 50–56, x̄ = 53.1 (38); dorsal-fin rays 128–154 (3); anal-fin rays 210–238 (3). Pores: SO 1 + 2; IO 4; M 6; B 1.

Proportions as % (n = 28, 77–209 mm TL) of TL: predorsal 53–63, x̄ = 60; preanal 42–47, x̄ = 45; head 12–13, x̄ = 12; depth at gill opening 26–48, x̄ = 39; depth at anus 30–49, x̄ = 38; width at gill opening 18–38, x̄ = 24; width at anus 16–29, x̄ = 23. Of HL: snout 11–16, x̄ = 14; upper jaw 29–44, x̄ = 34; lower jaw 28–42, x̄ = 32; eye 5.6–9.8, x̄ = 7.4; interorbit 2.6–10, x̄ = 5.2.

Head elongate; snout long and broad; jaws long, subequal. Eye moderate, above mid-jaw. Anterior nostril tubular; posterior nostril a small round pore above anterior margin of eye. Head pores as listed and illustrated; no variation noted. Teeth numerous, slender, and depressible. On upper jaw, an outer ring of small intermaxillary teeth, 8–15 on each side, enclosing three longitudinal rows of large canines, three teeth in median row and four in side rows. A single series of 10–14 small stout teeth on vomer, well separated anteriorly from intermaxillary patch. Maxillary teeth biserial; outer row of 16–30 small close-set teeth, inner row of 7–10 long, slender teeth. Dentary teeth biserial anteriorly, uniserial for most of their extent; outer row of 35–39 small teeth, inner row of 3–4 large teeth anteriorly.

Color. Head and tip of tail pale in preservative, remainder of body uniformly tan to medium brown. Pale area of head extends to margin of eye dorsally, to angle of gape ventrally; pale area on tail noticeable on large specimens, less obvious on juveniles. Anus in pale area. Head and body of smallest specimen (45 mm) entirely pale except for thin brown line from snout tip to center of front edge of eye. In life, the head is bright red or red-orange, the tip of the tail is yellowish, and the body is uniformly brown.

Size. A small species; the largest specimen we have seen is 209 mm TL. All individuals from

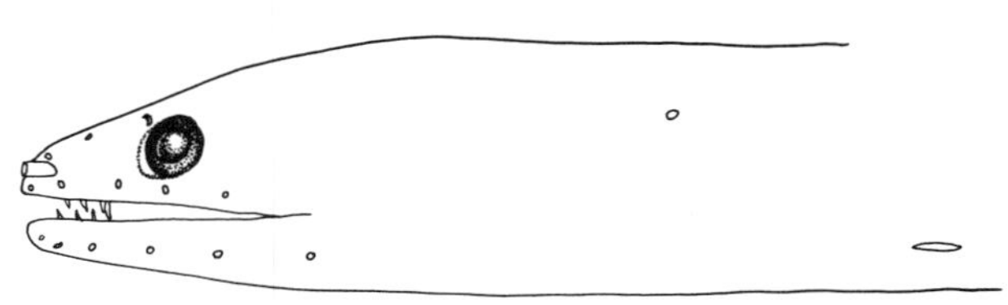

FIGURE 188. *Monopenchelys acuta*: ANSP 148102, 183 mm TL, Curaçao; head.

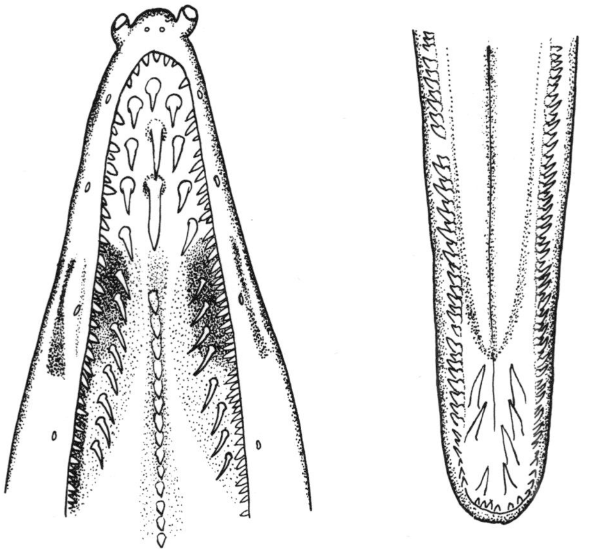

FIGURE 189. *Monopenchelys acuta*: ANSP 148103, 128 mm TL, Curaçao; dentition.

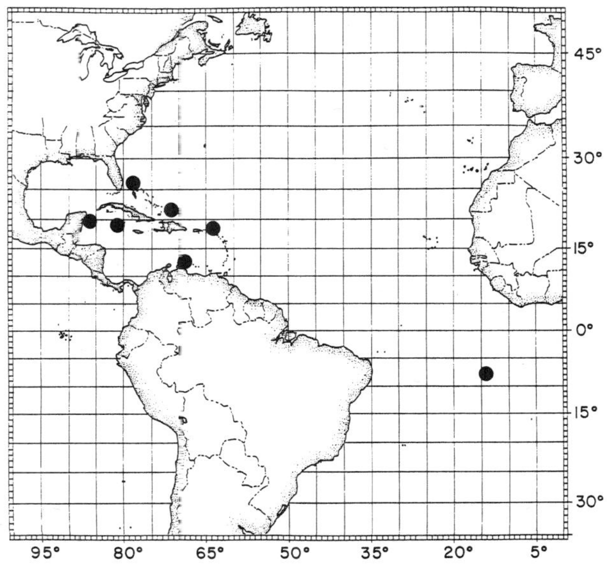

FIGURE 190. Distribution of *Monopenchelys acuta* in the Atlantic based on material examined.

Hawaii were much larger than those from the Indian and Atlantic oceans. Of six Hawaiian females (151–203 mm), three contained 0.9–1.0-mm eggs, and a fourth had exceptionally large 2.2-mm eggs; four specimens (196–209) were males. The largest Atlantic specimen is 183 mm; six females (95–155) had 0.5–0.8-mm eggs. Sex of the small Indian Ocean specimens (113–125) could not be determined.

Development. Leptocephalus described as Type 1 larvae by Keller (1976: 111) and as *Rabula acuta* by Smith (1979: 21) (see p. 905). Those described as *Rabula acuta* by Eldred (1969a) are not this species.

Behavior. This small moray is an inhabitant of coral heads and rocky ledges at moderate depths; it has usually been collected by ichthyocides, once by dredge.

Distribution (Fig. 190). In the western Atlantic, *Monopenchelys acuta* has been infrequently taken at islands in the Bahamas, Lesser Antilles, off Mexico, and at Grand Cayman and Ascension Islands. Acero et al. (1984: 53) reported three specimens off Colombia, and Lubbock (1980: 285) reported an 84.9-mm specimen from Ascension Island. In the Pacific, three collections were taken off Oahu, and one at Fiji; in the Indian Ocean, one specimen each from the Seychelles, Comores, and Agalega Islands. Depths were recorded from 13–45 m.

Etymology. From the Latin *acutus* (acute), in reference to the slender head and snout. An adjective.

Remarks. Monopenchelys acuta was known only from a few specimens from the western Atlantic until 1982, when additional material from Hawaii and the Indian Ocean was obtained. No differences were found in coloration, morphometric characters, or vertebral counts (Table 22).

Study Material. A total of 40 specimens, 45–209 mm TL, including the holotype.

Holotype: YPM 2604 (140); Caicos Islands, West Caicos Island, from the stomach of an *Epinephelus ascensionis* taken at 54 m; PAWNEE Sta. 45; 2 Apr. 1927.

Other material: ATLANTIC: *Bahamas:* ANSP 133684 (1, ca 140). UMML 22134 (1, 77). *Anegada Passage:* NMC 73-50 (1, ca 125). *Curaçao:* ANSP 126574 (1, ca 150, cleared and stained), 126575 (4, 140–155), 148102 (1, 183), 148103 (1, 128). *Cozumel Island:* ANSP 130017 (2, 95–112). *Grand Cayman Island:* UF 12470 (2, 83–148). *Ascension Island:* ANSP 149900 (6, 45–115). GMBL 80-50 (8, 70–111).

INDO-PACIFIC: *Hawaii:* ANSP 148940 (1, 190, cleared and stained). BPBM 15506 (4, 182–208), 17799 (2, 178–209). *Fiji:* USNM 259890 (1, 127). *Seychelles:* ANSP 134983 (1, 113). *Comoros:* CAS 35461 (1, 120). *Agalega Islands:* USNM 233946 (1, 125).

TABLE 22. Frequency distributions of vertebral counts in *Monopenchelys acuta.*

Area	Predorsal												Preanal						
	67	68	69	70	71	72	73	74	75	76	77	78	50	51	52	53	54	55	56
Western Atlantic							2	8*	2	2	-	1			3	4	4	3*	1
Ascension Island			1	1	2	4	1	1	1						2	9			
Hawaii	1	1	-	1	4	-	1	-	1	1			1	1	2	3	-	3	
Indian Ocean				1	-	1								1	1				

Area	Total											
	124	125	126	127	128	129	130	131	132	133	134	135
Western Atlantic	1	1	1	-	3*	3	4	2	1			
Ascension Island			1	-	-	2	5	4	1			
Hawaii						2	-	1	-	3	2	2
Indian Ocean							1	-	1			

* Includes holotype.

Genus *Muraena* Linnaeus, 1758

Muraena Linnaeus, 1758: 244 (type species *Muraena helena* Linnaeus, 1758, by subsequent designation by ICZN Opinion 77, 1922, Smithsonian Misc. Coll. 73(1): 71). Also spelled *Murena* by various authors.

Murenophis Cuvier, 1798: 54 (type species *Muraena helena* Linnaeus, 1758, by original designation and monotypy).

Muraenophis Lacepède, 1803: 627 (emendation of *Murenophis* Cuvier, 1798, type species *Muraena helena* Linnaeus, 1758, by subsequent designation of Jordan, 1917).

Limomuraena Kaup 1856a: 65 (type species *Muraena guttata* Solander in Richardson, 1848 = *Muraena pavonina* Richardson, 1848, by monotypy).

Limamuraena Kaup 1856b: 95 (emendation for *Limomuraena* Kaup, 1856).

Characters. Body elongate, laterally compressed posteriorly; tail tapering, head and trunk shorter than tail (41–50% TL in Atlantic species). Dorsal-fin origin on head, before gill opening. Snout moderate, subconical; jaws elongate, subequal and closing completely. Eye moderate, its center above mid-gape. Anterior nostril tubular, on tip of snout; posterior nostril tubular, above and before anterior margin of eye. Gill opening small and slit-like, at mid-body. Head pores developed, typically three supraorbital, four infraorbital, six mandibular, and two branchial pores. Teeth strong, pointed, with smooth edges, uniserial in adults.

Gill arches reduced (see Fig. 113, p. 112), limited to ossified ceratobranchials 1–4 and the upper and lower pharyngeal tooth plates. Peritoneum pale. Coloration variable, usually mottled and/or spotted, with the branchial opening usually surrounded by a large dark spot.

Size. Large eels, growing to over 1800 mm TL.

Distribution. Circumtropical and subtropical; juveniles inhabit shallow rock and coral areas, adults are found in deeper waters recorded to 50 m.

Etymology. From the Greek *Muraena*, the ancient name for *Muraena helena.* Feminine. Artedi (1738) proposed the name *Muraena* for all anguilliform fishes and, on that basis, Linnaeus (1758) described *M. helena* as the first eel in his catalogue.

Remarks. The genus *Muraena* at present is restricted to morays with both anterior and posterior nostrils tubular, the dorsal-fin origin before gill openings, and with strong caniniform dentition. Species of *Muraena* are distinguished from some of those in the *Gymnothorax* complex only by the posterior nostril condition (a character which may not deserve generic recognition); they are otherwise very similar to species in the subgenus *Taeniophis.* We herein refer *Muraena miliaris* (Kaup) (frequently placed in *Muraena* because its posterior nostril is in a short tube) to *Gymnothorax*, sensu lato, and the Indo-

Pacific *Muraena pardalis* Schlegel to *Enchelycore.* A thorough osteological study of the species of *Muraena* and *Gymnothorax,* as well as of other genera of the subfamily Muraeninae, is necessary before generic characters and limitations can be adequately determined.

Species. The total number of species of *Muraena* found world-wide remains uncertain until Indo-Pacific species are studied. In the eastern Pacific, the genus is represented by three species: *Muraena argus* (Steindachner, 1870), *M. clepsydra* Gilbert, 1898, and *M. lentiginosa* Jenyns, 1842. Blache (1967b) has adequately listed the voluminous literature and described the eastern Atlantic and Mediterranean species of *Muraena.* To that we add information based on our study, and somewhat modify his arrangement. We have examined specimens of *Muraena* (as currently defined) found on both sides of the Atlantic and in the Mediterranean, and have compared vertebral formulae as well as coloration, meristics and morphological data. We recognize six species of *Muraena* in the Atlantic: *Muraena robusta* Osorio, 1909, is present in both western and eastern Atlantic; also present in the western Atlantic are *pavonina,* Richardson, 1848, and *retifera,* Goode and Bean, 1882; and, in the eastern Atlantic we recognize three species, *augusti,* Kaup, 1860, *helena,* Linnaeus, 1758 (present also in the Mediterranean), and *melanotis,* Kaup, 1860. The species are separated by coloration and by vertebral counts (Table 19).

Muraena augusti was included in the synonymy of *M. helena* by Blache (1967b: 179); we consider the differences in coloration and vertebral counts to warrant species separation. *Muraena pavonina* is here redescribed as a valid species based on studies of the holotype (which is in very good condition) and other specimens, both old and recently collected. This species is very similar to *M. melanotis* from the eastern Atlantic (see *Remarks* under *M. pavonina*). *Muraena pavonina* and *M. melanotis* differ from the other *Muraena* species in possessing a different body shape (short and stubby in *melanotis* and *pavonina,* long and tapering in the others) and more complete biserial dentition; additional study and redefinition of the genera of the Muraeninae may result in the resurrection of the genus *Limomuraena,* created by Kaup (1856a: 65), for these two species.

KEY TO THE ATLANTIC SPECIES OF *MURAENA*

1a. Posterior nostril tubular, broader and larger than anterior nostril; large white spots or polygons on dark background; MVF 4-50-127 . . . . *pavonina,* p. 197
1b. Posterior nostril tubular, about equal to anterior nostril; coloration variously patterned 2
2a. Color dark with widely separated small white dots overall; MVF 5-53-138 . . *augusti* Kaup, 1860 (eastern Atlantic islands)
2b. Color pattern of large spots or rosettes 3
3a. Color pattern of large spots 4
3b. Color pattern of large rosettes, overlain with smaller spots, on dark background 5
4a. Body and fins with large dark spots, usually darker than background; MVF 4-64-154 . . . . *robusta,* p. 203
4b. Body and fins with large pale spots on dark honeycomb background; MVF 4-50-125 . . . . *melanotis* Kaup, 1860 (islands and coast of eastern Atlantic)
5a. Rosettes close together, background a dark reticulum; MVF 4-56-132 . . . . *retifera,* p. 200
5b. Rosettes far apart and separated, especially on tail; MVF 5-60-144 . . . . *helena* Linnaeus, 1758 (eastern Atlantic and Mediterranean)

TABLE 23. Frequency distributions of vertebral counts in Atlantic species of *Muraena*.

	Predorsal				Preanal																
	3	4	5	6	49	50	51	52	53	54	55	56	57	58	59	60	61	62	63	64	65
Western Atlantic																					
pavonina	10*	8	2		4	9	7*	2													
retifera																					
North		20*	4						1	-	9	8*	5	1							
Brazil		6	9	2				1	1	1	7	7									
robusta	1	1																2	-	-	1
Eastern Atlantic																					
augusti		2	6	1				4	1	3	-	1									
helena	2	4	7	2											2	8	5				
melanotis		8	4		4	4	3	1													
robusta	1	2																	1	1	

	Total																																			
	122	123	124	125	126	127	128	129	130	131	132	133	134	135	136	137	138	139	140	141	142	143	144	145	146	147	148	149	150	151	152	153	154	155	156	157
Western Atlantic																																				
pavonina	1	1	-	3	3	9	6*	2																												
retifera																																				
North							1	2	3	3	3	6	2*	3																						
Brazil							1	4	6	4	1																									
robusta																												1	-	1	-	1				
Eastern Atlantic																																				
augusti																3	3	1	1	-	1															
helena																1	-	-	-	-	2	3	1	4	3											
melanotis		3	2	3	3	1																														
robusta																														1	-	-	-	-	1	
Blache†																														1	2	2	3	4	5	3

* Includes holotype.
† From Blache, 1967: 211–212.

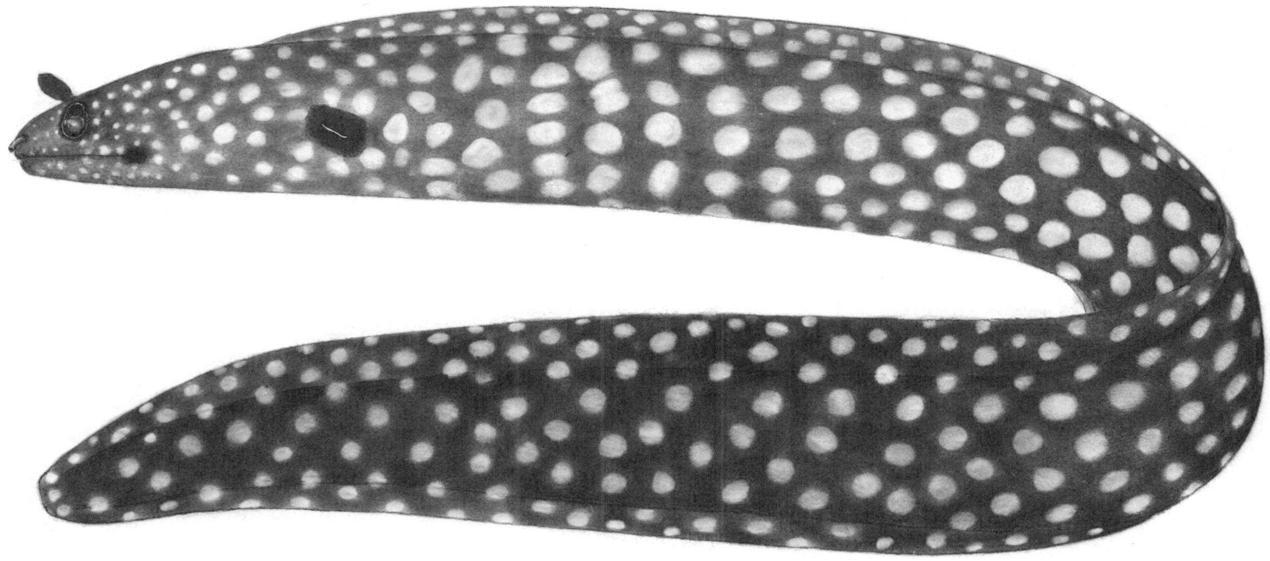

FIGURE 191. *Muraena pavonina*: ANSP 153460, 295 mm TL, Atol das Rocas.

Muraena pavonina Richardson, 1844
FLAGGED MORAY
Figures 191–194, Tables 9, 23

Muraena pavonina Richardson, 1844: 110, pl. 53, figs. 1–6 (original description, no locality given; holotype BMNH 1985.2.19.4).

Limamuraena guttata (not *Muraena guttata* of Solander in Richardson, 1848). Kaup, 1856a: 65. Kaup, 1856b: 96.

Muraenophis pavonina. Fowler, 1928: 51 (description based on Günther, 1870: 90; "in the [U.S.] National Museum, without locality, may have been obtained in Oceania").

Murenophis pavonina. Fowler, 1940: 754 (one 220-mm specimen obtained by Wilkes Expedition; "'Cape Verde Islands,' likely from Polynesia?").

Muraena melanotis (not of Kaup, 1856). Böhlke, 1977: 4 (listed only, from West Central Atlantic).

Discussion of Synonymy. Muraena pavonina was placed in the synonymy of *Muraena guttata* Solander in Richardson, 1848, by Kaup (1856a: 65, 1856b: 96). *Muraena guttata* was based on the illustration in the Banks and Solander manuscript collection of a moray from Brazil and Madeira. We have seen the illustration (at BMNH, courtesy of P. J. P. Whitehead) and determined that it is an accurate depiction of *Muraena augusti* from Madeira, the Brazil locality either erroneous or based on a specimen of *Muraena retifera* from that locale.

Distinctive Characters. A moderate-size, relatively stout, dark brown or black moray patterned with white spots, coloration uniform on head, body and fins; body short and stubby, tail not long and tapering; posterior nostril high and conspicuous, larger than anterior nostril; MVF 4-50-127.

Description. Total vertebrae 122–129, x̄ = 127 (n = 25); predorsal vertebrae 3–5, x̄ = 4.0 (20); preanal vertebrae 49–52, x̄ = 50 (26); dorsal-fin rays 270–300, x̄ = 284 (4); anal-fin rays 198–205, x̄ = 200 (4); kidney terminus 58–60, x̄ = 59 (3). Head pores: SO 1 + 2; IO 4; M 6; B 2.

Proportions as % (n = 15, 160–670 mm): Of TL: predorsal 7.4–10, x̄ = 8.5; preanal 41–47, x̄ = 44; head 11–14, x̄ = 13; depth at gill opening 6.1–11, x̄ = 7.5; depth at anus 5.9–10, x̄ = 8.0; width at gill opening 3.5–7.7, x̄ = 4.9; width at anus 4.0–9.0, x̄ = 5.5. Of HL: snout 15–20, x̄ = 17; upper jaw 34–48, x̄ = 40; lower jaw 34–47, x̄ = 38; eye 8.6–11, x̄ = 10; interorbit 10–14, x̄ = 11; length anterior nostril 2.5–4.9, x̄ = 4.1; length posterior nostril 8.5–14, x̄ = 11.

Snout short, its profile uniform in lateral aspect; jaws short; eye moderate, above mid-jaw. Anterior nostril narrowly tubular, reaching edge of snout when depressed; posterior nostril conspicuously large, fleshy, and erect, flag-like on preserved specimens, above and just before anterior margin of eye. Posterior nostril longer (2.8–

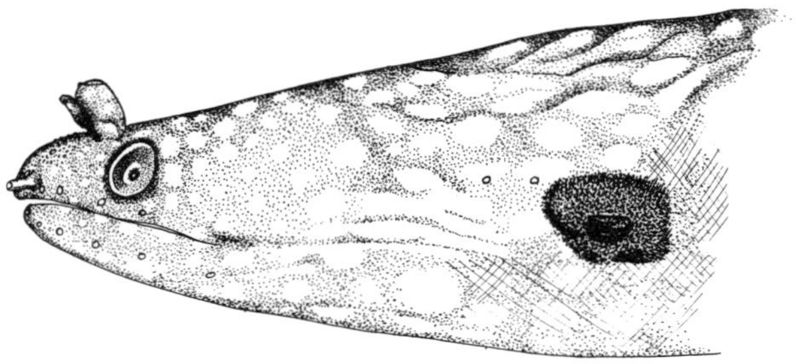

FIGURE 192. *Muraena pavonina*: data as for Figure 191; head.

3.8 times) and with broader base (2.4–3.2 times) than anterior nostril. Head pores as listed, with little variation; one specimen with seven mandibular pores on one side; two specimens with one branchial pore on one side, one with three on one side.

Dentition variable with size. In adults, on each side of upper jaw, 5–6 large, stout intermaxillary teeth (plus 3–5 small teeth between those of main series), continuous with maxillary teeth which are in one row of 10–20 widely-spaced, large, triangular teeth, decreasing in size posteriorly. On midline of upper jaw, 2–3 long, depressible fangs anteriorly (the posteriormost the longest tooth in the mouth), and a row of 6–11 small teeth on vomer, well separated from the anterior fangs and extending almost to level of posterior maxillary teeth in small specimens; fewer vomerine teeth visible in larger specimens. On lower jaw, 13–15 uniserial teeth, the anterior 2–4 the largest. In small specimens (less than 300 mm), the jaw teeth are biserial; an outer series of 16–22 small teeth and an inner row of 5–7 long slender sharp depressible teeth on maxillary, and on dentary a main series of 21–27 plus 3–4 large depressible inner teeth anteriorly (occasionally a few additional anterior tiny teeth).

Body shape short and stubby, tail tapering little, rounded at tip; depth and width of body at anus equal to or greater than that at gill opening. Grooves on dorsal surface of head between and behind eyes of large specimens, adorned with papillae and small leaf-like appendages. Lips of specimens larger than 300 mm papillose.

Color. Dark brownish black background with round white spots over head, body, and fins, coloration of uniform intensity all over, including belly. In small specimens the spots appear

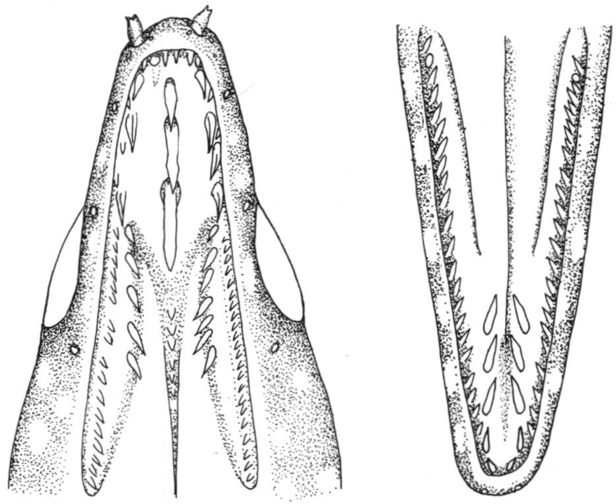

FIGURE 193. *Muraena pavonina*: data as for Figure 191; dentition.

to be in three lateral rows plus a row on edge of median fins; spots in four to five rows in medium (130–160 mm) specimens; spots randomly placed in larger specimens. Spots vary in size between specimens but usually remain distinct roundish spots, not appearing as a reticulated pattern. Gill opening dusky in small specimens, centered in large conspicuous black blotch in medium to large specimens, the patch remaining dark in old faded specimens. Mouth angle dusky. Both anterior and posterior nostrils with spotted color pattern.

Very small (less than 100 mm) specimens of *Muraena pavonina, M. retifera, Gymnothorax moringa, G. vicinus* and *Enchelycore anatina* have a similar coloration of small pale spots on a dark background. In *Gymnothorax moringa* and *G. vicinus* the spots are diffuse and the lower jaw is conspicuously pale; in *E. anatina* the dorsal-fin origin is behind the gill opening and the teeth are extremely long and sharp. The two species of *Muraena* are separable by vertebral counts (Table 19).

Size. The largest specimen studied is 670 mm TL, but no mature adults were found and the species may grow larger. Of 19 specimens checked for sex, four were females (295–300 mm) and four were males (434–670 mm).

Development. No leptocephalus in the western Atlantic has been identified as this species. Blache (1967b: 204, 1971: 215, 1977: 31) described leptocephali of *Muraena melanotis* from the eastern Atlantic, which could also include this species.

Behavior. Muraena pavonina inhabits shallow tide pools over coral and rocks. It has been collected at poison stations in groups of several individuals of various sizes (a recent collection, USNM 274346, of eight individuals, 135–490 mm TL, taken in tidepools on lithothamnion reef at Fernando Noronha Island). Large specimens have been speared from holes and crevices in coral habitats and among weeds at depths to 50 m (Lubbock and Edwards, 1981: 138).

Distribution (Fig. 194). Known only from islands of the mid-Atlantic and off Brazil. The exact locality for the holotype and for the Brazilian Trawling Survey specimens collected in 1944–1945 is not known. Recent, well-documented material has come from the islands of St. Paul's Rocks, Fernando-Noronha, Atol das Rocas, and from Ascension Island. Taken by ichthyocide at depths recorded to 6 m and by trawl and spear at unrecorded depths.

Etymology. Not stated; possibly from the Latin *pavo* (peacock) with an adjective ending, in reference to the eye-like spots. An adjective.

Remarks. Muraena pavonina was not known or recognized as a valid species of Atlantic moray until 1973 when the holotype and three additional specimens from Fernando Noronha were examined at the British Museum (by EBB and JEB). They were recognized to be conspecific with study specimens of a previously unknown species collected in islands off Brazil and from Ascension Island. No locality was given in the original description of *M. pavonina*; Richardson later (1848a: 90) listed it from "Southern Seas." Kaup (1856a: 65) incorrectly included *pavonina* in his synonymy for "*Limamuraena guttata*" (Solander in Richardson, 1848) from Madeira and Brazil; Günther (1870: 98) listed it as a valid species from "Southern Seas." Fowler (1928: 51) interpreted "Southern Seas" to refer to the Pacific Ocean and listed a specimen of *M. pavonina* in the USNM as "without locality, may have been obtained in Oceania."

Evidence strongly suggests that the holotype of *Muraena pavonina* came from Ascension Island. It was collected during the voyage of the H.M.S. SULPHUR during the years 1836–1842; the itinerary included Madeira, Teneriffe, and

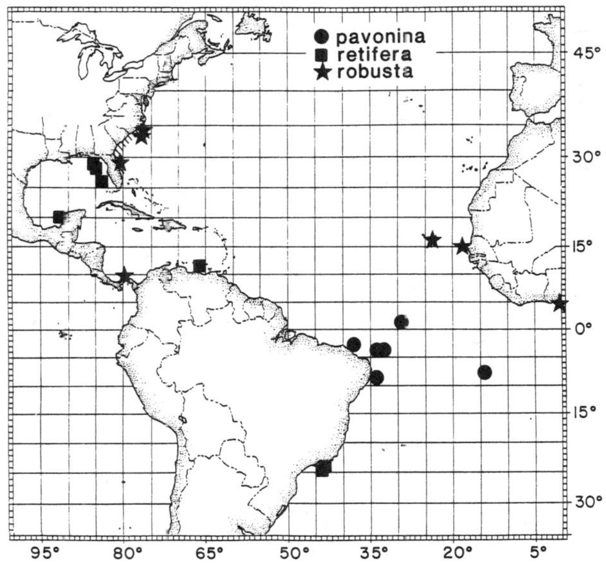

FIGURE 194. Distribution of western Atlantic species of *Muraena* based on material examined and records in Blache, 1967b.

the Brazilian coast during 1836, and St. Helena and Ascension Island on the return trip in the spring of 1842. Richardson noted that many of the specimens taken during the early part of the voyage were lost due to "the length of the voyage, and from our inadequate means of preserving them." The holotype is in excellent condition, well preserved and with good color, and most likely was taken during the last part of the voyage. It is remarkably similar in size and color to specimens recently collected at Ascension Island.

Muraena melanotis (Kaup, 1860) from the eastern Atlantic is very similar to *M. pavonina*, differing only in coloration and in the size of the posterior nostril. *M. melanotis* has a honeycomb-like pattern which is absent or indistinct on the belly, and the posterior nostril is in a tube which is about the same size or smaller than the anterior nostril. The differences are small but consistent, and we recognize both species at this time. Should future studies of additional material (we have seen no juveniles of *M. melanotis*) prove them to be conspecific, the older name of *pavonina* would be retained.

Study Material. A total of 47 specimens, 53–670 mm TL, including the holotype.

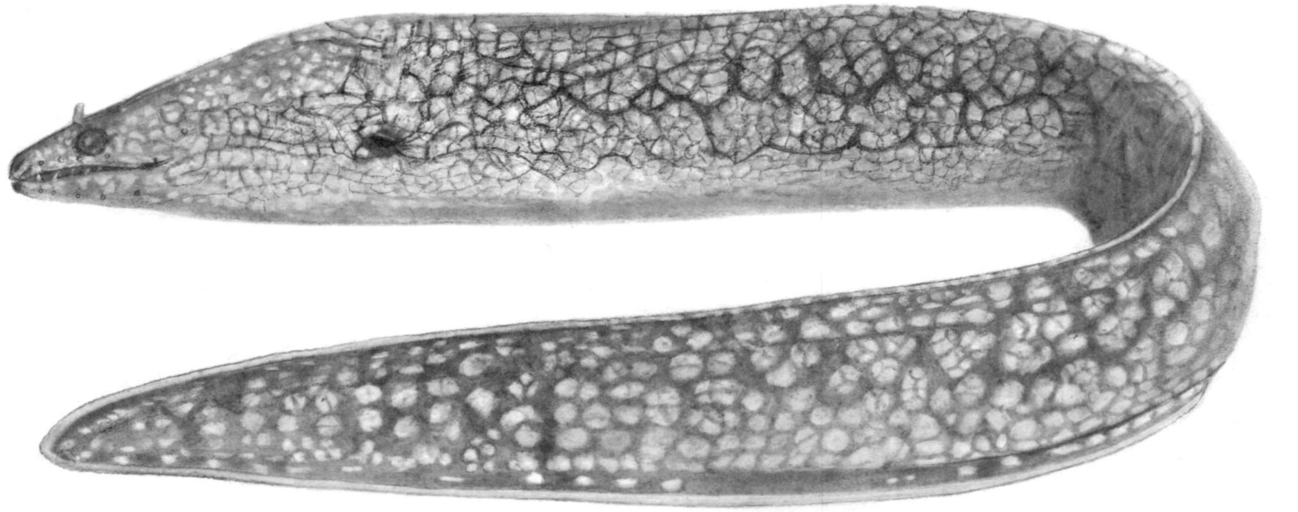

FIGURE 195. *Muraena retifera*: ANSP 106510, 432 mm TL, Bay of Campeche; typical coloration.

Holotype: BMNH 1985.2.19.4 (247.5); J. Richardson.

Other material: BRAZIL: *Forteleza* and *Recife* (taken by the Brazilian Trawling Survey, 1944–1945): ANSP 128018 (1, 362, cleared and stained), 152824 (2, 450–508), 152825 (1, 660), 152826 (1, 434). CAS 54313 (2, 450–508), 54314 (1, 660), 54315 (1, 434). CAS-SU 51720 (1, 456), 51722 (1, 484), 51723 (1, 670). MZUSP uncat. (1, 311). *Atol das Rocas:* ANSP 153460 (1, 295). MZUSP 11884 (1, 81.5). *St. Paul's Rocks:* UF 44624 (7, 56–285), 44627 (3, 186–565). *Fernando Noronha:* ANSP 142734 (1, 220). BMNH 1888.1.19.154-5 (3, 76–282). USNM 274346 (8, 135–490). ASCENSION ISLAND: ANSP 102797 (1, 660), 115276 (1, 410), 137870 (2, 68.5–160), 153460 (1, 295). GMBL 78-17 (4, 61–198), GMBL 80-54 (1, 58).

Comparative material: Muraena melanotis: EASTERN ATLANTIC: *Cape Verde Islands:* BMNH uncat. (1); Rev. R. J. Lowe. MNHN 1962-04 (1); Calypso sta. 25. USNM 83371 (1, 226); Wilkes Expedition. *Africa:* ANSP 127463 (1, 241, cleared and stained), 158140 (2, 177–528). CAS 55588 (1, 250). *Liberia:* USNM 193913 (1, 430). *Sénégal:* USNM 268509 (1, 326). NMW uncat. (1); Steindachner 1884-I-221. ZMC C.N.1 (1, 470); holotype of *Muraena albomarginata* Pfaff. *Annobon:* ANSP 158152 (5, 40–194).

Muraena retifera Goode and Bean, 1882

RETICULATE MORAY

Figures 194–198, Tables 9, 23

Muraena guttata Banks and Solander, MS., Icon. pict. Parkinson, II.1. Bib. Banks (in part, Brazil specimen).

Muraena guttata Solander in Richardson, 1848a: 90 (in part, Brazil specimen). Preoccupied by *Muraena guttata* Risso, 1826.

Murenophis punctatus Castelnau, 1855: 82 (original description, Rio de Janeiro, holotype MNHN B.2411). Preoccupied by *Muraena punctata* Bloch and Schneider, 1801.

Muraena helena (not of Linnaeus, 1758). Kaup, 1856a: 57. Kaup, 1856b: 55. Regan, 1903a: 60.

Muraena miliaris (not of Kaup, 1856). Günther, 1870: 100.

Muraena retifera Goode and Bean, 1882: 435 (original description, Charleston, South Carolina, holotype USNM 31393).

Gymnothorax conspersus (not of Poey, 1867). Jordan and Davis, 1891: 604. Howell Rivero, 1932: 16.

Lycodontis conspersus (not of Poey, 1867). Fowler, 1941: 137.

Discussion of Synonymy. "*Murenophis punctatus*" Castelnau (name preoccupied) has been synonymized with *M. helena, Gymnothorax miliaris* and *G. conspersus* by various authors as listed above. Our studies have found it to be a valid western Atlantic species, conspecific with *Muraena retifera* Goode and Bean, which is the next available name.

Distinctive Characters. A large moray with distinctive color pattern of brown rosettes overlaid with small bright white dots on a dark background which appears as a dark reticulum; anterior and posterior nostrils tubular, subequal; MVF 4-56-131.

Description. Total vertebrae 128–135, $\bar{x}$ = 131 (n = 39); predorsal vertebrae 4–6, $\bar{x}$ = 4.4 (41); preanal vertebrae 52–58, $\bar{x}$ = 56 (41); dorsal-fin

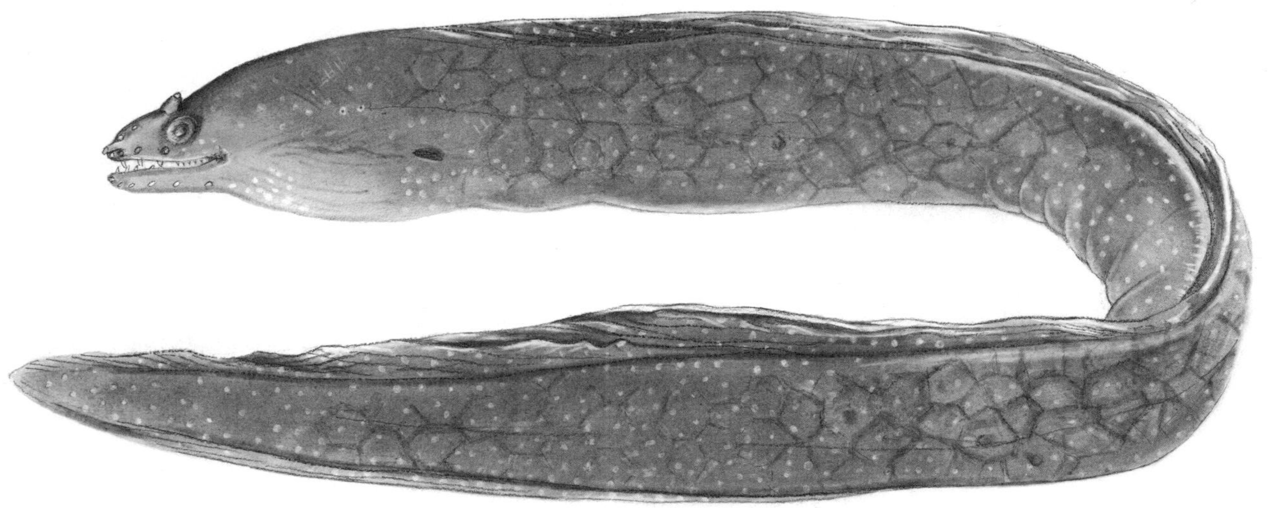

FIGURE 196. *Muraena retifera*: ANSP 158007, 468 mm TL, Brazil; faded coloration.

rays 282–323, x̄ = 300 (8); anal-fin rays 192–225, x̄ = 208 (8); kidney terminus 65–67, x̄ = 66 (3). Head pores: SO 1 + 2; IO 4; M 6; B 2.

Proportions as % (n = 21, 122–880 mm): Of TL: predorsal 7.0–13, x̄ = 10; preanal 43–50, x̄ = 47; head 12–16, x̄ = 14; depth at gill opening 5.3–11, x̄ = 7.5; depth at anus 5.2–10, x̄ = 7.0; width at gill opening 2.3–7.4, x̄ = 4.7; width at anus 2.6–6.5, x̄ = 4.5. Of HL: snout 16–22, x̄ = 20; upper jaw 38–50, x̄ = 44; lower jaw 37–48, x̄ = 42; eye 7.0–12, x̄ = 9.7; interorbit 8.6–15, x̄ = 12; anterior nostril 4.0–6.7, x̄ = 4.5; posterior nostril 2.4–7.6, x̄ = 5.0.

Snout and jaws moderate; eye moderate, above mid-jaw. Anterior nostril tubular, near tip of snout; posterior nostril in slightly shorter tube, above front margin of eye. Little variation in head pores; three specimens with one extra infraorbital pore on one side, one specimen with three branchial pores on one side and one with five on one side.

Dentition strong, uniserial in adults. On each side of upper jaw, an outer row of 5–6 large, stout intermaxillary teeth (plus 4–6 small teeth between or just outside this series in some specimens) continuous with a row of 7–14 widely-spaced, large, triangular maxillary teeth, directed back, decreasing in size posteriorly. A median row of 2–3 depressible intermaxillary fangs; vomerine series beginning at level of anterior margin of eye and extending back as far as posterior end of maxillary tooth row; 7–17 short sharp teeth, the anteriormost few biserial or staggered in some specimens. On lower jaw, 12–19 widely-spaced teeth, the anteriormost few largest, decreasing in size posteriorly, a few tiny teeth sometimes adjacent anteriorly. Jaw teeth of juveniles partly biserial; outer series of 15–21 teeth, inner series of 3–5 long slender depressible canines on maxillaries; dentaries with outer series of 19–27 teeth, inner row of 2–4 long, slender, depressible canines. Additional tiny intermaxillary teeth also sometimes present on juveniles.

Body elongate, tail tapering. Head of adults decorated with flaps and papillae on dorsal surface between eyes and on posterior nostrils, sometimes extending forward on snout before nostrils. Papillae increasing in number and size with increasing size of both sexes, on largest specimens appearing as small appendages of various sizes and shapes, from small rounded flaps of skin to tall, thin, branched or unbranched "tree-like" appendages. Lips of large specimens papillose.

Color. Typical coloration with close-set "rosettes" of light-brown blotches, each peppered with small bright white dots, on medium-brown to blackish background color which appears as a dark reticulum (Fig. 195). Head brown with white spots, the rosette pattern beginning at level

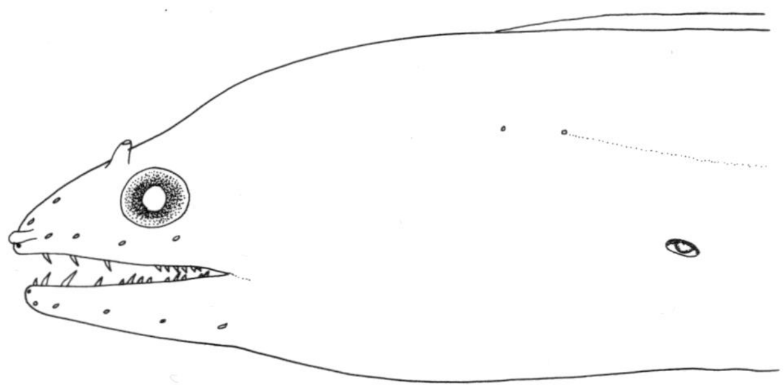

FIGURE 197. *Muraena retifera*: data as for Figure 196; head.

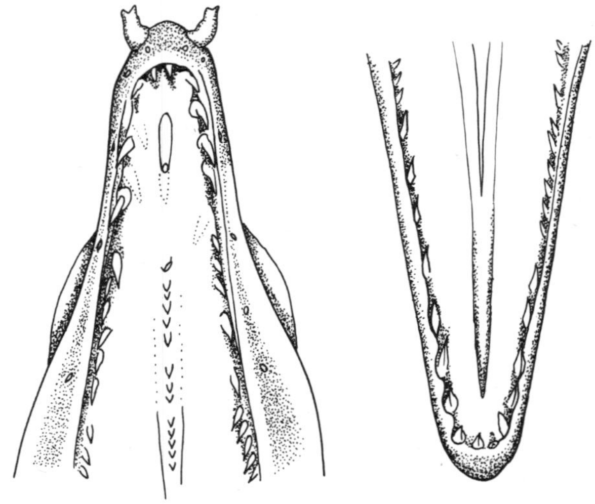

FIGURE 198. *Muraena retifera*: ANSP 107090, 395 mm TL, Florida Atlantic; dentition.

of gill opening. Gill opening surrounded by prominent black blotch; dark patch at mouth angle, eye ringed with dusky color. Edges of fins appear pale, usually with close-set small spots. Small individuals are dark with large white spots appearing in rows along body; fins with similar pattern, distal edges pale; tip of upper and lower jaws (but not entire chin) white in very small specimens.

Considerable variation in body coloration. In a number of old faded specimens (including the holotype of *Murenophis punctatus*, and the Brazilian Trawling Survey specimens) the body is medium brown overall with widely scattered small white dots, the rosette pattern not visible or very faint (Fig. 196). A recently captured specimen (ANSP 143073) had small scattered white dots visible on a brownish or blackish background (see discussion under *Remarks*). Another specimen (ANSP 134516, a 570-mm ripe female, taken with a 635-mm male which had typical coloration) had a color pattern on the trunk which resembled that of *Gymnothorax polygonius*, but exhibited typical coloration on the tail and had the nostrils, dentition, and vertebral counts of *Muraena retifera*.

A kodachrome taken by D. D. Flescher, NMFS-Woods Hole, just after capture of a 723-mm specimen off North Carolina at 37 m (specimen not seen) shows life colors very similar to those on preserved specimens; small pale spots on medium brown rosettes on a dark brown or black background; no other colors are visible.

Size. The largest specimen we have seen is 895 mm TL. Of some 30 specimens examined for sex, eight were females (ca 350–895 mm), four (535–895) ripe with 1.0–1.8-mm eggs, and eight were males (518–785).

Development. Leptocephalus not identified in the western Atlantic.

Behavior. Muraena retifera occupies a much different habitat than that of *M. pavonina.* It appears to be a mid- to shallow-water benthic species. All of the smaller specimens as well as many of the larger ones were obtained by trawls; it has not been collected at poison stations nor seen in any rock or reef habitats. It is neither uncommon nor abundant in collections.

Distribution (Fig. 194). Taken along the Atlantic coast of the United States from North Carolina (south of 34°51″) to Florida, from the Gulf of Mexico off Florida and Yucatan, off Venezuela, and off Brazil, recorded as far south as Rio de Janeiro (22°53″S). It has not been taken in the mid- or eastern Atlantic. Most often captured by trawl at depths of 21–91 m, on two occasions by hook and line, and twice in fish traps at 33 and 86 m.

Etymology. From the Latin *rete* (a net) and *fero* (to bear), in reference to its color pattern. A noun in apposition.

Remarks. Muraena retifera is part of a complex of species, including the eastern Atlantic *M. helena* and *M. augusti,* with a complicated nomenclatural history. Typical *M. helena* (Bloch, 1795: pl. 153; Blache, 1967b: fig. 1) and most specimens of *M. retifera* (Fig. 195) have similar color patterns of spotted areas, or rosettes, separated by dark interspaces; usually, the pale areas are more numerous and close-set in *retifera*, giving a net-like appearance to the dark background color, and the rosettes appear as widely-separated spotted blotches in *M. helena.* Specimens of *retifera* in which the rosettes are not obvious look much like *M. augusti,* which has small white spots on a uniformly dark brownish background. However, the three species are separable by vertebral counts (Table 19) and we consider them to be valid species.

Vertebral counts for *Muraena retifera* from the Carolinas to Venezuela are tabulated separately from those from southern Brazil. In the early stages of the study, the Brazilian specimens (all old and faded) were identified as Castelnau's *Murenophis punctatus,* which differed in vertebral counts and other characters from the various species with which it had been synonymized. The species was found to differ from *Muraena retifera* (with which it had not been compared) only in lacking the rosette color pattern. Closer inspection of some of the specimens revealed a faint pattern (seen in Fig. 196); subsequently, a fresh specimen taken off North Carolina (ANSP 143074) also showed the reticulum only faintly, superficially appearing to have a dark background color with small white dots overall (given another 100 years on a museum shelf, this individual would resemble the Brazilian ones). Counts (which are included in the description of *M. retifera* above) and proportions (which are not included above) for *M. "punctata"* were well within the range of those for *M. retifera,* and vertebral formulae were similar; Castelnau's *Murenophis punctatus* is here considered to be conspecific with *Muraena retifera,* and takes *retifera* as the next available name.

Study Material. A total of 57 specimens, 63–895 mm TL, including the holotype and the holotype of *Murenophis punctatus* Castelnau (22 specimens, 355–759 mm, were faded Brazilians ones).

Holotype: USNM 31393 (535 mm TL); South Carolina, Charleston; C. C. Leslie.

Other material: COASTAL U.S.: *North Carolina:* ANSP 143073 (1, 710), 143074 (1, 680). GMBL 73-60 (1, 820). *South Carolina:* CAS 19175 (1, 751). GMBL 71-10 (1, 763), 72-384 (1, 655). 79-147 (1, 78). UMML uncat. (1, 520). USNM 25380 (1, 710), 25396 (1, 570). *Florida Atlantic:* ANSP 94131 (1, 122), 103574 (1, 168), 106189 (1, 300+), 106515 (1, ca 210), 106516 (1, 159), 107090 (1, 395), 110414 (1, 242, cleared and stained), 126412 (1, 167), 138407 (1, 63), 138418 (1, 102), 138419 (1, 110), 138420 (2, 117–125). GULF OF MEXICO: USNM 157992 (1, 735). *Florida:* ANSP 103575 (1, 834), 158008 (1, 895). FSU 16369 (1, 725). UMML 28389 (1, 455). BAY OF CAMPECHE: ANSP 106510 (1, 432). VENEZUELA: ANSP 134516 (2, 570–635). MBUCV 3022 (1, 613).

BRAZIL: ANSP 152768 (1, 712); Brazilian Trawling Survey. ANSP 158007 (1, 468); Thayer Expedition. BMNH 1903.6.9.107-8 (1); Dr. Goeldi. BMNH 1923.7.30.64 (1, 469); Ternetz. MCZ 9066 (3, 489–535); Dr. Bourget, Thayer Expedition. MCZ 9070 (5, 410–530); L. Agassiz; Thayer Expedition. MNHN B.2411 (1, 759); Castelnau; holotype of *Murenophis punctatus.* MNHN B.2412 (1, 576+); Delelande. MNHN B.2504 (1, 355). MNRJ 9363 (1, 713); H. P. de Oliviera. MNRJ 9364 (1). NMW 4917 (1, 712); Steindachner. NMW 4953 (1, 527); M. Zurch. NMW 4954 (3, 500–566); Steindachner. NMW 4955 (1, 667); Steindachner. ZMB 4049 (1, 708); V. Alfers.

Muraena robusta Osorio, 1909

STOUT MORAY

Figures 194, 199–201, Tables 9, 23

Muraena robusta Osorio, 1909b: 75, pl. 2, fig. 1 (original description, Cape Verde Archipelago, holotype destroyed). Blache, 1967b: 206, figs. 15–19 (redescription and illustration; Congo, Senegal, Cape Verde Islands). Böhlke and Ross, 1981: 123 (Panama Atlantic, South Carolina).

Murenophis robustus. Fowler, 1936b: 1214 (after Osorio).

Muraena sp. aff *melanotis* (not of Kaup, 1860). Poll, 1953: 115 (listed in Blache, 1967b: 206; Angola).

Muraenophis robusta. Bassindale, 1961: 508.

Gymnothorax nevezi Roux, 1957: 164, fig. 17 (original description; Pointe-Noire, Congo, holotype unknown).

Gymnothorax galetae Rubinoff, 1966: 1, fig. 1 (original description, Panama-Atlantic, holotype MCZ 44035).

Distinctive Characters. A large robust moray with distinctive coloration of round dark spots on body and fins; anterior and posterior nostrils tubular, subequal; MVF 4-64-153.

Description. Total vertebrae 149–156, $\bar{x}$ = 153 (n = 5); predorsal vertebrae 3–4, $\bar{x}$ = 3.5 (5); preanal vertebrae 62–65, $\bar{x}$ = 64 (6); dorsal-fin rays 375–396, $\bar{x}$ = 384 (3); anal-fin rays 246–289,

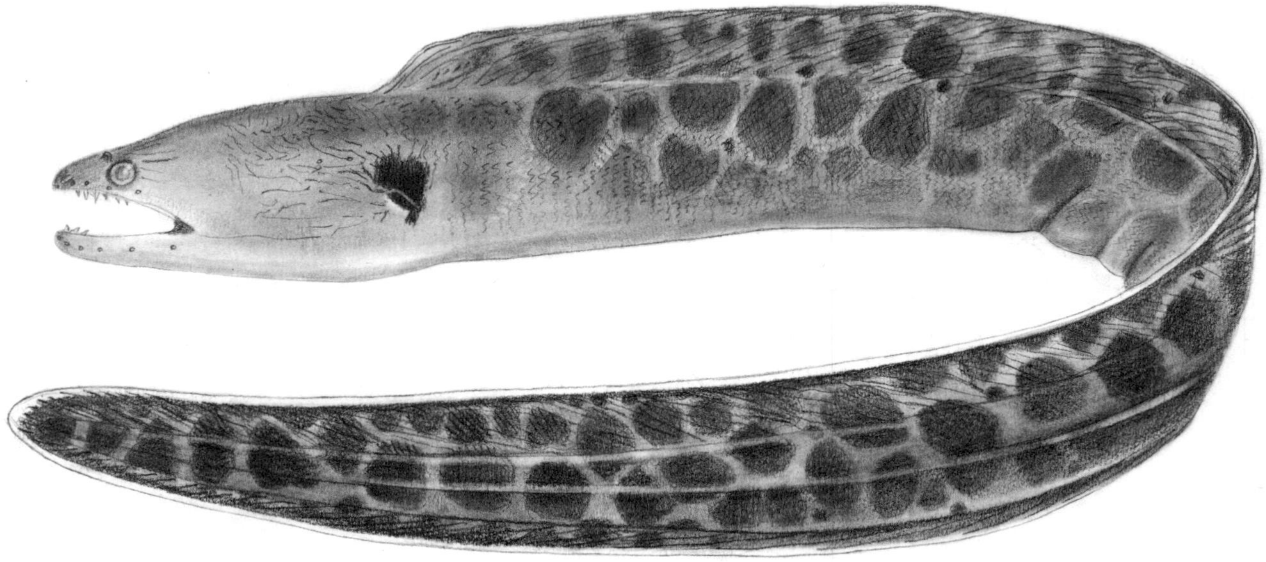

FIGURE 199. *Muraena robusta*: ANSP 138767, 459 mm TL, North Carolina; anterior portion of dorsal fin anomalous.

$\bar{x}$ = 263 (3). Head pores: SO 1 + 2; IO 4; M 6; B 2.

Proportions as % (given individually for four specimens, 128, 459, 1353, and 1680 mm TL): Of TL: predorsal 8.9, –, 9.9, 12; preanal 45, 47, 49, 49; head 13, 12, 14, 14; depth at gill opening 4.5, 6.7, 9.8, 13; depth at anus 3.4, 5.1, 5.8, 7.6; width at gill opening 2.0, 3.2, 4.1, 5.8; width at anus 2.6, 2.9, 5.0, 5.9. Of HL: snout 18, 17, 17, 18; upper jaw 38, 39, 44, 41; lower jaw 37, 38, 43, 41; eye 9.7, 8.3, 4.9, 4.8; interorbit 11, 12, 15, 20; anterior nostril 3.6, 7.5, 1.6, –; posterior nostril 'slightly raised,' 5.1, 1.5, –.

General characters similar to those of *Muraena retifera*. Anterior nostril tubular, directed forward; posterior nostril tubular, above and slightly in front of anterior margin of eye, its length about equal to that of anterior nostril. No variation seen in head pores.

Adult dentition uniserial. On upper jaw, six large intermaxillary teeth on each side, closely followed by row of 10–16 stout maxillary teeth; 2–3 median depressible intermaxillary teeth, the posteriormost the longest in the mouth; and a row of 6–11 small vomerine teeth, the first few biserial or staggered. On lower jaw, 22–25 large stout teeth, anteriormost 3–5 larger and flanked by 2–8 outer small teeth. Juvenile dentition unknown (only one small specimen available), jaw teeth probably partially biserial as in other species of *Muraena*.

Body size and shape of the two smaller specimens typically muraenid. Large specimens extremely robust, head and anterior body thick, muscles highly developed (see Blache, 1967b: fig. 15).

Color. *Muraena robusta* is readily identified by its bold and characteristic color pattern, which is highly contrasting in small individuals but which darkens and becomes less conspicuous with age. Head dusky to dark brown with darker reticulations; large round spots on body and fins behind gill openings, interspersed with smaller spots; spots appearing pale on a dark background before mid-body, gradually changing to dark spots on a pale background on tail. Gill opening in large dark blotch. Large old specimens appear dark overall, the spots barely discernible.

Life colors of the 459-mm specimen from off North Carolina (ANSP 138767, illustrated in Fig. 199, an abnormal specimen with the anteriormost part of the dorsal fin missing) were recorded as follows: Head golden brown dorsally with dark area from snout tip to interorbit and with darker longitudinal folds, and pale and grayish ventrally with some yellow invading back-

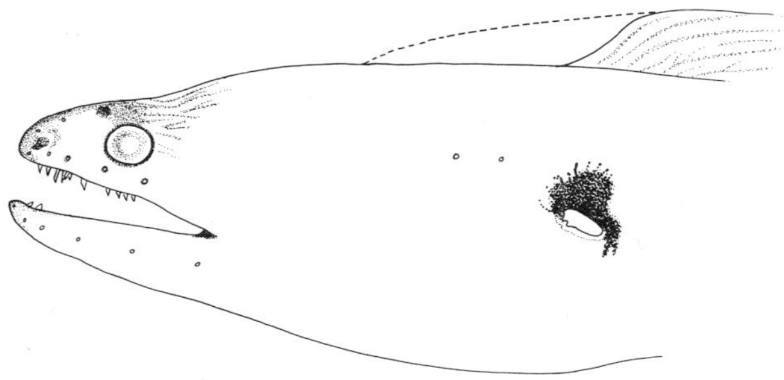

FIGURE 200. *Muraena robusta*: data as for Figure 199; head; dotted line shows normal dorsal-fin origin.

ground color and with darker reticulations present. Inner edges of lips and interior of mouth pale orange-gold; mouth angle dark. Eye and inside of gill opening bluish. Gill opening in conspicuous large dark blotch. Large round dark brown spots of varying sizes on body and fins posterior to gill opening, sizes of spots and intensity of color similar throughout length but somewhat slightly paler on belly; background color grayish-blue anteriorly to grayish-tan posteriorly, with dark reticulations throughout. Distal edge of posterior third of dorsal fin and of entire length of anal fin pale gray. Small golden patch on tip of tail. Color of the 1680-mm specimen (ANSP 152848, frozen) showed some variation: Head brown with dark orange reticulations; mouth interior bright orange, eyes blue. From gill opening to anus, body spots appeared dark orange on pale purple background, coloration paler ventrally; gradual change to dark brown spots on paler background on posterior half of tail. Anus bright orange. When preserved, orange and blue colors faded to light brown, color appearing brownish with darker brown spots and reticulations, fin margins pale.

Size. An 1860-mm specimen recently caught off North Carolina (USNM 267064, not seen) is the largest moray reported in the Atlantic. Our study material also included the smallest *Muraena robusta* known to us, the 128-mm holotype of *Gymnothorax galatae* from Panama. One of our specimens, 1353 mm, was a ripe female with 1.0-mm eggs, and the 1680-mm specimen was a male. Blache (1967b: 206) listed 41 specimens from the eastern Atlantic ranging in size from 625–1245 mm TL, none of which were sexually mature.

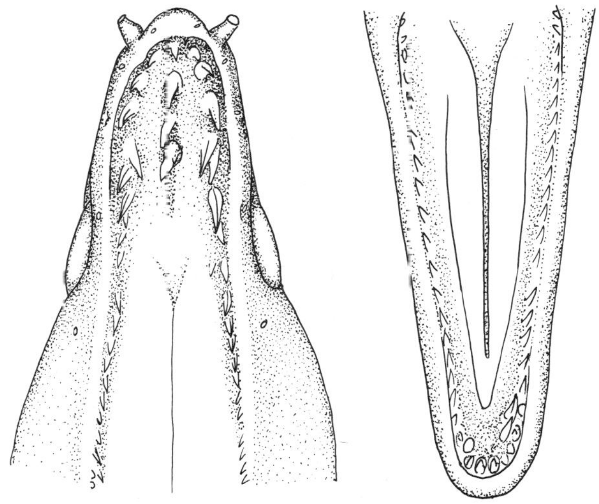

FIGURE 201. *Muraena robusta*: data as for Figure 199; dentition. Vomerine teeth not shown.

Development. Leptocephalus not known from the western Atlantic, although a ripe female was taken off the Atlantic coast of Florida. Blache (1967b: 213, fig. 19; 1977: 34) described nine larvae taken in the eastern Atlantic at 00°42′S, 07°00′E.

Behavior. Adults inhabit moderately deep waters; large specimens were observed and collected in rocky areas off North Carolina. Juveniles have been taken in shallow rock and coral areas. Said to eat crustaceans and various fishes (Blache, 1967b: 213).

Distribution (Fig. 194). A common eastern Atlantic species, *Muraena robusta* has been infrequently taken in the western Atlantic off North Carolina, Florida and Panama. Additional large specimens were observed during submersible dives on offshore North Carolina reefs on two occasions (reported in Böhlke and Ross, 1981: 124), and a ripe female (ANSP 152848) was later collected near the site. The two smaller specimens (MCZ 44035 and ANSP 138767) were taken by fish toxicants, the smallest at a depth of 6 inches, the other at 68 m. Large specimens were taken by hook and line at depths of 30–68 m. In the eastern Atlantic, *Muraena robusta* is common off the west coast of Africa from 21°N to 18°S, and has been taken from the islands of Cape Verde and the Bay of Biafra (Blache, 1967b).

Etymology. From the Latin *robustus* (robust), in reference to its body size and shape. An adjective.

Study Material. Six specimens, 128–1680 mm TL, including the holotype of *Gymnothorax galetae* Rubinoff.

WESTERN ATLANTIC: *North Carolina:* ANSP 138767 (1, 459); "30 mile rocks," 34°14.5′N, 76°35.5′W, depth 30 m; S. W. Ross, A. R. Emery; 6 July 1977. ANSP 152848 (1, 1680); off Cape Fear, 33°15.5′N, 77°16.5′W, depth 68 m; 8 Jan. 1984. *Florida:* UF 27970 (1, 1353); off Ponce de Leon Inlet, ca 13 mi SSE of Daytona Beach; 1980. *Panama:* MCZ44035(1,128);GaletaIsland,09°24′20″N, 79°52′18″W, depth 6 inches; 25 May 1962; I. Rubinoff; holotype of *Gymnothorax galetae*.

EASTERN ATLANTIC: *Sénégal:* USNM 221162 (1, 465+); Goree; 21 Sept. 1950. *Accra:* BMNH 1938.12.15.3 (1); Gold Coast; Irvine.

Family Synaphobranchidae

CATHERINE H. ROBINS *and* C. RICHARD ROBINS

Acknowledgments. In addition to those persons and grants mentioned in the general acknowledgment section of this volume and our fellow contributors to this volume, we wish to thank Jon C. Staiger for his help on many cruises where one or the other of us was senior scientist. Support for our work in the deep Bahamas basins was provided by NSF grants GA-38834 and OCE 7306639, C. Richard Robins, principal investigator. The Maytag chair of ichthyology has supported the junior author in this research. We have depended heavily on library resources of the Academy of Natural Sciences of Philadelphia. Kenneth J. Sulak, Nigel R. Merrett, and Luis Saldana have generously provided data based on their work in the American Bight and the eastern Atlantic. Peter H. J. Castle has especially been helpful in providing his unpublished information on *Synaphobranchus capensis.* Yukio Iwatsuki helped us greatly by providing translations of Japanese literature.

General Characters. Deep benthic eels. The best known synaphobranchids are among the most numerous and frequently collected of the deep benthic fishes, have scales, and are dark-colored, moderate (ca 40–50 cm) to large (70–100 cm) species. Most species, however, are small, pale-colored, scaleless, and infrequently collected. Gill openings small to moderate in size, low on body, at or below level of insertion of pectoral fin, sometimes opening on ventral or ventrolateral surface, and, in some species, through a common slit. Mouth terminal, gape extending at least to level of posterior margin of eye in all but one species. Gills four, the slit reduced behind the last gill. Eyes moderately large to much reduced. Anterior nostrils short, simple tubes near fleshy tip of snout, directed antero-ventrally. Posterior nostril anterior to and level with or slightly ventral to midpoint of eye. Dorsal, anal and caudal fins confluent. Caudal fin small, usually more darkly pigmented on lower half. Swimbladder present in all species examined. No pyloric caeca. Overall pigmentation somber: dark brown, purplish brown, blackish brown, gun-metal gray, or pinkish brown. Color varies in individuals of same species from different localities. Smaller species ordinarily pallid.

Leptocephalus larva distinguished from those of all other eels by possession of eyes which are vertically or diagonally elongated, usually referred to as tubular (e.g., Lea, 1913: 28), a term that seems inappropriate to us. Also the olfactory organ is far in front of the eye. Synaphobranchid leptocephali are treated by Smith (this volume) and generally are not discussed here except as relevant to nomenclatural, taxonomic, or other special matters.

Osteological Characters (Fig. 202). Frontals fused into one bony unit, with no trace of a median suture. Premaxillae fused to ethmoid complex or absent (see discussion below). Pterygoid variously present as complete arch between quadrate and vomer, a simple bony splint lacking bony connection, or completely absent. Basisphenoid present, occasionally fused with parasphenoid in small species with highly modified skulls. Sagittal otolith ovoid in profile. Caudal and pectoral fin supports reduced. Hypurals variable, greatest number seven (*Synaphobranchus*). Most branchiostegal rays (as many as 16) borne on epihyal, a few near the outer end of the ceratohyal. One to three pharyngobranchials. Third hypobranchials directed forward from midline, forming sharp angle with third ceratobranchials. Many vertebrae—from 110 to 205, most species between 120 to 140. Division of vertebrae into precaudal and caudal sections not sharp due to instability of formation of haemal arch. (Most workers, including us, have depended on radiographs, rather than on the determination of the

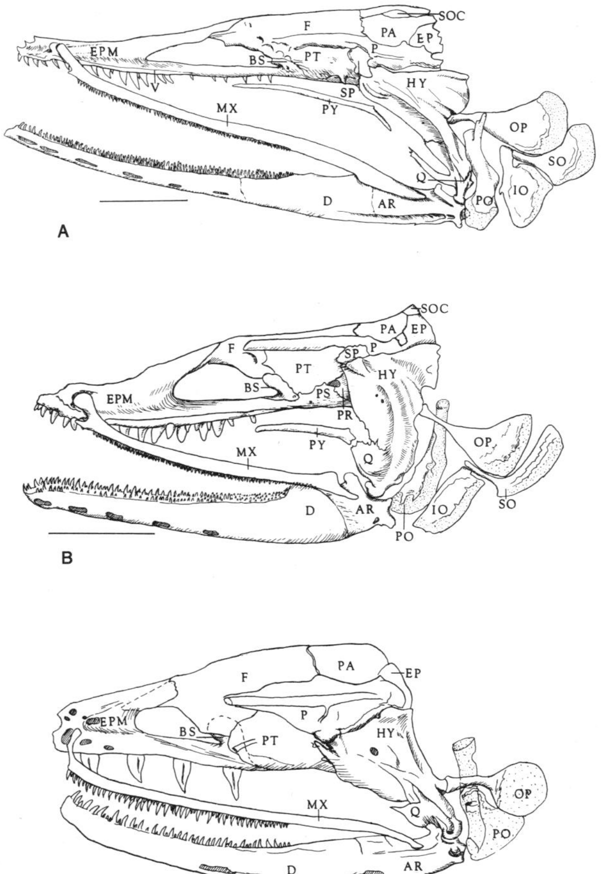

FIGURE 202. Lateral views of skull, suspensorium, jaws and opercular series of representative synaphobranchid eels, illustrating trend toward fusion of bones and streamlining within the family (from Robins and Robins, 1976). A. *Haptenchelys texis*: ANSP 133813, 460 mm TL. B. *Iliophis arx*: ANSP 133810, 401 mm TL. C. *Dysomma goslinei*: ANSP 133807, 108 mm TL. Abbreviations: AR, articular; BO, basioccipital; BS, basisphenoid; D, dentary; EP, epiotic; EPM, premaxillary-ethmoid complex; EX, exoccipital; F, frontal; HY, hyomandibular; IO, interopercle; MX, maxilla; OP, opercle; P, pterotic; PA, parietal; PO, preopercle; PR, prootic; PS, parasphenoid; PT, pterosphenoid; PY, pterygoid; Q, quadrate; SO, subopercle; SOC, supraoccipital; SP, sphenotic; V, vomer.

point at which the dorsal aorta enters the haemal canal, the latter a more reliable index, but one whose determination is time-consuming.) Ratio of dorsal-fin rays and their supports to vertebrae 3 to 2½:1. Ribs rather short.

Discussion. The Synaphobranchidae includes three subfamilies (Simenchelyinae, Synaphobranchinae, and Ilyophinae), eight genera (five monotypic), and twenty-three species. We have the front end of an undescribed species, and comparisons of descriptions of leptocephali and adults show that there are at least several other unrecognized species. The family, however, is a small one in terms of numbers of species (probably about 30), especially when compared to the speciose Muraenidae and Ophichthidae. The family is important, however, both biologically and systematically.

Synaphobranchus (especially *S. kaupi* and *S. brevidorsalis*), constitutes a significant part of the deep benthic fish fauna around the world in tropical and temperate latitudes, both in terms of biomass and numbers of individuals. Their role and that of their eggs and leptocephali in energy transfer between the deep ocean and surface layers is poorly understood at this time.

The Synaphobranchidae combines both primitive and derived characters, such that it is not easily related to other eel families. Nelson (1966a) recognized it as comprising one of his three eel lineages, the "synaphobranchoids," to which he assigned no definite taxonomic rank. Robins and Robins (1976) regarded the Synaphobranchidae as the sole family of the superfamily, Synaphobranchoidea, an action which we follow here. As in most eel groups, evolution within the group has featured consolidation and losses of structures. This makes characterization of the group as a whole difficult because the losses of some features result in convergences with eels of other families. Primitive features possessed by at least the primitive members of the Synaphobranchidae (and, hence, also of the Synaphobranchoidea) are: retention of scales of varying shapes and alignments, but mostly elongate and arranged in a basket-weave pattern on the body (*Simenchelys, Synaphobranchus,* some species of *Ilyophis*), retention of a complete palatopterygoid arcade (*Simenchelys*); presence of premaxillary teeth (in all except *Dysommina* and some species of *Dysomma*); presence of a more or less complete ring of bony ossicles around the eye; presence of four pairs of lower pharyngeal tooth plates (*Synaphobranchus*); retention of many branchiostegal rays (up to 16 in *Synaphobranchus*); only the outer two branchiostegal rays curve behind and above the opercle; hyomandibular inclined at least slightly caudally (*Synaphobranchus; Haptenchelys, Dysomma, Dysommina*); elongate body with fused median fins; retention of many caudal rays (as many as 22 in *Simenchelys*), and as many as seven hypurals (*Synaphobranchus affinis*). Having completely fused frontal bones is a derived character shared with the congroids and ophichthoids. Derived features unique to the Synaphobranchidae (and Synaphobranchoidea) are: leptocephalus larva with vertically elongate eyes; third hypobranchials directed forward from the midline, meeting the third ceratobranchials at a sharp angle. Gill slits placed very low on body, joined at midventral line in some (Fig. 203). Some eel genera in other families have the gill slits low on the side (although never so close to the midventral line), presumably a convergent condition.

Very little is known about the life history and ecology of any synaphobranchid, especially the more specialized ilyophines due to their rarity in collections and secretive nature. The more abundant species are in deep water and have been seen from submarines and in photographs, but these observations are few and rather random. The abundance of species of *Synaphobranchus* and the rarity of their easily recognized larvae, suggests that they have limited spawning periods and a short-lived leptocephalus stage.

Comparison of only some of the extreme types among the family might easily lead one to split the family and to recognize a high percentage of monotypic genera. A review of the section on Family Group Names and of the generic synonymies will show that this, in fact, has been a commonly supported viewpoint. With only 23 species allocated among eight genera, and with eight species in *Dysomma* and seven in *Synaphobranchus*, ours is scarcely a "lumper's" classification. This is a primitive group of eels, and the low ratio of species to genera perhaps reflects the relict nature of the group. The subfamilies are easily defined. Although recent data on *Ilyophis* (Saldanha and Merrett, 1982), compromised our earlier definition of the Ilyophinae, additional support has been found for this separation.

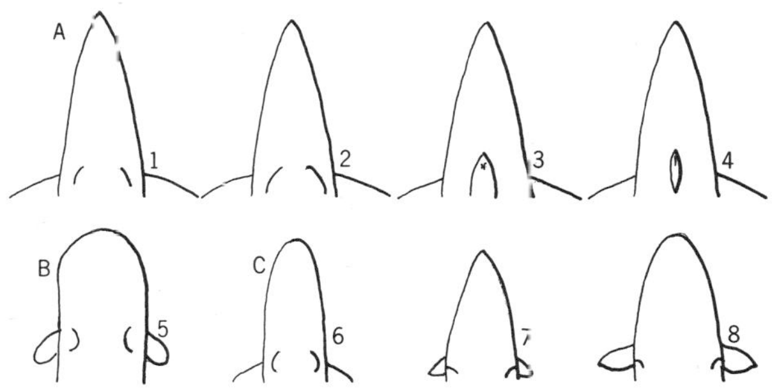

FIGURE 203. Condition of gill slits in the Synaphobranchidae. Ventral diagrammatic view of A: Synaphobranchinae. 1. *Haptenchelys texis.* 2. *Synaphobranchus capensis.* 3. *Synaphobranchus bathybius.* 4. *Synaphobranchus brevidorsalis.* B: Simenchelyinae. 5. *Simenchelys parasitica.* C: Ilyophinae. 6. *Ilyophis brunneus.* 7. *Dysomma anguillare.* 8. *Meadia abyssalis.*

Many ichthyologists would recognize the three subfamilies as families. We feel that to do so serves no useful purpose, and obscures the evolutionary trends exhibited within the Synaphobranchidae by making them artificially disjunct.

Family-group Names. Ichthyologists usually have not given synonymies of family group names of fishes and so are often unaware of priority of these names. In fact, the recent literature on the Synaphobranchidae is incorrect in this regard. At least seven generic names have been used as the bases for family-group names for eels of the family Synaphobranchidae. These names are discussed in chronological order below. In addition, various genera have at times been placed in other families, notably the Muraenidae, Nettastomatidae, Congridae, and Anguillidae.

Synaphobranchidae Johnson, 1862. Gill (1890b), while quoting Johnson's statement about the new family, failed to note Johnson's use of a family group name and was equivocal in attributing the name to Johnson. In fact, Johnson specifically noted that this new genus formed the type of a new eel family, the Synaphobranchidae, which he characterized. Gill (1890b: 161) discussed the history of this and other group names with this stem (Synaphobranchoidei Bleeker; Synaphobranchina Günther), and this need not be repeated. This is the oldest family-group name and is here used as the basis of the superfamily, family, and one of the subfamilies.

Simenchelyidae Gill in Goode and Bean, 1879. Gill, in a manuscript quoted by Goode and Bean, made his genus *Simenchelys* the type of a new family, the Simenchelyidae. This has been improperly emended to Simenchelidae (e.g., Regan, 1912b: 381). This is the oldest family group name for the monotypic subfamily Simenchelyinae.

Ilyophidae Gilbert, 1891. Gilbert described his genus *Ilyophis* and made it the type of what was then a monotypic family. Meanwhile, his manuscript was given to Jordan and Davis, who, in a paper in press at the same time, but which did not appear until 14 June 1892, desribed *Ilyophis* (quoting and crediting Gilbert) and making it the type of the family Ilyophididae. The latter name was not attributed by them to Gilbert except in the recapitulation of their eel classification (Jordan and Davis, 1891: 667). As Steyskal (1980: 171) noted, many names ending in *-is* do not have the *-id* stem. *Ilyophis* is such a name; hence, the correct name is Ilyophidae as Gilbert had originally proposed. This is the oldest group name for the subfamily previously called the "Dysomminae" (e.g., by Robins and Robins, 1976). Accordingly, this subfamily is here called the Ilyophinae.

Dysommidae Regan, 1912. Alcock, 1889, had placed his genus *Dysomma* (and most other eels) in the Muraenidae. Regan, 1912b: 387, recognized the family Dysommidae and included *Dysomma* and *Dysommopsis* in it. As noted by Steyskal (1980: 173) and followed by Nelson (1984: 106), the correct spelling of this group name is Dysommatidae and Dysommatinae. This name is not used here since it is predated by the Ilyophinae.

Todaridae Grassi, 1913. *Todarus* Grassi and Calandruccio, 1896, described in the Nettastomatidae, and preoccupied by *Todarus* Rafinesque, 1890 (a cephalopod), was placed by Grassi, 1913, in the monotypic family Todaridae. Whitley (1951b: 407) substituted *Nettodarus* and Nettodaridae for Grassi's names. Neither is used here since the type species (*brevirostre*) is placed in *Dysomma.*

Dysomminidae Böhlke and Hubbs, 1951. Ginsburg (1951: 450), described *Dysommina* in his catch-all family Congridae, and suggested (p. 412) that some ichthyologists might wish to divide the Congridae into three subfamilies, of which the Dysomminae (for *Dysomma* and *Dysommina*) would be one. Böhlke and Hubbs, 1951, placed *Dysommina* in their new family Dysomminidae. *Dysommina* is here placed in the Ilyophinae and no group name based on *Dysommina* is used.

To recapitulate, the family-group names as used in this paper are:

Synaphobranchidae Johnson, 1862
Synaphobranchinae Johnson, 1862
Simenchelyinae Gill, 1879
Ilyophinae Gilbert, 1891

The family-group names Dysommatidae (Dysommidae), Dysomminidae, and Nettodaridae (Todaridae) are all considered to be subfamilial synonyms of the Ilyophinae.

Relationships. As noted in our characterization of the family (p. 207) the Synaphobranchidae are easily defined by a number of shared specializations or synapomorphies, including the orientation of the third hypobranchials and the diagonally elongated eye possessed by the leptocephalus. These characters are such that Robins and Robins (1976) placed them as the sole family in a superfamily Synaphobranchoidea, a view which we continue to endorse.

The Synaphobranchoidea is one of three superfamilies which share the derived feature of the fused frontal bones, the others being the Ophichthoidea and the Congroidea. The nature of the frontal bones frequently has been described incorrectly (see discussion by Robins and Robins (1970: 312)). Fusion of the frontal bones could have occurred independently but this seems unlikely since comparison of primitive members of the three superfamilies show such an overall degree of similarity. We regard the fusion of the frontal bones as a shared derived character (see Robins, p. 19). The retention of such primitive features as a complete palatopterygoid arcade in *Simenchelys* (albeit specialized and expanded to support the muscles associated with its peculiar mode of feeding), scales (*Simen-*

chelys, Synaphobranchus, and some species of *Ilyophis*), and the posterior inclination of the hyomandibular suggest that the Synaphobranchoidea is a primitive lineage. There are no shared specializations yet demonstrated to suggest a sequence of splitting or interrelationship of the Synaphobranchoidea, Ophichthoidea, and Congroidea (see Robins, p. 19).

Infrafamilial Relationships. The Synaphobranchidae was divided into three subfamilies, Simenchelyinae, Synaphobranchinae, and Dysomminae by Robins and Robins (1976: 277–279). We continue to recognize these divisions except that, as noted earlier, we now call the "Dysomminae" by its earliest family group name, the Ilyophinae. Also we have altered our views with regard to some of the character states and details of the generic interrelationships. Some discussion of some of these characters follows. Saldanha and Merrett (1982) questioned the distinctiveness of the Synaphobranchinae and Dysomminae (= Ilyophinae).

Scales: The presence of scales is primitive. They are absent from all of the Ilyophine genera, except for some species of *Ilyophis,* and are present in all the other genera except for *Haptenchelys* which has independently lost them. Among those genera which have scales, elongate and regularly arranged scales is considered to be primitive, oval or rounded, less regularly arranged scales (e.g., as in *Synaphobranchus oregoni* and *S. brevidorsalis*) is considered to be derived. Elsewhere in the Anguilliformes, scales are retained only by the Anguillidae (Anguilloidea).

Palatopterygoid arcade: A complete palatopterygoid arcade is primitive and exists in the family only in *Simenchelys.* Retention of a pterygoid without bony connection is an intermediate state in the loss of this structure. This condition exists in varying degree in all other genera except *Dysomma,* in which the loss is complete.

Lower jaw: A projecting lower jaw is regarded as primitive (Simenchelyinae, Synaphobranchinae). A shorter lower jaw, overhung by the snout in varying degree is derived (Ilyophinae) and this may be seen in a progression of the shortening process seen through the series of species. The most projecting snout is found in *Dysomma.*

Dentition: The presence of small, simple teeth on the maxillae, dentary, and vomer (and very numerous at least on the first two), and with somewhat larger (but still numerous) sharp premaxillary teeth is primitive (Synaphobranchinae). *Simenchelys* has a few small chisel-like teeth on the premaxillae, maxillae, and dentary and may have one to three largely imbedded teeth, one behind the other, anteriorly on the vomer. This is judged to be an independent specialization associated with its peculiar mode of feeding. All ilyophines have enlarged vomerine teeth (often surrounded by complex pads of papillose tissue) and the anterior dentary teeth are enlarged. In *Atractodenchelys, Dysommina,* and *Dysomma,* the vomerine teeth are few in number, very much enlarged (by a factor of at least 10 to 1), and each is compound, apparently the result of an anterior-posterior fusion of two teeth. Reductions in numbers of premaxillary teeth occur in the ilyophines and they are lost entirely in *Dysommina* and some species of *Dysomma.* Maxillary teeth are largest in *Dysomma,* usually with one compound tooth anteriorly. The posterior 40 percent of the maxilla is toothless.

Maxilla: The front end of the maxilla articulates to the premaxilloethmoid region by a two-pronged head (each part forming an apophysis). In *Synaphobranchus, Ilyophis,* and *Atractodenchelys* (Robins, 1971: figs. 4–5; Robins and Robins, 1970: fig. 5; 1976: fig. 7B,C) both apophyses are well-developed, though the upper is distinctly larger. In *Haptenchelys* (Robins and Robins, 1976: fig. 7A) the lower apophysis is very small but the upper retains its strong forward slant. In *Dysomma* and *Dysommina,* the lower apophysis is absent or rudimentary and the orientation of the upper is vertical or even slightly to the rear of a vertical (as opposed to a forward inclination in all other genera). *Simenchelys* has a rudimentary lower apophysis and the upper is large, broad, and just slightly in front of vertical in its orientation (see Jacquet, 1920: pl. 2, fig. 43). Also its maxilla is short and broad and has a special process that abuts against the mandible. In most genera the maxilla is long and slender, with its rear end well-developed and reaching above the articular. The rearward extent of the dentition varies with the species. *Dysommina* has a narrow section just rearward of the midpoint and in *Dysomma* the entire rear part may be very slender, somewhat arched, and without teeth (see Robins and Robins, 1970: fig. 7).

Eye and bony socket: A relatively large eye and socket are primitive. These features are progressively reduced in the ilyophines. The bony socket is smallest in *Dysomma.*

Inclination of the hyomandibular: A posterior inclination of the hyomandibular is primitive. Apparently it has evolved in two directions. Only in *Simenchelys* is this orientation shifted to a vertical orientation, although it is close to vertical in *Ilyophis.* The extreme backward slope of the hyomandibular in *Dysomma* is surely a specialization. We disagree with our earlier interpretation (1976: 278) of this feature. The peculiar feeding mode of *Simenchelys* has resulted in a greatly modified head skeleton and massive musculature. The hyomandibular is much expanded to this end and is but one of many specializations in the head region of this genus.

Location of the anus: The primitive condition in the Synaphobranchidae, and probably early eels in general, is for the anus to be located at midbody as it is in *Simenchelys, Haptenchelys,* and some species of *Synaphobranchus.* There is a progressive forward movement of the anus and anal fin so that in some species of *Dysomma,* the anus is located under the base of the pectoral fin and only slightly behind the level of the gill slit.

Dorsal fin: A long-based dorsal fin is primitive and, in varying degree, this condition occurs in all three subfamilies. In *Synaphobranchus* (except for *S. bruuni* and *S. bathybius*), the dorsal fin begins farther back, usually well beyond the level of the anus.

The Synaphobranchidae appears to have had a long evolutionary history and has many surviving lineages which fall into the three definable units which we recognize as subfamilies.

Simenchelyinae. Although representing the most primitive lineage, the monotypic *Simenchelys* bristles with unique specializations (see p. 214) associated with its specialized feeding habits and the need to develop torque in twisting and pulling off pieces of flesh. It is the only synaphobranchid to retain the complete palatopterygoid arcade. It also has the anus at midbody and it is scaly. The supporting structure of its caudal region, although variable, is the best developed of any eel known to us, with broad epurals and hypurals (Jacquet, 1920: figs. 66–67), and this surely is another primitive state.

Synaphobranchinae. The cluster of species in *Synaphobranchus* presumably is a relict group with a mixture of characters that makes defining their interrelationships very difficult. For this reason we are unable to define subgenera unless one wished to propose a subgenus for each species. *Synaphobranchus bruuni* and *Synaphobranchus bathybius* form one pair united only by primitive features. *Synaphobranchus oregoni* shows some similarity in scale shape and arrangement to *S. brevidorsalis* but has greater similarity in head structure with *S. affinis. Synaphobranchus capensis* shares with *S. oregoni, S. affinis, S. brevidorsalis,* and *S. kaupi,* a forward shift of the anus and anal fin, but its separate gill slits are either a secondary specialization or a retention of a primitive feature. All synaphobranchines that we have seen from submarines or in underwater films and photographs swim with their head and body angled downward at about 30° with the anal fin broadly expanded to assist the undulating body in swimming. A forward shift of the anus and anal fin may be adaptive in this regard. The low dorsal fin is of little use in swimming and its reduction in height and in forward extent and the variability that one finds in the location of its insertion may reflect a lack of selective pressure. We have not found other species of eels to show so much individual variation in the location of the dorsal-fin structure.

Ilyophinae. This subfamily is more readily and substantially defined than we previously had indicated (Robins and Robins, 1976). The cluster of species in *Dysomma,* unlike *Synaphobranchus,* would seem to represent a more recent radiation as burrowers or hole-searchers in soft substrates of shelf and upper slope waters. The genus varies widely in its dentition, reduction of pectoral fin and girdle, sensory decorations of the snout in the form of papillae and plicae, elongation of the body, and vertebral number. Not surprisingly, ichthyologists who have emphasized such characters at the expense of shared specializations have recognized many monotypic genera. One cannot combine all elongate species, or all those without pectoral fins and end up with natural units. The alternative to our definition of *Dysomma* is a series of monotypic genera, several of which would require new names. Characters of the larvae and discovery of the undescribed species whose presence is indicated by unattributable larvae (see Smith, 1974, and this volume), may permit better resolution of the intrageneric relationships of *Dysomma*. With few exceptions, species of *Dysomma* are known from very few specimens so that studies of cleared and stained material, and extensive dissection to study soft parts has not been possible.

Zoogeography. Most of the synaphobranchid species which are known from many specimens have proved to be wide-ranging within certain surface temperature regimes. With regard to the Simenchelyinae and Synaphobranchinae, bottom temperatures at the depths at which the adults of these species live vary little and do not serve to limit geographic movement. Surface or near surface ecology essential for the survival of the leptocephalus apparently is the important factor in determining a species's geographic limit. Thus *Simenchelys* and *Synaphobranchus kaupi* occur under cold surface waters, *S. brevidorsalis* and *S. oregoni* under warmer surface waters. Where there are overlapping ranges as in the Tongue-of-the-Ocean, Bahamas, spawning may be restricted to different times of the year when surface waters are warmest or coldest (our unpublished data). *S. oregoni* is notable in that it is restricted to the Gulf of Mexico and nearby areas and seems to comprise part of the distinctive deepwater fauna of the Gulf of Mexico. In its early history the Gulf of Mexico opened westward and was then nearly, or possibly entirely, closed off for much of the time from the early Paleocene to the late Miocene (Salvador and Green, 1980). The origins of this fauna including the two synaphobranchines could date from this time. Additional important information on the early history of this area is provided by Anderson and Schmidt (1983) and especially, with regard to the isolation of the Gulf of Mexico, by Pindell and Dewey (1982) and Pindell (1985). Wide-ranging species such as *S. affinis* and *S. brevidorsalis* have not been studied in detail on a worldwide basis and possibly some populations will prove to be distinct.

Too little is known about most ilyophines to merit zoogeographic comment. The reasonably well-known and distinctive *Dysomma brevirostre*, long thought to be restricted to the Mediterranean Sea has been found as far afield as Hawaii and Florida. *Dysommina rugosa*, once thought to be restricted to the tropical western Atlantic, is now known from Hawaii and the Mozambique Channel. With this history, it is difficult to attach much significance to the fact that *Atractodenchelys* is known only from the Caribbean Sea and Straits of Florida. At the same time it is increasingly clear that there is an old, isolated, fauna centered in the Gulf of Mexico and this should be further studied with regard to eels.

Key to the Subfamilies of the Family Synaphobranchidae

1a. Head anterior to eyes extremely short, mouth a sphincterlike slit with gape not extending posterior to anterior nostril . . . . Simenchelyinae, p. 214

1b. Mouth extends to level of rear edge of eye or beyond 2

2a. Lower jaw longer than upper; all teeth numerous and very small, multiserial on dentary and maxilla, patch of teeth on premaxilloethmoid, and weak teeth in one or two rows on vomer . . . . Synaphobranchinae, p. 218

2b. Lower jaw shorter than upper; 3–30 moderate to much enlarged teeth on vomer, simple or, if compound, few and very large . . . . Ilyophinae, p. 234

Subfamily SIMENCHELYINAE

Characters. Head very short, blunt. Mouth a sphincterlike transverse slit. Body covered with deeply imbedded, elongate scales, arranged in a basket-weave fashion. Gill slits horizontal, crescentic, ventrolateral directly below pectoral-fin base. Anus at midbody. Complete bony pterygoid arch between quadrate and palatine region of vomer. Three pairs of pharyngobranchials. Premaxillae well-developed but fused to ethmoid. Maxilla short, broadly expanded, with posteroventral process; lower apophysis rudimentary, upper large, only slightly inclined anteriorly. One genus, one species: *Simenchelys parasitica* Gill.

Genus *Simenchelys* Gill, 1879

Simenchelys Gill in Goode and Bean, 1879a: 27 (type species *Simenchelys parasiticus* Gill, by monotypy).

Conchognathus Collett, 1889: 122–132 (type species *Conchognathus Grimaldii* Collett, by monotypy).

Gymnosimenchelys Tanaka, 1908: 2 (type species *Gymnosimenchelys leptosomus* Tanaka, by monotypy).

Simenchelis Pakhorukov, 1976: 325 (lapsus for *Simenchelys*).

Characters. (See also the definition of the subfamily above.) Head anterior to eyes very short, blunt. Upper and lower jaws short, massive. Mouth terminal, transverse. Snout tip blunt, not overhanging tip of lower jaw. Hyomandibular-quadrate complex aligned vertically. Pterygoid arch complete between quadrate and palatine region of vomer, strongly developed. Scales present in adults (absent in juveniles), set in groups of four or five, and aligned to each other at right angles in a basket-weave fashion. Scales deeply imbedded in skin, difficult to see, especially when they first appear in larger juveniles. Dorsal fin inserted slightly in front of level of appressed pectoral fin. Pectoral fins well-developed.

Size. Sexually mature at approximately 450 mm. To at least 610 mm.

Distribution. Essentially worldwide in tropical and temperate latitudes, along the continental slope, between 365 and 2620 m. The shallowest records are from regions with the coldest surface water, e.g., the Banquereau Bank off Nova Scotia.

Etymology. From the Greek *simos* (pugnosed) and *enchelys* (eel). Feminine.

Species. One, *Simenchelys parasitica*, but see discussion below.

Simenchelys parasitica Gill, 1879
PUGNOSE EEL
Figures 204–206

Simenchelys parasiticus Gill in Goode and Bean, 1879a: 27 (original description, type locality halibut grounds, Gulf of Maine, lectotype USNM 21669).

Conchognathus Grimaldii Collett, 1889: 123 (original description, type locality Azores).

Gymnosimenchelys leptosomus Tanaka, 1908: 2 (original description, type locality Sagami Bay, off Misaki, Japan).

Simenchelys dofleini Franz, 1910: 10 (original description, type locality Misaki, Japan).

Simenchelys leptosomus. Jordan and Thompson, 1914: 233 (large specimen from Owston collection).

Simenchelis parasitica. Pakhorukov, 1976: 325 (name erroneously given; record from southwestern Atlantic).

Description. The characters of the genus apply to the sole species, and are not repeated here. Vertebrae 115–117 (Solomon-Raju and Rosenblatt, 1971, recorded 121–125 myotomes from metamorphosing larvae and 121–125 vertebrae (45–49 precaudal) in 10 adults from the central Pacific). Precaudal vertebrae 46–53. Caudal vertebrae 64–69. Ratio of dorsal-fin rays and their supports to vertebrae varying from 2–3 to 1. Caudal-fin rays 15–17 (20–21 are indicated on the drawings by Jacquet, 1920: pl. 2, and given by him in text, p. 41, as 18–20, usually 19; Solomon-Raju and Rosenblatt, 1971, recorded 18 for 5 specimens; Castle, 1961, recorded 14 from 1 specimen from New Zealand).

Proportional measurements (expressed as percent of total length or head length*): predorsal length: 16.9 (16–19); preanal length (snout to origin of anal fin): 46.1 (44–49); head length (snout to front of gill opening): 9.02 (7.7–9.9); snout to front edge of eye: 28.6* (24–32); lower jaw tip to posterior bony end (angular) of jaw: 48.5*; hor-

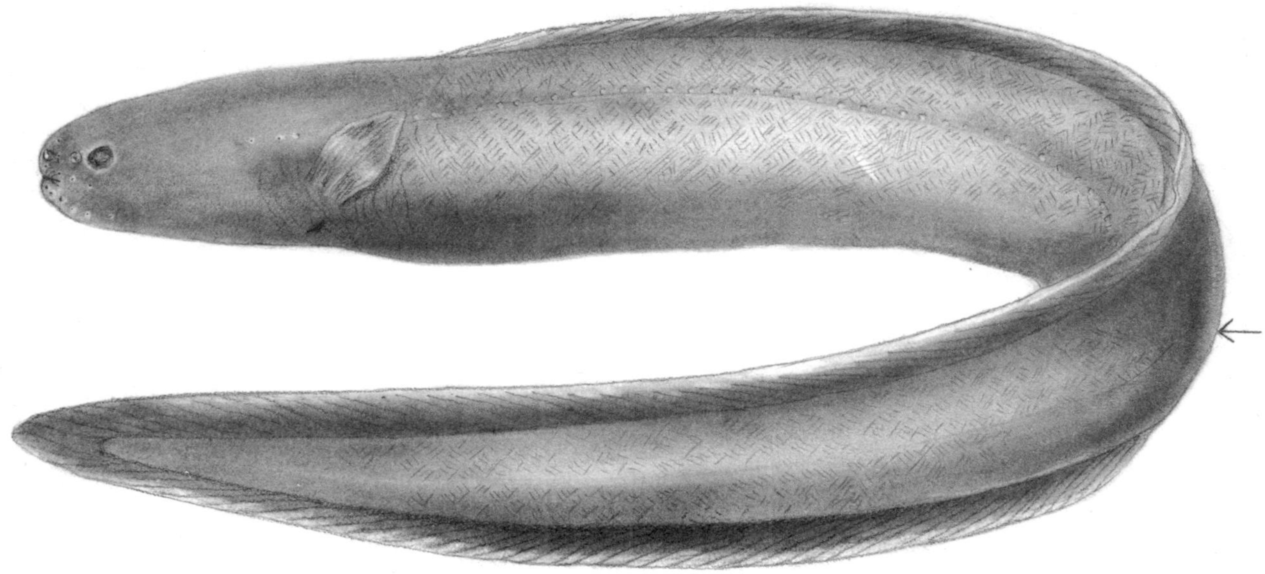

FIGURE 204. *Simenchelys parasitica*: ANSP 137428, 360 mm TL, Bahamas.

izontal diameter of eye: 15.8* (14–20); greatest postanal depth of body: 7.1 (6.1–7.9).

Dentition: Jacquet (1920) illustrated and described the dentition of his specimens as follows: There is a single row of contiguous, bluntly pointed teeth along the maxilla and dentary. These are shaped so as to make a closely fitted cutting edge. There are about 12 teeth in the maxillary row and about 17 on the dentary, the teeth in both rows becoming slightly smaller toward the rear. There are 3 large, conical teeth in a median row on the vomer. In material available to us the jaw teeth are smaller and less regular and the row continues around the premaxillary rim of the premaxilloethmoid. Also, we find one to three teeth on the vomer in intact specimens. These teeth apparently are easily broken off. Cephalic pores: supraorbital 3; infraorbital 6; preoperculomandibular: 6. There is some individual variation in these counts.

Body elongate, nearly cylindrical anterior to anus, laterally compressed posterior to anus, the dorsal and ventral profiles only gradually converging toward the tail. Head proportionally small even when short snout is taken into account. Eye small, but easily seen through covering skin.

Scales small, deeply embedded in skin which is slimy in life. Scales not readily apparent, closely set in very regular basketweave pattern (Fig. 204).

Pectoral fins present, oval, even somewhat paddle-shaped in outline. Dorsal-fin origin almost directly above tips of appressed pectoral

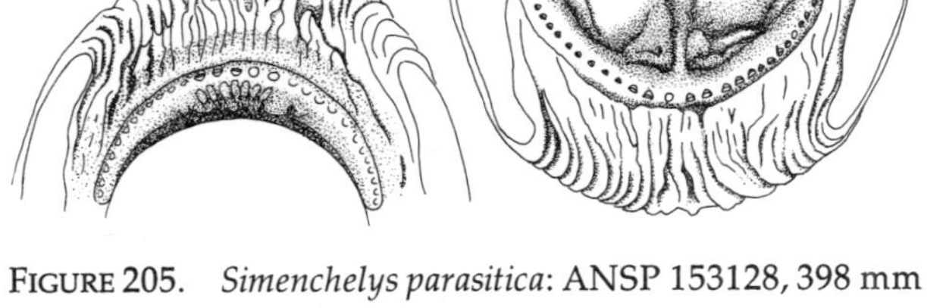

FIGURE 205. *Simenchelys parasitica*: ANSP 153128, 398 mm TL, off Virginia; dentition.

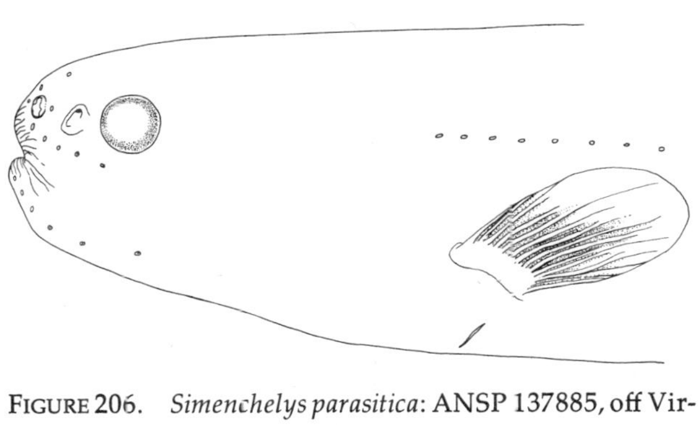

FIGURE 206. *Simenchelys parasitica*: ANSP 137885, off Virginia; lateral view of head.

fins, fleshy especially at the base, highest in the area above the anus, entirely confluent with caudal fin. Caudal fin approximately two-thirds size of pectoral fin. Anal-fin origin immediately posterior to anus, highest at midpoint in its length, entirely confluent with caudal fin.

Gill slits separate, horizontal, ventrolateral, directly below pectoral-fin base. Roughly crescentic in outline, the points of the crescent pointed dorsally. Branchiostegal rays usually about 10, curving dorsally in a semicircle, the most posterior few expanded laterally and curling around the upper posterior corner of the opercle (Jacquet, 1920; Castle, 1961). The branchiostegals can be seen best in cleared and stained specimens.

The head and mouth provide the most readily seen and distinctive features of the species. Mouth anterior, terminal, a transverse slit, the appearance sphincterlike due both to shrinkage in preservatives and the sensory papillae. Fleshy gape extremely short, the corner of mouth not reaching posterior to anterior nostril. Anterior nostril opening through a short, simple tube on anterior surface of snout. Posterior nostril immediately anterior to small, circular eye.

Lateral line complete on body, extending from lateral canal on head along the side of body nearly to base of caudal fin, pored throughout. Pores small, not tubular, and paler in color than surrounding tissues.

Color. Uniform, medium brown in preservation. In life (Castle, 1961), "uniform grey to greyish-brown, darker around vent, along dorsal and ventral mid-lines, on margins of fins, on head and along lateral line." Our experience agrees with Castle's account. Juveniles may be darker than adults. Local variation in color is to be expected.

Size. Five of fourteen specimens of *S. parasitica* from Tongue-of-the-Ocean and Exuma Sound in the Bahamas were ripe females, ranging in total length from 335–532 mm and in weight from 109–378 grams. Three were taken in late February and early March, two in late September and early October. No ripe males were collected. Grey (1956) lists the maximum size as 610 mm. The smallest juvenile seen by her was 95 mm.

Development. A newly metamorphosed juvenile or very late leptocephalus (elver) was described by Solomon-Raju and Rosenblatt (1971). Otherwise, the leptocephalus stage is unknown. See also the discussion of *Leptocephalus lachrymatus* Franz, below.

Food, Habits, and Behavior. In his description of *Simenchelys parasiticus*, Gill (in Goode and Bean, 1879a, and again in 1890b) stated that "individuals have been found burrowing into the flesh of the halibut," and that the first specimens seen by him were "obtained on the halibut trawls, on the off-shore banks." Bigelow and Schroeder (1953) noted that "it is partly parasitic in habit, burrowing into the bodies of halibut and other large fish, circumstances under which a considerable number of specimens have been brought in by fishermen. Very likely it was common inshore in the old days when halibut were plentiful there." Bigelow and Schroeder (1953) further noted that although there was no definite record of *Simenchelys* actually within the southern rim of the Gulf of Maine, "it has been recorded so often in water as shoal as 200 fathoms that it may be expected in the bottom of the Eastern Channel and in the southeastern deeps of the Gulf of Maine." They also noted ". . . it is as slimy as a hag and drips with sheets of mucus when drawn out of the water." Castle (1961) discussed the so-called "parasitic" nature of this species and noted that although records exist of the eels attacking living fish, most specimens are taken in traps and trawls, concluding that *Simenchelys* is probably more of a scavenger than a predator. Certainly the species is not parasitic in any precise definition of the term. In most parts of its range, *Simenchelys* occurs at much too great a depth (usually in excess of 550 m) to be more than a nuisance to commercial fishermen. The presumed mode of feeding is to rasp and twist at the flesh of dead or dying fishes, gradually excavating a path into the carcass. Examination of gut contents, largely a mince of fibers of flesh, with no discernible hard parts, tends to bear this out.

Distribution. Simenchelys parasitica is a component of the deep benthic fauna that is nearly cosmopolitan in cold, deep waters. The depth at which these fishes occur is in large part apparently dependent on the presence of waters of temperatures near 4–7°C over suitable substrate, coupled with suitably cool surface temperatures (less than 20°C). Consequently, *Simenchelys* (and many other synaphobranchids) are often collected at shallower depths in colder waters.

Simenchelys parasitica is listed by Grey (1956) from the "western Atlantic, 38–44°N., 56–73°W., common, 366–1645 m." We have also collected it from the deep Bahaman basins (TOTO, Exuma Sound) in 1280–1732 m. Vessels from the Virginia Institute of Marine Science collected 126 specimens of *S. parasitica* from 52 stations in the Middle Atlantic Bight; depth of capture ranged from 895–2125 m; temperatures were 3.5–9.2°C (K. J. Sulak, pers. comm.). Grey also reported it from the eastern Atlantic and western Pacific, while Castle (1961) recorded it from the South Pacific. Pakhorukov (1976: 325) recorded it from the Rio Grande Rise, southwestern Atlantic, 31°04°S, 35°34′W, in 1580 m.

Simenchelys parasitica is locally abundant. Castle recorded 17 from one trap, 25 and 22 from others. Bigelow and Schroeder (1953) commented on the large numbers collected in traps in 2 hours. Trawling is perhaps a less efficient means of collecting these eels for, in our use of a 41-foot semiballoon trawl in Tongue-of-the-Ocean and Exuma Sound, we collected only one or two at a time whereas as many as 45 specimens of *Synaphobranchus brevidorsalis* were collected. It is also likely that *S. parasitica* is near or at the edge of its range in the Bahaman basins, where surface temperatures commonly exceed 20°C, and is truly uncommon there. Presumably, local abundance is associated with areas in which dead or weakened fishes or other food items suitable for scavenging may settle.

Discussion. Simenchelys parasitica needs to be studied carefully on a worldwide basis. The dentition meticulously described by Jacquet (1920) is not like that seen in our material from the western North Atlantic. Also he, and Solomon-Raju and Rosenblatt (1971), recorded more caudal rays than we did (18–20, usually 19 vs 15–17; and 18, respectively). We have also noted variation in numbers of vertebrae (see above). We are not able to say at this time whether these differences are real, whether they are merely part of the usual variation, or whether they may be part of a suite of characters that define allopatric populational differences. The caudal skeleton of *S. parasitica* has the parhypural and the hypurals greatly expanded and variously fused (see Jacquet, 1920: pl. 2), a condition most like that in *Ilyophis brunneus* (see Robins, 1971: fig. 11). The characters of *S. parasitica* (and, indeed, of the Simenchelyinae) are a combination of primitive features and unique, derived characters (autapomorphies).

Etymology. From the Greek *parasitos* (parasite) and *ikos* (belonging to), in reference to its presumed parasitic nature. An adjective.

Study Material. Type specimens: Gill in Goode and Bean (1879a) did not designate type material in his original paper nor did he do so in subsequent treatments of the genus. He mentioned "several specimens" and the halibut grounds. In the introduction (1879a: 2), Goode and Bean mention specifically Georges, Le Have, Browns, and Sable Island banks. The following collections were entered in the catalogs of the United States National Museum between April and 2 December 1878: USNM 21669 (orig. no. 179) Banquereaux; USNM 21673 (orig. no. 226) near Sable I. Bank. (2 spec., ad. and juv.); USNM 21675 (orig. no. 191) southeast of Le Have, 300 fm. (this specimen was not found in the collections of the National Museum, nor has it been located elsewhere); USNM 21677 (skeleton), collected 1878; USNM 21872 (59°50′N, 43°25′W in 300 fm., 15 Oct. 1878). Additional material was catalogued in 1879 and may have been seen by Gill. However, Gill surely had the 1878 material. One (USNM 21677) had been fleshed and used for study of the skeleton. A second (USNM 21669) had its head split horizontally to expose the jaws and teeth. We here designate USNM 21669 as lectotype on the reasonable premise that the material listed constituted most or all of the original undesignated type material. All of the other material listed above is considered to be lectoparatypes. The following USNM material (also studied by us) was catalogued too late to be considered to be likely as type material: USNM 22791 (Earll 234) 42°47′N, 63°10′W, in 375 fm. (entered 2 May 1879) (now catalogued as MCZ 26773). USNM 24264 (orig. no. 519) 43°48′N, 59°W in 250 fm., Dec. 1879. USNM 24265 (orig. no. 526) 43°52′N, 59°09′W in 200 fm., Oct. 1879. USNM 24266 (orig. no. 542) 42°46′N, 65°18′W in 200 fm., Oct 1879. USNM 24267 (orig. no. 513) 42°46′N, 65°18′W, in 200 fm., Oct. 1879. FMNH 55614 (1 spec.) Owston sta. 156, Japan, Yeachigo, Hondo, March 1907.

We also examined all material in the collections of the United States National Museum (USNM), the Academy of Natural Sciences of Philadelphia (ANSP), and the University of Miami (UMML). We did not attempt a variational study of the species.

Status of Leptocephalus lachrymatus *Franz.* Franz (1910: 17) described and illustrated *Leptocephalus lachrymatus* based on two lots of 19 and 51 specimens from Sagami Bay, Japan. The number of myomeres was given as 140–142, but it is unknown whether this is based on the entire collection, a few specimens, or just an estimate based on 1 specimen. Franz also tentatively associated with this species, a "young eel," 13 cm long, with about 140–145 myomeres. The figured leptocephalus (pl. 2, fig. 3) shows a vertically elongate eye and the general shape of *Simenchelys.* If the counts of myomeres are even approximately correct, then *lachrymatus* cannot be *S. parasiticus.* Pairs of species with high-low vertebral counts exist in *Synaphobranchus* (e.g., *S. bathybius* and *S. bruuni, S. affinis* and *S. kaupi, S. brevidorsalis* and *S. oregoni;* the species with the low count is given first in each instance). If *Leptocephalus lachrymatus* is a *Simenchelys,* it is a species not currently recognized. Other synaphobranchids in the Sagami Bay region with myomere counts only slightly higher are *Synaphobranchus kaupi* and *Ilyophis brunneus.* At the present time, we cannot positively assign *lachrymatus* to any known species.

Subfamily SYNAPHOBRANCHINAE

CUTTHROAT EELS

Characters. Head conical; gape extending behind level of rear margin of eye; eyes large, approximately oval (the long axis horizontal); body covered with scales (*Synaphobranchus*) or naked (*Haptenchelys*); teeth numerous, small and simple, occurring on dentary, maxilla, premaxilloethmoid, and vomer, those on vomer uniserial or biserial never greatly enlarged, those on the premaxilloethmoid in a narrow patch and teeth slightly larger than those on other bones. Lower jaw longer than upper. Preopercle reasonably well-developed, its upper limit above the opercular condyle.

Gills slits very low on side and separated by only a small distance, or ventral and united, there being a common opening for the two sides (Fig. 203).

Fleshy tip of snout weakly pointed, proboscis-like and overhanging tip of lower jaw. Small papillae in tracts on snout, jaws and underside of head, and inside the mouth, but never large and conspicuously developed.

Pectoral fins present, tapered to a point. Dorsal and anal fins well-developed, not fleshy, confluent with small caudal fin. Dorsal-fin origin varying from far forward (originating above base of pectoral fin) to above a point well behind the level of the anus (almost at the midpoint of the body).

Body slender, laterally compressed, especially posterior to anus. Trunk region in cross-section elongate ovoid (the vertical axis longest) to simply oval.

Color varies among the species and with locality, but all mature individuals somberly colored: brown, purplish brown or dull reddish brown. The pigment is more intense on the trunk, branchial region, and around the anus. Vertical fins often paler distally than at base. Small, immature specimens may be quite pale. The dorsum of larger eels may appear whitish due to abrasion, an artifact of collection.

Pterygoid present as a slender bony splint, not connected by bone with any bony surface. Suspensorium long, posteriorly directed, concomitant with long gape. Basisphenoid present, not fused with parasphenoid. We recognize two genera and eight species, six of which are reported from the western North Atlantic.

Key to the Genera of the Subfamily Synaphobranchinae

1a. Body covered with small scales, most often elongate and arranged in basket-weave fashion, sometimes rounded or irregularly oval; dorsal-fin origin varies; gill slits united at midventral line or separate but low on body . . . . *Synaphobranchus*, p. 219

1b. Body naked, dorsal fin far foward; gill slits separate, low on body . . . . *Haptenchelys*, p. 232

Genus *Synaphobranchus* Johnson, 1862

Synaphobranchus Johnson, 1862: 169 (type species *S. kaupi* Johnson, 1862, by monotypy).

Nettophichthys Holt, 1891: 122 (type species *N. retropinnatus* Holt, 1891 = *Synaphobranchus kaupi* by monotypy).

Histiobranchus Gill, 1883: 255 (type species *H. infernalis* Gill, 1883 = *Synaphobranchus bathybius* Günther, 1877 by monotypy).

Diastobranchus Barnard, 1923: 441 (type species *D. capensis* Barnard, 1923, by monotypy).

Diactobranchus Pakhorukov, 1976: 325 (lapsus for *Diastobranchus*).

Characters. Body anguilliform, tapering from the anus toward either end; laterally compressed; covered with small elongate, oval, irregularly oval, or subcircular scales. Gill slits ventral, horizontal, opening through a single medially divided opening or separate but with anterior ends close together, diverging posteriorly, or separated by about the length of the gill slit and strongly diverging posteriorly.

Size. The smallest juvenile seen was 107 mm. The largest adult was 1175 mm. Species are sexually mature at 450 mm or larger, depending upon the species.

Distribution. Eels of the genus *Synaphobranchus* are important members of the deep benthic fauna of the continental slope. Essentially fishes of cold, deep waters, their depth range and distribution are dependent on temperature, substrate, and appropriate pelagic temperatures and currents to support growth and distribution of the distinctive leptocephalus larvae.

The genus is essentially worldwide except for the polar regions and certain areas in which conditions are presumably unfavorable for larval development. The distribution of the various species are better defined by surface or near surface requirements of the leptocephali than by conditions at or near the bottom. Currently, eight species are recognized, although at least that many more have been described at one time or another. Pending examination of fresh material, especially from the Pacific Ocean, using radiography and electron microscopy, it is assumed that such differences as exist in numbers of vertebrae and in body proportions, among samples from widely scattered areas of what we regard as a single species, are not at the species level of significance. It is common in the Synaphobranchidae for species to have an average difference in vertebral number of several vertebrae in a consistent, almost clinal way, from east to west (from mid Pacific to western Atlantic) without any evidence of concordant variation in other characters. Samples of *Synaphobranchus* species, from presumably identical stocks, vary in numbers of vertebrae by as many as ten. Worldwide study of variation in *Synaphobranchus* is clearly needed, but should not substantially affect the understanding of the species in the western North Atlantic as presented here.

Etymology. From the Greek words *synaphes* (united) and *branchia* (gills), in reference to the united midventral gill openings of some of the species. Masculine.

Key to the Species of *Synaphobranchus*

(Figure 207 provides a visual adjunct which facilitates identification of the most common western North Atlantic species of the genus: *S. kaupi, S. affinis, S. oregoni,* and *S. brevidorsalis.*)

1a. Gill slits united, midventral, opening through a common median slit 2

1b. Gill slits separate but close 5
2a. Scales elongate to oval 3
2b. Scales rounded 4
3a. Scales very elongate (3–4 times as long as wide), small, regularly arranged in aligned clusters of 4–5 set at right angles to other such groups; vertebrae usually 144–152; dorsal-fin origin behind level of anus; vomerine teeth uniserial but zig-zag anteriorly .... *S. kaupi,* p. 220
3b. Scales oval, not so regularly arranged; vertebrae 128–140 (rarely slightly fewer); dorsal-fin origin at or just posterior to level of anus; vomerine teeth uniserial .... *S. affinis,* p. 223
4a. Scales large, polygonal or rounded, epidermis between scales in preserved specimens often forming networklike raised surfaces; vertebrae 140–150 (rarely slightly fewer); dorsal-fin origin posterior to level of anus, about one snout length anterior to midpoint of body; vomerine teeth uniserial except for short anterior patch of a few teeth in a roughly biserial arrangement; larger specimens very deep through trunk region .... *S. oregoni,* p. 225
4b. Scales rounded or oval, irregularly arranged, small; vertebrae 130–140 (rarely fewer); dorsal-fin origin at least one head length posterior to level of anus, near midpoint of body; vomerine teeth irregularly biserial anteriorly, forming small expanded portion at tip; pectoral fin long, as much as 1.5–2 times as long as snout; soft fleshed; deep dwelling .... *S. brevidorsalis,* p. 227
5a. Gill slits separate, divergent toward bases of pectoral fins, separated by more than their length; scales elongate, irregularly arranged; vertebrae 164–174 (rarely a few more or less); dorsal-fin origin about one head length behind level of anus; a few anterior vomerine teeth much enlarged, others very small, biserial or irregularly triserial .... *S. capensis,* p. 229
5b. Gill slits separate, nearly parallel, separated by much less than their length, scales elongate-oval, deeply embedded, regularly arranged; dorsal-fin origin far anterior, above level of pectoral-fin base; vomerine teeth weakly biserial or triserial 6
6a. Vertebrae 125–135 .... *S. bathybius,* p. 230
6b. Vertebrae 160–170 .... *S. bruuni,* p. 232

Synaphobranchus kaupi Johnson, 1862

NORTHERN CUTTHROAT EEL

Figures 208–210

Synaphobranchus kaupi Johnson, 1862: 160 (original description, Atlantic Ocean off Madeira, syntype BMNH 1862.11.9.79).

Synaphobranchus pinnatus (not *Muraena pinnata* Gronovius). Günther, 1870: 23 (in part, those parts of the account based on *S. kaupi* Johnson). Murray and Hjort, 1912a: 395, 755, figs. 264, 549 (see discussion below). Matsubara, 1938 (variation in a collection from Japan). Matsubara and Ochiai, 1951: 258 (reidentification of previous reference to *S. kaupi;* variation, vertebral counts).

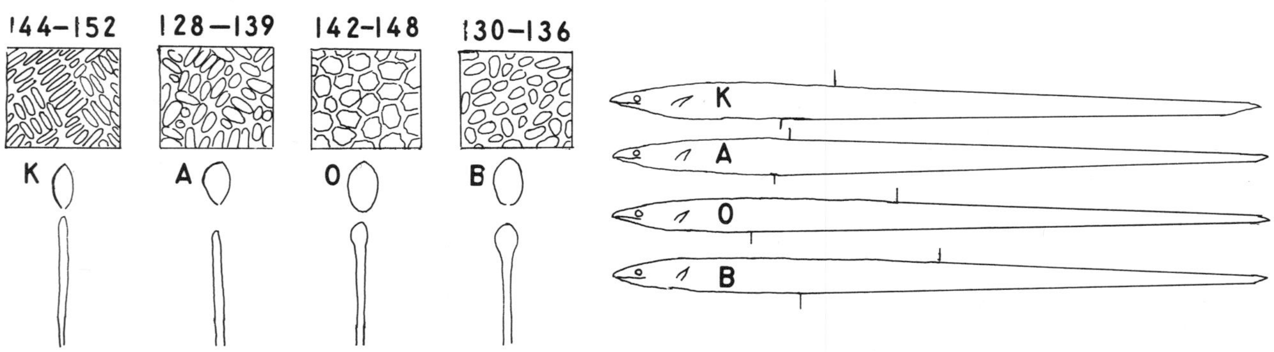

FIGURE 207. Diagrammatic guide to the identification of four western North Atlantic species of *Synaphobranchus* using the following characters: Number of vertebrae; scale pattern; shape of toothed areas of premaxilla and vomer; and relative positions of dorsal- and anal-fin origins. K: *Synaphobranchus kaupi.* A: *S. affinis.* O: *S. oregoni.* B: *S. brevidorsalis.*

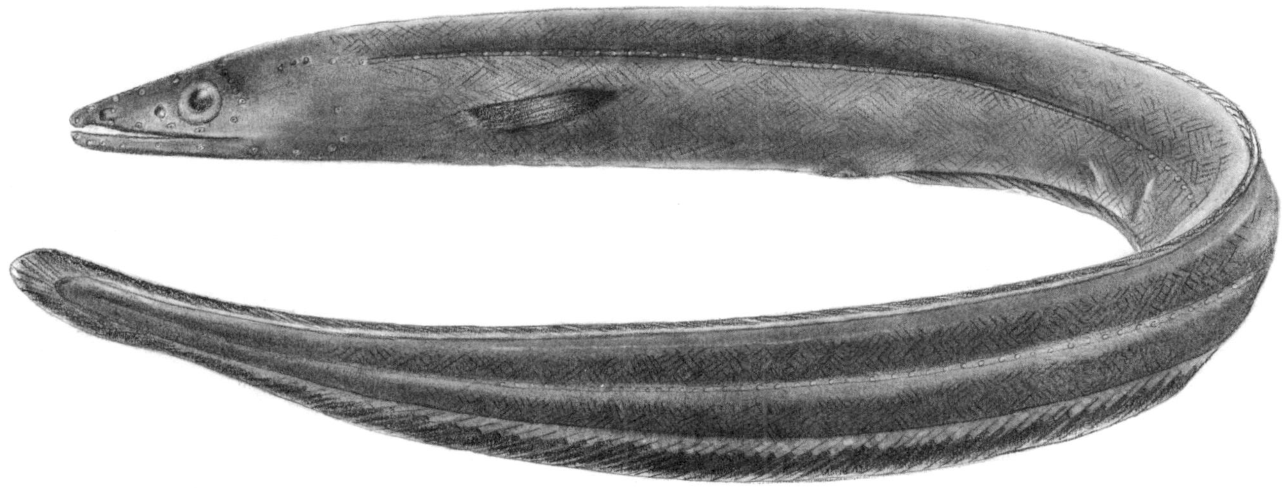

FIGURE 208. *Synaphobranchus kaupi*: ANSP 137425, 590 mm TL, Bahamas.

Nettophichthys retropinnatus Holt, 1891: 122 (original description, fishing grounds off west coast of Ireland).
Synaphobranchus iraconis Jordan and Snyder, 1901: 844 (original description, Japan, off the coast of Myiako, in Rikuchu, in 200 fm, holotype CAS-SU 6465).
Synaphobranchus pinnatus var. *parvipinnis* Roule 1916: 4 (original description, southeast of Portugal in 1500 m).

Discussion of Synonymy. Gray, 1854: 18, noted the following: "*Operculis branchialibus pinnisque pectoralibus utrinque.*" On p. 19, he continues "2. Muraena pinnata. Muraena unicolor; maxilla superiore longiore; pinnis pectoralibus aperturisque branchialibus utrinque. Muraena unicolor, maxilla superiore longiore. *Mus. Ichth.* 11, p. 11, n. 161." Gray gives no locality following the notation "Habitat." This indicates to us that there were no locality data, for such data are duly recorded for all other eel entries. Wheeler (1958) makes no mention of *Muraena pinnata* in his discussion of the Gronovius collection. We regard *Muraena pinnata* Gronovius (in Gray, 1854) as unidentifiable, a conclusion also reached by Bruun (1937: 4). The basis of Günther's referral of this species to *Synaphobranchus* is unclear to us since he makes no mention of any dried skin from the Gronovius collection in his list of material examined. *Leptocephalus Synaphobranchus pinnati* Murray and Hjort (1912a: 751, fig. 549) is not a proposal of a new name but merely a Latin notation for Leptocephalus of *Synaphobranchus pinnatus.* It is doubtful that the leptocephalus illustrated is correctly identified but the quality of reproduction of the photograph is too poor for a definite statement in this regard.

Description. Vertebrae 144–152, rarely one or two more or less. Precaudal vertebrae 75 and caudal vertebrae 71 in one specimen. Vertebral counts show little if any geographic variation in this species. Matsubara and Ochiai (1951: 254) gave counts for 53 specimens from Japan (142–150, mean 146.2). Caudal-fin rays 13. Pectoral-fin

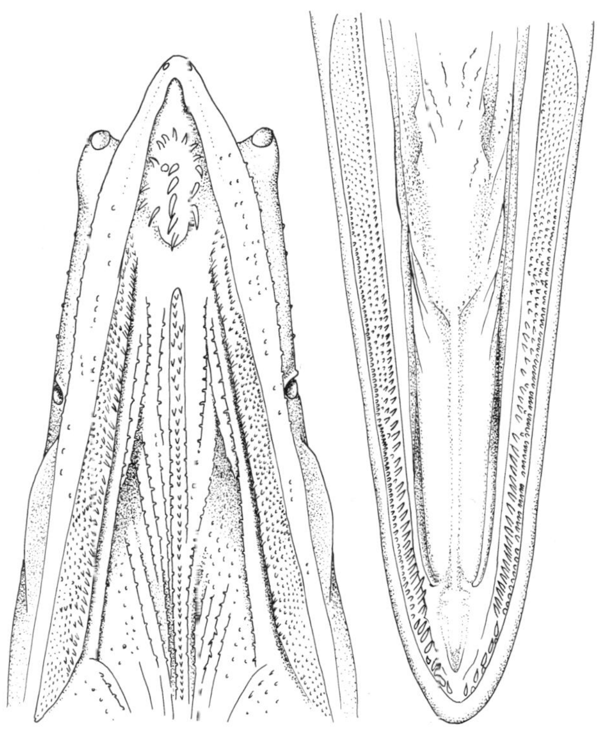

FIGURE 209. *Synaphobranchus kaupi*: dentition. Data as for Figure 208.

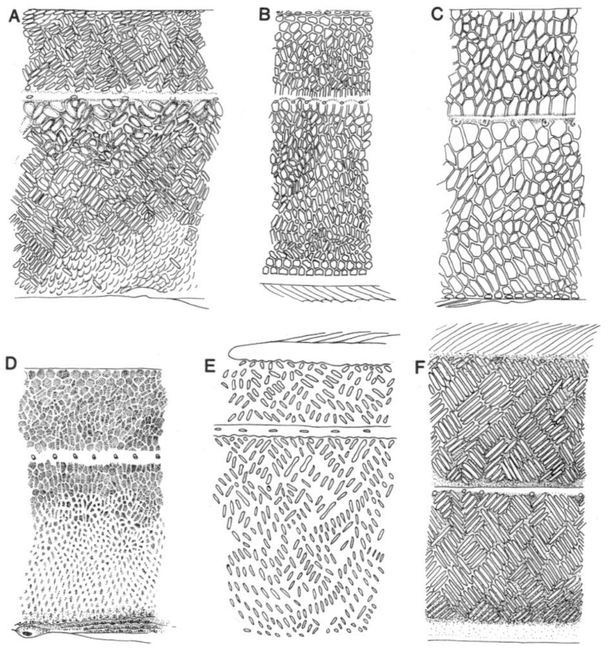

FIGURE 210. Scale patterns of synaphobranchines: A. *Synaphobranchus kaupi*. B. *S. affinis*. C. *S. oregoni*. D. *S. brevidorsalis*. E. *S. bathybius*. F. *Ilyophis brunneus*.

rays 14–17. Cephalic lateralis pores: infraorbital 10, supraorbital 5, preoperculomandibular 12. Pores difficult to count in small or older, ethanolic specimens. Some asymmetry and variation in pore counts fairly common. Proportional measurements (based on 20 specimens from the Middle Atlantic Bight and Tongue-of-the-Ocean, Bahamas) expressed as percent of total or head (*) length: predorsal length 31–42; head length 11–16; jaw length* 68–97; horizontal diameter of eye* 13–17; greatest postanal depth of body 3.5–7.5.

Small teeth present on dentary, maxilla, vomer, and premaxilloethmoid. All simple, pointed; many slightly recurved. Innermost teeth on front ends of lower and upper jaws larger than outer and more posterior teeth. Teeth on longitudinally oval patch on premaxilloethmoid slightly more enlarged and more recurved. Vomerine teeth small, deeply embedded in tissue and mucus. Those most anterior zigzag uniserial to roughly biserial, and slightly larger than those more posterior which are uniserial.

Body elongate, laterally compressed, especially posterior to anus. Depth through trunk region not great, even in large specimens, rounded to somewhat oval in cross-section. Scales small, elongate; width 3–5 times in length, closely set, arranged in clusters of 4–5 aligned scales set at right angles to other such clusters (Fig. 210). In fully adult specimens, small scales, less regularly arranged, extend between rami of lower jaw.

Dorsal-fin origin posterior to level of anus, but this feature variable, as noted by Matsubara and Ochiai (1938: fig. 1). Dorsal-fin origin may be nearly a head length posterior to level of anus or, less commonly, about directly above anus. Vertical fins confluent; caudal fin small, often slightly squared-off at tip due to erosion or artifact of collection and preservation. Anal fin well-developed; begins immediately posterior to anus. Pectoral fins present, well-developed, about as long as snout, tapering to a point. Lateral line complete from head to base of caudal fin.

Head laterally compressed (on average, more so than in other species of *Synaphobranchus*), depth from supraoccipital to ventral surface of lower jaw not great, so that eye appears exceptionally large in comparison with other species of *Synaphobranchus*. Gape large, fleshy crease extending to a point about equal to the horizontal diameter of eye below and behind the eye. Anterior nostrils tubular, short, directed anteroventrally, with base at level of articulation of front end of maxilla with the premaxilloethmoid. Rear nostril nearly circular, separated by its own diameter from midpoint of front margin of eye. Rim of rear nostril slightly thickened.

Color. In preservative pale brown to blackish brown, depending on locality. Most specimens from western North Atlantic dark brown. Abdomen, mouth, areas around dorsal and anal midlines, and caudal fin darker than other parts of body. Peritoneum black.

Size. Grey (1956) published a size range of 115–813 mm for *S. kaupi*. The largest of 1008 specimens from Tongue-of-the-Ocean was a ripe female, 640 mm TL. Sexually mature eels, the small ones usually male, may be about 430 mm or larger. The greatest number of reproductively active individuals occurred in February and April, with smaller numbers present in August and September. "Ripe" eels were collected in all months except May, June, December, and January.

Development and Growth. Grey (1956: 147) identified with *S. kaupi* leptocephali of 20–131 mm, but gave no details. Essentially nothing is known about the rate of growth of this or any other species of *Synaphobranchus.* The exact applicability of Bruuns's (1937) work on *Anguilla* to *Synaphobranchus* is questionable.

Food. Stomach contents of North Atlantic specimens of *S. kaupi* have often been squid, and it is common to observe circular superficial injuries, seemingly inflicted by squids, on specimens of *S. kaupi* from the southern Straits of Florida and Tongue-of-the-Ocean. Castle (pers. comm.) reported a small amphipod from the gut of a 266 mm specimen. The presence of injuries inflicted by squid may also indicate the use of *S. kaupi* as food by the squid. Saldanha (1980) described the digestive tract and reported on stomach contents from a series of specimens from the continental slope off Europe between 1080–2070 m. A broad diet of fishes, cephalopods, and crustaceans was indicated.

Distribution. Essentially worldwide, under cold surface waters; absent from the Mediterranean Sea. Deep, benthic inhabitant of continental slope with most records between 800–2000 m (Grey, 1956). Off New England, it has been collected as shallow as 131 m (off Cape Sable, Nova Scotia, USNM 143201) and it was first made known in the western North Atlantic by fishermen using deep longlines in 365–550 m (Goode and Bean, 1896). Jensen (1948: 58–59) recorded it from as far north as 66°37'N in the Davis Strait; bottom temperatures there ranged from 3.12–3.52°C. *S. kaupi* is the most abundant species of *Synaphobranchus* in the Middle Atlantic Bight. A total of 5392 individuals from 92 stations was collected by vessels from the Virginia Institute of Marine Sciences from a depth range of 316 (a single 1-gram specimen) to 2196 m; temperature range was 3.2–9.2°C. The largest number of specimens collected at one station was 588 from a depth of 1287 m (K. J. Sulak, pers. comm.). The deepest record is 2654 m (Grey, 1956). Jensen (1948: 58) reported it as occurring to 3465 m but the basis for this is unclear. The occurrence of *S. kaupi* in the Gulf of Mexico is doubtful. Robins (1971) did not record it from the southern Straits of Florida. At least some of the material listed by Grey (1956: 146) from Oregon Station 1303 is identified here as *S. brevidorsalis.* Martin (1978) also failed to collect it in his extensive work in the western Gulf of Mexico.

The distribution is clouded by the use of "*S. pinnatus*" for both *S. kaupi* and *S. affinis.* Its absence from regions under tropical surface waters (and possibly also the Mediterranean Sea) is apparently dictated by its larval life in surface or near surface waters.

Etymology. After J. J. Kaup who wrote the first major treatise on apodal fishes. A noun in the genitive case.

Study Material (in addition to that listed by Robins, 1971: 180). WEST ATLANTIC: USNM 21684 (2, 357–620), 24351 (1, 580), 28927 (1 female with eggs), 186166 (5), 21681 (1), 31598 (1, 377), 31608 (1, 457), 31612 (1, 362), 32655 (2, 308–325 + 4), 35596 (1), 38078 (1, 411), 143201 (1, 375), 177934 (1, 270), 273418 (2, 323–436), 273419 (ex 185657) (1, 397). WEST PACIFIC: USNM 49996 (2, 245–290), 160590 (6, 262–356), 160591 (1, 317), 273420 (2, 245–290). We have also examined all of the material in the collections of the University of Miami (including approximately 1000 uncatalogued specimens from deep waters of the Bahamas) and the Academy of Natural Sciences of Philadelphia.

Synaphobranchus affinis Günther, 1877
Figures 210–212

Synaphobranchus affinis Günther, 1877: 455 (original description, off Inosima, Japan, syntypes BMNH 1887.12.7.250). Matsubara and Ochiai, 1951: 255 (variation in Japanese specimens).

Synaphobranchus pinnatus (not of Gronow). Günther, 1887: 255 (Brazil, western Pacific).

Synaphobranchus brachysomus Gilbert, 1905: 583 (original description, Hawaiian Islands, holotype USNM 51591).

Synaphobranchus taketae Tanaka, 1916: 257 (original description, Japan, off Miyagi Prefecture, Sea of Japan).

Synaphobranchus indicus Bruun, 1937: 14 (original description based on leptocephalus, eastern Indian Ocean; on p. 15, Bruun divides this species into two subspecies, this being the nominate subspecies).

Synaphobranchus indicus occidentalis Bruun, 1937: 15 (original description based on leptocephalus, western Indian Ocean).

Description. Vertebrae 125–140. Precaudal 59; caudal 66 (n = 1). Matsubara and Ochiai (1951: 255) recorded 133–138 vertebrae for 6 specimens

FIGURE 211. *Synaphobranchus affinis*: ANSP 114619, 419 mm TL, NE Gulf of Mexico.

from Japan. Caudal-fin rays 14 (1). Pectoral-fin rays 14–17 (6). Cephalic lateralis system: Infraorbital pores 10; supraorbital pores 5; preoperculomandibular pores 12 (number may vary by 2–3 and the two sides of the same fish may differ).

Proportional measurements expressed as percent of total or head (*) length: predorsal length 27–37; preanal length (Norman and Trewavas, 1939) 27–31; snout to front margin of eye* 27–36; length of lower jaw* (to posterior bony corner) 47–59; horizontal diameter of eye* 9–17; greatest postanal depth 5–10. Small teeth present on dentary, maxilla, vomer, and premaxilloethmoid. All teeth simple, pointed. Those of inner rows on jaws larger than outer rows, especially anteriorly. Teeth on vomer very small, inconspicuous, essentially uniserial, although one or two teeth may appear out of line and, hence, biserial anteriorly. No expanded portion on anterior end of vomer.

Body elongate, more laterally compressed posteriorly than in trunk region. Trunk deep and robust. Scales small, oval-elongate, not so closely packed nor so regularly arranged as in *S. kaupi* (Fig. 210). No scales between jaws. Scales on dorsum may be nearly round, almost reptilian in appearance.

Dorsal-fin origin above anus or slightly posterior to this point, not so far posterior on the average as in *S. kaupi*, but individuals of *S. affinis* can have a more posterior dorsal-fin origin than those of *S. kaupi*. Median and paired fins essentially as described for *S. kaupi*. Lateral line complete.

Head less laterally compressed, blunter in outline, deeper in occipital region than in *S. kaupi*. Jaw and fleshy gape slightly shorter than in *S. kaupi*. Nostrils essentially the same.

Color. Variable in preservation depending on locality. Many specimens from Straits of Florida

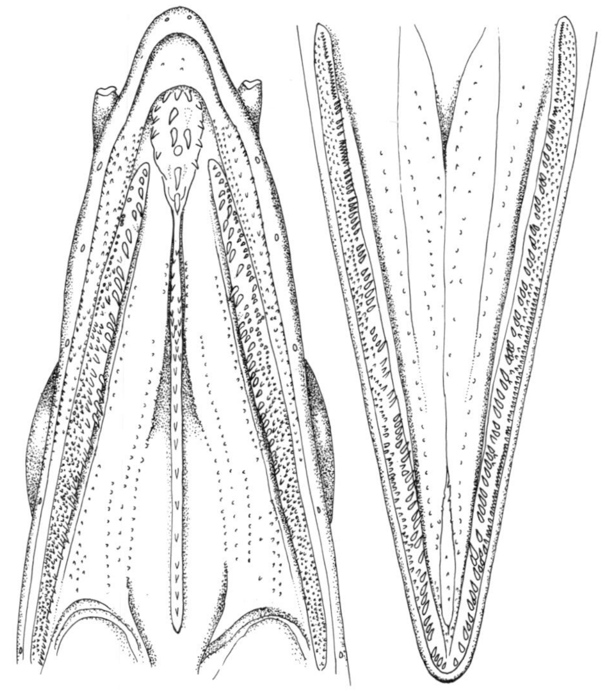

FIGURE 212. *Synaphobranchus affinis*: dentition. Data as for Figure 211.

grayish to brownish gray with pinkish dorsum. As in all species of *Synaphobranchus*, ventral surface, mouth (under congealed mucus), bases of fins, and caudal fin darker than other parts of body. Peritoneum black.

Size. In the western North Atlantic, *S. affinis* is possibly the smallest member of the genus. Ovigerous eels from the Straits of Florida ranged from 382–460 mm, and were collected in January, April, and September.

Food, Habits, and Habitat. Synaphobranchus affinis appears to be an opportunistic feeder upon crustaceans and small fishes, largely over rocky or rubble-strewn bottoms with some current. We have taken this species alive and from our observations, it presumably would be able to hold its own in moderate, local currents.

Distribution. Synaphobranchus affinis is the most common species of *Synaphobranchus* in the Straits of Florida, where it occurs principally between 640–825 m. The bottom of the Straits of Florida is subjected to north-south current reversals and bottom scouring. It is much rarer in collections north of Florida or about 30°N, in the Tongue-of-the-Ocean and Exuma Sound, Bahamas, and does not enter the deeper waters on the western part of the Gulf of Mexico. (Only 8 specimens were trawled in the Tongue-of-the-Ocean program which yielded more than a thousand specimens each of *S. kaupi* and *S. brevidorsalis.* Grey (1956) did not include *S. affinis* in her treatise on deep-dwelling fishes.) A total of 511 individuals from 27 stations (190 from a station in 636 m, 127 from a station in 809 m) were collected at depths ranging from 403–1190 m (data from K. J. Sulak, Virginia Institute of Marine Science, pers. comm.). The species is known from the Pacific, Atlantic, and Indian oceans, and presumably is of nearly worldwide distribution in temperate and tropical latitudes. Various populations have been studied in some detail and compared to other species in the region. No worldwide study of variation has been done. That variation exists is obvious. *Synaphobranchus affinis* from the Straits of Florida and Caribbean Sea is smaller and paler but otherwise seems similar to other Atlantic material. Japanese material is particularly troublesome, but that available to us is old and in rather poor condition. There may be two types in Japanese waters, one with a longer tail and very small scales. This problem cannot be resolved with material at hand.

Etymology. From the Latin meaning "related to" because of its perceived relationships to *S. brevidorsalis.* An adjective.

Study Material (excluding that reported by Robins, 1971: 175). EASTERN ATLANTIC: USNM 198618 (1, 490), 198623 (5, 329–350), 198637 (2). UMML 34394 (2, 285–298).

WESTERN ATLANTIC: USNM 132209 (3), 122327 (7), 273421 (3, 317–414), 273422 (14, 248–532), 185657 (1, 330), 197126 (1, 220), 211891 (10, 78–93).

INDIAN OCEAN: USNM uncat. (1, 497), USNM 273423 (1, 378), 273424 (1, 304), 273425 (1, 592).

INDONESIA-PHILIPPINES: USNM 135164 (1, 400), 135168 (1, 402), 135169 (1, 463), 175159 (2, 400–435), 175156 (2, 331–370), 135170 (1, 381), 175162 (1), 175152 (1, 560), 175145 (1, 385).

JAPAN: USNM 160594 (1, 253), 160586 (5, 328–493), USNM 160592 (3, 275–346), 160593 (1, 383), 160581 (375–470), 161451 (1, 378), 161449 (1, 410), 161452 (1, 387), 161450 (1, 350), 160587 (1, 440), 49955 (1, 375), 160598 (1, 540), 161453 (1, 185), 160597 (2, 355–375), 160595 (2, 312–340), 160596 (1, 300). FMNH 71760 (1, 578), 71761 (7, 460–687), 71758 (7, 300–400). USNM 164156 (1, 333), 117922 (2), 117985 (1), 117895 (2), 149896 (1).

HAWAII: BPBM uncat. (1, 508). USNM 51591 (1, 703; holotype of *S. brachysomus*). USNM 51696 (2, 499–540, paratypes of *S. brachysomus*).

Synaphobranchus oregoni Castle, 1960
Figures 210, 213, 214

Synaphobranchus oregoni Castle, 1960: 388 (original description, Gulf of Mexico, 29°01′N, 88°24′W, in 650–870 m, Oregon sta. 640, holotype USNM 185605). Robins, 1971: 153 (osteology, relationships). Martin, 1978 (distribution and ecology).

Description. Vertebrae 140–150 (usually 142–148). Precaudal vertebrae 55, 61; caudal vertebrae 82, 87. Caudal-fin rays 12–16 (4 specimens); branchiostegal rays 16 (1 specimen). Cephalic lateralis pores: infraorbital 10, supraorbital 5, preoperculomandibular 12, supraoccipital 5.

Proportional measurements as percent of total or head (*) length: predorsal length 31–34; preanal length 25–30 (Castle, 1960a); snout length* 24–36; length of jaw* 49–67; horizontal diameter

FIGURE 213. *Synaphobranchus oregoni*: ANSP 137423, 721 mm TL, Dominican Republic.

of eye* 12–19, mean 15.7; greatest body depth 5–12, mean 6.8.

Teeth present on dentary, maxilla, premaxilloethmoid, and vomer. Innermost teeth on front ends of jaws enlarged, but still small, simple, and pointed, as are all others. Teeth on premaxilloethmoid considerably enlarged, irregularly arranged in a longitudinally oval patch separated from the vomerine teeth by a variable distance. Vomerine teeth in most specimens irregularly biserial anteriorly, covering an expanded area at anteriormost end of vomer. The holotype and occasional other specimens have uniserial vomerine teeth.

Adults are laterally compressed, quite deep through the trunk; immatures much more slender, not so massive in the trunk region. Scales rounded or polygonal, large compared to those of other synaphobranchid eels (Fig. 210). Epidermis in some preservatives shriveled, raised above surface of scales into a network pattern. Scale condition of large adults clearly discernible; more difficult to see in immature eels.

Origin of dorsal fin posterior to a point above anus, ordinarily by a distance equal to about one head length. Preanal part of abdomen is second shortest in *Synaphobranchus* (as here defined). Median fins confluent as in all species of the family. Pectoral fins present, at least as long as the snout, often longer. Lateral-line canal complete.

Head broad at occiput, somewhat depressed. Eye large, fleshy gape extends at least a distance equal to the eye length behind the level of the rear margin of eye. Nostrils essentially as in *S. kaupi*, varying only in minor details, and then

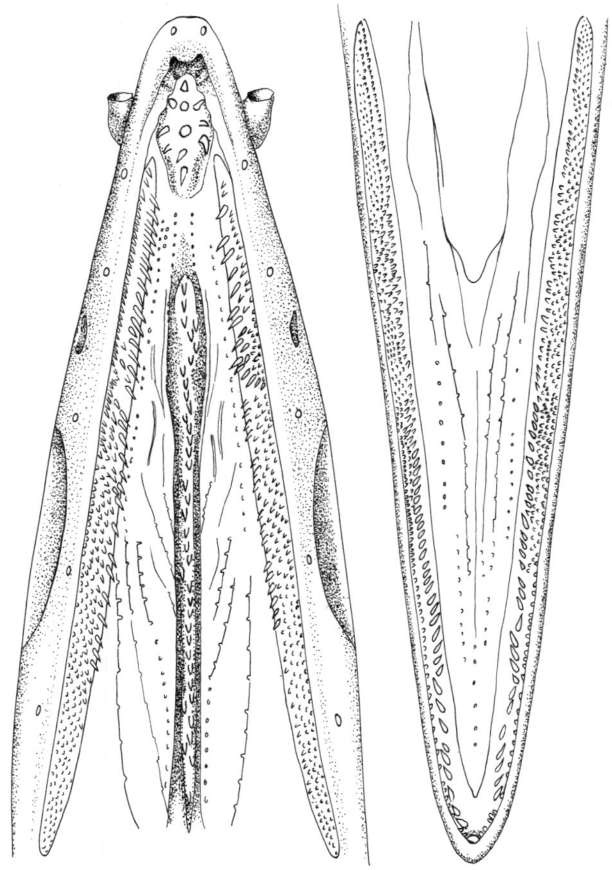

FIGURE 214. *Synaphobranchus oregoni*: ANSP 114618, 480 mm TL, Gulf of Mexico off Mississippi; dentition.

inconsistently. Gill slits ventral, united, almost directly below pectoral-fin base or slightly anterior to that point. No significant, consistent difference in size of gill slits of this species and those of *S. kaupi, S. affinis,* and *S. brevidorsalis.*

Color. Preserved specimens are brown, often with a reddish or orangish cast in larger individuals. Darker in region of abdomen, branchial region, bases of fins, and on caudal fin. Mouth dark, blackish; peritoneum black.

Size. Maximum size recorded, a female, 700 mm, from southern Straits of Florida (Robins, 1968: 180).

Food and Habitat. Like *S. affinis, S. oregoni* is an opportunistic feeder upon other fishes, crustaceans, and squids. Martin (1978) found that it frequented soft bottom areas of the Gulf of Mexico and was not taken in rocky or limestone shelf areas of the eastern Gulf of Mexico. This is corroborated by Robins (1968) who found a behavioral or ecological dependence on deeply sedimented, silty substrates.

Parasites. Overstreet and Martin (1974) recorded three families of digenetic trematodes from *S. oregoni.*

Distribution. Synaphobranchus oregoni is often locally abundant, but its geographic range appears to be limited to the Gulf of Mexico (where it is the most common synaphobranchid), Caribbean Sea (where it is rare), and the Straits of Florida. One small specimen was collected in 890 m in the Middle Atlantic Bight (36°36.2'N, 74°38.9'W) by the Virginia Institute of Marine Science (K. J. Sulak, pers. comm.). Presumably this is a stray.

Etymology. After the National Marine Fisheries Service's exploratory research vessel "Oregon." A noun in the genitive case.

Study Material (in addition to that reported by Robins, 1971: 175 and by Martin, 1978). WESTERN ATLANTIC: USNM 185605 (1, 340; holotype) and FMNH 47911 (6, 313–506, paratypes) 29°01'N, 88°24'W, in 631–869 m, 19 Sept. 1952, Oregon sta. 640. FMNH 69272 (1, 500), 71741 (1). USNM 188009 (1, 434), 157953 (1, 273), 186050 (1, 695), 195900 (1).

Status of Synaphobranchus sului *Bruun, 1937.* Bruun (1937: 12) distinguished *Synaphobranchus sului* based on 18 leptocephali from the Sulu and Mindanao seas. They have 124–129 myomeres and other features which Bruun felt were unique but which have been very little studied. He listed the material (1937: 39) under *Leptocephalus Synaphobranchus sului,* and illustrated one (56 mm TL) from Dana Station 3735 (1937: pl. 1, fig. 4) which he designated "type." We have adults of both *S. brevidorsalis* and *S. affinis* from this region. While the counts are more typical of vertebral numbers of *S. affinis,* we have counts as low as 124 for *S. brevidorsalis* from Japan. *Synaphobranchus sului* is most likely a junior synonym of one of these two species.

Synaphobranchus brevidorsalis Günther, 1887
Figures 210, 215, 216

Synaphobranchus brevidorsalis Günther, 1877: 445 (nomen nudum). Günther, 1887: 225 (original description, north of New Guinea, in 1957 m, syntypes BMNH 1887.12.7.256). Norman, 1939: 45 (off Zanzibar, in 1789 m; Maldive Is., in 914–1463 m). Robins, 1971 (osteology and relationships). Pakhorukov, 1976: 325 (recorded from the southwestern Atlantic). Martin, 1978 (distribution and ecology in Gulf of Mexico).

Synaphobranchus jenkinsi Jordan and Snyder, 1901: 845 (original description, Japan, Sagami Bay, off Enoshima, Albatross sta. 3696, holotype USNM 49727).

Synaphobranchus pinnatus var. *brevidorsalis* var. nov. Lloyd, 1909: 152 (Arabian Sea, Investigator sta. 338, in 1534 m; see comments below).

Synaphobranchus kaupi: Grey, 1956: 146 (in part, only that part dealing with *brevidorsalis,* "Of the several Indo-Pacific forms that have been considered synonyms of *S. kaupi* by some authors, only *S. brevidorsalis* is known from deep-abyssal waters").

Discussion of Synonymy. Lloyd's description of his new variety makes it clear that he regarded it as distinct from *Synaphobranchus brevidorsalis,* and, in fact, related to what he considered to be *S. pinnatus.* Names proposed as new varieties before 1960 are valid and available. His name is therefore both a junior homonym and a synonym of *S. brevidorsalis* Günther.

Description. Vertebrae 130–140 (140 is surely the upper extreme; Japanese specimens have as few as 124–125). Precaudal vertebrae 60–68; caudal vertebrae 64–78; caudal-fin rays 15–17; pec-

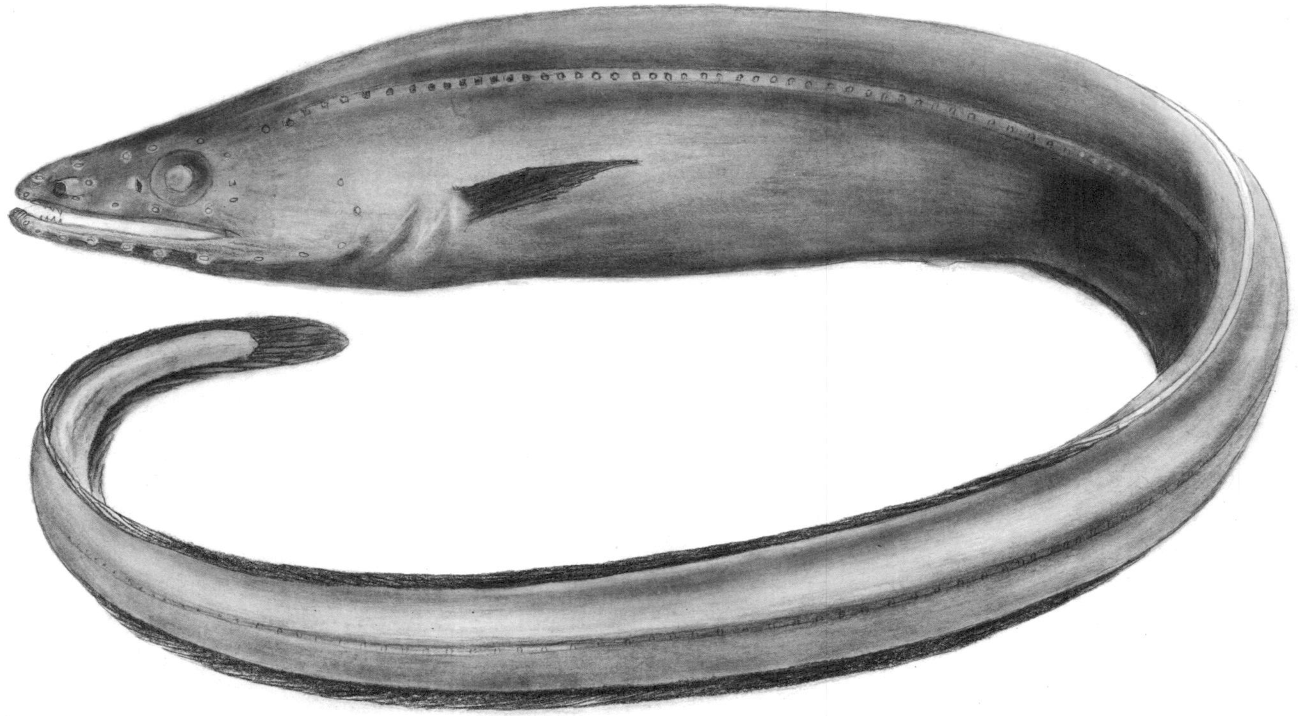

FIGURE 215. *Synaphobranchus brevidorsalis*: ANSP 137424, 635 mm TL, Bahamas.

toral-fin rays 14–17. Proportional measurements expressed as percent of total or head (*) length: predorsal length 45–52; head length 10–14; jaw length* 72–98; eye length* 11–16; depth of body 5.7–8.0.

Teeth on the dentary, maxilla, premaxilloethmoid, and vomer. All small, simple, pointed. Those of anterior innermost rows on jaws larger. Teeth in oval patch on premaxilloethmoid and at front end of vomer enlarged. Vomerine teeth irregularly biserial anteriorly, forming an expanded, rounded area much as in *S. oregoni*. Posterior vomerine teeth uniserial.

Body elongate, laterally compressed, slabsided, not deep through trunk even in large specimens. Flabby, soft and somewhat slimy or slick to the touch. Scales small, irregularly rounded or suboval, variable in shape over body from oval (somewhat resembling those of *S. affinis*) to rounded (but smaller on the average than those of *S. oregoni*); scales never so closely packed or regularly arranged as in *S. kaupi* (Fig. 210). Dorsal-fin origin far behind level of anus, only slightly anterior to or at midpoint of total length. Median fins confluent. Pectoral fin well-developed, one and a half to two times longer than snout length and tapered to a point. (Only *S. capensis*, not recorded from the western North Atlantic, has a longer pectoral fin.) Lateral line complete.

Head broad, deep through occipital region. Eye large, pronounced. Gape large, extending slightly more than one eye diameter posterior to rear margin of eye. Nostrils essentially as in previously discussed species in this genus.

Color. In preservative medium brown, occa-

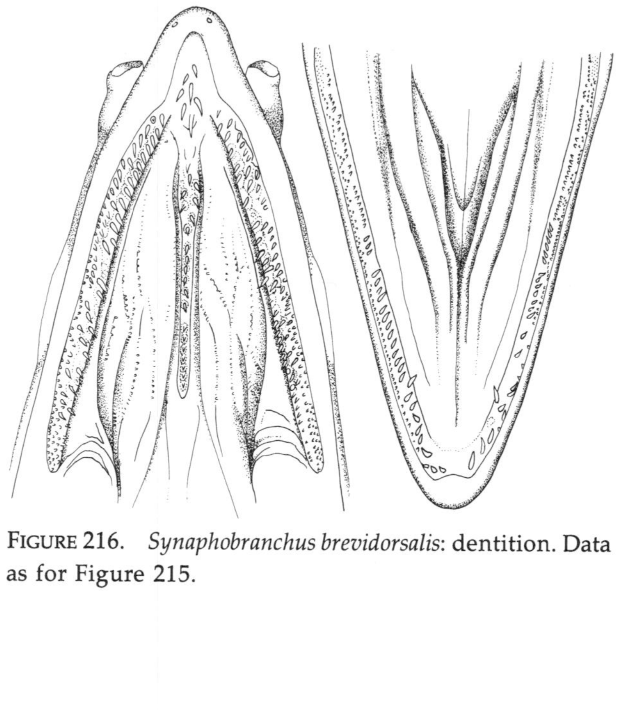

FIGURE 216. *Synaphobranchus brevidorsalis*: dentition. Data as for Figure 215.

sionally paler dorsally, with darker areas as in other species of *Synaphobranchus.*

Size. Synaphobranchus brevidorsalis is the largest of the genus in the western North Atlantic, reaching a meter in length, frequently slightly more. Extremes of length of more than 3000 specimens from Tongue-of-the-Ocean and Exuma Sound were 107 mm (3 g, sex unknown) and 1175 mm (2487 g, female). Despite the large numbers collected, very few were in "ripe" breeding condition. Those that were were collected in February, April, August, and September.

The rate of development of this species is unknown. It is the most flaccid of all the *Synaphobranchus* species; perhaps it matures more rapidly than others. However, it is the most consistently deep-dwelling, and the soft, poorly muscled bodies may not reflect rapid growth but the more energy-poor environment.

Distribution. Synaphobranchus brevidorsalis occurs in the western Atlantic from the Middle Atlantic Bight to southern Brazil. It is wideranging but rare in the Gulf of Mexico. Martin (1978) recorded 3 specimens from the northeastern Gulf of Mexico and 3 specimens from the southwestern Gulf of Mexico at sites with temperatures of 4.1–4.3°C. It seems to be limited to deep waters under tropical and warm temperature surface layers. Although the range of *S. brevidorsalis* in the western North Atlantic seems to be restricted to depths in excess of 1200 m, it has been recorded as shallow as 545 m in the Philippines (see material examined section below). The distribution is somewhat confused by earlier authors' inclusion of *S. brevidorsalis* in the synonymy of *S. kaupi* and mistakes of identification. Grey (1956) recorded it from the western North Atlantic, Indian, and western Pacific oceans. Pakhorukov (1976: 325) recorded one from the Rio Grande Rise, southwestern Atlantic, 31°22'S, 34°59'W, in 1410–1520 m. It can be locally very abundant; the largest sample from Tongue-of-the-Ocean, Bahamas was 130 (243–653 mm) for a tow of about 1 hour in duration. We have collected more than 35 at a station on many occasions. One small specimen was collected from the Middle Atlantic Bight from 1635 m, water temperature 3.0°C.

Etymology. From the Latin for short and dorsum and the suffix meaning having the condition of, in reference to the posterior origin of the dorsal fin. An adjective.

Study Material (in addition to that reported by Martin, 1978). ATLANTIC: 3345 specimens from the deep basins of the Bahamas, especially Tongue-of-the-Ocean and Exuma Sound are at UMML, and are not listed by individual collection. FMNH uncat. (1, 403). USNM 133208 (2, 462–488), 198659 (2, 460–607).

INDIAN OCEAN: USNM 273426 (1, 538; tail regenerated), 273427 (2, 690–710).

INDONESIA-PHILIPPINES: USNM 135166 (1, 252), 273428 (2, 462–488), 175163 (1, 482, tail regenerated), 135171 (1, 360), 175147 (1, 508), 175148 (1, 496), 175146 (1, 487), 175150 (1, 607), 175144 (1, 545; tail broken), 175154 (1, 475), 135167 (1, 220), 135165 (1, 195), 175143 (1, ca 475).

JAPAN: USNM 49727 (1, 412; holotype of *S. jenkinsi*) Sagami Bay, Albatross sta. 3696. USNM 160588 (1, 531), 117982 (1).

OTHER PACIFIC AREAS: SIO 68-443 (1, 780), SIO 68-460 (5, 425–835). USNM 135214 (8, 491–604).

Synaphobranchus capensis (Barnard, 1923)

Diastobranchus capensis Barnard, 1923: 441 (original description, South Africa, off Cape Point in 860 m). Norman and Trewavas, 1939: 359 (data on cotype, BMNH 1924.9.24.1; *Synaphobranchus danae* regarded as possible synonym).

Synaphobranchus danae Bruun, 1937: 13 (original description based on leptocephalus, Dana sta. 3644, east of New Zealand).

Ilyophis capensis: Smith, 1949: 386 (referred to genus *Ilyophis*, contrasted with *brunneus*).

Diastobranchus danae: Whitley, 1951a: 27–28, fig. 1 (new combination, description of a large gravid female from off Tasmania). Cowper and Downie, 1957: 5 (6 specimens recorded from 1006–1087 m, off Australia).

Diactobranchus capensis: Pakhorukov, 1976: 325–326 (lapsus for *Diastobranchus*, recorded from southwestern Atlantic).

Description. Vertebrae 164–173. The following measurements expressed as percent of total length or head length (*) are of one specimen 566 mm total length from the South Pacific: predorsal length 31; preanal length 21; head length 15; greatest postanal depth 7; snout length 28*; jaw length 68*; horizontal diameter of eye 13*; gill opening 11*; and pectoral-fin length 34*. Castle (1961) examined eleven specimens 327–1227 mm from Cook Strait and South Africa (type material). Measurements in percent of total length were as follows: head length 12.4–14.6; gape 6.6–7.6; preanal length 21.4–24.8; and predorsal length 29.8–37.5.

Teeth on dentary, maxilla, premaxilloethmoid, and vomer. Teeth on anterior parts of innermost rows of jaws larger than those of outer rows, and decidedly larger than those posteriorly in the rows. Teeth on premaxilloethmoid enlarged, recurved in oval or subcircular patch, visible in profile when mouth is opened wide. Anteriormost teeth of vomer similarly enlarged, with tissue around them forming a mound (much as in *S. brevidorsalis*, but more pronounced). Although these teeth are large, they are simple, acutely pointed, and not at all similar to those of the synaphobranchid eels of the subfamily Ilyophinae.

Body elongate, laterally compressed. Anus far forward, about one head length posterior to gill slits. Dorsal-fin origin posterior to a point above the anus by a distance equal to about one head length. Gill slits ventral but separate, about as long as horizontal eye diameter. The gill slits are separated in front by a distance about equal to their length, but diverge so that they are separated posteriorly by a distance equal to 1.5–2 times their length. (Not as widely divergent as illustrated by Castle, 1961: 26, fig. D.) Scales elongate, very similar to those of *S. kaupi*.

Castle (1975b) described the osteology of *S. capensis* in detail. Overall configuration of cephalic skeleton essentially as in other synaphobranchine eels, the principal difference being in slightly depressed dorsal surface of the cranium (concave in lateral view). Frontals fused. Pterygoid long, slender, more complete than in other synaphobranchines extending (according to Castle) from the vomer to the quadrate. Hyomandibular (suspensorium) inclined far posteriorly. Basisphenoid not fused with parasphenoid. Orbit large, not restricted.

Size. Longest species of *Synaphobranchus*, commonly exceeding 1 m in total length. Whitley (1951a) recorded a gravid female 4½ ft. long (1.4 m). Castle (1986: 189) states that it attains 1.8 m.

Distribution. South Pacific (Cook Strait, New Zealand; Tasmania, Victoria, and New South Wales, Australia), southwestern Indian Ocean (Cape Point, South Africa), near Antarctic convergence. Recently recorded by Pakhorukov (1976: 325) from Rio Grande Rise, southwestern Atlantic, at 31°04′S, 35°34′W, in 1580 m. Depth, about 750–2000 m. Said to be extremely common, in depths of about 1000 m, off Tasmania (Last et al., 1983: 179). No deep abyssal records.

Etymology. From the Latin for cape in reference to Cape Point, the type locality. An adjective.

Study Material. LACM 11340-1 (1, 566) 49°38–40′N, 178°56–57′E, Eltanin sta. 1852, 3 Jan. 1967.

Synaphobranchus bathybius Günther, 1877
Deepwater cutthroat eel
Figures 210, 217–219

Synaphobranchus bathybius Günther, 1877: 445 (original description, Pacific Ocean, 36°N, 178°E, in 3749 m, syntypes BMNH 1887.12.7.251; 1887.12.7.252-4).

Histiobranchus infernalis Gill, 1883: 255 (original description, off eastern United States, 38°N, 69°W, in 3166 m, holotype USNM 33279). Jordan and Davis, 1891: 673 (compiled, contrasted with *bathybius*).

Synaphobranchus infernalis: Günther, 1887: 254 (new combination). Bruun, 1937: 11 (nature and distribution of leptocephali).

Histiobranchus bathybius: Jordan and Davis, 1891: 673 (new combination, compiled). Castle, 1964a: 36 (synonymized *H. infernalis* with *H. bathybius*).

Synaphobranchus australis Regan, 1913: 235 (original description, Indian Ocean, 46°S, 45°E, in 2515 m, holotype BMNH).

Synaphobranchus (*Histiobranchus*) *bathybius:* Norman, 1939: 45 (off Zanzibar, in 2926 m).

Description. Vertebrae 126–140; precaudal 62–68 (3), caudal 60–69. Caudal-fin rays 14–15. Cephalic lateralis pores: infraorbital pores 6, supraorbital pores 7, preoperculomandibular pores 10. Proportional measurements as percent of total or head (*) length: predorsal length 13–16; preanal-fin length 44–47; head length 10–11; snout length* 82–91; eye diameter* 16–17; depth of body 8.3–9.0. Dentition essentially as in other species of *Synaphobranchus*. Premaxillary teeth numerous, small, recurved, arranged in a roughly oval patch. Vomerine teeth smaller than premaxillary

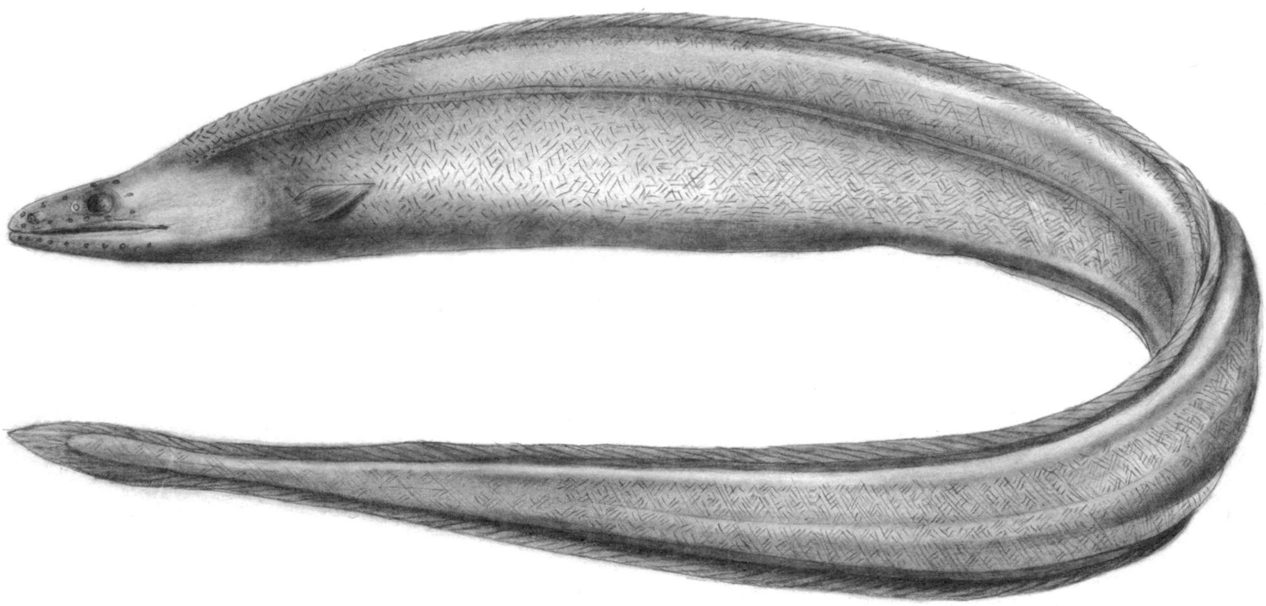

FIGURE 217. *Synaphobranchus bathybius*: ANSP 136743, 782 mm TL, Atlantic, Norfolk Canyon.

teeth, and becoming smaller posteriorly; irregularly biserial to multiserial. Lower jaws with teeth fewer and longer anteriorly than posteriorly, where they are cardiform. Dorsal-fin origin far forward, about over the pectoral-fin base. Dorsal fin well-developed, about as high as the anal fin. Head depressed.

Color. Generally rather uniform tan to brown. Robins et al. (1979: fig. 2) published a color photograph of a live individual from 2837–2873 m in Northeast Providence Channel, Bahamas.

Size. Reaches a length of at least 74 cm.

Distribution. Eleven specimens were collected by VIMS in 2125–2933 m from the Middle Atlantic Bight. Temperature ranged from 2.3–2.8°C. The individual illustrated by Robins et al. (1979: fig. 2) as *Histiobranchus* sp., from 2837–2873 m on the floor of Northeast Providence Channel Bahamas, is undoubtedly of this species. This is a large, widely distributed synaphobranchid of very deep, cold water. It may be locally abundant. It extends south almost to the Antarctic Convergence (DeWitt, 1971: 2, map 2).

Etymology. From the Greek for deep and life in reference to its habitat. An adjective.

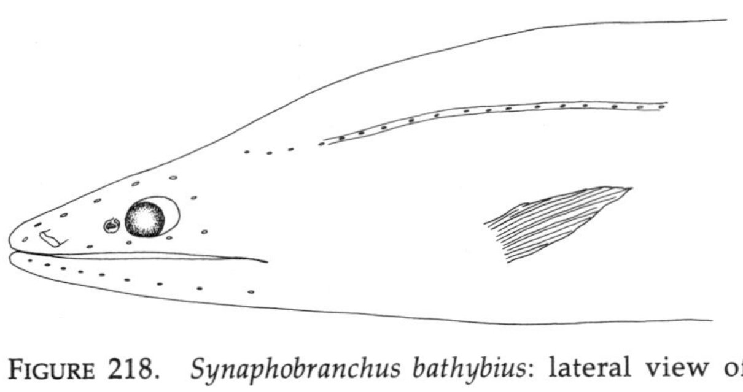

FIGURE 218. *Synaphobranchus bathybius*: lateral view of head. Data as for Figure 217.

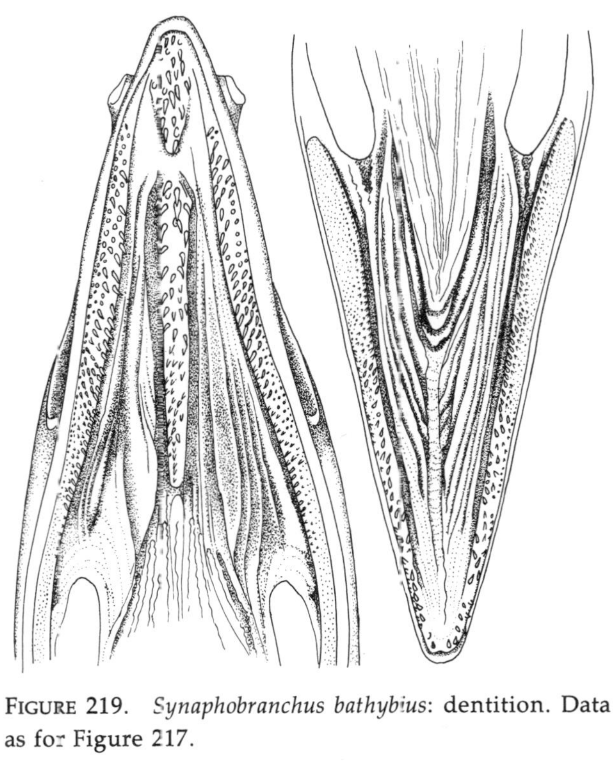

FIGURE 219. *Synaphobranchus bathybius*: dentition. Data as for Figure 217.

Study Material. WEST ATLANTIC: USNM 33279 (1), 38°N, 69'W, in 3166 m. EAST ATLANTIC: UMML 34395 (1, 417), 01°13–05'N, 07°46–43'E, in 2424 m. NEW ZEALAND: USNM 273429 (4, 330–456), 42°11'S, 175°05–11'E, in 2612 m. EAST PACIFIC: SIO 59-269-26A (1, 675), 22°11.2–25.5'N, 107°46.1–49.5'W. UMML 31350 (1, 735), 06°42–44.9'N, 78°56–54.5'W, in 3173–3208 m.

Synaphobranchus bruuni (Castle, 1964)

Histiobranchus bruuni Castle, 1964a: 34–36, fig. 1A, C, F (original description, Tasman Sea, Galathea Sta. 601, 45°51'N, 164°32'E, in 4400 m, holotype AMUC P 32444).

Comment. This species is not known from the Atlantic Ocean.

Distinctive Characters. Vertebrae: 162. Dorsal fin far forward, above a point about one-third out on the appressed pectoral fin; well-developed, as high as anal fin. Scales present, elongate-oval. Abdomen long; anus about at the midpoint of body. Gill slits ventral, horizontal, almost confluent anteriorly but separated by half their length posteriorly. Gape long.

Distribution. Known only from type locality in Tasman Sea (see details in synonymy above).

Etymology. For the late Danish biologist and oceanographer, Dr. Anton F. Bruun.

Genus *Haptenchelys* C. H. Robins and Martin, 1976

Haptenchelys C. H. Robins and Martin in Robins and Robins, 1976: 267 (type species *H. texis* C. H. Robins and Martin, by original designation).

Characters (modified from the original description). Naked synaphobranchine eels with dorsal-fin origin directly above base of pectoral fin. Preanal length of abdomen short, anus posterior to level of gill slits by a distance equal to about one and a half times the head length. Pectoral fins small. Dorsal and anal fins with fleshy bases. Median fins confluent. Gill slits separate, crescentic, oriented horizontally on the ventral surface of pharynx. Head depressed, fleshy tip of snout slightly overhanging tip of lower jaw. Eyes oval, well-developed, covered with skin, but not deeply subcutaneous. Gape extending to a point behind a vertical from the rear of eye about equal to half the eye diameter. Lateral-line canal pores present on anterior three-quarters of body. Number of pores in cephalic lateralis system somewhat variable; most common numbers: infraorbital pores 7, supraorbital pores 3, preoperculomandibular pores 8.

Dentition similar to that of species of *Synaphobranchus.* All teeth simple, small. Patch of teeth present on premaxilloethmoid. Vomerine teeth in single, irregular row, those toward the rear smallest. Teeth on dentary and maxilla in irregular series; innermost teeth somewhat enlarged.

Osteological features: Frontals fused; basisphenoid present, not fused to parasphenoid; suspensorium angled posteriorly; eye socket reduced; supraoccipital small, reduced, possibly variable in shape and placement. Two pairs of pharyngobranchials. Lower maxillary apophysis rudimentary, upper long, slender, sloping forward.

Etymology. From the Greek *haptos* (to join, fasten to) and *enchelys* (a feminine Greek noun for an eel), the name alluding to the fact that this species forms a link between what was once regarded as two families. Feminine.

Species. A single species, *Haptenchelys texis.*

Haptenchelys texis C. H. Robins and Martin, 1976
Figures 220–222

Haptenchelys texis C. H. Robins and Martin in Robins and Robins, 1976: 267 (original description; 25°16.75–18.75'N, 77°45.5–42.8'W, in 2780–2890 m; holotype ANSP 133811). Merrett and Saldanha, 1985: 273 (records from northeastern Atlantic). Anderson et al., 1986: 795 (recorded from the Caribbean Sea in 3993–4064 m).

Description. Vertebrae 125–132 (based in part on Merrett and Saldanha, 1985: table 2). Caudal-fin rays 15–18; pectoral-fin rays 15–17. Proportional measurements, expressed as a percent of total or head (*) length: predorsal length 14–17; preanal length 32–40; depth of body 4.1–5.5; snout to front edge of eye* 34–38; length of jaw* 73–

FIGURE 220. *Haptenchelys texis*: ANSP 133811, holotype, 431 mm TL; Bahamas.

84; horizontal diameter of eye* 9.4–15. Pores of cephalic lateralis system: infraorbital pores 6–8 (usually 7), supraorbital 3, preoperculomandibular 7–9 (usually 8).

Body subcircular in cross section anterior to anus, laterally compressed posteriorly. Head depressed, snout and jaws moderately pointed, with fleshy tip of snout overhanging tip of lower jaw.

All teeth simple, small, slender, sharp. About 12 teeth in irregular central patch on premaxilloethmoid, 16–17 on vomer in one irregular row and set in narrow folds of yellowish, granular tissue. Maxillary and dentary teeth in irregular rows, inner rows slightly enlarged. Irregular rows become more numerous and the teeth small toward the posterior end of the jaws.

Color. Most specimens evenly pigmented, dark brown in isopropanol, not sharply bicolored although bases of fins, abdomen, pharynx, and underside are darker than rest of body.

Size. The smallest known metamorphosed specimen is 183 mm (Merrett and Saldanha, 1985: 723). The largest adult is 511 mm. Of the original material reported by Robins and Martin, one female was ovigerous at 460 mm. A second ovigerous female, 445 mm TL, was reported by Merrett and Saldanha (1985: 730). A 358 mm male weighed 42.9 g.

Distribution. *Haptenchelys texis*, while rare, is widely distributed. The original material was from the Gulf of Mexico and Exuma Sound, Bahamas. Merrett and Saldanha (1985: 723) reported nine specimens from 17–25°N, 16–18°W, off northwestern Africa, and Anderson et al. (1986:

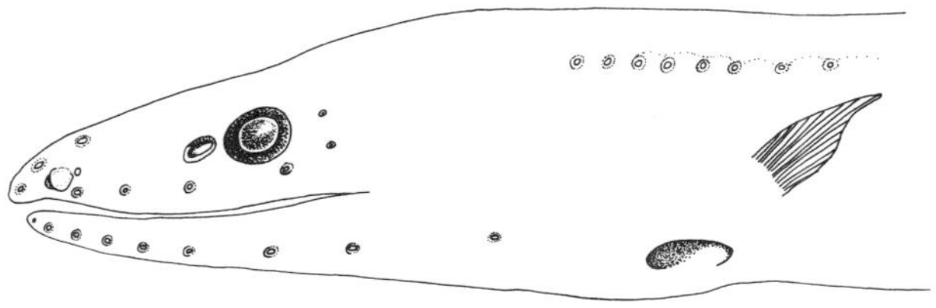

FIGURE 221. *Haptenchelys texis*: lateral view of head. Data as for Figure 220.

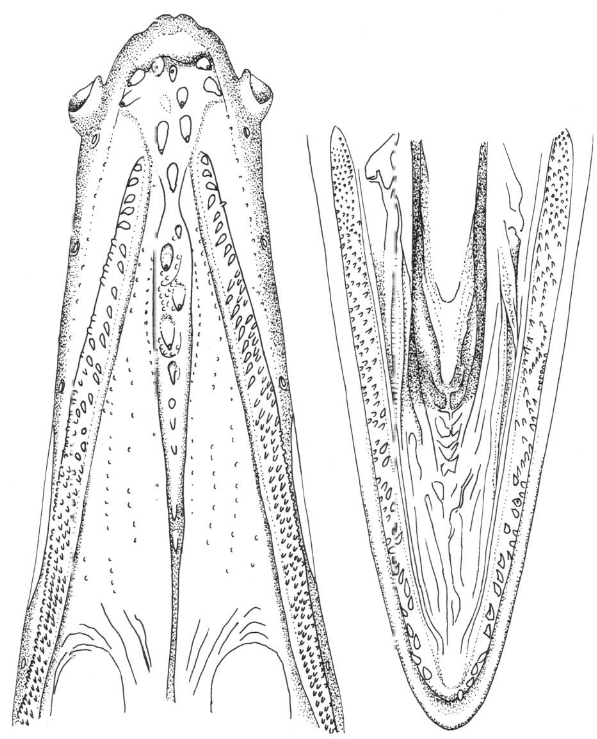

FIGURE 222. *Haptenchelys texis*: dentition. Data as for Figure 220.

795) recorded two additional specimens from the Caribbean Sea in 3993–4064 m. Depths of all collections range from 2160–4086 m. Obviously this species is a widely distributed inhabitant of the deep benthic community.

Discussion. Of the original material, one eel had a severely anomalous stricture of the abdomen causing displacement of the liver and other organs. The vomer of this specimen was also deformed. The supraoccipital in the cleared and stained specimens was absent in one and enclosed by the parietals in another. The number of pores in the cephalic lateralis system is more variable than we have found in other members of this family. The sample is too small to do other than note these variations, which may be a trait of the species.

Etymology. From the Greek *texis* (a wasting away or dissolution), again in reference to the artificial boundaries between the Ilyophinae and Synaphobranchinae. A noun in apposition.

Study Material (exclusive of that reported by Robins and Robins, 1976: 270). USNM 264638 (2, 245–378) 15°08′N, 69°13′W, 27 Oct. 1981. UMML 34396 (1, 500) 23°45.8′N, 77°14.6′W, in 1374 m, 6–7 April 1975, Columbus Iselin 374. UMML 34397 (2, 430–460) 25°15.7′N, 77°39.3′W, in 2977 m, 21 April 1975, Columbus Iselin 353.

Subfamily ILYOPHINAE

ARROWTOOTH EELS

Characters. Small to moderately large eels all of which have vomerine teeth which are enlarged, peglike or compound in some, and reduced in number from the condition found in the Synaphobranchinae. Gill slits separate, ventrolateral, variable in size according to species, but restricted in most. Most species naked, pallid in color except in deep-dwelling forms. Dorsal-fin origin and anus far forward in all species. Strong trend toward streamlining, shortening of snout, reduction of the bony orbit, and fusions in cranial osteology. Pterygoid present in those species possessing more completely ossified, less fused cephalic skeletons, absent in less well-ossified species; it never is connected by bone either in front or in back. Preopercle reduced in size, its dorsal extent about opposite the condyle of opercle and usually below that point. Pectoral fins present or absent; when present, usually small, broad-based, and not acutely tapering. Eye never large, varying from moderate in size with a normal skin covering, to very small and thickly covered with skin. Lower jaw always shorter than upper.

Only one deep benthic species locally abundant; most species are rare and from moderate depths. Most speciose subfamily of the Synaphobranchidae, the result of adaptive radiation of a deep-benthic group into shallower niches, and, perhaps, concomitant restrictions in larval distribution. Five genera and 14 species currently recognized (6 of them in western North Atlantic).

KEY TO THE GENERA OF THE ILYOPHINAE

1a. Pectoral fins absent . . . . *Dysomma* (in part), p. 244
1b. Pectoral fins present 2
2a. Body partly to rather fully scaled (scales imbedded, elongate, arranged in basket-weave pattern) . . . . *Ilyophis* (in part), p. 235
2b. Body naked 3
3a. Anus very far forward, distance from gill slit to anus much shorter than head 4
3b. Anus moderately far forward, distance from gill slit to anus about equal to distance from snout tip to gill slit, or longer 5
4a. 3–5 large teeth on midline of vomer . . . . *Dysomma* (in part), p. 244

4b. Teeth in anterior patch on premaxilloethmoid and in biserial row of about 12 teeth on vomer . . . . *Meadia*, p. 239
5a. Trunk relatively long, equal to at least 1.5 times the distance from tip of snout to gill opening . . . . *Ilyophis* (in part), p. 235
5b. Trunk shorter, about equal to distance from tip of snout to gill opening 6
6a. Distinctive patch of large teeth on premaxilloethmoid; body very slender . . . . *Atractodenchelys*, p. 240
6b. No premaxillary teeth; body stouter . . . . *Dysommina*, p. 242

Genus *Ilyophis* Gilbert, 1891

Ilyophis Gilbert, 1891b: 351 (type species *Ilyophis brunneus* Gilbert, by original designation and monotypy).

Characters. Ilyophine eels with moderately long abdomen (as opposed to the very short abdomen of *Dysomma*), moderately large, simple teeth in patch on premaxilloethmoid and in irregular, biserial rows on the front half of the vomer. Teeth surrounded by large, conspicuous papillae. Scales, when present, elongate, imbedded in the skin, and arranged in basket-weave pattern. Dorsal-fin origin above the posterior third of the small appressed pectoral fin or slightly posterior to a vertical through tip of appressed pectoral fin (to a distance about equal to the length of the pectoral fin). Anus about one and a quarter to two head lengths behind the level of the small, crescentic gill slits. Gill slits separate, set horizontally or slightly oblique, on ventral surface of body directly below base of pectoral fin. Pectoral fins small, not attenuate. Vertical fins confluent. Gape moderately long, extending at least to under rear margin of eye to a point about equal to one eye diameter farther back. Eyes moderate in size, round, subcutaneous. Lateral-line either complete (to caudal region) or restricted to anterior third of body. Cephalic lateralis pores: infraorbital 5–8, supraorbital 3–6, preoperculomandibular 7–11. Lateral-line pores pale.

Osteological features (based on *I. brunneus* and *I. arx*): Pterygoid present, suspensorium nearly vertical, basisphenoid present, not fused with parasphenoid. Branchiostegal rays 13–14.

Etymology. From the Greek *ilys* (ooze) and *ophis* (snake). Masculine.

Key to the Species of *Ilyophis*

1a. Head pores: infraorbital 7–8, supraorbital 5, preoperculomandibular 10–11; dorsal-fin origin behind level of tip of pectoral fin by a distance equal to length of the fin; gill slits oblique; vertebrae 179–188 . . . . *I. blachei*, p. 239
1b. Head pores: infraorbital 5, supraorbital 3, preoperculomandibular 7–9; dorsal-fin origin usually above posterior half of pectoral fin (occasionally slightly behind that point, but never as in A); gill slits horizontal; vertebrae 131–151 2
2a. Body always well scaled; lateral-line pores distinctly paler than body; body slender; preoperculomandibular pores 9; vertebrae 145–151 . . . . *I. brunneus*, p. 235
2b. Body naked (or partly scaled?); body stouter; preoperculomandibular pores 7; vertebrae 131–135 . . . . *I. arx*, p. 238

Ilyophis brunneus Gilbert, 1891

OOZE EEL

Figures 223–225

Ilyophis brunneus Gilbert, 1891b: 352 (original description; eastern Pacific near Galapagos Is., 00°36′S, 89°W; holotype USNM 44403). Gilbert in Jordan and Davis, 1891: 671 (see discussion below; description). Goode and Bean, 1896: 141, fig. 162 (description, illustration of the holotype). Jordan and Evermann, 1896a: 350 (description, compiled). Barnard, 1925: 182 (report of 2 specimens from off Cape Point, South Africa). Parr,

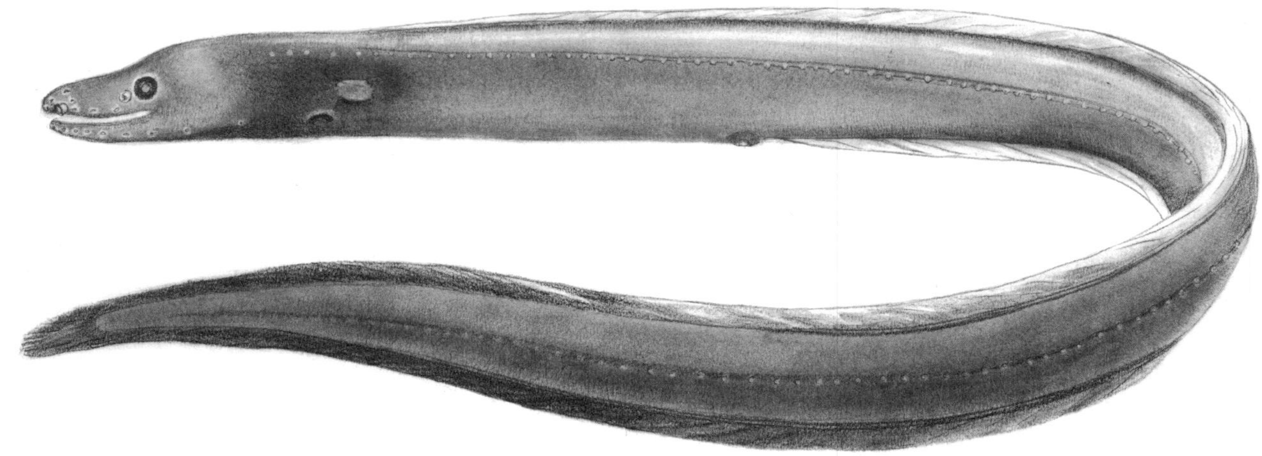

FIGURE 223. *Ilyophis brunneus*: ANSP 137446, 440 mm TL; Bahamas.

1932: 35, fig. 15 (data from a specimen from Pawnee st. 20, in the Bahamas, in about 1300 m; description of dentition of holotype). Smith, 1949: 385, fig. 1092 (compiled). Grey, 1956: 142–143 (synonymy; distribution, occurrence below 2000 m). Castle, 1964a: 39 (description; Galathea material). Castle, 1968d: 395 (recorded from South Pacific). Robins, 1971 (comparative osteology; occurrence in Straits of Florida and Caribbean Sea). Blache et al., 1973: 253 (in list). Pakhorukov, 1976 (reported from Rio Grande Rise, southwestern Atlantic). Robins and Robins, 1976: 264–265 (genus referred to Dysomminae in Synaphobranchidae; contrasted with *I. arx*). Karrer, 1982: 99–101, fig. 30 (description; summary of records from southwest Indian Ocean). Merrett and Saldanha, 1985: 735–739, fig. 8 (description based on 36 specimens from the northeastern Atlantic between 18–52°N).

Discussion of Synonymy. Gilbert provided Jordan and Davis with his description of *Ilyophis* and *I. brunneus* for inclusion in their work on apodal fishes. The result, Gilbert in Jordan and Davis (1891) has been cited as the original description of those taxa (e.g., by Grey, 1956: 42). In fact, that paper, which dates from 31 Dec. 1891, was predated by Gilbert's paper which was published 8 Sept. 1891. Lea (1913: 18) reported on a larva which he identified with *I. brunneus* because of agreement in myomere count with the vertebral count of 132 which he obtained from a specimen supposedly of *I. brunneus*. Smith (this volume) identified this leptocephalus as *Synaphobranchus* sp. since there are 3 species of that genus with distribution of numbers of vertebrae in this range. What adult Lea had, or whether he erred in his count of its vertebrae, is unclear. Koefoed's (1927: 61) reference of *I. brunneus* is based on Lea (1913).

Description. Vertebrae 145–151 (12 spec., extremes perhaps one or two more or less). Vertebral numbers from the northeastern Atlantic are lower, 139–148 (mean 143.5) based on counts of 36 specimens reported by Merrett and Saldanha (1985: table 2). Proportional measurements expressed as percent of total or head (*) length: predorsal length 11–12; abdominal length (posterior end of gill slit to fleshy protuberance around anus) 16–20; depth of body 3–5; head length 8–11; jaw length* 44–56 ($\bar{x}$ = 50.5); snout length* 31–37 ($\bar{x}$ = 35.4); length of gill slit* 6–14 ($\bar{x}$ = 8.9). These data based on 14 specimens except 12 for preanal length and depth of body. Data from the holotype are: predorsal length 11, preanal length 32, depth of body 4.

Patch of teeth on premaxilloethmoid, vomerine teeth in an irregularly biserial series (Parr, 1932: fig. 15, shows an irregular row for the holotype), and multiserial rows on maxilla and dentary. All teeth small, acute, some slightly recurved, and conical. Innermost premaxillary teeth, anterior vomerine teeth, and anterior and innermost teeth in jaws, largest. Oral epithelium covered with extensive system of complex papillae in which vomerine teeth appear imbedded.

Body completely covered with slender, rounded scales; head differentiated from trunk. Dorsal-

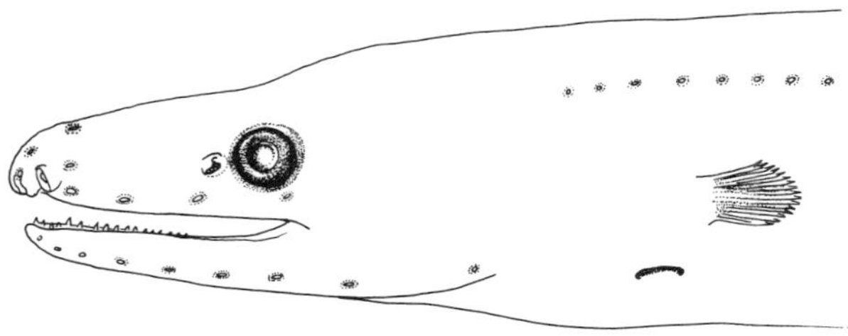

FIGURE 224. *Ilyophis brunneus:* lateral view of head. Data as for Figure 223.

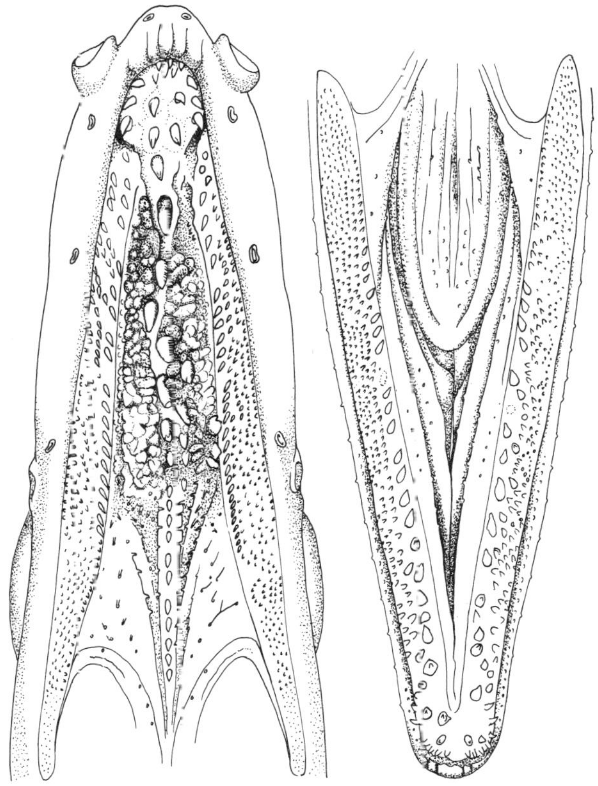

FIGURE 225. *Ilyophis brunneus*: dentition. Data as for Figure 223.

fin origin above tip of small pectoral fins (when appressed). Anus about 2 head lengths posterior to the small, crescentic gill slits. Gill slits ventral, separate, their axis horizontal. Median fins confluent. Gape extends below a point about equal to one eye diameter posterior to posterior margin of eye.

Cephalic lateralis pores (typically): infraorbital 5, supraorbital 3, preoperculomandibular 9. Lateral line long, to within a few millimeters of caudal fin. Scales elongate, narrow (as much as 5–6 times longer than wide), closely packed in a basket-weave pattern, and deeply imbedded in the skin (Fig. 210).

Color. In preservative ordinarily dark brown, especially along the lateral line, whose pores stand out as being conspicuously white, and along the bases of the pale dorsal and anal fins. Although all species of *Ilyophis* have pale lateral-line pores, those of *I. brunneus* are whitish and especially conspicuous against the dark background. Area immediately around anus very dark. Peritoneum black. Abraded areas may be quite pale.

Size. A total of 365 specimens of *I. brunneus* from Tongue-of-the-Ocean and Exuma Sound, Bahamas, included 222 females and 107 males. They were collected at 86 stations ranging in depth from 1234–2171 m. The largest collection (27 spec.) was trawled from a level area at 1360 m. The smallest specimen was 236 mm (10 g); the largest, a female, 569 mm (93 g). Ripe eels were collected in February through April and August through November. One female was ripe at 342 mm, but average mature size is probably close to 400 mm.

Food. No careful analysis of food items has been made as yet from these collections, but several specimens contained spiny bivalve shells. Robins (1968) examined gut contents of 5 specimens from depths of 732–1015 m in the Straits of Florida and found mostly crustacean parts (brachyuran crabs, some free-swimming shrimp, and galatheids). Martin (1978: 90) examined the stomachs of 8 specimens from a depth range of 700–1200 m and found an isopod, *Aega gracilipes* in one.

Distribution. Ilyophis brunneus was described from the eastern Pacific near the Galapagos Islands, and has subsequently been reported from the eastern and western North Atlantic, southwestern Atlantic, south Pacific, and southwestern Indian Ocean. Its known range in the eastern Atlantic was recently extended northward by Saldanha and Merrett (1982: 633) and Merrett and Saldanha (1985: 736) who reported on specimens from 49°N, 13°19.9′W to 51°20.8′N, 21°31.4′W in 1720–1789 m. The record from the southwestern Atlantic is by Pakhorukov (1976: 326), who reported on one specimen collected at

31°35'S, 34°39'W, in 1250 m. K. J. Sulak (pers. comm.) reported that a total of 108 specimens of *I. brunneus* were collected at depths of 1109–2745 m at 24 stations in the Middle American Bight (most northern record: 39°04'N, 72°32'W). Average weight per specimen was low, indicative of small size, presumably correlated with immature individuals. In the largest collection, 14 specimens weighed 300 g, an average of 21.4 g (a normal weight for a specimen of *I. brunneus* of less than 350 mm, at which size this species is normally sexually immature). A collection of 13 specimens from the Bahamas had an average weight of 64.5 g (49–88 g). Only three specimens from this station were listed as "immature," but, at their size, 445–486 mm, were also parenthetically noted as "spent?" (C. H. Robins, unpublished data). *Ilyophis brunneus* was reported by Martin (1978: 49–53) throughout the Gulf of Mexico, where he regarded it as the second most abundant synaphobranchid, occurring most frequently in the northern and eastern part of the Gulf between 700 and 1100 m. Martin stated that the temperature range for *I. brunneus* in the Gulf of Mexico was 6.5–4.3°C, with most specimens coming from a range of 5.2–4.5°C. Temperatures in the Middle Atlantic Bight (Sulak, pers. comm.) at stations from which this species was recorded were 2.9–4.3°C. Temperature range in the area of the Bahaman collections are 3.3–4.6°C. Only five of 24 stations in the Middle Atlantic Bight had temperatures in excess of 4.0°C whereas most from the Bahamas were in excess of 4.0°C.

That *I. brunneus* is an eel of soft or silty bottoms seems to have been concluded since its description by Gilbert (1891b). Since this species is not particularly slimy the reference seems to have been to the habitat. Robins (1968) and Martin (1978) both found this species to be associated with loose, soft, or silty bottom. The predominant sediment in the deep Bahaman basins is globigerina ooze.

Etymology. From the Latin *brunneus* (dark brown). An adjective.

Study Material (in addition to that listed by Robins, 1971: 182). Holotype: USNM 44403 (1, 377) Albatross sta. 2808, Chatham I., Galapagos, in 1152 m. UMML 23496 (1, 518) 10°16–10.4'N, 76°03–08.9'W in 818–1054 m, 15 July 1966, Pillsbury sta. 388. In addition, we have studied 365 specimens from the deep basins in the Bahamas, all of which are in the collections of the University of Miami.

Ilyophis arx C. H. Robins, 1976

Ilyophis arx C. H. Robins in Robins and Robins, 1976: 265–267, figs. 4, 6–8 (original description, eastern Pacific, 01°48'S, 90°19'W, in 3225 m, holotype ANSP 133808). ?Saldanha and Merrett, 1982: 635 (specimen from northwestern Atlantic, 51°04.4'N, 11°59.3'W, in 1533 m). ?Merrett and Saldanha, 1985: 730–735, figs. 6–7 (descriptions of specimens from northeastern Atlantic).

Description (modified from Robins and Robins, 1976). Naked ilyophine eel (see discussion below) with long abdomen (anus about one and two-thirds times the head length posterior to the level of the gill slits), and stout, conical teeth in a patch on premaxilloethmoid, in a roughly biserial arrangement on the vomer, these teeth surrounded by papillae which may equal the size of the teeth. Vertebrae 131–135.

Dorsal-fin origin far forward (above posterior third of small pectoral fin). Median fins confluent. Gill slits separate, very broadly crescentic, horizontally oriented and ventrally located on body. Fleshy tip of snout not overhanging tip of lower jaw to any significant extent. Gape moderate, extending only to rear margin of eye; cutaneous crease of gape about one eye diameter posterior to posterior border of eye. Eye well-defined, not deeply subcutaneous, although covered by skin. Lateral line short; open pores confined to front half of body. Cephalic lateralis pores: infraorbital pores 5, supraorbital 3, and preoperculomandibular 7.

Teeth not compound. Contiguous patches on premaxilla and on anterior half of vomer, followed by a single row of tiny teeth posteriorly on vomer. Premaxillary and anterior vomerine teeth set in bed of papillae which are about equal in size to the smaller teeth.

Distribution. Known with certainty only from the eastern Pacific. See discussion below.

Discussion. Known only from 5 (possibly 6) specimens, little can be said about the species aside from its physical characters and relationships based on character states. We have not seen the material described by Saldanha and Merrett (1982) and Merrett and Saldanha (1985) but the surface regimes, which seem to be very important in delineating distributions of many synaphobranchids, are very different in the northeastern Atlantic and the tropical eastern Pacific. This factor is much more important than mere geographic distance. In view of the differences in squamation noted by Saldanha and Merrett, there is reason to doubt the conspecificity of the two populations. The discussion by Merrett and Saldanha (1985: 734) of two Caribbean specimens with vertebral counts of 124 was based on *Haptenchelys texis.* Vertebral counts need to be assessed carefully. They often provide important species differences but sometimes distinct species of synaphobranchid eels have identical or similar counts. Identification of the northeastern Atlantic population as *Ilyophis arx* based on this character alone is unwarranted, especially in light of the differences in squamation. This remains an unresolved problem.

Ilyophis arx is no doubt part of the deep benthic, cosmopolitan fauna. It is rarely collected, but should be considered whenever naked ilyophine eels of unknown identity are collected.

Etymology. From the feminine Latin noun *arx* (a castle), an allusion to P. H. J. Castle. To be treated as an appositional noun.

Ilyophis blachei Saldanha and Merrett, 1982

?*Leptocephalus rostratus* Schmidt, 1909a: 5 (original description, 51°34′N, 11°50′W). Smith, 1974: 673, 680 (in key, discussion).

Ilyophis blachei Saldanha and Merrett, 1982: 623–636, figs. 2–5 (original description, northeastern Atlantic, 51°04′N, 11°59.3′W in 1494–1572 m, holotype BMNH 1981.3.17.1). Merrett and Saldanha, 1985: 739 (in discussion and key).

Distinctive characters (after Saldanha and Merrett, 1982). Vertebrae: 179–188. Proportional measurements expressed as percent of total or head (*) length: predorsal length 15.4–20.4; preanal length 26.7–31.9; head length 10.7–14.2; snout length* 32.8–38.6; horizontal diameter of eye* 10.2–12.7. Cephalic lateralis pores: infraorbital 7–8, supraorbital 5–6, preoperculomandibular 10–11. Lateral line essentially complete. Scales variably present: absent in some, some present in disjunct areas of others; scales ovoid-elongate, but not so slender or elongate as scales of *I. brunneus.*

Discussion. This species, specimens of which have not been examined by us, is included due to the cosmopolitan distribution of other species of synaphobranchid eels. Although described and known only from the eastern North Atlantic, its existence in the western North Atlantic in similar depths (1247–2070 m) is conceivable.

Etymology. For Jacques Blache, a notable student of eels. A noun in the genitive case.

Genus *Meadia* Böhlke, 1951

Meadia Böhlke, 1951: 6 (type species *Dysomma abyssale* Kamohara, by original designation).

Characters. Dorsal-fin origin far forward, above level of basal third of pectoral fin. Pectoral fins ovate, slightly less than snout length, set low on body. Anus far forward, less than one head length behind gill slit. Gill slits low on body, just ventral and anterior to base of pectoral fin. Eye well-developed, not deeply subcutaneous. Snout fleshy, slightly overhanging tip of lower jaw. Gape moderate; cutaneous crease of gape only moderately behind rear margin of eye. Cephalic lateralis system moderately well-developed: infraorbital pores 6, 3 supraorbital pores, and 7 preoperculomandibular pores. Lateral-line pores present on anterior third of body. Teeth not compound. Distinctive tooth pattern consisting of large canines (largest as long as tubule of anterior nostril) on the premaxilla in a patch, on the vomer in an irregular biserial series (becoming smaller and irregularly uniserial posteriorly). Maxillary teeth numerous, roughly triserial anteriorly; teeth on innermost row much enlarged.

Large dentary teeth biserial, those of inner row larger than those of outer row. When head is viewed in profile, with mouth slightly gaping, largest of the teeth in this series is visible. Pterygoid present. At least 14 branchiostegal rays.

Etymology. For Dr. Giles W. Mead. Feminine.

Species. One species, *Meadia abyssalis* (Kamohara).

Meadia abyssalis (Kamohara, 1938)

Dysomma abyssale Kamohara, 1938: 12, fig. 3 (original description, type locality Mimase Market, near Kochi City, Tosa Prov., Shikoku Island, Japan). Böhlke, 1949b: 36 (description, compiled). Kamohara, 1950: 50 (compiled). Kamohara, 1952: 20–21, fig. 13 (redescription, figure). Robins and Robins, 1970: 314 (discussion of history; *Meadia* referred to synonymy of *Dysomma*).

Meadia abyssale. Böhlke, 1951: 6 (made type of new genus *Meadia*). Kamohara, 1958: 15 (in catalog). Kamohara, 1961: 2 (in list of type specimens). Kamohara, 1964: 18 (compiled). Blache, 1975b: 805–810, figs. 1–3 (description of specimen from Reunion Island, Indian Ocean; genus resurrected).

Meadia abyssalis. Robins and Robins, 1976: 251–256 (redescription of genus and species, neotype designated FMNH 76869).

Distinctive Characters. Characters are those of the monotypic genus (see above); vertebrae 165–174.

Distribution. Known from the holotype (destroyed in the war), two specimens from Sagami Bay (including the neotype), and one specimen from near Reunion Island in the Indian Ocean. Available data indicate that this is an eel of moderate depths and one most likely to be collected in traps.

Relationships. Despite the very distinctive appearance of *Meadia abyssalis* and its strongly developed dentition, its relationships are much closer to *Ilyophis* than to *Atractodenchelys*, *Dysommina*, and *Dysomma*. All species of those three genera have the specialized or synapomorphic condition of having a reduced number of large compound teeth on the vomer, these arranged in a single row. In *Meadia* and *Ilyophis*, the vomerine teeth are in two rows, and, although large, all are simple.

Etymology. Adjectival form of the Greek word *abyssus* (of the deep sea, abyssal).

Genus *Atractodenchelys* C. H. Robins and C. R. Robins, 1970

Atractodenchelys Robins and Robins, 1970: 296–297 (type species *A. phrix* Robins and Robins, by original designation).

Characters (after Robins and Robins, 1970). Naked; pectoral fins present, moderately well-developed. Dorsal-fin origin far forward, above posterior half of pectoral fin. Anus far forward, slightly more than one head length behind level of gill opening. Caudal fin slightly emarginate, distinct, but connected by a low fold to dorsal and anal fins. Gill slits small, vertically oriented crescent, placed low on side of body. Lower jaw distinctly shorter than the fleshy and papillose snout. Lateral line well-developed, pores extending nearly to caudal-fin base. Cephalic lateralis pores: infraorbital 6, supraorbital 4, preoperculomandibular 10. A patch of medium-sized conical teeth present on premaxilloethmoid. Vomerine teeth consisting of five large conical teeth, each formed by fusion of two teeth, the tip of each fused pair slightly expanded. Maxillary and dentary teeth small, numerous. Osteological characters: pterygoid present, suspensorium nearly vertical, basisphenoid present, not fused to parasphenoid.

Etymology. From the Greek *atractos* (arrow), *odus* (tooth), and *enchelys* (an eel). Feminine, fide Opinion 915, International Commission of Zoological Nomenclature. The describers' declaration that it is masculine is invalid.

Species. One species, *Atractodenchelys phrix.*

Atractodenchelys phrix C. H. Robins and C. R. Robins, 1970

Figures 226–229

Atractodenchelys phrix Robins and Robins, 1970: 297–299 (original description, 11°36′N, 62°52′W, in 385–425 m, holotype ANSP 114437, immature female, 380 mm TL). Karrer, 1982: 99 (compared with *Ilyophis brunneus*).

Description. Inasmuch as there is only one species in the genus, the salient characters of *A. phrix* are those of the genus. The following mea-

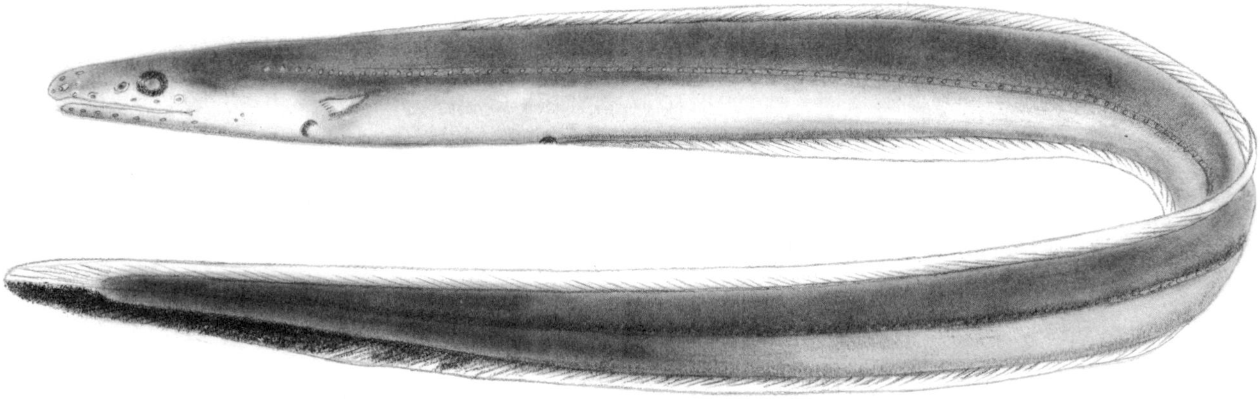

FIGURE 226. *Atractodenchelys phrix*: ANSP 114437, holotype, 380 mm TL; eastern Caribbean.

surements and counts are those of the type material. For detailed treatment of this rare eel, see Robins and Robins (1970). Vertebrae 168–173. Proportional measurements expressed as percent of total or head (*) length: predorsal length 11–12; preanal length 18–21; depth of body 3.5–5.0; snout length* 33–37; jaw length* 48–57; eye diameter* 13–16. Cephalic lateralis pores: infraorbital 6, supraorbital 4, preoperculomandibular 10 (corrected from original description, see Robins and Robins, 1976).

Body naked, elongate.

Five large compound vomerine teeth. Premaxillary teeth in patch of about 12 moderately large, grooved or compound teeth. Maxillary teeth numerous, irregularly biserial anteriorly and triserial posteriorly. Teeth of dentary slightly larger, in 3–4 very irregular rows. Innermost teeth in maxillary and dentary rows slightly larger than those in outer rows.

Color. Sharply bicolored for most of its length (dorsum dark brown, ventral surface much paler). Anal fin darkens posteriorly and is black in the section just anterior to caudal fin.

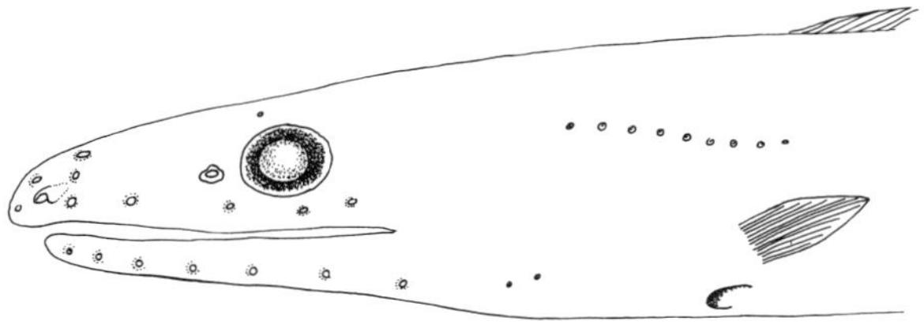

FIGURE 227. *Atractodenchelys phrix*: lateral view of head. Data as for Figure 226.

Distribution (Fig. 229). All records are from the Caribbean Sea and the eastern ccast of Florida.

Discussion. Atractodenchelys phrix, with its large compound teeth, is further along the line of ilyophine evolution toward reduction of number of vomerine teeth, coupled with an increase in

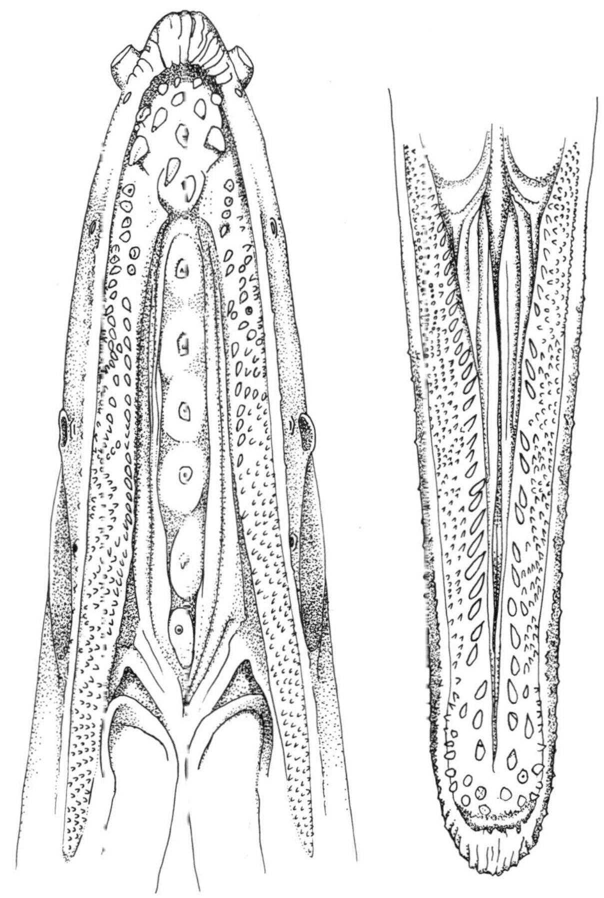

FIGURE 228. *Atrcctodenchelys phrix*: dentition. Data as for Figure 226.

their size. The extreme condition is seen in *Dysomma*. Like *Ilyophis* and *Meadia*, *Atractodenchelys* retains the premaxillary patch of teeth, but *Ilyophis* and *Meadia* lack the specialized compound vomerine teeth. *Atractodenchelys* has more numerous and better developed premaxillary teeth than *Dysomma* and *Dysommina*, which generally possess only two or lack them completely.

Little is known about *A. phrix*, other than the fact that it is a relatively shallow-water species (366–730 m) as would be inferred also from its pigmentation. A 564 mm paratype collected in October was ovigerous.

Etymology. From the feminine Greek noun *phrix* (a ruffling or a ripple), in allusion to the plicate snout. To be treated as an appositional noun.

Study Material (in addition to the material reported on by Robins and Robins, 1976). USNM 158938 (1, 607) 16°41′N, 81°02′W, Oregon sta. 1888, in 450 m, 23 Aug. 1957. USNM 197340 (2, 419–551) 24°48′N, 79°17′W, in 201 m, Silver Bay sta. 3475, 8 Nov. 1960. USNM 270794 (1, 346, incomplete) 09°N, 80°23′W, in 450 m, Oregon sta. 3599.

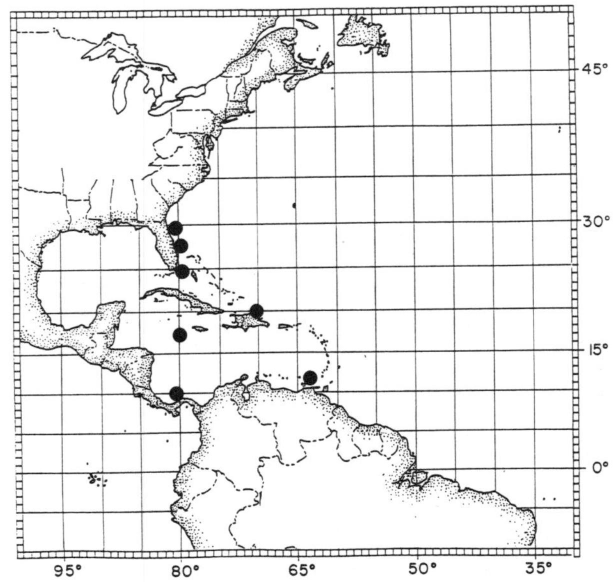

FIGURE 229. Distribution of *Atractodenchelys phrix*.

Genus *Dysommina* Ginsburg, 1951

Dysommina Ginsburg, 1951: 450 (type species *Dysommina rugosa* Ginsburg, by original designation and monotypy).

Characters. Naked ilyophine eels with small, broad-based pectoral fins, their tips not sharply pointed. Dorsal-fin origin forward, about one head length behind a vertical through the gill slits. Head not measureably larger than in *Dysomma*, but musculature of nape better developed so that head appears large. Head not laterally compressed. Fleshy, papillose snout slightly overhangs tip of lower jaw. Eye circular, covered by skin but not deeply subcutaneous. Gape short, extending only to vertical from posterior margin of eye or very slightly behind that point. Cutaneous crease of gape may extend one eye diameter farther posterior. Cephalic lateralis pores: infraorbital 5, supraorbital 3, preoperculomandibular 6. No open pores in lateral line on body but some shallow, slitlike depressions along lateral line. Gill slits separate, small (about equal to diameter of eye), crescentic, oriented nearly vertically on the ventrolateral surface.

Osteological characters: Pterygoid present, short, robust, near to quadrate but without bony connection (as in *Ilyophis* and *Atractodenchelys*). Frontals fused; parietals, supraoccipital, and epiotics strongly developed and elevated posteriorly, providing a large surface on top and rear of cranium for muscle attachment, this reflected in the bulging nape region of this genus. Eye socket reduced by posterior expansion of ethmoid. Basisphenoid fused in part with parasphenoid.

Etymology. Diminutive of *Dysomma*. Treated by Ginsburg as feminine.

Species. One species, *Dysommina rugosa*.

Dysommina rugosa Ginsburg, 1951
Figures 230–232

Dysommina rugosa Ginsburg, 1951: 450–451 (original description, off Cumberland Island, Georgia, 30°53′N, 79°04′30″W, in 500 m, holotype USNM 131594, 196 mm TL). Böhlke and Hubbs, 1951: 7–10, fig. 1 (type of new family Dysomminidae). Nelson, 1966a: 392 *et seq.* (osteology of gill arches, mostly attributed solely to *Dysommina* or Dysomminidae of which this is the sole species). Robins and Robins, 1970 (redescription; referred to family Dysommidae; comparative osteology).

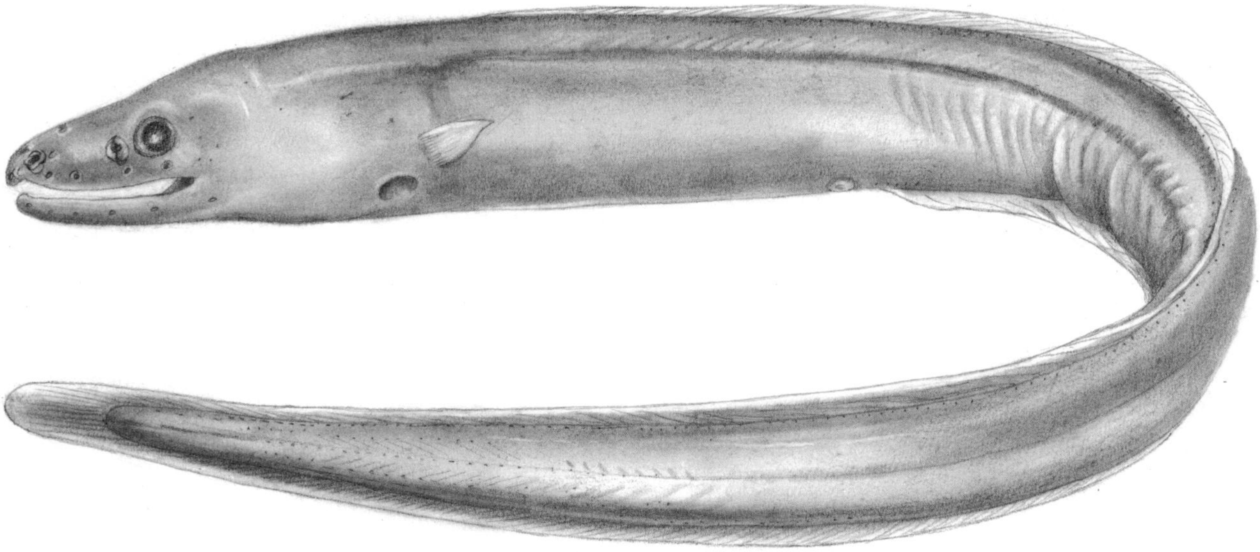

FIGURE 230. *Dysommina rugosa*: ANSP 137427, 296 mm TL; Colombian Caribbean.

Smith, 1974: 678 (discussion of leptocephali, see comments below). Robins and Robins, 1976: 264 (recorded from off Hawaii; placed in family Synaphobranchidae, subfamily Dysomminae). ?Karrer, 1982: 96–98, fig. 29 (description of material from the Mozambique Channel).

Leptocephalus proboscideus: (not of Lea, 1913) Smith, 1974: 678 (description of 2 metamorphosing leptocephali).

Discussion of Synonymy. Robins and Robins (1976: 264) agreed that the two metamorphosing leptocephali described by Smith (1974: 678) were *Dysommina rugosa,* but gave reasons not to identify them with *Leptocephalus proboscideus* Lea, a taxon which is not yet identifiable with certainty with any known adult, though it is likely a species of *Dysommina.* The possibility of a second species and small differences (see discussion below) between the description of the specimen from the Mozambique channel (Karrer, 1982: 96–98) and those from the western Atlantic and Hawaii causes us to question the identity of the specimen from the Mozambique Channel. No direct comparison has been made by Karrer or by us.

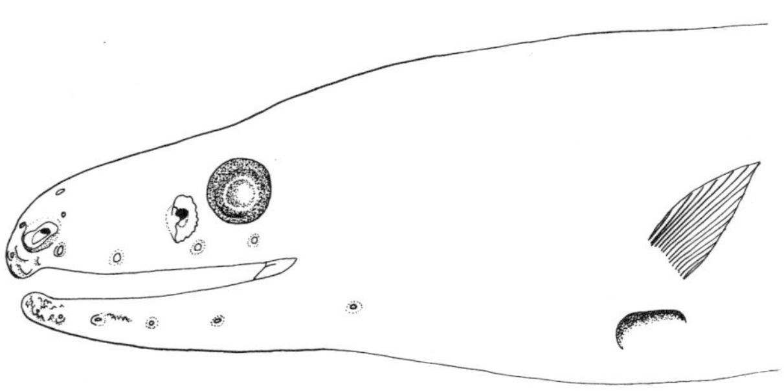

FIGURE 231. *Dysommina rugosa*: lateral view of head. Data as for Figure 230.

Description. Vertebrae 127–134 (138 in specimen from Mozambique Channel; see comments below). Measurements, based on Robins and Robins, 1970, 1976, expressed as percent of total or head (*) length: predorsal length 17–20; preanal length 26–33; depth of body 4.8–8.6; head length 12–15; snout length* 31–39; jaw length* 48–56; eye diameter* 9.5–14; length of gill slit* 3.6–15.

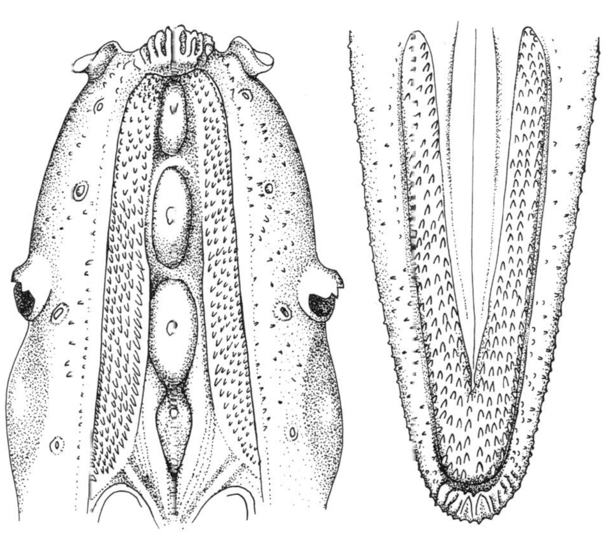

FIGURE 232. *Dysommina rugosa*: dentition. Data as for Figure 230.

Cephalic lateralis pores: infraorbital 5, supraorbital 3, preoperculomandibular 6.

Salient features are those of the generic diagnosis, since the genus is currently monotypic and all characters given are those of *D. rugosa.*

Dentition: Four large, compound vomerine teeth, aligned in one row, each tooth enclosed in a pad of papillose tissue. No premaxillary teeth (front part of ethmoid little developed). Maxillary and dentary teeth numerous, tiny, set in many, irregular rows.

Color. Plain brown in preservative. Dorsum and head subtly paler than remainder of body. Median fins and pectoral fins pale, except for lower half of caudal fin, which is very dark brown to blackish. Peritoneum black; gut pale, pinkish tan.

Size. Juveniles and adults range from 85–352 mm.

Distribution. Widespread in tropical latitudes in 260–775 m. In the western Atlantic it is known from the American Bight off the Carolinas and Georgia, and from localities along the Florida coast and in the Caribbean Sea. It is also known from Hawaii (Robins and Robins, 1976: 204) in 774 m, and (see comments above) from the Mozambique Channel, southwestern Indian Ocean, in 280–310 or 640–720 m (Karrer, 1982: 96).

Discussion. As previously noted by Robins and Robins (1976: 264) there seem to be two species represented by leptocephali that can be attributed to *Dysommina.* Both have rostral filaments (a feature almost certainly not unique to *Dysommina* in that Smith (this volume) recognizes 17 different leptocephali of this type), and similar myomere counts, but different pigment patterns. The one described by Smith (1974: 268) is *D. rugosa,* the other *Leptocephalus proboscideus* Lea (1913). The adult from the Mozambique Channel, at least as illustrated, appears to have a slightly different shape to the head, a smaller eye, a more anterior origin of the dorsal fin, a smaller and more paddleshaped pectoral fin, and slightly higher vertebral count (138). For these reasons we question whether it represents the same species as *D. rugosa.*

The similarities in dentition between the monotypic genus *Dysommina* and relatively speciose *Dysomma* might indicate congeneric relationship to some, but the osteological differences are striking, and the larvae are very different as well, at least from those that have been certainly identified with species of *Dysomma.*

Etymology. From the Latin adjective *rugosus* (full of wrinkles or folds) in reference to the condition of the snout. An adjective.

Study Material (exclusive of that listed by Robins and Robins, 1970: 316). USNM 131974 (1, 196; holotype) 30°51′N, 79°04′30″W, in 496 m, Albatross sta. 2667, 5 May 1886. USNM 270795 (1, 318) 15°45′N, 80°45′W, in 654 m, Oregon II sta. 10185, 18 Nov. 1968. MCZ 39483 (4, 133–167) and MCZ 39496 (9, 87–141) 30°21′N, 79°55′W, in 418–454 m, Atlantis sta. 3779, 24 Feb. 1940. MCZ 39486 (1, 167) Atlantis sta. 3781, 24 Feb. 1940.

Genus *Dysomma* Alcock, 1889

Dysomma Alcock, 1889: 459 (type species *D. bucephalus* Alcock, by monotypy).

Dysommopsis Alcock in Wood-Mason and Alcock, 1891: 137 (type species *Dysommopsis muciparus* Alcock, by monotypy).

Todarus Grassi and Calandruccio, 1896: 349 (type species *Nettastoma brevirostre* Facciola, by monotypy; generic name preoccupied).

Sinomyrus Lin, 1933: 93 (type species *S. angustus* Lin, by monotypy, species a synonym of *Dysomma anguillare* Barnard).

Nettodarus Whitley, 1951b: 407 (replacement name for *Todarus,* preoccupied; same type species).

Characters. Dorsal-fin origin far forward, from about one pectoral-fin length in front of level of the pectoral-fin base to a point about one-half head length behind level of gill slit. Pectoral fin present or absent. When present, base close to dorsal end of gill slit. Anus far forward (from a point directly below the gill slits to about one-half head length behind the gill slits). Gill slits small, restricted in all species except *D. bucephalus.* Snout in all species fleshy, plicate, and papillose, overhanging tip of lower jaw. Gape moderate to long; usually long. Cephalic lateralis pores few, often difficult to see because of fleshy and papillose nature of head. Premaxillary teeth (when present) and vomerine teeth compound-fused, each a pair fused front to back (as opposed to compound condition with the bases of the coalesced teeth still distinct, the external grooves obvious). Vomerine teeth large, 4–5, each set in individual pad of oral epithelium. Transversely oriented pair of premaxillary teeth (when present) slightly smaller than vomerine teeth.

Entire skull streamlined, frontals united (as in all synaphobranchids). Orbit reduced. Pterygoid absent. Basisphenoid fused with parasphenoid in all species examined. Suspensorium angled posteriorly.

Etymology. From the Greek *dys-* (bad) and *omma* (eye). Neuter.

Key to the Species of *Dysomma*

(Modified from Robins and Robins, 1976: 257.)

1a. Pectoral fins absent 2
1b. Pectoral fins present 5
2a. Snout tip and chin bulbous, ornamented with conspicuous papillae and ridges; vertebrae 195–205 . . . . *D. brevirostre*, p. 247
2b. Snout tip tapered, only modestly ornamented (papillae not obvious without magnification) or smooth; vertebrae 151–175 3
3a. Trunk shorter than head length; snout modestly ornamented; vertebrae 153–157 . . . . *D. muciparus*, p. 248
3b. Trunk 1.5 times length of head 4
4a. Snout smooth, no compressed premaxillary teeth projecting from snout; vertebrae 146–153 . . . . *D. dolichosomatum*, p. 249
4b. Snout rugose, with some papillae; 3 large, compressed premaxillary teeth projecting downward from tip of snout in front of lower jaw; vertebrae 175 . . . . *D. tridens*, p. 250
5a. Premaxillary teeth absent, dentary teeth multiserial, many and small (except those on inner row sometimes enlarged) 6
5b. Premaxillary teeth present, dentary teeth in one row, at least some of them enlarged 7
6a. Dorsal-fin origin in front of a vertical through pectoral-fin base; 4 large, compound vomerine teeth; lateral line short, confined to anterior fourth of body, with about 27 pores; vertebrae 130–131 . . . . *D. goslinei*, p. 249
6b. Dorsal-fin origin behind pectoral-fin base; 5 large, compound vomerine teeth; lateral line long, extending to within a head length of caudal fin, with about 110 inconspicuous pores; vertebrae 137 . . . . *D. melanurum*, p. 249
7a. Anus immediately behind a vertical from gill opening, under base or tip of pectoral fin; many dentary teeth 8
7b. Anus well behind a point below tip of appressed pectoral fin; about 10 very large dentary teeth; vertebrae 119–130 . . . . *D. anguillare*, p. 245
8a. Anus under base or anterior third of pectoral fin; vertebrae 107 . . . . *D. bucephalus*, p. 248
8b. Anus under tip of pectoral fin, vertebrae 107 . . . . *D. polycatodon*, p. 250

Only those species reported from the western North Atlantic will be described and discussed in detail. However, considering the occasionally unexpected, surprising distributions of other eels, and the great disjunctions of known distributions of some species in this genus, all other species are briefly diagnosed.

Dysomma anguillare Barnard, 1923
Figures 233–235

Dysomma anguillaris (sic) Barnard, 1923: 443 (original description, off Tugela R., Natal). Barnard, 1925: 195–196, pl. 8, fig. 8 (redescription). Chen and Weng, 1967: 82–83, fig. 62 (characters, synonymy; 7 specimens from Taiwan).

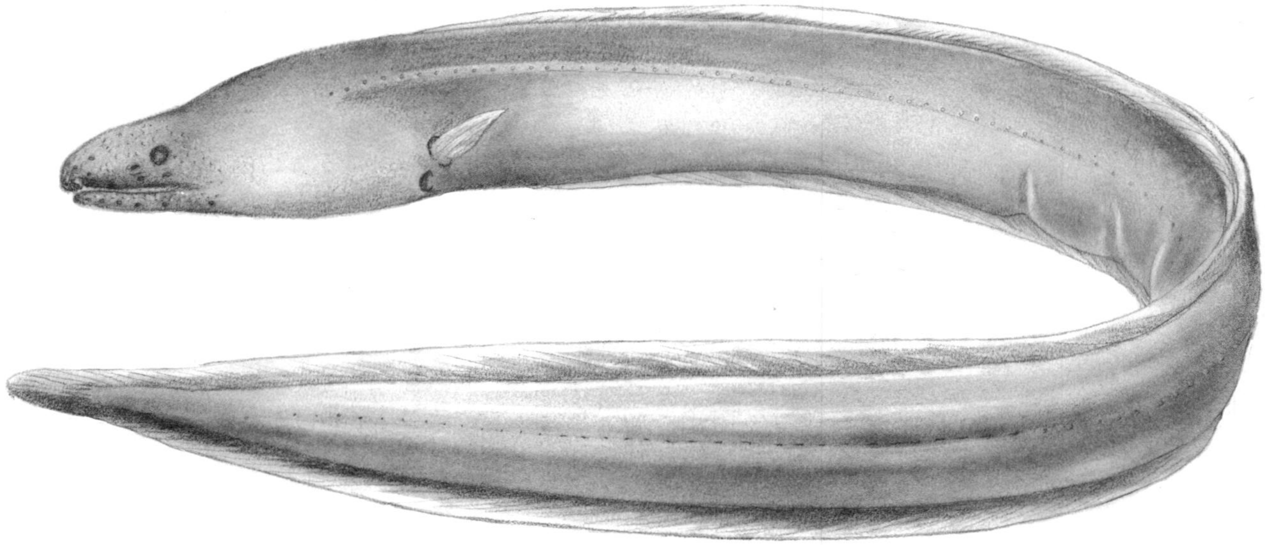

FIGURE 233. *Dysomma anguillare*: ANSP 114451, 227 mm TL; Suriname.

Sinomyrus angustus Lin, 1933: 93, fig. 1 (original description, Hoitow, Hainan). Böhlke, 1949b: 34–36 (placed in synonymy of *D. anguillare*).

Dysomma japonicus (sic) Matsubara, 1936: 900, fig. (original description, Japan, Kumano Nada, southeast of Kii Peninsula). Böhlke, 1949b: 34, 36 (placed in synonymy of *D. anguillare*).

Dysomma zanzibarensis (sic) Norman, 1939: 44, fig. 17 (original description, 5°38′54″S, 39°15′42″–17′36″E, in 183–194 m). Böhlke, 1949b: 34–36 (placed in synonymy of *D. anguillare*).

Dyssoma anguillaris. Smith, 1949: 394 (genus misspelled; compiled, stated to be possibly the same as *D. bucephalus*). Smith, 1961: 394 (repeat of earlier edition).

Dysomma anguillare: Böhlke, 1949b: 36–37 (redescription, synonymy; adjectival ending emended to agree with *Dysomma*, neuter). Nelson, 1966a: 392 *et seq.* (osteology of gill arches, mostly attributed simply to *Dysomma* but based solely on *D. anguillare*). Robins and Robins, 1970 (redescription, synonymy). Robins and Robins, 1976: 263–264 (summary of characters, relationships). Karrer, 1982: 93 (comparison with other Indian Ocean species). Gloerfelt-Tarp and Kailola, 1984: 56–67, 304 (collected at 08°27′S, 117°47′E; fig. listed as *Dysomma* cf *anguillare*).

Dysomma aphododera Ginsburg, 1951: 451–453, fig. 7 (original description; off Padre Island, Texas, in 91 m, 26°30′N, 96°26′W; holotype USNM 154992).

Dysomma angustum. Kamohara, 1950: 49, fig. 43 (includes *D. japonicus* in synonymy). Kamohara, 1952: 20 (characters; said to be possibly the same as *D. anguillaris*). Kamohara, 1958: 15 (in catalog; range). Kamohara, 1964: 18 (compiled; fig. 6, including insert figure of dentition).

Description (in part modified from Robins and Robins, 1976). Vertebrae 119–130; precaudal 58–

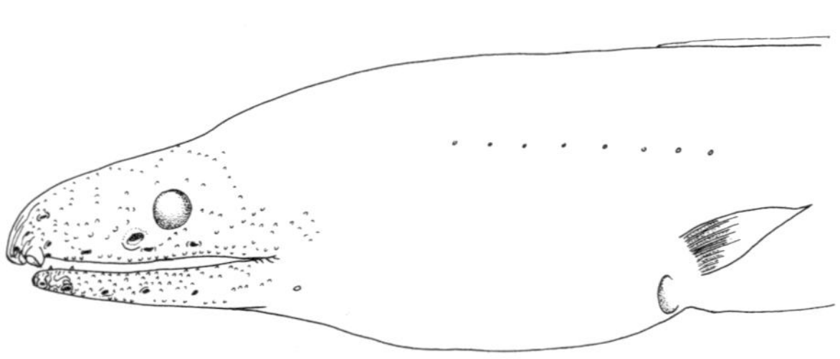

FIGURE 234. *Dysomma anguillare*: lateral view of head. Data as for Figure 233.

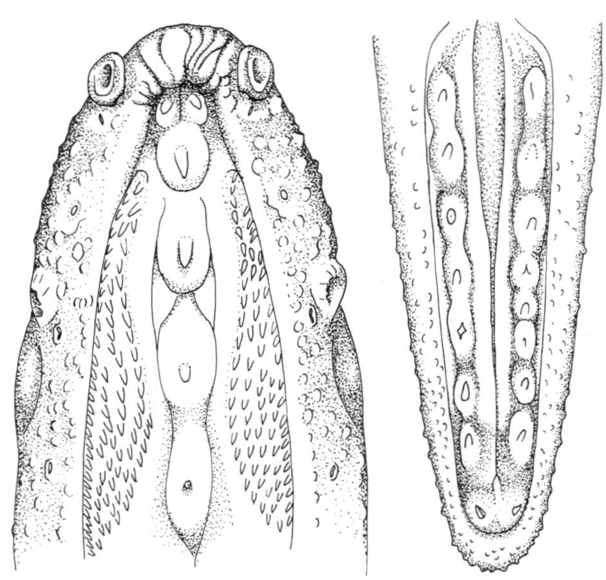

FIGURE 235. *Dysomma anguillare*: dentition. Data as for Figure 233.

63, caudal 59–66. Caudal-fin rays 10–12. Pectoral-fin rays 10–11. Ratio of dorsal-fin rays to vertebrae about 2.5–3.0 to 1. Proportional measurements expressed as percent of total or head (*) length: predorsal 11–14; preanal 16–22; head 12–16; snout* 21–28; jaw* 36–44.

Pair of premaxillary teeth, transversely oriented. Vomer with four compound teeth. Maxillary teeth tiny, irregularly multiserial; dentary teeth few (about 10), widely spaced, large. Cephalic lateralis pores: infraorbital pores 5, supraorbital pores 3, mandibular pores 6. Few inconspicuous ridges and papillae on snout. Dorsal-fin origin directly above base of pectoral fin.

Color. Pallid; tan dorsally, whitish below. Caudal fin quite dark on ventral half.

Size. 110–426 mm TL.

Distribution. Western North Atlantic (Gulf of Mexico, Florida, Caribbean Sea), western North Pacific (Philippines, Japan, China), usually in shallow warm waters, ordinarily 100 m or less in depth. Recently, Gloerfelt-Tarp and Kailola (1984: 304) reported on material from the Lesser Sunda Islands, Indonesia.

Etymology. From the adjective *anguillaris,* meaning eel-like. An adjective.

Study Material. All material was listed by Robins and Robins, 1970: 316–317.

Dysomma brevirostre (Facciolà, 1887)
BATNOSE EEL
Figure 236

Nettastoma brevirostre Facciolà, 1887: 166 (original description, Sicily, near Messina). Carus, 1893: 543 (characters summarized). Griffini, 1903: 171 (compiled).

Todarus brevirostris. Grassi and Calandruccio, 1896: 349 (placed in new genus *Todarus;* the generic and specific names not actually used together in this paper, but such a combination implied). Supino, 1905c: 255–259 (detailed description of specimen from Sicily). Grassi, 1913: 29, 170, pl. 10 (possible larval stages described and figured; development; caudal structure).

Nettodarus brevirostris. Whitley, 1951b: 407 (placed in new genus, *Nettodarus*). Böhlke and Robins, 1968: 477–480 (redescription; recorded from western Atlantic, off Florida; relationships). Robins and Robins, 1970: 315–316 (characters; referred to family Dysommidae). Blache et al., 1970 (description of larvae and adults from eastern Atlantic). Bini, 1971: 255–258 (description, color figure; status in Mediterranean Sea). Blache et al., 1973 (synonymy and bibliography for northeastern Atlantic and Mediterranean). Smith, 1974: 674 (discussion of larvae). Blache, 1977: 187–188 (leptocephalus from Gulf of Guinea).

Dysomma brevirostre. Robins and Robins, 1976: 256–257 (*Nettodarus* synonymized with *Dysomma,* placed in Synaphobranchidae). Smith and Castle, 1981: 460 (recorded from Hawaii).

Dysomma brevirostris. Karrer, 1982: 92 (in discussion of relationships).

Comment. This species has been fully discussed in the literature cited in the synonymy, especially that by Böhlke and Robins (1968), Robins and Robins (1970), and Bini (1971). No new information or metamorphosed material are available from the western Atlantic.

Description. Vertebrae 193–204. Cephalic lateralis pores: infraorbital pores four, mandibular pores six (none in preopercular portion of the preoperculomandibular canal). Lateral line short, with about 10 pores anteriorly on body. Proportional measurements expressed as percent of total length of the single western Atlantic specimen: predorsal 10, preanal 18, head 7.4, depth at gill opening 2.3, depth at anus 2.4, eye 0.6, snout 2.0, jaw 2.8. Pair of transversely oriented teeth on premaxilloethmoid followed by a median row of five large, compound teeth on the vomer. Lower jaw with three large teeth followed by a row of tiny teeth. No teeth on maxilla. Dorsal-fin origin behind a vertical from gill slits by a distance equal to slightly more than the length of the snout. Anus far forward, slightly more than one head length posterior to level of the gill slits. Body extremely compressed laterally, most slender and elongate species in the family Synaphobranchidae. Snout bulbous, adorned with ridges, papillae, and fleshy tabs; utterly distinctive. Pale tan.

Color. Body and head uniform brown, except mid-dorsum and dorsal fin pale, ventral edge of body near tail tip dusky.

Size. 123–245 mm.

Distribution. Mediterranean Sea, eastern and western Atlantic, and the central Pacific (Hawaii). Mostly between 200–1000 m.

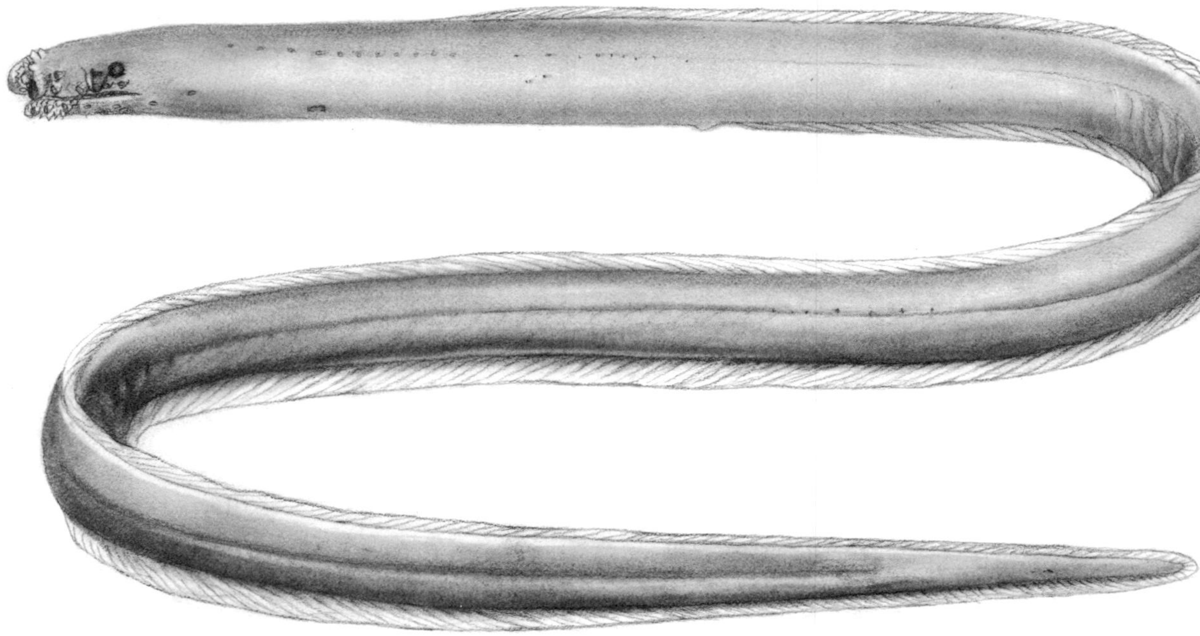

FIGURE 236. *Dysomma brevirostre*: ANSP 108623, 209 mm TL; Florida.

Etymology. From the Latin *brevis* (short) and an adjectival form of *rostrum* (snout). To be treated as a compound adjective.

Study Material. One specimen from the western Atlantic. ANSP 108623 (1, 209), off Miami, Florida, 25°31′30″N, 79°57–56′W, 351 m, by try net; Gerda sta. 67; 26 Sept. 1962.

Dysomma muciparus (Alcock, 1891)

Dysommopsis muciparus Alcock in Wood-Mason and Alcock, 1891: 137–138 (original description, Bay of Bengal, RIMS Investigator sta. 120, 440–505 m, lectotype ZSI 13107). Menon and Yazdani, 1968: 134 (list of syntypes). Robins and Robins, 1970: 315 (synonymy, compiled).

Dysomma muciparus: Robins and Robins, 1976: 259–261, figs. 1C, 2 (redescription; referred to genus *Dysomma;* lectotype designated). Smith and Castle, 1981: 460–461 (recorded from Hawaii).

Dysomma muciparum: Karrer, 1982: 91 (comparison with other Indian Ocean species: species name improperly modified, see below).

Distinctive Characters. An eel of the genus *Dysomma* with 5 large vomerine teeth, a transversely oriented pair of premaxillary teeth, and with 4 large teeth toward the front of the dentary followed by one row of small teeth. Snout moderately ornamented. Pectoral fins absent in the metamorphosed juvenile and adults. Vertebrae about 153–162.

Size. 164–257 mm.

Distribution. Not yet known from the western North Atlantic. Taken in the North Pacific (Hawaii) and Indian Ocean (Bay of Bengal) in less than 1000 m (known depth records: 440–505 m, 284 m). Robins and Robins (1976: 261) tentatively identified two leptocephali from off Gabon and Angola in the eastern Atlantic with this species.

Etymology. Robins and Robins (1976: 260) noted that Alcock did not explain the derivation of the species name but that they regarded it as coming from the Latin noun for mucus and the verb meaning to bear; it is thus a gerund or verbal noun and its ending is not to be modified.

Dysomma bucephalus Alcock, 1889

Dysomma bucephalus Alcock, 1889: 459 (original description, Bay of Bengal, 20°17′30″N, 88°51′W, holotype ZSI 11675). Alcock in Wood-Mason and Alcock, 1891: 137, fig. 5 (compiled). Alcock, 1892a: pl. 6, fig. 1 (figure only). Alcock 1899: 192–193 (description). Böhlke, 1949: 37–38 (description, synonymy, relationships). Menon and Yazdani, 1968: 134 (in list of type specimens). Robins and Robins, 1976: 257–259, fig. 1B,D (redescription, relationships).

Dysomma bucephalum: Karrer, 1982: 93 (descriptive comments on original series; comparison with other Indian Ocean species; specific name improperly modified, see below).

Distinctive Characters. An eel of the genus *Dysomma* with deeply subcutaneous, small eye, an extremely short abdomen (anus immediately behind the enormous gill slits), and with dorsal-fin origin near nape, in front of a vertical from the pectoral-fin base by a distance about equal to the length of that fin. A pair of transversely oriented teeth on the premaxilla; 4 enlarged compound teeth on the vomer; maxilla with tiny, irregularly arranged teeth along ventral surface; dentary teeth small, uniserial, widely spaced, and about 35–40 in number. Branchial region much enlarged, the gill slits huge. Vertebrae 107.

Size. 200–300 mm.

Distribution. Known only from 2 extant specimens from the Bay of Bengal. Alcock apparently had a larger series according to Karrer (1982: 93) who noted that Alcock cited 6 catalog numbers. The known depth range is 205–505 m.

Etymology. From the prefix *bu* (huge) and the Latinized noun *cephalus* (head). To be treated as an appositional noun, and, therefore, not to be modified. *Bucephalus,* meaning oxhead, is a classical name and was the one applied to the war horse of Alexander the Great. If such an allusion was intended, the word would still require treatment as a noun.

Dysomma goslinei C. H. Robins and C. R. Robins, 1976

Dysomma goslinei Robins and Robins, 1976: 261–262, figs. 3, 6–8 (original description, Indian Ocean, 14°52′N, 96°39′W, holotype ANSP 133805). Karrer, 1982: 93 (compared to *D. polycatodon*).

Distinctive Characters. No premaxillary teeth; dentary teeth all small; 4 large, conical, compound teeth on vomer; teeth on maxilla small, irregularly multiserial, inner row enlarged, but still small. Dorsal-fin origin in front of a vertical through pectoral-fin base. Anus far forward, abdomen short. Lateral line short. Pectoral fins present. Vertebrae 130–131, precaudal and caudal vertebrae about equal in number.

Distribution. Known from the original series of three specimens taken in the Indian Ocean on Anton Bruun Cruise I, sta. 39A.

Size. The types are newly transformed juveniles 106–121 mm.

Etymology. For Dr. William A. Gosline. A noun in the genitive case.

Dysomma melanurum Chen and Weng, 1967

Dysomma melanurum Chen and Weng, 1967: 84, fig. 63 (original description, Tungkong, Taiwan, holotype THUP 1687). Robins and Robins, 1970: 314 (discussion). Robins and Robins, 1976: 262–263, fig. 1 (redescription, relationships). Karrer, 1982: 93 (in discussion).

Distinctive Characters (from Robins and Robins, 1976: 262). Snout exceptionally sharp-pointed. No premaxillary teeth; 5 large vomerine teeth; dentary teeth tiny and numerous. Some bands of maxillary and dentary teeth visible when viewing the head from the side. Anus far forward, below tip of appressed pectoral fin. Dorsal-fin origin very slightly in advance of a vertical through anus. Snout slightly overhanging tip of lower jaw. Vertebrae 137.

Discussion. This species was described in detail by Robins and Robins (1976: 262–263). No new material has been collected. Karrer (1982: 93) emended our count of pores in the preoperculomandibular canal from 5 to 6 and recorded 120 lateral-line pores on the left side.

Distribution. Known only from the holotype, a 276 mm ovigerous female, from Taiwan.

Etymology. From the Greek *melanos* (black) and *oura* (tail). Treated by the describers as a compound adjective.

Dysomma dolichosomatum Karrer, 1982

Dysomma dolichosomatum Karrer, 1982: 93–96, fig. 28 (original description, Indian Ocean, Mozambique Channel, 22°25′S, 43°04.5′E, in 550–555 m, holotype MNHN 1979-4).

Distinctive Characters (after Karrer, 1982, but modified based on study of material listed below). Premaxilla with 2 compound teeth, one on each side, largely covered by soft tissue. Vomer with 4 large, median, compound teeth, shorter and more massive than in *D. polycatodon.* Maxilla with 3 rows of teeth, starting opposite 2nd vomerine tooth and ending in front of the corner of the gape; all conical, simple, and rather blunt. Dentary with a band of small teeth posteriorly but with 5 large, compound teeth, deeply set in soft tissue, anteriorly. Entirely pale. No pectoral fins. Dorsal fin begins above a point slightly nearer gill opening than anus. Head small, occipital length 4 times in predorsal distance. Preanal distance about 3½ times in total length. Vertebrae 146–153 (146 and 149 in the specimens from China, 151–153 in the types).

Distribution. Known from the Mozambique Channel off the coast of the Malagasy Republic and in the western Pacific (Lat. 14°01′N, Long. 120°22′E). The depth range is 174–555 m.

Etymology. From the Greek *dolichos* (long) and the adjectival form of *soma* (body). A compound adjective.

Study Material. USNM 270682 (2, 172–184) China, 23 Nov. 1969 (no precise locality data).

Dysomma polycatodon Karrer, 1982

Dysomma polycatodon Karrer, 1982: 89–93, fig. 27 (original description, Indian Ocean, Mozambique Channel, 15°21′S, 46°11′E in 170–175 m, holotype MNHN 1979-3).

Distinctive Characters (based on Karrer, 1982). Two transversely oriented premaxillary teeth, 5 large, uniserial vomerine teeth, surrounded by pads of papillose tissue. Premaxillary and vomerine teeth compound-fused. Anteriormost dentary teeth enlarged, followed by about 20–25 much smaller teeth. Maxillary teeth small, irregularly triserial and recurved. Pectoral fin present. Dorsal-fin origin far forward, on nape, about one snout length anterior to a vertical through pectoral-fin base. Vertebrae 140.

Distribution. Known only from the holotype collected northwest of the Malagasay Republic in the Mozambique Channel (see Synonymy for details).

Etymology. From the Greek words *poly* (many), *kata* (under, below), and *odon* (tooth). To be treated as a noun.

Dysomma tridens C. H. Robins, E. B. Böhlke, and C. R. Robins, new species
Figures 237–240

Diagnosis. A naked ilyophine eel of the genus *Dysomma* in which three large, laterally compressed premaxillary teeth of peculiar shape project below tip of fleshy snout. Three large median, compound teeth on vomer, many small teeth on maxillae, and four large compound teeth on front part of dentary followed by many markedly smaller, simple teeth. Dorsal-fin origin located about one snout length behind a vertical through the gill opening. Pectoral fin absent. Trunk short, body elongate (tail about 80% of total length). Head markedly convex in profile. Fleshy tip of snout (and the peculiar premaxillary teeth) overhang tip of lower jaw. Gape long (ca 1.5 times the snout length), extending beyond a vertical through the rear margin of the eye by a distance equal to about two times the eye diameter. Eye small, dark, rounded. Anterior nostrils tubular, posterior slitlike. Snout plicae and papillae moderately developed. Total vertebrae 175; predorsal vertebrae 14; preanal vertebrae 28.

Description. Measurements of the holotype, expressed as percent of total or head (*) length: predorsal 9.7; preanal 17.5; tail 82.5; head 6.8; depth of body at gill opening 2.3; depth of body at anus 2.1; width of body at gill opening 1.3; width of body at anus 1.2; snout to corner of gape* 37.4; snout tip to tip of lower jaw* 7.8; snout length* 22.8; eye diameter* 4.1; interorbital distance* 16.

The dorsal-fin origin is far forward, behind a vertical through the gill opening by a distance equal to one snout length. The anal-fin origin is behind the anus by a distance equal to about two times the eye diameter. The dorsal, caudal and anal fins are continuous. Pectoral fins are absent but long cleithra and supracleithra can be seen on the radiograph. All other fins are moderately

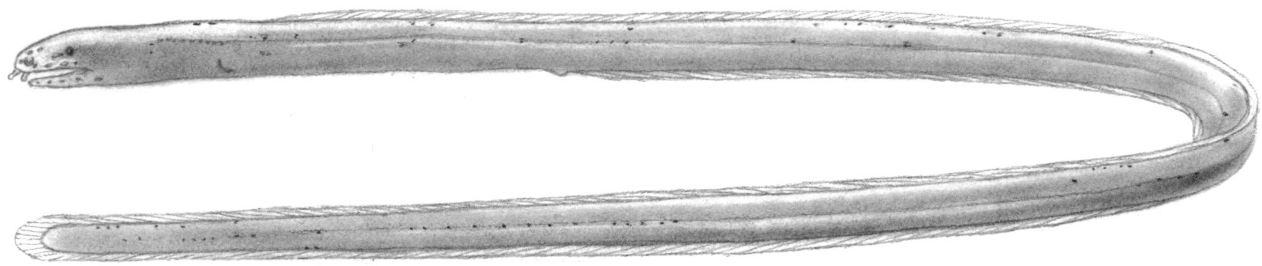

FIGURE 237. *Dysomma tridens*: USNM 193563, holotype, 217 mm TL; off Belize.

well-developed and somewhat fleshy at their bases. The gill openings are separate, low on body, more ventral than lateral. This region was damaged in the specimen. The gill opening, as reconstructed, is indicated by a broken line in Figure 238.

As is characteristic of most species of *Dysomma*, the fleshy tip of the snout overhangs the tip of the lower jaw, in this species by a distance slightly greater than the eye diameter. The snout is blunt, dorsally convex, and covered with small papillae. The tip of the snout has a few plicae. That area of the snout immediately adjacent to the enlarged premaxillary teeth is either smooth or so finely papillose as to appear smooth. The tip of the lower jaw is closely packed with a few, relatively large papillae. There is no free tongue. The radiograph shows the glossohyal low-set, deeply embedded in the tissue of the floor of the mouth.

The nostrils are simple. The anterior nostril opens through a short, ventrolaterally directed tube, placed at the same level as the eye. The tip of the tube neither reaches the gape nor the snout tip. The posterior nostril, angled slightly ventrally from the anterior margin of the eye, is elliptical, with a raised rim and an anterior, laterally projecting flap. The gape is large, but not exceptionally so for the genus, extending behind a vertical through the eye by a distance equal to two times the eye diameter. The suspensorium is posteriorly directed as in other species of *Dysomma*, but the peculiarities of the anterior end of the snout region result in a slightly shorter gape than in other species. (See below under dentition and osteological features.)

Placement of the cephalic lateralis pores is as follows. Three supraorbital pores: the first anterior to and level with the anterior nostril, the second dorsal to anterior surface of the tube of the anterior nostril, and the third dorsal to the posterior surface of that tube. Four infraorbital pores placed as follows: the first just ventral to the rear margin of the anterior nostril, the second just ventral and anterior to front margin of rear nostril, the third ventral to the rear margin of the rear nostril and the anterior margin of the eye, and the fourth ventral to the posterior margin of the eye. The first and second infraorbital pores are partly covered by an anterior flap. Preoperculomandibular pores four (perhaps five) along lower jaw and one preopercular pore posteroventral to the corner of gape by a distance equal to slightly more than one eye diameter. The ten inconspicuous lateral-line pores open at the end of short tubules and are confined to the branchial region; nine are anterior to a vertical through the gill opening.

The most salient feature of *Dysomma tridens* is the unique development of the three premaxillary teeth. One tooth is dorsal and medial to the other two, so that they form an inverted V or isosceles triangle. Each tooth is large (twice the size of the large vomerine teeth that typify the genus) and laterally compressed. The tips are expanded, but still compressed. The ventral pair of teeth are slightly twisted medially (toward

FIGURE 238. *Dysomma tridens*: lateral view of head and trunk. Data as for Figure 237. Illustration by C. H. Robins.

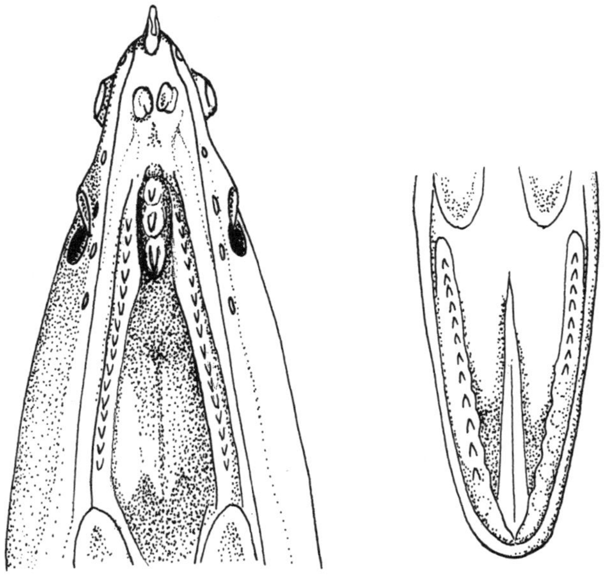

FIGURE 239. *Dysomma tridens*: dentition. Data as for Figure 237.

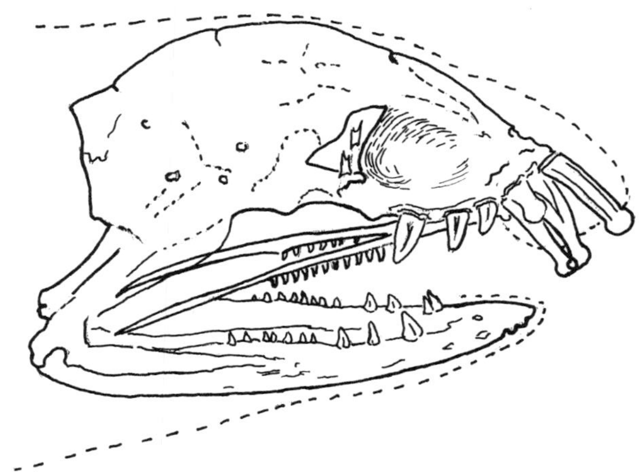

FIGURE 240. *Dysomma tridens*: camera lucida drawing from radiograph to show dentition and osteology (in part). Data as for Figure 237.

each other). All three project beyond the tip of the snout for about half or one-third of their length. The radiograph reveals a replacement tooth emerging in association with the right ventral tooth. Unlike the other two teeth, the right ventral tooth reveals no internal structure (pulp cavity). The median tooth is less dense to radiation through its entire central portion. The left ventral tooth shows a less dense area for half its length. The three large vomerine teeth increase in size posteriorly, and are set in individual hummocks of oral epithelium. The anterior four or five dentary teeth are uniserial, widely spaced, and large (about half the size of the most posterior (largest) vomerine tooth). The 8–9 more posterior dentary teeth, also uniserial, are much smaller. The vomerine teeth and larger dentary teeth are compound. There are about 20 small, uniserial and conical maxillary teeth.

Osteological features were deduced from the radiograph of the holotype. Concomitant with the convex head, the cranium bulges posteriorly, and is very rounded and streamlined. The most radiographically dense portions are in the area of the basisphenoid and prootic. There is no pterygoid. The suspensorium, much of which is weakly ossified, is strongly posteriorly inclined. The snout and eye region of the skull is modified in ways unique in the genus. Although species of *Dysomma* tend to have a restricted orbit due to ventral expansion of the frontal, in *D. tridens* the ethmoid portion (possibly including parasphenoid) of the premaxilloethmoid forms a vertical, median septum. Presumably this structure reinforces the snout region, an adaptation to the vigorous use of the strong, large premaxillary teeth. The premaxillary region of the premaxilloethmoid is distinct in the radiograph. Although there are no cephalic lateralis pores behind or above the eyes, there are small tubular ossifications in the soft tissue behind and above the eye. Despite the absence of pectoral fins, a cleithrum and supracleithrum are located at the level of the ninth vertebra.

Color. The single specimen is a pale, uniform tan, but retains a very few, small, irregular traces of larval pigment along the midline, dorsum, and ventral surfaces. All fins are pale yellowish cream. The peritoneum, gut, and mouth are pale.

Food. Remains that possibly are the chelae of a tiny shrimp are visible in the region of the stomach.

Leptocephalus. Smith (1974: 672, and this volume) described three types of ilyophine larvae from the western North Atlantic with myomere counts in the 170's. Two (B_7, B_{11}) have 173 and 175–177, respectively. The B group has no rostral filament and has midlateral melanophores. The traces of lateral pigment described above surely represent retention of the leptocephalus pattern. One of these two types may prove to be the leptocephalus of *D. tridens.*

Distribution. Known only from the holotype taken off Belize.

Etymology. From the Latin *tridens* (a trident), in reference to the premaxillary dentition. To be treated as an appositional noun.

Study Material. Holotype: USNM 193563, an immature specimen, 217 m TL; 16°44′N, 87°55′W (off Belize), in 348 m, 40 ft. shrimp trawl; Oregon sta. 3634, 9 June 1962.

Family Ophichthidae

SNAKE EELS and WORM EELS

JOHN E. McCOSKER, EUGENIA B. BÖHLKE, *and* JAMES E. BÖHLKE

Acknowledgments. See Acknowledgments for family Muraenidae.

Introduction

The snake eels and worm eels of the family Ophichthidae are the most diverse and speciose of true eels. The more than 250 living species distributed among 55 genera inhabit all tropical oceans and seas. Their habitats range from the intertidal zone to a depth of 750 m or more, from coral reefs to sand and mud substrates and rivers and estuaries, and even to the midwater realm. Because of their secretive, burrowing behavior, some species are known from but a single specimen, and it is likely that many more remain to be discovered. Ophichthids are of limited direct economic importance, although several large species are taken as a bycatch of trawl fisheries and consumed. Due to their abundance in many habitats, it is likely that they are of significance as both predator and prey of other commercially important fishes and invertebrates.

The systematics of ophichthid genera and species has had a checkered history. Numerous species from different families have been included with the ophichthids, and others belonging to the Ophichthidae were improperly assigned to other families. Recent authors have begun the cumbersome dissection of this entangled eel family, particularly Gosline (1951a), Rosenblatt and McCosker (1970), Castle (1972), Böhlke and McCosker (1975), McCosker (1972, 1974, 1977), McCosker and Böhlke (1982, 1984), and Leiby (1979a,b, 1981, 1982, 1984a,b) whose work concerned leptocephali and metamorphic forms.

No worldwide treatment of ophichthid species has been attempted since Günther (1870); however numerous regional studies have been achieved or are in preparation, including: northern Gulf of Mexico (Ginsburg, 1951); Bahamas (Böhlke and Chaplin, 1968); eastern Atlantic and Mediterranean (Blache, 1968b, 1971b, 1975a; Blache and Bauchot, 1972, 1976b; Blache and Cadenat, 1971); eastern Pacific (McCosker and Rosenblatt, in prep.); Hawaii (McCosker, 1979); western Australia (McCosker and Allen, in prep.); southeast Africa (McCosker and Castle, 1986); and Indian Ocean and Red Sea (Smith, 1962b). This study is the first comprehensive treatment of the Ophichthidae of the western Atlantic.

History of Ophichthid Classification. The first described ophichthid species, *Muraena ophis,* was the second apodal fish in Linnaeus' *Systema Naturae* (1758). Since that time, 116 nominal genera have been named, approximately 55 percent of which are valid today. The first generic name properly applied to an ophichthid was *Ophichthus* (Ahl, 1789), which should more properly have been written *"Ophichthys."* Bleeker, Günther, and other classicists emended Ahl's generic spelling (see Jordan and Gilbert, 1883: 648–651, and McCosker, 1977: 10–11), but Jordan and his later co-authors returned to Ahl's original spelling. The 1985 ICZN Code (Art. 29(b)(iii)) would indicate that Ahl's usage of *Ophichthus* is latinized with a change in termination, which would thereby disallow emendation.

The first attempt to combine species now

known to comprise the Ophichthidae was that of Duméril (1806: 153). He created several orders which independently included ophichthid species as well as numerous nonapodal fishes. Within his order Holobranches, sub-order Holobranches Apodes, was the family Peropteres, which included not only *Ophisure, Caecilie* and *Apteronote,* but also *Trichiure, Gymnote* and *Regalec* (pp. 110–113). His subsequent order, Ophichthyctes, contained genera now known to be muraenids as well as *Sphagebranchus* (pp. 152–153). Given the choice of either of Duméril's heterogeneous assemblages, and in the interest of nomenclatural stability, we have adopted his usage of Ophichthyctes as the basis of the family name. The family name for the snake eels is thus Ophichthidae. Duméril (p. 153) derived Ophichthyctes (more properly "Ophichthyes") from "de Oøis serpent, et de Ixovs poisson." We have recently discussed the correct form of the family name with C. Richard Robins, who stated (in litt.):

> A name proposed as an "order," if based on a generic name, and if subsequently used as a family group name is valid and dates from its original proposal. . . . Ophichthyctes proposed by Dumeril (1806) as an "order" but having the same basis as *Ophichthus* Ahl, 1789, is the earliest family group name, correctly emended to Ophichthidae, for the family of snake eels, because the Ophichthidae was later used as a family group name. Emendation of this name to Ophichthyidae is improper (despite example 41 in Table 2, of the 1985 Code) since *Ophichthys,* though classically correct, is not an allowable emendation.

This opinion was confirmed by W. D. L. Ride, chairman of the Editorial Committee of the ICZN.

The following is a summary of the major works, after Duméril (1806), in which a variety of family names were adopted. Rafinesque (1810b: 37, 42–43) named three "orders" which included ophichthid species: Anguillidae, Echelini and Dalofidini. Risso (1826) established the Ophisuridae, a family name that was variously followed for the next half century. Swainson's (1838) treatment of the eel-like fishes was based largely on Cuvier's (1817), and was the first clear attempt at an arrangement of the eels into family groups. Swainson (p. 215) designated the "*Muraenidae* (as) having two branchial spiracles in their ordinary position, and the *Sphagebranchidae,* or sea eels, where the branchial spiracles are either close together or united into one." Confusion ensues on the following page where the family name Gymnarchidae is apparently considered synonymous with the Symbranchidae, and further evidenced in his discussion (p. 218) of the gill openings, by his statement that "among the *Gymnarchidae,* or sea eels, for instance, they are close together and united under the throat as in *Sphagebranchus.*" Swainson divided the Muraenidae into two subfamilies, the Anguillinae and the Muraeninae, both of which contained species now known to be ophichthids. McClelland (1844) realigned the apodal classifications of Swainson and Cuvier and included within the Ophisuridae those eels with a rayless caudal. Kaup (1856a,b) disregarded most of McClelland's classification without comment, but retained the name Ophisuridae. Kaup divided the apodal fishes into two "sections"; these were the Cryptomycteres (containing only the Ophisuridae) which included those eels with labial nostrils, and the Phaneromycteres which contained all other apodal families. Duméril adopted the family name Ophisurides in 1856 and recognized three subfamilies: Myrines, Ophisurines and Sphagebranchines. Bleeker, in his *Systema Muraenorum Revisum* (1864a), recognized the family Ophisuroidei and considered the Myrophinae (as Myriformes) to be a subfamily of the family Congroidei. Günther (1870) considerably revised previous classification by placing the majority of the known eels into a single family, the Muraenidae, which he divided into ten "Groups." The Ophisuridae of earlier authors was divided into two groups, the Ophichthyina, containing those species with a rayless caudal, and the Myrina, containing those with a rayed caudal. Gill (1885) was perhaps the first author to use the family name Ophichthyidae, an action followed by Goode and Bean (1896) but ignored by most of his contemporaries. Jordan and Davis (1891) elevated Günther's groups to family rank; the Myrinae became the Echelidae (Jordan and Davis considered *Myrus* a synonym of *Echelus*) and the Ophichthyina of Günther (actually, the Ophisuroidei of Bleeker) became the Ophisuridae. Uncertainty concerning the synonymy of *Myrus* and *Echelus* resulted in the changing of the name Echelidae to Myridae by Jordan and Evermann (1896a) and by Jordan and Snyder (1901).

Familial Synonymy of the Ophichthidae. The Ophichthidae, as currently recognized, includes several families which were previously considered distinct. McCosker (1977: 12–13) presented the most recent familial synonymy, the results of which are summarized here. As mentioned above, the family names Ophisuridae and Sphagebranchidae are preempted by Duméril's earlier usage of Ophichthyctes and are thus relegated to the synonymy of Ophichthidae.

The family Neenchelidae (emended by subsequent authors to Neenchelyidae) was erected by Bamber (1915) to contain *Neenchelys microtretus*, a new genus and species from the Red Sea. It was considered to be closely related to the Muraenesocidae as defined by Regan (1912b). A second neenchelid, *N. buitendijki*, was described from the Indo-Australian archipelago. Böhlke (1960), on the basis of Bamber's description, suggested that *Neenchelys* may be related to species of *Pseudomyrophis* and thus properly be considered an ophichthid of the subfamily Echelinae (sensu Gosline, 1951a, 1952). Nelson's (1966b, 1967) examination of *N. buitendijki* and the holotype of *N. microtretus* confirmed the recognition of the Neenchelidae as ophichthids in the subfamily Echelinae (herein considered as Myrophinae). Dean (1972), McCosker (1977, 1982), and Smith and Böhlke (1983) supplemented those earlier findings.

The Acanthenchelyidae also belongs in the Ophichthidae. Family recognition was short-lived, consisting of Jordan, Evermann, and Clark's (1930) elevation of *Acanthenchelys* Norman (erroneously attributed to Regan) to family status, which was largely ignored by later authors. Randall and Robins (1966) relegated *Acanthenchelys* to the synonymy of *Ophichthus*, an action which we follow.

The Aoteidae, based upon *Aotea acus* Phillips (1926), was first recognized to be an ophichthid by Castle (1967). *Aotea acus* is now included in the synonymy of *Muraenichthys breviceps* of the subfamily Myrophinae.

The family Echelidae (= Myridae, Myrophinae, Myrophidae, and Muraenichthyidae) has been a catch-all group. The echelids were considered congrid or muraenesocid relatives by most nineteenth century authors. Bleeker (1865), for example, considered the Myriformes (containing *Myrophis, Echelys,* and *Muraenichthys*) to be a subfamily of the Congroidei. Kaup (1856a,b) was exceptional in placing considerable importance on the labial nostril condition and in allying the Myrophinae with the Ophisurinae as a single unit which excluded the congroids and relatives. Twentieth century authors considerably expanded the Myrophidae (which was to become known as the Echelidae, see Jordan and Evermann, 1896a) to include as many as 22 genera at various times (Schultz and Woods, 1949). The dissection of this large and cumbersome family was initiated by Myers and Storey (1939), and was followed by Schultz and Woods (1949) and by Gosline (1950, 1951a,b, 1952). Myers and Storey noted the presence of accessory branchiostegal rays (the "jugostegalia" of Parr, 1930a) in both ophichthids and echelids, but were hesitant to merge the families without an extensive anatomical examination. Gosline (1951a) established the similarities of the two families on an osteological basis and included the genera *Myrophis* and *Muraenichthys* in the Ophichthidae to comprise the subfamily Myrophinae. His subsequent (1952) osteological examination of *Echelus myrus* resulted in its inclusion into the family, and the replacement of the name Myrophinae with Echelinae. McCosker (1977: 75–76) demonstrated that Gosline was correct in considering the myrophines to be ophichthids, but that he erred in including *Echelus* with the Myrophinae. Although certain authors have continued to recognize a family Echelidae for the forms with caudal rays (Smith, 1962b; Blache, 1968b), no convincing arguments have been proposed which would merit familial separation.

In contrast to the above mentioned families, the Macrocephenchelyidae was incorrectly synonymized with the Ophichthidae, without comment, by McAllister (1968: 85). Robins and Robins (1971) re-erected the family on the basis of a thorough osteological examination of the paratype. Recent studies have shown its affinities to be with the Congridae and it is now referred to that family.

Based on extensive studies of adults, metamorphic individuals and leptocephali, the family Ophichthidae currently is divided into two subfamilies: the Myrophinae (containing the tribes Benthenchelyini and Myrophini) and the Ophichthinae (containing the tribes Bascanichthyini, Callechelyini, Ophichthini and Sphagebranchini).

Methods

In the following taxonomic treatments, the characteristics of each subfamily and tribe are defined, and all presently recognized genera are included in the generic keys. Valid western Atlantic species are described and illustrated, the species listed alphabetically within each tribe. Characters given for genera are not necessarily repeated in species accounts.

Osteological descriptions are based on specimens cleared and stained by the trypsin method of Taylor (1967). Rare specimens and holotypes were studied from radiographs and by gill arch removal and staining. Neurocrania were prepared by dissection, maceration in potassium hydroxide, and alizarin staining. Gill arch terminology is that of Nelson (1966a); bone terminology follows that of Asano (1962), with modifications. (See McCosker, 1977: 14, for additional details.)

Abbreviations and terms used are those given in the general introduction to this volume (pp. 2–6). Trunk length is measured from upper end of gill opening to mid-anus; depth of body is measured just behind the expanded branchial opening and does not include the median fins. For comparative purposes, ophichthids are described as extremely elongate (body depth less than 1% TL), elongate (body depth 1–2% TL), moderately elongate (2–3% TL), moderately stout (3–4% TL) and stout (depth greater than 4% TL). Color is described for preserved specimens, followed by life colors when known. Sex was determined by gross examination; egg size given is the greatest diameter of the largest egg observed.

The number of pores is expressed as follows:

SO—supraorbital pores, expressed as ethmoid pore + pores in supraorbital canal, i.e., 1 + 3.

IO—infraorbital pores, expressed as pores along upper jaw + those in vertical part of canal behind eye ("postorbital pores"), i.e., 4 + 2; frequently the last pore included along the upper jaw is part of the postorbital series.

POM—preoperculomandibular pores, expressed as pores along lower jaw + preopercular pores, i.e., 6 + 3.

F—pore in frontal commissure, present in all ophichthids; not listed for each species but position indicated by anteriormost line on head illustrations.

STC—total number of pores in supratemporal canal.

LL—pores in lateral line canal (preanal count sometimes given), infrequently included in the following accounts because vertebral counts are more accurately obtained from radiographs.

Synopsis of Characters

The paucity of adequate external morphological characteristics hampered earlier phylogenetic studies of the Ophichthidae. Patterns of relationship within the family are best exhibited by osteological characters, and it is on that basis that the modern family treatment is now recognized.

General Characters. Several character states, some unique to the family, may be used to define the Ophichthidae. They are: (a) branchiostegal rays numerous and broadly overlapping along the ventral midline; (b) supraorbital canals united by a transverse commissure through the fused frontals; (c) supratemporal canal present; (d) frontals of adults fused for their entire length and lacking an obvious suture; (e) first epibranchial connected by a continuous cartilaginous strap to the second infrapharyngobranchial; (f) no more than first basibranchial ossified; (g) third hypobranchial usually cartilaginous; (h) neural spines rudimentary or absent; (i) tongue adnate; (j) palatines absent; and (k) pterygoids well separated from vomer and generally free from hyomandibulars.

Osteological Characters. The history of osteological studies of the Ophichthidae was extensively treated by McCosker (1977: 11–12), who made the first attempt at a worldwide revision of the family on the basis of adult osteology. Genera then unavailable for study or later created have been treated in subsequent works, including: *Glenoglossa* and *Neenchelys* (McCosker, 1982); *Lethogoleos, Hyphalophis* and *Kertomichthys* (McCosker and Böhlke, 1982); *Gordiichthys* and *Ethadophis* (McCosker and J. E. Böhlke, 1984); and *Mixomyrophis* and *Asarcenchelys* (McCosker, 1985).

The osteology and functional anatomy of *Ophichthus zophochir* (Jordan and Gilbert), a generalized ophichthid, was illustrated, described in detail, and compared to other ophichthids by McCosker (1977: 17–48). *Ophichthus zophochir* appears most closely related to *O. gomesii* and its osteology does not demonstrably differ from that of *O. ophis,* the generic type. In the following section we summarize and illustrate McCosker's treatment in order to allow a comparison with eels of other families presented in this volume.

The neurocranium of *O. zophochir* is completely ossified and well fused along the cranial sutures. The skull is stout, rather elongate, and truncate posteriorly. The neurocranium of most ophichthids is small in relation to the total length of the eel. That of an adult *O. zophochir* occupies 5.5 percent of the total length, yet the neurocrania of more elongate ophichthids, such as *Phaenomonas cooperae* Palmer, occupy as little as 1.3 percent. Various aspects of the neurocranium of *O. zophochir* are illustrated in Figures 241 and 242. The ethmoid, vomer, and intermaxillary region are fused into a single complex commonly called the premaxilloethmovomer by most authors; most apodal fishes, however, lack the premaxillary bones and we therefore define this as the ethmovomer complex, but call the teeth the intermaxillary teeth. It articulates posterodorsally with the frontal and posteroventrally with the parasphenoid, and forms the anterior margin of the orbit. The anterior ethmoid portion of the ethmovomer is separate from the vomer in leptocephali, yet in adult ophichthids there are no distinct sutures separating the elements, and its precise limits can only be determined ontogenetically. The ethmoid portion is expanded in most ophichthids and is toothed (the teeth termed the intermaxillary teeth) in all genera except *Aprognathodon.* It narrows posteriorly to become the vomer, which is toothed in most ophichthids. The intermaxillary (ethmoid) dentition is continuous with that of the vomer in *O. zophochir* and in most ophichthines. A gap separates the intermaxillary and vomerine dentition of other ophichthid genera, and appears to be a useful character to indicate relationship. The vomerine dentition does not continue onto the parasphenoid. The ethmoid portion forms the anterior margin of the orbit and the medial margin of the nasals. The ethmoid is perforated anterolaterally by the lateral commissure of the first cranial nerve.

The nasals of *O. zophochir* are paired, thin, laminar, and cartilaginous along their external edges. The anterior portion of the supraorbital cephalic lateralis nerve tract passes through the canal along the median edge of the nasal. The nasals of the Myrophinae are either cartilaginous or absent. Nasal development among the Ophichthinae is variable.

The parasphenoid is a long, narrow, toothless bone, anteriorly overlying the vomer and forming the ventral margin of the orbit. Centrally, it is spread laterally to form the anterior floor of the cranium, narrows posteriorly, and splits into two short prongs.

The paired prootics combine with the paired basioccipitals and pterotics to form the otic bullae. They are small, nearly rectangular, and highly perforated with numerous openings for the passage of nerves and blood vessels. Through the most conspicuous foramen passes the hyoman-

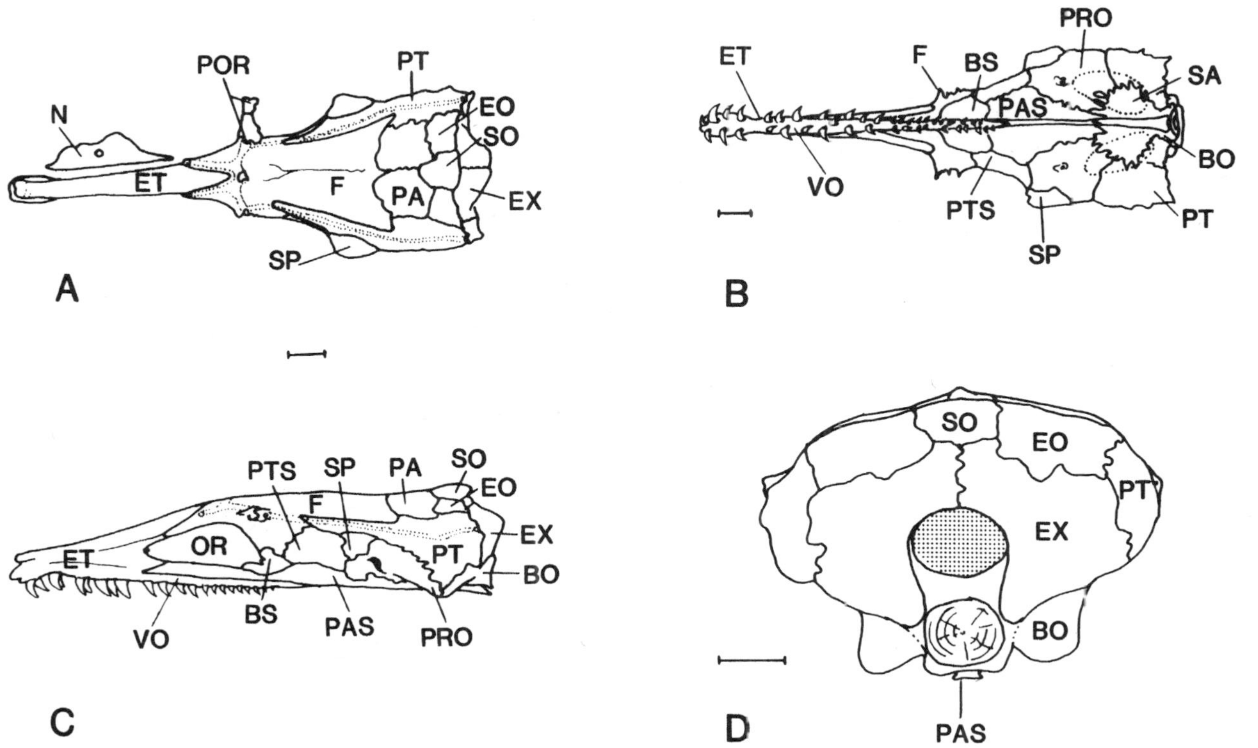

FIGURE 241. Neurocranium of *Ophichthus zophochir,* a typical ophichthine. A. Dorsal view. B. Ventral view. C. Left lateral view. D. Posterior view. Scale represents 1 mm. Abbreviations: BO, basioccipital; BS, basisphenoid; EO, epiotic; EX, exoccipital; ET, ethmoid portion of ethmovomer; F, frontal; N, nasal; OR, orbit; PA, parietal; PAS, parasphenoid; POR, postorbitals; PRO, protic; PT, pterotic; PTS, pterosphenoid; SA, sagittal otolith; SO, supraoccipital; SP, sphenotic; VO, vomer. Figures 241–251 from McCosker, 1977.

dibular trunk of the facial nerve (VII). Smaller foramina exist for the passage of the orbital artery and the jugular vein. The major axis of the prootic is horizontal.

The median basioccipital is a small, irregularly shaped bone which forms the posterior portion of the otic bullae. Its major axis, in contrast to the prootics, is vertical. It is bordered medially by the parasphenoid, anteriorly by the prootics, and dorsally by the pterotics and exoccipitals.

The paired pterotics are elongate, narrow anteriorly, and broadly flared laterally and posteriorly forming the lateral edge of the roof of the cranium. The cephalic lateralis nerve tract passes through the pterotic and opens anteriorly in the frontal and posteriorly at the posterior pterotic

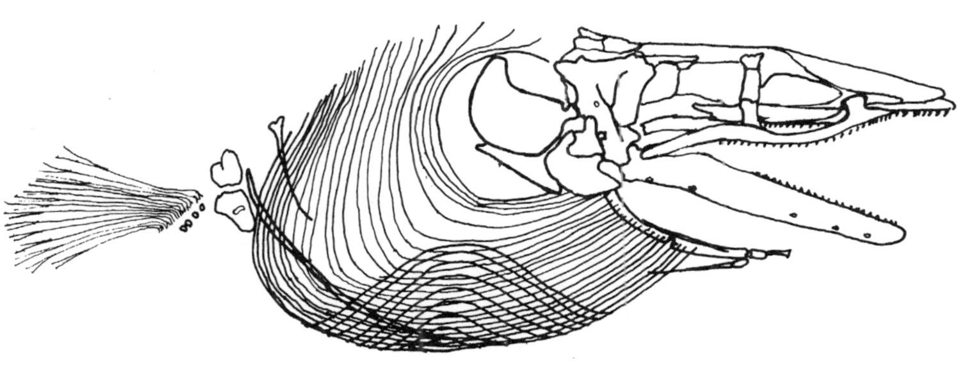

FIGURE 242. Head skeleton of *Ophichthus zophochir,* a typical ophichthine, displaying the relationship of the neurocranium, suspensorium, hyoid arch, and pectoral girdle.

margin. Anteromedially the pterotics are bordered by the frontal, followed medially by the parietals and epiotics, anterolaterally by the pterosphenoid and sphenotic, ventrally by the prootic and basioccipital, and posteriorly by the exoccipitals. Posterior to the sphenotics, the pterotic forms a sheet-like eave extending beyond the body of the cranium.

The paired exoccipitals form the dorsal and lateral margins of the foramen magnum. They are sutured along their dorsal midline, and extend posteriorly and laterally as a semicircular sleeve around the foramen magnum. Ventrally, they contact the basioccipital, and dorsally the supraoccipital. The foramen of the tenth cranial nerve opens posteriorly along the ventral exoccipital-basioccipital border, with the ninth opening lateroventrally from a foramen slightly anterolateral to that of the tenth.

The supraoccipital is single, small, square to subrectangular in shape, and lies along the posterodorsal cranial midline. In *O. zophochir* the narrow median crest of the supraoccipital extends posteriorly as a small point. The condition is typical of many ophichthids, although in certain generic groups, particularly *Callechelys* and related genera, the supraoccipital is rounded along the posterior supraoccipital margin.

The epiotics are paired, thin, and sub-rectangular, bordered anteriorly by the parietals, laterally by the pterotics, and medially by the supraoccipital. Posteriorly their major axis is changed from horizontal to vertical in forming the dorsolateral margin of the posterior cranial face bordering the exoccipitals. The epiotics, like the supraoccipital and the pterotics, form a narrow sheet-like eave along their posterior margins.

The paired, thin, sub-rectangular parietals overlay the posterior margin of the frontal and are fused medially. In certain ophichthids which have a prominent supraoccipital crest, the median ridge development begins along the parietal midline and gradually increases to its posterior projection.

The frontal is a single long element which, along with the epiotics and parietals, forms the roof of the cranium. Ontogenetically, the frontal is formed from the fusion of paired lateral elements, but in juveniles and adults there is no evident suture. In most *Ophichthus* species the frontal is ridged posteriorly along the dorsal midline. In *Aplatophis* this ridge is developed as a sharp crest. Several nerve tracts pass through the frontal, including the anterior tract of the cephalic lateralis nerve and the transverse frontal commissure, which is unique to the Ophichthidae. The frontal is deeply split anteriorly by the insertion of the ethmoid portion of the ethmovomer in some genera, and bordered anteroventrally by the orbit and orbitosphenoid, ventrally by the parasphenoid, laterally by the pterotic, and posteriorly by the parietals.

The basisphenoid (= orbitosphenoid of others) is a small, unpaired median bone with two lateral wings which forms the posteroventral margin of the orbit. It is bordered dorsally by the frontal, posteriorly by the pterosphenoids, and ventrally is supported by the parasphenoid (a myodome is not present). Robins (1971: 164–165) has noted that the use of the term "orbitosphenoid" in other eel studies (including Regan, 1912b; Trewavas, 1932; Gosline, 1950, 1951a, 1952; Robins and Robins, 1967; and others) actually pertained to the basisphenoid.

The small paired pterosphenoids (= alisphenoid) form the anterodorsal roof of the cranium. In *O. zophochir* they are concave, turning evenly from a longitudinal axis (along the margin of the frontal) to a nearly transverse axis which abuts the anterior margin of the sphenotic. The conspicuous foramen along the pterosphenoid-pterotic border is the anterior opening of the trigeminofacialis chamber.

Ophichthid otoliths, like those of most anguilliforms, are small, and hence have received little attention either on a descriptive or on a comparative basis. Studies are limited to those of Frost (1926), Kotthaus (1968), Böhlke and McCosker (1975) and McCosker (1977). Ophichthid sagittae (the largest of the three pairs; the asteriscus and lapillus are too small to be of comparative value) are ovate and biconvex, with a shallow sulcus on the medial surface. They are particularly distinctive in having a shallow ostial channel which opens anteriorly rather than turning

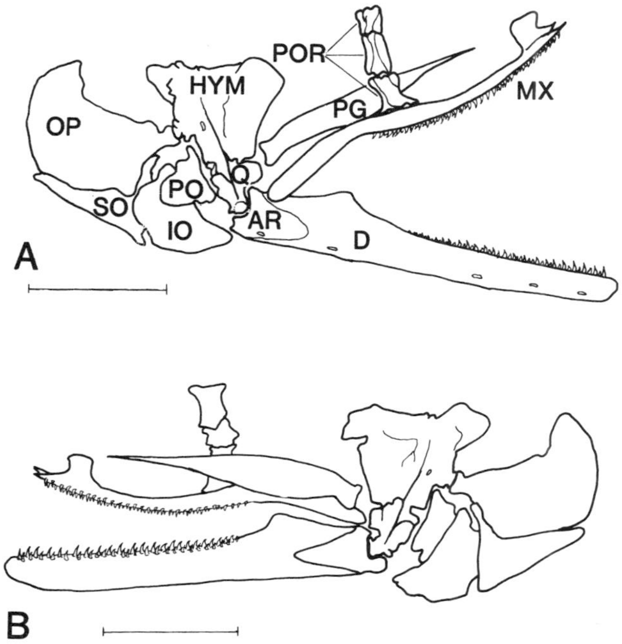

FIGURE 243. Suspensorium and jaws of *Ophichthus zophochir*, a typical ophichthine. A. Right outer face. B. Right inner face. Scale represents 5 mm. Abbreviations: AR, articular; D, dentary; HYM, hyomandibular; IO, interopercle; MX, maxilla; OP, opercle; PG, pterygoid; PO, preopercle; POR, postorbitals; Q, quadrate; SO, subopercle.

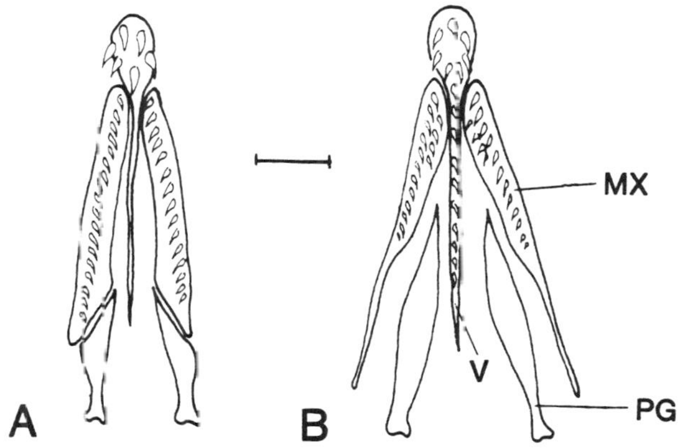

FIGURE 244. Vomer (V), maxillae (MX), and pterygoid (PG) of (A) *Ahlia egmontis* and (B) *Myrophis vafer*. Scale represents 1 mm.

dorsally and opening from the sulcus, as is typical of those of the Congridae (cf. Frizzel and Lamber, 1962).

The location and shape of teeth has often been used as a principal character to define and differentiate ophichthid genera. The dentition of *O. zophochir* represents the generalized ophichthid condition in being multiserial, conical, and on all tooth-bearing bones (vomer, ethmoid, maxilla, and dentary). Considerable variation exists within the family, including the elongate fang-like dentition of *Aplatophis*, the molariform or granular dentition of *Myrichthys*, *Pisodonophis* and certain species of *Muraenichthys*, the minute, nearly villiform dentition of *Schultzidia*, the multiserial congrid-like tooth bands of *Hyphalophis*, and the smooth, toothless vomer of *Leiuranus*, *Leuropharus*, *Phyllophichthus* and *Schultzidia*. Ontogenetic variation in tooth size has been demonstrated in certain ophichthins, particularly species of *Echiophis* and large *Ophichthus*.

The conditions of the dentition, suspensorium, and jaws (Fig. 243) are directly related to the feeding habits of the various genera. In ophichthids, differences in feeding habits and prey items are greater between the species of different genera than between congeners. All elements of the suspensorium and jaws are paired. The hyomandibular is stout, and shaped like an inverted right triangle. The hyomandibular of strong-jawed piscivorous genera is generally strongly ridged for the attachment of the massive adductor mandibularis muscle. A small irregular condyle along the antero-dorsal margin of the hyomandibular fits into a shallow socket formed along the sphenotic-pterotic suture. The large process on the posterior hyomandibular margin adjoins the anterior process of the opercle. The ventral portion of the hyomandibular contacts the quadrate, which is small, stout, and tightly sutured to the hyomandibular. The vertical ridge along the outer face of the hyomandibular is continuous along the quadrate. Ventrally, the quadrate bears a broad rounded concave process that contacts the articular bone of the mandible. The wedge-shaped articular is narrowed anteriorly and slides into a pocket within the dentary. A remnant of the coronomeckelian is present along the inner face of the articular. Posteriorly, a grooved socket in the articular meets the rounded socket of the quadrate. The dentary is the toothed bone of the mandible, joined by the articular posteriorly and adjoining its opposite member at the symphysis by a cartilaginous connection. The three postorbitals of most ophichthids are separate, weak ossicles that surround the nerve tract connecting

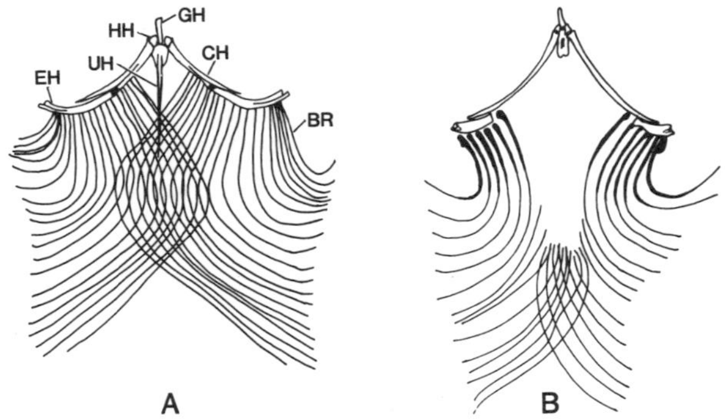

FIGURE 245. Hyoid arch and branchiostegals of a typical ophichthine (A, *Ophichthus zophochir*) and a typical myrophine (B, *Muraenichthys chilensis*, posterior-most branchiostegals not illustrated), shown in dorsal view. Abbreviations: BR, branchiostegal rays; CH, ceratohyal; EH, epihyal; GH, glossohyal; HH, hypohyals; UH, urohyal.

the supraorbital and infraorbital pore tracts. The postorbital of *Ophichthus zophochir, O. ophis,* and other piscivorous ophichthins (e.g., *Echiophis* and *Brachysomophis*) are specialized by enlargement and fusion to form a continuous strut bracing the mandible and neurocranium. Pterygoid development is variable with the family. The pterygoid is reduced to a narrow splint in most ophichthids. The pterygoid of all ophichthids tapers anteriorly, and often posteriorly, is largely cartilaginous in many species, and does not appear to serve any distinct purpose. *Ahlia* is exceptional in differing from the closely related species of *Myrophis* in the shape of its pterygoid and in lacking vomerine teeth (Fig. 244). A true palatine is not present in ophichthids, as shown by Robins and Robins (1970) in their discussion of the "palatopterygoid arcade." The maxilla of all ophichthids is toothed, elongate, and possesses an anterior dorsal process which articulates with the ethmovomer. The generalized ophichthid condition is that of a toothless, elongate, ossified or cartilaginous extension of the maxilla beyond and lateral to the articular. The location of maxillary articulation with the vomer is affected by the elongation of the snout and jaw, and is quite variable within the family.

In concluding this section on the neurocranium and suspensorium and jaws, certain comments are in order relating to their specializations and functional anatomy. Typical of the piscivorous adaptations of species of *Ophichthus, Brachysomophis, Echiophis, Kertomichthys* and related species are the strengthening provided by the cranial vault, the elongate pterotics, the broad junction of the hyomandibula along the neurocranium, and the bracing of the maxillae to the frontal by means of the fused postorbitals. An analagous condition exists in the muraenid species of *Gymnothorax* in which a postorbital strut strengthens the jaws and suspensorium (cf. Burton, 1956: fig. 6, and this volume, p. 110). Other ophichthid species, particularly among the Myrophinae and elongate ophichthines, are adapted to diets of minute invertebrate prey, and have extremely reduced neurocrania, suspensoria, jaws, and dentition.

The opercular series of ophichthids, and of anguilliforms in general, is greatly reduced. This reduction is apparently related to the increase in number and importance of the branchiostegal rays as supporting elements for the branchial cavity (Greenwood et al., 1966; McCosker, 1977). Gosline (1959) has correlated the reduction of the opercular series, posterior displacement of the gill arches and pectoral girdle, and the separation of the pectoral girdle from the neurocranium with the peculiar branchial pump and circulatory mechanism of anguilliforms. Opercular reduction and deossification to cartilage, and the increase of branchiostegals of ophichthids appear to be greatest in the smaller myrophines and elongate bascanichthyins and sphagebranchins. Several specializations in each subfamily are useful indicators of phylogeny. For example, the subopercle is produced posteriorly as a projection enclosing the ventral and posterior margins of the opercle in species of *Myrophis, Ahlia, Muraenichthys, Pseudomyrophis* and *Schismorhynchus.* This subopercle-opercle morphology is typical of other eel families, including certain Congridae (Asano, 1962), Moringuidae (Trewavas, 1932; Smith and Castle, 1972), Chlopsidae (Gosline, 1951b; Robins and Robins, 1967), and *Xenomystax atrarius* (Peden, 1972). The opercular series of *Callechelys* and related genera is reduced and has a conspicuously fringed appearance along the margin. The opercular series of *Stictorhinus,*

Apterichtus, Ichthyapus and related genera are quite reduced, with elements absent or greatly reduced.

The hyoid apparatus and the associated branchiostegals provide fundamental characters which help to unify the subfamilies within the Ophichthidae. In particular, the broad overlap along the ventral midline of the branchial basket was considered by McCosker (1977) to be a major unifying character of the family, not evidenced by homology or convergence in other eel families. The importance of this character later led to the combining of the Neenchelidae, Echelidae (in part), and the Ophichthidae (see McCosker, 1977: 28). The following description of the hyoid apparatus is based on that of *O. zophochir* (Fig. 245). The apparatus consists of the unpaired glossohyal and urohyal, and paired upper hypohyals, ceratohyals, and epihyals. The interhyal is absent in adults (although present in larvae). The outer posterodorsal margin of the epihyal is connected by cartilage to the inner face of the quadrate and provides support for the branchial basket. The ceratohyal and epihyal are connected by a stout cartilaginous strut affording little flexibility along the arch; further strengthening is provided by the flanking spike-like posterior extension of the ceratohyal along the outer edge of the epihyal. All the branchiostegal rays are inserted on the external face of the arch. The rays broadly overlap along the ventral midline in a characteristic manner. The distal rays of the left epihyal and cartilaginous interspace are overlain by all the rays of the right arch, which are in turn overlain by the remaining left ceratohyal rays. This pattern of the left ceratohyal rays overlaying the right ceratohyal and epihyal rays which overlap the left epihyal rays is consistent within the Ophichthidae.

The location and number of branchiostegal rays among the genera of ophichthids and the proximity of the branchiostegal rays to the hyoid arch differ markedly in the ophichthid subfamilies (Fig. 245). In the Myrophinae, which appear to follow the generalized anguilliform condition, the branchiostegal rays are attached to the outer face of the epihyal, often with a single ray on the ceratohyal. The remainder of the branchiostegal rays, which will hereafter be referred to as the "accessory branchiostegal rays," are unattached and basally lie well behind the hyoid arch. In the Myrophinae, these vary from as many as 13 to 42 pairs. In the Ophichthinae, all branchiostegal rays are attached to the outer face of the hyoid arch, although in some species the rays have secondarily become detached. The extreme development of this branchiostegal apparatus is obviously a means of strengthening the gill basket, particularly for those species with a diet of struggling prey, in contrast to the myrophine condition of free rays and their diet of comparatively weaker prey. The reduction of the opercular apparatus and the posterior displacement of the entire gill arch complex necessitates a supplementary skeletal framework to prevent the gill basket from collapsing during the normal burrowing activity. The posterior displacement of the gill arches among anguilliforms is extreme in the Ophichthidae and the Moringuidae (see Nelson, 1966a: fig. 58), both of which comprise predominantly sand and mud-burrowing forms.

The accessory branchiostegal apparatus of ophichthid and "echelid" eels was recognized by earlier workers, but Parr (1930) was the first to describe it and suggest its function. He created the term "jugostegalia" for the accessory skeleton of the gill cover in species of *Myrophis*. Because of their number he did not consider the attached rays to be homologous with the branchiostegals. McCosker (1977) demonstrated their homology and suggested that the term "jugostegalia" be discarded.

The urohyal of most ophichthines is produced posteriorly as an ossified spike. The urohyal of all myrophines is limited to an ossified basal plate with cartilaginous posterior filaments. The few ophichthines that lack the ossified spike are clearly derived from the generalized condition typified by species of *Ophichthus*, rather than from the somewhat similar myrophine condition. A major subfamilial difference is also evidenced in the ceratohyal. Without exception, the ceratohyal of the Ophichthinae is split into an elongate and pointed distal portion and a shorter, truncate

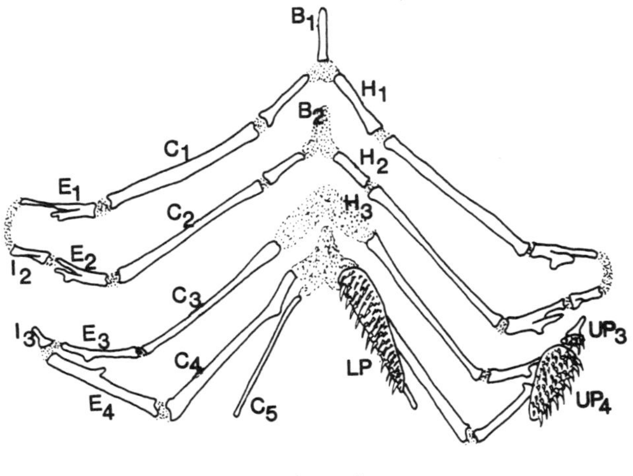

FIGURE 246. Gill arch skeleton of *Ophichthus zophochir*, a typical ophichthine. Dorsal view. Gill arches have been cut along the dorsal midline and spread laterally; left upper and lower pharyngeal tooth plates are removed to show underlying bones. Stippling indicates cartilage. Scale represents 1 mm. Abbreviations: B_1, first basibranchial; C_1, first ceratobranchial; E_1, first epibranchial; H_1, first hypobranchial; I_2, second infrapharyngobranchial; LP, lower pharyngeal tooth plate; UP_3, upper pharyngeal tooth plate.

medial portion which connects, by means of cartilage, with the epihyal. The myrophine ceratohyal is not split, but ends at the midpoint of the epihyal. The myrophine condition appears to involve a unique reduction, whereas the ophichthine condition is similar to that of the Congridae (Asano, 1962; Smith, 1971) and other eel families. The hypohyals of most ophichthines are like those of *Ophichthus.* In certain otherwise dissimilar genera the hypohyals are absent. Careful examination of the anterior end of the ceratohyal does not reveal a suture or line of fusion; it might be assumed that the hypohyals are lost altogether in those instances, or that they never develop, since the hypohyal comes from the same block of cartilage as the ceratohyal and epihyal.

Certain elements of the anguilliform gill arch skeleton have proven to be important indicators of relationship (Nelson, 1966a). The Ophichthidae differ from all other eels in a combination of gill arch characters, including: a cartilaginous connection between the proximal ends of the dorsal parts of the first and second gill arches (the condition peculiar to the Ophichthidae except for a few Myrophinae); first basibranchial either ossified or absent, all others cartilaginous, rudimentary, or absent; hypobranchials 1–2 ossified, and; second infrapharyngobranchial ossified. If one considers the anguillid or congrid gill arch conditions, that of numerous ossified elements with minor loss or reduction, to be primitive among the anguilliforms, then the ophichthids are considerably advanced in having several osseous elements replaced with cartilage, and in having others reduced or lost. The Ophichthinae, and in particular the Ophichthini, are even more primitive than the Myrophinae, which have lost the fifth ceratobranchial (present in larvae but fused with the lower pharyngeal plate in adults) and have reduced or lost certain basibranchials. The ophichthids are also specialized in having the gill arch skeleton displaced posteriorly in relation to the cranium. References to ophichthid gill arches include Popta (1904), Nelson (1966a,b), and the ophichthid papers of McCosker (adults) and Leiby (leptocephali).

The following description of the gill arch system of *Ophichthus zophochir* (Fig. 246) illustrates the presumably primitive condition within the Ophichthidae. The basibranchials are single elements, are not interconnected, and lie along the ventral midline; all other gill arch elements are paired. The first basibranchial is ossified, slender, and connected by cartilage to the first hypobranchials; basibranchials 2–4 are cartilaginous and connected to the adjoining hypobranchial pair. Hypobranchials 1–2 are ossified and stout; hypobranchials 3–5 are cartilaginous, with 4 and 5 fused. The third hypobranchial is cartilaginous in nearly all ophichthids. This specialization probably functions to increase the flexibility of the gill arch skeleton and, in particular, to allow further anterior movement of the lower pharyngeal tooth plates. For the same reason, we presume that hypobranchial 4 is never ossified. Ceratobranchials 1–4 are ossified and subequal. Ceratobranchial 5 is reduced to a slender filament which is fused for most of its length with the ventral surface of the ventral pharyngeal tooth plate. Epibranchials 1–4 are short, stout, and bear various processes for cartilaginous or ligamentous attachment. The first infrapharyn-

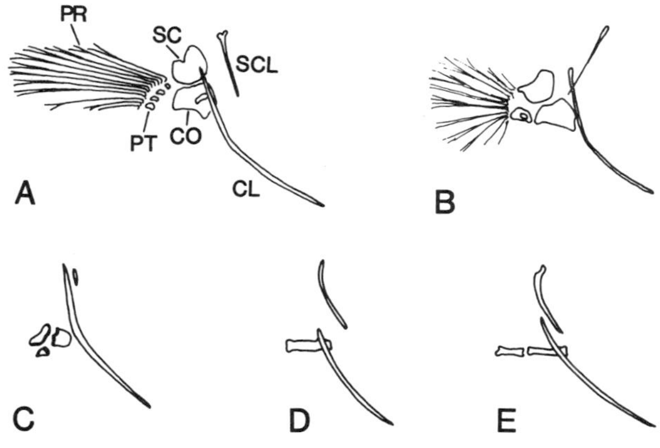

FIGURE 247. Pectoral girdles of various ophichthids, shown in right lateral view. A. *Ophichthus zophochir.* B. *Ahlia egmontis.* C. *Caralophia loxochila.* D. *Callechelys marmorata.* E. *Aprognathodon platyventris.* Abbreviations: CL, cleithrum; CO, coracoid; PR, pectoral rays; PT, pterygiophores; SC, scapula; SCL, supracleithrum.

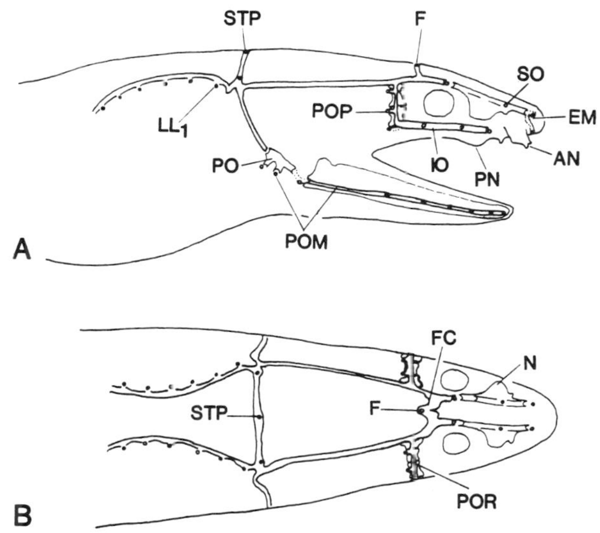

FIGURE 248. Cephalic lateralis system and associated bones of a typical ophichthid. A. Right side view. B. Dorsal view. Abbreviations: AN, anterior nostril; EM, ethmoid section of supraorbital canal; FC, transverse frontal commissure; IO, infraorbital pores; LL_1, anteriormost lateral line pore; N, nasal bone; POM, preoperculomandibular pores; PN, flap overlying posterior nostril; PO, preopercle; POP, postorbital pores; POR, postorbital bones; SO, supraorbital pore; STP, supratemporal pore.

gobranchial, as in all eels, is absent. The second connects to the first epibranchial by a cartilaginous strap. The third is "T" shaped and distally supports the third upper pharyngeal tooth plate. The upper pharyngeal plates are separated by a suture, the third being much smaller than the fourth. Each plate bears along one margin a single row of slightly retrorse conical teeth which grades to a fine-toothed pavement. In situ, the tooth plates overlie each other, and surround the esophageal canal so that the corresponding toothed areas of the upper and lower plates are aligned.

Nelson (1966a) and McCosker (1977) identified certain ophichthid lineages on the basis of the gill arch conditions. These comprised (a) those genera with a moderately well-developed series of basibranchials and an ossified ceratobranchial 5, (b) those with ceratobranchial 5 reduced or cartilaginous, and (c) those with very reduced basibranchials and lacking ceratobranchial 5, i.e., the Myrophinae. Subsequent ophichthid descriptions have fit within that pattern.

The pectoral girdle varies considerably within the Ophichthidae, grading from a well developed to a reduced condition in both subfamilies. The primitive condition, represented by *Ophichthus* (Fig. 247A) in the Ophichthinae and by *Ahlia* (Fig. 247B) in the Myrophinae, includes the retention of the cleithrum, supracleithrum, scapula (= hypercoracoid), and coracoid (= hypocoracoid), and in the Ophichthinae, the actinosts. The coracoid and scapula generally lie within a cartilaginous plate which provides a rigid support for pectoral fin movement. When present, the pectoral fin base lies in a dorsoventral plane and is broad-based in relation to the fin length, offering little rotational movement. The posttemporal is absent and the pectoral girdle is not attached to the cranium in anguilliforms (Gosline, 1971).

A peculiar coracoid and scapula condition exists in the Callechelyini (Fig. 247D,E) and several of the Bascanichthyini (Fig. 247C) representing a unique ophichthid specialization which is not seen in other apodal fishes (McCosker and Rosenblatt, 1972; McCosker, 1977). It is nearly universal among eels for the scapula to be a nearly round bone (flattened slightly on the posterodorsal edge) lying above the coracoid (also nearly round but slightly flattened on the posteroventral edge), one or both of which are fenestrated. This is also the generalized ophichthine and myrophine condition. The Callechelyini possess

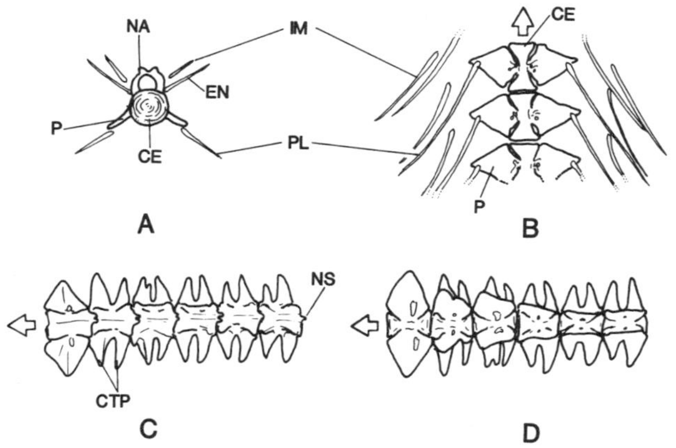

FIGURE 249. Trunk and caudal vertebrae of *Ophichthus zophochir,* a typical ophichthine. Arrows point anteriorly. A. Anterior view of 14th vertebra. Ribs appear foreshortened due to viewing aspect. B. Ventral view of 14th–16th vertebrae. C. Dorsal view of last precaudal (51st) and anterior five caudal (52nd–56th) vertebrae. Ribs and intramuscular bones not illustrated. D. Ventral view of vertebrae illustrated in C. Abbreviations: CE, centrum; CTP, caudal transverse processes; EN, epineural; IM, intramuscular bone; NA, neural arch; NS, neural spine; P, parapophysis; PL, pleural rib.

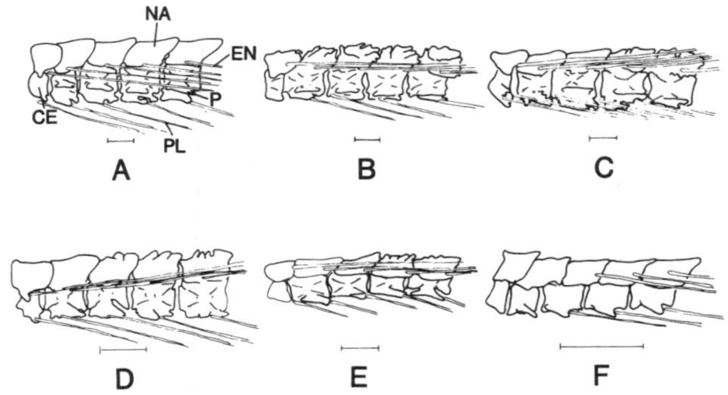

FIGURE 250. Anteriormost five vertebrae of representative species from the tribes of ophichthids. All in left lateral view. A. *Ophichthus zophochir.* B. *Stictorhinus potamius.* C. *Bascanichthys panamensis.* D. *Callechelys marmorata.* E. *Myrophis vafer.* F. *Benthenchelys cartieri.* Scales represent 1 mm. Abbreviations: CE, centrum; EN, epineural; NA, neural arch; P, parapophysis; PL, pleural rib.

either one or two small rod-shaped bones which are connected by cartilage and are oriented horizontally in the normal location of the coracoid.

The apodal lateralis system and associated bones have been shown to be useful indicators of relationships, particularly within the Congridae (Asano, 1962; Smith, 1971). The lateralis system within the Ophichthidae differs in a characteristic manner at the subfamilial and tribal level (McCosker, 1977). It is best developed in the Ophichthinae and reduced in the Myrophinae.

The ophichthid lateralis system (Fig. 248) is divisible into seven canals: lateral line, supraorbital, infraorbital, preoperculomandibular, temporal, supratemporal commissure, and frontal commissure. The canals lie either within certain cephalic bones (frontal, pterotic, nasal, preopercle, postorbital, and dentary) or within weakly ossified tubes (ossicles) which are broken at short intervals to provide flexibility. The Ophichthidae are distinctive in having the right and left sides of the cephalic lateralis system connected through the frontal and supratemporal canals. Suprageneric relationships among ophichthid genera are indicated by repeating pore patterns in certain cephalic canals, particularly in the preopercular portion of the preoperculomandibular canal and in the supratemporal canal. The generalized ophichthine condition consists of three preopercular pores and three pores in the supratemporal canal. The third preopercular pore is lost in certain Ophichthini and all Callechelyini and Bascanichthyini; certain Sphagebranchini are specialized in having a fourth preopercular pore and five pores in the supratemporal canal, as in *Apterichtus kendalli.*

The lateral line ossicles also indicate relationship in their degree of ossification, separation at each pore, and the pore position along the canal. Most pores lie below the midline of the lateral line canal, although some genera are specialized in having the pores located centrally within the canal. The number of preanal and postanal lateral line pores correlate well with vertebral number in that each structure develops from the same myomere. Lateral line pore development, however, is variable within the family and appears to be related to the substrate occupied by the juveniles and adults. Many shallow water forms have extensive development of prominent pores, extending nearly to within a head length of the tail tip. Others, such as those that occupy deep water soft benthic substrates, have their lateral lines reduced such that they disappear in the posterior trunk or within the anterior tail region.

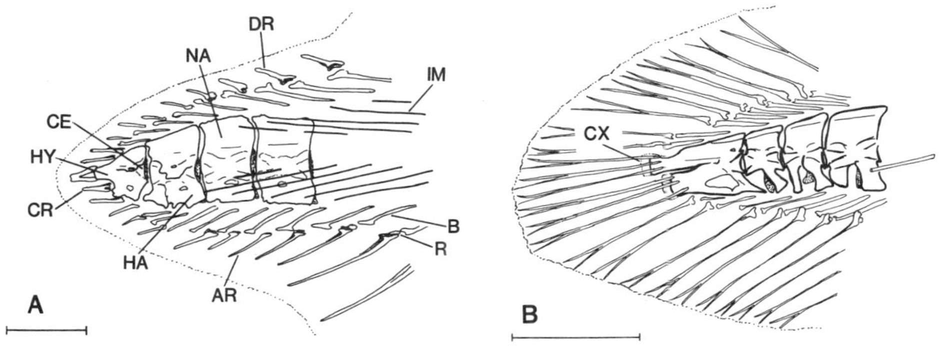

FIGURE 251. Caudal skeleton, right lateral view. A. A typical ophichthine. B. A typical myrophine. Scale represents 1 mm. Abbreviations: AR, anal ray; B, basal element of pterygiophore; CE, centrum; CR, caudal ray, CX, cartilaginous extension of terminal vertebrae; DR, dorsal ray; HA, haemal arch; HY, hypural; IM, intramuscular bone; NA, neural arch; R, radial element of pterygiophore.

This is apparently an adaptation to avoid pore clogging by soft muds. Lateral line ossicles are nearly solid structures in the Sphagebranchini and Callechelyini, less substantial in the Ophichthini, and most reduced in the Myrophinae.

The axial skeleton has been shown to be significantly different between and within apodal families (Regan, 1912b; Gosline, 1951a; McCosker, 1977). Typical vertebrae of *Ophichthus zophochir* are illustrated in Figure 249. Differences in neural arch condition and rib and intramuscular bone development are evident between genera and obviously relate to body strength and differences in life style. Differences in neural arch shape and sculpturing are evident in comparing the anteriormost five vertebrae of species of the type genus of each ophichthid tribe (Fig. 250). The neural arch of vertebrae 1–5 of members of the Callechelyini can be distinguished, at the tribal level, on that basis alone. Other characters of the axial skeleton are useful indicators of relationship (Fig. 250). The parapophyses of certain sphagebranchin genera, for example, possess a marginal process which is lacking in related genera (McCosker, 1977: fig. 33). Also, the caudal temporal processes are lacking in most myrophins, yet in *Muraenichthys* and related genera the anterior half of the column is similar to the ophichthine condition. Finally, the characteristics of the pleural ribs of *Asarcenchelys*, *Ahlia* and *Myrophis* differ from that of all other ophichthids in that they are limited to the anterior trunk vertebrae.

Differences in the myrophine and ophichthine caudal fins seemed important enough to most earlier authors to recognize the lineages as distinct families (see McCosker, 1977 for a review). The Ophichthidae of nineteenth century authors was indeed a unique and unified assemblage, primarily due to the conspicuously pointed tail tip which is associated with tail-first burrowing. Important members, most notably the species of *Echelus* were erroneously excluded from the Ophichthidae because they possess a weakly developed caudal fin. Gosline (1951a: 303) noted the similarity in the ophichthine and myrophine caudal skeletons, but felt that the continuous median fin condition merited subfamilial separation. His findings were clarified by McCosker (1977) who included species of *Leptenchelys* and *Echelus* within the Ophichthinae in spite of their caudal fin development. The homologies of ossified elements within the apodal caudal skeleton are difficult to ascertain, and especially so in the case of the sharp-tailed ophichthins and sphagebranchins that have undergone major modification in their adaptation to rapid burrowing. The caudal skeletons of a typical ophichthin and a typical myrophin are illustrated in Figure 251.

Vertebral numbers (counts of predorsal, preanal, and total vertebrae) have been shown to be useful characters for the separation of species and populations of apodal fishes (McCosker, 1977; Böhlke, 1982). Predorsal counts are characteristic for genera and can be utilized to separate genera within tribes of Ophichthidae. Trends in the number of total vertebrae are probably related to the mode of life and associated anatomical specializations of the species involved. The elongate species of *Phaenomonas, Allips,* and *Bascanichthys,* for example, tend to have increased vertebral numbers, primarily in the trunk region. The examination of radiographs and gut contents of specimens of those genera has disclosed the presence of copious sand and gravel particles in the gut and intestine. McCosker (1977) has suggested that these eels indiscriminately eat their way through the substrate, digesting any utilizable organic matter they encounter. In many ophichthids the length of the gut is increased by a loop that extends into the tail portion. The gut of species of *Phaenomonas* and *Bascanichthys* is straight, presumably an adaptation to prevent blockage by sand particles. The increased trunk length, typical of these genera, perhaps is an adaptive solution to achieve this feeding mode. Certain callechelyins exhibit a similar increase in trunk vertebrae and a comparable life style.

Internal Characters. The visceral anatomy of ophichthids has not been systematically analyzed. In general, the ophichthid digestive tract includes a gut diverticulum which branches off the anterior trunk region and extends posteriorly into the caudal region and then loops forward to the anus. Several genera, however, lack the posterior extension of the intestine into the tail region (see preceding section). The gasbladder connects anteriorly to the intestine through the pneumatic duct at approximately the mid-trunk level. The gasbladder is typically white or shiny in appearance and easily distinguishable from the pale stomach and intestine; its thickness and length is variable within the family, occupying between 6 and 45 percent of the total length of those species examined. It is unlikely that further analysis of visceral anatomy will be fruitful; variation within the family is broad and the genera and higher categories do not appear to be unique in their configurations.

Gonads of ophichthids are elongate paired structures on either side of the large intestine, which appear as thread-like tubes in juveniles. Sex of females can be determined at an earlier stage than that of males; the ovaries develop as highly convoluted yellowish structures in which first fat globules and later small eggs can be discerned by gross microscopic examination; eggs grow in size and increase in number until the peritoneum is completely filled and distended with eggs, the gut usually empty. As is apparently the case in many anguilliforms, the eggs of "ripe" females (the term used for those with the abdominal cavity filled with eggs) commonly reach a diameter of 1.0 mm (and larger) in all species; in very small species the eggs are few in number, appear very large, and are usually in one or two rows on either side of the intestine; large species contain thousands of what appear to be very small eggs (but which are the same 1–2 mm diameter) which fill and distend the abdomen. Ripe male gonads appear as milk-white lobes extending throughout and filling the length of the abdominal cavity. In some species (e.g. those of *Apterichtus*) the gonads extend posteriorly and fill the body cavity well behind the anus and kidney terminus. Little is known about spawning activities of ophichthids except as suggested by location and prevalence of leptocephali.

Distribution

The present zoogeographic patterns of living ophichthids allow an interpretation of the age of modern genera and the speciation patterns of many Atlantic genera and species. Presuming that modern ophichthids (now limited to tropical, subtropical and warm temperate seas) evolved from a lineage with similar environmental and physiological limitations, then it can therefore be assumed that the present distribution reflects the age of certain genera in relation to available seaways and the limitations of historical continental land masses.

Seven of the 25 genera found in the Atlantic are circumtropical. They are *Apterichtus, Bascanichthys, Callechelys, Ichthyapus, Myrichthys, Myrophis* and *Ophichthus.* They range from *Myrophis* with about seven species to *Ophichthus* (as broadly defined) with more than 50 species. Each of these circumtropical genera probably existed prior to the Miocene closure of the Tethyan Seaway (Phillips and Forsyth, 1972) and/or, alternatively, might have transgressed both oceans through the then open Central American Seaway (see McCosker, 1977: 96–100 for a review of ophichthid distribution). The earliest ophichthid known seems to be *Eomyrus dolloi*, an upper Eocene *Ophichthus*-like species from the Wemmelian Formation of western Europe (Storms, 1896). The genus *Phaenomonas*, although possessing single species in the Atlantic, eastern Pacific, and western Pacific, is probably post-Tethyan in origin, unlike the above-mentioned circumtropical genera, each having several species in each water mass.

Six Atlantic genera are limited to the New World and share their congeners with the eastern Pacific. They are *Echiophis, Ethadophis, Gordiichthys* (an undescribed eastern Pacific species exists), *Letharchus, Pseudomyrophis* and *Quassiremus*. Those genera presumably arose subsequent to the Tethyan Seaway closure and prior to the late Pliocene to Pleistocene uplift of the Panamic land mass which closed the Middle American Seaway (Whitmore and Stewart, 1965). Three of those genera (*Echiophis, Ethadophis* and *Pseudomyrophis*) also possess congeners and conspecifics in the eastern Atlantic, presumably reflecting an Atlantic transgression like that of many muraenid genera and species.

The ten monotypic Atlantic genera probably arose since the closure of the New World seaway. Admittedly, several of these very rare deepwater species are known from a single or very few specimens (*Asarcenchelys, Hyphalophis, Kertomichthys, Lethogoleos* and *Mixomyrophis*) and subsequent collections may change our understanding of their distribution patterns. However, it is very unlikely that the unique-appearing *Aplatophis* or the abundant shallow-water monotypic Atlantic genera (such as *Ahlia, Aprognathodon, Caralophia* and *Stictorhinus*) would contain species which have gone undetected in other oceans.

An odd distribution pattern exists within the genus *Herpetoichthys*. It is known from an eastern Pacific species, *H. fossatus* (Myers and Wade), and a mid-Atlantic species, *H. regius* (Richardson). Although its species are uncommon, it is curious that this shallow-water genus has not been discovered in the western Atlantic.

Within the Atlantic, several species appear to form separate populations which we conservatively consider to be conspecific. On the basis of mean vertebral differences, we identify northern and southern western Atlantic populations of *Echiophis intertinctus, Myrophis punctatus* and *Ophichthus gomesii*. We tentatively recognize *Bascanichthys paulensis, B. inopinatus* and *B. bascanium* as distinct species; should they prove conspecific, it too would show the above distributional pattern. *Myrophis punctatus* is further complicated in that it exhibits mean differences in myomere and vertebral numbers for western Gulf of Mexico, Caribbean and Brazilian specimens. Lacking more definitive data, we continue to deem them to be merely populations.

Key to the Subfamilies of the Family Ophichthidae

1a. Tail tip flexible, caudal fin rays conspicuous, confluent with dorsal and anal fins; gill opening midlateral, a constricted opening; accessory branchiostegal rays originate behind ends of epihyal, free rays more numerous than attached rays (visible by radiography or by dissection) ...... Subfamily Myrophinae, p. 270

1b. Tail tip a hard or fleshy finless point (except in two extralimital genera); gill opening midlateral to entirely ventral, unconstricted; all branchiostegal rays originate either in association with hyoid or before level of epihyal tips ...... Subfamily Ophichthinae, p. 296

Subfamily MYROPHINAE

WORM EELS

The subfamily Myrophinae contains those ophichthid eels most easily recognized by their externally visible caudal-fin rays. Other diagnostic characters of juveniles and adults include: gill opening mid-lateral, a constricted opening; dorsal fin mid-trunk or posterior in origin; ceratohyal not divided into a short median and a long distal portion; basal plate of urohyal ossified, its posterior extension cartilaginous; accessory branchiostegal rays originate behind tips of epihyal, free rays more numerous than attached; branchial skeleton reduced, fifth ceratobranchial absent; and coloration uniform or darkened dorsally.

The Myrophinae is divided into two tribes, the monotypic Benchenchelyini, and the Myrophini containing ten genera (McCosker, 1977).

Key to the Genera of Ophichthidae, Subfamily Myrophinae

(Genera entirely extralimital to the western Atlantic are included, but are not treated in the following generic diagnoses.)

1a. Tail tip flexible, caudal fin rays conspicuous, confluent with dorsal and anal fins; gill opening midlateral, a constricted opening; accessory branchiostegal rays originate behind ends of epihyal, free rays more numerous than attached rays (visible by radiography or by dissection) . . . . Subfamily Myrophinae, 2

1b. Tail tip a hard or fleshy finless point (except in two extralimital genera); gill opening midlateral to entirely ventral, unconstricted; all branchiostegal rays originate either in association with hyoid or before level of epihyal tips . . . . Subfamily Ophichthinae, p. 296

2a. Anterior nostril non-tubular, posterior nostril before eye; eye large, about 6 times in head length; pectoral fin moderately developed . . . . Tribe Benthenchelyini, *Benthenchelys* Fowler, 1934 (a single midwater species from the western Pacific and Indian Ocean)

2b. Anterior nostril tubular, posterior nostril before eye, along upper lip, or within mouth; eye small, 10 or more times in head; pectoral fin may be absent . . . . Tribe Myrophini, 3

3a. Posterior nostril before eye, above upper lip and not covered by a flap; pectoral fin present, but may be reduced to a small, barely noticeable flap in posterodorsal corner of gill opening 4

3b. Posterior nostril labial, either within lip and opening into mouth, or along lip and covered by a flap; pectoral fin either present and well developed or absent 5

4a. Dorsal-fin origin in anterior trunk region; snout conical; pectoral fin well developed, longer than snout; two preopercular pores . . . . *Neenchelys* Bamber, 1915 (several Indo-Pacific species)

4b. Dorsal-fin origin in posterior trunk region; snout broad, tumid; pectoral fin minute, smaller than eye; three preopercular pores (except in *P. atlanticus*) . . . . *Pseudomyrophis*, p. 289

5a. Pectoral fin well developed; pleural ribs absent behind 15th–20th vertebrae (visible in a radiograph) 6

5b. Pectoral fin minute or absent; pleural ribs present on all trunk vertebrae 8

6a. Dorsal-fin origin above or behind anus; maxilla stout and abutting pterygoid, not tapering posteriorly; vomerine teeth absent . . . . *Ahlia*, p. 272

6b. Dorsal-fin origin anterior to mid-trunk region; maxilla thin and tapering posteriorly, not closely associated with pterygoid; vomerine teeth present 7

7a. Body elongate, its depth more than 70 times in total length; teeth slender, slightly depressible: posterior nostril on lip and covered by a flap .... *Asarcenchelys*, p. 276
7b. Body stouter, its depth less than 50 times in total length; teeth stouter, not depressible; posterior nostril opens into mouth .... *Myrophis*, p. 279
8a. Pectoral fin minute but apparent; body moderately elongate, its depth 45 times or more in total length .... *Mixomyrophis*, p. 277
8b. Pectoral fin absent; body stouter, its depth less than 45 times in total length 9
9a. Tongue elongate, extending well beyond mouth and decorated with a fleshy appendage; inner edge of lips and palate decorated with fleshy lappets; teeth conical and uniserial .... *Glenoglossa* McCosker, 1982 (a single western Pacific species)
9b. Tongue not elongate, not extending outside of mouth, lacking a fleshy appendage at its tip; inner edge of lips and palate smooth; teeth either conical or blunt, uniserial or multiserial ... 10
10a. A prominent median toothed groove on ventral side of snout, bordered by dermal folds, extending anteriorly to anterior nostrils; anterior nostrils elongated tubes equal to eye in length .... *Schismorhynchus* McCosker, 1970 (a single Pacific species)
10b. Ventral side of snout without a prominent median groove bordered by dermal folds; anterior nostrils less than eye in length 11
11a. Teeth present on intermaxillary, maxillary, dentary and vomer; dorsal-fin origin either before or behind anus .... *Muraenichthys* Bleeker, 1853 (numerous species from the Indian and Pacific oceans)
11b. Teeth absent on vomer, absent or embedded on intermaxillary, those on maxillary and dentary minute or villiform; dorsal-fin origin behind anus .... *Schultzidia* Gosline, 1951 (two central and western Pacific species)

Tribe Benthenchelyini

The Benthenchelyini is monotypic, containing *Benthenchelys cartieri* Fowler, a pelagic species found over deepwater in the central Indo-Pacific (Castle, 1972). The species of this tribe differ most obviously from those of the Myrophini in possessing a short broad neurocranium and a non-tubular anterior nostril. The tribe is diagnosed as follows: body moderately elongate, compressed throughout; tail much longer than body; anterior nostril not tubular; posterior nostril lateral, before center of orbit; median fins elevated; pectoral fin moderately developed; head pores enlarged, a single preopercular pore; frontal commissure weakly developed; lateral-line ossicles nearly absent; neurocranium short, rounded; otic bulla and nasals absent; maxilla broad, not produced posteriorly, articulating beneath anterior margin of orbit; gill arches weakly ossified, hypobranchial 3 ossified; epipleural ribs limited to anterior 14–16 vertebrae; vertebrae distinctive, neural arches prominent; coloration uniform, although darker dorsally.

Tribe Myrophini

The Myrophini are relatively small species with plain coloration which live in a variety of habitats, from relatively fresh waters of rivers and tidal creeks to shallow sandy shores to waters as deep as 450 m. The tribe is diagnosed as follows: body short to elongate, compressed throughout; tail generally longer than body; anterior nostril tubular; posterior nostril either lateral or labial; gill opening lateral, a constricted opening; median fins low or elevated, dorsal-fin origin behind gill openings; pectoral fin present or absent; head pores variably developed; lateral-line canal weakly ossified; neurocranium not raised along frontal or parietal midline, supraoccipital crest developed in some species; gill arches reduced,

TABLE 24. Mean vertebral formulae (MVF) and ranges of vertebral counts in western Atlantic species of Myrophini.

		Predorsal		Preanal		Total	
Species	MVF	Range	n	Range	n	Range	n
*Ahlia egmontis**	68-64-159	65–72	28	60–66	30	155–167	33
*Asarcenchelys longimanus**	27-54-148	27	2	53–55	2	148	2
*Mixomyrophis pusillipinnis**	33-57-178	33	1	57	1	178	1
*Myrophis anterodorsalis**	16-47-141	14–17	15	45–49	15	135–147	15
*platyrhynchus**	20-47-143	18–22	18	46–48	18	140–146	16
*plumbeus**	27-54-150	25–30	17	53–57	19	146–154	17
punctatus	32-53-146	29–34	25	51–55	30	141–154	42
*Pseudomyrophis frio**	39-67-188	37–40	6	67–68	6	184–191	6
*fugesae**	30-52-157	27–33	2	51–53	2	155–159	2
*nimius**	49-73-214	47–53	14	71–75	15	211–217	15

* Includes holotype.

weakly ossified; epipleural ribs limited to anterior trunk vertebrae in some species; and coloration uniform, often darker dorsally.

The Myrophini includes ten genera, five of which (*) are present in the Atlantic, in three distinct lineages: *Myrophis**, *Ahlia**, *Asarcenchelys** and *Mixomyrophis**; *Neenchelys* and *Pseudomyrophis**; and *Muraenichthys, Glenoglossa, Schismorhynchus* and *Schultzidia.*

Genus *Ahlia* Jordan and Davis, 1891

Ahlia Jordan and Davis, 1891: 639 (type species *Myrophis egmontis* Jordan, 1884, by original designation).

Characters. Body moderately stout (depth 1.8–3.7% TL), laterally compressed throughout; head and trunk shorter than tail (41–46% TL). Dorsal-fin origin above or behind anus; median fins confluent with caudal; pectoral fin well devel-

TABLE 25. Vertebral counts of holotypes of valid species of western Atlantic myrophins and their synonyms.

Species	Catalog number	Predorsal	Preanal	Total
Ahlia egmontis (Jordan)	USNM 35086	67	63	159
Myrophis macrophthalmus Parr	BOC 2593	66	62	160
Myrophis microps Parr	BOC 2597	68	63	158
Asarcenchleys longimanus McCosker	MNHN 1968-215	27	53	148
Mixomyrophis pusillipinna McCosker	ANSP 152305	33	57	178
Pseudomyrophis frio Jordan & Davis	USNM 46760	38	68	188
Pseudomyrophis fugesae n. sp.	ANSP 153910	33	53	155
Pseudomyrophis nimius Böhlke	USNM 186274	50	72	213
Myrophis anterodorsalis n. sp.	ANSP 153670	16	48	146
Myrophis platyrhynchus Breder	BOC 30	20	47	142
Myrophis emmae Howell-Rivero	MCZ 33454	22	48	143
Myrophis plumbeus (Cope)	ANSP 22964	26	53	149
Myrophis punctatus Lütken	———			
Myrophis dolicorhynchus Parr	BOC 2602	33	52	143
Myrophis iguapensis Miranda-Ribero	———			
Myrophis longleyi Silvester	———			
Myrophis lumbricus Jordan & Gilbert	USNM 30896	32	52	143
Myrophis microstigmius Poey	MCZ 33440	33	53	147
Neoconger perlongus Poey	USNM 37478	31	51	144

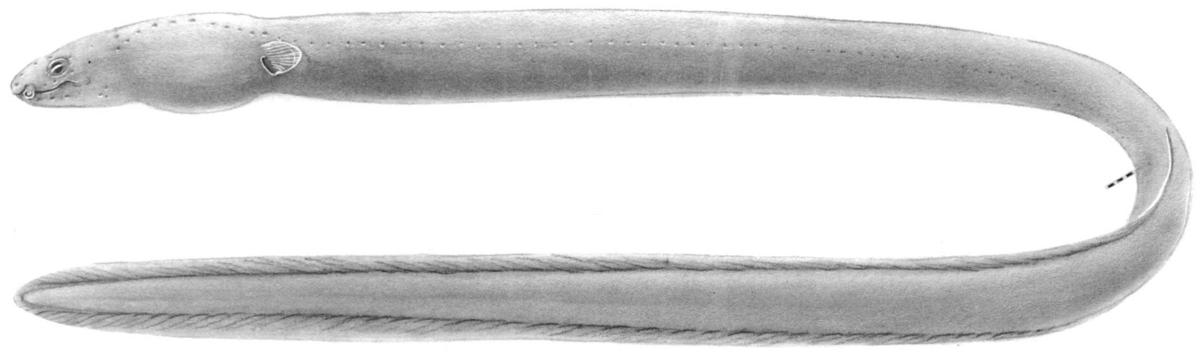

FIGURE 252. *Ahlia egmontis*: ANSP 111568, 348 mm TL, Bahamas. Illustration by S. P. Gigliotti.

oped, rounded and broad based. Snout subconical, broad, overhanging lower jaw, lacking a groove on its underside; lips without barbels; eye well developed, its size affected by sexual condition; anterior nostril tubular, with lateral interior projection; posterior nostril along edge of lip, beneath a flap and opening into mouth. Gill opening constricted. Head pores developed; three preopercular pores. Teeth conical, small, uniserial on jaws, absent on vomer.

Neurocranium subtruncate posteriorly; supraoccipital crest present; maxilla broad, not tapering posteriorly, closely abutting the short and broad pterygoid; otic bulla weakly developed; suspensorium nearly vertical; opercular series developed, their margins entire. Branchiostegal rays numerous, slender; 6–7 rays along the epihyal, the remainder lie behind epihyal tips; urohyal reduced to an ossified basal plate with a cartilaginous posterior filament. Gill arches less developed than those of *Myrophis*; first basibranchial ossified, 2–4 absent; hypobranchials 1–2 ossified, 3 absent; ceratobranchials 1–4 ossified, 5 absent; infrapharyngobranchial 2 absent, 3 ossified; upper pharyngeal tooth plates separate. Pectoral girdle well developed; cleithrum and supracleithrum slender, coracoid and scapula broad. Intramuscular bones, ribs, and caudal transverse processes developed; epipleural ribs limited to anteriormost 15–20 vertebrae; precaudal vertebrae less numerous than caudal. Gasbladder weakly developed. Peritoneum pigmented. Body coloration uniformly pale, darker dorsally.

Size. Maximum size 433 mm TL.

Distribution. Found only in the western Atlantic.

Etymology. Named for Jonas Nicolas Ahl, author of "De Muraena et Ophichtho" which "furnishes the beginning of our systematic arrangement of the eels" (Jordan and Davis, 1892: 639), with the noun suffix *-ia.* Neuter.

Remarks. Ahlia is easily recognized by its posterior dorsal-fin origin and its lack of vomerine teeth. Numerous authors have argued for and against *Ahlia* as a genus distinct from *Myrophis.* Recent authors (Myers and Storey, 1939; Wade, 1946; Nelson, 1966a; Dean, 1972; McCosker, 1977) have considered *Ahlia egmontis* to warrant generic status. The generic differences identified by McCosker (1977: 58) are clearly related to feeding specialization in *A. egmontis,* viz., tooth loss, maxillary-pterygoid bracing, and gill arch reduction, yet the universality of these characters among the species of *Myrophis* suggests that the species of *Myrophis* form a natural group from which *Ahlia* is a specialized offshoot.

Species. A single species, *Ahlia egmontis* (Jordan, 1884).

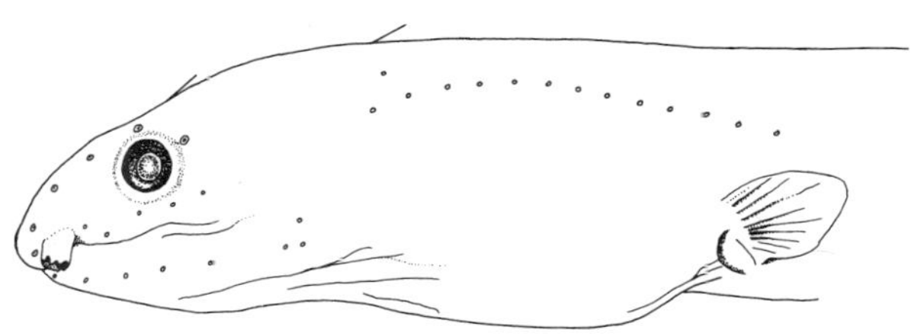

FIGURE 253. *Ahlia egmontis*: ANSP 105788, 271 mm TL, Lesser Antilles; head.

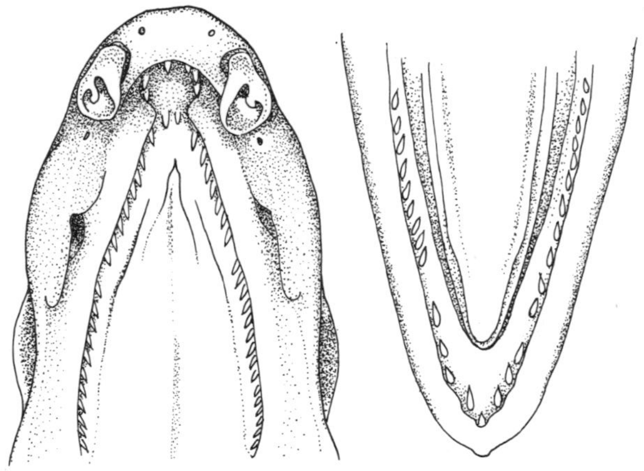

FIGURE 254. *Ahlia egmontis*: ANSP 110409, 305 mm TL, Santo Domingo; dentition.

Ahlia egmontis (Jordan, 1884)

KEY WORM EEL

Figures 252–255, Tables 24–26

Myrophis egmontis Jordan, 1884: 44 (original description, Egmont Key, Florida, lectotype USNM 35086).

Ahlia egmontis, Jordan and Davis, 1891: 639 (new genus). Jordan and Evermann, 1896a: 371 (lectotype designated).

Leptocephalus humilis Strömman, 1896: 29, pl. 2, figs. 7–9 (original description, 03°S, 30°W). Leptocephalus.

Leptocephalus crenatus Strömman, 1896: 32, pl. 3, figs. 4–5 (original description, off Georgia). Leptocephalus.

Leptocephalus hexastigma Regan, 1916: 141, pl. 7, fig. 6 (original description, 18°00′S, 31°45′W). Leptocephalus.

Myrophis macrophthalmus Parr, 1930b: 10, fig. 1, lower (original description, Green Cay, Bahamas, holotype YPM 2593).

Myrophis microps Parr, 1930b: 11, fig. 1, upper (original description, edge of Great Bahama Bank, holotype YPM 2597).

TABLE 26. Proportional differences, as percent of head length, in eye size and pectoral-fin length of immature and mature specimens of *Ahlia egmontis*.

	Immature	Ripe females		Ripe males	
	Range	Range	Mean	Range	Mean
Eye diameter	5.7–7.7	10–13	11	12–18	14
Pectoral fin length	10–17	20–21	21	18–26	22

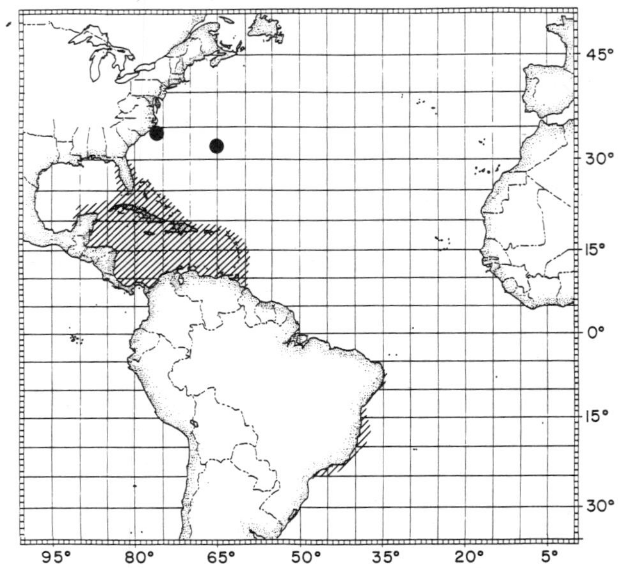

FIGURE 255. Distribution of *Ahlia egmontis* based on study material.

Misidentification:

Not *Ahlia egmontis*. Borodin, 1928: 9 (presumably a misidentification of *Myrophis vafer* Jordan and Gilbert, Galapagos Archipelago).

Distinctive Characters. A moderately stout myrophine with nearly uniform coloration or faintly bicolored; dorsal-fin origin above or just behind anus; vomerine teeth absent, jaw teeth uniserial; MVF 68-64-159.

Description. Total vertebrae 155–167, $\bar{x}$ = 159.3 (n = 33); predorsal vertebrae 65–72, $\bar{x}$ = 67.7 (28); preanal vertebrae 60–66, $\bar{x}$ = 64.2 (30). Pores: SO 1 + 4; IO 5 + 1; POM 5 + 3; ST 3.

Proportions as % of TL: head 7.9–10; trunk 33–36; tail 54–59; predorsal 44–49; depth behind gill openings 1.8–3.7. Of HL: snout 17–21; snout to rictus 23–30; eye 5.6–7.7; interorbital width 9.8–13; pectoral-fin length 10–17.

General characters as for the genus. Dentition as follows: an outer circle of 6–9 stout intermaxillary teeth, plus several smaller scattered inner teeth visible on stained specimens; vomerine teeth absent; jaw teeth small, conical and recurved, 11–25 teeth along maxillary, 12–19 highly recurved teeth along dentary.

Color. In life and in alcohol pale or tan with scattered melanophores dorsally, sometimes appearing partially bicolored. Sexually mature individuals darker overall, noticeably so posteriorly on tail and fins and in branchial area, with reddish-brown nuchal band encircling body at level of supratemporal canal, color continued back along lateral line; red-brown patch on lower jaw.

Size. The largest specimen examined was 433 mm TL. Of 18 females (272–433 mm), seven (276–387) were ripe with 0.3–0.6-mm eggs; of 15 males (142–402), 13 (142–343) were ripe.

Development. Leptocephalus identified (p. 795).

Cohen and Dean (1970) and McCosker (1977) have shown that with the onset of sexual maturity, this species moves offshore; the eye becomes large, and changes occur in coloration and in size of pectoral fin, males changing more than females (Table 22). Sexually mature individuals were found in collections from the southwestern Caribbean and from the Bahamas (some of the latter named *Myrophis macrophthalmus* by Parr). The Caribbean specimens were conspicuously darker and had greatly enlarged eyes; one collection (ANSP 126526) contained six ripe males and three ripe females, the body cavities of both sexes filled with gonads, the guts empty. The Bahamian specimens, all males, were small and pale but exhibited very large eyes and developed gonads. Females attain a considerably larger size than do males, and the bimodal distribution of sexes by length frequency (cf. Cohen and Dean, 1970: fig. 2) would suggest that the species might be a protandrous hermaphrodite.

Distribution (Fig. 255). A widespread species, known from North Carolina (Ross and Fast, 1977) to Brazil, including Bermuda, throughout the West Indies, the Caribbean, and off northern South America. An extremely common worm eel, found along sandy shores, tidal creeks and embayments, as well as inhabiting eel grass beds and the sand edge of coral reefs; usually found in shallow waters (13 m or less), recorded to 37 m.

Etymology. Named for Egmont Key, Florida, the type locality. An adjective.

Remarks. Parr (1930: 10–12) described two new species, *Myrophis macrophthalmus* and *M. microps*, which were similar to *Ahlia egmontis* in dorsal-fin placement but which differed in having teeth on the vomer. These nominal species are synonyms of *A. egmontis*; the "vomerine" teeth are the intermaxillary teeth, and the large eye of *macrophthalmus* was due to sexual condition.

Study Material. A total of 365 specimens, 55–433 mm TL, including the lectotype and the holotypes and paratypes of *Myrophis macrophthalmus* and *Myrophis microps*. [An additional 120 specimens from Brazil not examined.]

Lectotype: USNM 35086 (406); Florida, Egmont Key; Mr. E. Jewett.

Other material: BERMUDA: ANSP 133229 (1, 402), 134998 (1, 129), 148172 (1, 238), 148179 (1, 258). MCZ 34653 (1, 107). COASTAL UNITED STATES: ANSP 128387 (1, 246), 128388 (2, 128–185). *Florida Atlantic:* MCZ 64508 (1, 204). *Florida Keys:* ANSP 110433 (2, 62–101), 124020 (6, 75–158), 124547 (1, 100), 126195 (7, 196–386), 134506 (1, 135). USNM 33734 (6, 55–85), 167623 (6, 110–263). GULF OF MEXICO: *Florida:* ANSP 98512 (1, 242), 98515 (2, 338–346), 98516 (1, 236). USNM 116778 (1, 190), 117168 (1, 162).

BAHAMAS: ANSP 88484 (1, 58), 98305 (1, 163), 98306 (1, 96), 98307 (1, 192), 98308 (2, 94–225), 98309 (3, 57–72), 98310 (1, 42), 98311 (2, 184–207), 98312 (1, 169), 98313 (1, 208), 98314 (1, 104), 98315 (1, 192), 98316 (3, 134–345), 98317 (2, 142–341), 98318 (2, 146–217), 98319 (1, 195), 98320 (1, 275), 98321 (2, 178–184), 98322 (4, 124–151), 98323 (4, 103–270), 98324 (3, 228–280), 98325 (8, 88–168), 98326 (2, 91–154), 98327 (1, 139), 98328 (12, 93–256), 98329 (8, 134–303), 98330 (49, 54–290), 98331 (1, 402), 98332 (2, 115–323), 100198 (1, 273), 111568 (1, 348), 115004 (1, 156), 116140 (3, 182–308, cleared and stained), 117103 (2, 75–113), 128793 (1, 55), 128798 (6, 66–194), 128826 (1, 206, cleared and stained), 130555 (1, 185), 140511 (1, 186), 149254 (1, 94), 158802 (6, 56–88), 158803 (1, 102), 158804 (1, 87), 158805 (3, 82–111), 158806 (2, 85–139). USNM 272466 (9, 65–127). YPM 27 (1, 158), 29 (1, 276). YPM 2593 (1, 271); holotype of *Myrophis macrophthalmus*. YPM 2594 (1, 225) and 2595 (1, 146); paratypes of *M. macrophthalmus*. YPM 2597 (1, 163); holotype of *Myrophis microps*. YPM 2599 (1, 122) and YPM 2600 (1, 147); paratypes of *M. microps*.

WEST INDIES: USNM 108359 (1, 242). SANTO DOMINGO: ANSP 110409 (1, 305). HAITI: ANSP 114562 (2, 55–208), 119054 (1, 55). PUERTO RICO: ANSP 115551 (1, 174), 118705 (1, 110), 126500 (3, 275–433), 129900 (1, ca 305), 139831 (6, 79–258), 140581 (1, 167). USNM 184700 (1, 185). USNM uncat. (1, 148). VIRGIN ISLANDS: ANSP 123861 (2, 78–129), 124253 (2, 111–139), 124381 (1, 253), 124485 (3, 101–140), 126146 (4, 92–195). USNM 100460 (1, ca 55). LESSER ANTILLES: *Anguilla Island:* ANSP 120537 (1, 265). *St. Martin:* ANSP 105729 (4, 60–131). *St. Barthelemy:* ANSP 105700 (1, 138). MNHN 1974-139 (1). *Antigua:* ANSP 105776 (1, 168). BMNH 1934.9.10.62.67 (6, 60–87). *Dominica:* USNM 272463 (1, 345), 272464 (1,

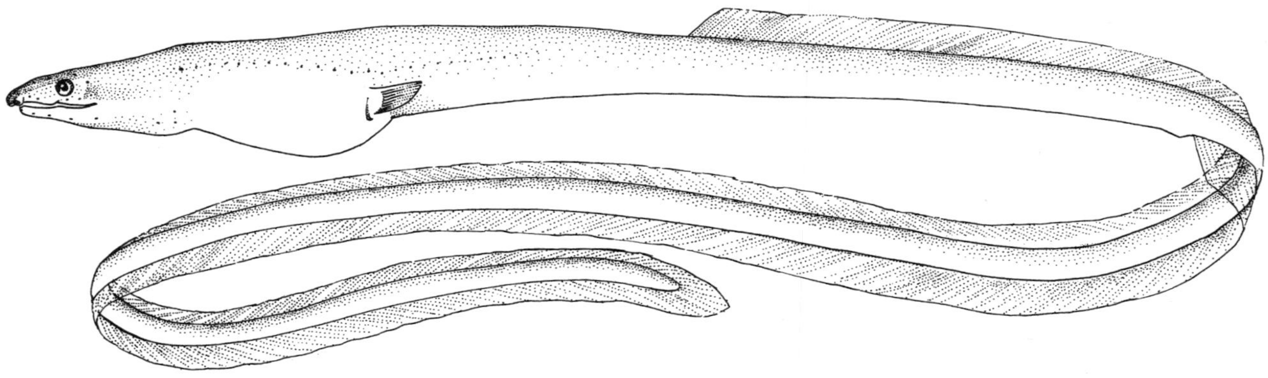

FIGURE 256. *Asarcenchelys longimanus*: MNHN 1968-215, holotype, 277 mm TL, Brazil. Illustration by A. Pertschuk.

111), 272465 (1, 192), 272467 (4, 111–153). *Martinique:* ANSP 105702 (2, 232–272), 106183 (7, 150–345). USNM 272464 (1, 231–272). *St. Lucia:* ANSP 105699 (7, 129–216), 105705 (1, 183), 105761 (1, 103). *Barbados:* BMNH 1933.8.8.28.31 (1). *Little St. Vincent:* ANSP 105745 (4, 89–131), 105760 (1, 177), 105781 (4, 111–154). *Union Island:* ANSP 105788 (3, 105–271). CAS 54463 (13, 109–270). *Grenada:* ANSP 105704 (3, 70–185), 105742 (1, 145). *Tobago:* ANSP 98474 (1, 127), 98535 (11, 179–285). USNM 178209 (1, 241). *Tortuga:* ANSP 75194 (1, 258).

COLOMBIA: ANSP 117562 (1, 100), 153662 (1, 85). BELIZE: USNM 278171 (1, 166). NICARAGUA: ANSP 126526 (9, 145–345). MEXICO: *Gulf of Campeche:* ANSP 98523 (3, 327–387). MCZ 47914 (7, 105–143). ST. ANDREWS ISLAND: ANSP 110418 (2, 113–210). COURTOWN KEY: ANSP 87670 (1, 246). OLD PROVIDENCE ISLAND: ANSP 87671 (2, 142–160), 128980 (7, 60–224). SERRANA BANK: ANSP 102449 (2, 133–191). GRAND CAYMAN: ANSP 141341 (1, 56).

VENEZUELA: *Los Roques:* MBUCV 315 (1), 857 (1), 1049 (1), 2738 (4). BRAZIL: MNRJ 9513 (3, 200–252). [MZUSP; 120 specimens, 21 lots from Pernambuco, Alagoas, Bahia, and São Paulo.]

Genus *Asarcenchelys* McCosker, 1985

Asarcenchelys McCosker, 1985: 12 (type species *Asarcenchelys longimanus* McCosker, 1985, by original designation).

Characters. Body elongate (depth 1.4% TL), laterally compressed throughout trunk and tail; head and trunk shorter than tail (38–44% TL). Dorsal-fin origin in anterior trunk region; median fins elevated, confluent at caudal; pectoral fins well developed, lanceolate. Snout subconical, tumid, not grooved ventrally, overhanging lower jaw; lips without barbels; eye large, behind middle of upper jaw; anterior nostril tubular; posterior nostril on outer edge of lip and covered by a flap. Gill opening constricted. Head pores developed; three preopercular pores. Teeth conical, large, not close set; biserial anteriorly on jaws and on vomer.

Gill arches appear to be well developed for a myrophin; first basibranchial ossified, the others appear to be cartilaginous or absent; ceratobranchials 1–4 ossified, 5 not apparent. Neurocranium stout, truncate posteriorly; supraoccipital crest developed; maxillae taper posteriorly; suspensorium posteriorly inclined. Pectoral girdle reduced to a stout cleithrum and thin supracleithrum. Epipleural ribs present only on anterior trunk vertebrae; caudal transverse processes apparently absent. Caudal vertebrae more numerous than precaudal. Peritoneum pallid. Body coloration uniformly pale, darker dorsally.

Size. To 277 mm TL.

Distribution. Found in the western Atlantic off Brazil.

Etymology. From the Greek *asarkos* (lean) and *enchelys* (eel), in reference to its emaciated appearance. Treated as feminine according to Opinion 915 of the Bulletin of Zoological Nomenclature, 1970.

Species. A single species, *Asarcenchelys longimanus* McCosker, 1985.

Asarcenchelys longimanus McCosker, 1985
EMACIATED WORM EEL
Figures 256–258, Tables 24, 25

Asarcenchelys longimanus McCosker, 1985: 12, figs. 3–5, 6c (original description, near Belem, Brazil, holotype MNHN 1968-215).

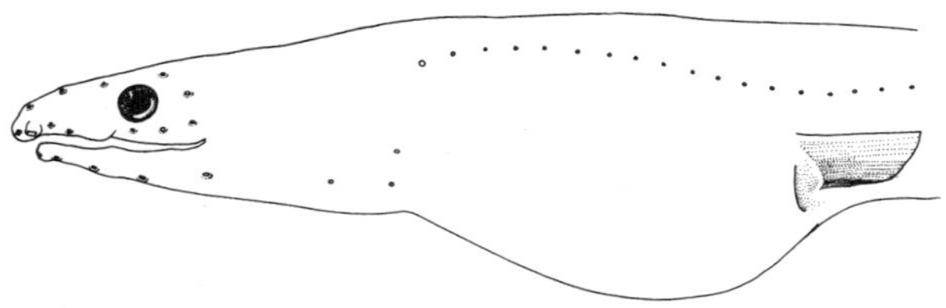

FIGURE 257. *Asarcenchelys longimanus*: MNHN 1968-215, holotype, 277 mm TL, Brazil; head. Illustration by A. Pertschuk.

Distinctive Characters. An elongate myrophin with pale coloration, darker dorsally; dorsal-fin origin in anterior trunk region; vomerine and jaw teeth biserial anteriorly and uniserial posteriorly; MVF 27-54-148.

Description. Total vertebrae 148 (n = 1); predorsal vertebrae 27 (2); preanal vertebrae 53–55, $\bar{x}$ = 54 (2). Pores: SO 1 + 3; IO 5 + 2; POM 5 + 3; ST 3.

Proportions as % of TL: head 9.7–12; trunk 28–31; tail 56–62; predorsal 20–23; depth behind gill opening 1.4. Of HL: snout 14–21; snout to rictus 24–32; eye 5.2–7.2; interorbital width 5.9–7.2; pectoral-fin length 15–18.

General characters as for the genus. Teeth conical, fairly large, not close set, subequal. A ring of 9 intermaxillary teeth; 24 vomerine teeth, 3 anterior uniserial teeth followed by 3–4 pairs, the remainder uniserial; maxillary teeth biserial anteriorly, uniserial posteriorly, 22–23 in outer row, 6 in inner row; 24–25 outer mandibular teeth plus 4 large anterior inner teeth.

Color. Cream to white, with numerous fine brown spots on snout, behind eye, and on dorsal surface of body. All fins transparent.

Size. The largest specimen is 277 mm TL.

Development. Leptocephalus tentatively identified (see p. 786).

Distribution. The two type specimens were captured off Brazil in 55 m.

Etymology. From the Latin *longus* (long) and *manus* (hand), in reference to the elongate pectoral fins. A noun in apposition.

Remarks. It should be noted that both type specimens of *A. longimanus* are damaged and thereby the total length measurements may be in error by a few mm. The holotype was broken during capture behind the fifteenth vertebra and is twisted in preservative; however, the head remains attached by the skin to the trunk region and allows measurements to be taken and characters to be analyzed. The paratype appears intact, but the radiograph indicates that the tail has been severed and regrown; it has 17 fewer vertebrae than the holotype. Records at the Paris Museum were examined by M.-L. Bauchot, but exact locality or means of capture of the type specimens could not be determined.

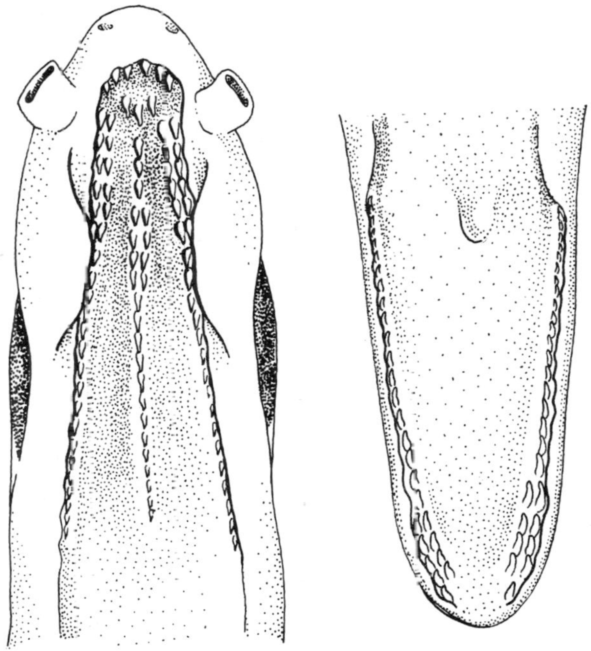

FIGURE 258. *Asarcenchelys longimanus*: MNHN 1968-215, holotype, 277 mm TL, Brazil; dentition. Illustration by A. Pertschuk.

Study Material. Holotype: MNHN 1968-215 (277 mm TL); Brazil, captured near Belem, 55 m; Sept. 1966. *Paratype:* MNHN B.2994 (147+); collected with the holotype.

Genus *Mixomyrophis* McCosker, 1985

Mixomyrophis McCosker, 1985: 10 (type species *Mixomyrophis pusillipinna* McCosker, 1985, by original designation).

Characters. Body moderately elongate (depth 2.2% TL), laterally compressed, particularly posteriorly; head and trunk shorter than tail (37% TL). Dorsal-fin origin in mid-trunk; median fins elevated; pectoral fins minute. Snout subconical, broad from above, not grooved ventrally, slightly overhanging lower jaw; lips without barbels; eye

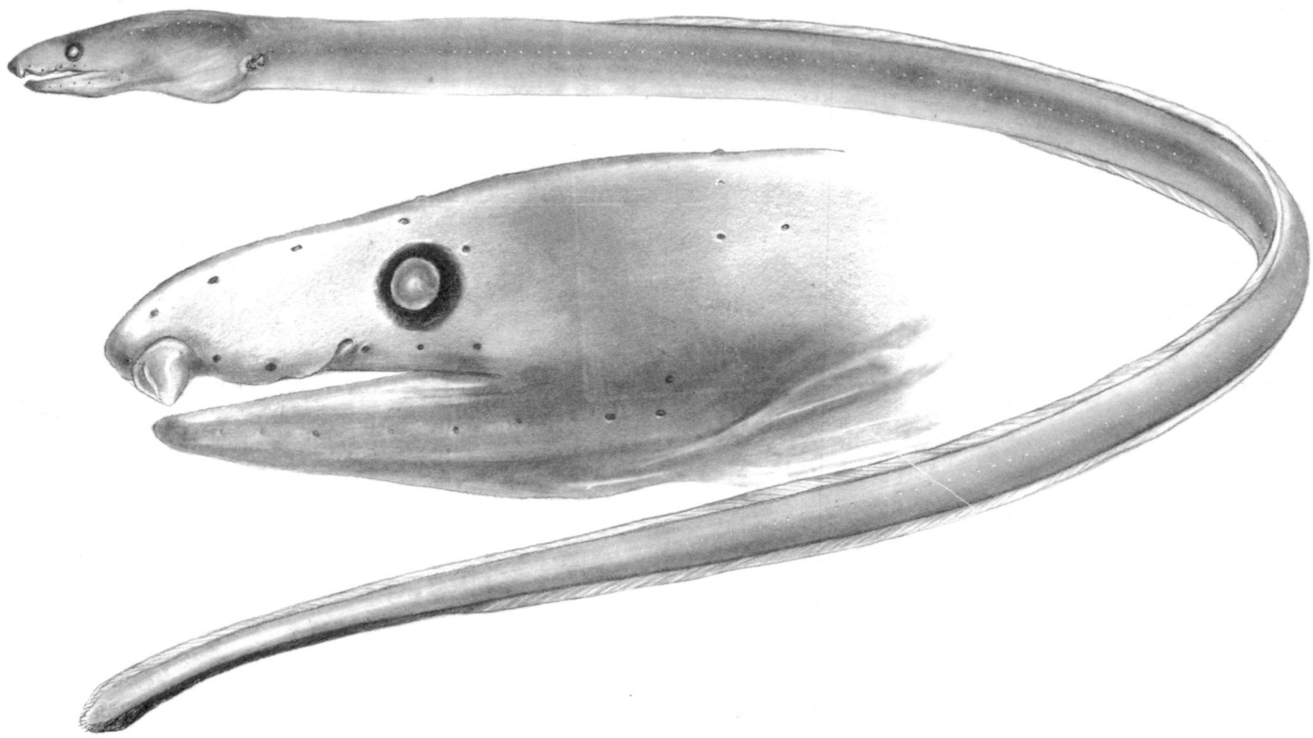

FIGURE 259. *Mixomyrophis pusillipinna*: ANSP 152305, holotype, 407 mm TL, Lesser Antilles. Insert: head.

large, behind middle of upper jaw; anterior nostril tubular; posterior nostril on outer edge of lip and covered by a flap. Gill opening constricted. Head pores developed; three preopercular pores. Teeth uniserial, conical, small and close-set.

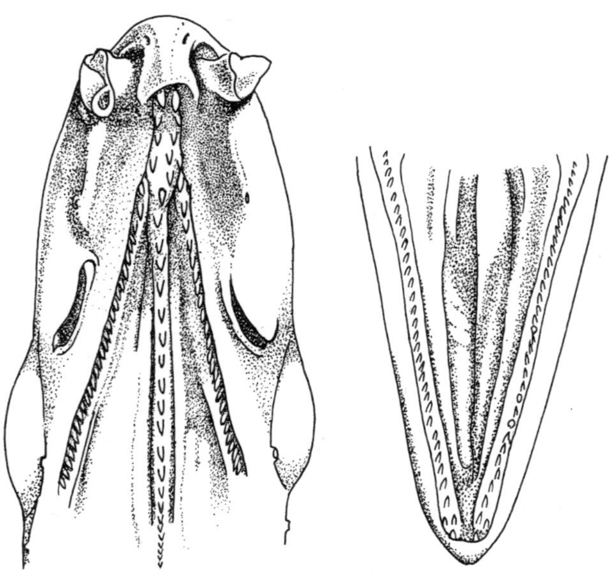

FIGURE 260. *Mixomyrophis pusillipinna*: ANSP 152305, holotype, 407 mm TL, Lesser Antilles; dentition.

Gill arches well developed for a myrophine; hypobranchials 1–2 ossified, the third cartilaginous; ceratobranchials 1–4 ossified, the fifth absent; upper pharyngeal tooth plate fused. Neurocranium stout, slightly sloping posteriorly; suspensorium anteriorly inclined; pterygoid stout, not bracing maxilla; maxillae elongate, tapering posteriorly; opercular series apparently moderately developed. Pectoral girdle reduced to a slender cleithrum and supracleithrum; epipleural ribs on all precaudal vertebrae; caudal transverse processes apparently absent; caudal vertebrae more numerous than precaudal. Peritoneum black. Coloration pale, peppered with small brown specks.

Size. To 407 mm TL.

Distribution. A single specimen from the western tropical Atlantic.

Etymology. From the Greek *mixis* (a mixing) and *Myrophis* (a related ophichthid), in reference to the combination of myrophine characters which this eel possesses. Masculine.

Species. A single species, *Mixomyrophis pusillipinna* McCosker, 1985.

Mixomyrophis pusillipinna McCosker, 1985

MOSAIC WORM EEL

Figures 259, 260, Tables 24, 25

Mixomyrophis pusillipinna McCosker, 1985: 10, figs. 1, 2, 6b (original description, Lesser Antilles, off Anguilla, holotype ANSP 152305).

Distinctive Characters. A moderately elongate myrophin with pale coloration; dorsal-fin origin in mid-trunk, pectoral fin minute; dentition uniserial; MVF 33-57-178.

Description. Total vertebrae 178 (n = 1); predorsal vertebrae 33 (1); preanal vertebrae 57 (1). Pores: SO 1 + 4; IO 5 + 1; POM 6 + 3; ST 3.

Proportions as % TL: head 9.0; trunk 28; tail 63; predorsal 23; depth behind gill openings 2.2. Of HL: snout 22; snout to rictus 33; eye 7.9; interorbital width 12; pectoral-fin length 2.7.

General characters as for the genus. Teeth small, conical, nearly uniform in size; a patch of 11 intermaxillary teeth continuous with uniserial row of 22 vomerine teeth; jaw teeth uniserial, 24–30 on maxillaries, 34 on each mandible plus 1–2 inner teeth at symphysis.

Color. Yellowish on head, chin, dorsal surface of trunk and tail; throat and belly whitish; body and tail finely peppered with small brown specks.

Size. The only known specimen is 407 mm TL, a female with ripening ovaries.

Development. Leptocephalus tentatively identified (p. 784).

Distribution. Taken off the northern Lesser Antilles by trawl in deep water.

Etymology. From the Latin *pusillus* (puny or insignificant) and *pinna* (fin), in reference to the minute pectoral fin. A noun in apposition.

Study Material. Holotype: ANSP 152305 (407 mm TL); LESSER ANTILLES: off Anguilla, 18°26.4′N, 63°12.6′W, 393–451 m; 22 July 1969.

Genus *Myrophis* Lütken, 1851

Myrophis Lütken, 1851: 14 (type species *Myrophis punctatus* Lütken, 1851, by monotypy).

Holopterura Cope, 1871: 482 (type species *Holopterura plumbea* Cope, 1871, by monotypy).

Leptoconger Poey, 1880: 250 (type species *Neoconger perlongus* Poey, 1876 = *Myrophis punctatus* Lütken, 1851, by original designation).

Hesperomyrus Myers and Storey, 1939: 157 (type species *Hesperomyrus fryi* Myers and Storey, 1939 = *Myrophis vafer* Jordan and Gilbert, 1882, by original designation).

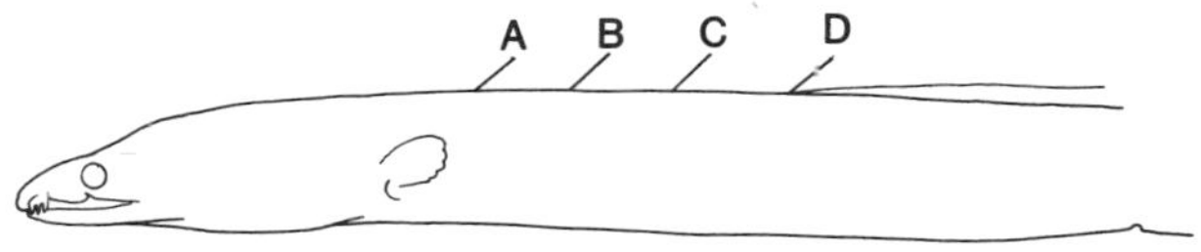

FIGURE 261. Diagram of position of dorsal-fin origin in Atlantic species of *Myrophis*. A. *M. anterodorsalis*. B. *M. platyrhynchus*. C *M. plumbeus*. D. *M. punctatus*.

Characters. Body moderately elongate to stout (depth 2.3–4.8% TL in western Atlantic species), laterally compressed throughout; head and trunk shorter than tail (36–42% TL). Dorsal-fin origin before mid-trunk; median fins confluent at caudal; pectoral fin well developed, generally rounded and broad based. Snout conical, overhanging lower jaw, lacking a groove on its underside; lips without barbels; eye well developed, its size affected by sexual condition; anterior nostril tubular; posterior nostril along edge of lip, beneath a flap and opening into mouth. Gill opening constricted. Head pores developed; three preopercular pores. Teeth conical, small, uniserial to multiserial on jaws and on vomer.

Neurocranium subtruncate posteriorly; supraoccipital crest present; maxilla elongate, slender posteriorly; pterygoid short, not bracing maxilla; otic bulla weakly developed; suspensorium nearly vertical; opercular series developed, their margins entire; subopercle produced posteriorly as a posteroventral border to opercle. Branchiostegal rays numerous, slender; 6–7 rays along the epihyal, the remainder lie behind epihyal tips; urohyal reduced to an ossified basal plate with a cartilaginous posterior filament. Gill arches well developed for a myrophin; first basibranchial ossified, 2–4 absent; hypobranchials 1–2 ossified, 3 cartilaginous; ceratobranchials 1–4 ossified, 5 absent; infrapharyngobranchials 2–3 ossified; upper pharyngeal tooth plates separate. Pectoral girdle well developed; cleithrum and supracleithrum slender, coracoid and scapula broad. Intramuscular bones, ribs, and caudal transverse

Table 27. Frequency distributions of vertebral counts in Atlantic species of *Myrophis*.

Predorsal	14	15	16	17	18	19	20	21	22	23	24	25	26	27	28	29	30	31	32	33	34	35
anterodorsalis	1	2	7*	4																		
platyrhynchus				2	3	6*	5	2														
plumbeus																						
Western Atlantic													2*	3	4	1	1					
Eastern Atlantic												2	2	2								
punctatus																						
Northwestern Atlantic																	2	3	4	2	4	1
Southwestern Atlantic																1	2	-	5	-	1	

Preanal	44	45	46	47	48	49	50	51	52	53	54	55	56	57
anterodorsalis		4	3	4	3*	1								
platyrhynchus	1	1	6	5*	5									
plumbeus														
Western Atlantic										2*	3	5	2	1
Eastern Atlantic										4	2	2		
punctatus														
Northwestern Atlantic								4	7	8	1			
Southwestern Atlantic									3	3	3	1		

Total	135	136	137	138	139	140	141	142	143	144	145	146	147	148	149	150	151	152	153	154
anterodorsalis	1	-	-	1	2	1	5	2	1	-	-	1*	1							
platyrhynchus						2	1	3*	8	-	2	1								
plumbeus																				
Western Atlantic														2	1*	-	1	3	2	3
Eastern Atlantic												1	1	1	2					
punctatus																				
Northwestern Atlantic							2	1	3	8	3	5	5	-	3					
Southwestern Atlantic											2	-	2	2	1	1	3	-	-	1

* Includes holotype.

processes developed; epipleural ribs limited to anteriormost 15–20 vertebrae; precaudal vertebrae less numerous than caudal. Gasbladder weakly developed. Peritoneum freckled. Body coloration uniformly pale, slightly darker dorsally.

Size. To about 500 mm TL.

Distribution. Known from all tropical seas, including seven or eight valid species.

Etymology. From the Greek *myrus* (eel) and *ophis* (snake). Masculine.

Species. Four species of *Myrophis* are currently recognized in the Atlantic: *M. anterodorsalis* new species, *M. platyrhynchus* Breder, 1927, *M. plumbeus* (Cope, 1871), and *M. punctatus* Lütken, 1851. They are similar in appearance, especially when young, and are most easily separated by the position of the dorsal-fin origin (Fig. 261) and by vertebral counts (Table 23). *Myrophis anterodorsalis* and *M. platyrhynchus* are small, pale, laterally compressed species which mature at small sizes (about 150 mm); *M. plumbeus* and *M. punctatus* grow larger and become robust and darker, maturing at larger sizes (greater than 350 mm), and they exhibit ontogenetic variation in dentition. *Myrophis platyrhynchus* differs from other species of the genus in having two postorbital pores rather than one, a difference sometimes considered deserving of generic or subgeneric rank.

Key to the Atlantic Species of *Myrophis*

1a. Two postorbital pores, one behind upper half of eye and one behind mideye; dorsal-fin origin about midway between snout tip and anus; 19–22 predorsal vertebrae .... *platyrhynchus*, p. 283
1b. One postorbital pore, behind upper half of eye 2
2a. Dorsal-fin origin greater than twice head length behind snout tip, about midway between gill opening and anus; 30–34 predorsal vertebrae .... *punctatus*, p. 286
2b. Dorsal-fin origin less than twice head length behind snout tip; 30 or fewer predorsal vertebrae 3
3a. Dorsal-fin origin behind gill opening by 1–2 times the pectoral fin length; 14–17 predorsal vertebrae, 45–49 preanal vertebrae .... *anterodorsalis* new species, p. 281
3b. Dorsal-fin origin behind gill opening by more than 3 times the pectoral fin length, before midpoint between gill opening and anus; 26–30 predorsal vertebrae, 51–57 preanal vertebrae .... *plumbeus*, p. 285

Myrophis anterodorsalis new species

LONGFIN WORM EEL

Figures 261–265,
Tables 24, 25, 27, 28

Distinctive Characters. A moderately elongate myrophine with nearly uniform pale coloration, although speckled dorsally; dorsal-fin origin 1–2 times pectoral-fin length behind gill opening; vomerine teeth uniserial; one postorbital pore; MVF 16-47-141.

Counts and Measurements (in mm) of the Holotype. Total vertebrae 146; predorsal vertebrae 16; preanal vertebrae 48. Pores: SO 1 + 4; IO 5 + 1; POM 5 + 3; ST 3; LL pores above branchial basket 10, to dorsal-fin origin 16, to mid-anus 48; total visible LL pores 55.

Total length 161; standard length 160; head length 17.5; trunk length 42.5; tail length 101; body depth at gill openings 7.4; body width at gill openings 3.6; body depth at anus 5.8; body width at anus 3.4; predorsal length 23.5; snout length 3.2; snout tip to rictus 7.6; eye diameter 1.4; fleshy interorbital distance 2.2; pectoral-fin length 3.6.

Description. See Table 28 for ranges of counts and proportions of the holotype and 14 paratypes.

Body moderately elongate, strongly compressed posteriorly. Snout bluntly rounded; low-

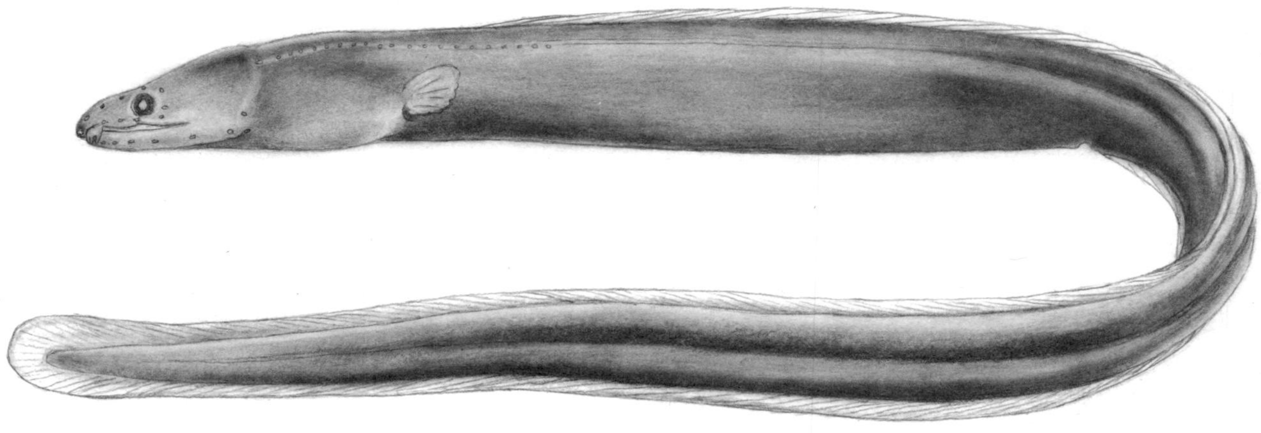

FIGURE 262. *Myrophis anterodorsalis*: ANSP 153670, holotype, 161 mm TL, Colombia.

er jaw included, its tip slightly before bases of anterior nostrils. Eye above mid-point of upper jaw. Anterior nostril in broad tube; posterior nostril along edge of upper lip, covered by a posteriorly-directed flap which projects laterally and appears as a slit at level of anterior edge of eye. Gill opening mid-lateral, a constricted opening.

Head pores visible, typically myrophin. Single median pore in frontal commissure located above and behind posterior margin of eyes. Ethmoid pore on underside of tip of snout; four supraorbital pores as illustrated, the posteriormost above posterior half of eye. Five pores in infraorbital canal, two between anterior and posterior nostrils and three below and behind eye, plus one postorbital pore at posterodorsal margin of eye. Five pores along mandible, the first minute, the last shortly behind rictus, plus three preopercular pores forming a sharp right angle. Three pores in supratemporal canal. Lateral-line pores visible in head and trunk region, only a few behind anus.

Teeth small, conical and slightly recurved. An outer ring of 9–11 intermaxillary teeth plus two large teeth between insertion of the maxillae, followed by a short gap and 12–16 uniserial, small vomerine teeth. Jaw teeth biserial; maxillary with 25–30 close-set teeth in outer row flanking inner row of 15–16 teeth; on mandible, several large subequal teeth at symphysis, followed by outer row of 26–30 small close-set teeth enclosing inner row of 18–20 teeth.

Median fins low anteriorly, lying partially within a groove, expanded in posterior half of body and confluent with extended caudal fin. Dorsal-fin origin in anterior trunk region, 1–2

TABLE 28. Counts and proportions (in thousandths of TL or HL) of the holotype and 14 paratypes of *Myrophis anterodorsalis* new species.

	Range	Mean
Total length (TL in mm)	92–161	125.6
Predorsal vertebrae	14–17	16.1
Preanal vertebrae	45–49	46.6
Total vertebrae	135–147	141.1
Proportion of TL		
Head	108–129	117
Trunk	245–271	256
Tail	612–644	626
Predorsal	146–176	160
Depth	28–48	36
Proportion of HL		
Snout	161–203	184
Upper jaw	356–448	389
Eye	65–86	74
Interorbital width	90–126	105
Pectoral-fin length	145–227	170

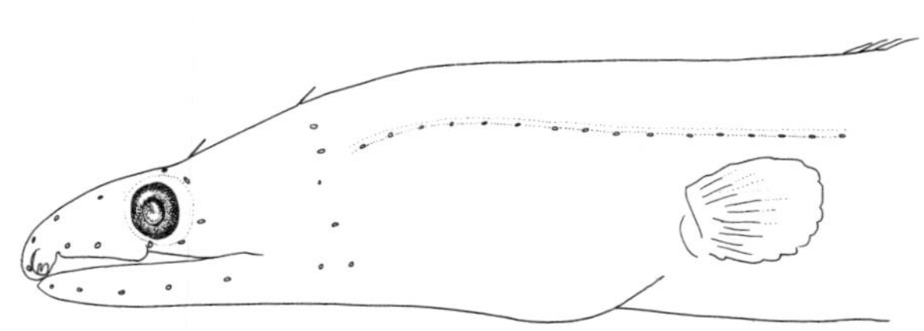

FIGURE 263. *Myrophis anterodorsalis*: ANSP 153670, holotype, 161 mm TL, Colombia; head.

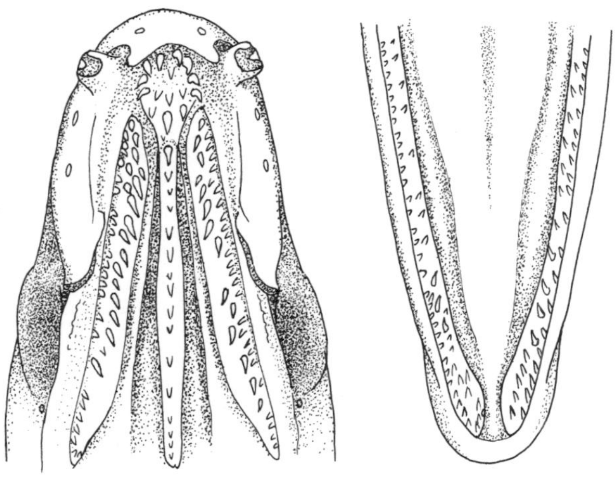

FIGURE 264. *Myrophis anterodorsalis*: ANSP 153671, paratype, 149 mm TL, Colombia; dentition.

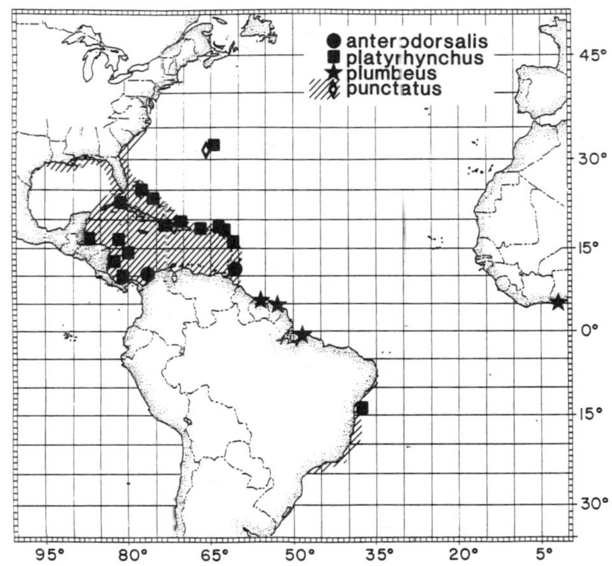

FIGURE 265. Distribution of Atlantic species of *Myrophis* based on study material.

times the pectoral-fin length behind the gill opening. Anal fin higher than dorsal fin. Pectoral fin paddle-shaped, its length about twice orbit.

Color. Pale yellow to tan, overlain by a fine speckling of brown spots on snout, nape, and on body above the lateral line. Fins pale.

Size. The largest specimen is the holotype, a female 161 mm TL. Eight paratypes (120–161 mm) are females, two (147–161) filled with 0.5-mm eggs.

Development. Leptocephalus not identified.

Distribution (Fig. 265). Known from islands off the Atlantic coast of Colombia and from Tobago, at depths to 1.3 m.

Etymology. From the Latin *antero* (anterior) and *dorsualis* [*dorsalis*] (of the back), in reference to the anterior dorsal-fin origin. An adjective.

Remarks. The new species is known only from shallow water and has been taken with *Ahlia egmontis* and with *Myrophis punctatus.* It is easily separable from its congeners on the basis of its anterior dorsal-fin position (more so than any other known *Myrophis*) combined with its head pore condition. It is closest to *Myrophis platyrhynchus,* which differs markedly in having two, rather than one, postorbital pores, and which appears to have fewer jaw teeth. The eastern Pacific *Myrophis vafer* Jordan and Gilbert differs primarily in its posterior dorsal-fin location, which is similar to that of *M. punctatus.* We suspect that the new species has sometimes been identified as *Myrophis platyrhynchus,* and that it will be found to be more widely distributed geographically.

Study Material. The holotype and 14 paratypes, 92–161 mm TL.

Holotype: ANSP 153670 (161); Colombia, Baru Island, outside entrance to Cartagena Bay; 10°12′35″N, 75°37′30″W, depth 0–1.3 m; L. K. Knapp, LK 69-36; 27 Sept. 1969.

Paratypes: COLOMBIA: ANSP 153671 (1, 149, cleared and stained) and ANSP 117563 (6, 104.5–147); taken with the holotype. ANSP 117574 (5, 92–135); Cabruna Island, 09°44.6′N, 75°41.1′W, depth 0–1.3 m; L. K. Knapp, LK 69-40; 1 Oct. 1969. CAS 55913 (1, 100.5); Ceycen Island, W side, 09°42.1′N, 75°51.1′W, depth 0–1.3 m; L. K. Knapp, LK 69-38; 30 Sept. 1969. TOBAGO ISLAND: ANSP 98472 (1, 119); Charlotteville, Man-o-War Bay, NE end of island, leeward (N) side; A. N. and D. M. Barringer, WA-20; 23 April 1962.

Myrophis platyrhynchus Breder, 1927

BROADNOSE WORM EEL

Figures 261, 265–268,
Tables 24, 25, 27

Myrophis platyrhynchus Breder, 1927: 8, figs. 3–4 (original description, Glover Reef, British Honduras, holotype YPM 30).

Myrophis emmae Howell Rivero, 1935: 342, fig. 2 (original description, Cuba, holotype MCZ 33454).

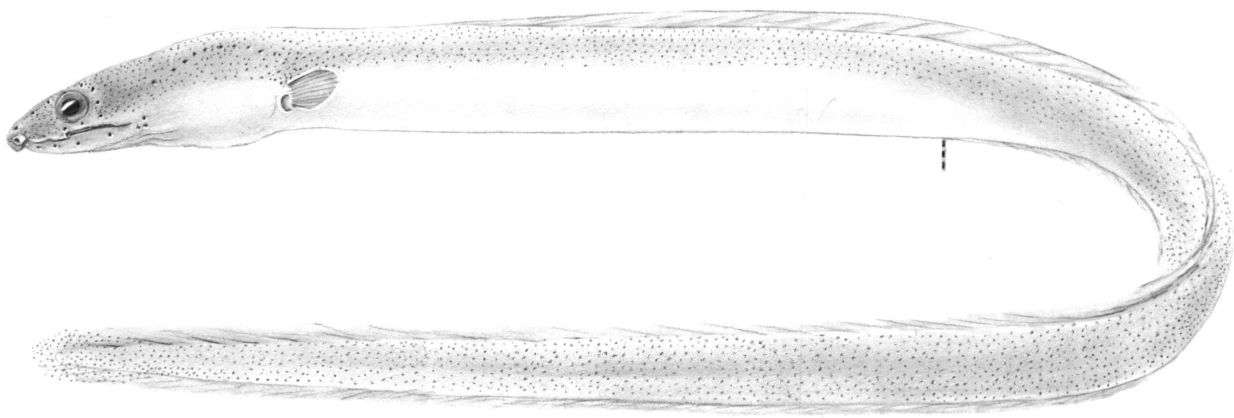

FIGURE 266. *Myrophis platyrhynchus*: ANSP 98287, 176 mm TL, Bahamas. Illustration by S. P. Gigliotti.

Distinctive Characters. A moderately elongate myrophine with nearly uniform coloration, although finely spotted dorsally; dorsal-fin origin about midway between snout tip and anus; vomerine teeth uniserial; two postorbital pores; MVF 20-47-143.

Description. Total vertebrae 140–146, $\bar{x}$ = 142.8 (n = 16); predorsal vertebrae 18–22, $\bar{x}$ = 20.1 (18); preanal vertebrae 46–48, $\bar{x}$ = 46.7 (18). Pores: SO 1 + 4; IO 5 + 2; POM 5 + 3; ST 3.

Proportions as % TL: head 10–11; trunk 25–29; tail 60–64; predorsal 17–20; depth behind gill openings 2.3–3.6. Of HL: snout 17–20; snout to rictus 31–36; eye 9.6–14; interorbital width 7.9–11; pectoral-fin length 17–24.

Snout broad, rounded, and flattened dorsally; eye large, its posterior margin above rictus. Anterior nostril tubular, with lateral projection; posterior nostril covered with a flap which projects laterally under anterior margin of eye. Head pores as for the genus, except this species has one additional pore in vertical (ascending) part of infraorbital canal, the additional pore just behind and close to eye.

An outer circle of 9–10 plus several median intermaxillary teeth, continuous with a mostly uniserial row of 14–18 teeth along vomer, a few staggered or biserial in middle of row. Maxillary teeth biserial, an outer series of 19–31 small close-set teeth enclosing an inner row of 11–14 taller widely-spaced teeth. Outer row of 23–36 small teeth on mandible, largest at symphysis, and an inner row of 6–10 larger teeth anteriorly.

Dorsal-fin origin about midway between snout tip and anus. Anal fin more highly developed than dorsal fin, both continuous with expanded caudal fin.

Color. Body pale, with small widely-spaced black pigment spots over the dorsal and lateral surfaces, visible microscopically. Fins colorless.

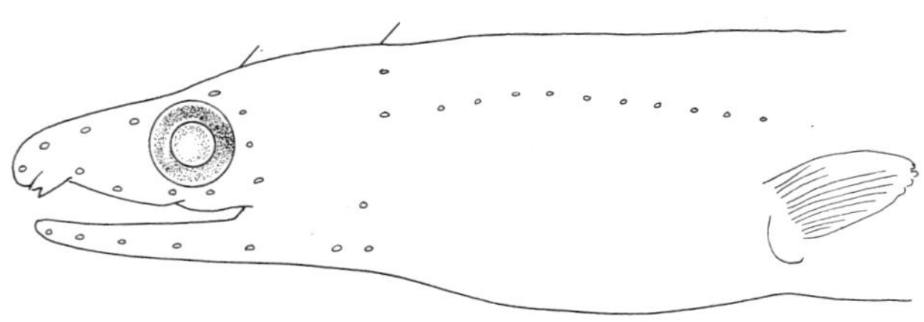

FIGURE 267. *Myrophis platyrhynchus*: ANSP 136523, 146 mm TL, Puerto Rico; head.

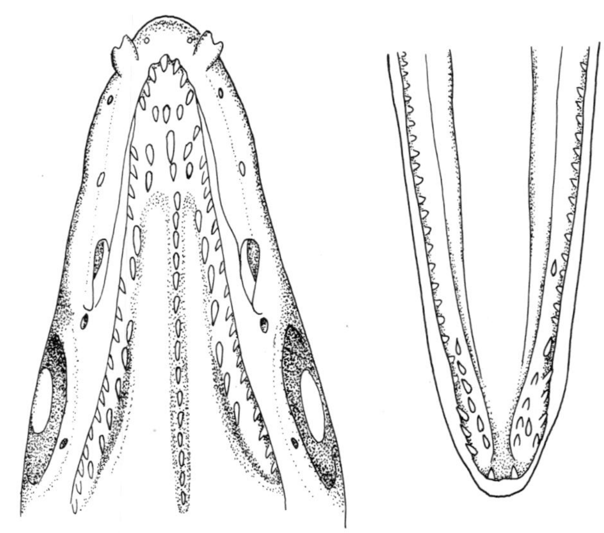

FIGURE 268. *Myrophis platyrhynchus*: ANSP 128813, 165 mm TL, Bahamas; dentition.

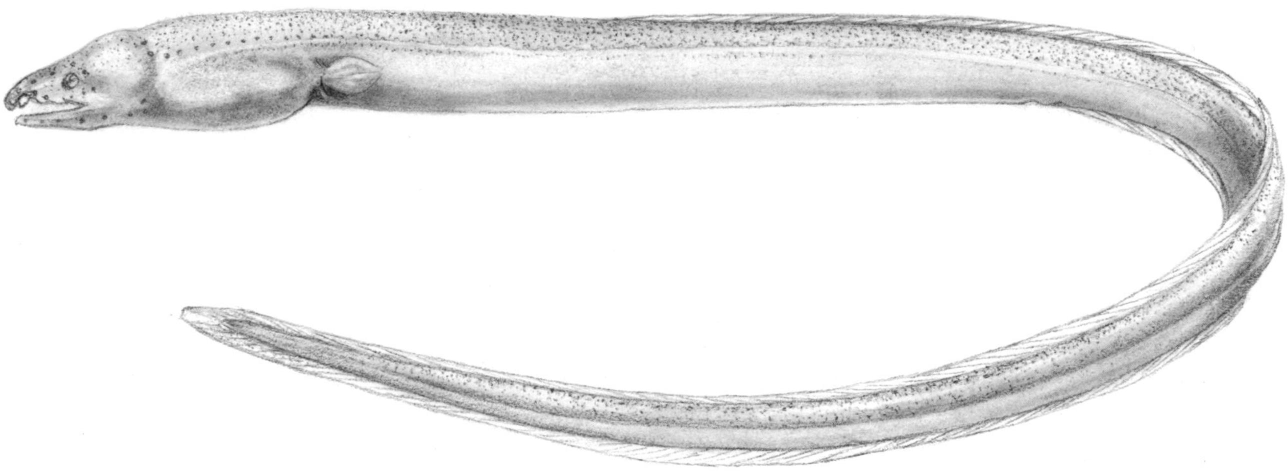

FIGURE 269. *Myrophis plumbeus*: ANSP 159107, 188 mm TL, French Guiana.

Size. A small species; the largest specimen examined was 208 mm TL. Of 15 females (146–208 mm), 12 (146–208) were filled with 0.3–0.8-mm eggs. Eight ripe males (142–186) were found; four were paratypes collected with the holotype (a ripe female with 0.6-mm eggs), the additional paratype a ripe female.

Development. Leptocephalus identified (p. 786).

Distribution (Fig. 265). Known from Bermuda, the Bahamas, the Antilles, Central America, and Brazil. Not common, generally collected in shallow water (recorded to 220 m but usually at less than 10 m), from semi-protected sandy beaches or embayments over sand and weeds, but not from exposed coastal zones. Collected with poison, speared, dip-netted at night, and once taken from the gills of a snapper. Several times collected with *Myrophis punctatus* and/or *Ahlia egmontis.*

Etymology. From the Greek *platy* (flat), and *rhynchos* (snout), named for the "peculiar broad flat snout" (Breder, 1927: 9). A noun in apposition.

Study Material. A total of 53 specimens, 53.5–208 mm TL, including the holotype and three of the paratypes, and the holotype of *Myrophis emmae* Howell Rivero.

Holotype: YPM 30 (193); British Honduras, Glover Reef, at night with light; PAWNEE; 18 April 1925.

Paratypes: YPM 31 (3, 142–158 [2 paratypes, 146–186, = *M. punctatus*]); taken with the holotype. YPM 32 (1, 121); Bahamas, Southwest Harbor, at night with light; 16 Mar. 1925. YPM 33 (2, 113–158); Bahamas, Saddle Rock; 23 Mar. 1925.

Other material: BERMUDA: ANSP 148369 (1, 147). BAHAMAS: ANSP 98282 (2, 56–140), 98283 (3, 112–148), 98284 (2, 79–94), 98285 (1, 125), 98286 (2, ca 170–208), 98287 (1, 176), 98288 (1, 56), 116338 (1, 54), 128813 (1, 165), 158807 (1, 143), 158808 (5, 133–156), 158809 (1, 146), 158810 (3, 118–171), 158811 (2, 78–108). YPM 2603 (1, 108).

ANTILLES: *Cuba:* MCZ 33454 (1, 103); Havana; L. Howell Rivero; 1934; holotype of *Myrophis emmae* Howell Rivero. USNM uncat. (1, 210). *Haiti:* ANSP 111371 (1, 55). *Santo Domingo:* ANSP 110073 (1, 179). *Puerto Rico:* ANSP 136523 (1, 146). *Virgin Islands:* ANSP 140577 (1, 166). CAS 54475 (2, 98–102). *St. Barthelemy:* ANSP 106070 (2, 89–125), 106177 (3, 149–178). *Antigua:* BMNH 1934.9.10.57 (1).

PANAMA: ANSP 113757 (1, 199), 116137 (1, 179, cleared and stained). HONDURAS: USNM 268508 (1, 118). NICARAGUA: USNM 268507 (1, 127). SERRANA BANK: ANSP 153663 (1, 145). GULF OF CAMPECHE: MCZ 64509 (1, 169). BRAZIL: MZUSP 12045 (6), 12046 (1, 136), 12047 (1).

Myrophis plumbeus (Cope, 1871)

LEADEN WORM EEL

Figures 261, 265, 269, Tables 24, 25, 27

Myrophis longicollis (not of Cuvier, 1817). Kaup, 1856a: 30; 1856b: 70.

Holopterura plumbea Cope, 1871: 482 (original description, West Africa, holotype ANSP 22964).

Myrophis vafer (not of Jordan and Gilbert, 1882). Pietschmann, 1913: 177.

Myrophis punctatus (not of Lütken, 1851). Monod, 1927: 678.
Paramyrus plumbeus. Fowler, 1936a: 290.

Distinctive Characters. A moderately elongate to stout myrophine with tan to brown coloration with small spots dorsally; dorsal-fin origin before midpoint between gill opening and anus; vomerine teeth numerous and mostly biserial; one postorbital pore; MVF 27-55-150.

Description. Total vertebrae 146–154, $\bar{x}$ = 150.5 (n = 14); predorsal vertebrae 25–30, $\bar{x}$ = 27.0 (17); preanal vertebrae 53–57, $\bar{x}$ = 54.3 (19). Pores: SO 1 + 4, IO 5 + 1; POM 5 + 3; ST 3.

Proportions as % of TL: head 10–13; trunk 27–30; tail 58–63; predorsal 20–24; depth behind gill opening 2.8–4.6. Of HL: snout 13–19; snout to rictus 25–31; eye 5.3–8.3; interorbital width 9.9–14; pectoral-fin length 12–19.

Head and nostrils as for the genus. Ring of 8–10 outer intermaxillary teeth, continuous with 22–34 vomerine teeth, which are tri- to biserial for most of their extent. Maxillary teeth in two rows; 21–30 small close-set teeth in outer row, 5–7 larger spaced teeth in inner row. Mandibular teeth partly biserial, outer series of 22–30 teeth flanking anterior inner series, which is sometimes a short patch of 3–4 and sometimes a row of 4–11.

Body of small specimens laterally compressed posteriorly; large specimens cylindrical and muscular. Dorsal fin arises before midpoint between gill opening and anus, less than two times head length from snout tip. Dorsal fin low anteriorly, higher posteriorly, confluent with caudal fin which is developed and continuous with higher anal fin.

Color. Small specimens pale tan, larger ones dark brown or grey, appearing uniform but peppered with small spots dorsally on body and overall on tail. Sometimes bicolored, darker dorsally, as described for the holotype. Fins colorless.

Size. The largest specimen studied was 465 mm. Three females (358–457 mm), all from the eastern Atlantic, were ripe with 0.2–0.4-mm eggs. Specimens from the western Atlantic are smaller, the sex not determinable.

Development. Leptocephalus identified (p. 790).

Distribution (Fig. 265). *Myrophis plumbeus* was previously considered restricted to the eastern Atlantic, where adults are exceedingly abundant. Blache (1968b: 1521) studied over 1900 adults, but had collected only nine larvae. In the western Atlantic the larvae are very common but our study material includes only 79 postmetamorphic specimens, all from off northern South America (Suriname, French Guiana, and Brazil). Recent collections of numerous small specimens from French Guiana (ANSP 136645, 136647 and 136650) were taken in brackish or fresh waters of the Mahury River. We could find no differences between western and eastern Atlantic specimens, nor do leptocephali from the two areas differ.

Etymology. From the Latin *plumbeus* (leaden), in reference to the ventral coloration described for the holotype. An adjective.

Study Material. A total of 78 specimens, 44–452 mm TL, including the holotype.

Holotype: ANSP 22964 (224); West Africa; Dr. Goheen.

Other material: WESTERN ATLANTIC: SURINAME: BMNH 1870.3.10.66 (1, 287). MNHN B.2143 (1, 376). FRENCH GUIANA: ANSP 136650 (61, 56–181), 136647 (1, 122), 136645 (1, 78+), 159107 (1, 188). BRAZIL: MZUSP 12059 (6, 87–262). EASTERN ATLANTIC: DAHOMEY: USNM 274269 (4, 230–457). AFRICA: USNM 48488 (1, 354).

Myrophis punctatus Lütken, 1851

SPECKLED WORM EEL

Figures 261, 265, 270–272,
Tables 24, 25, 27

Myrophis punctatus Lütken, 1851: 15 (original description, West Indies, holotype unknown). Jordan, 1883: 282 (redescription of Lütken's type specimen).
Myrophis microstigmius Poey, 1867b: 250, pl. 3, fig. 4 (original description, Cuba, holotype MCZ 33440).
Neoconger perlongus Poey, 1876b: 67, pl. 9, figs. 3–4 (original description, Matanzas, Cuba, holotype USNM 37478).
Leptoconger perlongus. Poey, 1880: 250 (new genus).
Myrophis lumbricus Jordan and Gilbert, 1882c: 261 (original description, Galveston, Texas, holotype USNM 30896).
Myrophis iguapensis Miranda Ribeiro, 1908: 4, text-fig. (original description, Brazil, holotype unknown).
Myrophis longleii Silvester, 1916: 214 (original descrip-

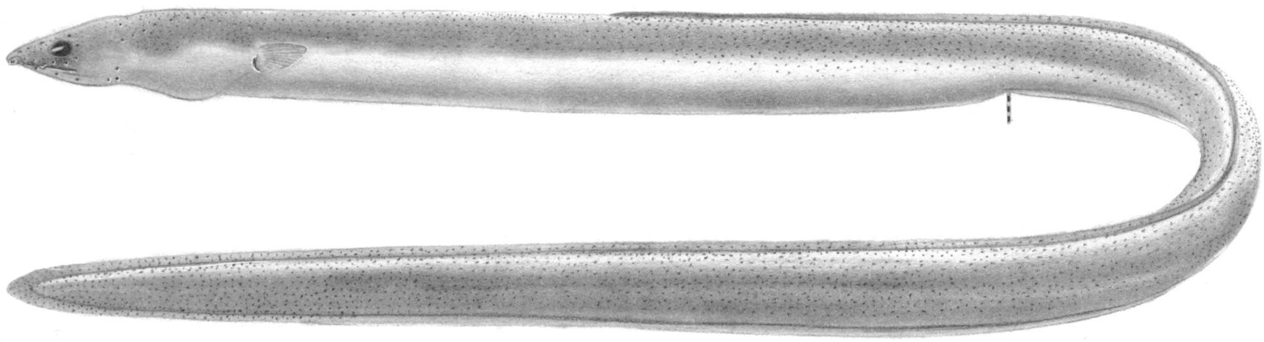

FIGURE 270. *Myrophis punctatus*: ANSP 111572, 188 mm TL, Bahamas. Illustration by S. P. Gigliotti.

tion, Guanica Harbor, Puerto Rico, holotype PU 3083, not extant).

Myrophis platyrhynchus (not of Breder, 1927). Breder, 1927: 8 (two of the five paratypes of *M. platyrhynchus*).

Myrophis dolichorhynchus Parr, 1930: 13, fig. 1, center (original description, Washerwoman, Bahamas, holotype YPM 2602).

Leptocephalus mollis Fowler, 1944: 126, fig. 47 (original description, Old Providence, Caribbean Sea, holotype ANSP 70118). Leptocephalus.

Discussion of Synonymy. "*Myrophis longicollis*" has long been included in the synonymy of *M. punctatus*, based on a specimen from Suriname in the Paris Museum. It was studied by Kaup, identified by him as *M. longicollis*, and synonymized by him with *M. punctatus* (1856b: 30). However, Cuvier's name *longicollis* (1817: 231) was based on "Murene Myre" of Lacepède (1800: plate 3, number 3), which depicts a recognizable specimen of *Echelus myrus*, a valid species. We have examined Kaup's specimen (MNHN B.2143) and identify it as *M. platyrhynchus* on the basis of measurements and vertebral counts.

Distinctive Characters. A moderately elongate myrophine with pale tan to brown coloration overlain with square freckles dorsally on body and on tail; dorsal-fin origin behind midpoint between gill opening and anus; vomerine teeth tri- to uniserial; one postorbital pore; MVF 32-53-146.

Description. Total vertebrae 141–154, x̄ = 146.1 (n = 42); predorsal vertebrae 29–34, x̄ = 32.0 (25); preanal vertbrae 51–55, x̄ = 52.6 (30). Pores: SO 1 + 4; IO 5 + 1; POM 5 + 3; ST 3.

Proportions as % of TL: head 9.0–12; trunk 27–31; tail 59–63; predorsal 23–28; depth behind gill openings 2.4–3.5. Of HL: snout 11–16; snout to rictus 22–32; eye 5.9–9.2; interorbital width 5.2–9.2; pectoral-fin length 10–21.

Head, nostrils and pores as for genus; one postorbital pore. Dentition consisting of outer circle of 9–10 plus several median intermaxillary teeth, continuous with vomerine teeth; the latter an anterior patch followed by a row of 24–29 bi- to uniserial teeth, usually uniserial posteriorly; maxillary teeth biserial, an outer row of 23–25

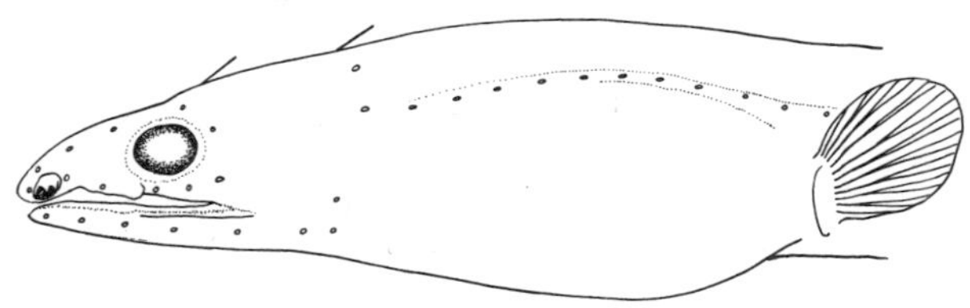

FIGURE 271. *Myrophis punctatus*: ANSP 111572, 188 mm TL, Bahamas; head.

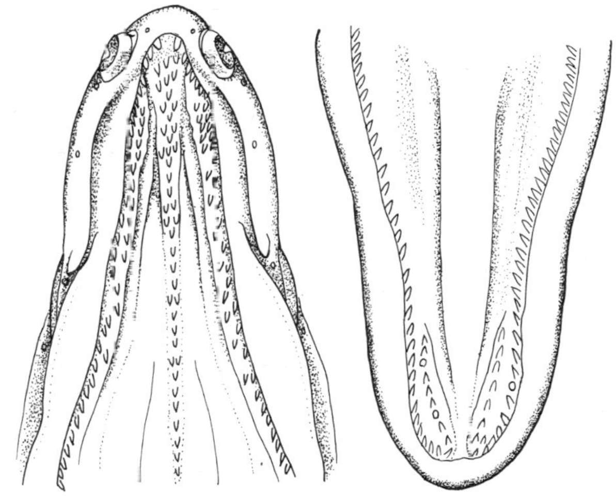

FIGURE 272. *Myrophis punctatus*: ANSP 114540, 353 mm TL, Puerto Rico; dentition.

small close-set teeth and an inner row of 6–12 longer, sharp, spaced teeth; mandibular teeth partially biserial, an outer row of 26–28 small teeth, an anterior inner row or patch of 6–7 subequal teeth. Considerable variation in teeth of larger specimens: vomerine teeth become biserial throughout, 3–4 across anteriorly; maxillary teeth tri- or quadriserial anteriorly, and anterior mandibular patch larger, partially triserial.

Dorsal-fin origin about midway between gill opening and anus, greater than two times head length behind gill opening. Dorsal fin low anteriorly, expanded posteriorly; anal fin higher than dorsal fin, both continuous with caudal fin.

Color. Pale tan to dark brown, with a peppering of squarish brown freckles over the dorsal and lateral surfaces of the head and body and on chin; spots denser and extending over entire body and fins posterior to anus. Large specimens have more pigment and appear darker brown overall. Sometimes appearing bicolored; one specimen (ANSP 114540) from Puerto Rico strongly bicolored, yellowish ventrally.

Size. The largest specimen examined was 353 mm TL, probably growing to larger sizes (no ripe adults found in study material examined). Of 22 females, 147–353 mm, only two contained measurable eggs (0.1 mm), the specimens not ripe.

Development. Leptocephalus well known, common in the western Atlantic, described by Leiby (1979b, and this volume, p. 791).

Distribution (Fig. 265). The most abundant myrophine in the western Atlantic. Known from the Carolinas south to Brazil, including the Gulf of Mexico and throughout the West Indies. Generally collected in shallow water, frequently found in brackish waters of tidal creeks and protected bays to a depth of 7 m; also collected at the surface offshore by night-light.

Etymology. From the Latin *punctatus* (speckled), in reference to its coloration. An adjective.

Remarks. Böhlke (1968: 100) suggested that this species, like *Ahlia egmontis,* undergoes an enlargement of its eye and pectoral fin with the onset of sexual maturity. A difference in eye size accounts for some of the nominal species within its synonymy. Vertebral counts for *Myrophis punctatus* from the Gulf of Mexico, the Caribbean, and Brazil show some geographical variation (Table 23), and Leiby (this volume, p. 000) has shown mean differences, but broadly overlapping ranges, in leptocephalus myomere counts for *M. punctatus* from the Gulf of Mexico and the Caribbean. However, the populations are considered conspecific at this time.

Blache (1968b: 1520–1521) suggested that eastern Atlantic records of *Myrophis punctatus* (Fowler, 1936a: 291; Cadenat, 1950: 297, 314; Poll, 1953: 138; and Gras, 1961: 579) were misidentifications of *M. plumbeus.* Since we have found specimens of the latter in the western Atlantic, it is quite possible that *M. punctatus* also exists on both sides, and the specimens reported above should be reexamined with this in mind.

Study Material. A total of 530 specimens, 44–353 mm TL, including holotypes of *Myrophis dolichorhynchus, M. lumbricus, M. microstigmius,* and *Neoconger perlongus.*

BERMUDA: ANSP 127811 (1, 56), 133567 (1, 206), 133568 (16, 121–222). USNM 164752 (1, 267). VIRGINIA: ANSP 158462 (1, 157, from stomach of tuna). NORTH CAROLINA: ANSP 110415 (2, 204–200+). SOUTH CAROLINA: ANSP 128403 (1, 240). FLORIDA ATLANTIC: ANSP 71024–71025 (2, 264–283), 115296 (1, 174), 126196 (32, 45–202). FLORIDA KEYS: ANSP 113044 (1, 80), 124022 (5, 44–62), 124023 (3, 51–61), 124025 (10, 50–293), 124026 (2, 143–175), 124378 (6, 52–144), 124544 (2, 150+–179). UMML 30535 (1, 94), 31691 (1, 324), 32030 (1, 311). USNM 77749 (1, 137), 203586 (2, 104–107), 274257 (4, 135–180). GULF OF MEXICO: *Florida:* ANSP 74885 (1, 125), 98514 (1, 283), 103180 (1, 285), 110432 (1, 60), 110434 (1, 118). Escambia River: ANSP 73074 (10, 104–186), 73092 (5, 92–133), 79020 (3, 90–134), 79029 (7), 73015 (1). *Mississippi Delta:* ANSP 98513 (4, 101–257). MNHN 1974-145 (2, 97–179). *Louisiana:* Calcasieu River: ANSP 55611 (1, 214). *Texas:* Sabine River: ANSP 73817 (2, 52–81), 73842 (12, 47–55). Neches River: ANSP 73914 (1, 259), 85229 (5, 50–60), 95068 (2, 116–116), 99055 (15, 81–191), 99086 (5, 59–98), 99119 (2, 70–78), 99160 (7, 66–136), 99193 (8, 62–73), 141222 (1, 55).

BAHAMAS: ANSP 98292 (17, 51–180), 98293 (62, 56–305), 98294 (3, 98–136), 98295 (16, 58–124), 98296 (1, 156), 98297 (1, 156), 98298 (12, 102–183), 98299 (1, 80), 98300 (3, 69–116), 98301 (2, 50–54), 98302 (53, 53–156), 98303 (7, 110–156), 98304 (38, 62–242), 111572 (1, 188), 112006 (1, 234), 113289 (8, 62–158), 116138 (2, 165–173, cleared and stained), 122683 (22, 56–149), 128827 (1, 110, cleared and stained), 158812 (1, 58), 158813 (1, 63). YPM 2602 (1, 141); holotype of *Myrophis dolichorhynchus.* BMNH

1986.10.14.1 (1, 76). CUBA: ANSP 90769 (1, 159). MCZ 33440 (1, 324); holotype of *Myrophis microstigmius*. USNM 37478 (1, 320); holotype of *Neoconger perlongus*. USNM 37567 (2, 365–380). PUERTO RICO: ANSP 114540 (1, 353), 134326 (1, 121). USNM 147623 (2, 176–177), 147630 (8, 88–231). VIRGIN ISLANDS: ANSP 124024 (1, 121). LESSER ANTILLES: *St. Barthelemy:* ANSP 106176 (2, 98–143). *Martinique:* ANSP 106185 (11, 59–263). CAS 54476 (3, 145–242). *Trinidad:* MBUCV 70 (1). *Curaçao:* USNM 142878 (1, 325). VENEZUELA: MBUCV 7547 (4, to 244). COLOMBIA: ANSP 117561 (1, 224), 117573 (1, 217), 117575 (1, 119). PANAMA: ANSP 113748 (1, 64), 113750 (10, 98–218). NICARAGUA: USNM 178987 (1, 113). BRITISH HONDURAS: YPM 31 (2, 146–186, 2 of 5 paratypes of *M. platyrhynchus*). MEXICO: Yucatan: ANSP 130080 (1, 229). UANL P1785 (1, 367). GRAND CAYMAN: ANSP 126502 (6, 154–187). BRAZIL: ANSP 121206 (2, 157–189), 121332 (6, 170–265). MCZ 9165 (1, 255). MNRJ 1926 (1, 398), 2243 (1), 9530 (1, 320). MNRJ uncat. (2, 238–256). MNRJ uncat. (2, 257–307). MZUSP 410 (1, 220). SMF Exp. 2125 (1).

Genus *Pseudomyrophis* Wade, 1946

Pseudomyrophis Wade, 1946: 199 (type species *Pseudomyrophis micropinna* Wade, 1946, by original designation).

Characters. Body moderately to extremely elongate, (depth 1.0–2.4% TL in western Atlantic species), laterally compressed throughout; head and trunk shorter than tail (35–39% TL). Dorsal-fin origin behind mid-trunk; median fins confluent at caudal; pectoral fin minute, smaller than eye. Snout broad, tumid, overhanging lower jaw, lacking a groove on its underside; lips without barbels; eye small to moderate; anterior nostril tubular; posterior nostril an elongate slit before lower margin of orbit. Gill opening round and constricted. Head pores developed; three preopercular pores. Teeth small, conical, sharp; dentition tri- to uniserial on vomer, mostly uniserial on jaws.

Neurocranium rounded posteriorly; supraoccipital crest absent; maxilla elongate, slender posteriorly; pterygoid short, not produced anteriorly; otic bulla weakly developed; suspensorium nearly vertical; opercular series moderately developed; subopercle produced posteriorly, as in *Myrophis*, as a posterodorsal border to opercle. Branchiostegal rays numerous, slender; 7 rays along the epihyal, the remainder lie behind the epihyal tips. Gill arches weakly developed; first basibranchial ossified, 3–4 absent; hypobranchials 1–2 ossified, 3 cartilaginous; ceratobranchials 1–4 ossified, 5 absent; upper pharyngeal tooth plates separate. Pectoral girdle reduced to a slender cleithrum and supracleithrum. Epipleural ribs present on all precaudal vertebrae; precaudal less numerous than caudal vertebrae. Peritoneum lightly pigmented. Body coloration uniformly brown.

Size. To 450 mm TL.

Distribution. Found in the western and eastern Atlantic and the eastern Pacific.

Etymology. From the Greek *pseudos* (false), and *Myrophis* (a related ophichthid). Masculine.

Remarks. The western Atlantic species differ from the type species in possessing one postorbital pore rather than the three pores shown by both Pacific congeners (*P. micropinna* and an undescribed species), and were thought to represent a new genus at one time (Dean, 1972: 53). The eastern Atlantic species differs in having two preopercular pores. They are all provisionally retained in *Pseudomyrophis* pending further study and description of the new Pacific species. In addition, generic relationships of *Pseudomyrophis* and the related Old World genus *Neenchelys* Bamber, 1915 are in need of further study (McCosker, 1982; Smith and Böhlke, 1983).

Species. About eight species, three of them undescribed: in the western Atlantic, *Pseudomyrophis frio* (Jordan and Davis, 1892), *P. fugesae* new species, *P. nimius* Böhlke, 1960, and one species known only from the leptocephalus (77 specimens) described by Leiby (p. 793); two species in the eastern Atlantic, *P. atlanticus* Blache, 1975 and an undescribed species (the subadult described by Blache, 1977: 202 as *P. nimius*, M. Leiby, pers. comm., 1986); and two in the eastern Pacific, *P. micropinna* Wade, 1946 and an undescribed species. The three described western Atlantic species are extremely similar in general appearance; they are readily separable by vertebral counts (Table 25), and less easily by the position of the dorsal-fin origin, head length and comparative body depth. *Pseudomyrophis nimius* is extremely elongate, its head proportionately shorter and its snout longer, and is taken over deeper waters; *Pseudomyrophis fugesae* has the deepest body, the longest head, and is the smallest species.

KEY TO THE ATLANTIC SPECIES OF *PSEUDOMYROPHIS*

1a. Body extremely elongate, its depth 1–1.5% TL; head 5.2–6.1% TL; dorsal-fin origin behind mid-trunk; total vertebrae 213–217 . . . . . *nimius*, p. 295
1b. Body elongate, its depth greater than 1.5% TL; head length greater than 6.9% TL 2
2a. Dorsal-fin origin before mid-trunk; head length 7.5–8.7% TL; body depth 1.6–2.8% TL; total vertebrae 170–174 . . . . . *atlanticus* Blache, 1975
(an eastern Atlantic species from Congo and Senegal)
2b. Dorsal-fin origin about at mid-trunk 3
3a. Head length 6.9–7.8% TL; body depth 1.5–2.2% TL; total vertebrae 184–191 . . . . . *frio*, p. 290
3b. Head length 9.5–9.7% TL; body depth 2.1–2.4% TL; total vertebrae 155–159 . . . . . *fugesae*, p. 292

Pseudomyrophis frio (Jordan and Davis, 1891)
SLENDER WORM EEL
Figures 273–276, Tables 24, 25, 29

Myrophis frio Jordan and Davis, 1891: 640 (original description, Cape Frio, near Rio de Janeiro, Brazil, holotype USNM 46760).

Distinctive Characters. An elongate brownish myrophin with dorsal-fin origin at mid-trunk, and pectoral fin a small lappet at upper gill opening; body depth 1.5–2.2% TL; head length 6.9–7.8% TL; MVF 39-67-188.

Description. Total vertebrae 184–191, x̄ = 187.5 (n = 6); predorsal vertebrae 37–40, x̄ = 39 (6); preanal vertebrae 67–68, x̄ = 67.3 (6). Pores: SO 1 + 4; IO 5 + 1; POM 6 + 3; ST 3.

Proportions as % of TL: head 6.9–7.8; trunk 30–31; tail 61–62; predorsal 22–23; depth behind gill openings 1.5–2.2. Of HL: snout 18–19; snout to rictus 27–31; eye 4.3–4.9; interorbital width 9.4–11.5.

Snout blunt, rugose; eye moderate, above posterior third of jaw. Anterior nostrils tubular, flared at edge with short lateral barbel; posterior nostril an irregular elongate slit in front of lower margin of eye. Head pores conspicuous, as listed above. Two or three pairs of intermaxillary teeth continuous with teeth on vomer which are in an elongate tri- to biserial or staggered patch anteriorly and uniserial posteriorly. Jaw teeth uniserial, slender, directed back, slightly larger anteriorly; 9–18 teeth on maxillary, about 23 along mandible.

Median fins low, confluent with expanded

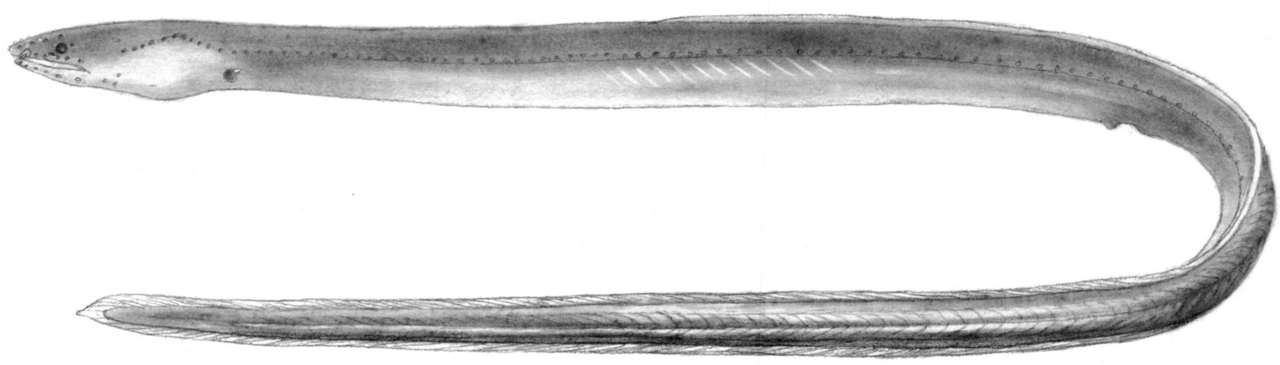

FIGURE 273. *Pseudomyrophis frio*: ANSP 110423, 215 mm TL, Venezuela.

TABLE 29. Frequency distributions of vertebral counts in Atlantic species of *Pseudomyrophis*.

	Predorsal																										
	27	28	29	30	31	32	33	34	35	36	37	38	39	40	41	42	43	44	45	46	47	48	49	50	51	52	53
atlanticus							1*	1	-	-	-	1															
frio											1	1*	2	1	1												
fugesae	1	-	-	-	-	-	1*																				
nimius																					1	4	5*	3	1	-	2

	Preanal																								
	51	52	53	54	55	56	57	58	59	60	61	62	63	64	65	66	67	68	69	70	71	72	73	74	75
atlanticus												1*	-	-	2										
frio																	4	2*							
fugesae	1	-	1*																						
nimius																					2	1*	6	6	2

	Total																													
	155	156	157	158	159	160	...	169	170	171	172	173	174	...	183	184	185	186	187	188	189	190	191	...	211	212	213	214	215	216	217
atlanticus									1*	1	-	-	1																		
frio																2	-	-	-	1*	2	-	1								
fugesae	1*	-	-	-	1																										
nimius																									2	-	4*	2	5	1	3

* Includes holotype.

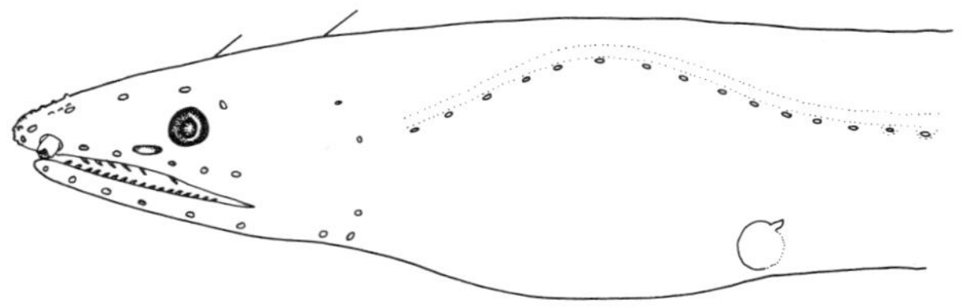

FIGURE 274. *Pseudomyrophis frio*: ANSP 110423, 215 mm TL, Venezuela; head.

caudal fin; dorsal-fin origin at or slightly behind mid-trunk. Pectoral fin greatly reduced, appearing as a small lappet at upper end of gill opening.

Color. Uniformly brown to reddish brown, with small freckles and spots visible microscopically. Fins colorless.

Size. The largest specimen examined is the holotype, 308 mm TL, a ripe female with 0.6-mm eggs. Sex could not be determined for other specimens.

Development. Leptocephalus identified (p. 774).

Distribution (Fig. 276). A deep-water species, known only from off northern South America, from Venezuela to Brazil (to 33°10'S); captured by trawl between 100–420 m depth (only one capture recorded below 200 m).

Etymology. Named for Cape Frio, the locality of the holotype. A noun in apposition.

Study Material. Four specimens, 157–308 mm TL, including the holotype. [An additional 36 specimens are recorded from Brazil but were not included in this study.]

Holotype: USNM 46760 (308); Brazil, from Cape Frio, near Rio de Janeiro, 59 fm (108 m); ALBATROSS Station 2762; 30 Dec. 1887.

Other material: VENEZUELA: ANSP 110423 (1, 215); 11°39'N, 69°38'W; 420 m; OREGON 4407. BRAZIL: MZUSP 12305 (1, 158), and MZUSP 12306 (1, 213); 32°48'S, 50°27'W, 197 m; R/V PROF. W. BESNARD, sta. 1758; 22 April 1972. [MZUSP uncataloged material: 36 specimens from off Rio Grande do Sul, 23°20'–33°10'S, 42°37'–50°50'W, 100–197 m.]

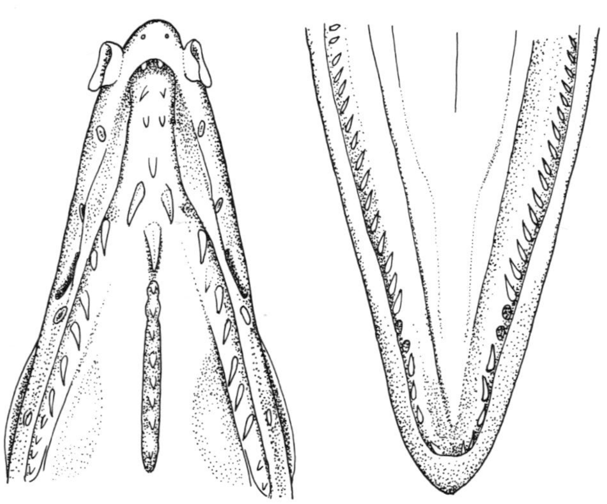

FIGURE 275. *Pseudomyrophis frio*: ANSP 110423, 215 mm TL, Venezuela; dentition.

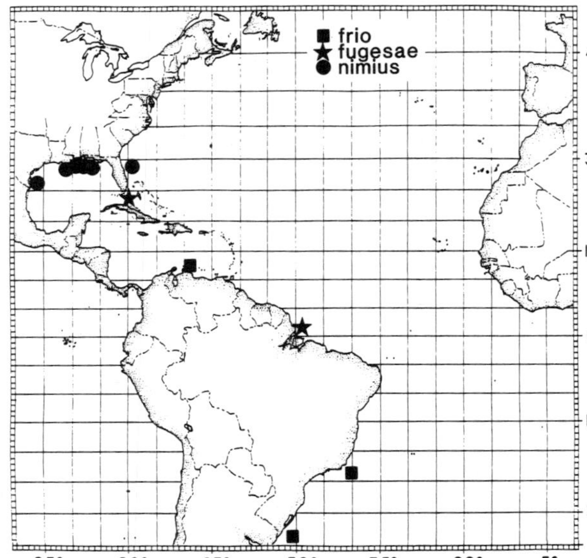

FIGURE 276. Distribution of western Atlantic species of *Pseudomyrophis* based on study material.

Pseudomyrophis fugesae new species

DIMINUTIVE WORM EEL

Figures 276–279, Tables 24, 25, 29

Distinctive Characters. A very small, elongate myrophin with dorsal-fin origin at or behind mid-trunk, and pectoral fin a minute broad-based flap; body depth 2.1–2.4% TL; head length 9.5–9.7% TL; MVF 30-52-157.

Counts and Measurements. (Those of the holotype parenthetically followed by those of the paratype, all in mm.) Total length 94 (95); head length 9.1 (9.0); trunk length 27.9 (24); tail length 57 (62); body depth at gill openings 2.3 (2.0); body width at gill openings 1.4 (1.3); body depth at anus 1.8 (1.5); body width at anus 1.5 (1.3); predorsal length 24 (21.5); snout length 1.6 (1.7); snout tip to rictus 2.5 (2.7); eye diameter 0.6 (0.5); fleshy interorbital distance 0.8 (1.0); length gill opening 0.3 (0.5); width isthmus 0.9 (1.5); length

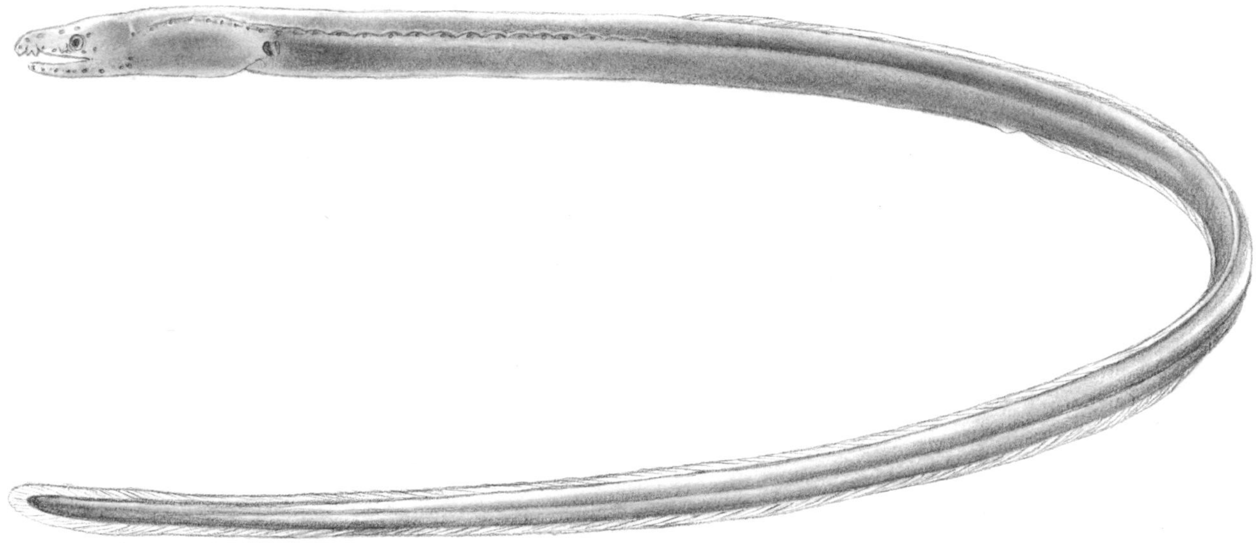

FIGURE 277. *Pseudomyrophis fugesae*: ANSP 153910, holotype, 94 mm TL, Florida.

pectoral fin 0.1 (–); length pectoral-fin base 0.4 (–). Total vertebrae 155 (159); predorsal vertebrae 33 (27); preanal vertebrae 53 (51); 9 lateral line pores before gill opening.

Description. Vertebrae as above. Pores: SO 1 + 4; IO 5 + 1; POM 6 + 3; ST 3.

Proportions as % TL: head 9.5–9.7; trunk 25–30; tail 61–65; predorsal 23–26; depth at gill opening 21–24; depth at anus 16–19; width at gill opening 14–15; width at anus 14–16. Of HL: snout 18–19; snout to rictus 28–30; eye 5.6–6.6; interorbital width 8.8–11; length gill opening 33–56; length isthmus 99–170; length pectoral fin 11 (for holotype).

Snout rugose, slightly tumid. Eye moderate, posterior margin of eye slightly before rictus. Anterior nostril tubular; posterior nostril an elongate slit in front of and below eye (hidden by fold of skin on holotype). Pores typical for genus; ethmoid pore under tip of snout; four supraorbital pores, the posteriormost above posterior half of eye; median pore in frontal commissure slightly behind posterior margin of eye; five infraorbital pores along upper lip, two between anterior and posterior nostrils and three below eye, plus one pore behind dorsal margin of eye; six pores along mandible plus three preopercular pores forming an angle greater than 90°; three pores in supratemporal canal.

Dentition typical for the genus. A patch or ring of 6–7 intermaxillary teeth, closely followed by vomerine series which consists of 6–7 biserially staggered teeth, then a uniserial row of 5–8. Maxillary teeth 13–16; mandibular teeth 24–27, plus 2 larger inner teeth anteriorly.

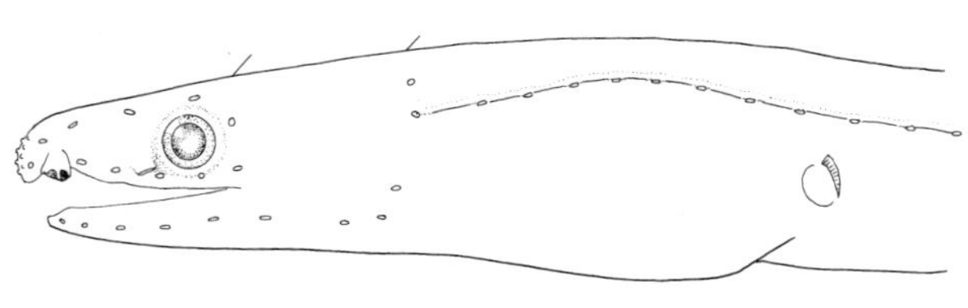

FIGURE 278. *Pseudomyrophis fugesae*: ANSP 153910, holotype, 94 mm TL, Florida; head.

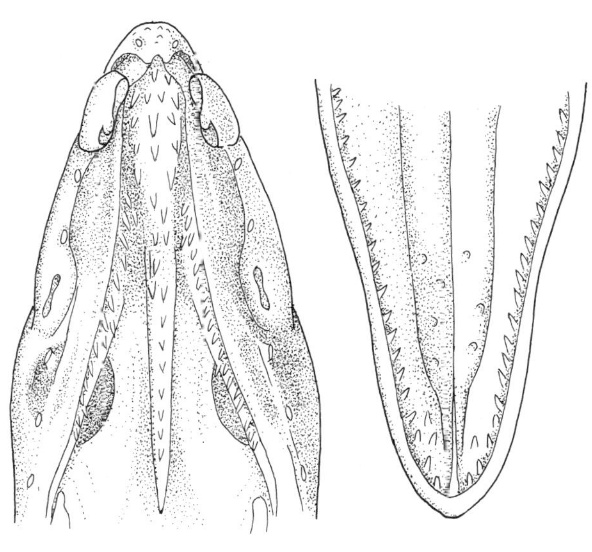

FIGURE 279. *Pseudomyrophis fugesae*: ANSP 153910, holotype, 94 mm TL, Florida; dentition.

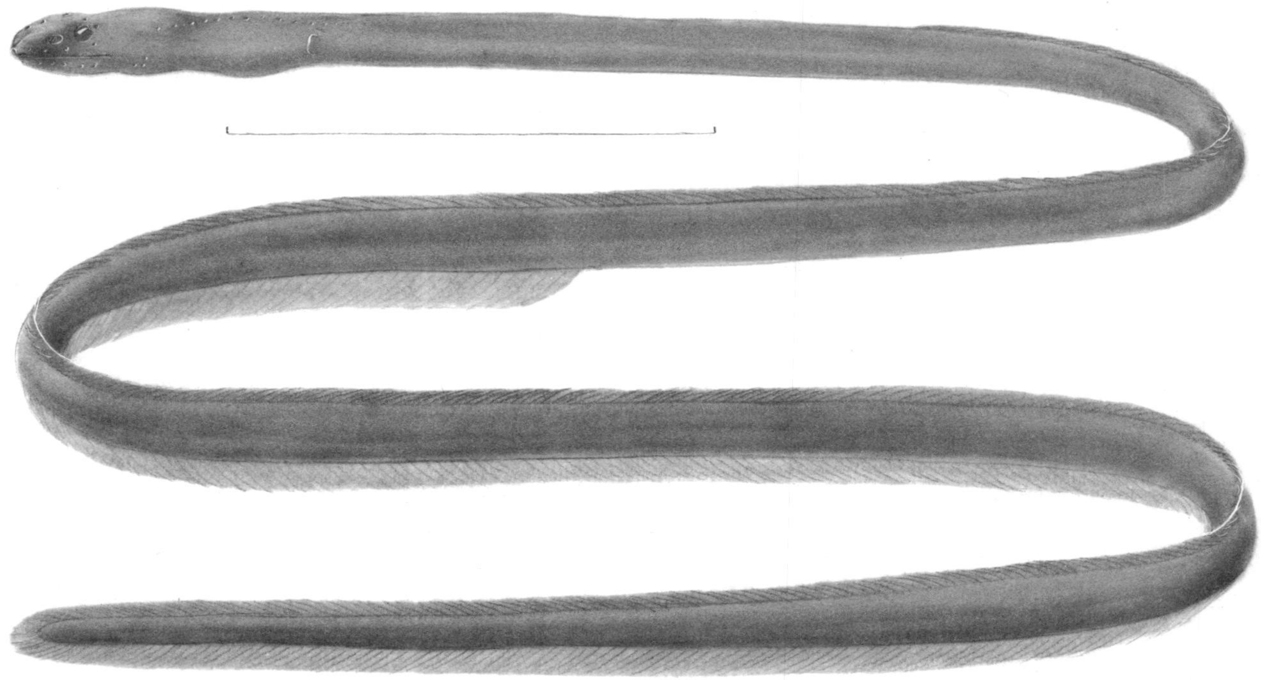

FIGURE 280. *Pseudomyrophis nimius*: ANSP 91939, paratype, 248 mm TL, Gulf of Mexico off Mississippi. Illustration by S. P. Gigliotti.

Dorsal-fin origin at mid-trunk or slightly behind that point, the fin continuous with expanded caudal fin. Pectoral fin minute, appearing as a broad-based flap extending along gill opening, its upper edge just above upper end of gill opening.

Color. Reddish-brown overall, the skin transparent in head region. Fins colorless, appearing as light streaks along midline.

Size. A small species; the type specimens are 94 and 95 mm TL, the body cavity of the larger filled with 0.5-mm eggs.

Development. Leptocephalus identified (p. 778).

Distribution (Fig. 276). Known from two specimens; the holotype from Florida, taken by dredge at 119 m, and the paratype from Brazil, captured at 56 m.

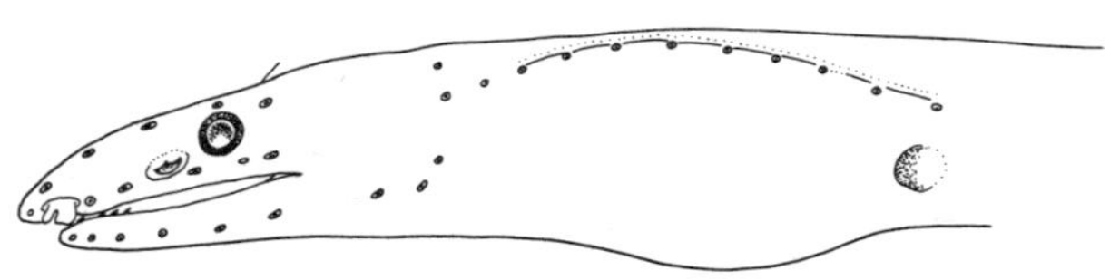

FIGURE 281. *Pseudomyrophis nimius*: ANSP 114878, 291 mm TL, Gulf of Mexico off Florida; head.

Remarks. This species is best separated from its congeners by vertebral count. Its rarity is probably due to its very small size and deep-water habitat, which enable it to evade capture by trawls and dredges.

Etymology. Named for Mary H. Fuges, artist and illustrator of numerous eels for this volume and for many publications during the past 14

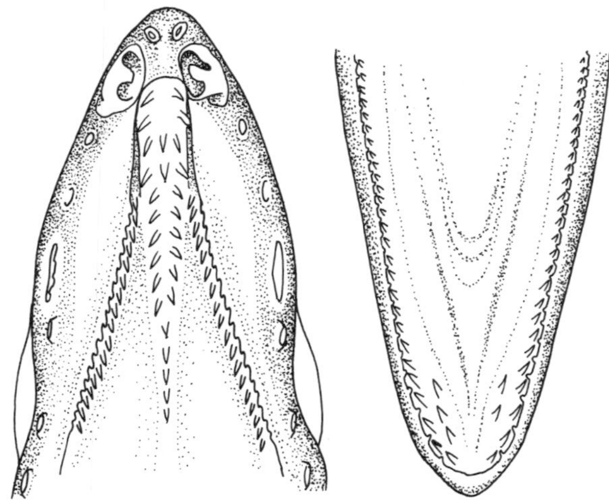

FIGURE 282. *Pseudomyrophis nimius*: ANSP 114878, 291 mm TL, Gulf of Mexico off Florida; dentition.

years; in recognition of her artistic ability, her meticulous attention to detail (well represented in the rendition of the dentition and head pores of this diminutive species), and her patience and encouragement throughout the preparation of this volume.

Study Material. The holotype and paratype, 94–95 mm TL.

Holotype: ANSP 153910 (94, a ripe female); Florida, off Government Cut, dredge at 119 m; R. Chesher on the TURSIOPS; 7 Jan. 1966.

Paratype: MZUSP 35644 (1, 95); Brazil, 01°45'N, 48°18'W, 56 m; R/V ALMIRANTE SALDANHA, sta. 1892; 1 May 1968.

Pseudomyrophis nimius Böhlke, 1960

ELONGATE WORM EEL

Figures 276, 280–282, Tables 24, 25, 29

Pseudomyrophis nimius Böhlke, 1960: 1, fig. 1 (original description, Gulf of Mexico, holotype USNM 186274).

New genus *nimius*, Dean 1972: 53 (description and illustrations of osteology).

Distinctive Characters. An extremely elongate myrophin with dorsal-fin origin behind mid-trunk and pectoral fin minute or not apparent; body depth 1–1.5% TL; head length 5.2–6.1% TL; MVF 49-73-214.

Description. Total vertebrae 211–217, $\bar{x}$ = 214 (n = 11); predorsal vertebrae 47–53, $\bar{x}$ = 49.3 (10); preanal vertebrae 71–75, $\bar{x}$ = 73.2 (11). Pores: SO 1 + 4; IO 5 + 1; POM 6 + 3; ST 3; total visible LL pores 143–160.

Proportions as % TL: head 5.2–6.1; trunk 29–33; tail 61–65; predorsal 24–28; depth behind gill openings 1–1.5. Of HL: snout 20–25; snout to rictus 29–33; eye 5.8–7.2; interorbital width 10–15.

Head elongate; snout tumid and papillose; skin transparent, especially over head. Eye above posterior half of upper jaw, its margin in advance of rictus. Head pores conspicuous, with raised pale rims, the number and arrangement as in *Pseudomyrophis frio.* A patch or ring of 7–8 intermaxillary teeth, continuous with tri- to uniserial vomerine teeth, about 7 pairs anteriorly and about 7 uniserial teeth posteriorly. Jaw teeth uniserial except anteriorly on mandible; maxillary with 17–20 teeth, 5 inner teeth on one specimen only; mandible with 24–27 teeth plus 2–4 larger inner teeth anteriorly.

Dorsal fin low, its origin behind mid-trunk, continuous with caudal and anal fins. Pectoral fin minute, frequently not readily visible, a small fringe or flap at upper end of gill opening.

Color. Uniform brown or reddish-brown overall, with a silvery-bluish cast, slightly darker along each side of lateral line; scattered pigment spots visible microscopically. Fins colorless. Anus in pale area.

Size. The largest specimen examined is 369 mm TL. Two females (297–348 mm) were ripe with 0.4–0.5-mm eggs. Four of five males (248–303) were ripe.

Development. Leptocephalus identified (p. 781).

Distribution (Fig. 276). A relatively rare species taken over deep waters of the northern Gulf of Mexico, from Florida to Texas. Taken by trawl over recorded depths of 320–755 m.

Etymology. From the Latin *nimius* (excessive), in reference to its length. An adjective.

Study Material. A total of 28 specimens, 208–369 mm TL, including the holotype and nine paratypes.

Holotype: USNM 186274 (319); Gulf of Mexico, east of the delta of the Mississippi River, 29°05'N, 88°22'W, 250 fm [457 m], with 40-ft flat trawl; SILVER BAY 1203; 11–12 June 1959.

Paratypes: ANSP 91937 (7, 208–324), 91938 (1, 288, cleared and stained), and 91939 (1, 248), taken with the holotype.

Other material: FLORIDA ATLANTIC: FDNR 16290 (1, 307); 28°20.4'N, 79°46.9'W, 411 m; 1 June 1984. FDNR 16291 (1, 369); 28°50.3'N, 79°55.7'W, 320 m; 30 May 1984. GULF OF MEXICO: *Florida:* ANSP 114859 (2, 297–348) and CAS 54480 (1, 330); 29°17'N, 87°03'W, 755 m; 27 Feb. 1969. ANSP 114878 (1, 291); 29°10'N, 87°00'W, 731 m; 7 Feb. 1970. ANSP 114950 (1, ca 295); 28°59'N, 87°01'W, 731 m; 10 Feb. 1970. *Mississippi:* ANSP 110150 (3, 248–297); 29°10'N, 87°57'W, 731 m; 27 July 1962 USNM 193589 (1, 245); 29°03.5'N, 88°32'W, 366–384 m; 26 Oct. 1962. USNM 193590 (3, 257–315); 29°04.5'N, 88°26.5'W, 366–384 m; 27 Oct. 1962. USNM 272468 (1, 280); 28°59.5'N, 88°39.5'W, 338–348 m; 12 Aug. 1962. *Louisiana:* ANSP 152306 (1, 208); 27°45'N, 91°18.5'W, 548 m; 23 Feb. 1964. *Texas:* ANSP 105677 (1, 282); 26°30'N, 96°16.5'W, 457 m; 24 Jan. 1964. UF 40278 (1, 314); 26°13.5'N, 96°18.9'W, 529 m; 23 Aug. 1983.

Subfamily OPHICHTHINAE

SNAKE EELS

Most juvenile and adult ophichthids can be recognized as members of the subfamily Ophichthinae by their hard-pointed tail tip which lacks externally visible fin rays (the species of *Echelus* and *Leptenchelys* are exceptions). Other diagnostic characters of juveniles and adults include: gill opening condition variable, mid-lateral to entirely ventral and longitudinal, generally elongate and unconstricted; origin of dorsal fin, if present, from nape to behind anus; ceratohyal divided into a short median and a longer distal portion, the median section connecting to the epihyal by a cartilaginous splint (Fig. 245); urohyal generally continues posteriorly from its basal plate as a slender ossified spike; all branchiostegal rays originate either in association with the hyoid or before the level of the epihyal tips; gill arch skeleton variably developed, generally more ossified than that of Myrophinae, fifth ceratobranchial present in several genera; and coloration variably uniform, freckled, spotted, saddled, banded or striped.

The Ophichthinae has been divided by McCosker (1977) into four tribes: Bascanichthyini, Callechelyini, Ophichthini, and Sphagebranchini. Characters are summarized and western Atlantic species treated under tribal headings.

Key to the Genera of Ophichthidae, Subfamily Ophichthinae

(Genera entirely extralimital to the western Atlantic are included in the key but not in the generic diagnoses.)

1a. Tail tip flexible, caudal-fin rays conspicuous, confluent with dorsal and anal fins; gill opening midlateral, a constricted opening; accessory branchiostegal rays originate behind ends of epihyal, free rays more numerous than attached rays (visible by radiography or by dissection) . . . . Subfamily Myrophinae, p. 270

1b. Tail tip a hard or fleshy finless point (except in two extralimital genera); gill opening midlateral to entirely ventral, unconstricted; all branchiostegal rays originate either in association with hyoid or before level of epihyal tips . . . . Subfamily Ophichthinae, 2

2a. Dorsal fin present, generally elevated, its origin on nape above supraoccipital; pectoral fin absent; gill openings inferior, parallel or converging forward, isthmus narrower than gill opening length . . . . Tribe Callechellyini, 3

2b. If present, dorsal fin arises behind nape; pectoral fin present or absent; gill openings inferior or lateral 7

3a. Anal fin absent 4

3b. Anal fin present 5

4a. Anterior nostril a hole, without a raised rim; four supraorbital pores; head brown or black, with white markings (not spotted), dorsal fin white . . . . *Letharchus*, p. 313

4b. Anterior nostril tubular; three supraorbital pores; head and dorsal fin spotted or mottled . . . . *Paraletharchus* McCosker, 1974 (two eastern Pacific species)

5a. Median groove on underside of snout extends to and beyond anterior nostril bases; vomerine and intermaxillary teeth present . . . . *Callechelys*, p. 303

5b. Underside of snout not incised anteriorly; vomerine and/or intermaxillary teeth absent . . . 6

6a. Vomerine teeth absent, intermaxillary teeth present; snout and nape with numerous papillae . . . . *Leuropharus* Rosenblatt & McCosker, 1970 (a single eastern Pacific species)

6b. Vomerine teeth present, intermaxillary teeth absent; snout and nape not heavily papillose . . . . *Aprognathodon,* p. 300
7a. Pectoral fin present, generally as large as or larger than eye (a small flap in two genera); median fins generally elevated; coloration variable, may be spotted or striped . . . . Tribe Ophichthini, 23
7b. Pectoral fin absent or vestigial; median fins reduced or absent; coloration without large spots 8
8a. Gill openings entirely ventral, or nearly so; head pores developed, second supratemporal pore and third preopercular pore generally present (Fig. 283); all fins may be absent . . . . Tribe Sphagebranchini, 9
8b. Gill openings low lateral, crescentic; head pores reduced, three supratemporal and two preopercular pores; at least a dorsal fin present . . . . Tribe Bascanichthyini, 16
9a. All fins absent 10
9b. Median fins present, although sometimes low and difficult to observe 12
10a. Upper lip with numerous, obvious cirri . . . . *Cirricaecula* Schultz, 1953 (a single central Pacific species)
10b. Upper lip without a row of obvious cirri 11
11a. Posterior nostril opening outside mouth, with a flap; anterior nostril tubular; eye moderately developed . . . . *Apterichtus,* p. 317
11b. Posterior nostril opening inside mouth, with or without a flap; anterior nostril not tubular; eye minute . . . . *Ichthyapus,* p. 322
12a. Lateral head profile, from above, narrows sharply from epiotics to orbit, then extends evenly to a narrow, pointed snout; body stout, its depth less than 30 times in its length; vomerine teeth enlarged, pointed and recurved . . . . *Lamnostoma* Kaup, 1856 (several Indo-Pacific species)
12b. Lateral head profile narrows evenly from epiotics to snout; body moderately elongate, its depth more than 40 times in total length; vomerine teeth conical, not enlarged 13
13a. Anterior nostril not tubular, a hole with lateral fleshy projections into it; eye minute, greater than 5 times in snout; dorsal-fin origin in mid-trunk region . . . . *Stictorhinus,* p. 325
13b. Anterior nostril either tubular, or with a short but noticeable posterior rim; eye larger, less than 3 times in snout; dorsal-fin origin above, before, or behind gill opening 14
14a. Gill openings with an anterolateral duplication forming a deep pouch; head nearly flat across region of parietals and epiotics . . . . *Caecula* Vahl, 1794 (one or two western Pacific species)
14b. Gill openings without a deep pouch; head raised along midline 15
15a. Anterior nostril not tubular, but with a groove surrounding it; snout without a groove on its underside . . . . *Hemerorhinus* Weber & deBeaufort, 1916 (two species, in western Pacific and eastern Atlantic oceans)
15b. Anterior nostril tubular; snout grooved on its underside, intermaxillary teeth exposed . . . . *Yirrkala* Whitley, 1940 (several Indo-Pacific species)
16a. The only fin a short dorsal, originating mid-head and ending about 1–3 head lengths behind it; trunk extremely elongate; tail very short, about 30% of total length . . . *Phaenomonas,* p. 351
16b. Median fins continuous to near end of tail; trunk moderately to extremely elongate; tail 35% or more of total length 17
17a. A small, lappet-like pectoral fin present in upper gill opening corner 18
17b. Pectoral fin absent 19

18a. Dorsal-fin origin on head . . . . *Bascanichthys*, p. 329
18b. Dorsal-fin origin more than a head length behind head . . . . *Allips* McCosker, 1972 (a single western Pacific species)
19a. Median fins continuous around caudal, caudal-fin rays evident . . . . *Leptenchelys* Myers & Wade, 1941 (a single eastern Pacific species)
19b. Caudal tip blunt, finless 20
20a. Underside of snout not grooved, intermaxillary teeth inconspicuous 21
20b. Underside of snout with grooved, intermaxillary teeth protruding 22
21a. Anterior nostril a pore-like opening, without raised rim . . . . *Caralophia* p. 338
21b. Anterior nostril tubular, projecting below upper lip . . . . *Ethadophis*, p. 341
22a. Tail longer than head and trunk; dorsal-fin origin behind gill openings . . . . *Dalophis* Rafinesque, 1810 (several eastern Atlantic and Mediterranean species)
22b. Tail shorter than head and trunk; dorsal-fin origin before gill openings . . . *Gordiichthys*, p. 343
23a. Caudal fin present, confluent with dorsal and anal fins; frontal, postorbital, and supratemporal pores absent . . . . *Echelus* Rafinesque, 1810 (two eastern Atlantic and European species)
23b. Tip of tail a finless point; frontal, postorbital, and supratemporal pores present 24
24a. Upper lip conspicuously fringed (Fig. 284) 25
24b. Upper lip entire, although narial barbels may be present 26
25a. Lower lip fringed; body and tail subequal; jaws subequal; canine teeth in jaws and on vomer . . . . *Brachysomophis* Kaup, 1856 (an eastern Atlantic and several Indo-Pacific species)
25b. Lower lip entire; tail much longer than head and trunk; lower jaw included; no canine teeth . . . . *Cirrhimuraena* Kaup, 1856 (several Indo-Pacific species)
26a. Anterior teeth of both jaws long fang-like canines extending far outside mouth when it is closed; lower jaw extends beyond snout . . . . *Aplatophis*, p. 354
26b. Anterior teeth in jaws not fangs extending beyond snout tip 27
27a. Teeth molariform or granular; pectoral fin broad-based (Fig. 285) 28
27b. Teeth pointed; pectoral fin base restricted, opposite upper half of gill openings (Fig. 286) . . . 29
28a. Dorsal-fin origin above or behind gill openings; three preopercular pores usually present . . . . *Pisodonophis* Kaup, 1856 (an eastern Atlantic and several Indo-Pacific species)
28b. Dorsal-fin origin well in advance of gill openings; two preopercular pores . *Myrichthys*, p. 372
29a. Pectoral fin rudimentary, smaller than or equal to eye; head and trunk slightly longer than tail 30
29b. Pectoral fin developed, larger than eye 31
30a. Maxillary teeth biserial; two preopercular pores . . . . *Evips* McCosker, 1972 (a single western Pacific species)
30b. Maxillary teeth uniserial; three preopercular pores . . . . *Quassiremus*, p. 409
31a. Conspicuous leaf-like appendages on anterior nostrils; vomerine teeth absent . . . . *Phyllophichthus* Gosline, 1951 (a single Indo-Pacific species)
31b. No leaf-like appendages on anterior nostrils; vomerine teeth usually present 32
32a. Eye before middle of upper jaw; some teeth long and fanglike 33

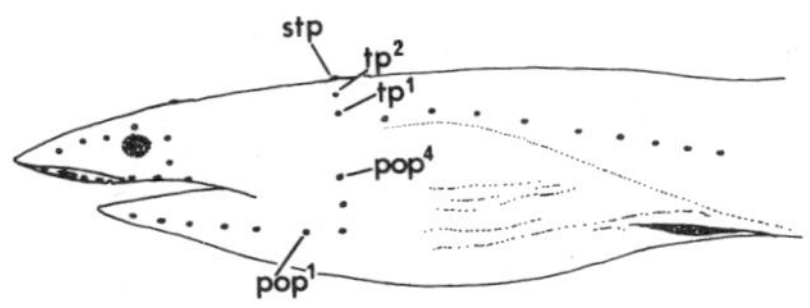

FIGURE 283. Diagrammatic view of a typical sphagebranchin with well-developed head pores.

FIGURE 285. Diagrammatic view of an ophichthin with a broad-based pectoral fin.

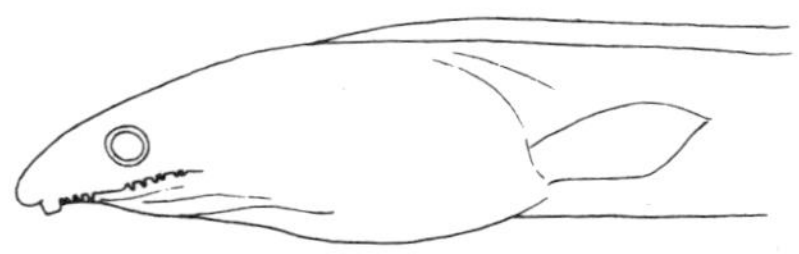

FIGURE 284. Diagrammatic view of an ophichthin with a conspicuously fringed upper lip.

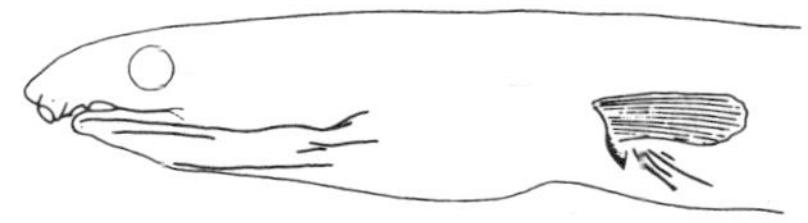

FIGURE 286. Diagrammatic view of an ophichthin with a narrow-based pectoral fin.

32b. Eye over middle of upper jaw; teeth not long and fanglike 36
33a. Snout short, its profile depressed, narrow posteriorly and clavate anteriorly; vomerine teeth numerous and small, mostly uniserial . . . . *Kertomichthys*, p. 368
33b. Snout longer and linear; vomerine teeth larger, either biserial or uniserial 34
34a. Maxillary teeth in 3 rows; vomerine teeth large and uniserial; preopercular pores absent . . . . *Lethogoleos*, p. 370
34b. Maxillary teeth in 1 or 2 rows; vomerine teeth either biserial or uniserial; at least 2 preopercular pores present, although possibly difficult to observe 35
35a. Snout short, 7–12 times in head length; secondary cephalic papillae absent; vomerine teeth biserial . . . . *Echiophis*, p. 357
35b. Snout longer, 6 or less times in head length; secondary cephalic papillae well developed; vomerine teeth uniserial . . . . *Mystriophis* Kaup, 1856 (two eastern Atlantic species)
36a. Vomerine teeth absent or, at most, 1–3 small embedded teeth; tail shorter than head and trunk . . . . *Leiuranus* Bleeker, 1853 (two or three Indo-Pacific species)
36b. Vomerine teeth developed, either uniserial or in multiserial rows; tail longer or shorter than head and trunk 37
37a. Tail shorter than head and trunk; pectoral fin reduced, 6 or more times in head length . . . 38
37b. Tail longer than head and trunk; pectoral fin developed, less than 6 in head length 39
38a. Pectoral fin minute, more than 8 times in head length; tail much shorter than head and trunk . . . . *Scytalichthys* Jordan & Davis, 1892 (a single eastern Pacific species)
38b. Pectoral fin better developed, about 7 times in head length; tail slightly shorter than head and trunk . . . . *Xyrias* Jordan & Snyder, 1901 (a single western Pacific species)
39a. Snout very long, attenuate, clavate at tip, its length greater than ¼ of head length; jaws of adults slender and elongate, incapable of closing completely . . . . *Ophisurus* Lacepède, 1800 (three eastern Atlantic and Indo-Pacific species)
39b. Snout moderate or short, ¼ of head length or less; jaws not slender and elongate 40
40a. Teeth minute, conical, in four rows on maxillary; lateral line abruptly arched and abutting its partner in head region . . . . *Hyphalophis*, p. 367
40b. Teeth in 3 rows or less on maxillary; lateral line not abruptly arched in head region 41

41a. Jaws equal; snout tip blunt; narial barbels pendant and conspicuous ... *Herpetoichthys*, p. 364
41b. Snout overhanging lower jaw; snout tip rounded to pointed; barbels on upper lip present or absent, not pendant and conspicuous ... 42
42a. Dorsal-fin origin well before gill openings ... *Malvoliophis* Whitley, 1934 (a single western Pacific species)
42b. Dorsal-fin origin above or behind gill openings ... *Ophichthus*, p. 379

Tribe Callechelyini

The Callechelyini are small to moderately large, elongate species, often with distinctive striped or spotted coloration. They live primarily in shallow sandy habitats and are collected by fish toxicants and occasionally by trawl or dredge. The tribe may be diagnosed as follows: moderately elongate, laterally compressed; head and trunk longer than tail; snout acute, rounded; lower jaw included; eye small; posterior nostril opens into mouth; gill openings low lateral to entirely ventral, converging forward, their length much greater than isthmus; dorsal-fin origin on nape; anal fin present or absent; pectoral fin absent; head pores reduced, two preopercular pores; teeth conical, jaw teeth uniserial; skull short, sloping posteriorly; supraoccipital rounded; suspensorium nearly vertical; branchiostegal rays numerous; gill arches reduced, ceratobranchial 5 fused to lower pharyngeal tooth plate; precaudal vertebrae more numerous than caudal; and coloration variable.

The tribe Callechelyini contains five genera, three of which (*) are found in the western Atlantic: *Aprognathodon**, *Callechelys**, *Letharchus**, *Leuropharus* and *Paraletharchus*.

Genus *Aprognathodon* Böhlke, 1967

Aprognathodon Böhlke, 1967: 99 (type species *Aprognathodon platyventris* Böhlke, 1967, by original designation).

Characters. Body moderately elongate (depth 2.5–3.0% TL), laterally compressed; head and trunk longer than tail (65–67% TL); tail tip a hard finless point. Dorsal-fin origin on nape, above supraoccipital; anal fin present; pectoral fins absent. Snout moderate, rounded at tip, overhanging lower jaw, not grooved on underside; lips without barbels; eye small; anterior nostril tubular; posterior nostril opens into mouth. Gill openings low lateral, converging forward, length greater than isthmus width. Head pores reduced; three supraorbital pores, two preopercular pores and three pores in supratemporal canal. Teeth few, conical, small to moderate, uniserial on jaws and on vomer; intermaxillary teeth absent.

Neurocranium short, sloping posteriorly; supraoccipital crest absent; maxilla elongate, slen-

Table 30. Mean vertebral formulae (MVF) and ranges of vertebral counts in western Atlantic species of Callechelyini.

		Predorsal		Preanal		Total	
Species	MVF	Range	n	Range	n	Range	n
*Aprognathodon platyventris**	H‡-96-150	H‡	11	93–99	9	148–152	14
*Callechelys bilinearis**	H-97-160	H	12	95–101	12	155–165	20
guineensis	H-115-178	H	10	111–118	9	172–182	10
*muraena**	H-83-142	H	7	80–85	7	139–144	7
*springeri**	H-108-169	H	3	107–108	2	167–171	3
Letharchus aliculatus	H-99†-159	H-1	7	96–103†	5	155–164	8
*velifer**	1-84-140	1–2	15	80–87†	19	136–144	22

* Includes holotype.
† Number of vertebrae before anus.
‡ H = on head.

der posteriorly; pterygoid slender, elongate, free and tapering posteriorly; otic bulla developed; suspensorium nearly vertical; opercular series weakly developed. Branchiostegal rays numerous, along arch; rays along epihyal broadened basally; hyoid arch stout, inflexible; urohyal a single slender filament posteriorly. Gill arches reduced; basibranchial 1 ossified, 2 cartilaginous; hypobranchials 1–2 ossified (and 3 in some specimens); ceratobranchials 1–4 ossified, 5 absent; upper pharyngeal tooth plates separated. Pectoral girdle reduced to cleithrum, supracleithrum and two rod-shaped elements. Intramuscular bones, ribs, and caudal transverse processes developed. Precaudal vertebrae more numerous than caudal. Gasbladder weakly developed. Peritoneum pale. Body coloration strongly striped longitudinally.

Size. To 447 mm TL.

Distribution. Known only from the western Atlantic.

Etymology. From the Greek *a* (without), *pro* (forward), *gnathos* (jaw), and *odontos* (tooth); in reference to the lack of intermaxillary dentition. Masculine.

Species. A single species, *Aprognathodon platyventris* Böhlke, 1967.

Aprognathodon platyventris Böhlke, 1967

STRIPE EEL

Figures 287–290, Table 30

Aprognathodon platyventris Böhlke, 1967: 100, fig. 3 (original description, Great Bahama Bank, holotype ANSP 106303).

Distinctive Characters. An elongate callechelyin with a broad lateral stripe on body from gill opening to tip of tail, and narrow stripes dorsally along margin of dorsal fin and ventrally along ventral ridges and anal fin; intermaxillary teeth absent, vomerine and dentary teeth uniserial; MVF H-96-150.

Description. Total vertebrae 148–152, $\bar{x}$ = 150 (n = 14); dorsal-fin origin on head before first vertebra (H); preanal vertebrae 93–99, $\bar{x}$ = 96 (9). Pores: SO 1 + 3; IO 4 + 2; POM 4 + 2; ST 3.

Proportions as % of TL: head 7–9.8; trunk 56–60; tail 33–35; predorsal 2.1–2.8; depth behind gill openings 2.5–3. Of HL: snout 10–14; snout to rictus 21–26; eye 5–8; interorbital width 7–10.

Head moderate, snout short, lower jaw very short, reaching mid-point of upper jaw; eye moderate, above tip of lower jaw. Anterior nostril tubular, thick and prominent, with fleshy projections at tip; posterior nostril opening into mouth, the opening covered by a flap of skin, its position marked externally by a short groove in upper jaw. Deep lateral grooves paralleling posterior two-thirds of both jaws. Head pores as illustrated. Teeth few; intermaxillary teeth absent; 4–7 vomerine teeth, some staggered or paired; jaw teeth slender, sharp, strongly canted posteriorly and mesially, 3–8 teeth on maxillaries, 8–18 teeth on dentaries.

Dorsal fin high, beginning far forward on head; anal fin low. Venter flattened between grooved lateral ridges on each side which begin below posterior ends of gill openings and extend back to level of anus.

Color. Ground color pale, overlain by three dark stripes. A broad stripe along mid-body, extending from head or level of gill openings almost to tail tip, its width from below lateral line nearly to dorsal fin; a narrow stripe along distal edge of dorsal fin, which is pale at base; and a narrow stripe along ventral lateral ridges which degrades behind anus to two parallel lines or an interrupted line of spots in tail region. Head and anterior dorsal-fin region variously mottled or spotted; diagonal streaks along dorsal-fin rays at regular intervals. Specimens smaller than 100-mm pale and with remnants of larval pigmentation, the blotches and stripes appearing at larger sizes. In life, the median body stripe brown or black, the dorsal stripe varying from orange to brown or black.

Size. The largest specimen examined was 447 mm TL. Other than three small metamorphosing individuals, most of the study specimens examined were mature adults. Of 51 for which sex was recorded, 31 (118–447 mm) were females, 21 (213–447) ripe with eggs of 0.9–1.2-mm diameter; and 20 (196–289) were ripe males. Males were generally smaller than females. Large collections

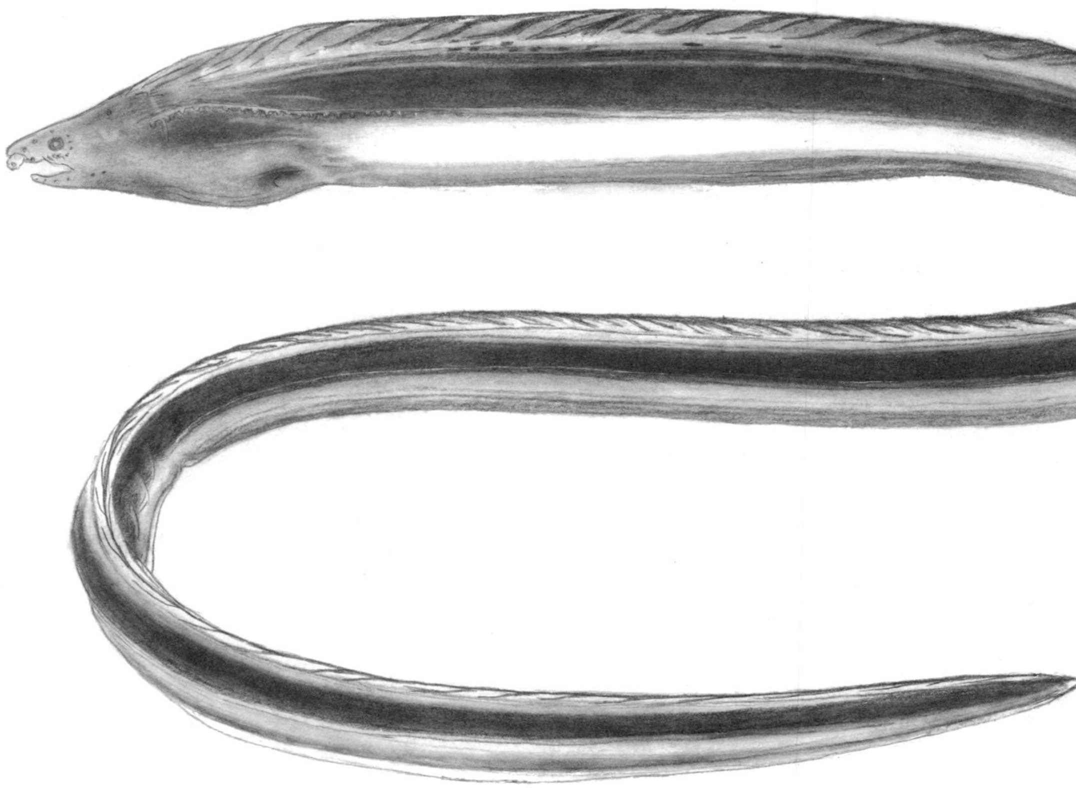

FIGURE 287. *Aprognathodon platyventris*: ANSP 106301, paratype, 332 mm TL, Bahamas.

of ripe specimens were obtained in May, July and August.

Development. Leptocephalus described by Leiby (1984b: 401, and this volume, p. 881).

Distribution (Fig. 290). An insular species, taken in the Florida Keys and throughout the West Indies, including the islands off the coast of Venezuela. This species apparently prefers shallow habitats where it burrows in sand; all specimens were collected by use of ichthyocides over shallow sand or turtle grass bottom, from the shoreline to a depth of 17 m.

Etymology. From the Greek *platy* (flat) and *ventris* (belly), in reference to the flattened ventrum. A noun in the genitive case.

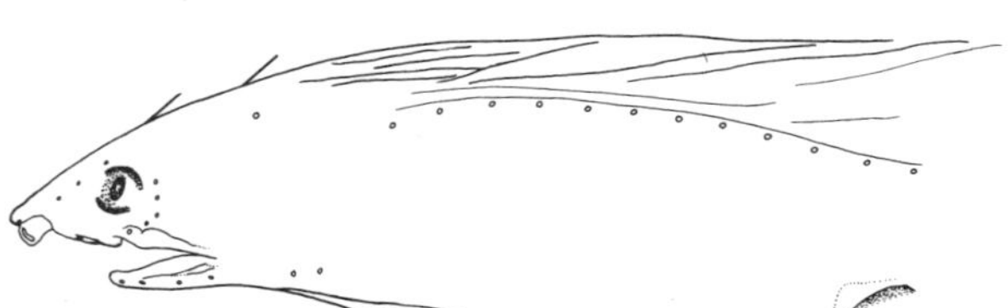

FIGURE 288. *Aprognathodon platyventris*: ANSP 106307, paratype, 222 mm TL, Grand Cayman; head.

Study Material. A total of 134 specimens, 52–447 mm TL, including the holotype and 109 paratypes.

Holotype: ANSP 106303 (233.5 mm TL); Great Bahama Bank, ca ¼ mi N of Green Cay (N of Rose Island), Nassau vicinity, surface to 10 m; J. E. Böhlke et al., sta. 295; 7 May 1956.

Paratypes: BAHAMAS: AMNH 20616 (1, 269). ANSP 106298 (1, 447), 106302 (3, 242–368), 106301 (1, 332), 106387 (1, 337, cleared and stained), 106296 (57, 56–315),

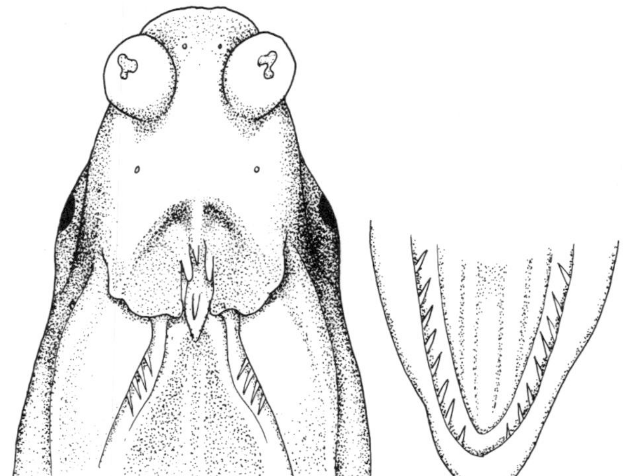

FIGURE 289. *Aprognathodon platyventris*: ANSP 138770, 244 mm TL, Bahamas; dentition.

106299 (3, 236–293), 106300 (3, 166–272), 106297 (1, 54). FLORIDA: UMML 21754 (3, 338–358). UMML 21753 (1, 330). GRAND CAYMAN ISLAND: ANSP 106307 (1, 222). UF 12469 (3, 52–223). PUERTO RICO: ANSP 106308 (1, 118). VIRGIN ISLANDS: ANSP 106306 (2, 308–328). UMML 6661 (1, 185). UPR 1117 (1, 381). ANGUILLA ISLAND: ANSP 106305 (3, 93–263). ST. BARTHELEMY: ANSP 106304 (21, 103–402). VENEZUELA: UMML 12520 (1, 351).

Other material: BAHAMAS: ANSP 130554 (18, 53–260), 138770 (1, 244). CAS 54464 (4, 64–335).

Genus *Callechelys* Kaup, 1856

Callechelys Kaup, 1856a: 51 (type species *Callechelys guichenoti* Kaup, 1856 = *Dalophis marmorata* Bleeker, 1853, by monotypy).

Cryptopterygium Ginsburg, 1951: 482 (type species *Cryptopterygium holochroma* Ginsburg, 1951, by original designation).

Characters. Body moderately elongate (depth 1.7–3.7% TL in Atlantic species), laterally compressed; head and trunk much longer than tail (60–72% TL); tail tip a hard finless point. Dorsal-fin origin on nape, above supraoccipital; anal fin present; pectoral fins absent. Snout acute, rounded at tip, overhanging lower jaw, deeply grooved on underside; lips without barbels; eye moderately developed; anterior nostril tubular; posterior nostril opens on underside of snout. Gill openings low lateral to entirely ventral, converging forward, length much greater than isthmus. Head pores reduced; three supraorbital pores, two preopercular pores, and three pores in supratemporal canal. Teeth conical, slender, small to moderate, uniserial on jaws, those of intermaxillary separated from vomer by a gap.

Neurocranium short, sloping posteriorly; supraoccipital crest absent; maxilla elongate, slender posteriorly; pterygoid slender, elongate, free and tapering posteriorly; otic bulla developed; suspensorium nearly vertical; opercular series weakly developed, their margins irregularly ossified. Branchiostegal rays numerous, along arch; hyoid arch stout; urohyal either a single slender filament posteriorly or split into two divergent rays. Gill arches reduced; basibranchial 1 ossified, 2 cartilaginous; hypobranchials 1–2 ossified, 3 cartilaginous; ceratobranchials 1–4 ossified, 5 ab-

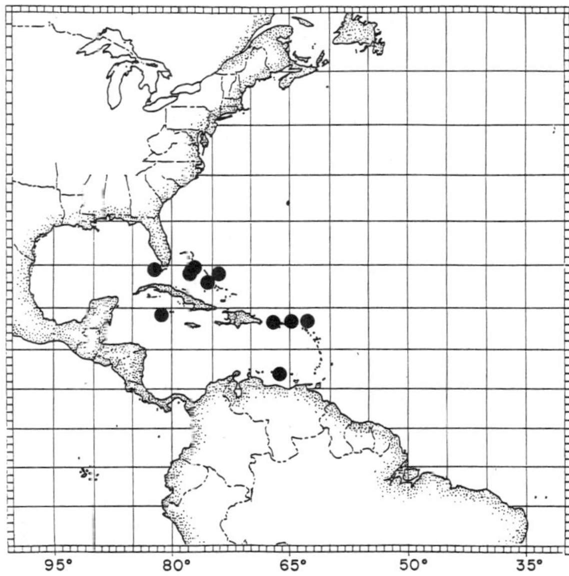

FIGURE 290. Distribution of *Aprognathodon platyventris* based on study material.

sent; upper pharyngeal tooth plates separate. Pectoral girdle reduced to cleithrum, supracleithrum and one or two rod-shaped elements (McCosker's (1977: table 8) observation that the holotypes of *Callechelys perryae* and *Cryptopterygium holochroma* lacked a second pectoral girdle element is erroneous; the examination of subsequent radiographs indicates that it is present in both specimens). Intramuscular bones, ribs, and caudal transverse processes developed. Precaudal vertebrae more numerous than caudal. Gasbladder weakly developed. Peritoneum pale. Body coloration variable: striped, barred, spotted, mottled or uniform.

Size. The largest western Atlantic callechelyin, reported to 1724 mm TL.

Distribution. Known from all tropical and subtropical seas.

Etymology. From the Greek *kallos* (beauty) and *enchelys* (eel). Treated as feminine according to Opinion 915 of the Bulletin of Zoological Nomenclature, 1970.

Species. There are approximately 15 species of *Callechelys* worldwide. We recognize four species in the western Atlantic: *C. bilinearis* Kanazawa, 1952, *C. guineensis* (Osorio, 1894), *C. muraena* Jordan & Evermann, 1897, and *C. springeri* Ginsburg, 1951. *Callechelys bilinearis* is an insular species, easily identified by its distinctive coloration of alternating dark and light stripes along the body and fins. The other three species have similar color patterns of dark spots of varying sizes and shapes on a pale background, and are best separated by vertebral count. *Callechelys guineensis* and *C. springeri* are more elongate and have fewer jaw teeth than *C. muraena; C. guineensis* and *C. springeri* are separated by differences in vertebral counts and subtle color variations. The single eastern Atlantic species, *Callechelys leucoptera* (Cadenat, 1954), has uniform dark coloration.

We have followed Blache, Bauchot, and Saldanha (1979) in placing the western Atlantic species previously known as *Callechelys perryae* in the synonymy of *C. guineensis* from the eastern Atlantic, based on their conclusions and our examination of the neotype of *guineensis*. It is possible that study of additional material will result in the validation of *C. perryae*, or may, as well, result in the synonymy of *C. springeri* with this species (see *Remarks*, p. 312).

Key to the Atlantic Species of *Callechelys*

1a. Body uniformly colored or conspicuously striped, not spotted 2
1b. Body covered with numerous brown to black spots 3
2a. Body coloration uniform brown to black, fins colorless; total vertebrae 162–165 . . . . *leucoptera* (Cadenat, 1954)
(an eastern Atlantic species, from Senegal to the Ivory Coast)
2b. Body coloration strongly delineated with alternating black and white stripes running the length of the body; total vertebrae 155–165 . . . . *bilinearis*, p. 304
3a. Body moderately elongate, depth 2.6–3.7% TL; jaw teeth numerous, 7–11 on maxillary, 11–17 on mandible; head 7.0–8.8% TL; total vertebrae 141–144 . . . . *muraena*, p. 310
3b. Body elongate, depth 1.7–2.2% TL; jaw teeth few, 4–5 on maxillary, 9–11 on mandible; head 5.6–7.1% TL; total vertebrae 169–186 4
4a. Preanal vertebrae 111–121, total vertebrae 171–186; dorsal spots of large individuals not coalesced into saddles . . . . *guineensis*, p. 308
4b. Preanal vertebrae 107–108, total vertebrae 169–171; dorsal spots of large individuals coalesced into dark saddles . . . . *springeri*, p. 311

Callechelys bilinearis Kanazawa, 1952

TWOSTRIPE SNAKE EEL

Figures 291–294, Tables 30, 31

Callechelys bilinearis Kanazawa, 1952: 72, fig. 9 (original description, Bermuda, holotype FMNH 48973).

Discussion of Synonymy. Fowler (1944: 130, 163, fig. 49) described and illustrated *Leptocephalus caribbaeus* from Old Providence Island in the Caribbean. These leptocephali are questionably referred to this species (p. 891); if this assignation is correct, *caribbaeus* is the senior synonym of *bilinearis*. However, there are several discrepancies between Fowler's description and illustration and the type specimens (ANSP 70122–70130) which make it impossible to assign them to one species with certainty. Therefore, *L. caribbaeus* is considered a *nomen dubium* and the name *Callechelys bilinearis* is retained for the species.

Distinctive Characters. A moderately elongate callechelyin, boldly colored with two dark lateral stripes along body and one stripe along dorsal fin; body depth 2.8–3.5% TL; head 7.0–8.0% TL; maxillary teeth 8–12, dentary teeth 12–19; MVF H-97-160.

Description. Total vertebrae 155–165, $\bar{x}$ = 160.4

TABLE 31. Frequency distributions of vertebral counts in western Atlantic species of *Callechelys*.

	Preanal vertebrae																																
	80	81	82	83	84	85	86	...	95	96	97	98	99	100	101	102	...	106	107	108	...	111	112	113	114	115	116	117	118	119	120	121
bilinearis									2	4	7	2[1]	-	-	1																	
guineensis																																
W. Atlantic																						1	-	1	2[2]	-	2	2	1			
E. Atlantic																														1[3]	1	
muraena	1[1]	-	2	2	1	1																										
springeri																			1	2[1,5]												

	Total vertebrae																													
	139	140	141	142	143	144	...	155	...	158	159	160	161	162	163	164	165	166	167	168	169	170	171	172	...	176	177	178	179	180	181	182	...	186
bilinearis								1	-	2	4	5	2	5	3[1]	-	1	1	1															
guineensis																																		
W. Atlantic																								1		1	1[2]	1	-	4	1	1		
E. Atlantic																													1[4]	-	-	-	-	1
muraena	1	-	1[1]	1	3	1																												
springeri																				1[5]	1	-	1[1]											

[1] Holotype of species.
[2] Holotype of *C. perryae.*
[3] Holotype of *C. guineensis,* counts from Blache et al. (1979: 97), preanal count from lateral line pores.
[4] Holotype of *C. pantherina,* counts from Blache (1971: 189), preanal count from lateral line pores.
[5] Holotype of *C. holochroma.*

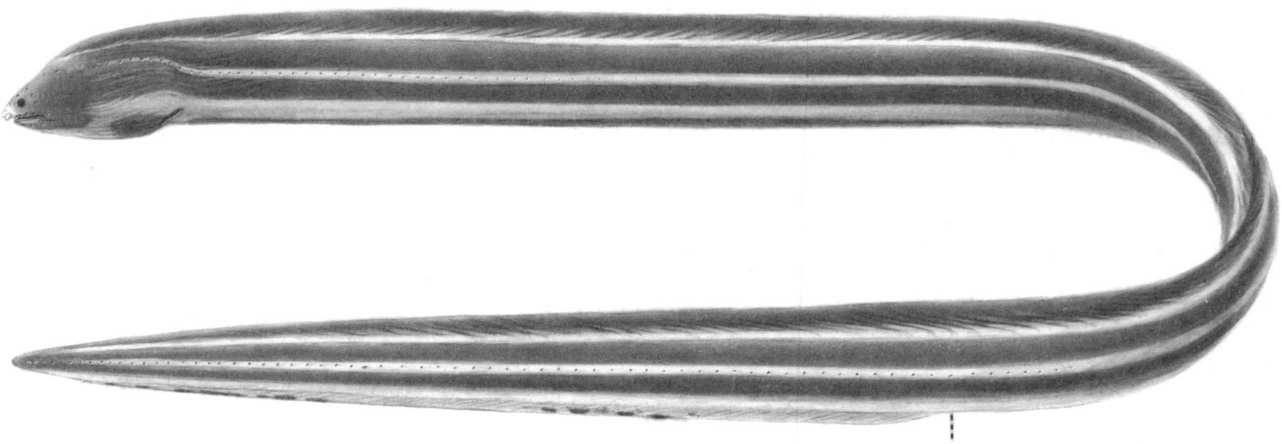

FIGURE 291. *Callechelys bilinearis*: ANSP 98336, 760 mm TL, Bahamas. Illustration by S. P. Gigliotti.

(n = 20); preanal vertebrae 95–101, x̄ = 96.9 (12); dorsal-fin origin before first vertebra (H). Pores: SO 1 + 3; IO 4 + 2; POM 5 + 2; ST 3.

Proportions as % of TL: head 7.0–8.0; trunk 54–64; tail 28–38; predorsal 2.3–3.8; depth behind gill openings 2.8–3.5. Of HL: snout 10–14; snout to rictus 24–30; eye 3.8–6.3; interorbital width 5.5–9.6.

Head short; snout short and acute, a deep groove on underside of snout extending forward between anterior nostrils; lower jaw short, reaching midpoint of upper jaw. Eye moderate, located about midway between snout tip and rictus, above tip of lower jaw. Anterior nostril tubular, directed down, the opening small. Posterior nostril on underside of snout, lying outside lower jaw and exposed when mouth is closed, covered by an anterior flap which is visible in lateral aspect as a vertical slit in upper lip just below eye. Head pores as illustrated, with some variation: of 92 specimens, one has four supraorbital pores on one side and one has four on both sides; one has an extra infraorbital pore below eye on both sides; three specimens have six pores along each mandible, the holotype has six pores on right side and five on left.

Teeth long, slender and sharp. Two or three stout intermaxillary teeth in groove on underside of snout, these the largest teeth in the mouth, exposed when mouth is closed but partially obscured by papillae-like folds of skin on either side. A large gap followed by 7–12 teeth on vomer, biserially staggered except far posteriorly. Jaw teeth uniserial; 8–12 on maxillary, 12–19 teeth along mandible.

Dorsal fin high and prominent, its origin forward on head. Anal fin moderate. Oblique longitudinal grooves in skin on head and body. Body slightly flattened ventrally between grooved lateral ridges from just behind gill openings to anus.

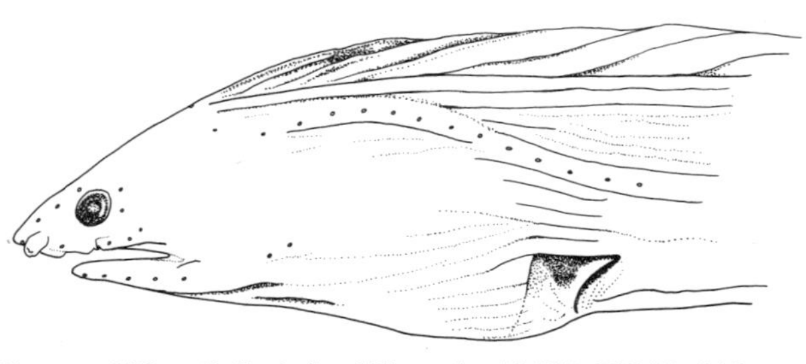

FIGURE 292. *Callechelys bilinearis*: ANSP 98347, 368 mm TL, Bahamas; head.

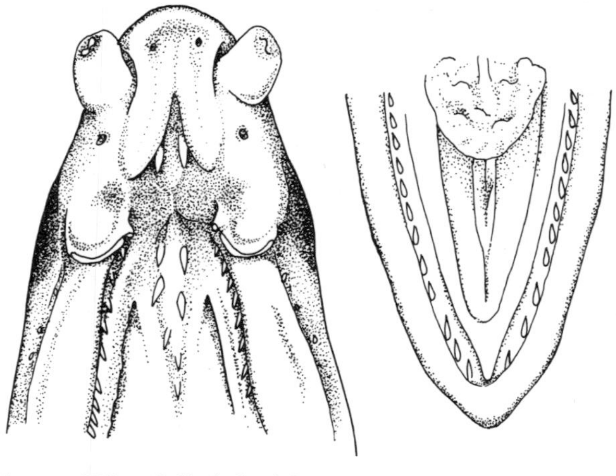

FIGURE 293. *Callechelys bilinearis*: ANSP 98345, 275 mm TL, Bahamas; dentition.

Color. Three prominent dark stripes alternating with three narrow pale stripes along the length of the eel, two of the dark stripes on the body (the basis of the common name), plus a broad stripe distally along dorsal fin, the first pale stripe basally along that fin. An additional diffuse dark ventral stripe from gill opening to anus between body ridges. Head and chin variously mottled to dark; tip of snout sometimes pale, a thin pale band encircling snout just before eyes sometimes present. Anal fin of small specimens pale, variously spotted or with irregular dark margin on larger specimens. Pale areas on fins and head yellowish in life.

Size. Cervigón (1973: 11) reported a 1724-mm specimen poisoned in Archipelago de Los Roques off Venezuela; the largest specimen we examined is 871 mm TL. Many of the specimens larger than 200 mm were maturing and a large percentage of them were ripe; of 31 females (221–871 mm), 10 (314–834) were ripe with 0.4–1.0-mm eggs; of 21 males (210–460), nine (240–466) were ripe.

Development. Leptocephalus described by Leiby (1984b: 407, and this volume, p. 890).

Behavior. A sand-dwelling eel, most often found in waters shallower than 4 m. "It lives on open beaches and beaches that are semi-protected from the open ocean, but seems particularly to occur near the ends of sand-spits . . ." (Böhlke and Chaplin, 1968: 109). Large ripe specimens have been taken in deeper waters (to 22 m).

Distribution (Fig. 294). An insular species, widespread throughout tropical western Atlantic, and the most common species of *Callechelys* in the Atlantic. It has been taken from Bermuda, throughout the West Indies, from islands off Central and South America, and from St. Helena and Ascension islands in the mid-Atlantic. Not reported from American coastal waters, nor from the eastern Atlantic. Our study material came from rotenone stations to depths of 22 m, most often from shallow waters to 4 m.

Etymology. From the Latin *bilinearis* (two-lined), in reference to the stripes along the body. An adjective.

Remarks. This species is unmistakable in its coloration and might only be confused with *Aprognathodon platyventris* which has only one lateral black stripe along the body and which differs in the structure of the upper jaw and the anterior dentition. We note that the specimen from St. Helena has a slightly higher vertebral count (167) than that of the others (155–165).

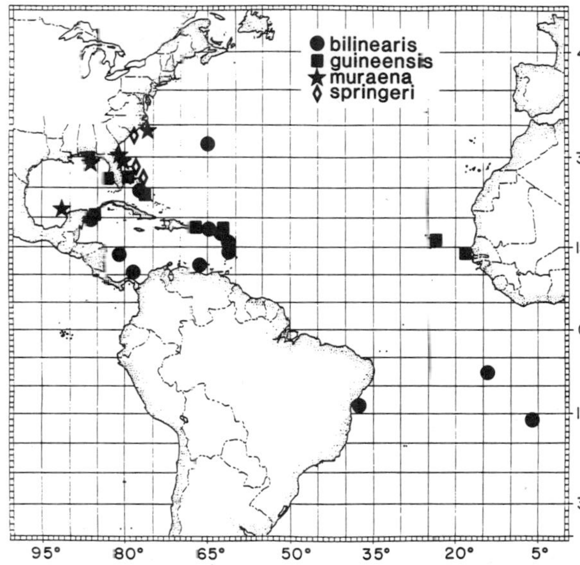

FIGURE 294. Distribution of western Atlantic species of *Callechelys* based on study material and records in Blache et al., 1979.

Study Material. 92 specimens, 86–871 mm TL, including the holotype and paratype.

Holotype: FMNH [CNHM] 48973 (597); Bermuda, Cooper's Island and St. Davids; Mr. F. Lamb; June 1938.

Paratype: USNM 157363 [formerly CNHM 48974] (571); Bermuda, taken from stomach of a seven-pound porgy, which was taken at depth of 9 fm; Mr. Astwood; 17 Dec. 1937.

Other material: BERMUDA: ANSP 133588 (1, 376), 133559 (1, 543), 133560 (1, 377), 148173 (4, 175–625), 148174 (2, 314–314), 148175 (4, 260–607). BAHAMAS: ANSP 74703 (3, 210–266), 98336 (1, 760), 98337 (6, 257–617), 98338 (15, 198–436), 98339 (4, 285–428), 98340 (10, 220–460), 98341 (3, 293–333), 98342 (1, 239), 98343 (2, 299–341), 98344 (1, 130), 98345 (1, 275), 98346 (1, 371), 98347 (1, 368), 113401 (2, 348–374), 115511 (2, 344–355, cleared and stained), 115533 (1, 320, cleared and stained), 119814 (5, 178–232), 137929 (1, 303). ANTILLES: *Virgin Islands, St. Croix:* ZMC P.3210 (1, 402). *St. Barthelemy:* ANSP 105856 (1, 371). *St. Lucia:* CAS 54467 (1, 393). *Dominica:* USNM 270501 (1, 315). MEXICO: *Cozumel:* ANSP 115291 (1, 834). PANAMA: ANSP 113768 (1, 668). CAS 29116 (1, 284). PROVIDENCIA: UF 18807 (1, 282). VENEZUELA:

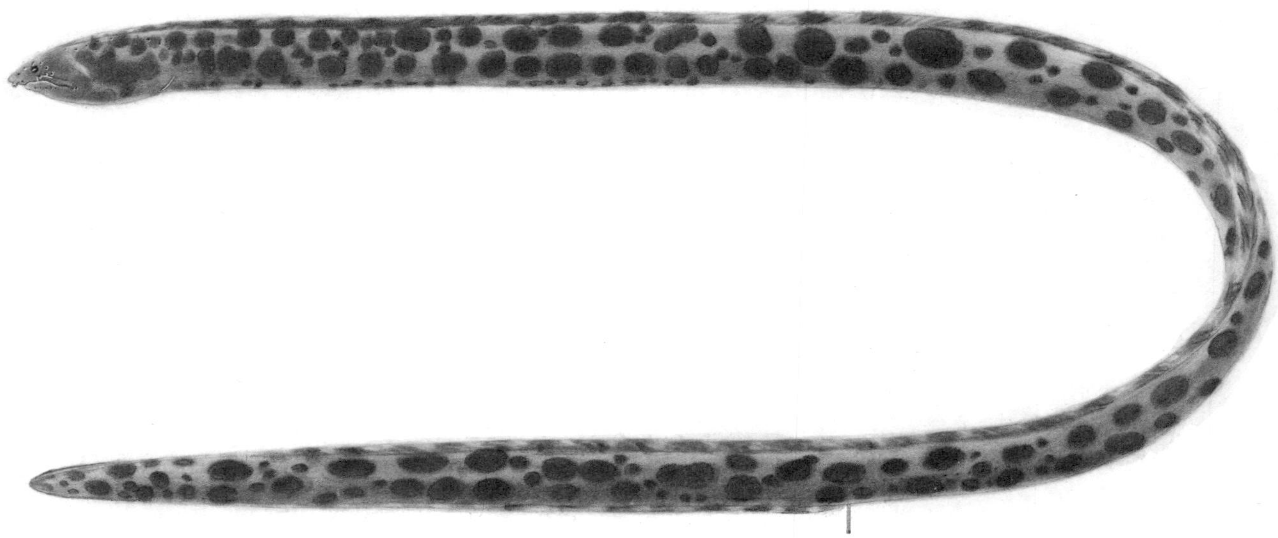

FIGURE 295. *Callechelys guineensis*: ANSP 98333, 722 mm TL, Bahamas. Illustration by S. P. Gigliotti.

Los Roques: MBUCV 984 (2, 676–701). BRAZIL: Bahia: GCRL 9408 (2, 188–221), 10770 (1, 154), 10883 (1, 375).

MID-ATLANTIC: *St. Helena:* BMNH 1965.12.1.10 (1, 257). *Ascension:* USNM 270466 (2, 86–165).

Callechelys guineensis (Osorio, 1894)

SHORTTAIL SNAKE EEL

Figures 294–297, Tables 30, 31

Ophichthys (*Sphagebranchus*) *guineensis* Osorio, 1894a: 179 (original description, Ile de São Tome, Cape Verde Islands, neotype MNHN B.1635).

Callechelys perryae Storey, 1939: 71, figs. 2, 4 (original description, Sanibel Island, Florida, Gulf of Mexico, holotype CAS-SU 33027).

Caecula pantherina Cadenat, 1956: 1267, fig. 1 (original description, Sénégal, holotype MNHN B.1635).

Callechelys pantherina. Cadenat, 1961: 235.

Callechelys perryae. Blache and Cadenat, 1971: 186 (redescription and illustration of type of *Caecula pantherina*, synonymized with *Callechelys perryae*).

Callechelys guineensis. Blache, Bauchot, and Saldanha, 1979: 97 (*perryae* and *pantherina* synonymized with *guineensis*: designation of the holotype of *Caecula pantherina* as the neotype of *Ophichthys guineensis*).

Distinctive Characters. An elongate callechelyin with cream to tan coloration overlain with numerous dark brown spots and blotches; body depth 1.7–2.2% TL; head 5.6–6.7% TL; jaw teeth few, 4–5 on maxillary, 9–11 on mandible; MVF H-115-178.

Description. Total vertebrae 172–182, $\bar{x}$ = 177.9 (n = 10); preanal vertebrae 111–118, $\bar{x}$ = 114.8 (9); dorsal-fin origin on head (H). Pores: SO 1 + 3; IO 4 + 2; POM 5 + 2; ST 3.

Proportions as % of TL: head 5.6–6.7; trunk 59–62; tail 32–34; predorsal 2.0–2.8; depth behind gill openings 1.7–2.2. Of HL: snout 11–17; snout to

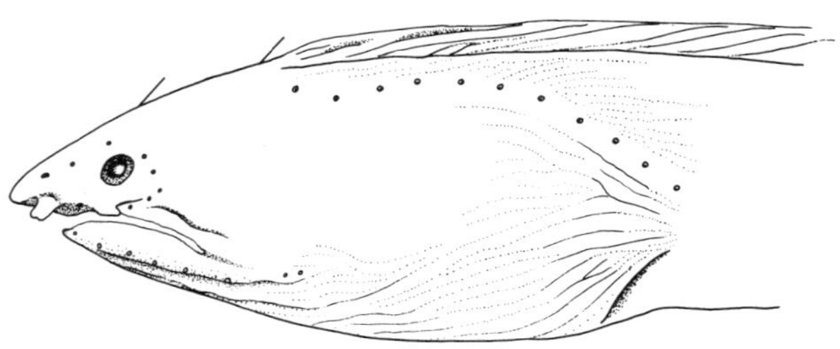

FIGURE 296. *Callechelys guineensis*: ANSP 98333, 722 mm TL, Bahamas; head; posterior (sixth) POM pore usually not present.

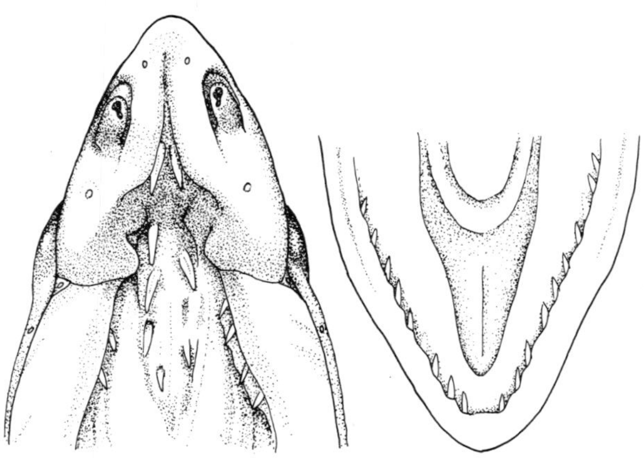

FIGURE 297. *Callechelys guineensis*: ANSP 98333, 722 mm TL, Bahamas; dentition.

rictus 23–33; eye 3.9–7.7; interorbital width 7.7–11.

General characters as for *C. bilinearis.* Body very elongate; ventral ridges on trunk not pronounced. One specimen (ANSP 151696, head illustrated) with six pores along mandible. Teeth few in number; typically two large intermaxillary teeth, partially exposed when mouth is closed; 6–9 stout teeth on vomer in two widely divergent rows, sometimes appearing as an elongate patch; 4–5 widely spaced teeth along maxillary; 9–11 teeth along mandible.

Color. Variably cream to tan body overlain with numerous brown to black spots; belly pale, spots between ventral ridges pale and few in number or absent, spots behind anus extend ventrally onto body and anal fin. Spots on snout, nape and chin small, about equal to eye; those on posterior head, body and median fins are larger and vary in size. A row of uniform spots distally on dorsal fin.

Life colors of the 601-mm specimen from Puerto Rico (CAS 54469) were recorded by J. E. Randall (in litt.): "Body pale blue, pink on abdomen, dark brown spots indistinctly edged with dusky yellow; head pale yellow with smaller dark brown spots; anterior roof of mouth yellow-white, remaining pink; iris brassy." Cadenat described life colors of the holotype of *Caecula pantherina* as orange dorsally, fading to cream ventrally, with irregular chocolate brown spots.

Size. The largest specimen examined is 1072 mm TL. Blache et al. (1979: 97) reported that the type of *Ophichthys guineensis* was 1080 mm. Four of the study specimens (795–1072) were female, none of them ripe; one 722-mm ripening male.

Development. Leptocephali of this species were described as *Callechelys perryae* by Leiby (1984b: 407); see p. 892.

Distribution (Fig. 294). Apparently widely distributed throughout the eastern and western Atlantic, but infrequently taken. In the western Atlantic, known from eastern and western Florida, Puerto Rico, the Bahamas and St. Barthelemy. Cervigón (1966: 193) tentatively identified a 54-mm specimen taken off Venezuela as this species or *C. muraena.* In the eastern Atlantic, known from Sénégal and the Cape Verde Islands. Our large specimens (624–1072) were taken in the Bahamas by rotenone over sand or a mixture of sand, turtle grass and rock, at depths to 15 m; the smallest study specimen was trawled off Yucatan over 36 m, and one specimen was collected by clam dredge at 10–13 m.

Etymology. Named for its type locality, a former Portuguese colony. An adjective.

Remarks. The identity of Bahaman specimens of this species as "*Callechelys perryae*" was questioned by Böhlke and Chaplin (1968: 108), and two color phases were described. Study of additional material and comparison of vertebral counts of all specimens confirms their identity with the holotype of *C. perryae.*

Blache et al. (1979) synonymized the western Atlantic *Callechelys perryae* with the eastern Atlantic *C. guineensis* and *C. pantherina.* They published data on the type of *Ophichthys guineensis,* which had been destroyed in the fire at the Museu Bocage in March, 1978, and made the holotype of *Caecula pantherina* the neotype of *Ophichthys guineensis.* The holotype of *C. pantherina* has 179 vertebrae, and they reported 186 vertebrae and 119 preanal lateral line pores for the holotype of *Ophichthys guineensis;* western Atlantic specimens have 172–182 ($\bar{x}$ = 178) total vertebrae and 111–118 ($\bar{x}$ = 115) preanal vertebrae. (This appears to be analagous to the condition of *C. bilinearis,* wherein the St. Helena specimen has slightly more vertebrae.) Body proportions of eastern Atlantic specimens, as measured by Blache et al. (1979) and Blache and Cadenat (1971), do not differ significantly from those of western Atlantic specimens. We follow them in retaining *perryae* as a junior synonym of *guineensis,* but note the difference in vertebral counts.

We cautiously recognize *Callechelys springeri* (and its junior synonym, *Cryptopterygium holochroma*) as distinct from *C. guineensis* on the basis of its lower total and preanal vertebral counts.

Study Material. Twelve specimens, 136–1072 mm TL, including the neotype (which is also the holotype of *Caecula pantherina*) and the holotype of *Callechelys perryae.*

Neotype: MNHN B.1635 (460); Coast of Sénégal, S of

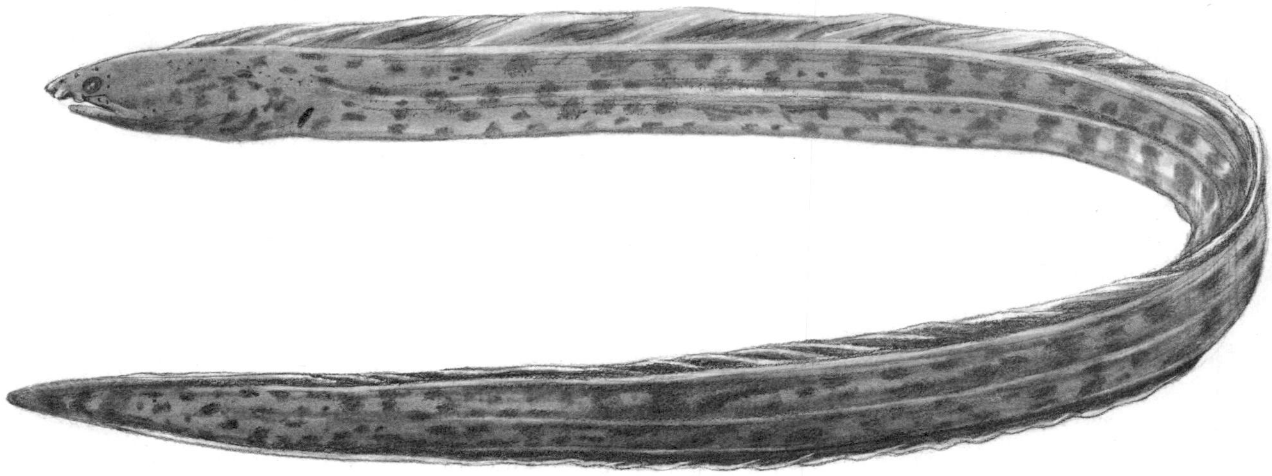

FIGURE 298. *Callechelys muraena*: ANSP 151696, 232 mm TL, Gulf of Campeche.

Dakar, 13°50′N, 35 m; 5 May 1956; holotype of *Caecula pantherina*.

Other material: BAHAMAS: ANSP 98333 (1, 722); Green Cay, N of Rose Island, 10 m; J. E. Böhlke et al., sta. 299; 13 May 1956. ANSP 98334 (1, 624) and ANSP 98335 (1, 775); Grand Bahama Island, W end, 1 m; J. E. Böhlke et al., sta. 416; 19 Feb. 1958. ANSP 115290 (1, 1072); Green Cay, N of Rose Island, 15 m; J. E. Böhlke et al., sta. 652; 27 Aug. 1969. FLORIDA ATLANTIC: UF-FSU 11632 (1, 320); Jupiter Inlet, Palm Beach; N. T. and G. W. Christiansen; 28 Feb. 1959. GULF OF MEXICO: *Florida:* CAS 33027 (1, 734); holotype of *Callechelys perryae*); a mile off S shore of Sanibel Island, 2–3 fm (3.6–5.5 m), mud and broken shell bottom; M. H. Storey; June 1935. FDNR 5988 (1, cleared and stained); off Laguna Beach and Destin, 10–13 m, clam dredge; Jolly and Irby; 15 Jan. 1970. PUERTO RICO: CAS 54469 (1, 601); Crashboat Basin, Aguadilla, 13 m, rotenone; J. E. Randall; 24 July 1956. ST. BARTHELEMY: ANSP 103290 (2, 310–395); Fourche Island, 6–7 m; J. C. Tyler et al., TE-50; 12 July 1965. OFF YUCATAN: GCRL 21391 (1, 136); 20°40′30″N, 87°37′00″W, 36 m, BELLOWS; 29 May 1974.

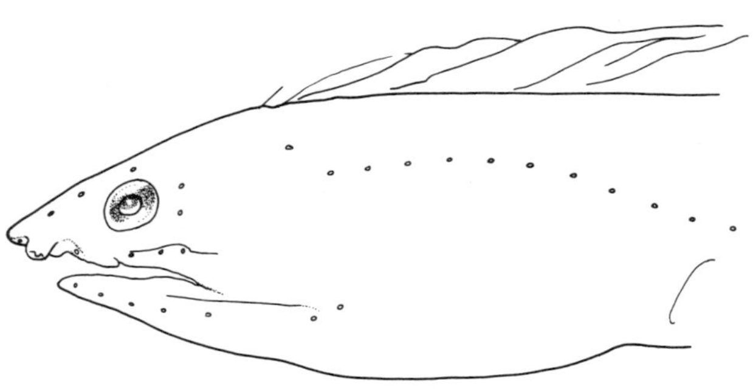

FIGURE 299. *Callechelys muraena*: ANSP 151696, 232 mm TL, Gulf of Campeche; head.

Callechelys muraena Jordan and Evermann, 1887

BLOTCHED SNAKE EEL

Figures 294, 298-300, Tables 30, 31

Callechelys muraena Jordan and Evermann, 1887: 466 (original description, Snapper Banks, Florida, holotype USNM 37966).

Negative reference:

Callechelys muraena (not of Jordan and Evermann). Ginsburg, 195: 481 (in part, one specimen, USNM 39652, = *Bascanichthys scuticaris*).

Distinctive Characters. A moderately elongate callechelyin with brown spots and mottlings on pale body; body depth 2.6–3.7% TL; head 7.0–8.8 % TL; maxillary teeth 7–11, dentary teeth 11–19; MVF H-83-142.

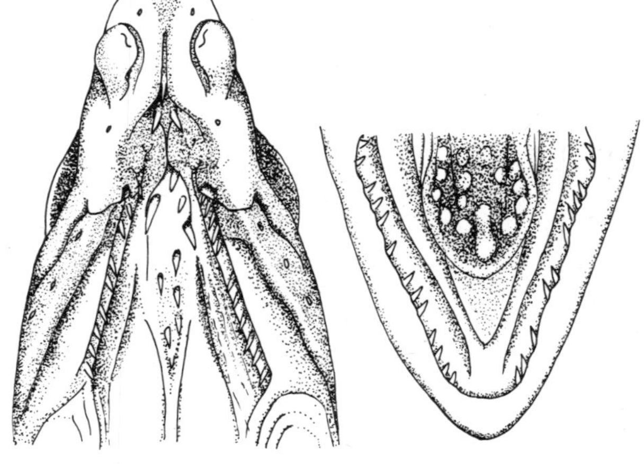

FIGURE 300. *Callechelys muraena*: ANSP 138411, 317 mm TL, Florida Atlantic; dentition.

Description. Total vertebrae 139–144, $\bar{x}$ = 142.1 (n = 7); preanal vertebrae 80–85, $\bar{x}$ = 82.7 (7); dorsal-fin origin on head (H). Pores: SO 1 + 3; IO 4 + 2; POM 5 + 2; ST 3.

Proportions as % of TL: head 7.0–8.8; trunk 48–55; tail 37–40; predorsal 2.9–3.5; depth behind gill openings 2.6–3.7. Of HL: snout 13–17; snout to rictus 24–32; eye 5.8–9.7; interorbital width 9.1–12.

General characters of head, body and fins as for *C. bilinearis*. Dorsal fin high; ventral ridges from gill opening to anus not as pronounced as those of *bilinearis* (possibly due to smaller size of specimens). Teeth similar to those of *C. bilinearis*; two pairs of intermaxillary teeth; 8–12 teeth on vomer, biserially staggered anteriorly; 7–11 teeth on maxillary; and 11–17 teeth on mandible.

Color. Pale tan ground color on head and anterior trunk, grading to darker brown posteriorly, covered by numerous brown eye-sized spots and mottling, except cheeks, lower jaw and snout which are pale yellowish. Median fins brown at their bases, pale or yellow at margin.

Size. A smaller species than its congeners, the largest specimen examined is 590 mm TL. Two specimens (317–590 mm) were females, the smaller one ripe with 0.8-mm eggs; three specimens (258–379+) were males, the largest one ripe.

Development. Leptocephalus described and illustrated by Leiby (1984b: 404, and this volume, p. 894).

Distribution (Fig. 294). A rare species in collections, infrequently taken along the east coast of the United States off North Carolina and Florida, from the Gulf of Mexico off northern Florida, and from the northern Yucatan peninsula. Taken once by trawl over 115 m and three times by dredge at depths of 27–42 m; the specimen recorded from off North Carolina was found in the stomach of a *Carcharhinus plumbeus* captured at 44 m. To date not taken by poison.

Etymology. From the Latin *muraena* (moray); the authors compared the species with "*Callechelys*" *scuticaris* and "C" *bascanium* and said "... much stouter and more like *Muraena* in form and color." To be treated as a noun in apposition.

Study Material. Seven specimens, 232–590 mm TL, including the holotype.

Holotype: USNM 37996 (334.5); snapper banks off Pensacola, Florida; Mr. Silas Stearns; 1885.

Other material: NORTH CAROLINA: ANSP 153538 (1, 379+); 34°06.5′N, 76°33′W, from stomach of *Carcharhinus plumbeus*, 44 m; J. Dodrill; 25 Sept. 1984. FLORIDA ATLANTIC: ANSP 126148 (1, 258); 28°07′N, 80°27′W, 27 m; SILVER BAY 2284; 11 Oct. 1960. ANSP 138411 (1, 317); 28°39′N, 80°13′E, 42 m; OREGON 17772; 23 June 1975. CAS 54468 (1, 355); 30°09.5′N, 80°46.5′W, 31 m; OREGON 7918; 14 Apr. 1968. GULF OF MEXICO: *Florida:* USNM 44651 (1, 590); Pensacola Harbor; J. E. Kauser. GULF OF CAMPECHE: ANSP 151696 (1, 232); 22°18′N, 91°21.5′W, 115 m; OREGON 2189; 22 May 1958.

Callechelys springeri (Ginsburg, 1951)

RIDGEFIN EEL

Figures 294, 301–303, Tables 30, 31

Gordiichthys springeri Ginsburg, 1951: 484, fig. 16 (original description, from the stomach of a shark caught off Salerno, Florida, holotype USNM 121604).

Cryptopterygium holochroma Ginsburg, 1951: 482, fig. 15 (original description, off Cape Fear, North Carolina, holotype USNM 154994).

Callechelys springeri. Rosenblatt and McCosker, 1970: 494. Robins et al., 1980: 16, 71 (*holochroma* synonymized with *springeri*).

Callechelys holochromus. McCosker and Rosenblatt, 1972: 22.

Distinctive Characters. An elongate callechelyin with cream coloration, entire body overlain with dark brown to black spots and blotches; body depth 1.9–2.2% TL; head 5.6–7.1% TL; jaw teeth few, 4–5 on maxillary, 9–10 on mandible; MVF H-108-169.

Description. Total vertebrae 167–171, $\bar{x}$ = 169 (n = 3); preanal vertebrae 107–108, $\bar{x}$ = 107.7 (3); dorsal-fin origin on head (H). Pores: SO 1 + 3; IO 4 + 2; POM 5 + 2; ST 3.

Proportions as % TL: head 5.6–7.1; trunk 58–60; tail 34–35; predorsal 2.3–2.6; depth behind gill openings 1.9–2.2. Of HL: snout 12–14; snout to rictus 24–26; eye 4.4–6.9; interorbital width 6.5–10.

General characters as for *Callechelys bilinearis*. Dentition similar to that of *C. guineensis*; two large intermaxillary teeth, 5–7 teeth in patch on vomer, 4–5 maxillary teeth, and 9–10 on mandible.

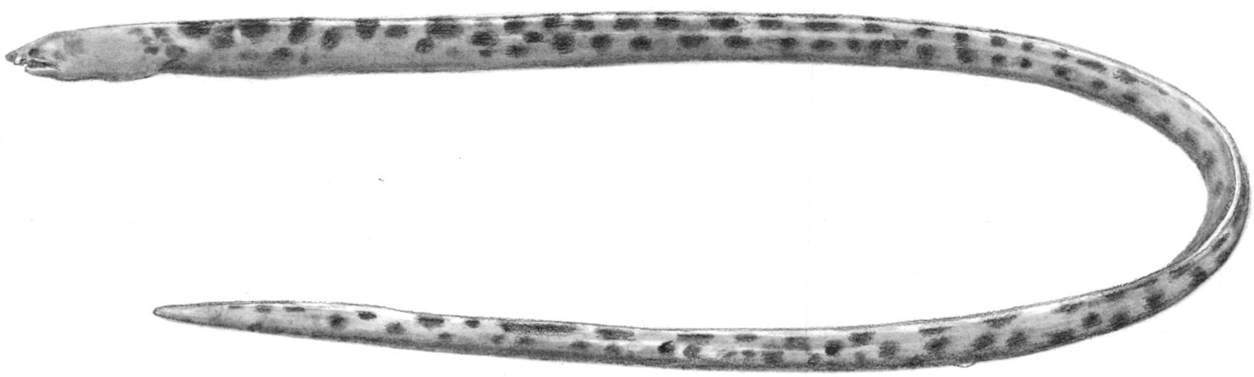

FIGURE 301. *Callechelys springeri*: ANSP 151698, 367 mm TL, Florida Atlantic.

Color. Similar to that of *C. guineensis*; the spots are somewhat larger and more numerous, and the spotted color pattern extends evenly around the body, including the preanal ventral area which is pale in *C. guineensis.* The largest specimen (the holotype of *Cryptopterygium holochroma*) is very heavily spotted, the dorsal spots coalescing into large saddles which meet across the fin. We have observed large dorsal spots and blotches on large specimens of *C. guineensis,* but none approach the condition of this specimen.

Size. The largest specimen is 797 mm TL, a female with 0.6-mm eggs.

Development. Leptocephali not known. Those described as "*Callechelys* species cf. *springeri*" by Leiby (1984b: 412), despite marginal agreement of some adult and larval meristics, differed in several tribal characters. They have been tentatively reidentified as *Gordiichthys randalli.*

Distribution (Fig. 294). Known from three specimens from the Atlantic coast; one off Cape Fear, North Carolina and two off central Florida. Two specimens were taken by trawl over 22–36 m; the holotype was found in the stomach of a shark captured at 33 m.

Etymology. The species was named after Mr. Stewart Springer, who obtained the type specimen. A noun in the genitive case.

Remarks. The three specimens of *C. springeri* are in poor condition, having come from fish stomachs or otherwise damaged. Ginsburg's inability to determine the exact position of the anus and the features of the gill openings (including the presence or absence of pectoral fins) in the holotype of *Cryptopterygium holochroma* prompted him to erect a new genus; however, radiographs of the holotype provide reliable vertebral counts and other data which confirm its generic status and its identity with *Callechelys springeri.*

We provisionally recognize *Callechelys springeri* as distinct from *C. guineensis,* but they are very similar. Vertebral counts differ slightly; there are subtle differences in coloration; and there is a slight difference, but significant overlap, in the tail lengths. In addition, the scarcity of material

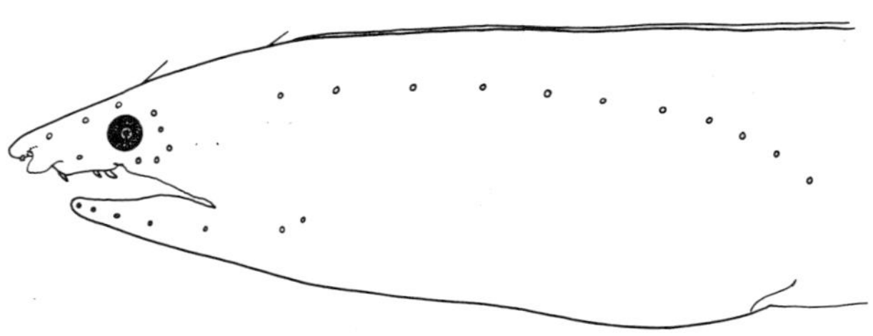

FIGURE 302. *Callechelys springeri*: ANSP 151698, 367 mm TL, Florida Atlantic; head.

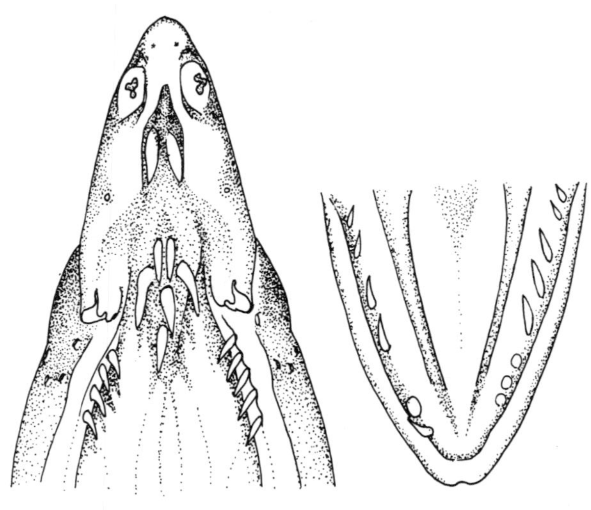

FIGURE 303. *Callechelys springeri*: ANSP 151698, 367 mm TL, Florida Atlantic; dentition.

could indicate this species to be a deep-water inhabitant not readily collected by usual methods.

Study Material. Three specimens, 367–801 mm TL, including the holotype and the holotype of *Cryptopterygium holochroma* Ginsburg.

Holotype: USNM 121604 (372); from the stomach of *Carcharhinus milberti* taken off Salerno, Florida, in 18 fathoms [33 m]; Stewart Springer; 1–3 July 1943.

Other material: SOUTH CAROLINA: USNM 154994 (1, 801); off Cape Fear, 33°30′N, 78°13′30″W, 22 m; PELICAN 183-9; 13 Feb. 1940; holotype of *Cryptopterygium holochroma.* FLORIDA ATLANTIC: ANSP 151698 (1, 367); 27°42′N, 80°04.5′W, 36 m; SILVER BAY 5533; 3 Mar. 1964.

Genus *Letharchus* Goode and Bean, 1882

Letharchus Goode and Bean, 1882: 437 (type species *Letharchus velifer* Goode and Bean, 1882, by original designation).

Characters. Body moderately elongate (depth 1.9–2.6% TL in Atlantic species), laterally compressed; head and trunk longer than tail (58–64% TL); tail tip a hard finless point. Dorsal-fin origin on nape, above epiotics; anal and pectoral fins absent. Snout acute, rounded at tip, overhanging lower jaw, not grooved on underside; lips without barbels; eye small; anterior nostril a hole, its rim not raised; posterior nostril opens into mouth. Gill openings entirely ventral, converging forward, length greater than isthmus; a pocket formed by an obvious anterolateral membrane. Head pores reduced; four supraorbital pores, two preopercular pores and three pores in supratemporal canal. Teeth slender and pointed; basically uniserial on jaws and on vomer, those of intermaxillary a biserial patch.

Neurocranium short, depressed, sloping posteriorly; supraoccipital crest absent; maxilla elongate, slender posteriorly; pterygoid slender, elongate, free and tapering posteriorly; otic bulla developed; suspensorium nearly vertical; opercular series weakly developed. Branchiostegal rays numerous, along arch; hyoid arch stout; urohyal a single slender filament posteriorly. Gill arches reduced; basibranchial 1 ossified, 2 cartilaginous; hypobranchials 1–2 ossified, 3 cartilaginous; ceratobranchials 1–4 ossified, 5 absent; upper pharyngeal tooth plates separate. Pectoral girdle reduced to cleithrum, supracleithrum and 2 rod-shaped elements. Intramuscular bones, ribs, and caudal transverse processes developed. Precaudal vertebrae more numerous than caudal. Gasbladder weakly developed. Peritoneum pale. Body coloration uniformly dark or longitudinally striped.

Size. Maximum known size for western Atlantic species is 578 mm.

Distribution. A New World genus, known from one eastern Pacific and two western Atlantic species.

Etymology. From the Greek *lethos* (to forget) and *archos* (anus), referring to the absent anal fin. Masculine.

Species. Three species: *Letharchus velifer* Goode and Bean, 1882 and *L. aliculatus* McCosker, 1974 in the western Atlantic, and *L. rosenblatti* McCosker, 1974 in the eastern Pacific. They are separable by coloration and by vertebral counts.

Key to the Species of *Letharchus*

1a. Trunk and tail dark brown to black, dorsal fin pale; vertebrae 135–151 2
1b. Trunk and tail cream-colored, with a broad brown mid-lateral stripe; total vertebrae 157–164 . . . . *aliculatus*, p. 314
2a. Supraorbital pores linear when viewed from above; total vertebrae 136–144 . . . . . *velifer*, p. 315
2b. Second supraorbital pore displaced medially when viewed from above; total vertebrae 144–151 . . . . *rosenblatti* McCosker, 1974 (eastern Pacific, Gulf of California to Ecuador)

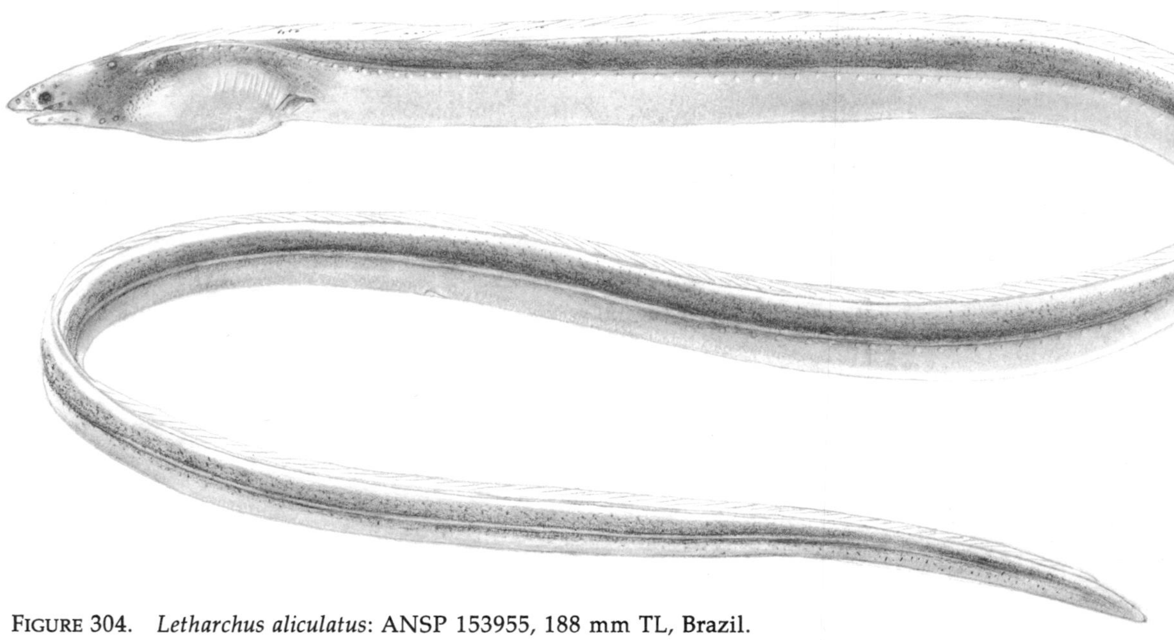

FIGURE 304. *Letharchus aliculatus*: ANSP 153955, 188 mm TL, Brazil.

Letharchus aliculatus McCosker, 1974

STRIPED SAILFIN EEL

Figures 304–307, Table 30

Letharchus aliculatus McCosker, 1974: 626, fig. 6 (original description, Bahia, Brazil, holotype MZUSP 9984).

Distinctive Characters. An elongate pale callechelyin with a light brown stripe along body; dorsal-fin origin on nape, anal fin absent; teeth slender and sharp, mostly uniserial on jaws and vomer; MVF H-[99]-159.

Description. Total vertebrae 155–164, $\bar{x}$ = 159.1 (n = 8); dorsal-fin origin before or over first vertebra, H-1, $\bar{x}$ = 0.4 (7); vertebrae before anus 96–103, $\bar{x}$ = 99.4 (5). Head pores: SO 1 + 4; IO 4 + 2; POM 4 + 2; ST 3.

Proportions as % of TL: head 7.4–9.6; trunk 53–56; tail 36–39; depth behind gill openings 1.9–2.6. Of HL: predorsal 36–44; snout 11–18; snout to rictus 21–27; eye 3.3–5.7.

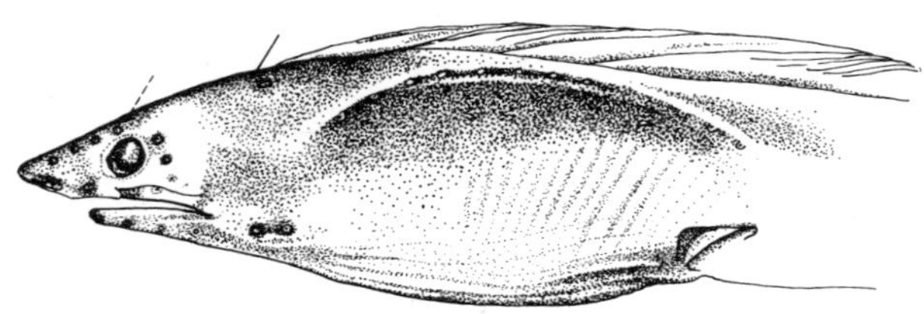

FIGURE 305. *Letharchus aliculatus*: ANSP 153955, 188 mm TL, Brazil; head.

Head moderate, snout short, overhanging lower jaw. Eye small, about midway between snout tip and rictus. Anterior nostril a hole with lateral projections into opening; posterior nostril broad, opening inside mouth in front of maxillary tooth patch, partially covered by flap anteriorly, and visible externally as a slit in upper lip beneath eye. Gill opening ventral, converging forward, its length greater than isthmus width, with a deep pocket anteriorly. Head pores as listed above, except one specimen (GCRL 22311) with five pores along each side of lower jaw. Median

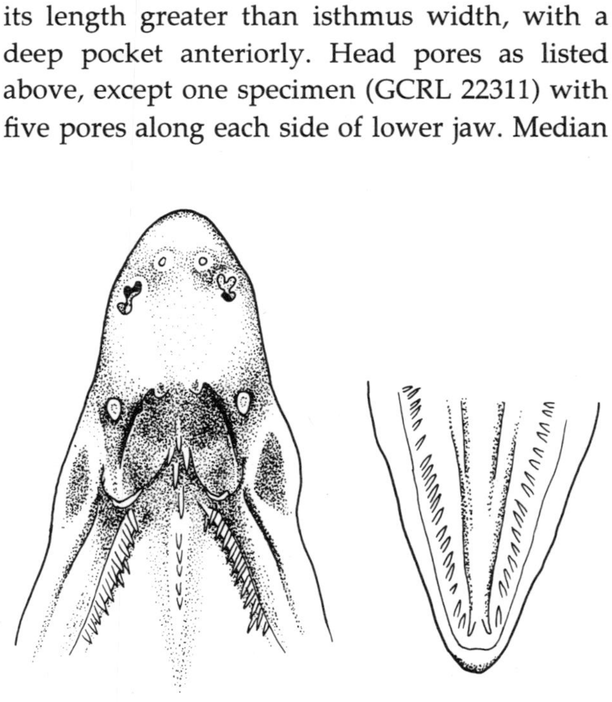

FIGURE 306. *Letharchus aliculatus*: ANSP 153955, 188 mm TL, Brazil; dentition.

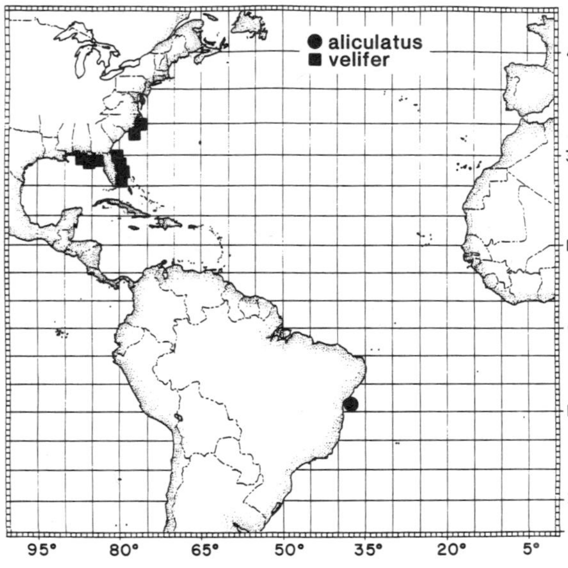

FIGURE 307. Distribution of Atlantic species of *Letharchus* based on study material.

pore of frontal commissure frequently difficult to see; pores in supratemporal commissure close together on dorsum; lateral line pores visible to tip of tail.

Teeth slender, transparent, difficult to see. No anterior intermaxillary teeth; a patch of 4–6 teeth at ethmovomer fusion, followed by 4–6 uniserial teeth along shaft of vomer, this series extending posteriorly only to anterior maxillary teeth on largest (258-mm) specimen. Nine to eleven teeth along maxillary, headed by a pair; 18–21 teeth along mandible, canted inward strongly and not readily visible.

Dorsal fin moderately elevated, originating behind nape and ending well before tail tip. Anal fin absent, but pterygiophores present, visible in cleared and stained specimens and on radiographs.

Color. Body and fins cream to pale tan, overlain by a broad brown longitudinal stripe of medium to coarse freckles, extending from anterior head region nearly to tail tip, the lower edge along lateral line. Body stripe expanded on head above opercle to adjoin its opposite across nape; a few spots below lateral line posterior to anus. Inside of mouth pigmented. Faint brown slashes on anterior part of dorsal fin of largest specimens.

Size. The largest specimen examined is 258 mm TL, an immature female. No mature adults have been found; the species probably grows to a larger size.

Development. Leptocephalus described as *Callechelys* sp. by Leiby (1984b: 410); see p. 885.

Distribution (Fig. 307). All known specimens are from Bahia, Brazil, collected in shallow water (recorded to 1 m), from sand and rock tidepool habitats using ichthyocides.

Etymology. From the Latin *alicula* (a light upper garment), and the noun suffix *-atus* (provided with), in reference to the coloration. An adjective.

Remarks. The coloration of this species resembles that of *Aprognathodon platyventris.* The two are easily separated by generic characters; *Aprognathodon* possesses an anal fin, the anterior nostrils are tubular, and it has only three pores dorsally on the snout.

Study Material 32 specimens, 110–258 mm TL, including the holotype and 27 paratypes.

Holotype: MZUSP 9984 (215); Brazil, Bahia, Salvador, Perto de Tibras, off the south end of Parque Interlagos, taken using rotenone during ebb tide, between 0–1 m over a rock tidepool and sand bottom.

Paratypes: BRAZIL: Bahia: ANSP 124011 (3, 148–193). CAS 23820 (2, 160–176), 23821 (1, 169, cleared and stained). GCRL 10307 (16, 110–204). MZUSP 9985–9986 (2, 120–156). USNM 210600 (3, 121–202).

Other material: BRAZIL: Bahia: ANSP 153955 (1, 188). GCRL 22311 (1, 115), 22313 (2, 199–258).

Letharchus velifer Goode and Bean, 1882

SAILFIN EEL

Figures 307–310, Table 30

Letharchus velifer Goode and Bean, 1882: 437 (original description, west Florida, lectotype USNM 31458).

Distinctive Characters. An elongate callechelyin with dark brown coloration on body and contrasting white dorsal fin; dorsal-fin origin on nape, anal fin absent; teeth slender and sharp, mostly uniserial on jaws and vomer; MVF 1-[84]-140.

Description. Total vertebrae 136–144, $\bar{x}$ = 140.3 (n = 22); predorsal vertebrae 1–2, $\bar{x}$ = 1.4 (15); vertebrae before anus 80–87, $\bar{x}$ = 84.2 (19). Pores: SO 1 + 4; IO 4 + 2; POM 4 + 2; ST 3.

Proportions as % of TL: head 7.4–8.4; trunk 50–56; tail 37–42; depth behind gill openings 1.9–

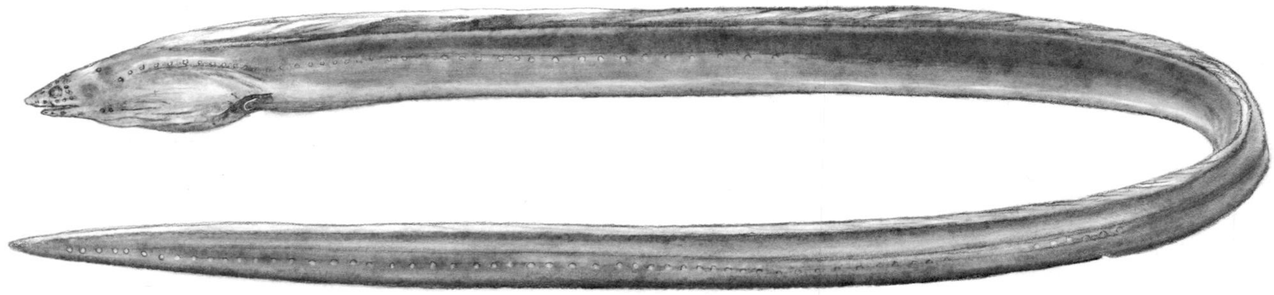

FIGURE 308. *Letharchus velifer*: ANSP 110431, 292 mm TL, Florida Atlantic.

2.9. Of HL: predorsal 32–44; snout 8.9–13; snout to rictus 18–27; eye 2.5–6.2.

General characters as for *Letharchus aliculatus.* Head pores as listed above; pores in supratemporal canal situated dorsally and close together.

Teeth slender and pointed; anteriormost intermaxillary teeth absent; 6–13 biserial or staggered teeth followed by 0–6 uniserial teeth on the shaft of the vomer. An outer row of 8–13 teeth on maxillary plus 2–8 inner teeth which appear as an anterior patch in small specimens or as a biserial row on anterior two-thirds of jaw of large specimens. A row of 17–23 teeth on mandible, canted inward and not readily visible. Larger specimens have the greater number of teeth; however, the teeth on the vomerine shaft and the posterior maxillary series are missing from the largest specimen studied (ANSP 102471), the dentition appearing as three small patches in the upper jaw, the mandibular teeth normal in number and extent.

Color. Body dark brown overall with darker mottlings, sharply contrasted with the dorsal fin, which is white or pale in small specimens, and which has a dark margin in larger ones. Pale areas on head; head pores (especially the postorbital pores) in conspicuous dark areas, the pores pale.

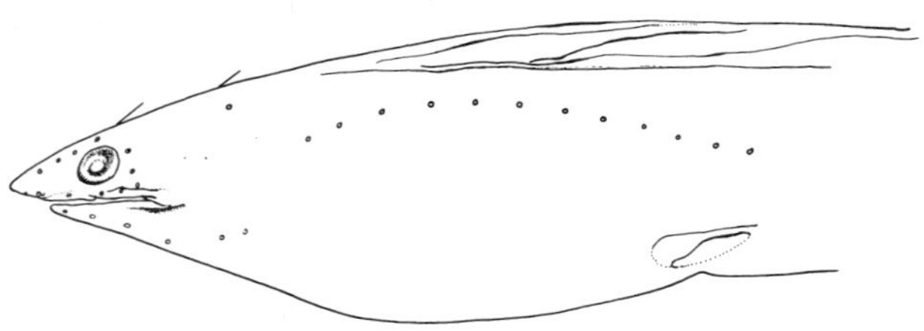

FIGURE 309. *Letharchus velifer*: ANSP 110431, 292 mm TL, Florida Atlantic; head (drawn from right side).

Size. Letharchus velifer is the largest species of the genus; the largest specimen examined is 578 mm TL. Of five females (192–578), two were ripe (192–482) with 0.3–0.7-mm eggs. Four specimens (281–477) were males, one (477) ripe.

Development. Leptocephalus described by Fahey and Obenchain (1978: 473) and by Leiby (1984b: 415, and this volume, p. 887).

Distribution (Fig. 307). Known from coastal United States from North Carolina to the northern Gulf of Mexico off Florida. Most specimens taken by dredge at depths ranging from 5–90 m, usually below 20 m. Also taken from fish stomachs and by ichthyocide collecting to a depth of 20 m.

Etymology. From the Latin *velum* (sail) and *fero* (I bear), in reference to the dorsal fin. A noun in apposition.

Remarks. A holotype was not designated in the original description. Ginsburg (1951: 482) re-

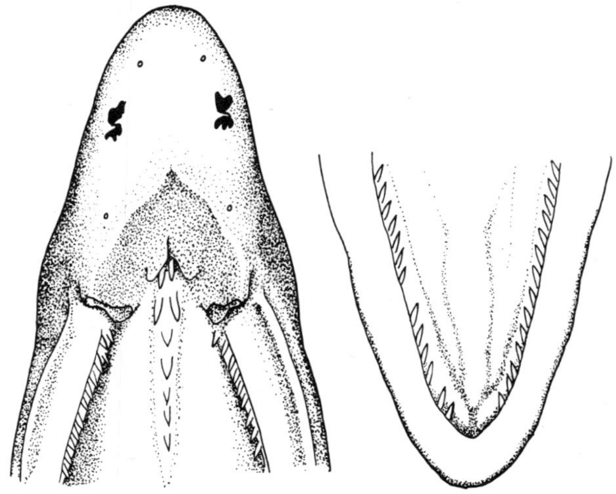

FIGURE 310. *Letharchus velifer*: ANSP 126139, 436 mm TL, Florida Atlantic; dentition.

TABLE 32. Mean vertebral formulae (MVF) and ranges of vertebral counts in western Atlantic species of Sphagebranchini.

Species	MVF	Predorsal Range	Predorsal n	Preanal Range	Preanal n	Total Range	Total n
*Apterichtus ansp**	—54†-128	—	-	52-56†	6	123-132	25
*kendalli**	—62†-142	—	-	61-63†	6	137-145	14
*Ichthyapus ophioneus**	—48†-132	—	-	43-51†	4	123-139	75
*Stictorhinus potamius**	27-64-140	26-28	13	62-67	13	135-143	14

* Includes holotype.
† Vertebrae before anus.

ferred to the "types," and McCosker later (1974: 625) incorrectly listed a holotype. We herein designate his "holotype," USNM 31458, 396 mm TL, as lectotype; the three paralectotypes have been recatalogued as USNM 278182.

Study Material. 73 specimens, 118–578 mm TL, including the lectotype, here designated, and three paralectotypes.

Lectotype: USNM 31458 (396); West Florida; Kaiser and Martin, no. 5762.

Paralectotypes: USNM 278182 (3, 408–483); taken with the holotype.

Other material: NORTH CAROLINA: ANSP 109977 (1, 248). MCZ 62071 (1, 445). SOUTH CAROLINA: CAS-SU 3932 (1, 424). FLORIDA ATLANTIC: ANSP 94215 (1, 482), 94216 (1, 477), 102468 (1, 281), 102471 (1, 578), 110416 (1, 251), 110420 (1, 180), 110425 (1, 290, cleared and stained), 110426 (1, 231), 110430 (1, 227), 110431 (1, 292), 126139 (1, 436), 138410 (1, 192), 138426 (1, 216), 138427 (1, 172), 138429 (2, 118–121). CAS 54471 (1, 390). UF 10238 (2, 270–308). UF-FSU 12016 (2, 296–344). GULF OF MEXICO: *Florida:* ANSP 105683 (2, 319–436), CAS 27478 (1, 362). FDNR 5972 (1, 242, cleared and stained). LACM 2455 (1, 162). UF 5384 (1, 352), 5592 (1, 312), 18132 (2, 204–238). UF-FSU 3264 (1, 314), 15122 (1, 473), 15697 (1, 390), 17298 (9, 323–435), 17560 (2, 348–379), 17989 (4, 227–376), 18480 (3, 383–475), 18703 (1, 254), 18893 (6, 227–450), 19122 (2), 19607 (2, 240–261), 20230 (3, 303–352), 23571 (1, 399).

TRIBE SPHAGEBRANCHINI

The tribe Sphagebranchini contains small, thin, moderately elongate, uniformly-colored species; they appear pointed at both ends, the median fins are absent or very low and inconspicuous, and the anus is before midbody. They live buried in the sand and are seldom seen unless disturbed; they are captured by toxicants and by dredge. Species of two genera inhabit fresh waters as juveniles and adults. The tribe may be diagnosed as follows: moderately elongate, cylindrical, often compressed posteriorly; head and trunk either nearly equal to or shorter than tail; snout pointed, often broad and depressed; posterior nostril opens into mouth in most species; gill openings entirely ventral; median fins absent or, if present, low; pectoral fin absent; tail tip sharply pointed; head pores well developed; teeth conical, not caniniform, and generally uniserial; neurocranium elongate, generally depressed and truncate posteriorly; orbit reduced; gill arches reduced, ceratobranchial 5 absent in most species; and coloration generally uniform, or darker dorsally.

The tribal name of Sphagebranchini is derived from the former family-group name based on the genus *Sphagebranchus* Bloch, now considered a junior synonym of *Caecula* Vahl (see discussion in McCosker, 1977: 65).

The tribe Sphagebranchini contains eight genera, three of which (*) are present in the western Atlantic: *Apterichtus**, *Caecula*, *Cirricaecula*, *Hemerorhinus*, *Ichthyapus**, *Lamnostoma*, *Stictorhinus** and *Yirrkala*.

Genus *Apterichtus* Duméril, 1806

Caecilia Lacepède, 1800: 134 (type species *Caecilia branderiana* Lacepède, 1800 = *Muraenea caeca* Linnaeus, 1758, by monotypy; preoccupied by *Caecilia* Linnaeus, 1758, a genus of Amphibia).

Apterichtus Duméril, 1806: 331 (type species *Muraena cae-*

ca Linnaeus, 1758, by monotypy). Also spelled *Apterichthys, Apterichthus, Apterichthe,* and *Apterichtes,* by other authors.

Typhlotes Fischer, 1813: 81 (replacement name for *Caecilia* Lacépède, preoccupied).

Branderius Rafinesque, 1815: 93 (replacement name for *Caecilia* Lacépède, preoccupied).

Ophisurapus Kaup, 1856a: 52 (type species *Ophisurapus gracilis* Kaup, 1856, by monotypy).

Ophisuraphis Kaup, 1856b: 29 (emendation for *Ophisurapus* Kaup, 1856a).

Verma Jordan and Evermann, 1896a: 374 (type species *Sphagebranchus kendalli* Gilbert, 1891, by original designation).

Characters. Body elongate (depth 1.5–2.3% TL in Atlantic species), cylindrical, pointed at both ends and entirely finless; head and trunk slightly shorter than tail (42–48% TL in Atlantic species). Snout pointed, subconical, overhanging lower jaw, its underside flattened and grooved; lips without barbels; eye moderately developed; anterior nostril tubular; posterior nostril a horizontally ovate slit along edge of lip, or entirely outside of mouth immediately preceding a labial flange. Gill openings ventral, converging forward, their isthmus narrow. Head pores developed, numerous, variable; three or four preopercular pores and three or five pores in supratemporal canal. Teeth pointed, uniserial on jaws and on vomer, intermaxillary teeth the largest and separated from those of vomer by a gap.

Neurocranium slightly depressed, subtruncate to rounded posteriorly; orbit reduced; nasals moderately and nasal cartilage well-developed; supraoccipital crest short and blunt posteriorly; maxilla elongate and pointed posteriorly; suspensorium nearly vertical, or anteriorly inclined; pterygoid slender, pointed and reduced; opercular margins entire; hyomandibular broad; otic bulla moderately developed. Hyoid arch slender; hypohyals separated from ceratohyals by a gap; branchiostegal rays closely associated with hyoid; urohyal cartilaginous posteriorly. Gill arches typically sphagebranchin; fifth ceratobranchial absent; upper pharyngeal tooth plates separate. Pectoral girdle reduced to a broad cleithrum and a thin supracleithrum. Intramuscular bones, ribs, and caudal transverse processes well developed; posterior trunk parapophyses with an anterior marginal projection; precaudal vertebrae nearly equal to or fewer than caudal. Gasbladder moderately developed. Peritoneum not blackened. Body coloration in life and in preservative pale, although some species have faint spotting on their head and dorsal surfaces.

Size. Maximum size of western Atlantic species 542 mm TL.

Distribution. Found in all tropical and subtropical seas, with ten valid described species and two undescribed Indo-Pacific species.

Etymology. From the Greek *apteron* (without fins) and *ichtus* (correctly written *ichthys,* fish). Masculine.

Remarks. The genus *Verma* was established on the basis of a difference in posterior nostril location, and was recognized by Böhlke (1968) and by Blache and Bauchot (1972). However, McCosker (1977: 66) concluded that species referred to *Verma* were congeneric with the type species of *Apterichtus, A. caecus* (Linnaeus, 1758) from the Mediterranean.

Species. Two species of *Apterichtus* are found in the western Atlantic: *A. ansp* (Böhlke, 1968) and *A. kendalli* (Gilbert, 1891), separable by vertebral counts (Table 28) and by number of supratemporal pores. Blache and Bauchot (1972) reported four species in the eastern Atlantic and Mediterranean: *A. caecus, A. gracilis* (Kaup, 1856), *Verma anguiformis* (Peters, 1877), and *V. monodi* (Roux, 1966). They differ from western Atlantic species in nostril and pore condition and in vertebral counts. A worldwide generic revision is in preparation (McCosker, MS).

Key to Western Atlantic Species of *Apterichtus*

1a. Supratemporal pores 5, preopercular pores usually 4 (occasionally 3); total vertebrae 123–132; preanal lateral-line pores 53–58, total lateral-line pores 123–131 . . . . *ansp,* p. 319

1b. Supratemporal pores 3, preopercular pores 3; total vertebrae 137–144; preanal lateral-line pores 61–65, total lateral-line pores 135–142 . . . . *kendalli,* p. 321

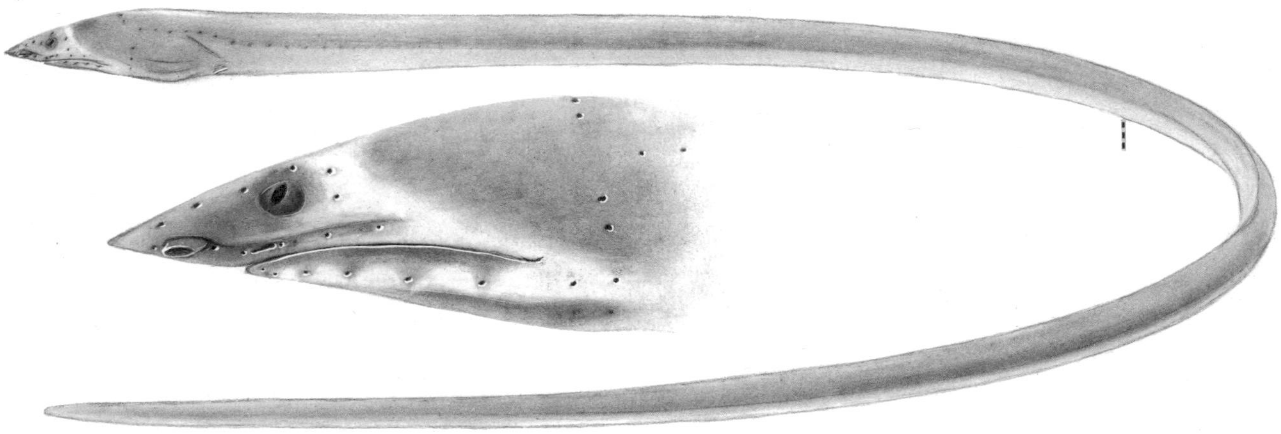

FIGURE 311. *Apterichtus ansp*: ANSP 98366, holotype, 222.5 mm TL, Bahamas. Insert: head. Illustration by S. P. Gigliotti.

Apterichtus ansp (Böhlke, 1968)

ACADEMY EEL

Figures 311–313, Table 32

Verma ansp Böhlke, 1968: 3, fig. 1 (original description, Green Cay, Bahamas, holotype ANSP 98366).
Apterichtus ansp. McCosker, 1977: 66.

Distinctive Characters. An elongate finless sphagebranchin with almost uniform coloration; three or four preopercular pores, five pores in supratemporal canal; preanal lateral-line pores 53–58; mean total vertebrae 128, mean vertebrae before anus 54.

Description. Total vertebrae 123–132, $\bar{x}$ = 128 (n = 25); vertebrae before anus 52–56, $\bar{x}$ = 54 (6); kidney terminus at vertebrae 62–64, $\bar{x}$ = 63 (4). Pores: SO 1 + 4; IO 5 + 2; POM 5-6 + 4; ST 5.

Proportions as % of TL: head 6.8–8.1; trunk 35–36; tail 55–58; preanal 42–45; depth behind gill openings 1.9–2.3. Of HL: eye 4.9–6.9; snout 18–20; snout to rictus 38–42; interorbital width 4.9–6.9.

Head short; snout pointed, with short groove on underside, and projecting well beyond lower jaw. Eye moderate, above midpoint of upper jaw and tip of lower jaw. Anterior nostril in a short tube with lateral projections, a groove around base; posterior nostril an oval pore on upper lip below eye. Head pores numerous and prominent: ethmoid pore at tip of underside of snout, four supraorbital pores as illustrated, the third noticeably ventrad to the others; two infraorbital pores before posterior nostril, three behind nostril, and two pores in ascending canal behind eye; mandibular pores 4–5 plus 3–4 (usually four) preopercular pores, the extra (fourth) pore the highest in the ascending canal. Usually five supratemporal pores; one specimen with four (the median pore missing) and one with six pores.

Dentition uniserial, teeth conical and slightly recurved at tips. Intermaxillary teeth the largest, typically five, arranged as an inverted V and situated in groove anterior to tip of lower jaw.

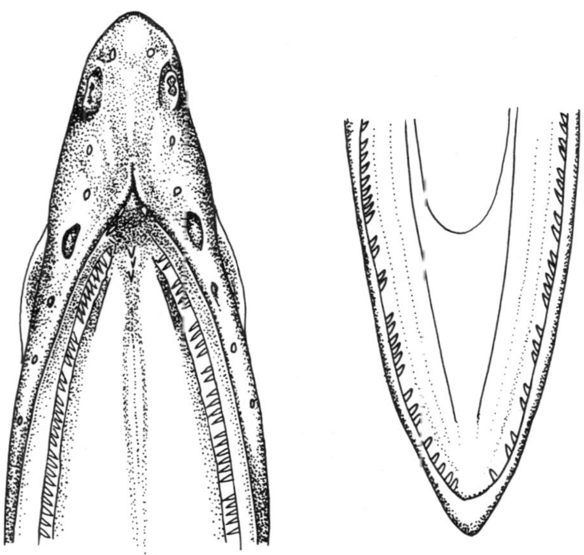

FIGURE 312. *Apterichtus ansp*: ANSP 130692, paratype, 219 mm TL, St. Barthelemy; dentition.

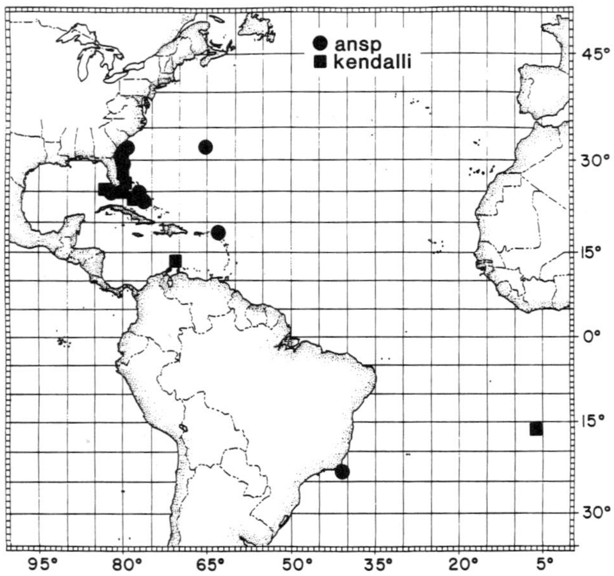

FIGURE 313. Distribution of western Atlantic species of *Apterichtus* based on study material.

Vomerine teeth 3–4, sometimes staggered. Number of jaw teeth varies with growth; 31–36 on maxillaries and ca 35 on dentaries of 243-mm stained paratype, 27–30 and 27–28, respectively, on 276-mm specimen.

Coloration. Nearly uniform pale to tan. Freshly preserved and live specimens display a fine sprinkling of melanophores over the dorsal surface of the head and body, with pale areas behind eye and along lower jaw as illustrated. Lateral line canal visible as a pale stripe.

Size. The largest specimen reported is 373 mm (M. M. Leiby, pers. comm.). Of 14 females (159–373 mm), six were ripe with eggs to 1.0 mm; none of the three males (173–344) were ripe.

Development. Leptocephalus described by Leiby (1981: 68, 1982: 220, and this volume, p. 851).

Distribution (Fig. 313). Not common, taken infrequently off the east coast of the United States from the Carolinas to the Florida Keys, from Bermuda, the Bahamas, the Lesser Antilles, and off Brazil. *Apterichtus ansp* was considered to be a shallow-water species, usually taken in shoreline collections over bare sand bottom at depths to 15 m, and was frequently in company with *Ichthyapus ophioneus.* Recent collections of large ripe specimens were obtained by trawl at depths of 20–38 m.

Etymology. "The name *ansp* should be considered an arbitrary combination of letters" (Böhlke, 1968: 3), the abbreviation for the Academy of Natural Sciences of Philadelphia. A noun in apposition.

Remarks. Apterichtus ansp is very similar in appearance to *A. kendalli.* The most useful external features distinguishing the two are the number of supratemporal pores (five in *A. ansp,* three in *A. kendalli*) and the number of lateral-line pores before the anus (53–58 for *A. ansp,* 61–65 for *A. kendalli*). The number of mandibular pores (mentioned as a diagnostic character by Böhlke, 1968: 5–6, and given as five for *A. ansp* versus four for *A. kendalli*), and the number of preopercular pores (usually four for *A. ansp* and three for *A. kendalli*) cannot be used due to variation in counts for both species. In lateral aspect, *Apterichtus ansp* has a pointed snout and smaller eye situated above and just behind the tip of the lower jaw; *A. kendalli* has a more rounded snout and larger eye located well behind the lower jaw tip. A difference in dentition is apparent when comparing specimens of similar size. *Apterichtus ansp* has more jaw teeth: e.g., a 243-mm specimen of *A. ansp* has 36 and 35 upper and lower jaw teeth, respectively, a 235-mm specimen of *A. kendalli* has 23 and 27; a 276-mm *A. ansp* has 27 and 28 maxillary and mandibular teeth, versus 17 and 18 for a 260-mm *A. kendalli.*

Apterichtus ansp is very close to *A. gracilis* in the eastern Atlantic (coast of Guinea), which differs in posterior nostril condition (Blache and Bauchot, 1972: 714–717).

Study Material. A total of 41 specimens, 58–276 mm TL, including the holotype and 19 paratypes.

Holotype: ANSP 98366 (222.5 mm TL); Bahamas, Green Cay (N of Rose Island, Nassau vicinity), small patch reef ca ¼ mi. N of center of cay, depth 50 ft (15 m); J. E. Böhlke et al., sta. 513; 14 Nov. 1959.

Paratypes: BAHAMAS: ANSP 111662 (1, 243.5), 98367 (4, 57.7–165.5), 98368 (4, 78–94), 98369 (2, 100–102). UF 10304 (1, 118). ST. BARTHELEMY: ANSP 111663 (5, 158–248), 111664 (1, 243, cleared and stained), 130692 (1, 219).

Other material: SOUTH CAROLINA: GMBL 77-33 (1, 276). USNM 181873 (2, ca 200–270+), 197895 (1, 226). GEORGIA: ANSP 158162 (1, 285), 158163 (1, 264), 158165 (1, 233). FDNR 12613 (1, 239), 12624 (1, 173). FLORIDA:

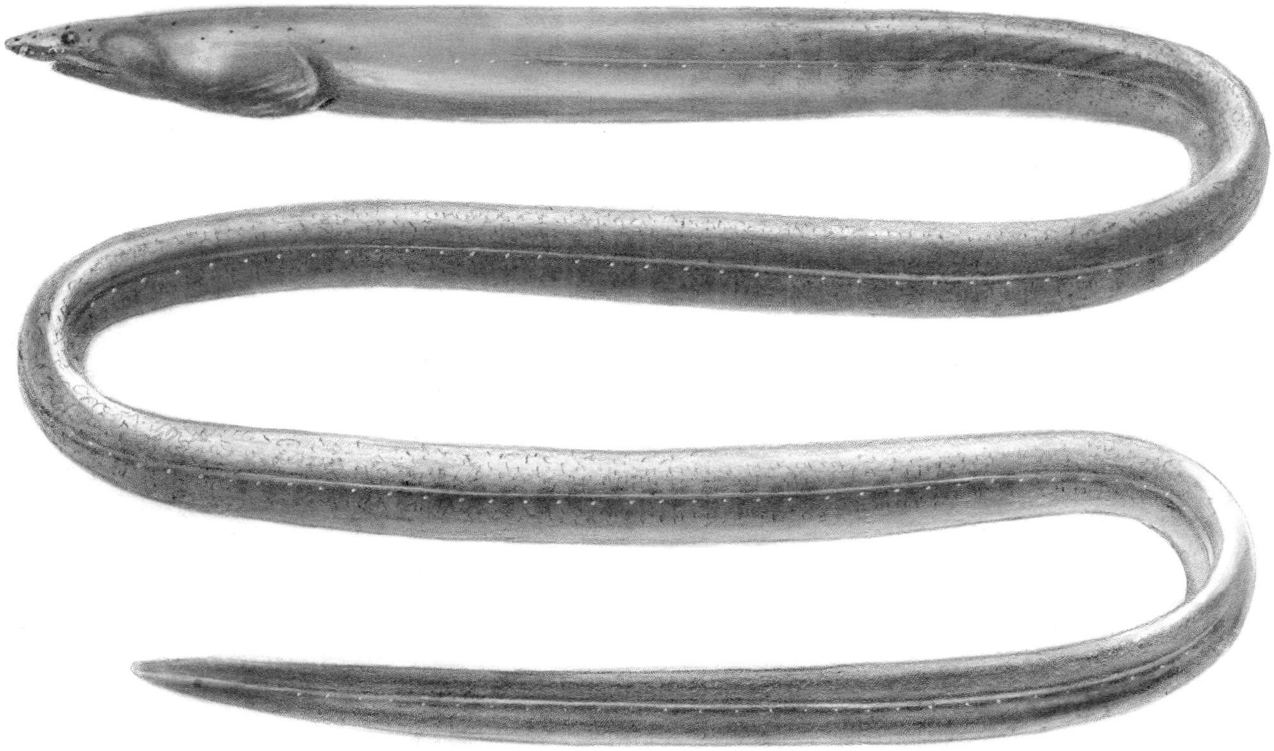

FIGURE 314. *Apterichtus kendalli*: ANSP 98923, 542 mm TL, Florida Atlantic.

ANSP 158164 (1, 344). FDNR 12608 (1, 293), 12612 (1, 233), 12800 (1, 373). UMML 18144 (1, 283). BERMUDA: ANSP 148178 (2, 126–140). CAS 54478 (2, 91–152). BAHAMAS: ANSP 137931 (1, 97). BRAZIL: MNRJ 10442 (1, 120).

Apterichtus kendalli (Gilbert, 1891)

FINLESS EEL

Figures 313–316, Table 32

Sphagebranchus kendalli Gilbert, 1891a: 310, text-fig. (original description, west coast of south Florida, 25°34′N, 82°50′W, holotype USNM 44304).
Verma kendalli. Jordan and Evermann, 1896a: 375.
Caecula kendalli. Myers and Wade, 1941: 75.
Apterichtus kendalli. McCosker, 1977: 66.

Distinctive Characters. An elongate, finless and colorless sphagebranchin with three preopercular pores and three pores in supratemporal canal; preanal lateral-line pores 61–65; mean total vertebrae 142, mean vertebrae before anus 62.

Description. Total vertebrae 137–145, $\bar{x}$ = 142 (n = 14); vertebrae before anus 61–63, $\bar{x}$ = 62 (6). Pores: SO 1 + 4; IO 5 + 2; POM 4–5 + 3; ST 3.

Proportions as % of TL: head 5.3–7.1; trunk 38–43; tail 52–56; preanal 44–48; depth behind gill openings 1.5–1.9. Of HL: eye 5.7–9.4; snout 19–21; snout to rictus 38–40; interorbital width 7.5–13.

General appearance and characters similar to those for *Apterichtus ansp*. Snout more rounded, eye larger, and fewer jaw teeth in specimens of *A. kendalli* of comparable size (see *Remarks* under *A. ansp*).

Color. Nearly uniform pale yellow to tan after long periods in preservative. Freshly preserved and living specimens display a fine sprinkling of melanophores dorsally which can take on a mottled or freckled appearance. Two brown blotches behind the eye surround a pale patch

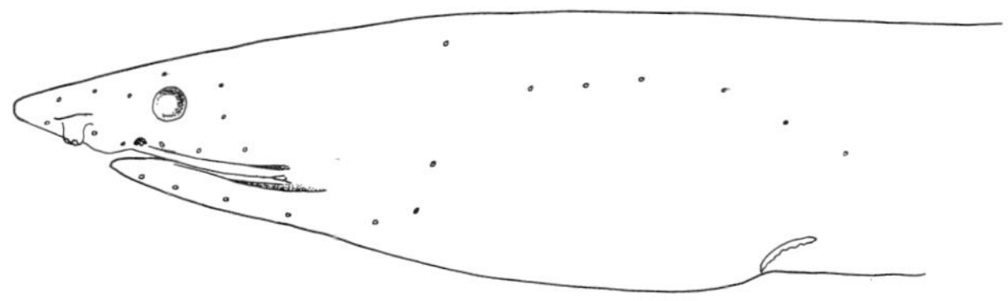

FIGURE 315. *Apterichtus kendalli*: ANSP 98923, 542 mm TL, Florida Atlantic; head.

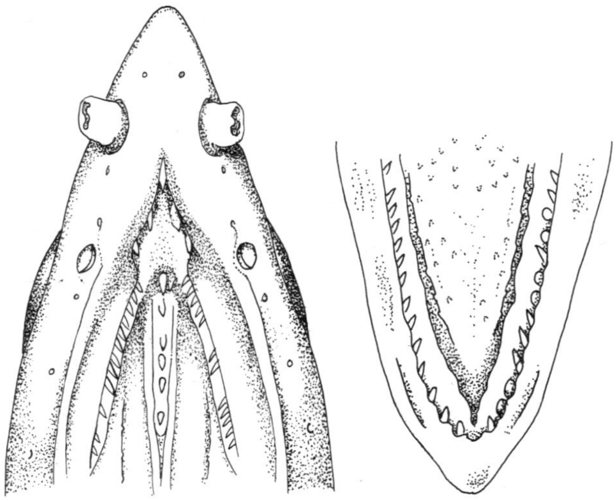

FIGURE 316. *Apterichtus kendalli*: ANSP 98923, 542 mm TL, Florida Atlantic; dentition.

about equal to the eye, another fainter pale patch behind the second is faintly edged with pigment. Lateral line canal visible as narrow pale stripe.

Size. The largest specimen examined was 542 mm TL. Four females ranged in size from 225–317 mm; three (225–246) were ripe, with eggs to 0.9 mm diameter filling the body cavity anteriorly and posteriorly to well behind anus and beyond the kidney terminus.

Development. Leptocephalus described by Leiby (1981: 68, 1982: 123, and this volume, p. 853).

Distribution (Fig. 313). Not common; taken off Florida, the Bahamas, Venezuela and St. Helena. Collected by poison and by dredge at depths recorded from 6 to 401 m; most often taken at moderate depths between 30–80 m where it lives burrowed in sand.

Etymology. "Named in honor of Mr. W. C. Kendall, naturalist of the GRAMPUS at the time the fish was taken" (Gilbert, 1891a: 310). A noun in the genitive case.

Study Material. A total of 26 specimens, 78–542 mm TL, including the holotype.

Holotype: USNM 44304 (ca 169 mm TL); Florida, west of the southern tip of the state, 25°34′N, 82°50′W, by dredge over sand at 25 fm (46 m); GRAMPUS sta. 5080; 2 Mar. 1889.

Other material: FLORIDA ATLANTIC: ANSP 98923 (1, 542), 111666 (1, 318), 138425 (1, 278). UMML 4824 (1, 78), 5124 (2, 82–91), 23949 (1, 333), 23950 (1, 246), 23951 (1, 299). USNM 73272 (4, 140–306), 93782 (1, ca 246), 101601 (1), 153160 (1, 204), 158890 (1, 308), 161341 (1, 268). BAHAMAS: ANSP 111667 (2, 226, and one incomplete), 156882 (1, 175). VENEZUELA: CAS 54479 (1, ca 225). ST HELENA: ZMUC P32854–56 (3, 211–261).

Genus *Ichthyapus* de Barneville, 1847

Ichthyapus de Barneville, 1847: 219 (type species *Ichthyapus acutirostris* de Barneville, 1847, by monotypy).

Rhinenchelys Blache and Bauchot, 1972: 718 (type species *Sphagebranchus ophioneus* Evermann and Marsh, 1902, by original designation).

Characters. Body elongate (depth 1.4–2.8% TL in western Atlantic species), cylindrical, pointed at both ends and entirely finless; head and trunk shorter than tail (36–38% TL). Snout pointed, depressed and broad dorsally, overhanging lower jaw, its underside flattened and grooved; lips without barbels; eye small; anterior nostril flush with snout; posterior nostril opens into mouth. Gill openings entirely ventral, with a thin medial membrane, converging forward, their isthmus narrow. Head pores developed, numerous; three or four preopercular pores and three or five pores in supratemporal canal. Teeth pointed, uniserial on jaws and on vomer, intermaxillary teeth the largest and separated from those of vomer by a gap. Body coloration in life and in preservative uniformly pale or gray; peritoneum not blackened.

Neurocranium depressed, broad, and truncate posteriorly; orbit extremely reduced; nasals moderately, and nasal cartilage well-developed; supraoccipital crest moderately projecting posteriorly; maxilla elongate and pointed posteriorly; suspensorium nearly vertical; opercular margins entire, preopercle reduced; hyomandibular broad, expanded anteriorly and posterodorsally; otic bulla moderately developed; pterygoid elongate and rectangular posteriorly, with a slender projection from the anterodorsal corner. Hyoid thickened; hypohyals separated from ceratohyal by a narrow gap; branchiostegal rays not numerous, slender and generally unbranched, closely associated with hyoid, outermost rays along epihyal broadened basally; urohyal cartilaginous for posterior two-thirds. Gill arches developed for a sphagebranchin; fifth cerato-

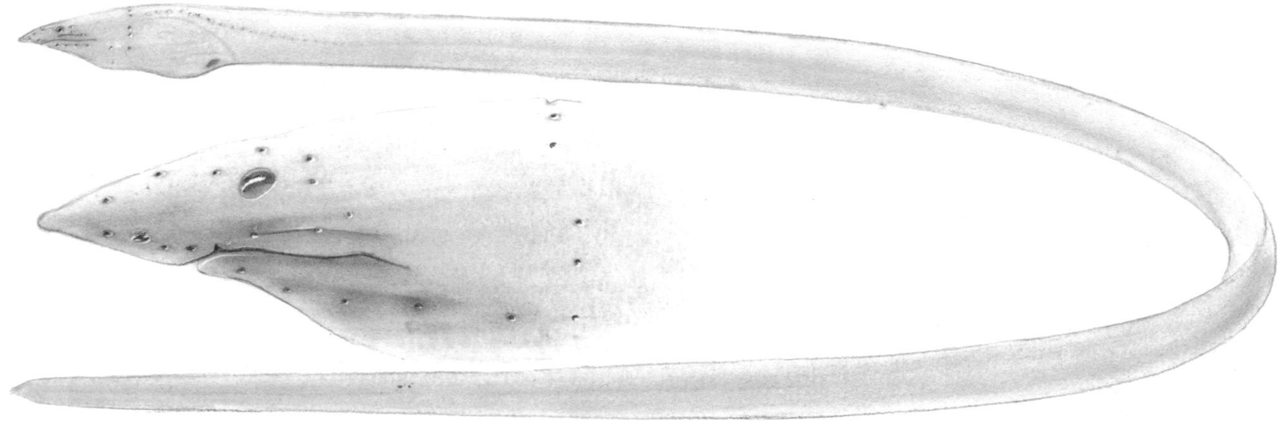

FIGURE 317. *Ichthyapus ophioneus*: ANSP 111591, 390 mm TL, Bahamas. Illustration by S. P. Gigliotti.

branchial present, cartilaginous or ossified; upper pharyngeal tooth plates separate. Pectoral girdle reduced to a broad cleithrum and, if present, a thin supracleithrum. Intramuscular bones, ribs, and caudal transverse processes well developed; posterior trunk parapophyses with an anterior marginal projection; precaudal vertebrae nearly equal to or fewer than caudal vertebrae.

Size. Maximum size recorded for western Atlantic species 479 mm TL.

Distribution. Known from all tropical seas.

Etymology. From the Greek *ichthys* (fish) and *apous* (without foot), likely in reference to the lack of fins. Masculine.

Remarks. The genus *Sphagebranchus* was a catchall for most finless or nearly-finless ophichthins for some time. Recent studies showed the type species, *Sphagebranchus rostratus* Bloch, to be a junior synonym of *Caecula pterygera* Vahl, a species generically distinct from many of those previously placed in *Sphagebranchus* (see discussions in Böhlke and McCosker, 1976: 1, and in McCosker, 1977: 68). The next available name for the genus is *Ichthyapus* de Barneville, 1847. Blache and Bauchot (1972: 718) created a new genus, *Rhinenchelys*, for the species *ophioneus* on the basis of minor differences from the type species in nostril condition and intermaxillary tooth location, features that are considered undeserving of generic rank.

Species. The genus *Ichthyapus* as here defined contains four valid described species and several undescribed species. A worldwide generic revision is in preparation (McCosker, MS). In the Atlantic, one described species, *Ichthyapus ophioneus*, and an undescribed species from Ascension Island.

Ichthyapus ophioneus
(Evermann and Marsh, 1902)
SURF EEL
Figures 317–320, Table 32

Sphagebranchus ophioneus Evermann and Marsh, 1902: 73, fig. 7 (original description, off Mayaguez, Puerto Rico, holotype USNM 49526).

Sphagebranchus conklini Eigenmann, 1916: 55, pl. 9 (original description, New Providence, Bahamas, holotype FMNH 57731).

Sphagebranchus sp., vel *acutirostris* (in part). Cadenat and Marchal, 1963: 1240 (St. Helena).

Rhinenchelys ophioneus. Blache and Bauchot, 1972: 718 (type species of new genus *Rhinenchelys*).

Ichthyapus ophioneus. McCosker, 1977: 67 (*Rhinenchelys* synonymized with *Ichthyapus*).

Distinctive Characters. An elongate, finless, colorless sphagebranchin; tail length 62–64% TL; four preopercular pores, five pores in supratemporal canal; mean total vertebrae 133, mean vertebrae before anus 48.

Description. Total vertebrae 125–139, $\bar{x}$ = 132.6 (n = 77); vertebrae before anus 43–51, $\bar{x}$ = 48 (4). Pores: SO 1 + 4; IO 5 + 2; POM 4 + 4; ST 5.

Proportions as % TL: Head 8–9.5; trunk 27–30; tail 62–64; depth behind gill openings 1.4–2.8. Of HL: snout 13–19; snout to rictus 25–38; eye 3.2–8.4; interorbital width 3.8–13.

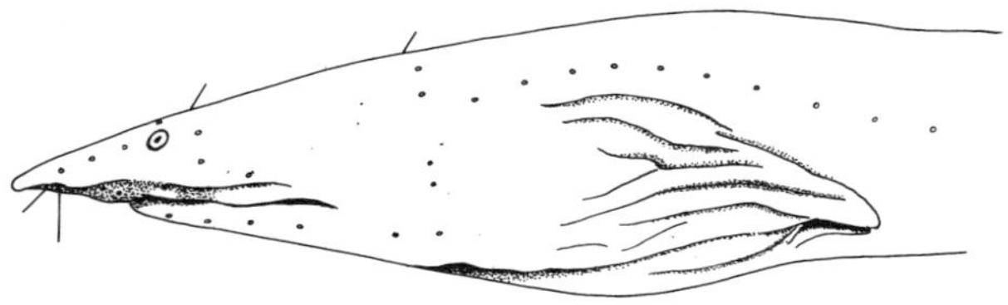

FIGURE 318. *Ichthyapus ophioneus*: ANSP 136748, 394 mm TL, Bahamas; head.

Head short; snout sharply pointed, its underside grooved; lower jaw short, included. Eye small, above anterior part of lower jaw. Anterior nostril non-tubular, a convoluted opening on underside of snout, about midway between tip of snout and tip of lower jaw when mouth is closed. Posterior nostril with irregular rim, opening inside mouth. Head pores conspicuous, some variation in numbers; preoperculomandibular pores 5 + 4 on one side in three specimens, 4 + 3 on both sides in two specimens; three specimens had 6 pores in supratemporal canal. Branchial openings ventral, low, with median fold forming a pouch; branchial region expanded, forming a dorsal pouch in largest specimens.

Teeth uniserial, conical and sharp. A chevron of 5–7 intermaxillary teeth, a short gap, then about 19 teeth on vomer, biserial anteriorly. Jaw teeth uniserial, small and close-set, of uniform size, 24–30 teeth on both jaws.

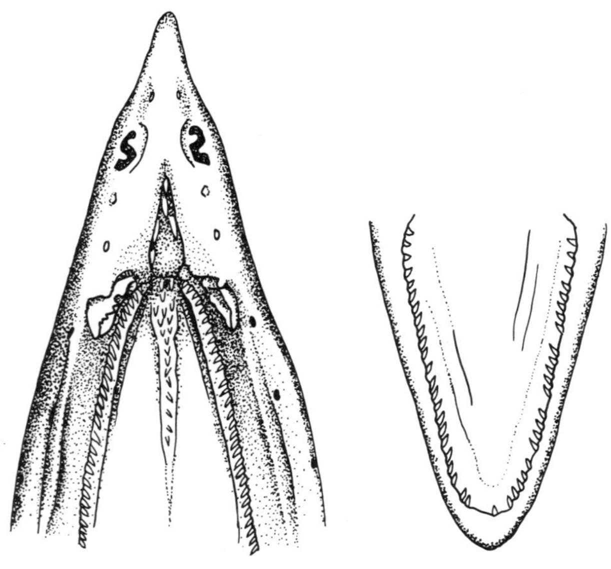

FIGURE 319. *Ichthyapus ophioneus*: 119125, 280 mm TL, Belize; dentition.

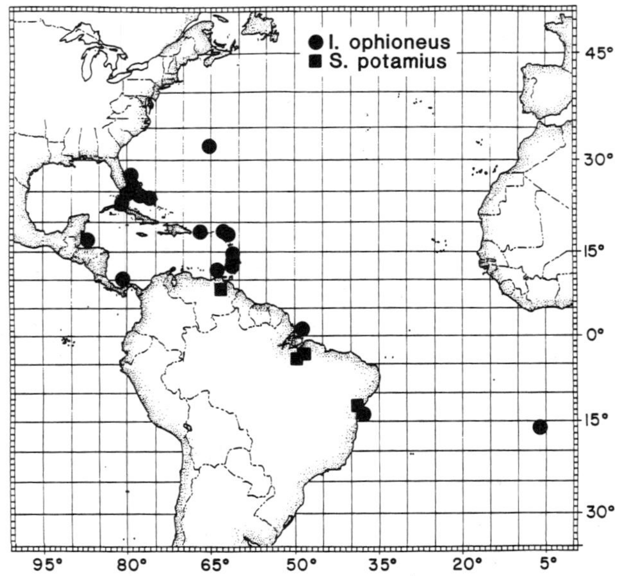

FIGURE 320. Distribution of *Ichthyapus ophioneus* and *Stictorhinus potamius* based on study material.

Color. Uniformly colored, pale in preservative; no obvious markings on head or body, although subcutaneous pigment along lateral line canal is visible microscopically, and some specimens have faint pigment on head. The holotype was described as pale-green overall, but numerous specimens collected subsequently have been described as flesh-colored to pallid (Böhlke and Chaplin, 1968: 102).

Size. The largest specimen examined is 479 mm TL. A disproportionately large number of specimens were ripening females; of 42 females (178–479 mm), 26 (178–479) were ripe with 0.5–1.3-mm eggs; none of the seven males (204–395) contained mature gonads. Many of the largest females did not contain measurable eggs (perhaps spent?); mean size of 26 individuals with eggs was 291 mm, mean size of 16 nongravid females was 335 mm.

Development. Leptocephalus described by Leiby (1982: 225, and this volume, p. 848).

Distribution (Fig. 320). Known from Bermuda, the Bahamas, Florida, throughout the Greater Antilles and from St. Helena. *Ichthyapus ophioneus* is locally abundant, large collections (as many as 258 specimens) containing individuals of varying sizes. Most often taken by ichthyocides in very shallow waters, usually less than 3 m, recorded once to 15 m and one collection from Brazil (USNM 214065) from 50 m. Single specimens occasionally taken by dredge, the holotype at 7-11 m, other depths unrecorded. This sand-burrowing eel was reported by Böhlke and Chaplin (1968: 102) to occupy semi-protected bays and, more abundantly, bare sand bottom at the surf zone on the seaward side of Bahamian islands.

Remarks. Specimens of *Ichthyapus ophioneus* have been taken at St. Helena over several decades. However, those from Ascension Island (listed under *Comparative material*) appear to be an undescribed species, which differs from *I. ophioneus* in number of total vertebrae (122–126 vs 125–139), number of vertebrae before anus (44–47 vs 43–51) and body length (head and trunk 41–43% TL vs 36–40% TL); although there is some overlap in these values, the means are significantly different.

Etymology. From the Greek *ophioneus* (snake-like), presumably in reference to the body shape. A noun in apposition.

Study Material. A total of 1074 specimens, 50–479 mm TL, including the holotype.

Holotype: USNM 49526 (286); Porto Rico, Mayaguez, dredged in 4–6 fm [7–11 m], coral bottom; Fish Hawk Sta. 6065; 20 Jan. [1899].

Other material: BERMUDA: ANSP 133573 (2, 225–275), 133574 (9, 177–308), 133621 (8, 182–323), 133622 (23, 81–306), 148176 (2, 125–334), 148177 (19, 130–332). MCZ 29284 (1, 107), 37345 (1, 291). FMNH 48153 (1, 353), 48418 (1, 334), 48596 (1, 204). USNM 164782 (1, 187), 164838 (1, 250). BAHAMAS: ANSP 72301 (1, 392), 72302 (1, 320), 74304 (31, 150–333), 98370 (258, 76.5–274), 98371 (54, 66–332), 98374 (97, 118–307), 98373 (1, 138), 98374 (2, 50–66), 98375 (1, 188), 98376 (1, 149), 98377 (5, 224–342), 98378 (1, 242), 98379 (3, 209–284), 98380 (2, 222–252), 98381 (3, 220–305), 98382 (10, 158–407), 98383 (2, 241–262), 98384 (89, 62–397), 98385 (48, 162–346), 98386 (10, 108–322), 98387 (11, 167–386), 98388 (1, 63), 98389 (21, 57–273), 98390 (1, 254), 111571 (1, 164), 111591 (1, 350), 113264 (9, 226–322), 113429 (38, 106–458), 115517 (12, 106–468, cleared and stained), 119815 (24, 163–325), 128794 (2, 249–283), 128802 (6, 222–331), 129925 (1, 223, cleared and stained), 136748 (1, 394), 137930 (79, 84–364), 148096 (27, 82–318), 148986 (4, 224–264). CAS 54477 (5, 183–280). MCZ 64506 (1, 248). USNM 53080 (1, ca 167).

FLORIDA: ANSP 55946 (1, 348), 113694 (3, 336–395), 134502 (1, 286), 134503 (1, 275), 134504 (1, 295). UMML 19381 (1, 283). USNM 8585 (1, ca 455). WEST INDIES: ANSP 105727 (1, 158). ZMB 4045 (1). CUBA: USNM 82355 (1, ca 53). PUERTO RICO: ANSP 115553 (1, 203), 136521 (1, 302). LESSER ANTILLES: *Anguilla:* ANSP 110096 (9, 243–373). *St. Martin:* ANSP 105703 (1, 273). *St. Barthelemy:* ANSP 105313 (16, 135–338). *Dominica:* USNM 274243 (2, 105–125), 274245 (19, 69–266). *St. Lucia:* ANSP 106180 (4, 165–395). *Grenadines:* ANSP 105731 (1, 150). PANAMA: ANSP 113752 (4, 279–324), 113760 (22, 164–376), 113763 (9, 103–387). BELIZE: ANSP 119125 (1, 280). VENEZUELA: MBUCV 3247 (3, 450–479). BRAZIL: MZUSP 12100 (1), 12101 (2). USNM 214065 (1, 150). ST. HELENA: ANSP 126915 (5, 178–260). BMNH 1932.8.3.1 (1, 161). USNM 267893 (2, 127–164). ZMC uncat. (1, 120). ZMC uncat. (19).

Comparative material: Ichthyapus sp.: ASCENSION ISLAND: ANSP 158903 (2, 166–248). BMNH 1922.11.7.13 (1, 208). USNM 214480 (10, 227–422).

Genus *Stictorhinus* Böhlke and McCosker, 1975

Stictorhinus Böhlke and McCosker, 1975: 5 (type species *Stictorhinus potamius* Böhlke and McCosker, 1975, by original designation).

Characters. Body moderately elongate (depth 1.9–2.3% TL), cylindrical, pointed at both ends; head and trunk shorter than tail (44–49% TL). Median fins low but distinct, dorsal-fin origin approximately mid-trunk; pectoral fin absent. Snout pointed, sub-conical, overhanging lower jaw, its underside flattened and grooved; lips without barbels; eye small; anterior nostril without a tube, a hole with lateral fleshy projections into it; posterior nostril within mouth, its position not marked externally. Gill openings entirely ventral, converging forward, without an anterolateral pouch, their isthmus narrow. Head pores developed and numerous; two or three preopercular pores, five pores in supratemporal canal. Teeth pointed, biserial to uniserial, intermaxillary teeth the largest and separated from those of vomer by a gap.

Neurocranium depressed, broad and truncate posteriorly; basisphenoid not elongate as in *Ichthyapus*; orbit reduced; nasals moderately and nasal cartilage well-developed; supraoccipital crest reduced; maxillae elongate and pointed posteriorly; pterygoid thin, elongate and rectangular posteriorly, tapering to a narrow point anterior-

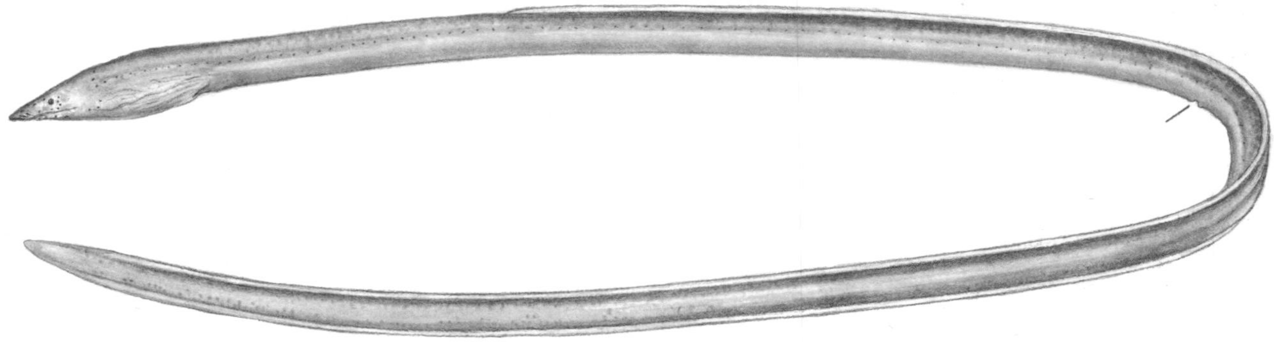

FIGURE 321. *Stictorhinus potamius*: ANSP 128520, paratype, 316 mm TL, Brazil.

ly; suspensorium nearly vertical; opercle well-developed; preopercle minute, interopercle apparently absent; hyomandibular broad; otic bulla moderately-developed. Hypohyals separated from ceratohyal by a gap; branchiostegal rays numerous, branched basally in some individuals, closely associated with hyoid, outermost rays along epihyal broadened basally; urohyal a slender ossified filament posteriorly. Gill arches developed, typically sphagebranchin; fifth ceratobranchial absent. Pectoral girdle reduced to a broad cleithrum which is expanded anteriorly; supracleithrum absent; scapula and coracoid reduced. Intramuscular bones, ribs, and caudal transverse processes well-developed; trunk parapophyses lack an anterior projection; precaudal vertebrae less numerous than caudal. Body coloration pale; peritoneum pale.

Size. To about 345 mm TL.

Distribution. Known only from fresh-waters of northeastern South America.

Etymology. From the Greek *stiktos* (pricked) and *rhinos* (nose), in reference to the nature of the anterior nostrils. Masculine.

Species. A single species, *Stictorhinus potamius* Böhlke and McCosker, 1976.

Stictorhinus potamius Böhlke and McCosker, 1975
FRESHWATER SAND EEL
Figures 320–323, Table 32

Stictorhinus potamius Böhlke and McCosker, 1975: 5, figs. 1–7 (original description, Rio Tocantins, Para, Brazil, holotype MZUSP 9248).

Distinctive Characters. An elongate spagebranchin with pale coloration; median fins present, low, dorsal-fin origin at mid-trunk; pectoral fin absent; 2–3 preopercular pores, five supratemporal pores; MVF 17-64-140.

Description. Total vertebrae 135–143, $\bar{x}$ = 139.9 (n = 14); predorsal vertebrae 26–28, $\bar{x}$ = 27.5 (13); preanal vertebrae 62–67, $\bar{x}$ = 63.8 (13). Pores: SO 1 + 3; IO 3 + 3; POP 4 + 2–3; ST 5.

Proportions as % of TL: head 7.5–8.5; trunk 36–40; tail 51–56; preanal 44–49; predorsal 20–22; depth behind gill openings 1.9–2.3. Of HL: snout 20–22; snout to rictus 25–28; eye 2.4–3; interorbital width 8.3–9.7.

Snout sharply pointed, flattened ventrally; a median groove on underside extending to or beyond anterior nostrils. Eye small, situated over posterior half of lower jaw. Anterior nostril without a tube or delineated by a groove, with three fleshy projections into opening; one specimen with two pairs of anterior nostrils (ANSP 128522, teeth illustrated) and one with supernumerary nostril on left side only. Posterior nostril entirely inside mouth, the opening covered by a thin fleshy flap, its presence not indicated externally by any groove or notch in upper lip. Head pores

FIGURE 322. *Stictorhinus potamius*: ANSP 128520, paratype, 316 mm TL, Brazil; head.

small but visible; considerable variation in placement. Typically four pores in supraorbital canal as illustrated; three infraorbital pores along maxillary plus three pores in vertical canal just behind eye; four pores along mandible plus two or three preopercular pores, the third or upper pore frequently absent (in 50 specimens, 34 had two pores on both sides, 10 had 2 on one side and 3 on the other, and 6 had 3 pores on both sides). Five pores in supratemporal canal.

Teeth small, conical, embedded and difficult to see. Two to five intermaxillary teeth, typically one anterior median tooth plus two pairs. Vomerine teeth beginning before maxillary series, biserial anteriorly, uniserial posteriorly, about 19 teeth in series. Maxillary teeth about 14, mostly uniserial except biserial posteriorly at lateral curve. Mandibular teeth uniserial, about 18–20 teeth.

Median fins present, low; dorsal-fin origin at mid-trunk. Pectoral fins absent.

Color. Pale, slightly darker dorsally and on areas of head and snout due to speckling of melanophores. Darker regions on top of head, along body on either side of dorsal fin, and a transverse band in area of temporal canal. Median fins pale.

Size. The largest specimen examined is 345 mm TL. Nine specimens (201–339 mm) were females, the two largest (337–339) with developing 0.5-mm eggs; no male gonads were observed.

Development. Leptocephalus described by Leiby (1982: 230, and this volume, p. 855).

Distribution (Fig. 320). The types were collected from shallow, freshwater pools in the states of Para and Bahia, Brazil, to a depth of 20 m. Subsequent collections were obtained from the Rio Orinoco, Venezuela, the largest collection of 41 specimens (USNM 233791) from "117 mi upstream [the Orinoco] from sea buoy."

Remarks. This species is the only Atlantic member of the tribe Sphagebranchini known to inhabit freshwater. Böhlke and McCosker (1975) provided an extensive description of the osteology of this species.

Etymology. From the Greek, *potamios* (of a river), in reference to its freshwater habitat. A noun in apposition.

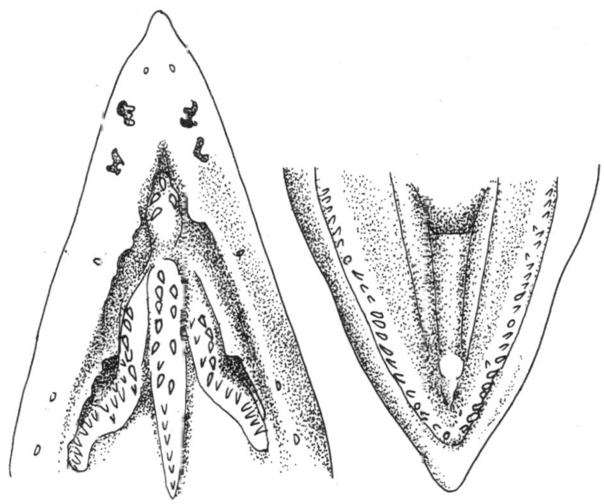

FIGURE 323. *Stictorhinus potamius*: ANSP 128522, paratype, 273 mm TL, Brazil; dentition; this specimen with two sets of anterior nostrils.

Study Material. A total of 102 specimens, 117–345 mm TL, including the holotype and 57 paratypes plus 44 additional specimens.

Holotype: MZUSP 9248 (330); Brazil, Para, freshwater pools of Rio Tocantins near Tucurui, 5 m; N. A. Menezes et al.; 14 Sept. 1970.

Paratypes: BRAZIL: *Para:* Collected with the holotype: ANSP 128520 (1, 316.5); ANSP 128521 (8, 237.5–341.5); CAS 30473 (1, 289.5, cleared and stained); CAS 30472 (8, 227–335); USNM 212060 (2, 273–290); MZUSP 9269–9281 (13, 219–300). ANSP 128522 (1, 312). MZUSP 8959–8979 (21, 151–345). *Bahia:* GCRL 13871 (1, 288).

Other material: VENEZUELA: Rio Orinoco: USNM 233791 (41, 108–333). USNM 268510 (1, 116). USNM 268511 (2, 103–201).

TRIBE BASCANICHTHYINI

The Bascanichthyini includes moderately small, very elongate, plain-colored or striped species; they have overhanging bluntly pointed snouts, and the anus is at or behind mid-body. They live buried in the sand at shallow to moderate depths, and are most often collected by fish toxicants in shallow water, but are also captured by trawl or dredge in deep water. The tribe may be diagnosed as follows: moderately to extremely elongate, generally cylindrical, compressed posteriorly in some species; head and trunk either equal to or longer than tail; posterior nostril opens into mouth; gill openings low lateral, crescentic, never entirely ventral; median fins generally low;

TABLE 33. Mean vertebral formulae (MVF) and ranges of vertebral counts in western Atlantic species of Bascanichthyini.

Species	MVF	Predorsal		Preanal		Total	
		Range	n	Range	n	Range	n
*Bascanichthys bascanium**	1-100-185	1–2	9	95–102	10	182–189	10
*inopinatus**	1-103-201	1	4	101–105	4	198–205	4
*paulensis**	1-101-192	1–2	7	100–103	8	190–196	8
*scuticaris**	2-86-163	1–2	31	83–89	31	159–167	31
*Caralophia loxochila**	2-75-145	2–3	28	68–79	22	138–155	30
*Ethadophis akkistikos**	4-96-165	3–4	4	92–98	4	161–173	4
*Gordiichthys ergodes**	1-98-186	1	1	98	1	186	1
*irretitus**	2-107-197	2	4	105–111	4	193–206	4
*leibyi**	2-95-171	2	5	92–98	4	168–176	5
*randalli**	2-91-181	1–2	8	89–95	8	176–186	8
Phaenomonas longissima	2-138†-212	2	4	135–140†	2	206–216	5

* Includes primary type.
† Number of vertebrae before anus.

dorsal-fin origin on head of most species; pectoral fin generally absent, if present, a minute flap in upper gill opening corner; head pores reduced, two preopercular pores; teeth conical, not caniniform; neurocranium variable in proportions and posterior shape; orbit generally reduced; branchiostegal rays numerous; gill arches reduced, ceratobranchial 5 absent in most species; precaudal vertebrae generally more numerous than caudal; and coloration nearly uniform or darker dorsally.

The tribe Bascanichthyini contains eight genera, five of which (*) are found in the western Atlantic: *Allips, Bascanichthys**, *Caralophia**, *Dalophis, Ethadophis**, *Gordiichthys**, *Leptenchelys* and *Phaenomonas**.

Remarks. The Tribe Bascanichthyini comprises a complex of worm-like burrowing eels that are perhaps the most difficult of ophichthids to identify, even to the generic level. The collection and examination of abundant Atlantic material allows an expanded discussion of McCosker's (1977: 94) original analysis of the tribe. (We lack adequate material and biological information for species of *Dalophis* and therefore exclude them for this discussion.)

We have identified character states that support the separation of distinct lineages within the tribe; however, the question remains in several cases as to the delineation of species groups as genera or as subgenera. These differences between groups are apparently related to the feeding behavior and the substrate composition inhabited by each species. The species of *Bascanichthys* are the most primitive of bascanichthyins and inhabit shallow-water sand substrates that are coarser than the fine sand, mud and silt bottoms within which the deeper-water, trawl-caught Atlantic species of *Ethadophis, Gordiichthys* and *Phaenomonas* were collected. Species of *Bascanichthys* and *Allips* are moderately elongate and possess a tubular anterior nostril, minute pectoral fin, and a grooved snout from which the anterior intermaxillary teeth protrude. The monotypic *Allips concolor* has a further reduced gill arch condition and median fin translocation. The sister group to *Bascanichthys*/*Allips* contains all other bascanichthyins, which are more elongate eels with reduced gill arches and lacking any vestige of a pectoral fin. The remaining specialized lineages contain species of *Phaenomonas* and *Gordiichthys,* which are extremely elongate and retain the plesiomorphic split-snout character state, and those of *Ethadophis* and *Caralophia,* whose upper lip remains as a continuous frenum. *Caralophia* is further advanced in having its anterior nostril modified to become a pore-like opening within a smooth snout. The *Phae-*

nomonas lineage possesses three species, represented in all oceans, which retain the primitive snout and anterior nostril condition, but have lost nearly all fins and undergone extreme elongation with the trunk region.

Within the Bascanichthyini there has been an adaptation toward a vermiform existence. Median fin reduction has been achieved in all advanced genera, either by shrinkage of the fins to a skin-covered ridge or complete loss of the fin rays and the supporting pterygiophores. The extreme elongation of bascanichthyins, particularly within the genera *Gordiichthys* and *Phaenomonas,* has occurred principally within the trunk region. Postanal vertebrae within the Atlantic bascanichthyins are similar in number, ranging from 69 (*Ethadophis*) to 91 (*Bascanichthys paulensis*), whereas preanal vertebrae range broadly from 75 (*Caralophia*) to 138 (*Phaenomonas*). As suggested elsewhere (p. 000), elongation in the trunk region allows a lengthy, linear gut, and avoids clogging due to the ingestion of indigestible particulates.

Genus *Bascanichthys* Jordan and Davis, 1891

Bascanichthys Jordan and Davis, 1891: 621 (type species *Caecula bascanium* Jordan, 1885, by original designation).

Characters. Body elongate (depth 1.2–2.4% TL for Atlantic species), cylindrical and compressed far posteriorly; head and trunk longer than tail (52–57% TL); tail tip blunt. Dorsal-fin origin on head; median fins low; pectoral fin minute. Head markedly rugose; snout short and blunt, overhanging lower jaw, its underside grooved; jaws short; lips without barbels; eye small; anterior nostril tubular; posterior nostril opens into mouth. Gill opening low lateral to horizontal, crescentic, length about equal to isthmus. Head pores reduced; two preopercular pores and three pores in supratemporal canal. Teeth conical, small; intermaxillary teeth exposed when mouth is closed, separated from those of vomer by a short gap; vomerine teeth biserial; jaw teeth mostly uniserial.

Neurocranium short, sloping posteriorly; orbit

TABLE 34. Frequency distributions of vertebral counts in western Atlantic species of *Bascanichthys*.

	Predorsal		Preanal																			
	1	2	83	84	85	86	87	88	89	...	95	96	97	98	99	100	101	102	103	104	105
bascanium	6	3*									1	-	1	1	-	3	-	4*			
inopinatus	4*																2	-	-	1*	1
paulensis	5	2*														4	3	-	1*		
scuticaris	18*†	15	4	5*	7†	7	5	3	1												

	Total									
	159	160	161	162	163	164	165	166	167	...
bascanium										
inopinatus										
paulensis										
scuticaris	1	3	5	4†	6	5*	3	4	2	

	Total (cont.)																					
	182	183	184	185	186	187	188	189	190	191	192	193	194	195	196	197	198	199	200	201	...	205
bascanium	2	2	-	1	1	1	2	1*														
inopinatus																	1	1	-	1*		1
paulensis									3	1	2*	-	1	1								
scuticaris																						

* Includes primary type.

† Lectotype of *Sphagebranchus teres*.

reduced; nasals and nasal cartilage moderately developed; supraoccipital crest absent; maxilla weak, slender, pointed posteriorly; pterygoid pointed anteriorly; otic bulla moderately developed; suspensorium nearly vertical; opercular series weakly developed. Branchiostegal rays numerous, often branched basally, all along arch, the distal rays not broadened basally; urohyal either ossified or cartilaginous posteriorly. Gill arches reduced; basibranchial 1 ossified, 2 cartilaginous or absent; hypobranchials 1–2 ossified, 3 cartilaginous; ceratobranchials 1–4 ossified, 5 ossified or absent; upper pharyngeal tooth plates separated. Pectoral girdle consists of a reduced cleithrum, supracleithrum, scapula and coracoid. Intramuscular bones, ribs, and caudal temporal processes developed. Precaudal vertebrae more numerous than caudal. Gasbladder moderately developed. Peritoneum pale. Body coloration uniform or bicolored, darker dorsally; color frequently obscured by heavy mucus secretion.

Size. To about 1000 mm TL.

Distribution. Known from all tropical seas, with approximately 12 valid species.

Etymology. From *Bascanion,* the generic name for the black snake, from the Greek *baskanos* (malignant), and *ichthys* (fish). Masculine.

Species. We recognize four species in the western Atlantic: *Bascanichthys bascanium* (Jordan, 1884), *B. inopinatus* new species, *B. paulensis* Storey, 1939, and *B. scuticaris* (Goode and Bean, 1880). They are very similar in external appearance, and morphological characters broadly overlap. They are separable by vertebral counts (Tables 29, 30) and by subtle differences in coloration (*B. scuticaris* is marked by conspicuous white spots on head and body along the lateral line, and *inopinatus* and *paulensis* are more definitively bicolored), in size of pectoral-fin base (that of *B. bascanium* is broader than that of the others), in head length (*B. scuticaris* a slightly longer head), and in maxillary dentition (*B. scuticaris* usually partly biserial, in other species usually uniserial, briefly biserial in a few specimens only). The pectoral-fin base differences appear to be significant and also reflect differences between eastern Pacific species (McCosker and Rosenblatt, in prep.).

Blache and Cadenat (1971: 168–185) listed three species of *Bascanichthys* present in the eastern Atlantic: *B. congoensis* Blache and Cadenat, 1971, is here synonymized with *B. paulensis;* "*B.*" *longissimus* Cadenat and Marchal, 1963, has been referred to the genus *Phaenomonas* (p. 000); and *B. ceciliae* Blache and Cadenat, 1971 (which we have not examined) appears to be a valid species, characterized by higher counts of 225–226 total vertebrae and 135–140 preanal lateral-line pores (= preanal vertebrae).

Key to the Western Atlantic Species of *Bascanichthys*

1a. A series of small, pale spots above each lateral-line pore along body, tail and in branchial region (may be diffuse along body and tail of larger specimens); pectoral fin minute, its base 13–32% length of gill opening; total vertebrae 159–167 . . . . *scuticaris,* p. 336

1b. No obvious pale spotting along lateral line or in branchial region; pectoral-fin base greater than 25% length of gill opening; total vertebrae more than 177 2

2a. Pectoral fin small, broad-based, its width 44–77% length of gill opening; total vertebrae 177–190 . . . . *bascanium,* p. 331

2b. Pectoral fin minute, narrow-based, its width 25–37% length of gill opening; total vertebrae 190 or more 3

3a. Pectoral-fin base 25–33% length of gill opening; total vertebrae 190–196 . . . . *paulensis,* p. 335

3b. Pectoral-fin base 36–37% length of gill opening; total vertebrae 198–205 . . . . . . *inopinatus,* p. 333

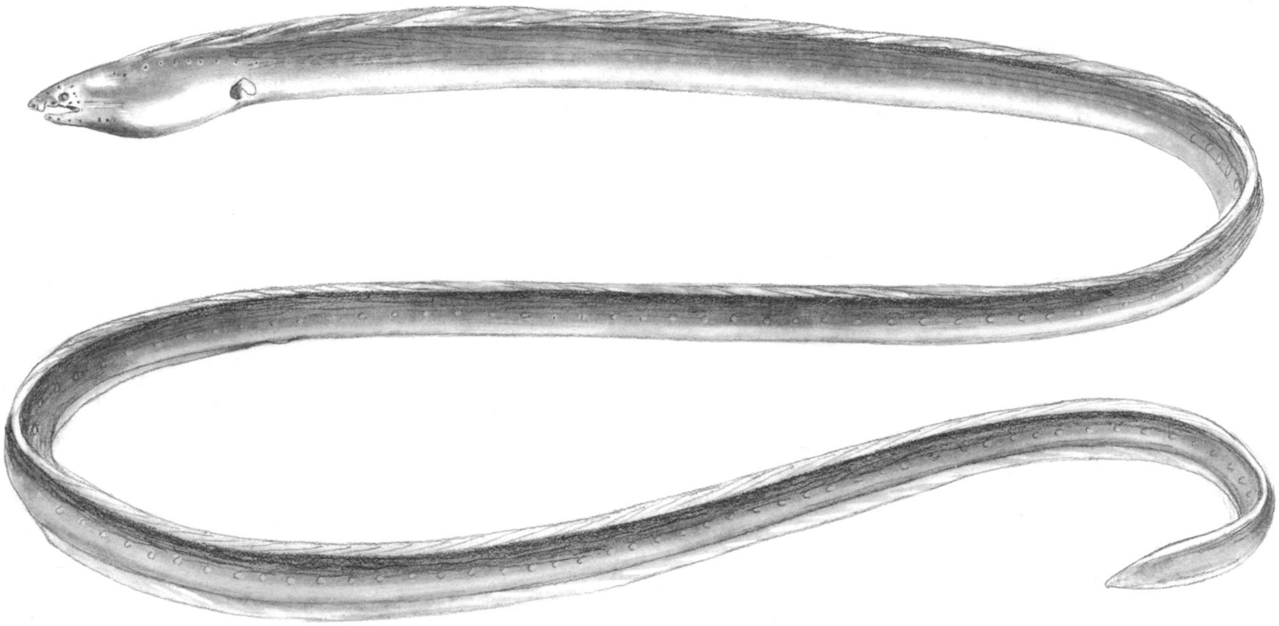

FIGURE 324. *Bascanichthys bascanium*: ANSP 131487, 701 mm TL, Gulf of Mexico off Florida.

Bascanichthys bascanium (Jordan, 1884)
SOOTY EEL
Figures 324–327, Tables 33, 34

Sphagebranchus teres (not of Goode and Bean, 1882). Goode and Bean, 1882: 436 (in part, paralectotype USNM 220010 = *bascanium*).
Caecula bascanium Jordan, 1884: 43 (original description, Egmont Key, Florida, neotype USNM 219832).
Callechelys bascanium. Jordan, 1887a: 840.
Bascanichthys bascanium. Jordan and Davis, 1891: 621 (designated type species of new genus, *Bascanichthys*).
Leptocephalus gilberti Eigenmann and Kennedy, 1902: 92, fig. 14 (original description, 34°57′N, 75°43′30″W, holotype USNM 49766). Leptocephalus.
Bascanichthys teres. Ginsburg, 1951: 479 (synonymized *bascanium* with *teres*).
Bascanichthys bascanium. Leiby and Yerger, 1980: 405 (revision of western north Atlantic species, *bascanium* resurrected, designation of neotype, USNM 219832).

Distinctive Characters. An elongate dark bascanichthyin, slightly darker dorsally, without conspicuous pale spots above lateral-line pores; pectoral fin small, broad-based, its base 44–77% length of gill opening; jaw teeth uniserial; MVF 1-100-185.

Description. Total vertebrae 182–189, $\bar{x}$ = 185.3 (n = 10); predorsal vertebrae 1–2, $\bar{x}$ = 1.3 (9); preanal vertebrae 95–102, $\bar{x}$ = 99.8 (10). Pores: SO 1 + 3; IO 4 + 2; POM 4 + 2; ST 3. Leiby and Yerger (1980: 406) reported 177–190 total vertebrae, 94–101 preanal vertebrae, 163–174 total lateral-line pores, and 18–19 branchiostegal rays on epihyal.

Proportions as % of TL: head 4.6–6.3; trunk 50–52; tail 43–46; predorsal 2.0–2.8; depth behind gill openings 1.2–1.8. Of HL: snout 13–15; snout to rictus 22–27; eye 4.0–5.2; interorbital width 8.8–13; pectoral-fin length 3.2–8.3. Pectoral-fin base 44–77% of gill-opening length.

Head and jaws short; snout rugose; eye small, over posterior half of mid-jaw. Anterior nostril tubular; posterior nostril on under side of upper jaw, visible externally as a slit in upper lip. Head pores as illustrated.

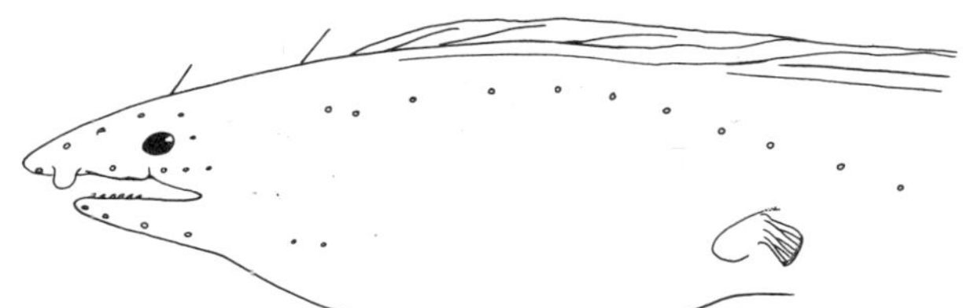

FIGURE 325. *Bascanichthys bascanium*: ANSP 131487, 701 mm TL, Gulf of Mexico off Florida; head.

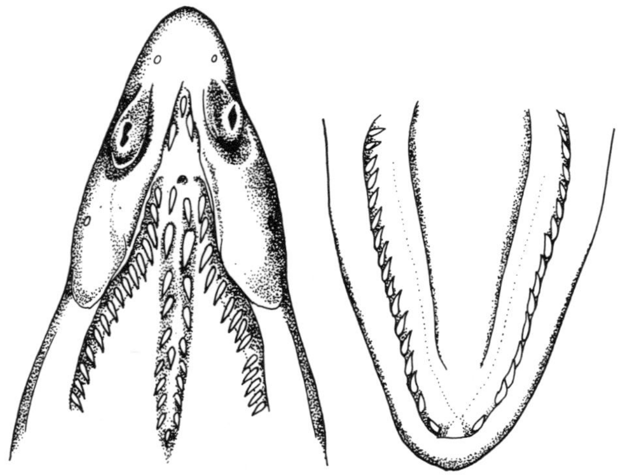

FIGURE 326. *Bascanichthys bascanium*: ANSP 131487, 701 mm TL, Gulf of Mexico off Florida; dentition.

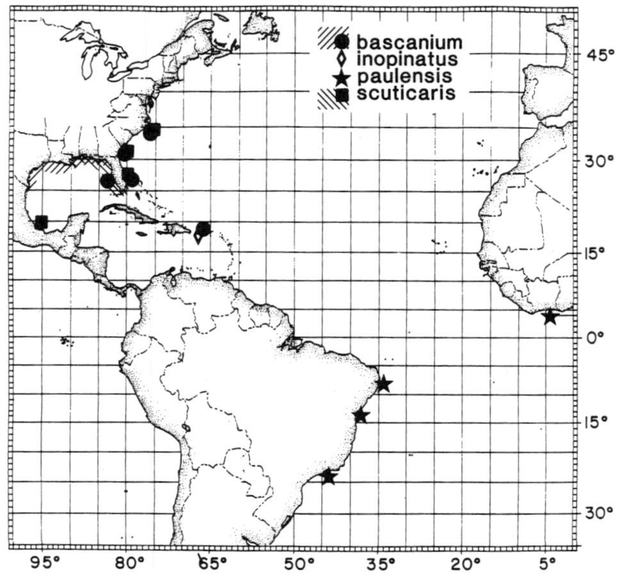

FIGURE 327. Distribution of western Atlantic species of *Bascanichthys* based on study material and records in Leiby and Yerger (1980).

Teeth moderately slender, recurved. Three large, stout, anterior intermaxillary teeth in an inverted V, separated from row of 21 irregularly biserial teeth on vomerine shaft; 15–16 uniserial maxillary teeth, and 15–18 long slender dentary teeth which decrease in size from front to rear.

Body muscular and round; diagonal longitudinal folds on head and gular region, extending back on body and fins. Median fins low, developed; pectoral fin small, with moderate base about half the width of gill opening, the fin squarish, its length equal to or slightly longer than length of base.

Color. Dark dorsally (darkest along base of dorsal fin), pale ventrally, but not sharply bicolored, some pigment extending below mid-side. No pale spots along lateral line or in branchial region as in *Bascanichthys scuticaris.* Head paler than body, some areas with brown pigment. Fins pale, dorsal fin appearing as a white stripe contrasting with dark dorsum. Coloration frequently obscured by mucus.

Size. The largest specimen we examined is 701 mm TL, a ripe female; two specimens (535 and 570 mm) were ripe males. Leiby and Yerger (1980) examined a 786-mm specimen of this species, and reported one ripe female.

Development. Leptocephalus described by Leiby (1981: 49, and this volume, p. 859).

Behavior. Leiby and Yerger discussed the feeding behavior of *Bascanichthys bascanium* and *B. scuticaris.* They found large quantities of sand and shell, as well as polychaete worms, sipunculids, small crustaceans, and sizeable pieces of unidentifiable animal tissue in stomachs examined. This, plus observations of live specimens, suggested to them that *Bascanichthys* actively seeks and captures live prey.

Distribution (Fig. 327). Fairly common in the western Atlantic off coastal United States from North Carolina to Florida and in the northern Gulf of Mexico as far west as Texas; also taken off Mexico and Puerto Rico. Leiby and Yerger examined 66 specimens and noted that *Bascanichthys bascanium* is found with *B. scuticaris* from Cape San Blas, Florida to Mobile Bay, Alabama, but is more abundant from Mobile Bay to Brownsville, Texas where *B. scuticaris* has not been collected (see *Distribution* for *B. scuticaris,* p. 338). There appears to be no difference in habitat preference between the two species; most specimens of *bascanium* were taken from shallow sandy beaches, one specimen by trawl in 24 m.

Etymology. From *Bascanion,* the black snake, which the body much resembles, its name from the Greek *baskanos* (malignant). A noun in apposition.

Study Material. A total of 17 specimens, including the neotype.

Neotype: USNM 219832 (560); Florida, tidal stream near

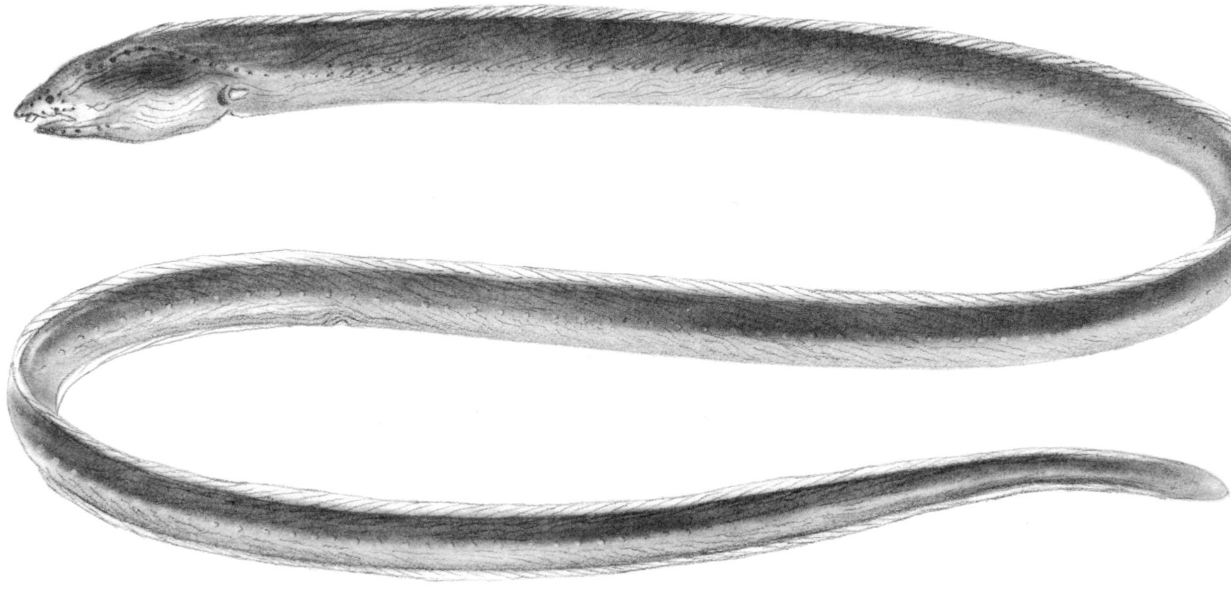

FIGURE 328. *Bascanichthys inopinatus*: ANSP 158885, holotype, 658 mm TL, Puerto Rico.

old Florida State University Marine Laboratory in Alligator Harbor; R. W. Yerger et al., RWY 856; 2 Oct. 1971; formerly FSU 21239.

Other material: NORTH CAROLINA: USNM 119349 (1, 253), 155005 (1, 530). GEORGIA: ANSP 128354 (1, 429). FLORIDA: ANSP 128404 (1, 570). GULF OF MEXICO: *Florida:* ANSP 127982 (2, 465–535), 131487 (1, 701). CAS 54465 (1). USNM 214502 (1, 624). USNM 220010 (1, 470); paralectotype of *Bascanichthys teres* Goode and Bean. *Alabama:* MCZ 9167 (1). *Texas:* USNM 125910 (1, 793), 156855 (1, 367), 278180 (1, 637). PUERTO RICO: ANSP 118632 (2, 189–199), 129903 (1, 414).

Bascanichthys inopinatus new species

PUERTO RICO SAND EEL

Figures 327–330, Tables 33–35

Distinctive Characters. An elongate bicolored bascanichthyin, darker dorsally, with lateral-line pores dark anteriorly and pale posteriorly; head 4.8–5.4% TL; pectoral fin small, its base 36–67% length of gill opening; jaw teeth uniserial, vomerine teeth irregularly biserial; MVF 1-103-201.

Counts and Measurements (in mm) of the Holotype. Total vertebrae 199; predorsal vertebrae 1; preanal vertebrae 101. Pores: SO 1 + 4; IO 4 + 2; POM 4 + 2; ST 3; LL pores above branchial basket 10, to dorsal-fin origin 1; to mid-anus 101.

Total length 658; head length 34.9; trunk length 325.1; tail length 298; body depth at gill openings 12.2; body width at gill openings 7.8; body depth at anus 8.4; body width at anus 8.3; predorsal length 16.6; snout length 4.8; snout tip to rictus 8.3; eye diameter 1.5; fleshy interorbital distance 3.5; length gill opening 4.0; length isthmus 4.8;

TABLE 35. Counts and proportions (in thousandths of TL or HL) of the holotype and three paratypes of *Bascanichthys inopinatus*..

	Range	Mean
Total length (TL in mm)	373–813	618
Predorsal vertebrae	1	1
Preanal vertebrae	101–105	103
Total vertebrae	198–205	201
Proportion of TL		
Head	47–54	51
Trunk	463–494	482
Tail	453–483	467
Predorsal	20–25	23
Depth at gill openings	16–18	17
Depth at anus	13–14	13
Proportion of HL		
Snout	133–148	138
Upper jaw	153–247	214
Eye	41–54	46
Interorbital width	89–100	95
Isthmus width	113–138	127
Pectoral-fin length	79–93	86
Length pectoral-fin base	38–88	54

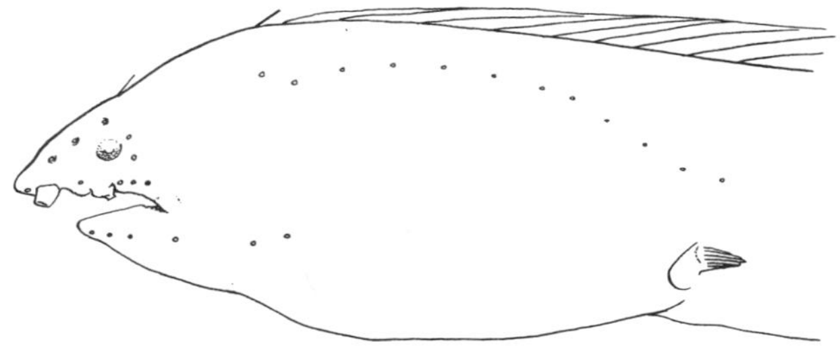

FIGURE 329. *Bascanichthys inopinatus*: ANSP 158885, holotype, 658 mm TL, Puerto Rico; head.

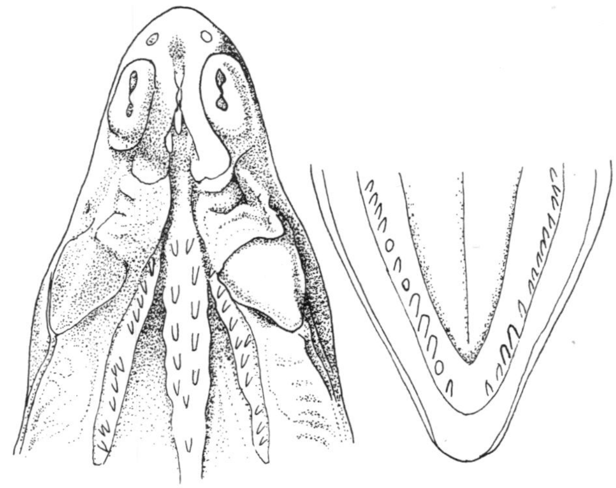

FIGURE 330. *Bascanichthys inopinatus*: ANSP 158886, paratype, 628 mm TL, Puerto Rico; dentition.

pectoral-fin length 3.0; length pectoral-fin base 1.4.

Description. See Table 31 for ranges of counts and proportions of the holotype and three paratypes.

Body elongate, muscular and rounded, little compressed posteriorly; anus shortly behind midbody. Diagonal longitudinal folds on head and gular region, extending back on body and fins. Median fins developed but low, ending shortly before tip of tail; pectoral fin reduced, its base one-third to two-thirds the length of gill opening, the fin moderately long for *Bascanichthys* (1–2.5 times the length of its base), somewhat triangular, the posterior edge fleshy.

Head and jaws short; snout short, pointed, and rugose, overhanging lower jaw, with a short groove on underside. Eye small, above posterior half of lower jaw. Anterior nostril tubular, directed anterolaterally; posterior nostril on underside of snout, covered by a large flap, visible externally as a slit in upper lip. Head pores small, well-defined, as listed above and as illustrated in Figure 329. Lateral-line pores noticeable, the pores large and obvious in large specimens. Gill opening moderate in size, vertical and low-lateral on side.

Dentition typical for the genus. Three large, stout, anterior intermaxillary teeth in an inverted V in groove on underside of snout, exposed when mouth is closed. Teeth on vomerine shaft irregularly biserial anteriorly, uniserial posteriorly, a total of 9–15 teeth, well-separated from intermaxillary teeth. Maxillary teeth conical and slender, uniserial, 11–14 in number; 11–15 uniserial teeth on mandible.

Color. Sharply bicolored, brown dorsally, pale below lateral line, with some pigment extending below mid-side; pigment visible microscopically as fine brown dots very close together and overlapping. Lateral-line pores dark anteriorly, in pale spots posteriorly, noticeable on all specimens except the smallest (373 mm, faded). Head and snout dark, some pigment on chin. Fins pale; scattered brown pigment along base of dorsal fin, the fin appearing as a pale stripe contrasting with body. Coloration partially obscured by mucus.

Size. The largest specimen is 813 mm TL. Three specimens (628–813 mm) are females, none ripe.

Development. Leptocephalus described by Leiby (see p. 863).

Distribution. The four type-specimens of *Bascanichthys inopinatus* were all taken in Puerto Rico with ichthyocides; the one depth recorded was 40 m. The holotype and one paratype were taken with one *B. bascanium.*

Etymology. From the Latin *inopinatus* (unexpected), because the authors did not initially expect to describe this as a new species (see *Remarks*).

Remarks. The four specimens of *Bascanichthys inopinatus* were originally considered by us to be abberrant or geographically variant forms of *B. paulensis.* However, Leiby found a uniquely pigmented leptocephalus whose meristics matched those of these specimens; he wrote "I have little

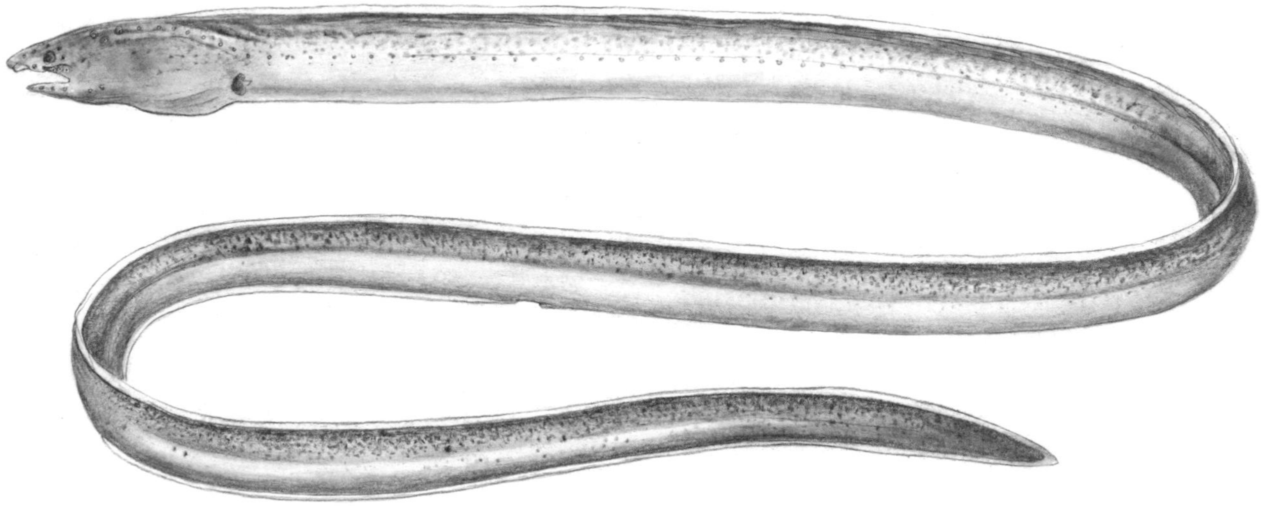

FIGURE 331. *Bascanichthys paulensis*: ANSP 158821, 303 mm TL, Brazil.

doubt . . . the four specimens you listed and the one [leptocephalus] I have are a valid, undescribed, species." Our subsequent analysis appears to have proven him to be correct.

Study Material. The holotype and three paratypes, 373–813 mm TL.

Holotype: ANSP 158885 (658); Puerto Rico, Puerto Yabucoa, 1¾ mi SSE of Playa de Guayanes, Municipio de Yabucoa; J. J. Loos; 24 July 1973.

Paratypes: PUERTO RICO: ANSP 158886 (1, 628); taken with the holotype. UF 18131 (1, 373); Crashboat Basin, Aguadilla, 22 m; J. E. Randall, G. Cooke; 24 July 1965. UPRM 3654 (1, 813); Puerto Rico, Humacao Beach; Eger and Storkus; 6 July 1969.

Bascanichthys paulensis Storey, 1939

BICOLORED SAND EEL

Figures 327, 331–333, Tables 33, 34

Bascanichthys paulensis Storey, 1939: 73, figs. 3, 5 (original description, São Paulo, Brazil, holotype CAS 5586).

Bascanichthys congoensis Blache and Cadenat, 1971: 179, figs. 16–20 (original description, Pointe-Noire Bay, Congo, holotype MNHN 1965-660).

Distinctive Characters. An elongate, markedly bicolored bascanichthyin without conspicuous light spots along lateral line pores; pectoral fin minute, its base 25–33% length of gill opening; jaw teeth uniserial; MVF 1-101-192.

Description. Total vertebrae 190–196, x̄ = 192 (n = 8); predorsal vertebrae 1–2, x̄ = 1.3 (7); preanal vertebrae 100–103, x̄ = 100.8 (8). Pores: SO 1 + 3; IO 4 + 2; POM 4 + 2; ST 3.

Proportions as % TL: head 4.8–5.6; trunk 47–50; tail 45–48; predorsal 2.3–3.0; depth behind gill openings 1.6–1.9. Of HL: snout 12–16; snout to rictus 22–24; eye 3.7–5.8; interorbital width 7.6–13; pectoral-fin length 3.0–8.1. Pectoral-fin base 25–33% of gill-opening length.

General characters similar to those of *Bascanichthys bascanium.* This species differs from its congeners in vertebral counts, and the pectoral fin is very small and rounded with a narrow base. Dentition similar to that of *B. bascanium* and *B. inopinatus*; 3 intermaxillary teeth, well separated from about 13 teeth on vomerine shaft; 11–14 maxillary teeth, 13–16 dentary teeth. Maxillary dentition of our study material is uniserial; Blache and Cadenat (1971: 181) said of their larger specimens of *B. congoensis*: ". . . dents maxillaires . . . avec tendance à la biseriation au milieu. . . ."

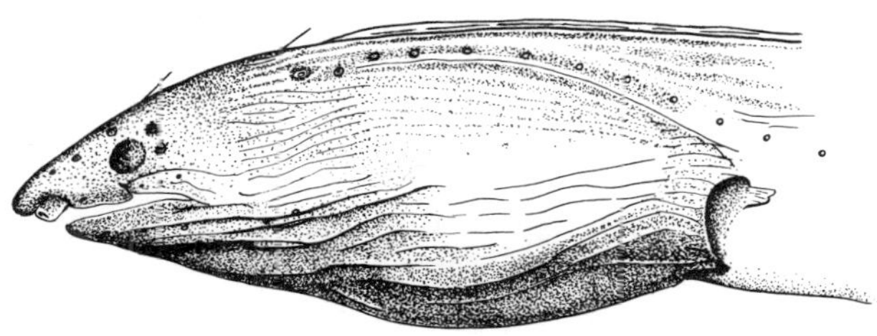

FIGURE 332. *Bascanichthys paulensis*: ANSP 158821, 303 mm TL, Brazil; head.

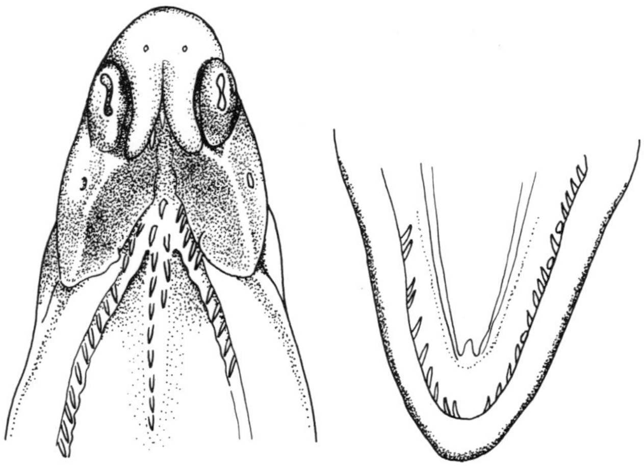

FIGURE 333. *Bascanichthys paulensis*: ANSP 158821, 303 mm TL, Brazil; dentition.

Color. Markedly bicolored; brown dorsally, pigment spots large and stellate, visible to naked eye; abruptly paler below lateral line. No pale spots marking lateral-line pores along body or in branchial region as in *B. scuticaris.* Head and chin pigmented. Fins pale, dorsal fin contrasting sharply with brown body color. Inside of mouth pigmented. Body color frequently obscured by mucous secretions.

Size. The largest western Atlantic specimen examined is 510 mm, although none of the specimens are sexually mature. The single eastern Atlantic specimen (UMML 21789) is a very ripe 570-mm female.

Distribution (Fig. 327). All western Atlantic specimens seen are from Brazil. In the eastern Atlantic recorded from the Gulf of Guinea and Pointe Noire (Blache, 1971). Western Atlantic study material came from shallow water sand beaches and was taken by poison; eastern Atlantic specimens were taken by line and by trawl.

Etymology. Named for São Paulo, the capture site of the type specimen. An adjective.

Remarks. Bascanichthys congoensis, described from the eastern Atlantic by Blache and Cadenat (1971: 179), appears to be conspecific with *B. paulensis.* Data obtained on our one study specimen (UMML 21789, a ripe female) agrees with that presented by Blache and Cadenat for the holotype and two paratypes; the two nominal species share a similar pectoral fin condition, coloration, morphometry, and vertebral numbers (UMML 21789 has a vertebral formula of 1-101-190; Blache and Cadenat (1971: 185) reported 189–190 total vertebrae and 99–103 preanal lateral-line pores for the type-specimens of *B. congoensis*).

Study Material. A total of 18 specimens (213–570 mm TL), including the holotype.

Holotype: CAS 5586 (635); Brazil, São Paulo, "probably taken from the stomach of a larger fish." The holotype is partially digested and broken, but intact enough to allow accurate measurements.

Other material: BRAZIL: *Recife:* USNM 214068 (1, 361). *Bahia:* ANSP 156906 (1, 247), ANSP 158821 (1, 303) and MZUSP 12008 (3, 321–406); Isla Itaparica, Caixa Pregos, 13°07′S, 38°47′W; 25 Aug. 1972. ANSP 156907 (1, 311); Salvador, Playa Arembepe, ca 1 km S of town, 12°45′S, 38°10′W, 0–1.3 m; 26 Aug. 1972. GCRL 9529 (1, 210); Salvador, Arembepe, ca 0.9 km S of town; 12°45′S, 38°10′W; 23 Aug. 1972. GCRL 9564 (1, 235); Isla Itaparica, Barra do Gil, 13°00′S, 38°37′W; 24 Aug. 1972. GCRL 10884 (1, 318); Salvador, Playas Arembepe, ca 1 km S of town, 12°45′S, 38°10′W; 30 Apr. 1973. MZUSP 12007 (2, 277–303); Praia de Arembepe; 23 Aug. 1972. MZUSP 12009 (2, 494–510); Praia de Arembepe; 26 Aug. 1972. MZUSP 12010 (1, 372); Barra do Gil, Ilha de Itaparica; 28 April 1973. NMW 65355 (1); Rio?; F. Steindachner, 1874-I.

EASTERN ATLANTIC: UMML 21789 (1, 567); Gulf of Guinea, 04°06′N, 05°58′–06°04′E, 24 m; 14 May 1965.

Bascanichthys scuticaris (Goode and Bean, 1880)

WHIP EEL

Figures 327, 334–336, Tables 33, 34

Sphagebranchus scuticaris Goode and Bean, 1880: 343 (original description, Cedar Key, Florida, holotype USNM 23636).

Sphagebranchus teres Goode and Bean, 1882: 436 (original description, west Florida, lectotype USNM 31457).

Caecula scuticaris. Jordan and Gilbert, 1882a: 358.

Caecula teres. Jordan and Gilbert, 1882a: 897.

Callechelys scuticaris. Jordan and Gilbert, 1883: 651.

Callechelys teres. Jordan and Gilbert, 1883: 651.

Bascanichthys scuticaris. Jordan and Davis, 1891: 621 (new genus, *teres* synonymized with *scuticaris*).

Bascanichthys teres. Storey, 1939: 74 (*teres* resurrected). Ginsburg, 1951: 480 (recognized as valid).

Bascanichthys scuticaris. Leiby and Yerger, 1980: 403 (revision of western north Atlantic species of *Bascanichthys, teres* synonymized with *scuticaris*).

Distinctive Characters. An elongate dark bascanichthyin with pale spots marking lateral-line pores on head above branchial basket and along

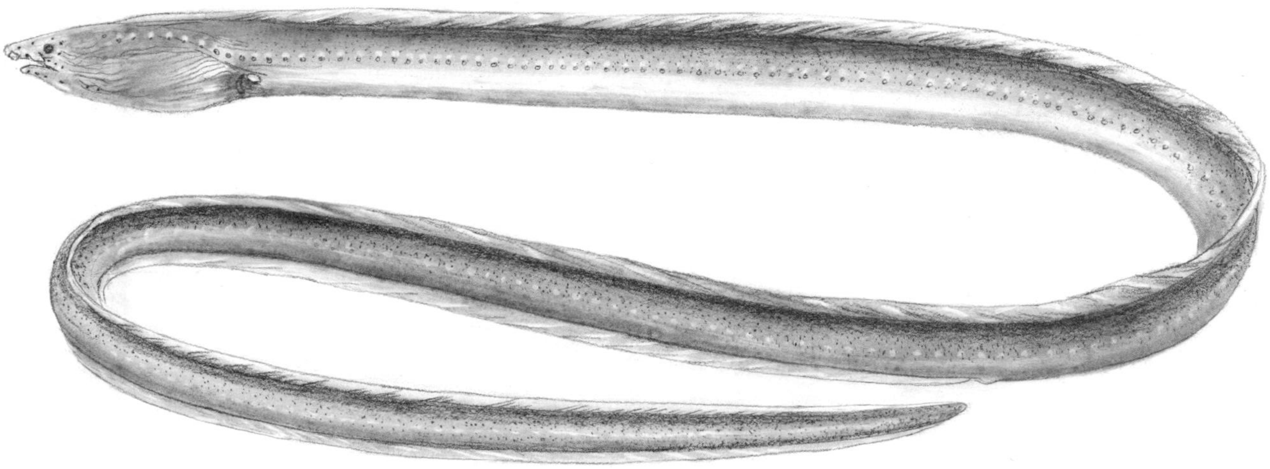

FIGURE 334. *Bascanichthys scuticaris*: ANSP 117782, 479 mm TL, Gulf of Mexico off Florida.

body; pectoral fin minute, its base 13–32% of gill-opening length; upper jaw teeth partly biserial; MVF 2-86-163.

Description. Total vertebrae 159–167, x̄ = 163.2 (n = 31); predorsal vertebrae 1–2, x̄ = 1.5 (31); preanal vertebrae 83–89, x̄ = 85.5 (31). Pores: SO 1 + 3; IO 4 + 2; POM 4 + 2; ST 3. Leiby and Yerger (1980: 404) reported 151–173 total vertebrae, 79–89 preanal vertebrae, 141–159 total LL pores and 16–17 branchiostegal rays on epihyal.

Proportions as % of TL: head 5.4–6.5; trunk 47–50; tail 44–46; predorsal 2.6–3.4; depth behind gill openings 1.6–2.4. Of HL: snout 14–15; snout to rictus 23–28; eye 3.3–6.6; interorbital width 11–13; pectoral-fin length 2.4–6.0. Pectoral-fin base 13–32% of gill-opening length.

Characters similar to those of *Bascanichthys bascanium*. As compared to its western Atlantic congeners, *B. scuticaris* has a slightly longer head and partly biserial maxillary dentition. Three to five large outer intermaxillary teeth in an inverted V (plus 1–2 median teeth on largest specimens), separated from vomerine row of 11–18 irregularly biserial teeth (uniserial far posteriorly); an outer row of 16–23 small teeth on maxillary, an inner row of 6–12 teeth at mid-section of outer row, sometimes extending to posterior end of main row. Dentary with 16–19 longer slender teeth, decreasing in size from front to rear. The pectoral fin is a minute narrow-based flap at the upper edge of the gill opening and frequently appears fleshy due to mucous secretion. Head pores as illustrated.

Color. Brown dorsally, melanophores appearing microscopically as very small overlapping dots; pale ventrally, except posteriorly on tail where color extends ventrally; not sharply bicolored. Distinctive pale spots marking lateral-

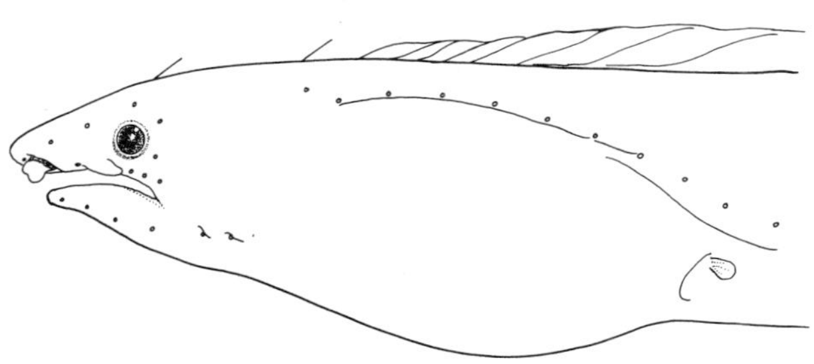

FIGURE 335. *Bascanichthys scuticaris*: ANSP 117782, 479 mm TL, Gulf of Mexico off Florida; head.

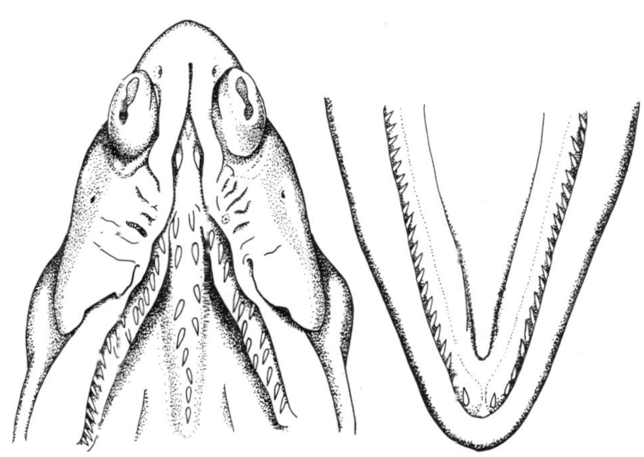

FIGURE 336. *Bascanichthys scuticaris*: ANSP 117782, 479 mm TL, Gulf of Mexico off Florida; dentition.

line pores in branchial region and on body, the spots just above the pore openings; body spots may be absent in specimens less than 175 mm and diffuse on those longer than 600 mm, but always visible in branchial region. Head paler than body, with variable pigmentation; inside of mouth pigmented. Fins pale, some pigment basally along dorsal fin. Body color frequently obscured by mucus.

Size. The largest specimen examined is the holotype, 756 mm. Of the study material, 9 specimens (244–539 mm) were females, none ripe; one 534-mm specimen was a ripe male. Leiby and Yerger (1980) examined a 770-mm specimen and reported others at least 1000 mm found on the beach after a red-tide kill off Sarasota, Florida. Their data suggested that females grow to greater sizes than males; no males were larger than 600 mm, but 6% of the females were 600–800 mm TL. They found seven ripe females, the smallest 471 mm; the smallest ripening male was 356 mm.

Development. Leptocephalus described by Leiby (1981: 50, and this volume, p. 864).

Distribution (Fig. 327). A common species, taken off coastal United States from North Carolina to the northern Gulf of Mexico; also taken off Mexico. Leiby and Yerger (1980) studied 374 specimens of *B. scuticaris,* and noted: "*B. scuticaris* is more common in collections from Cape San Blas, Florida, to the Florida Keys with *B. bascanium* being taken occasionally in that area. Collections from Mobile Bay to Cape San Blas yielded nearly equal numbers of both species. Although both species have been taken in Mexican waters, the scarcity of specimens from those waters makes it impossible to determine any distributional trends . . . [or] the southernmost limit of either species. . . . The small number of collections from Florida to North Carolina makes it impossible to draw a conclusion as to the more common species in that area." Most of their specimens were from exposed low tide flats and sandy beaches, always from sand bottom near grass beds; one specimen was captured by trawl at 46 m, and one was taken by night light over 31 m. Several of our study specimens were captured by trawl, one at a recorded depth of 27 m.

Etymology. From the Latin *scutica* (a whip), in reference to its shape. An adjective.

Study Material. A total of 52 specimens, 85–756 mm TL, including the holotype, and the lectotype and one paralectotype of *Sphagebranchus teres* Goode and Bean.

Holotype: USNM 23636 (756); Florida, Cedar Key; J. W. Velie.

Other material: NORTH CAROLINA: MCZ 9225 (1). USNM 132516 (1, 643). FLORIDA ATLANTIC: UMML 6208 (2, 244–334), 13941 (1, 426), 18749 (1, 300+), 31687 (1, 416), 32216 (1, 481). GULF OF MEXICO: *Florida:* ANSP 69939 (1, 539), 101021 (6, 340–534), 117782 (1, 479), 136518 (10, 224–408). CAS 31397 [previously IU 3242] (1, ca 400, broken and desiccated), 54466 (1, 545). MCZ 34169 (1). USNM 31457 (1, 525); lectotype of *Sphagebranchus teres* Goode and Bean. USNM 39652 (1, 160), 43116 (1, 409+), 104328 (4, 350–488), 104334 (9, 260–568), 114762 (1, 480), 152252 (1, 85), 152575 (1, 482); paralectotype of *Sphagebranchus teres.* USNM 157941 (1, 530), 278174 (1, 504), 278181 (1, 470).

Genus *Caralophia* Böhlke, 1955

Caralophia Böhlke, 1955: 1 (type species *Caralophia loxochila* Böhlke, 1955, by original designation).

Characters. Body elongate (depth 1.6–2.3% TL) and cylindrical, round in cross-section and not compressed posteriorly; head and trunk longer than tail (52–55% TL); tail tip blunt. Dorsal-fin origin on head; median fins low; pectoral fins absent. Head rugose; snout subconical, blunt from above, overhanging lower jaw, its underside not grooved; jaws short; lips without barbels; eye moderate; anterior nostril not tubular, a hole with interior lateral projections; posterior nostril opens into mouth. Gill openings ventral, converging forward, longer than isthmus. Head pores reduced; two preopercular pores and three pores in supratemporal canal. Teeth bluntly conical, moderate, uniserial on jaws and vomer; intermaxillary teeth separated from those of vomer by a short gap.

Neurocranium sloping posteriorly, depressed dorsally; orbit reduced; nasals stout, nasal cartilage well developed; supraoccipital crest moderately developed; maxilla weak, slender, pointed posteriorly; pterygoid broad centrally, closely associated with maxilla anteriorly; otic bulla

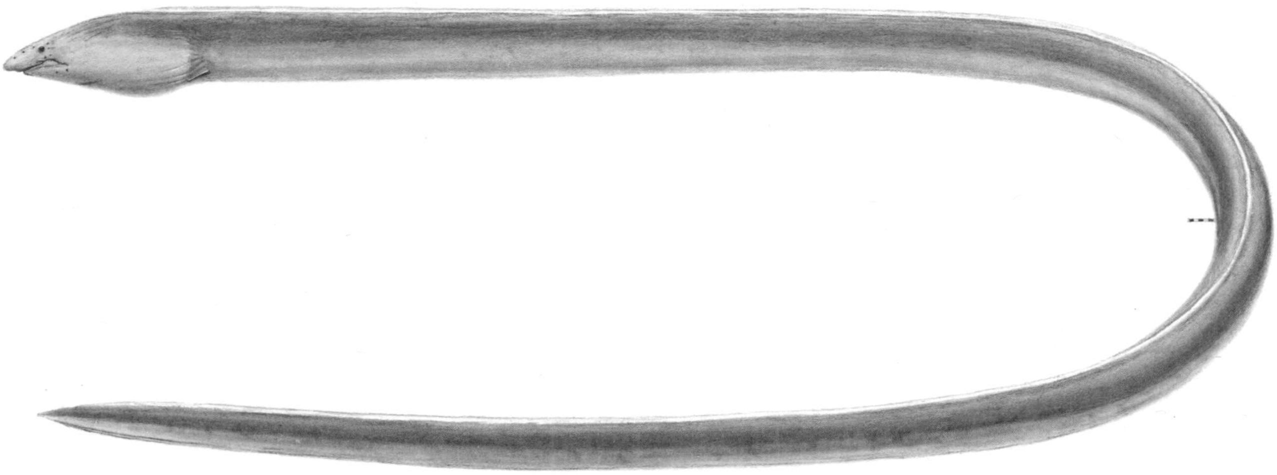

FIGURE 337. *Caralophia loxochila*: ANSP 73433, holotype, 317 mm TL, Bahamas. Illustration by S. P. Gigliotti.

moderately developed; suspensorium posteriorly inclined; opercular series developed. Branchiostegal rays numerous, often branched basally, all along arch, the distal rays broadened basally; hyoid stout; urohyal a cartilaginous filament posteriorly. Gill arches reduced; basibranchial 1 ossified, 2–4 cartilaginous; hypobranchials 1–2 ossified, 3 cartilaginous; ceratobranchials 1–4 ossified, 5 absent; upper pharyngeal tooth plates fused. Pectoral girdle reduced to a cleithrum and fragments of a supracleithrum, scapula and coracoid. Intramuscular bones, ribs, and caudal transverse processes developed. Precaudal vertebrae more numerous than caudal. Gasbladder moderately developed. Peritoneum pale. Body coloration uniform, darker dorsally.

Distribution. A single western Atlantic species.

Etymology. From the Greek *kara* (head) and *lophia* (mane), in reference to the anterior dorsal-fin origin. Feminine.

Caralophia loxochila Böhlke, 1955

SLANTLIP EEL

Figures 337–340, Table 33

Caralophia loxochila Böhlke, 1955: 2, fig. 1 (original description, Rose Island, Bahamas, holotype ANSP 73433).

Distinctive Characters. An elongate bascanichthyin with brown coloration, darker dorsally; anterior nostrils non-tubular, on underside of snout; teeth few, stout, blunt, and uniserial; MVF 2-75-145.

Description. Total vertebrae 138–155, $\bar{x}$ = 145.2 (n = 30); predorsal vertebrae 2–3, $\bar{x}$ = 2.1 (28); preanal vertebrae 68–79, $\bar{x}$ = 74.6 (22). Pores: SO 1 + 3; IO 4 + 2; POM 5 + 2; ST 3.

Proportions as % of TL: head 6.5–8.7; trunk 43–47; tail 45–48; predorsal 4.3–5.2; depth behind gill openings 1.6–2.3. Of HL: snout 15–18; snout to rictus 23–30; eye 3.2–4.4; interorbital width 6.6–8.8.

Head short; snout sharply pointed when viewed laterally, bluntly pointed from above. Eye small and embedded, above and just behind midpoint between snout tip and rictus. Anterior nostril on underside of snout and flush with snout surface, a convoluted hole with interior lateral projections; posterior nostril opening inside mouth, covered by an anterior flap and marked externally by a short inconspicuous lateral groove. A flange on lower jaw below posterior nostril, bordered by a ventral groove. Ethmoid pores close

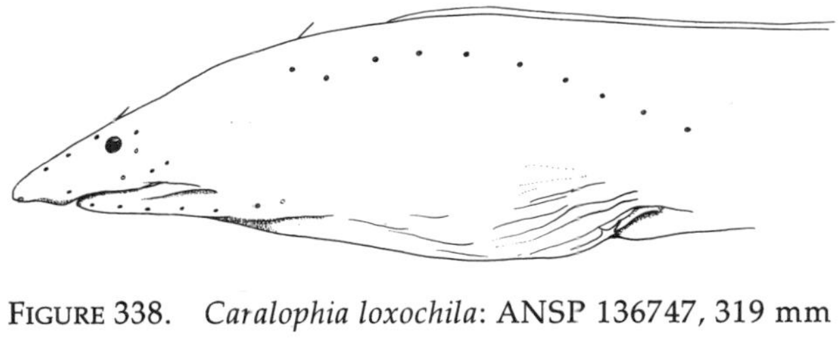

FIGURE 338. *Caralophia loxochila*: ANSP 136747, 319 mm TL, Bahamas; head.

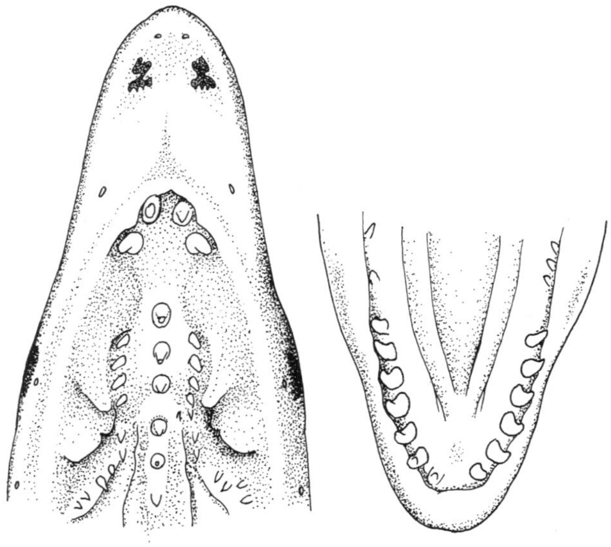

FIGURE 339. *Caralophia loxochila*: ANSP 105489, 274 mm TL, Martinique; dentition.

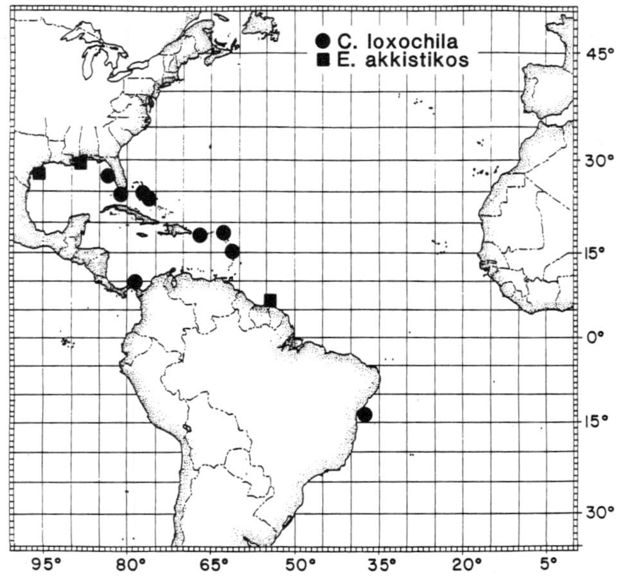

FIGURE 340. Distribution of *Caralophia loxochila* and *Ethadophis akkistikos* based on study material.

together on underside of snout; third pair of supraorbital pores between eyes.

Teeth stout, the anteriormost intermaxillary and vomerine teeth the largest. Four intermaxillary teeth in broad curve, two median ones usually exposed when mouth is closed, well separated from uniserial row of 5–8 teeth along vomer. Row of 8–12 teeth along maxillary, closely paralleling the vomerine row anteriorly and flaring outward posteriorly. Mandibular teeth 8–11, largest anteriorly.

Color. Almost uniformly brown, darker dorsally, pigment spots visible microscopically on small specimens (88 mm). Fins pale, the low dorsal fin appearing as a pale streak along the dorsal midline. Three large (mature) specimens from Puerto Rico sharply bicolored, the head and snout darker (see *Remarks*).

Size. The largest specimen examined is 456 mm TL. Sex could be determined only for three large (350–456 mm) specimens which were ripe males.

Development. Leptocephalus identified (see p. 878).

Distribution (Fig. 340). Primarily an insular species, known from the Gulf of Mexico off Florida, and from the Bahamas south to Brazil, including Panama, Puerto Rico and the Lesser Antilles. Found in coves and bays rather than on open beaches, collected by using ichthyocides over sand and turtle grass bottoms at depths usually less than 3 m.

Etymology. From the Greek *loxos* (oblique) and *chilos* (a lip), referring to the flange on either side of the lower jaw. A noun in apposition.

Remarks. Five of the 30 specimens x-rayed had an unusually high number of total vertebrae (151–155 vs 140–148). One of these (GCRL 22312) was from Brazil, one was from the Bahamas (one specimen in ANSP 98349), and three were from Puerto Rico (ANSP 136519). The latter three, all ripe males collected at 12 m, exhibited several other differences: they were strongly bicolored, the snout appeared shorter, the lower jaw profile was distended downward, and the anterior nostril had a slight rim. The range of 140–155 total vertebrae is not unduly wide, and other differences could be due to sexual maturity. However, study of additional material may support the existence of a southern species with a higher vertebral count.

Study Material. A total of 65 specimens, 52–456 mm TL, including the holotype and paratype.

Holotype: ANSP 73433 (317); Bahamas, bay on the north shore near the west end of Rose Island, 25°05′47″N, 77°11′56″W; J. E. Böhlke et al.; 8 April 1955.

Paratype: ANSP 73434 (1, 230.5); same data as holotype.

Other material: FLORIDA: ANSP 134505 (1, 166). FSBC 3644 (1, 206, cleared and stained). BAHAMAS: ANSP

98348 (7, 82–197), 98349 (7, 226–405), 98350 (3, 88–186), 98351 (1, 254), 98352 (1, 52), 98353 (2, 221–427), 98354 (5, 154–265), 115534 (3, 183–356, cleared and stained), 136747 (1, 319). PUERTO RICO: ANSP 136519 (3, 137–350), 138777 (2, 155–456). LACM 6522-7 (2, 328–396). LESSER ANTILLES: *St. Barthelemy:* ANSP 115520 (1, 309, cleared and stained). CAS 54470 (7, 74–299). *Martinique:* ANSP 105312 (6, 240–274), 105489 (1, 274). PANAMA: CAS 29118 (2, 213–362). SIO 70-228 (1, 445), 70-376 (1, 238). BRAZIL: GCRL 9561 (1, 214), 22312 (1, 225). MZUSP 12014 (3, 94.5–283).

Genus *Ethadophis* Rosenblatt and McCosker, 1970

Ethadophis Rosenblatt and McCosker, 1970: 498 (type species *Ethadophis byrnei* Rosenblatt and McCosker, 1970, by original designation).

Microrhynchus Blache and Bauchot, 1972: 728 (type species *Sphagebranchus foresti* Cadenat and Roux, 1964, by original designation; *Microrhynchus* preoccupied by earlier authors for a lepidopteran (1823), and subsequently for a mammal, a crustacean, a bird, an hemipteran, and a worm).

Characters. Body moderately elongate (depth 1.5–2% TL in western Atlantic species), cylindrical anteriorly, compressed posteriorly; head and trunk longer than tail (58–62% TL); tail tip blunt. Dorsal-fin origin on head; median fins low; pectoral fins absent. Head rugose. Snout short and rounded, overhanging lower jaw, its underside not grooved; jaws short; lips without barbels; eye moderate; anterior nostril tubular; posterior nostril opens into mouth. Gill openings crescentic, low on body, their major axis vertical, less than isthmus in length. Head pores reduced; two preopercular pores and three pores in supratemporal canal. Teeth conical, small, uniserial on jaws and vomer; intermaxillary teeth separated from those of vomer by a short gap.

Neurocranium subtruncate posteriorly, less so than in *Gordiichthys*; orbit reduced; nasals and nasal cartilage moderately developed; supraoccipital crest absent; maxilla slender, pointed posteriorly; pterygoid pointed anteriorly; otic bulla moderately developed; suspensorium nearly vertical; opercular series weakly developed. Branchiostegal rays numerous, all along arch; urohyal ossified, tapering to a thin filament posteriorly. Gill arches reduced; basibranchial 1 ossified, 2 cartilaginous, 3–4 absent; ceratobranchials 1–4 ossified, 5 absent; upper pharyngeal tooth plates separated. Pectoral girdle reduced to a slender cleithrum and supracleithrum, and a scapula and coracoid (?). Intramuscular bones, ribs, and caudal transverse processes developed. Precaudal vertebrae more numerous than caudal. Peritoneum pale. Body coloration uniformly pale, slightly darker dorsally.

Size. Grows to 530 mm TL in the western Atlantic.

Distribution. A tropical and subtropical genus, known from one western Atlantic species, two eastern Atlantic and two eastern Pacific species. One additional larval species is referred to *Ethadophis* (p. 876), the adult unknown.

Etymology. From the Greek *ethas* (ordinary) and *ophis* (snake), named for its lack of outstanding characteristics. Masculine.

Species. One recently described species, *Ethadophis akkistikos* McCosker and Böhlke, 1984 is found in the western Atlantic; two species, *Sphagebranchus foresti* Cadenat and Roux, 1964, and, provisionally, *Microrhynchus epinepheli* Blache and Bauchot, 1972, are in the eastern Atlantic; and *Ethadophis byrnei* Rosenblatt and McCosker, 1970 and *E. merenda* Rosenblatt and McCosker, 1970 are in the eastern Pacific. We note that the holotype and only known specimen of *E. epinepheli* was from the stomach of a grouper and is partially digested, disallowing verification of the snout and nostril conditions. We provisionally retain it in *Ethadophis*; however, study of intact material may disclose the species to be a *Gordiichthys*.

Key to the Species of *Ethadophis*

1a. Body extremely elongate, its depth 71–91 times in TL; 4–6 uniserial vomerine teeth 2
1b. Body elongate, its depth 54–66 times in TL; 10 or more uniserial vomerine teeth 3

2a. Body depth 91 times in TL; total vertebrae 187; preanal length 1.41 in TL .... *foresti* (Cadenat and Roux, 1964) (eastern Atlantic, Cape Verde Archipelago)

2b. Body depth 71 times in TL; total vertebrae 192; preanal length 2.02 in TL .... *epinepheli* (Blache and Bauchot, 1972), incertae sedis (eastern Atlantic, Sénégal)

3a. Vertebrae 189; dorsal-fin origin slightly in advance of gill openings .... *byrnei* Rosenblatt and McCosker, 1970 (eastern Pacific, Gulf of California)

3b. Vertebrae 159–173; dorsal-fin origin mid-head 4

4a. Vertebrae 159; head 13.7 in TL .... *merenda* Rosenblatt and McCosker, 1970 (eastern Pacific, Baja California)

4b. Vertebrae 161–173; head 15.2–17.2 in TL .... *akkistikos*, p. 342

Ethadophis akkistikos
McCosker and Böhlke, 1984
INDIFFERENT EEL

Figures 340–343, Table 33

Ophichthidae, new genus, new species A. Böhlke, 1977b: 4, 6.

Ethadophis akkistikos McCosker and J. E. Böhlke, 1984: 41, figs. 12–14 (original description, Gulf of Mexico, 27°55′N, 95°40′W, holotype ANSP 149884).

Distinctive Characters. A small, elongate, pallid bascanichthyin which lacks distinguishing features; dorsal-fin origin at mid-head; anterior nostril tubular; dentition uniserial; MVF 4-96-165.

Description. Total vertebrae 161–173, $\bar{x} = 165.2$ (n = 4); preodorsal vertebrae 3–4, $\bar{x} = 3.5$ (4); preanal vertebrae 92–98, $\bar{x} = 96.2$ (4). Pores: SO 1 + 3; IO 4 + 2; POM 5 + 2; ST 3.

Proportions as % TL: head 5.8–6.6; trunk 52–55; tail 38–42; predorsal 4.4; depth behind gill openings 1.5–2. Of HL: snout 16–18; snout to rictus 26–29; eye 3.4–4; interorbital width 8–10.

Body elongate and cylindrical, becoming laterally compressed posteriorly and tapering to a blunt finless tip. Snout rounded, underside not grooved; lower jaw included, its tip well behind anterior nostrils, anteriormost pair of intermaxillary teeth exposed when mouth is closed. Center of eye slightly before midpoint of upper jaw. Anterior nostril tubular; posterior nostril on underside of snout, visible laterally as a minute slit in lip at level of eye. Median fins low, dorsal-fin origin at mid-head; pectoral fins absent. Gill openings low on body, on vertical axis. Head pores reduced; lateral line pores apparent but not countable.

Teeth uniserial, small and difficult to count. An intermaxillary semicircle of four teeth separated from 8–14 uniserial vomerine teeth which are largest anteriorly; 11–16 teeth on maxillary; mandibular teeth 12–16, the anterior four large and stout.

Color. Body coloration pale yellow to tan, with a few minute brown punctations dorsally.

Size. The largest specimen examined was 316 mm TL. The 242-mm specimen (RMNH 29318) collected in April off Suriname had eggs of varying sizes to 0.6 mm diameter.

Development. Leptocephalus identified (p. 875).

Distribution (Fig. 340). This rare eel has been collected by trawl, dredge, and bottom grab between 30–60 m depth, from the Gulf of Mexico and off Suriname.

Etymology. From the Greek *akkistikos* (pretending indifference), "in reference to the coy appearance of this eel's face" (McCosker and J. E. Böhlke, 1984: 42). A noun in apposition.

Remarks. Species of *Ethadophis* possess few distinguishing characters. *E. akkistikos* can be distinguished from its eastern Atlantic congeners by its vertebral count; it also appears to have more vomerine teeth.

Study Material. The five type specimens, 237–316 mm TL.

Holotype: ANSP 149884 (239.8 mm); Gulf of Mexico,

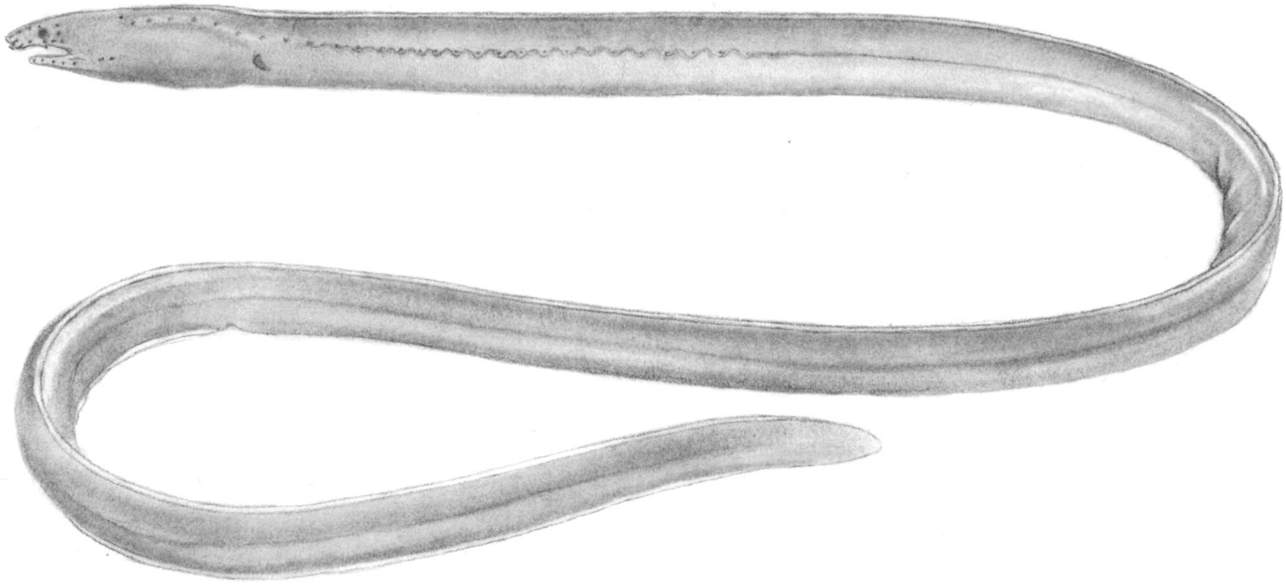

FIGURE 341. *Ethadophis akkistikos*: ANSP 149884, holotype, 240 mm TL, Gulf of Mexico off Texas.

off Texas, 27°55′N, 95°40′W, 5 ft tumbler dredge at 55–58 m; OREGON sta. 3878; 21 Sept. 1962.

Paratypes. GULF OF MEXICO: *Alabama:* USNM 263570 (1, 237); 29°58′N, 87°57′W, 33 m; 27 June 1977. *Mississippi:* ANSP 135526 (1, 316, cleared and stained); 29°58.5′N, 88°21′W, depth 30 m; 27 June 1977. *Texas:* CAS 52672 (1, 260); 27°55.5′N, 95°34.5′W, 60 m; OREGON 3875; 20 Sept. 1962. SURINAME: RMNH 29318 (1, 242); 06°50.7′N, 53°55.2′W, 51.5 m; Van Veen grab, O.C.P.S.II; 12 Apr. 1969.

Genus *Gordiichthys* Jordan and Davis, 1891

Gordiichthys Jordan and Davis, 1891: 644 (type species *Gordiichthys irretitus* Jordan and Davis, 1892, by original designation).

Characters. Body moderately to extremely elongate (depth 0.8–1.6% TL), cylindrical, compressed far posteriorly; head and trunk longer than tail (52–59% TL); tail tip blunt. Dorsal-fin origin on head; median fins low; pectoral fins absent. Head markedly rugose. Snout short and blunt, overhanging lower jaw, its underside grooved; jaws short, lower jaw included, not reaching posterior base of anterior nostrils, anteriormost intermaxillary teeth exposed; lips without barbels; eye small; anterior nostril tubular or inset, partly set off by a groove from underside of snout; posterior nostril opens into mouth, visible as a minute vertical slit in lip at level of eye. Gill opening crescentic, low on body, the main axis vertical,

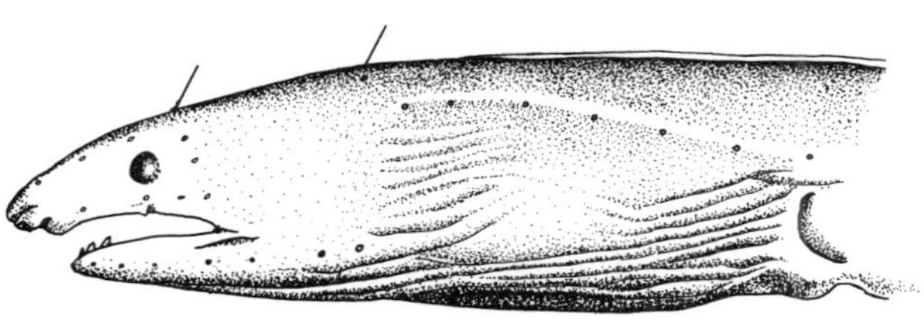

FIGURE 342. *Ethadophis akkistikos*: ANSP 149884, holotype, 240 mm TL, Gulf of Mexico off Texas; head.

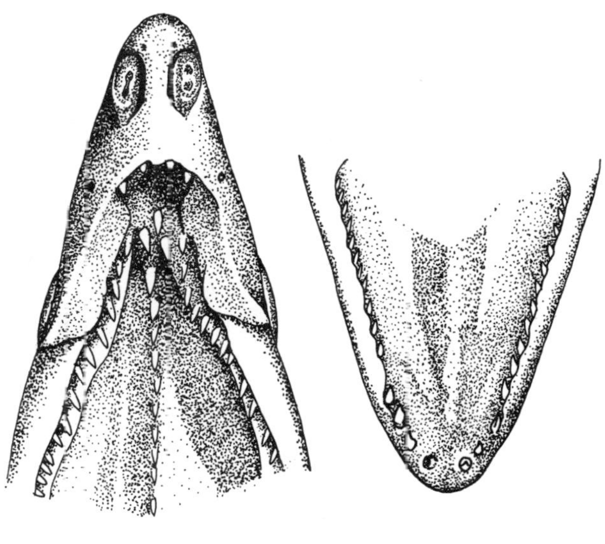

FIGURE 343. *Ethadophis akkistikos*: ANSP 149884, holotype, 240 mm TL, Gulf of Mexico off Texas; dentition.

less than isthmus in length. Head pores reduced; two preopercular pores and three pores in supratemporal canal. Teeth conical, small, uniserial on jaws; intermaxillary teeth separated from those of vomer by a short gap.

Neurocranium subtruncate posteriorly; orbit reduced; nasals and nasal cartilage moderately developed; supraoccipital crest absent; maxilla weak, slender, pointed posteriorly; pterygoid pointed anteriorly; otic bulla moderately developed; suspensorium nearly vertical; opercular series weakly developed. Branchiostegal rays numerous, all along arch; urohyal ossified, tapering to a thin filament posteriorly. Gill arches reduced; basibranchial 1 ossified, 2 cartilaginous, 3–4 absent; hypobranchials 1–2 ossified, 3 cartilaginous; ceratobranchials 1–4 ossified, 5 absent; upper pharyngeal tooth plates separated. Pectoral girdle consists of a reduced cleithrum and supracleithrum. Intramuscular bones, ribs, and caudal transverse processes developed. Precaudal vertebrae more numerous than caudal. Gasbladder moderately developed. Body coloration pale or darker dorsally; peritoneum pale.

Size. To 790 mm TL.

Distribution. Four known species, all from the western Atlantic.

Etymology. From *Gordius,* the horsehair worm (named after Gordius, the king whose complicated knot was cut by Alexander), and *ichthys* (fish). Masculine.

Species. The species are *Gordiichthys ergodes* new species; *G. irretitus* Jordan and Davis, 1892; *G. leibyi* McCosker and Böhlke, 1984; and *G. randalli* McCosker and Böhlke, 1984. They are very similar in general appearance. *Gordiichthys ergodes* differs from all others in possessing tubular anterior nostrils, those of the other species set in grooves but not projecting downward. *Gordiichthys randalli* is a larger, more robust fish which is bicolored (darker dorsally), with a rounded snout, partially biserial vomerine dentition, and typically three intermaxillary teeth and four mandibular pores. *Gordiichthys leibyi* is elongate and uniformly pallid, with a rounded snout, uniserial vomerine teeth, and typically four intermaxillary teeth and four mandibular pores. *Gordiichthys irretitus* is even more elongate with a more pointed snout, also uniformly pallid, with uniserial vomerine teeth, and typically with three intermaxillary teeth and five mandibular pores. All are separable by vertebral counts (Table 32).

KEY TO SPECIES OF *GORDIICHTHYS*

1a. Anterior nostrils within a short tube, extending noticeably beneath underside of snout .... *ergodes* new species, p. 344
1b. Anterior nostrils not tubular, partly set off by a groove from underside of snout 2
2a. Body moderately elongate, its depth 62–79 times in TL; bicolored in life and in preservative, dark dorsally and pale ventrally; vomerine teeth biserial anteriorly .... *randalli,* p. 349
2b. Body extremely elongate, wormlike, its depth 80–128 times in TL; coloration in preservative uniform, not darker dorsally; vomerine teeth uniserial throughout 3
3a. Body depth 80–91 times in TL; preanal vertebrae 92–98 and total vertebrae 168–176; lateral line pores 161–166 .... *leibyi,* p. 348
3b. Body depth 100–128 times in TL; preanal vertebrae 105–111 and total vertebrae 193–206; lateral line pores 186–189 .... *irretitus,* p. 347

Gordiichthys ergodes new species

IRKSOME EEL

Figures 344–347, Tables 33, 36

Distinctive Characters. An elongate bascanichthyin with brown coloration; depth behind gill opening 1.6% TL; head 5.8% TL; anterior nostrils tubular; five mandibular pores; uniserial vomerine teeth; MVF 1-98-186.

Counts and Measurements (in mm) of the Holotype. Total vertebrae 186; predorsal vertebrae 1;

preanal vertebrae 98. Pores: SO 1 + 3; IO 4 + 2; POM 5 + 2; ST 3. Total length 339; head length 19.6; trunk length 168.4; tail length 151; body depth at gill opening 5.5; body width at gill opening 4.0; predorsal length 10.0; snout length 3.5; snout tip to rictus 5.8; length gill opening 1.35; isthmus 2.0; eye diameter 0.9; interorbital distance 1.6.

Description. Proportions as % of TL: head 5.8; trunk 50; tail 44; predorsal 3.0; depth behind gill openings 1.7. Of HL: snout 18; snout to rictus 30; eye 4.6; interorbital width 8.2.

Body elongate, depth at gill openings 61.6 in TL, tapering posteriorly to blunt finless tip; body and tail nearly cylindrical, becoming laterally compressed posteriorly. Anus shortly behind mid-body. Head and trunk 1.8 and head 17.3 in TL. Median fins low; dorsal fin begins above first vertebra, about mid-head. Pectoral fins absent. Gill openings crescentic, low on body, their main axis about 20° from vertical, less than isthmus in length.

Head and jaws short; snout rounded at tip, but not blunt, conical when viewed from above; lower jaw included, not reaching posterior base of anterior nostrils, anteriormost intermaxillary teeth barely exposed. Snout flat on underside, median groove extends to middle of anterior nostril bases. Center of eye behind midpoint of upper jaw. Anterior nostrils tubular, their tips well below underside of snout. Posterior nostrils within upper lip, visible as minute vertical slits in lip below level of eye. Head pores reduced, small but visible; lateral-line pores difficult to discern.

Teeth conical, short, rounded at tips, uniserial on jaws and vomer. Intermaxillary chevron of 4 teeth, followed by a gap and 11 vomerine teeth. Nine to eleven teeth on maxillaries and 9 mandibular teeth on each side. Anterior vomerine teeth and those at mandibular symphysis the largest.

Color. Body coloration in preservative reddish-brown; small freckles and punctations visible microscopically. Median fins colorless.

Size. The only known specimen is an immature female, 339 mm TL.

Development. The leptocephalus of this species

TABLE 36. Frequency distributions of vertebral counts in species of *Gordiichthys*.

	Predorsal		Preanal																						
	1	2	89	90	91	92	93	94	95	96	97	98	99	100	101	102	103	104	105	106	107	108	109	110	111
ergodes		1*										1*													
irretitus		4*															1*	-	1	1	-	-	-	-	1
leibyi		4*				1	-	1	-	-	1*	1													
randalli	1*	7	1*	3	-	-	3	1																	

	Total																										
	168	169	170	171	172	173	174	175	176	177	178	179	180	181	182	183	184	185	186	...	192	193	194	195	196	197	...	206
ergodes																			1*									
irretitus																						1	2	-	-	1*		1
leibyi	1	-	1	2*	-	-	-	-	1																			
randalli									1	-	2	-	2*	-	2	-	-	-	1									

* Includes holotype.

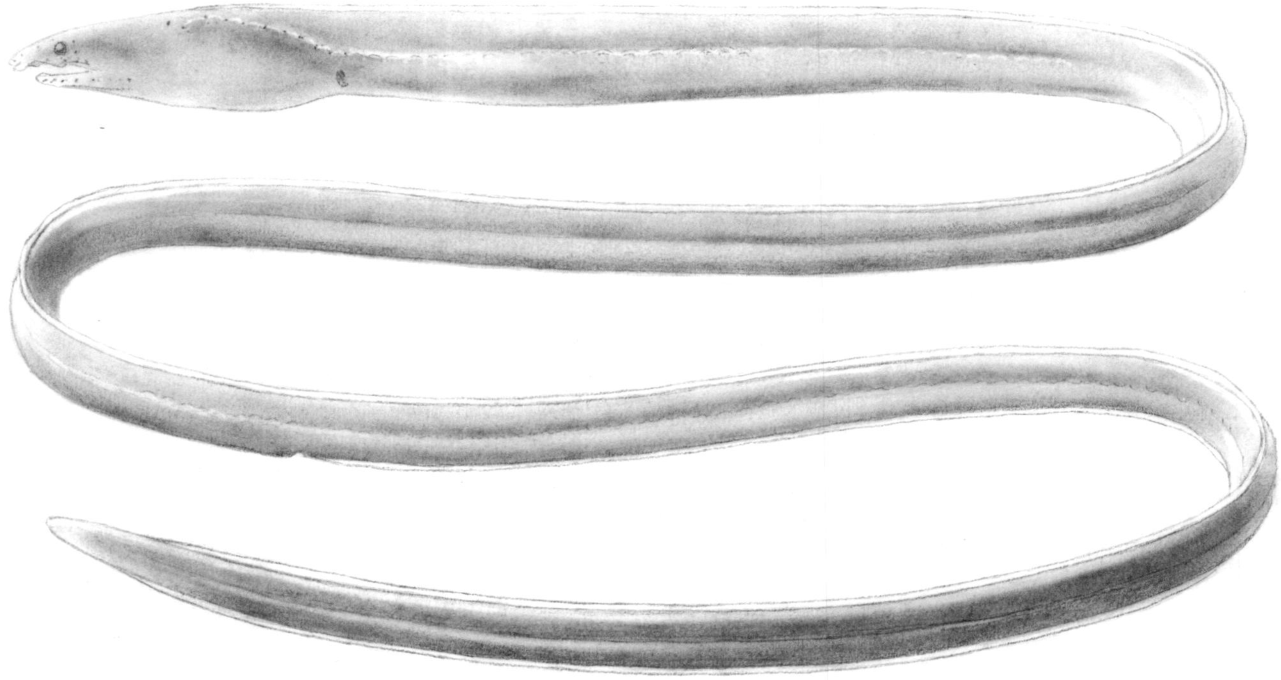

FIGURE 344. *Gordiichthys ergodes*: USNM 272462, holotype, 339 mm TL, northeastern Gulf of Mexico.

was previously described as *Callechelys* sp. by Fahay and Obenchain (1978: 471) and as "Tribe Bascanichthyini genus and species unknown" by Leiby (1984b: 419); see p. 865.

Etymology. From the Greek *ergodes* (irksome or troublesome), in reference to the difficulties this new species has created with our previous concept of generic limits. To be treated as a noun in apposition.

Remarks. Subsequent to our review of *Gordiichthys* and *Ethadophis* (McCosker and J. E. Böhlke, 1984), this specimen was brought to our attention by Kenneth Tighe of the USNM. It had been received at USNM in a box of polychaete worms collected by core samples taken during the MAFLA program (1975–1978), the station number not known. Its combination of tubular anterior nostrils as well as a well-formed groove along the underside of the snout has caused us to question the generic differences of these as well as the closely-related genera *Allips* and *Phaenomonas.* With hesitation, we continue to recognize the species of *Ethadophis* as distinct from those of *Gordiichthys,* and conservatively recognize this new species within the latter genus. The anterior nostril condition of *G. irretitus, leibyi* and *randalli* differs from that of *G. ergodes* in that, although

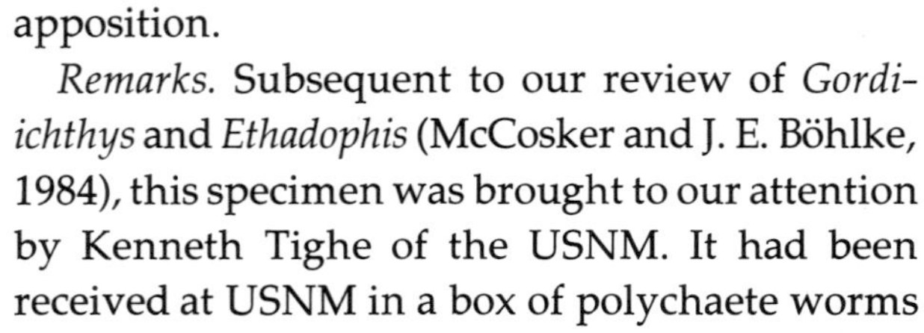

FIGURE 345. *Gordiichthys ergodes*: USNM 272462, holotype, 339 mm TL, northeastern Gulf of Mexico; head.

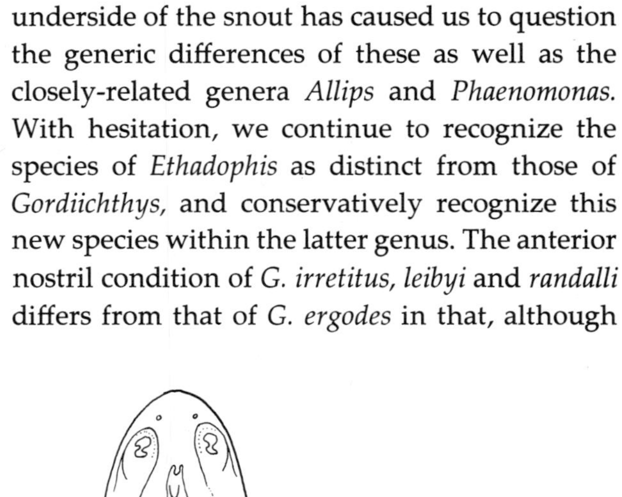

FIGURE 346. *Gordiichthys ergodes*: USNM 272462, holotype, 339 mm TL, northeastern Gulf of Mexico; dentition.

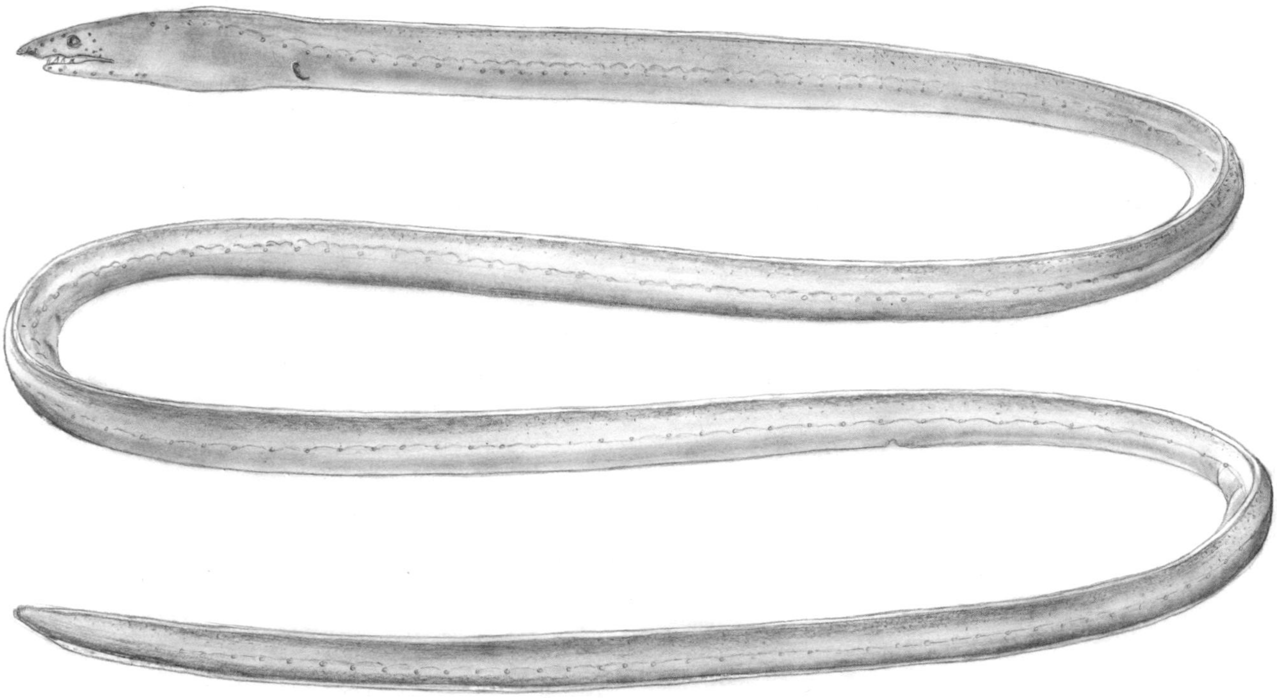

FIGURE 347. *Gordiichthys irretitus*: ANSP 115297, 370 mm TL, Gulf of Mexico off Florida.

tubular, they are depressed within the surface of the underside of the snout. The snout groove exposing the anterior intermaxillary teeth is similar in all species of *Gordiichthys.* The anterior nostril condition of *G. ergodes* is more like that of the species of *Ethadophis*; however this plesiomorphic condition is easily modified to the character state of all other species of *Gordiichthys.* The snout groove, present in all *Gordiichthys* but absent in *Ethadophis,* appears to be a more significant transition, and on that basis we identify the species of *Ethadophis* as a separate lineage. The unique character states separating the species of *Phaenomonas, Allips,* and *Caralophia* concern the dorsal-fin origin and anterior nostril conditions; we continue to recognize these as separate lineages in that intermediate forms do not exist. The differences observed in bascanichthyin fin condition, body elongation, and general physiognomy appear to be related to the burrowing behavior, habitat preference, and feeding strategy of these forms (see discussion under Tribe Bascanichthyini, p. 328).

Study Material. The holotype: USNM 272462 (339 mm TL); Gulf of Mexico, continental shelf between Naples, Florida and Biloxi, Mississippi (exact location not recorded); Reinek box core sample from sand, mud or rocky bottom; depth 10–189 m; MAFLA Program; 1975–1978.

Gordiichthys irretitus Jordan and Davis, 1891
HORSEHAIR EEL
Figures 347–350, Tables 33, 36

Gordiichthys irretitus Jordan and Davis, 1891: 644 (original description, northeastern Gulf of Mexico, from "the Snapper Banks at Pensacola," holotype USNM 44303).

Distinctive Characters. An extremely elongate bascanichthyin with uniform pallid coloration; depth behind gill opening 0.8–1.0% TL; head 4.5–5.4% TL; anterior nostril set in a groove; vomerine teeth uniserial; five mandibular pores; MVF 2-107-197.

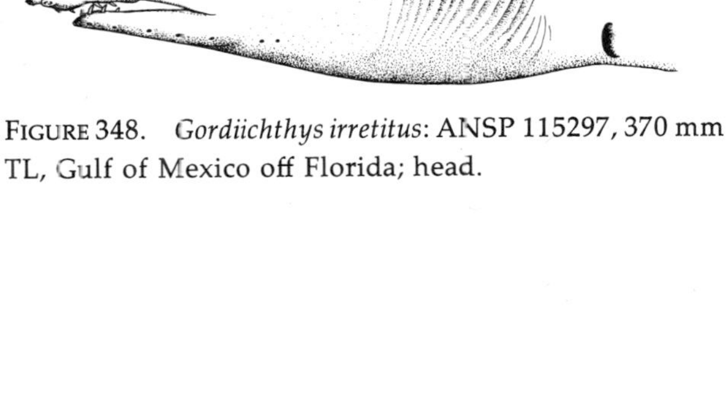

FIGURE 348. *Gordiichthys irretitus*: ANSP 115297, 370 mm TL, Gulf of Mexico off Florida; head.

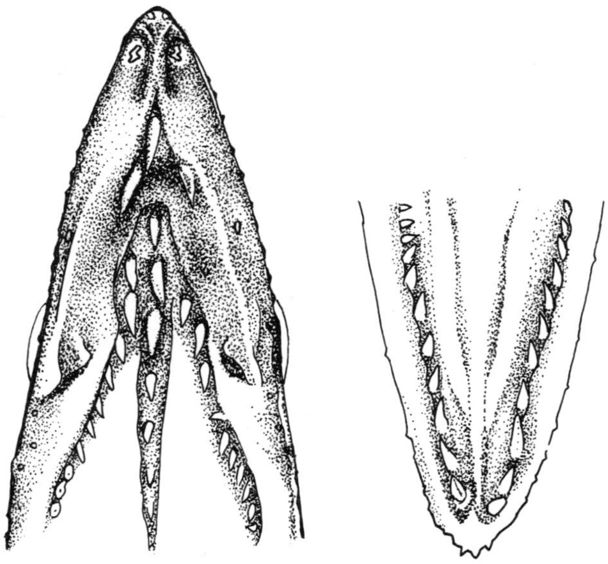

FIGURE 349. *Gordiichthys irretitus*: ANSP 115297, 370 mm TL, Gulf of Mexico off Florida; dentition.

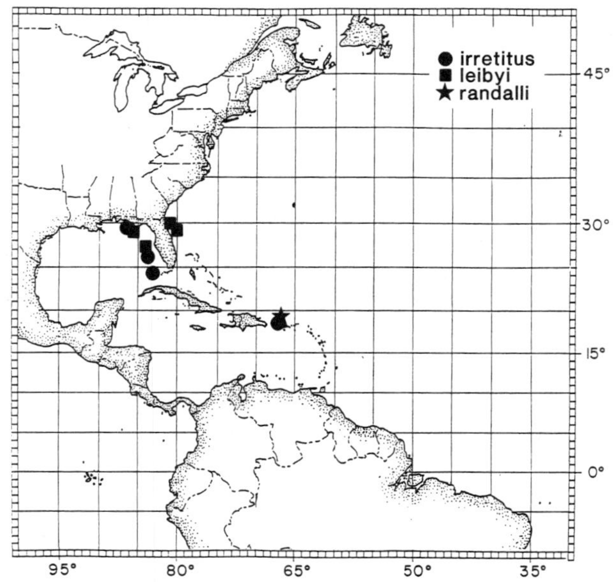

FIGURE 350. Distribution of species of *Gordiichthys* based on study material.

Description. Total vertebrae 193–206, x̄ = 196.3 (n = 4); predorsal vertebrae 2 (4); preanal vertebrae 105–111, x̄ = 107 (4). Pores: SO 1 + 3; IO 4 + 2; POM 5 + 2; ST 3.

Proportions as % TL: head 4.5–5.4; trunk 50–53; tail 42–46; depth behind gill openings 0.8–1.0. Of HL: snout 19–21; snout to rictus 30–35; eye 4–6; interorbital width 8–10.

Characters as for the genus. Snout pointed; five mandibular pores. Three intermaxillary teeth, separated by a short gap from 7 uniserial vomerine teeth, the second the longest tooth in the mouth; 9–10 maxillary teeth; 11 teeth on dentary, the second and third enlarged.

Color. Body coloration in alcohol yellow to tan with minute brown punctations above the midline. Median fins transparent.

Development. Leptocephalus identified (see p. 868). There are two distinct populations of leptocephali, a northern form from the Gulf of Mexico and the Caribbean and a southern population from Brazil; they differ in preanal and total myomere counts. We have no adult material referable to the southern population.

Size. The largest specimen examined was 790 mm TL. The 370-mm specimen (ANSP 115297) was a female with eggs of 0.5-mm diameter.

Distribution (Fig. 350). Rarely taken, known only from the Gulf of Mexico and Puerto Rico. It has been collected by trawl, try-net, fish trap, and "from the spewings of snappers," at depths between 90–200 m over sand and mud bottoms.

Etymology. From the Latin *irretitus* (entangled), probably referring to the appearance of the partly digested and broken holotype. An adjective.

Study Material. The holotype and four specimens, 310–790 mm TL.

Holotype: USNM 44303 (ca 790, partly digested and in two pieces); Gulf of Mexico, on the Snapper Banks at Pensacola.

Other material: GULF OF MEXICO: Off Florida: ANSP 115297 (1, 370) and CAS 52674 (1, 310); Key West, 8 km SE of Sand Key, caught with try-net digging into sand at 91–183 m; R. T. Abbott; March 1968. ANSP 135748 (1, ca 560, cleared and stained); 26°26′N, 83°48′W, 90 m; 19 Dec. 1975. PUERTO RICO: CAS 54472 (1, 391); 19 km W of Mayaguez, 18°12′N, 67°22′W, fish trap in 100 m; D. W. Arenson; 6 July 1977.

Gordiichthys leibyi McCosker and Böhlke, 1984

STRING EEL

Figures 350–352, Tables 33, 36

Gordiichthys species A. Böhlke, 1977b: 5.

Gordiichthys leibyi McCosker and J. E. Böhlke, 1984: 36, figs. 7–8 (original description, Florida, 30°05′N, 80°41.5′W, holotype ANSP 110417).

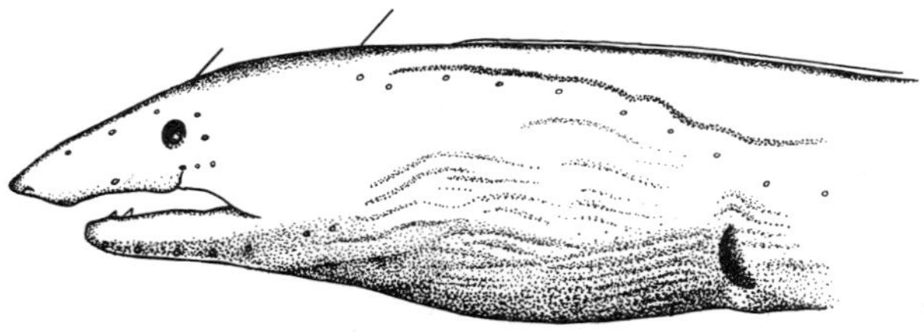

FIGURE 351. *Gordiichthys leibyi*: ANSP 110417, holotype, 336 mm TL, Florida Atlantic; head.

Distinctive Characters. A very elongate bascanichthyin with uniform pallid coloration; depth behind gill openings 1–1.2% TL; head 5–5.9% TL; four mandibular pores; vomerine dentition uniserial; MVF 2-95-171.

Description. Total vertebrae 168–176, x̄ = 171.2 (n = 5); predorsal vertebrae 2 (5); preanal vertebrae 92–98, x̄ = 95.2 (4). Pores: SO 1 + 3; IO 4 + 2; POM 4 + 2; ST 3.

Proportions as % TL; head 5–5.9; trunk 51–53; tail 41–44; depth behind gill openings 1–1.2. Of HL: snout 18–21; snout to rictus 28–32; eye 4–5.3; interorbital width 10–11.

General characters as for the genus. Mandibular pores four, except five on left side of holotype (the illustrated specimen). Teeth on jaws and on vomer uniserial; anterior vomerine teeth and those at mandibular symphysis the largest. Four intermaxillary teeth; 6–8 vomerine teeth; 8–10 maxillary teeth; 10–12 dentary teeth.

Color. Body coloration yellow to tan with minute brown punctations above the midline. Median fins transparent.

Size. The largest specimen examined was 424 mm TL. Two (216–274 mm) were females, the 274-mm specimen (GCRL 21392) with one row of 0.9-mm eggs on each side.

Development. Leptocephalus identified (see p. 869).

Distribution (Fig. 350). This rare eel has been collected by trawl and by dredge between 37–72 m depth, from the Atlantic coast of Florida and the Gulf of Mexico; it apparently lives in sand and mud.

Etymology. Named in honor of Mark M. Leiby, "friend and apodal ichthyologist" (McCosker and J. E. Böhlke, 1984: 37). A noun in the genitive case.

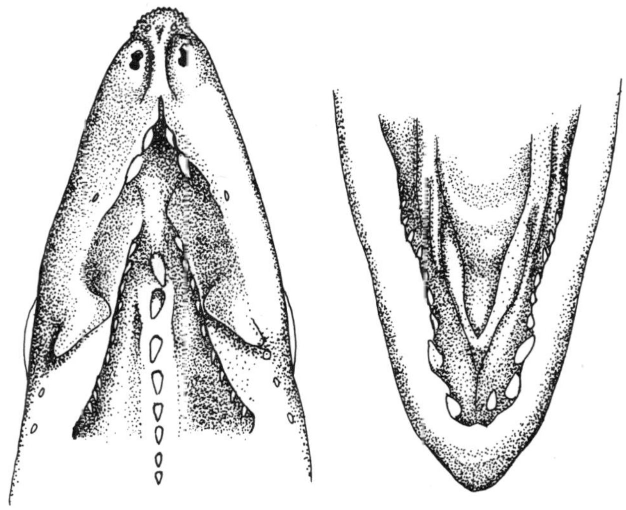

FIGURE 352. *Gordiichthys leibyi*: ANSP 110417, holotype, 336 mm TL, Florida Atlantic; dentition.

Study Material. The holotype and five paratypes, 89–424 mm TL.

Holotype: ANSP 110417 (336); Florida Atlantic, 30°05′N, 80°41.5′W, depth 36 m, 6-ft tumbler dredge; SILVER BAY sta. 4843; 4 April 1963.

Paratypes: FLORIDA ATLANTIC: USNM 263571 (1, 424); 28°50′N, 80°10′W, 49 m; DELAWARE II sta. 88; 25 June 1982. GULF OF MEXICO: *Florida.* Off St. Petersburg: ANSP 151694 (1, 216); 27°37′N, 84°13′W, 72 m; HOURGLASS MM 67-1054; 4 Apr. 1967. ANSP 151695 (1, 89, cleared and stained); 27°37.5′N, 83°59.3′W, 54.8 m; IC-76-56. GCRL 21392 (1, 274); 28°14′N, 84°17.5′W, 37.5 m; BELLOWS BLM Cruise 4; 18 June 1974. Off Panama City: CAS 52671 (1, 337); 29°48′N, 86°03.5′W, 40 m; 4 June 1974.

Gordiichthys randalli
McCosker and Böhlke, 1984
ROPE EEL
Figures 350, 353–355, Tables 33, 36

Gordiichthys species B. Böhlke, 1977b: 5.
Gordiichthys randalli McCosker and J. E. Böhlke, 1984: 38, figs. 9–11 (original description, Puerto Rico, holotype ANSP 151690).

Distinctive Characters. An elongate bicolored bascanichthyin, darker dorsally; depth behind gill openings 1.3–1.6% TL; head 5.6–6.4% TL; vomerine dentition biserial anteriorly; four mandibular pores; MVF 2-91-181.

Description. Total vertebrae 176–186, x̄ 181.1 (n = 8); predorsal vertebrae 1–2, x̄ = 1.9 (8);

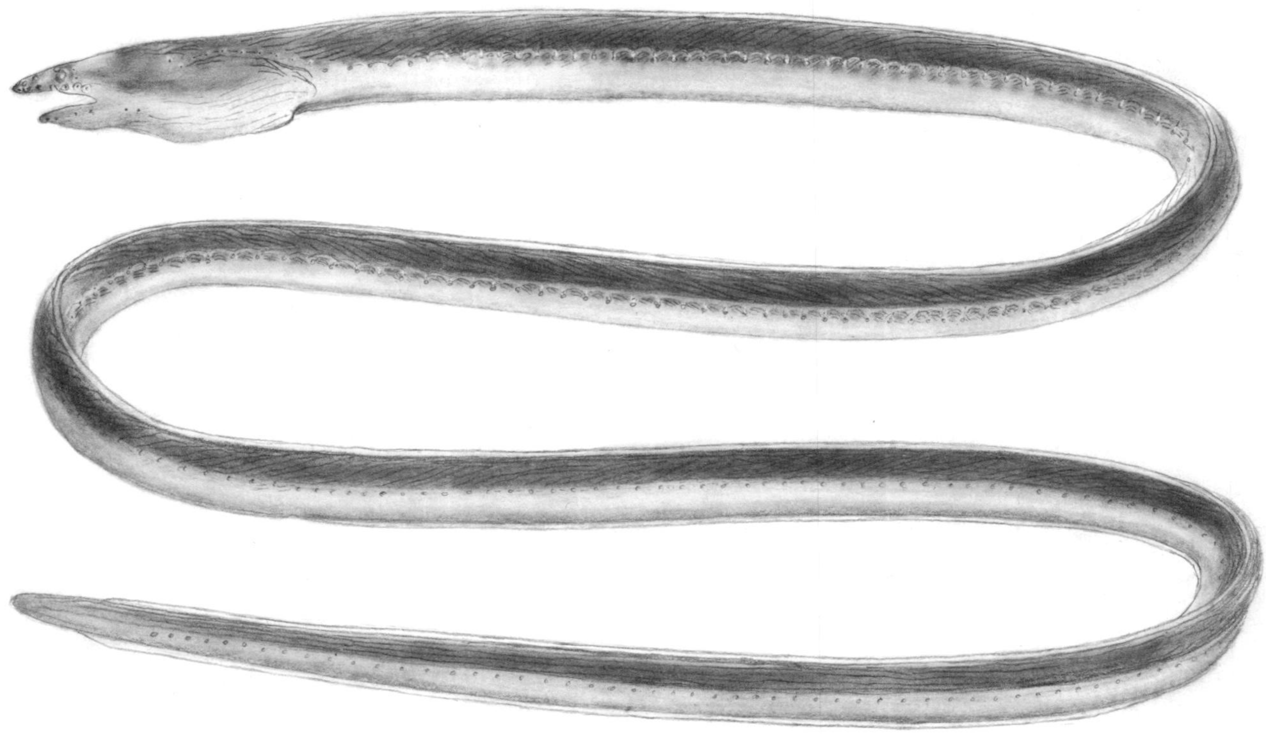

FIGURE 353. *Gordiichthys randalli*: ANSP 151690, holotype, 537 mm TL, Puerto Rico.

preanal vertebrae 89–95, x̄ = 91.3 (8). Pores: SO 1 + 3; IO 4 + 2; POM 4 + 2; ST 3.

Proportions as % TL: head 5.6–6.4; trunk 46–48; tail 45–48; depth behind gill openings 1.3–1.6. Of HL: snout 13–15; snout to rictus 26–32; eye 3.2–3.7; interorbital width 6.4–9.2.

General characters as for the genus. Typically four mandibular pores, one specimen with five pores on one side; one specimen had an extra supraorbital pore on one side. Teeth in jaws uniserial, vomerine teeth partially biserial; anterior vomerine teeth and those at the mandibular symphysis the largest. Three intermaxillary teeth followed by a gap, then two to three pairs of irregularly biserial teeth followed by a uniserial row of three to four teeth on vomer. Seven to eight teeth on each maxillary; 10–12 teeth on mandible, the second to fourth larger and stouter than others.

Color. Body uniformly brown above the midline and yellowish ventrally; lateral line canal visible as a pale irregular stripe. Snout and jaw tip brown, with paired white patches above and behind eyes and laterally before eye; pores pale. Median fins colorless. In life, head ridges brown

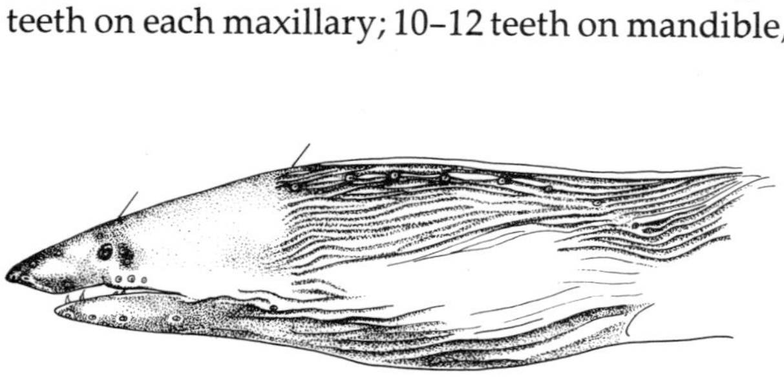

FIGURE 354. *Gordiichthys randalli*: ANSP 151690, holotype, 537 mm TL, Puerto Rico; head.

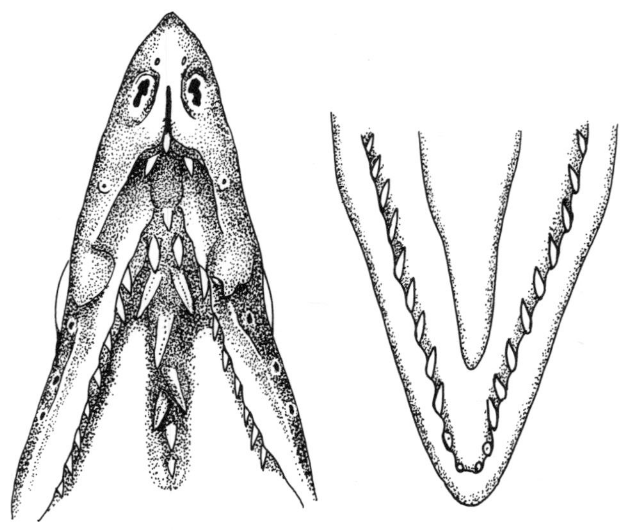

FIGURE 355. *Gordiichthys randalli*: ANSP 151690, holotype, 537 mm TL, Puerto Rico; dentition.

and grooves noticeably yellow; head patches are obvious, and underside of body is yellowish-white, except for abdomen, which is pinkish.

Size. The largest specimen examined was 669 mm TL. Of five females (537–669 mm), two (537–598 mm) were ripe with 0.8–0.9-mm eggs. One 662-mm specimen was a ripe male.

Development. The leptocephalus described as *Callechelys* sp. cf. *springeri* by Leiby (1984b: 412), is tentatively identified as this species (p. 871).

Distribution (Fig. 350). All known specimens were captured in Puerto Rico using ichthyocides over sand and algal beds of *Halophyla* between 7.5–12 m depth.

Etymology. Named in honor of John E. Randall, fellow ichthyologist and collector of the type specimens. A noun in the genitive case.

Remarks. This species, the most robust and colorful of *Gordiichthys,* differs markedly in general appearance from its slender congeners. Upon cursory examination, it might be mistaken for a species of *Bascanichthys* (which possess a small pectoral fin and a blunter snout) or *Caralophia* (which differs markedly in its nostril condition). The difference in appearance of *G. randalli* from its congeners is probably related to its different habitus and life style; the other species are from deeper water and presumably more uniform bottom conditions.

Study Material. The holotype and seven paratypes, 537–669 mm TL.

Holotype: ANSP 151690 (537 mm TL); Puerto Rico, Aguadilla, Crashboat Basin, depth 7.5 m over sand and *Halophyla;* J. E. Randall; 12 June 1965.

Paratypes: PUERTO RICO: Aguadilla, Crashboat Basin: ANSP 138370 (1, 555, cleared and stained), ANSP 151691 (1, 614), ANSP 151692 (1, 669), CAS 52673 (6, 614), and USNM 263572 (1, 598); ca 30 m off NW corner of main pier, 10.5 m; J. E. Randall; 11 Nov. 1965. ANSP 151689 (1, 662); ca 40–45 m NE of easternmost outer docks, 12 m; J. E. Randall, G. Cooke; 24 July 1965. BPBM 29361 (1, 540); ca 30–45 m NE of easternmost outer dock, 12 m; J. E. Randall, G. Cooke; 24 July 1965.

Genus *Phaenomonas* Myers and Wade, 1941

Phaenomonas Myers and Wade, 1941: 77 (type species *Phaenomonas pinnata* Myers and Wade, 1941, by original designation).

Characters. Body extremely elongate (depth 0.7–1.0% TL), cylindrical; head and trunk much longer than body (67–77% TL). The only fin a short low dorsal, arising about mid-head and ending in anterior trunk. Snout blunt, overhanging lower jaw, a short groove on its underside; lips without barbels; eye minute; anterior nostril tubular; posterior nostril opens into mouth, its location not marked externally. Gill openings low lateral, oblique and elongate, about equal to isthmus. Head pores reduced; two preopercular pores, three pores in supratemporal canal. Teeth stout, conical and uniserial, largest at intermaxillary, separated from those of vomer by a gap.

Neurocranium sloping posteriorly, orbital foramen a narrow slit; supraoccipital crest weakly developed; maxillae taper posteriorly; opercular series reduced, all but opercle serrate at margin; suspensorium anteriorly inclined; otic bulla well developed; pterygoid free and tapering anteriorly. Hypohyal separated from ceratohyal by a gap; branchiostegal rays numerous, slender and along hyoid; urohyal club-shaped, cartilaginous posteriorly. Gill arches typically bascanichthyin; third basibranchial absent; fifth ceratobranchial absent; upper pharyngeal tooth plates separate. Pectoral girdle reduced, only a cleithrum, supracleithrum, and a thin coracoid. Intramuscular bones, ribs, and caudal transverse processes weakly developed; precaudal vertebrae more numerous than caudal. Coloration uniform.

Size. To about 590 mm TL.

Distribution. Circumtropical, known from one eastern Pacific species, one wide-ranging Indo-Pacific species, and one western Atlantic species (McCosker, 1975; McCosker and J. E. Böhlke, 1984).

Etymology. From the Greek *phaneros* (visible) and *monas* (single), apparently in reference to the single dorsal fin. Feminine.

Species. The three species, *Phaenomonas cooperae* Palmer, 1970, *P. longissima* (Cadenat and Marchal, 1963) and *P. pinnata* Myers and Wade, 1941, are found in different oceans, and are separable by vertebral counts.

KEY TO THE SPECIES OF *PHAENOMONAS*

1a. Vertebrae 243–270; body depth less than 1% TL .... *cooperae* Palmer, 1970 (widespread in Indo-Pacific)
1b. Vertebrae less than 240 2
2a. Vertebrae 175–194; body depth 1.2–1.7% TL .... *pinnata* Myers and Wade, 1941 (widespread in eastern Pacific)
2b. Vertebrae 206–216; body depth 0.74–1.05% TL .... *longissima*, p. 352

Phaenomonas longissima (Cadenat and Marchal, 1963)

SHORT-MANED SAND EEL

Figures 356–358, Table 33

Callechelys longissimus Cadenat and Marchal, 1963: 1243, fig. 6 (original description, St. Helena, lectotype MNHN 1963-209).

Bascanichthys longissimus. Blache and Cadenat, 1971: 174 (referred to *Bascanichthys* and designation of lectotype).

Phaenomonas longissima. McCosker and J. E. Böhlke, 1984: 33 (referred to *Phaenomonas*).

Distinctive Characters. An extremely elongate, pale bascanichthyin lacking all fins but a short section of the dorsal fin above head and anterior trunk; tail 23–33% TL; uniserial dentition; MVF 2-[138]-212.

Description. Total vertebrae 206–216, $\bar{x}$ = 211.8 (n = 5); predorsal vertebrae 2 (4); vertebrae before anus 135–140, $\bar{x}$ = 137.5 (2). Pores: SO 1 + 3; IO 3 + 2; POM 4 + 2; ST 3.

Proportions as % TL: head 3.8–4.3; trunk 63–73; tail 23–33; predorsal 22–42; depth behind gill openings 0.7–1.0. Of HL: snout 19–20; snout to rictus 25–30; eye 2.6–3.2; interorbital width 12–13.

Snout blunt, lower jaw included; eye small and embedded, above posterior quarter of upper jaw. Groove on underside of snout, bordered by two folds of skin which appears as two backward-directed papillae between anterior nostrils. Anterior nostril large and tubular, within a groove and with interior lateral projections; posterior nostrils opening within mouth, lateral to maxillary teeth and covered by a flap, the opening beneath eye and not marked externally. Anterior edge of gill opening bordered with short striations; a flap of skin forming a slight pouch posteriorly.

Teeth strong, stout, largest anteriorly, directed back. Anterior intermaxillary pair at median edge of papillae between anterior nostrils, followed by a second pair of slightly smaller teeth; a short gap, then 4–6 teeth on vomer, decreasing in size posteriorly. Jaw teeth uniserial; 7–9 teeth on maxillary, 7–12 along mandible.

Dorsal-fin origin on head, the fin extending back about three head-lengths behind gill openings; last fin-rays visible on radiographs above vertebrae 29–31 (3), the last pterygiophores above vertebrae 34–40. Additional dorsal and anal pterygiophores visible far posteriorly on tail, above and below the last 9–12 vertebrae (excluding the hypural).

Color. Uniform tan, slightly darker dorsally and on sides. Life color for the specimen from Ascension Island (ANSP 149899) was recorded as "rose-colored, somewhat darker anteriorly" (W. D. Anderson, in litt.).

Size. The largest specimen examined was 553 mm TL, a ripe female filled with 0.9-mm eggs.

Development. Leptocephalus identified (see p. 873).

Distribution. Rarely caught, known from St. Helena and Ascension islands and from Brazil. Collected by dredge to depths of 60 m and by ichthyocide in tidepool and on beach.

Remarks. The two Indo-Pacific species of this bizarre genus were reviewed by McCosker (1975). The Atlantic species was ignored because earlier descriptions suggested it possessed typical dorsal and anal fins. After study of additional specimens and reexamination of the lectotype, McCosker and J. E. Böhlke (1984: 33) referred *longissimus* to *Phaenomonas*.

Etymology. From the Latin *longissimus* (longest), in reference to its extreme length. An adjective.

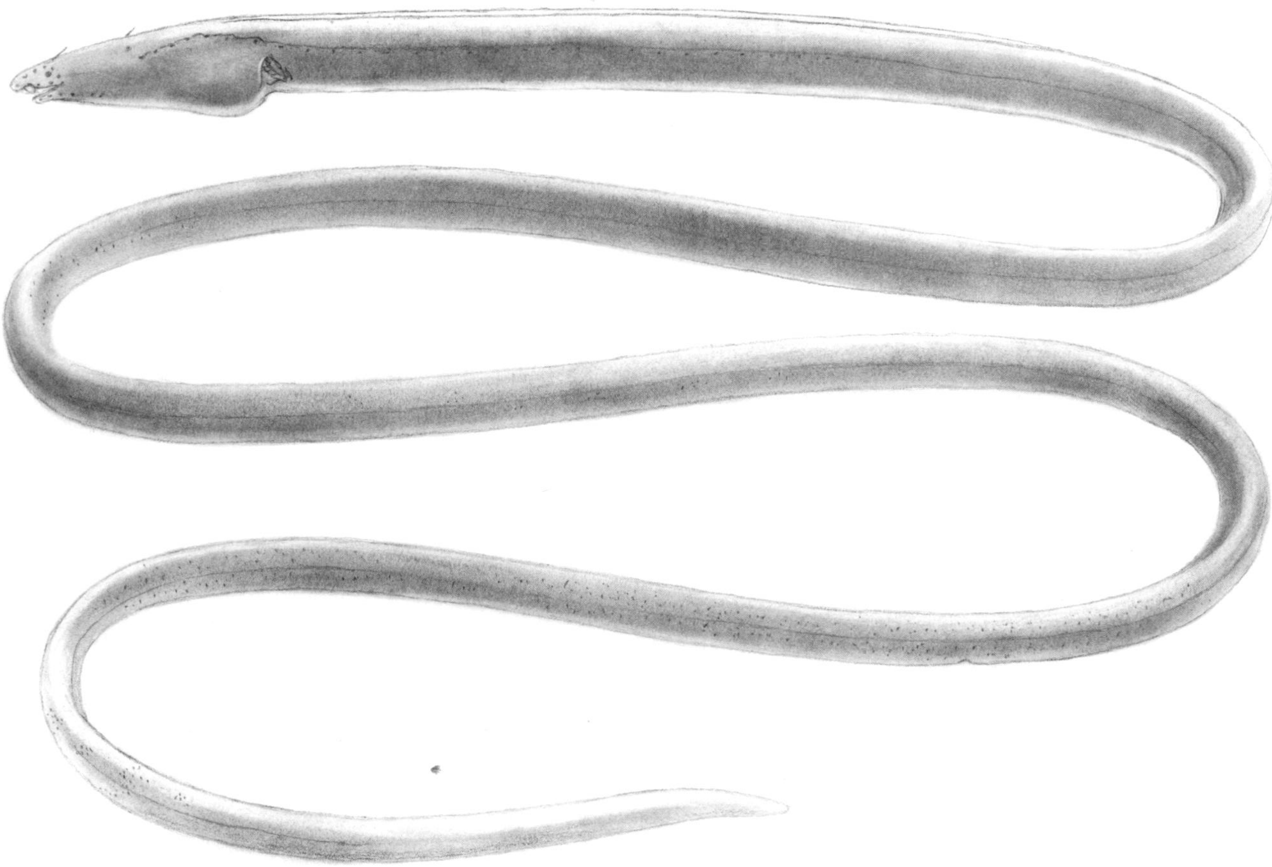

FIGURE 356. *Phaenomonas longissima*: ANSP 149899, 444 mm TL, Ascension Island.

Study Material. Six specimens, 332–553 mm TL, including the lectotype.

Lectotype: MNHN 1963-209 (498); St. Helena, Lemon Valley, by dredge at 35 m; La REINE-POKOU; 25 Feb. 1962.

Other material: BRAZIL: MZUSP 12099 (1, 347); Bahia, beach at Arembepe; N. A. Menzes, C. E. Dawson; 26 Aug. 1972. ASCENSION ISLAND: ANSP 149899 (1, 444); NE edge of English Bay, in tide pools, to 1 m; W. D. Anderson et al.; 30 July 1980. ST. HELENA: ZMUC P.32862 (1, 332); Flagstaff Bay, dredge at 60 m; Th. Mortensen, sta. 83; 24 Feb. 1930. ZMUC P.32863 (1, 485) and ZMUC P.32864 (1, 553); off Egg Island, dredge; Th. Mortensen; 5–6 Mar. 1930.

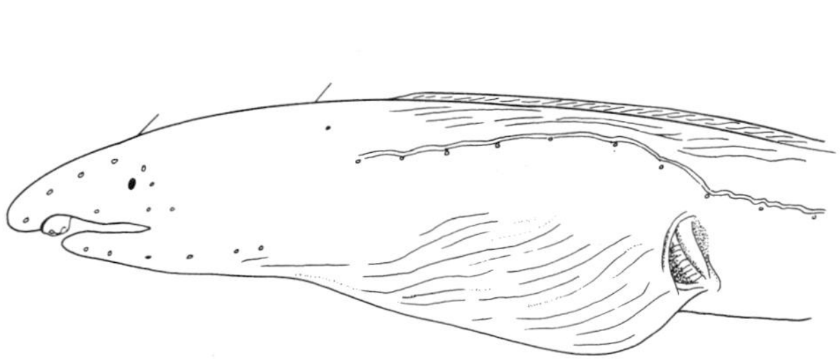

FIGURE 357. *Phaenomonas longissima*: ANSP 149899, 444 mm TL, Ascension Island; head.

TRIBE OPHICHTHINI

The Ophichthini include moderate-sized to the largest species of ophichthids; they are generally cylindrical and robust, and often have distinctive

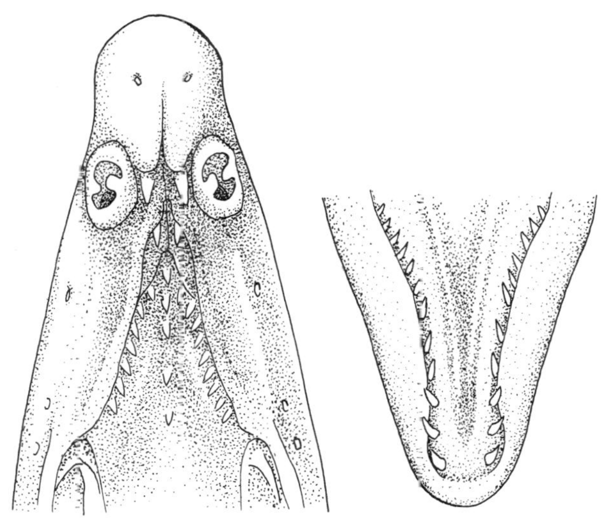

FIGURE 358. *Phaenomonas longissima*: ANSP 149899, 444 mm TL, Ascension Island; dentition.

TABLE 37. Mean vertebral formulae (MVF) and ranges of vertebral counts in western Atlantic species of Ophichthini.

		Predorsal		Preanal		Total	
Species	MVF	Range	n	Range	n	Range	n
*Aplatophis chauliodus**	23-54-112	20–26	9	53–56	9	110–115	10
*Echiophis intertinctus**	16-56-129	14–18	44	53–58	54	125–130	54
punctifer	18-58-136	16–19	36	54–61	39	136–143	44
*Herpetoichthys regius**	15-77-160	15–16	3	76–78	4	157–161	5
*Hyphalophis devius**	23-52-144	23	1	52	1	144	1
*Kertomichthys blastorhinos**	22-61-143	22	1	61	1	143	1
*Lethogoleos andersoni**	17-58-132	17	1	58	1	132	1
*Myrichthys breviceps**	3-63-170	2–3	15	61–65	15	165–175	25
ocellatus	2-59-168	2–3	10	57–61	10	164–173	24
*Ophichthus brasiliensis**	23-57-153	23	1	57	1	153	1
*cylindroideus**	14-43-145	11–15	36	39–45	37	140–151	40
*cruentifer**	16-58-149	14–19	33	56–61	31	144–155	48
*gomesii**	14-47-145	12–18	100	44–52	111	138–153	142
*hyposagmatus**	12-54-136	12–13	6	52–55	6	134–139	6
*melanoporus**	12-52-183	11–12	10	50–54	10	177–186	10
*menezesi**	19-56-153	16–23	14	55–58	15	150–156	15
*omorgmus**	14-57-141	14	2	56–57	2	141+	1
ophis	14-61-164	13–17	11	57–64	8	161–167	12
puncticeps	14-53-134	12–17	38	50–58	39	127–141	77
*rex**	16-46-118	15–18	19	45–49	25	115–121	25
*spinicauda**	20-61-145	18–21	7	61–62	7	144–146	7
Quassiremus ascensionis	9-67-134	9–11	15	65–69	14	129–136	15

* Includes holotype.

coloration of spots, bands or saddles. Some ophichthins are found in shallow coral reef areas but most are inhabitants of coastal waters at moderate to fairly great depths, and are obtained by hook and line, trawls, and in traps. The tribe may be diagnosed as follows: moderately elongate to stout, cylindrical anteriorly, generally compressed posteriorly; head and trunk shorter than tail or subequal; snout conical or subconical; jaws subequal; anterior nostril generally tubular; posterior nostril opens into mouth in most species; gill openings lateral, crescentic, their major axis vertical; median and pectoral fins present; caudal fin usually absent, tail tip sharply pointed or blunt; teeth variable, often multiserial, including caniniform and molariform conditions; neurocranium moderately elongate; orbit well developed; gill arches well developed; ceratobranchial 5 present in many species; intramuscular bones, ribs, and caudal transverse processes well developed; and coloration variable.

The Ophichthini are the most speciose of the ophichthids; the tribe contains 21 genera, nine of which are found in the western Atlantic (*): *Aplatophis**, *Brachysomophis*, *Cirrhimuraena*, *Echelus*, *Echiophis**, *Evips*, *Herpetoichthys** (including *Pogonophis*), *Hyphalophis**, *Kertomichthys**, *Leiuranus* (including *Elapsopis*), *Lethogoleos**, *Malvoliophis*, *Myrichthys**, *Mystriophis*, *Ophichthus**, *Ophisurus*, *Phyllophichthus*, *Pisodonophis*, *Quassiremus**, *Scytalichthys* and *Xyrias*.

Genus *Aplatophis* Böhlke, 1956

Aplatophis Böhlke, 1956b: 1 (type species *Aplatophis chauliodus* Böhlke, 1956, by original designation).

Characters. Body stout (depth 3.6–6.4% TL), not elongate, laterally compressed posteriorly; head and trunk slightly longer than tail (51–54% TL). Dorsal-fin origin behind pectoral tips; tail tip blunt; pectoral fins small, lanceolate and narrow-based. Snout short; jaws elongate, the lower jaw

FIGURE 359. *Aplatophis chauliodus*: ANSP 116366, 439 mm TL, Gulf of Mexico off Texas.

projecting beyond upper; one short barbel below eye; eye small; anterior nostril tubular and just before posterior nostril which is in a short broad tube before and beneath eye. Tongue developed and enlarged as a fleshy lure. Gill opening vertical, lateral and elongate, less than isthmus. Head pores reduced, minute and difficult to observe; median pores absent although frontal commissure and supratemporal canal are present; two preopercular pores; lateral-line pores inconspicuous. Anterior teeth of both jaws excessively developed as long fang-like canines extending far outside mouth when closed; jaw teeth biserial; vomerine teeth uniserial, few, and continuous with intermaxillary teeth.

Neurocranium subtruncate posteriorly; supraoccipital crest absent; supraoccipital anteriorly contacts frontals and separates parietals; frontal midline elevated anteriorly forming a sharp ridge extending to supraoccipital; maxilla elongate, rounded posteriorly; pterygoid broad posteriorly, free and terete anteriorly; orbit large, postorbital ossicles weak, not fused as a strut to maxilla; nasals extremely reduced; nasal cartilage rudimentary or absent; otic bulla well developed; suspensorium posteriorly inclined; opercular series developed; suborbital scythe-shaped; preorbital enlarged. Branchiostegal rays few, slender, those of epihyal joined basally; urohyal a spike posteriorly. Gill arches well developed, *Ophichthus*-like; fifth ceratobranchial ossified; upper pharyngeal tooth plates separate. Pectoral girdle well developed; cleithrum and supracleithrum stout, coracoid and scapula large, actinosts absent. Intramuscular bones, ribs, and caudal transverse processes well developed. Precaudal about equal to caudal vertebrae. Gasbladder stout, muscular, about 9% TL. Body coloration pale and freckled; peritoneum pale.

Size. To 815 mm TL.

Distribution. A single western Atlantic species.

Etymology. From the Greek *aplatos* (unapproachable) and *ophis* (snake), referring to its threatening appearance. Masculine.

Species. One species, *Aplatophis chauliodus* Böhlke, 1956.

Aplatophis chauliodus Böhlke, 1956

TUSKY EEL

Figures 359–363, Table 37

Aplatophis chauliodus Böhlke, 1956b: 3, fig. 1 (original description, Mayaguez Pier, Puerto Rico, holotype FMNH 61772).

Distinctive Characters. A robust ophichthin with tan coloration with small spots and freckles on head and dorsally on body; snout short, jaws elongate, anterior teeth excessively developed into fangs projecting far outside of closed mouth;

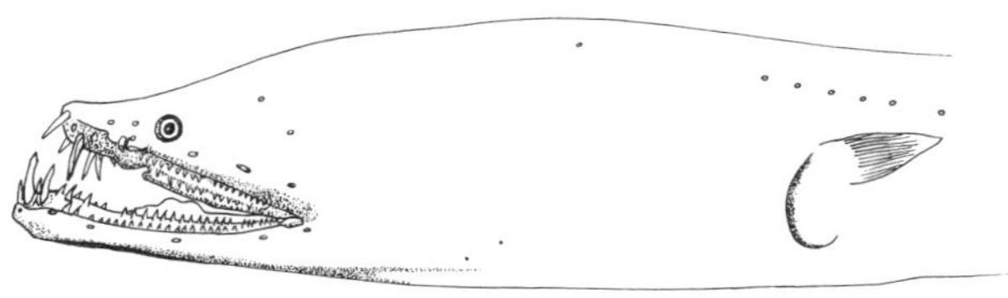

FIGURE 360. *Aplatophis chauliodus*: ANSP 116366, 439 mm TL, Gulf of Mexico off Texas; head.

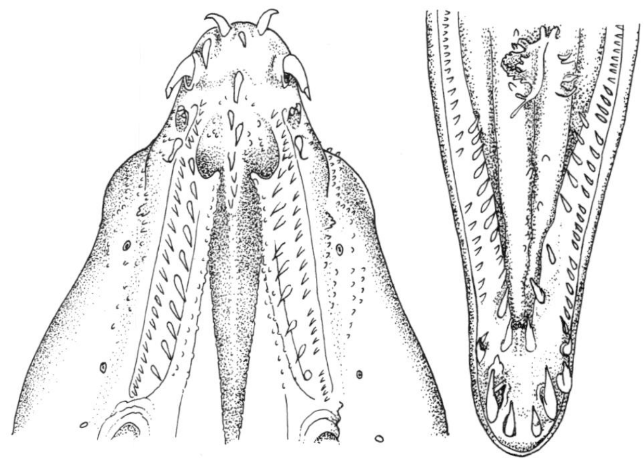

FIGURE 361. *Aplatophis chauliodus*: ANSP 151697, 190 mm TL, Gulf of Mexico off Texas; dentition.

dorsal-fin origin behind pectoral tips; MVF 23-54-112.

Description. Total vertebrae 110–115, $\bar{x}$ = 111.9 (n = 10); predorsal vertebrae 20–26, $\bar{x}$ = 23.2 (9); preanal vertebrae 53–56, $\bar{x}$ = 54.1 (9); kidney terminus 57–58 (2). Pores: SO 1 + 2; IO 4 + 2; POM 6 + 2.

Proportions as % of TL: head 14–16; trunk 34–39; tail 46–49; predorsal 25–28; depth behind gill openings 3.6–6.4. Of HL: snout 13–17; snout to rictus 32–40; eye 4–6.3; interorbital width 13–18; pectoral-fin length 13–17.

Dorsal profile of head concave; snout short, lower jaw projecting beyond upper; eye small, located just before mid-gape and close to or bordering upper lip. Anterior nostril a small narrow tube just in front of and below eye; posterior nostril a short broad tube opening ventrally just behind anterior nostril. Cephalic pores small, reduced in number and difficult to see. Supraorbital pores consisting of ethmoid pore placed ventrally on snout and two pores situated low-lateral on snout and before eye. Four infraorbital pores along upper jaw and two postorbital pores behind eye as illustrated. Six mandibular pores near edge of lower jaw; two minute preopercular pores. Median frontal pore and pores in supratemporal commissure apparently absent. Single papilla on upper lip projecting down below eye; small papillae on top and sides of head.

Dentition highly developed and prominent. An outer ring of large fang-like intermaxillary teeth which project outside mouth, four or five extremely elongate teeth alternating with shorter ones, plus two median teeth (three in smallest specimen). Two to seven small conical teeth on vomer, the series continuous with last median intermaxillary tooth. Maxillary teeth in two widely separated rows on a fleshy ridge which has a conspicuous anterior projection; the outer row of 17–28 shorter, conical teeth, the inner row of 11–13 widely-spaced, long, depressible teeth. Three or four long fangs anteriorly on each mandible, followed by two rows of smaller teeth; the outer row of 19–38 small fixed teeth, the inner row of 9–20 larger depressible ones. Number of teeth increasing with growth.

Tongue conspicuous and complex, located posteriorly in mouth (Fig. 362), a highly structured fleshy appendage, supported by cartilage. It is decorated with papillae and projections, including one or more long forward-directed "lures"; the entire structure possibly capable of extending forward and out of the mouth.

Color. Uniform tan overlain with brown speckling; fine spots on head, larger freckles on body; belly pale. Dorsal and pectoral fins pale or with scattered spots; distal edge of dorsal and anal fins sometimes dark.

Size. The largest specimen seen was 804 mm TL; Cervigón (1966: 194) reported an 815-mm specimen. Three of the study specimens (190–439 mm) were females, the 439-mm one filled with 0.6-mm eggs; three (173–466) were males, the 334-mm specimen ripe.

Development. Leptocephalus identified (p. 834).

Distribution (Fig. 363). Not common, known from the northern Gulf of Mexico, Puerto Rico, Panama and northern South America. Captured by trawl or dredge over mud bottom at depths recorded from 33–91 m; once taken by "beach seine."

Etymology. From the Greek *chauliodus* (tusky),

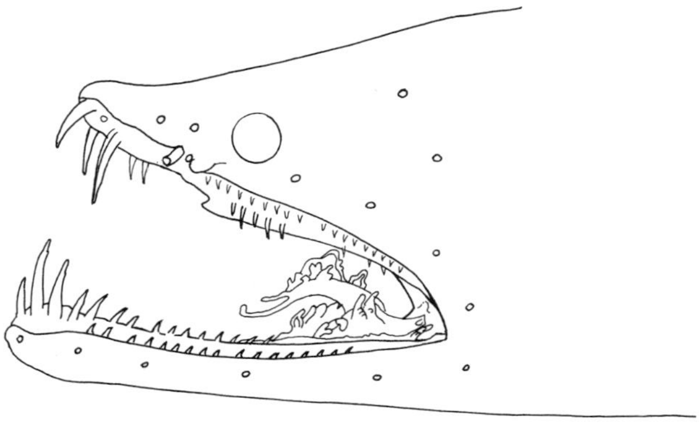

FIGURE 362. *Aplatophis chauliodus*: ANSP 151697, 190 mm TL, Gulf of Mexico off Texas; tongue.

"for obvious reasons" (Böhlke, 1956: 8). An adjective.

Remarks. The unique structure of the tongue has not previously been reported. We hypothesize that *Aplatophis chauliodus* might lie within a burrow with mouth open and tongue extended to entice incautious prey into its tusky maw. An analogous, even more elongate and extensible structure exists within the subfamily Myrophinae in the monotypic Indo-Pacific *Glenoglossa wassi* (McCosker, 1982).

Study Material. Ten specimens, 168–804 mm TL, including the holotype and two paratypes.

Holotype: FMNH 61772 (466); Puerto Rico, brought up in a dredge at Mayaguez pier; V. Biaggi, Jr.; April 1934.

Paratypes: PUERTO RICO: ANSP 73871 (1, 225); Cano Corazones, opening into Bahia de Mayaguez just N of Punta Guanajibo; V. Biaggi, Jr.; May 1952. PANAMA ATLANTIC: LACM 21708 (1, 173); outside Caledonia Bay, 2 mi E of Isla de Oro, 08°54′43″N, 77°38′25″W, dredged in 51–53 m; VELERO III; 4 April 1939 [formerly AHF 2464].

Other material: GULF OF MEXICO: GCRL 15078 (1, 333.5); probably between Mobile Bay and SW Pass, Louisiana, to 91 m; July 1976. *Texas:* ANSP 116366 (1, 439); 26°18′N, 96°28′W, 68 m; OREGON 10455; 21 Mar. 1969. ANSP 151697 (1, 189.5); S of Galveston, 28°10′N, 94°38′W, 62 m; MBI 942; 23 Feb. 1978. PUERTO RICO: UPRM 3653 (1, 381+); Mayaguez, Mani Beach, beach seine; R. T. Ying; 19 July 1970. PANAMA: ANSP 156880 (1, 168, cleared and stained); Gulf of Darien, 08°57.5′–09°00.3′N, 76°33.6′–76°30.5′W, 55–64 m, 10-ft otter trawl; PILLSBURY 362; 12 July 1966. BRITISH GUIANA: BMNH 1961.8.31.100 (1, 647); 08°32′N, 59°15′W, 33–35 m, in mud, shells and weeds; CAPE ST. MARY sta. 935; 6 Jan. 1959. SURINAME: BMNH 1985.11.27.1 (1, 804); shrimp trawl; OH-B6A.

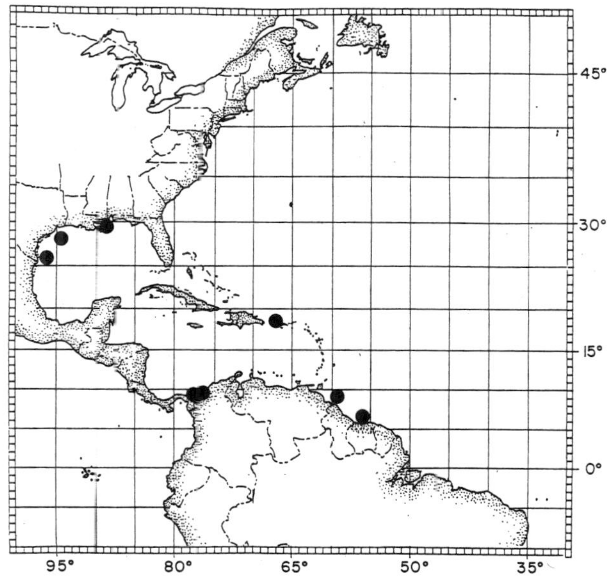

FIGURE 363. Distribution of *Aplatophis chauliodus* based on study material.

Genus *Echiophis* Kaup, 1856

Echiophis Kaup, 1856a: 46 (type species *Ophisurus intertinctus* Richardson, 1848, by monotypy).

Echiopsis Kaup, 1856b: 13 (emendation of *Echiophis* Kaup, 1856).

Crotalopsis Kaup, 1860: 12 (type species *Crotalopsis punctifer* Kaup, 1860, by monotypy). Also spelled *Crotalopis* or *Crotalophis* by other authors.

Macrodonophis Poey, 1867b: 251 (type species *Conger mordax* Poey, 1860, by monotypy).

Notophtophis Castro-Aguirre and Suarez de los Cobos, 1983: 113 (type species *Notophtophis brunneus* Castro-Aguirre and Suarez de los Cobos, 1983, by monotypy).

Characters. Body stout (depth 2.9–5.0% TL), cylindrical, moderately elongate; head and trunk shorter than tail (43–48% TL). Dorsal-fin origin behind pectoral tips; tail tip blunt; pectoral fin developed, rounded. Snout short, subconical; jaws elongate and subequal; lips without barbels; eye moderate, situated far forward on head; anterior nostril tubular; posterior nostril in a short tube before and beneath eye. Gill opening vertical, lateral and elongate, about equal to isthmus. Head pores developed; two or three preopercular pores. Teeth strong, pointed and biserial, largest anteriorly on intermaxillary and on jaws; vomerine teeth nearly continuous with intermaxillary teeth.

Neurocranium subtruncate posteriorly; supraoccipital crest absent; preorbital region reduced, orbit large; postorbitals fused to skull and maxilla, forming a stout strut; maxilla elongate, rounded posteriorly, toothed along most of its length; pterygoid broad posteriorly, free and terete anteriorly; otic bulla well developed; suspensorium posteriorly inclined; opercular series developed, their margin entire, supraorbital scythe-shaped. Branchiostegal rays not numerous, all slender, unbranched and along hyoid; urohyal a spike posteriorly. Gill arches well developed, *Ophichthus*-like; fifth ceratobranchial ossified; upper pharyngeal tooth plates separate. Pectoral girdle well developed; cleithrum and supracleithrum strong; coracoid and scapula large; actinosts present. Intramuscular bones, ribs, and caudal transverse processes well developed. Precaudal about equal to caudal vertebrae. Gasbladder not well developed, about 7% of TL. Peritoneum pale. Coloration in life and in preservative pale ventrally, darker dorsally, overlain with small to large dark spots.

Size. Grows to about 1200 mm in the western Atlantic.

Distribution. A largely New World genus, with one or two eastern Pacific species and two western Atlantic species, one of which is also in the eastern Atlantic.

Etymology. From the Greek *echis* (viper) and *ophis* (snake), in reference to its snake-like appearance. Masculine.

Remarks. Species in the closely-related genera *Mystriophis, Echiophis,* and *Kertomichthys* have been included under *Mystriophis* Kaup, 1856, in the past. *Echiophis* was resurrected by McCosker (1977) for the species treated below and *Kertomichthys* was created for the single species *blastorhinos* by McCosker and Böhlke (1984), leaving two eastern Atlantic species (*crosnieri* and *rostellatus*) in *Mystriophis.*

Juveniles of this genus have sometimes been misidentified due to ontogenetic changes in head profile and in dentition. Adults have broad heads with bulbous cheeks, with eyes much narrower than interorbital width; the head profile of juveniles is steep and narrow, the eyes equal to or larger than the interorbital width. Jaws of adults have an outer row of widely-spaced, large fang-like teeth and an inner row of much smaller uniformly sized teeth, but in small specimens the two rows are similar in size and shape. Vomerine teeth of adult *Echiophis* are typically biserial; in smallest specimens they are in one row. See additional comments under species descriptions.

Species. Five nominal Atlantic species have been described which belong in the genus *Echiophis*: *Ophisurus intertinctus* Richardson, 1848; *Crotalopsis punctifer* Kaup, 1860; *Conger mordax* Poey, 1860; and *Ophichthys Schneideri* Steindachner, 1880 in the western Atlantic; and *Mystriophis creutzbergi* Cadenat, 1956 in the eastern Atlantic. Previous authors have considered the western Atlantic forms to be either a single species, *intertinctus* (Jordan and Davis, 1891), or three species, *intertinctus, mordax,* and *punctifer,* which are separable only by coloration (Ginsburg, 1951). (*Ophichthys Schneideri* was not previously identified with the genus.) The eastern Atlantic form was synonymized with the single species *intertinctus* by Blache (1971b). We recognize two species or species groups, *intertinctus* and *punctifer* (including *mordax, Schneideri* and *creutzbergi* as synonyms), which are separable by preopercular pore condition, vertebral numbers, and coloration. There is some overlap in each of these characters, and, in addition, overlapping differences exist in vertebral numbers and coloration between the nominal species *mordax* and *punctifer* and between northern and southern populations of *E. intertinctus.* Further discussion is presented under each species heading. A comprehensive analysis of characters based on adequate samplings from all localities (we lack material from localities other than the northwestern Atlantic) could alter the conclusions presented here.

We note that the similarities between species and the problems in determining species, as well as the distribution of the *Echiophis* species complex, closely parallel those of the *Gymnothorax ocellatus* complex in the family Muraenidae.

Key to the Atlantic Species of *Echiophis*

1a. Color pattern of large brown spots on pale background; two inconspicuous preopercular pores; dorsal-fin origin behind pectoral tips by length less than pectoral-fin length .... *intertinctus*, p. 359
1b. Small brown spots on tan background; three preopercular pores set in dark spots; dorsal-fin origin behind pectoral tips by length greater than pectoral-fin length .... *punctifer*, p. 362

Echiophis intertinctus (Richardson, 1848)

SPOTTED SPOON-NOSE EEL

Figures 364–367, Tables 37, 38

Ophisurus intertinctus Richardson, 1848a: 102 (original description, West Indies, holotype BMNH 1982.5.7.4).
Echiophis intertinctus. Kaup, 1856a: 46 (type of new genus).
Echiopsis intertinctus. Kaup, 1856b: 13 (emendation of *Echiophis* Kaup, 1856).
Ophichthys intertinctus. Günther, 1870: 57.
Mystriophis intertinctus. Jordan and Davis, 1891: 635.
Leptocephalus caudomaculatus Eigenmann and Kennedy, 1902: 87, fig. 5 (original description, 34°57′00″N, 75°43′30″W, syntypes USNM 49758–49759). Leptocephalus.

Distinctive Characters. A stout ophichthin with large brown spots along the pale body; dorsal-fin origin behind pectoral tips by length less than pectoral-fin length; pectoral-fin length 20–28% HL; two inconspicuous preopercular pores; MVF 16-56-129.

Description. Total vertebrae 125–138, $\bar{x}$ = 129.4 (n = 54); predorsal vertebrae 14–18, $\bar{x}$ = 15.6 (44); preanal vertebrae 53–58, $\bar{x}$ = 55.9 (54). Pores: SO 1 + 3; IO 4 + 2; POM 6 + 2; ST 3.

Proportions as % TL: head 11–12; trunk 32–37; tail 52–57; predorsal 13–17; depth behind gill openings 2.9–4.4. Of HL: snout 11–14; snout to rictus 40–48; eye 5.2–9.2; interorbital width 4.4–13 (see generic remarks); pectoral-fin length 20–28.

Snout short, jaws elongate and subequal, or lower jaw slightly longer (in large specimens); eye moderate, located far forward over anterior third of jaw, slightly dorsad. Anterior nostril in a small tube about midway between tip of snout and anterior margin of eye; posterior nostril a short distance behind anterior nostril, just before and below eye, covered by a short tubular flap which projects below lip line. Cephalic pores small, some pores difficult to see; supraorbital pores consisting of ethmoid pore on tip of snout, two pores on snout before eye, and the last pore above and behind posterior margin of eye; first infraorbital pore between anterior and posterior nostrils pale, three pores below eye and two well behind eye in small dark spots; mandibular pores in dark spots, the distance between pores increasing from snout tip to rictus; two preopercular pores below and behind rictus minute and difficult to find, slightly pigmented only on largest specimens. One specimen with an extra supraorbital pore over mid-eye on left side.

Considerable variation in dentition, which is basically biserial, with some large stout fang-like teeth anteriorly. An outer semi-circle of five intermaxillary teeth flanking two very stout median teeth on a fleshy ridge, these the largest teeth in the mouth. A short gap followed by a chevron of 20–43 vomerine teeth which are biserial, widely separated anteriorly, the last 1–3 teeth sometimes uniserial. Maxillary dentition uniserial anteriorly (4–5 teeth increasing in size posteriorly, becoming very large and fang-like), biserial for most of the remainder; an outer row of 14–25 teeth (large stout spaced fangs with or without smaller teeth between followed by a posterior series of uniform small close-set teeth) and a shorter inner row of 8–14 uniform small sharp teeth partially embedded in fleshy ridge. Mandibular teeth similar to those on maxillary; anteriorly 4–6 increasingly larger teeth, followed by an outer row of 14–22, the last 9–12 small and uniform in size, and a shorter inner row of 14–26 small teeth embedded in fleshy ridge and sometimes hidden by tongue.

The above description is of dentition of moderate to large specimens (greater than 300 mm). In smaller specimens all teeth are moderately large and sharp and the rows appear fairly uniform in size; the anterior 4–6 graduated teeth

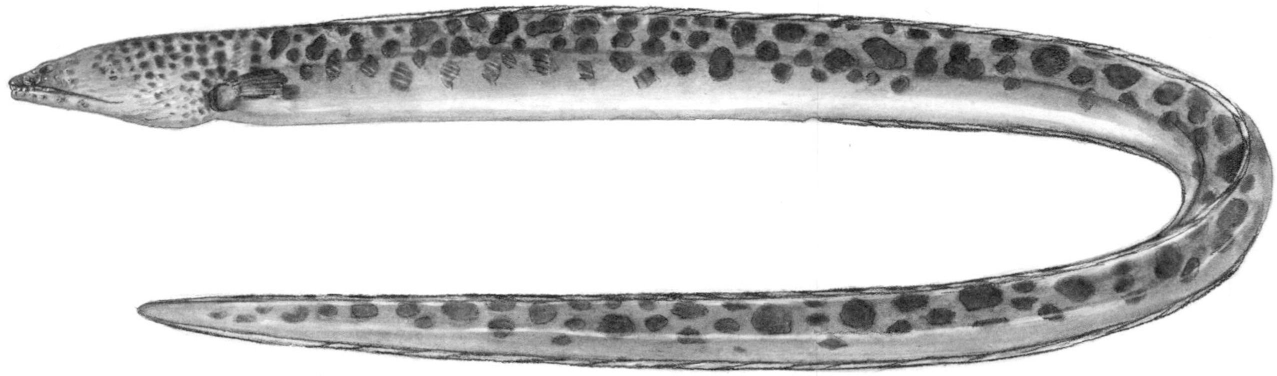

FIGURE 364. *Echiophis intertinctus*: ANSP 116134, 737 mm TL, Florida Atlantic.

(uniserial in large specimens) are part of the inner row, and are flanked by smaller outer teeth which are not found in larger specimens. With growth, some of the anterior teeth become large and fang-like, and the inner teeth appear very small because they become embedded in fleshy folds.

Dorsal-fin origin behind pectoral-fin tips by a length less than pectoral-fin length. Pectoral fin lanceolate or roundly so, its base beginning above and extending about one-third the length of gill opening. Gill opening semicircular, large and pouch-like, equal to or larger than isthmus.

Color. Background color cream, slightly darker above lateral line; faint yellow to tan stripe at bases of dorsal and anal fins. Body overlain on and above the lateral line with 1–3 irregular rows of large brown spots, the majority of spots (measuring 10–34% TL) larger than eye and some as large as pectoral fin; smaller spots in anterior trunk region and on posterior half of head. Nape and chin dusky, with dark speckling; spots marking jaw pores (except anteriormost pores), preopercular pores pale and not visible to naked eye. Dorsal fin pale basally, with diffuse spots and intermittent dark margin; anal fin pale basally, with prominent dark margin; pectoral fin dark except for base.

Size. The largest specimen examined was 1030 mm TL. Thirteen of the study specimens were females (420–985 mm), seven (420–855) ripe with 0.8-mm eggs; two ripe males measured 478 and 863 mm. All ripe specimens were taken between April and August.

Development. Leptocephalus described as *Leptocephalus caudomaculatus* by Eigenmann and Kennedy (1902: 811); see p. 830.

Distribution (Fig. 367). Taken along the Atlantic coast from North Carolina to Florida, and in the northeastern Gulf of Mexico off Florida and west to Mobile Bay area. Isolated specimens taken in the West Indies, off Mexico, and off southern Brazil. Usually captured by trawl at recorded

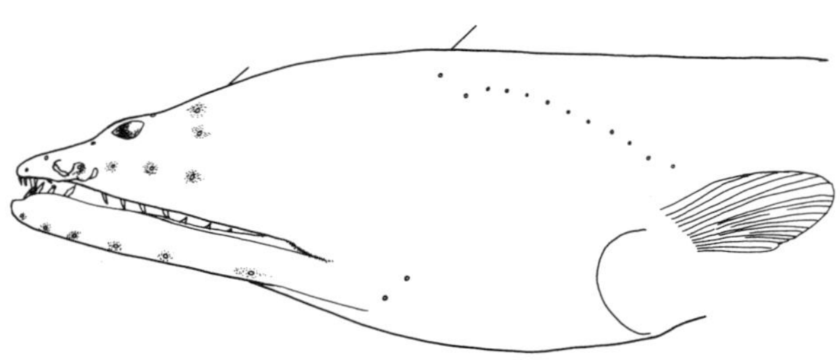

FIGURE 365. *Echiophis intertinctus*: ANSP 116134, 737 mm TL, Florida Atlantic; head.

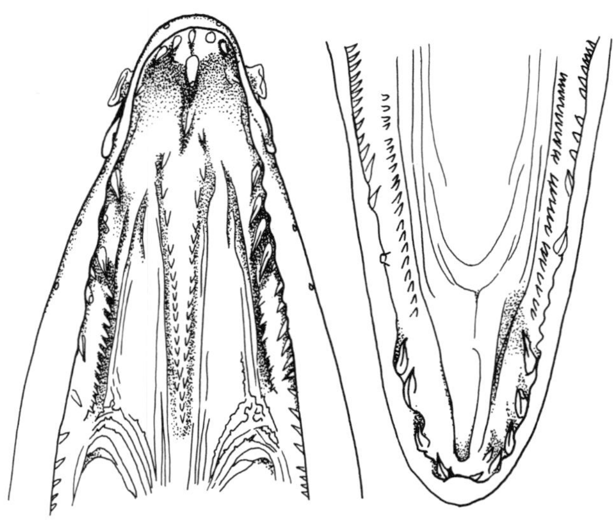

FIGURE 366. *Echiophis intertinctus*: ANSP 116134, 737 mm TL, Florida Atlantic; dentition.

depths of 20–64 m; reported taken by hook and line off Brazil (Menezes, 1975, pers. comm.). Apparently a continental shelf species which lives over rock and sand bottom; its abundance in the northeastern Gulf of Mexico and relative scarcity throughout the Caribbean may reflect bottom habitat preference.

Etymology. From the Latin *inter* (between) and *tinctus* (colored), the application not stated. A noun in apposition.

Remarks. Our data support the separation of *Echiophis intertinctus* from the other nominal forms, although there is some overlap in vertebral counts and gradation in the numbers of rows and sizes of spots (Table 38). The most reliable character is the preopercular pore condition; two inconspicuous pores in *intertinctus* and three readily visible pores in dark spots in *punctifer*;

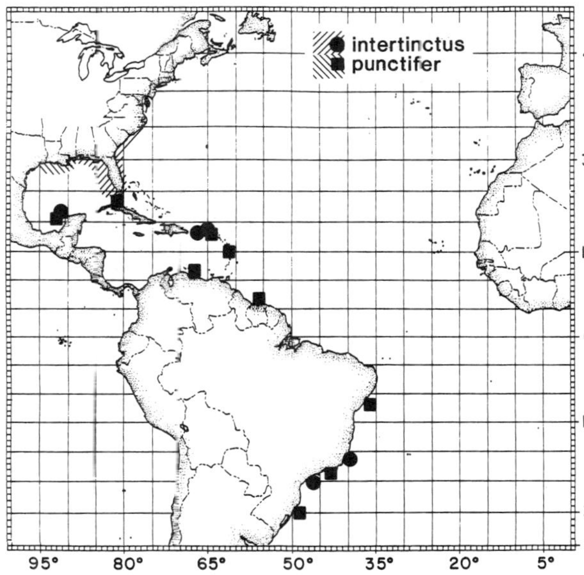

FIGURE 367. Distribution of western Atlantic species of *Echiophis* based on study material.

TABLE 38. Geographic variation in spots and vertebral counts in Atlantic species of *Echiophis*.

	Spots		Predorsal vertebrae							Preanal vertebrae								
	Rows	Size	14	15	16	17	18	19	20	53	54	55	56	57	58	59	60	61
intertinctus																		
Coastal U.S.	1–4	10.7–38.2	1	13	25	1	1			2	6	13	16	8	4			
Caribbean	1–4	15.3	1	-	1*								1	-	2*			
Brazil	2–3	25.6–32.4		1	1										2			
punctifer																		
Coastal U.S.	5–12	1.3–11.3		1	1	4	9	5	1		1	1	1	3	9	7	2	1
Caribbean	4–8	6.0–11.8					2	3							2	1	1	1
Brazil	4–10	5.5–10.3				4	3	3			1	-	1	1	4	3		
E. Atlantic†	ca 5	———								X	X	X	X	X	X	X	X	X

	Total vertebrae																			
	124	125	126	127	128	129	130	131	132	133	134	135	136	137	138	139	140	141	142	143
intertinctus																				
Coastal U.S.		1	3	3	15	7	7	8	2	2	1									
Caribbean						2	-	-	1*											
Brazil												1	-	-	1					
punctifer																				
Coastal U.S.							1	-	2	3	6	5	3	6	2					
Caribbean											1	-	-	-	-	2	2	1		
Brazil									1	-	-	1	-	-	-	2	2	2	1	1
E. Atlantic†											X	X	X	X	X	X	X	X	X	

* Includes holotype.

† Data from Blache 1971b; range only given.

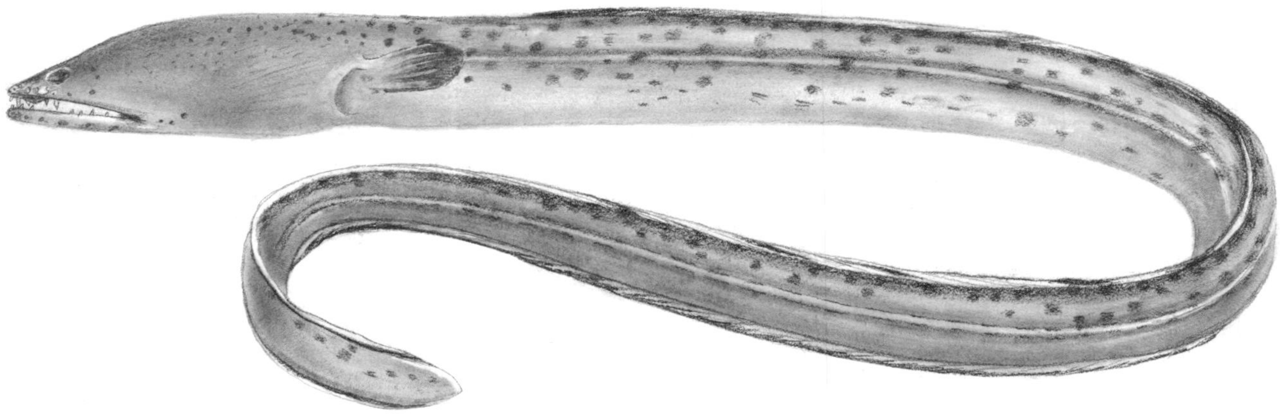

FIGURE 368. *Echiophis punctifer*: ANSP 116367, 590 mm TL, Gulf of Mexico off Mississippi.

only three out of 157 specimens of the two species showed variation in this character.

The status of Brazilian *E. intertinctus* is not clear, due to incomplete data for the few study specimens; they appear to have larger body spots and more total vertebrae than do northwestern Atlantic specimens (Table 38), but show typical preopercular pore condition. The specimens reported by Lema et al. (1979) are *E. punctifer*, based on the figures.

Richardson's holotype is a small specimen with one row of very large spots and a high (132) vertebral count; its locality is given only as "West Indies."

Study Material. A total of 99 specimens, 171–1030 mm TL, including the holotype.

Holotype: BMNH 1982.5.7.4 (316); West Indies; Scrivener.

Other material: COASTAL UNITED STATES: *North Carolina:* USNM 151927 (1, 1010). ZMC 20110 (1, 408). *South Carolina:* ANSP 144111 (1, 925). GMBL 73-308 (1, 558). UMML SB 3657 (2, 835–1030). USNM 151981 (1, 985). *Georgia:* GMBL 74-223 (1, 233). UMML Com 339 (1, 482). UMML SB 5719 (1, 526). *Florida:* ANSP 116134 (1, 737). CAS 54483 (1, 620). USNM 158331 (1, 825). GULF OF MEXICO: *Florida:* ANSP 78432 (1, 209), 78540 (3, 240–309), 78742 (1, 612), 87977 (1, 835), 87978 (1, 420), 96663 (1, ca 350), 101431 (1, 734), 116124 (1, 855), 116125 (1, 820), 116126 (1, 713), 116128 (1, 550), 117791 (1, 178), 136002 (2, 682–730), 157632 (1, 335, cleared and stained). BMNH 1883.12.14.252 (1). CAS 29292 (1, 876). FMNH 9708 (2, 171–254), 34981 (1, 633), 34982 (1, 295), 34983 (1, 610), 35586 (1, 238). MCZ 9228 (3, 645–800), 9330 (2, 687–727), 31432 (1, 444). UF 9149 (11, 296–430), 17718 (1, 572), 19973 (1, 663). UF-FSU 1511 (1, 508), 2554 (1, 310), 15130 (3, 222–543), 18882 (1, 601), 20332 (1, 486), 20776 (2, 498–628), 20909 (1, 596), 21484 (1, 288). UMML 1957 (1, 732), 4979 (1, 448). UMML FHB 65-2 (1, 440). USNM 5984 (1, 666), 22865 (1, 732), 23635 (1, 230), 31891 (3, 425–ca 700), 32758 (1, 303), 37663 (1, pieces), 43553 (1, 61), 49797 (1, 239), 152251 (1, 353), 156811 (1, ca 517), 158558 (1, ca 520), 158748 (1, 534), 161433 (1 head), 171826 (1, 855), 188035 (1, 900), 274249 (1, 580). *Alabama:* TCWC 0550.1 (2, 478–708).

WEST INDIES: *Puerto Rico:* LACM 6522–23 (2, 508–770). *St. Thomas:* USNM 6956 (1, 372). YUCATAN: ANSP 116122 (1, 960). BRAZIL: MNRJ 2293 (1, 875), 10440 (1, 863). MZUSP 10062 (1, 808).

Echiophis punctifer (Kaup, 1860)

STIPPLED SPOON-NOSE EEL

Figures 367–370, Tables 37, 38

Crotalopsis punctifer Kaup, 1860: 13, pl. 1, fig. 3 (original description, Puerto Cabello, holotype ZMH 392).

Conger mordax Poey, 1860: 319 (original description, Cuba, holotype MCZ 9220).

Macrodonophis mordax. Poey, 1867b: 252 (type of new genus).

Ophichthys punctifer. Günther, 1870: 56.

Crotalopsis mordax. Poey, 1876a: 153.

Ophichthys Schneideri Steindachner, 1880: 184 (original description, Brazil, holotype NMW 65096).

Mystriophis mordax. Ginsburg, 1951: 470.

Mystriophis punctifer. Ginsburg, 1951: 474.

Mystriophis creutzbergi Cadenat, 1956b: 857 (original description, Sierra Leone, holotype unknown).

Echiopsis intertinctus (not of Richardson). Blache, 1971b: 219 (synonymized *creutzbergi* with *intertinctus*, placed in "*Echiopsis*").

Echiophis punctifer. McCosker, 1977: 77.

Distinctive Characters. A stout ophichthin with small brown spots on tan background; dorsal-fin

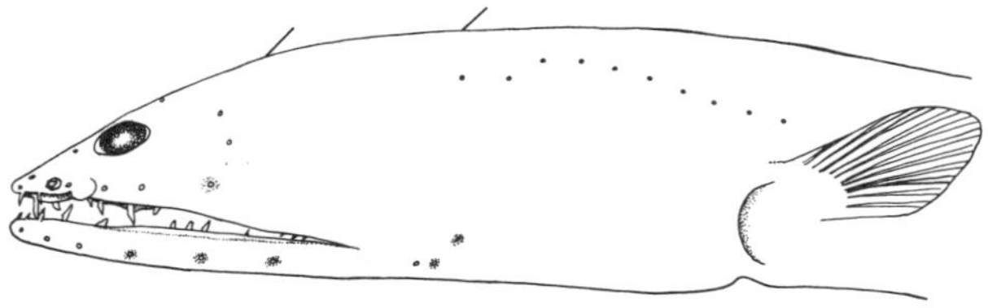

FIGURE 369. *Echiophis punctifer*: ANSP 116367, 590 mm TL, Gulf of Mexico off Mississippi; head.

origin behind pectoral tips by length greater than pectoral-fin length; pectoral fin 25–32% TL; three preopercular pores set in dark spots; MVF 18-58-136.

Description. Total vertebrae 130–143, x̄ = 136.4 (n = 44); predorsal vertebrae 16–19, x̄ = 18.0 (36); preanal vertebrae 54–61, x̄ = 58.2 (39). Pores: SO 1 + 3; IO 4 + 2; POM 6 + 3; ST 3.

Proportions as % TL: head 11–13; trunk 32–35; tail 52–55; predorsal 16–19; depth behind gill openings 3.2–5. Of HL: snout 11–14; snout to rictus 42–52; eye 4.8–7.9; interorbital width 4.5–9.2 (see generic remarks); pectoral-fin length 25–32.

General characters as for *Echiophis intertinctus,* from which it differs in the following: Three preopercular pores surrounded by dark pigment, readily visible to the naked eye (with the exception of three specimens which had three preopercular pores on one side and two on the other side, all of these from the Mobile Bay area of transition); dorsal-fin origin behind pectoral-fin tips by a length equal to or greater than pectoral-fin length; pectoral fin generally more rounded and somewhat longer than that of *E. intertinctus.* Dentition similar to that of *E. intertinctus.*

Color. Background tan; a brown stripe dorsally on either side of dorsal fin, a faint brown or yellow stripe along lateral line, and a reddish-brown ventral stripe along base of anal fin; anus in yellowish area. Numerous small brown spots on body equal to or less than size of eye (1.3–11% TL), forming 6–8 indistinct rows. Spots on head numerous and smaller; pores along jaws, including preopercular pores, set in prominent dark areas; lips black-edged; inside of mouth pigmented. Chin, throat and belly generally unspotted. Dorsal fin pale basally, with few large

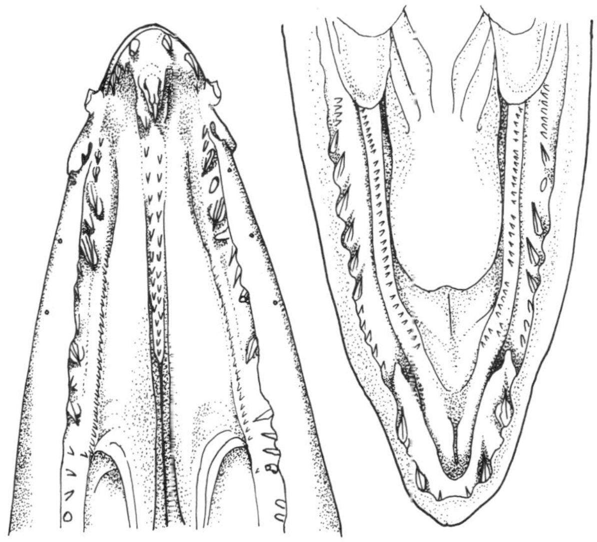

FIGURE 370. *Echiophis punctifer*: ANSP 116367, 590 mm TL, Gulf of Mexico off Mississippi; dentition.

diffuse spots, margin black for most of extent; anal fin with prominent black margin; pectoral fin dark except basally.

Size. The largest specimen examined is 1191 mm TL. Of 16 females (332–1191 mm), five (620–1191 mm) were ripe with eggs to 1.6 mm; two ripe males measured 590 and 868 mm.

Distribution (Fig. 367). *Echiophis punctifer* is commonly taken in the northwestern Gulf of Mexico from Mobile Bay to Texas; occasionally taken in the West Indies, off Central America, and off northern South America; reported off southern Brazil (as *E. intertinctus*) by Lema et al. (1979). In the eastern Atlantic, reported off Sierra Leone (as *Mystriophis creutzbergi*). Taken by trawl and seine, at depths recorded to 100 m, usually less than 43 m.

Etymology. From the Latin *punctifer* (dotted), referring to the coloration; treated as a noun in apposition.

Remarks. Ginsburg, presenting data on a total of 18 specimens, recognized three species of *Echiophis* in the Gulf of Mexico: the large-spotted *intertinctus, punctifer* with very small spots, and Poey's species *mordax* with spots intermediate in size. At the beginning of our study (of a total of 154 specimens) we attempted to separate the same three groups on the basis of coloration, and found it possible to do so with some degree of success, especially if spot-size was correlated with locality; *E. intertinctus* along the Atlantic coast of the United States and in the northeastern Gulf of Mexico to the Mobile Bay area; *mordax* from the northwestern Gulf of Mexico from Mobile Bay to Texas, and *punctifer* from the southern Caribbean and northern South America. Other differences (in preopercular pores, vertebral counts, and dorsal and pectoral fins) were found to distinguish *intertinctus*, but none were found to separate the nominal form *mordax* from *punctifer*. We therefore synonymize *mordax* with *punctifer*, but note subtle differences in spot size and vertebral counts.

Blache (1971b) synonymized *M. creutzbergi* with *E. intertinctus* (all western Atlantic specimens considered to be one species); *creutzbergi* agrees with *E. punctifer* in preopercular pore condition and in vertebral numbers (Table 34) and is here included in the synonymy for this species.

We did not examine the holotype of *Crotalopsis punctifer*, but the excellent illustration that accompanied its description leaves little doubt that it is the species we identify as *E. punctifer*. MCZ 9220 has long been considered to be the holotype of *Conger mordax*; however, Ginsburg (1951: 473) called attention to differences between that specimen and Poey's description of spot size ("le plus grandes ne depassant pas cinc millimetres" in Poey, measured 10 mm by Ginsburg and by us) and the size of the specimen (990 mm according to Poey, 940 mm now, a difference too great to be due to shrinkage). Regardless of the status of the "holotype," the name remains in synonymy based on the description. The holotype of *Ophichthys Schneideri* was examined in Vienna (by EBB and JEB) and found to be conspecific with *O. punctifer*.

Study Material. A total of 59 specimens, 125–1191 mm TL, including the "holotype" of *Conger mordax* Poey and the holotype of *Ophichthys Schneideri* Steindachner.

GULF OF MEXICO: *Mississippi:* ANSP 116367 (1, 590). GCRL 4035 (1, 885). USNM 125911 (1, 1191), 152990 (1, 1185), 157930 (1, 690), 157954 (2, 587–620). *Louisiana:* CAS 54482 (1, 755). FSC 5404 (1, 830), 18777 (1, 625). FSC sta. 8 (1, 563). GCRL 255 (1, 563), 1175 (1, 589), 1385 (1, 125), 2789 (1, 788). TU 4181 (1, 355), 5389 (1, 607), 10763 (1, 725), 57979 (1, 495), 69540 (2, 727–755), 80651 (1, 856). *Texas:* ANSP 116129 (1, 606, cleared and stained), 157633 (1, ca 225, cleared and stained), 157634 (1, ca 340, cleared and stained). TCWC 0551.1 (1, 332), 0552.1 (3, 625–735), 0553.1 (2, 522–690), 4209.2 (1, 710), 4210.1 (1, ca 780). USNM 152240 (2, 872–924), 154995 (1, 464), 154996 (1, 665), 157935 (1, 495).

CUBA: ANSP 92926 (1, 868). MCZ 9217 (1, 431). MCZ 9220 (1, 940); holotype of *Conger mordax*. VIRGIN ISLANDS: ANSP 150012 (1, 874). MARTINIQUE: MNHN B.2138 (1, 722). GULF OF CAMPECHE: ANSP 116123 (1, 620). FMNH 45951 (1, 777). SURINAME: MNHN B.2743 (1). VENEZUELA: MBUCV 1633 (1, 948). BRAZIL: MCZ 4712 (3, 948–1159), 9208 (1, 985), 9209 (1, 1090). MNHN 4808 (1, 724). MNRJ 3195 (1, 1189). MZUSP 8878 (3, ca 710–765). NMW 65096 (1, 718); holotype of *Ophichthys Schneideri*. ZMC J.128 (1, 845).

Genus *Herpetoichthys* Kaup, 1856

Herpetoichthys Kaup, 1856a: 44 (type species *Herpetoichthys ornatissimus* Kaup, 1856 = *Ophisurus regius* Shaw in Richardson, 1848, by Jordan, 1919b, as first reviser).

Pogonophis Myers and Wade, 1941: 78 (type species *Pogonophis fossatus* Myers and Wade, 1941, by original designation).

Characters. Body moderately stout (depth 3.9–4.9% TL in Atlantic species), cylindrical, not laterally compressed; head and trunk slightly longer than tail (53–55% TL). Dorsal-fin origin behind gill opening; tail tip pointed; pectoral fins small to minute, not broad based. Snout blunt; jaws subequal; a single pendulous barbel on upper lip behind anterior nostril; eye large, closer to snout than to rictus; anterior nostril tubular; posterior nostril opens into mouth. Gill opening vertical, semicircular. Head pores developed; two preopercular pores. Teeth pointed, uniserial except on maxillae and anterior vomer where biserial; intermaxillary teeth the largest and continuous with those of vomer.

Neurocranium subtruncate posteriorly; supraoccipital crest weakly pointed; maxilla elongate, slender posteriorly; pterygoid broad posteriorly, free and terete anteriorly; orbit large; otic bulla moderately developed; nasals and nasal cartilage reduced; opercular series developed. Branchiostegal rays numerous, all slender and associated with hyoid; urohyal broad anteriorly, short and slender posteriorly. Gill arches moderately de-

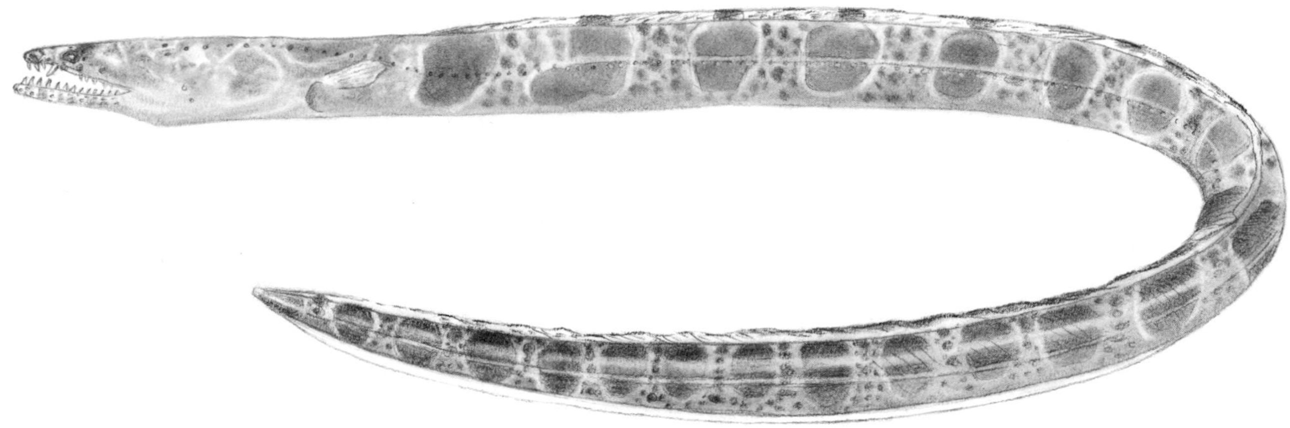

FIGURE 371. *Herpetiochthys regius*: BMNH 1867.10.8.44, 826 mm TL, St. Helena.

veloped, *Ophichthus*-like, except fifth ceratobranchial absent; upper pharyngeal tooth plates separate. Pectoral girdle reduced; actinosts absent. Intramuscular bones, ribs and caudal transverse processes developed. Precaudal about equal to caudal vertebrae. Gasbladder short, 6.7% TL. Coloration boldly patterned and spotted; peritoneum pale.

Distribution. Known from one Atlantic and one eastern Pacific species; the locality of a purported Indian species is in doubt.

Remarks. McCosker (1977: 83) described and analyzed the osteology of *Pogonophis fossatus.* Blache and Bauchot (1976b) resurrected the genus *Herpetoichthys* and showed *Pogonophis* to be a junior synonym of *Herpetoichthys;* they erred in synonymizing the eastern Pacific and Atlantic species (see *Remarks* below).

Etymology. From the Greek *herpeton* (snake) and *ichthys* (fish), in reference to its snake-like appearance. Masculine.

Species. Two species: *Herpetoichthys regius* (Shaw, 1848), from the mid- and eastern Atlantic; and *H. fossatus* (Myers and Wade, 1941) in the eastern Pacific.

Herpetoichthys regius (Shaw, 1848)

REGAL SNAKE EEL

Figures 371–373, Table 37

Ophisurus regius Shaw in Richardson, 1848a: 106 (original description, no locality given, holotype BMNH 1985.2.19.5 (Old Collection)).

Herpetoichthys ornatissimus Kaup, 1856a: 44 (original description, Malabar, holotype MNHN B.2757). Kaup, 1856b: 7, fig. 4 (not fig. 5 as listed).

Herpetoichthys regius. Blache and Bauchot, 1976b: 441 (synonymy of *Pogonophis* and the species of *Herpetoichthys*).

Distinctive Characters. A moderately stout ophichthin with distinctive color pattern of brown markings on pale background; conspicuous barbel projecting down between anterior and posterior nostrils; dorsal-fin origin behind pectoral tips; pectoral fin reduced and narrow-based; MVF 15-77-160.

Description. Total vertebrae 157–161, $\bar{x}$ = 159.6 (n = 5); predorsal vertebrae 15–16, $\bar{x}$ = 15.3 (3); preanal vertebrae 76–78, $\bar{x}$ = 76.8 (4). Pores: SO 1 + 3; IO 4 + 2; POM 5 + 2; ST 3.

Proportions as % of TL: head 12–13; trunk 41–42; tail 45–47; predorsal 16; depth behind gill openings 3.9–4.9. Of HL: snout 14–18; snout to rictus 41–44; eye 6.2–10; interorbital width 8.9–13; pectoral-fin length 12–18.

Head wide and compressed, widest at rictus, narrowing to pointed snout, appearing triangular when viewed from above; mouth horizontal, lower jaw projecting beyond snout. Eye moderately small, dorsad, above anterior half of jaw. Anterior nostril in a short tube on side of snout; posterior nostril an oval opening into mouth below and slightly before margin of eye, covered by an anterior flap and not discernable externally. A prominent fleshy papilla projecting down at edge of upper jaw between nostrils, its

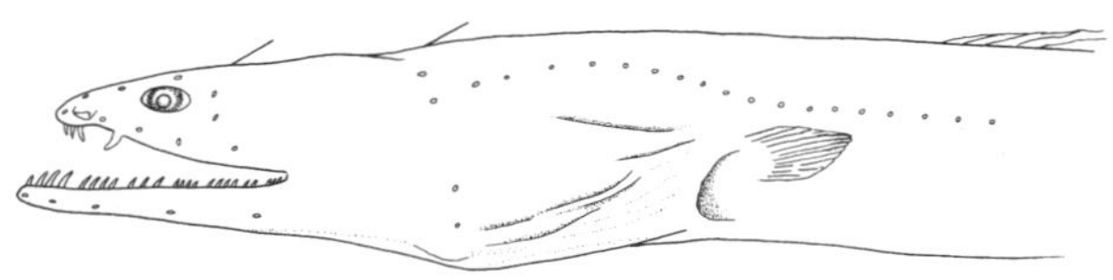

FIGURE 372. *Herpetiochthys regius*: BMNH 1867.10.8.44, 826 mm TL, St. Helena; head.

length about one and one-half times the length of anterior nostril. Head pores as listed above; dorsal head pores and anterior lateral-line pores ringed with black pigment and conspicuous, ethmoid pore and preoperculomandibular pores pale and difficult to see.

Teeth conical and sharp. A semicircle of 5–7 intermaxillary teeth continuous with vomerine teeth which are irregularly biserial anteriorly, uniserial posteriorly, totaling 17 on illustrated specimen. Maxillary teeth biserial; an outer row of about 23–25 beginning just behind posterior nostril and an inner row of an equal number beginning 3–4 teeth anteriorly at the level of the barbel. Mandibular teeth uniserial, about 32 moderately sharp strong canines, slightly larger anteriorly.

Body markedly cylindrical, muscular, and snake-like in appearance, body width equal to or greater than depth; tail tapering to a blunt point. Dorsal and anal fins developed, ending just before tail tip; dorsal-fin origin slightly behind pectoral-fin tips. Pectoral fin small, lanceolate, its base beginning slightly above and extending less than half the distance of the gill opening.

Color. Background color pale cream. A series of 19–23 large brown, pale-bordered saddle-like spots along the dorsal flank, beginning above pectoral fin and extending to tail tip; interspaces patterned with central medium spots bordered by smaller spots on each side. Small spots on head, chin, throat and tail, one or two larger median spots on nape and head; pattern lines of small white papillae on head between frontal and supratemporal commissures. Dorsal fin variously spotted, body pattern sometimes extending onto fin; few spots on anal fin; pectoral fin with small pale spots.

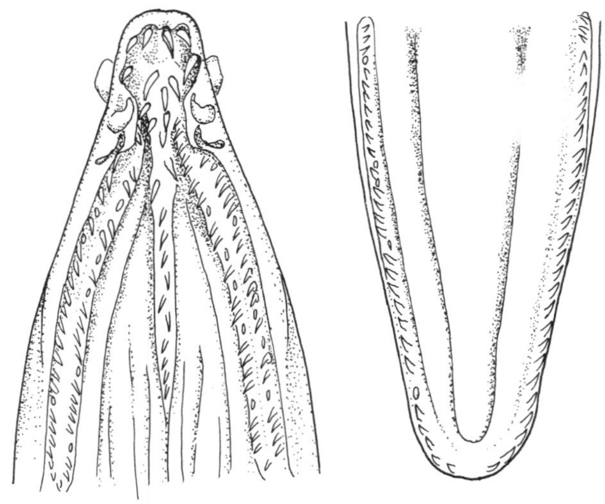

FIGURE 373. *Herpetoichthys regius*: BMNH 1867.10.8.44, 826 mm TL, St. Helena; dentition.

Size. The largest specimen examined is the holotype, 850 mm TL. The 826-mm illustrated specimen is a female, filled with 1.0-mm eggs.

Development. Leptocephalus not identified in the western Atlantic. Three larvae described as Ophichthini sp. 4 (p. 841) may possibly be this species.

Distribution. A rare eel, known from the holotype from an unknown locality, four specimens from St. Helena, one from Mauritania (Blache and Bauchot, 1976b), and the holotype of *H. ornatissimus* of which Kaup said "Malabar (found by M. Dussumier)." We are unaware of subsequent Indian Ocean specimens and consider the latter locality questionable. Edwards and Glass (1987) commented on the likelihood of a St. Helena origin of the holotype.

Remarks. Blache and Bauchot (1976b) showed *Pogonophis* to be a junior synonym of *Herpetoichthys*, and synonymized *H. ornatissimus* with *H. regius*, based on their study of the holotype of *H. ornatissimus* (MNHN B.2757). However, they erred in synonymizing the eastern Pacific *H. fossatus* (and its synonym *Ophichthus afuerae* Hildebrand, 1946) with the eastern Atlantic *regius*. *Herpetoichthys fossatus* differs in having a smaller pectoral fin, more vertebrae, and in its coloration.

Etymology. From the Latin *regius* (royal), presumably in reference to its striking coloration. An adjective.

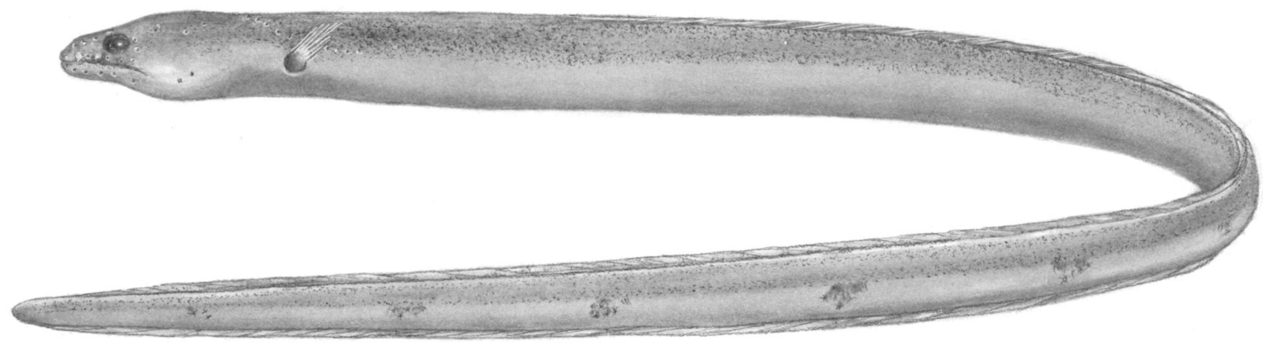

FIGURE 374. *Hyphalophis devius*: ANSP 115294, holotype, 286 mm TL, Lesser Antilles.

Study Material. Five specimens, 110–850 mm TL, including the holotype.

Holotype: BMNH 1985.2.19.5 (850); "Old Collection."

Other material: ST. HELENA: BMNH 1867.10.8.44–45 (2, 698–826); J. C. Melliss. BMNH 1869.3.18.7 (1, 110); J. C. Melliss. BMNH 1910.9.9.4 (1, 702); J. T. Cunningham.

Genus *Hyphalophis* McCosker and Böhlke, 1982

Hyphalophis McCosker and Böhlke, 1982: 116 (type species *Hyphalophis devius* McCosker and Böhlke, 1982, by original designation).

Characters. Body moderately elongate (depth 3.0% TL), cylindrical, laterally compressed posteriorly; skin loose and flabby; head and trunk shorter than tail (40% TL). Dorsal-fin origin well behind end of pectoral fins; tail tip blunt; pectoral fins well developed. Snout bluntly rounded; jaws subequal; lips without barbels; eye well developed, over center of jaw; anterior nostril tubular; posterior nostril at edge of lip, covered by a flap. Gill openings semicircular. Head pores small; two preopercular pores; lateral-line canal and pores arch abruptly in head region, abutting but not combining dorsally. Teeth minute, conical and numerous; in four rows on maxillary, uniserial on mandible and vomer.

Neurocranium short for an ophichthin, subtruncate, sloping posteriorly; supraoccipital crest apparently absent, condyle a short upturned knob posteriorly; orbit large; maxilla not tapering posteriorly; otic bulla well developed, otoliths large; nasals reduced; suspensorium nearly vertical. Branchiostegal rays numerous, slender, all closely apposed to hyoid. Gill arches reduced for an ophichthine; fifth ceratobranchial absent; upper pharyngeal tooth plates separate. Pectoral girdle moderately developed; cleithrum, supracleithrum, scapula and coracoid visible on radiograph. Intramuscular bones and ribs visible on radiograph, caudal transverse processes apparently not developed. Caudal vertebrae more numerous than precaudal. Body coloration uniform.

Size. To 286 mm TL.

Distribution. Known only from the western Atlantic.

Etymology. From the Greek *hyphalos* (submerged under the sea) and *ophis* (snake), in reference to the extreme depth of capture of the type specimen. Masculine.

Species. The single species is *Hyphalophis devius* McCosker and Böhlke, 1982.

Hyphalophis devius McCosker and Böhlke, 1982

FLABBY SNAKE EEL

Figures 374–376, Table 37

Ophichthidae, new genus, new species B. Böhlke, 1977b: 4, 6.

Hyphalophis devius McCosker and Böhlke, 1982: 117, figs. 4–6 (original description, Lesser Antilles, SW of Grenada, 11°34′N, 62°49′W, holotype ANSP 115294).

Distinctive Characters. A moderately elongate, pallid ophichthin with lateral line arching over head; dorsal-fin origin well behind pectoral tips; maxillary teeth quadriserial, vomerine and dentary teeth uniserial; MVF 23-52-144.

Description. Total vertebrae 144 (n = 1); pre-

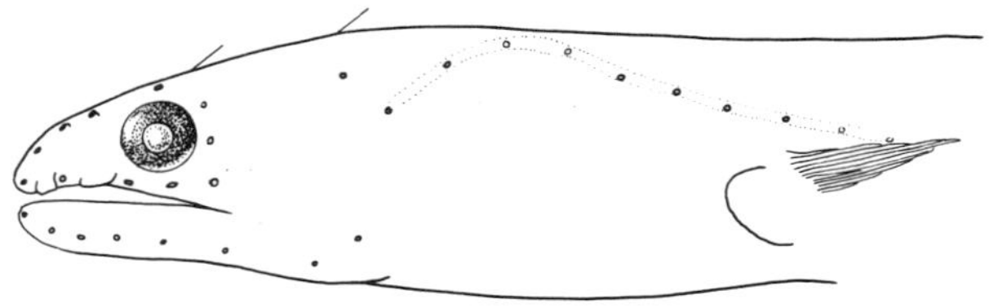

FIGURE 375. *Hyphalophis devius*: ANSP 115294, holotype, 286 mm TL, Lesser Antilles; head.

dorsal vertebrae 23; preanal vertebrae 52. Pores: SO 1 + 4; IO 4 + 2; POM 6 + 2; ST 3.

Proportions as % of TL: head 8.4; trunk 31; tail 60; predorsal 18; depth behind gill openings 3. Of HL: snout 17; snout to rictus 31; eye 12.5; interorbital width 8.3; pectoral fin length 26.

Body moderately elongate, nearly cylindrical, becoming laterally compressed posteriorly; tail a blunt finless point. Head broad, interorbital area flat. Snout bluntly rounded, lower jaw subequal; eyes large, lateral, centered above upper jaw. Anterior nostril tubular; posterior nostril at edge of lip in front of eye, entirely outside of mouth and covered by small flap. Head pores small and difficult to see; lateral line arching abruptly in head region, the two sides abutting but not joining dorsally.

Teeth minute, conical and numerous. Two pairs of intermaxillary teeth, posterior pair larger, separated by short gap from uniserial row of eight teeth on vomer. Maxillary teeth in four rows, outermost row of about 50 small close-set teeth separated from inner three rows by broad gap. Mandibular teeth uniserial, about 50 small teeth similar to those on maxillary.

Median fins low; dorsal-fin origin more than a head length behind gill opening. Pectoral fin lanceolate.

Color. Uniformly pale, with a fine peppering of melanophores dorsally, visible only with microscope. Six irregular dusky splotches along tail. Fins colorless.

Size. The single known specimen is 286 mm TL. It is a female with eggs of varying sizes to 0.5 mm diameter.

Development. Leptocephalus has not been identified.

Distribution. Presumably a deepwater species, the holotype obtained by trawl off the Lesser Antilles below 300 m.

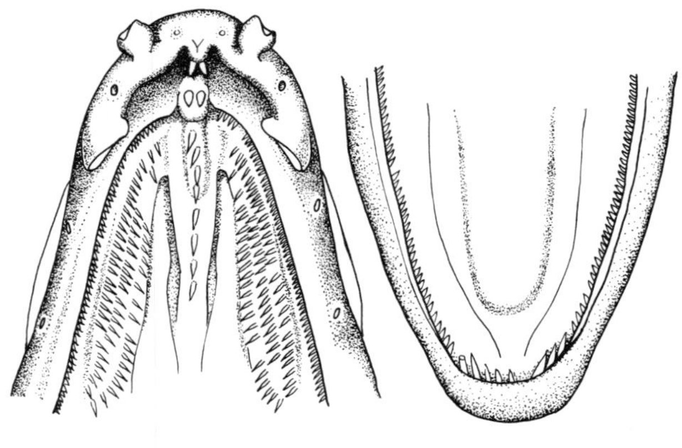

FIGURE 376. *Hyphalophis devius*: ANSP 115294, holotype, 286 mm TL, Lesser Antilles; dentition.

Etymology. From the Latin *devius* (off the highway), in reference to its specialized dentition and lateral line condition. An adjective.

Study Material. The holotype: ANSP 115294 (286 mm, female); Lesser Antilles, SW of Grenada, 11°34′N, 62°49′W, 40-ft bottom trawl at 293–366 m; R/V OREGON sta. 2781; 20 April 1960.

Genus *Kertomichthys* McCosker and Böhlke, 1982

Kertomichthys McCosker and Böhlke, 1982: 119 (type species *Mystriophis blastorhinos* Kanazawa, 1963, by original designation).

Characters. Body moderately elongate (depth 3.3% TL), cylindrical, laterally compressed posteriorly; head and trunk shorter than tail (44% TL). Dorsal-fin origin well behind pectoral fins; tail tip blunt; pectoral fins well developed. Snout short, its profile depressed, narrow posteriorly and clavate anteriorly; jaws elongate; lips without barbels; eye well developed, near tip of snout; anterior nostril in a short tube; posterior nostril at edge of lip, covered by a flap. Gill opening crescentic, its major axis vertical. Head pores developed; three preopercular pores. Teeth pointed, conical; jaw teeth biserial, the outer row enlarged and widely spaced; vomerine teeth small and mostly biserial.

Neurocranium subtruncate posteriorly; supra-

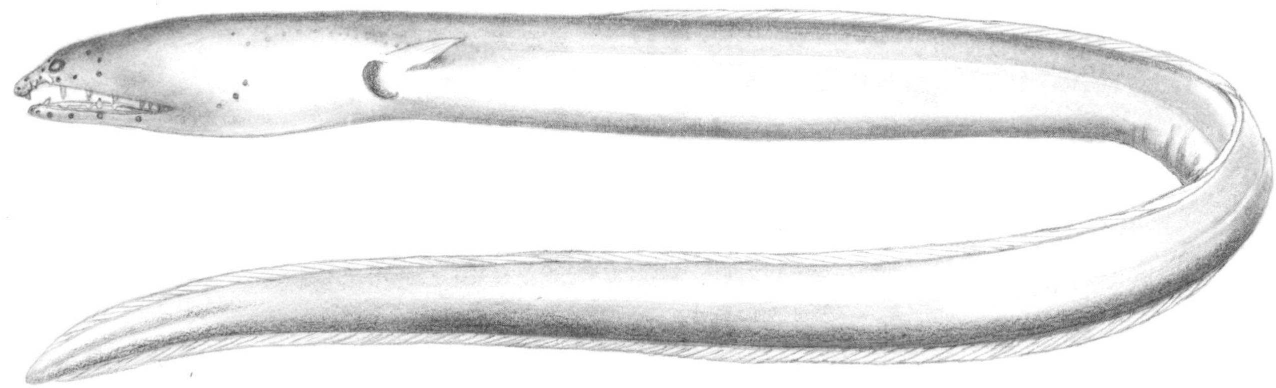

FIGURE 377. *Kertomichthys blastorhinos*: USNM 158960, holotype, 418 mm TL, French Guiana.

occipital crest and condyle well developed; orbit moderately large, braced posteriorly by a stout strut formed by postorbitals; maxilla stout, not slender posteriorly; otic bulla developed; suspensorium nearly vertical. Branchiostegal rays numerous, slender and along arch. Gill arches developed, *Ophichthus*-like; fifth ceratobranchial ossified; upper pharyngeal tooth plates fused. Pectoral girdle developed; cleithrum and supracleithrum slender, coracoid and scapula broad. Intramuscular bones, ribs, and caudal transverse processes developed. Precaudal about equal to caudal vertebrae. Body coloration pale with minute spots.

Size. To 418 mm TL.

Distribution. Known from a single western Atlantic species.

Etymology. From the Greek *kertomios* (mocking) and *ichthys* (fish), "in reference to the facial expression of the type species, as well as the difficulty that this eel has caused for the ichthyologists who try to understand its relationships" (McCosker and Böhlke, 1982). Masculine.

Species. The single species is *Kertomichthys blastorhinos* (Kanazawa, 1963).

Kertomichthys blastorhinos (Kanazawa, 1963)

SNUB-NOSE SNAKE EEL

Figures 377–379, Table 37

Mystriophis blastorhinos Kanazawa, 1963: 282, figs. 1, 2A, pl. 1, left (original description, French Guiana, 07°18′N, 53°32′W, holotype USNM 158960).

Kertomichthys blastorhinos. McCosker and Böhlke, 1982: 119.

Distinctive Characters. A robust ophichthin with pale coloration; dorsal-fin origin well behind pectoral tips; jaws elongate, snout spatulate, eye near tip of snout; jaw teeth biserial, outer teeth fang-like; three preopercular pores; MVF 22-61-143.

Description. Total vertebrae 143 (n = 1); predorsal vertebrae 22; preanal vertebrae 61. Pores: SO 1 + 3; IO 4 + 2; POM 6 + 3.

Proportions as % of TL: head 12; trunk 31; tail 56; predorsal 19; depth behind gill openings 3.3. Of HL: snout 8.8; snout to rictus 44; eye 6.4; interorbital width 4.8; pectoral-fin length 22.

Characters as for the genus. Head pores small and inconspicuous, difficult to discern. Last two pores along upper jaw raised above jaw, two postorbital pores far behind eye; preopercular pores inconspicuous, low and mostly lateral; no pores visible in supratemporal canal. Lateral-line canal apparent, pores barely visible as raised papillae.

The holotype has three widely spaced intermaxillary teeth, slightly separated from about 30 small, mostly biserial teeth on vomer. Maxillary teeth biserial: an outer row of about 10–12 enlarged and widely spaced teeth, the largest strongly fixed and alternating with smaller slender depressible fangs; the inner row of 30–38 small teeth. Mandibular teeth biserial and similar to those on upper jaw; 12–16 in outer row and 36–42 in inner row.

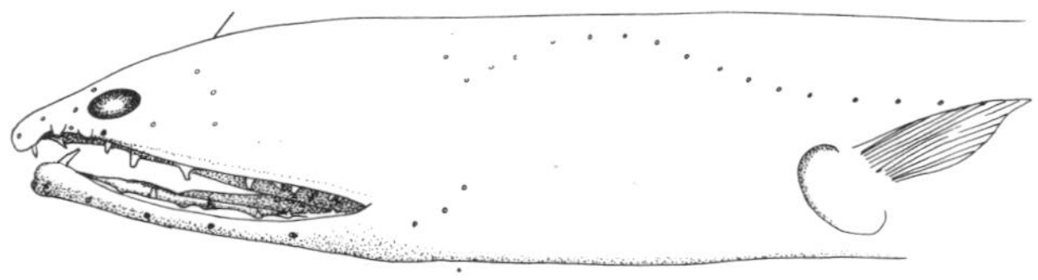

FIGURE 378. *Kertomichthys blastorhinos*: USNM 158960, holotype, 418 mm TL, French Guiana; head.

Color. Pale tan background, darker dorsally, overlain with fine speckling. A few head pores outlined with dark borders. Fins completely pale, although the original description stated that the distal edge of the dorsal fin was dusky.

Size. The single specimen is 418 mm TL, the sex not determinable.

Development. No larvae have been identified as this species.

Distribution. The holotype, the only known specimen, was collected by trawl in 183 m off French Guiana.

Etymology. From the Greek *blastos* (bud) and *rhinos* (nose), in reference to its clavate snout. A noun in apposition.

Remarks. The physiognomy of this eel suggests behavior and habitat similar to its closest relatives. McCosker and Böhlke (1982: 120) speculated: "We presume, from the appearance of the down-turned snout and forward eye position, that *Kertomichthys blastorhinos* inhabits a burrow with only its snout and eyes exposed. It would therefore feed on passing prey much like species of *Brachysomophis* or *Echiophis* . . . which rapidly dart out of their burrows to seize passing prey."

Study Material. The holotype: USNM 158960 (418 mm TL); FRENCH GUIANA, 07°18′N, 53°32′W, 40-ft shrimp trawl at 183 m; OREGON sta. 2021; 8 Nov. 1957.

Genus *Lethogoleos* McCosker and Böhlke, 1982

Lethogoleos McCosker and Böhlke, 1982: 114 (type species *Lethogoleos andersoni* McCosker and Böhlke, 1982, by original designation).

Characters. Body moderately elongate (depth 3–3.3% TL), cylindrical, laterally compressed posteriorly; head and trunk shorter than tail (46% TL). Dorsal-fin origin behind end of pectoral fins; tail tip blunt; pectoral fins well developed. Snout acute, overhanging lower jaw, underside not grooved; jaws elongate; lips without barbels; eye well developed, nearer snout tip than rictus; anterior nostril tubular; posterior nostril entirely outside of mouth, at edge of lip. Gill openings crescentic, their major axis vertical. Head pores minute; one preopercular pore, anterior pores of lateral line absent. Teeth elongate, conical, sharply pointed and depressible; maxillary teeth triserial, mandibular teeth biserial, vomerine teeth large and uniserial.

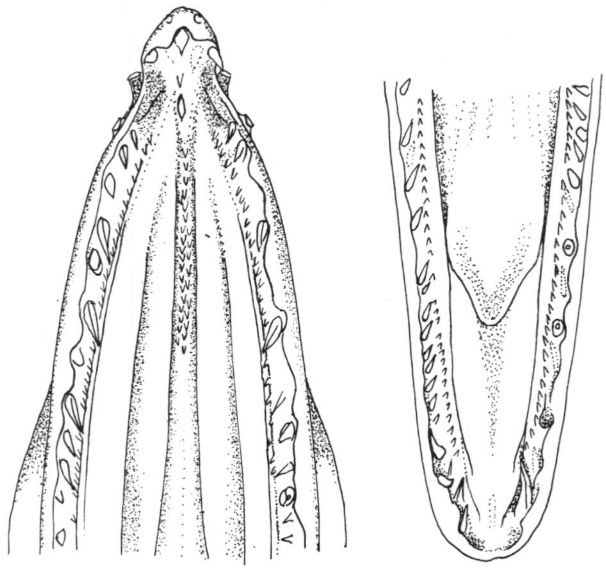

FIGURE 379. *Kertomichthys blastorhinos*: USNM 158960, holotype, 418 mm TL, French Guiana; dentition.

Neurocranium truncate posteriorly; orbit large, a moderately stout strut formed from postorbitals; supraoccipital crest developed; suspensorium nearly vertical; jaws elongate. Gill arches developed, *Ophichthus*-like; fifth ceratobranchial ossified; upper pharyngeal tooth plates separate. Pectoral girdle developed. Intramuscular bones, ribs, and caudal transverse processes developed. Caudal vertebrae more numerous than precaudal. Gasbladder well developed, 18% of TL. Body coloration pallid, with small spots; peritoneum lightly pigmented.

Size. To 318 mm TL.

Distribution. Known from a single western north Atlantic species.

Etymology. From the Greek *lethos* (to forget) and *goleos* (a hole), in reference to the unique absence of several pores. Masculine.

Species. The single species is *Lethogoleos andersoni* McCosker and Böhlke, 1982.

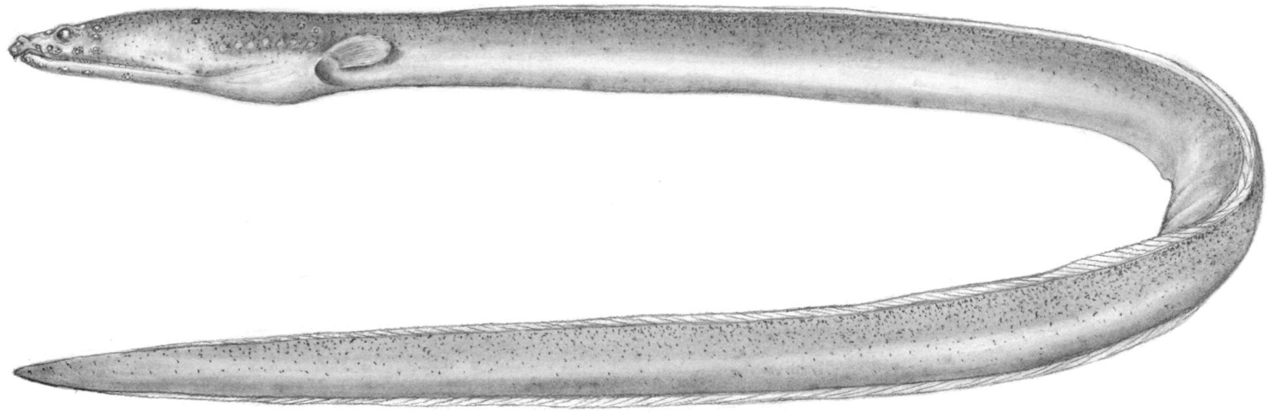

FIGURE 380. *Lethogoleos andersoni*: CAS 50931, paratype, 274 mm TL, South Carolina.

Lethogoleos andersoni
McCosker and Böhlke, 1982
FORGETFUL SNAKE EEL
Figures 380–382, Table 37

Ophichthidae, new genus, new species C. Böhlke, 1977b: 4, 6.

Lethogoleos andersoni McCosker and Böhlke, 1982: 118, figs. 1–3 (original description, South Carolina, 32°29.3′N, 78°42.2′W, holotype ANSP 149062).

Distinctive Characters. A moderately elongate, pale ophichthin with brown freckling dorsally and on snout; dorsal-fin origin well behind pectoral-fin tips; one preopercular pore; dentition triserial on maxillary, biserial on mandible, and uniserial on vomer; MVF 17-58-132.

Description. Total vertebrae 130–134, $\bar{x}$ = 132 (n = 3); predorsal vertebrae 16–18, $\bar{x}$ = 17 (3); preanal vertebrae 57–59, $\bar{x}$ = 58 (3). Pores: SO 1 + 3; IO 4 + 2; POM 5 + 1; ST 3; preanal LL pores incomplete, 47–49.

Proportions as % of TL: head 11.5–11.6; trunk 35; tail 54; predorsal 16.4–17; depth behind gill openings 3–3.3. Of HL: snout 13–15; snout to rictus 46–50; eye 7.3–7.6; interorbital width 3.8–4.1; pectoral-fin length 17–19.

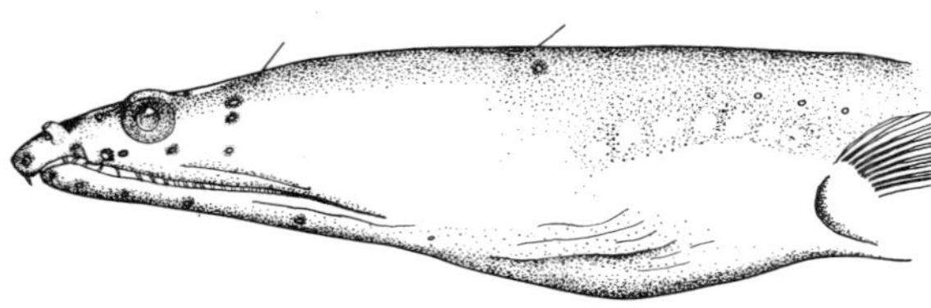

FIGURE 381. *Lethogoleos andersoni*: CAS 50931, paratype, 274 mm TL, South Carolina; head.

Head and jaws elongate; snout spatulate and downcurved, under side not grooved. Eyes superior, protruding, entirely over anterior half of gape. Anterior nostril tubular, about midway between snout tip and eye; posterior nostril outside mouth, closer to eye than to anterior nostril, covered by a flap that extends down below upper lip margin. Lower jaw included, intermaxillary teeth exposed when mouth is closed.

Head pores small and difficult to see, reduced in number and displaced posteriorly. Last pair of supraorbital pores between eyes, above pos-

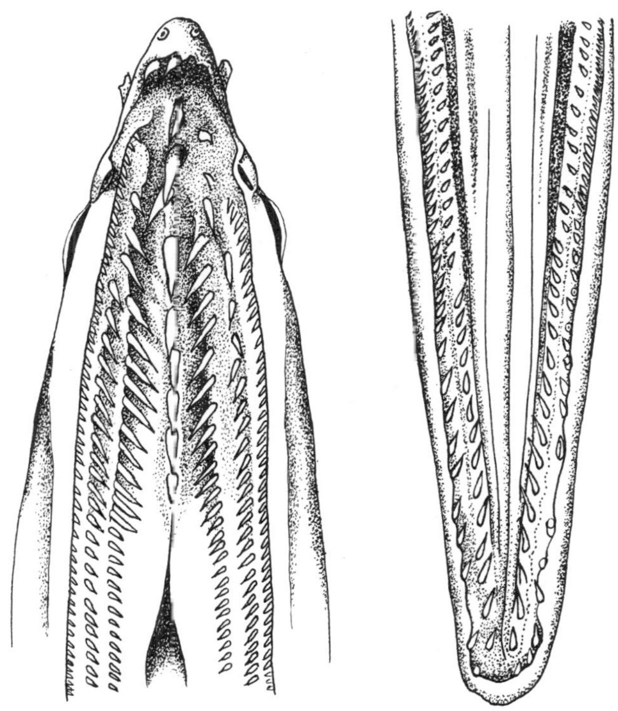

FIGURE 382. *Lethogoleos andersoni*: CAS 50931, paratype, 274 mm TL, South Carolina; dentition.

terior margin of eyes; median frontal pore more than an eye diameter behind eye; postorbital pores well behind eye and postorbital strut. Posteriormost pores along lower jaw ringed by dark pigment; the single preopercular pore visible is small and inconspicuous, located just behind and below rictus. Three pores in supratemporal commissure. Lateral-line pores difficult to discern, only two to four pores visible on each side before gill opening.

Teeth numerous, conical and sharply pointed. An outer semicircle of four large intermaxillary teeth, closely followed by a row of nine vomerine teeth equal in size to those of intermaxillary. Maxillary teeth in three rows; outer row of over 50 close-set small fixed teeth, inner rows of longer depressible teeth, about 22–30 in middle row and 16–27 in innermost row. Mandibular teeth biserial; outer row of 39–48 small close-set teeth, inner row of 21–30 longer ones, the anterior teeth the largest.

Median fins low. Dorsal-fin origin well behind pectoral tips. Pectoral fins elongate; dorsal base of pectoral fin slightly above gill opening, lower base above mid-line of gill opening. Gill opening crescentic, its dorsal end slightly anterior to ventral end.

Color. Pale ventrally, peppered with small brown spots laterally and dorsally, the spots increasing in density toward the dorsum. Chin dark. Most head pores ringed with dark color. Fins pale.

Size. The largest specimen is the paratype, 317.9 mm TL, an immature female. The 274.4-mm holotype is a developing female with 0.6-mm eggs.

Development. Leptocephalus not identified, although larvae listed as *Ophichthus hyposagmatus* could possibly be this species.

Distribution. The two type specimens were captured off South Carolina by bottom trawl in deep water. A third very small specimen (USNM 149652) captured off Cuba many years ago is tentatively identified as this species; the preanal and total vertebral counts agree as does the general appearance of the head and dentition. Probably a bottom-dwelling species, with habits similar to those of *Echiophis, Kertomichthys* and *Mystriophis* which it closely resembles.

Etymology. Named in honor of William D. Anderson, Jr., who made the type-specimens available to the authors. A noun in the genitive case.

Study Material. The two types, 274.4–317.9 mm TL, plus one 59-mm specimen tentatively identified.

Holotype: ANSP 149062 (274.4); South Carolina, ca 97.5 km from Charleston Light, 32°29.3′N, 78°42.2′W, bottom trawl at 141–146 m; DOLPHIN; 2 April 1977.

Paratype: CAS 50931 (1, 318); South Carolina, 32°18′N, 78°50′W, bottom trawl at 192 m; DOLPHIN Cruise 5; 8 Nov. 1973.

Other material: USNM 149652 (1, 59); Cuba, 23°10′54″N, 81°18′W; ALBATROSS 2321; 17 Jan. 1885.

Genus *Myrichthys* Girard, 1859

Myrichthys Girard, 1859b: 58 (type species *Myrichthys tigrinus* Girard, 1859, by monotypy).

Chlevastes Jordan and Snyder, 1901: 867 (type species *Muraena colubrina* Boddaert, 1781, by original designation).

Characters. Body elongate (depth 2.5–3.0% TL in Atlantic species), cylindrical, laterally compressed far posteriorly; head and trunk shorter than tail (37–42% TL). Dorsal-fin origin on head well before gill opening; tail tip blunt; pectoral fin short, broad based. Snout short, broad from above, underside grooved; jaws short, the upper overhangs the lower; lips without barbels; eye moderate to large; anterior nostril tubular; posterior nostril opens into mouth. Gill opening vertical, semicircular. Head pores developed, two preopercular pores, three pores in supratemporal canal. Teeth molariform or granular, multiserial on jaws and vomer; intermaxillary and vomerine teeth the largest.

Neurocranium subtruncate posteriorly; supraoccipital crest weakly developed; maxilla elongate, slender posteriorly; pterygoid broad posteriorly, free and terete anteriorly; orbit large; otic bulla well developed; nasals and nasal cartilage moderately developed; opercular series developed. Branchiostegal rays numerous, slender, all or most along hyoid; urohyal a spike posteriorly. Gill arches well developed, *Ophichthus*-like, except teeth of pharyngeal plates, which are generally small and pavement-like; fifth ceratobranchial ossified; upper pharyngeal tooth plates separate. Pectoral girdle reduced; coracoid reduced, scapula and actinosts absent. Intramuscular bones, ribs, and caudal transverse processes developed. Precaudal vertebrae less numerous than caudal. Gasbladder poorly developed. Coloration spotted, banded, or both; peritoneum pale.

Species of the genus *Myrichthys* are easily identifiable to the generic level by the combination of the multiserial molariform teeth, dorsal-fin origin on the head, and the presence of the small pectoral fin.

Size. Grows to just over 1000 mm TL in the western Atlantic.

Distribution. A circumtropical genus, found in all tropical oceans.

Etymology. From the Greek *myrus* (eel) and *ichthys* (fish). Masculine.

Remarks. As a result of our studies, the names of both species of *Myrichthys* found in the western Atlantic must be changed. *Ophisurus breviceps* Richardson, 1848, was found to be a senior synonym of *Muraena acuminata* Gronow, published in Gray, 1854, based on examination of Richardson's holotype in the British Museum (see additional remarks under species description); therefore the species known as *Myrichthys acuminatus* becomes *M. breviceps* (Richardson).

The name change for the species known as *Myrichthys oculatus* (Kaup) is more complicated. The fish described by Lesueur as *Muraenophis ocellatus* (1825: 108, pl. 4, fig. 3, holotype unknown) from Barbados is not the western Atlantic species commonly known as *Ophichthus ocellatus* (Lesueur) (see *Remarks* under *O. puncticeps*, p. 402), but quite certainly a species of *Myrichthys*. It was described as an elongate, cylindrical, spotted eel with tubular anterior nostrils, dorsal fin beginning far forward on the head, with no pectoral fin evident; the illustration is readily recognizable as a species of *Myrichthys*. In addition, the behavior and habitat described, "... more agile; its cylindric and attenuated body enables it to pass readily between the rocks and corals of the shores of Barbadoes ...," is characteristic for only *Myrichthys* among the spotted snake eels (species of *Ophichthus* and *Quassiremus* usually observed buried in sand, seldom seen swimming in the manner described). However, the species identification is not clear; the illustration, with a series of pale spots on a dark body and smaller pale spots on the head, looks much like the species known as *M. acuminatus*; the coloration described in the text would more readily fit that previously known as *M. oculatus* ("*Body* ... of a rufous colour, ornamented with round and distant spots of orange, edged with blue; dorsal fin pale, edged with blue, united to the anal; *head* short, covered with small round orange spots"). It is entirely possible that Lesueur had seen both species. In order to avoid future nomenclatorial contention, and to prevent the phonetic confusion that might arise from assigning the name *ocellatus* to the species currently known as *acuminatus* (which would result in two western Atlantic species of *ocellatus* and *oculatus* in the genus *Myrichthys*), we declare *Muraenophis ocellatus* to be a senior synonym of *Pisodonophis oculatus* Kaup, and designate ANSP 107304 as neotype, the species to be known as *Myrichthys ocellatus* (Lesueur).

Species. Currently nine valid species of *Myrichthys*; the Pacific species might be more numerous, however, should populations of insular forms of *Myrichthys maculosus* prove to be unique. Three species present in the Atlantic, separable by subtle differences in vertebral counts (Table 35), the two western Atlantic species also separable by coloration; *M. breviceps* (= *M. acuminatus*) has diffuse pale spots on a dark body, *M. ocellatus* (= *M. oculatus*) has dark spots, usually with bright centers, on a pale background. *Myrichthys pardalis* of the eastern Atlantic is similar in appearance to *ocellatus* but differs in number of total vertebrae.

TABLE 39. Frequency distributions of vertebral counts in Atlantic species of *Myrichthys*.

	Predorsal		Preanal								
	2	3	57	58	59	60	61	62	63	64	65
breviceps	6	9					2	2	7	3	1
ocellatus	6	5	2	1	1	2	5				
pardalis†					X	X	X				

	Total																					
	151	152	153	154	155	156	157	158	159	...	164	165	166	167	168	169	170	171	172	173	174	175
breviceps												1	-	1	3	6	3	5	2	2	1	1
ocellatus											1	4	5	1	2	2	5	3	1	1		
pardalis†	1	-	2	-	-	-	-	1*	1													

* Includes holotype.

† From Blache and Cadenat (1971: 166–167); X = lateral-line pores to anus.

KEY TO THE ATLANTIC SPECIES OF *MYRICHTHYS*

1a. Coloration of round pale spots on a dark background of brown or green ...... *breviceps*, p. 374

1b. Coloration of diffuse dark spots with bright centers (gold in life) on a pale background 2

2a. Total vertebrae 151–159 .... *pardalis* (Valenciennes, 1835) (an eastern Atlantic species)

2b. Total vertebrae 164–173 .... *ocellatus*, p. 377

Myrichthys breviceps (Richardson, 1848)

SHARPTAIL EEL

Figures 383–386, Tables 37, 39

Ophisurus breviceps Richardson, 1844a: 99 (original description, locality unknown, holotype BMNH uncat.).

Muraena acuminata Gronow in Gray, 1854: 21 (original description, "in insula Divi Eustachii in America," holotype unknown).

Pisodonophis guttulatus Kaup, 1856a: 49 (original description, Martinique, syntypes MNHN B.2693, B.2694).

Pisoodonophis [sic] *guttulatus*. Kaup, 1856b: 21, fig. 10 (redescribed, head illustrated).

Pisoodonophis coronata Kaup, 1860: 14, pl. 2, fig. 1 (original description, St. Thomas, holotype ZMH 391).

Ophisurus longus Poey, 1867b: 254 (original description, Cuba, holotype MCZ 9155).

Ophichthys acuminatus. Günther, 1870: 83.

Pisodontophis longus. Poey, 1876a: 196.

Ophichthys longus. Poey, 1880: 253.

Ophichthys pisavarius Poey, 1880: 253 (original description, Cuba, holotype unknown).

Ophisurus acuminatus. Jordan, 1887a: 841.

Myrichthys acuminatus. Jordan and Davis, 1891: 619.

Leptocephalus undulatus Strömman, 1896: 27, pl. 2, figs. 4–6 (original description, 29°N, 72°W). Leptocephalus.

Discussion of Synonymy. In 1848 Richardson described a single specimen of *Ophisurus breviceps* present in the British Museum, originally from the College of Surgeons, habitat unknown. Kaup (1856a: 48) placed this species in his genus *Pisodonophis* (*Pisoodonophis* in 1856b: 20) and later (1860: 14) described and illustrated *Pisoodonophis coronata* from the island of St. Thomas in the West Indies. Günther (1870: 82) listed two specimens of *Ophichthys breviceps* (Richardson's type and a second specimen from the Pacific coast of Central America) and included Kaup's species *coronata* in synonymy. Richardson's holotype has been examined and x-rayed and found to be conspecific with the species commonly known as *Myrichthys acuminatus* Gronow, 1854, which must therefore take the name *breviceps*. While we have not examined the holotype of *Pisoodonophis co-*

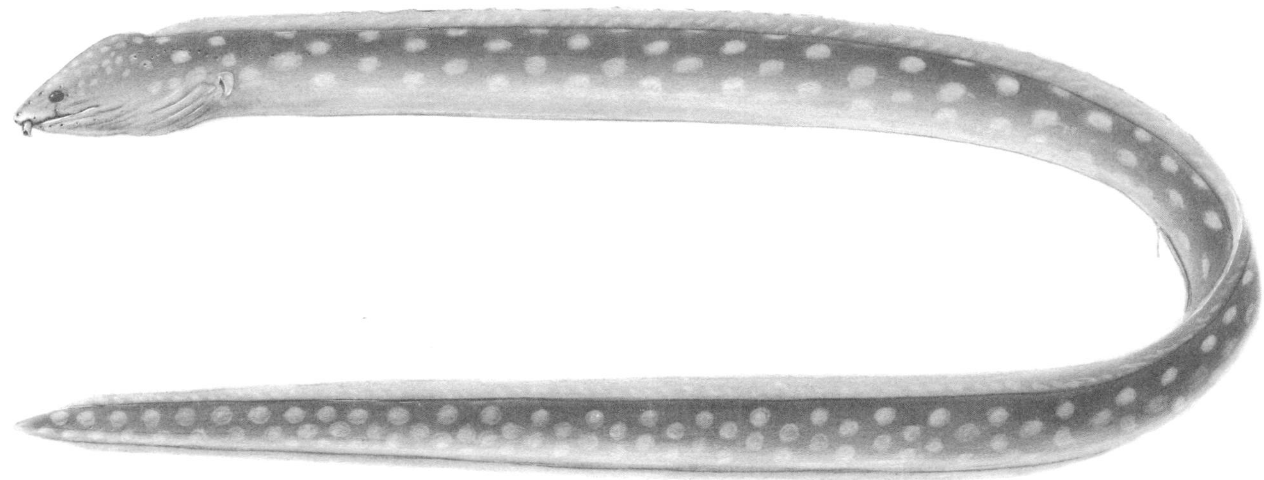

FIGURE 383. *Myrichthys breviceps*: ANSP 98359, 714 mm TL, Bahamas. Illustration by S. P. Gigliotti.

ronata, its description and illustration leave little doubt that Günther was correct in placing it in the synonymy of *breviceps.*

Distinctive Characters. An elongate ophichthin with two to four rows of pale round spots overlying brown or green background; dorsal-fin origin on head; pectoral fin small and broad-based; teeth molariform and multiserial; MVF 3-62-170.

Description. Total vertebrae 165–175, x̄ = 170.2 (n = 25); predorsal vertebrae 2–3, x̄ = 2.6 (15); preanal vertebrae 61–65, x̄ = 62.9 (15). Pores: SO 1 + 3; IO 4 + 2; POM 5 + 2; ST 3.

Proportions as % TL: head 7.5–8.6; trunk 30–34; tail 58–61; predorsal 4.8–6.3; depth behind gill openings 2.5–3.0. Of HL: snout 16–22; snout to rictus 27–33; eye 6.4–8.4; interorbital width 13–18; pectoral-fin length 5–7.9.

Head and jaws short, snout overhanging lower jaw, with short groove on underside. Eye moderate, above middle of upper jaw. Anterior nostril in prominent tube, longer posteriorly, with lateral projections which sometimes extend as small papillae. Posterior nostril opening within mouth, covered by a flap, and not marked externally.

Teeth stout and molariform, mostly biserial in small specimens, becoming multiserial with growth. In small specimens, 7–8 biserial intermaxillary teeth followed by a short gap; in larger specimens, a patch of 14–16 intermaxillary teeth which are continuous with vomerine teeth. Vomerine teeth 29–42, biserial anteriorly, becoming multiserial and extending far posteriorly. Maxillary tooth patch short, extending about half the length of vomerine patch; 18–35 teeth, biserial anteriorly to triserial posteriorly. Mandibular tooth patch about the length of vomerine patch; 33–70+ teeth, biserial anteriorly, multiserial posteriorly, becoming six rows across at posterior end in largest specimens. Intermaxillary and anterior vomerine and mandibular teeth the largest.

Color. Brown dorsally on head and trunk and overall on tail, venter pale; 30–35 pale spots, roughly in pairs, along each side of dorsal fin, plus one to three rows of alternating spots below. Numerous smaller spots on snout, cheek, and nape; a pair of pale spots in advance of dorsal fin origin. Dorsal fin variously spotted, the edge

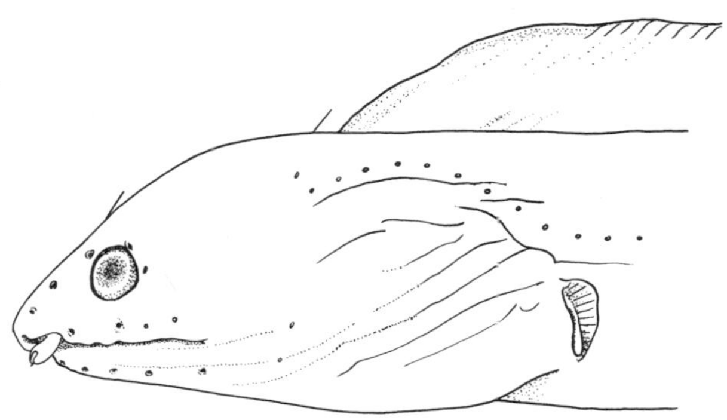

FIGURE 384. *Myrichthys breviceps*: ANSP 98360, 730 mm TL, Bahamas; head.

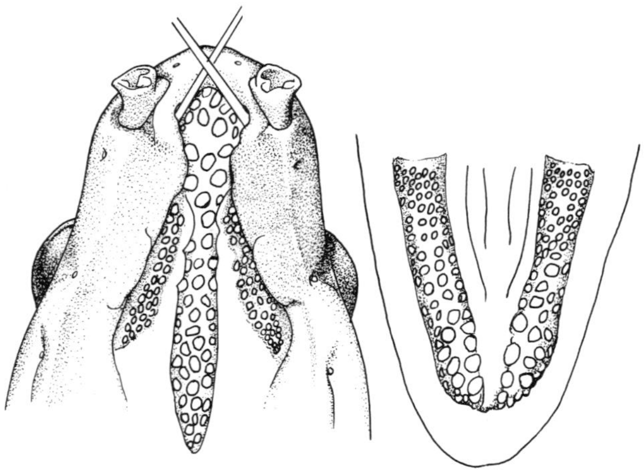

FIGURE 385. *Myrichthys breviceps*: ANSP 98360, 730 mm TL, Bahamas; dentition.

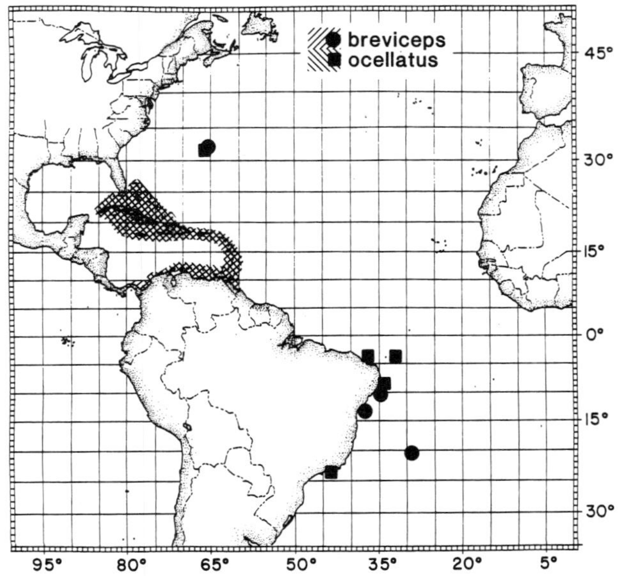

FIGURE 386. Distribution of western Atlantic species of *Myrichthys* based on study material.

usually appearing dark; anal and pectoral fins colorless. Small specimens (less than 150 mm) with only one row of spots along dorsum; larger specimens with two or more rows of spots. Very small specimens (less than 80 mm) have no spots and cannot be identified to species. In life, the background color is green or greenish-brown, the spots pale, yellowish, or with yellow centers.

Size. The largest specimen examined is 783 mm TL. Of 13 females (360–783 mm), the two largest (730–783) were ripe with 0.8-mm eggs; none of the four males (280–550) were ripe.

Development. Leptocephalus described as *Leptocephalus undulatus* by Strömman (1896: 27); see p. 827.

Behavior. Myrichthys species live in shallow-water grass beds or coral areas, and are seen threading their way through these areas during daylight hours (the only spotted snake eels likely to be seen). This behavior accounts for the occasional and incorrect reports in the popular press of "sea snakes" (family Hydrophiidae) in the western Atlantic. Randall (1968: 32) reported that *Myrichthys* feed mainly on crabs. Thresher (1985: 14) reported large aggregations of what were tentatively identified as *Myrichthys* near the surface well off the Atlantic coast of Panama, and suggested that they migrate to deeper waters to spawn.

Distribution (Fig. 386). An insular species, known from Bermuda, the Bahamas, the Florida Keys, throughout the West Indies and offshore islands south to Brazil. Collected over sand and turtle grass beds at depths recorded to 9 m, usually less than 3 m; depth recorded by Böhlke and Chaplin (1968: 106) for one specimen as "deeper than 180 m off the northern edge of the Great Bahama Bank" is questionable; it was collected by the OREGON during a cruise in the Bahamas, the station (and the exact locality) unknown.

We have examined and reidentified as *Myrichthys breviceps* a specimen (ROM 19448) reported as *Omochelys cruentifer* by Scott & Crossman (1959: 344), later illustrated with the coloration of *M. breviceps* in Leim and Scott (1966: 161). It was found in the stomach of a swordfish which was taken in the Laurentian Channel (45°30′N, 57°10′W). The specimen, although partially digested, exhibits the typical color pattern and has the vertebral formula of *M. breviceps.* However, we do not consider this record to constitute a range extension; most likely it was ingested at a more southern locality.

Etymology. From the Latin *brev* (short) and *ceps* (head) in reference to its short head. A noun in the genitive case.

Remarks. The two western Atlantic species of *Myrichthys* overlap broadly in meristic and morphological characteristics; they are separable primarily by coloration and by preanal vertebral counts (Table 35). The leptocephali are separable by differences in pigmentation and number of nephric myomeres. Juveniles and adults of both sexes of the two species have been taken throughout the study area, although never in the same collection.

Study Material. A total of 92 specimens, 60–941 mm TL, including the holotype, and the holotype of *Ophisurus longus* Poey and the syntypes of *Pisoodonophis guttulatus* Kaup. [Eight small specimens which cannot be identified to species are also listed.]

Holotype: BMNH uncat. (ca 880); habitat unknown; College of Surgeons No. 812.

BERMUDA: ANSP 124406 (1, 366), 133861 (1, 510). FLORIDA KEYS: ANSP 136002 (1, 550), 159275 (1, 357, cleared and stained). USNM 31456 (1, 450), 61116 (1, 660), 89691 (1, ca 510). BAHAMAS: ANSP 98359 (1, 714), 98360 (1, 730), 98361 (1, 249), 98362 (2, 254–360), 98363 (2, 198–280), 98364 (6, 67–253), 98365 (1, 783), 101724 (1, 742).

ANTILLES: *Cuba:* MCZ 9155 (1, 753); holotype of *Ophisurus longus.* USNM 24935 (1, 747), 89594 (1, 723). *Haiti:* ANSP 126670 (1, 343), 126678 (1, 687). *Jamaica:* LACM 5054 (1, 294). *Puerto Rico:* ANSP 118633 (2, 139–311), 126413 (2, 365–600), 139839 (1, 117–541). *Virgin Islands:* ANSP 153908 (1, 555). *Barbuda:* USNM 190494 (1, 120). *Dominica:* ANSP 105772 (1, 410). USNM 274246 (1, 364), 274250 (6, 143–585), 274251 (2, 239–292), 274255 (1, 157), 274256 (1, 269). *Martinique:* ANSP 105498 (1, 398). MNHN B.2693 (1) and MNHN B.2694 (1), syntypes of *Pisoodonophis guttulatus. St. Lucia:* ANSP 106071 (2, 119–161). *Barbados:* ANSP 106149 (1, 151). BMNH 1866.6.7.11.12 (1). CAS 54473 (2, 345–415). *Grenadines:* ANSP 105482 (1, 397), 105697 (1, 600), 105728 (1, 94), 105770 (1, 274), 106178 (1, 498). *Grenada:* ANSP 105771 (1, 382). *Curaçao:* CAS 59205 (1, 635). USNM 143183 (1, 626).

PANAMA: ANSP 113754 (1, 193). GRAND CAYMAN: ANSP 104904 (7, 84–211), 104912 (1, 260), 104982 (1, 155), 115519 (3, 206–242, cleared and stained). COLOMBIA: ANSP 117586 (1, 409). USNM 274244 (1, 287). VENEZUELA: MBUCV 775 (2), 1432 (1), 1497 (1), 2570 (1, 746). BRAZIL: MHNG 665-29 (1, 941). MNRJ 3476 (1, 626). MZUSP 12021 (1), 12022 (1), 12023 (1).

LOCALITY UNKNOWN: MNHN B.2695 (1), MNHN B.2697 (1). ROM 19448 (1, ca 600).

Species unknown: Vertebral counts are inconclusive and the color pattern on the following small specimens is not sufficiently developed to permit species identification: FLORIDA KEYS: USNM 73271 (1, 75). PUERTO RICO: ANSP 158800 (1, 74). USNM 184701 (1, 76), 287585 (1, 60). VIRGIN ISLANDS: UMML 29569 (1, 78). BRAZIL: GCRL 1569 (1, 66), 1579 (1, 68), 22314 (1, 65).

Myrichthys ocellatus (Lesueur, 1825)

GOLDSPOTTED EEL

Figures 386, 387, Tables 37, 39

Muraenophis ocellatus Lesueur, 1825: 108, pl. 4, fig. 3 (original description, Barbados, neotype ANSP 107304).
Pisodonophis oculatus Kaup, 1856a: 49 (original description, Curaçao, holotype unknown).
Pisoodonophis oculatus. Kaup, 1856b: 22.
Ophisurus latemaculatus Poey, 1867b: 252, pl. 3, fig. 1 (original description, Cuba, holotype MCZ 27223).
Pisodontophis latimaculatus. Poey, 1876a: 195.
Ophichthys latimaculatus. Poey, 1880: 252.
Myrichthys oculatus. Jordan and Davis, 1891: 618.
Myrichthys keckii Silvester, 1916: 214 (original description, Puerto Rico, Guanica Harbor, holotype PU 3082, not extant).

Discussion of Synonymy. As previously stated, *Muraenophis ocellatus* Lesueur is not the species commonly known as *Ophichthus ocellatus,* but is a species in the genus *Myrichthys.* For reasons presented above, we identify it as a senior synonym of the species previously known as *Myrichthys oculatus,* and designate ANSP 107304 as neotype.

Distinctive Characters. An elongate ophichthin with bold coloration of diffuse dark spots with small bright centers on a pale body; dorsal-fin origin on head; pectoral fin short and broad-based; teeth molariform and multiserial; MVF 2-59-168.

Description. Total vertebrae 164–173, $\bar{x}$ = 168.2 (n = 24); predorsal vertebrae 2–3, $\bar{x}$ = 2.4 (10); preanal vertebrae 57–61, $\bar{x}$ = 59.5 (10). Pores: SO 1 + 3; IO 4 − 2; POM 5 + 2; ST 3.

Proportions as % of TL: head 7.6–9.4; trunk 29–32; tail 60–63; predorsal 2.3–6.1; depth behind gill openings 2.5–3. Of HL: snout 17–20; snout to rictus 27–34; eye 5.9–8.2; interorbital width 13–16; pectoral-fin length 4.7–7.4.

General characters and dentition similar to those of *M. breviceps.* Two specimens with six pores along mandible.

Color. Body pale tan or brown, overlain with 25–30 pairs of ocellated spots along body on either

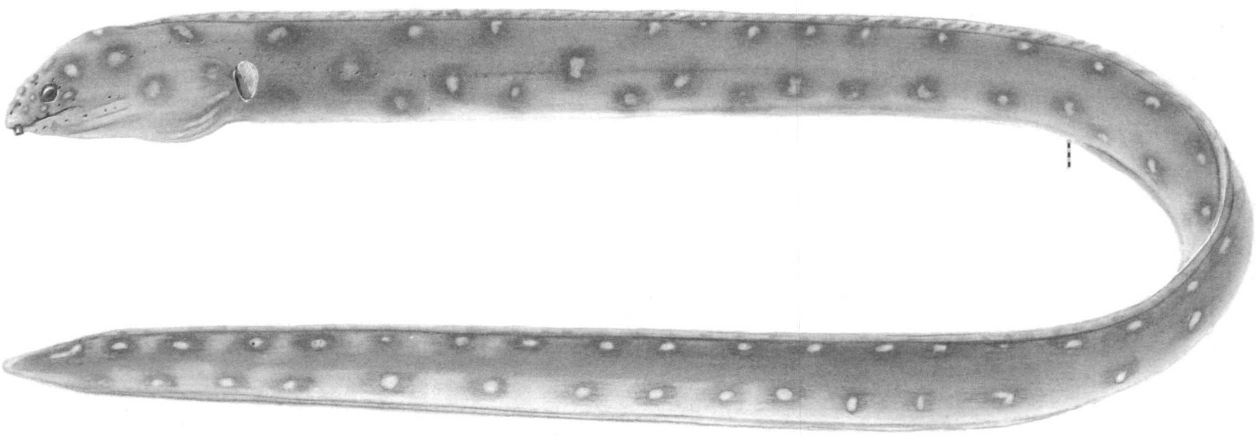

FIGURE 387. *Myrichthys ocellatus*: ANSP 98410, 494 mm TL, Bahamas. Illustration by S. P. Gigliotti.

side of dorsal fin, and a second row of spots alternating in interspaces and lying just below lateral line. Dorsal spots have bright well-defined centers surrounded by diffuse dark rings; the centers may contain small amounts of dark pigment in largest specimens. Spots in lower row sometimes diffuse and without bright centers, or sometimes missing (juveniles and smallest specimens). Snout and postocular region with smaller ocellated spots; a single large ocellated spot on nape just in front of dorsal-fin origin. A series of spots on dorsal fin, distal edge of fin usually pale; anal and pectoral fins colorless. In life, the background color varies from pale to deep greenish brown dorsally, becoming pale greenish ventrally; central spots bright yellow to orange, surrounded by dark rings.

Size. The largest specimen examined is 1080 mm TL. Of 13 females (234–585 mm), one was ripe with 0.7-mm eggs; of eight males (424–955 mm), five (525–955 mm) were ripe.

Development. Leptocephalus identified (see p. 829).

Distribution (Fig. 386). An insular species, known from Bermuda, the Florida Keys, throughout the West Indies, and south to Brazil. A common species, taken more often than *Myrichthys breviceps* (except in the Florida Keys where it is rare); found in similar habitats of shallow grass beds or coral areas, at depths to 7 m. Not infrequently seen swimming through grass beds or among corals. (See *Behavior* for preceding species.)

Remarks. The western Atlantic goldspotted eel has been recognized as *Myrichthys oculatus* (= *ocellatus*) in recent literature, although Günther (1870: 82) and Fowler (1936a: 295) both synonymized it with the eastern Atlantic *M. pardalis* (Valenciennes, 1835). Although meristic and morphometric data as well as coloration are similar for the two nominal forms, vertebral counts given for *M. pardalis* by Blache and Cadenat (1971: 167) are considerably outside the range found for *M. ocellatus* (151–159 vs 164–173). We continue to recognize the two forms as distinct species.

Etymology. From the Latin *ocellatus* (spotted as with little eyes), in reference to the eye-like spots. An adjective.

Study Material. A total of 184 specimens, 85–1080 mm TL, including the neotype and the holotype of *Ophisurus*

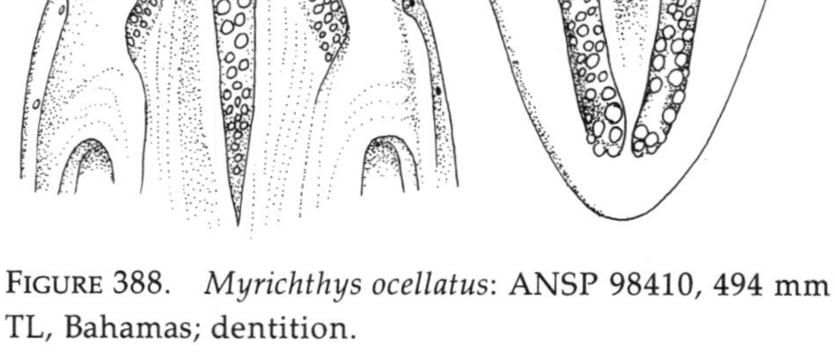

FIGURE 388. *Myrichthys ocellatus*: ANSP 98410, 494 mm TL, Bahamas; dentition.

latemaculatus Poey. [An additional 70 specimens in 19 lots from Brazil not included.]

Neotype: ANSP 107304 (451); Lesser Antilles, Martinique, 15°23′N, 61°26′W, 15 m, by fish pump; OREGON 5983; 20 Mar. 1966.

BERMUDA: ANSP 151999 (posterior half, 485+). USNM 274247 (1, 848). FLORIDA KEYS: UMML 532 (1, 790+). UMML 1264 (3, 146–170). BAHAMAS: ANSP 72248 (1, 549), 98398 (1, 210), 98399 (1, 342), 98400 (3, 154–173), 98401 (1, 276), 98402 (1, 544), 98403 (11, 155–534), 98404 (4, 174–424), 98405 (1, 593), 98406 (28, 85–590), 98407 (23, 100–685), 98408 (1, 668), 98409 (3, 378–502), 98410 (1, 494), 98411 (1, 597), 98412 (1, 606), 98413 (1, 955), 115516 (3, 218–445, cleared and stained), 116318 (5, 292–410), 120713 (1, 175, cleared and stained).

CUBA: MCZ 27223 (1, 738); holotype of *Ophisurus latemaculatus.* USNM 33092 (1, 703), 57788 (1, 672). SANTO DOMINGO: ANSP 110577 (1, 614). PUERTO RICO: ANSP 115552 (5, 198–506), 115645 (1, 234), 118634 (1, 326), 118706 (1, 163), 129905 (6, 266–510), 139840 (1, 327). CAS 54474 (3, 243–465). FMNH 74763 (13, 190–554). USNM 126437 (1, ca 605), 274242 (1, 535), 274252 (2, 420–565). VIRGIN ISLANDS: ANSP 97935 (1, 510), 124484 (3, 136–355). LESSER ANTILLES: *St. Martins:* ANSP 1054 (1, 391). *Antigua:* USNM 170115 (1, 476). *Guadeloupe:* MNHN B.2703 (1). *Martinique:* ANSP 105486 (1, 270+). *St. Lucia:* ANSP 110088 (3, 278–654). *Curaçao:* CAS 59205 (1, 637). ZMA 113.015 (1); Boeke. *Grand Cayman:* ANSP 104907 (7, 143–318), 104916 (1, 298), 104921 (1, 525), 104981 (1, 167).

BELIZE: ANSP 159276 (1, 230, cleared and stained). COLOMBIA: ANSP 117583 (10, 328–582). VENEZUELA: MBUCV 773 (1, 520), 775 (1). BRAZIL: CAS-SU 22126 (1, 530), 51728 (1, 413). MNHG 665-30 (1, 521). MNHN B.2722 (1, 822). MNRJ 9513 (1, 174), 9844 (1, 281). MNRJ uncat. (1, 1080). *Fernando Noronha:* USNM 274248 (4, 147–400). [MZUSP—19 lots, 70 specimens, not studied.]

LOCALITY UNKNOWN: UF 17662 (1, 190, cleared and stained). USNM 112038 (1, ca 710); from stomach of swordfish.

Comparative material: Myrichthys pardalis from the eastern Atlantic: *Cape Verde Islands:* BMNH 1935.7.10.1 (1). BMNH 1937.2.10.2 (1).

Genus *Ophichthus* Ahl, 1789

Innominado Parra, 1787: 96 (a junior synonym of *Muraena ophis* Linnaeus, 1758, non-binomial).

Ophichthus Ahl, 1789: 5 (type species *Muraena ophis* Linnaeus, 1758, by original designation). Improperly emended to *Ophichthys* by other authors.

Ophis Turton, 1807: 87 (type species "*O. maculata* . . . Spotted Serpent. Shaw Zool., iv. p. 22 . . . Bloch t. 154," = *Muraena ophis* Linnaeus, 1758, by monotypy).

Cogrus Rafinesque, 1810a: 62 (type species *Cogrus maculatus* Rafinesque, 1810, by monotypy).

Ophithorax McClelland, 1844: 212 (type species *Ophisurus ophis* Lacepède, 1800 = *Muraena ophis* Linnaeus, 1758, by Jordan, 1919, as first reviser).

Centrurophis Kaup, 1856a: 42 (type species *Ophisurus spadiceus* Richardson, 1844 = *Ophichthys cephalazona* Bleeker, 1864, by Jordan, 1919, as first reviser).

Poecilocephalus Kaup, 1856a: 43 (type species *Poecilocephalus bonaparti* Kaup, 1856, by monotypy).

Microdonophis Kaup, 1856a: 43 (type species *Microdonophis altipinnis* Kaup, 1856, by monotypy).

Coecilophis Kaup, 1856a: 44 (type species *Ophisurus compar* Richardson, 1844 = *Ophisurus apicalis* Bennett, 1830, by monotypy).

Scytalophis Kaup, 1856a: 46 (type species *Scytalophis magnioculis* Kaup, 1856 = *Ophisurus gomesii* Castelnau, 1855, by Jordan, 1919a, as first reviser).

Leptorhinophis Kaup, 1856a: 46 (type species *Ophisurus gomesii* Castelnau, 1855, by Jordan 1919a, as first reviser).

Cryptopterus Kaup, 1860: 11 (type species *Cryptopterus puncticeps* Kaup, 1860, by monotypy).

Uranichthys Poey, 1867b: 256 (type species *Muraena hauannensis* Bloch and Schneider, 1801 = *Muraena ophis* Linnaeus, 1758, by Jordan and Davis, 1892, as first revisers).

Paramyrus Günther, 1870: 51 (type species *Conger cylindroideus* Ranzani, 1839, by Jordan and Davis, 1892, as first revisers).

Oxyodontichthys Poey, 1880: 254 (type species *Ophichthys macrurus* Poey, 1867 = *Ophisurus gomesii* Castelnau, 1855, by original designation).

Omochelys Fowler, 1918: 3 (type species *Pisodonophis cruentifer* Goode and Bean, 1895, by original designation; described as a subgenus of *Pisodonophis* Kaup, 1856).

Syletor Jordan, 1919c: 343 (type species *Pisodonophis cruentifer* Goode and Bean, 1895, by original designation).

Styletor Jordan, 1919 in Jordan, Evermann, and Clark, 1930: 86 (lapsus for *Syletor* Jordan, 1919).

Acanthenchelys Norman, 1922b: 296 (type species *Acanthenchelys spinicauda* Norman, 1922, by original designation).

Cryptopterenchelys Fowler, 1925a: 1 (substitute name for *Cryptopterus* Kaup, 1860, preoccupied; described as a subgenus of *Ophichthus* Ahl, 1789).

Zonophichthus Whitley, 1930: 250 (type species *Ophichthys cephalazona* Bleeker, 1864, by original designation).

Gisenchelys Fowler, 1944: 188 (type species *Ophichthys zophochir* Jordan and Gilbert, 1881, by original designation; described as a subgenus of *Ophichthus* Ahl, 1789).

Syletophis Whitley, 1950: 44 (substitute name for *Syletor* Jordan, 1919, preoccupied).

Antobrantia Pinto, 1970: 13 (type species *Antobrantia ri-*

beiroi Pinto, 1970 = *Muraena ophis* Linnaeus, 1758, by original designation).

Characters. Body moderately elongate to stout (depth 1.5–7.7% TL in Atlantic species), laterally compressed posteriorly; head and trunk usually shorter than tail (29–52% TL in Atlantic species). Dorsal-fin origin above or behind gill openings; tail tip blunt; pectoral fin developed, rounded or lanceolate, not extremely broad-based. Snout moderately developed, subconical, overhanging lower jaw; jaws moderately elongate; lips usually without barbels, or one or two between nostrils; eye moderate; anterior nostril tubular; posterior nostril opening along lower edge of lip or into mouth, usually partially covered by a flap. Gill opening ventral, elongate and crescentic. Head pores developed; two or three preopercular pores, three pores in supratemporal canal. Dentition variable, from numerous, small, and multiserial to few, large, and uniserial or biserial, never caniniform; largest teeth the intermaxillary and vomerine teeth.

Neurocranium subtruncate posteriorly; supraoccipital crest moderately to well developed; orbit large; postorbital strut developed in some species; nasals and nasal cartilage moderately developed; maxilla elongate, slender or rounded posteriorly; pterygoid broad posteriorly, free and terete anteriorly; otic bulla well developed; suspensorium nearly vertical to posteriorly inclined; opercular series well developed, their margins entire. Branchiostegal rays slender, often numerous, unbranched and along arch; urohyal a spike posteriorly. Gill arches well developed; first basibranchial ossified, 2–4 cartilaginous, rudimentary or absent; hypobranchials 1–2 ossified, 3 cartilaginous; ceratobranchials 1–5 ossified; infrapharyngobranchials 2–3 ossified; upper pharyngeal tooth plates ossified and separate. Pectoral girdle well developed; cleithrum and supracleithrum stout, scapula and coracoid broad, actinosts present. Intramuscular bones, ribs, and caudal transverse processes developed. Caudal vertebrae more numerous than precaudal vertebrae. Gasbladder developed, elongate, about 20–25% of TL. Coloration variably uniform, bicolored, saddled, banded or spotted; peritoneum pale.

Size. Considerable variation in size of species; includes the largest known ophichthin species. In the western Atlantic, maximum size varies from 281 mm for *Ophichthus menezesi* to over 2.4 m for the giant snake eel, *Ophichthus rex*.

Distribution. Ophichthus (sensu lato) is the most speciose of ophichthid genera, with approximately 50 tropical and subtropical species worldwide.

Etymology. From the Greek *ophis* (snake) and *ichthus* (more correctly written *ichthys*) (fish), in reference to its snake-like appearance. Masculine.

Remarks. Several subgenera are identifiable within *Ophichthus* (McCosker, 1977); however a worldwide revision has not been attempted. In the western Atlantic, most species would be in the subgenus *Ophichthus.*

Species. Twelve species of *Ophichthus* are found in the western Atlantic; they are highly variable in coloration. *Ophichthus brasiliensis* (Kaup, 1856), *O. cruentifer* (Goode and Bean, 1896), *O. cylindroideus* (Ranzani, 1839), and *O. gomesii* (Castelnau, 1855) are uniformly colored or bicolored; *O. melanoporus* Kanazawa, 1963, and *O. omorgmus* McCosker and Böhlke, 1984, are pale with small dark spots marking head pores and lateral-line pores; *O. punctifer* (Kaup, 1860) (*ocellatus* of authors) has a row of white spots along a darker body; *O. menezesi* McCosker and Böhlke, 1984, is pale with dark blotches along midbody; *O. ophis* (Linnaeus, 1758) has large dark spots on a pale body; and *O. hyposagmatus* McCosker and Böhlke, 1984, *O. spinicauda* (Norman, 1922) and *O. rex* Böhlke and Caruso, 1980, have dark saddles on a pale background. Species are further separated by pore and fin conditions and by differences in dentition and vertebral counts (Tables 37 and 40).

Status of Ophichthus pauciporus (Poey, 1867). Ophichthys pauciporus was briefly described and the head illustrated by Poey (1867b: 255, pl. 3, fig. 5). To date, no specimen is known which can be identified to this species, and the 760 mm holotype is apparently lost. After an exhaustive

TABLE 40. Summary of distinguishing characters for western Atlantic species of *Ophichthus*. (See Table 37 for vertebral counts.)

Coloration of species	Distinctive characters	Proportions (% TL) Tail	Depth	DFO†	Head pores SO*	POP	Jaw teeth
Pale or bicolored							
cylindroideus	anterior nostril barbel	63–68	2.4–5.4	13–17	3	7 + 3	biserial
gomesii	DFO before pectoral tips	62–65	2.7–3.7	13–17	3	6 + 2	biserial
cruentifer	DFO behind pectoral tips, biserial dentition	57–62	2.3–3.0	10–16	4	5–6 + 2–3	biserial
brasiliensis	DFO behind pectoral tips, uniserial dentition	56	2.7	19	3	6 + 2	uniserial
Pale with dark pore spots							
melanoporus	slender uniserial teeth	69–71	1.5–1.8	8.2–9 4	4	6 + 3	uniserial
omorgmus	stout biserial teeth	55	3.1	14	3	6 + 3	biserial
Large spots or blotches							
puncticeps	one row of white spots on gray body	51–56	3.7–5.0	15–18	3	6 + 3	biserial
ophis	several rows of irregular black spots	57–61	2.9–3.9	11–14	4	6 + 3	biserial
menezesi	one row dark blotches on pale body	58–62	2.4–3.0	14–18	4	5 + 2	biserial, uniserial
Dark saddles or bands							
hyposagmatus	DFO above pectoral tips	53–57	2.9–4.5	14–16	3	6–7 + 2–3	biserial, uniserial
rex	DFO behind pectoral tips	50–54	4.5–7.7	17–19	4	7 + 3	multiserial
spinicauda	DFO far behind pectoral tips	48–53	4.5–6.4	17–21	3	6 + 3	biserial

* Supraorbital pores, excluding first (ethmoid) pore.
† Dorsal-fin origin = predorsal length.

analysis of all other Atlantic taxa, we are unable to apply Poey's name with certainty.

Jordan and Davis (1891: 634) placed *pauciporus* in the synonymy of *Ophichthus parilis* (= *O. cylindroideus*) based only on the descriptions of both species. However, Poey did not mention the anterior nostril tentacle, which is characteristic for *O. parilis*, nor is it depicted in his illustration, and he stated that the dorsal-fin origin was behind the pectoral tips rather than before that point. The latter difference was noted but discounted by Jordan and Davis, who stated that "in *pauciporus* the insertion of the dorsal is said to be behind (*mas atras*) the end of the pectoral. It is before the tip in *parilis*, but this is likely either an error or an individual variation." We find their conclusion unlikely.

The species to which *O. pauciporus* is most similar are those with biserial dentition and a posterior dorsal-fin origin, viz. *cruentifer*, *rex*, and *spinicauda*. *Ophichthus cruentifer* is unknown from waters approaching Cuba, and the maximum known size of 467 mm TL would make it an unlikely candidate. The large *Ophichthus* species, *rex* and *spinicauda*, agree in size, dorsal-fin origin, and dentition, but have a somewhat shorter tail than that of *pauciporus*, dark saddles or bands along the body, and share a peculiar "spinous ray" condition in the posterior portion of the dorsal fin, also not mentioned by Poey. We therefore cannot identify any extant specimens as *Ophichthus pauciporus* and, lacking Poey's type specimen, consider it to be a *nomen dubium*.

Key to the Western Atlantic Species of *Ophichthus*

1a. Anterior nostril with a tentacle from its inner rim longer than length of nostril tube; small barbel along lip between anterior and posterior nostrils; coloration uniform, fins with black-edges; teeth on jaws and on vomer biserial .... *cylindroideus,* p. 386
1b. Anterior nostril tubular, lacking an obvious tentacle from its inner rim 2
2a. Teeth of upper jaw uniserial (small juveniles may have biserial teeth posteriorly) throughout their length; body elongate, not robust 3
2b. Teeth of upper jaw biserial for most of their length; body moderately elongate to stout ... 4
3a. Tail 70% or more of TL; pores of head and lateral line dark and conspicuous, in striking contrast to head and body .... *melanoporus,* p. 394
3b. Tail less than 60% of TL; coloration uniform, pores of head and lateral line not dark .... *brasiliensis,* p. 383
4a. Dorsal-fin origin well behind tips of pectoral fins 5
4b. Dorsal-fin origin above, slightly behind, or in advance of tips of pectoral fins 8
5a. Body dark (although median fins may be lighter), lacking bands or spots; teeth small, biserial to multiserial in jaws .... *cruentifer,* p. 384
5b. Body banded or spotted; teeth stout, upper jaw teeth biserial or multiserial; teeth of vomer uniserial or biserial 6
6a. Body with dark blotches equal to or larger than eye along flank; tail longer than 60% of TL .... *menezesi,* p. 396
6b. Body without spots, but with 13–15 dark saddles; tail less than 60% of TL 7
7a. Dorsal-fin origin behind pectoral-fin tip by a distance greater than length of pectoral fin; vomerine teeth of adults uniserial; total vertebrae 144–147 .... *spinicauda,* p. 407
7b. Dorsal-fin origin behind pectoral-fin tip by a distance less than the length of pectoral fin; vomerine teeth of adults biserial to triserial; total vertebrae 115–119 .... *rex,* p. 405
8a. Vomerine teeth biserial to triserial; tail longer than 60% of TL; body without dark bands or light or dark spots .... *gomesii,* p. 389
8b. Vomerine teeth uniserial (although sometimes a pair anteriorly); tail shorter than 60% of TL; body with dark bands or dark or light spots 9
9a. Body with brown bands (although faint or lost in preservative in many specimens), dorsal fin edged with brown .... *hyposagmatus,* p. 393
9b. Body with dark or pale spots, lacking broad bands 10
10a. Body grey or bicolored, dark dorsally, with pale eye-size spots along midline; head pores in dark spots .... *puncticeps,* p. 402
10b. Body pale, with irregular large dark blotches or with small dark spots outlining lateral-line and head pores 11
11a. Large dark spots and blotches along pale body, often meeting along the back; head, chin and throat freckled with numerous small dark spots .... *ophis,* p. 399
11b. Lateral-line pores and head pores set in small dark spots, sharply contrasting with pale body color; dorsal fin with black edge .... *omorgmus,* p. 397

Ophichthus brasiliensis (Kaup, 1856)

BRAZILIAN SNAKE EEL

Figures 389–392, Tables 37, 40

Centrurophis brasiliensis Kaup, 1856a: 43 (original description, "Rio Janeiro durch Quoy und Gaim," holotype MNHN A.8674). Kaup, 1856b: 4.
Ophichthys brasiliensis. Günther, 1870: 73.
Ophichthus brasiliensis. Jordan and Davis, 1892: 628.

Distinctive Characters. An elongate ophichthin with uniform dark coloration; tail 56% of TL; dorsal-fin origin well behind pectoral-fin tips; pectoral fin spatulate, its length 22% of HL; three

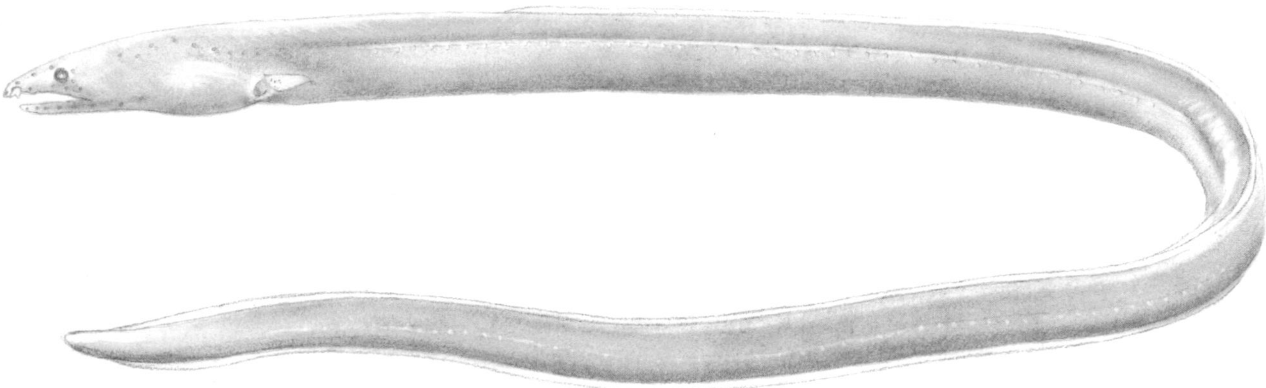

FIGURE 389. *Ophichthus brasiliensis*: MNHN A.8679, holotype, 405 mm TL, Brazil.

dorsal supraorbital, six mandibular and two preopercular pores; dentition uniserial except anteriorly on mandible; MVF 23-57-153.

Description. Total vertebrae 153 (n = 1); predorsal vertebrae 23; preanal vertebrae 57. Pores: SO 1 + 3; IO 4 + 2; POM 6 + 2; ST 3; lateral-line pores terminating far (38 mm) in advance of tail tip.

Proportions as % of TL: head 8.8; trunk 35; tail 56; predorsal 19; depth behind gill openings 2.7. Of HL: snout 21; snout to rictus 33; eye 8.4; interorbital width 12; pectoral-fin length 22.

Jaws short; eye small, over posterior half of upper jaw. Anterior nostril tubular; posterior nostril opening within mouth, not marked externally. Head pores small, as listed above. Teeth conical, stout and subequal, basically uniserial. A patch of eight intermaxillary teeth, a short gap followed by four staggered and 11 uniserial teeth on vomer. Maxillary teeth 17; mandibles with outer row of 19–22 teeth plus three larger inner teeth anteriorly.

Color. After many years in preservative, the holotype is uniformly brown, the median fins pale. Originally described as yellowish brown with black dots.

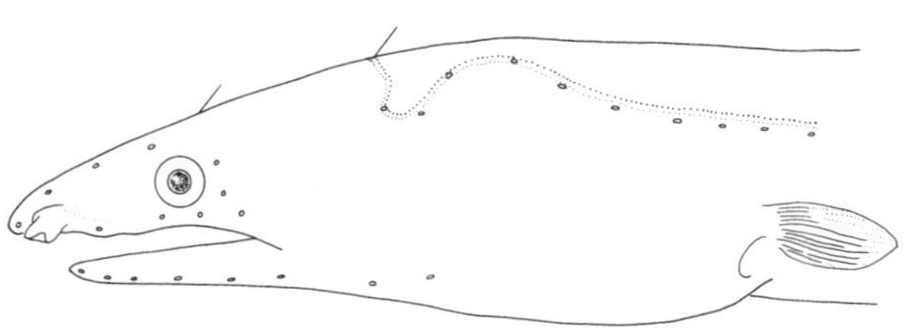

FIGURE 390. *Ophichthus brasiliensis*: MNHN A.8679, holotype, 405 mm TL, Brazil; head.

Size. The single known specimen is 405 mm TL.

Distribution (Fig. 392). "Rio Janeiro durch Quoy und Gaim." is the locality data given for the holotype (see *Remarks*).

Etymology. Named for the supposed country of its origin. An adjective.

Remarks. This species is similar to *Ophichthus menezesi* in several characters (general appearance, dorsal-fin placement, vertebral number, and body elongation); it differs in having uniserial rather than biserial maxillary dentition, a shorter tail, and three rather than four (as has *O. menezesi*) pores on the dorsal surface of the snout. In that no subsequent specimens of *O. brasiliensis* have been taken since its description well over a century ago, it seems possible that the locality

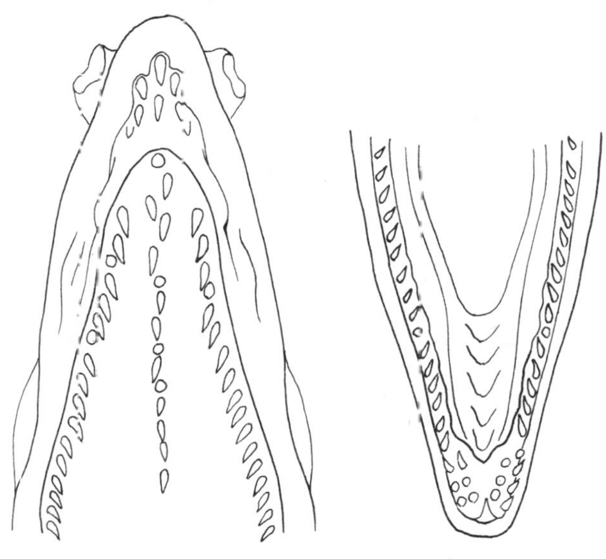

FIGURE 391. *Ophichthus brasiliensis*: MNHN A.8679, holotype, 405 mm TL, Brazil; dentition.

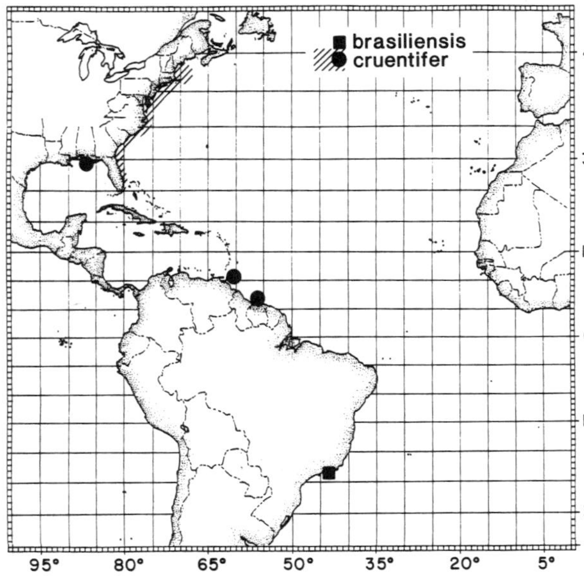

FIGURE 392. Distribution of *Ophichthus brasiliensis* and *O. cruentifer* based on study material.

may be erroneous. Examination of the records, field notes, and letters relating to this specimen and to Quoy and Gaimard's expedition (M. L. Bauchot, in litt.) did not clarify this matter.

Study Material. The holotype: MNHN A.8674 (405 mm TL); BRAZIL, Rio de Janeiro; Quoy and Gaimard.

Ophichthus cruentifer (Goode and Bean, 1896)

MARGINED SNAKE EEL

Figures 392–395, Tables 37, 40

Pisoodonophis [sic] *cruentifer* Goode and Bean, 1896: 147, fig. 166 (original description, Gulf Stream, lectotype USNM 28938).

Omochelys cruentifer. Fowler, 1918: 3 (referred to *Omochelys* new genus).

Leptocephalus conger (not of Linnaeus) Eigenmann, 1902: 37, figs. 1–14 (original description, eggs hatched in laboratory). Leptocephalus.

Leptocephalus mucronatus Eigenmann and Kennedy, 1902: 90, fig. 11 (original description, 41°07′N, 65°26′W and 38°25′N, 72°40′W, syntypes USNM 49746-49747). Leptocephalus.

Omochelys marginatus Fowler, 1948: 1, figs. 1–5 (original description, New Jersey, holotype ANSP 72103).

Omochelys cruentifer. Fowler, 1952: 110 (*O. marginatus* synonymized with *O. cruentifer*).

Pisodonophis cruentifer. Wenner, 1976: 656 (biology and morphology of eggs and adults).

Ophichthus cruentifer. McCosker, 1977: 80 (*Omochelys* synonymized with *Ophichthus*).

Misidentification:

Not *Omochelys cruentifer.* Scott and Crossman, 1959: 344; Leim and Scott, 1966: 161 (= *Myrichthys breviceps* (Richardson)).

Distinctive Characters. An elongate ophichthin with nearly uniform greyish-brown coloration; tail 57–62% TL; dorsal-fin origin well behind pectoral-fin tips; pectoral fin broad, spatulate, its length 20–30% HL; four dorsal supraorbital, five or six mandibular and two or three preopercular pores; teeth small, dentition mostly biserial; MVF 16-58-149.

Description. Total vertebrae 144–155, $\bar{x}$ = 149.1 (n = 48); predorsal vertebrae 14–19, $\bar{x}$ = 16.3 (33); preanal vertebrae 56–61, $\bar{x}$ = 57.9 (31). Pores: SO 1 + 4; IO 4 + 2; POM 5–6 + 2–3; ST 3.

Proportions as % of TL: head 6.9–8.7; trunk 31–34; tail 57–62; predorsal 9.5–16; depth behind gill openings 2.3–3. Of HL: snout 20–23; snout to rictus 29–37; eye 5.6–9.2; interorbital width 9.6–16; pectoral-fin length 20–30.

Snout and jaws short; eye moderate, posterior margin above and just anterior to rictus. Anterior nostril tubular, with dorsal groove; posterior nostril opening outside mouth, entirely covered by fleshy flap which projects below lip-line. Head pores small, inconspicuous, as listed above; variation in number of pores along mandible (usually six, occasionally five) and preopercular pores (usually three, occasionally two).

Teeth small, conical and pointed, mostly biserial. Usually an anterior median tooth followed by two pairs of intermaxillary teeth (the anterior a pair also in one cleared and stained specimen), continuous with vomerine teeth or only a short space separating them. On vomer, 3–7 pairs (sometimes headed by a median anterior tooth), followed by a row of 9–15 staggered to uniserial teeth; the largest specimen had 27 teeth on the vomer, triserial anteriorly tapering to staggered uniserial posteriorly. Maxilla with 15–20 teeth in outer row, 10–18 in inner row, plus 2–5 forming a short triserial patch anteriorly in large specimens. Dentary with 17–25 outer teeth, 3–13 in inner row (three anterior teeth on smallest specimens, a triserial patch on largest specimens).

Median fins low, folded into groove. Dorsal-

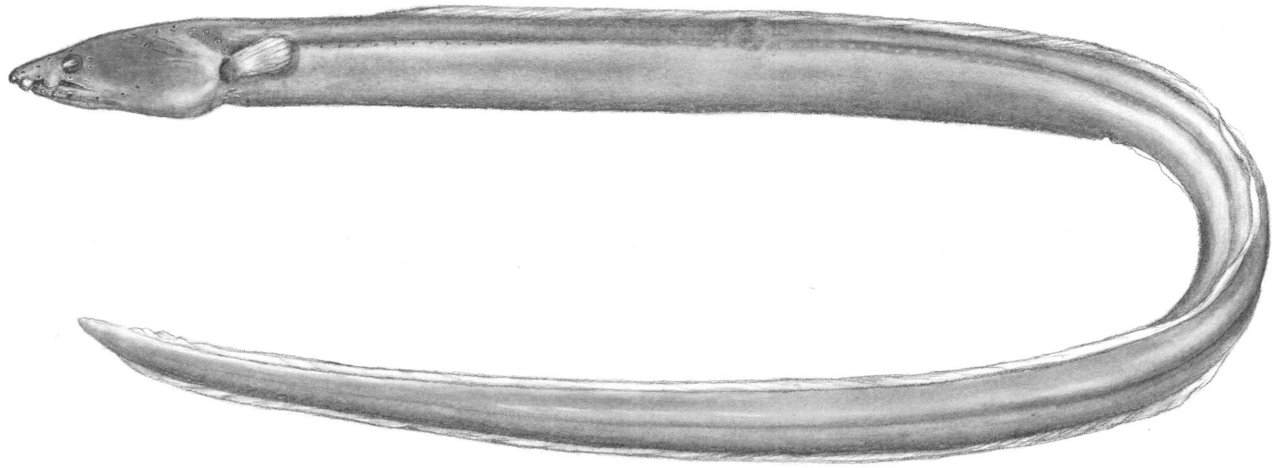

FIGURE 393. *Ophichthus cruentifer*: ANSP 117711, 338 mm TL, Delaware.

fin origin always well behind tips of pectoral, although considerable variation in distance between pectoral-fin tips and dorsal-fin origin. Pectoral fin variably small and rounded to broad and spatulate.

Color. Uniform tan to greyish-brown body, pigment widely scattered on small (tan) specimens and confluent on larger ones, coloration uniform, not bicolored. Fins colorless in small specimens, body pigment extending onto dorsal fin and pectoral base in larger ones, giving the illusion of a dark margin to those areas. The dark margins on the dorsal and pectoral fins described for the holotype of *Omochelys marginatus* are not evident, although both fins are pigmented throughout; other specimens show overall pigmentation only.

Size. The largest specimen examined was 467 mm TL. Of 18 females (177–467 mm), eight (200–467) were ripe, with eggs of 0.4–1.0-mm diameter. Of three males (255–301), only the largest was ripe.

FIGURE 394. *Ophichthus cruentifer*: ANSP 117711, 338 mm TL, Delaware; head.

Development. Leptocephalus described as *Leptocephalus conger* Eigenmann, 1902, as *Leptocephalus mucronatus* Eigenmann and Kennedy, 1902, and subsequently studied by various authors (see p. 809). Richardson (1974) described the eggs and larvae of *O. cruentifer* from plankton tows off the continental shelf off Virginia. Wenner (1976) described and illustrated the sexual condition of 84 ripe adults from Norfolk Canyon; the sex ratio of mature individuals did not deviate significantly from equality, and mature females were significantly larger (227–423 mm TL, $\bar{x}$ = 278) than mature males (197–317 mm TL, $\bar{x}$ = 263).

Behavior. The discovery of "mummified" specimens of *Ophichthus cruentifer* and other

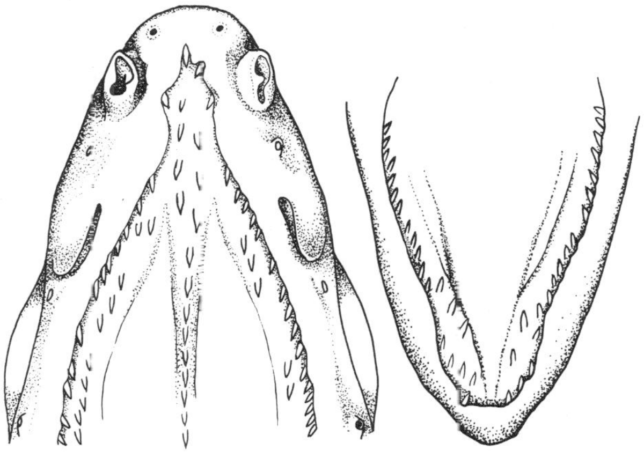

FIGURE 395. *Ophichthus cruentifer*: ANSP 98521, 376 mm TL, Mississippi; dentition.

ophichthids in the coelomic cavity of larger predatory fishes suggested to earlier authors that these species were parasitic borers, not unlike *Simenchelys parasiticus* (Goode and Bean, 1896; Deraniyagala, 1932; Breder and Nigrelli, 1934; Breder, 1953). Walters (1955) dispelled that notion and suggested that if swallowed entire, the tail-burrowing ophichthids would penetrate the alimentary tract and die within the coelom. This species seems particularly adept at this behavior and has repeatedly been found in the peritoneal cavity of fish.

Distribution (Fig. 392). A common and the northern-most occurring ophichthid of the western Atlantic, taken off the east coast of North America from off St. George's Bank and the Gulf of Maine to Florida. Only three of the study specimens were from other localities: one from the northern Gulf of Mexico, two from off northern South America, one of these likely from the stomach of another fish. Usually captured by trawl over sandy and muddy substrates between 36–1350 m depth. Wenner (1976) reported this species having been observed in the Norfolk Canyon from the submersible *Alvin* with only their heads exposed from the sandy substrate or resting on the sediment with their bodies in S-shaped curves; they were most abundant between 250–350 m depth.

Etymology. From the Latin *cruentifer* (bearing blood), presumably in reference to "the savage physiognomy of this fish" as suggested by its descriptors. A noun in apposition.

Remarks. Because of its small teeth and broad-based pectoral fins, many authors have placed *cruentifer* in the genus *Pisodonophis*. The Indo-Pacific species of *Pisodonophis*, however, are robust, shallow water forms, often associated with rivers, swamps and tidal creeks, a habitat extremely different than that of the deepwater, oceanic *Ophichthus cruentifer*. Although *O. cruentifer* is somewhat unlike the robust species of *Ophichthus* (sensu stricto) in general appearance, its osteological condition (McCosker, 1977) and leptocephalus (Leiby, 1981: 69) support its retention in *Ophichthus*. A world-wide revision of the genus and separation of the many species of *Ophichthus* may result in the recognition of *Omochelys* as a distinct genus or subgenus.

Study Material. A total of 80 specimens, 67–467 mm TL, including the lectotype and paralectotype, herein designated, and the holotype of *Omochelys marginatus* Fowler.

Lectotype: USNM 28938 (415); New Jersey, 39°57′N, 69°28′W, 120 fm (219 m); FISH HAWK Sta. 1035; 14 Sept. 1881.

Paralectotype: USNM 278183 (1, 360); taken with the lectotype.

Other material: GEORGE'S BANK: USNM 116410 (1, ca 330). GULF OF MAINE: USNM 133872 (1, 325+). MASSACHUSETTS: USNM 169867 (2, 375–467); from stomach of shark. USNM 30233 (1, 290); from stomach of ray. USNM 83838 (1, 257). CONNECTICUT: USNM 82094 (1, ca 300). NEW YORK: USNM 118948 (1, 225). NEW JERSEY: ANSP 72103 (1, 260); holotype of *Omochelys marginatus*. ANSP 84434 (1, 255), 95622 (1, 320+), 96652 (1, 233), 137877 (1, 404). MCZ 27800 (1, 266), 34527 (1, 217), 34528 (1, ca 67). USNM 31711 (1, 232), 35694 (1, 301). DELAWARE: ANSP 117711 (1, 338), 124778 (1, 255), 124779 (1, 239, cleared and stained). MCZ 34529 (10, 193–275). MARYLAND: ANSP 137875 (1, 344), 137876 (1, ca 250), 137881 (2, 386–415), 137882 (15, 225–312, cleared and stained), 137878 (2, 322–420). MCZ 34526 (2, 202–203), 62059 (1, 274), 62075 (1, 298). VIRGINIA: MCZ 34524 (1, 242), 34525 (1, 296), 38494 (1, 254). NORTH CAROLINA: ANSP 141180 (1, 274), 153178 (1, 218), 153177 (1, anterior part), 153643 (1, 242); from stomach of grouper. FLORIDA: UMML 27182 (1, 246), 27303 (1, 240), 27294 (1, 151), 13885 (1), 30642 (1, 280), 17677 (1, 200). USNM 148498 (1, 213). EAST COAST: USNM 67639 (4, 283–320), 67640 (1, 410); from stomach of a tilefish. GULF OF MEXICO: MISSISSIPPI: ANSP 98521 (1, 376). SOUTH AMERICA: VENEZUELA: UMML 27451 (1, 177). SURINAME: ANSP 119166 (1, anterior part).

Ophichthus cylindroideus (Ranzani, 1839)

TENTACLE-NOSE EEL

Figures 396–399, Tables 37, 40, 41

Conger cylindroideus Ranzani, 1839: 18, pl. 6, fig. 2a–b (original description, Brazil, holotype at Bologne Museum, uncatalogued). Ranzani, 1840: 80, pl. 13, fig. 2a–b.

Ophisurus parilis Richardson, 1848a: 105 (original description, West Indies, holotype BMNH 1985.11.27.2). Also spelled *parilus* by subsequent authors.

Scytalophis parilis. Kaup, 1856a: 46; Kaup, 1856b: 14, fig. 8.

Paramyrus cylindroideus, Günther, 1870: 51 (type of new genus, *Paramyrus*).

Ophichthys parilis. Günther, 1870: 59.

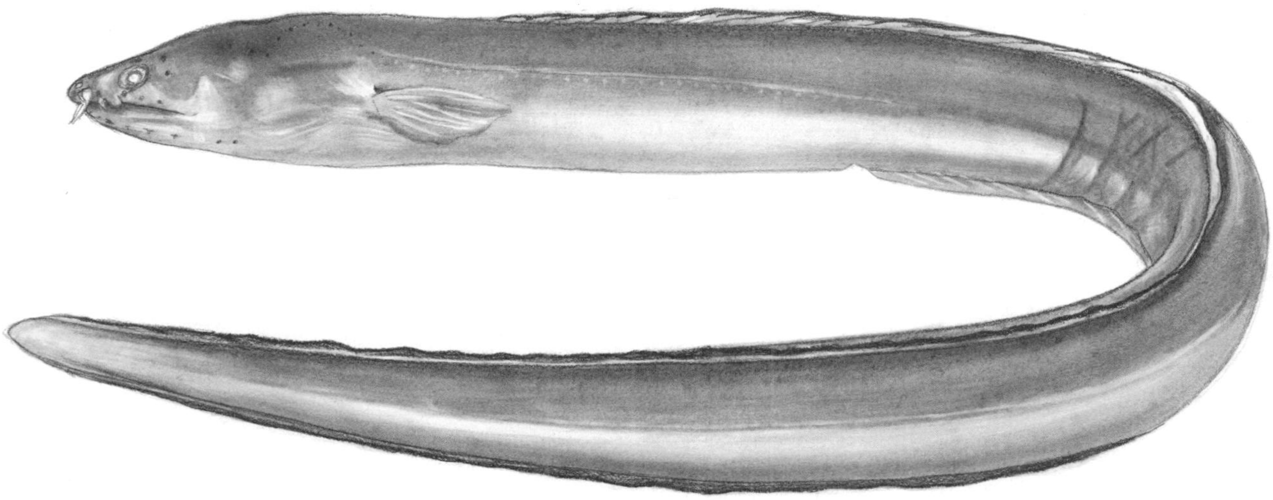

FIGURE 396. *Ophichthus cylindroideus*: ANSP 157320, 520 mm TL, French Guiana.

Discussion of Synonymy. Jordan and Davis (1891: 634) placed *Ophichthys pauciporus* Poey (1867b: 255) in the synonymy of *Ophichthus parilis* (= *O. cylindroideus*). We find that most unlikely and consider *O. pauciporus* to be a *nomen dubium* (see additional remarks, p. 380).

Distinctive Characters. A robust ophichthin, uniformly gray dorsally, pale ventrally, median fins black-edged; anterior nostril with conspicuous mesial tentacle; tail length 63–68% TL; dorsal-fin origin above mid-pectoral; pectoral fin long, spatulate, its length 34–50% HL; three dorsal supraorbital, seven mandibular and three preopercular pores; dentition biserial; MVF 14-43-145.

Description. Total vertebrae 140–151, x̄ = 144.8 (n = 40); predorsal vertebrae 11–15, x̄ = 13.4 (36); preanal vertebrae 39–45, x̄ = 42.4 (37). Pores: SO 1 + 3; IO 4 + 2; POP 7 + 3; ST 3.

Proportions as % of TL: head 11–13; trunk 21–23; tail 63–68; predorsal 13–17; depth behind gill openings 2.4–5.4. Of HL: snout 17–20; snout to rictus 33–44; eye 5.5–9; interorbital width 10–17; pectoral-fin length 34–50.

Snout broad, not pointed. Jaws short; eye moderate, over midpoint of upper jaw. Anterior nostril tubular, with conspicuous fleshy tentacle which is longer than the nostril tube extending from its posterior rim. Posterior nostril a large opening with irregularly rimmed flap on underside of upper jaw, the opening mostly outside of mouth; a small papilla between the two nostrils just behind anteriormost infraorbital pore. Head pores small; ethmoid pore on underside of snout, first two supraorbital pores far anteriorly on snout, third pore above posterior rim of eye. Preoperculomandibular pores usually 7 + 3; one specimen with 6 + 3 on both sides, two with 6 + 3 on one side only, and one (from Trinidad) with 6 + 2 on one side.

Teeth biserial, sharp, moderate in size. An outer row of 5–7 intermaxillary teeth, the anteriormost sometimes visible when mouth is closed, usually covered by folds of upper lip. Vomerine teeth beginning after a short gap; 19–35 mostly biserial teeth (an anterior broad patch sometimes briefly triserial), the two rows narrowing posteriorly. Outer maxillary teeth numerous and close-set, 18–30 teeth; inner row of 13–21 slightly larger teeth, canted inward, the row beginning slightly behind outer row. Mandible with 20–30

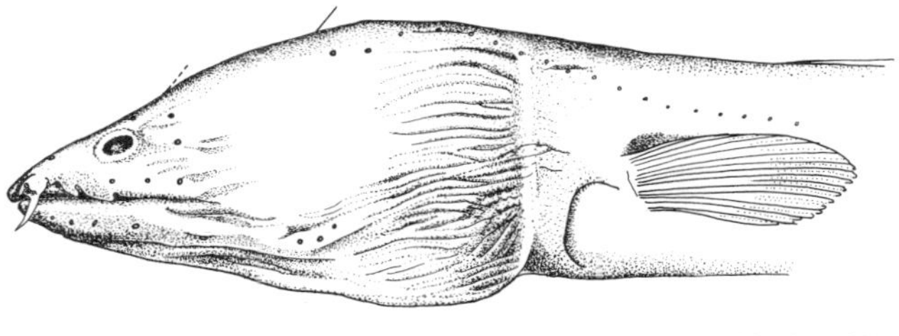

FIGURE 397. *Ophichthus cylindroideus*: ANSP 49840, 625 mm TL, Panama; head.

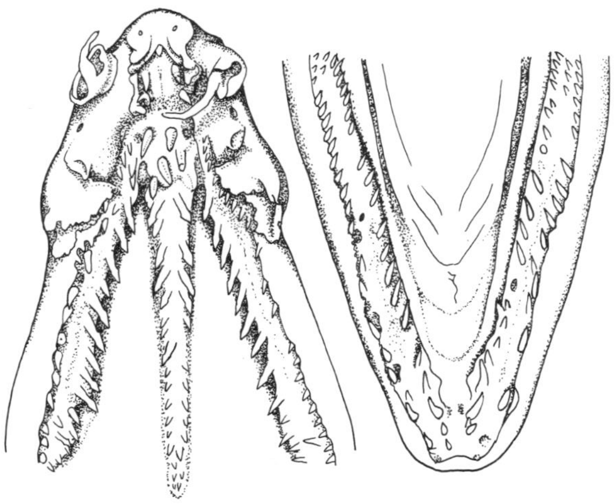

FIGURE 398. *Ophichthus cylindroideus*: ANSP 130221, 486 mm TL, Brazil; dentition.

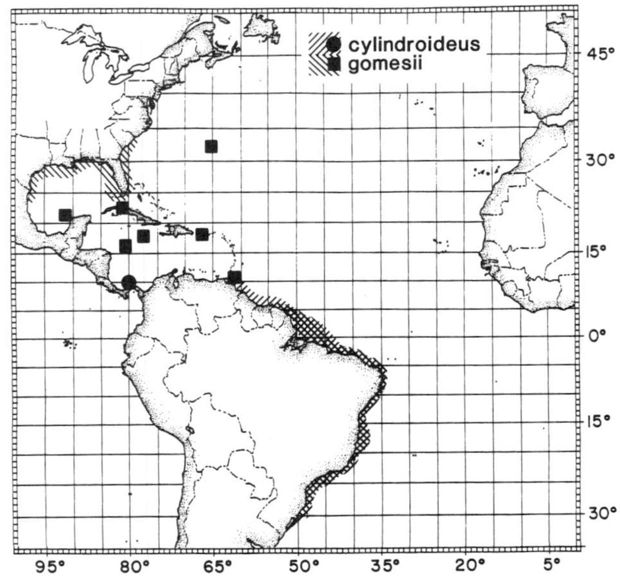

FIGURE 399. *Distribution of Ophichthus cylindroideus* and *O. gomesii* based on study material.

teeth in each row, the teeth similar in size and number.

Dorsal fin low, the origin variable, between mid-pectoral and pectoral-fin tips. Pectoral fin long and spatulate.

Color. Bicolored, uniformly gray dorsally, pigment diminishing below lateral line, lower third of body pale; body darkest along dorsal-fin base; on some specimens a pinkish-brown stripe along base of anal fin, darkest at end of tail. Head colored as body except throat, chin, and cheeks white. Dorsal fin pale basally with dark margin, darkest near tip of tail; anal fin coloration similar to that of dorsal fin; pectoral fin colored as dorsum.

Size. The largest specimen examined is 933 mm TL; Figueiredo and Menezes (1978: 18) reported a specimen of 1 m. Of 15 females (419–737 mm), six (486–737) were ripe with 0.3–0.9-mm eggs; of eight males (428–626), four (428–626) were ripe.

Development. Leptocephalus described (see p. 813).

Distribution (Fig. 399). Known from Trinidad, Panama, and the South American coast from Guyana to southern Brazil. Most frequently taken by trawl at recorded depths of 24–46 m. Richardson gave the locality for *Ophisurus parilis* as "West Indies"; Günther (1870: 59) stated that the holotype was from Cuba, "purchased of Mr. Scrivener." We have not seen any other specimens from north of Trinidad, but larvae have been taken off Puerto Rico.

Remarks. On the basis of Ranzani's original description and illustration, *Conger cylindroideus* has been variously referred to the Congridae, Echelidae, and the Ophichthidae. The holotype was examined and x-rayed at the Paris museum (by EBB and JEB where it was on loan to C. Karrer in 1978) and found to agree in all characters with Richardson's *O. parilis.* We note also that Ranzani's illustration (pl. 6, fig. 2 in the 1839 privately-printed publication, pl. 13, fig. 2 in the commonly cited 1840 work) depicts a very recognizable specimen of this species, including the nostril tentacle. *Ophichthus cylindroideus* is thus the correct name for this species.

Etymology. From the Greek *cylindros* (cylinder) and *oidos* (like), in reference to its body shape. A noun in apposition.

Study Material. A total of 68 specimens, 287–933 mm TL, including the holotype and the holotype of *Ophisurus parilis.*

Holotype: Bologne uncat. (485); Brazil.

Other material: WEST INDIES, CUBA: BMNH 1985.11.27.2 (1, 469); Mr. Scrivener; holotype of *Ophisurus parilis.* TRINIDAD: ANSP 76302 (1, 527), 76387 (2, 480–504), 96617 (1, 420). BMNH 1931.12.5.123 (1, 517). PANAMA: ANSP 49840 (1, 625). USNM 38522 (1, 522).

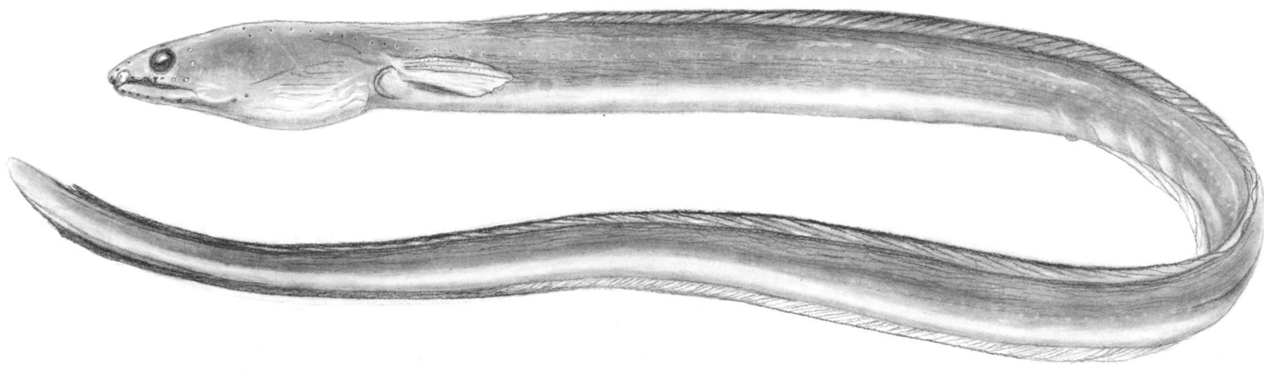

FIGURE 400. *Ophichthus gomesii*: ANSP 137567, 450 mm TL, Bermuda.

COLOMBIA: USNM 94755 (1, 622). GUYANA: BMNH 1959.3.17.117 (1, 530). BMNH 1961.8.31.101 (1, 419). USNM 185622 (1, 540). FRENCH GUIANA: ANSP 138435 (1, 590, cleared and stained), 157320 (1, 520). FMNH 69182 (1, 491). MNHN 1974-171 (1, 433). UMML 32483 (1, 486). SURINAME: ANSP 126487 (1, 480+). UMML 32484 (1, 615). MNHN uncat. (1, 366). MNHN uncat. (1, 507). FMNH 69183 (1), 69184 (1, 517). RMNH 21429 (1, 410), 21437 (1, 415), 21450 (1, 314). RMNH Sta. A9 (1, 428). RMNH Sta. N78 (1, 327). RMNH uncat. (1, 344). UF uncat. (1, 561). USNM 196001 (1, 287).

BRAZIL: ANSP 130221 (1, 486), 136634 (1, 554), 136656 (1, 562), 136657 (2, 575–580), 136658 (1, 694), 138433 (1, 720), 138434 (1, 674). CAS 54321 (1, 429), 54322 (1, ca 527), 59178 (1, 578), 59179 (1, 887), 59180 (1, 488). FMNH 69185 (1, 455). MCZ 4715 (2, 626–737), 9207 (2, 520–607), 9210 (1, 794). MNHN uncat. (2, 533–615). MNHN uncat. (1). MNRJ 4131 (4, 474–594), 4132 (1, 580), 4133 (1, 518), 7776 (1, 887), 8251 (1, 488). MZUSP 8879 (1, 598), 10063 (1), 12098 (1, 933). USNM 38522 (1, 523), 158912 (1, 428), 214061 (1, 455). ZMC J.122 (1, 410).

Ophichthus gomesii (Castelnau, 1855)

SHRIMP EEL

Figures 399–402, Tables 37, 40, 41

Ophisurus gomesii Castelnau, 1855: 84, pl. 44, fig. 2 (original description, Rio de Janeiro, holotype MNHN B. 2745).

Scytalophis magnioculis Kaup, 1856a: 46 (original description, St. Croix and Brasil, syntype MNHN B.2749). Kaup, 1856b: 13, fig. 7 (not fig. 8 as listed).

Leptorhinophis Gomesi. Kaup, 1856a: 47 (referred to new genus *Leptorhinophis*). Kaup, 1856b: 14.

Ophisurus chrysops Poey, 1860: 321 (original description, Havana, holotype MCZ 27212).

Ophichthys magnioculis. Kner, 1865: 376.

Ophichthys macrurus Poey, 1867b: 256 (original description, Havana, holotype unknown).

Ophichthys brachyurus Poey, 1868b: 426 (original description, Havana, holotype unknown).

Ophichthys magnoculus (emendation for *magnioculis*). Günther, 1870: 59.

Ophichthys gomesii. Günther, 1870: 60.

Oxyodontichthys limbatus Poey, 1880: 254 (substitute name for *O. brachyurus*). New genus created for *chrysops*, *limbatus*, *macrurus*, and *nova species*.

Ophichthus gomesi. Jordan and Davis, 1891: 632.

Ophichthus sp. Böhlke and Chaplin, 1968: 104 (Bahamas, the "Pallid snake eel").

Discussion of Synonymy. We follow previous authors in placing Poey's species *Ophichthys macrurus* (Günther, 1870: 59; Jordan and Davis, 1891: 632; Howell Rivero, 1932: 12) and *O. brachyurus* (Jordan and Davis, loc. cit.; Howell Rivero, loc. cit.) in the synonymy of *O. gomesii*. No type specimens are known, but the characters given in the brief descriptions agree with those for *O. gomesii*. The holotype of *Ophisurus chrysops* was examined and found to be conspecific with *O. gomesii*.

Distinctive Characters. A robust bicolored ophichthin, gray dorsally, pale ventrally, with median fins dark-edged; tail length 62–65% TL; dorsal-fin origin above posterior third of pectoral fin; pectoral fin spatulate, its length 37–45% HL; three dorsal supraorbital, six mandibular and two preopercular pores; dentition biserial to quadriserial; MVF 14-47-145.

Description. Total vertebrae 138–153, $\bar{x}$ = 144.6 (n = 142); predorsal vertebrae 12–18, $\bar{x}$ = 14.5 (100); preanal vertebrae 44–52, $\bar{x}$ = 47.2 (111). Head pores: SO 1 + 3; IO 4 + 2; POM 6 + 2; ST 3.

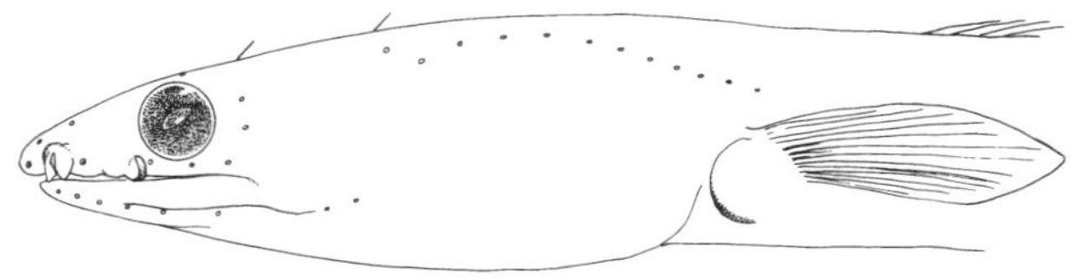

FIGURE 401. *Ophichthus gomesii*: ANSP 137567, 450 mm TL, Bermuda; head.

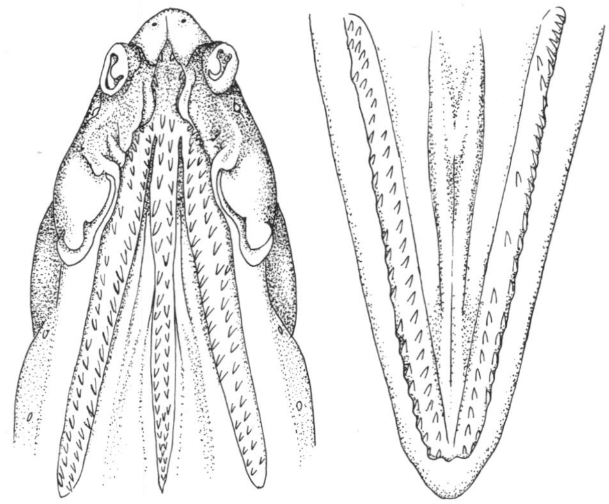

FIGURE 402. *Ophichthus gomesii*: ANSP 137567, 450 mm TL, Bermuda; dentition.

Proportions as % TL: head 9.6–14; trunk 23–28; tail 62–65; predorsal 13–17; depth behind gill openings 2.7–3.7. Of HL: snout 15–20; snout to rictus 37–53; eye 8.6–12; interorbital width 8–15; pectoral-fin length 37–45.

Snout moderate, jaws short; eye large, over posterior half of upper jaw. Anterior nostrils tubular, with convoluted margin but without projecting barbel; posterior nostril in underside of upper lip, opening partly inside mouth, with flap. Supraorbital pores as illustrated, two dorsal pores on anterior third of snout, the third pore above mid-eye. Preoperculomandibular pores typically 6 + 2; two specimens had 7 pores along mandible, and four had 3 preopercular pores on one side.

Dentition biserial in young, becoming tri- to quadriserial in larger specimens, the teeth moderate, sharp, and similar in size. An outer ring or inverted V of 5–7 intermaxillary teeth, mostly exposed when mouth is closed. A short gap followed by vomerine teeth which are irregularly bi- to triserial and which extend as far as or posterior to rows on maxillaries; in large specimens headed by a wide patch of slightly larger teeth. Maxillary teeth of smallest specimens in two parallel rows which are equal in length and number of teeth, 23–27 in each row; larger specimens with many more teeth, becoming triserial posteriorly to tri- or quadriserial throughout length of jaw. Teeth on lower jaw similar to those on upper jaw.

Dorsal-fin origin variably over posterior third of pectoral fin, sometimes above the tips of fin. Pectoral fin long, lanceolate or rounded.

Color. Grayish-brown to slate-gray dorsally, pale ventrally, pigment extending ventrally on body; sometimes appearing sharply bicolored, but pale or faded specimens appearing almost uniformly colored. Head dark, pigment extending onto underside, sometimes a dark patch on throat. Head pores variously pale and inconspicuous or set in small dark spots. Dorsal and anal fins pale basally, and with dark margins. Pectoral fin as body color or darker. Small specimens are pale or yellowish, with diffuse dorsal pigment.

Size. The largest specimen examined is 665 mm TL. Of 41 females (182–665 mm), 13 (335–577) were ripe with eggs to 0.9 mm. Of 16 males (319–570), 12 (330–570) were ripe.

Development. Leptocephalus identified (see p. 814), abundant from the Scotian Shelf to northern Brazil. Leiby (1979b) presented an extensive description of the leptocephali and their development from just-hatched to metamorphosis. He found some variation in pigmentation between those spawned in the Gulf of Mexico and those spawned from the Lesser Antilles to Brazil.

Distribution (Fig. 399). Reported to be widely distributed from New England to southern Brazil; however, we have seen specimens of *Ophichthus gomesii* from only as far north as South Carolina (we have found specimens of *O. cruentifer* misidentified as *O. gomesii*). Taken in the Atlantic from South Carolina to Florida, and throughout the northern Gulf of Mexico, where it is abundant and frequently taken as a bycatch of shrimp trawl fisheries. It is infrequently collected in the West Indies, but often taken off Brazil. Obtained by trawl to depths of 90 m (one catch recorded from 411–457 m); also taken by hook and line. Both northern and southern populations are found in bays and estuaries, as well as in offshore waters.

Variation. We have noted considerable variation in dentition, which seems generally related to growth, and previous authors have noted differences in eye size (*O. magnioculis* was named for its large eye), a change observed in other eels and found to be related to sexual maturity. There is geographic variation in both larvae and adults. The adult Brazilian population differs considerably from that of the northwestern Atlantic and Gulf of Mexico in some characters, notably in number of vertebrae (Table 41) and in head size (head length 9.7% TL in Brazilian specimens vs 11.0 for others). However, other characters (pore condition, dorsal-fin placement, tail length) are consistent and distinctive for the species, and we consider all forms to be conspecific.

Etymology. Named for Dr. Ildefonso Gomes, the physician who cured Castelnau of a malady in Rio de Janeiro. A noun in the genitive case.

Remarks. This species differs from many other western Atlantic *Ophichthus* in possessing two rather than three preopercular pores and in its multiserial dentition. In general appearance, morphology and osteology, it is similar to *O. cylindroideus* (which possesses a distinctive conspicuous tentacle on the anterior nostril and which has seven mandibular plus three preopercular pores), and to *O. cruentifer* (which has the dorsal-fin origin well behind the pectoral-fin tips and which has six mandibular plus three preopercular pores). The latter is commonly found off northern New England, and specimens of *O. gomesii* recorded from that area, such as that reported by Backus (1957: 61) from Massachusetts (not seen), are probably *O. cruentifer.*

We include *Scytalophis magnioculis* Kaup in the synonymy of *Ophichthus gomesii,* based on study of the holotype in the Paris museum, and have found most specimens previously identified as *magnioculis* to be *gomesii.* We note, however, that USNM 38522 from Aspinwall (= Colon), Panama, referred to *magnioculis* by Jordan and Davis (1892: 633), is *O. cylindroideus.*

Study Material. A total of 240 specimens, 74–665 mm TL, including the holotype and the holotypes of *Ophisurus chrysops* Poey and *Scytalophis magnioculis* Kaup.

Holotype: MNHN B.2745 (456); Rio de Janeiro; Castelnau.

Other material: BERMUDA: ANSP 124410 (1, 540), 137567 (1, 450). BAHAMAS: ANSP 111566 (1, 182). COASTAL UNITED STATES: *South Carolina:* GMBL 73-115 (1, 552). USNM 29970 (1, 465). *Georgia:* GMBL 70-317 (1, 537). TU 21349 (1, 368). *Florida:* ANSP 126141 (1, 292), 126142 (1, 460), 126153 (2, 374–390), 126436 (1, 538), 128381 (1, 345, cleared and stained), 128382 (1, 578), 143075 (1, 613), 152295 (1, 488). UF 10248 (6, 346–476). UF uncat. (1, 502). UF uncat. (1, 441). UMML 1511 (1, 424), 5070 (1, 570), 8533 (1). USNM 152579 (1, 239), 155007 (1, 535). *Florida Keys:* UMML 56 (1), 946 (1, 378), 2078 (1), 2688 (1, 495), 4694 (2), 8641 (1, 440), 8975 (1, 382), 9135 (1), 13124 (3, 356–408). UMML uncat. (1). USNM 152254 (1, 420).

GULF OF MEXICO: *Florida:* ANSP 70876–70877 (2, 235–408), 87974 (1, 183), 126435 (2, 411–437), 157374 (1, 375). FMNH 34963–34967 (5, 343–485), 34984–34987 (4), 38366 (1). SCMRD 70-296 (1, 292). TU 2102 (1, 574), 20328 (1, 270). UF 2047 (2, 411–468), 3267 (1, 409), 10297 (1, 214). UF-FSU 540 (1, 294), 1244 (2, 366–473), 7810 (1, 340), 8144 (1, 472), 10950 (1, 442), 16376 (2, 437–478), 16951 (3, 379–509), 18300 (8, 213–404), 18626 (1, 309), 19868 (1, 281), 20081 (2, 227–331), 23472 (2, 346–356). USNM 26578 (1, 364), 30707 (3, ca 565–580), 36995 (1, 112), 43177 (11, 385), 68420 (1, 457), 157931 (1, 547), 158742 (2, 356–410), 158743 (1, 332), 158744 (2, 371–550), 274254 (1, 385). *Alabama:* USA 04324 (1, 215). USNM 157929 (1, 465), 157933 (1, 355+). *Mississippi:* FMNH 31785 (1, 480), 31786 (1, 477). GCRL 423 (2, 350–415). MNHN 1974-170 (1). UMML uncat. (1). USA 03685 (1, 640). USNM 143216 (2, 417–515), 157928 (1, 317), 157936 (1, 354). *Louisiana:* ANSP 126143 (2, 374–391). FSC 815 (1, 440), 3309 (1, 457), 3829 (1, 510), 6727–6729 (3, 407–535), 18778 (1, 529). GCRL 254 (1, 586), 1178 (2, 362–400), 5766 (1, 507). TU 1892 (1, 477), 4190 (1, 517), 4395 (1, 557), 5368 (2, 514–528), 5492 (2, 490–545), 5500 (1, 511), 5561 (1, 461), 6056 (5, 447–550), 8230 (1, 276), 8411 (1, 463), 8923 (1, 614), 9057 (1, 455), 10022 (1, 541), 10040 (9, 472–555), 11191 (5, 501–528), 17592 (1, 458), 18766 (1, 472), 21349 (1, 368), 28166 (3, 408–435), 51380 (2, 447–493), 87132 (1, 394), 94506 (2, 577–591). USNM 125920 (1, 497), 133652 (1, 320), 155008 (1, 615), 157804 (2, 363–380), 162581 (1, 480). *Texas:* ANSP 152355 (1, 319). FMNH 11000 (1, 540), 38662 (1), 40300 (2, 395–416), 45060 (1). USNM 147809 (4, 370–ca 500), 152253 (1, 550), 152581 (1, 365), 155306 (1, 360). *Bay of Campeche:* FMNH 45377 (1).

WEST INDIES: CUBA: AMNH 12298 (1, 453). ANSP

TABLE 41. Geographic variation in vertebral counts in *Ophichthus cylindroideus* and *O. gomesii.*

	Predorsal							
	11	12	13	14	15	16	17	18
cylindroideus								
W. Indies		2	3	3[c]				
N. South America		1	3	3	1			
Brazil	1	2[a]	5	9	1			
gomesii								
U.S. Atlantic		1		6	4	3		
Gulf of Mexico		1	7	30	15	13		
West Indies[b]			3	6	3[e]			
Brazil				5	3	4[a,d]	2	1

	Preanal													
	39	40	41	42	43	44	45	46	47	48	40	50	51	52
cylindroideus														
W. Indies			2	5[c]	–	1								
N. South America			2	2	3	1								
Brazil	1	1	1	5[a]	7	4	1							
gomesii														
U.S. Atlantic						1	–	2	5	3	1			
Gulf of Mexico						2	3	16	23	12	10			
West Indies[b]							1	2	2	5	2[e]	2		
Brazil									1	3[d]	6[a]	6		

	Total															
	138	139	140	141	142	143	144	145	146	147	148	149	150	151	152	153
cylindroideus																
W. Indies				1	1	–	2	1	–	1	1[c]					
N. South America				1	3	2	1	1	1							
Brazil			1	2	–	2	4	3	2	2[a]	1	2	–	1	1	
gomesii																
U.S. Atlantic					2	6	4	1	1	–	2					
Gulf of Mexico	1	2	2	6	14	18	22	8	13	6	3					
West Indies[b]			1	–	–	2	5[e]	3	1	4	–	1				
Brazil								1	–	1	2[a,d]	3	1	5	3	2

[a] Includes holotype.
[b] Including Bermuda and the Bahamas.
[c] Holotype of *O. parilis.*
[d] Holotype of *O. magnioculis.*
[e] Holotype of *O. chrysops.*

126151 (1, 391). MCZ 9157 (2, 530–720), 9213 (1, 374), 9215 (1, 423), 27212 (1, 720; holotype of *Ophisurus chrysops* Poey). PUERTO RICO: ANSP 134328 (1, 630), 153334 (1, 74). MNHN 1974-169 (1). USNM 125507 (1, 210). JAMAICA: FMNH 2711 (1, ca 435). TRINIDAD: USNM 142879 (1, ca 228). ROSALIND BANK: FMNH 71764 (1, 385).

SOUTH AMERICA: FRENCH GUIANA: ANSP 126488 (1, 433). USNM 185629 (1, 416), 185660 (1, 330). BRAZIL: BMNH 1886.1.21.38–40 (2, 460–665). BMNH 1890.1.3.18 (1, 456). CAS 59181 (1, 470). MCZ 974 (1, 385), 4713 (1, 580), 9211 (1, 510), 9212 (1, 593), 9221 (3, 485–552). MNHN B.2749 (1, 488); holotype of *Scytalophis magnioculis* Kaup. MNHN B.2762 (1, 469). MNRJ 1659 (1, 342), 5758 (2, 335–

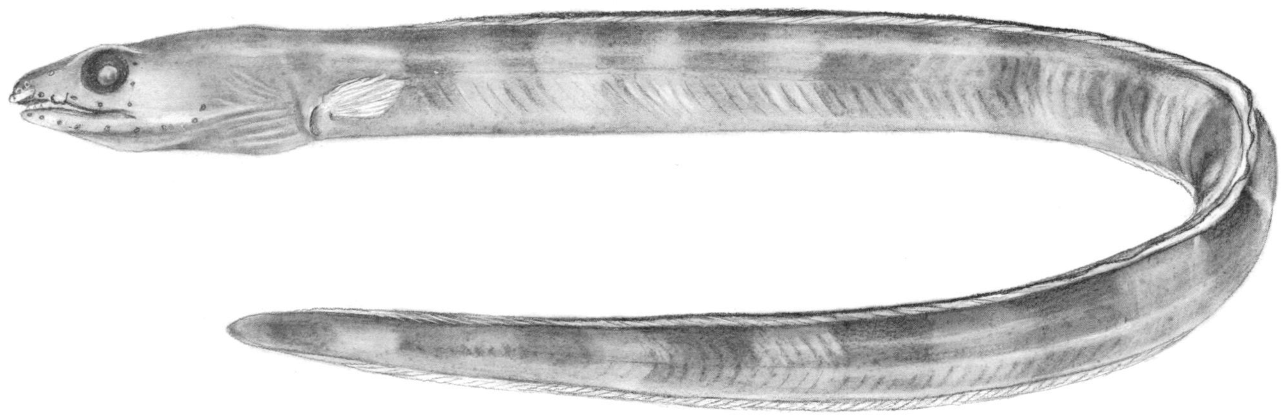

FIGURE 403. *Ophichthus hyposagmatus*: ANSP 152300, holotype, 357 mm TL, Gulf of Mexico off Florida.

378), 5825 (3, 265–511), 6642 (1, 470), 9365 (1, 483), 9529 (1, 395), 9530 (2, 320–321), 9744 (1). MNRJ uncat. (1). MZUSP 5190 (3, 465–489), 10064 (2, 458–490), 12075 (2), 12076 (1, 128), 12077 (1, 88.5), 12078 (1, 286), 12079 (1, 103), 12080 (1, 729), 12081 (1), 12082 (2, 468–694), 12083 (1, 484), 12084 (1, 500), 12085 (1, 133), 12086 (1, 437), 12087 (1, 517), 12088 (1, 522), 12089 (1, 333), 12090 (1, 476), 12091 (2, 471), 12092 (1, 130). NMW 61608 (2, 367–392), 61971 (2, 474–497), 65108 (2, 363–392+), 65358 (2, 283–300), 65359 (1, 592), 65378 (1, 462), 65379 (1, 453), 65380 (1, 516), 65408 (1, 516), 65409 (1, 540), 65410 (2), 65470 (2, 450–657). USNM 83383 (1, 414).

Ophichthus hyposagmatus McCosker and Böhlke, 1984

FAINT-SADDLED SNAKE EEL

Figures 403–406, Tables 37, 40

Ophichthus species B. Böhlke, 1977b: 5.
Ophichthus hyposagmatus McCosker and E. B. Böhlke, 1984: 24, figs. 1–3 (original description, Florida, 29°56′N, 86°38′W, holotype ANSP 152300).

Distinctive Characters. A robust pale tan ophichthin with 16 faint saddles along body; tail 53–57% TL; dorsal-fin origin above pectoral tips; pectoral-fin length 25–31% HL; three dorsal supraorbital, 6–7 mandibular, and 2–3 preopercular pores; jaw teeth biserial, vomerine teeth uniserial; MVF 12-54-136.

Description. Total vertebrae 134–139, x̄ = 135.7 (n = 6); predorsal vertebrae 12–13, x̄ = 12.5 (6); preanal vertebrae 52–55, x̄ = 53.5 (6). Pores: SO 1 + 3; IO 4 + 2; POM 6–7 + 2–3; ST 3.

Proportions as % of TL: head 11–12; trunk 31–35; tail 53–57; predorsal 14–16; depth behind gill openings 2.9–4.5. Of HL: snout 20–33; snout to rictus 35–42; eye 12–14; interorbital width 7–14; pectoral-fin length 25–31.

Anterior half of head conical, interorbital region flat. Jaws short; eye large and bulbous, its posterior margin just above rictus. Anterior nostril narrowly tubular; posterior nostril at edge of lip, entirely outside mouth, covered anteriorly by a small flap. Head pores minute, not conspicuous; variation in preoperculomandibular pores.

Teeth small, conical, subequal. Intermaxillary teeth an inverted V of four teeth, separated by short gap from vomerine teeth which begin with a pair, followed by a row of 12–18 teeth. Maxillary teeth biserial, 24–36 in outer row, 18–26 in inner row. Mandibular teeth partly biserial; an outer row of 25–32 close set teeth, an inner row of four teeth anteriorly and 4–10 far posteriorly.

Median fins low, elevated posteriorly; dorsal-fin origin just before or behind pectoral-fin tips. Pectoral fins moderately elongate.

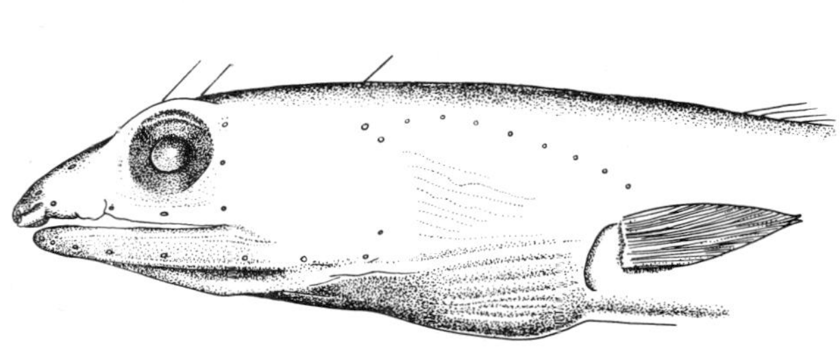

FIGURE 404. *Ophichthus hyposagmatus*: ANSP 152301, paratype, 267 mm TL, Dominican Republic; head. Anteriormost line indicates SO_4.

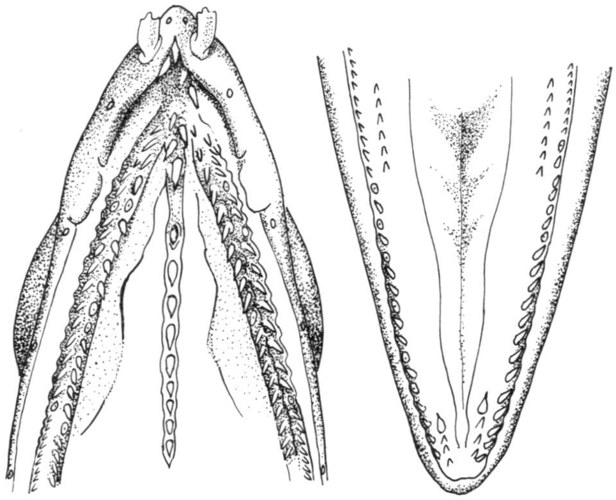

FIGURE 405. *Ophichthus hyposagmatus*: ANSP 152301, paratype, 267 mm TL, Dominican Republic; dentition.

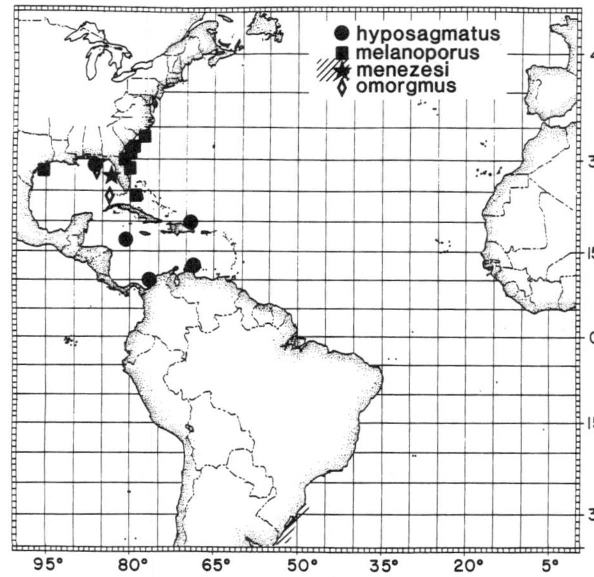

FIGURE 406. Distribution of *Ophichthus hyposagmatus, O. melanoporus, O. menezesi,* and *O. omorgmus* based on study material.

Color. Body color yellow to pale tan, slightly darker dorsally. A tan triangle between eyes extending slightly beyond them; snout tip brown. Sixteen faint tan saddles along body, the first in advance of gill openings. Fins colorless except dorsal fin dark-edged. Smallest specimens have faded and become colorless in preservative.

Size. The largest specimen examined was 385 mm TL. Four specimens (218–385) are females, the 385-mm one ripe with 1.0-mm eggs; the 357-mm specimen is a ripe male.

Development. Leptocephalus not identified.

Distribution (Fig. 406). The six known specimens were taken off Florida, the Dominican Republic, between Honduras and Jamaica, Venezuela, and Colombia, by trawl and dredge between 88 and 293 m depth.

Etymology. From the Greek *hypo* (somewhat) and *sagmatus* (saddled), in reference to the faint body pigmentation. An adjective.

Remarks. Ophichthus hyposagmatus is similar in appearance, morphometry and dentition to *O. spinicauda,* which differs in its more posterior dorsal-fin origin and more numerous vertebrae (144–147). It is also similar in appearance to the comparably saddled *O. rex,* which possesses multiserial dentition, a posterior dorsal-fin origin, and fewer vertebrae (115–119). *Ophichthus spinicauda* and *O. rex* are large species (1–2 m TL), whereas *O. hyposagmatus* is relatively small (a ripe male and female are both less than 400 mm). Faded specimens of *O. hyposagmatus* which lack the dorsal saddles might be mistaken for *O. gomesii,* but are easily separable by the biserial vomerine dentition of the latter.

Study Material. The holotype and five paratypes, 178–385 mm TL.

Holotype: ANSP 152300 (357); Gulf of Mexico, S of Destin, Florida, 29°56′N, 86°38′W, by trawl at 108 m; R. L. Shipp et al.; 29 Aug. 1976.

Paratypes: DOMINICAN REPUBLIC: ANSP 152301 (1, 267); 18°21.3′N, 69°12.6′W, 165–167 m; PILLSBURY sta. 1395; 10 July 1971. CAS 53292 (1, 234); 18°21′N, 69°14.3′W, 170–176 m; PILLSBURY sta. 1303; 21 July 1970. CARIBBEAN SEA: Between Honduras and Jamaica: FMNH 71764 (1, 385); 16°45′N, 81°28′W, 293 m; OREGON sta. 1882; 22 Aug. 1957. COLOMBIA: USNM 263573 (1, 178); 09°46.2′N, 76°10.9′W, 88 m; PILLSBURY sta. 393; 16 July 1966. VENEZUELA: UMML 31629 (1, 214.5); 11°39′N, 69°22′W, 161–186 m; PILLSBURY sta. 757; 27 July 1968.

Ophichthus melanoporus Kanazawa, 1963

BLACKPORED EEL

Figures 406–409, Tables 37, 40

Ophichthus melanoporus Kanazawa, 1963: 284, figs. 2B, 3, pl. 1 right (original description, Straits of Florida, west of Andros Island, Bahamas, holotype USNM 191118).

Distinctive Characters. A very elongate, pale ophichthin with conspicuous dark spots marking head pores and lateral-line pores; tail very

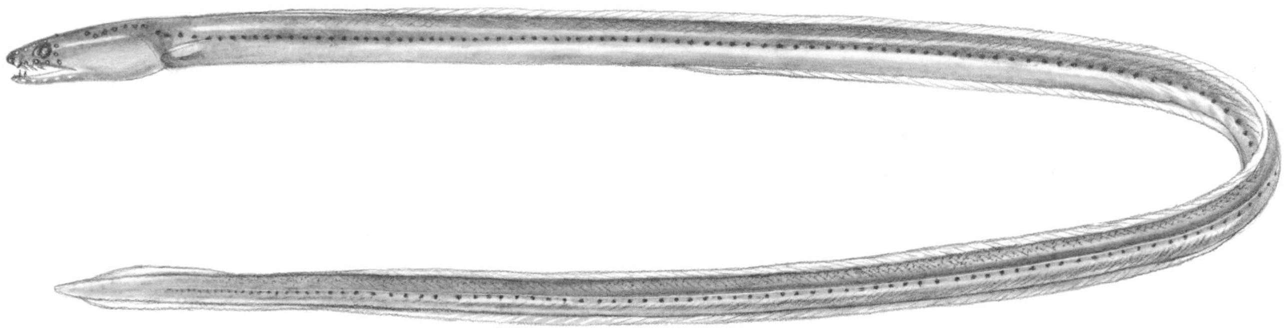

FIGURE 407. *Ophichthus melanoporus*: ANSP 128380, 470 mm TL, South Carolina.

long, 69–71% TL; dorsal-fin origin above tips of pectoral fins; pectoral fin lanceolate, its length 20–25% HL; four dorsal supraorbital, six mandibular, and three preopercular pores; teeth few, widely spaced, uniserial; MVF 12-52-183.

Description. Total vertebrae 177–186, $\bar{x}$ = 183.2 (n = 10); predorsal vertebrae 11–14, $\bar{x}$ = 12.0 (10); preanal vertebrae 50–54, $\bar{x}$ = 51.8 (10). Pores: SO 1 + 4; IO 4 + 2; POM 6 + 3; ST 3.

Proportions as % of TL: head 6.3–7; trunk 22–24; tail 69–71; predorsal 8.2–9.4; depth behind gill openings 1.5–1.8. Of HL: snout 14–18; snout to rictus 36–44; eye 8.4–11; interorbital width 5.2–8.4; pectoral-fin length 20–25.

Head short; jaws elongate, narrow, and pointed from above; eye above mid-gape. Anterior nostrils in narrow tube; posterior nostril on upper lip, covered by a flap, opening outside mouth. Head pores as listed; three preopercular pores low and difficult to see, sometimes one pore apparently missing.

Teeth uniserial, widely spaced, fine and needlelike, frequently broken off or missing. Four small intermaxillary teeth, all exposed when mouth is closed, closely followed by 8–11 stouter teeth on vomer, the second the longest in the mouth. Typically 12–17 teeth on maxillary; 20–26 teeth on mandible.

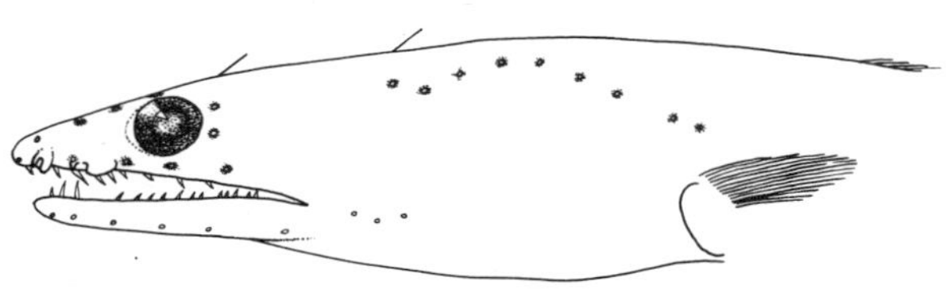

FIGURE 408. *Ophichthus melanoporus*: ANSP 128380, 470 mm TL, South Carolina; head.

Color. Background color yellow to tan, overlain with a fine brown spotting dorsally. Head pores (except those of lower jaw and preopercle) and lateral line pores lie in conspicuous brown spots, spots paler and less obvious posteriorly on tail. Fins colorless.

Size. The largest specimen examined is 693 mm TL. Two specimens (487–609 mm) were females, and two (470–578) were males, none of them ripe. Kanazawa (1963: 284) reported the holotype to be female, the three paratypes males.

Development. Leptocephalus described by Leiby (1981: 52, and this volume, p. 817).

Distribution (Fig. 406). Known from North Carolina to the Florida Straits, and the northern Gulf of Mexico west to Corpus Christi, Texas. Captured infrequently by trawl in 51–460 m depth.

Etymology. From the Greek *melano* (black) and

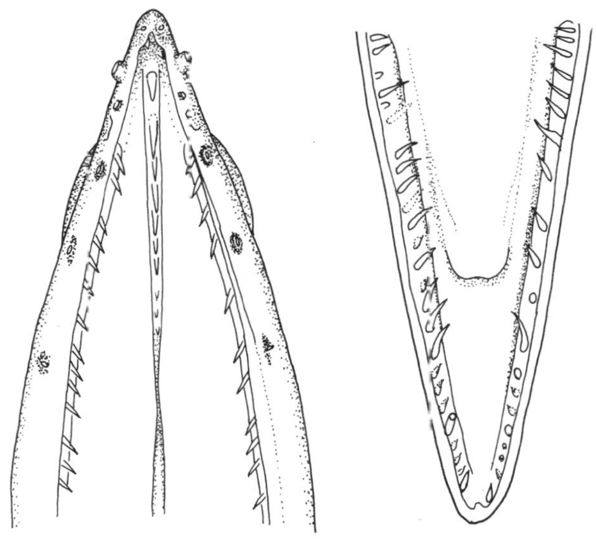

FIGURE 409. *Ophichthus melanoporus*: ANSP 127792, 609 mm TL, North Carolina; dentition.

porus (pore), referring to the conspicuous black pores on head and lateral line. An adjective.

Remarks. Ophichthus melanoporus differs considerably from other western Atlantic species in the genus (with the exception of *O. brasiliensis*) in its slender body, elongate tail, and distinctive uniserial dentition. The closest relative to this species appears to be an undescribed eastern Pacific *Ophichthus*, known from Costa Rica to Panama (McCosker and Rosenblatt, in prep.).

Study Material. A total of 13 specimens (411–693 mm TL), including the holotype and three paratypes.

Holotype: USNM 191118 (690 mm TL); Straits of Florida, west of Andros Island, Bahamas, 24°04′N, 79°15′W, 40-ft flat net in 250 fm (457 m); COMBAT sta. 448; 24 July 1957.

Paratypes: USNM 191119 (3, 542–672); collected with the holotype.

Other material: NORTH CAROLINA: ANSP 127792 (1, 609); 33°04′N, 77°42′W, 155 m; DOLPHIN; 15 Nov. 1973. SOUTH CAROLINA: ANSP 128380 (1, 470); 32°19′N, 79°09′W, 51 m; DOLPHIN; 12 Apr. 1974. GEORGIA: ANSP 128378 (2, 487–578); 31°30′N, 79°40.2′W, 88 m; DOLPHIN; 11 April 1974. GMBL 75-55 (1, 587); 31°44.5–42.5′N, 79°30.0–31.0′W, 95–104 m; 1975. FLORIDA: ANSP 128379 (1, ca 680, cleared and stained); 30°20.3′N, 80°56.5′W, 27 m; DOLPHIN; 7 April 1974. GMBL 75-56 (1, 693); 28°58.7–58.0′N, 80°08.6–07.5′W, 70–75 m; DOLPHIN; 29 Jan. 1975. UF 10245 (1, 411); Flagler Co., beach at Marineland; C. R. Gilbert et al.; 17 Sept. 1962. TEXAS: ANSP 126197 (1, ca 580); 27°55′N, 95°40′W, 55–58 m; OREGON 3878; 21 Sept. 1962.

Ophichthus menezesi McCosker and Böhlke, 1984

BLOTCHSIDE SNAKE EEL

Figures 406, 410–412, Tables 37, 40

Ophichthus menezesi McCosker and E. B. Böhlke, 1984: 29, figs. 7–9 (original description, Brazil, 28°46′S, 47°48′W, holotype MZUSP 12166).

Distinctive Characters. A moderately stout, pale ophichthin with 20–25 dark blotches along midbody; tail 58–62% TL; dorsal-fin origin well behind pectoral-fin tips; pectoral fin rounded, its length 14–22% HL; four dorsal supraorbital, five mandibular and two preopercular pores; maxillary teeth biserial, vomerine and mandibular teeth uniserial; MVF 19-56-153.

Description. Total vertebrae 150–156, $\bar{x}$ = 153.3 (n = 15); predorsal vertebrae 16–23, $\bar{x}$ = 19 (14); preanal vertebrae 55–58, $\bar{x}$ = 56.2 (15). Pores: SO 1 + 4; IO 4 + 2; POM 5 + 2; ST 3.

Proportions as % of TL: head 7.5–8.9; trunk 30–32; tail 58–62; predorsal 14–18; depth behind gill openings 2.4–3. Of HL: snout 23–26; snout to rictus 34–38; eye 10–13; interorbital width 8.5–14; pectoral-fin length 14–22.

Anterior half of head conical, interorbital region flat; jaws moderate; eye moderate, above posterior half of upper jaw. Anterior nostril tubular; posterior nostril at edge of lip, entirely outside mouth, covered anteriorly by a small flap. Head pores minute, not conspicuous. Five supraorbital pores, posteriormost above center of eye.

Teeth small, conical, nearly subequal. An inverted chevron of 5–7 intermaxillary teeth, well separated from vomerine teeth which are partially biserial, one to three pairs anteriorly followed by a row of 9–14 teeth. Maxillary teeth biserial; an outer row of 11–23, a shorter inner row of 3–14. Dentary teeth uniserial and close set, 19–25 in row.

Median fins low. Dorsal-fin origin well behind gill opening, by a distance nearly equal length of head. Pectoral fins moderately short, rounded.

Color. Uniformly pale yellowish body overlain with 20–25 round dark blotches, equal to or larger than eye in size, their centers near or on the lateral line. All fins colorless. This species apparently retains the larval pigmentation of dark lateral blotches into maturity, a unique condition; all specimens exhibited the coloration, including the ripe female.

Size. The largest specimen examined is 281 mm TL. Six specimens (117–281 mm) were females, the largest ripe with 0.8-mm eggs; eight (123–235) were males, none of them ripe.

Development. Leptocephalus identified (p. 818).

Distribution (Fig. 406). The 13 type specimens are all from Brazil, off the state of Rio Grande do Sul, and were collected by trawl between 169–209 m depth. One additional specimen is from the Gulf of Mexico off Florida; it is a ripe female, and it was collected with a try-net at 1200–1400 m depth over sand and rubble bottom.

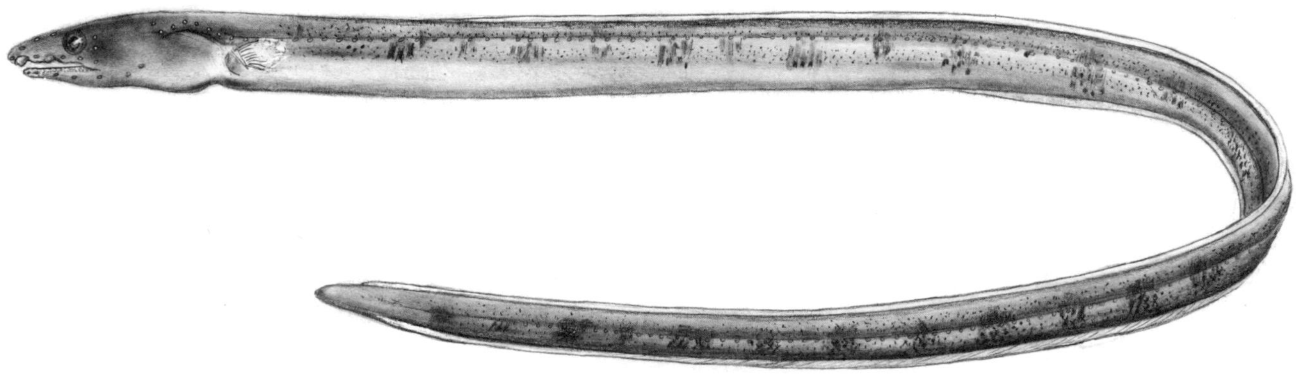

FIGURE 410. *Ophichthus menezesi*: MZUSP 12166, holotype, 346 mm TL, Brazil.

Remarks. Ophichthus menezesi is similar in general appearance and vertebral counts to *O. brasiliensis,* but differs in having biserial maxillary dentition and five rather than four pore pairs between the snout tip and the frontal pore. The Brazilian locality given for *O. brasiliensis* has been questioned (p. 383).

Etymology. Named in honor of Naercio A. Menezes who made the specimens available to the authors. A noun in the genitive case.

Study Material. A total of 14 specimens, 119–281 mm TL, including the holotype and 12 paratypes.

Holotype: MZUSP 12166 (235); Brazil, Rio Grande do Sul, 28°46′S, 47°48′W, captured by trawl at 177 m; PROF. W. BESNARD, sta. 1702; 31 Jan. 1972.

Paratypes: BRAZIL: Rio Grande do Sul (all taken by trawl by the R/V PROF. W. BESNARD): ANSP 152303 (1, 200, cleared and stained); 31°33′S, 49°52′W, 209 m; 25 Jan. 1972. ANSP 152304 (1, 197); 34°28′S, 51°50′W, 169 m; 19 Apr. 1972. CAS 53291 (1, 158), MZUSP 12168 (1, 168) and MZUSP 12169 (1, 119.5); 31°31′S, 49°52′W, 200 m; 11 Apr. 1972. MNHN 1983-679 (1, 167.5); 32°21′S, 50°13′W, 180 m; 21 Aug. 1972. MZUSP 12162 (1, 166); 33°38′S, 51°04′W, 215 m; 19 Jan. 1972. MZUSP 12164 (1, 128) and MZUSP 12165 (1, 123.5); 30°20′S, 48°37′W, 194 m; 30 Jan. 1972. MZUSP 12172 (1, 245); 30°42′S, 49°03′W, 192 m; 6 Aug. 1972. MZUSP 12173 (1, 148); 31°03′S, 49°46′W, 144 m; 6 Aug. 1972. USNM 263574 (1, 153); 29°53′S, 48°19′W, 194 m; 3 Aug. 1972.

Other material: USNM 197737 (1, 281); Gulf of Mexico, off Florida, 250° course from Egmont Key, 1200–1400 m; J. H. Finucane; 23 May 1963.

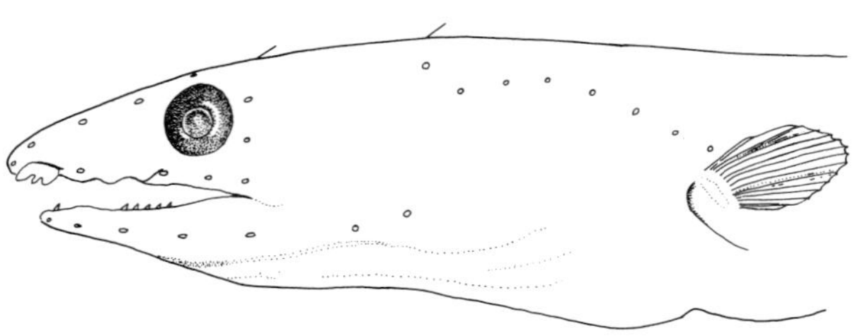

FIGURE 411. *Ophichthus menezesi*: MZUSP 12166, holotype, 346 mm TL, Brazil; head.

Ophichthus omorgmus McCosker and Böhlke, 1984

DOTTED-LINE SNAKE EEL

Figures 406, 413–415, Tables 37, 40

Ophichthus omorgmus McCosker and E. B. Böhlke, 1984: 26, figs. 4–6 (original description, Gulf of Mexico, off Northern Florida, holotype ANSP 152303).

Distinctive Characters. A moderately elongate, pale tan ophichthin marked by conspicuously dark head pores and lateral-line pores; tail 55% TL; dorsal-fin origin slightly behind pectoral tips; pectoral fin rounded, its length 30% HL; three

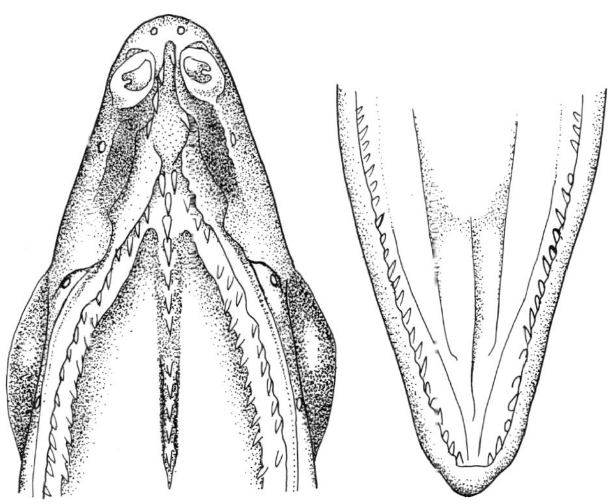

FIGURE 412. *Ophichthus menezesi*: MZUSP 12166, holotype, 346 mm TL, Brazil; dentition.

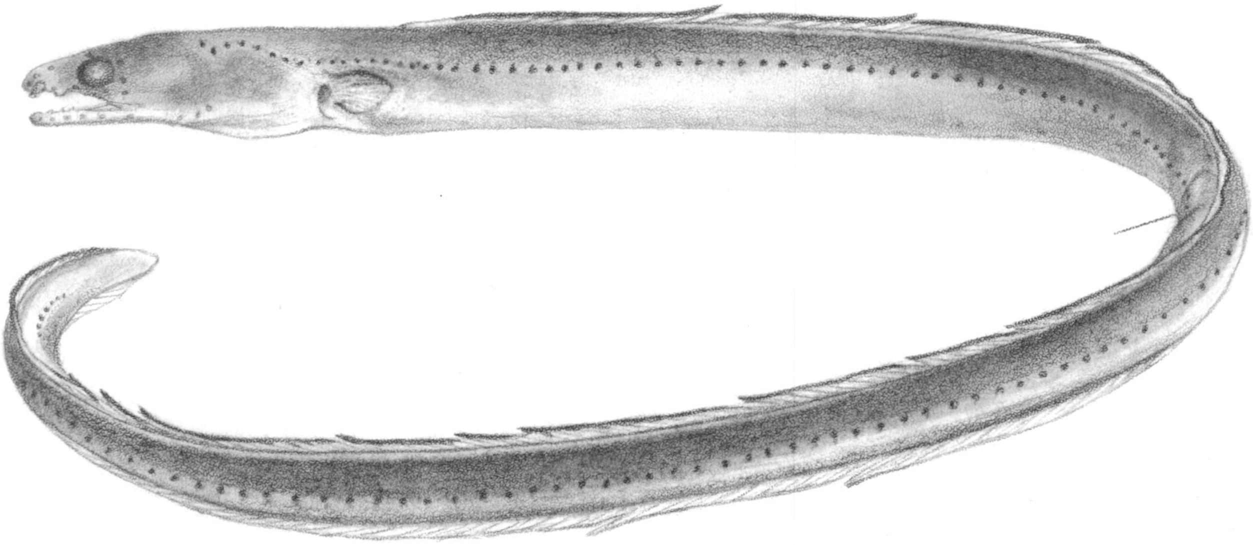

FIGURE 413. *Ophichthus omorgmus*: ANSP 152302, holotype, 453 mm TL, Gulf of Mexico off Florida.

dorsal supraorbital, six mandibular, and three preopercular pores; jaw teeth biserial, vomerine teeth uniserial; MVF 14-56-141+.

Description. Total vertebrae 141+ (n = 1); predorsal vertebrae 14 (n = 2); preanal vertebrae 56–57 (n = 2). Pores: SO 1 + 3; IO 4 + 2; POM 6 + 3; ST 3.

Proportions as % of TL (although the tip of the tail of the holotype has been broken off, it appears to be a nearly complete specimen; body proportions are approximate): head 9.7; trunk 35; tail 55; predorsal 14; depth behind gill openings 3.1. Of HL: snout 19; snout to rictus 36; eye 15; interorbital width 12; pectoral-fin length 30.

Head conical, flattened across interorbital region. Jaws short; eye large, its posterior margin just above rictus. Anterior nostril tubular; posterior nostril at edge of lip, entirely outside of mouth, covered anteriorly by a small flap. Head pores developed, as listed above; the holotype with six pores along left mandible and five on right side.

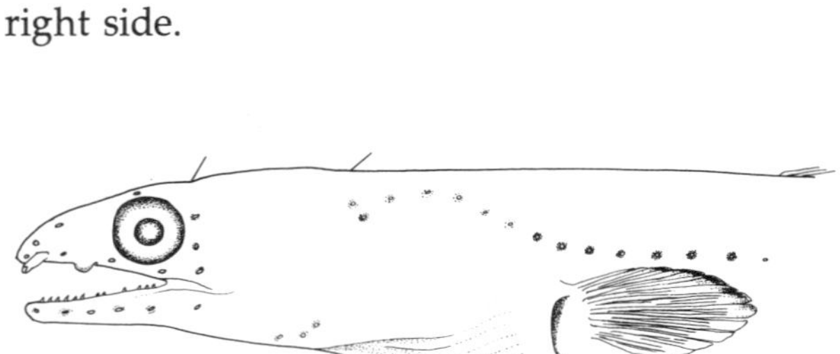

FIGURE 414. *Ophichthus omorgmus*: ANSP 152302, holotype, 453 mm TL, Gulf of Mexico off Florida; head.

Teeth small and conical, subequal. Intermaxillary teeth an outer chevron of four, closely followed by a pair of vomerine teeth heading a uniserial row of 14. Maxillary teeth biserial, 21 in each row, the outer slightly larger. Mandibular teeth similarly biserial, approximately 20 in each row on holotype.

Median fins elevated; dorsal-fin origin above and immediately behind pectoral-fin tips; middle third of anal fin elongate. Pectoral fin moderately elongate and rounded.

Color. Body pale, darker dorsally; anterior nostrils, head pores and lateral-line pores in con-

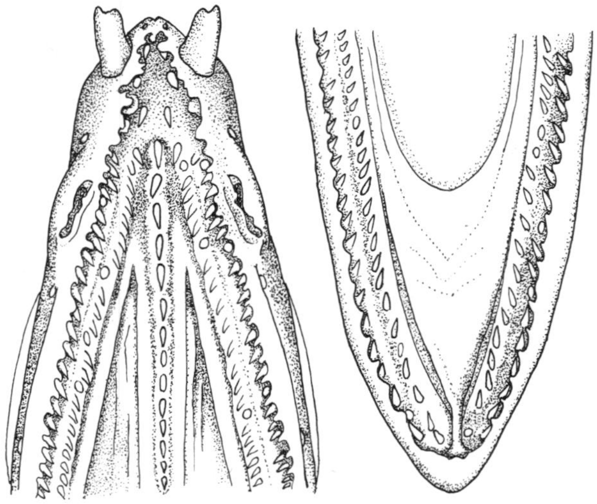

FIGURE 415. *Ophichthus omorgmus*: ANSP 152302, holotype, 453 mm TL, Gulf of Mexico off Florida; dentition.

spicuous black spots. Top of head noticeably dark. Dorsal fin pale with a sharp black edge; anal fin colorless; pectoral fins colorless except for black edge on posterior dorsal margin.

Size. The holotype is a male, 453 mm TL, 203 mm preanal. The second specimen is a ripe male, 411 mm TL, 177 mm preanal, which has been broken off at the 133rd vertebra. Size at maturity would thus appear to be about 500 mm.

Development. Leptocephalus not identified.

Distribution. The holotype was captured by otter trawl in the Gulf of Mexico off northern Florida at 183 m depth. A second specimen (UF 45173) was recently captured off the Florida Keys at 271 m.

Etymology. From the Greek *omorgmus* (spotted), in reference to the spots around the lateral line. An adjective.

Remarks. The coloration of *Ophichthus omorgmus* is similar to that of *O. melanoporus*, and the two might be confused on the basis of coloration alone. However, *O. melanoporus* differs from the above species, and from most western Atlantic species of *Ophichthus*, in its elongate body and very long tail (70–76% TL), its slender uniserial teeth, and in possessing four pairs of dorsal supraorbital pores.

Study Material. The holotype: ANSP 152302 (453 mm TL); Gulf of Mexico off northern Florida, 29°16.6'N, 86°05.5'W, 16-ft otter trawl at 183 m; TURSIOPS cruise 7110, sta. 21; 12 April 1971.

Other material: UF 45173 (1); Gulf of Mexico off Florida Keys, 24°24.38'N, 83°17.56'W, 271 m; R/V CHAPMAN, Cr. 19, haul 10; Aug. 1987.

Ophichthus ophis (Linnaeus, 1758)

SPOTTED SNAKE EEL

Figures 416–419, Tables 37, 40

Serpens Marinus, Maculosus Lister in Willughbeii, 1686: App. 19, pl. G.9 (pre-Linnaean, no locality given; non-binomial).

Muraena ophis Linnaeus, 1758: 244 (original description, "in Ocean Europaeo," holotype unknown). Bloch, 1786: 35, pl. 154.

Innominado Parra, 1787: 96, pl. 37, fig. 2 (original description, Cuba; non-binomial).

Ophichthus ophis. Ahl, 1789: 9.

Ophisurus ophis. Lacepède, 1800: 196, pl. 6, fig. 2 (in part, portion copied from Bloch, 1786, 1787).

Muraena hcuannensis Bloch and Schneider, 1801: 491 (name based on the figure of Parra, 1787; the specific name a typographical error, corrected by later authors to *havannensis* and by Poey to *havanensis*).

Anguilla ophis. Shaw, 1803: 22 (after Bloch, 1787: pl. 154; synonymized with *Anguilla serpens* Shaw).

Ophisurus guttatus Cuvier, 1817: 232 (original description, Surinam, evidently based on Bloch, 1787: pl. 154).

Herpetoichthys sulcatus Kaup, 1856a: 44 (original description, locality not stated, holotype MNHN B.2758). Kaup, 1856b: 8, pl. 1, fig. 5 [not fig. 6 as listed].

Ophisurus havanensis. Poey, 1860: 320 (redescription of the species on the basis of another Cuban specimen).

Herpetoichthys callisoma Abbott, 1861: 475 (original description, type locality unknown, holotype ANSP 38148).

Uranichthys havanensis. Poey, 1867b: 257.

Uranichthys brachycephalus Poey, 1867b: 257 (original description, type locality Cuba, holotype unknown).

Ophichthys havannensis. Günther, 1870: 67. Osorio, 1898: 200 (Ilha Cabo Verde).

Ophichthys triserialis (not of Kaup, 1856). Günther, 1870: 58 (in part, the Caribbean and Bahia specimens and *Herpetoichthys callisoma*). Osorio, 1894b: 184 (Bissau, West Africa). Osorio, 1898: 188, 200 (Ilha Cabo Verde, Bissau, Ilha S. Tomé, America).

Ophichthus ophis. Jordan and Evermann, 1898: 2804 (*Muraena hcuannensis* placed in the synonymy of *Ophichthus ophis*). Hubbs, 1932: 26 (*Ophis* Turton, 1870 synonymized with *Ophichthus* Ahl, 1789, and discussion of synonymy of *Ophis maculata* (Shaw)). Böhlke and Robins, 1959: 1 (extensive synonymy and description of western Atlantic specimens). Blache and Saldanha, 1972: 135 (description and illustrations of eastern Atlantic specimens). Böhlke and Menezes, 1977: 786 (*Antobrantia ribeiroi* synonymized with *O. ophis*).

Antobrantia ribeiroi Pinto, 1970: 13, figs. 1–2 (original description, Rio de Janeiro, holotype MNRJ 9367).

Discussion of Synonymy. The synonymy listed above has been well documented with the exception of *Herpetoichthys callisoma* Abbott. Abbott suggested that the holotype of *Herpetoichthys callisoma* might come from the Pacific, ". . . various circumstances induce me to suspect it is a species belonging to the Pacific fauna." However it agrees in all characters, including vertebral counts, with *Ophichthus ophis*; we add it to the synonymy for that species, and consider the suggested locality to be incorrect.

Distinctive Characters. A robust ophichthin with large dark spots along the pale body and a dark nuchal band; tail 57–61% TL; dorsal-fin origin

FIGURE 416. *Ophichthus ophis*: ANSP 94848, 825 mm TL, St. Barthelemy.

above middle to tips of pectoral fins; pectoral fin spatulate, its length 23–27% HL; four dorsal supraorbital, six mandibular and three preopercular pores; jaw teeth biserial, vomerine teeth uniserial; MVF 14-61-164.

Description. Total vertebrae 161–167, x̄ = 164.4 (n = 12); predorsal vertebrae 13–17, x̄ = 14.2 (11); preanal vertebrae 57–64, x̄ = 60.6 (8). Pores: SO 1 + 4; IO 4 + 2; POM 6 + 3; ST 3.

Proportions as % of TL: head 9.9–12; trunk 28–32; tail 57–61; predorsal 11–14; depth behind gill openings 2.9–3.9. Of HL: snout 16–20; snout to rictus 33–42; eye 7.9–11; interorbital width 10–14; pectoral-fin length 23–27.

Snout moderate, lower jaw included, intermaxillary teeth exposed. Eye large, horizontally elongate or elliptical, over mid-gape, protruding above head profile in lateral aspect. Anterior nostrils tubular, with conspicuous barbel from mesial rim. Posterior nostril large, opening outside mouth and entirely bordered by irregular flange-like lip. Head pores small and black-margined. Numerous pits and grooves on head, fine papillae on lips and chin.

Teeth stout and conical. Six outer intermaxillary teeth, slightly separated from vomerine teeth which are headed by a pair followed by 12–17 uniserial, progressively smaller teeth. Maxillary teeth biserial; outer row of 8–15 diminishing in size posteriorly, and inner row of 10–19 longer teeth which begins before outer row. Dentaries with outer row of 14–23 larger teeth and inner row of 14–19, the three anterior teeth in inner

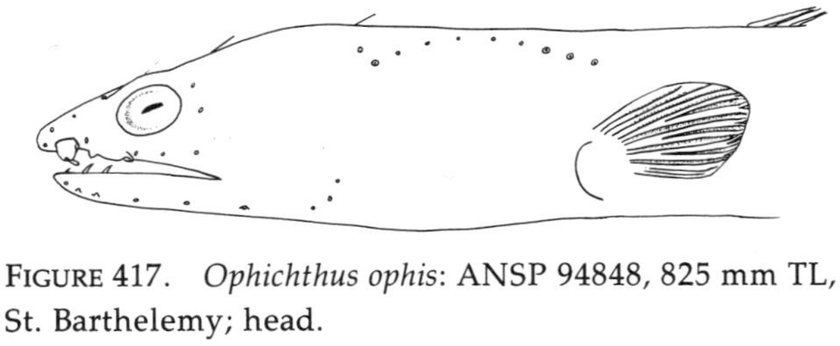

FIGURE 417. *Ophichthus ophis*: ANSP 94848, 825 mm TL, St. Barthelemy; head.

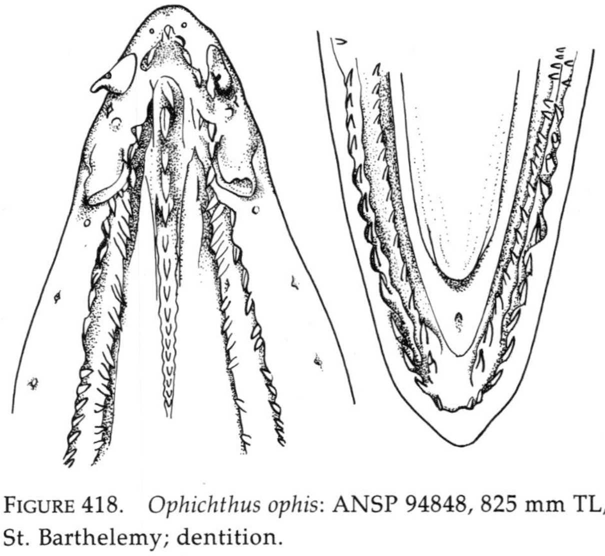

FIGURE 418. *Ophichthus ophis*: ANSP 94848, 825 mm TL, St. Barthelemy; dentition.

row large and stout, sometimes followed by a gap.

Median fins well developed. Pectoral fin spatulate.

Color. Body pale, overlain with about 20 large dark spots along midline and 25–35 pairs of smaller spots abutting base of dorsal fin; ventrum pale with few spots anteriorly. A single dark spot at dorsal fin insertion, preceded by numerous freckles on nape, snout, chin and pectoral fins. A wide nuchal band behind the mouth, fading at level of jaws. Fins pale, overlain with numerous spots and slashes. Dark spots surrounding minute papillae dorsally on head, similar to white spots on *O. puncticeps* but not as well developed or obvious. An underwater photograph of the head of one specimen shows life colors of very dark brownish-black spots on a white background.

Size. Reported to almost 2 m. The largest specimen we examined is 1400 mm TL; Hurdis (1897: 355) reported an eel from Bermuda, identified by Goode as *O. triserialis* (= *O. ophis*), to be 64 inches (ca 1620 mm) and Cervigón (1973: 13) reported one to be 1922 mm. Of 14 females examined (692–1400 mm), three (977–1241) contained eggs 0.4–0.7 mm in diameter. Only one 904-mm specimen had identifiable male gonads.

Development. Leptocephalus in the eastern Atlantic described by Blache (1977: 259), in the western Atlantic described by Leiby (1981: 53, and this volume, p. 820).

Behavior. Reported to be benthic on the continental shelf, living burrowed in sand at 10–50 m (Bauchot, 1986: 582). Many of our specimens were obtained by trawls, none recorded taken by poison. A color photograph sent from Puerto Rico (courtesy of D. Arnesen) shows a specimen of *Ophichthus ophis* protruding from a fine sand bottom, only the head anterior to the pectoral fins exposed (size of specimen unknown).

Distribution (Fig. 419). Found on both sides of the Atlantic. In the western Atlantic, known from southern Florida to Brazil, including Bermuda, Cuba, the Lesser Antilles and northern South America, the greater number of specimens from off Brazil. In the eastern Atlantic recorded from the islands of Cape Verde and San Tome and off the African coast from Sénégal to Angola. West African records of *O. triserialis* (Kaup) were based on *O. ophis* according to Blache and Saldanha (1972: 135). A fairly shallow-water species, depths recorded in the western Atlantic of 1 m (ANSP 103864) and 10 m (the photographed specimen); in the eastern Atlantic recorded depths were 10–50 m.

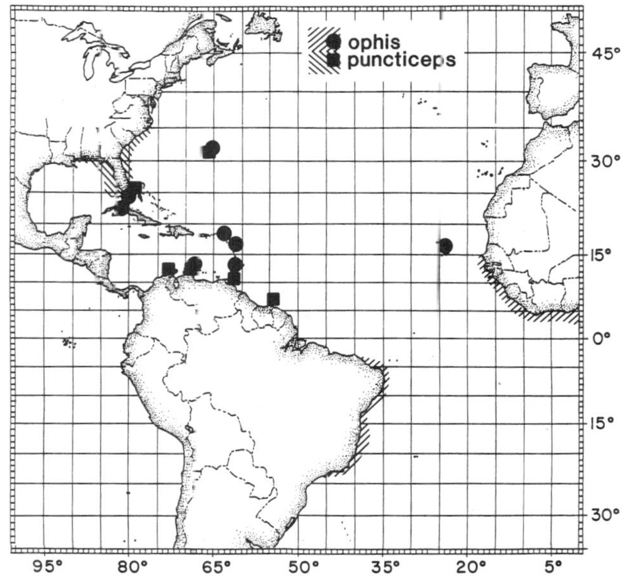

FIGURE 419. Distribution of *Ophichthus ophis* and *O. puncticeps* based on study material and records in Blache and Saldanha, 1972.

Etymology. From the Greek *ophis* (snake), referring to its similarity to a snake. A noun in apposition.

Remarks. Ophichthus ophis is closely related to its eastern Pacific congener, *O. triserialis,* and many specimens of *O. ophis* were initially identified as that species. However, the species differ in dentition and meristics; *O. triserialis* has a mean vertebral formula of 14-56-152. *Ophichthus triserialis* is not uncommon in the eastern Pacific, generally found in shallow water, and juveniles are often collected in intertidal and shallow subtidal rotenone stations. Juveniles of *O. ophis* are not found in similar habitats; no specimens have been collected by poison, and we have seen no specimens smaller than 578 mm TL.

Study Material. A total of 36 specimens, 578–1400 mm TL (five additional specimens from Brazil unlisted).

BERMUDA: FMNH 48605 (1, 700). USNM 23812 (1, ca 1400). FLORIDA: ANSP 82329 (1, 1220) and ANSP 82330 (1, skull and skin). WEST INDIES: BMNH 1933.6.13.1 (1, ca 721). CUBA: MCZ 9219 (1, 855). ZMB 6362 (1, 578). LESSER ANTILLES: *St. Barthelemy:* ANSP 94848 (1, ca 825). *Guadeloupe:* MNHN B.2759 (1, 809). *St. Vincent:* ANSP 103864 (1). *Curaçao:* UMML 32125 (1, incomplete). CARIBBEAN SEA: BMNH 1855.12.26.445 (1, ca 1241).

BRAZIL: *Coast:* MNHN B.2758 (1, 797); holotype of *Herpetoichthys sulcatus* Kaup. MCZ 9171 (1, 977). *Forteleza* (all from Brazilian Trawling Survey, 1945): CAS 54318 (1, ca 1040), 54319 (2, 633–ca 900), 54320 (1, ca 810). CAS-SU 51724 (2, 597–962), 51725 (2, 661–806), 51726 (2, 750–904), 51727 (1, 823), 67820 (1, 692), MNRJ 10441 (1, ca 1010). *Pernambuco:* MZUSP 12096 (2, 786–993). *Bahia:* MHNG 665-31 (1, 847). *Rio de Janeiro:* BMNH 1923.7.30.65 (1, 926). MNRJ 3037 (1, 1016). MNRJ 9367 (1, ca 970); holotype of *Antobrantia ribeiroi* Pinto. NMW uncat. (1, 740).

LOCALITY UNKNOWN: ANSP 38148 (1, 995); holotype of *Herpetoichthys callisoma.*

Ophichthus puncticeps (Kaup, 1860)

PALESPOTTED EEL

Figures 419–422, Tables 37, 40

Ophisurus ocellatus (not of Lesueur). Richardson, 1848a: 104.

Muraenopsis ocellata (not of Lesueur). Kaup, 1856a: 46; 1856b: 11.

Cryptopterus puncticeps Kaup, 1860: 11, pl. 1, fig. 2 (original description, Caribbean Sea, at Puerto Cabello, holotype ZMH 192).

Herpetoichthys ocellatus (not of Lesueur). Goode and Bean, 1879b: 155.

Ophichthys guttifer Bean and Dresel, 1884a: 99 (original description, Pensacola, Florida, holotype USNM 32647).

Ophichthys retropinnis Eigenmann, 1887: 116 (original description, Snapper Banks, off Pensacola, Florida, holotype USNM 38054).

Acanthenchelys ocellatus (not of Lesueur). Norman, 1922b: 296 (referred to new genus *Acanthenchelys* Norman).

Ophichthus ocellatus (not of Lesueur). Günther, 1870: 68. Schroeder, 1941: 45 (*guttifer* synonymized with *ocellatus*). Randall and Robins, 1966: 610 (*guttifer* and *retropinnis* synonymized with *ocellatus*).

Discussion of Synonymy. This species has long been known as *Ophichthus ocellatus* (Lesueur). However, both the description and the figure published by Lesueur (1825: 108–109, pl. 4, fig. 3) clearly depict a species of *Myrichthys*—a long slender eel with dorsal-fin origin on head, minute pectorals, and "rufous color, ornamented with round and distant spots of orange, edged with blue, head covered with small round orange spots." We therefore consider *Muraenophis ocellatus* as described by Lesueur to be a senior synonym of the species previously known as *Myrichthys oculatus*, its name now *Myrichthys ocellatus* (Lesueur, 1825) (see additional Remarks under *Myrichthys*, p. 000).

The species from the West Indies described by Richardson (1848a: 104) as *Ophisurus ocellatus* and that described from the Atlantic coast of Tropical America by Günther (1870: 68) and subsequent authors as *Ophichthys ocellatus* define the species "*Ophichthus ocellatus*" as we know it; this species must now take the next available name, *Ophichthus puncticeps* (Kaup).

Distinctive Characters. A moderately robust, gray ophichthin with about 20 white eye-size spots along the lateral line; tail 51–56% TL; dorsal-fin origin above or just before pectoral-fin tips; pectoral fin lanceolate, its length 30–45% HL; three dorsal supraorbital, six mandibular and three preopercular pores; jaw teeth biserial, vomerine teeth uniserial; MVF 14-53-134.

Description. Total vertebrae 127–141, $\bar{x}$ = 134 (n = 77); predorsal vertebrae 12–17, $\bar{x}$ = 13.9 (38); preanal vertebrae 50–58, $\bar{x}$ = 53.4 (39). Pores: SO 1 + 3; IO 4 + 2; POM 6 + 3; ST 3.

Proportions as % of TL: head 11–14; trunk 30–35; tail 51–56; predorsal 15–18; depth behind gill openings 3.7–5. Of HL: snout 17–21; snout to rictus 35–48; eye 9.4–12; interorbital width 7.4–16.5; pectoral-fin length 30–45.

Snout broad, flattened, triangular when viewed from above. Jaws moderate; eye large, over mid-gape. Anterior nostril tubular, longer posteriorly than anteriorly; posterior nostril opening outside mouth well in advance of anterior margin of eye, covered with a short flap. Head pores small, as listed above.

Teeth conical, sharp, subequal. Typically three intermaxillary pairs, followed by a short gap; two to six paired or staggered teeth followed by a row of 11–14 progressively smaller teeth along vomer. Jaw teeth biserial, the rows roughly equal in size and number of teeth; 17–25 outer teeth

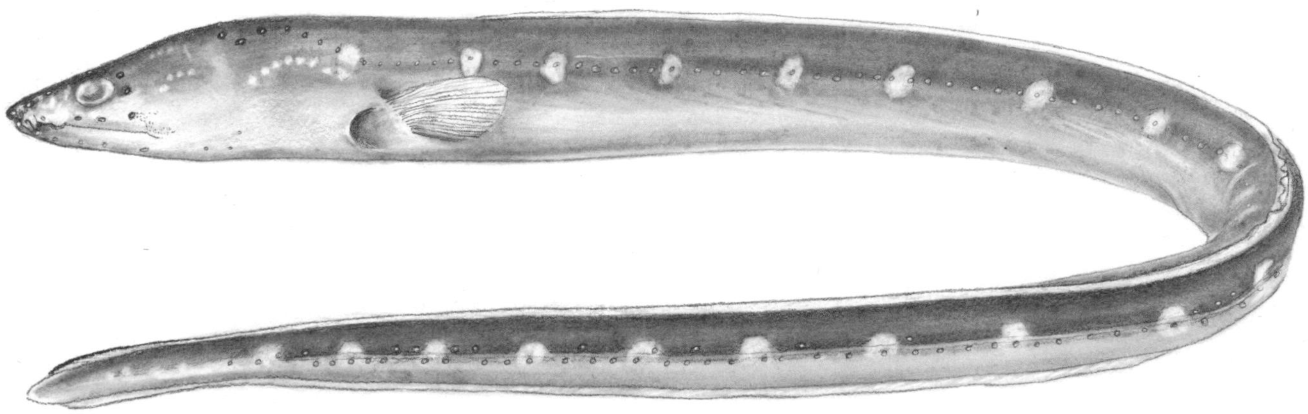

FIGURE 420. *Ophichthus puncticeps*: ANSP 128353, 464 mm TL, South Carolina.

and 13–24 slightly larger inner teeth on maxillaries. Dentaries with outer row of 17–29 teeth that are larger and stouter than those of inner row; on small specimens only the anteriormost three or four large teeth visible on inner row of dentaries.

Median fins low, sometimes within groove on preserved specimens. Position of dorsal-fin origin varies considerably, from before to over to just behind pectoral-fin tips. A posterior portion of the dorsal fin with short erect "spinous" rays (this the character for which the genus *Acanthenchelys* was erected), followed by a short expanded section of fin with conspicuous black coloration.

Color. Small specimens gray dorsally fading to white ventrally; a series of 18–22 evenly-spaced white or pale spots, about equal in size to eye, along lateral line. Larger specimens darker grey to brown, the spots not as obvious, and with a dark reddish brown stripe along lateral line, the stripe appearing as dark slashes between the pale spots; additional reddish-brown color on body along dorsal and anal fins. Dorsal fin with dark edge and with conspicuous black area at posterior end; anal and pectoral fins colorless. Head pores (except anteriormost snout pores and preopercular pores) and branchial lateral-line pores outlined with dark pigment. A characteristic pattern of white dots or dashes on snout and head, enclosing small papillae which are raised and observable on large specimens. A kodachrome of a fresh specimen showed life colors of greenish-brown dorsally and white from midside down, the lateral spots white.

Size. The largest specimen examined is 927 mm TL. Of 17 females, 233–927 mm TL, seven (383–927) were ripe with 0.7–1.1-mm eggs. Of nine males (221–498), six (387–498) appeared ripe.

Development. Leptocephalus described as *Ophichthus ocellatus* by Fahey and Obenchain (1978:

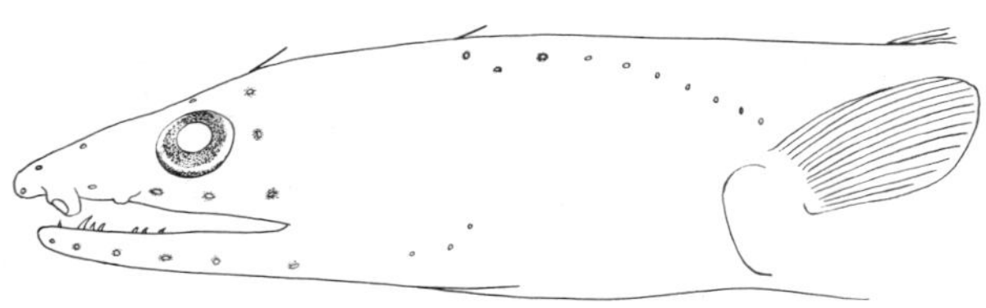

FIGURE 421. *Ophichthus puncticeps*: ANSP 128353, 464 mm TL, South Carolina; head.

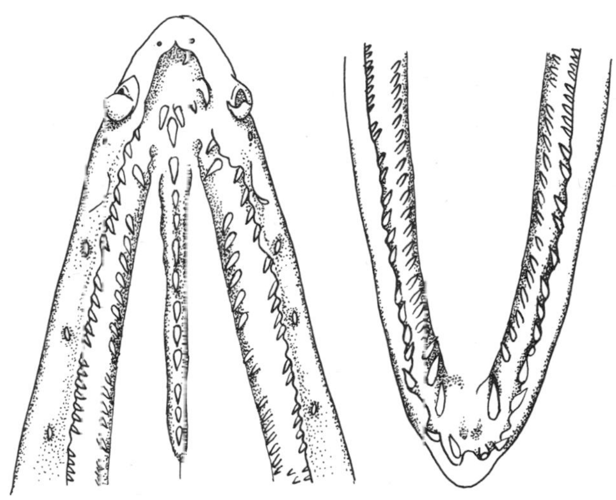

FIGURE 422. *Ophichthus puncticeps*: ANSP 128353, 464 mm TL, South Carolina; dentition.

460) and by Leiby (1981: 69), and in this volume, p. 821.

Variation. The largest specimen (ANSP 140306, 927 mm), a ripe female from off North Carolina at recorded depth of 165 m, showed some variation: color dark brown, lateral line streaks observable but spots not visible; five supraorbital pores; snout excessively short and blunt. Five specimens from one other collection (ANSP 126531, 236–498 mm) from the Gulf of Mexico off Florida, recorded as taken by longline over 3300 m, showed similar coloration and had five SO pores. Two specimens from off Colombia (ANSP 116147) also had five SO pores, but showed typical spots. Vertebral counts for all are within the range for *Ophichthus puncticeps,* and they are retained within this species, the variation noted.

Distribution (Fig. 419). Frequently taken along coastal United States from North Carolina to Florida, and in the northern Gulf of Mexico off Florida, Alabama and Texas; infrequently taken in the West Indies and off northern South America to Suriname (not as far south as Brazil—see below). Taken most often by trawl, also by dredge and by longline; depths of capture recorded from surface to 219 m, most often between 20–200 m.

The distribution previously given for "*Ophichthus ocellatus*" [= *puncticeps*] has included "South America" or "Brazil" ever since Kaup (1856a: 46) included *Ophisurus ramiger* Valenciennes, 1839 (changed to *remiger* by subsequent authors), no locality given, in the synonymy of "*Muraenopsis ocellatus*" Lesueur. Valenciennes' illustration of *O. ramiger* is more similar to the eastern Pacific *Ophichthus pacifici* (Günther, 1870), known from Chile to Costa Rica, which also has a row of spots along the lateral line, the spots generally smaller than those of *O. puncticeps.* In addition, *O. pacifici* and *O. puncticeps* are separable by vertebral counts; ten specimens of *O. pacifici* have 18–20 predorsal, 53–58 preanal, and 150–158 total vertebrae. The six syntypes of *Ophisurus ramiger* (MNHN B.2748 and MNHN A.9955) were x-rayed (courtesy of M. L. Bauchot, MNHN) and found to have 18–20 predorsal, 54–58 preanal, and 152–157 total vertebrae, indicating *pacifici* to be conspecific with *ramiger,* the type-locality for *O. ramiger* likely the west coast of South America rather than the east coast. There are no records of *Ophichthus puncticeps* from any area south of Suriname.

Etymology. From the Latin *puncticeps* (a spot), in reference to the coloration. An adjective.

Study Material. A total of 152 specimens, 221–927 mm TL, including the holotypes of *Ophichthys guttifer* Bean and Dresel, and *Ophichthys retropinnis* Eigenmann.

NORTH CAROLINA: ANSP 112003 (1, ca 395), 140306 (1, 927). GMBL 73-67 (1, 555). UMML 2321 (1, 260). USNM 143051 (1, 580). SOUTH CAROLINA: ANSP 128353 (1, 464), 126503 (2, 584–602). GMBL 55-7 (1, 444), 71-444 (1, 612), 74-227 (1, 525). MNHN 2137 (1, 392). UMML uncat. (1). USNM 25359 (1, 583), 25585 (1, 463), 129911 (1, 520). GEORGIA: ANSP 128383 (1, 423). FMNH 64299 (1). GMBL 73-287 (1, 444). GMBL uncat. (1, 442). UMML 422 (1). USNM 118152 (1, 450). FLORIDA: *Atlantic coast:* ANSP 82434 (3, 468–533), 115515 (1, 437, cleared and stained), 116119 (1, 387), 116120 (1, 630), 126152 (1, 383), 128384 (1, 311), 138430 (1, 233), 138431 (1, 221). GMBL 73-281 (1, 410), 74-224 (2, 264–270), 74-225 (1, 500). UF 715 (1, 584), 10246 (4, 386–494), 13040 (1, 505), 19493 (1, 473), 29827 (1, 367), 29828 (2, 225–389). UMML 1071 (1), 3293 (1, 515), 7027 (1, 402), 8386 (2, 259–271), 12318 (1, 432), 13468 (1), 17671 (1, 346), 20992 (1, 260), 26971 (1, 294), 30581 (2). UMML uncat. (1). UMML uncat. (1, 470). USNM 48877 (1, 292), 153157 (1, anterior half), 153158 (1, 208), 153229 (1, 234). *Keys:* TU 35811 (5, 350–595). UMML 13509 (1).

GULF OF MEXICO: *Florida:* ANSP 39951 (1, ca 410), 79422 (1, 237), 126531 (5, 236–498). CAS-IUM 8576 (1, 669), 6989 (1, 729). CAS-SU 2119 (3, 615–707). FMNH 34988 (1, 253). MCZ 35109 (9, 502–592). TU 6827 (1, 620), 12965 (3, 319–393), 83858 (1, 231). UF 3582 (1, 474), 21766 (1, 312). UF-FSU 3277 (1, 521), 18881 (1, 468), 20670 (2, 415–422), 20716 (1, 439), 21405 (3, 326–433), 21475 (1, 252). UMML uncat. (1). USNM 22289 (1, 570). USNM 32647 (1, 570); holotype of *Ophichthys guttifer.* USNM 30160 (1, 535), 36995 (1, 112), 37662 (1, 547). USNM 38054 (1, 518); holotype of *Ophichthys retropinnis.* USNM 106542 (2, 340–390), 114763 (1, 645), 134072 (1, 158), 157931 (1, 477), 157934 (1, 458), 157938 (1, 520), 158451 (4, 410–626), 158745 (1, 508), 158746 (1, 538), 198766 (3, 460–545). *Alabama:* GCRL 2526 (1, 433), 2799 (1, 438), 3554 (1, 489). USNM 134198 (2, 390–470+), 158747 (1, 608). *Mississippi:* USNM 157949 (1, 257).

BAHAMAS: Bimini: UMML uncat. (2, 435–510). USNM 48877 (1, 292). LESSER ANTILLES: ANSP 153651 (1, 395). *Tobago:* BMNH 1922.6.22.3-4 (2, 473–559). COLOMBIA: ANSP 116147 (2, 510–543). USNM 198755 (1, 527). VEN-

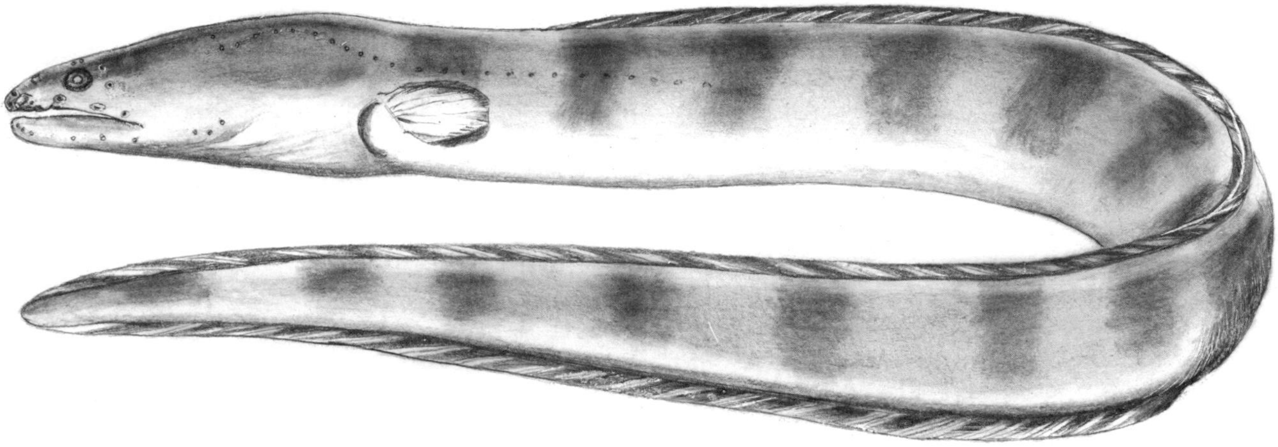

FIGURE 423. *Ophichthus rex*: ANSP 135555, paratype, 1422 mm TL, Gulf of Mexico off Louisiana.

EZUELA: MBUCV 168 (2, incomplete–706). UMML 31325 (1, 373), 31628 (1, 276). SURINAME: USNM 158897 (1, 478).

Ophichthus rex Böhlke and Caruso, 1980

GIANT SNAKE EEL

Figures 423–426, Tables 37, 40

Ophichthus n. sp. Gallaway, Parker, and Moore, 1972: 37 (in key to species of estuarine and marine fishes of Texas).

Ophichthus sp. Hoese and Moore, 1977: 132 (northwestern Gulf off Louisiana and from Port Aransas).

Ophichthus species A. Böhlke, 1977b: 5.

Ophichthus rex Böhlke and Caruso, 1980: 239, figs. 1–3 (original description, northern Gulf of Mexico, off Louisiana, holotype ANSP 136745).

Distinctive Characters. The largest western Atlantic ophichthid, robust, pallid, with 14–15 dark saddles along body; tail 50–54% TL; dorsal-fin origin behind pectoral-fin tips by length less than pectoral-fin length; pectoral fin spatulate, its length 26–34% HL; four dorsal supraorbital, seven mandibular and three preopercular pores; biserial to quadriserial dentition; MVF 16-46-118.

Description. Total vertebrae 115–121, x̄ = 118 (n = 25); predorsal vertebrae 15–18, x̄ = 16.6 (19); preanal vertebrae 45–49, x̄ = 46.6 (25). Pores: SO 1 + 4; IO 4 + 2; POM 7 + 3; ST 3.

Proportions as % of TL: head 11–13; trunk 35–37; tail 50–54; predorsal 17–19; depth behind gill openings 4.5–7.7. Of HL: snout 17–21; snout to rictus 35–44; eye 6–10; interorbital width 16–22; pectoral-fin length 26–34.

Snout short, jaws moderate, lower jaw included, slightly shorter than upper, only one or several intermaxillary teeth exposed when mouth is closed. Shallow median groove on underside

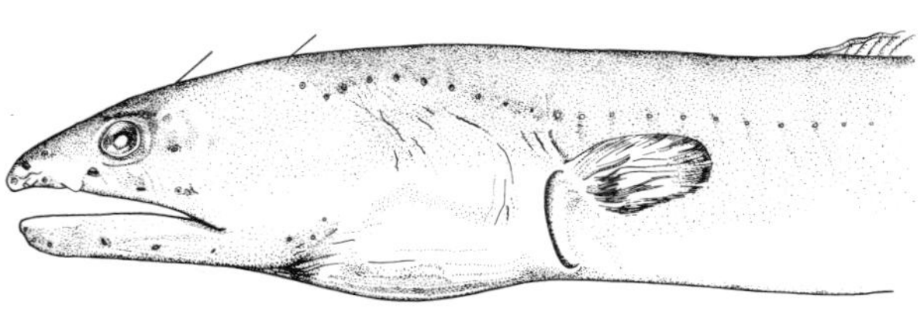

FIGURE 424. *Ophichthus rex*: ANSP 135555, paratype, 1422 mm TL, Gulf of Mexico off Louisiana; head.

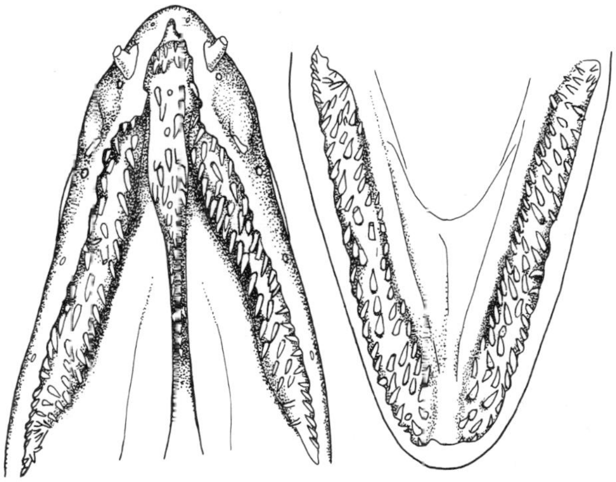

FIGURE 425. *Ophichthus rex*: ANSP 135555, paratype, 1422 mm TL, Gulf of Mexico off Louisiana; dentition.

of snout extending vertically up and around tip of snout of large specimens. Eye moderate, about equidistant from tip of snout and rictus. Anterior nostril tubular, with notch in dorsal margin and ridge inside ventral margin. Posterior nostril opening outside mouth, opening completely covered by posteroventrally-directed flap. Head pores generally small and difficult to see; the first dorsal supraorbital and second infraorbital pores enlarged. The illustrated specimen (ANSP 135555) had a double pore over the eye, and six mandibular pores on one side and eight on the other. UF 28451 had only two preopercular pores visible. Lips, snout and chin finely papillose.

Teeth stout, conical, increasing in number with growth. An outer semicircle of intermaxillary teeth in small individuals, median teeth added with growth until ethmovomer bone is covered with teeth in large specimens. Vomerine teeth continuous with intermaxillary teeth; biserial anteriorly to uniserial posteriorly in small individuals, becoming tri- and quadriserial anteriorly in largest specimens. Maxillary teeth of males and small females mostly biserial, becoming tri- to quadriserial in largest females. Mandibular teeth of small individuals uni- and biserial, becoming quadriserial to sometimes briefly six rows across in largest females. Dentition of the males examined was not as well developed as that of the largest females; however, no males were as large, and the dentition of males and females of similar size was the same.

Dorsal-fin origin well behind tips of pectorals. Pectoral fin spatulate, short and fleshy; pectoral fin missing on left side of ANSP 143064. Pits and pockets in skin, especially dorsally and on top of head.

Color. Body brown or slate gray, darker dorsally, with 14–15 faint to dark saddles along body, hardly visible on some specimens. Vertical fins striped at base; pectoral fins darker on dorsal half. In life, pale yellowish-brown dorsally, with an abrupt midline differentiation to a white ventral surface, overlain with dark, olive-brown rectangular saddles.

Size. The largest specimen we examined is 1801 mm TL, although individuals larger than 2.1 m have been reported (Böhlke and Caruso, 1980), and a recent publication (Burgess et al., 1987) stated that the species is said to reach 11 feet. Of 12 females examined (945–1793 mm TL), only one (1687) was ripe with 0.9-mm eggs; of seven males (720–965), five (722–896) were ripe, this suggesting some degree of difference in size at maturation. Burgess et al. studied 25 specimens, 568–2080 mm TL, including 10 mature females (1605–2080); fecundity studies indicated that large females (1740–2080) may produce 1.8–4.7 million eggs per year.

Development. Leptocephalus identified (see p. 825).

Behavior. Ophichthus rex has most often been taken at sites of offshore oil rigs in the Gulf of Mexico and has become widely known by sport and commercial fishermen. Its existence was known for several years before it was described, but few specimens were collected; its large size and threatening appearance made most fishermen throw specimens overboard as quickly as possible when they appeared. Recent interest in developing a market for edible fillets, prepared and sold in Florida under the name "Keoghfish," prompted a study of its biology, ecology, and food value by Burgess et al. (1987). They found it to be an obligate mud-dweller, preferring blue mud (sandy or calcareous mud bottoms were marginal habitat choices) where it lives in burrows or tubes. It was almost always caught on the bottom-most hook of longlines, suggesting that the eel takes bait dropped on the bottom near its burrow. They recommended that the species be utilized as a food-source, and reported that the flesh had been well-received when prepared as steaks, fillets, or in chowder, the single disadvantage being the large number of intramuscular bones which must be removed.

Distribution (Fig. 426). The type specimens were from the northern Gulf of Mexico off Louisiana and Texas, with one specimen from off Florida, taken by hook and line at depths exceeding 15 m, by shrimp trawls, and by bottom longlines at 22–366 m depth. Additional specimens were collected by bottom line on the Campeche Bank. Burgess et al. collected 25 specimens in the fish-

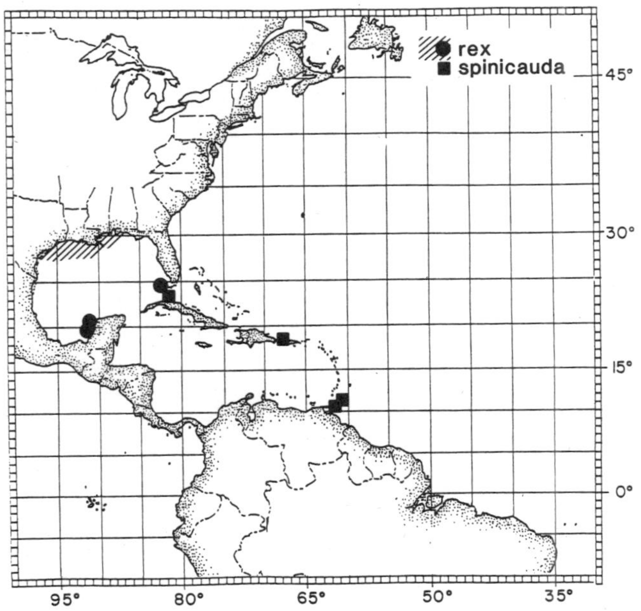

FIGURE 426. Distribution of *Ophichthus rex* and *O. spinicauda* based on study material.

ing grounds southwest of Mobile, Alabama during a four-month period, reporting most captures at depths from 16–150 m.

Etymology. From the Latin *rex* (king), alluding to the preeminently large size of this species. A noun in apposition.

Remarks. Coloration and external appearance of *Ophichthus rex* and *O. spinicauda* are almost identical. They are distinguishable primarily by vertebral counts which differ greatly (MVF for *rex* 16-46-118 versus 20-61-145 for *spinicauda*). They differ in number of pores along the lower jaw (usually seven in *O. rex* and six in *O. spinicauda*), but there is some variation in number and the pores are difficult to see. The two distinguishing characters given in the key are questionably useful; the relationship of the dorsal-fin origin to the length of the pectoral fin cannot be compared unless specimens of the same size are at hand, and even then variability in both measurements can be expected; the greater number of rows of teeth in *rex* could be due to the larger size of the specimens studied rather than a difference between species (the partially triserial dentition on the lower jaw of the largest specimens of *spinicauda* would support this).

Study Material. A total of 33 specimens, 720–2100 mm TL, including the holotype and 26 paratypes.

Holotype: ANSP 136745 (1594 mm TL); northern Gulf of Mexico, off Ship Shoal, Louisiana, 28°17′N, 90°30′W, 64 m; A. M. Landry; 17 Dec. 1962.

Paratypes: FLORIDA KEYS: USNM 133826 (1, 1793). GULF OF MEXICO: *Mississippi:* ANSP 142201 (1, 1208). *Louisiana:* ANSP 133550 (1, 1624), 135555 (1, 1422), 136744 (1, 1634), 142968 (1, 966), 143071 (1, 1687). GCRL 3691 (2, 1565+–1638), 4745 (2, 1222–1398). TU 82453 (1, 1783), 88498 (1, 1801), 88499 (1, 1589), 88500 (1, 1487), 88501 (1, 1469), 88502 (1, 1698), 88505 (1, 945), 88506 (1, 1507), 88507 (1 skull and skeleton parts), 88508 (1, 722), 88510 (1, 1239). USNM 200495 (1, ca 2100). *Texas:* ANSP 142969 (1 head of specimen ca 1900), 143064 (1, 720). FSU 23592 (1, 1429).

Other material: GULF OF MEXICO: *Texas:* USNM 287583 (skull and anterior vertebrae). *Campeche Bank:* ANSP 145827 (1, 896), 145828 (1, 879). UF 28445 (2, 965–1072), 28451 (1, 858), 28458 (1, 843).

Ophichthus spinicauda (Norman, 1922)

SPINEFIN SNAKE EEL

Figures 426–429, Tables 37, 40

Acanthenchelys spinicauda Norman, 1922b: 296, text-fig. (original description, Tobago, holotype BMNH 1922.6.22.5). Jordan, Evermann, and Clark, 1930: 89 (erroneously attributed to Regan, elevated to family status). Ginsburg, 1951: 475 (suggested that *Acanthenchelys* is a synonym of *Ophichthus*).

Ophichthus zonatus Howell Rivero, 1932: 11 (original description, off Cojimar, Cuba, holotype MCZ 32788).

Ophichthus spinicauda. Randall and Robins, 1966: 610 (*Acanthenchelys* synonymized with *Ophichthus*, holotype of *A. spinicauda* redescribed).

Distinctive Characters. A large robust ophichthin with 14 dark saddles on tan or grey body; tail 48–53% TL; dorsal-fin origin behind pectoral-fin tips by length greater than pectoral-fin length; pectoral fin spatulate, its length 23–30% HL; three dorsal supraorbital, six mandibular and three preopercular pores; dentition biserial, uniserial posteriorly on vomer; MVF 20-61-145.

Description. Total vertebrae 144–146, $\bar{x}$ = 145.3 (n = 7); predorsal vertebrae 18–21, $\bar{x}$ = 19.6 (7); preanal vertebrae 61–62, $\bar{x}$ = 61.3 (7). Pores: SO 1 + 3; IO 4 + 2; POM 6 + 3; ST 3.

Proportions as % of TL: head 11–14; trunk 34–38; tail 48–53; predorsal 17–21; depth behind gill openings 4.5–6.4. Of HL: snout 17–22; snout to rictus 36–44; eye 5.4–11; interorbital width 14–17; pectoral-fin length 23–30.

FIGURE 427. *Ophichthus spinicauda*: MCZ 32788, holotype of *Ophichthus zonatus*, 842 mm TL, Cuba.

Snout short, jaws moderate; lower jaw shorter than upper, only one or two intermaxillary teeth exposed in shallow median groove when mouth is closed. Eye moderate, about equidistant from tip of snout and rictus. Anterior nostril narrow and tubular; posterior nostril opening outside mouth, covered by flap, with flange around entire opening. Head pores small and difficult to see.

Teeth stout, conical and strong, increasing in number with growth. An outer semicircle of six intermaxillary teeth in small specimens, 1–4 additional median teeth in larger ones. Vomerine teeth continuous with intermaxillary teeth, 1–3 sometimes staggered pairs followed by a row of 10–12 teeth. Maxillary teeth biserial, 15–24 teeth in each row, the inner teeth larger. Dentary teeth biserial, 19–25 teeth in each row; a few additional teeth form short triserial row in largest specimens.

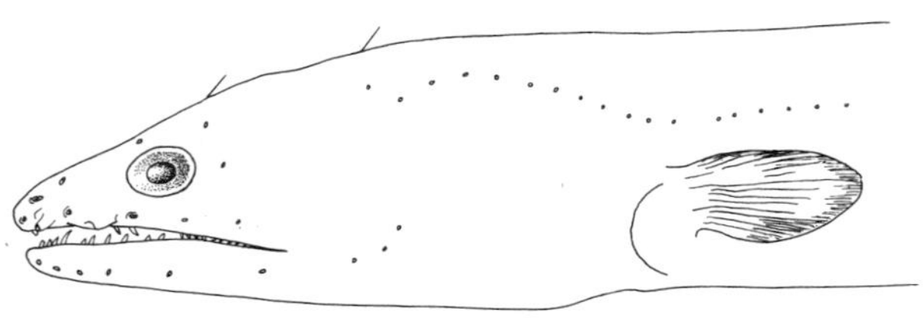

FIGURE 428. *Ophichthus spinicauda*: ANSP 138143, 1000 mm TL, Puerto Rico; head.

Pits and pockets in skin, especially dorsally. Dorsal-fin origin well behind tips of pectorals; posteriorly, a short section of the dorsal fin erect and conspicuous due to thickened rays, this section particularly noticeable on the holotype, less so on other specimens. Pectoral fins spatulate and fleshy.

Color. Background color tan or grey, overlain with 14 darker saddles or bands, separated by pale interspaces; body slightly lighter ventrally. Snout dark, head with a broad diffuse band dorsally. Dorsal fin dark-edged with occasional dark smudges; posterior part of anal fin variously dark-edged. Pectoral fins pale.

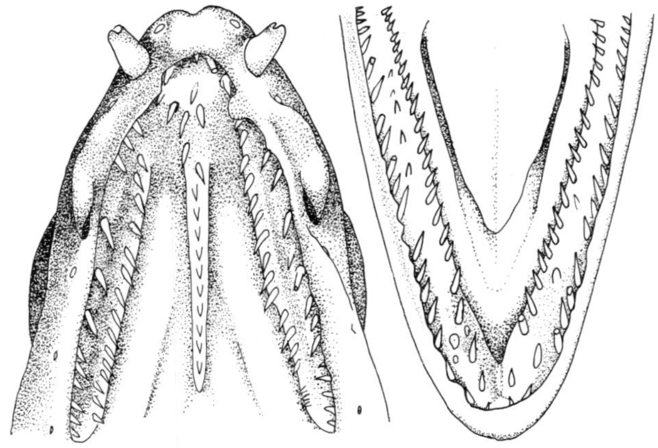

FIGURE 429. *Ophichthus spinicauda*: MCZ 32788, holotype of *Ophichthus zonatus*, 842 mm TL, Cuba; dentition.

Size. The largest specimen examined is 1077 mm TL. Six specimens (745–1077 mm) were females, none of them ripe.

Development. Leptocephalus not found for this species.

Distribution (Fig. 426). A rare species, taken off Tobago, Trinidad, Cuba and Puerto Rico, and reported from Isla Margarita off Venezuela (Cervigón, 1966: 192). Howell Rivero reported the holotype of *Ophichthus zonatus* was taken in "very deep water." The four specimens recently collected off Puerto Rico (ANSP 138143 and ANSP 138145) were obtained in fish traps at depths of 235–310 m. Probably a deep-water species, although Cervigón reported his specimen was captured at a "depth probably not greater than 60–70 fm [110–128 m]."

Remarks. Ophichthus spinicauda is very similar in external appearance to *O. rex*, differing primarily in vertebral counts (see *Remarks* for *O. rex*).

Norman (1922b) erected the genus *Acanthenchelys* on the basis of a single anatomical characteristic, the region of low "spinous" rays in the posterior segment of the dorsal fin. This condition is shared by *O. gomesii* and *O. puncticeps*, and felt by us and by others (Ginsburg, 1951; Randall and Robins, 1966; McCosker, 1977) to be variable, possibly related to sexual condition, and insignificant as a generic character.

Etymology. From the Latin *spina* (thorn) and *cauda* (tail), in reference to the posterior dorsal fin. A noun in apposition.

Study Material. Seven specimens, 745–1077 mm TL, including the holotype and the holotype of *Ophichthus zonatus* Howell Rivero.

Holotype: BMNH 1922.6.22.5 (847 mm TL); Tobago; Guppy.

Other material: CUBA: MCZ 32788 (1, 842); off Cojimar, near Havana, in very deep water; holotype of *Ophichthus zonatus* Howell Rivero. PUERTO RICO: ANSP 138143 (2, 1000–1077); west coast, 3.7 km NW of Mayaguez, 18°15′N, 67°15′W, trap at 310 m; 12 May 1977. ANSP 138145 (2, 991–1062); off San Juan, fish pots at 235 m; R/V AUGUSTIN STAHL; 12 Aug. 1976. TRINIDAD: BMNH 1891.9.9.8 (1, 745); G. Turner Wayman.

Genus *Quassiremus* Jordan and Davis, 1891

Quassiremus Jordan and Davis, 1891: 622 (type species *Ophichthus evionthas* Jordan and Bollman, 1889, by original designation).

Characters. Body stout (depth 3.2–4.3% TL in Atlantic species), laterally compressed posteriorly; head and trunk longer than tail (53–56% TL in Atlantic species). Dorsal-fin origin behind gill opening; tail tip blunt; pectoral fin minute, no longer than eye diameter. Snout conical, overhanging lower jaw, a short groove on its underside; lips without barbels; eye well developed; anterior nostril tubular; posterior nostril opens into mouth. Gill opening crescentic, its major axis vertical, shorter than isthmus. Head pores developed; three preopercular pores. Teeth pointed, uniserial, largest the intermaxillary teeth which are continuous with those of vomer.

Neurocranium truncate posteriorly; supraoccipital crest developed, pointed posteriorly; maxilla elongate, slender posteriorly; pterygoid broad posteriorly, free and terete anteriorly; orbit large, postorbital strut absent; otic bulla well developed; suspensorium anteriorly inclined. Branchiostegal rays numerous, slender, and associated with hyoid. Gill arches reduced, but *Ophichthus*-like; fifth ceratobranchial absent; upper pharyngeal tooth plates separate. Pectoral girdle reduced to a slender cleithrum and supracleithrum. Intramuscular bones, ribs, and caudal transverse processes developed. Gasbladder elongate, thin walled, 24% of TL. Body coloration spotted; peritoneum pale.

Size. To 700 mm TL.

Distribution. A New World genus, found only in the eastern Pacific and western Atlantic.

Etymology. From the Latin *quassus* (obliterated) and *remus* (oar), in reference to the minute pectorals. Masculine.

Species. One species in the western Atlantic, *Quassiremus ascensionis* (Studer, 1889); two eastern Pacific species, *Quassiremus evionthas* (Jordan and Bollmann, 1889) and *Quassiremus nothochir* (Gilbert, 1890). The three species are separable by vertebral counts and by coloration.

KEY TO THE SPECIES OF *QUASSIREUMS*

1a. Pectoral-fin length equal to or longer than eye diameter; upper half of body covered with numerous oblong spots, each slightly smaller than eye; total vertebrae 149 .... *evionthas* (Jordan and Bollmann, 1889) (eastern Pacific, Galapagos Island endemic)

1b. Pectoral-fin length less than eye diameter; upper half of body overlain with large brown spots, mostly larger than eye; total vertebrae 129–142 2

2a. Sides of body overlain with 14–16 large hourglass-shaped pale spots, ringed with brown, meeting at the dorsal midline as large ocelli; total vertebrae 138–142 .... *nothochir* (Gilbert, 1890) (eastern Pacific)

2b. Body dark dorsally, pale ventrally, with 16–20 dark spots or saddles which meet at the dorsal midline, an alternating row of dark spots or blotches below the lateral line; total vertebrae 129–136 .... *ascensionis*, p. 410

Quassiremus ascensionis (Studer, 1889)

BLACKSPOTTED SNAKE EEL

Figures 430–433, Table 37

Ophichthys Ascensionis Studer, 1889: 48, pl. 19, fig. 2 (original description, Ascension Island, holotype unknown).

Quassiremus productus Seale, 1917: 82 (original description, Nassau, Bahamas, holotype MCZ 9150).

Quassiremus goslingi Beebe and Tee-Van, 1932: 110 (original description, Bermuda, holotype USNM 170563).

Comments on Synonymy. Ophichthys Ascensionis, described by Studer in 1889, has been either ignored or placed in synonymy by previous authors. The holotype is unknown, and its Ascension Island locality placed it outside most faunal studies. Blache and Bachot (1976b: 451) suggested it might be a species of *Callechelys,* while Lubbock (1980: 287) tentatively placed it in the synonymy of *Ophichthus ophis.* However, the original description and the illustration (describing the color, the number of spots, the dorsal-fin origin behind the gill openings, the very small pectoral flap, and the dentition) depict a readily recognizable specimen of what has been known as *Quassiremus productus.* This species has recently been observed and collected at Ascension Island. We therefore adopt Studer's name and place *productus* as well as *goslingi* in the synonymy of *Quassiremus ascensionis.*

Distinctive Characters. A moderately elongate ophichthin, darker dorsally, with a row of dark spots or saddles along dorsal midline and an alternating series of spots or blotches below the lateral line; dorsal-fin origin behind gill opening; pectoral fin minute, less than ½ eye diameter; teeth conical and sharp, uniserial on jaws and vomer; MVF 9-67-134.

Description. Total vertebrae 129–136, $\bar{x}$ = 134.1 (n = 15); predorsal vertebrae 9–11, $\bar{x}$ = 9.5 (15); preanal vertebrae 65–69, $\bar{x}$ = 67.4 (14). Pores: SO 1 + 3; IO 4 + 2; POM 5 + 3; ST 3.

Proportions as % of TL: head 8.7–10; trunk 43–47; tail 44–47; predorsal 10–12; depth behind gill openings 3.2–4.3. Of HL: snout 20–25; snout to rictus 32–40; eye 7.2–10; interorbital width 10–14; pectoral fin minute, less than ½ eye.

Snout short, pointed down; eye moderate, above posterior half of upper jaw. Anterior nostril tubular, with several interior lateral projections; posterior nostril an oval slit opening into mouth, hidden by underside of snout and not marked externally. Head pores conspicuous, rounded, with rim; two postorbital pores high and close together; three preopercular pores forming an angle greater than 90°; frontal pore above posterior margin of eye.

Teeth uniserial, conical, pointed and sharp. Six anterior intermaxillary teeth forming an inverted V, followed by two smaller slender teeth; series of 10–13 teeth on vomer, the anterior few teeth the largest in the mouth. Jaw teeth uniserial, 16–19 teeth on maxillary and 15–19 on mandible, all similar in size and shape.

Color. In alcohol, dark dorsally, pale ventrally,

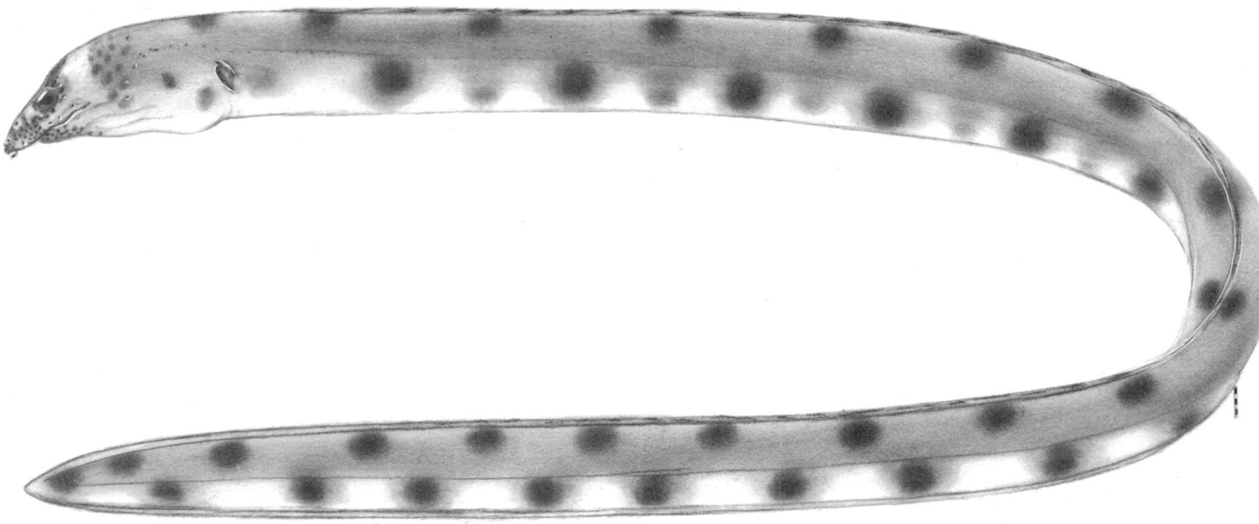

FIGURE 430. *Quassiremus ascensionis*: ANSP 98355, 360 mm TL, Bahamas. Illustration by S. P. Gigliotti.

overlain with two rows of dark spots; considerable variation in size, shape and intensity of spots. Typically, one row of large spots or saddles on either side of the dorsal midline, abutting each other, and one row of smaller, less obvious spots along and below the lateral line, these alternating with dorsal spots, one or two smaller spots between each dorsal spot. Various small spots on snout, nape, and chin; pale areas before and behind eyes; head pores in conspicuous light areas. Dorsal and anal fins variously spotted or pale; pectoral fin colorless. Some specimens in preservative appear almost uniformly bicolored, very dark dorsally and pale ventrally, the large dorsal spots appearing very faintly as bands or not visible at all. The holotype of *Quassiremus productus* (collected in 1861 but not described until 1917) was described as having wide black bands down to the lateral line, alternating with spots on the sides below the lateral line; the specimen is now very dark, the black bands hardly visible.

Life coloration was well described by Beebe and Tee-Van (1932: 111), based on the three types of *Quassiremus goslingi*, all large ripe males:

> General color of sides above the lateral line salmon-buff, changing abruptly beneath the lateral line to creamy white. Seventeen saddle-like markings of salmon-orange along the dorsal surface, these spots becoming darker centrally, this darkening most pronounced on the three last markings A series of large light orange spots which shade to deep chrome in the center, on the midsides, one between each of the saddle-like markings. Between the upper saddles and the

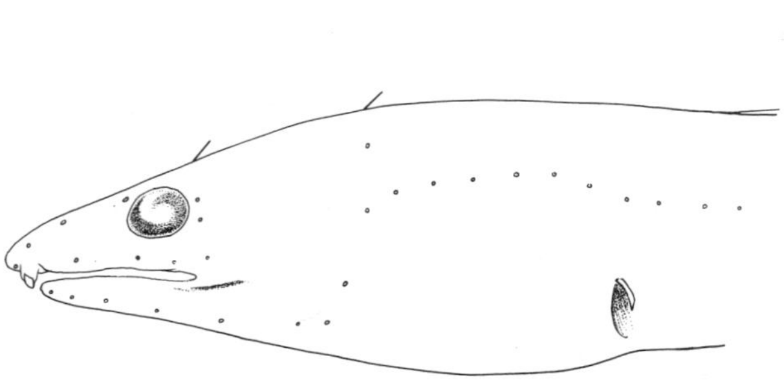

FIGURE 431. *Quassiremus ascensionis*: ANSP 98358, 471 mm TL, Bahamas; head.

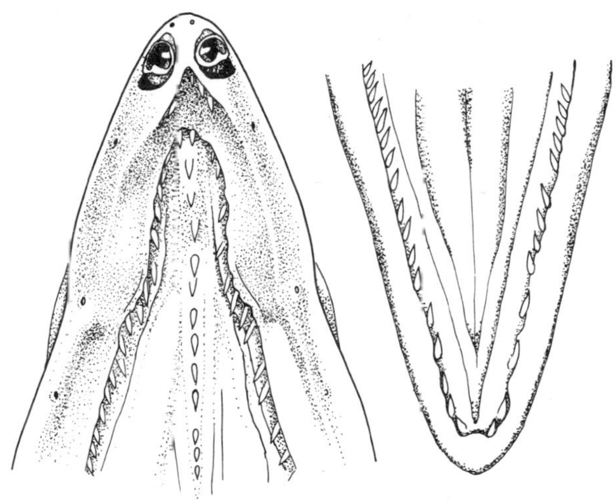

FIGURE 432. *Quassiremus ascensionis*: ANSP 98358, 471 mm TL, Bahamas; dentition.

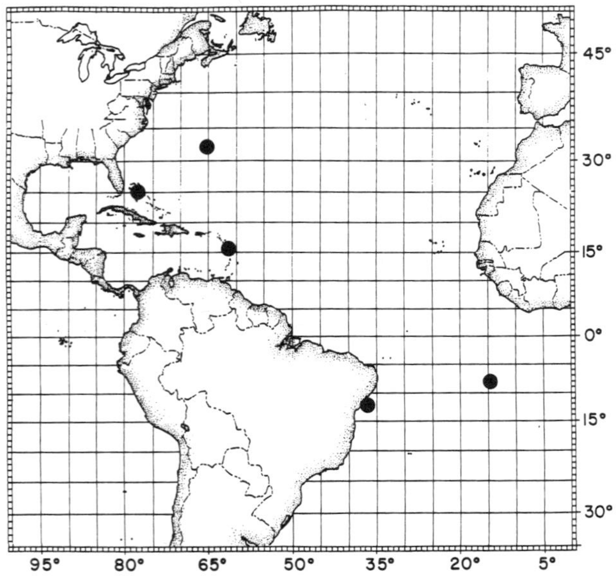

FIGURE 433. Distribution of *Quassiremus ascensionis* based on study material.

lateral spots are smaller, similarly-colored spots, the majority with dark centers. Snout with a pinkish buff background and with small irregular blackish spots. Back of the eye these spots enlarge and change to antique brown. Iris pale gold with dark markings anteriorly and posteriorly. Lateral line pores white, conspicuous from head to tail. Dorsal fin marked with pale orange spots, somewhat like those on the body, the fin margined with white. Anal fin transparent, with a few orange spots posteriorly.

Recently-collected specimens from Ascension Island (ANSP 158084, 158789, and USNM 268539), speared at night, lacked body spots and appeared darkly bicolored when preserved, although the heads showed typical spots and the pale areas before and behind the eyes, and both head pores and lateral line pores were in conspicuous white spots. The collector, M. McDowell, wrote (of USNM 268539): "Body color of salmon orange faded to brown quickly after putting in formalin"; kodachromes taken at time of capture showed typical orangish coloration with large spots.

Size. The largest specimen examined is 700 mm TL. Three specimens (229–470 mm) were females, none ripe; five (250–690) were males, three (457–690) ripe.

Development. Leptocephalus tentatively identified (p. 835).

Behavior. A fossorial species. McDowell (in litt., 1984) observed three specimens on several occasions while diving at 7–10 m off Ascension Island. He stated that they remained buried with only the spotted triangle of the head exposed, and, when disturbed, backed under and appeared to swim through the sand, although one swam some 10 m through the water before disappearing under the sand. In the eastern Pacific, adult *Quassiremus evionthas* and *Q. nothochir* are occasionally seen swimming in a snakelike manner over the open sand bottom.

Distribution (Fig. 433). A rare species, known from Bermuda, the Bahamas, the Lesser Antilles, Brazil, and Ascension Island. Collected in shallow water (to 12 m), most often with ichthyocides over mixed patches of sand and turtle grass; two specimens speared at 10–12 m.

Etymology. From Ascension Island, the type locality. A noun in the genitive case.

Study Material. A total of 17 specimens, 143–700 mm TL, including the holotypes of *Quassiremus goslingi* Beebe and Tee-Van and *Quassiremus productus* Seale.

BERMUDA: USNM 170563 (1, 646); holotype of *Quassiremus goslingi* Beebe and Tee-Van. USNM 170564 (2, 608–690); paratypes of *Q. goslingi.* BAHAMAS: ANSP 98355 (1, 360), ANSP 115518 (1, 363, cleared and stained), and CAS 54481 (1, 172); Silver Cay, 1–5 m; J. E. Böhlke et al.; 24 July 1955. ANSP 98357 (1, 143); Hog Island, 3.6 m; J. E. Böhlke et al.; 10 May 1959. ANSP 98358 (1, 471); Rose Island, 0.3–0.6 m; H. R. & H. Roberts; 1 April 1953. MCZ 9150 (1, 595); Nassau; J. D. Sargent; 10 Nov. 1861; holotype of *Quassiremus productus* Seale. LESSER ANTILLES: Dominica: USNM 216546 (3, 231–457); to 7.5 m; V. G. Springer; 10 Nov. 1964. BRAZIL: MNRJ 9366 (1, 689); Salvador, Bahia. ZMB 12556 (1, 398); Santos; Salmin. ASCENSION ISLAND: ANSP 158084 (1, 570); English Bay, 12 m; M. McDowell; Aug. 1985. ANSP 158789 (1, 722); English Bay, 55 m; June 1986. USNM 268539 (1, 700); Boatswain Bird Island, 10 m; M. McDowell; 5 Feb. 1984.

Family Colocongridae

DAVID G. SMITH

Acknowledgments. See family Anguillidae.

Introduction. The family Colocongridae was established by Smith (1971: 83) to contain the single genus *Coloconger* Alcock, which had rested uncomfortably for many years in the Congridae. *Coloconger* is a short, stubby, blunt-nosed eel, the least elongate of any of the Anguilliformes. At first glance it hardly looks like an eel at all. The species of *Coloconger* live on the mid to upper continental slope of all oceans except the eastern and central Pacific. They seem to favor open terrain over muddy bottoms and are fairly commonly collected within their depth range. Nevertheless, four of the five nominal species were described only in the last thirty years. Virtually nothing is known of their biology.

Characters. Body short, stubby, deepest shortly behind head, tapering toward tail, strongly compressed posteriorly; anus well behind midlength; tip of tail flexible. Dorsal and anal fins well developed, confluent with caudal fin; dorsal fin begins slightly behind base of pectoral fin; dorsal- and anal-fin rays segmented. Pectoral fin well developed. Gill opening a vertical slit, upper corner opposite middle to upper end of pectoral-fin base. Lateral line complete, pores open through low tubes. Scales absent.

Head wide and deep, snout short and bluntly rounded. Snout projects slightly beyond lower jaw, fleshy tip of snout extends anteriorly beyond intermaxillary teeth. Mouth moderate, rictus under posterior margin of pupil or posterior margin of eye; teeth fully concealed when mouth closed. No flange on upper or lower lip. Eye large. Anterior nostril a short tube near tip of snout, directed anterolaterally. Posterior nostril a large, round opening with a low, raised rim, about midway between tip of snout and eye, at mid-eye level.

Head pores numerous; except for few anteriormost pores in supraorbital, infraorbital, and preoperculomandibular canals, all pores open through fleshy tubes; additional series of small dermal papillae, presumably sensory, behind eye, across supratemporal region, parallel to preopercular canal, before dorsal fin, and at scattered locations elsewhere on head. Adnasal pore present. At least some species with pore in temporal canal.

Teeth small, conical. Intermaxillary teeth in two transverse series; posterior series complete, continuous with maxillary teeth; anterior series develops with growth, often incomplete, with teeth on outer part of bone but not medially. Maxillary teeth in one or two series; teeth of outer series larger and forming a continuous cutting edge; teeth of inner series smaller, concealed by flesh. Mandibular teeth in two or three rows. No teeth on shaft of vomer.

Osteology (based on specimens of *Coloconger meadi,* UMML 22964, and *C. raniceps,* ANSP 115508). Neurocranium (Fig. 434) relatively short and broad, with a large orbit and a greatly shortened antorbital section. Ethmovomer short, thin, extending little beyond orbit, with prominent palatine processes; no lateral ethmoid process. Frontals fused, with a broad lateral shelf on each side over orbit. Parietals fused. Supraoccipital present, with a crest. Temporal canal not completely enclosed within pterotic.

Jaws (Fig. 435): Maxilla with a pedicel at anterior end; posterior end bent down slightly. Mandible moderately elongate.

Suspensorium moderately short, slightly in-

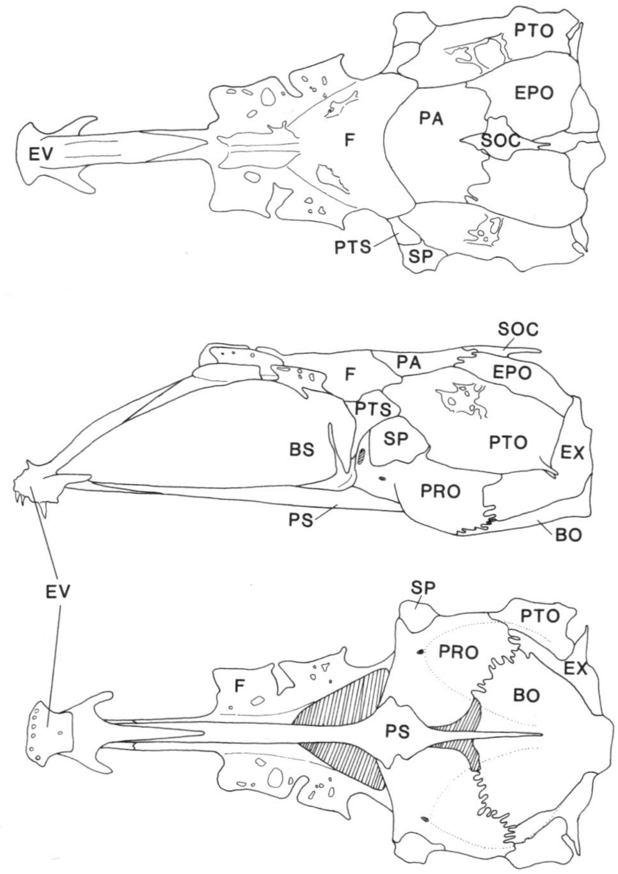

FIGURE 434. Neurocranium of *Coloconger meadi* in dorsal view (top), lateral view (middle), and ventral view (bottom). Abbreviations on pp. 5–6.

clined forwardly (Fig. 436). Pterygoid fairly well developed, closely connected to palatine process of ethmovomer.

Hyoid arch consists of paired hypohyals, ceratohyals, epihyals, and unpaired glossohyal and urohyal (Fig. 437). Hypohyals short, closely articulated to, but clearly separate from, proximal end of ceratohyals. Branchiostegal rays moderately numerous (13 in specimen of *Coloconger*

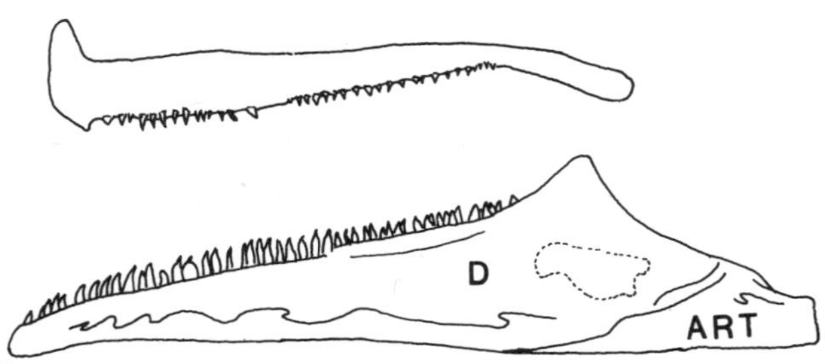

FIGURE 435. Jaws of *Coloconger meadi* in lateral view: maxilla (top) and mandible (bottom).

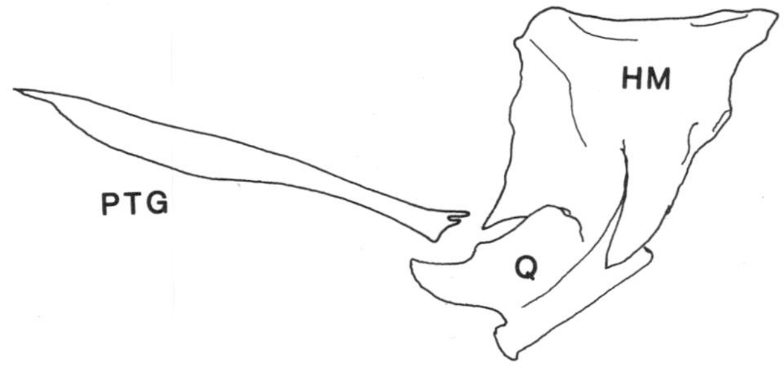

FIGURE 436. Suspensorium of *Coloconger meadi.*

meadi), first four associated with ceratohyal, remainder with epihyal. Glossohyal short, slightly depressed, somewhat triangular in shape in dorsal or ventral view, expanded laterally toward anterior end. Urohyal slender, with an abrupt, paddle-like expansion at anterior end and a slight bifurcation at posterior end.

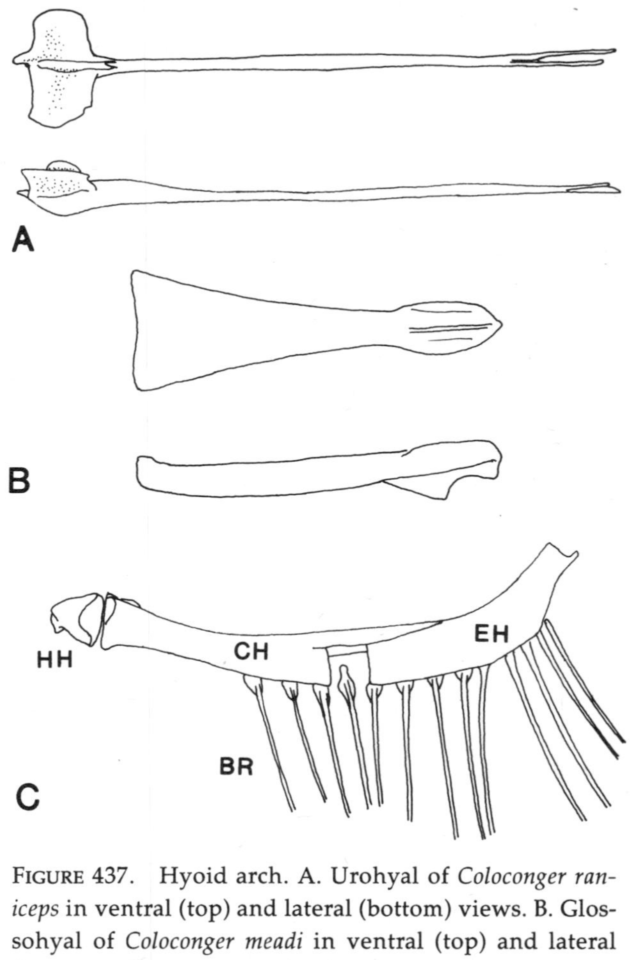

FIGURE 437. Hyoid arch. A. Urohyal of *Coloconger raniceps* in ventral (top) and lateral (bottom) views. B. Glossohyal of *Coloconger meadi* in ventral (top) and lateral (bottom) views. C. Hypohyal, ceratohyal, epihyal, and branchiostegal rays of *Coloconger meadi*, left side, dorsal view.

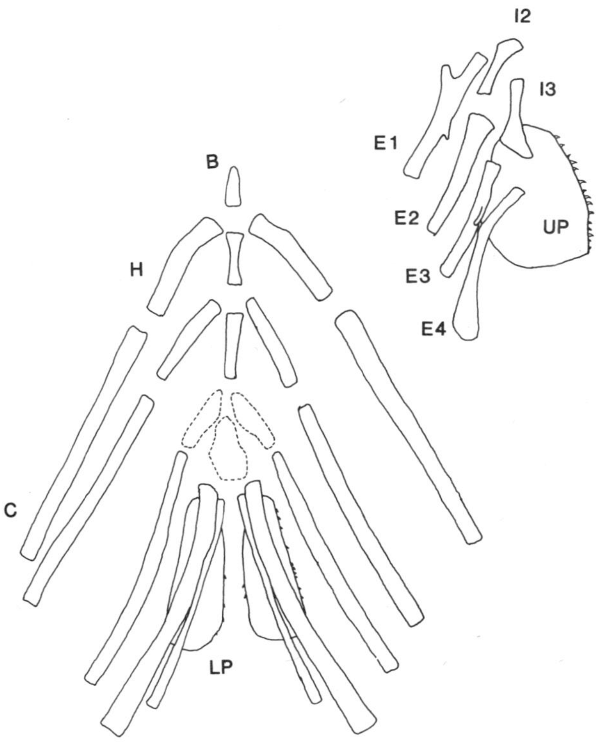

FIGURE 438. Gill arches of *Coloconger meadi*. Ventral elements in ventral view, dorsal elements in dorsal view.

Gill arches complete (Fig. 438). First three basibranchials ossified, fourth cartilaginous; first basibranchial short, not articulated with glossohyal. First two hypobranchials ossified, third cartilaginous. Five ceratobranchials, first four stout, fifth slender and somewhat shorter than fourth. Four epibranchials, two infrapharyngobranchials. Pharyngeal tooth plates large and undivided; lower one broadly rectangular in shape, upper one very broad, almost round.

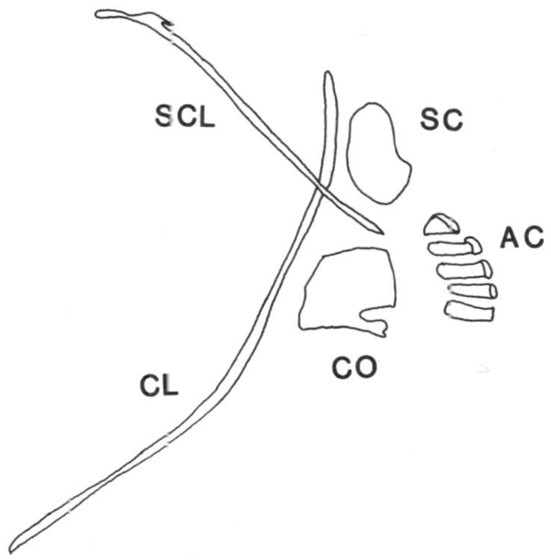

FIGURE 440. Pectoral girdle of *Coloconger meadi*.

All four opercular elements present (Fig. 439). Opercle broadly fan-shaped, posterior margin smooth. Subopercle sickle-shaped, but relatively broad, middle part of posterior margin serrate. Preopercle larger than opercle; dorsal part forms a semi-enclosed tube carrying preopercular canal; ventral part expanded, fan-shaped. Interopercle as large as opercle, roughly triangular with apex at anterior end.

Pectoral girdle complete (Fig. 440). Cleithrum and supracleithrum slender, cleithrum longer. Scapula and coracoid present. Five actinosts in specimen examined.

Circumorbital bones (Fig. 441): Nasal elongate, poorly ossified. Supraorbital not apparent in specimen examined. Preorbital elongate, ossified mainly in posterior half of length, occupies entire section of infraorbital canal below eye; ventral margin without projections. Three suborbitals,

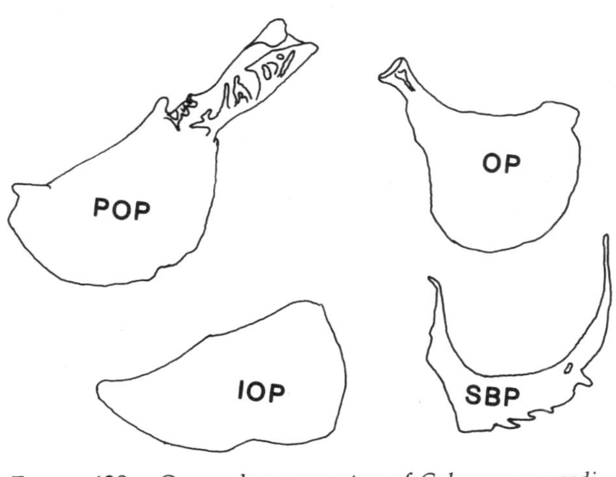

FIGURE 439. Opercular apparatus of *Coloconger meadi*.

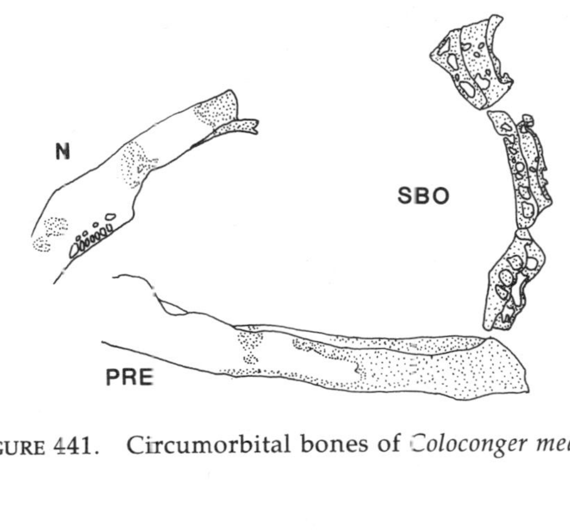

FIGURE 441. Circumorbital bones of *Coloconger meadi*.

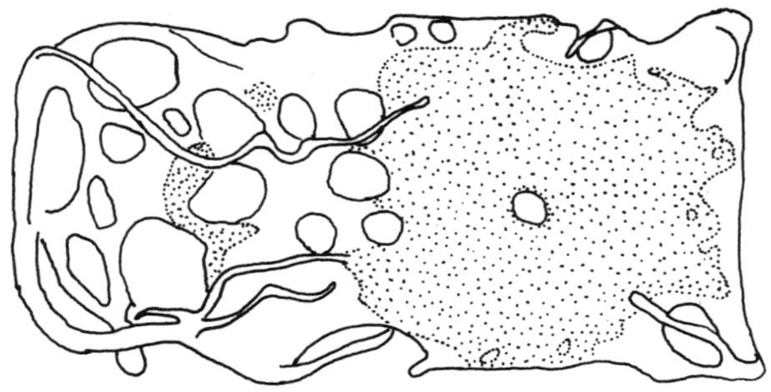

FIGURE 442. Lateral-line ossicle of *Coloconger meadi.*

well ossified, all in vertical section of canal behind eye.

Lateral-line ossicles stout, somewhat elongate horizontally; posterior half rather solidly ossified, anterior half largely cartilaginous and heavily fenestrated (Fig. 442).

Vertebrae slender, foreshortened, centra with large central opening (Figs. 443, 444). Myorhabdoi absent. Ribs absent. Caudal vertebrae without epicentral processes. Dorsal- and anal-fin rays segmented.

Remarks. In his treatment of the anatomy of *Coloconger scholesi,* Chan (1967) made a few points that deserve comment. I am unable to interpret the cartilaginous structure on the cranium that he called the lateral ethmoid process. It was not present in the specimens of *C. meadi* and *C. raniceps* examined by me, and it does not appear to be the true lateral ethmoid process found in the bathymyrine congrids. He described the premaxilla as being "distinctly demarcated from both maxillary and vomer" as seen ventrally. In my specimens, the ethmovomer is without any trace of sutures. He described and illustrated three cir-

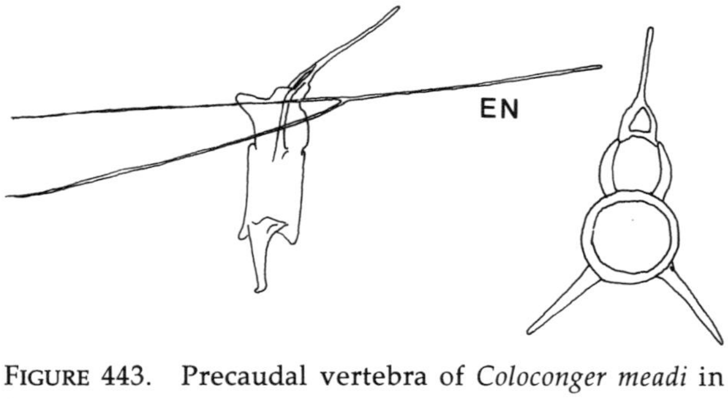

FIGURE 443. Precaudal vertebra of *Coloconger meadi* in lateral and frontal views.

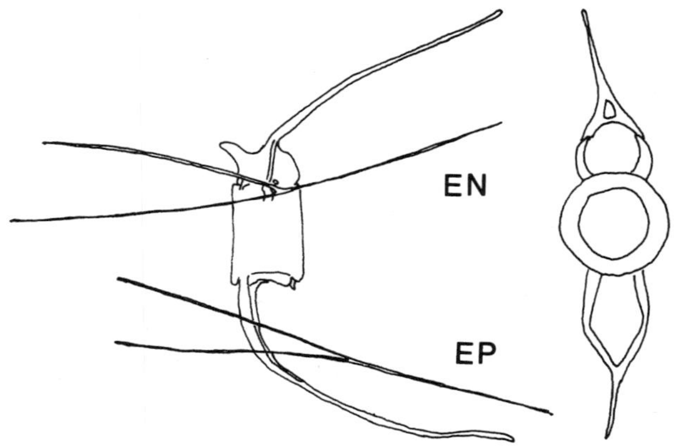

FIGURE 444. Caudal vertebra of *Coloconger meadi* in lateral and frontal views.

cumorbital bones below the eye and a nasal bone in three pieces. In my specimen of *C. raniceps* there was a single elongate element below the eye, and the nasal was in one piece. Finally, the structure he called the air bladder, located between the liver and the esophagus, seems to be the gall bladder instead.

Some of Chan's observations may be simply erroneous, such as his misidentification of the gall bladder. Others may be real. If so, it is uncertain whether the differences between his specimen and mine are characteristic of the species or reflect the range of individual variation.

Discussion. Matsubara (1955) was apparently the first to question the traditional assignment of *Coloconger* to the Congridae. Asano (1962: 70) followed Matsubara and excluded *Coloconger* from his treatment of Japanese congrid eels. Smith (1971: 83) compared *Coloconger* to the other congrid genera and concluded that it deserved recognition as a family. Blache and Bauchot (1976a: 427) summarized this evidence and agreed with the conclusion.

Coloconger shows no obvious relationship to any of the true congrid genera or, indeed, to any other eel. Most of the characters it shares with the Congridae, such as the large eye, well developed fins, complete lateral line, complete gill arches, etc., are primitive. About the only derived character it shares with the Congridae is the fusion of the frontals, a feature found in several other families as well. On the whole, *Coloconger*

TABLE 42. Distinguishing characters of the five species of *Coloconger*. Characters are: 1, number of supratemporal pores; 2, HL in % TL; 3, number of pores over posterior nostril; 4, dorsal-fin rays; 5, anal-fin rays; 6, pectoral-fin rays; 7, rictus under posterior margin of eye or pupil; 8, dorso-pectoral distance in % pectoral-fin length; 9, diameter of eye in relation to length of snout; 10, number of vertebrae. Dorso-pectoral distance is predorsal minus HL. Data from Kanazawa (1957, 1961b), Castle (1968c, 1969a), Blache and Bauchot (1976a), Okamura and Kitajima (1984), and personal observation.

	Character									
Species	1	2	3	4	5	6	7	8	9	10
raniceps	6	20–22	2	210–214	115–119	20–22	pupil	50–54	eye larger	144–145
scholesi	3	18–20	2	184–241	72–110	20–23	pupil	50–71	snout larger	142–157
japonicus	3–4	17–18	?	203–220	97–109	19–22	?	31–46	snout larger	149–153
cadenati	3	17–21	1–2	235–263	133–142	24–25	eye	10–30	eye larger	147–153
meadi	3	17–18	1	209–232	102–124	19–23	pupil	20–59	eye larger	147–163

does not fit the prevailing trends of character evolution in congrids. For example, in congrids the loss of the upper labial flange is accompanied by the development of projections from the ventral margin of the preorbital, which remains stout and well ossified. *Coloconger* lacks the flange, but the preorbital is elongate, poorly ossified, and has no ventral projections. *Coloconger* has discrete hypohyals, a primitive feature lacking in almost all congrids. On the other hand, the fused parietals are a specialization found in no congrid. The stubby body and the very short tail are also very uncongrid-like. Among congrids, only some species of *Ariosoma* have the tail slightly shorter than the body, but there the tip of the tail is stiffened and used for burrowing. *Coloconger*, by contrast, has a flexible tail and shows no evidence of being a burrower. The lateralis system of *Coloconger* is also unusual, with pores more numerous than in congrids and opening through short tubes. The absence of vomerine teeth is another character that separates *Coloconger* from the congrids.

For all these reasons *Coloconger* is best treated as a distinct family, the Colocongridae.

Distribution. Tropical eastern and western Atlantic and Indo-West Pacific from east Africa to Japan. Not reported from the eastern Pacific or the islands of the central Pacific. Depth about 300–900 m.

Genera. A single genus, *Coloconger* Alcock.

Genus *Coloconger* Alcock, 1889

Coloconger Alcock, 1889: 456 (type species *Coloconger raniceps* Alcock, 1889, by monotypy).

Characters. Those of the family.

Size. To approximately 500–600 mm TL.

Distribution. That of the family.

Etymology. From the Greek *kolos* (clipped, shortened, truncated) and *Conger* (the conger eel), in reference to the stubby body and short tail. Masculine

Species. Five very similar species have been described: *Coloconger raniceps* Alcock, 1889, from the Indo-West Pacific, *C. scholesi* Chan, 1967, from the Indo-West Pacific, *C. japonicus* Machida, 1984 (in Okamura and Kitajima, 1984), from the western Pacific, *C. cadenati* Kanazawa, 1961, from the eastern Atlantic, and *C. meadi* Kanazawa, 1957, from the western Atlantic. They are distinguished by a combination of meristic and morphometric characters (Table 42).

Coloconger meadi Kanazawa, 1957
Figures 445–448, Tables 42, 43

Coloconger raniceps (not of Alcock, 1889). Mead and Nicholson, 1956: 62.

Coloconger meadi Kanazawa, 1957: 234, fig. 1 (original description, 24°16′N, 83°22′W, holotype USNM 157926); 1961b: 110 (additional characters given to separate *C. meadi* from *C. raniceps* and *C. cadenati*).

Distinctive Characters. Coloconger meadi is distinguished from the other species of *Coloconger*

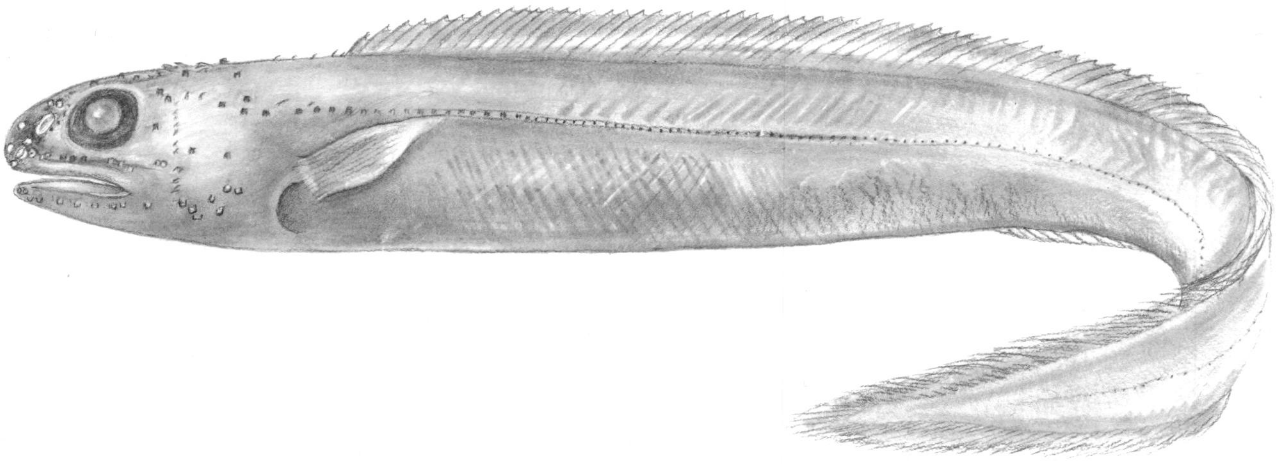

FIGURE 445. *Coloconger meadi*: ANSP 126969, 260 mm TL.

by the characters given in Table 42. In the western Atlantic it is recognized by the characters of the genus.

Description (based on 36 specimens plus data from Kanazawa, 1961b). Total vertebrae 147–163 (n = 35), predorsal vertebrae 11–17 (29), preanal vertebrae 75–86 (29). Branchiostegal rays 11–13 (2). Dorsal-fin rays 209–232 (7), anal-fin rays 102–124 (7), pectoral-fin rays 19–23 (27). Pores: LL to anus 64–74 (26), POM 14–22 (8), IO 7–12 (8) + 3–6 (7) + 2–5 (7), SO 1 + 5–7 (9), temporal 1 (8), STC 3 (8).

Proportions as % TL: preanal 58–65 (16), predorsal 20–23 (13), head 17–21 (14), depth at anus 8–10 (10), pectoral fin 7–8 (10). Of HL: snout 19–27 (15), eye 24–30 (15), interorbital 22–32 (9), snout-rictus 38–47 (11), gill opening 10–15 (9), interbranchial 27–39 (6), pectoral fin 40–54 (10). Of pectoral-fin length: dorso-pectoral (predorsal minus HL) 22–59 (9).

Stomach short, about one-third to one-fourth length of intestine. Gas bladder rather short, free of vertebrae, begins slightly behind junction of stomach and intestine, ends at about midpoint of intestine (Fig. 448).

Color. In preservative gray or brown, sensory pores and papillae blackish. Stomach and intestine black.

Size. The largest specimen examined was 505 mm TL (USNM 193614).

Development and Growth. The leptocephalus is unknown. Castle (1969a: 7, fig. 2) described a metamorphic specimen of *Coloconger raniceps.* It had a series of melanophores below the gut but none on the side of the body.

Variation. There is some indication of a slight geographic variation in total, predorsal, and preanal vertebrae (Table 43).

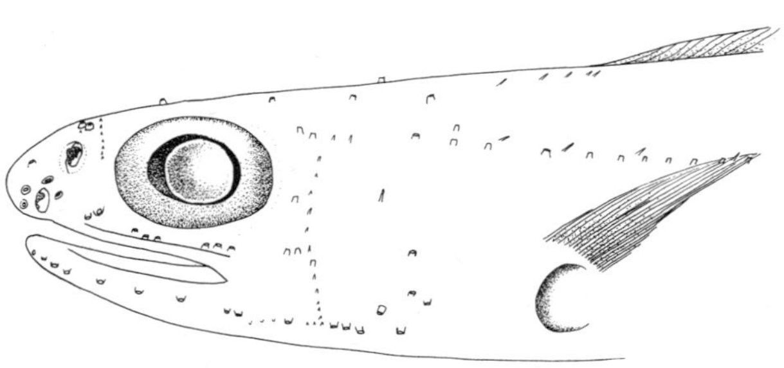

FIGURE 446. *Coloconger meadi*: ANSP 126969, 260 mm TL; head.

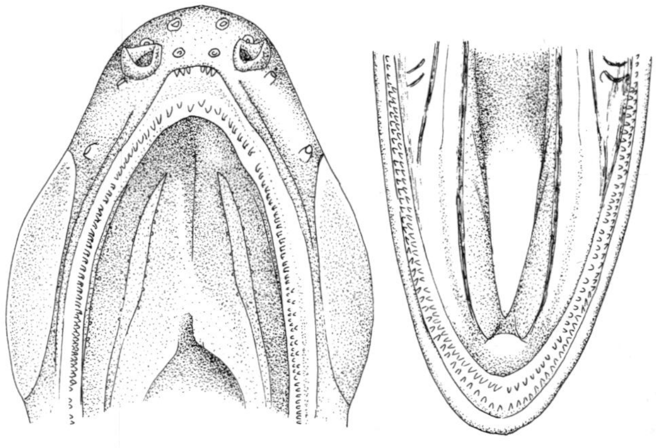

FIGURE 447. *Coloconger meadi*: ANSP 126969, 260 mm TL; dentition.

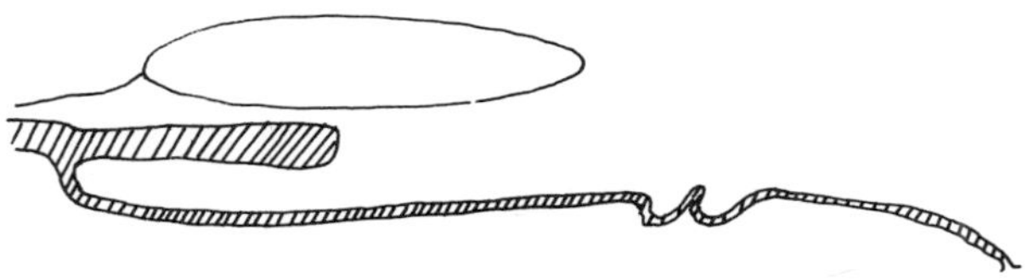

FIGURE 448. Diagrammatic view of viscera of *Coloconger meadi*: gas bladder (top); stomach (middle), and intestine (bottom). Shaded areas are black.

Distribution. Gulf of Mexico, West Indies, and the coast of Central and South America to the Guianas. Depth 366–915 m (mean 678 m). Common.

Etymology. Named for Dr. Giles W. Mead, who provided the type specimens to Kanazawa.

Study Material. 231 specimens, 177–505 mm TL.

Holotype: USNM 157926 (282 mm TL), Oregon 1019, southeastern Gulf of Mexico, 24°16′N, 83°22′W, 686 m, 16 April 1954.

Other material: FLORIDA GULF: ANSP 126184 (1, 367), 114995 (3, 324–410), 117320 (1, 385). LOUISIANA: ANSP 109612 (1, 410). MEXICO GULF: ANSP 117003 (1, 270), 117093 (1, 517). USNM 198747 (2, 445–460). LESSER ANTILLES: ANSP 103988 (5), 109609 (1, 455), 109610 (5, 278–401), 114143 (1, 385), 114144 (1, 341), 114935 (1, 280), 114945 (1, 250), 117068 (1, 350), 126969 (1, 260). USNM 268868 (27, 345–468). VENEZUELA: UMML 22643 (1, 221). USNM 198743 (3, 247–271), 198749 (3, 328–380), 198752 (1, 375), 198753 (1, 306), 198761 (7, 340–384). COLOMBIA: ANSP 109613 (4, 287–383), 114157 (1, 270), 126133 (1, 310), 126173 (2, 269–335). UMML 22964 (1, 197). USNM 198757 (2, 250–257), 268806 (1, 371). PANAMA: USNM 193561 (2, 177–261), 193573 (5, 190–327), 193614 (1, 505), 193615 (15, 337–470), 268805 (1, 318). NICARAGUA: ANSP 109603 (1, 390), 109614 (2, 386–396). USNM 158880 (1, 190), 179242 (5, 155–244), 187809 (1, 227), 193543 (2, 186–205), 193557 (1, 190), 193616 (3, 320–464), 268807 (1, 310). HONDURAS: USNM 193570 (2, 229–265), 193572 (5, 236–300), 268808 (1, 315). BELIZE: USNM 193608 (2, 280–285). SURINAME: ANSP 114842 (4, 310–400), 114845 (2, 305–412), 114850 (1, 352), 114871 (3, 335–358), 114956 (1, 295), 114960 (1, 326), 114961 (1, 352), 114998 (5, 280–329), 117028 (3, 315–450). USNM 158884 (1, 222), 158928 (2, 267–276), 158940 (6, 295–350), 158941 (3, 190–276), 179222 (4, 347–421), 179225 (1, 468). FRENCH GUIANA: ANSP 114852 (2, 295–302), 114872 (2, 319–320), 114833 (3, 285), 114943 (1, 353), 114946 (2, 278–284), 114997 (2, 260–289), 117010 (2, 290–320), 117016 (2, 305–317). USNM 158939 (3, 200–292), 179221 (15, 282–443), 179223 (4, 256–355), 179224 (14, 295–379), 198686 (1, 320), 268804 (2, 260–291).

TABLE 43. Numbers of vertebrae in specimens of *Coloconger meadi*.

	Total vertebrae														
Area	147	148	149	150	151	152	153	154	155	156	157	158	159	n	x̄
Gulf of Mexico						1	1	–	–	1	3			6	155.3
Caribbean				1	1	–	2	2	2	1	1			10	153.8
Guianas	1	1	–	2	1	1	–	2	2	–	–	–	1	11	152.3

	Preanal vertebrae												
Area	75	76	77	78	79	80	81	82	83	84	85	86	n	x̄
Gulf of Mexico						1	–	1	3	1			6	82.5
Caribbean				2	2	–	1	5	1	1			12	81.0
Guianas	1	–	1	4	2	1	–	1	–	–	–	1	11	79.1

	Predorsal vertebrae								
Area	11	12	13	14	15	16	17	n	x̄
Gulf of Mexico			1	1	3	–	1	6	14.8
Caribbean		1	3	4	2	1		11	13.9
Guianas	1	2	3	2	4			12	13.5

Family Derichthyidae

CATHERINE H. ROBINS

Acknowledgments. In addition to those persons and grants listed in the general acknowledgments section of this volume, I wish to thank Karsten E. Hartel (Museum of Comparative Zoology, Harvard), William H. Krueger (Department of Zoology, University of Rhode Island) and John E. McCosker (Steinhart Aquarium) for making available specimens of derichthyids. Much of the material at the Museum of Comparative Zoology was collected by James Craddock and Richard H. Backus (Woods Hole Oceanographic Institution). As always, discussions with colleagues, especially C. Richard Robins, who also examined material for me, have been both instructive and thought-provoking.

General Characters. Adults mesopelagic to bathypelagic; larvae pelagic. Small (to about 600 mm TL), lacking scales and plain-colored as adults (mostly brownish with paler fins). Dorsal-fin origin in anterior third of body, posterior to a point above tip of pectoral fins; anus and anal-fin origin far posterior. Tail of adults equal to or shorter than remainder of total length. Branchial region rounded, more slender than remainder of body which is expanded dorsoventrally and compressed. Gill openings small, ventral slits immediately anterior to level of pectoral-fin base. Gill region not obviously extensible or expanded. Mouth terminal, or moderately oblique, gape extending either to a point under anterior margin of eye or to posterior margin. Upper jaw projecting beyond tip of lower jaw either slightly or by a distance equal to eye diameter so that the premaxilloethmoid teeth are at least partially exposed when the smaller, narrower lower jaw is closed against the upper. Nostrils simple, porelike. Dorsal and anal fins present, confluent with caudal fin. Dorsal and anal fins abruptly much lower in tail region. Pectoral fins present, about 4–5 times as long as pectoral-fin base or approximately 1.5–2.0 times the eye diameter. Eye large, well-developed.

Lateral line complete to within head length of tail tip. Pore system of head exceptionally developed; pores large, some as conspicuous as the nostrils. Additional system of parallel cicatrix-like lines, arranged in groups associated with the head pores.

Leptocephali fairly small (50–90 mm), unspecialized. Distinctive features: deep-bodied with minute midlateral melanophores in caudal region. Myomeres ca 125–180 (see Smith, this volume).

Osteological Characters (Figs. 449–452). Frontals fused into a single bony unit. Premaxillae present, joined as single bone but separate from ethmoid in *Derichthys serpentinus*, and fused to form a premaxilloethmoid in *Nessorhamphus.* Pterygoid present as small, bony splint. Basisphenoid present. Otoliths (sagitta) oval with large sulcus. Pectoral-fin supports reduced, small; four radials, slender cleithrum. Caudal-fin supports much reduced to two or three elements. Branchiostegals six or seven on each side, most borne on ceratohyal. Branchial elements as follows: four basibranchials (the fourth cartilaginous), three hypobranchials (the third cartilaginous, short, *not* curved posteromedially), five ceratobranchials, three or four epibranchials and two pharyngobranchials plus upper and lower tooth plates. Fused hyomandibular and quadrate (= suspensorium) angled sharply forward to articulate with lower jaw. Vertebrae numerous, neural and haemal spines not heavily sculptured or elaborated. Vertebral number: 125–160 (presumed extremes); actual counts of limited number of specimens of the three species range from 125–155; precaudal

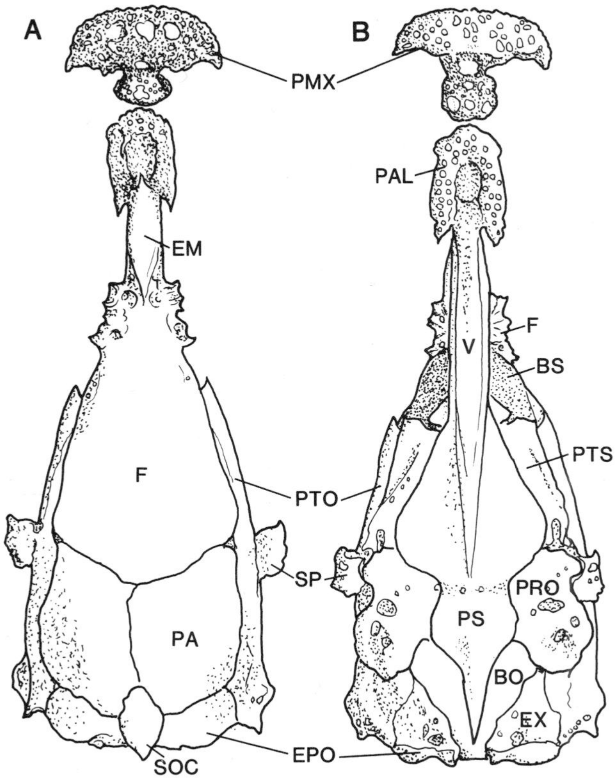

FIGURE 449. Neurocranium of *Derichthys serpentinus* (premaxilla shown disarticulated); UMML 34397, NW Atlantic. A. Dorsal view of skull. B. Ventral view of skull. Abbreviations on pp. 5–6.

vertebrae 53–74. Ratio of dorsal-fin rays (and supports) to vertebrae roughly 2:1. Simple epipleural ribs present; no pleural ribs.

Discussion. The odd little eels of the family Derichthyidae are distinguished by an unexpanded branchial region followed by a dorsoventrally expanded, compressed body, a dorsal (and occasionally anal) fin that becomes abruptly much reduced in the caudal region, an elaborate cephalic sensory system, peculiarities of the snout region, and similar leptocephali (Smith, this volume). The affinities of the Derichthyidae with the Congroidea were discussed by Robins and Robins (1971).

Gill (1884), in his description of *Derichthys serpentinus,* was so struck by the unusual dentition of the upper jaw that, in his expanded description of *Derichthys* (Gill in Goode and Bean, 1896), he erected a separate order, the Carenchelyi which he characterized as follows: "Teleost fishes with the intermaxillaries and supramaxillaries developed and united by suture, and immoveably connected with the cranium; branchial apparatus as in Apodes; scapular arch remote from the skull and the body anguilliform."

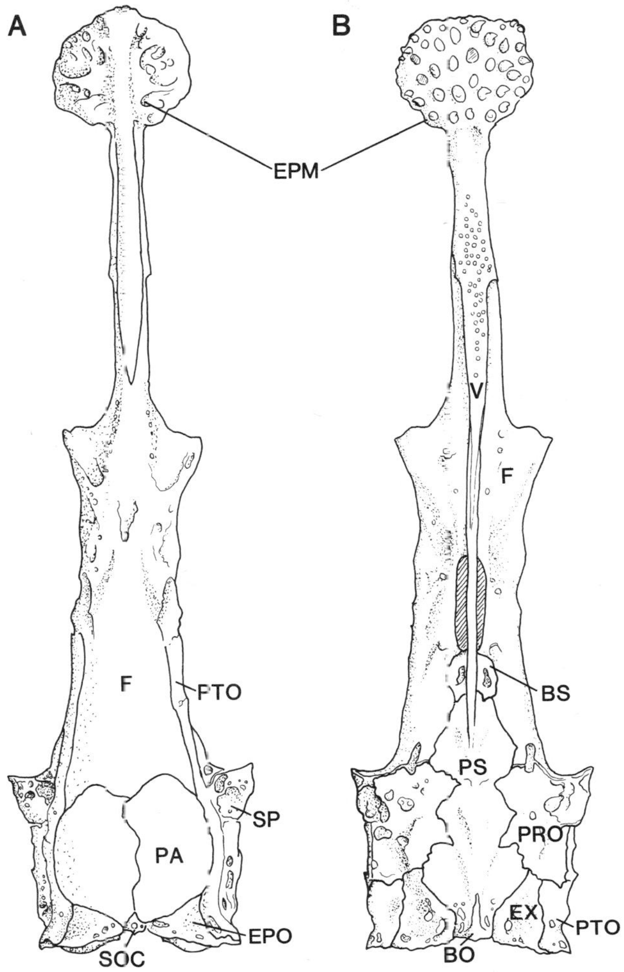

FIGURE 450. Neurocranium of *Nessorhamphus ingolfianus;* UMML 20883, NW Atlantic. A. Dorsal view of skull. B. Ventral view of skull.

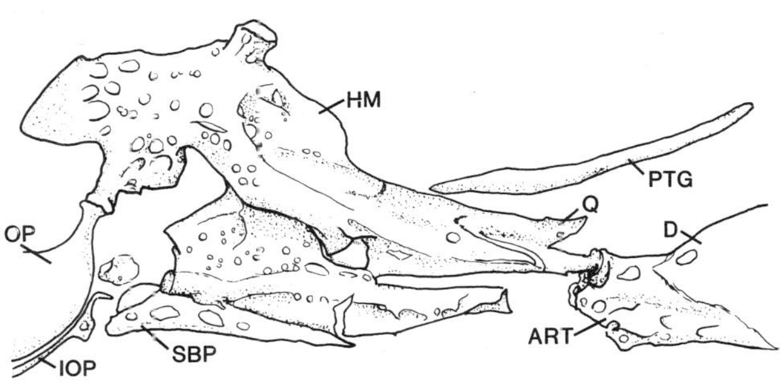

FIGURE 451. Right suspensorium, opercular bones, pterygoid, and posterior end of lower jaw of *Nessorhamphus ingolfianus* (data as for Fig. 450).

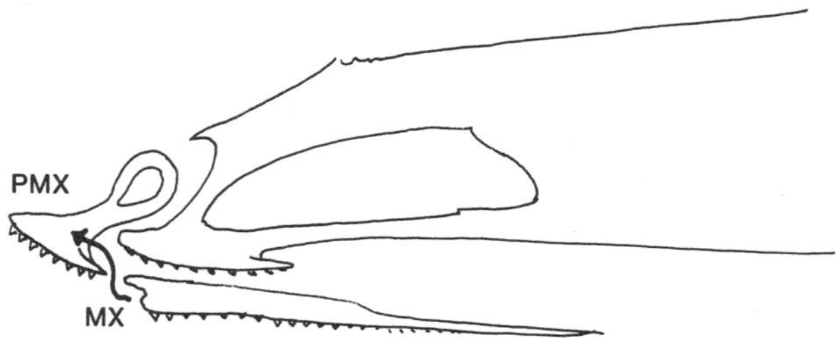

FIGURE 452. Diagram of relationship of free premaxilla to ethmoid and maxilla of *Derichthys serpentinus*.

There are four tooth-bearing bones in the upper part of the mouth of *Derichthys serpentinus*: the two maxillae, a premaxilla (a fused, single, median unit) which (unlike other eels) is not fused with the ethmoid, and a V-shaped or horseshoe shaped bone which is fused with the parasphenoid. The last complex presumably is Gill's fusion of the "intermaxillaries and supramaxillaries," since the premaxilla is occasionally slightly movable in an intact specimen and readily disarticulates in a cleared and stained preparation (Figs. 449–452).

A premaxilla that is separable from the ethmoid complex is unique in the Anguilliformes in which the trend has been toward fusion of the premaxillae with the ethmoid to form a single unit, the premaxilloethmoid. Additionally, there has been extreme reduction of the premaxillae or even complete loss of the premaxillae and concomitant loss of the ethmoid. Few to many vomerine teeth (the vomer often simply a thin lamella of bone easily peeled from the parasphenoid) in one or more rows are also common conditions in eels. The V-shaped pattern of teeth in *Derichthys* is unusual especially since the teeth are confined to ventrolateral projections from the fused ethmoid and parasphenoid, and the median portion (where vomerine teeth might be expected) is concave and toothless. Something suggestive of this can be seen in the congrids *Rhyncoconger* sp. (Asano, 1962) and *Macrocephenchelys* (Robins and Robins, 1971). In *Macrocephenchelys* and many other congrids the pterygoid is fused with or at least articulated with the quadrate. In *Macrocephenchelys*, the jaws, suspensorium and pterygoid are curiously modified so that the pterygoid forms a strut between the suspensorium and a section of the ethmoid and vomerine portion of the skull, a part labeled a fused palatine by Robins and Robins (1971).

The pterygoid of *Derichthys* is a simple splint suspended in the connective tissue of the suborbital region by ligaments to the suspensorium and parasphenoid. The pterygoid does not directly abut on any bone, including the peculiar V-shaped structure fused with the parasphenoid. Nonetheless, considering the toothed area's location and the congroid affinities of the Derichthyidae, it seems reasonable to suggest that this represents fused palatines as in *Macrocephenchelys.*

The premaxillae and ethmoid of the other derichthyid genus, *Nessorhamphus,* are completely fused (as in many congroids and synaphobranchoids) to form a single unit, termed the premaxilloethmoid (Fig. 450). The entire snout region of *Nessorhamphus* is very elongate. Uniserial to multiserial teeth are present on the vomer, which is unremarkable in shape from that of many other congroid vomers. The vomer is widest at its midpoint, completely toothed (about 6 teeth wide), then abruptly much narrower.

Despite the great difference in the snout region, *Derichthys* and *Nessorhamphus,* the only two derichthyid genera, share many other important osteological characters: the suspensorium is angled sharply forward, the bones are extensively perforated by many foramina, the shapes and relationships of the opercular bones are nearly identical, the preopercle is fused to the ventral edge of the suspensorium and the parasphenoid is large and bilobate. The branchial arches of both genera are essentially identical except for a difference of one pair of epibranchials and in configuration (the branchial and hyal elements of *Derichthys* are less elongate, especially the glossohyal and basibranchial, than in *Nessorhamphus,* a reflection of the overall prolongation of the head of *Nessorhamphus*).

Relationships. That the relationships of the Derichthyidae are with the Congroidea is clear. The family's exact position within this assemblage remains to be delineated.

The Derichthyidae are included by Robins (this volume) in the superfamily Congroidea and the intraordinal relationships of the Derichthyidae are discussed and amplified by him. In concert with the principal features of the superfamily, the Derichthyidae possess the following components of the suite of characters associated with the Congridae. A palatine process is present in *Derichthys* (as in *Macrocephenchelys* and various congrid genera described by Asano, 1962, who termed this process "lateral ethmoid"). Further ontogenetic study is needed to clarify the nature of this process. A complete bracing pterygoid arch is absent, however, both in *Derichthys* and *Nessorhamphus.* The suspensorium is acutely inclined forward in both genera. The tubular axis of the preopercle is almost horizontal in *Nessorhamphus,* with its dorsal and ventral flanges extending from its central portion. The opercle of *Nessorhamphus* has approximately a half dozen posterior spinelike projections. There are six or seven branchiostegals on each side; long, slender and curved around the opercular bones in *Nessorhamphus,* shorter and less curved in *Derichthys.* The suspensorium, lower jaw, preopercle, premaxilla and some ventral cranial bones are deeply sculptured by large foramina, depressions and ridges. Whether this feature, shared by some congrids, is merely an adaptation to mesopelagic life, or is indicative of relationships with the Congridae is uncertain.

As noted earlier, the premaxilla of *Derichthys* (but not *Nessorhamphus*) is a separate, unpaired bony fusion which is attached to the ethmoid by flesh and ligaments, rather than fused to the ethmoid bone as in all other eels. Evolutionarily, this feature may be neotenic, but the cleared and stained individuals, although not the largest specimens seen, were both sexually mature, so that retention of a separate premaxilla is not a feature of juveniles only. In contrast, the premaxillae of a 100 mm TL juvenile of *Synaphobranchus affinis,* cleared and stained by C. Wenner, are completely fused with the ethmoid although sutures of some of the cranial bones are not complete.

The exact provenance of the dentition at the anterior end of the anguilliform upper jaw has been discussed extensively, but most authors have followed Regan (1912b) in considering the commonly subcircular patch of teeth to be the result of fusion of the premaxillae with ethmoid components. Regan included the vomer in this fusion, but most eels have a clearly distinct or even separable, toothed vomer. Torlitz (1922), Norman (1926) and Trewavas (1932) all demonstrated the presence of paired, separate premaxillae in very small specimens of *Anguilla* (ca 40–78 mm TL). Presence of a separate, unpaired premaxilla in adults, is, as far as known, unique to *Derichthys* among the Anguilliformes. The condition in *Nessorhamphus ingolfianus* and *N. danae* is typical of eels; that is, complete fusion of the premaxillae to each other and to the ethmoid.

Trewavas (1932) and Beebe (1935a,b) each illustrated some points of the osteology of *Derichthys* and *Nessorhamphus.* Based on better, more numerous specimens and having the advantage of the last twenty years of advances in anguilliform research, the following comments and emendations can be made. *Derichthys*: the frontals are fused as a single unit (not paired). The preopercle, although held to the suspensorium by connective tissue, is not fused to it. The articular is present, as is a coronomeckelian cartilage (both genera). The pterygoid, although not fused to any bone, is ligamentously connected to the vomer-parasphenoid region as well as to the quadrate as is typical of many eels in which the pterygoid does not form a bony arcade between the suspensorium and the skull. The epihyal and ceratohyal are not fused. Four basibranchials are present. *Nessorhamphus*: Trewavas (1932) and Beebe (1935a,b) both illustrated a "symplectic" with completely delineated margins dorsolateral on the suspensorium, or crossing the junction between the hyomandibular and quadrate. In material examined in this study, the hyomandibular does enter deeply into the quadrate with a long splint of bone in this position, but there is no free dorsal end to this splint or evidence of a suture as shown by Beebe and Trewavas. Norman (1926) regards the symplectic as a "ventral prolongation" of the hyomandibular which provides for a strong articulation of the hyomandibular and quadrate. Presumably, the anteroventral process of the hyomandibular seen in *Nessorhamphus* (Fig. 451) represents the fused symplectic carrying out this function. The cartilaginous element at the connection between the border of the hyomandibular and quadrate possibly is the interhyal.

Key to Genera and Species of Adult Derichthyids

1a. Snout and lower jaw short, snout approximately equal to eye diameter; head pores conspicuous; many cicatrix-like ridges on head in association with head pores . . *Derichthys serpentinus*, p. 424

1b. Snout long, three to four times eye diameter; head pore system less conspicuous; cicatrix-like ridges less well developed 2

2a. Snout long, about four times eye diameter, very slender (less than one-half eye diameter at narrowest point); no midventral line of dark pigment; vertebrae 147–159 . . . . *Nessorhamphus ingolfianus*, p. 427

2b. Snout shorter, about three times eye diameter, less slender (about four-fifths eye diameter at narrowest point); conspicuous stripe of dark, midventral pigment from lower jaw to anus; vertebrae 134–145 . . . . *Nessorhamphus danae*, p. 430

Genus *Derichthys* Gill, 1884

Derichthys Gill, 1884: 433 (type species *Derichthys serpentinus* Gill, 1884, by monotypy; new family). Gill in Goode and Bean, 1896: 161 (amplified familial and generic diagnosis, species description; placed in new order).

Derichtys Borodin, 1929: 110 (name misspelled; new species described).

Grammatocephalus Norman, 1930: 358 (type species *Grammatocephalus kempi* Norman, 1930).

Characters (see also general characters of family, above). Snout and lower jaw short; snout approximately equal to horizontal diameter of eye. Lower jaw shorter and narrower than upper. Cleft of mouth horizontal, extending to posterior margin of eye or slightly posterior to this point. Premaxillary teeth in advance of dentary teeth. Median, V-shaped patch of teeth in roof of mouth (pointed base of "V" directed anteriorly). Head with many large pores (some equal in size to the simple, circular nostrils) and conspicuous cicatrix or weal-like ridges regularly arranged in parallel with groups of ridges set at angles to other groups. Premaxillae present as a single fused, toothed bone, articulated with, but not fused to, the ethmoid. Median V-shaped patch of teeth fused with ethmoid and parasphenoid. Maxillae with anterodorsally directed process articulating with grooves on premaxilla. Glossohyal and basibranchial relatively short; four epibranchials.

Distribution. Widely distributed in Atlantic, Indo-West and southcentral Pacific. No records from Caribbean Sea or Gulf of Mexico.

Etymology. From the Greek words for neck and fish; gender masculine.

Species. One, *Derichthys serpentinus* Gill.

Derichthys serpentinus Gill, 1884

Narrowneck Eel

Figures 453–455

Derichthys serpentinus Gill, 1884: 433 (brief original description, establishment of family name, no specimen or locality given). Gill in Goode and Bean, 1896: 161–162, fig. 169 (amplified description, 33°44′30″N, 71°04′W, in 1022 fm., holotype USNM 33523).

Leptocephalus anguilloides Schmidt, 1916b: 17, figs. 12–14 (original description, North Atlantic, type material lost).

Derichtys iselini Borodin, 1929: 110 (generic name misspelled, original description, 50°41′N, 27°17′W, holotype MCZ 31670, based on misinterpretations of cephalic pores).

Grammatocephalus kempi Norman, 1930: 339, fig. 34 (original description, off S. Africa, placed in Congridae).

Derichthys kempi. Trewavas, 1932: 639, 641 (redescription).

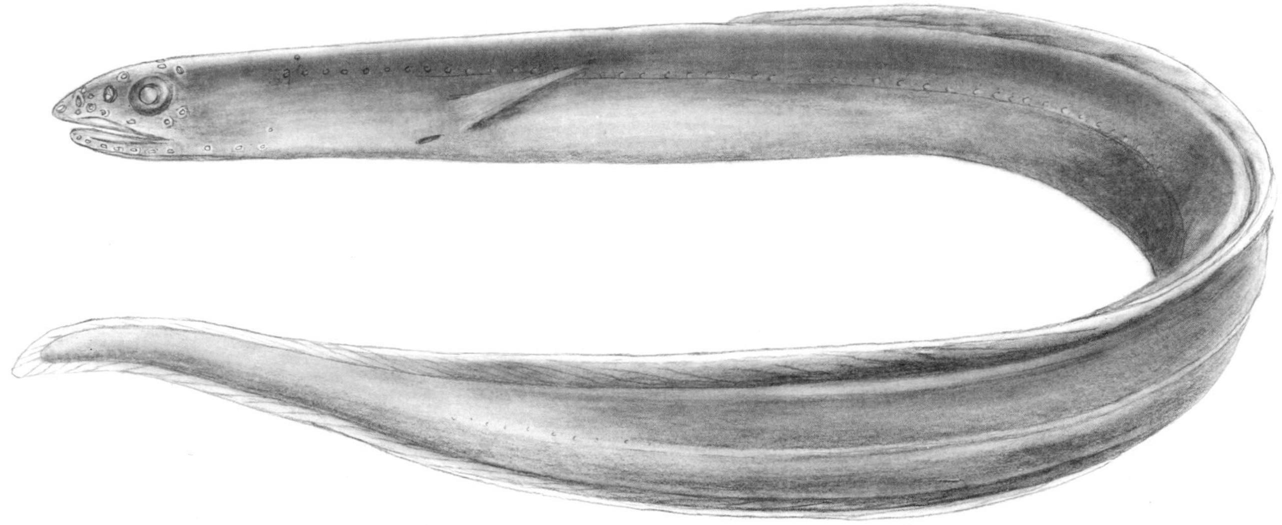

FIGURE 453. *Derichthys serpentinus*: ANSP 124921, off Chile.

Distinctive Characters. Characters of the genus apply to the sole species.

Description. Vertebrae 128–134 (53–57 precaudal). Caudal-fin rays 9–10, pectoral rays 13 (Beebe, 1935a; Castle, 1970b; radiograph). Ratio of dorsal-fin rays (and supports) to vertebrae, 2:1. Branchiostegal rays 7 (5 on ceratohyal; 2, loosely, on epihyal).

Proportional measurements (in % of TL or HL*, based on 16 specimens, 85–353 mm TL): predorsal 24–28, preanal (snout to vent) 48–56, head 9–15, snout* (snout tip to anterior margin of eye) 15–28, jaw* (tip of lower jaw to angular) 44–51, horizontal diameter of eye* 13–19, greatest depth of body 4.4–8.4.

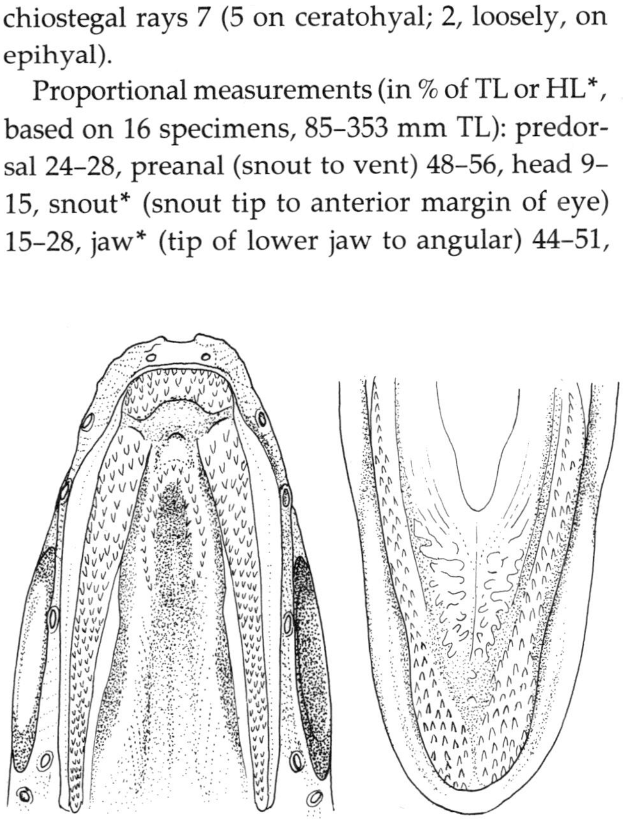

FIGURE 454. *Derichthys serpentinus*: ANSP 124921, off Chile; dentition.

Many small conical or slightly recurved teeth on maxillae, premaxillae (fused as one bone), palatine region of parasphenoid, and dentaries. Viewed ventrally, most of the premaxilla and the anterior ends of the maxillae are visible, projecting beyond the smaller lower jaw. Toothed portion of lower jaw fitting between maxilla and premaxilla, surrounding and forming tight closure with the V-shaped patch of palatine teeth. Small, partially free tongue fits into median concavity formed by palatine tooth patch.

Cephalic sensory system (based on several individuals, one specimen illustrated, Fig. 455, and described below). Many pores present, conspicuous. Simple, very slightly tubular anterior nostril (the rim slightly elevated) located dorsolaterally on snout at base of large, tubular, anteriorly

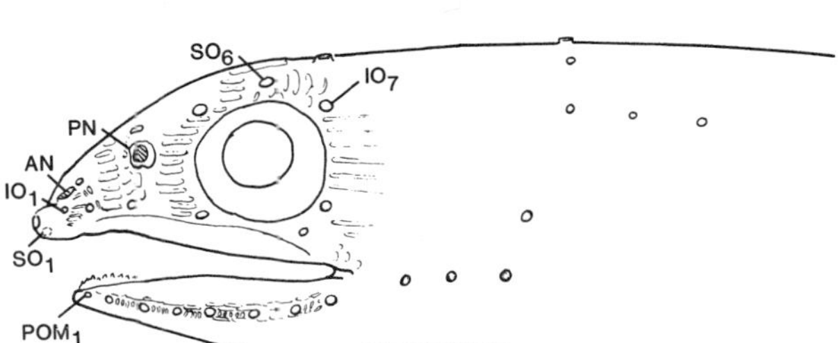

FIGURE 455. *Derichthys serpentinus*: MCZ 60629, NW Atlantic; diagram of head pores.

directed head pore. Posterior nostril nearly equidistant between eye and anterior nostril, at level slightly above horizontal midline of eye. Six pores in supraorbital series (designated S and followed by appropriate numeral). Pores located as follows: S_1 - small, at tip of snout on lower, medial surface of large, anteriorly directed tubule of pore S_2; S_3 - dorsoposterior to anterior nostril, small; S_4 - mediodorsal to posterior nostril; S_5 - at level of anterodorsal quadrant of eye; S_6 - mediodorsal to dorsal margin of eye.

Seven infraorbital pores: I_1 - immediately ventral to anterior nostril; I_2 - ventral and slightly posterior to anterior nostril; I_3 - ventral to posterior nostril; I_4 - ventral and slightly posterior to anterior margin of eye; I_5 - ventral and slightly anterior to posterior margin of eye; I_6 - posterodorsal to I_5, close to posterior margin of eye; I_7 - dorsal to I_6, posterior to eye at level of dorsal margin of eye. Two pores on dorsal surface of head at level of posterior margin of eye, shown by Karmovskaya (1985), belong to the preoperculomandibular series and lateral-line pore canal rather than to the infraorbital series.

Preoperculomandibular series: P_1–P_8 - spaced at slightly increasing intervals in a specialized channel of sensory tissue along the lower jaw from close to the tip of the jaw to the crease of the gape; P_9–P_{12} - continue posteriorly from gape, with P_{12} curving dorsoposteriorly from P_{11} toward lateral line.

Associated with the head-pore system are many straight, short, slightly raised or inflated ridges, parallel within each group, the groups set at right angles to each other. They give the appearance of cicatrices or weals.

Lateral-line pores extend forward to very slightly posterior to level of last preoperculomandibular pore. Lateral line of each side connected by three pores in supratemporal commissure. Open pores occur along the lateral line to within one or two head lengths of tail's tip. Cicatrix-like ridges occur ventral to anterior portion of lateral line on trunk but are easily visible only in larger, well-preserved specimens.

The dorsal fin is abruptly much lower about one and one-half head lengths from tip of tail.

Color. Brown. Beebe (1935) described color of fresh specimens as "tawny olive to mouse gray" and also noted an "oval, opaque, whitish patch, possibly luminous" in each web of the "dorsal and anal" fins. This coloration is lost in preservative.

Size. Largest specimen examined, MCZ 60629, was 353 mm TL, a nearly ripe female. UMML uncat., 229 and 279 mm TL, were also carrying maturing or nearly mature eggs. Beebe (1935a) counted 4100 eggs, about 0.75 mm diameter in a 268 mm specimen, which also had "an equal number" of very small eggs.

Leptocephalus. See Castle, 1970b, and Smith, this volume.

Distribution. As for the genus.

Etymology. From the Latin *serpentinus,* snaky or snakelike; an adjective.

Study material: USNM 280353 (3, 185–204) 31°42′–46′N, 63°51′–57′W; 280354 (1) 32°16′N, 64°18′W; 280355 (1) Acre 12-83; 280356 (1, 115) Acre 11-7-B. ANSP 153305 (1, 85) 10°59′N, 41°38′W; 126098 (1, 197) 47°23.6′–28.0′S, 116°53.5′–49.8′E. UMML 22919 (1, 97) 24°29.5′–32.1′N, 77°39.4′–39.3′W; 34397 (2, 280–395, cleared and stained) 41°29′–26′N, 60°14′–08.5′W. URI uncat. (3, 124–185) 31°42′N, 63°51′W. URI uncat. (1, 116) 31°54′N, 64°19′W. URI uncat. (1, 140) 32°16′N, 64°18′W.

MCZ 31670 (1, 161; holotype of *Derichtys iselini* Borodin) 50°40′48″N, 27°17′W, in 1463–1829 m, 28 Aug. 1928; 35611 (1) Albatross (?) sta. 3031; 39502 (1, 182) R/V Atlantis 3106B; 41553 (1, 216) 15°56′–16°06′N, 36°13′–26′W; 47841 (1, 183) 33°03′–10′S, 77°03′–76°59′W; 47842 (1, 175) 33°44′–53′S, 75°53′–48′W; 47847 (2, 199–210) 33°20′S, 73°41′–42′W; 53012 (2, 126) 35°28′–23′N, 67°35′–27′W; 53013 (1, 162) 40°55′–58′N, 65°27.1′–10.1′W; 53014 (1, 173) 36°06′–03′N, 62°32′–19′W; 53015 (1, 203) 36°05′–35°52.2′N, 62°26′–27.2′W, in 0–1000 m; 53016 (3, 167–202) 38°08′–37°59′N, 70°10′–69°58′W, in 0–1000 m; 53017 (1, 118) 35°49′–55′N, 71°20′–30′W; 53018 (1, 109) 35°25′–28′N, 61°00′–13′W; 53019 (2, 199–210) 39°38′–47′N, 69°48′–68°52′W; 53020 (1, 165) 35°25′–24′N, 66°23′–36′W; 53022 (2, 128–240) 38°52′–39°09′N, 69°19′–40′W; 53023 (2, 280–350) 39°33′–47′N, 70°56′–71°03′W; 53024 (2, 160–189) 39°38′–27′N, 70°03′–69°50′W; 55326 (3, 185–241) 39°46′N, 70°30′W; 60629 (4, 84–353) 38°49.38′N, 71°40.47′W; 62160 (1) 39°42′N, 67°23′W; 63268 (2, 216–245) 34°33′N, 75°25′W; 63269 (1, 215) 33°44′N, 74°23′W; 64361 (2, 190–210) 34°33′N, 75°25′W; 64362 (1, 207) 35°42.6′N, 67°21.0′W; 64363 (2, 200–230) 40°04′–07′N, 68°07′–18′W; 64364 (6, 190–271) 34°33′N, 75°25′W; 64365 (1, 216) 35°49′S, 02°16′–20′W; 63268 (2, 220–250) 39°48′N, 70°27′W.

Genus *Nessorhamphus* Schmidt, 1930

Nessorhamphus Schmidt, 1930: 372 (type species *Leptocephalus ingolfianus* Schmidt, 1912, by monotypy).

Characters (see also General Characters of family, above). Snout long (three to four times eye diameter), spatulate or duck-like. Premaxillary teeth in advance of lower jaw's tip. Mouth oblique, not extending posterior to level of midpoint of eye. Tail length less than half total length of body. Premaxillary teeth in subcircular or oval patch; premaxillae fused with ethmoid to form premaxilloethmoid. Vomerine teeth irregularly multiserial (three to six rows) anteriorly, gradually becoming uniserial posteriorly. Maxillae toothed, slender with small ascending process anteriorly connected with ethmoid rather than premaxillae. Glossohyal and basihyal long.

Distribution. Widely distributed in pelagic to bathypelagic zones of Atlantic, Indian and western Pacific oceans as larvae and adults.

Etymology. From the Greek words for duck and beak; gender masculine.

Species. Two, *Nessorhamphus ingolfianus* Schmidt, and *N. danae* Schmidt.

Nessorhamphus ingolfianus (Schmidt, 1912)
SPOONBILL EEL
Figures 456–458

Leptocephalus ingolfianus Schmidt, 1912a: 49, pl. 3, fig. 8 (original description, Atlantic, southwest of Azores, 32°03′N, 39°00′W, syntypes ZMUC P 32520–23).

Avocettina scapulorostris Borodin, 1929: 109 (original description, 41°29′N, 47°48′W, holotype MCZ 31671). Borodin, 1931: 74, pl. 3, figs. 1–3 (placed in Nemichthyidae).

Nessorhamphus ingolfianus: Schmidt, 1930: 373, pls. 4–5 (formal description based largely on adult 248.5 mm, see comment below). Schmidt, 1931a: 487, fig. 4 (contrasted with *N. danae*).

Platuronides danae (not of Schmidt, 1931). Beebe and Crane, 1937a: 332, text-fig. 2 (misidentification of larvae of *N. ingolfianus* with larval serrivomerid).

Platuronides acutus (not of Schmidt, 1930). Beebe and Crane, 1937a: 332, text-fig. 2 (misidentification of larvae of *N. ingolfianus* with larval serrivomerid).

Discussion of Synonymy. Schmidt (1930) used the designation *Nessorhamphus ingolfianus* n. sp. incorrectly since he had already (unintentionally or not) described the species (1912a) in a valid manner. The adult specimen from 45°24′N, 41°58′W, which formed the basis of the redescription, cannot be considered type material since it was not part of the original description.

Distinctive Characters. Vertebrae 147–159 (precaudal 70–74). Snout long (about four times eye diameter) and narrow (less than one-half eye diameter). Overall color various shades of brown; no midventral line of dark pigment.

Description. Caudal-fin rays 10–12; pectoral-fin rays 12–13 (Beebe, 1935a; Trewavas, 1932). Branchiostegal rays 7 (5 on ceratohyal; 2, loosely, on epihyal). Proportional measurements (based on 10 specimens from the western North Atlantic, 128–597 mm TL, expressed as percent of total length or head length*): predorsal 19–26, preanal 54–62, head (snout to anterior edge of gill slit) 13–16, tip of snout to eye* 37–43, jaw* (jaw tip to angular, based on 4 specimens) 35–45, eye diameter* 9.0–10.7, greatest depth of body 1.6–4.2.

Teeth present on premaxilloethmoid, maxillae, vomer and dentaries. Premaxillary teeth in subcircular patch, longer than those on other toothbearing bones, conical or conical-recurved. No regular rows, as many as 6–8 teeth across widest part of bone. Maxillae closely studded with small, bluntly conical teeth which steadily decrease in size posteriorly. Vomerine teeth in irregular series, three or four teeth across at anterior end of vomer, decreasing slightly in size to an irregular uniserial row posteriorly. Dentary completely toothed, anterior teeth largest in cluster two or three rows wide anteriorly, followed by two rows of slightly smaller teeth posteriorly. Posterior three-quarters of dentary covered with teeth of decreasing size arranged in irregular rows of three, then uniserial at posterior end.

Cephalic sensory pores of *Nessorhamphus* are of two types: large pores (as in *Derichthys*) plus more numerous, smaller pores in series with the large pores. Presence of this secondary series is perhaps associated with prolongation of snout region of *Nessorhamphus*. Secondary series considerably variable in number and asymmetry.

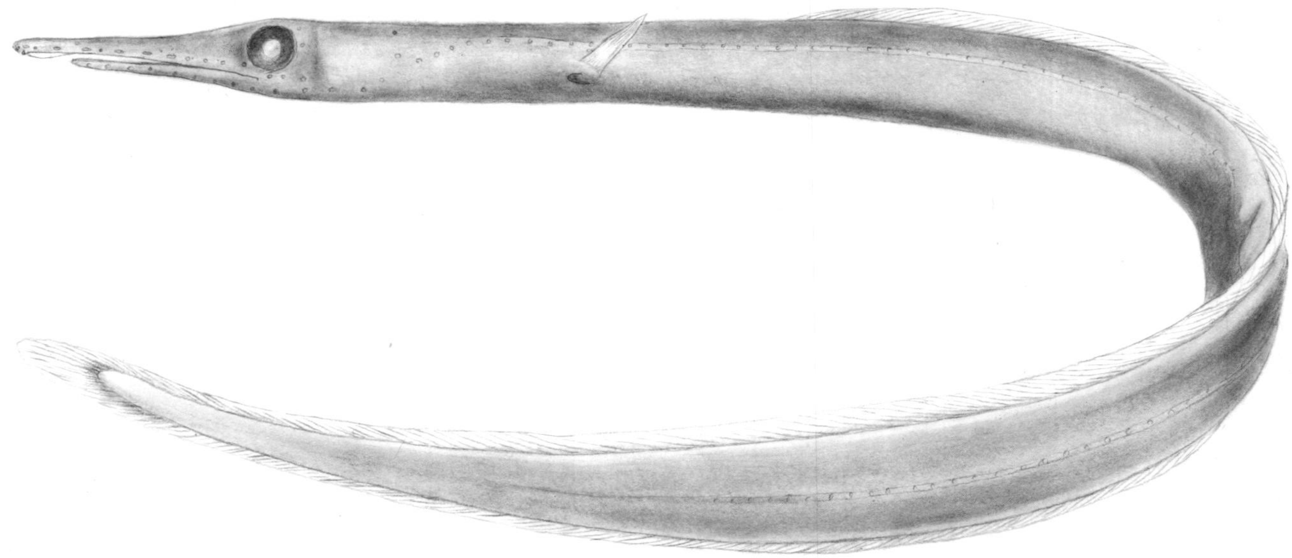

FIGURE 456. *Nessorhamphus ingolfianus*: ANSP 117076, off North Carolina.

In the description below and in the illustration (Fig. 458), only the primary pores are numbered. Study of nostrils and cephalic pores is largely based on three large specimens (445–597 mm TL); the description below and the illustration are based on the single best specimen.

Simple, porelike anterior nostril ventrolateral near tip of fleshy, spatulate snout. Posterior nostril larger, also simple, but dorsolateral.

Five supraorbital pores as follows: S_1 – very small, ventromedial at tip of snout; S_2 – lateral, much larger, tubular; S_3 – small, terminal on snout; S_4 – larger, mediodorsal on snout tip; S_5 – largest, simple pore with short raised rim, medial and posterior to posterior nostril. Fourteen much smaller secondary pores in series extending posteriorly from S_5 to distance of about one eye diameter from anterior margin of eye.

Six infraorbital pores placed as follows: I_1 – on upper lip at level of tip of lower jaw (followed by two smaller secondary pores); I_2 – posterior to I_1 (followed by series of seven secondary pores); I_3 – immediately anterior to anteroventral quadrant of eye margin (followed by seven secondary pores); I_4 – immediately posterior to posteroventral quadrant of eye margin (two secondary pores dorsal to I_4); I_5 – approximately at midpoint of posterior margin of eye (one small pore dorsal); and I_6 – immediately posterior to posterodorsal quadrant of eye margin. Dorsal to I_6, the large pore shown by Karmovskaya (1985) is shown by clearing and staining to be part of the posttemporal series connecting with the lateral line.

On the dorsal surface of the head at the level of the eyes, there is a median unpaired series of 5 secondary pores, set in a ridge of sensory tissue.

Thirteen preoperculomandibular pores: P_1–P_{10} are along the lower jaw from close to the tip of jaw to the crease of the gape; P_{11}–P_{13} are more widely spaced, curling dorsally and posteriorly toward the lateral line. The lateral line is connected by the supratemporal commissure with three pores to that of the other side.

Karmovskaia (1985) described and illustrated two infraorbital pores far anterior on the snout in association with the anterior nostril (as in *Derichthys*). These pores are not present in any of the three large western North Atlantic specimens examined by me and C. R. Robins. Nor does the infraorbital canal pass anterior to the first pore of the infraorbital series (Fig. 458). Our counts are rather widely at variance with those of Merrett and Saldanha (1985), in part due to their counting primary and secondary pores together and, surely, in part due to difficulties inherent in finding sufficiently large, intact material of *Nessorhamphus*. If one overlooks the elongation of the snout in *Nessorhamphus*, then the infraor-

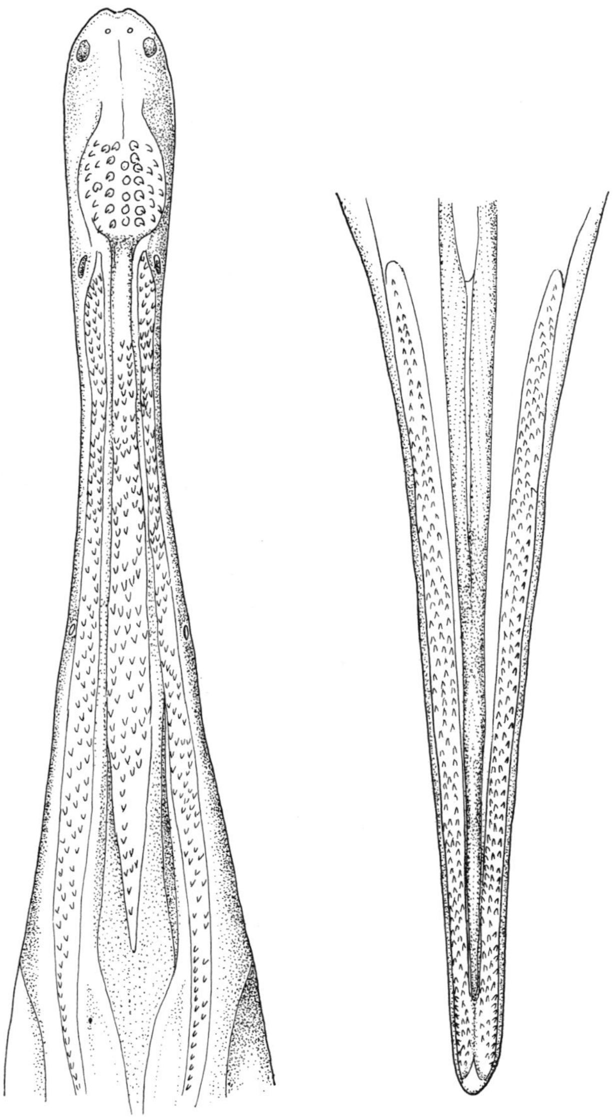

FIGURE 457. *Nessorhamphus ingolfianus*: ANSP 117076, off North Carolina; dentition.

bital canal as described above ends at the same point in *Nessorhamphus* and *Derichthys*.

In addition to pores, many papillae and plicae are present on the jaws and snout of *Nessorhamphus*. Cicatrix-like ridges are also present, especially in the postorbital region, but are much less strongly developed than in *Derichthys*.

The dorsal and anal fin are much lower about one head length from tip of tail; the caudal fin, when undamaged, is truncate to emarginate.

Color. Smaller specimens are ordinarily pale in preservative; larger specimens are dark or darkish brown. Body cavity and organs pale.

Size. To about 600 mm TL (Merrett and Saldanha, 1985 reported 502 mm TL; MCZ 63271 is 597 mm), perhaps larger. Size at sexual maturity unknown, but both sexes are identifiable at about 300–400 mm TL.

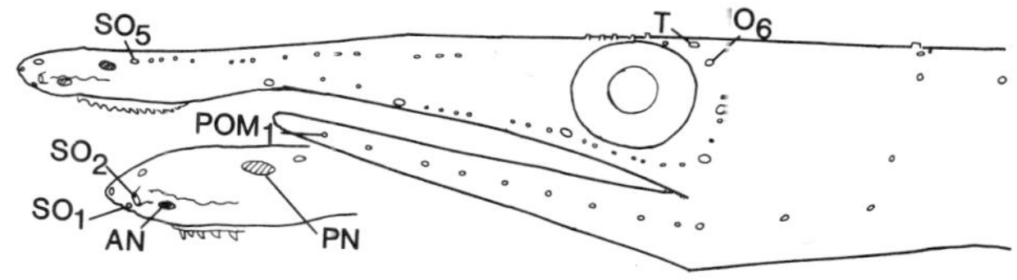

FIGURE 458. *Nessorhamphus ingolfianus*: MCZ 53003, NW Atlantic; diagram of head pores and detail of snout.

Leptocephalus. Long recognized (Schmidt described the species from leptocephali). See Smith, this volume.

Distribution. North and South Atlantic, Indian, and central Pacific oceans. Not known from the Gulf of Mexico or the Caribbean Sea. Karmovskaya (1985: fig. 8) mapped the distribution of both adults and leptocephali. Castonguay and McCleave (1987) summarized the occurrence of the leptocephali in the western Sargasso Sea relative to physical factors in the environment.

Etymology. For the Danish ship Ingolf.

Study Material. USNM 280357 (1, 155) 33°33′N, 64°04′W; 280358 (2, 132–139) 32°N, 64′W; USNM 280359 (2, 128–139) 32°N, 64′W; USNM 280360 (1, 103) 32°N, 64′W; ANSP 130810 (1, 228) 36°40.7′N, 74°30.5′W; UMML 20883 (1, 355, cleared and stained) 05°57′–56′N, 03°59′–04°03′E; UMML 22920 (2, 168–?) 24°29.5′–32.1′N, 77°39.4′–39.3′W.

MCZ 40954 (4) 39°28′N, 71°27′–21′W; 40955 (2) 39°47′–40′N, 70°36′–43′W; 40956 (1) 39°43′–42′N, 70°19′–09′W; 40957 (9) 39°45′N, 71°08′W; 40958 (3) 39°45′N, 71°08′W; 40959 (4) 39°45′N, 71°08′W; 40961 (1) 41°53′–50′N, 64°23′–27′W; 41282 (12) 39°45′N, 71°08′W; 41283 (24) 39°45′N, 71°08′W; 41284 (33) 39°44′–48′N, 70°39′–28′W; 41285 (5) 39°43′–41′N, 70°04′–69°54′W; 41286 (21) 39°48′–47′N, 70°21′–36′W; 41446 (3) 41°38′–24.5′N, 65°41.5′–44′W; 41556 (1) 02°54′–03°02′N, 21°12′–19′W; 44207 (2, 103–191) 38°34′–33′N, 71°08′–11′W; 44223 (1) 32°50′–38.5′N, 64°04′–23′W; 47800 (1) 04°12′–03°56′N, 60°08′E; 53000 (1) 35°33′N, 67°46′W; 53001 (1, 158) 39°40′–27′N, 70°33′–32′W; 53002 (1, 132) 36°17′N, 61°53′W; 53003 (1, 510) 36°43′N, 64°23′W; 53004 (1, 155) 39°38′–47′N, 69°48′–68°52′W; 53005 (1) 35°25′N, 66°23′W; 53006 (1) 37°52′N, 66°18′W; 53007 (1, 92) 37°51′–38°04′N, 66°39′–36′W; 53008 (2, 220–234) 38°52′–39°09′N, 69°19′–10′W; 53009 (1, 175) 39°09′–38°56′N, 69°10′–14′W; 53010 (1, 215) 39°22′N, 70°49′W; 53011 (1, 200) 39°38′–27′N, 70°03′–69°50′W; 63270 (1, 435) 40°23′–24′N, 55°48′W; 63271 (2, 213–510) 35°42.6′N, 67°21.0′W;

64393 (1, 187) 17°48′-42′N, 18°28′W; 64394 (1, 187) 15°06′-02′N, 24°20′-17′W; 64395 (2, 150–154) 33°03′-05′N, 17°40′-45′W; 64396 (1, 150) 31°53′-56′N, 16°10′-16′W; 64397 (3, 150–155) 33°06′-08′N, 17°46′-49′W; 64398 (1), 39°71′N, 41°10′W; 64399 (1, 137) 31°52′N, 77°05′W; 64400 (1, 263) 34°29.0′N, 71°54.8′W; 64401 (1, 193) 15°34′-28′N, 24°39′-33′W; 64402 (1, 158) 33°21′N, 73°58′W; 64403 (1, 148) 39°20′N, 66°45′W; 64404 (1, 152) 34°10′-05′N, 14°10′-07′W; 64405 (1, 124) 30°07′N, 78°03′W; 64406 (2, 204–215) 36°46.5′N, 69°30.0′W; 64407 (1, metamorphosing leptocephalus) 37°44′N, 67°00′W; 64408 (1) 27°11′-13′N, 43°46′W; 64409 (1, 155) 36°35′N, 70°31′W; 64410 (2) 28°17′N, 45°21′-26′W; 64411 (1) 26°37′-38′N, 41°35′-32′W; 64412 (1, 343) 38°53.8′N, 70°48.0′W; 64413 (1) 40°10′N, 39°57′W; 64414 (1, 104) 24°46′N, 35°44′W; 64415 (1, 54) 26°34′N, 40°50′W; 64416 (1, 109) 08°59′N, 53°05′W; 64417 (1, 100) 22°16′N, 45°27′W; 64418 (1, 89) 23°15′N, 44°57′W; 64419 (1, 129) 22°57′N, 64°12′W; 64420 (1, 158) 36°39.5′N, 68°27.0′W; 64421 (1, 144) 36°37.3′N, 66°41.0′W; 64422 (1, 147) 39°24.6′-31.0′W; 64423 (1, 170) 39°36.8′N, 65°49.0′W; 64424 (1, 172) 39°36.8′N, 65°49.0′W; 64425 (1, 285) 34°27.0′N, 71°18.5′W; 64426 (1, 264) 38°30.5′N, 71°59.3′W; 64427 (1, 177) 39°09.0′-09.3′N, 71°22.0′-20.5′W; 64428 (1, 292) 32°22′-09′N, 59°45′-59′W; 64429 (1, 235) 43°00′-03′N, 21°42′-46′W; 64430 (6, 143–229) 39°29.6′N, 69°42.2′W; 64431 (2, 147–232) 38°19.0′N, 69°08.0′W; 64432 (3, 185–260) 40°04′-07′N, 68°07′-18′W.

Nessorhamphus danae Schmidt, 1931

Nessorhamphus danae Schmidt, 1931a: 487, fig. 4 (original description, Indian Ocean, west of Sumatra, also recorded from elsewhere in Indian, southwestern Pacific, and Atlantic oceans). Karmoskaia, 1985: 896 (synonymy, designation of lectotype, 102 mm TL, from Indian Ocean west of Sumatra).

Nessorhamphus ingolfianus (not of Schmidt, 1912). Fourmanoir, 1971: 269 (misidentification).

Distinctive Characters. Snout shorter than that of *Nessorhamphus ingolfianus* (approximately three times eye diameter in *N. danae*), much broader (at narrowest point, about four-fifths eye diameter). Dark pigment in streak along ventral midline from just posterior to a point directly under the eye to anus in smaller individuals.

Description. Vertebrae 134–145 (precaudal 50–55), based on material examined and literature. Pectoral-fin rays 11–12 in two specimens. Proportional measurements (four specimens, 161–178 mm TL) expressed as percent of total or head* length. Predorsal 22–23, preanal 60–61, head 15–16, snout tip to eye* 38–40, jaw* (tip to angular) 36–45, eye diameter* 12, greatest depth of body 3.7–5.0.

Teeth are present on premaxilloethmoid, maxillae, vomer, and dentaries. Premaxilloethmoid teeth in oval patch. All teeth are slightly more sparse than in *N. ingolfianus* in this small sample of material.

The nostrils are far anterior on the spatulate snout, above a point just posterior to premaxilloethmoid tooth patch. Anterior nostril larger, dorsoventral on snout over olfactory organ which is visible through the opening. Posterior nostril smaller, placed about one-third eye diameter posterodorsal to anterior nostril. Relationship of nostrils to olfactory organ clearly seen in semitranslucent small specimens.

Cephalic sensory system (based on four specimens 161–178 mm TL). There are many papillae, presumably sensory, arranged in series around the open pores; these are not described here. Five open supraorbital pores are located as follows: S_1 - small, on anteroventral tip of snout, S_2 - at tip of snout, S_3 - about equidistant between S_2 and anterior nostril, S_4 - on dorsolateral surface of snout between posterior nostril and eye, and S_5 - on dorsal surface of head at level of posterior margin of eye. A median pore anterior to pore S_5 on dorsal surface of head at level of midpoint of eye. Eight infraorbital pores located as follows: I_1 - in front of anterior nostril (posteroventral to S_3), I_2 - anteroventral of I_1, I_3 - on lip anterior to level of anterior nostril, I_4 - on lip slightly anterior to level of posterior nostril, I_5 - in front of anteroventral quadrant of eye, I_6 - ventral to eye, approximately level with gape, I_7 - at midpoint of posterior margin of eye, and I_8 - posterior to eye at level of dorsal margin. Twelve preoperculomandibular pores are located from the tip of the lower jaw along the jaw to the preopercular region.

A few small, parallel, sensory ridges are associated with the head pore system, especially in the postorbital and nape area; none so conspicuous or well-developed as in *Derichthys*. Many small papillae and plicae present on tip of snout. First lateral line pore dorsal to most posterior preoperculomandibular pore. Lateral-line canal of each side connected by supratemporal canal with 3 dorsal pores. Lateral line elements or a lightly pigmented line extends to within one head length of tail tip.

Dorsal, anal and caudal fin shaped as in *Nessorhamphus ingolfianus.*

Color. Adults medium brown in preservative, the dark midventral streak conspicuous; head somewhat paler than body. Fins quite pale to clear.

Size. Karmovskaya (1985) listed material as large as 286 mm TL. The 178 mm specimen listed above is immature, possibly female.

Distribution. Atlantic records of *Nessorhamphus danae* are of leptocephali except for three records of adults from the South Atlantic reported by Karmovskaya (1985: fig. 8) and the two transformed specimens noted below. Leptocepahali of *N. danae* are distributed equatorially in the Atlantic and Indo-Pacific.

Leptocephalus. See Smith, this volume

Etymology. For the research vessel Dana.

Study Material. 7 specimens, all from Indo-Pacific. SIO 73-133 (1, 169) 21°20′–30′N, 158°20′–30′W, 6 Sept. 1971. SIO 73-122 (1, cleared and stained). SIO 73-134 (1, 178) 21°20′–30′N, 158°20–30′W, 8 Sept. 1971. SIO uncat. (1, 161) 04°47.4′S, 129°51.8′E, in 0–1500 m, Alpha Helix sta. 110, 13 May 1975. SIO uncat. (1, 163) 00°18.05′S, 129°00.8′E, in 700–1000 m, Alpha Helix sta. 150, 21 May 1975. MCZ 47800 (1) 04°12′N, 60°08′E. MCZ 41556 (1) 03°00′N, 21°15′W.

Family Muraenesocidae

DAVID G. SMITH

Acknowledgments. See family Anguillidae.

Introduction. The Muraenesocidae is a small family of somewhat uncertain composition and affinity, possibly related to the Congridae. Muraenesocids are heavy-bodied, predatory eels with strong teeth and jaws. The family includes some of the largest of all eels, and a significant fishery exists for them in the Far East. The family name dates from Günther (1870: 20), in the form Muraenesocina. Cope (1872) recognized it as the Muraenesocidae. Gill (1890d) reviewed the family and gave a morphological diagnosis; he recognized the same three genera that I do in this volume. The family was later expanded and became something of a catchall (see below).

Characters. Large to moderately large eels, relatively heavy bodied, preanal length approximately 40% TL. Tail moderate, tip flexible. Dorsal and anal fins well developed, confluent with caudal fin; dorsal fin begins slightly before base of pectoral fin; dorsal- and anal-fin rays segmented. Pectoral fin well developed. Gill opening moderately large, crescentic, located on lower side. Lateral line complete, but with a complex system of branches or diverticula with multiple pores. Scales absent.

Snout and jaws moderate to moderately elongate, upper jaw projects slightly beyond lower. Fleshy tip of snout extends anteriorly beyond intermaxillary teeth. Upper and lower labial flange absent. Teeth sometimes exposed when mouth closed. Eye well developed. Anterior nostril tubular, located in a variable distance behind tip of snout, directed laterally or anterolaterally. Posterior nostril elliptical, located at mid-eye level a variable distance in front of eye. Head pores complex. No frontal commissure.

Teeth strong. Intermaxillary tooth patch short, slightly separated from vomerine and maxillary teeth, anterior teeth enlarged. Vomer with a median series of large, strong, fang-like teeth, variable in form, flanked on each side by a row of smaller teeth. Maxillary teeth in narrow bands, those of one row sometimes distinctly enlarged. Mandibular teeth generally triserial, those of one row generally distinctly enlarged; anteriormost mandibular teeth large, fang-like.

Osteology (based on *Cynoponticus savanna*, ANSP 114155 and ANSP uncat. Oregon 20242). Neurocranium massive and moderately elongate (Fig. 459). In lateral view, ventral outline noticeably convex near anterior part of orbit where ethmovomer bears large median fangs. Vomerine portion of ethmovomer extends posteriorly well beyond posterior margin of orbit. Immediately in front of orbit near posterior end of ethmovomer is a structure that seems identical in form and location to lateral ethmoid process of bathymyrine congrids. Supraoccipital present and bears a crest. Otic bullae fairly well developed.

Maxilla with a well developed pedicel at anterior end; posterior end straight and not bent downward as it is in many congrids (Fig. 460). Mandible moderately elongate, consisting of dentary, articular, and coronomeckelian (Fig. 461). In lateral view, articular covered by dentary except at posterior end. Sensory canal runs through dentary and opens through several pores.

Suspensorium stout and nearly vertical (Fig. 462). Hyomandibula considerably larger than quadrate. Pterygoid heavy and well developed, but without teeth.

Small hypohyal present, clearly separate from head of ceratohyal (Fig. 463). Ceratohyal and

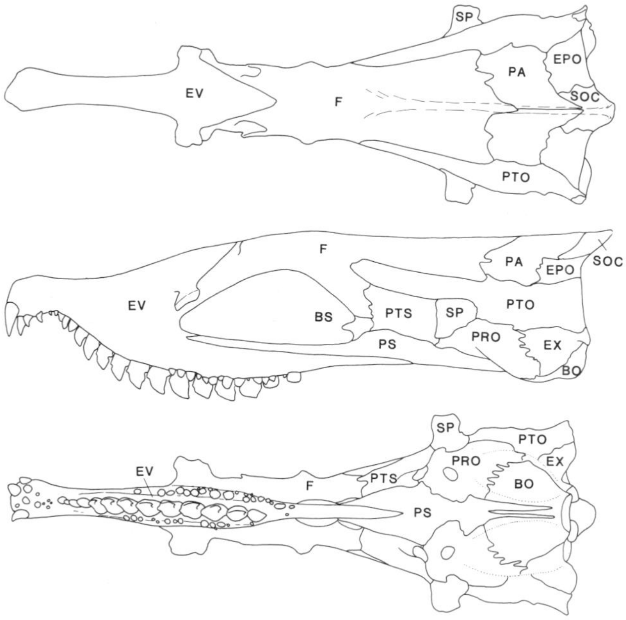

FIGURE 459. Neurocranium of *Cynoponticus savanna* in dorsal view (top), lateral view (middle), and ventral view (bottom). Abbreviations on pp. 5–6.

epihyal rather slender and about equal in length (Fig. 463). Branchiostegal rays numerous; first two associated with ceratohyal, remainder with epihyal (Fig. 463). Glossohyal simple, cylindrical, not articulated with basibranchials (Fig. 464, bottom). Urohyal simple, somewhat broadened and depressed at anterior end, simple and needle-like posteriorly (Fig. 464, top).

Gill arches complete (Fig. 465). Four basibranchials, first three ossified, fourth cartilaginous. Three hypobranchials, first two ossified, third cartilaginous. Five ceratobranchials, fifth smaller and more slender than others. Four epibranchials and two infrapharyngobranchials. Upper and lower pharyngeal tooth plates large, undivided. Pectoral girdle complete.

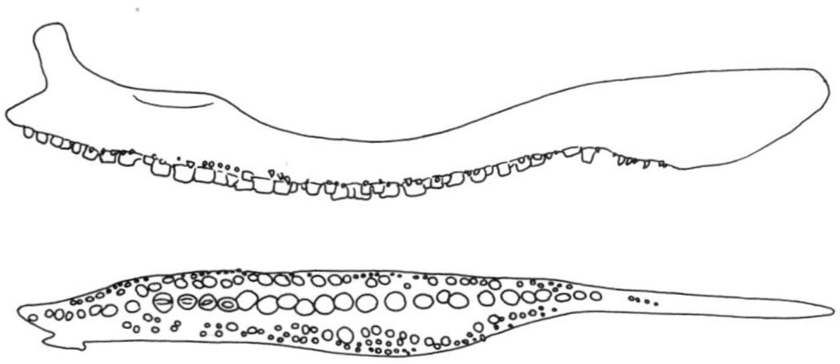

FIGURE 460. Maxilla of *Cynoponticus savanna* in lateral view (top) and ventral view (bottom).

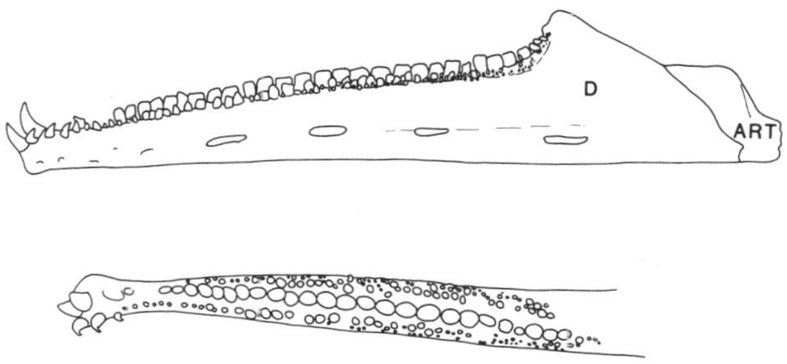

FIGURE 461. Mandible of *Cynoponticus savanna* in lateral view (top) and dorsal view (bottom).

Circumorbital bones (Fig. 466A): Nasal large, partially ossified. Preorbital elongate, poorly ossified, without projections from ventral edge. A small supraorbital and four suborbitals. Lateral-line ossicles heavy, elongate, trough-like (Fig. 466B).

All four opercular elements present (Fig. 467). Opercle broad, posterior margin serrate. Subopercle sickle-shaped, lies under ventral and posterior edge of opercle. Preopercle somewhat smaller than opercle, houses part of preoperculomandibular canal. Interopercle slightly smaller than preopercle, squarish in shape.

Pectoral girdle well developed (Fig. 468). Supracleithrum slightly longer than cleithrum. Cleithrum with short fork ventrally. Scapula and coracoid broad, well developed. Four actinosts.

Vertebrae (Figs. 469, 470): Myorhabdoi absent. Ribs present. Neural arches of first eight vertebrae articulated with centrum rather than fused. Epineurals of anterior few vertebrae fused to neural arch. Three extra epineurals and ribs at anterior end of vertebral column, connected to

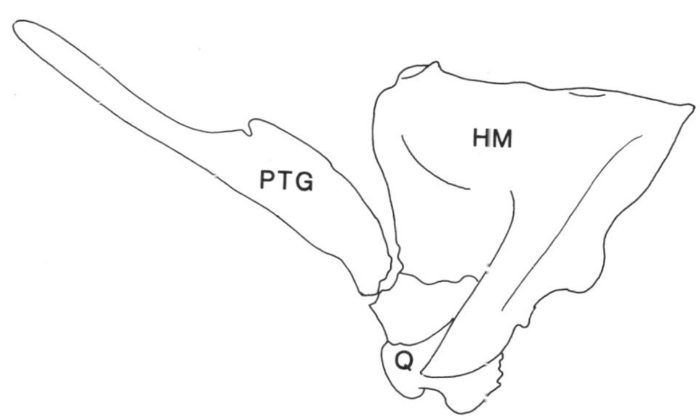

FIGURE 462. Suspensorium of *Cynoponticus savanna*.

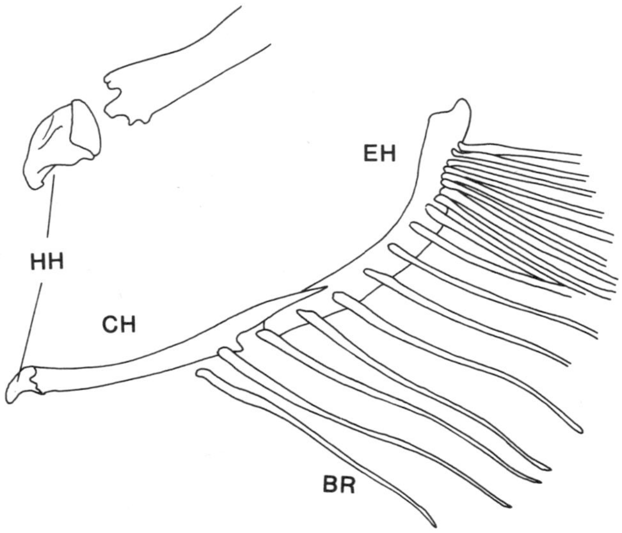

FIGURE 463. Hyoid arch of *Cynoponticus savanna*: hypohyal, ceratohyal, epihyal, and branchiostegal rays. Inset shows hypohyal and tip of ceratohyal, separated.

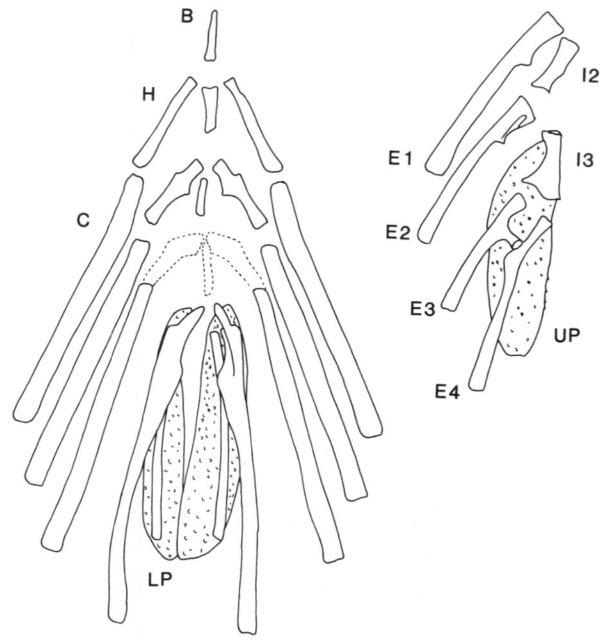

FIGURE 465. Gill arches of *Cynoponticus savanna*, ventral elements in ventral view, dorsal elements in dorsal view.

skull (Fig. 469). Caudal vertebrae without epicentral processes (Fig. 470B).

Discussion. The Muraenesocidae has never been adequately defined. It was long a dumping ground for a variety of eels that did not fit elsewhere. Jordan and Evermann (1896a: 358) included representatives of five different families. In recent years the family has been restricted to the genera *Muraenesox* McClelland, *Congresox* Gill, *Cynoponticus* Costa, *Sauromuraenesox* Alcock, *Oxyconger* Bleeker, *Hoplunnis* Kaup, *Xenomystax* Gilbert, *Paraxenomystax* Reid, and *Gavialiceps* Alcock. *Hoplunnis* has now been shown to be a nettastomatid (Smith and Castle, 1982; this volume, p. 577), and *Xenomystax* (*Paraxenomystax* is a synonym) is related to the *Uroconger* group within the Congridae. *Gavialiceps* is closely related to *Xenomystax.*

Muraenesox, Congresox, and *Cynoponticus* are closely related and constitute the core of the family Muraenesocidae. They resemble *Hoplunnis* and *Xenomystax* in some respects, especially the den-

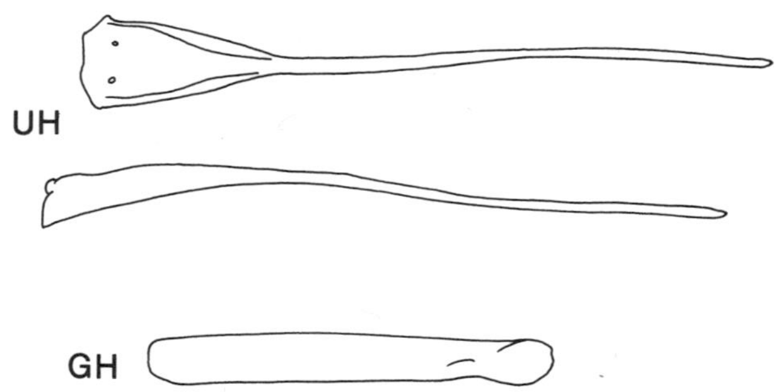

FIGURE 464. Hyoid arch of *Cynoponticus savanna.* Urohyal in ventral (top) and lateral (middle) views, and glossohyal in lateral view (bottom).

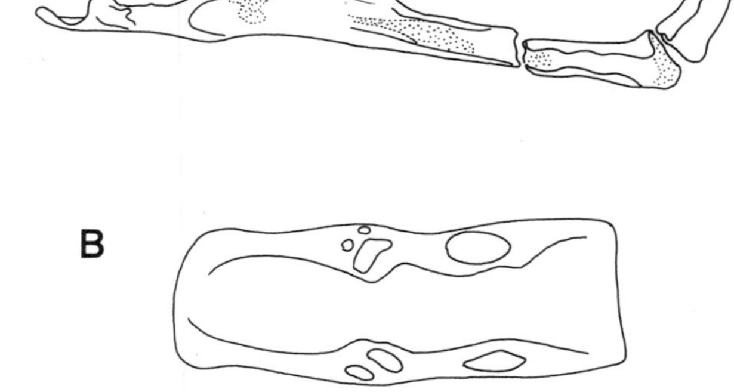

FIGURE 466. Bones associated with lateralis system of *Cynoponticus savanna.* A. Circumorbital bones. B. Lateral-line ossicle.

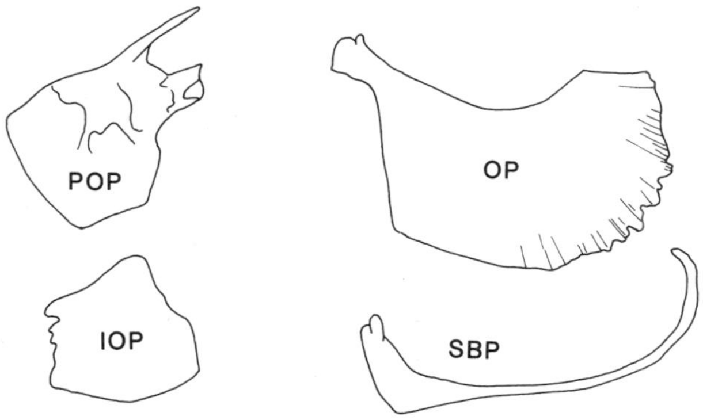

FIGURE 467. Opercular apparatus of *Cynoponticus savanna.*

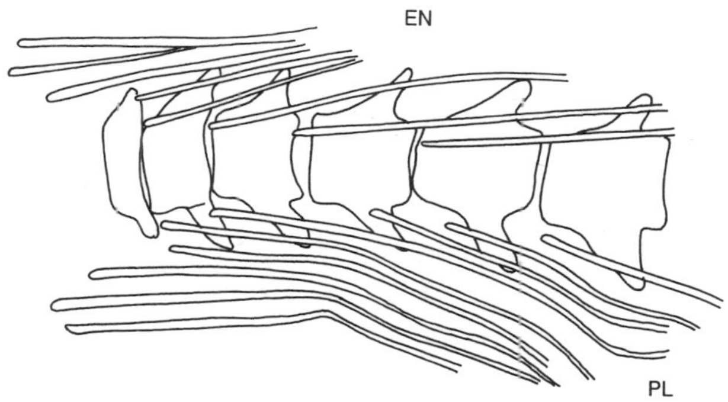

FIGURE 469. Anterior few vertebrae of *Cynoponticus savanna.*

tition, but in other ways they are peculiar. The lateralis system is unique. The complex branching canals and multiple pores are unlike those seen in any other eel. The lateral-line ossicles are heavy and horizontally elongate, quite unlike the slender, vertically elongate structures found in *Hoplunnis* and *Xenomystax. Cynoponticus,* at least, has discrete hypohyals, a primitive feature absent in *Hoplunnis* and *Xenomystax* as well as in almost all other congrids. The tail, though flexible, is not attenuate as it is in *Hoplunnis* and *Xenomystax.* In the latter two genera the fleshy lips are reduced, leaving the teeth exposed when the mouth is closed; *Muraenesox* and its relatives have relatively heavy lips that cover the teeth from the outside. *Hoplunnis* has a number of specialized nettastomatid characters, such as the close articulation between the glossohyal and the first basibranchial, which *Muraenesox* and its allies lack.

Sauromuraenesox has not been examined in detail, but externally it shows many similarities to *Muraenesox* et al. Its most distinctive feature is the sharp division between the stout trunk and the narrow tail. *Oxyconger* remains a question. It lacks the median vomerine teeth characteristic of the *Muraenesox* group, though it shows other resemblances. Its relationships will remain uncertain until it is studied in more detail.

If the Muraenesocidae is restricted to *Muraenesox, Congresox, Cynoponticus,* and, probably, *Sauromuraenesox,* it remains to determine where the family stands in relation to other eels. Its fused frontals place it among the Congridae, Colocongridae, Nettastomatidae, Synaphobranchidae, Ophichthidae, Derichthyidae, Serrivomeridae, and Nemichthyidae. The Muraenesocidae has none of the diagnostic characters of the Syn-

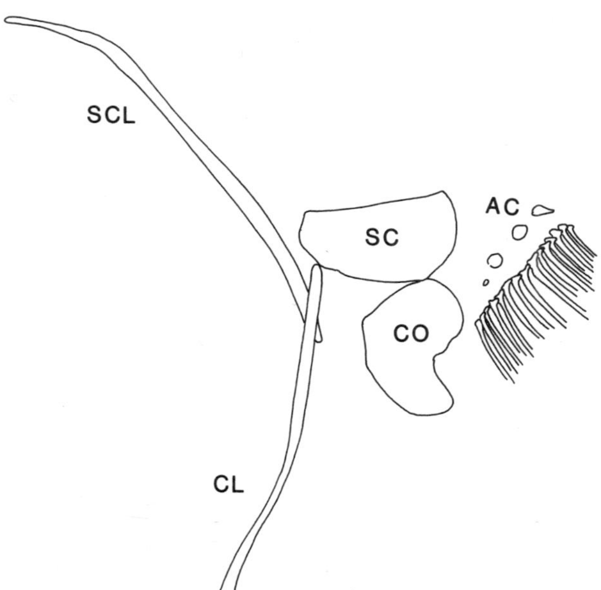

FIGURE 468. Pectoral girdle of *Cynoponticus savanna.*

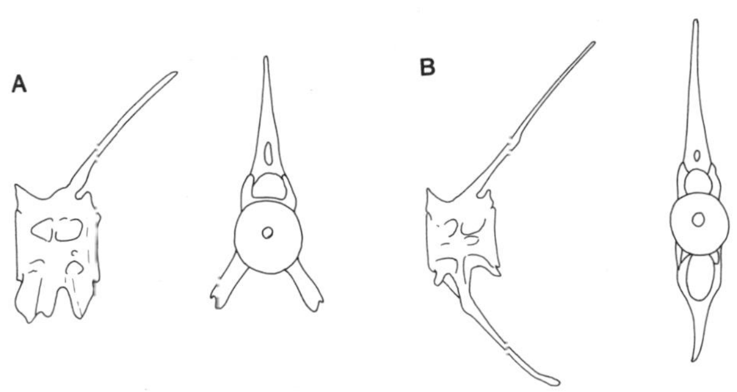

FIGURE 470. Vertebrae of *Cynoponticus savanna,* in lateral and frontal views. A. Precaudal. B. Caudal.

aphobranchidae or the Ophichthidae and cannot be considered close to those families. Nor is there any reason to ally it with the Serrivomeridae and Nemichthyidae, which are specialized midwater eels. The derichthyids are more generalized midwater eels, but there is no reason to believe that they are especially close to the muraenesocids. The Colocongridae shows an entirely different set of specializations, none of them indicating a special relationship with the Muraenesocidae. The superficial resemblances to the nettastomatid genus *Hoplunnis* were already discussed.

That leaves the Congridae, more or less by default, but the shared characters are mainly primitive. Muraenesocids do not fit the trends of character modification found in the various congrid genera. For example, muraenesocids lack a flange on the upper lip, a feature found in all the more primitive or basal members of the Congridae. The loss of the flange in congrids is accompanied by a modification of the preorbital bone, which sends three projections ventrally into the lip. Muraenesocids have no flange, but their preorbital is weak and without ventral projections. Those congrids and congrid derivatives, such as *Xenomystax* and the nettastomatids, that have neither a flange nor ventral projections on the preorbital have an additional suite of characters such as an attenuate tail and reduced lateral-line ossicles. These characters are not found in muraenesocids. Congrids almost universally lack hypohyals and have relatively few branchiostegal rays (12 or fewer). Muraenesocids have hypohyals and relatively numerous branchiostegals (17 in *Cynoponticus savanna*). Both the latter characters can be considered primitive. On the other hand, the lateralis system is highly specialized and unlike anything found in the Congridae. The skull of *Cynoponticus savanna* has a structure that, although not as well developed, seems identical to the lateral ethmoid process found in the bathymyrine genera and the heterocongrine *Gorgasia* but absent in all other congrids. *Cynoponticus,* however, shows none of the special characters of the bathymyrines or heterocongrines. The lateral ethmoid process of *Cynoponticus* is either a convergence or a primitive character that has been lost in the more advanced congrids. Because of their peculiar combination of primitive and derived characters, muraenesocids cannot be derived from any particular group within the Congridae. If relationships within the Congridae are visualized as the diagram on page 481 (family Congridae, Fig. 522) then the Muraenesocidae must arise somewhere before the first major branching point.

Distribution. Circumtropical, including the Mediterranean, but apparently absent from the islands of the central Pacific.

Genera. As recognized here, the Muraenesocidae includes the three genera *Muraenesox* McClelland, 1843, *Congresox* Gill, 1890, and *Cynoponticus* Costa, 1845. *Sauromuraenesox* Alcock, 1889 probably belongs here as well, but it has not been studied in detail.

The differences between *Muraenesox, Congresox,* and *Cynoponticus* are primarily in the degree of elongation of the snout and the form of the teeth. In *Congresox* the snout and jaws are markedly elongate. The anterior nostril is located a substantial distance behind the tip of the snout. The teeth are long and slender, round in cross section and lanceolate at the tips. *Muraenesox* has a somewhat shorter snout. The median vomerine teeth are massive, compressed, and triangular, with small basal cusps; the larger teeth on the mandible are similar in form. *Cynoponticus* has the shortest snout of the three genera. Its median vomerine teeth are also massive, compressed, and roughly triangular, but the edges are rounded in profile. The larger teeth on the mandible and maxilla are rather blunt.

Muraenesox and *Congresox* each contain two species, all found in the Indo-West Pacific. *Cynoponticus* has three species, one each in the eastern Atlantic, western Atlantic, and eastern Pacific. See Castle and Williamson (1975) for a summary of these three genera and their species.

Genus *Cynoponticus* Costa, 1845

Cynoponticus Costa, 1845: 1 (type species *Cynoponticus ferox* Costa, 1845, by monotypy).

Brachyconger Bleeker, 1864b: 22 (type species *Muraena savanna* Bancroft, 1831, by original designation).

Phyllogramma Pellegrin, 1934: 46 (type species *Phyllogramma regani* Pellegrin, 1934, by monotypy).
Cheiromuraenesox Fowler, 1944: 187 (type species *Muraenesox coniceps* Jordan and Gilbert, 1881, by original designation).

Characters. Body moderately stout, round in cross section anteriorly, compressed posteriorly; tail flexible but not attenuate; anus near anterior two-fifths of TL. Dorsal fin begins slightly ahead of base of pectoral fin. Pectoral fin large, well developed. Lateral line complete, but without simple pores; instead a series of short branches or diverticula, some of them simple pouches, others with secondary lobes, each with one or several small pores.

Snout moderately short, rather blunt, projecting beyond lower jaw. Fleshy tip of snout projects slightly beyond intermaxillary teeth. Mouth relatively large, rictus behind posterior margin of eye. Lips heavy, teeth largely concealed when mouth closed; no flange on upper or lower lip. Anterior nostril tubular, near anterolateral corner of snout, directed anterolaterally. Posterior nostril elliptical, in front of mid-eye. Eye well developed. Gill opening crescentic, nearly vertical, on lower half of side, greater than interbranchial space.

Teeth strong, heavy. Median vomerine teeth greatly enlarged, closely set, compressed, and blade-like, each with a large central cusp and smaller anterior and posterior cusps; teeth in middle of row largest, those at anterior and posterior ends smaller. Maxillary and mandibular teeth bluntly conical.

Size. To at least 2 m TL.

Distribution. Eastern Atlantic from the western Mediterranean to Angola; western Atlantic from the West Indies to Brazil; tropical eastern Pacific.

Etymology. From the Greek *kyon* or *kynos* (dog), *pontos* (the open sea), and *-icus* (belonging to). Masculine.

Species. Cynoponticus contains three species: *C. ferox* Costa, 1845 from the eastern Atlantic, *C. savanna* (Bancroft, 1831) from the western Atlantic, and *C. coniceps* (Jordan and Gilbert, 1881) from the eastern Pacific. They are distinguished by the number of preanal lateral-line pores (or diverticula), the number of vertebrae, and the length of the pectoral fin.

Cynoponticus savanna (Bancroft, 1831)

Figures 471–474

Muraena savanna Cuvier, 1829: 350 (nomen nudum).
Conger savanna Bancroft, 1831: 135 (original description, Jamaica, holotype unknown).
Conger brasiliensis Ranzani in Alessandrini, 1838: 433 (nomen nudum). Ranzani, 1839: 17, pl. 6, fig. 1a–b (original description, Brazil, holotype unknown); 1840: 79, pl. 13, fig. 1a–b.
?*Congrus curvidens* Richardson, 1844: 111 (original description, no locality given, holotype BMNH 1843.2.1.14).
Conger limbatus Castelnau, 1855: 83, pl. 43, fig. 3 (original description, Rio de Janeiro, Brazil, holotype MNHN 8458).
Muraenesox savanna. Kaup, 1856a: 75; 1856b: 117, pl. 15, fig. 74.
Brachyconger savanna. Bleeker, 1864b: 20.
Cynoponticus savanna. Castle and Williamson, 1975: 3.

Misidentifications:

Not *Muraenesox savanna* (Bancroft). Fowler, 1936: 277 (*Cynoponticus ferox*).

Discussion of Synonymy. The name *Muraena savanna* was given by Cuvier (1829: 350) in a footnote ("la *savanne* de la Martinique (*M. savanna*, N.)") as an example of "congres etrangers" with the dorsal fin beginning over or in advance of the pectoral-fin base. He gave no other characters, and the brief description is insufficient to distinguish the "savanne" from numerous other eels. Bancroft (1831: 135) described the "common *Sea-Eel* or *Conger* of Jamaica" and suggested that it was "perhaps identical to the *Savanne* of Martinique (*Muraena Savanna*, Cuv.)." Bancroft's description of the distinctive dentition leaves no doubt about the identity of his fish, and he must be considered the author of the name.

Ranzani's name *Conger brasiliensis* appeared three times within a three-year period. The first appearance (in Alessandrini, 1838) was simply in a list of names and must be considered a nomen nudum. The 1839 and 1840 references are essentially identical, and the description must date from 1839.

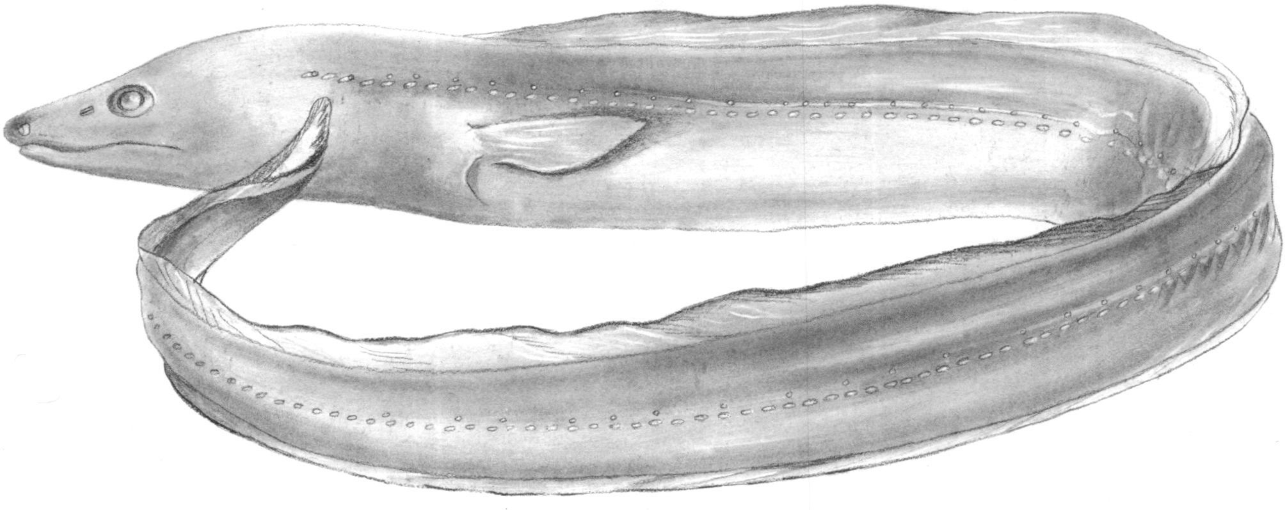

FIGURE 471. *Cynoponticus savanna*: ANSP 86695, 413 mm TL.

Distinctive Characters. According to Castle and Williamson (1975: 5), *Cynoponticus savanna* has fewer vertebrae than *C. ferox* (131–136 vs 140–147), more preanal lateral-line (LL) pores or diverticula (42–46 vs 36–40), and a shorter pectoral fin (29–37% HL vs 38–50). It differs from *C. coniceps* by having slightly more LL pores (42–46 vs 41–42) and a shorter pectoral fin (29–37% HL vs 40–67). The LL-pore character seems unreliable; the counts obtained here for *C. savanna* (ca 38–44) overlap those of the other two species. In the western Atlantic, *Cynoponticus savanna* is best recognized by the characters of the genus.

Description. Total vertebrae 132–135 (n = 6), predorsal vertebrae 9–10 (3), preanal vertebrae 44–46 (4). Branchiostegal rays 17 (1). LL pores ca 38–44 (11). Because of complex nature of lateralis system, counts of head pores cannot be given.

Proportions as % of TL: preanal 40–43 (6), predorsal 13–15 (5), head 15–16 (5), depth at anus 6–7 (4). Of preanal length: predorsal 30–36 (14), head 34–38 (14), depth at anus 11–17 (13), pectoral fin 11–15 (12). Of HL: snout 18–25 (13), eye 8–13 (14), interorbital 10–17 (10), snout-rictus 32–44 (12), tip of snout to tip of lower jaw 3–6 (5), gill opening 10–16 (14), interbranchial 6–10 (11), pectoral fin 29–41 (12).

Skin on head just before dorsal-fin origin with a reticular pattern of raised ridges, superficially almost scale-like in appearance; these ridges also prominent on top of snout, giving a sponge-like texture. Lips especially thick between anterior and posterior nostrils, bulging downward and inward near anterior end of maxillary teeth.

Intermaxillary teeth in a short patch, the anterior ones enlarged and conical, the posterior ones small; intermaxillary teeth separated from vomerine teeth. Vomer with a median series of large, heavy, compressed, blade-like teeth, largest in middle of series, smaller at anterior and posterior end; each tooth roughly triangular in profile, anterior edge convex, posterior edge more or less straight, with a large central cusp and smaller cusps anteriorly and posteriorly; the auxilliary cusps developed best on middle teeth, less apparent on anterior and posterior teeth. Main series of vomerine teeth flanked on each side by an irregular series of smaller, rather blunt teeth beginning behind anterior end of main series and extending slightly behind posterior end of main series. Maxillary teeth blunt, in roughly three series, those of middle series much the largest and arranged in a single row; teeth of outer

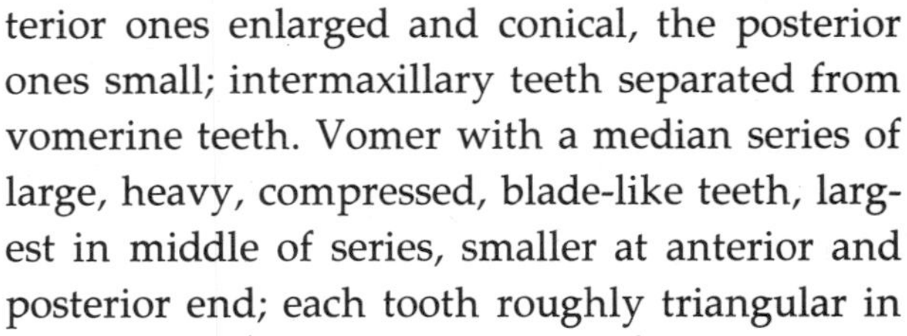

FIGURE 472. *Cynoponticus savanna*: ANSP 86695, 413 mm TL; head.

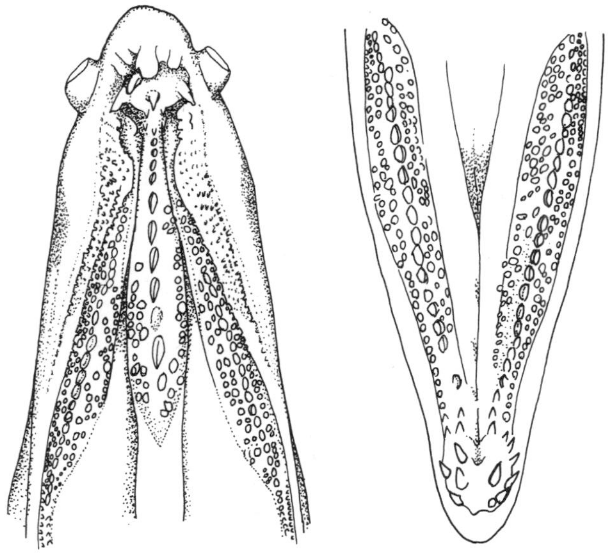

FIGURE 473. *Cynoponticus savanna*: ANSP 86695, 413 mm TL; dentition.

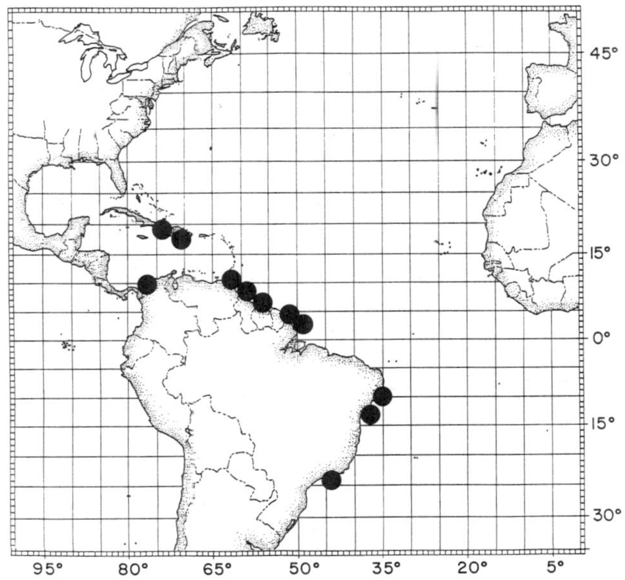

FIGURE 474. Distribution of *Cynoponticus savanna*.

series variable in size, in places arranged in two irregular rows; teeth of inner series also variable in size, arranged in two or three irregular rows. Mandibular teeth similar in form and arrangement to those of maxilla, with a middle series of moderately large, blunt teeth and an inner and outer series of smaller, irregularly bi- or triserial teeth; several teeth at anterior tip of mandible enlarged, conical, and recurved.

Stomach long, ends a short distance behind anus. Gas bladder slender, thick-walled, white; begins near junction of stomach and intestine, extends ⅔–¾ of distance to anus; broadest near anterior third, tapers to pointed posterior end.

Color. Gray or brown, lighter below. Vertical fins edged in black. Middle and distal part of pectoral fin dark. Stomach and intestine pale.

Size. Maximum size 1–2 m TL.

Growth and Development. The leptocephalus is unknown. Nair (1947) and Nair and Mohamed (1960a, 1960b) described the leptocephali of the related *Muraenesox cinereus* (Forsskål), *Congresox talabonoides* (Bleeker), and *Congresox talabon* (Cuvier) from Indian waters. These leptocephali are congrid-like, with a moderately elongate body and a rather long, convex snout; melanophores are present ventrally but not laterally. Such leptocephali have not been found in the western Atlantic.

Habits and Behavior. Judging by its formidable dentition, *Cynoponticus savanna* must be an efficient predator. Blache (1968a: 701) reported that the related *Cynoponticus ferox* feeds on small and medium-sized fishes, crustaceans, and even molluscs.

The species of *Muraenesox* and *Congresox* are important food fishes in the Far East. *Cynoponticus savanna* is less important in this respect, although it is caught and consumed locally. According to Castle (1978) it is taken mainly with bottom longlines and handlines at night, and also with bottom trawls. Robins (1978: 72) reported that it is marketed in the vicinity of the Gulf of Uraba, Colombia under the name "zafiro."

Remarks. The three Brazilian specimens (MNHN 8458, MZUSP 11927, MZUSP 11928) were examined by J. E. Böhlke at their respective institutions. Counts and measurements for these specimens were taken from his notes and included in the description above.

Four specimens for which counts and measurements were taken have since disappeared (Oregon 10242, Oregon 14476–14477, TABL 68-231, and GTS 1-11-2). At this point, I cannot explain what happened to them.

Distribution (Fig. 474). Through the West Indies from Cuba to Trinidad, and along the coast of Central and South America from Panama to Brazil at least as far as Rio de Janeiro. Depth less than 100 m.

Etymology. Latinization of *savanne,* the vernacular name at Martinique.

Study Material. 15 specimens, 288–569 mm TL.

WEST INDIES: USNM 33730 (1, 485), no other data. HISPANIOLA: ANSP 77109 (1, 396), Santo Domingo, 1928. ANSP 91759 (1, 515), Port-au-Prince, Haiti, 1949. VENEZUELA: ANSP 86695 (1, 413), Trinidad, La Brea Market, 13 May 1930. COLOMBIA: ANSP 114155 (1, cleared and stained), 9°37.5′N, 75°49′W, 37 m, 26 May 1964. UMML Oregon 10242 uncat. (1, 438, cleared and stained), 8°54′N, 76°33′W, 37 m, 29 Nov. 1968. GUYANA: USNM 185641 (1, 539), 8°10′N, 58°28′W, 51–53 m, 20 Sept. 1958. SURINAME: USNM 185626 (1, 539), 46–55 m, Sept. 1958, no other data. FRENCH GUIANA: ANSP 138432 (1, 510), 4°19′N, 50°51′W, 55 m, 23 May 1976. ANSP or UMML uncat. Oregon 14476–14477 (1, 353, missing), 5°00′N, 51°58′W, 37 m, 2 June 1974. BRAZIL: ANSP 136642 (1, 492), 3°15′N, 49°38′W, 77 m, 9 May 1975. MNHN 8458 (1, 340, holotype of *Conger limbatus* Castelnau), Rio de Janeiro. MZUSP 11927 (1, 366), Itapoa, Salvador, Bahia, 27 Jan. 1971. MZUSP 11928 (1, 288), Lagoa Managuaba, Maceio, 17 Oct. 1973. LOCALITY UNCERTAIN: TABL 68-231 (1, 569, missing).

Comparative material: Cynoponticus ferox: USNM 198741 (1, 530), Liberia. USNM uncat. GTS 1, 11-2 (1, 545, missing). *Cynoponticus coniceps:* USNM 268825 (1, 464), Panama, Bay of Guarachine, "7/12/67." USNM 268826 (2, 544–595), Panama, Gulf of San Miguel, Rio Congo, "7/12/67."

Family Nemichthyidae

SNIPE EELS

DAVID G. SMITH *and* JØRGEN G. NIELSEN

Acknowledgments. In our previous paper (Nielsen and Smith, 1978) we listed the people to whom we owe thanks in our studies of this group. To that list we add Drs. A. Post and F.-W. Tesch for the opportunity to join the cruise of the ANTON DOHRN from the Sargasso Sea to Europe, March–May, 1979.

Introduction. The Nemichthyidae is a small family of strongly modified midwater eels. They are characterized by their elongate body and the long, non-occlusible beak formed from the ethmovomer and the dentaries. The common name snipe eel derives from the resemblance of this beak to that of the long-billed shore bird of the same name. Curiously, the fully mature males lack the characteristic beak.

The family name Nemichthyidae dates from the Nemichthyina of Günther (1870: 21).

Methods and Terminology. Because of the nature of the jaws and frequent damage to the tail, total length and those characters associated with it are considerably less useful for snipe eels than for most other fishes. For this reason, we have used an artificial standard length that is not affected by broken jaws and regenerated tails. In *Nemichthys* this is the SL_{200}, defined as the distance from the posterior margin of the eye to the posterior end of the 200th vertebra. D_{200} and A_{200} are the dorsal- and anal-fin rays, respectively, anterior to this point. For *Labichthys* and *Avocettina,* which have many fewer vertebrae, the point chosen was the 100th vertebra, and the corresponding values are SL_{100}, D_{100}, and A_{100}. Because the lateral-line pores of *Labichthys* and *Avocettina* are easily counted, and because they are related to vertebrae on a one-to-one basis, we have used the 100th lateral-line pore as a more convenient measuring point for the SL_{100} in these genera. Predorsal and preanal measurements are taken from the posterior margin of the eye to the dorsal-fin origin and the anal-fin origin, respectively. Postorbital length is the distance from the posterior margin of the eye to the upper end of the pectoral-fin base. Infraorbital pores are listed as those below the eye plus those behind the eye (for example IO 7 + 3).

Characters. Body elongate to very elongate, moderately to strongly compressed. Tail moderately attenuate to filiform. Anus far forward, either under pectoral fin or less than one head length behind. Dorsal and anal fins long, confluent with caudal fin, anal fin higher than dorsal; dorsal fin begins over or slightly before pectoral fin; pectoral fin well developed. Scales absent.

Upper and lower jaws in females and immature males produced into a long, slender beak, the two jaws curving away from each other in the vertical plane toward their tips and becoming non-occlusible; upper jaw longer than lower; jaws short in fully mature males. Rictus under eye or slightly behind posterior margin of eye. Eye large and well developed. Anterior and posterior nostrils close together immediately in front of eye, anterior nostril anteroventral to posterior nostril. Anterior nostril without a tube or with a low tube in females and immature males, strongly tubular and anteriorly directed in mature males.

Lateralis system well developed on head and body (Fig. 475). Supraorbital canal extends some distance in front of anterior nostril, but first pore still much closer to nostril than to tip of snout. Adnasal pore present.

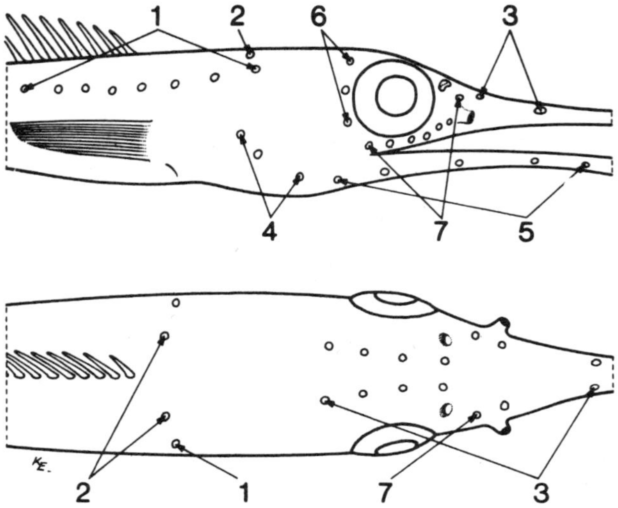

FIGURE 475. Head pores in Nemichthyidae (from Nielsen and Smith, 1978: fig. 3). 1. Lateral-line pores. 2. Supratemporal pores. 3. Supraorbital pores. 4. Preopercular pores. 5. Mandibular pores [4+5 = preoperculomandibular pores]. 6. Postorbital pores. 7. Suborbital pores [6+7 = infraorbital pores].

Teeth small with recurved tips, close-set in diagonal rows on maxilla, dentaries, and ethmovomer. Most of length of upper jaw consists of greatly elongate ethmovomer, which forms upper half of beak. Maxillary tooth band less than half length of toothed portion of ethmovomer. Dentaries produced to form lower half of beak, toothed for nearly entire length.

Osteology (from Trewavas, 1932; Beebe and Crane, 1937b; plus observations on an alizarin-stained specimen of *Nemichthys scolopaceus*, ANSP 115526). Neurocranium (Figs. 476–479) slender and delicate but relatively complete; ethmovomer greatly prolonged, anterior tip expanded into a small, spatulate knob in *Labichthys* and *Avocettina* but not in *Nemichthys*; orbit large. Frontals completely fused in specimen examined (Fig. 479); in literature, frontals described as fused anteriorly but divided posteriorly (Figs. 476–478, top).

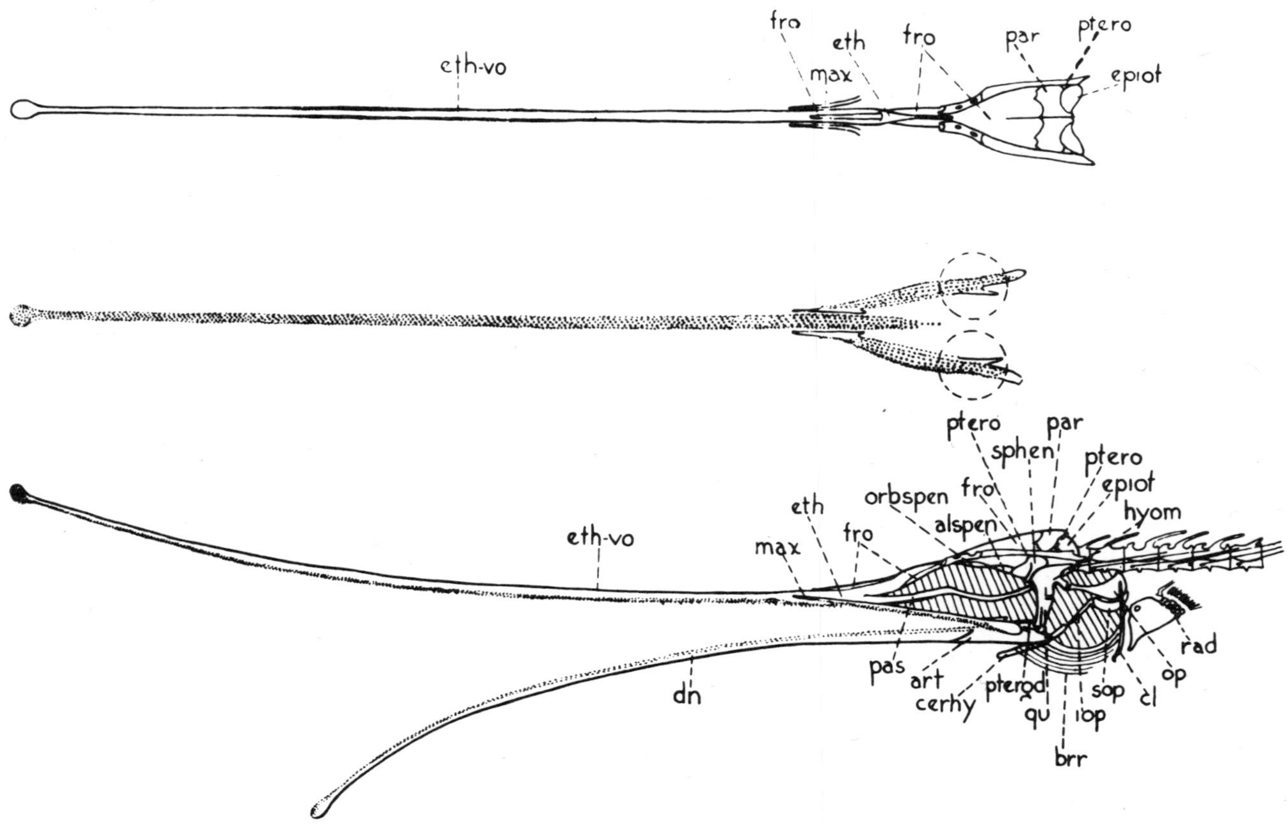

FIGURE 476. Skull of *Avocettina infans.* Top, dorsal view of neurocranium; middle, ventral view of ethmovomer and maxillae; bottom, lateral view of anterior skeletal elements in situ (from Beebe and Crane, 1937b: text-figs. 12–14). Abbreviations: alsphen = alisphenoid (basisphenoid), art = articular, brr = branchiostegal rays, cerhy = ceratohyal, cl = cleithrum, dn = dentary, epiot = epiotic, eth = ethmoid, eth-vo = ethmovomer, fro = frontal, hyom = hyomandibula, iop = interopercle, max = maxilla, op = opercle, orbsphen = orbitosphenoid, par = parietal, pas = parasphenoid, ptero = pterotic, ptergd = pterygoid; qu = quadrate, rad = radial (actinost), sop = subopercle, sphen = sphenotic.

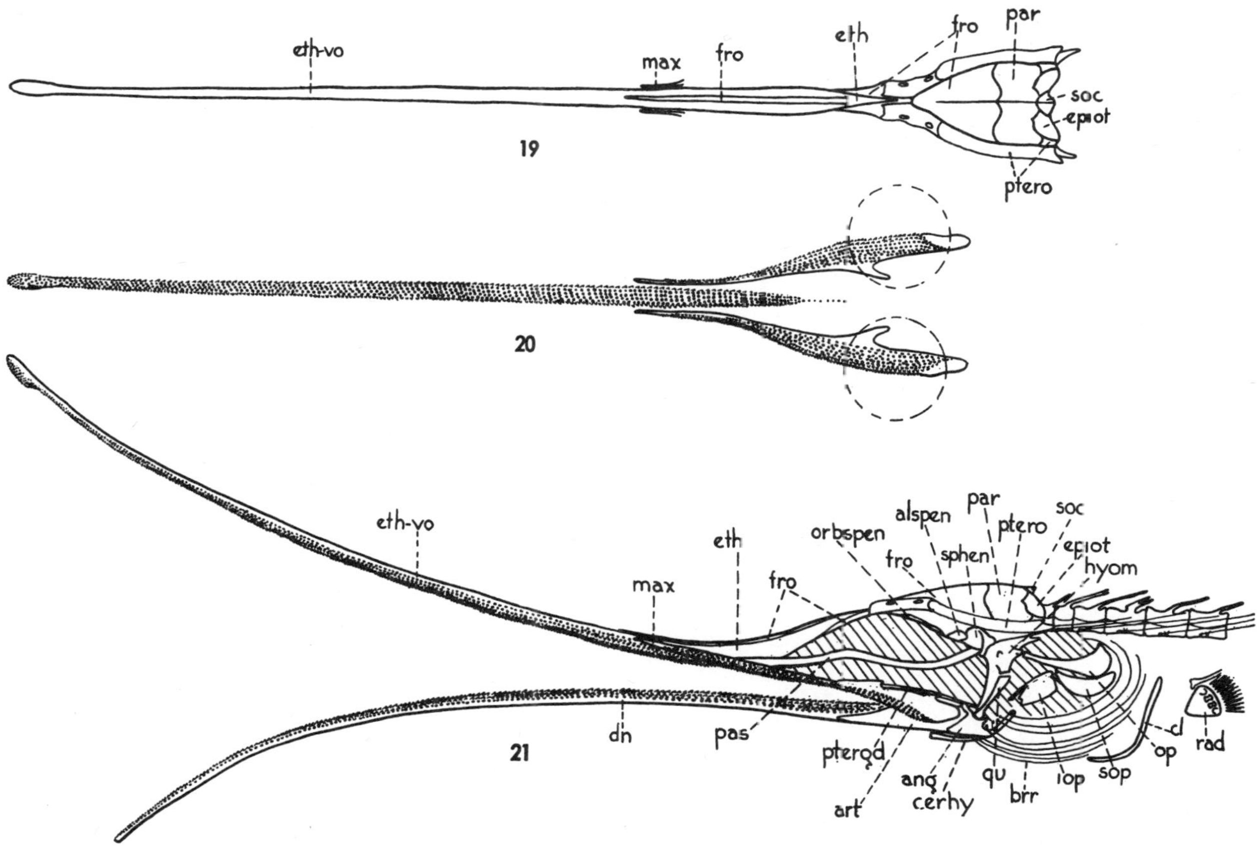

FIGURE 477. Skull of *Labichthys carinatus*. Top, dorsal view of neurocranium, middle, ventral view of ethmovomer and maxillae; bottom, lateral view of anterior skeletal elements in situ (from Beebe and Crane, 1937b: text-figs. 19–20). Abbreviations as in Figure 476.

Parietals fused in specimen examined (Fig. 479); Trewavas (1932: pl. 2) showed parietals fused in young specimen of *Nemichthys*, but Beebe and Crane (1937b: figs. 6, 12, 19) showed divided parietals in all three genera (Figs. 476–478, top). Supraoccipital present in *Labichthys* but not in *Avocettina* and *Nemichthys*. Basisphenoid present in *Avocettina* and *Labichthys* (Beebe and Crane, 1937b: text-figs. 14, 21, as alisphenoid) but absent in *Nemichthys* (Beebe and Crane, 1937b: text-fig. 8; also in specimen examined).

Maxilla (Fig. 480) relatively short, not reaching midpoint of ethmovomer, less than half length of mandible (Figs. 476–478, bottom); a shelf-like extension from medial side with postero-medial projection from posterior end; anterior end drawn out very finely, closely appressed to ethmovomer, toothed surface rotated outward anteriorly. Specimen of *Nemichthys* examined has a tendinous ossification between ethmovomer and maxilla (Fig. 481). Mandible slender and decurved, articular restricted to posterior extremity; anterior tip expanded into a small spatulate knob in *Labichthys* and *Avocettina* but not in *Nemichthys*; coronomeckelian appears absent.

Suspensorium (Fig. 482) vertical or inclined slightly forward; pterygoid small and weak, splint-like, approaching postero-medial projection of maxilla.

Opercle well developed in all three genera (Figs. 476–478, bottom; Fig. 482). According to literature, subopercle and interopercle well developed in *Labichthys*, present but somewhat reduced in *Avocettina*, absent in *Nemichthys*. Specimen of *Nemichthys* examined also lacks sub- and interopercle. Preopercle claimed to be absent in all genera, but specimen of *Nemichthys* examined has a slender, tubular structure in position of preopercle and carrying part of preopercular canal; this structure presumed to be preopercle (Fig. 482).

Hyoid arch reduced (Figs. 483–485). A forked

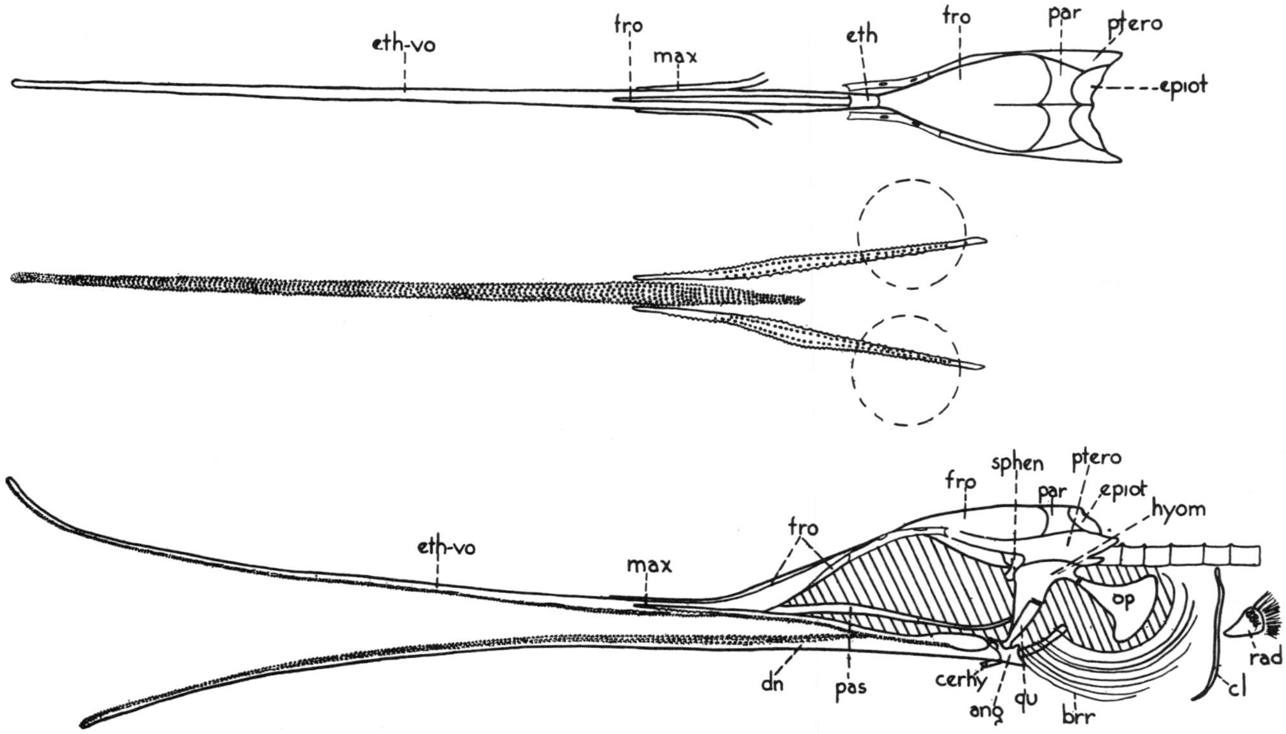

FIGURE 478. Skull of *Nemichthys scolopaceus*. Top, dorsal view of neurocranium; middle, ventral view of ethmovomer and maxillae; bottom, lateral view of anterior skeletal elements in situ (from Beebe and Crane, 1937b: text-figs. 6–8). Abbreviations as in Figure 476.

urohyal illustrated in *Labichthys* and *Avocettina*, and a glossohyal in *Avocettina* (Figs. 483, 484). Hypohyals also indicated in both genera, closely sutured to ceratohyal. In specimen of *Nemichthys* examined, glossohyal and urohyal absent. Ceratohyal and epiphyal slender, poorly ossified (Fig. 485A). Anterior end of ceratohyal cartilaginous except for ossified piece at anteriormost tip, which extends forward as a filamentous rod (Fig. 485B). Alternatively, this anteriormost piece might represent hypohyal, but no suture evident.

Gill arches relatively complete (Figs. 483, 484, 486). Five ceratobranchials present, usually ossified. Specimen of *Nemichthys* examined appears to lack basibranchials and hypobranchials, but Nelson (1966a: 394) listed one basibranchial and three hypobranchials (two ossified) for the genus. Beebe and Crane (1937b: text-figs. 15, 22) showed one basibranchial and two hypobranchials for *Avocettina*, two hypobranchials for *Labichthys* (Figs. 483, 484). Dorsal elements all present and well developed (Fig. 486).

Pectoral girdle with a slender cleithrum, broad scapula and coracoid, four actinosts (Figs. 476–478, bottom; Fig. 486). Scapula and coracoid embedded in a large cartilaginous plate. Supracleithrum absent.

Circumorbital bones in *Nemichthys* poorly ossified (Fig. 487). In specimen examined, nasal long and slender with numerous separate areas

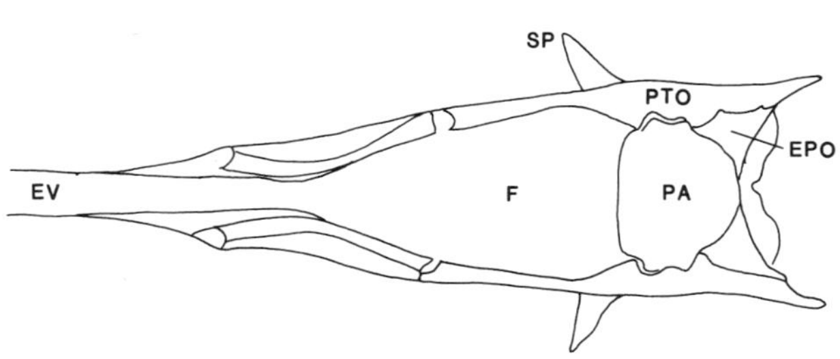

FIGURE 479. Neurocranium of *Nemichthys scolopaceus* in dorsal view, showing undivided frontals and parietals.

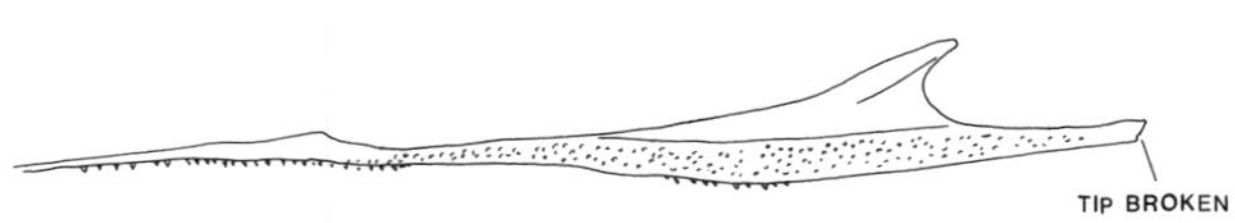

FIGURE 480. Maxilla of *Nemichthys scolopaceus*, left side in dorsal view.

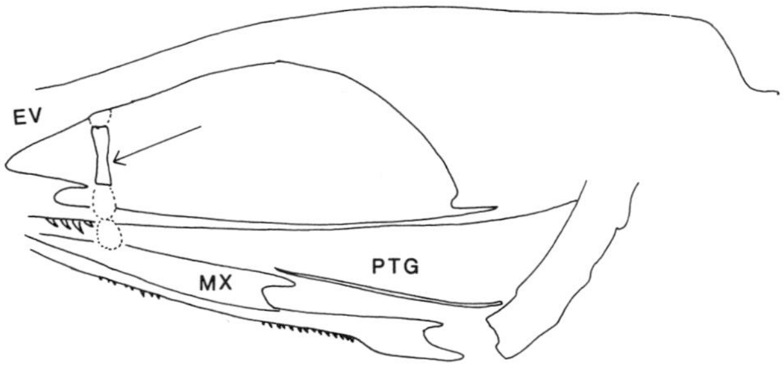

FIGURE 481. Orbital portion of skull of *Nemichthys scolopaceus,* showing tendinous ossification between ethmovomer and maxilla.

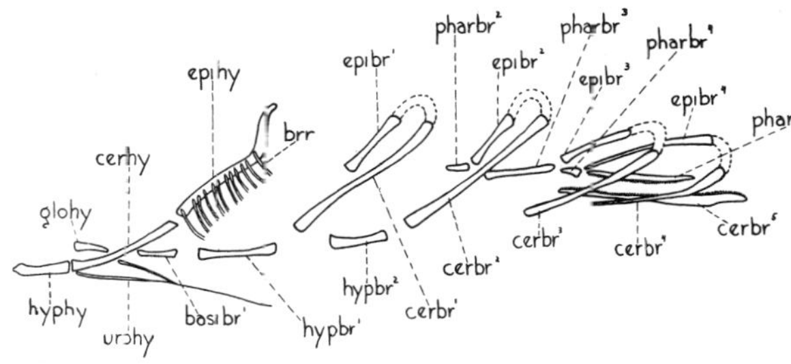

FIGURE 483. Hyoid and gill arches of *Avocettina infans* (from Beebe and Crane, 1937b; text-fig. 15). Abbreviations: basibr = basibranchial, brr = branchiostegal ray, cerbr = ceratobranchial, cerhy = ceratohyal, epibr = epibranchial, epihy = epihyal, glohy = glossohyal, hypbr = hypobranchial, hyphy = hypohyal, pharbr = pharyngobranchial (infrapharyngobranchial), urohy = urohyal.

of ossification. Several poorly ossified suborbitals behind eye. Adnasal bone not evident, but canal clearly visible, extending anterodorsally between anterior and posterior nostrils. Information on condition in *Avocettina* and *Labichthys* not available.

Vertebrae (Fig. 488) rather elongate; neural spines arise from middle of vertebra in *Avocettina* and *Labichthys,* from posterior end of vertebra in *Nemichthys.* Complex terminal vertebra with hypural plates in *Labichthys* and *Avocettina* but not in *Nemichthys.* In specimen of *Nemichthys* examined, two epineurals associated with skull, before first vertebra.

Sexual Dimorphism. Male snipe eels undergo a remarkable transformation when they reach full maturity. The characteristic beak disappears and the teeth are lost. The anterior nostrils enlarge into forwardly directed tubes, the pectoral fin is displaced posteriorly, the eyes enlarge, and in some species the color darkens. These changes are so great that for many years the ripe males were thought to belong to different genera and subfamilies from the females and immatures of the same species; some authors even placed them in different families. The reasons for these modifications are not completely clear, but their degenerative nature suggests that snipe eels die after spawning. The enlarged tubular nostrils indicate that olfaction is important in helping the male locate the female at spawning time. Many animals are known to give off chemical attractants, or pheromones, which influence the behavior of other individuals of their species. If female snipe eels produce such attractants, the males may be able to locate them by following scent trails or chemical gradients in the ocean. The loss of teeth is undoubtedly related to a cessation of feeding when sexual maturation begins. Freshwater eels (*Anguilla* spp.) are known to stop feeding when they begin their spawning migration.

Female snipe eels retain the beak and generally resemble the immature forms even at full maturity. Nevertheless, some modifications oc-

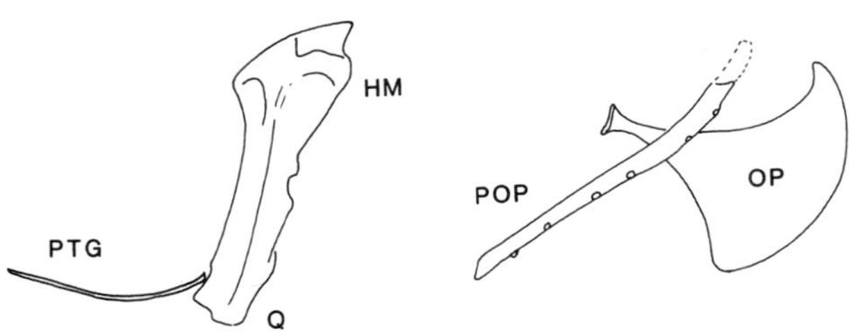

FIGURE 482. Suspensorium (left) and opercle and preopercle (right) of *Nemichthys scolopaceus.*

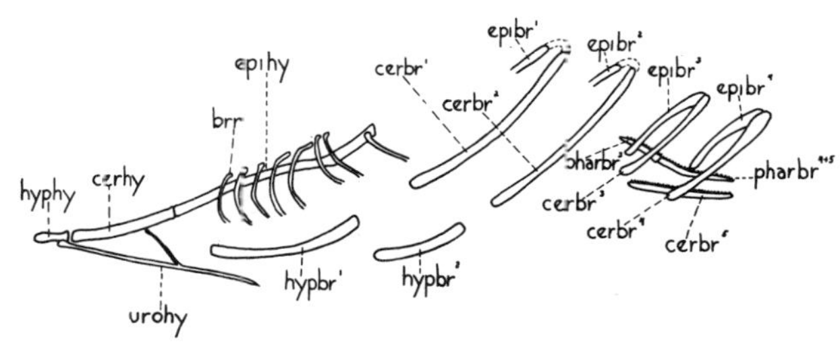

FIGURE 484. Hyoid and gill arches of *Labichthys carinatus* (from Beebe and Crane, 1937b: text-fig. 22). Abbreviations as in Figure 483.

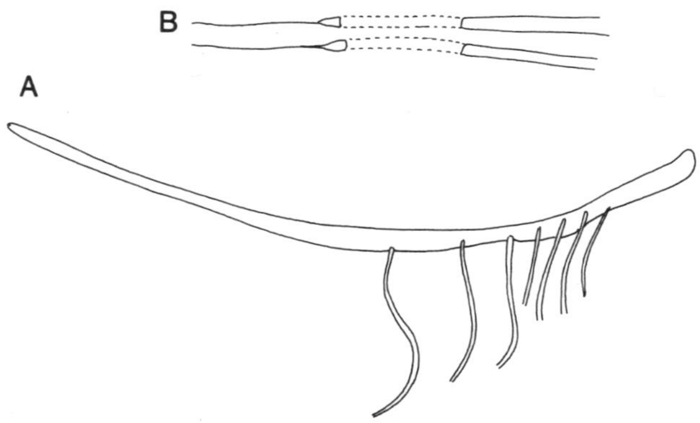

FIGURE 485. Hyoid arch of *Nemichthys scolopaceus*. A. Ceratohyal, epiphyal, and branchiostegal rays. B. Anterior end of ceratohyals in situ.

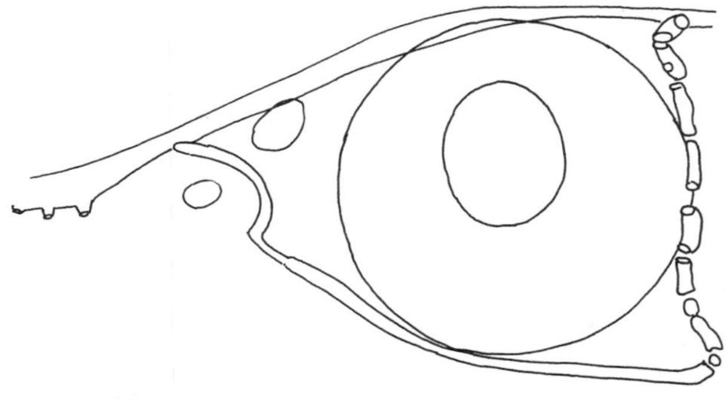

FIGURE 487. Circumorbital bones of *Nemichthys scolopaceus*.

cur. A running-ripe female *Nemichthys scolopaceus* (ISH 3176/79) seen by us on the ANTON DOHRN in the Sargasso Sea was toothless and uniformly black in color. Immatures of this species are countershaded, with the dorsal aspect relatively pale.

Habits. Snipe eels are wholly pelagic fishes, living their entire lives in the midwaters of tropical and temperate seas. Leptocephali are found relatively close to the surface, generally in the upper 100 m or so. After metamorphosis they descend to greater depths, although precise depth preferences are unknown. Adults have been captured from the surface down to 2000 m or more. Snipe eels have been observed from submersibles and are usually reported as hanging vertically and showing little movement.

From examination of stomach contents, it seems that snipe eels feed mainly if not exclusively on crustaceans. Frequently a bulge will appear in the ventral outline of the eel, caused by a large shrimp in the stomach (Nielsen and Smith, 1978: fig. 9). It has been suggested (Mead and Earle, 1970) that the long, non-occlusible beak serves to entangle the antennae of these shrimp. Since feeding has not been directly observed, this theory remains unproved. In a more general sense, the jaws are admirably adapted for grasping and swallowing small to moderate-sized prey. The backwardly directed teeth serve to hold the prey securely, allowing it to slide down the gullet but not to slide out.

Remarks. Some discrepancies were noticed between the osteology as described in the literature and in the specimen of *Nemichthys scolopaceus* examined here. Both Trewavas (1932) and Beebe

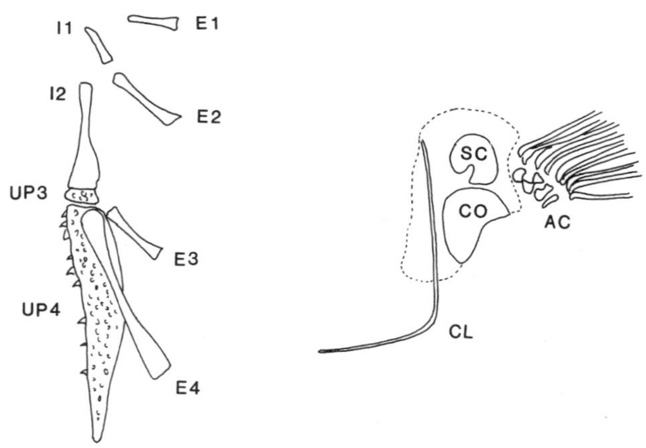

FIGURE 486. Gill arches, dorsal elements in dorsal view (left), and pectoral girdle (right) of *Nemichthys scolopaceus*.

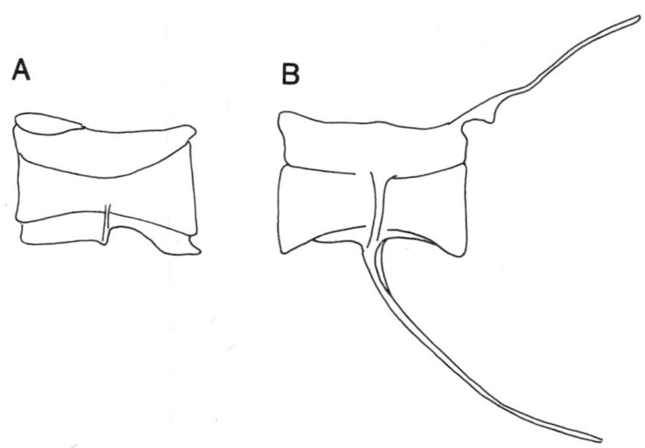

FIGURE 488. Vertebrae of *Nemichthys scolopaceus*. A. Precaudal vertebra. B. Caudal vertebra.

and Crane (1937b) showed the frontals united anteriorly but separated by a suture posteriorly. The specimen examined here, however, has completely fused frontals without a hint of a suture. Trewavas described the parietals of a young *N. scolopaceus* as fused, as they are in the specimen examined, but Beebe and Crane showed them separate in all three genera. The preopercle seems to be present in the specimen examined, but neither of the references cited mentioned it. The tendinous bone between the ethmovomer and the maxilla has not been described before.

Distribution. All tropical and temperate seas.

Genera. The Nemichthyidae is clearly divisible into two groups, one comprising the genera *Avocettina* Jordan and Davis, 1891 and *Labichthys* Gill and Ryder, 1883, and the other containing only *Nemichthys* Richardson, 1848. *Avocettina* and *Labichthys* have a moderate number of vertebrae, about 200, and the tail is not drawn out into a filament. The lateral line consists of a single series of pores, one per segment, and a series of elongate dermal ridges, presumably sensory in nature, is present on the head. *Nemichthys* has many more vertebrae, up to 750 or more, and the tail is drawn out into a slender, thread-like filament. The lateral line contains three rows of pores, and there are no dermal ridges on the head. All three genera are represented in the western North Atlantic.

Eight other genera have been recognized in the Nemichthyidae at one time or another, but all are synonyms of the three genera mentioned above.

KEY TO THE GENERA OF NEMICHTHYIDAE

1a. Anus well behind pectoral fin . . . . *Avocettina*, p. 447
1b. Anus below pectoral fin 2
2a. Caudal region not filamentous; one row of LL pores; dermal ridges on head . . . *Labichthys*, p. 450
2b. Caudal region filamentous; three rows of LL pores; no dermal ridges on head . . . . *Nemichthys*, p. 452

Genus *Avocettina* Jordan and Davis, 1891

Avocettina Jordan and Davis, 1891: 655 (type species *Nemichthys infans* Günther, 1878, by original designation).
Avocettinops Roule and Bertin, 1924: 63 (type species *Avocettinops schmidti* Roule and Bertin, 1924, by original designation).
Borodinula Whitley, 1931: 334 (substitute name for *Avocettina* Jordan and Davis, 1891).

Discussion of Synonymy. We previously (Nielsen and Smith, 1978: 22) discussed the confusing nomenclatural history of this genus and stated our reasons for believing that *Avocettina* Mulsant, Verreaux, and Verreaux, 1866 (a hummingbird) was a lapsus without nomenclatural standing and should not be regarded as preoccupying *Avocettina* Jordan and Davis, 1891. We follow that treatment here but acknowledge, as we did then, that the case is ambiguous and could be interpreted either way. Hubbs et al. (1979: 6) chose the alternative interpretation and used *Avocettinops*, the next available name.

Characters. Body elongate, strongly compressed, posterior end not drawn out into a long, slender filament; anus well behind pectoral fin. Total vertebrae about 176–216, precaudal vertebrae about 60–70. Dorsal fin originates behind pectoral-fin base in females and immature males, slightly before that point in mature males; bases of dorsal-fin rays at midbody relatively weak; anteriormost few dorsal-fin rays more close-set than remainder. Lateral-line pores in a single row, one per segment, equal in number to vertebrae. Upper jaw much longer than lower; tips of jaws expanded slightly to form a small, spatulate knob. Dermal ridges, possibly sensory in nature, behind eye, on occiput, behind anterior nostril, on snout, and dorsally on nape anterior to dorsal fin; most of these are longitudinally oriented, but a series of transverse ridges occurs on each side of occiput; dermal ridges longer than those of *Labichthys*. Anterior nostril in immature specimens with a small, laterally directed tube. Neural arches originate on middle of vertebral centra; intermuscular bones present. Color brown or black.

Size. To about 800 mm TL.

Distribution. All oceans.

Etymology. From *Avocetta* (the avocet, a bird with a long, curved bill) and *-ina* (having the nature of), in reference to the long, curved jaws. Feminine.

Species. This is an extremely difficult genus to understand taxonomically. In our previous paper (Nielsen and Smith, 1978) we recognized four species: *Avocettina bowersi* (Garman, 1899), *A. infans* (Günther, 1878), *A. acuticeps* (Regan, 1916), and *A. paucipora* Nielsen and Smith, 1978. *Avocettina bowersi*, an eastern and east-central Pacific endemic, is easily distinguished by its low eye/postorbital ratio, few close-set anterior dorsal-fin rays, and more posterior dorsal-fin origin. The other three species present severe problems. *Avocettina acuticeps* and *A. paucipora* are southern hemisphere forms. They are sympatric and are completely separable from each other by the number of dorsal- and anal-fin rays, lateral-line pores, and the predorsal length as a % of preanal length. Unfortunately, the northern hemisphere *A. infans* bridges the gap in each of these characters. Furthermore, *A. infans* varies substantially over its range. Specimens from the northeastern Pacific (Oregon to Alaska) have fewer lateral-line pores than specimens from other areas (although the overlap is considerable) and have at times been called a separate species, *A. gilli* (Bean). Specimens from the Indonesian area also have slightly fewer lateral-line pores, on the average, than specimens from other areas.

Karmovskaya (1982) synonymized *A. paucipora* with *A. gilli* and recorded the species from across the North Pacific. She has since recanted on the synonymization (pers. comm.), but she has recorded what she calls *A. paucipora* from the North Pacific. She has kindly allowed us to examine some of the specimens on which these records are based. Of particular interest is a specimen from the Bering Sea ("Equator" 17/III/69). In the number of lateral-line pores (189) and the predorsal/preanal ratio (30%), the specimen agrees with *A. infans*, but the number of dorsal- and anal-fin rays (ca 265 and 219, respectively) are typical of *A. paucipora*. We make no attempt to solve the problem here. More study is obviously needed, especially of specimens from the north-central and northwestern Pacific, which were not available to us for our earlier paper. In addition, the transitional zones between the northern and southern hemisphere species must be sampled more completely.

In the meantime, we note that the problems do not seem to involve the western North Atlantic, and we therefore refer all our material to *Avocettina infans*.

Avocettina infans (Günther, 1878)
Figures 489–491

Nemichthys infans Günther, 1878: 251 (original description, 5°48'N, 14°20'W, holotype BMNH 1887.12.7.259).

Labichthys elongatus Gill and Ryder, 1883a: 262 (original description, 39°22'N, 68°34'W, holotype USNM 33577).

Labichthys gilli Bean, 1890: 45 (original description, 55°20'N, 136°20'W, holotype USNM 44239).

Avocettina infans. Jordan and Davis, 1891: 655.

Avocettina gilli. Jordan and Davis, 1891: 655.

Avocettina elongata. Jordan and Evermann, 1898: 2802.

Leptocephalus oxycephalus (not of Pappenheim, 1914). Pappenheim, 1914: 190, pl. 9, figs. 3, 5, part (one or two North Atlantic paratypes only). Leptocephalus.

Avocettinops schmidti Roule and Bertin, 1924: 63 (original description, 23°13'N, 82°21'W, holotype ZMUC P31193). Mature male; see discussion below.

Borodinula infans. Whitley, 1931: 334.

Avocettina exophthalma Parr, 1932: 16, fig. 8 (original description, 23°42'N, 76°43'W, holotype YPM 2656).

Avocettinops normani Bertin, 1947: 55 (original description, western Indian Ocean, holotype BMNH 1939.5.24.581). Mature male.

Avocettinops sp. Parin et al., 1977: 137. Mature male.

Borodinula gilli. Karmovskaya, 1977: 196, part (some northern hemisphere specimens only).

Misidentifications:

Not *Avocettina infans* (Günther). Roule and Bertin, 1929: 22, part (specimens from Gulf of Panama are *A. bowersi*).

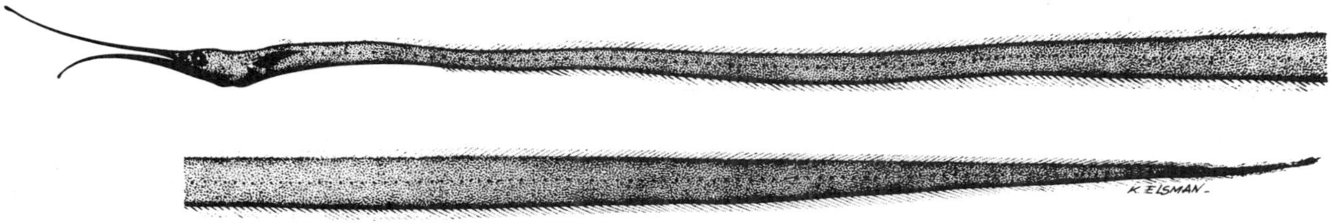

FIGURE 489. *Avocettina infans*, female: ZMUC P311076, 400 mm TL (from Nielsen and Smith, 1978: fig. 12).

Not *Borodinula infans* (Günther). Castle, 1961: 11 (species unidentifiable); 1964c: 73 (*A. acuticeps*).

Discussion of Synonymy. Roule and Bertin (1929: 30) presented *Avocettinops schmidti* as a new genus and species, but the names were actually introduced earlier (Roule and Bertin, 1924). In the 1924 paper, they gave a diagnosis of the genus *Avocettinops* but simply listed the species *schmidti* without a separate specific description. As *schmidti* was the only species included, the description of the genus must apply to the species as well. The name actually appeared as *Avocettina schmidti,* but this was most likely an error, because it was presented under the genus *Avocettinops* (Böhlke and Cliff, 1956: 95).

The holotype and four paratypes of *Leptocephalus oxycephalus* Pappenheim, 1914 came from the South Atlantic and are either *Avocettina paucipora* or *Labichthys.* The remaining two paratypes are from the North Atlantic; one of these is definitely *Avocettina infans* and the other is uncertain (Nielsen and Smith, 1978: 34).

Distinctive Characters. As the only species of *Avocettina* found in the North Atlantic, *A. infans* is best recognized by the characters of the genus. The following combination of characters separate it from its three extralimital congeners: predorsal lateral-line pores 5-8, close-set anterior dorsal-fin rays 9-15, eye 24-47% of postorbital, total lateral-line pores 181-201, predorsal 21-39% of preanal.

Description. Total vertebrae 187-202 (n = 13), preanal vertebrae 17-22 (6), predorsal vertebrae 5-9 (7), precaudal vertebrae 62-70 (19), prepectoral vertebrae 3-9 (7). Dorsal-fin rays 279-432 (91), anal-fin rays 240-372 (86), D_{100} 130-210 (154), A_{100} 103-176 (150), close-set anterior dorsal-fin rays 9-15 (94), first anal-fin ray below dorsal-fin ray number 23-43 (156), pectoral-fin rays 14-18 (96). Pores: total LL 181-201 (215), preanal LL 16-26 (239), predorsal LL 5-8 (175), prepectoral LL 3-8 (153), IO 6-10 (131) + 2-4 (135), SO 1 + 4-7 (127), POP 2-3 (103), STC 1 (111).

Proportions as % of TL: SL_{100} 47-58 (146). Of preanal length: predorsal 21-39 (241). Of postorbital length: eye 24-47 (194).

Color. Brown, with an intensification of pigment along lateral line. Mature males somewhat darker.

Size. To about 800 mm TL.

Development and Growth. The leptocephalus has been identified (see p. 929).

Variation. The holotype and only known specimen of *Avocettina exophthalma* Parr has an unusually large eye and interorbital space (Parr, 1932: fig. 8; Nielsen and Smith, 1978: pl. 1, figs. 3B, 4B), but we do not believe that it represents a distinct species. The specimen falls within the range of *A. infans* in all its meristic and mor-

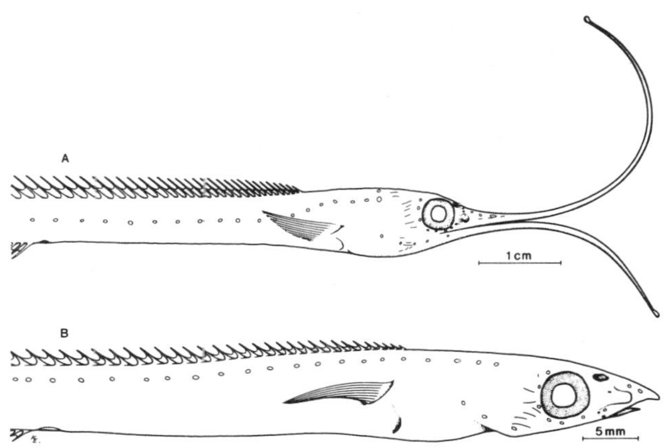

FIGURE 490. *Avocettina infans,* head and anterior part of body (from Nielsen and Smith, 1978: fig. 13). A. Female, ZMUC P311086, 420 mm TL. B. Mature male, ISH 790/68.

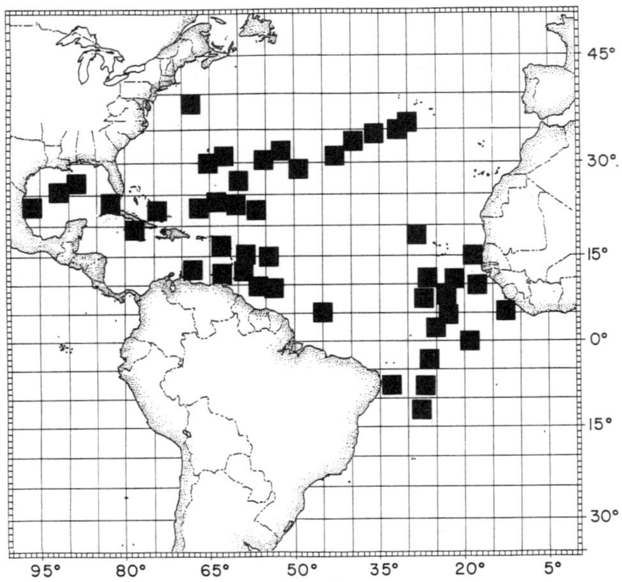

FIGURE 491. Distribution of *Avocettina infans.*

phometric characters, and aside from the bulging eyes there seem to be no other differences. In examining more than 300 specimens of *A. infans* we never found another specimen comparable to *A. exophthalma,* but we did find a specimen of *Nemichthys curvirostris* (Strömman) (ISH 473/74) that differs from typical *N. curvirostris* in the same way and to almost the same extent that *Avocettina exophthalma* differs from *A. infans* (Nielsen and Smith, 1978: pl. 1, figs. 5A, 6A). For these reasons we consider *A. exophthalma* to be simply an aberrant specimen of *A. infans.*

Prepectoral pores vary as follows: immatures 3–5 (127), mature females 4–6 (8), mature males 4–8 (18) (Nielsen and Smith, 1978: table 10).

Distribution (Fig. 491). Eastern and western Atlantic from about 40°N to 10°S, including the Caribbean and Gulf of Mexico but not the Mediterranean. Northern Indian Ocean. Western Pacific from about 30°N to 10°S; central Pacific around Hawaii from 30°N to 10°N; eastern Pacific from the coast of British Columbia to central Mexico, including the Gulf of California. Common.

Precise information on depth preference is difficult to obtain, as most of our study material was collected in open nets. In the Sargasso Sea in 1979 *Avocettina infans* was collected predominantly in tows that fished between 1200 and 2000 m. Only two specimens were collected at a lesser depth, 170 m (ISH 1336/79).

Etymology. Latin *infans* (immature). The significance is not clear; perhaps the fish had a larval or undeveloped appearance to Günther. The word apparently can be used as a noun or an adjective; the form is invariable.

Study Material. 364 specimens, 100–745 mm TL: 323 from Nielsen and Smith (1978: 62) plus 41 additional specimens listed below.

Holotype: BMNH 1887.12.7.259 (female, 280), off West Africa, 5°48′N, 14°20′W, 0–4572 m, Challenger 101, 19 Aug. 1873.

Other material: GULF OF MEXICO: ANSP 151955 (1). SARGASSO SEA: ISH 267/79 (2), 357/79 (3), 762/79 (1), 851/79 (1), 1139/79 (3), 1259/79 (3), 1336/79 (2), 1566/79 (3). CENTRAL ATLANTIC: ISH 1973/79 (8), 2113/79 (7), 2245/79 (4). ZMUC P311703 (1), P311704 (1), P311705 (1).

Genus *Labichthys* Gill and Ryder, 1883

Labichthys Gill and Ryder, 1883a: 261 (type species *Labichthys carinatus* Gill and Ryder, 1883, by subsequent designation of Jordan and Davis, 1892: 655).

Characters. Body elongate, strongly compressed, posterior end not drawn out into a long, slender filament; anus under pectoral fin. Total vertebrae 172–191, precaudal vertebrae 67–75; total lateral-line pores 170–191. Dorsal fin originates over pectoral-fin base in females and immature males, considerably before pectoral-fin base in mature males; basal portion of dorsal-fin rays at midbody relatively weak; anteriormost few dorsal-fin rays more close-set than remainder. Lateral-line pores in a single row, one per segment, equal in number to vertebrae. Upper jaw much longer than lower; tips of both jaws expanded to form a small, spatulate knob. Dermal ridges, possibly sensory in nature, behind eye, on occiput, medial to nostrils, behind anterior nostril, on snout, and dorsally on nape anterior to dorsal fin; these ridges shorter than those of *Avocettina.* Anterior nostril in immature specimens with a small, laterally directed tube. Neural arches originate on middle of vertebral centra; intermuscular bones present. Color uniform brown.

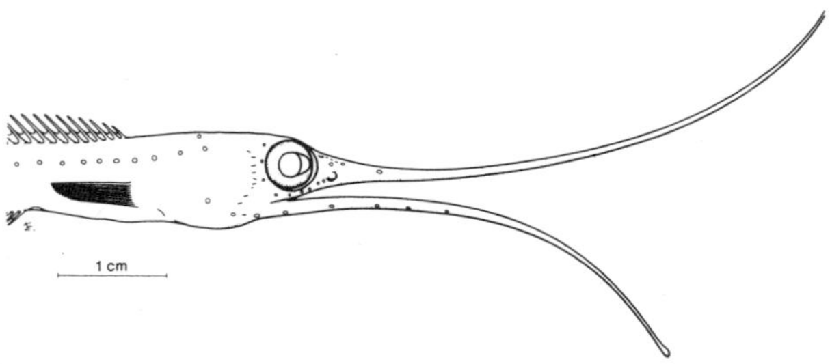

FIGURE 492. *Labichthys carinatus*, female: ZMUC P311002, 505 mm TL; lateral view of head (from Nielsen and Smith, 1978: fig. 6).

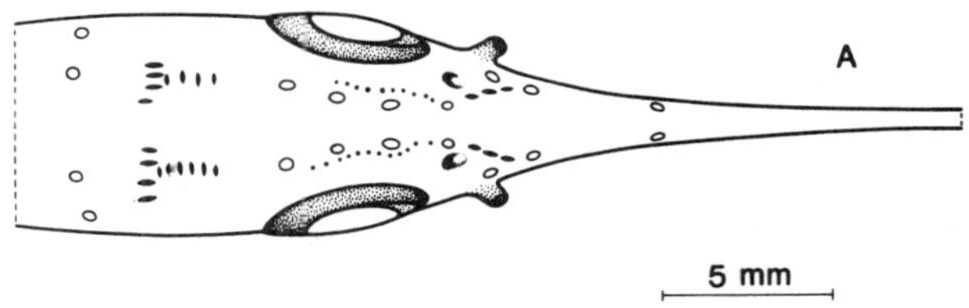

FIGURE 493. *Labichthys carinatus*, female: ZMUC P311002, 505 mm TL; dorsal view of head (from Nielsen and Smith, 1978: fig. 7A).

Labichthys closely resembles *Avocettina*, from which it is most readily distinguished by the more anterior anus.

Size. To about 800 mm TL.

Distribution. Irregularly distributed in all oceans. Records from Indian and North Pacific widely scattered; probably absent from eastern Pacific except extreme southern part.

Etymology. From the Greek *labis* (forceps) and *ichthys* (fish), in reference to the long, slender jaws. Masculine.

Species. Two species are recognized. *Labichthys carinatus* Gill and Ryder, 1883 occurs in the Atlantic, Indian, and central Pacific north of about 20°S, and *L. yanoi* (Mead and Rubinoff, 1966) lives in all oceans south of about 25°S.

Labichthys carinatus Gill and Ryder, 1883
Figures 492–494

Labichthys carinatus Gill and Ryder, 1883a: 261 (original description, 41°13′N, 66°00.5′W, holotype USNM 33369). Nielsen and Smith, 1978: 17.

Leptocephalus hjorti Blegvard, 1913: 136, figs. 1–3 (original description, 21°34′N, 29°50′W, holotype unknown). Leptocephalus.

Distinctive Characters. As the only species of *Labichthys* in the North Atlantic, *L. carinatus* is best recognized by the characters of the genus. The following characters distinguish it from the southern hemisphere *L. yanoi*: lateral-line pores without tubes; eye with aphakic space (asymmetrical extension of pupil in anterior direction, see Fig. 492); eye 33–44% of postorbital length; anterior nostril tube dark; head pores small (see Nielsen and Smith, 1978: 17, table 3).

Description. Total vertebrae 172–181 (n = 2), preanal vertebrae 9–11 (11), predorsal vertebrae 5–6 (11), prepectoral vertebrae 4–6 (11), precaudal vertebrae 67–70 (11). Dorsal-fin rays 239–361 (10), anal-fin rays 272–357 (9), D_{100} 114–180 (15), A_{100} 134–181 (14), first anal-fin ray below dorsal-fin ray no. 8–17 (12), pectoral-fin rays 12–14 (15). Pores: total LL 170–191 (9), preanal LL 8–13 (14), predorsal LL 4–6 (15), IO 7–8 (15) + 3–5 (17), SO 1 + 4–5 (13), POP 2 (16).

Proportions as % of SL_{100}: preanal 8–11 (14), predorsal 5–6 (15), postorbital 4–6 (13), eye 2 (15). Of postorbital: eye 33–44 (13).

Eye with an aphakic space anteriorly (an extension of pupil). Pores on head small; all pores without tubes.

Color. Uniformly light brown. Tube of anterior nostril dark.

Size. To about 800 mm TL.

Development and Growth. The leptocephalus has been identified (see p. 931).

Variation. The two specimens collected south of the equator (ISH 1227/68 and ZMUC P311001) have more total LL pores (185–191) than those from the northern hemisphere (170–181).

Distribution (Fig. 494). In the Atlantic, *Labichthys carinatus* has been recorded from the Gulf Stream off Georges Bank south through Bermuda, the Sargasso Sea, the Bahamas, and the West Indies to about 20°S off Brazil, and east through the central Atlantic to the northwestern coast of Africa. It occurs in the Caribbean and the eastern Gulf of Mexico but not in the Mediterranean.

Outside the Atlantic, records are much scarcer. We have seen one specimen from the Indian Ocean north of Madagascar (ZMUC P311001) and

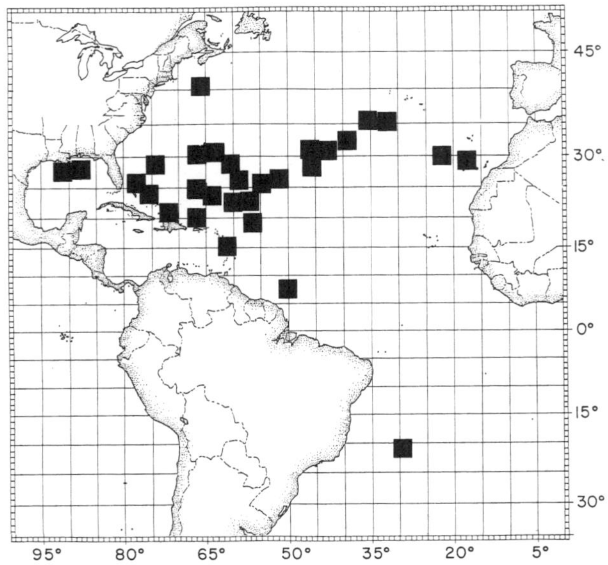

FIGURE 494. Distribution of *Labichthys carinatus.*

two from off Hawaii (USNM 219563 and FIDO IV; the latter cannot be located).

Labichthys carinatus has always seemed to be the rarest of all the snipe-eel species, but during the ANTON DOHRN cruise in the Sargasso Sea and central North Atlantic in 1979 it was one of the two most common species taken (along with *Avocettina infans*): 44 out of 138 adult and subadult specimens. By contrast, Bengtson (1973) found only one *L. carinatus* out of 1048 nemichthyid specimens collected in the Bermuda Ocean Acre. Perhaps *L. carinatus* has more precise ecological requirements than the other species and occurs in numbers only in certain areas. In the Sargasso Sea, this species was collected predominantly in tows that went below 1200 m.

Etymology. Latin *carinatus* (keeled), referring to a ridge or keel on the dorsal surface of the rostrum in the holotype. The ridge was caused by the skin shrinking and pulling tight around the posterolateral edges of the ethmovomer; it has no morphological or taxonomic importance. An adjective.

Study Material. 60 specimens (160–765 mm TL): 17 from Nielsen and Smith (1978: 61) plus 43 additional specimens listed below.

Holotype: USNM 33369 (1, 415), off Georges Bank, 41°13′N, 66°00′50″W, 0–1657 m, Albatross 2076, 4 Sept. 1883.

Other material: GULF OF MEXICO: ANSP 94206 (1), 94212 (1). BERMUDA-AZORES: ISH 268/79 (1), 356/79 (2), 472/79 (4), 599/79 (1), 671/79 (5), 673/79 (1), 852/79 (2), 954/79 (1), 1033/79 (1), 1140/79 (8), 1260/79 (3), 1438/79 (2), 1824/79 (1), 1923/79 (1), 1974/79 (1), 2246/79 (1), 2360/79 (1), 3017/79 (2). ZMUC P311706 (1), P311707 (1), P311708 (1).

Genus *Nemichthys* Richardson, 1848

Nemichthys Richardson, 1848b: 16 (type species *Nemichthys scolopacea* Richardson, 1848, by monotypy).

Leptorhynchus Lowe, 1852: 54 (type species *Leptorhynchus leuchtenbergi* Lowe, 1852, by monotypy), preoccupied by *Leptorhynchus* Clift, 1829 (Reptilia).

Belonopsis Brandt, 1854: 174 (substitute name for *Leptorhynchus* Lowe, 1852).

Investigator Goode and Bean, 1896: 518 (type species *Nemichthys acanthonotus* Alcock, 1894, by original designation).

Nematoprora Gilbert, 1905: 587 (type species *Nematoprora polygonifera* Gilbert, 1905, by original designation).

Cercomitus Weber, 1913: 54 (type species *Cercomitus flagellifer* Weber, 1913, by monotypy).

Paravocettinops Kanazawa and Maul, 1967: 2 (type species *Paravocettinops trilinearis* Kanazawa and Maul, 1967, by original designation).

Characters. Body greatly elongate, slightly compressed, posterior end drawn out into a long, thread-like filament; anus under pectoral fin. Vertebrae extremely numerous, 400–750 or more. Dorsal fin originates before pectoral-fin base; bases of dorsal-fin rays near midbody short and spine-like, although in undamaged specimens the rays continue distally in a delicate extension, increasing height of fin. Lateral-line pores in three longitudinal rows, the dorsal and ventral rows containing two pores for every pore in median row, resulting in a pattern of five pores per segment, in form of a rectangle with fifth pore in middle (Fig. 495). Upper jaw very slightly longer than lower; tips of jaws not expanded to form a knob. No dermal ridges on head. Neural arches originate on posterior part of vertebral centrum; intermuscular bones absent. Color variable, from dark brown or black through various degrees of countershading to completely pale.

Discussion. Nemichthys is the most specialized of the three nemichthyid genera and stands somewhat apart from *Labichthys* and *Avocettina.* The caudal filament and the three-rowed lateral line can be regarded as derived from the more typical caudal structure and single-rowed lateral line of *Labichthys* and *Avocettina. Nemichthys* lacks the dermal ridges on the head that are found in *Labichthys* and *Avocettina.* In *Nemichthys* the upper and lower jaws are nearly equal in length and the tips are not noticeably expanded, whereas in *Labichthys* and *Avocettina* the upper jaw is much longer and the tip of each jaw has a small, knob-like expansion. In *Nemichthys* the neural arch originates on the posterior part of the vertebral centrum, and intermuscular bones are absent; in *Labichthys* and *Avocettina* the neural arch originates on the middle of the centrum, and intermuscular bones are present. Finally, the basal portions of the dorsal-fin rays at mid-body are distinctly stronger and more spine-like in *Nemichthys* than in the other two genera.

Size. Specimens of *Nemichthys* species are known with total lengths in excess of 1 m, but much of this length consists of the caudal filament. The greatest SL_{200} we measured was 732 mm (*N. scolopaceus,* LACM 30379).

Distribution. All oceans. In the western North Atlantic, *Nemichthys* occurs nearly everywhere except the Arctic region.

Etymology. From the Greek *nema* (thread) and *ichthys* (fish), referring, no doubt, to the thread-like caudal region of the fish. Masculine.

Species. We recognize three species of *Nemichthys. Nemichthys scolopaceus* Richardson, 1848 is found in all oceans. *Nemichthys curvirostris* (Strömman, 1896) is found in the Atlantic, Indian, and South Pacific Oceans but seems to be absent from the North Pacific. *Nemichthys larseni* Nielsen and Smith, 1978 is restricted to the northeastern Pacific and Hawaii.

These are highly variable species, and identification is not always easy. The two species occurring in the western North Atlantic can be distinguished most easily by the color and the arrangement of pores on the head and lateral line. *Nemichthys scolopaceus* is either uniformly dark or shows a reverse countershading, dark ventrally and pale dorsally (Fig. 496). *Nemichthys curvirostris* is usually completely pale with a cluster of black spots below the stomach; a series of

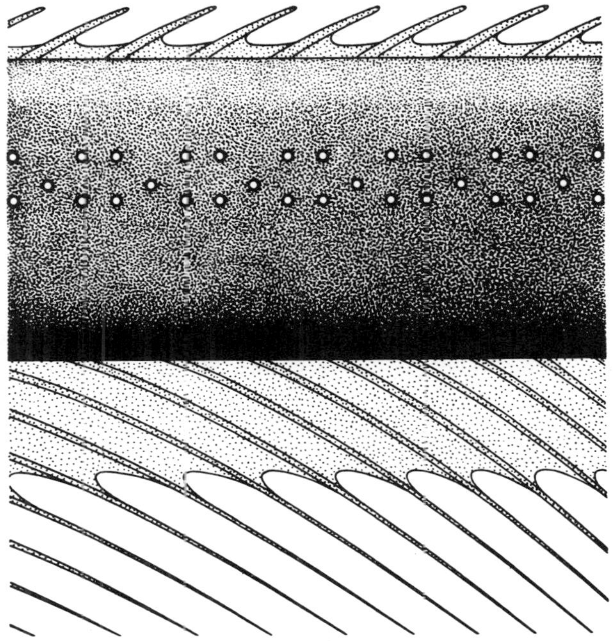

FIGURE 495. Semi-diagrammatic view of midbody of *Nemichthys scolopaceus,* showing pattern of lateral-line pores (from Nielsen and Smith, 1978: fig. 4A).

dark, subcutaneous, vertical bars between the vertebrae is evident the length of the body, and the end of the tail is black (Fig. 500A). At maturity, however, males of *N. curvirostris* darken. One semi-mature male of *N. curvirostris* observed in the Sargasso Sea (ISH 26/79) had the countershaded pattern typical of *N. scolopaceus.* In the Indo-West Pacific, *N. scolopaceus* frequently occurs in a pale color phase strongly reminiscent of *N. curvirostris,* but this pattern does not seem to occur in the Atlantic. In *N. scolopaceus* the rectangle formed by the lateral-line pores is nearly square (Fig. 495), whereas in *N. curvirostris* it is considerably longer than high. Specimens of *N. scolopaceus* from the Atlantic usually have more than 10 postorbital pores, whereas *N. curvirostris* usually has fewer than 10, although there is some overlap. As a consequence of their greater number, the postorbital pores in *N. scolopaceus* are arranged in a staggered row (Fig. 497) as opposed to an even row in *N. curvirostris* (Fig. 501). The number of preopercular pores shows almost no overlap in western North Atlantic populations of the two species; *N. scolopaceus* has 6–13 (Fig. 497) and *N. curvirostris* has 2–6 (Fig. 501). In addition, the teeth of *N. curvirostris* are distinctly larger than those of *N. scolopaceus.*

The mature male of *N. curvirostris* (Figs. 500B, 502) is smaller and more slender than that of *N. scolopaceus* (Fig. 498) and is pale brown rather than dark brown. In *N. curvirostris* the head has a more squared-off appearance and is differentiated more sharply from the slender body than in *N. scolopaceus.*

Nemichthys scolopaceus metamorphoses at a smaller size than *N. curvirostris* but grows larger at maturity.

Key to the Species of *Nemichthys* in the Western North Atlantic

1a. Jaws produced into a long, non-occlusible beak (females and immature males 2

1b. Jaws short and toothless (mature males) 3

2a. Body dark or countershaded; no dark, subcutaneous vertical bars; postorbital pores 6–17 (usually more than 10), arranged in a staggered row; preopercular pores 6–13; teeth smaller .... *scolopaceus*, p. 454

2b. Body pale with a cluster of black spots below stomach (semi-mature males may be countershaded like *N. scolopaceus*); dark, subcutaneous vertical bars between vertebrae; postorbital pores 5–14 (usually fewer than 10), arranged in an even row; preopercular pores 2–6; teeth larger .... *curvirostris*, p. 457

3a. Body dark brown; head more rounded in profile, not as sharply set off from body; postorbital pores 11–19, arranged in a staggered row; preopercular pores 6–13 .... *scolopaceus*, p. 454

3b. Body light brown; head more squared-off in profile, set off more distinctly from body; postorbital pores 5–8, arranged in an even row; preopercular pores 2–3 .... *curvirostris*, p. 457

Nemichthys scolopaceus Richardson, 1848
Figures 496–499, Table 44

Nemichthys scolopacea Richardson, 1848b: 25, pl. 10, figs. 1–3 (original description, South Atlantic off Brazil, holotype BMNH 1871.7.16.1).

Leptorhynchus leuchtenbergi Lowe, 1852: 54 (original description, Madeira, holotype lost).

Belonopsis leuchtenbergi. Brandt, 1854: 174.

Nemichthys avocetta Jordan and Gilbert, 1881: 409 (original description, Puget Sound, Washington, holotype USNM 27399).

Nemichthys acanthonotus Alcock, 1894: 136 (original description, Bay of Bengal, holotype ZSI F-13643).

Investigator acanthonotus. Goode and Bean, 1896: 518.

Nemichthys fronto Garman, 1899: 324, pl. 65, fig. 1 (original description, 7°31.5′N, 79°14.0′W, lectotype USNM 153591).

Nemichthys mediterraneus Ariola, 1904: 126 (original description, Gulf of Genoa, holotype MSNG 6822).

Nematoprora polygonifera Gilbert, 1905: 587, fig. 234 (original description, Hawaii, holotype USNM 51589).

Leptocephalus andreae Schmidt, 1912a: 50, pl. 3, fig. 9 (original description, North Atlantic, holotype ZMUC P31510). Leptocephalus.

Leptocephalus canaricus Lea, 1913: 40, pl. 6, no. 1 (original description, Canary Islands, syntypes BNM 2506, 5112). Leptocephalus.

Cercomitus flagellifer Weber, 1913: 54, pl. 6, fig. 1 (original description, 9°03′S, 119°01′E, lectotype ZMA 112625).

"Leptocéphale A." Roule and Bertin, 1929: 68 (North Atlantic). Leptocephalus.

"Tilurelle A." Roule and Bertin, 1929: 85, part (North Atlantic). Early metamorphic forms of both *Nemichthys scolopaceus* and *N. curvirostris.*

"Tilurelle B." Roule and Bertin, 1929: 89, part (North Atlantic). Late metamorphic forms of both *Nemichthys scolopaceus* and *N. curvirostris.*

Tilurella nemichthydis infantis (not of Günther, 1878). Borodin, 1931: 73.

Nemichthys infans (not of Günther, 1878). Borodin, 1931: 73.

Serrivomer sector (not of Garman, 1899). Borodin, 1931: 74.

Paravocettinops trilinearis Kanazawa and Maul, 1967: 3, figs. 1–3, pls. 1–2 (original description, Madeira, holotype MMF 21163). Mature male.

?*Tilurella gaussiana* Pappenheim, 1914: 187, pl. 5, fig. 3 (original description, Atlantic, holotype ZMB 19332). Metamorphic *Nemichthys*, probably *N. scolopaceus*, but description inadequate for positive identification and holotype in poor condition.

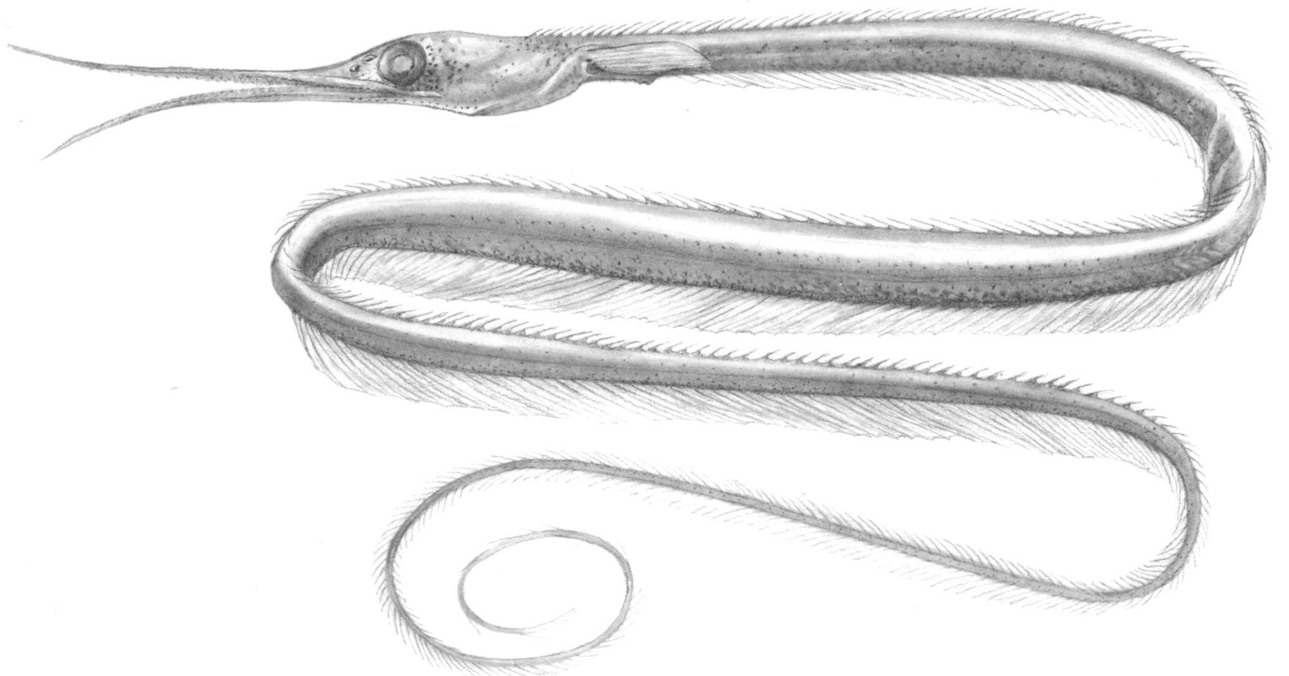

FIGURE 496. *Nemichthys scolopaceus,* female: ANSP 102474.

?*Nemichthys gaussiana.* Fowler, 1936a: 286 (after Pappenheim).

Distinctive Characters. Characters distinguishing *Nemichthys scolopaceus* from *N. curvirostris* are given in the account of the genus.

Description. Precaudal vertebrae 77–105 (n = 95), predorsal vertebrae 2–5 (81), preanal vertebrae 9–15 (79), prepectoral vertebrae 5–12 (89). D_{200} 170–253 (124), A_{200} 186–273 (121), pectoral-fin rays 10–14 (56). Branchiostegal rays 7–15 (76). Pores: IO 10–23 (44) + 3–20 (426), SO 1 + 6–21 (92), POP 2–18 (445), STC 2–8 (41).

Proportions as % SL_{200}: preanal 4–9 (85), predorsal 2–5 (116), eye 1–2 (120), postorbital 2–7 (116). Of postorbital: eye 18–51 (463), predorsal 37–96 (368).

Color. Immature specimens are generally countershaded, light above and dark below, with an intensification of pigment on the ventral midline below the stomach. The intensity of the color varies widely, from nearly uniform brown to very pale; Atlantic specimens are generally well pigmented, with prominent countershading. At full maturity both males and females become uniformly dark brown or black.

Size. To 1 m TL or more. The greatest SL_{200} measured was 732 mm (LACM 30379).

Development and Growth. The leptocephalus has been identified (see p. 926).

Variation. This is a highly variable species and several populations have been accorded species rank at one time or another. We have discussed this in detail elsewhere (Nielsen and Smith, 1978: 43). The North Atlantic population is relatively homogeneous with slight differences from east to west in a few characters.

The metamorphosis undergone by mature males of this and all other nemichthyid species has been described above. Females undergo similar though less extreme modifications. A running-ripe female collected in the Sargasso Sea in

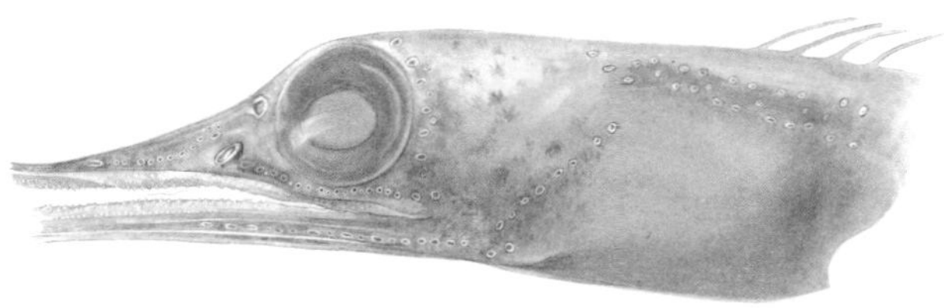

FIGURE 497. *Nemichthys scolopaceus,* female: ANSP 87975; head.

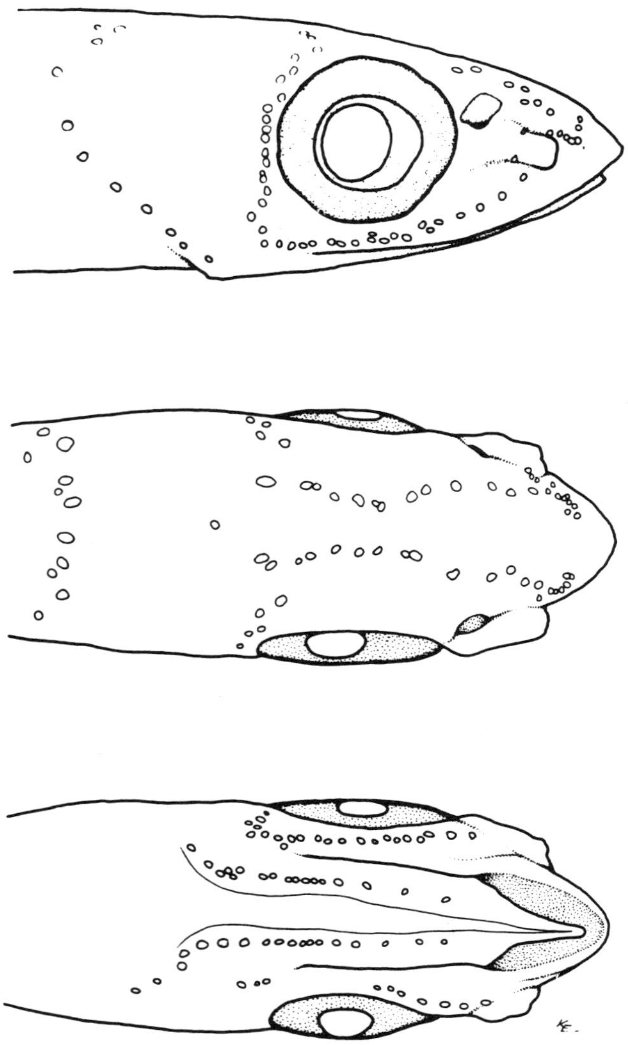

FIGURE 498. *Nemichthys scolopaceus,* mature male: ISH 1698/71; head in lateral (top), dorsal (middle), and ventral (bottom) views (from Nielsen and Smith, 1978: fig. 5B).

1979 (ISH 3176/79) illustrated this. The color was uniform velvety black. The jaws were still prolonged into a beak, but teeth were absent. The anterior nostril was tubular and directed anteriorly but was not as large as it is in the males.

Ontogenetic variations occur in several characters (Table 44).

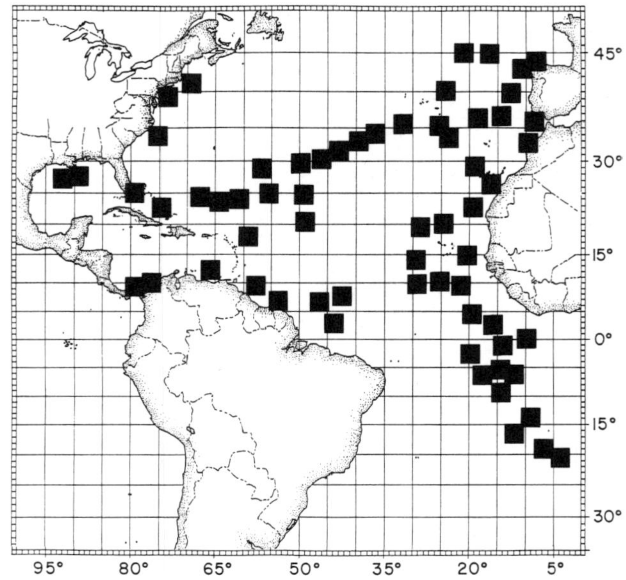

FIGURE 499. Distribution of *Nemichthys scolopaceus.*

Remarks. Ariola published two papers in 1904 concerning new fishes from the Mediterranean. Both papers were cited by Roule and Bertin (1929) and Beebe and Crane (1937b). The description of *Nemichthys mediterraneus* appeared in the second of these, published in the *Ann. Aquicoltura Lombarda.* Unfortunately, in our earlier paper (Nielsen and Smith, 1978: 59) we erroneously cited the other one (*Annali del Museo Civico di Storia Naturale di Genova*). The correct reference is given at the end of this volume.

Distribution (Fig. 499). Eastern and western North Atlantic to about 55°N, including the Caribbean, eastern Gulf of Mexico, and Mediterranean. Equatorial and South Atlantic to about 50°S, predominantly in the eastern part. Indian Ocean from the Arabian Sea and Bay of Bengal to about 50°S; western Pacific from Japan to New Zealand. Central Pacific from about 40°N to 20°S. Eastern Pacific from British Columbia to Peru. This is the most common and widespread snipe-eel species.

TABLE 44. Numerical values for four characters that show ontogenetic variation in *Nemichthys scolopaceus.*

	Prepectoral vertebrae	Postorbital as % SL_{200}	Predorsal as % postorbital	Eye as % postorbital
Females and immature males	5–9	3–6	60–95	25–50
Mature males	7–11	5–6	48–63	22–33

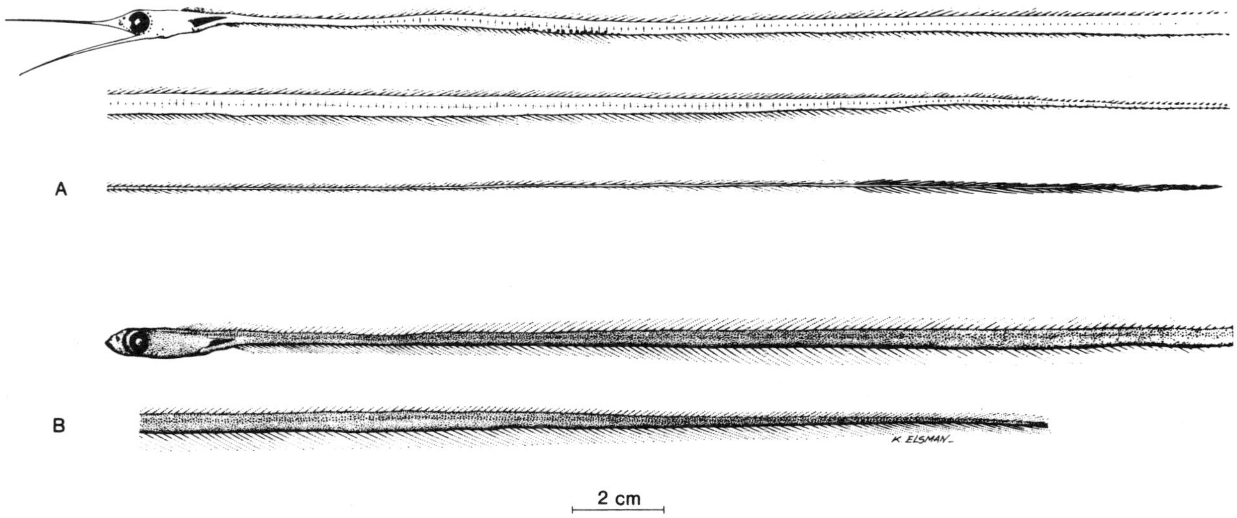

FIGURE 500. *Nemichthys curvirostris* (from Nielsen and Smith, 1978: fig. 30). A. Female, ISH 173/68. B. Mature male, ISH 228/68.

Nemichthys scolopaceus seems to occur at shallower depths than the other Atlantic nemichthyids. In the Sargasso Sea and the central and northeastern Atlantic in 1979, 11 of the 20 stations at which the species was collected were made at less than 1000 m (closing nets were not used). This compares to six of 17 for *Nemichthys curvirostis*, three of 22 for *Labichthys carinatus*, and one of 17 for *Avocettina infans.*

Etymology. From *Scolopax* (the snipe, a long-billed bird) and *-eus* (having the nature of), in reference to the long jaws. An adjective.

Study Material. 579 specimens (108–732 mm SL_{200}): 553 from Nielsen and Smith (1978: 66) plus 26 additional specimens listed below.

Holotype: BMNH 1871.7.16.1, South Atlantic off Brazil, HMS Samarang.

Other material: GULF STREAM: ANSP 87975 (1). SARGASSO SEA: ISH 96/79 (1), 200/79 (1), 850/79 (1), 1032/79 (1 mature male), 1138/79 (2). ZMUC P311722 (1). COLOMBIA: ANSP 115526 (1). CENTRAL ATLANTIC: ISH 1924/79 (1), 2114/79 (1), 2170/79 (1), 2248/79 (1), 3176/79 (1 mature female). ZMUC P311711–12 (2), P311713–14 (2), P311723–24 (1). NORTHEASTERN ATLANTIC: ZMUC P311715–16 (2), P311717–18 (2), P311719 (1), P311720–21 (2).

Nemichthys curvirostris (Strömman, 1896)
Figures 500–503

Leptocephalus curvirostris Strömman, 1896: 41, pl. 5, figs. 1–2 (original description, 22°N, 65°W, ZMUU 410); Bertin, 1936a: 9 (redescription with myomere counts). Leptocephalus.

Leptocephalus polymerus Lea, 1913: 27, fig. 17, pl. 3, fig. 5 (original description, North Atlantic, two syntypes ZMUB 5097). Leptocephalus.

Tilurus curvirostris. Pappenheim, 1914: 188.

Nemichthys scolopaceus (not of Richardson, 1848). Roule and Bertin, 1929: 9, part.

"Leptocéphale B." Roule and Bertin, 1929: 81 (North Atlantic). Leptocephalus.

"Tilurelle A." Roule and Bertin, 1929: 85, part (North Atlantic). Early metamorphic form of both *Nemichthys scolopaceus* and *N. curvirostris.*

"Tilurelle B." Roule and Bertin, 1929: 89, part (North Atlantic). Late metamorphic form of both *Nemichthys scolopaceus* and *N. curvirostris*).

Nemichthys sp. Parin et al., 1977: 136.

Nemichthys curvirostris. Nielsen and Smith, 1978: 47.

Distinctive Characters. Characters distinguishing *Nemichthys curvirostris* from *N. scolopaceus* are given in the account of the genus.

Description. Precaudal vertebrae 84–101 (n = 12), predorsal vertebrae 3–5 (17), preanal vertebrae 10–14 (13), prepectoral vertebrae 4–9 (22).

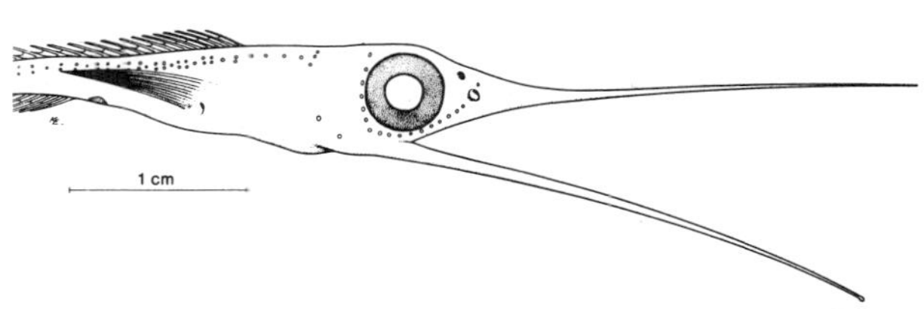

FIGURE 501. *Nemichthys curvirostris,* female: ISH 173/68; head (from Nielsen and Smith, 1978: fig. 31).

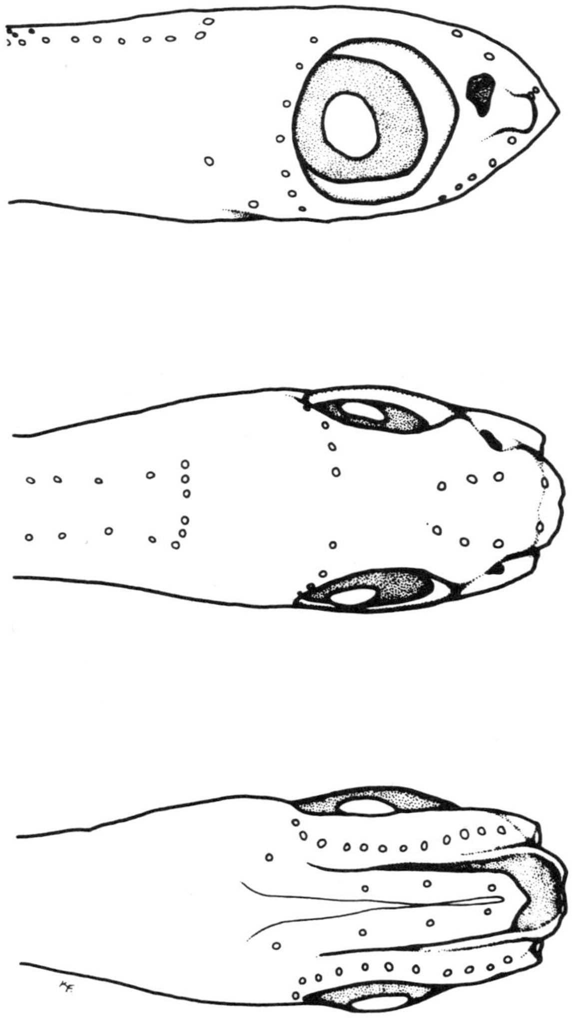

FIGURE 502. *Nemichthys curvirostris,* mature male: ISH 228/68; head in lateral (top), dorsal (middle), and ventral (bottom) views (from Nielsen and Smith, 1978: fig. 32).

D_{200} 157–210 (36), A_{200} 151–213 (33), pectoral-fin rays 8–12 (33). Branchiostegal rays 7–12 (20). Pores: IO 8–15 (14) + 5–14 (94), SO 1 + 4–13 (15), POP 2–6 (85), STC 1–4 (16).

Proportions as % of SL_{200}: preanal 10–26 (23), predorsal 4–17 (85), eye 2–7 (88), postorbital 4–18 (88). Of postorbital: predorsal 58–117 (106), eye 33–75 (108).

Color. Immatures: in preservative pale yellow to white with a cluster of large black spots beneath skin on abdomen; a variable number of small superficial black spots scattered along ventral part of body; brown pigment on jaws; dark vertical bars between vertebrae, located internally and not composed of discrete melanophores; posteriormost dorsal- and anal-fin rays black. Mature females: body grayish, ventral pigment more intense. Mature males: uniform light brown; vertical bars and ventral spots not evident. Semi-mature males: pale dorsally, flank covered with closely spaced brown chromatophores, giving a countershaded pattern similar to that of *N. scolopaceus.*

The following color notes were made from a freshly caught specimen (female) in the Sargasso Sea (ISH 3306/79). Caudal part velvety black, with long black fin rays connected by black skin. Portion of body immediately anterior to black caudal part pink from visible blood vessels; most of remainder of body translucent; head behind eyes and anteriormost part of body pink from underlying blood vessels. Black abdominal spots small and scattered. Vertical bars indistinct. Iris with metallic luster.

Size. The greatest TL we observed was 1430 mm (SAM 27474), the greatest SL_{200} was 508 mm (ANSP 126347). Although *Nemichthys curvirostris* metamorphoses at a larger size than *N. scolopaceus* (350 mm TL or greater vs ca 250 mm), it is smaller at maturity. It is a more slender species than *N. scolopaceus* and the caudal filament may constitute a greater percentage of the TL.

Development and Growth. The leptocephalus has been identified (see p. 927).

Variation. Nemichthys curvirostris shows less variation over its range than does *N. scolopaceus.* The population off the coast of Chile lacks the abdominal spots and has more anal-fin rays than other populations. Specimens from the eastern South Atlantic have more postorbital pores and slightly more preopercular pores than those from the North Atlantic (Nielsen and Smith, 1978: tables 34, 35). Dorsal-fin rays, the predorsal/postorbital ratio, and the eye/postorbital ratio also vary somewhat from place to place (Nielsen and Smith, 1978: tables 33, 36, 37), but on the whole *N. curvirostris* is relatively homogeneous.

Ontogenetic variation occurs in the following characters. Prepectoral vertebrae 4–7 (17) in females and immature males, 8–9 (5) in mature males. Predorsal length as % of postorbital 80–

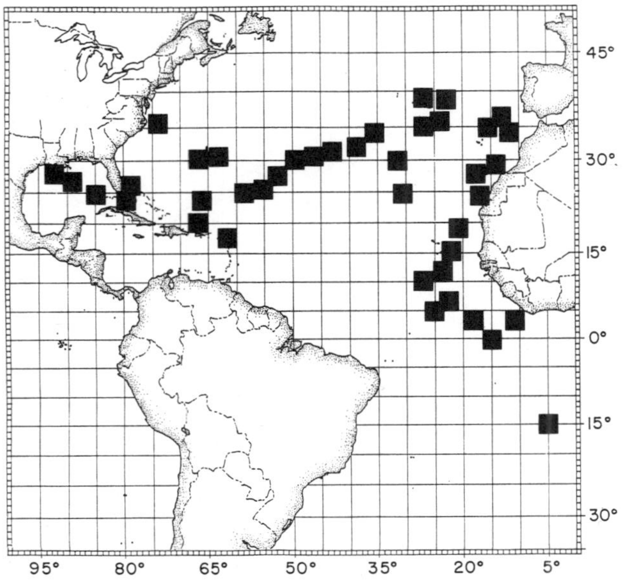

FIGURE 503. Distribution of *Nemichthys curvirostris.*

115 (55) in females and immature males, 70–80 (3) in mature males (Nielsen and Smith, 1978: table 32).

Distribution (Fig. 503). Eastern and western Atlantic from about 40°N to 40°S, including the Gulf of Mexico but not the Mediterranean; not recorded from the Caribbean or the coast of South America. Indian Ocean from about 10°N to 30°S. Indonesian area. Western and central Pacific from the equator to 50°S. Coast of Chile, about 30–35°S. Absent from the North Pacific and eastern tropical Pacific. Common in places but less abundant overall than *N. scolopaceus.*

In the Sargasso Sea and central Atlantic in 1979, this species was collected at somewhat lesser depths than were *Labichthys carinatus* and *Avocettina infans* but at greater depths than *Nemichthys scolopaceus.* Of 17 stations, six were made at less than 800 m, 11 at more than 1300 m. The extremes were 100 m and 2000 m, although closing nets were not used.

Etymology. From the Latin *curvis* (curved) and *rostrum* (beak, snout), referring to the slightly curved snout of the leptocephalus. An adjective.

Study Material 179 specimens (188–508 mm SL_{200}): 161 from Nielsen and Smith (1978: 69), plus 18 additional specimens listed below.

Holotype: ZMUU (leptocephalus, 238 mm SL), north of Virgin Islands, 22°N, 65°W, coll. Capt. Eckman.

Other material: SARGASSO SEA: ISH 26/79 (1 semimature male), 902/79 (1), 1137/79 (3), 1261/79 (1), 1755/79 (1), 2998/79 (1), 3074/79 (2). CENTRAL ATLANTIC: ISH 2247/79 (1), 2457/79 (3), 3305/79 (1), 3306/79 (1). ZMUC P311709 (1), P311710 (1).

Family Congridae

CONGER EELS

DAVID G. SMITH

Acknowledgments. Robert H. Kanazawa provided many of the counts and measurements used in the accounts of *Rhechias dubia* and *Xenomystax congroides.* I have elsewhere expressed my gratitude to him and to others (see Acknowledgments, family Anguillidae).

Introduction. The Congridae is one of the largest and most diverse of the eel families. Only the Ophichthidae may exceed it in number of genera and species. It has also been one of the most difficult to define and classify, largely because of the uniformity in external appearance of most members. With few strongly marked characters available, workers in the past were forced to base their classifications on such trivial features as fin position, dentition, and proportional measurements. This resulted in much confusion.

The first serious attempt to organize the Congridae was that of Ogilby (1898), who divided it into five genera: *Conger* Oken, *Uroconger* Kaup, *Congermuraena* Kaup, *Congrellus* Ogilby, and *Bathycongrus* Ogilby. Jordan and Hubbs (1925), studying Japanese species, added seven new genera based on differences in dentition. Parr (1932) demonstrated the weakness of the generic definitions when he tried unsuccessfully to identify a single specimen of a new congrid species. He showed that the criteria used to characterize genera were unreliable and subject to different interpretation. After an exhaustive analysis of the literature, he concluded that the existing classification was worthless and provisionally lumped a whole series of nominal genera into an expanded genus *Ariosoma* Swainson. The resolution of this group had to await the discovery of more fundamental characters.

A step in this direction was made by Reid (1934). Based on the presence or absence of a flange on the upper lip, the nature of the bony support of the infraorbital canal, and the dentition, he divided Parr's *Ariosoma* into five genera: *Ariosoma, Promyllantor* Alcock, *Congrina* Jordan and Hubbs, *Bathyuroconger* Fowler, and *Uroconger.* Wade (1946) expanded upon Reid's system and recognized four additional genera: *Chiloconger* Myers and Wade, *Alloconger* Jordan and Hubbs, *Thyreoconger* Wade, and *Rhynchocymba* Jordan and Hubbs. Wade misinterpreted *Ariosoma,* however, falsely attributing to it the presence of bony projections from the lower edge of the infraorbital canal. Matsubara (1955) also used these characters in revising the Japanese congrids and recognized eleven genera. Classification of congrid eels, however, had gone about as far as it could using only external features.

The foundation for the modern classification of the Congridae was laid by Asano (1962), who examined the osteology and other internal as well as external structures. He found numerous significant characters that allowed for the first time a natural arrangement of genera.

The family name Congridae was used by Kaup (1856a: 71). I am unaware of an earlier use, although I have not made an exhaustive search. Rafinesque (1810a) placed *Conger* in his "ordine" Anguillidi. The only congrid generic names introduced before 1856 were Swainson's (1838 and 1839) *Ariosoma* and *Ophisoma,* and I know of no family names ever based on them. Thus it seems impossible that there could be an earlier family name than Congridae.

Methods and Material. Counts and measurements follow Smith and Kanazawa (1977). All measurements are straight lines. Preanal length is measured from the tip of the snout to the posterior margin of the anus. Head length (HL) is from the tip of the snout to the upper end of the pectoral-fin base; in those species without a pectoral fin, it is measured to the upper corner of the gill opening. Eye diameter is measured horizontally across the clear "spectacle" covering the eye, and snout length is measured from the tip of the snout to the anterior margin of the spectacle. Depth is measured at the anus to reduce the effects of stomach contents. The gill opening is measured from corner to corner, and interbranchial distance between the lower corners of the gill openings across the isthmus.

Total vertebrae include the hypural as a single vertebra. The first vertebra with a closed hemal arch is considered the first caudal vertebra; because the division between precaudal and caudal vertebrae is difficult to determine from radiographs, precaudal vertebral counts were usually determined only for the cleared and stained specimens.

Unless otherwise stated, lateral-line (LL) pores are those in front of a vertical line through the anterior edge of the anus.

For taxonomic purposes, the two small pores at the tip of the snout are treated as supraorbital (SO) pores, although the first one really belongs to a separate ethmoidal canal and the second appears to connect with both the ethmoidal and supraorbital canals (Kanazawa, 1958). The SO pores are expressed in the formula 1 + X, where the 1 represents the ethmoidal pore proper and X the remainder. Individual SO pores are numbered as follows. Numbers 1 (the ethmoidal pore) and 2 (joint ethmoidal and SO) are located anterior to the anterior nostril at the tip of the snout. Number 3 is above or slightly behind the anterior nostril and is the first to belong exclusively to the SO canal. Number 4 is between the anterior and posterior nostrils, number 5 is above the anterior margin of the eye, and number 6 is above the posterior margin of the eye. Numbers 5 and 6 will often be referred to as interorbital pores.

The infraorbital (IO) pores are expressed as the number along the upper jaw, plus the number in line with these behind the rictus, plus the ones in the section of the canal that curves up behind the eye (for example, "4 + 1 + 3"). Individual IO pores are numbered as follows. Number 1 is the adnasal pore (see below under comparative osteology, lateralis system); in *Paraconger* it is small and located on the posterior side of the tube of the anterior nostril; in *Conger* it is larger and located immediately adjacent to the third SO pore; most other genera lack it. Number 2 is located on the lip just behind the anterior nostril. Numbers 3 and 4 are located on the lip between number 2 and the level of the posterior nostril. Number 5 is located under the anterior part of the eye, number 6 is on a line with the preceding pores but behind the rictus (except in *Xenomystax,* which has a long jaw), and numbers 7–9 are located behind the eye. Numbers 7–9 will often be referred to as postorbital pores.

The preoperculomandibular (POM) and supratemporal (STC) pores are given as the total number in the canal.

The terminology of bones is basically that of Asano (1962), with some exceptions. His orbitosphenoid is here considered a basisphenoid, and his alisphenoid is here called a pterosphenoid. The palatopterygoid is treated here as a pterygoid. In the pectoral girdle, the terms scapula and coracoid are used instead of hyper- and hypocoracoid, and cleithrum and supracleithrum are used instead of clavicle and supraclavicle. Myorhabdoi are the intermuscular bones located above the epineurals and below the epipleurals. Terminology of gill arches follows Nelson (1966a).

Representatives of the following species were cleared and stained with alizarin for osteological studies. *Ariosoma balearicum* (UMML uncatalogued, Pillsbury 687), *Ariosoma selenops* (UMML 20964), *Ariosoma anale* (UMML 29605), *Parabathymyrus oregoni* (UMML 22297), *Paraconger caudilimbatus* (UMML 9422), *Conger oceanicus* (UMML 5218), *Conger triporiceps* (UMML 9243), *Gnathophis bracheatopos* (UMML 32627), *Gnathophis bathytopos* (UMML 22591), *Gnathophis tritos* (UMML 29613), *Rhynchoconger flavus* (from

UMML 28787 and 28788), *Rhynchoconger gracilior* (UMML 9971), *Rhechias dubia* (UMML uncat. Gerda 1016), *Rhechias thysanochila* (UMML uncat. Pillsbury 739), *Rhechias vicinalis* (UMML uncat. Silver Bay 1184), *Bathyuroconger vicinus* (UMML 16945 and 22582), *Uroconger syringinus* (UMML 22296), *Japonoconger caribbeus* (from UMML 28785), *Pseudophichthys splendens* (from UMML 16079 and 16729), *Acromycter perturbator* (from UMML 13901), *Heteroconger halis* (ANSP 116443 and 156858), *Xenomystax bidentatus* (UMML uncat. Oregon 10831), *Xenomystax congroides* (ANSP 156859), and *Blachea xenobranchialis* (MNHN 1979-2). Information on the osteology of *Gorgasia punctata* was obtained from Rosenblatt (1967).

Characters. Relatively generalized eels, most without unusual modifications. Body variable, moderately stout to elongate, most species moderately elongate. Anus usually between about midbody and anterior quarter or fifth of TL. Tip of tail variable, from stout and stiffened to almost filamentous. Dorsal and anal fins present, confluent with caudal fin; dorsal fin begins near level of gill opening, sometimes slightly ahead or behind, but always closer to gill opening than to anus. Caudal fin usually present, but greatly reduced in some heterocongrines. Pectoral fin usually present, but reduced in heterocongrines. Gill opening on side of head, in contact with pectoral fin when latter present. Scales absent.

Head and snout variable, from long and slender to short and pug-nosed, but usually moderate. Jaws equal or upper jaw longer; in heterocongrines, lower jaw longer. Most species with well developed lips that conceal maxillary and mandibular teeth when mouth closed; intermaxillary teeth often exposed. Upper lip usually divided by a groove into inner and outer portions; outer portion more or less smooth; inner portion with papillose ventral edge, lying flap-like along lateral side of maxillary tooth patch. Lower lip undivided. An upturned flange on outside of upper lip of some genera; a similar flange on lower lip of many genera. Eye large, well developed. Anterior nostril tubular, near tip of snout; posterior nostril round or elliptical, in most species located on side of head in front of eye, but sometimes high on head or low near edge of lip.

Teeth variable in size and form, from granular patches to fang-like canines. Teeth present on ethmovomer, maxilla, and dentary.

Lateral line complete to tail; usually a single row of simple pores, one per vertebra, but modifications occur in some genera. On head, pores present in all canals except temporal, although some species lack supratemporal pores. Frontal commissure absent.

Usually drab in color, brown or gray, dorsal and anal fins commonly edged in black. Some heterocongrines boldly patterned.

Comparative Osteology. Neurocranium (Figs. 504–506): The front of the neurocranium is formed by the massive ethmovomerine complex, which varies in form among the various genera, from short and deep to long and slender. The genera *Ariosoma, Bathymyrus, Parabathymyrus, Paraconger,* and *Gorgasia* have a strut-like lateral ethmoid process just in front of the orbit, against which the maxilla abuts. The sides of the vomer bear a pair of short processes that are closely connected to the anterior end of the pterygoid. Dorsally, the posterior portion of the ethmoid overlaps the massive fused frontal, which forms the entire dorsal boundary of the orbit and the anterior part of the brain case. The frontal is grooved above the posterior portion of the orbit to receive the supraorbital canal. The parietals are large and paired. The posterodorsal edge of the skull is formed by the epiotics and, usually, the median supraoccipital. The supraoccipital, however, is commonly reduced and is absent in *Ariosoma, Bathymyrus, Parabathymyrus, Congriscus,* and *Pseudophichthys.* There is usually a transverse ridge on the epiotics that forms the posterior edge of the skull. The posterior surface of the skull is composed of the epiotics dorsally, the pterotics laterally, the exoccipitals medially, and the basioccipital ventrally. The foramen magnum is enclosed by the exoccipitals and the basioccipital.The floor of the skull is formed by the basioccipital and the elongate parasphenoid, which extends anteriorly over the dorsal side of the vomer about to the anterior corner of the orbit. Otic bullae are present, but in some genera, such as *Conger,* they are greatly reduced. The lateral wall of the skull is formed by the pterosphenoid, sphenotic, prootic, and pterotic. The elongate pterotic is hollow and houses the temporal canal. *Gnathophis* has the pterosphenoid, sphenotic, and prootic fused into a unit. In *Uroconger* the sphenotic and prootic are fused, but the pterosphenoid remains separate. The orbit is large in congrids; at the front is an irregularly shaped cartilage which apparently aids in supporting the eyeball.

Jaws (Figs. 507, 508): The maxilla forms most of the border of the upper jaw. A well developed pedicel is present anteriorly, which articulates with the skull just behind the intermaxillary area. The maxilla of *Heteroconger* is arched outward posteriorly. The lower jaw consists of the dentary, articular, and a small coronomeckelian visible on the medial side.

Suspensorium (Fig. 509): The hyomandibula and the quadrate are closely articulated into a strong, forwardly-inclined, trapezoidal structure. Anteriorly it fits into a socket formed by the sphenotic and prootic, posteriorly into a socket in the pterotic. The opercle articulates with the hyomandibula just below the posterodorsal corner. The epihyal is connected to the posterior part of the hyomandibula by a strong ligament. The pterygoid is attached to the anterodorsal corner of the quadrate; sometimes it articulates with part of the hyomandibula as well. Anteriorly it is closely connected to the lateral process of the vomer by a strong ligament. The forward inclination of the suspensorium is greatest in those genera with a small mouth, such as *Pseudophichthys, Acromycter,* and especially *Heteroconger.*

Hyoid arch (Figs. 510, 511): The hyoid apparatus consists of a median glossohyal and urohyal, paired ceratohyals and epihyals, and about 7–12 branchiostegal rays. Usually one or two branchiostegals are joined to the ceratohyal and the rest to the epihyal. The branchiostegals of the two sides do not overlap ventrally.

In *Ariosoma,* the urohyal is trifurcated posteriorly into two long lateral processes and a short median one. In *Parabathymyrus,* the two lateral processes are greatly reduced, much shorter than the median process. In some genera, the posterior part of the urohyal is compressed to form a keel-like expansion. In the specimen of *Gnathophis bracheatopos* examined, the posterior part of the urohyal appeared to be expanded into median and lateral keels, T-shaped in cross section. Other genera have a simple urohyal.

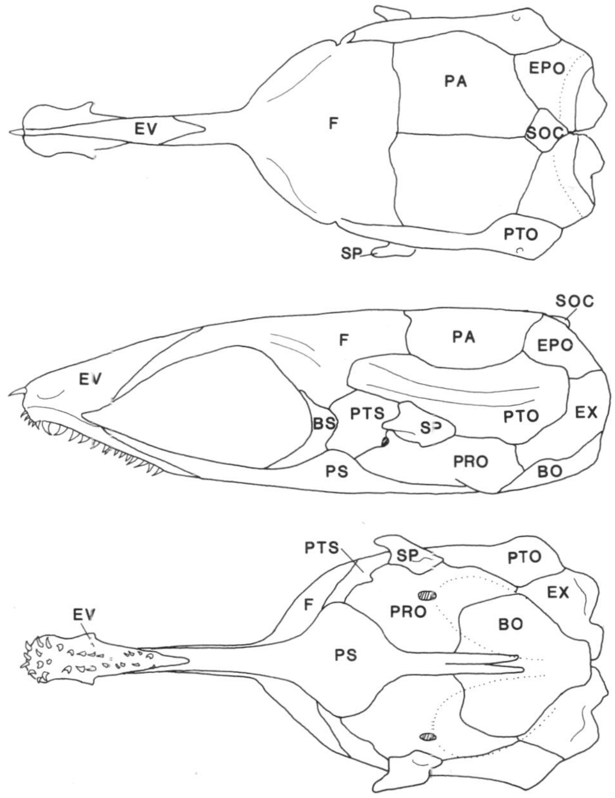

FIGURE 504. Neurocranium of *Heteroconger halis* in dorsal view (top) lateral view (middle), and ventral view (bottom). Abbreviations on pp. 5–6.

Hypohyals were not found in any of the species examined here, nor did Asano (1962) find them in any of his species. Rosenblatt (1967) reported them in *Gorgasia punctata.* In the same location in *Heteroconger halis,* Böhlke (1957: 74, fig. 4A) illustrated, without comment, structures which he called basihyals. In the specimen examined here, the proximal tip of the ceratohyal appeared at first to be separated by a suture. Closer examination revealed no suture, however; the tip of the ceratohyal is undivided.

Gill arches (Fig. 512): The gill arches in con-

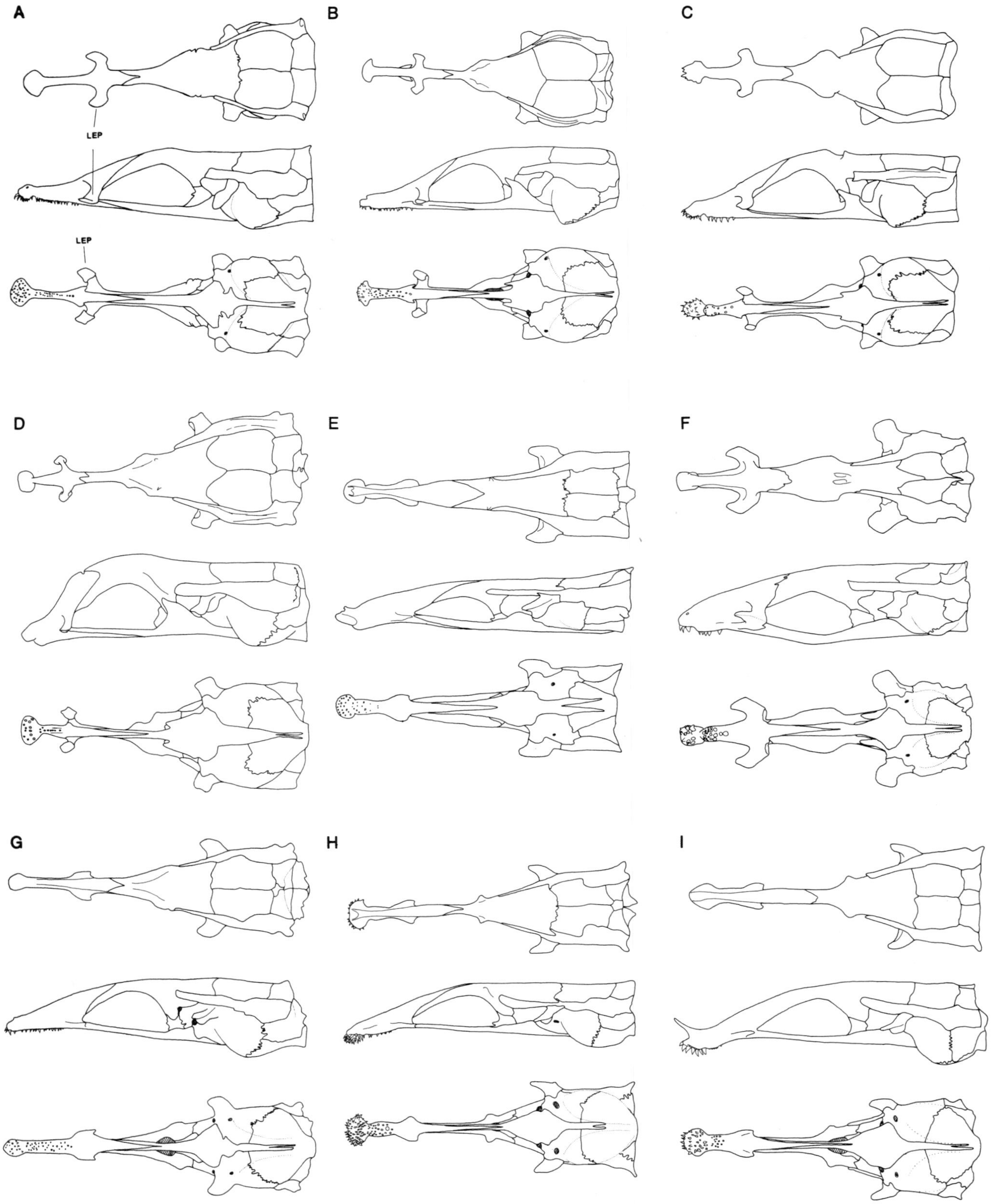

FIGURE 505. Neurocrania (see Fig. 504 for explanation). A. *Ariosoma balearicum*. B. *Ariosoma selenops*. C. *Ariosoma anale*. D. *Parabathymyrus oregoni*. E. *Conger oceanicus*. F. *Paraconger caudilimbatus*. G. *Gnathophis bathytopos*. H. *Rhynchoconger gracilior*. I. *Rhynchoconger flavus*.

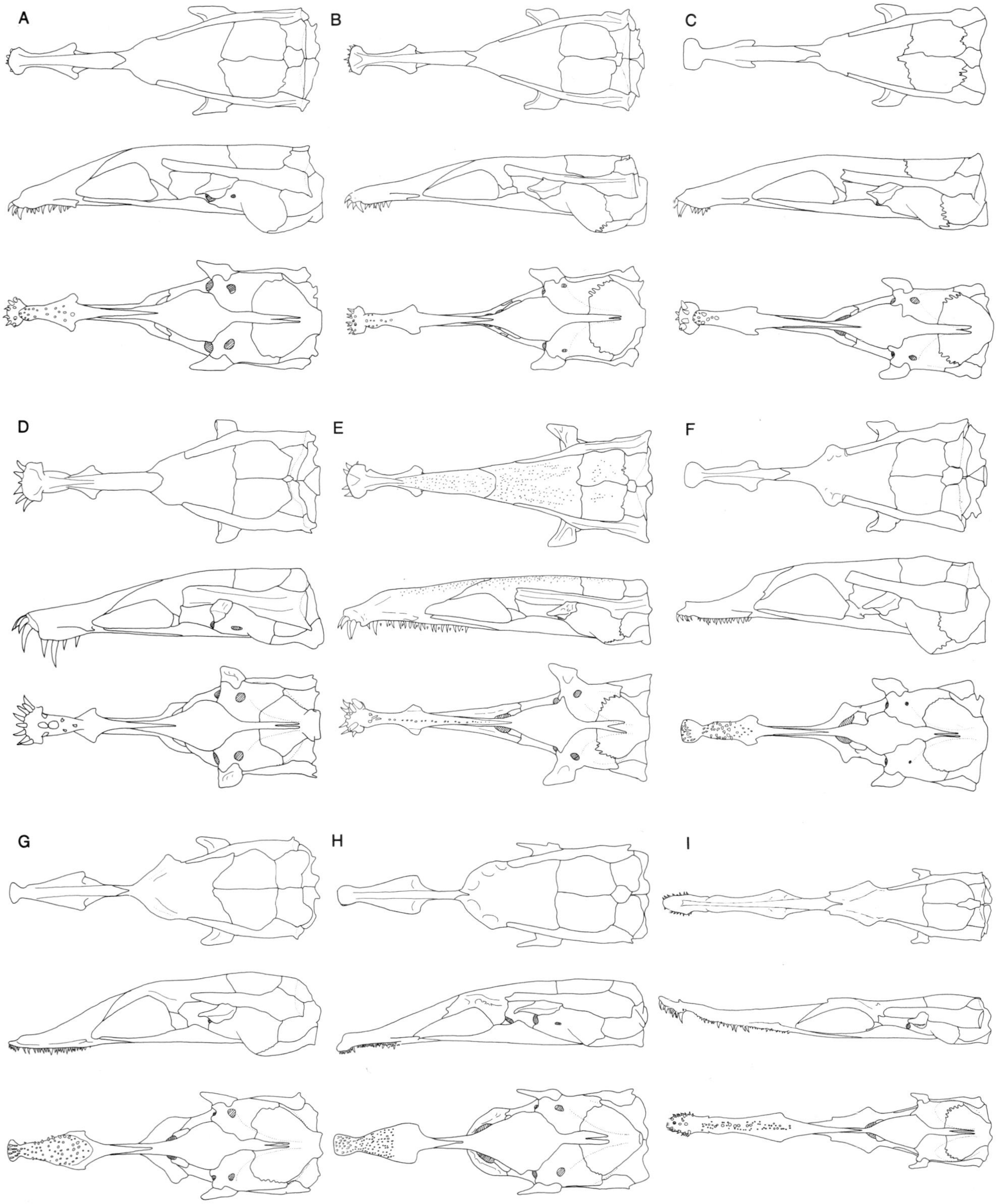

FIGURE 506. Neurocrania (see Fig. 504 for explanation). A. *Rhechias dubia*. B. *Rhechias thysanochila*. C. *Rhechias vicinalis*. D. *Bathyuroconger vicinus*. E. *Uroconger syringinus*. F. *Japonoconger caribbeus*. G. *Pseudophichthys splendens*. H. *Acromycter perturbator*. I. *Xenomystax bidentatus*.

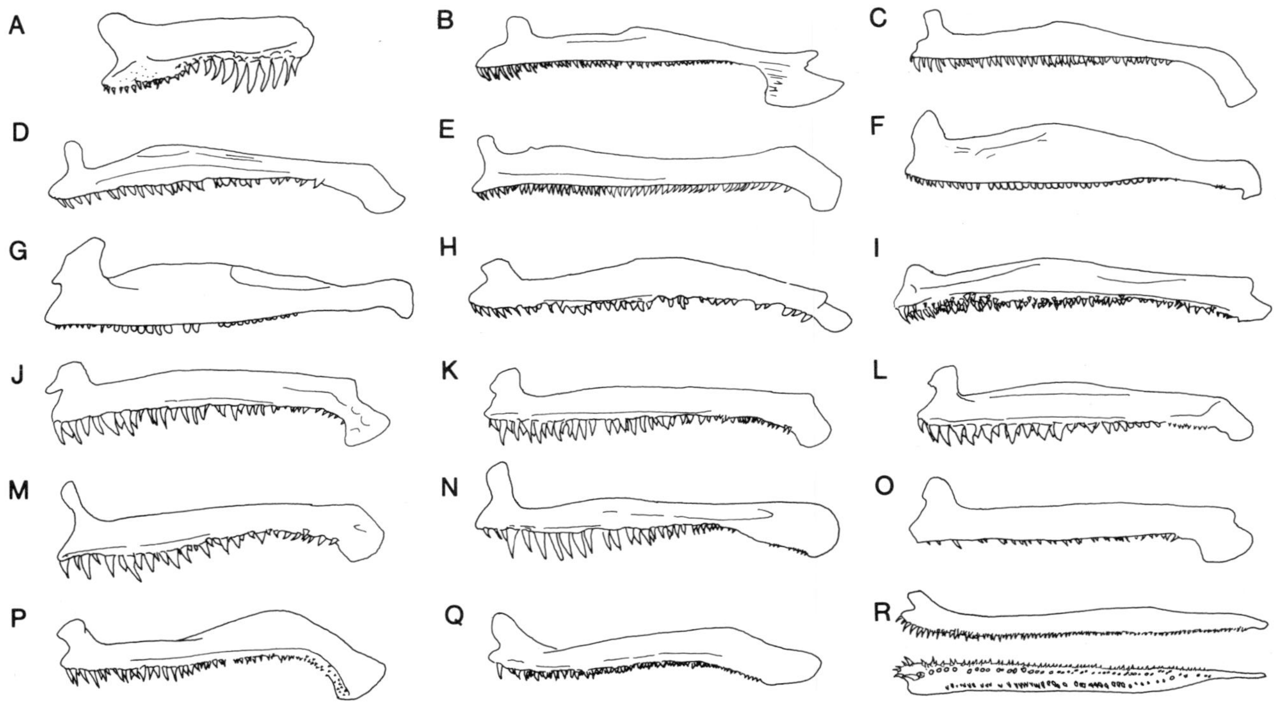

FIGURE 507. Maxilla. A. *Heteroconger halis.* B. *Ariosoma balearicum.* C. *Ariosoma selenops.* D. *Ariosoma anale.* E. *Parabathymyrus oregoni.* F. *Paraconger caudilimbatus.* G. *Conger oceanicus.* H. *Gnathophis bathytopos.* I. *Rhynchoconger gracilior.* J. *Rhechias dubia.* K. *Rhechias thysanochila.* L. *Rhechias vicinalis.* M. *Bathyuroconger vicinus.* N. *Uroconger syringinus.* O. *Japonoconger caribbeus.* P. *Pseudophichthys splendens.* Q. *Acromycter perturbator.* R. *Xenomystax bidentatus,* lateral view (top) and ventral view (bottom).

grids are complete. Unlike some eel families, congrids show no tendency to lose elements. Four basibranchials are present, the first three ossified and the fourth cartilaginous. There are three pairs of hypobranchials, the first two ossified and the third cartilaginous. There are five pairs of ceratobranchials; all are ossified, but the fifth is reduced in size. Usually there is a single pharyngeal tooth plate on each side, but *Conger, Rhynchoconger,* and *Xenomystax* have two; in *Conger* and *Rhynchoconger* the anterior element becomes fused to the proximal portion of the fifth ceratobranchial. Nelson (1966a: 399) stated that the two lower pharyngeals of *Conger* fuse during ontogeny. Asano (1962: fig. 12) showed them divided in *Conger cinereus* but undivided in two other species. The lower pharyngeals are reduced in *Japonoconger* and *Acromycter* and greatly reduced in *Heteroconger.*

The dorsal elements consist of the first through fourth epibranchials, the second and third infrapharyngobranchials, and the upper pharyngeal tooth plates. In *Conger* the first epibranchial bears an anterior process which, as Nelson (1966a) speculated, may be a primitive feature relating to the missing first infrapharyngobranchial. *Pseudophichthys* has a prominent medially directed process on the second infrapharyngobranchial, a feature not found elsewhere. The upper pharyngeals are undivided in all but three genera. *Conger* and *Rhynchoconger* each have two upper pharyngeals, representing the third and fourth arches, with the anterior one much the smaller. *Japonoconger* has the two elements more nearly equal in size.

Opercular apparatus (Fig. 513): The opercular apparatus is well developed for eels. The fan-shaped opercle articulates with the hyomandibula. In the bathymyrines and in *Conger, Congriscus, Gnathophis,* and *Xenomystax* the distal margin is smooth; in the other congrine genera, it is serrate. The sickle-like subopercle follows the distal margin of the opercle. The preopercle lies along the posterior edge of the hyomandib-

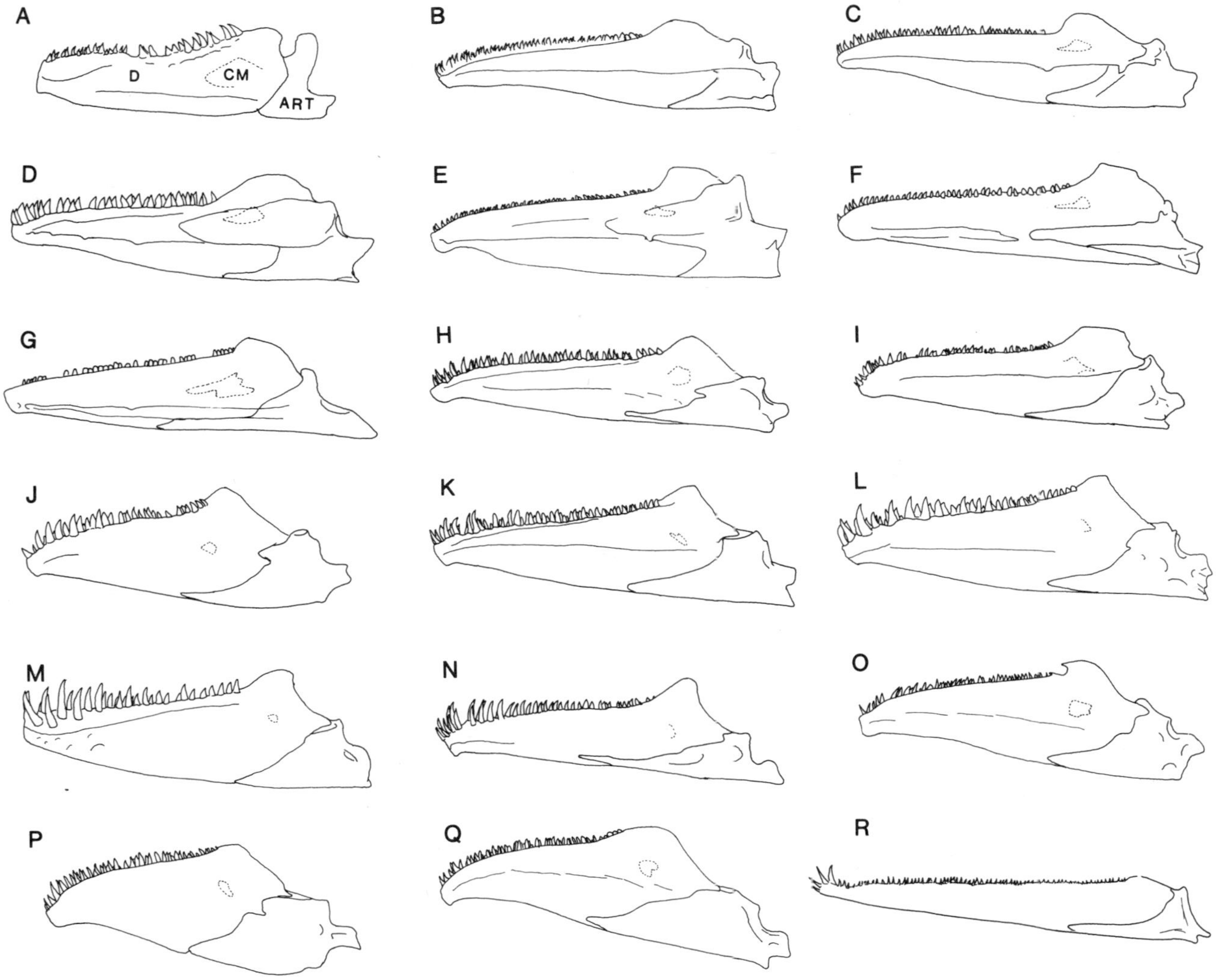

FIGURE 508. Mandible. A. *Heteroconger halis.* B. *Ariosoma balearicum.* C. *Ariosoma selenops.* D. *Ariosoma anale.* E. *Parabathymyrus oregoni.* F. *Paraconger caudilimbatus.* G. *Conger oceanicus.* H. *Gnathophis bathytopos.* I. *Rhynchoconger gracilior.* J. *Rhechias dubia.* K. *Rhechias thysanochila.* L. *Rhechias vicinalis.* M. *Bathyuroconger vicinus.* N. *Uroconger syringinus.* O. *Japonoconger caribbeus.* P. *Pseudophichthys splendens.* Q. *Acromycter perturbator.* R. *Xenomystax bidentatus.* (Position of coronomeckelian on medial side indicated by dotted line.)

ula; it does not have a well developed bony structure to enclose the preopercular canal as it does in some other eels. The interopercle is a roughly triangular plate that is partially concealed by the preopercle.

Pectoral girdle (Fig. 514): The pectoral girdle is complete in most genera and contains the cleithrum, supracleithrum, scapula, coracoid, and about four actinosts. In *Heteroconger* the cleithrum and supracleithrum are well developed, but the scapula and coracoid are greatly reduced. Böhlke (1957: 75) was uncertain whether the upper element represented the scapula or an actinost. In other genera examined, all elements are well developed.

Vertebrae (Figs. 515–518): The form of the vertebrae varies considerably over the length of the body. The first vertebra is shorter than the others, and its centrum articulates with the basioccipital. The anterior few vertebrae have more elaborately developed neural spines than the others. The neural arches of the anterior few vertebrae are fused to the centra in most genera, but in *Japonoconger, Pseudophichthys,* and *Xenomystax* they are articulated. The typical abdominal or precaudal vertebrae have broad parapophyses and a

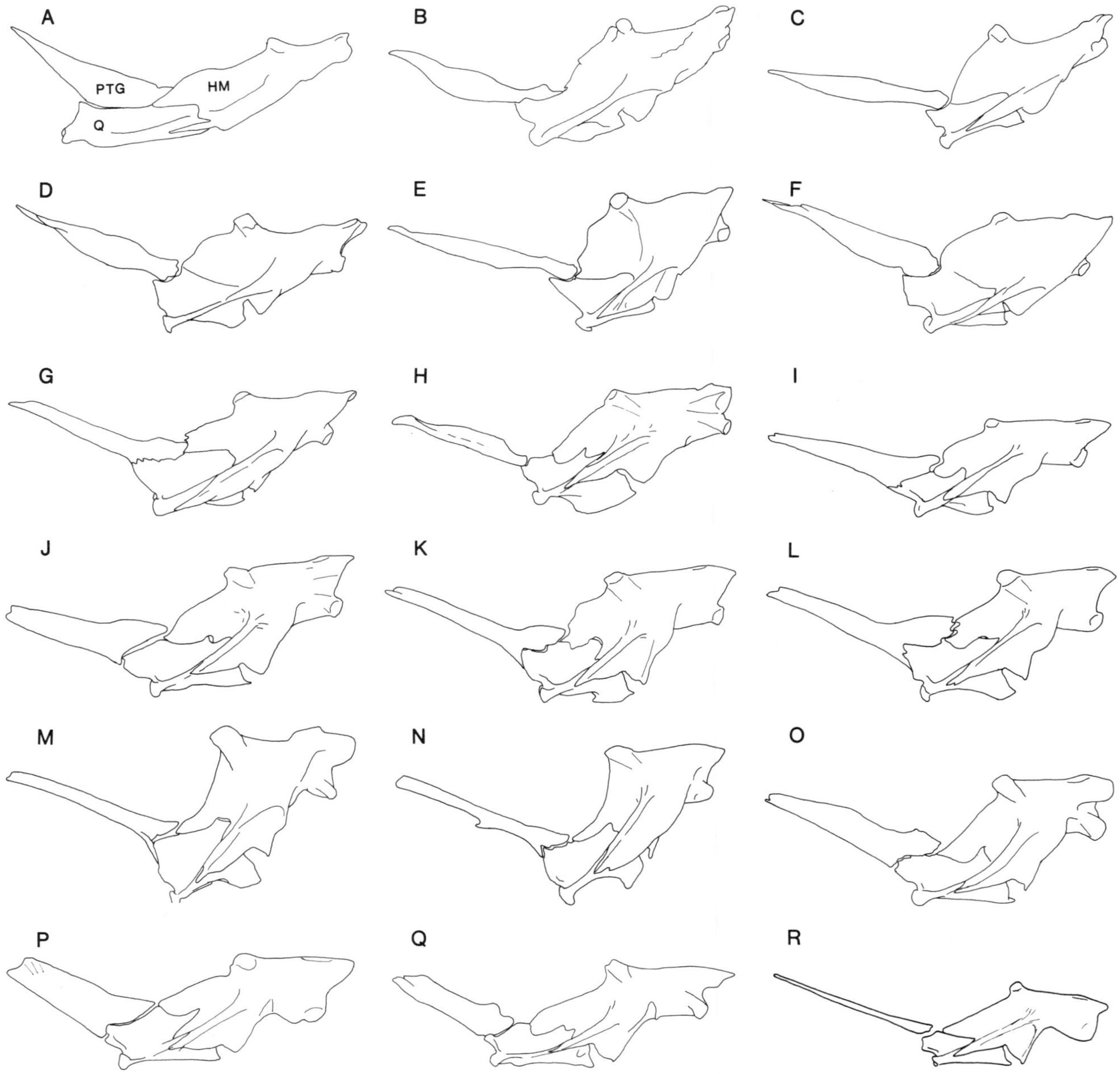

FIGURE 509. Suspensorium. A. *Heteroconger halis*. B. *Ariosoma balearicum*. C. *Ariosoma selenops*. D. *Ariosoma anale*. E. *Parabathymyrus oregoni*. F. *Paraconger caudilimbatus*. G. *Conger oceanicus*. H. *Gnathophis bathytopos*. I. *Rhynchoconger gracilior*. J. *Rhechias dubia*. K. *Rhechias thysanochila*. L. *Rhechias vicinalis*. M. *Bathyuroconger vicinus*. N. *Uroconger syringinus*. O. *Japonoconger caribbeus*. P. *Pseudophichthys splendens*. Q. *Acromycter perturbator*. R. *Xenomystax bidentatus*.

slender neural spine. In the posteriormost few abdominal vertebrae the parapophyses are bent increasingly downward until a complete hemal arch is formed. The caudal vertebrae have a slender neural and hemal arch. Epicentral processes are well developed on the caudal vertebrae of *Ariosoma balearicum*, *Paraconger*, *Conger*, *Gnathophis*, *Rhynchoconger*, *Uroconger*, and *Gorgasia*. *Heteroconger* has weak epicentral processes, and the remaining genera lack them. In *Ariosoma*, *Parabathymyrus*, *Paraconger*, *Heteroconger*, and *Blachea* the abdominal vertebrae constitute more than 40% of the total. In the other genera abdominal vertebrae are 20–40% of the total.

Epineurals are present on all vertebrae except those involved in the caudal skeleton. The anterior few epineurals are fused to the neural arch in *Paraconger*, *Conger*, *Rhynchoconger*, *Japonocon-*

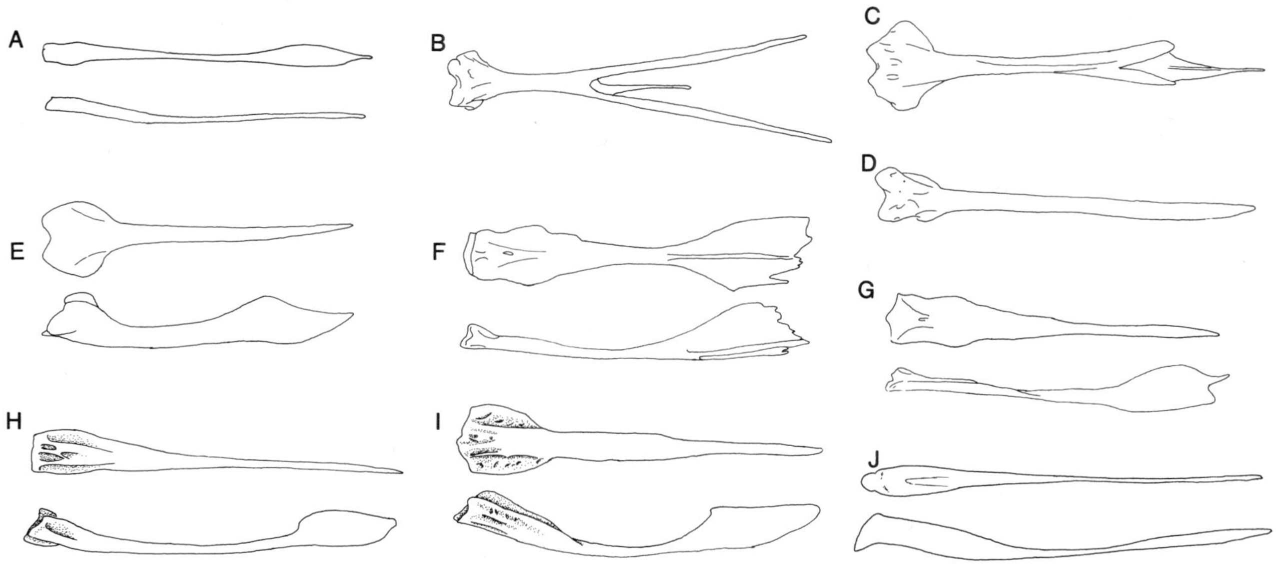

FIGURE 510. Urohyal (anterior end to left). A. *Heteroconger halis,* ventral (top) and lateral (bottom) views. Ventral views: B. *Ariosoma balearicum.* C. *Parabathymyrus oregoni.* D. *Paraconger caudilimbatus.* Ventral (top) and lateral (bottom) views: E. *Conger oceanicus.* F. *Gnathophis bracheatopos.* G. *Gnathophis bathytopos.* H. *Rhechias thysanochila.* I. *Uroconger syringinus.* J. *Xenomystax bidentatus.*

ger, Pseudophichthys, and *Xenomystax.* Except for the anterior and posterior few, the epineurals of most genera are bifurcated. *Gnathophis* and *Blachea* are exceptions, having undivided epineurals; in *Gnathophis* the distal end of the epineural is expanded and flattened. In some genera there is an extra epineural anteriorly associated with the neurocranium. Epipleurals are present on the last few abdominal vertebrae and the caudal vertebrae. Like the epineurals, the epipleurals are bifurcated in most genera, but undivided in *Gnathophis* and *Blachea. Parabathymyrus* and some species of *Ariosoma* have an additional set of intramuscular bones above the epineurals and below the epipleurals. They are present in *Ariosoma selenops, A. anale* and *A. coquettei; A. balearicum* lacks them. Asano (1962) termed these structures myorhabdoi. Pleural ribs are present on the abdominal vertebrae of *Ariosoma, Parabathymyrus, Paraconger, Japonoconger, Pseudophichthys, Gorgasia,* and *Heteroconger.*

The caudal skeleton of congrid eels, like that of all eels, is rather specialized and reduced. The terminal vertebra bears two hypurals; the penultimate vertebra also contributes a posteroventral projection that serves as a third hypural. *Ariosoma, Parabathymyrus, Paraconger,* and *Heteroconger* have the terminal vertebra and the caudal-fin rays shortened, in association with the stiffening of the tail and the tail-first burrowing habit. In *Gnathophis* the caudal structure is short

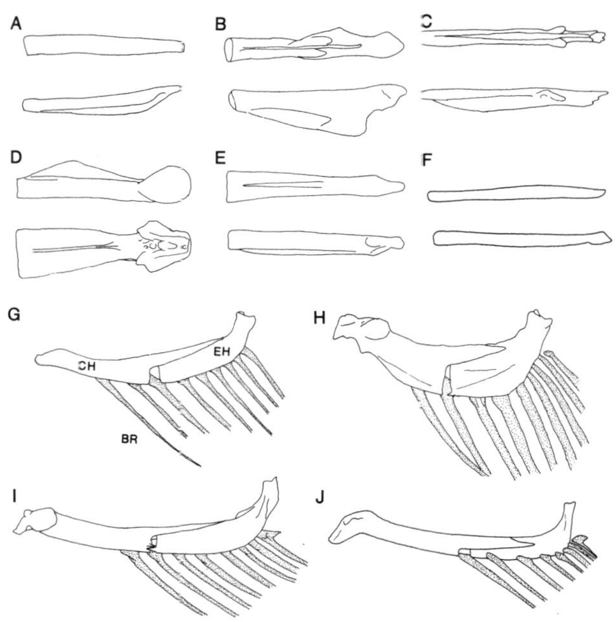

FIGURE 511. Hyoid arch. Glossohyal (anterior end to left): A. *Heteroconger halis.* B. *Ariosoma balearicum.* C. *Ariosoma anale.* D. *Parabathymyrus oregoni.* E. *Rhechias thysanochila.* F. *Xenomystax bidentatus.* Paired hyoid elements: G. *Heteroconger halis.* H. *Ariosoma balearicum.* I. *Rhechias thysanochila.* J. *Xenomystax bidentatus.*

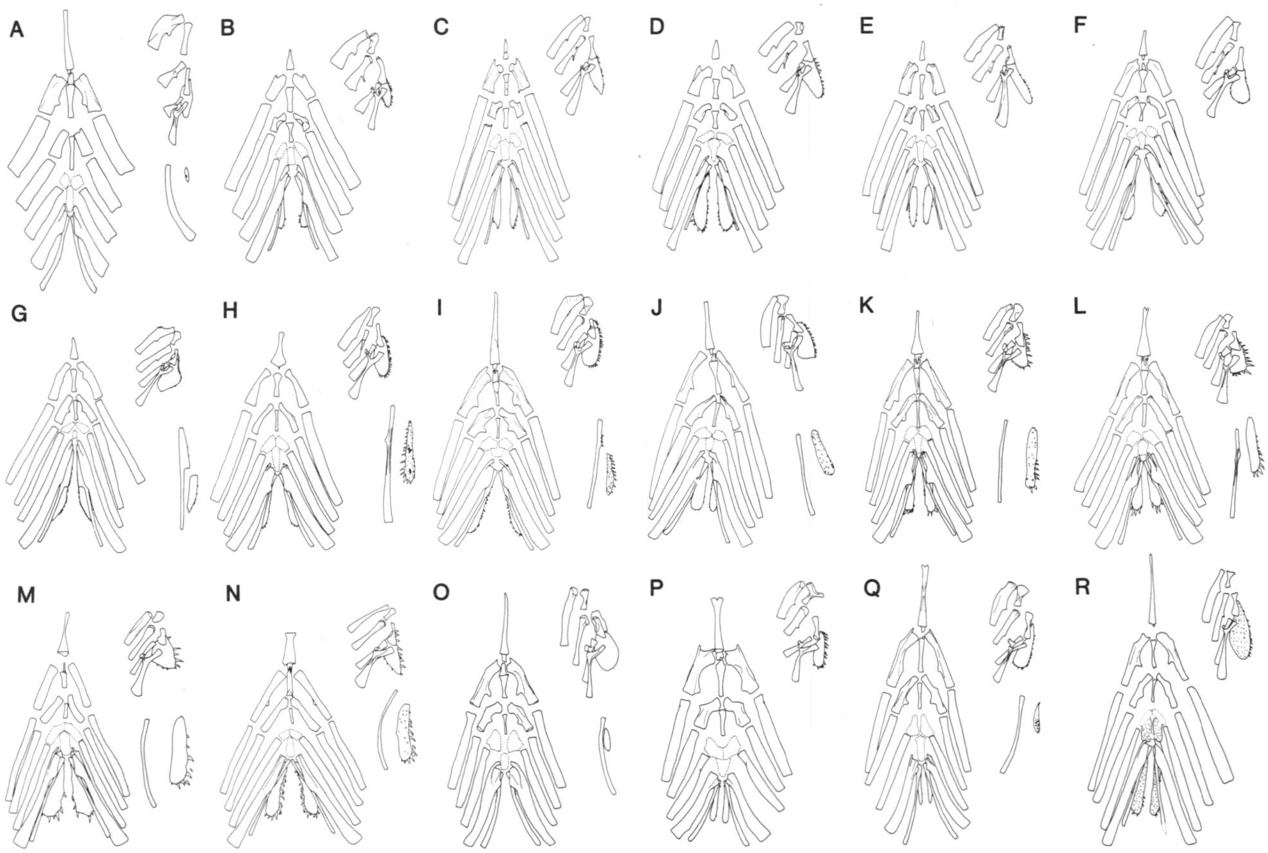

FIGURE 512. Gill arches, ventral view of ventral elements (left), dorsal view of left side of dorsal elements (upper right). A. *Heteroconger halis* (with separate view of fifth ceratobranchial and lower pharyngeal tooth plate on lower right). B. *Ariosoma balearicum*. C. *Ariosoma selenops*. D. *Ariosoma anale*. E. *Parabathymyrus oregoni*. F. *Paraconger caudilimbatus*. G. *Conger oceanicus*. H. *Gnathophis bathytopos*. I. *Rhynchoconger gracilior*. J. *Rhechias dubia*. K. *Rhechias thysanochila*. L. *Rhechias vicinalis*. M. *Bathyuroconger vicinus*. N. *Uroconger syringinus*. O. *Japonoconger caribbeus*. P. *Pseudophichthys splendens*. Q. *Acromycter perturbator*. R. *Xenomystax bidentatus*.

and deep, but the rays are not as stout as in the preceding genera. Several genera have slender, attenuate tails that are frequently damaged. When this happens a pseudocaudal fin can be regenerated which superficially looks much like the normal structure (Fig. 518F). It can usually be distinguished by the more numerous caudal rays and by the fact that the upper hypural is smaller than the lower. In addition, the penultimate vertebra is not modified in the regenerated tail as it is in the normal structure.

Lateralis system: sensory canals and associated bones (Figs. 519, 520): The lateralis system is generally divided into seven canals: the lateral line, preoperculomandibular, infraorbital, ethmoidal, supraorbital, temporal, and supratemporal. The only connection between the right and left sides is the supratemporal commissure. Congrids have no frontal commissure, and there seems to be no ethmoidal commissure, either, although the latter needs further study.

The ethmoidal canal is the remnant of the rostral commissure and derives from the infraorbital rather than the supraorbital canal. In congrids it always contains two pores, both near the tip of the snout. From the first pore, the canal penetrates deeply into the fleshy snout, passes through cartilaginous supports on the anterodorsal surface of the ethmovomer, and returns to the surface at the second pore. The ethmoidal and the supraorbital canals are connected through the second ethmoidal pore, which serves jointly as the first true supraorbital pore (in taxonomic convention, as mentioned above, all three are

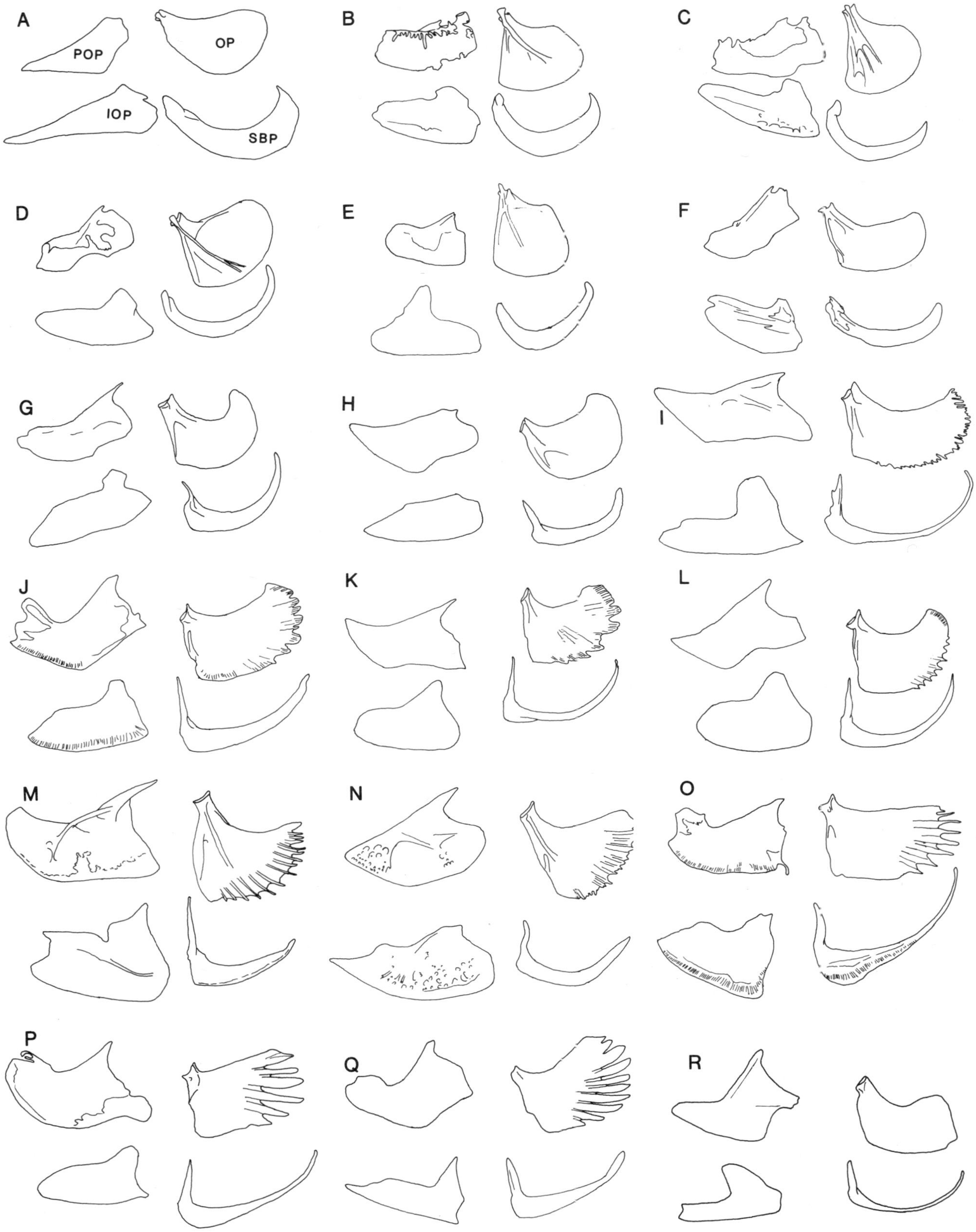

FIGURE 513. Opercular apparatus. A. *Heteroconger halis.* B. *Ariosoma balearicum.* C. *Ariosoma selenops* D. *Ariosoma anale.* E. *Parabathymyrus oregoni.* F. *Paraconger caudilimbatus.* G. *Conger oceanicus.* H. *Gnathophis bathytopos.* I. *Rhynchoconger gracilior.* J. *Rhechias dubia.* K. *Rhechias thysanochila.* L. *Rhechias vicinalis.* M. *Bathyuroconger vicinus.* N. *Uroconger syringinus.* O. *Japonoconger caribbeus.* P. *Pseudophichthys splendens.* Q. *Acromycter perturbator.* R. *Xenomystax bidentatus.*

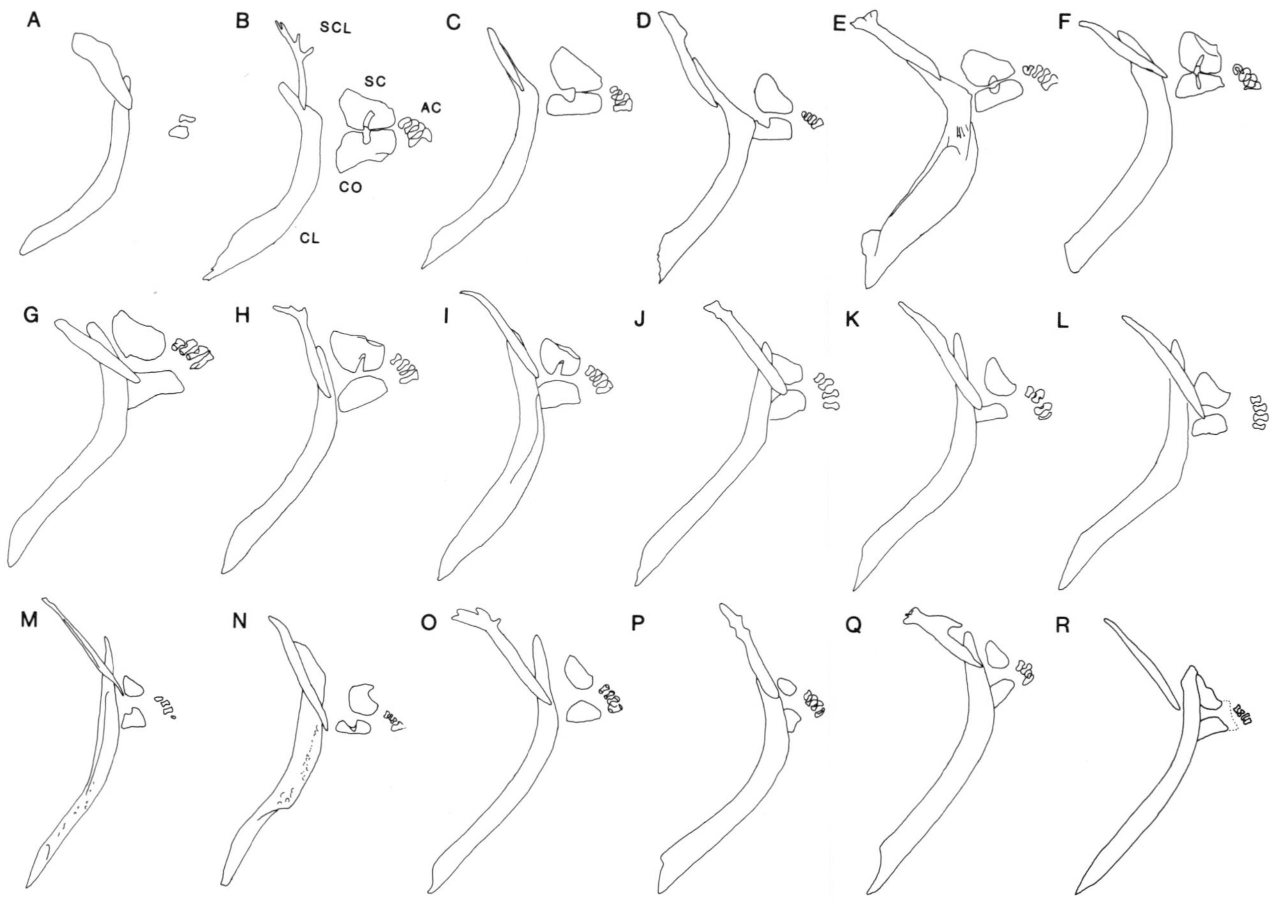

FIGURE 514. Pectoral girdle. A. *Heteroconger halis.* B. *Ariosoma balearicum.* C. *Ariosoma selenops.* D. *Ariosoma anale.* E. *Parabathymyrus oregoni.* F. *Paraconger caudilimbatus.* G. *Conger oceanicus.* H. *Gnathophis bathytopos.* I. *Rhynchoconger gracilior.* J. *Rhechias dubia.* K. *Rhechias thysanochila.* L. *Rhechias vicinalis.* M. *Bathyuroconger vicinus.* N. *Uroconger syringinus.* O. *Japonoconger caribbeus.* P. *Pseudophichthys splendens.* Q. *Acromycter perturbator.* R. *Xenomystax bidentatus.*

referred to as supraorbital pores). The connection between the ethmoidal canal and the infraorbital canal has been lost in congrids as in all other eels. This is perhaps a consequence of the loss of the premaxilla and the rostral ossicles, which bear a portion of the ethmoidal canal.

Behind the joint ethmoidal–supraorbital pore, the supraorbital canal always bears one pore above the anterior nostril. This pore is markedly enlarged in some genera, small in others. Three additional supraorbital pores may or may not be present, one between the anterior and posterior nostrils, one near the anterior margin of the eye, and one near the posterior margin of the eye.

The infraorbital canal begins near the base of the anterior nostril, where a pore is always located. From this point or slightly behind it, a short branch ascends dorsally behind the anterior nostril. A few genera, including *Conger, Congriscus,* and *Paraconger,* have a pore at the end of this branch. In *Paraconger* the pore is small and barely visible, but in *Conger* it is quite conspicuous and lies immediately adjacent to the third supraorbital pore. The small ossicle supporting this branch is termed the supraorbital by Asano (1962: 58), but it appears to be identical to the adnasal bone of other elopomorphs (McDowell, 1973: figs. 1–3). The presence of an adnasal canal is apparently unique to elopomorphs among living teleosts. Asano (1962: 57) stated that the infraorbital and supraorbital canals are connected near their anterior ends. Smith (1971: 20) re-

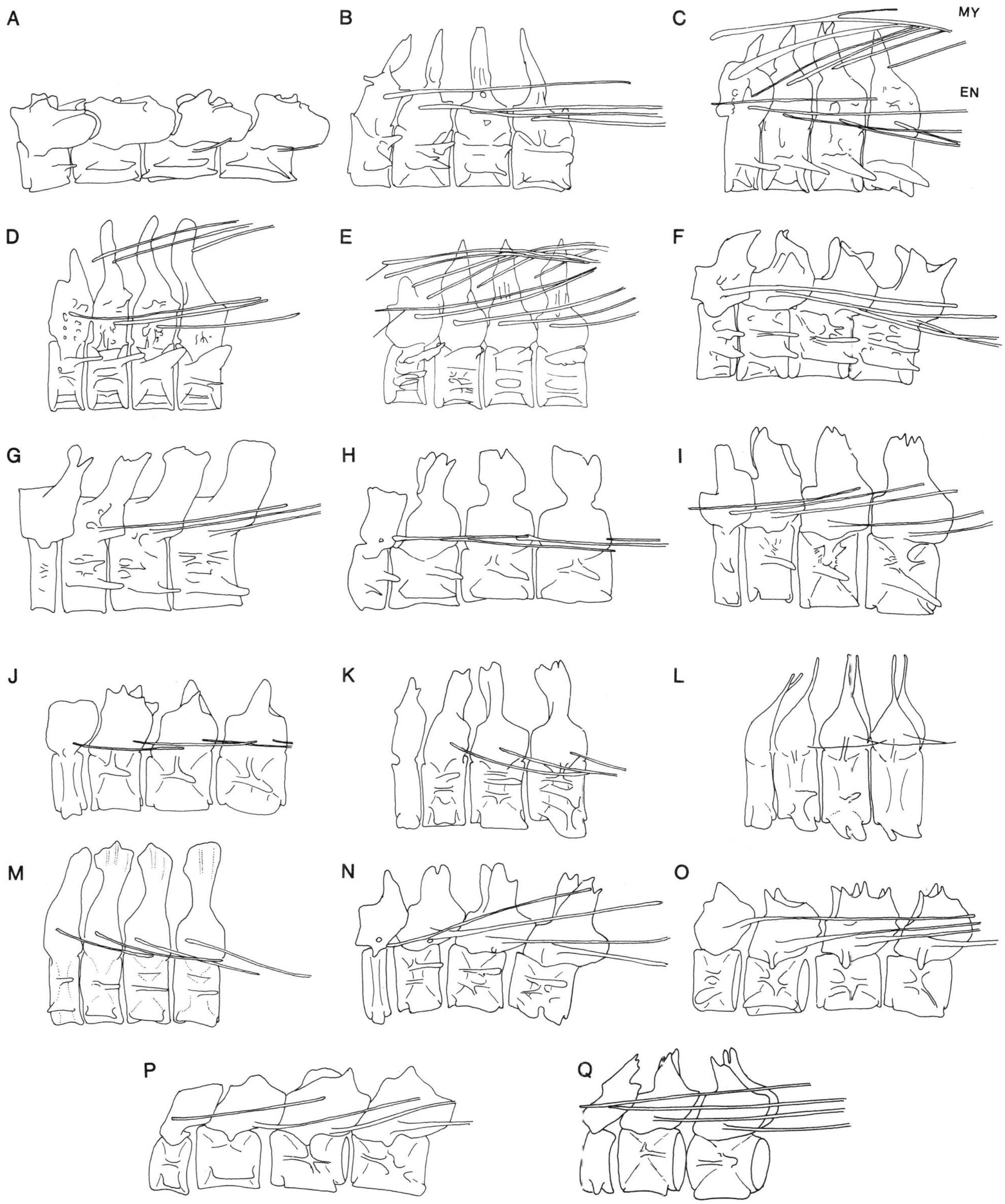

FIGURE 515. Anteriormost few vertebrae. A. *Heteroconger halis.* B. *Ariosoma balearicum.* C. *Ariosoma selenops.* D. *Ariosoma anale.* E. *Parabathymyrus oregoni.* F. *Paraconger caudilimbatus.* G. *Conger oceanicus.* H. *Gnathophis bathytopos.* I. *Rhynchoconger flavus.* J. *Rhechias dubia.* K. *Rhechias thysanochila.* L. *Bathyuroconger vicinus.* M. *Uroconger syringinus.* N. *Japonoconger caribbeus.* O. *Pseudophichthys splendens.* P. *Acromycter perturbator.* Q. *Xenomystax bidentatus.*

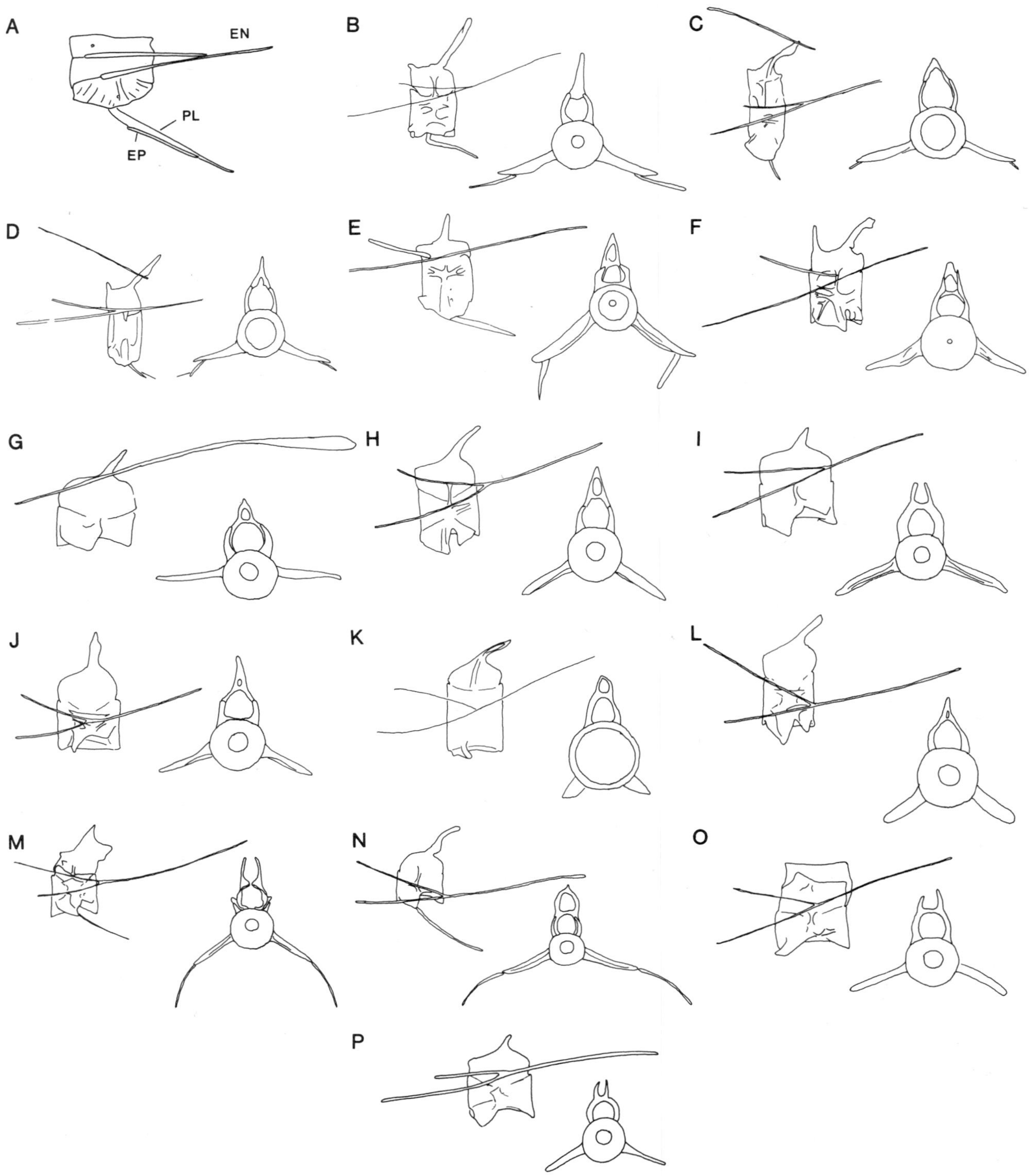

FIGURE 516. Precaudal vertebrae, left side and front view. A. *Heteroconger halis.* B. *Ariosoma balearicum.* C. *Ariosoma anale.* D. *Parabathymyrus oregoni.* E. *Paraconger caudilimbatus.* F. *Conger oceanicus.* G. *Gnathophis bathytopos.* H. *Rhynchoconger flavus.* I. *Rhechias dubia.* J. *Rhechias thysanochila.* K. *Bathyuroconger vicinus.* L. *Uroconger syringinus.* M. *Japonoconger caribbeus.* N. *Pseudophichthys splendens.* O. *Acromycter perturbator.* P. *Xenomystax bidentatus.*

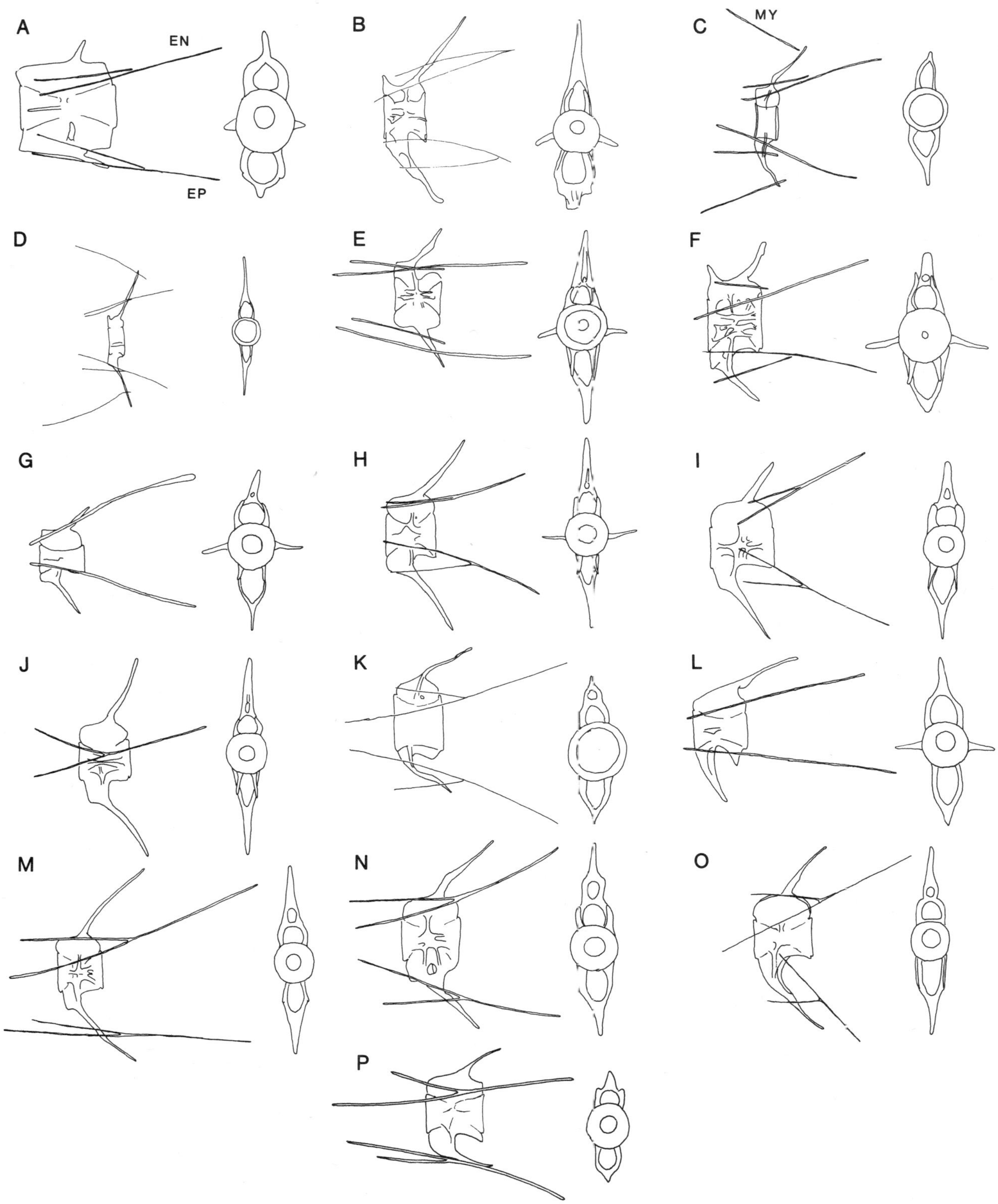

FIGURE 517. Caudal vertebrae, left side and front view. A. *Heteroconger halis.* B. *Ariosoma balearicum.* C. *Ariosoma anale.* D. *Parabathymyrus oregoni.* E. *Paraconger caudilimbatus.* F. *Conger oceanicus.* G. *Gnathophis bathytopos.* H. *Rhynchoconger flavus.* I. *Rhechias dubia.* J. *Rhechias thysanochila.* K. *Bathyuroconger vicinus.* L. *Uroconger syringinus.* M. *Japonoconger caribbeus.* N. *Pseudophichthys splendens.* O. *Acromycter perturbator.* P. *Xenomystax bidentatus.*

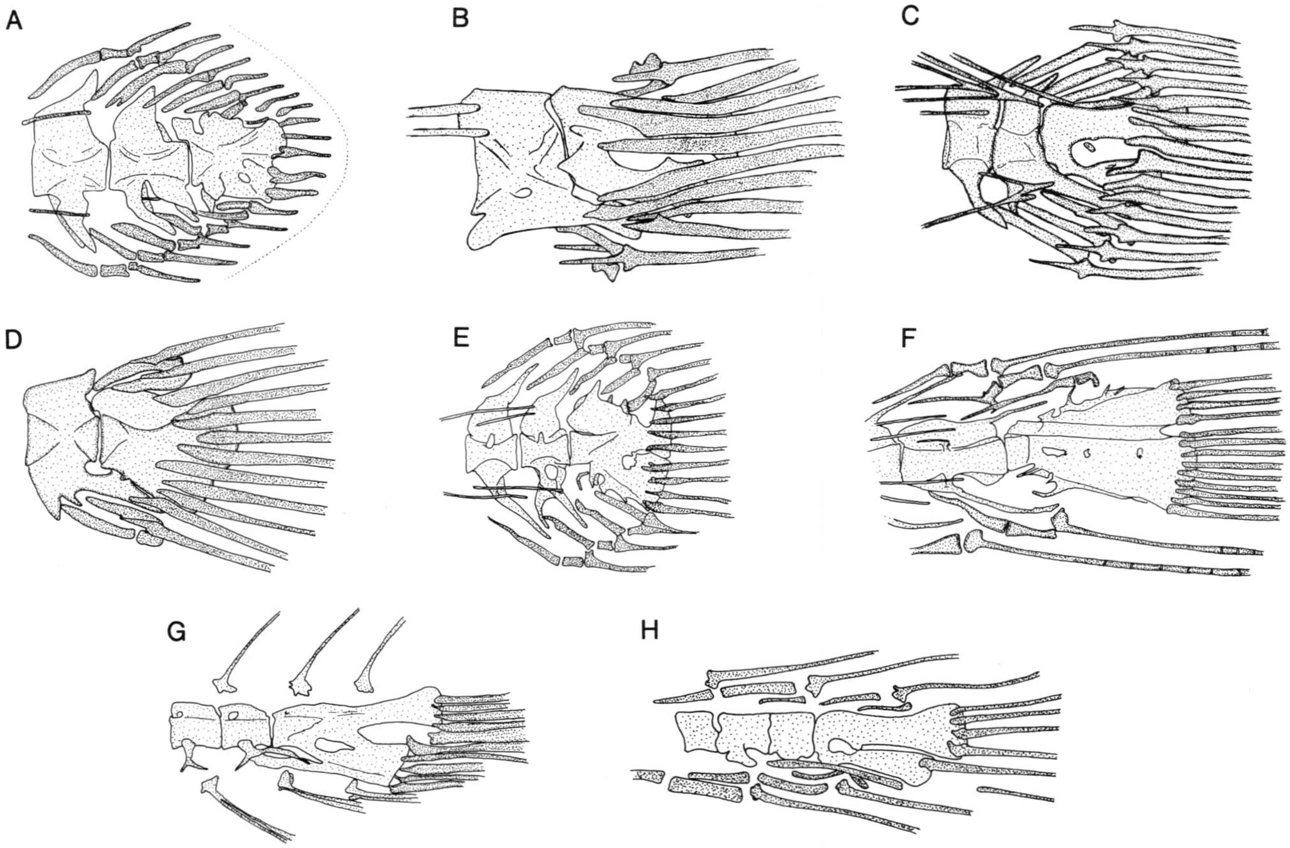

FIGURE 518. Caudal skeleton. A. *Heteroconger halis.* B. *Ariosoma balearicum.* C. *Parabathymyrus oregoni.* D. *Paraconger caudilimbatus.* E. *Gnathophis bathytopos.* F. *Rhynchoconger flavus,* regenerated. G. *Bathyuroconger vicinus.* H. *Xenomystax bidentatus.*

peated this claim, but in fact no connection exists. The main section of the infraorbital canal lies along the upper jaw and then curves dorsally immediately behind the eye. Most congrids have four pores in the anterior part of this canal between the anterior nostril and the rictus, another pore in line with these behind the rictus, and up to three pores behind the eye. The infraorbital pores can thus be subdivided into four groups or series: the adnasal, maxillary, postrictal, and postorbital. *Paraconger* is unusual in having only two maxillary pores (IO 3 and 4 are missing). *Heteroconger,* as a consequence of its truncated snout, has only two maxillary pores and no postrictal pore. In *Xenomystax* the gape is enlarged and there is no postrictal pore; all five pores along the upper jaw are before the rictus.

The preoperculomandibular canal contains about 8–11 pores. The two anteriormost pores are often smaller than the remainder, and the third one is sometimes enlarged.

The temporal canal is usually enclosed within the pterotic bone and thus without pores. The only known exception is *Blachea,* in which the temporal canal is exposed and bears pores.

The supratemporal commissure is sometimes without pores, sometimes with a single median pore, and sometimes with a median pore and a

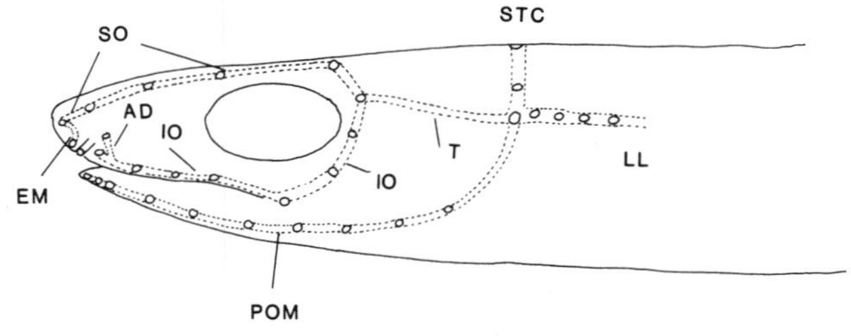

FIGURE 519. Diagrammatic view of sensory pores and canals on the head of a congrid eel.

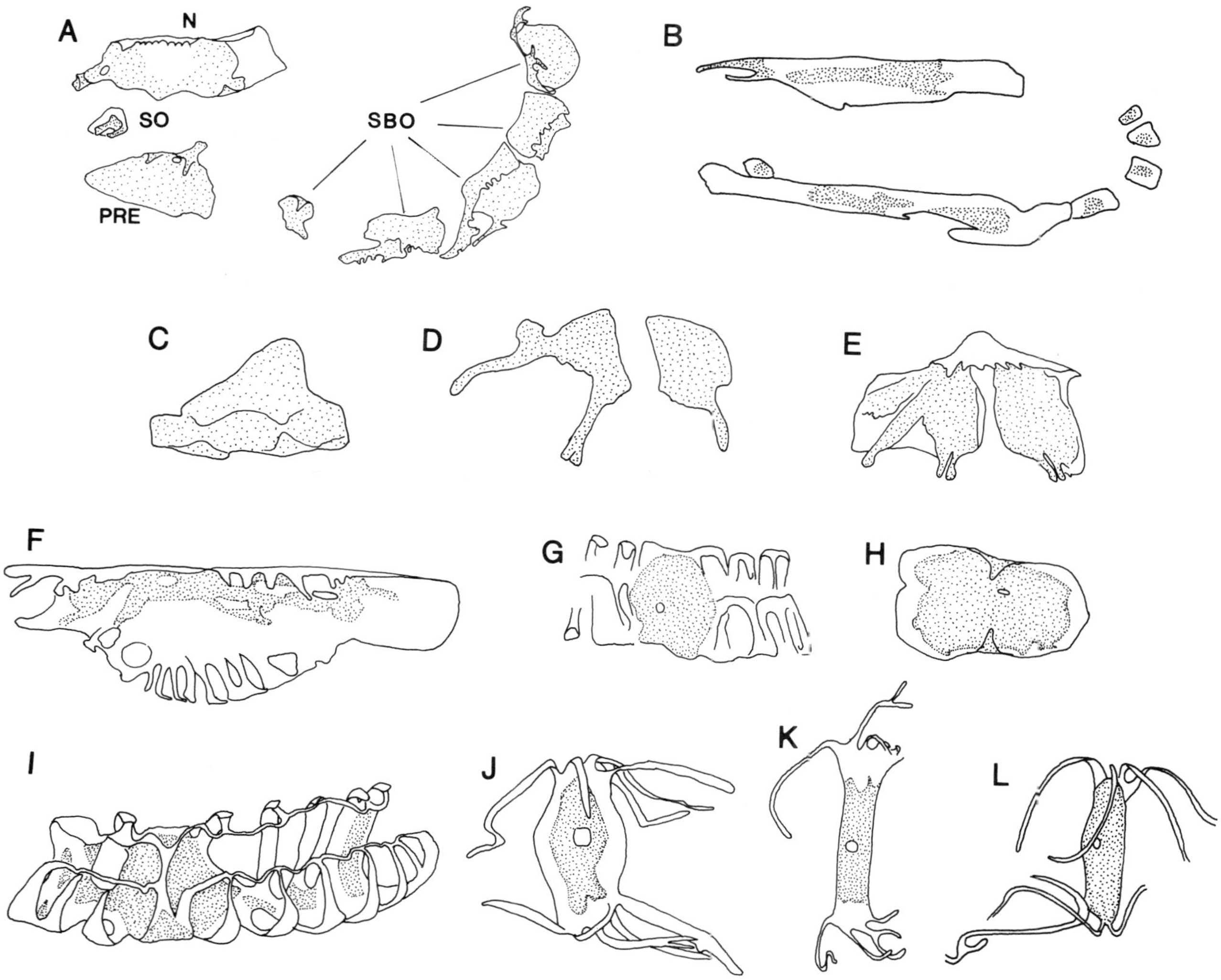

FIGURE 520. Bones associated with lateralis system (shaded areas are ossified). Circumorbital bones: A. *Ariosoma balearicum.* B. *Xenomystax bidentatus.* Preorbital bones: C. *Conger oceanicus.* D. *Gnathophis bathytopos.* E. *Rhynchoconger gracilior.* Nasal bones: F. *Rhechias thysanochila.* Lateral-line ossicles: G. *Heteroconger halis.* H. *Ariosoma balearicum.* I. *Conger oceanicus.* J. *Rhechias thysanochila.* K. *Bathyuroconger vicinus.* L. *Xenomystax bidentatus.*

flanking pore on each side. *Japonoconger caribbeus* usually has two flanking pores on each side, for a total of five.

The lateral line is complete to the tail, usually with one pore per segment except sometimes near the end of the tail. Some heterocongrines have significantly fewer pores than vertebrae. *Diploconger* has two pores per segment, at the end of short dorsal and ventral branches from the main canal. In *Scalanago* the canals are proliferated into an intricately branched network covering much of the body. In *Blachea* the canal is greatly enlarged; the pores are reduced in number but enlarged in size.

The supraorbital and infraorbital canals are partially enclosed by a series of trough-like circumorbital bones, closed on the medial side and open outwardly. The largest of these is the nasal bone, associated with the anterior part of the supraorbital canal. The small adnasal bone, the supraorbital of Asano (1962), has already been mentioned and supports the short ascending branch of the infraorbital canal. The anterior end of the maxillary section of the infraorbital canal is enclosed by a relatively large bone called the preorbital by Asano (1962) and the lachrymal by McDowell (1973) and others. In *Ariosoma, Parabathymyrus, Paraconger, Conger,* and *Blachea,* the genera with a well developed labial flange, the preorbital is stout and its ventral margin is smooth. *Heteroconger* has a well developed flange, but the preorbital is weak. The remaining congrid genera have a reduced flange or none at all, and in all but *Xenomystax* the ventral margin of the preorbital sends three projections into the lip. In *Gnathophis* the preorbital is divided into two elements. In *Xenomystax* the preorbital is a long, weakly ossified structure that extends under most of the orbit; it has no ventral projections. Posterior to the preorbital are three to five smaller ossicles, which Asano (1962) termed suborbitals. These begin under the eye and extend into the postorbital section of the infraorbital canal.

The lateral-line ossicles are solid and fairly well ossified in *Ariosoma, Parabathymyrus,* and *Paraconger,* with the upper and lower margins bent outwardly to form a trough. In *Conger, Gnathophis,* and *Heteroconger* the ossicles are rather elongate and somewhat less substantial, appearing more like a framework than a solid structure; less of the structure is ossified. In the remaining genera the lateral-line ossicles are slender and vertically elongate, consisting of a partly ossified central body with the upper and lower margins drawn out into a series of finger-like processes. In *Blachea* these processes are greatly enlarged and extended. In all lateral-line ossicles there is a small central opening in the inner face for the nerve. The pores open between adjacent ossicles.

Digestive Tract and Gas Bladder (Fig. 521). The digestive tract consists of an esophagus, stomach and intestine. The stomach is a blind sac, with the only entrance at the anterior end. The length of the stomach varies among genera. The digestive tract may be completely or partially black or completely pale.

The gas bladder is always present but varies in form and length among various genera and species. In most species it is relatively long and elliptical, beginning near junction of the stomach and intestine and ending near the anus. In *Bathyuroconger,* however, it is relatively short, reaching less than halfway to the anus. *Japonoconger* and *Pseudophichthys* also have a short gas bladder, but here it is located posteriorly, beginning well behind the stomach and ending well behind the anus. In *Uroconger* it is long but constricted in the middle. *Ariosoma, Parabathymyrus,* and *Paraconger* have the dorsal surface of the gas bladder attached to the vertebrae. Rosenblatt (1967: 96) stated that in the heterocongrines (garden eels) the gas bladder is free of the vertebrae, but in the specimen of *Heteroconger halis* examined by me, it appeared to be attached. In the remaining genera the gas bladder is free. A slender pneumatic duct connects the gas bladder to the esophagus, but it is greatly reduced and probably not functional.

Relationships of the Congridae. The Congridae is one of nine families of true eels with fused frontal bones, the others being the Ophichthidae, Synaphobranchidae, Muraenesocidae, Colocongridae, Nettastomatidae, Derichthyidae, Serrivomeridae, and Nemichthyidae. The Serrivomeridae and Nemichthyidae are highly modified midwater eels, in which the fusion could be secondary; in any event they show no obvious relationship to the Congridae. The remaining families are more generalized, and it seems most likely that the fusion of the frontals occurred in an ancestor common to all of them.

The Ophichthidae is distinguished from the Congridae by the more numerous branchiostegal rays (from 15 to more than 40) that overlap on the ventral midline, reduced gill arches, and the presence of a frontal commissure in the cephalic lateralis system. In addition there is a strong tendency among ophichthids toward a reduction of the caudal fin and a stiffening of the tail tip for burrowing.

Although the Synaphobranchidae includes some of the most primitive living eels, even these have certain specializations not found in the Congridae. *Simenchelys* Gill, on the whole the most primitive of the synaphobranchids, has

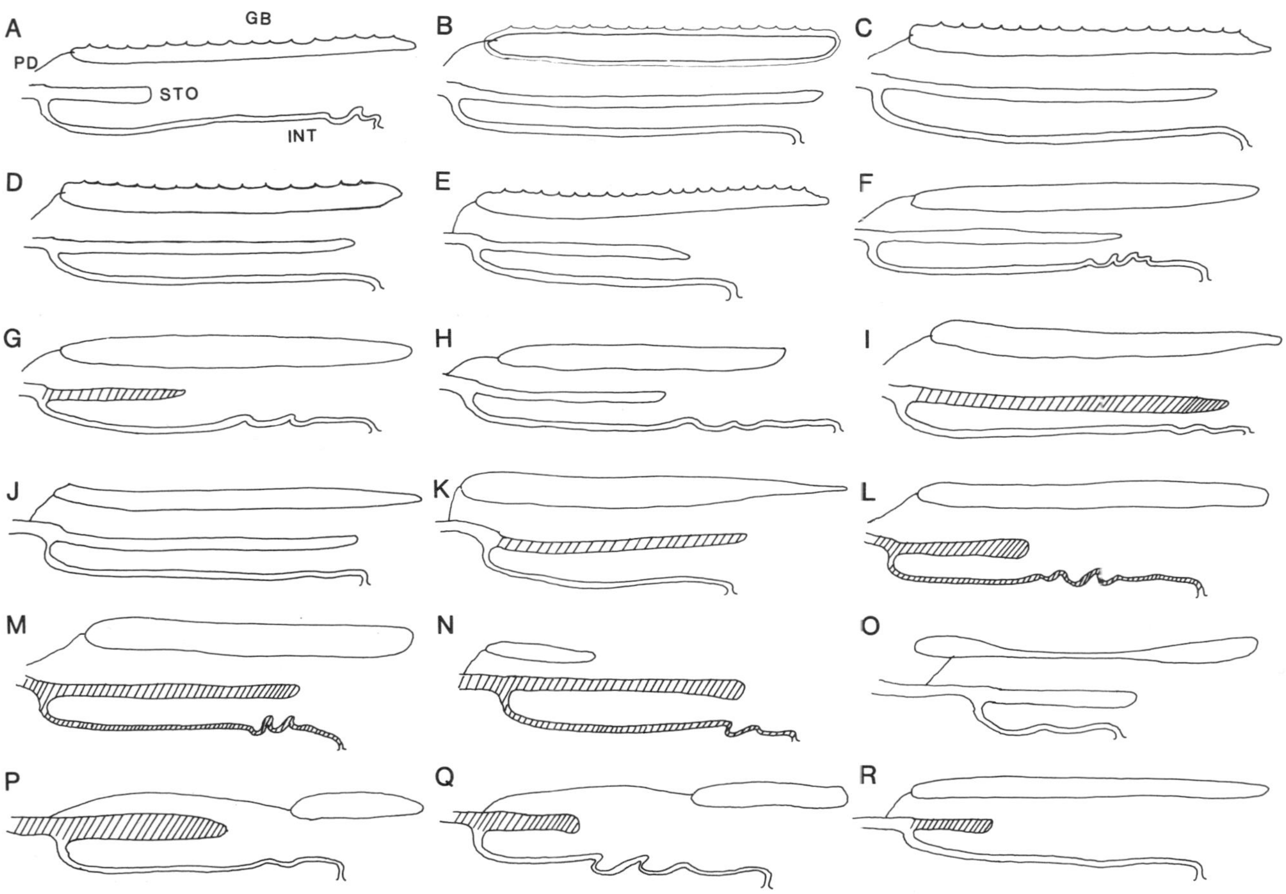

FIGURE 521. Diagrammatic view of digestive tract and gas bladder (shaded areas are black). A. *Ariosoma balearicum.* B. *Ariosoma selenops.* C. *Ariosoma anale.* D. *Parabathymyrus oregoni.* E. *Paraconger caudilimbatus.* F. *Gnathophis bracheatopos.* G. *Gnathophis bathytopos.* H. *Gnathophis tritos.* I. *Rhynchoconger gracilior.* J. *Rhynchoconger flavus.* K. *Rhynchoconger guppyi.* L. *Rhechias dubia.* M. *Rhechias thysanochila.* N. *Bathyuroconger vicinus.* O. *Uroconger syringinus.* P. *Japonoconger caribbeus.* Q. *Pseudophichthys splendens.* R. *Acromycter perturbator.* GB = gas bladder, INT = intestine, PD = pneumatic duct, STO = stomach.

greatly modified the anterior end of the head in response to its peculiar feeding habits and can hardly be considered ancestral to anything. The synaphobranchines, although less modified superficially than *Simenchelys,* show other specializations such as a posteriorly directed suspensorium, ventrally displaced gill openings, and reduced pterygoid. In addition all synaphobranchids have the third hypobranchial reversed in orientation, that is the distal end is anterior to the proximal.

The Muraenesocidae, here restricted to the genera *Muraenesox, Congresox, Cynoponticus,* and *Sauromuraenesox,* is probably the most congrid-like, at least superficially, of all the other fused-frontal families. Nevertheless, it has a peculiar combination of primitive and specialized characters that do not seem to fit within the Congridae as recognized here.

Coloconger has been placed in the Congridae in the past, but it clearly does not belong there. Its resemblances to the Congridae lie entirely in primitive characters, and its unique specializations are quite unlike anything found in congrids.

The Derichthyidae contains two genera of relatively unspecialized midwater eels. They are not clearly related to any of the other fused-frontal families, and their resemblance to the Congridae lies mainly in primitive characters.

The Nettastomatidae is the one family that does seem to be related to the Congridae, specifically to the *Uroconger* and *Xenomystax* groups. If the inferences drawn here are correct, a case could be made for ranking it as a subfamily of the Congridae. I have followed the conservative course here and retained it as a family because of its morphological distinctiveness.

Relationships Within the Congridae. Asano (1962) found that his congrid eels could readily be divided into two major groups, which he called subfamilies. One, which he called the Anagoinae, comprised the genera *Anago* Jordan and Hubbs and *Alloconger* Jordan and Hubbs (both synonyms of *Ariosoma*), and was characterized as follows: lateral ethmoid process present on the skull; gas bladder attached to vertebrae; dorsal- and anal-fin rays unsegmented; caudal vertebrae about as numerous as precaudal; caudal fin reduced and tip of tail stiffened; upturned flange present on upper lip; five suborbital bones; lateral-line ossicles well formed, without cartilaginous loops or "fingers." He also stressed the absence of the supraoccipital bone, but this structure shows a tendency toward reduction throughout the Congridae and has disappeared in at least three independent lines. Asano's other subfamily was the Congrinae and was characterized as follows: lateral ethmoid process absent; air bladder free from vertebrae; dorsal- and anal-fin rays segmented; caudal vertebrae considerably more numerous than precaudal; caudal fin not reduced and tip of tail not stiffened; three or four suborbital bones; lateral-line ossicles less substantial than in the preceding group, with cartilaginous loops or finger-like processes.

Asano's system was fine as far as it went, but he treated only the nine genera present in Japanese waters. He did not consider the Heterocongrinae, which had been recognized as a subfamily by Böhlke (1957), nor did he consider the problem of related groups such as the Muraenesocidae and Nettastomatidae. Robins and Robins (1971) redescribed *Macrocephenchelys* Fowler and studied its osteology. They provisionally retained it in its own family but emphasized its congrid affinities. Finally, Karrer and Smith (1980) described *Blachea*, a new genus that did not fit either of Asano's subfamilies.

The classification presented here is based on Asano's but modified to accommodate the forms not treated by him. It is based on those genera and species I have personally examined and those that have been reported in the literature. Several distinctive Indo-Pacific forms have not been studied in detail, and their eventual inclusion may lead to modifications of the scheme. Three subfamilies are recognized here, although they are not all exactly equivalent. They are the Bathymyrinae (= Anagoinae), Congrinae, and Heterocongrinae. The hypothesized relationships are represented in Figure 522.

The Bathymyrinae is basically as defined by Asano under the name Anagoinae, with the addition of *Bathymyrus* Alcock, *Parabathymyrus* Kamohara, *Paraconger* Kanazawa, and *Chiloconger* Wade. Since Böhlke (1949a) had already established a family-group name based on *Bathymyrus,* the group must be known as the Bathymyrinae. *Alloconger* and *Anago,* as defined by Asano, seem ends of a continuum rather than two discrete groups and are here treated as synonyms of *Ariosoma. Bathymyrus* and *Parabathymyrus* are closely related to *Ariosoma,* particularly the *A. anale* (Poey) group, and represent the most derived of the bathymyrines. These three genera have a distinctive leptocephalus with a very long gut, short dorsal fin, and pigment in a series of short diagonal lines outlining the myosepta just below the midlateral line. *Paraconger* seems somewhat more primitive, having retained the supraoccipital and having a leptocephalus more like that of *Conger* than of *Ariosoma. Chiloconger* has not been studied in detail, but externally it resembles the bathymyrines and may be close to *Paraconger.* Another character that seems common to all bathymyrines is a ventral displacement of the posterior nostril below mid-eye level.

The Heterocongrinae is a highly specialized group that shows certain resemblances to the Bathymyrinae. The caudal fin is reduced and the tip of the tail is stiffened for burrowing, a trend

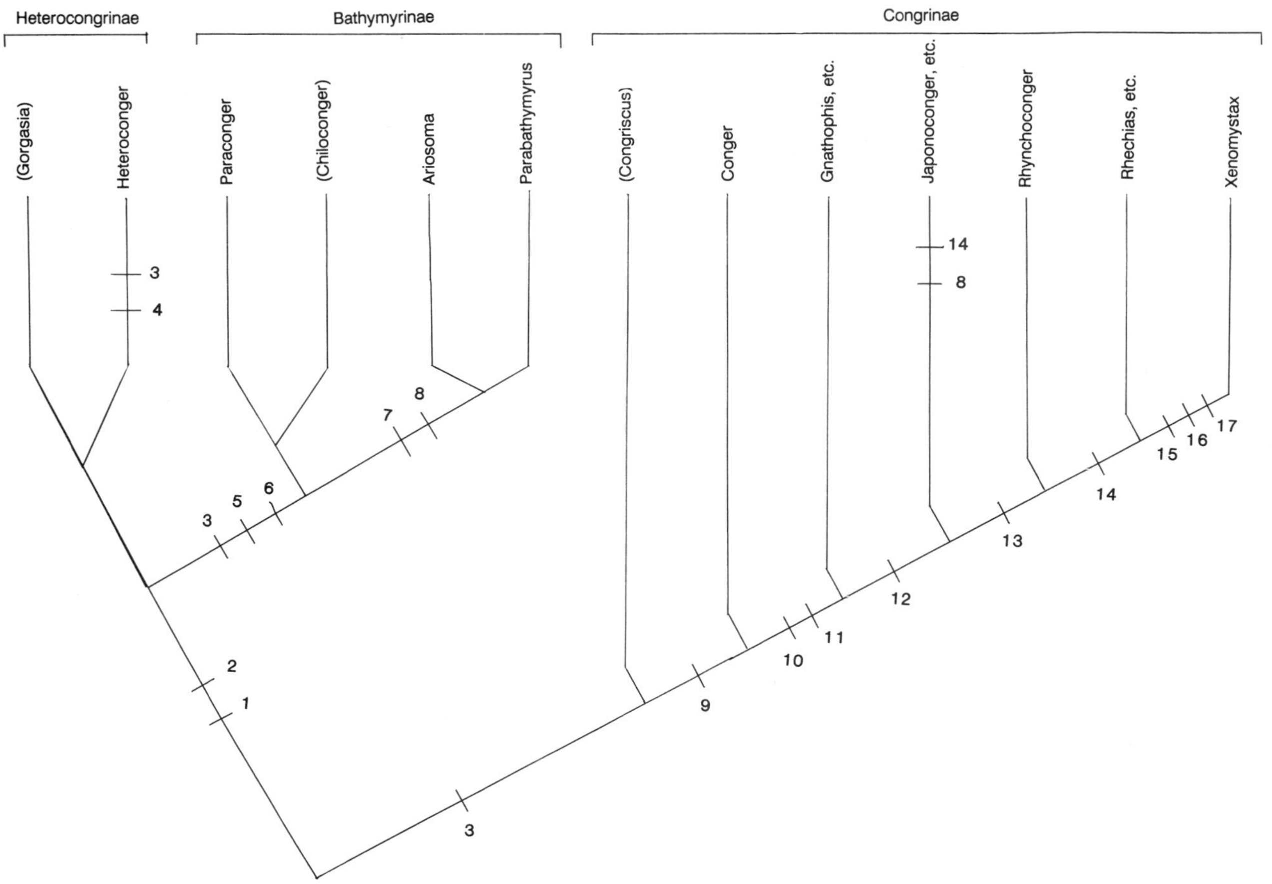

FIGURE 522. Hypothesized phylogenetic relationships among the genera of Congridae (genera in parentheses are not found in the western North Atlantic). The following represent presumed derived character states: (1) lateral ethmoid process; (2) unsegmented fin rays; (3) loss of hypohyals; (4) loss of lateral ethmoid process; (5) abdominal vertebrae equal in number or more numerous than caudal vertebrae; (6) gas bladder attached to vertebrae; (7) loss of supraoccipital; (8) modified leptocephalus; (9) reduction of lateral-line ossicles to *Conger* type; (10) preorbital with three ventrally directed projections; (11) reduction of upper labial flange; (12) reduction of lateral-line ossicles to *Uroconger* type; (13) attenuate tail; (14) loss of upper labial flange; (15) teeth enlarged; (16) upper lip reduced, teeth exposed; (17) reduction of preorbital.

clearly begun in the Bathymyrinae. The unsegmented fin rays and the well developed labial flange are also characteristic of the Bathymyrinae, as is the presence of a lateral ethmoid process in at least one species (Rosenblatt, 1967). Nevertheless, the group is so divergent and highly specialized that, following current practice, it is here recognized as a subfamily. The leptocephalus is similar to that of *Paraconger* but has a shorter gut and a longer dorsal fin.

The Congrinae is the largest and most diverse of the subfamilies. Indeed, it could be said that the Congrinae represents the main stem of congrid evolution. Whereas the Bathymyrinae is characterized by a suite of characters that are uniformly present in all its members (lateral ethmoid process, unsegmented fin rays, gas bladder attached to vertebrae), the Congrinae is characterized by trends in the evolution of several characters among its members. For example the lateral-line ossicles become less substantial, the labial flange is reduced and lost, the preorbital sends three projections ventrally into the lip, and the tail becomes slender and attenuate. *Conger* and *Uroconger*, near the two ends of the series of genera, differ in many of these characters, but they are connected by intermediate stages in the other genera.

Although *Blachea* shares some characters with the Congrinae, it has several peculiarities that make it difficult to place. Karrer and Smith (1980: 648) suggested that it might deserve subfamilial status, but they hesitated to take such a step until a complete osteological study was made of all congrid genera. Since *Blachea* does not occur in the western Atlantic, a formal decision on its status will be deferred.

Distribution. World wide, tropical and temperate.

KEY TO THE SUBFAMILIES OF THE CONGRIDAE

1a. Body extremely long and slender; pectoral and caudal fins reduced or absent; mouth short, oblique; lower jaw projecting beyond upper .... Heterocongrinae, p. 482
1b. Body moderately stout to moderately slender; pectoral and caudal fins present; mouth horizontal; jaws equal or upper jaw projecting 2
2a. Preanal length usually greater than 40% TL; caudal fin short and tip of tail stiff; posterior nostril below mid-eye level; dorsal- and anal-fin rays unsegmented; flange present on upper lip ... . Bathymyrinae, p. 490
2b. Preanal length usually less than 40% TL; posterior nostril at or above mid-eye level; dorsal- and anal-fin rays segmented; flange on upper lip present or absent .... Congrinae, p. 511

Subfamily HETEROCONGRINAE

GARDEN EELS

Characters. Body greatly elongate, slightly compressed throughout, anus near anterior third of TL. Tip of tail stiff; caudal fin reduced or absent externally, although rays always present. Dorsal and anal fins low, sometimes confluent with short caudal fin, sometimes terminating externally slightly before tip of tail; dorsal fin begins over, before, or slightly behind gill opening. Pectoral fin variously reduced or absent. Lateral line complete, but some species have fewer pores than vertebrae.

Head short, about 4–8% TL, with longitudinal folds or grooves in skin of throat. Mouth short, oblique, rictus near level of anterior margin of eye or pupil. Lower jaw projects beyond upper. Upper and lower labial flanges well developed, those of left and right sides continuous around tip of snout and lower jaw in *Heteroconger* but not in *Gorgasia.* Teeth concealed when mouth closed. Eye large, pupil elongate anteriorly. Anterior nostril a small tube near tip of snout, enclosed in labial flange in *Heteroconger,* free in *Gorgasia.* Posterior nostril on snout in front of eye. Head pores variously reduced in some species. Teeth small, pattern of dentition variable. Pterygoid teeth present in at least two species.

Lateral ethmoid process present or absent. Supraoccipital present. Pterygoid stout. Suspensorium inclined steeply forward. Hypohyals possibly present in at least one species. Lower pharyngeals reduced. Posterior edge of opercle smooth. Preorbital weak, without projections from ventral edge. Pectoral girdle complete, although scapula and coracoid sometimes reduced. Myorhabdoi absent. Ribs present. Caudal vertebrae with small epicentral processes. Lateral-line ossicles substantial, though with some fenestration and reduction in ossification, somewhat elongate horizontally, central portion ossified, dorsal and ventral edges curved around to form a trough. Dorsal- and anal-fin rays unsegmented.

Habits and Behavior. Heterocongrines live in colonies on sandy bottoms in clear, tropical waters in places where a slight to moderate current brings a continuous supply of small, planktonic food organisms. Each individual lives in a burrow. Typically the head and anterior part of the body protrude while the tail remains inserted. The eel faces into the current, waving back and forth snapping at small prey that drifts by. A colony behaving in this way resembles a garden of eels, hence the common name. Should an individual be approached by a diver or other large object, it will gradually withdraw into its burrow. One of the best descriptions of this phenomenon is that given by Beebe (1938: 60) and quoted by Böhlke (1957: 59) for the eastern Pacific species *Heteroconger digueti* (Pellegrin).

> Before me, covering the considerable expanse of sand within view, was a garden of eels. From the pale surface, in every direction, there arose perhaps fifty forms. They resembled iron rods as much as anything, slightly bent above the middle of their length, sticking up from the sand, but, unrodlike, swaying very slightly. As I crawled slowly toward them, those nearest me, without effort, and so slowly that the motion was scarcely observable, sank gently into the sand, until only the heads were visible. When even the latter disappeared the sand closed over the small holes and the eels were as if they had never been. There was a perfect gradation from the vanished ones nearest me to those twenty feet away which projected two feet or more. They were all in profile, facing up-current, and paid no attention to me other than to sink from view at the deadline of about six feet.

Such gardens may be as small as a few square meters, or they may extend for several kilometers. Depths range from a few meters to perhaps 50 m or more. The lower end of the depth range is difficult to determine. Most specimens are collected by scuba divers, who seldom venture below 50–60 m. The type of *Heteroconger herrei* (Wade) was dredged from 229–275 m.

A typical colony contains individuals of all sizes. Fourteen specimens of *Heteroconger halis* (Böhlke) (ANSP 138793, 138794) collected from a single colony near West End, Roatan, Honduras ranged from 78 to 346 mm TL. The smallest specimen still had traces of the pigmentation of the leptocephalus, indicating that these eels take up their characteristic way of life soon after metamorphosis. The variety of sizes indicates that colonies are being settled continuously. Unlike the species observed by Beebe (1938), *Heteroconger halis* lives in permanent burrows; when the eel retreats fully, the opening is still visible.

Garden eels are particulate plankton feeders. The principal adaptation is a shortening of the snout. The large eyes are thus brought close to the tip of the snout, aiding close-up binocular vision (Rosenblatt, 1967: 96). The pupil is somewhat elongate anteriorly, providing a better line of sight. The mouth is small and oblique, a modification that has taken place in a number of other fishes (for example, the serranid genus *Paranthias* and the embiotocid *Brachyistius*) and is an adaptation for seizing small planktonic organisms (Rosenblatt, 1967: 96). Davis and Birdsong (1973) illustrated this trend in several groups of coral-reef fishes. By opening the small mouth and at the same time expanding the buccal cavity, a suction is created, which draws the prey into the mouth. The grooves or folds on the skin of the throat of garden eels suggest that the pharyngeal cavity can be expanded, causing the necessary suction. Ecologically, garden eels resemble certain shorter-bodied burrow dwellers such as opistognathids, chaenopsids, and some gobies (L. Kaufman, pers. comm.). These fishes live in burrows and limit themselves to short forays into the water column to capture plankton that drifts by. Garden eels differ in being in constant physical contact with the burrow. Such fishes all require a dependable current to bring a continuous supply of food within range and can be regarded as almost semi-sessile. Randall (1967: 682) found that copepods formed 66.3% of the stomach contents by volume of *Heteroconger halis* collected at Puerto Rico and the Virgin Islands. Of the remaining third, the largest component was tunicates with 18.6%, followed by pteropods, ostracods, shrimp larvae, unidentified eggs, and gastropod larvae with less than 5% each.

Garden eels must be relatively secure from predation. They apparently never leave their burrows except possibly to spawn. A large predator could seldom if ever approach a garden eel without the latter retreating into its burrow. The only predators that would seem able to feed on garden eels are those capable of going into the sand after them, such as certain ophichthids. This, however, is speculation.

Little is known about the reproductive habits of heterocongrines. Fricke (1970) theorized that they might lack a free-swimming leptocephalus stage and develop directly in the sand. Like Herald's (1961: 132) similar suggestion about morays, however, this has been proved wrong, and heterocongrines are now known to have a typical leptocephalus larva (Raju, 1974b; Smith, 1979). There is no evidence that mature individuals become pelagic or undertake long spawning migrations. It seems most likely that spawning takes place close to the area of residence. Clark (1972) described purported courtship behavior in a colony of *Gorgasia sillneri* Klausewitz in the Red Sea. The male and female occupied burrows some 15 cm apart and engaged in an intertwining behavior. Males were stated to maintain harems of several females. It is unclear, however, whether the author actually saw the spawning act.

Although garden eels have no direct economic value, their unique way of life lends great interest to them, and in places they have become something of an underwater tourist attraction. With sport diving increasing in popularity, more and more people will have the opportunity to see these remarkable eels.

Distribution. Circumtropical, wherever suitable habitat is found.

Genera. The Heterocongrinae can be divided into two groups. The first consists of the genus *Gorgasia* Meek and Hildebrand, 1923. Here the upper labial flange is not continuous around the tip of the snout and does not enclose the anterior nostril and the ethmoidal pore. The teeth are in narrow bands or uniserial, and at least one species has a lateral ethmoid process (Rosenblatt, 1967).

The second group comprises those species that have been described under *Heteroconger* Bleeker, 1868, *Taenioconger* Herre, 1923, *Nystactichthys* Böhlke, 1958, and *Xarifania* Klausewitz and Eibl-Eibesfeldt, 1959. In this group the labial flange is continuous around the tip of the snout and encloses the anterior nostril and the ethmoidal pore. Teeth are in broad bands and there is no lateral ethmoid process. These genera were distinguished from each other by color and by the relative development of the pectoral and caudal fins. *Heteroconger* had a caudal fin but lacked pectoral fins. *Taenioconger* had both caudal and pectoral fins. *Nystactichthys* lacked an external caudal fin, although rays were present beneath the skin. *Xarifania* was said to lack caudal rays entirely. *Heteroconger* has a banded color pattern, whereas the other genera were plain. Rosenblatt (1967: 91, footnote) disposed of *Xarifania* by showing that the supposedly absent caudal-fin rays were in fact present. The distinctions between the remaining genera have been further compromised in recent years by the discovery of additional species with new characters and combinations of characters. Böhlke and Randall (1981) showed that the traditional arrangement of species into three genera was no longer tenable. When all these species have been studied in sufficient detail, some natural generic groupings may appear. Until then, they are all provisionally placed in *Heteroconger*.

Böhlke (1957) and Rosenblatt (1967) gave reasons for considering *Gorgasia* more primitive overall than the species of *Heteroconger*. Certainly the divided labial flange of *Gorgasia* is more generalized than the continuous flange of *Heteroconger* and more typical of other congrids. The presence of a lateral ethmoid process in *Gorgasia* takes on added significance in this view. It suggests that the structure was present in the common ancestor of heterocongrines and bathymyrines and was secondarily lost in the more advanced heterocongrines, possibly in association with the further shortening of the snout.

Heteroconger is the only genus present in the western North Atlantic.

Genus *Heteroconger* Bleeker, 1868

Heteroconger Bleeker, 1868: 331 (type species *Heteroconger polyzona* Bleeker, 1868, by monotypy).

Taenioconger Herre, 1923: 152 (type species *Taenioconger chapmani* Herre, 1923, by original designation).

Nystactes Böhlke, 1957: 68 (type species *Nystactes halis*

Böhlke, 1957, by original designation). Preoccupied by *Nystactes* Gloger, 1827 (Aves), *Nystactes* Kaup, 1829 (Mammalia).

Nystactichthys Böhlke, 1958: 59 (substitute name for *Nystactes* Böhlke).

Xarifania Klausewitz and Eibl-Eibesfeldt, 1959: 138 (type species *Xarifania hassi* Klausewitz and Eibl-Eibesfeldt, 1959, by original designation.

Characters. Those of the subfamily, with the following restrictions. Upper labial flange continuous across tip of snout, enclosing anterior nostril and ethmoidal pore. Pectoral fin reduced to various degrees, absent in *H. polyzona* Bleeker. Caudal fin variously reduced, rays clearly visible in some species, concealed by flesh in others.

Teeth small, pattern of dentition variable. Intermaxillary teeth continuous with those of vomer; in some species an additional tooth on midline above and anterior to main patch (called an "ethmoid tooth" by Böhlke, 1957). Vomerine tooth patch usually rather long, sometimes wider anteriorly. Maxillary and mandibular teeth in bands, at least anteriorly; in some species posterior maxillary teeth uniserial, sometimes enlarged. Pterygoid teeth sometimes present, usually absent.

Size. Probably to about 700 mm TL, but most species are smaller.

Distribution. Circumtropical in clear, relatively shallow water.

Etymology. From the Greek *heteros* (different) and *Conger* (the conger eel). Masculine.

Species. Fourteen species are currently recognized in *Heteroconger*. They are *Heteroconger polyzona* Bleeker, 1868, Indonesia and Philippines; *H. longissimus* Günther, 1870, eastern Atlantic; *H. chapmani* (Herre, 1923), eastern Pacific; *H. digueti* (Pellegrin, 1923), eastern Pacific; *H. herrei* (Wade, 1946), eastern Pacific; *H. halis* (Böhlke, 1957), western Atlantic; *H. hassi* (Klausewitz and Eibl-Eibesfeldt, 1959), Indo-West Pacific; *H. obscurus* (Klausewitz and Eibl-Eibesfeldt, 1959), Indian Ocean; *H. canabus* (Cowan and Rosenblatt, 1974), eastern Pacific; *H. camelopardalis* (Lubbock, 1980), Ascension Island and Fernando de Noronha, South Atlantic; *H. cobra* Böhlke and Randall, 1981, Indo-West Pacific; *H. lentiginosus* Böhlke and Randall, 1981, western Pacific; *H. perissodon* Böhlke and Randall, 1981, Indonesia and Philippines; and *Heteroconger luteolus* n. sp., described below from the Gulf of Mexico.

KEY TO THE SPECIES OF *HETEROCONGER* IN THE WESTERN NORTH ATLANTIC

1a. Large, conspicuous spots . . . . . . *camelopardalis*, p. 489
1b. Solid color, no spots 2
2a. Brown anteriorly, grading into pale yellow posteriorly; vertebrae 158–168 . . . . . . *halis*, p. 485
2b. Yellow anteriorly; vertebrae 139–148 . . . . . . *luteolus*, p. 488

Heteroconger halis (Böhlke, 1957)

BROWN GARDEN EEL

Figures 523–526

Nystactes halis Böhlke, 1957: 68, figs. 1a, 2–7 (original description, Green Cay, Bahamas, holotype ANSP 75156).

Nystactichthys halis. Böhlke, 1958: 59 (replaced *Nystactes*, preoccupied, with *Nystactichthys*).

Taenioconger halis. Randall, 1967: 665, 682.

Heteroconger halis. Böhlke and Randall, 1981: 368.

Misidentifications:

Not *Nystactichthys halis* (Böhlke). Klausewitz, 1971 (probably *Heteroconger longissimus*, see below).

Not *Nystactichthys halis* (Böhlke). Smith, 1979: 31, fig. 46 (*Heteroconger luteolus*).

Distinctive Characters. Heteroconger halis, H. longissimus, and *H. luteolus* have the posterior maxillary teeth markedly enlarged and pointed forward. In other species of the genus, the posterior maxillary teeth are not enlarged or only slightly enlarged. *Heteroconger halis* differs from *H. luteolus* in having more vertebrae (157–169 vs 137–148) and in being dark brown anteriorly instead of yellow. It differs from *H. longissimus* apparently by having slightly fewer vertebrae (157–169 vs 166–175) and preanal lateral-line pores

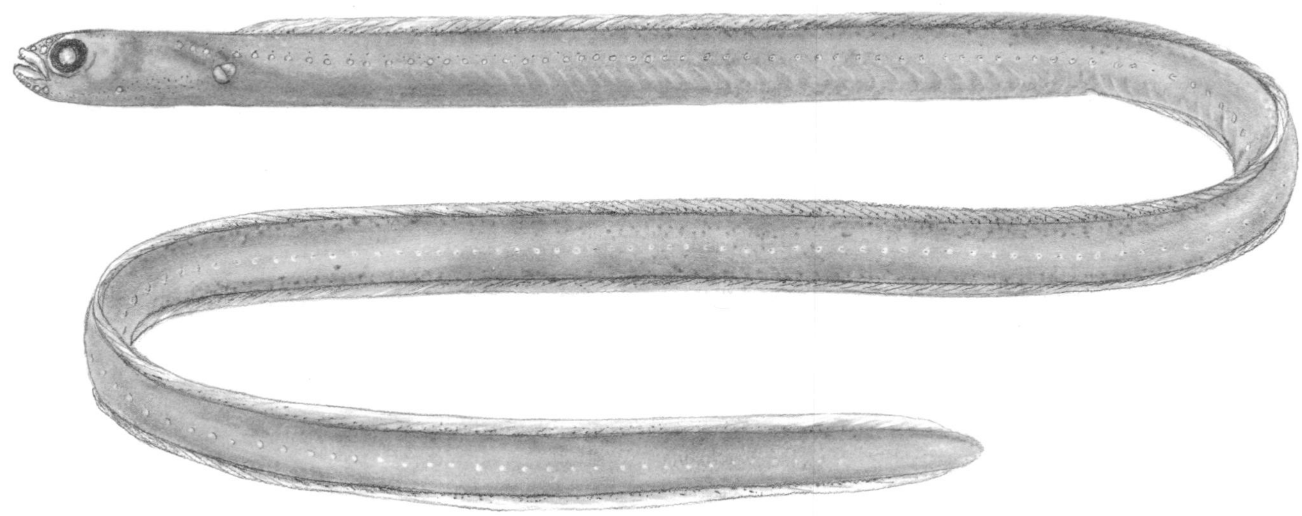

FIGURE 523. *Heteroconger halis*: ANSP 157599, 272 mm TL.

(45–51 vs 51–54), lacking postorbital pores, and lacking an external caudal fin.

Description (based on 68 specimens plus data from Böhlke, 1957). Total vertebrae 157–169 (n = 64), predorsal vertebrae 7–9 (48), preanal vertebrae 51–58 (49) precaudal vertebrae 59 (1). Branchiostegal rays 8–9 (2). Dorsal-fin rays 469 (1), anal-fin rays 311 (1), pectoral-fin rays 11–12 (2), caudal-fin rays 6–8 (2). Pores: LL to anus 45–51 (11), POM 4–6 (10), IO 2 (10) + 0 (10) + 0 (10), SO 1 + 2 (9), STC 0 (10).

Proportions as % TL: preanal 30–34 (22), predorsal 6–7 (22), head 5–6 (22), depth at anus 1–2 (22). Of preanal length: predorsal 18–21 (12), head 15–18 (12). Of HL: snout 11–14 (22), eye 16–24 (22), interorbital 11–17 (14), snout-rictus 16–24 (21), gill opening 7–12 (22), interbranchial 20–22 (3).

Body greatly elongate, slightly compressed, the degree of compression uniform throughout length. Dorsal fin begins above or slightly behind gill opening. Tip of tail rather hard, fleshy, without external caudal fin, although buried caudal-fin rays present. Pectoral fin small, almost vestigial. Gill opening crescentic, at mid-side, oblique, upper corner anterior to lower corner. Lateral line nearly complete, but slightly fewer pores than vertebrae.

Head short, depth about equal to that of body behind gill opening. Skin of throat and branchial region with longitudinal folds or grooves ventrally. Mouth small and oblique, lower jaw projects beyond upper, rictus near level of anterior margin of eye; well developed flange on upper and lower lips, continuous around tip of snout. Snout very short, less than eye diameter. Eye large, somewhat elongate in an anteroventral to posterodorsal direction; pupil also elongate with a space in front of lens. Anterior nostril tubular, near tip of snout, those of right and left sides close together, directed anteriorly, enclosed by labial flange. Posterior nostril round, just in front of eye at mid-eye level. Head pores reduced in number. Preoperculomandibular canal with four or five pores before rictus and one pore behind rictus (one specimen lacks the pore behind rictus on one side). Infraorbital canal with only two pores, one beneath labial flange near tip of snout and one above flange before rictus. Ethmoidal pore opens through labial flange just above and

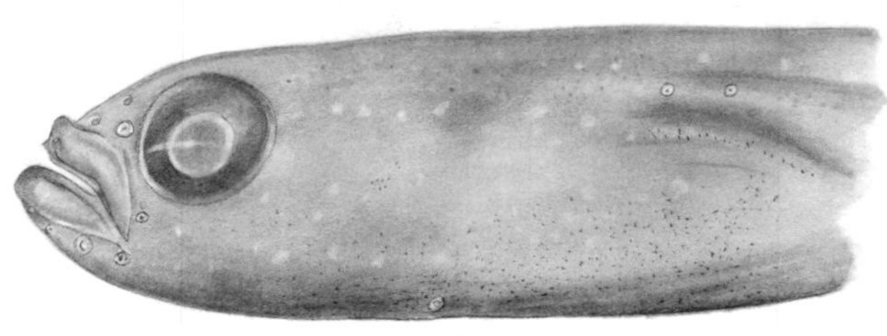

FIGURE 524. *Heteroconger halis*: ANSP 157599, 272 mm TL; head.

medial to anterior nostril. Supraorbital canal with two pores, one just above labial flange and one posterior to first pore. No supratemporal pores.

Intermaxillary teeth confluent with vomerine teeth, the whole forming an elongate patch widest anteriorly, narrowing at about midlength, not quite reaching level of posterior end of maxillary teeth. A single enlarged tooth in front of main intermaxillary patch, at anterior tip of ethmovomer, directed anteriorly (the ethmoid tooth of Böhlke, 1957). Maxillary teeth in bands anteriorly, narrowing to a single series posteriorly, the posterior teeth enlarged and curved forward; maxillary tooth patches of each side flare abruptly outward slightly behind their midpoint and then turn in again to become parallel to each other; the enlarged canine teeth are in this outwardly displaced section. Mandibular teeth in bands, widest anteriorly, narrowing to a single series at posterior end; posteriormost teeth somewhat enlarged, but not as much as maxillary teeth; mandibular tooth patch not flared outward like maxillary patch. Anterior teeth in all series curved posteriorly, posterior teeth curved anteriorly. No teeth on pterygoid.

Stomach very short, about one-sixth distance from gill opening to anus. Gas bladder long, thin-walled, appears to be attached to vertebrae.

Color. Anterior two-thirds of body and head dark brown, flecked with numerous small, pale spots; abdomen and throat paler; a gradual transition posteriorly to a pale yellowish color, tip of tail practically without pigment. No blotches, bands, or lines anywhere on body. Stomach and intestine pale. A 78-mm specimen (ANSP 138794) has some larval melanophores still visible.

Size. The largest specimen examined was 481 mm TL (ANSP 75162). Five females of 258–274 mm TL (ANSP 138794) had relatively large eggs in the body cavity. A male of 294 mm TL (ANSP 138794) had enlarged testes.

Growth and Development. The leptocephalus has been identified (see p. 737).

Remarks. The distinction between *Heteroconger halis* and the eastern Atlantic *H. longissimus* is slight and somewhat uncertain. Radiographs of

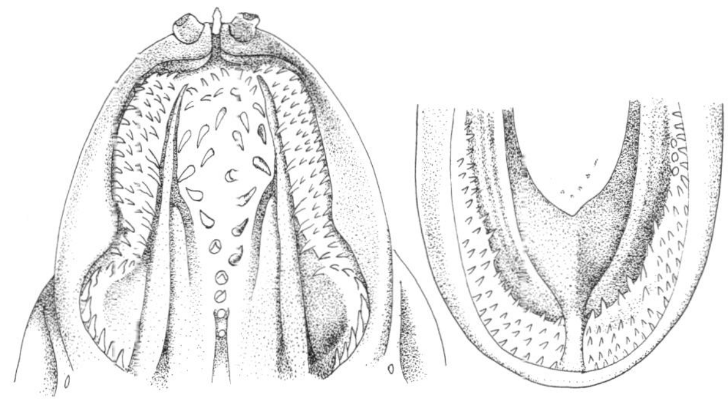

FIGURE 525. *Heteroconger halis*: ANSP 152111, 321 mm TL; dentition.

seven specimens of *Heteroconger* from the eastern Atlantic (see study material) showed 166–175 vertebrae, slightly more than in *H. halis* (157–169). Preanal vertebrae were also more numerous in these specimens (54–64) than in *H. halis* (51–58). I examined two of the specimens (ANSP 84096 and CAS-SU 62417, both from Madeira) and could find no differences from *H. halis* other than the vertebral counts. Both specimens had 51 preanal LL pores, within the range of *H. halis* although at its upper end, and there were no postorbital pores. Pectoral fins were present as small flaps, but there was no external caudal fin. The dentition was identical to that of *H. halis.* Blache and Bauchot (1976a: 423), however, described the two specimens from Senegal (MNHN 1975-1171 and 1975-1172) as having two postorbital pores and an external caudal fin, features not found in *H. halis.* Both specimens had 54 preanal LL pores, more than in *H. halis* and the two Madeira specimens. The status of these specimens and the relationship between *Heteroconger halis* and *H. longissimus* cannot be determined without further study. For the time being, the two species will be retained.

Klausewitz (1971) reported *Heteroconger halis* from the Canary Islands, the type locality of *H. longissimus.* Assuming that the two species are distinct, Klausewitz's fish was probably *H. longissimus* rather than *H. halis.*

Distribution (Fig. 526). Florida Keys, Bahamas, and throughout the Caribbean. Absent from the Gulf of Mexico. Depth approximately 10–50 m, but the lower end of the range is uncertain. Common.

Etymology. From the Greek *hals* or *halos* (sea, salt) and *-is* (pertaining to, belonging to). An adjective.

Study Material. 131 specimens, 78–481 mm TL.

Holotype: ANSP 75156 (285 mm TL), 25°07′00″N, 77°11′15″W, ca 0.4 km north of east end of Green Cay, Bahamas, fine white sand off isolated coral head, depth ca 10–11 m (35 feet), 16 May 1956.

Paratypes: BAHAMAS: ANSP 75157 (9, 257–303), 75158 (1, stained), 75159 (1, 230), 75160 (3, 158–271), 75161 (1, cleared and stained), 119074 (1, 276). PUERTO RICO: ANSP 75162 (1, 481).

Other material: FLORIDA KEYS: UMML 6876 (1, 322). BAHAMAS: ANSP 102892 (2, 135–210). JAMAICA: ANSP 126970 (2, 248–261), 157599 (1, 272). LESSER ANTILLES: ANSP 103295 (89, 260–327), 108494 (1, 139+), 116443 (1, cleared and stained), 152111 (1, 321). HONDURAS: ANSP 138793 (3, 308–346), 138794 (10, 78–299), MBI 14 (1, 274, specimen missing). LOCALITY UNKNOWN: ANSP 156858 (1, cleared and stained).

Comparative material (*Heteroconger longissimus*): MADEIRA: ANSP 84096 (1, 481). MNHN 56-62 (2). CAS-SU 62417 (1, 558). CANARY ISLANDS: BMNH (holotype). SENEGAL: MNHN 1975-1171, 1975-1172 (2, 243–345).

Heteroconger luteolus new species

YELLOW GARDEN EEL

Figures 526, 527

Nystactichthys halis (not of Böhlke, 1957). Smith, 1979: 31, fig. 46 (leptocephalus).

Nystactichthys sp. Williams and Shipp, 1980: 18.

Distinctive Characters. Heteroconger luteolus resembles *H. halis* in most of its characters, including the presence of an ethmoid tooth and enlarged, anteriorly directed posterior maxillary teeth. It has distinctly fewer vertebrae than *H. halis* (137–148 vs 157–169), however, and a greater predorsal length (23% preanal vs 18–21). In addition, *H. luteolus* is yellow anteriorly instead of brown.

Description. Total vertebrae unknown, but total myomeres in leptocephali 137–148 (n = 44). Predorsal vertebrae 8–9 (2), preanal vertebrae 52 (2). Pores: LL to anus 46 (1), POM 5 (1), IO 2 (1) +0 (1) + 0 (1), SO 1 + 2 (1), STC 0 (1).

Proportions as % of preanal length: predorsal 23 (2), head 18 (2), depth at anus 5 (1). Of HL: snout 12 (2), eye 17–18 (2), snout-rictus 16 (1), gill opening 6 (1), interbranchial 23 (1).

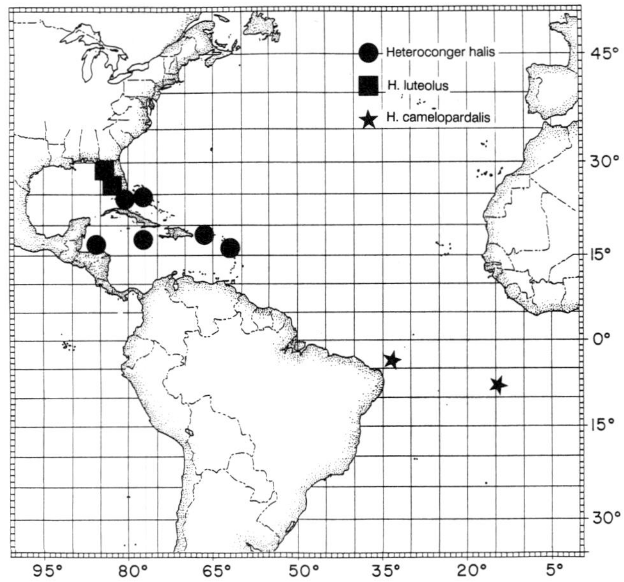

FIGURE 526. Distribution of *Heteroconger halis, H. luteolus,* and *H. camelopardalis.*

Otherwise like *H. halis.*

Color. Color in the freshly caught holotype was described by Williams and Shipp (1980: 19): "The specimen was bright yellow dorsally, with numerous fine scattered melanophores. Yellow extended laterally to just beneath the eye and to the upper level of the gill opening. Lower flanks and belly were white. Lips were heavily pigmented distally. Distinct spots or other markings were absent on anterior portions of the body." In preservative, the color is pale yellow with scattered tiny, dark spots.

Size. The two specimens are 148 mm and 105 mm TL, but both are incomplete. Preanal lengths are 74 mm and 66 mm, respectively. How representative of their species these two specimens are is unknown, but they are relatively small by the standards of *Heteroconger halis.*

Development and Growth. For a description of the leptocephali see p. 738.

Remarks. Heteroconger luteolus is closely related to *H. halis.* The two species share such specialized features as the median ethmoid tooth and the enlarged, forwardly directed posterior maxillary teeth. They differ mainly in the color and in the

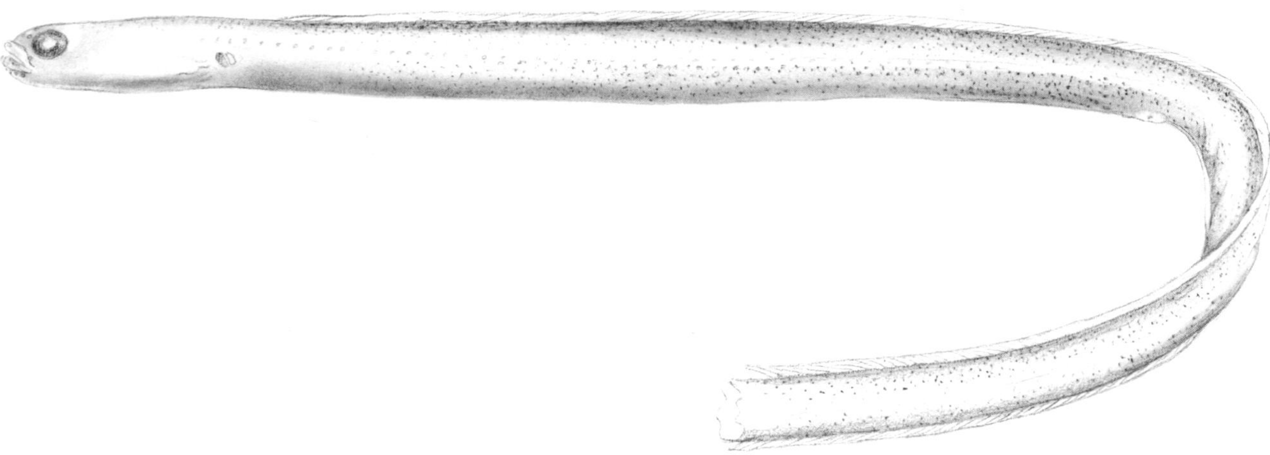

FIGURE 527. *Heteroconger luteolus*: ANSP 153941, holotype, 148+ mm TL.

number of vertebrae. In addition they seem to have mutually exclusive distributions. *Heteroconger luteolus* is found only in the Gulf of Mexico; *H. halis* occurs nearly everywhere in the tropical western Atlantic except the Gulf of Mexico.

Based on the abundance of leptocephali, this is a fairly common species, yet only two adult specimens are known. Why the adults should be so rarely collected is a matter of speculation. Perhaps they live in deeper water than *H. halis*, or perhaps they simply inhabit places not frequented by divers.

Distribution (Fig. 526). *Heteroconger luteolus* appears to be restricted to the Gulf of Mexico, where it replaces *H. halis.* The two adults were collected off the west coast of Florida in depths of 33 and 37 m. In addition, leptocephali have been collected just north of Campeche Bank, in Campeche Bay, and off the lower Texas coast, indicating that the species is widely distributed in the Gulf of Mexico. A few leptocephali have been collected in the Straits of Florida and the Gulf Stream off North Carolina, but these could easily have been carried from the Gulf by the currents. The one adult *Heteroconger* examined from the Florida Keys (UMML 6876) proved to be *H. halis.*

Etymology. From the Latin *luteus* (yellow) and *-olus* (diminutive), in reference to the color. An adjective.

Study Material. Two specimens, both incomplete.

Holotype: ANSP 153941 (1, 148+), eastern Gulf of Mexico off Florida, 28°42′00.3″N, 84°20′00.7″W, 33 m, by box core, 30 Aug. 1977.

Paratype: ANSP 153942 (1, 105+), eastern Gulf of Mexico off Florida, 26°24′N, 82°38′W, 37 m, by dredge, date "9/4/66."

Heteroconger camelopardalis (Lubbock, 1980)
Figure 526

Taenioconger camelopardalis Lubbock, 1980: 285, fig. 1 (original description, Ascension Island, holotype BMNH 1979.1.5.17).

Distinctive Characters. Heteroconger camelopardalis is distinguished from all other species by its distinctive color pattern of giraffe-like spots posteriorly and irregular diagonal bands anteriorly. Otherwise similar to *H. halis.*

Description (from Lubbock, 1980). Total vertebrae 171–173 (n = 2), preanal vertebrae 60 (2). Dorsal-fin rays 546–595 (2), anal-fin rays 312–378 (2), pectoral-fin rays 14 (2), caudal-fin rays ca 8 (2). Total LL pores 155–157 (2).

Proportions as % of TL: preanal 35.1–36.8 (2), predorsal 4.7–4.9 (2), head 5.3–5.4 (2), depth at anus 1.5–1.6 (2), depth at gill opening 1.7–1.9 (2), last LL pore to tail tip 2.4 (2). Of HL: snout 11.8–12.2 (2), eye 14.4–17.3 (2), interorbital 11.8–12.7 (2), upper jaw 26.2–27.7 (2), width of upper jaw 19.1–21.7 (2), longest pectoral ray 10.0–10.7 (2), gill opening 7.4–10.0 (2), body width at gill opening 22.5–25.9 (2), body width at anus 23.6–30.0 (2).

Dentition similar to that of *H. halis* but posterior maxillary teeth only slightly enlarged. Ethmoid tooth absent. Head pores as in *H. halis.* Pectoral fin small but more than just a fleshy tag; when folded forward covers gill opening. External caudal fin present on paratype (pers. obs.), but stated to be absent by Lubbock (1980: 287).

Color. "In life, body of holotype with a reticulate pattern of greyish brown spots on a whitish background, the spots becoming smaller and less conspicuous anteriorly. Anterior fifth of body with distinctive dark brownish olive markings on dorsal contour and flanks; markings irregularly shaped posteriorly, but forming two diagonal bands anteriorly which are joined dorsally; anteriormost band extends over gill opening and pectoral fin base on to posterior part of operculum. Head mottled light greyish brown, becoming whitish ventrally; a vertical row of four small white spots just posterior to eye, with another vertical row of three larger white spots about half way from eye to gill opening; lips mostly black, iris beige, darker posterodorsally; fins hyaline."

"In alcohol, pattern remains clearly visible but colours fade to shades of brown." (Lubbock, 1980: 287)

Size. The three known specimens are 385–501 mm TL.

Growth and Development. Leptocephali of the *Leptocephalus inferomaculatus* type collected off equatorial Brazil have a myomere count similar to the vertebral count of *H. camelopardalis* (p. 739), but confirmation of such an identification is lacking. Indeed, it seems unlikely in view of the general similarity of *H. camelopardalis* to *H. halis.*

Variation. The single specimen from Fernando de Noronha is stated to differ slightly in coloration from the Ascension Island specimens (Lubbock, 1980: 287).

Remarks. Lubbock (1980: 287) described the vertical fins as "terminating (externally) before tail tip" and the tail tip itself as having a "rather hard, fleshy covering." In the paratype examined by me (USNM 219406), however, the caudal fin, though reduced, was clearly present.

Distribution (Fig. 526). Ascension Island and Fernando de Noronha. Although both localities are in the South Atlantic, the latter is within four degrees of the equator. "Found on sand plains or occasionally rubble at 20–>50m depth" (Lubbock, 1980: 287).

Etymology. Latin *camelus* (camel) and *pardus* (leopard), together constituting the ancient name of the giraffe, and *-alis* (having the nature of), in reference to its giraffe-like spots. An adjective.

Study Material. One specimen, 408 mm TL.

Paratype: USNM 219406 (1: 408), Ascension Island, Hummock Point, depth 20 m on open sandy slope, date "10.1.1978."

Subfamily BATHYMYRINAE

Characters. Body moderate to moderately stout, preanal length more than 40% TL. Caudal fin reduced, tip of tail stiff. Pectoral fin well developed. Mouth moderate, rictus below eye. Lips well developed, maxillary and mandibular teeth concealed when mouth closed. Well developed flange on upper lip. Teeth small, in granular patches. Posterior nostril below mid-eye level. Dorsal- and anal-fin rays unsegmented. Air bladder generally thick-walled, attached to vertebrae dorsally. Lateral ethmoid process present; pterygoid stout; five suborbital bones; posterior edge of opercle smooth; preorbital bone solid, lower edge without ventral projections; lateral-line ossicles solid and well formed, without cartilaginous loops and processes; ribs present; precaudal vertebrae more than 40% of total; myorhabdoi present or absent.

Distribution. Worldwide, tropical to warm temperate, continental shelf and upper slope.

Genera. The Bathymyrinae can be divided into two principal groups, one containing *Paraconger* Kanazawa, 1961 and *Chiloconger* Myers and Wade, 1941, and the other containing *Ariosoma* Swainson, 1838, *Bathymyrus* Alcock, 1889, and *Parabathymyrus* Kamohara, 1938. The former seems more primitive in several characters than the latter.

TABLE 45. Selected character states in species of *Ariosoma* and *Parabathymyrus*.

Species	PO pores	SO pores	STC pores	Pre-caudal verte-brae as % total	Myo-rhabdoi	Stom-ach	Gut of leptocephalus	Epicen-tral proc-esses	Trifur-cations of uro-hyal	Posterior nostril
A. balearicum	3	6	3	ca 50	absent	short	sessile	present	long	exposed
A. selenops	2	4	0	ca 54	present	long	short free section	absent	long	exposed
A. coquettei	0	4	0	ca 48	present	short	long free section	?	?	exposed
A. anale	0	4	0	ca 58	present	long	long free section	absent	long	exposed
P. oregoni	0	4	0	ca 40	present	long	long free section	absent	short	covered by flap

Paraconger retains the supraoccipital bone and the adnasal pore and has a simple urohyal. Its leptocephalus has the ventral and midlateral pigment in the form of single, moderate-sized melanophores. The ventral melanophores are superficial and lie along the ventral or lateral aspect of the gut. This leptocephalus resembles those of the *Conger* group and the heterocongrines more than it does the more specialized leptocephalus of the *Ariosoma* group. *Chiloconger* has not been studied in detail, but it has the external characters of the Bathymyrinae. Larval pigment retained in a small specimen of *C. labiatus* (CAS 38794) resembles that of *Paraconger* rather than *Ariosoma*, and on that basis *Chiloconger* would be referred to the *Paraconger* branch of the subfamily.

The species of *Ariosoma*, *Bathymyrus*, and *Parabathymyrus* have lost the supraoccipital bone and the adnasal pore, and the posterior end of the urohyal is trifurcate. Their leptocephali are large with a long gut and a very short dorsal fin. The lateral pigment is in the form of short oblique rows of tiny melanophores along the myosepta just below the midlateral line. The ventral melanophores are very small and lie internally on the surface of the gut rather than externally. In the leptocephali of some species the gut is longer than the body and trails free posteriorly. In several characters the species of these genera form a graded series with *Ariosoma balearicum* at one end and *Parabathymyrus* at the other. Some of the advanced characters found in *Parabathymyrus* occur in some of the other species of *Ariosoma* as well (Table 45).

Hypothesized relationships of the genera of Bathymyrinae and Heterocongrinae are shown in Figure 528.

KEY TO THE GENERA OF BATHYMYRINAE IN THE WESTERN NORTH ATLANTIC

1a. Posterior nostril covered by a flap .... *Parabathymyrus*, p. 503
1b. Posterior nostril exposed 2
2a. Upper end of gill opening at middle of pectoral-fin base; jaw teeth in bands . . *Ariosoma*, p. 491
2b. Upper end of gill opening at or above upper end of pectoral-fin base; jaw teeth in one or two series forming a cutting edge .... *Paraconger*, p. 506

Genus *Ariosoma* Swainson, 1838

Ariosoma Swainson, 1838: 220 (type species *Ophisoma acuta* Swainson, 1839 = *Muraena balearica* Delaroche, 1809, by subsequent designation of Bleeker, 1864b: 20).

Ophisoma Swainson, 1839: 334 (type species *Ophisoma acuta* Swainson, 1839 = *Muraena balearica* Delaroche, 1809, by subsequent designation of Bleeker, 1864b: 20).

Helmichthys Costa, 1846: 1 (type species *Helmichthys diaphanus* Costa, 1846, by monotypy). Leptocephalus.

Congermuraena Kaup, 1856a: 71; 1856b: 108 (type species *Muraena balearica* Delaroche, 1809, by subsequent designation of Bleeker, 1864a: 116).

?*Diaphanichthys* Peters, 1864: 399 (type species *Leptocephalus brevicaudus* Peters, 1864, by monotypy). Leptocephalus.

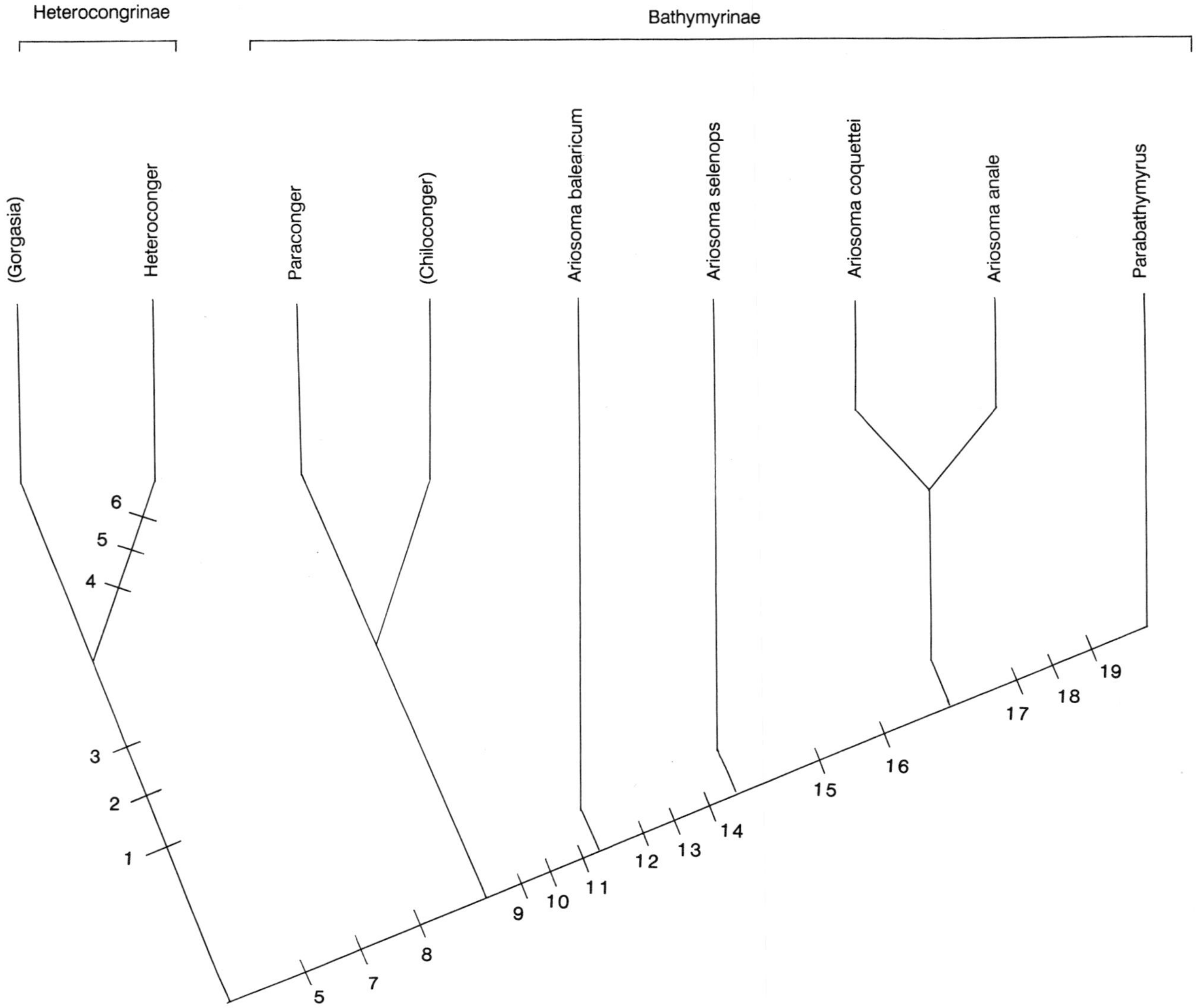

FIGURE 528. Hypothesized relationships among the genera of Heterocongrinae and Bathymyrinae. The following are presumed derived character states: (1) reduction of circumorbital bones; (2) reduction of pectoral fin; (3) "garden eel" habits; (4) loss of lateral ethmoid process; (5) loss of hypohyals; (6) inclusion of anterior nostril and ethmoidal pore by upper labial flange; (7) number of abdominal vertebrae more than 40% of total; (8) short dorsal fin in leptocephalus; (9) *Ariosoma*-type leptocephalus; (10) trifurcate urohyal; (11) loss of supraoccipital; (12) loss of one postorbital pore, two supraorbital pores, and all supratemporal pores; (13) short trailing gut in leptocephalus; (14) myorhabdoi present; (15) loss of remaining postorbital pores; (16) long trailing gut in leptocephalus; (17) forks on urohyal reduced; (18) ventral displacement of posterior nostril; (19) short snout.

Congromuraena Günther, 1870: 40 (emendation of *Congermuraena*).

Congrellus Ogilby, 1898: 286 (type species *Muraena balearica* Delaroche, 1809, by original designation).

Anago Jordan and Hubbs, 1925: 191 (type species *Conger anago* Temminck and Schlegel, 1846, by original designation).

Alloconger Jordan and Hubbs, 1925: 192 (type species *Leptocephalus flavirostris* Snyder, 1908, by original designation).

Nesocongrus Whitley, 1935: 219 (type species *Congermuraena howensis* McCulloch and Waite, 1916, by original designation).

Thyreoconger Wade, 1946: 189 (type species *Thyreoconger hemiaspidus* Wade, 1946, by original designation).

Discussion of Synonymy. Ariosoma has a most confusing nomenclatural history. Swainson (1838: 220) established *Ariosoma* in a brief description without including any species. He repeated this treatment early in the second volume of this work (1839: 196). Later in the same volume (p. 334) he omitted *Ariosoma* but introduced *Ophisoma,* with essentially the same description and position relative to the other genera as *Ariosoma.* No one seriously questions the fact that *Ophisoma* was intended as a replacement name for *Ariosoma.* For *Ophisoma* Swainson included two species, *acuta* and *obtusa,* without designating a type species. According to Article 69 (a) (i) (1) of the ICZN (1985), *acuta* and *obtusa* must be recognized as the originally included species of *Ariosoma.* Kaup (1856a: 71) described *Congermuraena* and included, without designating a type species, three species: *Congrus habenatus* Richardson, 1848, *Muraena balearica* Delaroche, 1809, and *Muraena mystax* Delaroche, 1809. Bleeker (1864a: 116) designated *Muraena balearica* Delaroche as the type species of *Congermuraena.* Shortly thereafter (1864b: 20) he synonymized *Congermuraena* with *Ariosoma* and *Ophisoma,* recognizing *Ophisoma* as the valid genus. He then stated "spec. typ. *Ophisoma balearicum* Blkr. = *Ophisoma acuta* Swns. ?" In other words, he synonymized *O. acuta* with *Muraena balearica* and designated that species as the type of *Ophisoma,* and therefore of *Ariosoma.* Ginsburg (1951: 439) rejected this for two reasons. First, *balearica* was not one of the originally included species. Article 69 (a) (v) of the ICZN (1985) permits it, however, because Bleeker synonymized *balearica* with *acuta,* which was an originally included species. Ginsburg's second objection was that Bleeker's use of a question mark in his synonymy invalidated the type-designation. Article 67 (c) states: "the term 'designation' in relation to the fixation of a type species must be rigidly construed; the following are not designations under the Code: . . . (3) one made in an ambiguous or conditional manner." Whether the rule applies in this case is debatable. Bleeker made it quite clear that he was designating a type species. Viewed another way, he designated *acuta* as the type while suggesting that *balearica* was synonymous.

Accepting this, a major question remains: what is the identity of *Ophisoma acuta*? The Mediterranean Sea, the type locality, contains only three species of congrid eels. *Conger conger* (Linnaeus) can be eliminated because it is almost certainly the basis of *Ophisoma obtusa.* Swainson's description of *obtusa* (dorsal fin commencing behind the end of the pectoral, fins margined in black) could fit only *Conger.* Remaining are *Muraena balearica* and *M. mystax* Delaroche. Swainson mentioned the silvery sides and blood-red pectoral fin in his account of *acuta,* features characteristic of *balearica.* If the measurements given by Swainson are converted to percentages, *acuta* has a preanal length of 45% TL, a value characteristic of *balearica* but greater than that of *mystax.* His illustration (Swainson, 1839: fig. 111a) also seems to resemble *balearica* more than *mystax.* It seems likely that *Ophisoma acuta* is indeed a synonym of *Muraena balearica.* The interest of nomenclatural stability will be served best by recognizing Bleeker's designation of *balearica* (= *acuta*) as the type species of *Ariosoma. Ariosoma* has been universally applied in recent years (except by Ginsburg), with *Congermuraena* disappearing except for occasional use as a catch-all.

Characters. Body moderately stout, not extremely elongate; preanal length more than 40% of total; tip of tail blunt and stiff, caudal fin reduced. Dorsal fin begins near level of pectoral-fin base. Pectoral fin well developed. Snout rounded, projecting slightly beyond lower jaw; upper and lower lip with well developed flange. Anterior nostril tubular, near tip of snout, directed anterolaterally; posterior nostril small, elliptical, in front of eye somewhat below mid-eye level. Head pores small; pores along upper jaw located on side of head rather than on labial flange; adnasal pore absent.

Teeth small, acute, in patches or bands. Intermaxillary teeth not sharply separated from maxillary teeth; vomerine tooth patch elongate, from a third to half the length of maxillary patch; maxillary and mandibular teeth in narrow bands, wider anteriorly.

Lateral ethmoid process present; lower edge of preorbital bone without projections; five suborbital bones; supraoccipital absent; posterior edge of opercle smooth; otic bulla present; urohyal trifurcate, the two lateral branches longer than median one; myorhabdoi present or absent; ribs present; precaudal vertebrae more than 40% total; epicentral processes on caudal vertebrae present or absent; anterior epineurals not fused to neural arches; lateral-line ossicles well formed, solid, rectangular or elliptical structures, slightly elongate horizontally, without cartilaginous loops or processes; dorsal- and anal-fin rays unsegmented. Digestive tract pale; stomach long or short.

Discussion. Ariosoma is one of the most abundant and speciose of congrid genera. The small size and secretive habits of most species make them unfamiliar to most people, but the abundance of their leptocephali in plankton collections indicates that they form an important part of the fauna.

Asano (1962) divided the Japanese species into two genera, *Anago* Jordan and Hubbs and *Alloconger* Jordan and Hubbs, based on the number of head pores, the relative number of precaudal and caudal vertebrae, and the presence or absence of myorhabdoi. *Anago* lacked postorbital, supratemporal and interorbital pores; had more precaudal than caudal vertebrae; and had myorhabdoi. *Alloconger* had a full complement of head pores, had about as many precaudal as caudal vertebrae, and lacked myorhabdoi. Among the western Atlantic species, *Ariosoma balearicum* fits the definition of *Alloconger* and *A. anale* agrees with *Anago*. To the characters given by Asano it can be added that *Alloconger* has a short stomach, reaching less than halfway to the anus, whereas *Anago* has a long stomach, reaching nearly to the anus. As substantial as these differences are, the two groups seem more like end points of a graded series than discrete genera (Table 45). *Ariosoma selenops* resembles *Anago* in having myorhabdoi, more precaudal than caudal vertebrae, and a long stomach, but it has postorbital and interorbital pores. *Ariosoma coquettei* has myorhabdoi and the pore pattern of *Anago*, but it also has a short stomach and slightly fewer precaudal than caudal vertebrae, both characteristic of *Alloconger*. *Ariosoma gilberti* from the eastern Pacific has a short stomach and lacks myorhabdoi (*Alloconger*) but lacks interorbital and supratemporal pores (*Anago*). It thus does not seem possible to divide the species of *Ariosoma* into more than one genus.

There is some doubt whether the species treated by Asano (1962: 72) as *Anago anago* (Temminck and Schlegel) is really the species described by those authors as *Conger anago*. Temminck and Schlegel (1846: 260) described the anus of *C. anago* as opening "sur la fin du quatrieme novieme de la longueur totale" (over the end of the fourth ninth of the TL); that is, equal to 44.4% of the TL. Asano gave a preanal length of 47–51% TL for the species he treated as *Anago anago*, 43–46% TL for *Alloconger* (= *Ariosoma*) *anagoides*, and 44–46% TL for *Alloconger shiroanago*. Thus the only diagnostic character retrievable from Temminck and Schlegel's description places their species with one of the "*Alloconger*" group. The next available name for Asano's species is *Congrellus meeki* Jordan and Snyder, 1900. If a distinct genus were recognized for this species, no name would be available for it.

In several characters, certain species of *Ariosoma* approach the condition found in *Parabathymyrus*. Indeed, of the eight characters in Table 45, *Ariosoma anale* actually shares more with *Parabathymyrus* than it does with *A. balearicum*. Although the decision is somewhat arbitrary, *Parabathymyrus* is recognized as a genus here because of its distinctive appearance with its short snout and its concealed posterior nostril.

Size. These are small eels, probably never exceeding 400–500 mm TL.

Distribution. Circumtropical. Most species live on the continental shelf, from the shoreline to about 200 m, but at least one species, *A. selenops* Reid, occurs on the upper continental slope from 220–549 m.

Etymology. From the Greek *aris* (auger) and *soma* (body). Neuter.

Species. Further study is needed before a reliable figure can be given for the number of species contained in *Ariosoma*. Critical comparisons have not been made between similar species described from geographically distinct areas. Some additional species probably await discovery. The following provisional list includes 17 species with their known ranges. *Ariosoma balearicum* (Delaroche, 1809), North Atlantic and Indian Ocean; *A. anale* (Poey, 1858), tropical Atlantic; *A. coquettei*

Smith and Kanazawa, 1977, Suriname and French Guiana; *A. selenops* Reid, 1934, western Atlantic; *A. opistophthalmus* (Ranzani, 1839), western South Atlantic; *A. mellissi* (Günther, 1870), St. Helena, South Atlantic; *A. scheelei* (Strömman, 1896), Indo-West Pacific; *A. mauritianum* (Pappenheim, 1914), Indo-West Pacific; *A. nigrimanus* Norman, 1939, Indian Ocean; *A. bauchotae* Karrer, 1982, Indian Ocean; *A. obud* Herre, 1923, Philippines; *A. meeki* (Jordan and Snyder, 1900), Taiwan to Japan; *A. anagoides* (Bleeker, 1864), Indonesia to Japan; *A. anago* (Temminck and Schlegel, 1846), Taiwan to Japan (based on distribution, *Alloconger major* Asano, 1958 is probably a junior synonym); *A. shiroanago* (Asano, 1958), Japan; *A. marginatum* (Vaillant and Sauvage, 1875), Hawaii; and *A. gilberti* (Ogilby, 1898), eastern Pacific. *Ariosoma howensis* (McCulloch and Waite, 1916) was considered a synonym of *A. mauritianum* by Castle (1964b: 12), and *A. neoguinaicus* (Bleeker, 1859) and *A. fijiensis* (Ogilby, 1898) were tentatively synonymized with *A. scheelei* by Castle (1968c: 692). *Ariosoma somaliense* Kotthaus, 1968 was synonymized with *A. balearicum* by Bauchot and Blache (1979).

KEY TO THE SPECIES OF *ARIOSOMA* IN THE WESTERN NORTH ATLANTIC

1a. Three supratemporal pores; interorbital pores present; three postorbital pores . . *balearicum*, p. 495
1b. Supratemporal and interorbital pores absent; zero to two postorbital pores 2
2a. Two postorbital pores; interbranchial shorter than gill opening . . . . *selenops*, p. 499
2b. No postorbital pores (may occasionally be present in *A. anale*); interbranchial longer than gill opening 3
3a. Preanal length 43–45% TL; preanal lateral-line pores 47–53 . . . . *coquettei*, p. 501
3b. Preanal length 49–52% TL; preanal lateral-line pores 55–59 . . . . *anale*, p. 502

Ariosoma balearicum (Delaroche, 1809)
Figures 529–531, Tables 45, 46

Muraena balearica Delaroche, 1809: 314, 327, fig. 3 (original description, Ibiza, Balearic Islands, Mediterranean, holotype MNHN 8434).

Muraena cassini Risso, 1810: 91 (original description, Nice, Mediterranean, holotype lost according to Blache et al., 1973: 241).

Echelus ciuciara Rafinesque, 1810a: 65, pl. 15, fig. 3 (original description, Sicily, Mediterranean, holotype unknown).

Conger cassini. Risso, 1826: 205.

Ophisoma acuta Swainson, 1839: 396, fig. 111a (original description, Sicily, Mediterranean, holotype unknown).

Conger auratus Costa, 1844–1845, fasc. 48–49: 6, pl. 29 (original description, Naples, Mediterranean, holotype unknown).

Conger balearicus. Bonaparte, 1845: 38.

Helmichthys diaphanus Costa, 1846: 2, pl. 31 (original description, Naples, Mediterranean, holotype unknown). Leptocephalus.

Congermuraena balearica. Kaup, 1856a: 71; 1856b: 110.

Leptocephalus taenia Kaup, 1856b: 151, pl. 19, fig. 18 (original description, Atlantic, Indian, and western Pacific Oceans, 15 syntypes, MNHN), part, Atlantic specimens only. Leptocephalus.

Leptocephalus marginatus Kaup, 1856b: 152, pl. 19, fig. 19 (original description, Atlantic and Indian Oceans, several syntypes, MNHN), part, Atlantic specimens only. Leptocephalus.

Conger impressus Poey, 1860: 318 (original description, Cuba, holotype unknown).

Ophisoma balearicum. Bleeker, 1864b: 20.

Ophisoma impressus. Poey, 1867a: 248.

Ophiosoma impressus. Poey, 1868b: 424.

Congromuraena balearica. Günther, 1870: 41.

Congromuraena impressa. Poey, 1876a: 152.

Leptocephalus affinis Facciolà, 1883a: 4, pl. 4, fig. 1 (original description, Strait of Messina, Mediterranean, holotype unknown). Leptocephalus.

Leptocephalus inornatus Facciolà, 1883a: 5, pl. 5, fig. 2 (original description, Strait of Messina, Mediterranean, holotype unknown). Leptocephalus.

Leptocephalus eckmani Strömman, 1896: 16, pl. 1, figs. 1–3 (original description, central and western Atlantic, lectotype ZMUU). Leptocephalus.

Leptocephalus scheelei Strömman, 1896: 21, pl. 1, figs. 6–7 (original description, Timor Sea and western Atlantic, lectotype ZMUU), part, Atlantic specimens only. Leptocephalus.

?*Leptocephalus dentex* Collett, 1896: 153 (original description, North Atlantic, holotype unknown). Leptocephalus.

Congrellus balearicus. Ogilby, 1898: 286.

Leptocephalus rex Eigenmann and Kennedy, 1902: 86, fig. 3 (original description, Bahamas, two syntypes, USNM 49764). Leptocephalus.

Leptocephalus microphthalmus Beebe and Tee Van, 1928: 58, fig. (original description, Hispaniola, West Indies, holotype USNM 170906). Leptocephalus.

Ariosoma minor Howell Rivero, 1935: 339, fig. 1 (original description, Havana, Cuba, holotype MCZ 33452).

Ariosoma balearica. Fowler, 1936a: 273, fig. 127.

Congermuraena impressa. Ginsburg, 1951: 43.

Ariosoma somaliense Kotthaus, 1968: 27, figs. 121–122 (original description, Somalia, Indian Ocean, holotype ZMH IOES 52 a).

Misidentifications:

Not *Ophisoma balearicum* (Delaroche). Jordan and Davis, 1891: 661, part (records from eastern Pacific should be referred to *Ariosoma gilberti* and records from Brazil to *A. opistophthalmus*).

Not *Congermuraena balearica* (Delaroche). Jordan and Evermann, 1896a: 356, part (same comments as for *Ophisoma balearicum*, Jordan and Davis).

Not *Ariosoma balearica* (Delaroche). Fowler, 1941: 146 (Cape Frio, Brazil, = *Ariosoma opistophthalmus*).

Not *Arisoma* [sic] *balearica* (Delaroche). Miranda Ribeiro, 1961c: 3 (Brazil, = *Ariosoma opistophthalmus*).

Discussion of Synonymy. Like many early authors, Strömman (1896) did not designate holotypes for his newly described species. Bertin (1936a) assumed that he did, however, and some of Bertin's comments appear to have fulfilled the requirements for lectotype designation (ICZN, 1985: Article 74). For *Leptocephalus eckmani*, Strömman illustrated one of the 451 syntypes and gave a series of measurements for it. Bertin assumed that this specimen was the holotype and by so stating (. . . "le type figure par Strömman"), apparently selected it as the lectotype. With 126–131 myomeres, *L. eckmani* is clearly *Ariosoma balearicum.*

Leptocephalus scheelei was based on five specimens, one from the Timor Sea, three from the western North Atlantic, and one from the coast of Brazil. It is unclear how Strömman distinguished the North Atlantic specimens from his own *L. eckmani*, and he did not designate a holotype. Bertin (1936a: 3) decided that the three North Atlantic specimens belonged to *L. eckmani* and only the single Indo-Malayan specimen belonged with certainty to *L. scheelei* (the Brazilian specimen could not be found). Bertin thus left only a single specimen in *scheelei* and subsequently referred to it as "le type de Strömman." The latter statement again seems to be a valid lectotype selection. The lectotype was from the Timor Sea, and on this basis Castle (1964b: 3) applied the name to a species of *Ariosoma* from the Indo-West Pacific. Bertin assigned the three western Atlantic specimens to Strömman's *Leptocephalus eckmani*, which is the larva of *Ariosoma balearicum.*

Bertin (1935) also re-examined the types of Kaup (1856b), who did not designate holotypes either, but in this case Bertin said nothing whatever about types. For *Leptocephalus taenia*, he found that the five Atlantic specimens had 125–130 myomeres, hence must be *Ariosoma balearicum.* For *Leptocephalus marginatus*, only four of the several syntypes still existed when Bertin examined them. Only two of these, both from the Atlantic, were sufficiently well preserved to yield useful data. The myomere counts of ca 135 indicate that they are *Ariosoma balearicum.*

Distinctive Characters. In the western North Atlantic, *Ariosoma balearicum* is the only species of the genus with three supratemporal pores, three postorbital pores, and six supraorbital pores. It also has fewer vertebrae (120–136) than the other species.

Description. Total vertebrae 120–136 (n = 170), predorsal vertebrae 7–10 (15), preanal vertebrae 49–54 (15), precaudal vertebrae 62–66 (3). Pores: LL to anus 46–52 (64), POM 10 (33), IO 4 + 1 + 3 (33—one specimen had 5 in maxillary section), SO 1 + 5 (30), STC 3 (33—one specimen had a single pore here). Branchiostegal rays 7–10 (23). Pectoral-fin rays 10 (1).

Proportions as % of TL: preanal 45–49 (72), predorsal 15–19 (72), head 15–19 (72), depth at anus 4–8 (38), pectoral fin 4–7 (58). Of HL: snout 19–25 (69), eye 16–24 (69), snout-rictus 26–32 (39), gill opening 8–15 (68), interbranchial 12–24 (63), pectoral fin 28–40 (59).

Anus slightly before midlength. Dorsal fin be-

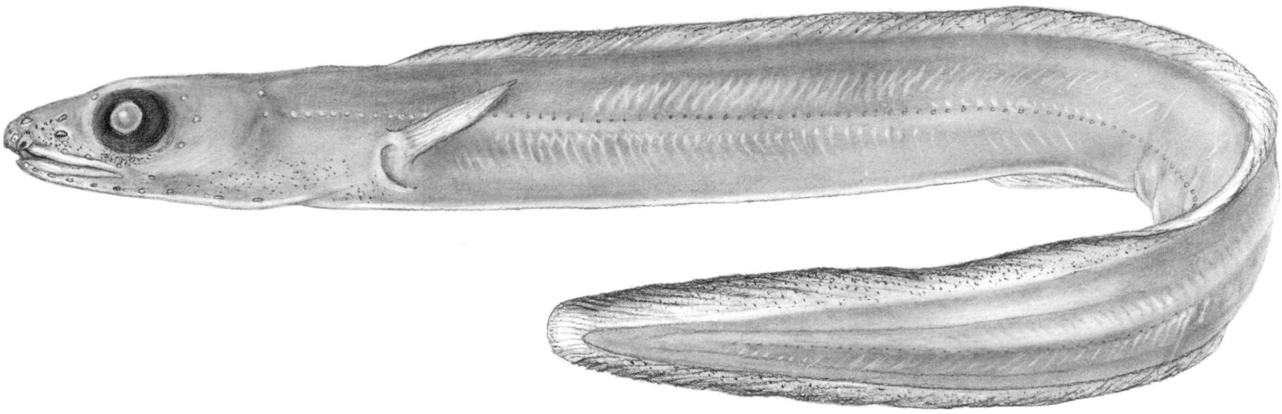

FIGURE 529. *Ariosoma balearicum*: ANSP 126967, 191 mm TL.

gins over or very slightly ahead of pectoral-fin base. Full complement of head pores present: five supraorbital pores, including two between eyes; eight infraorbital pores, including three behind eye; three supratemporal pores.

Intermaxillary tooth patch about as broad as long, indistinctly separated from maxillary teeth. Vomerine tooth patch elongate, separated from intermaxillary teeth, extending posteriorly to about midpoint of maxillary patch. Maxillary teeth in narrow bands, somewhat wider anteriorly. Mandibular teeth in bands, wider anteriorly.

Precaudal vertebrae about half total; epicentral processes on caudal vertebrae; myorhabdoi absent. Stomach short, ending less than half distance from base of stomach to anus.

Color. In life brown with silvery or golden reflections on lower half of flank; eye with crescentic orange area above pupil; pectoral fin red; margins of vertical fins dark. In preservative uniform brown with dark-edged vertical fins. Stomach and intestine pale.

Size. A rather small eel; the largest specimen examined was 336 mm TL (UMML 31032).

Development and Growth. The leptocephalus has been identified (see p. 726).

Habits and Behavior. Little direct information is available. The structure of the tail, stiff at the tip and with the caudal rays shortened, suggests that the fish burrows tail-first, and it probably spends a good part of its time buried in the substratum. Asano (1962: 136) observed a related species, *Ariosoma meeki* (Jordan and Snyder), in an aquarium in Japan and reported that during the day it entered the sand tail-first leaving only the head exposed. At night it emerged and swam about looking for food.

Radiographs show the gut frequently filled with an opaque, gravel-like material, suggesting that *Ariosoma balearicum* feeds largely on soft-bodied invertebrates picked off the bottom. Fish bones were absent and only one large crustacean, a stomatopod, appeared in some 170 radiographs examined.

Spawning appears to be widespread in *A. balearicum.* Small leptocephali less than 10 mm SL have been collected in the Gulf of Mexico, off the Bahamas, and in the Mediterranean. This

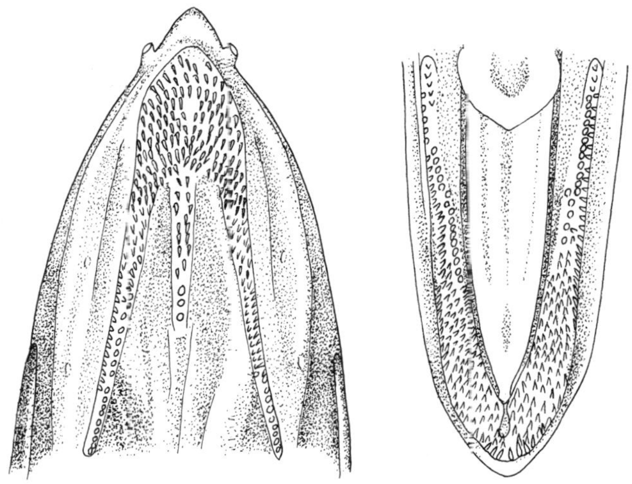

FIGURE 530. *Ariosoma balearicum*: ANSP 128394, 210 mm TL; dentition.

TABLE 46. Total vertebrae in *Ariosoma balearicum* from various locations. Areas: 1, U.S. Atlantic coast north of Palm Beach, Florida; 2, Biscayne Bay, Florida; 3, Florida Keys; 4, Gulf of Mexico; 5, Bahamas; 6, Bermuda; 7, Antilles; 8, Western Caribbean, Yucatan-Panama; 9, Colombia; 10, Guianas-Brazil; 11, Lebanon; 12, Western Atlantic composite low count; 13, Western Atlantic composite high count; 14, Western Atlantic northern high count; 15, Western Atlantic southern high count.

Area	120	121	122	123	124	125	126	127	128	129	130	131	132	133	134	135	136	137	n	Mean
1									1	-	6	9	6	2	1	-	2		27	131.5
2		1	-	2	2														5	123.0
3					1	-	1	-	1										3	126.0
4			2	-	1	2	-	-	-	1									6	124.5
5		3	-	1	7	3	3	3	-	1	1								22	124.9
6			1	-	-	-	1												2	124.0
7				1															1	123.0
8	1	-	-	-	3	5	1	1	1										12	124.8
9							1	-	-	1	-	1							3	128.7
10				1	-	-	-	2	5	4	6	-	-	2	1				21	129.2
11					1	5	11	15	14	12	7	2	1						68	127.7
12	1	4	3	4	14	10	6	4	2	2	1								51	124.6
13				1				2	6	4	12	9	6	4	2	-	2		47	130.6
14									1	-	6	9	6	2	1	-	2		27	131.5
15				1	-	-	-	2	5	4	6	-	-	2	1				21	129.2

plus the existence of geographically distinct populations indicates that this species does not undertake long spawning migrations.

Variation. It has been noted several times (Schmidt, 1912a; Lea, 1913; Castle, 1966a; Smith, 1971; Blache, 1977) that when the myomere counts of large numbers of leptocephali of *Ariosoma balearicum* are plotted as a frequency distribution, the result is a bimodal curve. Schmidt noted that specimens from the Gulf of Mexico and the Mediterranean had two or three fewer myomeres than those from the central and western Atlantic. Castle found that specimens collected off Mauritania had an average of five fewer myomeres (123–132) than Lea's sample from the central Atlantic (129–137). Blache found an additional difference in pigmentation between Lea's specimens and his own from the Gulf of Guinea.

Table 46 compares vertebral counts of adult *A. balearicum* from various places in the western Atlantic and one locality in the Mediterranean (Lebanon). The result shows two high-count populations, one off the U.S. east coast and the other off South America. The counts from these two areas, though close, differ by an average of about two vertebrae and seem certainly to represent distinct populations. Specimens from the western Caribbean, the Gulf of Mexico, the Florida Keys (including Biscayne Bay), the Bahamas, the West Indies, and Bermuda have a distinctly lower count, although there is overlap. The three specimens from Colombia seem to fall with the Guianas-Brazil group. The Lebanese specimens have a slightly lesser predorsal and head length than the western Atlantic specimens (15–17 vs 16–19% TL and 15–17 vs 15–19% TL, respectively), but the vertebral counts fall in the middle of the range of the western Atlantic populations.

The sharpest boundary between populations seems to lie along the lower east coast of Florida. Specimens from north of Jupiter Inlet all have high vertebral counts, whereas those from Biscayne Bay have low counts. No specimens from intermediate locations were available. It is difficult to explain how such a sharp gradient can be maintained by a species with a long-lived pelagic larva, unless the animals are responding to ecological factors such as bottom type, and leptocephali carried beyond the limits of their adult range cannot settle out and survive. This would imply that the populations in question are full species separated by a biological isolating mechanism. Further study may confirm this, but at the present state of knowledge it seems impractical to recognize as species populations whose only difference, the number of vertebrae, overlaps so broadly with neighboring populations. *Ariosoma balearicum* will be provisionally treated as a single widespread species which contains a number of more or less distinct geographic races.

Some variation also occurs in the number of branchiostegal rays. Of six specimens examined from the Guianas, five had nine branchiostegals and one had eight. Of eight specimens from the Caribbean and Gulf of Mexico, all had eight branchiostegals. Of six specimens from the U.S. east coast, three had eight branchiostegals, two had nine, and one had ten. Of three specimens from Lebanon, one had seven branchiostegals and two had eight.

Distribution (Fig. 531). Western North Atlantic from North Carolina to northern Brazil, including the Gulf of Mexico and the Caribbean. Eastern Atlantic from the Mediterranean to the Gulf of Guinea. Also recorded from the Red Sea and the Indian Ocean (Bauchot and Blache, 1979). Depth 1–732 m, although the great majority of specimens is collected on the continental shelf and most of those in less than 100 m. The species seems to prefer sandy bottoms and clear water. Based on the abundance of leptocephali, this must be one of the commonest eels in the western Atlantic.

Etymology. Named for the Balearic Islands in the Mediterranean. An adjective.

Study Material. 205 specimens, 72–336 mm TL. BERMUDA: ANSP 133631 (2, 86–114). NORTH CAROLINA: UMML 31033 (3, 195–324). USNM 155002 (1, 221), 195995 (2, 201–242). SOUTH CAROLINA: ANSP 114147 (1, 242), 128395 (1, 208). UMML 31032 (1, 336). GEORGIA: ANSP 127793 (3, 156–240), 128394 (1, 210), 128397 (1, 250), 128398 (1, 242). UMML 31026 (1, 249). USNM 179246 (1, 208). FLORIDA ATLANTIC: ANSP 128396 (1, 207). UMML 2909 (1, 75), 3399 (2, 181–214), 8383 (1, 152), 15582 (1, 144), 31024 (1, 157), 31025 (1, 117), 31027 (1, 236), 31028 (1, 172), 31029 (1, 198), 31030 (1, 202), 31031 (2, 159–216), uncat. (1, 200). USNM 155003 (1, 176), 185638 (1, 200), 197117 (1, 130). FLORIDA STRAITS: UMML 4451 (1, 100), 13662 (1, 89), 19620 (1, 140). FLORIDA GULF: ANSP 109608 (1, 214). USNM 158253 (1, 247), 185631 (1, 171). ALABAMA: ANSP 112047 (1, 196). USNM 174924 (1, 269), 198737 (1, 292). MEXICO GULF: USNM 158187 (2, 194–230). BAHAMAS: ANSP 102894 (2, 72–120), 102903 (1, 139), 111565 (1, 150). USNM 181346 (24, 72–147). CUBA: MCZ 9218 (1, 252). USNM 37568 (1, 212). PUERTO RICO: USNM 114659 (1, 115). COLOMBIA: UMML 28784 (3, 181–265). PANAMA: ANSP 126967 (1, 191). NICARAGUA: USNM 179237 (1, 78). MEXICO QUINTANA ROO: ANSP 114931 (1, 106), 114932 (11, 90–130). UMML 9733 (1, 93). GUYANA: UMML uncat. (1, 261, cleared and stained). SURINAME: RMNH 27397 (1, 215), 27401 (1, 188). UMML 29596 (1, 186). USNM 185632 (1, 193), 185655 (1, 203). FRENCH GUIANA: UMML 5030 (9, 136–188), 29593 (1, 178). USNM 158916 (1, 205), 185662 (1, 184). BRAZIL: USNM 218913 (1, 94). MADEIRA: ANSP 91116 (1, 292). ITALY: USNM 29747 (1, 324), 48283 (1, 265). LEBANON: USNM 196299 (2, 180–244), 198859 (7, 163–203), 218911 (76, 149–301), 218912 (1, 220).

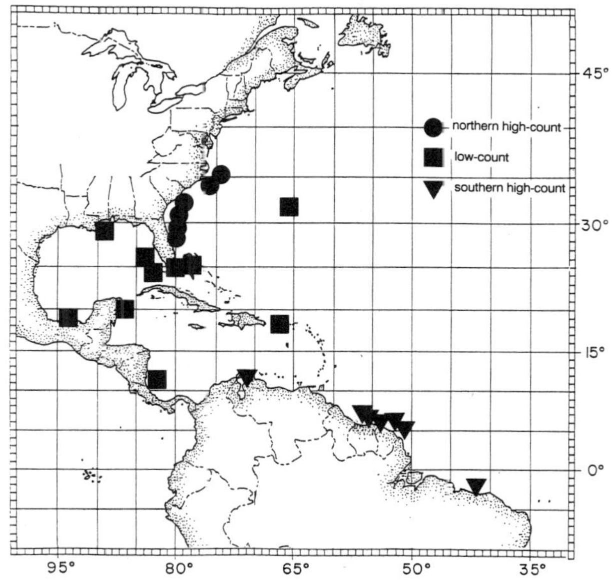

FIGURE 531. Distribution of *Ariosoma balearicum*.

Ariosoma selenops Reid, 1934
Figures 532, 533, Table 45

Ariosoma selenops Reid, 1934: 4, fig. 1 (original description, 18°40.5′N, 64°50.0′W, holotype USNM 93310).

Distinctive Characters. Ariosoma selenops is the only species of the genus in which the gill opening is wider than the interbranchial space. Like *A. anale* and *A. coquettei*, *A. selenops* has myorhabdoi, but unlike them it has two postorbital

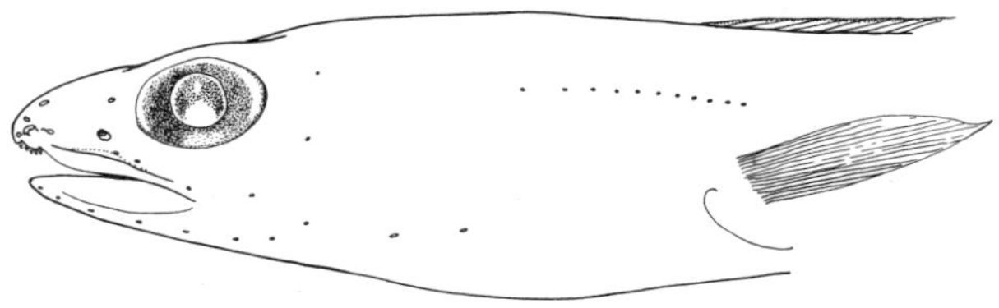

FIGURE 532. *Ariosoma selenops*: ANSP 116132, 383 mm TL; head.

pores. It has more vertebrae than the other species, and it lives in deeper water.

Description. Total vertebrae 158–174 (n = 50), predorsal vertebrae 7–12 (21), preanal vertebrae 57–62 (20), precaudal vertebrae 94 (1). Branchiostegal rays 10 (1). Pectoral rays 13–16 (4). Pores: LL to anus 51–59 (37), POM 10–11 (18), IO 4 + 1 + 2 (28), SO 1 + 3 (24), STC O (24).

Proportions as % of TL: preanal 41–47 (37), predorsal 16–19 (37), head 17–19 (37), depth at anus 5–7 (36), pectoral fin 5–6 (26). Of HL: snout 18–24 (28), eye 18–22 (28), snout-rictus 27–32 (28), gill opening 12–19 (28), interbranchial 8–17 (28), pectoral fin 26–35 (26).

Anus distinctly before midlength. Dorsal fin begins over pectoral-fin base. Head pores somewhat reduced: four supraorbital pores, none between eyes; seven infraorbital pores, two behind eye; no supratemporal pores.

Dentition similar to that of *A. balearicum*, but intermaxillary patch somewhat broader than long.

Precaudal vertebrae ca 54% of total; caudal vertebrae without epicentral processes; myorhabdoi present. Stomach long, reaching level of anus.

Color. In preservative brown or yellowish-brown with dark-edged vertical fins. Sometimes a dusky bar below eye. Stomach and intestine pale.

Size. This is one of the larger species of *Ariosoma.* The largest specimen examined was 537 mm TL (USNM 187627).

Development and Growth. The leptocephalus has been identified (see p. 730).

Variation. Of the seven specimens from Brazil, six had 158–162 vertebrae and the seventh had 169. The 43 specimens from the Caribbean and the Bahamas that were x-rayed all had 167–174 vertebrae. This situation is a bit difficult to interpret. A difference in vertebral number between the northern and southern populations would not be unusual in itself, but the single count of 169 from Brazil is difficult to explain. Additional material would be helpful, but no specimens are available from the entire area between the mouth of the Amazon and Aves Island in the east-central Caribbean. A decision on the taxonomic status of the northern vs southern populations is therefore deferred.

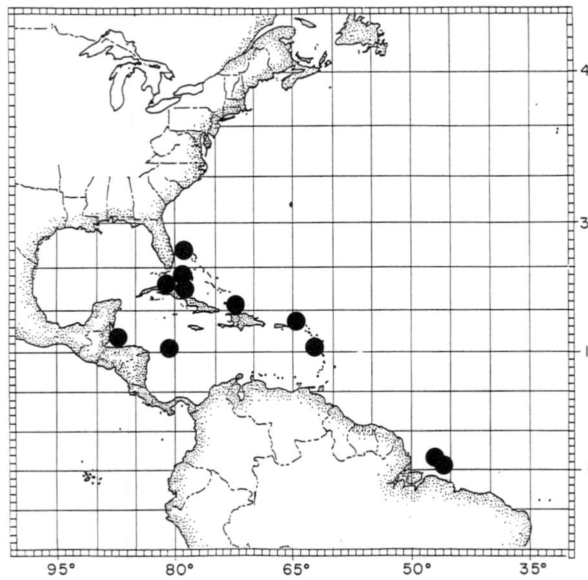

FIGURE 533. Distribution of *Ariosoma selenops.*

Most of the specimens examined had six preoperculomandibular (POM) pores before the rictus, but two specimens had six on one side and seven on the other. Seventeen out of 18 specimens counted had five POM pores behind the rictus, one specimen had four.

Distribution (Fig. 533). Known from the Bahamas, the northern coast of Cuba, Belize, the Honduran banks, the Virgin Islands, Aves Island in the eastern Caribbean, and off the mouth of the Amazon. In the Straits of Florida it seems restricted to the insular side. There are no records from the mainland North American coast. Depth 220–549 m.

Etymology. From the Greek *selene* (moon) and *ops* (eye), in reference to the large eye. Apparently a noun in apposition.

Study Material. 139 specimens, 166–580 mm TL.

Holotype: USNM 93310 (female, 475), Caroline 101, Virgin Islands, 18°40′30″N, 64°50′00″W to 18°45′40″N, 64°48′00″W, 438–549 m, 4 March 1933.

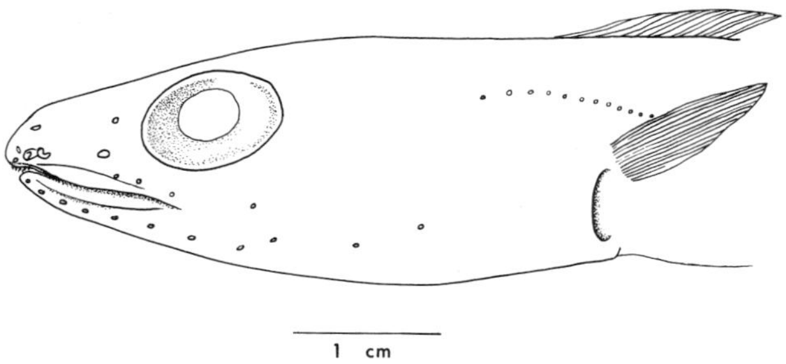

FIGURE 534. *Ariosoma coquettei*: ANSP 134203, holotype, 233 mm TL; head (from Smith and Kanazawa, 1977: fig. 4).

Other material: BAHAMAS: ANSP 109600 (1, 580), 109606 (4, 272–430), 114164 (1, 290). UMML 13850 (1, 338), 13876 (1, 316), 17481 (1, 462), 17722 (1, 326), 18651 (1, 362), 20964 (2, 215–230), 26845 (1, 316), 29590 (1, 432), 29598 (1, 349), 29671 (2, 348–362). UF 38485 (2, 229–331), 38486 (2, 317–320), 38490 (2, 330–363). USNM 158920 (2, 210–313), 187627 (11, 244–537), 185633 (2, 248–254), 197141 (2, 283–408), 197327 (4, 239–400), 197328 (19, 215–361), 197329 (3, 268–362), 197330 (5, 285–420), 197331 (6, 310–408), 197332 (8), 197333 (5, 320–395), 197334 (1, 447). TURKS AND CAICOS: UMML 29654 (5, 166–345). CUBA: ANSP 116132 (1, 383), 116133 (1, 460). MCZ 38791 (1, 350), 38793 (1, 459), 38794 (2, 407–438), 38795 (2, 384–461), 38796 (1, 457), 38798 (3, 431–490), 38799 (1, 361), 38806 (1, 480), 38810 (1, 403), 38812 (1, 405), 38813 (1, 447), 38815 (2, 315–350), 38821 (1, 432). USNM 157878 (1, 298), 157888 (2, 455–462), 157960 (2, 355–357), 157961 (2, 312–425). LESSER ANTILLES: USNM 198687 (1, 360). HONDURAS: USNM 158878 (1, 220). BELIZE: ANSP 116130 (4, 359–448). USNM 268823 (1, 385). BRAZIL: USNM 158914 (5, 326–387), 158918 (1, 293), 158919 (1, 328).

Ariosoma coquettei Smith and Kanazawa, 1977
Figures 534, 535, Table 45

Ariosoma sp. B. Smith, 1971: 49.
Ariosoma coquettei Smith and Kanazawa, 1977: 533 (original description, 07°10′N, 53°36′W, holotype ANSP 134203).

Distinctive Characters. Ariosoma coquettei differs from *A. balearicum* in lacking interorbital, postorbital and supratemporal pores. It differs from *A. selenops* in lacking postorbital pores. It differs from *A. anale* in having a more anterior anus, more vertebrae, and a shorter stomach.

Description. Total vertebrae 155–160 (n = 5), predorsal vertebrae 7–10 (5), preanal vertebrae 54–56 (5), precaudal vertebrae ca 75 (1). Pectoral-fin rays ca 15–16 (2). Pores: LL to anus 47–53 (6),

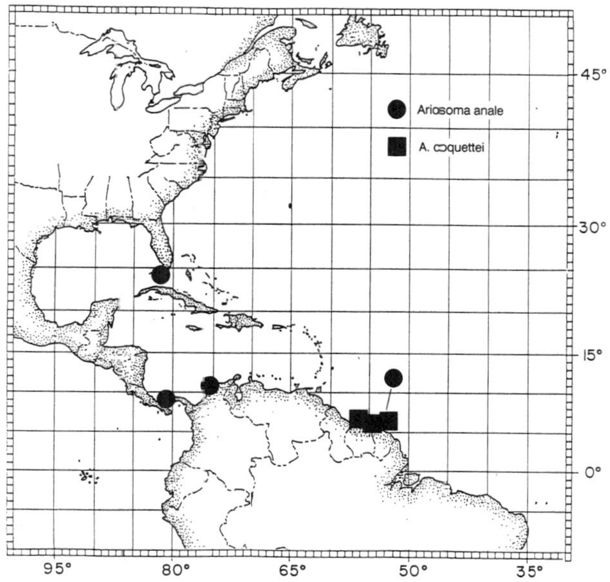

FIGURE 535. Distribution of *Ariosoma coquettei* and *A. anale*.

POM 11 (6), IO 4 + 1 + 0 (6), SO 1 + 3 (6), STC 0 (6).

Proportions as % of TL: preanal 43–45 (6), predorsal 16–18 (6), head 17–19 (6), depth at anus 6–7 (6), pectoral fin 4–7 (6). Of HL: snout 20–24 (6), eye 20–23 (6), snout-rictus 27–30 (6), gill opening 8–19 (6), interbranchial 12–17 (6), pectoral fin 23–38 (6).

Anus distinctly before midlength. Dorsal fin begins over or very slightly before pectoral-fin base. Head pores reduced: three supraorbital pores, none between eyes; five infraorbital pores, none behind eye; no supratemporal pores.

Precaudal vertebrae ca 48% of total; myorhabdoi present. Stomach short, reaching about midpoint between pectoral-fin base and anus.

Note: The vertebral count of "152 (1)" given by Smith and Kanazawa (1977: 533) is in error. It should be 156.

Color. In preservative brown or yellowish-brown with dark-edged vertical fins. Stomach and intestine pale.

Size. The largest specimen was 284 mm TL (RMNH 27428).

Development and Growth. The leptocephalus has been identified (see p. 733).

Distribution (Fig. 535). This species has been collected only off the coasts of Suriname and French Guiana. Depth 53–135 m.

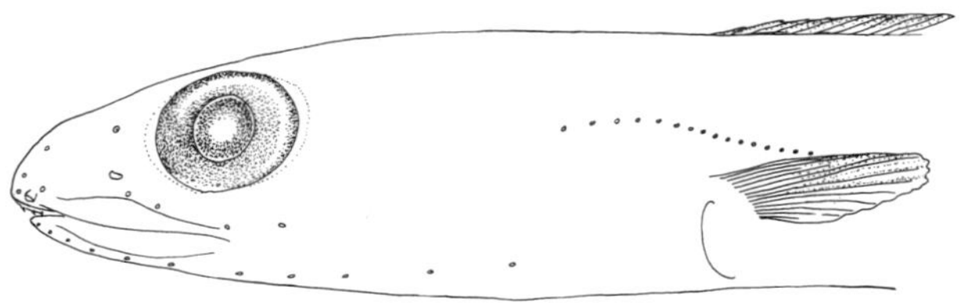

FIGURE 536. *Ariosoma anale*: ANSP 134334, 228 mm TL; head (from Smith and Kanazawa, 1977: fig. 3).

Etymology. Named for the U.S. Bureau of Commercial Fisheries Research Vessel COQUETTE, which collected the first specimen.

Study Material. Six specimens, 173–284 mm TL.

Holotype: ANSP 134203 (male, 233), Pillsbury 658, French Guiana, 07°10′N, 53°36′W, 126–135 m, 9 July 1968.

Paratypes: SURINAME: RMNH 27391 (2, 173–261), 07°11.3′N, 54°23′W, 81 m, 10 Apr. 1969. RMNH 27428 (2, 283–284), 07°10.7′N, 54°04′W, 90–100 m, 15 Apr. 1969. USNM 158917 (1, 185), 06°50′N, 55°24′W, 53 m, 20 July 1957.

Ariosoma anale (Poey, 1860)
Figures 535, 536, Table 45

Conger analis Poey, 1860: 318 (original description, Cuba, holotype MCZ 9325).
Ariosoma sp. A. Smith, 1971: 48.
Ariosoma sp. C. Smith, 1971: 51.
Ariosoma analis. Smith and Kanazawa, 1977: 533.

Distinctive Characters. Ariosoma anale is distinguished from the other western Atlantic species of *Ariosoma* by the relatively posterior position of the anus, at or slightly behind midlength. The precaudal vertebrae are correspondingly numerous. It further differs from *A. balearicum* and *A. selenops* by lacking interorbital, postorbital, and supratemporal pores. It has fewer vertebrae and a longer stomach than *A. coquettei.*

Description. Total vertebrae 146–150 (n = 10), predorsal vertebrae 10 (3), preanal vertebrae 60–63 (3), precaudal vertebrae 83–90 (5). Branchiostegal rays 10 (1). Pectoral-fin rays ca 13–15 (2). Pores: LL to anus 55–59 (9), POM 9–11 (7), IO 4 +1 + 0–2 (7), SO 1 + 3 (7), STC 0 (7).

Proportions as % of TL: preanal 49–52 (10), predorsal 18–21 (8), head 19–21 (8), depth at anus 6–9 (8), pectoral fin 6–7 (7). Of HL: snout 19–22 (8), eye 20–24 (8), snout-rictus 30–34 (7), gill opening 12–16 (8), interbranchial 14–22 (8), pectoral fin 31–34 (7).

Anus at or slightly behind midlength. Dorsal fin begins over pectoral-fin base. Head pores reduced: four supraorbital pores, none between eyes; five infraorbital pores, none behind eye; no supratemporal pores.

Teeth slightly larger and fewer than those of other species of *Ariosoma.* Intermaxillary patch about as long as broad, indistinctly separated from maxillary teeth. Vomerine patch relatively short, not reaching midpoint of maxillary patch, separated from intermaxillary teeth. Maxillary teeth in a narrow band, somewhat wider anteriorly. Mandibular teeth in bands, wider anteriorly.

Precaudal vertebrae ca 58% of total; caudal vertebrae without epicentral processes; myorhabdoi present. Stomach long, reaching nearly to anus.

Color. In preservative brown or yellowish-brown with dark-edged vertical fins. Some dark pigment on pectoral fin. Stomach and intestine pale.

Size. The largest specimen examined was 363 mm TL (USNM 200321).

Development and Growth. The leptocephalus has been identified (see p. 731).

Variation. Specimens from the Gulf of Guinea differ in no significant way from those from the western Atlantic.

One specimen (USNM 185634) differs from the others by having two postorbital pores and a distinct black patch on the pectoral fin. The possibility was briefly considered that this specimen represents a distinct species, but its close agreement in all other characters with *A. anale* makes it more likely to be simply a variant of this species. *Ariosoma anale* does show a tendency toward variation in head pores, and the pectoral fin frequently contains some dark pigment. The specimen in question is a nearly mature male and might be expected to show a modification such as an intensification of pigment in the pectoral fin.

Remarks. Among western Atlantic species, *Ariosoma anale* is closest to *A. coquettei,* sharing with that species the lack of interorbital, postorbital, and supratemporal pores, and the presence of myorhabdoi. Among extralimital species, it seems especially close to *A. meeki* from the western Pacific.

Distribution (Fig. 535). Known in the western Atlantic from Cuba, the Florida Keys, and the northern coast of South America from Panama to the Guianas. In the eastern Atlantic from the Gulf of Guinea. Depth 11–55 m.

Ariosoma anale is a rather uncommon species, and this makes it difficult to define the range precisely. The trans-Atlantic distribution suggests that the species probably occurs off Brazil as well.

Etymology. From the Latin *anus* (anus) and *-alis* (pertaining to), probably referring to the posterior position of the anus. An adjective.

Study Material. Ten specimens, 119–363 mm TL.

Holotype: MCZ 9325 (female, 315), Cuba.

Other material: FLORIDA STRAITS: UMML 31018 (1, 330), 24°34′N, 81°41′W, 55 m, 28 Oct. 1960. COLOMBIA: UMML 28783 (1, 119), 10°50′N, 75°22′W, 11–13 m, 30 July 1968. PANAMA: ANSP 134334 (1, 228) and UMML 29605 (1, 127), 08°49.6′N, 81°13′W, 18 m, 21 July 1966. SURINAME: USNM 185634 (1, 221), 06°53′N, 55°59′W, 42–44 m, 2 Sept. 1958. FRENCH GUIANA: UMML 32623 (1, 301), 06°38′N, 53°58′W, 46 m, 2 Feb. 1974. GULF OF GUINEA: UMML 16431 (1, 167), 04°56′N, 00°47.5′W, 35–37 m, 28 May 1964. USNM 200318 (1, 300), 04°52.5′N, 05°57.5′W, 73 m, 4 March 1964. USNM 200321 (1, 363), 07°32.5′N, 03°20′W, 73 m, 5 June 1964.

Genus *Parabathymyrus* Kamohara, 1938

Parabathymyrus Kamohara, 1938: 14 (type species *Parabathymyrus macrophthalmus* Kamohara, 1938, by original designation); 1952: 24 (redescription and illustration).

Characters. Body relatively stout, tip of tail stiff, caudal fin reduced. Dorsal fin begins over base of pectoral fin, sometimes slightly ahead or slightly behind; pectoral fin well developed. Snout short, bluntly rounded, projecting slightly beyond lower jaw; upper and lower labial flange present, upper flange well developed. Anterior nostril tubular, near tip of snout, directed obliquely anteroventrally; posterior nostril low, near lip close behind anterior nostril, covered by a flap from above. Eye large. Head pores small; pores along upper jaw on side of head rather than on flange; adnasal pore absent.

Teeth small, acute. Intermaxillary tooth patch slightly longer than broad, separated from maxillary and vomerine teeth. Vomerine teeth in a short median row. Maxillary and mandibular teeth uniserial for most of length, expanding to two or more series at anterior end.

Lateral ethmoid process present. Five suborbital bones; lower edge of preorbital bone smooth. Supraoccipital absent. Posterior edge of opercle smooth. Urohyal trifurcate, but the two lateral branches much shorter than median ventral one. Ribs present. Myorhabdoi present. Precaudal vertebrae more than 40% of total; no epicentral processes on caudal vertebrae. Lateral-line ossicles well formed, without cartilaginous loops or processes. Dorsal- and anal-fin rays unsegmented.

Digestive tract pale; stomach long, reaching nearly to anus. Gas bladder attached to vertebrae dorsally, as long as or slightly longer than stomach, extending posteriorly to anus.

Discussion. This genus is very close to *Bathymyrus* Alcock. The extraoral extension of the intermaxillary tooth patch varies considerably in extent among the three known species of *Bathymyrus*. As Castle (1968b: 6) pointed out, this character forms a graded series from *B. simus* J. L. B. Smith through *B. echinorhynchus* Alcock and *B. smithi* Castle to *Parabathymyrus* and cannot be used to diagnose genera. The only character remaining is the nature of the posterior nostril. In *Parabathymyrus* the nostril is covered by a flap; in *Bathymyrus* the nostril, although near the lip, does not have a flap. Castle's (1968b) description of the osteology of *Bathymyrus smithi* does not show anything clearly identifiable as myorhabdoi. He also shows the anterior pair of branchiostegals connected at their tips across the midline, a feature not found in *Parabathymyrus oregoni*. *Parabathymyrus* and *Bathymyrus* will be maintained here as distinct genera. Despite the similarities, they do form two distinct series without intermediate forms.

On the whole, *Parabathymyrus* (and *Bathymyrus* as well) does not differ greatly from *Ariosoma*.

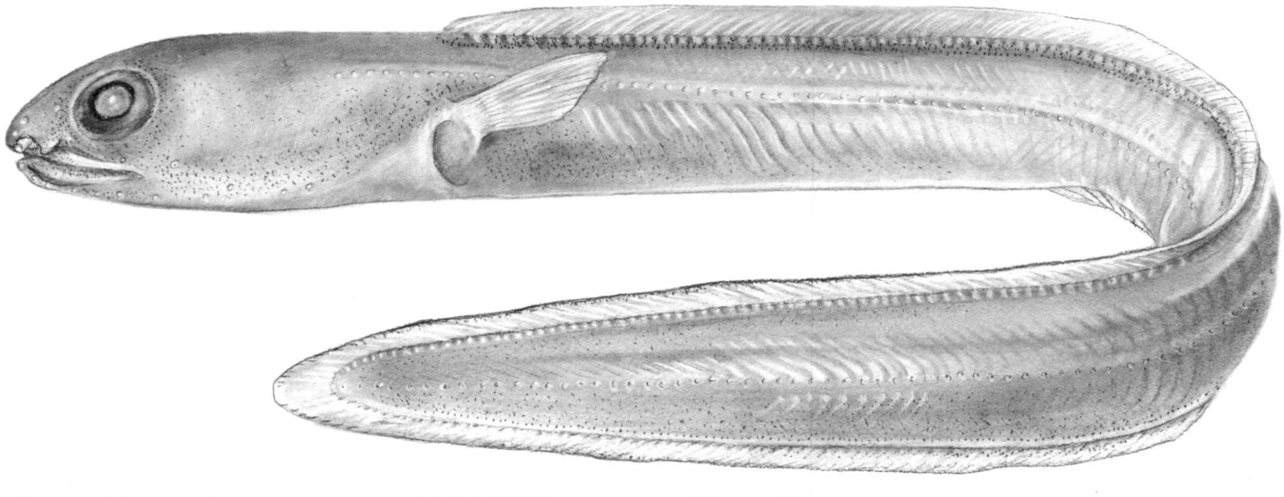

FIGURE 537. *Parabathymyrus oregoni*: ANSP 126974, paratype, 252 mm TL.

It seems closest to the *Ariosoma anale-meeki* group. The principle differences are a shortening of the snout in *Parabathymyrus* and an anteroventral displacement of the posterior nostril with the development of a nasal flap. The trifurcation of the urohyal is much reduced in *Parabathymyrus.* The maxillary and mandibular teeth are uniserial in *Parabathymyrus* rather than in the bands characteristic of *Ariosoma. Ariosoma, Parabathymyrus,* and *Bathymyrus* form a closely knit group within the Congridae.

Size. Probably not more than 500 mm TL; most specimens less than 400 mm.

Distribution. Recorded from Japan, Taiwan, the Philippines, the Mozambique Channel, and the western North Atlantic. In addition, leptocephali have been reported from the Gulf of Guinea (Blache, 1977).

Etymology. From the Greek *para* (near, beside) and *Bathymyrus* (a related genus). Masculine.

Species. Three nominal species of *Parabathymyrus* are currently recognized, differing mainly in the number of preanal lateral-line (LL) pores and vertebrae. *Parabathymyrus macrophthalmus* Kamohara, 1938, known from Taiwan and Japan, has 39 LL pores and 133 vertebrae; *P. brachyrhynchus* (Fowler, 1934) from the Philippines has 50–52 LL pores and 166–168 vertebrae; *P. oregoni* Smith and Kanazawa, 1977, the western Atlantic species, has 45–48 LL pores and 149–155 vertebrae. *Parabathymyrus oregoni* and *P. brachyrhynchus* have the maxillary and mandibular teeth in a single row for most of the length. Kamohara described uniserial teeth in his description of *P. macrophthalmus,* but Chen and Weng (1967: 42) described and illustrated the teeth in narrow bands in their specimens from Taiwan.

The specimens described by Karrer (1982) from the western Indian Ocean had the same number of LL pores as *P. oregoni* but more vertebrae (157–160). Karrer was unwilling to establish a new species based solely on a slight difference in vertebral number and suggested that *P. oregoni* and *P. brachyrhynchus,* along with her Indian Ocean specimens, might represent a single species that shows a west-to-east cline in vertebral numbers. The leptocephali described by Blache (1977: 102) from the Gulf of Guinea (misidentified as *Ariosoma mellissi*), however, had 141–151 myomeres and do not fit this clinal pattern. There would seem, then, to be five more or less distinct populations of *Parabathymyrus* worldwide.

Parabathymyrus oregoni
Smith and Kanazawa, 1977
Figures 537–539, Tables 45, 47

Parabathymyrus sp. Smith, 1971: 43.
Parabathymyrus oregoni Smith and Kanazawa, 1977: 532, fig. 2 (original description, 07°15′N, 53°25′W, holotype USNM 158900).

Distinctive Characters. Parabathymyrus oregoni is distinguished from all other western Atlantic congrids by the nature of the posterior nostril,

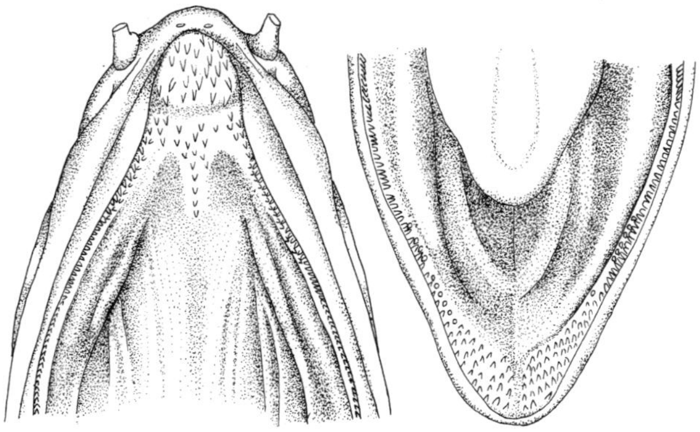

FIGURE 538. *Parabathymyrus oregoni*: ANSP 151864, 280 mm TL; dentition.

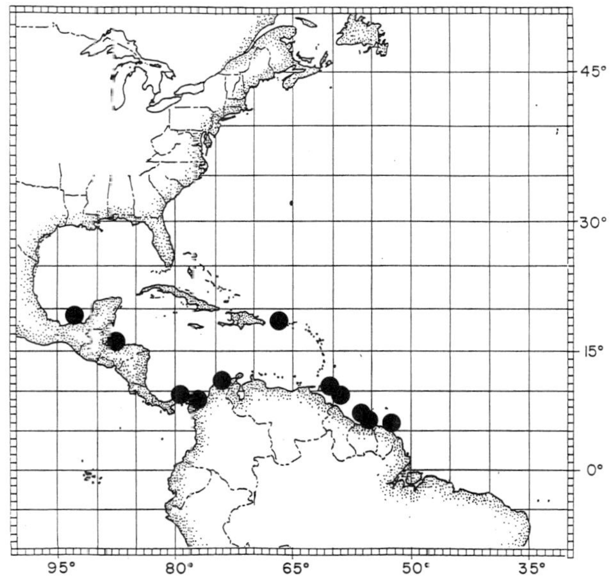

FIGURE 539. Distribution of *Parabathymyrus oregoni*.

low on the side of the head and covered by a flap. It is distinguished from its extralimital congeners by the number of preanal lateral-line pores and vertebrae.

Description. Total vertebrae 149–155 (n = 17), predorsal vertebrae 10–12 (4), preanal vertebrae 50–54 (5), precaudal vertebrae ca 67–70 (2). Branchiostegal rays 10 (1). Pectoral-fin rays ca 13–16 (4). Pores: LL to anus 45–48 (22), POM 9–11 (9), IO 4 (7) + 1 (10) + 0 (10), SO 1 + 3 (9), STC 0 (10).

Proportions as % of TL: preanal 41–45 (23), predorsal 17–21 (23), head 17–19 (23), depth at anus 5–8 (14), pectoral fin 6–7 (14). Of HL: snout 17–19 (14), eye 15–20 (14), snout-rictus 29–34 (14), gill opening 12–16 (14), interbranchial 14–25 (14), pectoral fin 33–43 (14).

Color. In preservative brown with dark-edged vertical fins. Stomach and intestine pale.

Size. The largest specimen examined was 330 mm TL (USNM 158898).

Development and Growth. The leptocephalus has been identified (see p. 733).

Variation. Specimens from the Guianas (including Venezuela east of Trinidad) have slightly fewer total vertebrae than those from the Caribbean and Gulf of Mexico (Table 47). The two specimens from the Gulf of Mexico have more predorsal (12) and preanal (54) vertebrae than do those from elsewhere (10 and 50–51, respectively).

The variation in the number of POM pores occurs mainly in posterior part of the canal, behind the rictus. One specimen had three pores in this section, four specimens had four, and five specimens had five. All specimens examined had six pores ahead of the rictus except one, which had six on one side and seven on the other.

Distribution (Fig. 539). Southern Gulf of Mexico, Caribbean, and coast of South America at least to French Guiana.

Etymology. Named for the National Marine Fisheries Service Research Vessel OREGON. A noun in the genitive case.

Study Material. 22 specimens, 111–330 mm TL.

Holotype: USNM 158900 (female: 322), Oregon 2022, French Guiana, 07°15′N, 53°25′W, 210 m, 9 Nov. 1957.

TABLE 47. Total vertebrae in *Parabathymyrus oregoni* from three areas.

Area	Total vertebrae 149	150	151	152	153	154	155	n	x̄
Gulf of Mexico		1	–	–	–	1		2	152.0
Caribbean			1	2	1	1	1	6	152.8
Guianas	2	2	4	1	1			10	150.7

Paratypes: MEXICO GULF: ANSP 136755 (1, 268). PUERTO RICO, USNM 190466 (1, 280). VENEZUELA: UMML 29600 (1, 241). COLOMBIA: ANSP 126974 (1, 252), 134335 (1, 214). UMML 22298 (2, 115–153). PANAMA: UMML 29594 (3, 111–156). HONDURAS: UMML 29697 (1, 289). GUYANA: USNM 158899 (1, 261). SURINAME: RMNH 27394 (1, 257). UMML 29603 (2, 168–250). USNM 158898 (2, 274–330). FRENCH GUIANA: USNM 158883 (3, 290–307).

Other material: MEXICO GULF: ANSP 151864 (1, 280).

Genus *Paraconger* Kanazawa, 1961

Paraconger Kanazawa, 1961a: 4 (type species *Echelus caudilimbatus* Poey, 1867, by original designation).

Characters. Body moderately stout, tail blunt-tipped with reduced caudal fin. Dorsal fin begins over pectoral fin but behind its base. Pectoral fin well developed. Upper corner of gill opening at or above upper edge of pectoral-fin base.

Snout moderately tapered, projecting slightly beyond lower jaw. Flange present on upper and lower lips, upper flange wide. Eye large, diameter about equal to snout length. Anterior nostril tubular, near tip of snout, directed anterolaterally; posterior nostril in front of eye, slightly below mid-eye level. Adnasal pore present, though small, on posterior side of tube of anterior nostril; third and fourth infraorbital pores absent.

Teeth small, acute. Intermaxillary tooth patch square, separated from maxillary and vomerine teeth. Vomerine teeth in a short patch, about as long as broad. Maxillary teeth in two irregular rows, the outer row forming a cutting edge. Mandibular teeth in one or two rows posteriorly, several rows anteriorly, the outer row forming a cutting edge.

Lateral ethmoid process present. Supraoccipital present. Five suborbital bones; lower edge of preorbital bone smooth. Posterior edge of opercle smooth. Ribs present. Myorhabdoi absent. Anterior few epineurals fused to neural arch; well developed epicentral processes on caudal vertebrae. Dorsal- and anal-fin rays unsegmented.

Gas bladder attached to vertebrae, extending posteriorly well beyond anus. Digestive tract pale; stomach fairly long, reaching nearly to anus.

Discussion. The head pores in *Paraconger* are very small and difficult to see, reducing their value as a taxonomic character. The adnasal pore is present, but the third and fourth infraorbital pores are absent. *Paraconger* is thus unusual among congrids in having only three pores along the upper jaw. In the supraorbital series the fifth pore is absent and the sixth is variable.

Paraconger is unique among congrids in having the base of the pectoral fin enclosed by the gill opening. The upper edge of the gill opening is at or above the level of the upper edge of the pectoral-fin base.

Size. To about 500–600 mm TL.

Distribution. Eastern and western Atlantic, eastern Pacific, tropical to warm temperate. Not recorded from the Indo-West Pacific.

Etymology. From the Greek *para* (near, beside) and *Conger* (a genus of congrid eels). Masculine.

Species. Six species of *Paraconger* are recognized: *P. californiensis* Kanazawa, 1961 and *P. similis* (Wade, 1946) from the eastern Pacific, *P. caudilimbatus* (Poey, 1868) and *P. guianensis* Kanazawa, 1961 from the western Atlantic, and *P. macrops* (Günther, 1870) and *P. notialis* Kanazawa, 1961 from the eastern Atlantic.

Key to the Species of *Paraconger* in the Western North Atlantic

1a. Preanal lateral-line pores 40–43; postorbital pores present .... *caudilimbatus*, p. 506
1b. Preanal lateral-line pores 31–36; postorbital pores absent .... *guianensis*, p. 509

Paraconger caudilimbatus (Poey, 1867)
Figures 540–541

Echelus caudilimbatus Poey, 1867b: 249 (original description, Cuba, holotype MCZ 9324).

Ophiosoma caudilimbatus. Poey, 1868b: 424.

Conger macrops (not of Günther, 1870). Günther, 1870: 40, part.

Conger caudilimbatus. Poey, 1876a: 152.

Conger caudicula Bean in Goode and Bean, 1882: 435 (original description, Pensacola, Florida, holotype USNM 30709).

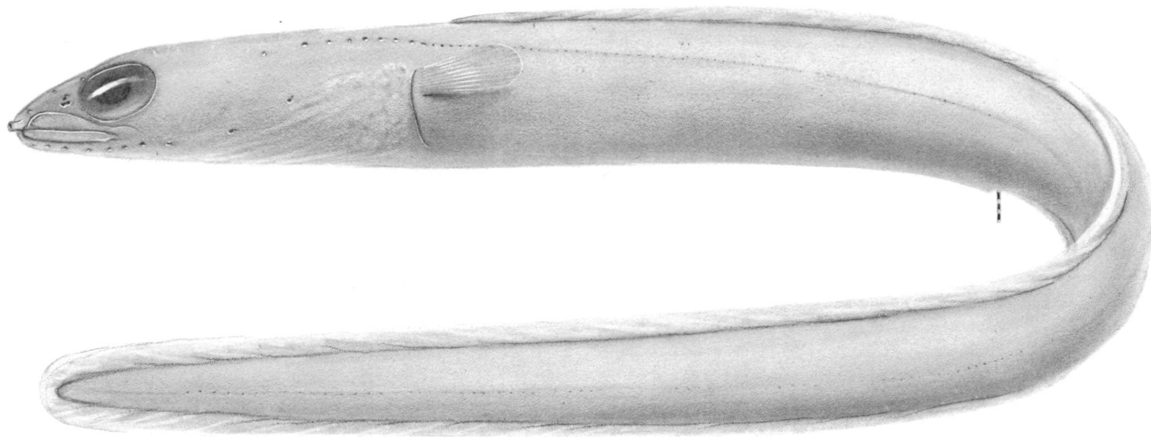

FIGURE 540. *Paraconger caudilimbatus*: ANSP 74958, 159 mm TL. Illustration by S. P. Gigliotti.

Leptocephalus caudilimbatus. Jordan and Davis, 1891: 666.
Leptocephalus harringtonensis Mowbray, 1931: 1 (original description, Bermuda, holotype FMNH 48442).
Conger harringtonensis. Beebe and Tee Van, 1933b: 46.
Paraconger caudilimbatus. Kanazawa, 1961a: 11.
Paraconger harringtonensis. Kanazawa, 1961a: 12.

Discussion of Synonymy. Günther (1870: 40) described *Conger macrops* from two specimens, one from Madeira and one from the Bahamas. Kanazawa, 1961a: 8, 11, determined that the Bahamian specimen was *P. caudilimbatus* and restricted *macrops* to the Madeira specimen.

Distinctive Characters. Total vertebrae 121–127. Preanal lateral-line pores 40–43; postorbital pores present.

Description. Total vertebrae 121–127 (n = 11), predorsal vertebrae 11–13 (6), preanal vertebrae 44–45 (6), precaudal vertebrae 52 (1). Branchiostegal rays 9–10 (2). Pectoral-fin rays ca 12 (1), dorsal-fin rays 181–240 (3, plus data from Kanazawa, 1961a), anal-fin rays 140–172 (4, plus data from Kanazawa, 1961a). Pores: LL to anus 40–43 (6), POM 9–10 (3), IO 3 (1) + 1 (2) + 2–3 (3), SO 1 + 4 (3), STC 0–3 (3).

Proportions in % of TL: preanal 39–46 (14), predorsal 18–22 (14), head 16–18 (13), depth at anus 4–6 (11), pectoral fin 4–5 (11). Of HL: snout 18–22 (11), eye 18–22 (11), snout-rictus 30–38 (10), gill opening 16–23 (10), interbranchial 10–17 (8), pectoral fin 23–29 (10).

Fifth SO pore absent, sixth present.

Color. In preservative brown or gray, lighter ventrally. Vertical fins dark-edged. Stomach and intestine pale.

Size. The largest specimen examined was 461 mm TL (UMML 7403).

Development and Growth. Leptocephali of *Paraconger* are common in the western Atlantic, and many of these are certainly *P. caudilimbatus* (see p. 736). Because the vertebral counts of *P. caudilimbatus* and *P. guianensis* overlap, however, identification of single specimens is not always possible. Furthermore, as discussed below, two species appear to be present in the Gulf of Mexico. The latter have not yet been distinguished as adults, and this adds another difficulty to the specific identification of larval *Paraconger*.

Variation. There is some evidence that two species may be involved here, one confined to the Gulf of Mexico and the other occurring in both the Gulf and the Caribbean. Tables 48 and 49 show the distribution of total and LVBV myomere counts of leptocephali of *Paraconger* collected in the Yucatan Channel and the western Gulf of Mexico during the MBI cruises (see p. 737). Within the western Gulf two groups of leptocephali can be distinguished. One group has 123–129 myomeres (n = 27, mean 126.1) and the last vertical blood vessel (LVBV) at myomere 57–61 (28, mean 59.1). The other group has 119–126 myomeres (29, mean 122.3) and the LVBV at myomere 51–56 (33, mean 53.4). The high-count leptocephali were present in the Gulf in the summer (July–August), the low-count leptocephali in the fall and winter (November and February). Two significant exceptions to this temporal distribution occurred. In the extreme southern Gulf

TABLE 48. Total myomeres in leptocephali of *Paraconger* from the Yucatan Channel and the Gulf of Mexico. 1, Yucatan Channel; 2, Gulf of Mexico except Veracruz, summer, >34 mm SL; 3, Gulf of Mexico, fall–winter; 4, Gulf of Mexico except Veracruz, summer, <34 mm SL; 5, Veracruz, summer.

	Myomeres												
Area	119	120	121	122	123	124	125	126	127	128	129	n	x̄
1						2	1	–	2	1	1	8	126.8
2					2	3	7	5	4	4	2	27	126.1
3	4	1	–	8	5	3	2	1				24	122.3
4			1	–	–	1						2	122.5
5		1	0	0	1	0	1					3	122.7

of Mexico, represented by the stations off Veracruz, low-count leptocephali were present in the summer as well, replacing the high-count form there. Elsewhere the summer leptocephali differed in myomere count according to size. Specimens smaller than 34 mm SL had 121–124 myomeres (2), those larger than 34 mm had 123–129 (27). The most reasonable explanation for this pattern is that the low-count form begins spawning in the summer and its leptocephali then predominate in most of the Gulf of Mexico through the fall and winter. In the southern Gulf, however, spawning is extended and the leptocephali are found throughout the year. The high-count form spawns in the spring and summer; its leptocephali predominate in the summer and metamorphose by fall. The clear sympatry of the two forms plus an apparent difference in the length of larval life indicate that they are distinct species. The high-count form is found in the Caribbean as well and probably represents the true *P. caudilimbatus.* The low-count form has been found only in the Gulf of Mexico. Corresponding differences have not been documented in the adults; this must await the collection of more material from the Gulf of Mexico.

Remarks. Paraconger harringtonensis (Mowbray) is known only from the holotype, which remains the only specimen of *Paraconger* known from Bermuda. Kanazawa (1961a: 11) distinguished it from *P. caudilimbatus* on the basis of three characters: lack of a supratemporal pore, more dorsal- and anal-fin rays, and higher vertical fins.

Among the specimens of *P. caudilimbatus* examined here, one had three STC pores, one had one STC pore, and one lacked STC pores. This character seems of no value in separating species.

Kanazawa (1961a: 11) recorded 181–213 dorsal-fin rays for *P. caudilimbatus* as opposed to 241 in *P. harringtonensis* (a recount by E. B. Böhlke yield-

TABLE 49. Position of the last vertical blood vessel (LVBV) in leptocephali of *Paraconger* from the Yucatan Channel and the Gulf of Mexico. 1, Yucatan Channel; 2, Gulf of Mexico except Veracruz, summer, >34 mm SL; 3, Gulf of Mexico, fall–winter; 4, Gulf of Mexico except Veracruz, summer, <34 mm SL; 5, Veracruz, summer.

	LVBV myomeres												
Area	51	52	53	54	55	56	57	58	59	60	61	n	x̄
1								1	3	5		9	59.4
2							3	5	8	10	2	28	59.1
3	1	5	7	8	3	1						25	53.4
4	1	–	3	–	1							5	53.0
5				2	1							3	54.3

ed 240 in the latter). Two specimens examined by me (USNM 218909), however, had 218–221, counts that are actually closer to that of *P. harringtonensis* than to the lower extreme of *P. caudilimbatus.* Similarly, Kanazawa recorded 140–152 anal-fin rays in *P. caudilimbatus* as opposed to 176 in *P. harringtonensis* (172, according to E. B. Böhlke). Three specimens examined by me (UMML 7403 and USNM 218909) had 159–163, again bridging the gap. Fin rays thus seem to be of little value as a diagnostic character.

The third character, the greater height of the vertical fins, is probably an artifact of preservation. The holotype of *P. harringtonensis* appears to have been preserved with its fins pinned and spread out; in most preserved eels the vertical fins are closely appressed to the body, making it difficult to determine the true height. Other counts and measurements of *P. harringtonensis* are as follows: preanal 46% TL, predorsal 19% TL, HL 18% TL, snout 20% HL, eye 19% HL, snout-rictus 35% HL, gill opening 16% HL, interbranchial 10% HL, depth at anus 4% TL, pectoral fin 26% HL, preanal LL pores 42, total vertebrae 124. Of all these characters, only the interbranchial distance falls outside the range shown by specimens of *P. caudilimbatus.* The weight of evidence does not support the existence of a distinct species of *Paraconger* at Bermuda, and *P. harringtonensis* is here considered a synonym of *P. caudilimbatus.*

Distribution (Fig. 541). Gulf of Mexico, Florida, Bermuda, Bahamas, Cuba, Caribbean coast of Mexico, and eastern Colombia. Depth 0–70 m.

Etymology. From the Latin *cauda* (tail) and *limbatus* (bordered), presumably referring to the dark borders of the dorsal and anal fins. An adjective.

Study Material. 15 specimens, 75–461 mm TL.

Holotype: MCZ 9324 (female, 296), Cuba.

Other material: BERMUDA: FMNH 48442 (1, 387, holotype of *Leptocephalus harringtonensis* Mowbray), Harrington Sound, 13 m, 20 July 1929. FLORIDA ATLANTIC: UMIM (UMML) 4318 (1, 387), Dade Co., off Baker's Haulover, 38 m. UMML 7403 (1, 461), Dade Co., off Baker's Haulover, Aug. 1961. FLORIDA STRAITS: USNM 131526 (1, 128), Florida, Key West, on reefs at low tide, Jan. 1885. FLORIDA GULF: USNM 33000 (1, 280), Pensacola. TEXAS: USNM 158903 (1, 279), 28°03′N, 94°57′W, 60–70 m, 2 Oct. 1957. BAHAMAS: ANSP 74958 (1, 159), north of Green Cay, 0.4 km off east end, just north of coral head, 13 May 1956. ANSP 103453 (1, 75), Green Cay, coral head ca 0.4 km north of east end, 11 m, 13 May 1956. UMML 9422 (1, 412), Bimini, Aug. 1961. CUBA: MCZ 62372 (1, 329). COLOMBIA: USNM 218909 (2, 353–372), 11°53′N, 72°29′W, 35 m, 12 Oct. 1965. MEXICO QUINTANA ROO: ANSP 116131 (2, 303–405), Cozumel, off San Miguel, 8 m, 10 Oct. 1968.

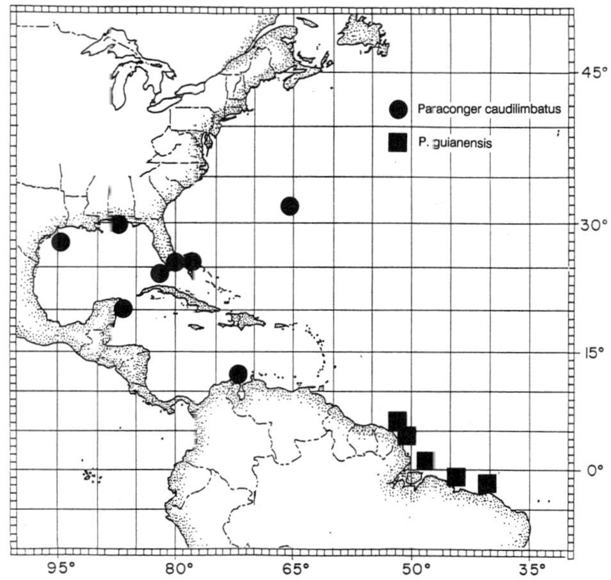

FIGURE 541. Distribution of *Paraconger caudilimbatus* and *P. guianensis.*

Paraconger guianensis Kanazawa, 1961
Figures 541–544

Paraconger guianensis Kanazawa, 1961a: 8 (original description, 5°52′N, 52°03′W, holotype USNM 158902).

Paraconger caudilimbatus (not of Poey, 1867). Randall et al., 1978: 5 (synonymized *guianensis* with *caudilimbatus*).

Distinctive Characters. Total vertebrae 126–131. Preanal lateral-line pores 31–36; postorbital pores absent.

Description. Total vertebrae 126–131 (n = 7), predorsal vertebrae 12–14 (7), preanal vertebrae 39–40 (7). Pectoral-fin rays 16–18 (5—from Kanazawa, 1961a). Pores: LL to anus 31–36 (7), POM 9–10 (2), IO 3 (1) + 1 (2) + 0 (2), SO 1 + 3–4 (1), STC 0 (3).

Proportions as % of TL: preanal 41–43 (11), predorsal 21–26 (11), head 18–20 (11), depth at

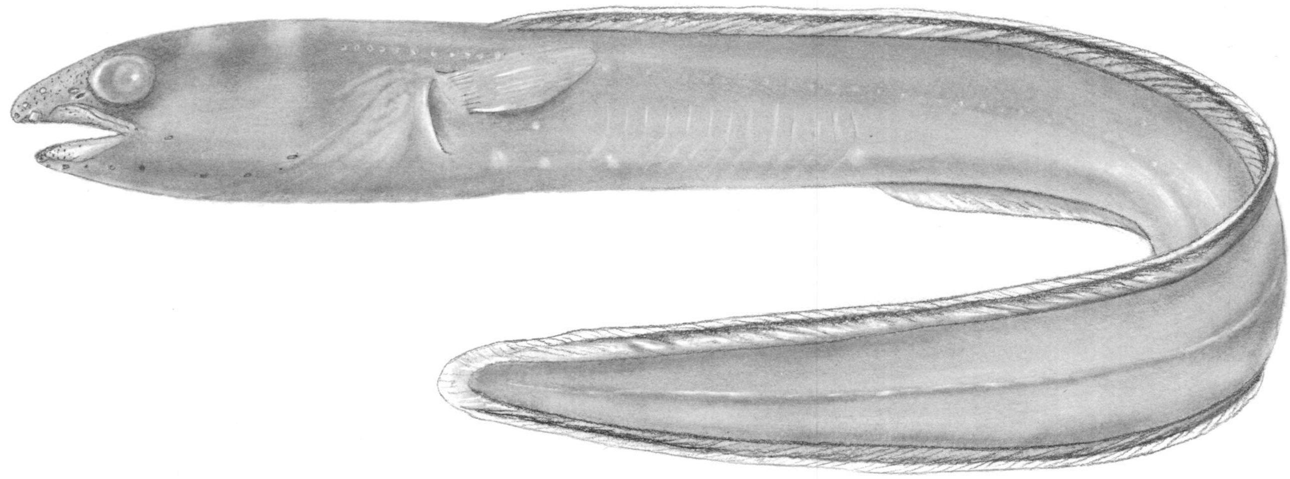

FIGURE 542. *Paraconger guianensis*: ANSP 126975, 335 mm TL.

anus 5–7 (11), pectoral fin 6–8 (11). Of HL: snout 20–22 (11), eye 14–21 (11), snout-rictus 30–37 (11), gill opening 17–22 (11), interbranchial 16–22 (11), pectoral fin 30–41 (11).

Fifth SO pore absent, sixth usually absent.

Color. In preservative light brown, lighter ventrally. Dark pigment above, below, and behind eye, sometimes forming an ill-defined streak; also dusky pigment on labial flange, densest anteriorly. Dark pigment near tip of lower jaw. Vertical fins dark-edged.

Size. The largest specimen examined was 538 mm TL (USNM 198768).

Development and Growth. Leptocephali of *Paraconger* collected off the Guianas and Brazil probably belong to this species (see p. 736).

Distribution (Fig. 541). French Guiana and northern Brazil. Depth 35–73 m (mean 60 m).

Etymology. From Guiana, the region on the northeast coast of South America. An adjective.

Study Material. 12 specimens, 230–538 mm TL.

Holotype: USNM 158902 (female, 460), Oregon 2044, French Guiana, 05°52′N, 52°03′W, 73 m, 12 Nov. 1957.

Paratypes: USNM 158901 (3, 280–400), French Guiana, 05°39′N, 51°56′W, 68 m, 12 Nov. 1957.

Other material: BRAZIL: ANSP 126975 (1, 335), 04°30′N, 50°52′W, 66 m, 27 Feb. 1963. ANSP 136633 (1, 332), 04°46′N, 50°59′W, 70–71 m, 6 May 1975. ANSP 136637 (1, 345), 01°25′N, 47°48′W, 48 m, 12 May 1975. ANSP 136644 (1, 230), 01°40′S, 404°45′W, 35–40 m, 18 May 1975. ANSP 136649 (2, 343–385), 04°31′N, 50°49′W, 73 m, 7 May 1975. USNM 198678 (1, 538), 02°20′S, 40°24′W, 40 m, 12 March 1963. Locality uncertain: GCRL 22467 (1,530).

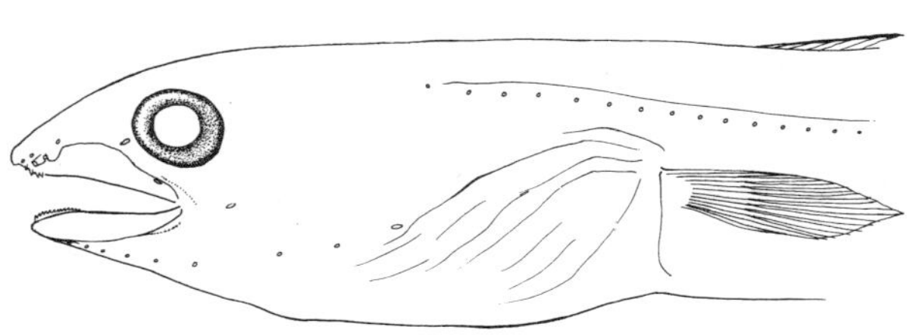

FIGURE 543. *Paraconger guianensis*: ANSP 126975, 335 mm TL; head.

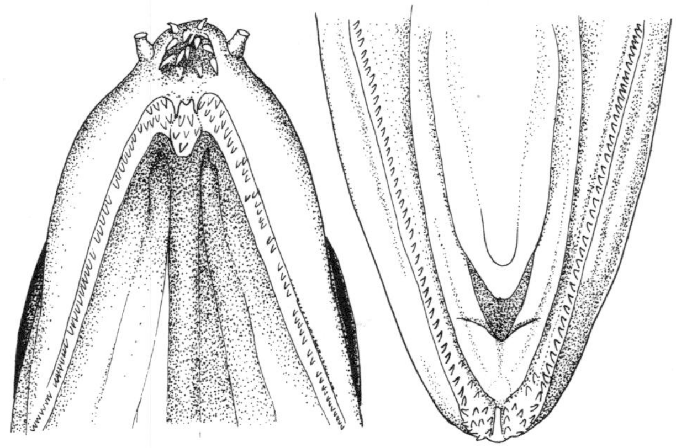

FIGURE 544. *Paraconger guianensis*: ANSP 136644, 230 mm TL; dentition.

Subfamily CONGRINAE

Characters. Body moderately elongate, preanal length usually less than 40% TL. Caudal fin usually not reduced (slightly reduced in *Gnathophis*); tip of tail flexible (slightly stiffened in *Gnathophis*). Pectoral fin well developed. Mouth small to moderately large, position of rictus varies from below anterior margin of eye to behind level of posterior margin of eye. Lips usually well developed, maxillary and mandibular teeth concealed when mouth closed (except in *Xenomystax*). Upper labial flange present or absent. Posterior nostril usually at mid-eye level, sometimes elevated. Teeth variable, from small and granular to large and fang-like. Gas bladder thin-walled, free of vertebrae. Dorsal- and anal-fin rays segmented. Lateral ethmoid process absent; pterygoid stout to moderately slender; preorbital usually solid, ventral margin entire or with three ventrally directed projections; three or four suborbital bones; posterior margin of opercle smooth or serrate; lateral-line ossicles variable, from elongate and trough-like to short and narrow; ribs absent in most genera; precaudal vertebrae less than 40% of total; myorhabdoi absent.

Distribution. Tropical to temperate seas worldwide, from the shoreline to depths of about 2000 m.

Genera. This is a much larger and more heterogeneous group than either the Bathymyrinae or the Heterocongrinae. Most of the genera included in this subfamily are similar in overall appearance, but they can be arranged in a series showing progressive modification in several characters.

Congriscus Jordan and Hubbs, 1925 appears to be the most primitive of the congrine genera and not far removed from the bathymyrine stem. The tip of the tail is relatively stiff, an upper labial flange is present, and the lateral-line ossicles are stout and well ossified. *Conger* Oken, 1817 is also primitive, with a well developed labial flange. The tip of the tail is flexible but not attenuate. The lateral-line ossicles are large, horizontally elongate, and in most species are incompletely ossified. The leptocephalus is moderately elongate with a series of moderately small melanophores along the intestine and a crescentic patch of pigment under the eye.

Gnathophis Kaup, 1860 seems close to *Conger*, although it shows a few modifications that foreshadow conditions in the more advanced congrines. The labial flange is reduced, and the preorbital (actually divided into two elements) sends three processes ventrally into the upper lip. The lateral-line ossicles are horizontally elongate as in *Conger* but somewhat less substantial. In one feature *Gnathophis* resembles the Bathymyrinae: the tip of the tail is slightly stiffened and the caudal fin is reduced. According to Asano (1962: 11), *Gnathophis* burrows tail-first but not quite as efficiently as *Ariosoma*. The genera *Poeciloconger* Günther, 1871, *Diploconger* Kotthaus, 1968, and *Scalanago* Whitley, 1935 appear to be derived from a *Gnathophis*-like ancestor. *Poeciloconger* differs primarily in its strongly banded or mottled color pattern. *Diploconger* and *Scalanago* are distinguished by the modified lateral line. In some species of *Gnathophis* several anterior lateral-line pores are elevated to the dorsal margin of the canal. In *Diploconger* this is carried one step further. There are two pores per segment, at the end of short branches extending dorsally and ventrally from the main canal. In *Scalanago* these branches have proliferated into a complex network of secondary canals covering much of the body surface. The leptocephalus of *Gnathophis* closely resembles that of *Conger*.

Beyond *Gnathophis* on the congrine family tree several evolutionary trends are apparent. The upper labial flange is reduced or lost, and the preorbital sends three processes ventrally into the upper lip. The lateral-line ossicles are reduced to small, vertically elongate structures with slender, cartilaginous, finger-like processes from the dorsal and ventral margins. The tip of the tail is soft and flexible, and the posterior edge of the opercle is serrate.

Two evolutionary lines seem discernible at this level of the Congrinae. One begins with *Japonoconger* Asano, 1958 and proceeds to *Pseudoph-*

ichthys Roule, 1915 and *Acromycter* Smith and Kanazawa, 1977. The peculiar form described as *Rhynchoconger brevirostris* Chen and Weng, 1967 (not a species of *Rhynchoconger*) also seems to belong here, and it in turn may form a bridge to *Macrocephenchelys* Fowler, 1934. In this group the anteriormost neural arches articulate with the centra rather than fuse. The lower pharyngeals are small, and pleural ribs are present in *Japonoconger* and *Pseudophichthys*. The tail is slender but not filiform, and its tip is pliable. The teeth are small and in granular patches, and the lining of the mouth and gill area is black. *Pseudophichthys* and *Acromycter* have swollen, bulbous snouts; small, inferior mouths; and relatively small eyes. In all but *Japonoconger* the dorsal fin begins behind the base of the pectoral fin. "*Rhynchoconger*" *brevirostris* resembles *Pseudophichthys* and *Acromycter* as follows: short, bulbous snout; small, inferior mouth; small eye; mouth and gill cavity black; dorsal fin begins behind pectoral-fin base; granular teeth. It differs most notably in its unusually long pectoral fin. *Macrocephenchelys* has a very short snout and jaws with modification of the associated skeletal elements (Robins and Robins, 1971). The snout is bulbous and the lips are elaborated into a series of folds and ridges. Despite these peculiarities, some resemblances to *Pseudophichthys* and *Acromycter* can be noticed. The eye is small and the snout is swollen. The gill cavity is black, and the dorsal fin begins behind the pectoral fin. The posterior margin of the opercle is strongly serrate, and the lower pharyngeals are absent. Despite its specializations, there are enough similarities to suggest that *Macrocephenchelys* originated from the *Japonoconger-Pseudophichthys-Acromycter* group of congrines. The known leptocephali of this group differ markedly from those of the *Conger-Gnathophis* group and the *Uroconger* group. The leptocephalus of *Pseudophichthys* has melanophores scattered over the entire lateral surface of the body. That of *Acromycter* has a series of ophichthid-like loops in the gut. Neither has the subocular pigment.

The other evolutionary line includes *Rhynchoconger* Jordan and Hubbs, 1925, *Rhechias* Jordan, 1921, *Bathyuroconger* Fowler, 1934, and *Uroconger* Kaup, 1856. Here the tail becomes extremely slender, almost filiform, and is frequently regenerated. The eye is large, and in *Rhechias, Bathyuroconger,* and *Uroconger* the teeth are enlarged. There is also a tendency toward enlargement of certain head pores, especially the anterior infraorbital and preoperculomandibular pores and the third supraorbital pore. The leptocephali of this group resemble those of the *Conger-Gnathophis* group, including the subocular pigment.

Xenomystax Gilbert, 1891, despite its distinctive appearance, does not seem far removed from the *Uroconger* group. The principal modification is an elongation of the snout and jaws. The teeth are enlarged, especially the intermaxillary and anterior maxillary and mandibular teeth, and the lips are drawn back away from the outer maxillary and mandibular teeth, leaving them exposed when the mouth is closed. The innermost series of teeth on the maxilla, and to a lesser extent on the mandible, is separated from the others by a distinct edentulous groove. When the mouth is closed the tip of the lower jaw fits into a gap between the intermaxillary teeth and the maxillary and vomerine teeth. The preorbital is elongate and poorly ossified, a condition undoubtedly related to the reduction of the upper lip. The anteriormost neural arches articulate with the centra. Despite these substantial differences, *Xenomystax* shows some similarities to the *Rhynchoconger-Uroconger* group. The tail is attenuate, the teeth are enlarged, and the lateral-line ossicles are like those of *Rhynchoconger*. Leptocephali of *Xenomystax* are large and moderately elongate, with the body deepening abruptly behind the head. The gut is long and simple, like that of most congrids. Ventral pigment is limited to a few large melanophores, concentrated near the anterior and posterior ends, and lateral melanophores are large and stellate. There is no subocular pigment.

Extralimital genera that appear to belong in the Congrinae include *Bathycongrus* Ogilby, 1898, *Congrosoma* Garman, 1899, *Microcephalocongrus* Fowler, 1934, *Silvesterina* Fowler, 1934, *Uranoconger* Fowler, 1934, *Bassanago* Whitley, 1948, and *Lumiconger* Castle and Paxton, 1984.

Key to the Genera of Congrinae in the Western North Atlantic

1a. Inner row of maxillary and mandibular teeth separated from outer rows by endentulous groove; maxillary and mandibular teeth exposed when mouth closed .... *Xenomystax*, p. 558
1b. Inner row of maxillary and mandibular teeth not separated from outer rows by edentulous groove; maxillary and mandibular teeth concealed when mouth closed 2
2a. Maxillary and mandibular teeth in one or two rows, the outer row forming a cutting edge; upper labial flange well developed .... *Conger*, p. 513
2b. Maxillary and mandibular teeth in bands or in two rows, not forming a cutting edge; upper labial flange reduced or absent 3
3a. Tip of tail slightly stiffened; narrow upper labial flange present .... *Gnathophis*, p. 521
3b. Tip of tail soft and flexible; upper labial flange rudimentary or absent 4
4a. Vomerine teeth in a single row, reaching posterior end of maxillary tooth patch ....
.... *Uroconger*, p. 545
4b. Vomerine teeth in more than two rows, at least anteriorly, ending before posterior end of maxillary tooth patch 5
5a. Vomerine tooth patch either with a few enlarged teeth or, if teeth granular, patch ends less than halfway to posterior end of maxillary tooth patch 6
5b. Vomerine teeth in a more or less broad patch, reaching at least halfway to posterior end of maxillary tooth patch 8
6a. Teeth small; vomerine tooth patch round or slightly elongate; rudimentary upper labial flange present; pores along upper jaw small .... *Rhynchoconger*, p. 525
6b. Vomerine tooth patch with some enlarged teeth, sometimes forming a short row posteriorly; upper labial flange absent; pores along upper jaw enlarged 7
7a. Upper jaw projects beyond lower; teeth moderately large .... *Rhechias*, p. 532
7b. Jaws nearly equal; anterior teeth very strong, fang-like .... *Bathyuroconger*, p. 541
8a. Intermaxillary tooth patch separated from maxillary and mandibular teeth by a distinct gap; supratemporal pores present; dorsal fin begins before pectoral-fin base .... *Japonoconger*, p. 549
8b. Intermaxillary tooth patch confluent with maxillary and vomerine teeth; supratemporal pores absent; dorsal fin begins behind pectoral-fin base 9
9a. Posterior nostril at mid-eye level .... *Pseudophichthys*, p. 551
9b. Posterior nostril opposite upper margin of eye .... *Acromycter*, p. 554

Genus *Conger* Oken, 1817

Les Congres. Cuvier, 1817: 231 (non-Latin).

Conger Oken, 1817: 1181 (misprinted as 1781) (type species *Muraena conger* Linnaeus, 1758, by absolute tautonymy). Cuvier's name Latinized.

Isognatha DeKay, 1842: 313 (type species *Anguilla oceanica* Mitchill, 1818, by original designation).

Congrus Richardson, 1844: 105 (emendation of *Conger*).

Veternio Snyder, 1904: 516 (type species *Veternio verrens* Snyder, 1904, = *Conger cinereus marginatus* Valenciennes, 1841, by original designation).

Microconger Fowler, 1912a: 9 (type species *Leptocephalus caudalis* Fowler, 1912, = *Conger cinereus* Rüppell, 1828, by original designation). Described as a subgenus of *Leptocephalus* Gronow, and based on a specimen erroneously reported from the west coast of Mexico.

Astroconger Jordan and Hubbs, 1925: 194 (type species *Anguilla myriaster* Brevoort, 1856, by original designation).

Forskalichthys Whitley, 1935: 219 (type species *Conger cinereus* Rüppell, 1828, by original designation).

Discussion of Synonymy. DeKay (1842: 313) stated in his account of *Anguilla oceanica* Mitchill that: "Some years since, I arrayed this as a representative of a new genus *Isognatha* in a catalogue of New York fishes.... It may possibly be included under *Conger*." I have been unable to locate this earlier reference, and indeed there seems to be no catalog of New York fishes by DeKay apart from his work on the New York

State Natural History Survey, which culminated in the publication cited (DeKay, 1842). Preliminary reports prepared for the Survey, at least those I have been able to examine, contain no mention of *Isognatha.* If the earlier reference does not exist or exists in an unpublished form, then the 1842 reference must be considered the description of *Isognatha.* Here DeKay presented the name and clearly associated it with a species, *Anguilla oceanica.* This would appear to be the indication required by Article 12 (b) (5) of the ICZN (1985) to make the name available.

Leptocephalus Gronow, 1763 had been used for many years to refer to the peculiar larval form of eels and related groups. When the true nature of these larvae was discovered and the type species of *Leptocephalus, L. morrisii* Gmelin, 1788, turned out to be the larva of *Conger conger* (Linnaeus), many ichthyologists, led by Jordan, adopted *Leptocephalus* in place of *Conger.* The International Commission on Zoological Nomenclature dealt with the matter twice. Initially (Opinion 44, 1912) it recognized *Leptocephalus* as the valid genus of conger eels but later (Opinion 93, 1926) reversed itself and suppressed *Leptocephalus* in favor of *Conger* Cuvier, 1817 (later changed in Direction 87 to *Conger* Oken, 1817). *Leptocephalus* has been used continuously as a term for larval eels, but for a time, beginning with Jordan and Davis (1891), it was also used as a substitute for *Conger.*

Castle (1969c: 9) considered *Helmictis* Rafinesque, 1810 a possible synonym of *Conger.* It was based on *Helmictis punctatus* Rafinesque, 1810, a larva or postlarva. Since Opinion 93 and Direction 87 established *Conger* as the valid name, the status of *Helmictis* is probably an academic question. In any case, the uncertainty over the identity of *Helmictis punctatus* would prohibit consideration of *Helmictis* as a senior synonym. *Helminctis* Bonaparte, 1841 and *Helminthichthys* Agassiz, 1846 are emendations of *Helmictis.*

Characters. Body moderately elongate, anus distinctly before midlength. Tail moderately slender, neither blunt-tipped nor attenuate, caudal fin flexible. Dorsal fin begins behind base of pectoral fin, usually over tip of appressed pectoral fin. Pectoral fin well developed.

Jaws nearly equal, fleshy part of snout extending slightly in front of intermaxillary teeth. Flange present on upper and lower lips, that of upper lip wide. Head pores small, those along upper jaw located on side of head rather than on flange; adnasal pore well developed, immediately adjacent to third supraorbital pore, the two forming what appears to be a double pore.

Teeth small, acute. Intermaxillary tooth patch rounded, not sharply separated from maxillary and vomerine teeth. Vomerine tooth patch short, triangular. Maxillary and mandibular teeth in one or two rows, the outer row forming a cutting edge.

Lateral ethmoid process absent. Supraoccipital present. Otic bulla not developed. Lower edge of preorbital bone without projections; four suborbitals. Posterior edge of opercle smooth. First epibranchial with short anterior process; upper pharyngeals divided, the anterior element much the smaller; lower pharyngeals divided, but may fuse with growth. Anterior few epineurals fused to neural arch; well developed epicentral processes on caudal vertebrae; ribs absent; myorhabdoi absent. Lateral-line ossicles horizontally elongate, incompletely ossified with many gaps in structure. Dorsal- and anal-fin rays segmented.

Digestive tract pale. Stomach rather long, not quite reaching anus posteriorly. Gas bladder reaching anus posteriorly.

Size. This genus includes the largest of all congrid eels. *Conger conger* (Linnaeus) of Europe is said to reach 3 m in length and 73 kg in weight, and the American *Conger oceanicus* reaches 2 m and 10 kg (Bigelow and Schroeder, 1953: 155). *Conger verreauxi* Kaup of Australia is reported to reach 16–18 kg (Castle, 1964b: 26).

Distribution. Eastern and western Atlantic, Indo-West Pacific. Tropical to cool temperate.

There are only two records of the genus from the eastern Pacific, and both appear to be erroneous. Jordan and Gilbert's (1882d: 378) *Leptocephalus conger* from Baja California was really an *Ariosoma* (Jordan and Davis, 1891: 665). Kanazawa (1958: 228) gave convincing evidence that the other record, Fowler's (1912a) *Leptocephalus caudalis*, a misidentified *Conger cinereus*, was also erroneous.

Etymology. From the Greek *gongros*, the ancient name of the conger eel. Masculine.

Species. Kanazawa (1958) recognized 14 species of *Conger* worldwide: *C. oceanicus* (Mitchill, 1818), *C. esculentus* Poey, 1861, and *C. triporiceps* Kanazawa, 1958, all western North Atlantic; *C. orbignyanus* Valenciennes, 1847, western South Atlantic; *C. conger* (Linnaeus, 1758), eastern North Atlantic; *C. cinereus* Rüppell, 1828, Indo-West Pacific; *C. wilsoni* (Bloch and Schneider, 1801), Australia, New Zealand, and South Africa; *C. verreauxi* Kaup, 1856, Australia and New Zealand; *C. philippinus* Kanazawa, 1958, and *C. macrocephalus* Kanazawa, 1958, both Philippines and Indonesia; *C. myriaster* (Brevoort, 1856), Japan and China; *C. jordani* Kanazawa, 1958, Japan and South Africa; *C. erebennus* (Jordan and Snyder, 1901), Japan; and *C. oligoporus* Kanazawa, 1958, Hawaii. Asano (1962) resurrected *Conger japonicus* Bleeker, 1879, which Kanazawa had considered a synonym of *C. myriaster*. He further suggested that *C. erebennus* and *C. jordani* might be synonyms of *C. japonicus*. Castle (1968c: 697) presented evidence that *C. wilsoni* and *C. jordani* may be synonymous. From my own observations, I would question the distinction between *C. macrocephalus* and *C. philippinus*. The species composition of *Conger* remains unsettled.

Key to the Species of *Conger* in the Western North Atlantic

1a. Three supratemporal pores; one or two postorbital pores .... *triporiceps*, p. 519
1b. One supratemporal pore; no postorbital pores 2
2a. Preanal lateral-line pores 32–36; total vertebrae 132–137 .... *esculentus*, p. 519
2b. Preanal lateral-line pores 37–44; total vertebrae 143–147 .... *oceanicus*, p. 515

Conger oceanicus (Mitchill, 1818)

CONGER EEL

Figures 545–548, Tables 50, 51

Anguilla conger (not of Linnaeus, 1758). Mitchill, 1814: 360.
Anguilla oceanica Mitchill, 1818: 407 (original description, Long Island Sound, Block Island, holotype unknown).
Isognatha oceanica. DeKay, 1842: 313.
Conger oceanica. DeKay, 1842: 313.
Conger occidentalis DeKay, 1842: 314 (a new name intended to distinguish the American from the European conger; no type locality or type specimen).
Leptocephalus gracilis Storer, 1846: 524 (original description, Cherryfield, Maine, holotype unknown); 1848: 76. Leptocephalus.
Echelus conger (not of Linnaeus, 1758). Jordan, 1888: 90.
Conger conger (not of Linnaeus, 1758). Bigelow and Welsh, 1925: 86.
Conger oceanicus. Schmidt, 1931b: 602.

Distinctive Characters. One supratemporal pore. Preanal lateral-line pores 37–44. Total vertebrae 143–147.

Description. Total vertebrae 143–147 (n = 14), predorsal vertebrae 14–19 (13), preanal vertebrae 41–46 (13), precaudal vertebrae 50 (2). Branchiostegal rays 9 (5). Pectoral-fin rays 16–18 (from Kanazawa, 1958). Pores: LL to anus 37–42 (16), POM 9 (11), IO 4–5 (12) + 1 (14) + 0–2 (14), SO 1 + 2–3 (14), STC 1–2 (15).

Proportions as % of TL: preanal 38–43 (13), predorsal 19–24 (12), head 14–17 (12), depth at anus 4–6 (11), pectoral fin 4–6 (18). Of HL: snout 21–28 (18), eye 12–19 (18), interorbital 16–21 (17), snout-rictus 34–40 (18), gill opening 9–16 (18), interbranchial 21–35 (18), pectoral fin 26–36 (18).

Color. In preservative brown or gray, lighter ventrally. Vertical fins black-edged.

Size. Said to reach 2 m and 10 kg (Bigelow and Schroeder, 1953: 155). Females grow larger than males.

Development and Growth. The leptocephalus has been identified (see p. 742).

Habits and Behavior. Evidence suggests that *Conger oceanicus* undertakes a long spawning migration. Although adults live along the east coast

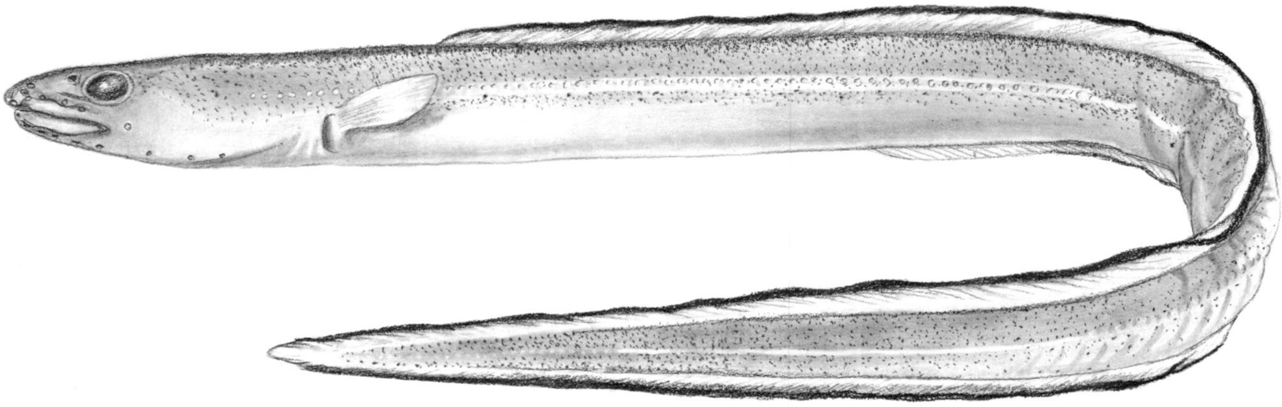

FIGURE 545. *Conger oceanicus*: ANSP 126968, 159 mm TL.

of the United States, small leptocephali have never reliably been reported from there. The eggs and young leptocephali described by Eigenmann (1902: 37) as those of the conger belong rather to the ophichthid *Ophichthus cruentifer* (Goode and Bean) (Naplin and Obenchain, 1980). Schmidt (1931b: 603) reported finding "quite tiny larvae" in the West Indies and concluded that the species breeds there. According to Bigelow and Schroeder (1953: 155) ripe congers are never caught on hook and line, indicating that as they approach maturity they stop feeding. Observations on the European species, *Conger conger* (Linnaeus), have shown that certain changes take place as the individual ripens, including enlargement of the eyes, loss of the teeth, and a softening of the bones (Cunningham, 1891–1892). These changes are similar to those that occur in the freshwater eels, *Anguilla,* as they begin their migration. Bigelow and Schroeder speculated that the American conger ripens in the summer, but positive data are lacking. Unlike *Anguilla rostrata,* leptocephali of *Conger oceanicus* seem to be collected throughout the year.

Like all eels, the conger is a predator. Bigelow and Schroeder (1953: 155) reported that it feeds chiefly on fishes (butterfish, herring, and eels, among others) but also takes shrimps, crabs, and small molluscs.

Able et al. (1982) reported *Conger oceanicus* associated with tilefish (*Lopholatilus chamaeleonticeps*) burrows off the coast of New Jersey.

Relationship to Man. Next to the freshwater eel, *Anguilla rostrata,* the conger is the most familiar eel to Americans. In the New England and Middle Atlantic states it lives in quite shallow water and is frequently caught by anglers during the warm months. Although the European conger is a food fish of some importance, the American species is not utilized to any significant extent, at least today. This is probably due to custom

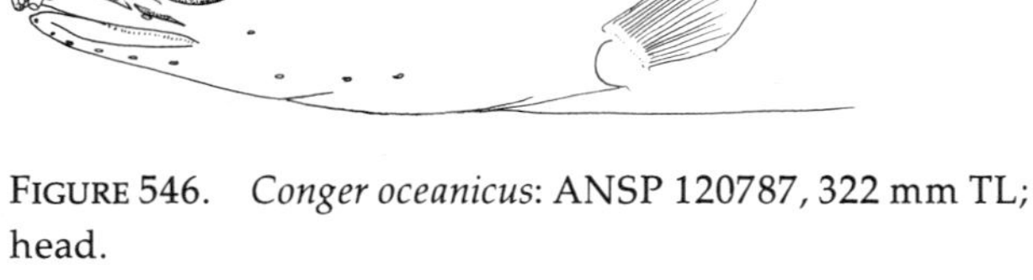

FIGURE 546. *Conger oceanicus*: ANSP 120787, 322 mm TL; head.

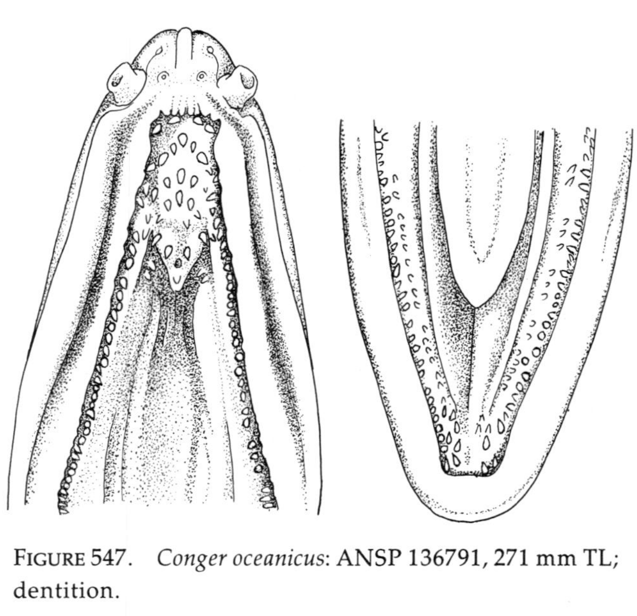

FIGURE 547. *Conger oceanicus*: ANSP 136791, 271 mm TL; dentition.

TABLE 50. Relationship between HL and TL in 12 specimens of *Conger oceanicus.*

HL in % TL	TL in mm					$\bar{x}$
14	159	287	311	320	—	269.3
15	264	275	302	318	389	309.6
16	—	—	—	—	—	—
17	328	442	328	—	—	366.0

rather than an inherent inferiority of the American species. Hildebrand and Schroeder (1928: 117) quoted Uhler and Lugger (1876: 181) as saying that at that time the conger was common in the lower Potomac and frequently brought to market. By 1928, however, congers were rare in Chesapeake Bay and of no commercial importance.

Variation. Eleven of 12 specimens examined had five infraorbital pores along the upper jaw, one specimen had four. Fourteen specimens had no postorbital pores, one specimen had two. Twelve specimens had three supraorbital pores on both sides, one specimen had four, and one had three on one side and four on the other. Fourteen specimens had one supratemporal pore, one specimen had two.

Kanazawa (1958: 259) reported a clinal difference in the relative length of the head and snout from north to south. In both characters, specimens from New England had lower values than those from New Jersey to Virginia; specimens from North Carolina to Florida had higher values still, and those from the Gulf of Mexico had the highest of all. The problem here is twofold. First, both characters change with growth, a fact acknowledged by Kanazawa (1958: 224). Second, some specimens lose a portion of their tails. Although the regenerated tails form a new caudal fin and are often nearly indistinguishable from the undamaged condition, the reduced length changes all characters based on total length. My own examination of these two characters at first showed the same condition found by Kanazawa, i.e., lower values for northern specimens. When the total length of the specimens was considered, however, the variation in head and snout length appeared to reflect size more than locality. Table 50 shows the total lengths of specimens with unregenerated tails (determined from radiographs) related to head length. Although the overlap is considerable, the specimens with the greater head length are clearly larger. A similar relationship exists with snout length (Table 51). I also re-examined the four specimens from the Gulf of Mexico reported by Kanazawa (USNM 164239), which had the greatest head and snout lengths. Three of them clearly showed regenerated tails, and the fourth had an unusually low vertebral count (139), indicating that it, too, was probably incomplete. Proportional measurements based on TL of these specimens would thus be unnaturally high. Geographic variation in morphometric characters of *Conger oceanicus* must therefore be considered unproved. If this species does indeed migrate long distances to spawn, and if the leptocephali must rely on large-scale current systems for transport back to the continent, then it is difficult to see how clinal variations could come about.

Distribution (Fig. 548). *Conger oceanicus* is found mainly, if not exclusively, along the Atlantic coast of the United States and the eastern Gulf of Mexico. In the north it seldom occurs north of Cape Cod. Although metamorphic leptocephali appear regularly in the Gulf of Maine, the only records of adults from there are two specimens from North Truro and Provincetown in Cape Cod Bay (Bigelow and Schroeder, 1953: 156). The species seems most common from Cape Cod south

TABLE 51. Relationship between snout length and TL in 16 specimens of *Conger oceanicus.*

Snout in % HL	TL in mm								$\bar{x}$
25 or less	159	264	275	287	311	318	320	328	282.8
26 or more	335	374	389	442	471	533	559	710	476.6

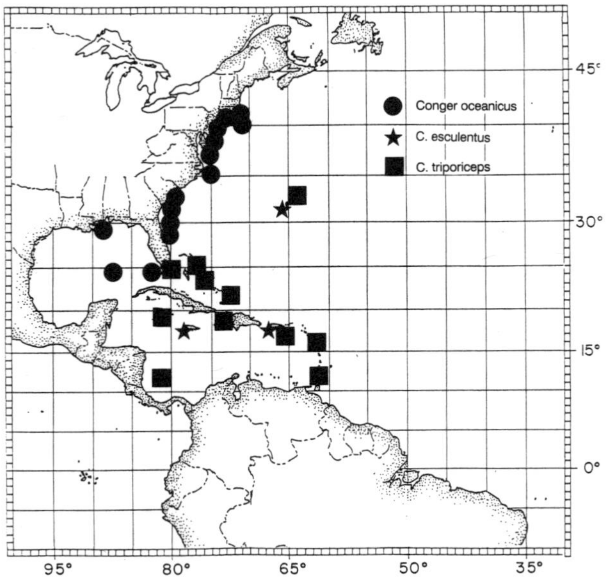

FIGURE 548. Distribution of *Conger oceanicus, C. esculentus,* and *C. triporiceps.*

to New Jersey and Delaware, or at least it is most frequently encountered in shallow water there. From Virginia to Florida most of the records are from offshore. Hildebrand and Schroeder (1928: 117) stated that it was "only a straggler" in Chesapeake Bay. In the Gulf of Mexico it has been found off Pensacola, Florida (UMML 8895) and off the mouth of the Mississippi (USNM 164239). There do not appear to be reliable records from the western Gulf of Mexico. Ginsburg (1951) did not include it in his account of the eels of the northern Gulf of Mexico. Walls (1975) also omitted it, and Hoese and Moore (1977: 129) included only a brief reference to Kanazawa (1958). The MBI cruises in the western and southern Gulf of Mexico failed to collect any leptocephali of *Conger oceanicus.*

The southern limit of its range is uncertain. I have seen a specimen (USNM 179226) taken just west of the Dry Tortugas and two others (ANSP 114150, 126178) from just off Campeche Bank north of the Yucatan Peninsula. The latter are the southernmost records and the only clear records not from the continental shelf or slope of the United States. In the Caribbean *Conger oceanicus* appears to be replaced by *C. esculentus* Poey. Only two characters reliably separate these two species: preanal lateral-line pores and total vertebrae. The former very nearly overlaps and the latter is affected by regenerated tails. The few specimens of *Conger* other than *C. triporiceps* examined from the Caribbean have all proved to be *C. esculentus.*

Although adult *Conger oceanicus* have not been reported from the Caribbean, leptocephali do occur there. This would seem to support Schmidt's (1931b) conclusion that the species spawns in the West Indies, presumably after a long migration from the North American coast.

Günther (1870: 39) recorded a specimen from the South Atlantic island of St. Helena as *Conger vulgaris* (= *C. conger*). Kanazawa (1958: 253) provisionally reidentified it as *Conger esculentus* Poey, largely because it had fewer preanal lateral-line (LL) pores than *C. conger* (37 vs 44–47). Dr. Alasdair Edwards (pers. comm.) recently collected a second specimen from St. Helena and obtained counts and measurements from both it and Günther's specimen. The specimens have 39–42 LL pores and 142–145 vertebrae, values characteristic not of *C. esculentus* but of *C. oceanicus.* None of the other characters differ significantly from those of *C. oceanicus.* Whether the St. Helena population is conspecific in a biological sense with *Conger oceanicus* is a question that cannot be answered. They may simply share the few characters that are useful in distinguishing the various species of *Conger.*

Conger oceanicus inhabits a wide range of depths, from near shore to beyond the edge of the continental shelf. The deepest reliable record is 577 m in the Gulf of Mexico (ANSP 126178). Southern records tend to be deeper and farther offshore. Bigelow and Schroeder (1953: 157) suggested that, at least in northern areas, the conger moves offshore in winter.

Etymology. From the Latin *oceanus* (sea, ocean) and *-icus* (belonging to), referring to the marine habitat of this eel, apparently intended by Mitchill to distinguish it from the freshwater eel, *Anguilla rostrata.* An adjective.

Study Material. 26 specimens, 159–710 mm TL.

NEW JERSEY: ANSP 120402 (1, 311), 120641 (1, 287), 120750 (1, 302), 120760 (1, 264), 126541 (1, 275), 127212 (2, 275–295). DELAWARE: ANSP 120787 (1, 322), 128770

(2, 318–320), 136791 (1, 271), 152118 (1, 357). NORTH CAROLINA: ANSP 114849 (1, 389), 114953 (1). SOUTH CAROLINA: ANSP 127794 (1). FLORIDA ATLANTIC: ANSP 126968 (1, 159). UMML 5218 (1, 374), 15009 (1, 685). GULF OF MEXICO: FLORIDA: UMML 8895 (1, 710). USNM 179226 (1, 328). ALABAMA: USNM 164239 (4, 442–559). MEXICO: ANSP 114150 (1, 335), 126178 (1, 442).

Conger esculentus Poey, 1861
Figure 548

Conger esculentus Poey, 1861: 346 (original description, Cuba, holotype MCZ 9330).

Conger niger (not of Risso, 1810). Bean and Dresel, 1884b: 169 (Jamaica).

Distinctive Characters. One supratemporal pore. Preanal lateral-line pores 32–36. Total vertebrae 136–137.

Description. Total vertebrae 136–137 (n = 5), predorsal vertebrae 14–16 (5), preanal vertebrae 36–37 (5). Branchiostegal rays 9 (1). Pectoral-fin rays 18–19 (from Kanazawa, 1958). Pores: LL to anus 32–35 (6), POM 9 (4), IO 5 (4) + 1 (4) + 0 (4), SO 1 + 2 (4), STC 1 (4).

Proportions as % of TL: preanal 39–41 (6), predorsal 21–24 (6), head 16–18 (6), depth at anus 6–8 (5), pectoral fin 6–7 (5). Of HL: snout 25–27 (5), eye 13–15 (5), interorbital 16–18 (4), snout-rictus 32–38 (5), gill opening 12–16 (5), interbranchial 31–38 (5), pectoral fin 32–41 (5).

Color. In preservative brown or gray, lighter ventrally. Vertical fins dark-edged.

Size. The largest specimen examined was 1330 mm TL (ANSP 128124).

Development and Growth. The leptocephalus has been identified (see p. 744).

Relationship to Man. The name chosen by Poey for this species and the fact that he gave a common name for it (congrio) suggest that it was known to the local population and was used as food. Randall et al. (1978) reported that it is taken by longlines, hook and line, and traps around Cuba and Jamaica but that there is no special fishery for it.

Remarks. Information from Karsten E. Hartel of the Museum of Comparative Zoology, Harvard University indicates that the holotype of *Conger esculentus* is probably MCZ 9330 rather than MCZ 9328 as generally believed (Howell Rivero, 1938: 173; Kanazawa, 1958: 253). Poey gave a TL of 1025 mm for his type. MCZ 9330 measures 1015 mm with the caudal fin broken off where it meets the body; this could add 10–20 mm to the TL, making it very close to Poey's original measurement. MCZ 9328 measures 965 mm. When these two specimens were originally catalogued into the MCZ collection, no note was made on the type status of either specimen. The word "type" was entered in the catalog many years later, probably by Borodin. This might have been done because there was an original Poey number (581, the number that appears in the original description) in the bottle containing MCZ 9328, but not with MCZ 9330. Hartel points out, however, that both specimens were originally received by the MCZ under Poey's number 581. Based on this evidence, I consider MCZ 9330 as the holotype of *Conger esculentus.*

Distribution (Fig. 548). *Conger esculentus* is known from Bermuda and the Greater Antilles. Metzelaar (1919: 14) recorded "*Conger vulgaris*" from Curaçao. Kanazawa (1958: 254) did not see the specimen but, based on the locality, conditionally referred it to *C. esculentus.*

Kanazawa (1958: 254) tentatively identified a specimen from the island of St. Helena in the South Atlantic as *Conger esculentus.* It is now known that the specimen is closer to *Conger oceanicus* in its diagnostic characters than to *C. esculentus* (see account of *Conger oceanicus* above).

Etymology. Latin *esculentus* (edible). An adjective.

Study Material. Six specimens, 700–1330 mm TL.

Holotype: MCZ 9330 (female, 1015), Cuba.

Other material: BERMUDA: ANSP 128124 (1, 1330), south shore, 165 m, 12 June 1973. ANSP 137866 (1, 855). CUBA: MCZ 9328 (1, 965). PUERTO RICO: ANSP 139682 (1, 700), off Mayagüez, 1978. ANSP 144593 (1, 684), Cabo Rojo, east of buoy "2," 229 m, 27 Feb. 1976.

Conger triporiceps Kanazawa, 1958
Figures 548, 549

Conger brasiliensis Kaup, 1856a: 73 (original description, Brazil, holotype MNHN 8427); 1856b: 115. Preoccu-

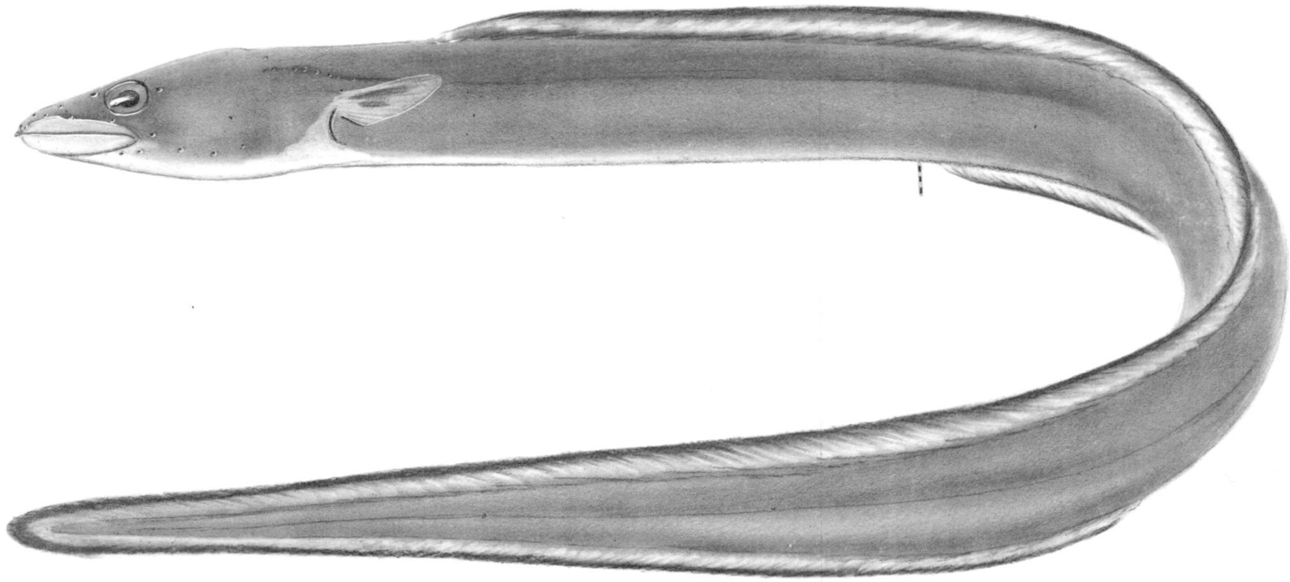

FIGURE 549. *Conger triporiceps*: ANSP 108711, 542 mm TL. Illustration by S. P. Gigliotti.

pied by *Conger brasiliensis* Ranzani, 1839, = *Muraena savanna* Bancroft, 1831.

Conger occidentalis (not of DeKay, 1842). Kaup, 1856a: 72; 1856b: 114 (Martinique).

Leptocephalus conger (not of Linnaeus, 1758). Nichols, 1921: 22 (Bahamas).

Conger triporiceps Kanazawa, 1958: 243 (substitute name for *Conger brasiliensis* Kaup).

Distinctive Characters. Three supratemporal pores. Postorbital pores present. Total vertebrae 156–160.

Description. Total vertebrae 156–160 (n = 11), predorsal vertebrae 12–14 (7), preanal vertebrae 38–40 (6). Branchiostegal rays 8 (5). Pectoral-fin rays 14–17 (from Kanazawa, 1958). Pores: LL to anus 35–39 (8), POM 11 (7), IO 5 (8) + 1 (8) + 1 (6), SO 1 + 4 (8), STC 3 (8).

Proportions as % of TL: preanal 32–34 (15), predorsal 15–19 (9), head 12–15 (9), depth at anus 4–6 (9), pectoral fin 3–4 (5). Of HL: snout 24–29 (9), eye 13–18 (9), interorbital 13–18 (7), snout-rictus 36–41 (8), gill opening 7–13 (8), interbranchial 21–26 (8), pectoral fin 25–29 (5).

Color. Head and body brown; lower part of head and belly pale; lateral-line pores pale; vertical fins black-edged, pale brown basally; pectoral fin pale in young individuals, upper part becoming dark in larger specimens (Böhlke and Chaplin, 1968: 92).

Size. To about 1 m TL.

Development and Growth. The leptocephalus has been identified (see p. 743).

Habits and Behavior. Böhlke and Chaplin (1968: 92) reported collecting this species from fine, white sand surrounding patch reefs at depths of 4–15 m. *Conger triporiceps* seems to prefer clear, relatively shallow water and sandy bottoms. This and the Indo-Pacific *Conger cinereus* are the only species of *Conger* commonly found around coral reefs.

Distribution (Fig. 548). Recorded from southern Florida, Bermuda, Bahamas, West Indies, western Caribbean, and Brazil. Kanazawa (1958: 244) reported a specimen from Pensacola, Florida, and leptocephali have been collected in the Gulf of Mexico. *Conger triporiceps* probably occurs throughout the tropical western Atlantic wherever conditions are favorable.

Etymology. From the Latin *tri* (three), *porus* (pore), and *caput* (head), referring to the three pores in the supratemporal commissure. A noun in apposition.

Study Material. 17 specimens, 80–737 mm TL. BERMUDA: ANSP 133875 (2, 134–163), 133876 (1, 176),

134299 (1), 152005 (1, 309). FLORIDA ATLANTIC: UMML 9243 (1, 280). BAHAMAS: ANSP 103049 (2, 80–84), 103050 (2, 87–104), 108711 (1, 542), 112276 (1, 93). HAITI: ANSP 114165 (1, 207). LESSER ANTILLES: ANSP 105751 (1, 103), 110090 (1, 737). PROVIDENCIA: ANSP 128808 (1, 595). CAYMAN ISLANDS: ANSP 95621 (1, 615).

Genus *Gnathophis* Kaup, 1860

Gnathophis Kaup, 1860: 7 (type species *Myrophis heterognathos* Bleeker, 1859, by original designation). Castle, 1963: 16 (definition and history of *Gnathophis*, with synonymy).

Rhynchocymba Jordon and Hubbs, 1925: 194 (type species *Leptocephalus nystromi* Jordan and Snyder, 1901, by original designation).

Poutawa Griffin, 1936: 15 (type species *Congrus habenatus* Richardson, 1848, by original designation).

Fimbriceps Whitley, 1946: 60 (type species *Fimbriceps umbrellabia* Whitley, 1946, by original designation). Castle, 1977c: 581 (synonymized *Fimbriceps* with *Gnathophis*).

Lemkea Kotthaus, 1968: 32 (type species *Lemkea heterolinea* Kotthaus, 1968, by original designation).

Characters. Body moderately elongate, anus at approximately two-fifths TL; tail moderately blunt, tip slightly stiffened, caudal fin short. Dorsal fin begins slightly behind base of pectoral fin. Pectoral fin well developed.

Snout projects beyond lower jaw, fleshy part of snout projects anteriorly well beyond intermaxillary teeth; longitudinal fleshy keel on underside of tip of snout. Flange on upper lip narrow. Anterior nostril tubular, on underside of snout near tip, directed anteroventrally, rim incised to form a flap or scroll-like structure. Posterior nostril elliptical, in front of mid-eye. Head pores small. Adnasal pore absent; pores along upper jaw located on flange. Lateral supratemporal pores low on canal, at end of short, posteriorly directed branches.

Teeth moderately small, conical. Intermaxillary tooth patch longer than broad, mostly excluded from closed mouth. Vomerine tooth patch elongate, about a third the length of maxillary tooth patch. Maxillary and mandibular teeth in narrow bands.

Lateral ethmoid process absent; supraoccipital present; pterosphenoid, sphenotic, and prootic fused into a single element. Preorbital divided, with bony projections from lower edge; four suborbitals. Posterior edge of opercle smooth. Precaudal vertebrae less than 40% of total; ribs absent; prominent epicentral processes on caudal vertebrae. Lateral-line ossicles like those of *Conger*.

Intestine pale; stomach black or pale, not reaching anus posteriorly. Gas bladder long.

Size. Less than 500 mm TL.

Distribution. World-wide, tropical and warm temperate.

Etymology. From the Greek *gnathos* (jaw) and *ophis* (snake). Masculine.

Species. This is a large but poorly known genus that could rival *Ariosoma* and *Conger* in number of species. The following list of 14 species should be considered provisional. *Gnathophis bracheatopos* Smith and Kanazawa, 1977, *G. bathytopos* Smith and Kanazawa, 1977, and *G. tritos* Smith and Kanazawa, 1977, all western North Atlantic; *G. mystax* (Delaroche, 1809), Mediterranean and adjacent northeastern Atlantic; *G. codoniphorus* Maul, 1972, Meteor Seamount, North Atlantic; *G. capensis* (Kaup, 1856), South Africa; *G. leptosomatus* Karrer, 1982 and *G. heterolinea* (Kotthaus, 1968), both Indian Ocean; *G. habenatus* (Richardson, 1848), South Africa to New Zealand; *G. heterognathos* (Bleeker, 1859), *G. nystromi* (Jordan and Snyder, 1901), *G. ginanago* (Asano, 1958) (either *ginanago* or *nystromi* is most likely a junior synonym of *G. heterognathos*, see Castle, 1963: 18), and *G. xenicus* (Matsubara and Ochiai, 1951), all Japan; *G. umbrellabia* (Whitley, 1946), Australia; *G. catalinensis* (Wade, 1946), eastern Pacific.

The principal characters distinguishing the various species are the number of vertebrae, the number and arrangement of sensory pores, the color of the stomach, the form of the anterior nostril, certain proportional measurements, and the form of the teeth. The nature of these characters is not known for all species, and critical comparisons cannot be made from the literature alone. Further study is needed before the number of species of *Gnathophis* is known.

KEY TO THE SPECIES OF *GNATHOPHIS* IN THE WESTERN NORTH ATLANTIC

1a. Second and about seventh to thirteenth lateral-line pores elevated; stomach black .... *bathytopos*, p. 522
1b. Only second and at most one other lateral-line pore elevated; stomach pale 2
2a. Three postorbital pores; preanal lateral-line pores 26–31 .... *bracheatopos*, p. 523
2b. Two postorbital pores; preanal lateral-line pores 34–36 .... *tritos*, p. 524

Gnathophis bathytopos
Smith and Kanazawa, 1977
Figures 550–552

Gnathophis sp. B. Smith, 1971: 60.
Gnathophis bathytopos Smith and Kanazawa, 1977: 535, fig. 6 (original description, 25°13′N, 84°15′W, holotype USNM 190597).

Distinctive Characters. All *Gnathophis* have the second lateral-line pore elevated, but *G. bathytopos* is the only western Atlantic species with five or six additional pores elevated. It is the only species in the area with a black stomach.

Description. Total vertebrae 128–133 (n = 8), predorsal vertebrae 9–10 (2), preanal vertebrae 34–36 (2), precaudal vertebrae 40–43 (2). Branchiostegal rays 8 (1). Pectoral-fin rays 10–13 (7), dorsal-fin rays 186–215 (3), anal-fin rays 124–159 (3). Pores: LL to anus 27–37 (11), POM 10 (9), IO 4 (10) + 1 (11) + 3 (11), SO 1 + 5 (11), STC 3 (11).

Proportions in % TL: preanal 38–40 (11), predorsal 18–20 (11), head 15–18 (11), depth at anus 3–6 (12), pectoral fin 3–6 (15). Of HL: snout 27–32 (12), eye 18–22 (12), snout-rictus 36–37 (10), gill opening 8–14 (11), interbranchial 19–36 (8), pectoral fin 18–30 (15).

Dorsal fin begins over or slightly behind base of pectoral fin. Precaudal vertebrae about 32% total. In addition to second lateral-line pore, the 7th–8th through 12th–16th pores also elevated. Gas bladder terminates behind anus.

Color. In preservative brownish. Stomach black, intestine pale.

Size. The largest specimen examined measured 351 mm TL (UMML 30226), but it had a damaged tail with an estimated 20–30 vertebrae missing.

Development and Growth. Leptocephali of *Gnathophis* have been identified in the western North Atlantic. Some of these probably belong to *G. bathytopos*, but the overlap in vertebral count between it and *G. bracheatopos* makes positive identification difficult (see p. 746).

Remarks. Gnathophis bathytopos is similar to the eastern Atlantic *G. mystax*. Both have a black stomach and extra elevated lateral-line pores. They differ primarily in the number of vertebrae, 128–134 in the former, 134–141 in the latter.

Distribution (Fig. 552). East coast of North America from off Chesapeake Bay to Florida,

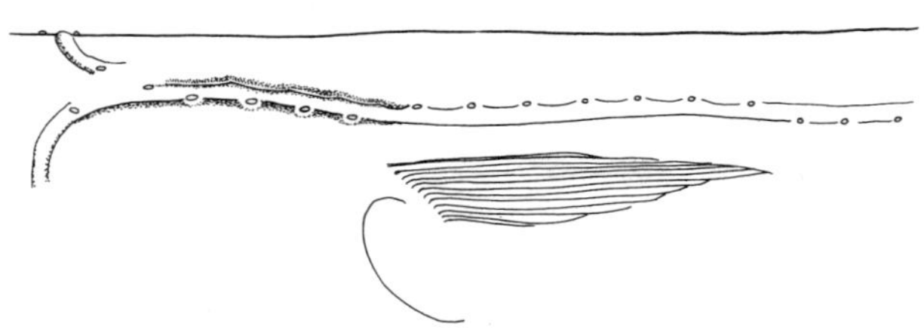

FIGURE 550. *Gnathophis bathytopos*: ANSP 127791, paratype, 342 mm TL; anterior part of lateral line showing elevated pores (from Smith and Kanazawa, 1977: fig. 6).

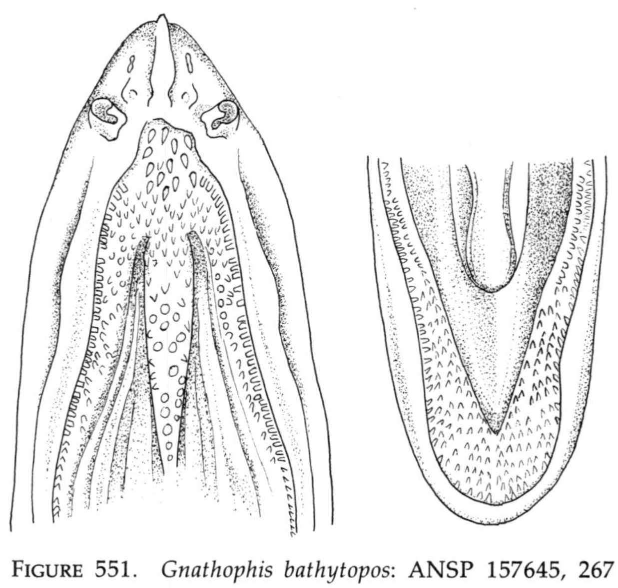

FIGURE 551. *Gnathophis bathytopos*: ANSP 157645, 267 mm TL; dentition.

southeastern Gulf of Mexico, and Yucatan Channel. Depth 90–366 m.

Etymology. From the Greek *bathys* (deep) and *topos* (place), referring to the relatively deep-water habitat of this species. A noun in apposition.

Study Material. 13 specimens, 111–351 mm TL.

Holotype: USNM 190597 (female, 340), Silver Bay 1200, southeastern Gulf of Mexico, 25°13′N, 84°15′W, 183 m, 9–10 June 1959.

Paratypes: SOUTH CAROLINA: ANSP 127791 (1, 342), 32°18′N, 78°50′W, 187–188 m, 8 Nov. 1973. FLORIDA ATLANTIC: UMML 17085 (1, 111), 25°02′N, 80°12′W, 185 m, 22 Jan. 1965. UMML 22591 (1, 174), 24°52′N, 80°20′W, 187–192 m, 16 Aug. 1966. USNM 197113 (2, 218–240), 25°29′N, 79°19′W, 366 m, 9 Nov. 1960. FLORIDA GULF: USNM 213558 (2, 314–330), same data as holotype. MEXICO QUINTANA ROO: UMML 30226 (1, 351), 20°11′N, 86°30′W, 165–311 m, 28 Jan. 1968. USNM 213557 (1, 339), 21°54′N, 86°28′W, 238 m, 22 Jan. 1967.

Other material: VIRGINIA: AMNH 40229 (2, 263–331), 36°40′N, 74°45′W, 90 m, 22 Nov. 1978. SOUTH CAROLINA: ANSP 157645 (1, 267), 33°07.0′N, 77°29.0′W, 156–170 m, 15 Jan. 1976.

Gnathophis bracheatopos Smith and Kanazawa, 1977

Figures 552, 553

Gnathopis sp. A. Smith, 1971: 59.

Gnathophis bracheatopos Smith and Kanazawa, 1977: 534, fig. 5 (original description, 28°50′N, 85°06′W, holotype USNM 163523).

Distinctive Characters. Among the western Atlantic species of *Gnathophis, G. bracheatopos* is distinguished from *G. bathytopos* by having fewer preanal lateral-line pores, having only the second lateral-line pore elevated, and having a pale stomach. It differs from *G. tritos* in having fewer vertebrae, fewer preanal lateral-line pores, more postorbital pores, and a longer gas bladder.

Description. Total vertebrae 125–130 (n = 23), predorsal vertebrae 7–9 (21), preanal vertebrae 30–34 (21), precaudal vertebrae 40 (1). Branchiostegal rays 8 (3). Pectoral-fin rays 10–14 (12), dorsal-fin rays 180–194 (4), anal-fin rays 129–151 (4). Pores: LL to anus 26–31 (21), POM 9–10 (7), IO 4 (9) + 1 (10) + 3 (14), SO 1 + 4–5 (14), STC 1–4 (14).

Proportions as % of TL: preanal 36–40 (20), predorsal 17–19 (21), head 16–18 (21), depth at anus 4–6 (14), pectoral fin 5–6 (15). Of HL: snout 31–35 (16), eye 19–21 (17), snout-rictus 39–42 (7), gill opening 7–15 (16), interbranchial 22–29 (15), pectoral fin 25–39 (15).

Dorsal fin begins over or very slightly behind base of pectoral fin. Only second lateral-line pore elevated. Precaudal vertebrae about 32% of total. Gas bladder terminates behind anus.

Color. In preservative brownish, paler ventrally. Dorsal- and anal-fin rays dark; distal portion of caudal fin and posteriormost dorsal and anal fin black. Stomach and intestine pale.

Size. The largest specimen examined was 352 mm TL (UMML 31019).

Development and Growth. Leptocephali of *Gnathophis* have been identified in the western North Atlantic. Some of these probably belong

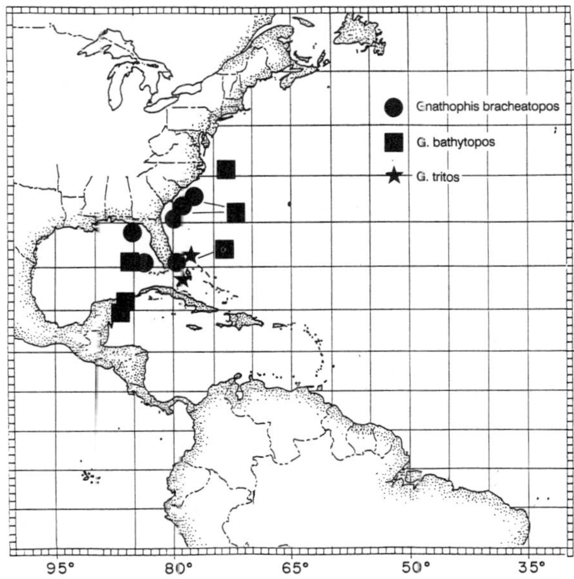

FIGURE 552. Distribution of *Gnathophis bathytopos, G. bracheatopos,* and *G. tritos.*

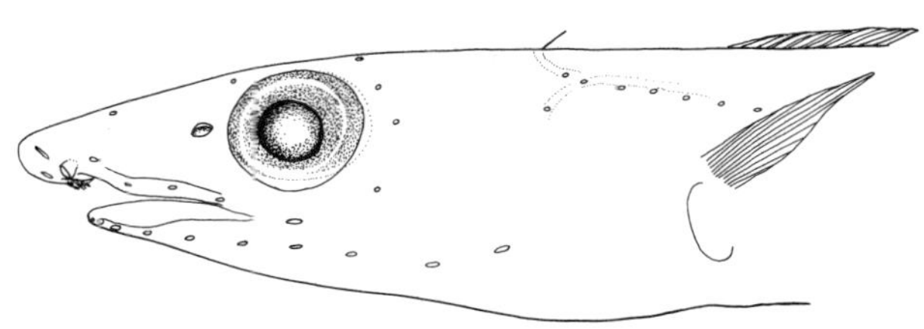

FIGURE 553. *Gnathophis bracheatopos:* ANSP 130817, paratype, 245 mm TL; head (from Smith and Kanazawa, 1977: fig. 5).

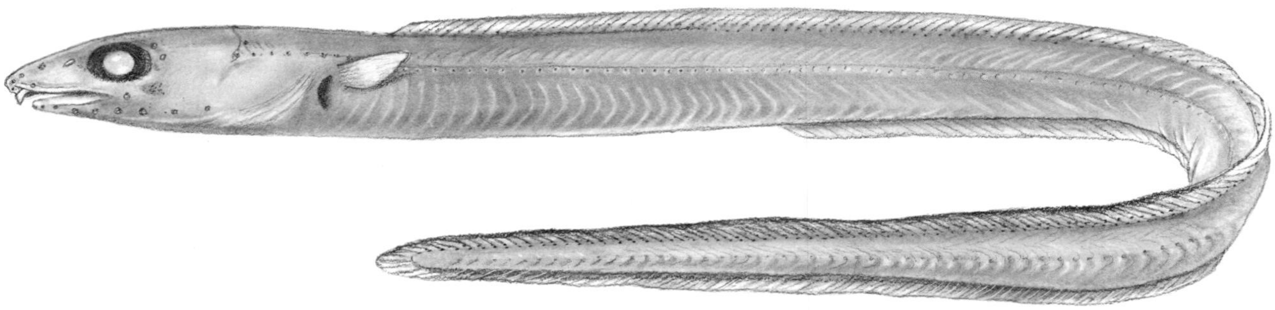

FIGURE 554. *Gnathophis tritos*: ANSP 126972, holotype, 181 mm TL.

to *G. bracheatopos*, but the overlap in vertebral count between it and *G. bathytopos* makes positive identification difficult (see p. 746).

Variation. Most specimens have three supratemporal pores; one specimen had four, with an extra pore on the left side, and another specimen had only the median pore. Six of seven specimens had 10 preoperculomandibular pores, the other had nine. Thirteen specimens had five supraorbital pores, one had four. It should be noted that the same specimen accounted for all but one of the above anomalies: it had one STC pore, nine POM pores, and four SO pores. The specimen was cleared and stained and cannot be rechecked.

Distribution (Fig. 552). Known only from the lower east coast of the United States and the eastern Gulf of Mexico, from South Carolina to northwestern Florida. In the Straits of Florida, it has been collected only on the mainland side. Depth 55–110 m.

Etymology. From the Greek *brachos* (shallow) and *topos* (place), referring to the relatively shallow-water habitat of this species. A noun in apposition.

Study Material. 30 specimens, 198–360 mm TL.

Holotype: USNM 163523 (male, 325), Oregon 896, northeastern Gulf of Mexico, 28°50′N, 85°06′W, 64 m, 7 March 1954.

Paratypes: SOUTH CAROLINA: ANSP 123583 (1, 325), 128391 (1, 303), 128392 (1, 305), 128393 (1, 290). UMML 31019 (1, 352). GEORGIA: ANSP 128390 (2, 259–312). FLORIDA ATLANTIC, ANSP 109602 (2, 246–267), 128389 (7, 209–321), 130817 (1, 245). UMML 32627 (1, 236). USNM 163485 (2, 198–318), 197128 (1, 269).

Other material: SOUTH CAROLINA: GMBL 75-25 (7, 192–360). GMBL 75-106 (1, 273).

Gnathophis tritos Smith and Kanazawa, 1977
Figures 552, 554

Gnathophis sp. C. Smith, 1971: 61.
Gnathophis tritos Smith and Kanazawa, 1977: 536, fig. 7 (original description, 23°25′N, 79°24′W, holotype ANSP 126972).

Distinctive Characters. Gnathophis tritos appears to be the only species in the genus with fewer than three postorbital pores. It has a shorter gas bladder and more vertebrae than the other western Atlantic species, and fewer supraorbital and preoperculomandibular pores. It further differs from *G. bracheatopos* by having more lateral-line pores and from *G. bathytopos* by having a pale stomach.

Description. Total vertebrae 136–138 (n = 3), predorsal vertebrae 10 (1), preanal vertebrae 38 (1), precaudal vertebrae 42 (1). Pectoral-fin rays 10–11 (3). Pores: LL to anus 34–36 (2), POM 9 (3), IO 4 (1) + 1 (2) + 1–2 (2), SO 1 + 4 (2), STC 3 (1).

Proportions in % of TL: preanal 37–39 (3), predorsal 18–19 (3), head 15–16 (3), depth at anus 3–4 (3), pectoral fin 4 (2). Of HL: snout 26–28 (2), eye 22–23 (2), snout-rictus 36 (1), gill opening 12–15 (3), interbranchial 20 (1), pectoral fin 26–27 (2).

Dorsal fin begins slightly behind base of pectoral fin. In one specimen only second lateral-line pore elevated, in another seventh pore also elevated. Missing supraorbital pore is number 5, near level of anterior margin of eye. Gas bladder terminates before anus.

Color. In a freshly caught specimen, body uniformly pale, anterior dorsal-fin rays dark, at least basally. Supratemporal commissure outlined in dark brown. Top of head between eyes dark brown. An iridescent green patch on each opercle, and a patch of irridescent blue on dorsal surface of snout from anterior nostril to about midway between top of snout and over brain. Anterior nostril darker brown. In preservative, green and blue disappear, brown areas remain.

Size. A small eel, based on the limited material available. The largest specimen was the holotype, only 181 mm TL.

Development and Growth. Leptocephali of *Gnathophis* from the western Atlantic have 126–141 myomeres, a range that suggests that more than one species is involved. A peak in the distribution of these counts occurs around 137–138. This corresponds closely to the vertebral count of *Gnathophis tritos,* but the rarity of *G. tritos* makes such an identification difficult to confirm. See p. 746.

Variation. Each of two specimens that were examined for the character (ANSP 126972 and UMML 15758) had one postorbital pore on one side and two on the other.

Distribution (Fig. 552). The three known specimens were taken from off the northwestern corner of the Little Bahama Bank and the southern entrance to the Santaren Channel. Depth 458–567 m.

Etymology. From the Greek *tritos* (third). This was the third species of *Gnathophis* discovered in the western Atlantic. An adjective.

Study Material. Three specimens, 173–181 mm TL.

Holotype: ANSP 126972 (male, 181), Pillsbury 1171, Santaren Channel, 23°25′N, 79°24′W, 516–521 m, 27 June 1970.

Paratypes: UMML 15758 (female, 173), Bahamas, 27°07′N, 79°21′W, 567 m, 5 Feb. 1964 (cleared and stained). UMML 29613 (female, 173), same data as holotype.

Genus *Rhynchoconger* Jordan and Hubbs, 1925
Figures 555, 556

Rhynchoconger Jordan and Hubbs, 1925: 196 (type species *Leptocephalus ectenurus* Jordan and Richardson, 1919, by original designation).

Hildebrandia Jordan and Evermann, 1927: 502 (type species *Congermuraena flava* Goode and Bean, 1896, by original designation).

Characters. Body moderately elongate, anus near first quarter or third of TL; tail slender, attenuate, frequently damaged. Dorsal fin begins over or slightly ahead of base of pectoral fin; pectoral fin well developed.

Snout projects well beyond lower jaw; fleshy part of snout extends anteriorly well beyond intermaxillary teeth; longitudinal fleshy keel on underside of tip of snout. Anterior nostril tubular, near anteroventral corner of snout, directed anteroventrally; posterior nostril elliptical, in front of mid-eye or slightly higher. Flange on upper and lower lip, that of upper lip narrow. Head pores small; pores along upper jaw located on flange; adnasal pore absent.

Teeth small, conical. Intermaxillary tooth patch rounded, slightly separated from maxillary and vomerine teeth, completely exposed when mouth closed. Vomerine tooth patch short. Maxillary and mandibular teeth in narrow bands, somewhat wider anteriorly.

Lateral ethmoid process absent. Supraoccipital absent. Preorbital bone with projections from lower edge; four suborbitals. Posterior edge of opercle serrate. Lower pharyngeals divided, anterior element fused to fifth ceratobranchial; upper pharyngeals divided, anterior element much the smaller. Myorhabdoi absent. Ribs absent. Anterior few epineurals fused to neural arch; prominent epicentral processes on caudal vertebrae; precaudal vertebrae less than 40 percent of total. Lateral-line ossicles weak, oval, vertically elongate structures with cartilaginous finger-like projections from upper and lower edges. Dorsal- and anal-fin rays segmented.

Stomach long, reaching or nearly reaching anus. Gas bladder long, terminating posterior to anus.

Discussion. The differences between the Atlantic and eastern Pacific species heretofore treated as *Hildebrandia* (Smith and Kanazawa, 1977: 536) and the Indo-West Pacific *Rhynchoconger ectenurus* (Jordan and Richardson) are so slight that there seems no reason to separate them generically. Asano (1962) gave a thorough account of the osteology of *R. ectenurus*. The only difference apparent between this species and the Atlantic species is that Asano described and illustrated both the upper and lower pharyngeals as being undivided. A cleared and stained specimen from the Arabian Sea (ANSP 127466), however, had both pharyngeals divided exactly as described above. It is possible that this character is influenced by growth and that fusion occurs in larger specimens. The specimens examined by Asano (1962: 101) were all considerably larger than those available to me. Nevertheless, even if the difference in the pharyngeals were real, it seems insufficient for separating genera. Three intact specimens of *R. ectenurus* from the South China Sea (CAS 40909) were available to me for study, and they showed no important differences from the Atlantic species.

Size. Probably to 1 m TL, although most specimens are considerably smaller.

Distribution. Indo-West Pacific, eastern Pacific, western Atlantic, and probably eastern Atlantic. Tropical and subtropical.

Etymology. From the Greek *rhynchos* (snout, muzzle) and *Conger* (the conger eel), perhaps in reference to the strongly projecting snout. Masculine.

Species. Five nominal species are recognized here: *Rhynchoconger ectenurus* (Jordan and Richardson, 1919), Indo-West Pacific; *R. nitens* (Jordan and Bollman, 1889), eastern tropical Pacific; *R. flavus* (Goode and Bean, 1896), *R. gracilior* (Ginsburg, 1951), and *R. guppyi* (Norman, 1925), all western North Atlantic. In addition, Blache (1977: 128) reported leptocephali from the Gulf of Guinea. The fact that three species are present in the western Atlantic while only one is currently recognized from the entire Indo-West Pacific region suggests that the latter area should be examined more carefully. Asano (1962: 102) gave the vertebral count for Japanese *R. ectenurus* as 155–159. Kotthaus (1968) gave a count of 165 for a specimen from the Arabian Sea, and the cleared and stained specimen from the same area mentioned above (ANSP 127466) had 169. Some of the species described by Alcock from the INVESTIGATOR cruises, such as *Congermuraena squaliceps*, *C. musteliceps*, and *C. macrocercus* resemble *Rhynchoconger*. *Rhynchoconger brevirostris* Chen and Weng, 1967 is a very different eel and does not belong in *Rhynchoconger*.

The principal characters separating the species of *Rhynchoconger* are the presence or absence of the fourth supraorbital pore (Fig. 555), the shape of the intermaxillary and vomerine tooth patches (Fig. 556), the color of the stomach, and the number of vertebrae. The frequency of damage to the slender tail reduces the value of the last character; *Rhynchoconger flavus* seems particularly susceptible to this problem.

Key to the Species of *Rhynchoconger* in the Western North Atlantic

1a. Fourth supraorbital pore present between anterior and posterior nostril .... *gracilior*, p. 526
1b. Fourth supraorbital pore absent 2
2a. Vomerine tooth patch as broad as long; stomach pale .... *flavus*, p. 529
2b. Vomerine tooth patch slightly longer than broad; stomach black .... *guppyi*, p. 531

Rhynchoconger gracilior (Ginsburg, 1951)
Figures 555A, 556A, 557

Congermuraena flava (not of Goode and Bean, 1896). Goode and Bean, 1896: 138, part.
Congrina gracilior Ginsburg, 1951: 445, fig. 3 (original description, 28°36′N, 85°33′W, holotype USNM 44617).
Hildebrandia gracilior. Smith, 1971: 63. Smith and Kanazawa, 1977: 536.

Discussion of Synonymy. The original type series of *Congermuraena flava* Goode and Bean contained two species. Ginsburg (1951: 445) selected the illustrated specimen as the lectotype of *C. flava* and described the other species as *Congrina gracilior*.

Distinctive Characters. Fourth supraorbital pore present. Intermaxillary tooth patch broader than

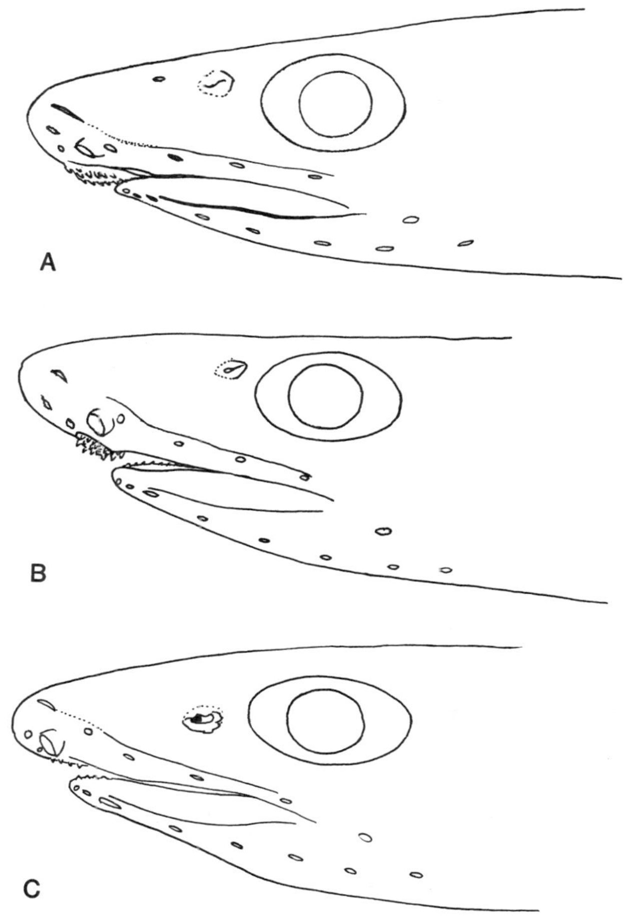

FIGURE 555. Heads of three species of *Rhynchoconger*. A. *R. gracilior*, ANSP 114149, 340 mm TL, note fourth SO pore in front of posterior nostril. B. *R. flavus*, ANSP 114882, 305 mm TL. C. *R. guppyi*, USNM 213812.

long; vomerine tooth patch longer than broad. Stomach black. Vertebrae 175–181.

When the two species are compared directly, *Rhynchoconger gracilior* is seen to be more slender than *R. flavus*. Despite this, *R. gracilior* does not seem to suffer tail damage as often as *R. flavus*, and intact specimens are more frequently found. The presence of the fourth supraorbital pore distinguishes *R. gracilior* at once from the other western Atlantic species of *Rhynchoconger*.

Description. Total vertebrae 175–182 (n = 8), predorsal vertebrae 6–7 (4), preanal vertebrae 29 (4), precaudal vertebrae 41–42 (3). Branchiostegal rays 8–10 (14). Pectoral-fin rays 9–13 (12). Pores: LL to anus 23–27 (14), POM 10 (11), IO 4 (8) + 1 (11) + 0 (11), SO 1 + 3 (27), STC 1 (10).

Proportions as % of TL: preanal 27–29 (5). Of preanal length: predorsal 45–53 (20), head 44–54 (21), depth at anus 9–16 (16), pectoral fin 13–17 (15). Of HL: snout 23–28 (15), eye 17–20 (14), snout-rictus 34–40 (11), gill opening 8–15 (15), interbranchial 16–25 (14), pectoral fin 27–32 (15).

Intermaxillary tooth patch broader than long; vomerine tooth patch longer than broad.

Color. In preservative brown or yellowish-brown, vertical fins edged in black posteriorly. Stomach black, intestine pale.

Size. The largest specimen examined was 434 mm TL, 127 mm preanal (USA 03275). Several specimens had greater total lengths but lesser preanal lengths. Because of the frequency of regenerated tails, preanal length is a better indicator of size than is total length.

Development and Growth. The leptocephalus has been identified (see p. 749).

Remarks. Ginsburg (1951: 445) chose the holotype and paratype of this species from among the type material of *Congermuraena flava* Goode and Bean, which contained both species. The paratype, MCZ 37165, cannot be found. A note on the type card says that it was loaned to Ginsburg in 1951 and never returned. MCZ 28080 is present, however, and it bears the same Louis Agassiz collection number (CCLXIV) as MCZ 37165. Although Ginsburg (1951: 445) listed MCZ 28080 as *R. flavus*, the specimen is in fact *R. gracilior*. MCZ 28080 originally contained two specimens, although only one is now in the jar. It seems likely that MCZ 37165 was taken from MCZ 28080 and made a paratype of *R. gracilior*. Remaining unexplained are what happened to MCZ 37165 and why Ginsburg did not recognize that both specimens were *gracilior*.

Distribution (Fig. 557). From Georgia, the east coast of Florida, and the eastern Gulf of Mexico south through the Caribbean to South America at least as far as Suriname. Depth 82–458 m (mean 211 m).

Whereas *Rhynchoconger flavus* prefers areas of high runoff and mud bottoms, *R. gracilior* is found predominantly in areas of low runoff and clear water. This preference is reflected in their distributions. In the northern Gulf of Mexico, *R. gracilior* is found east of Mobile Bay, whereas *R.*

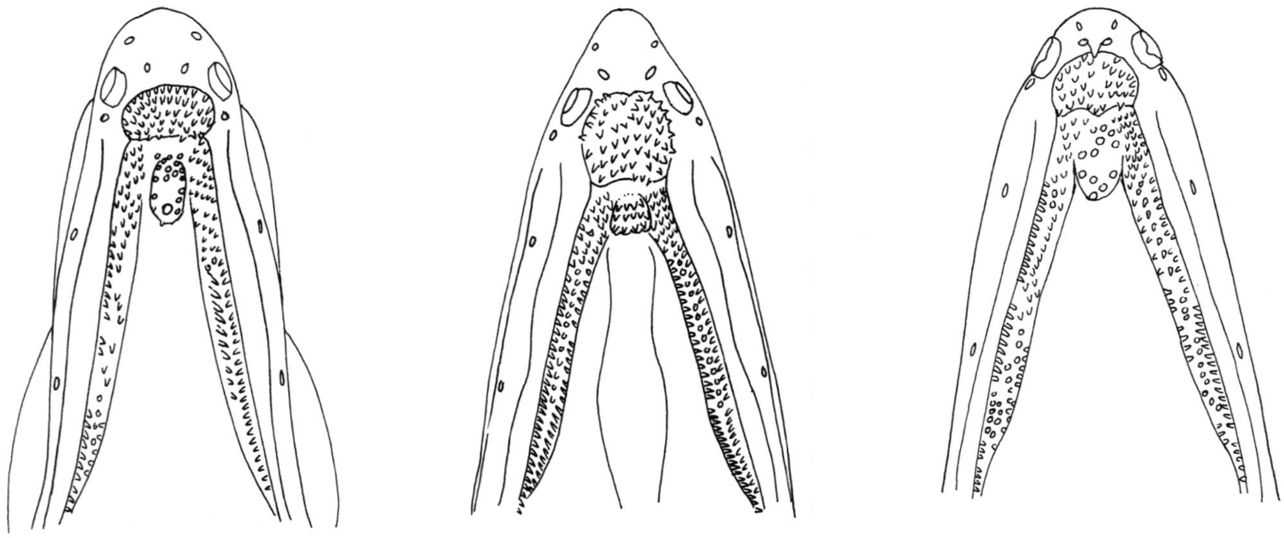

FIGURE 556. Upper dentition of three species of *Rhynchoconger*. A. *R. gracilior*, ANSP 114149, 340 mm TL. B. *R. flavus*, ANSP 114882, 305 mm TL. C. *R. guppyi*, USNM 213812.

flavus is found mainly to the west. Although *R. gracilior* occurs on the outer part of the banks off Honduras and Nicaragua, it has not been recorded from the narrower continental shelf from Costa Rica to the Gulf of Venezuela, where *R. flavus* is common. Only from about Caracas, Venezuela to the Guianas do *R. flavus* and *R. gracilior* coexist to any great extent. In the West Indies both species have been recorded, but *R. gracilior* predominates.

Etymology. Latin *gracilior*, comparative of *gracilis* (slender), in reference to the fact that this species is more slender-bodied than *R. flavus*. An adjective.

Study Material. 100 specimens, 97–560 mm TL.

Holotype: USNM 44617 (194 mm TL), Albatross 2402, northeastern Gulf of Mexico, 28°36′N, 85°33′W, 203 m, 14 March 1885.

Other material: GEORGIA: GMBL 80-137 (2, 97–195). FLORIDA ATLANTIC: UMML 14751 (3, 148–233), 29647 (1, 560), 33395 (1, 420), 33396 (2, 421–425). USNM 195996 (6, 290–354), 195997 (1, 296). FLORIDA STRAITS: UMML 9971 (1, 282). USNM 197119 (1, 256), 197974 (2, 166–214). FLORIDA GULF: UMML 17423 (3, 192–217), 20520 (1, 193). USA 03275 (10, 150–434), 03277 (2, 232–305), 03494 (1, 173), 03495 (1, 285). USNM 156009 (5, 309–397), 163497 (8, 201–387). ALABAMA: ANSP 114160 (1, 300). MEXICO GULF: MCZ 28080 (1, 364). BAHAMAS: USNM 157880 (1, 445), 157957 (1, 385), 197121 (1, 334), 197336 (2, 440–445). CUBA: MCZ 38804 (2, 438–470). PUERTO RICO: USNM 190598 (1, 283). VIRGIN ISLANDS: ANSP 114149 (1, 340). LESSER ANTILLES: USNM 221431 (1, 334), 268820 (5, 280–430). VENEZUELA: UMML 29646 (2, 232–249), 29649 (1, 581). USNM 218910 (2, 294–460). NICARAGUA: UMML 29241 (1, 335). USNM 268821 (2, 275–366). HONDURAS: UMML 29645 (1, 109). USNM 158909 (2, 383–395), 158910 (2, 374–383), 193550 (1, 288), 193560 (1, 417), 193593 (6, 285–347), 268822 (5, 315–442). GUYANA: USNM 158904 (1, 313), 158906 (1, 402). SURINAME: USNM 185628 (2, 318–345). FRENCH GUIANA: UMML 29648 (1, 540).

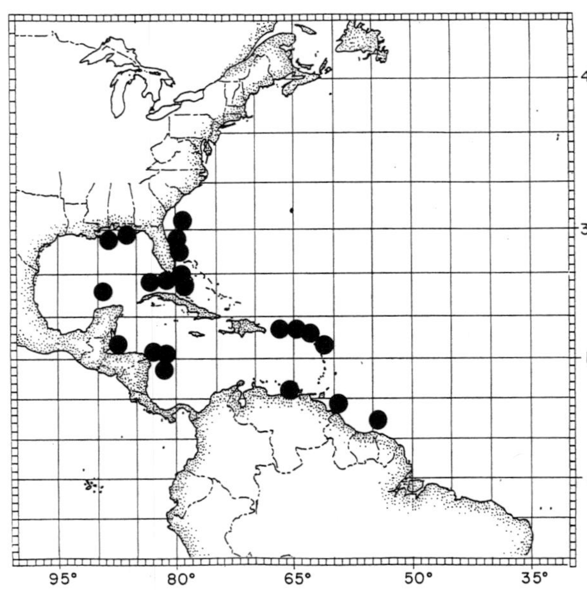

FIGURE 557. Distribution of *Rhynchoconger gracilior*.

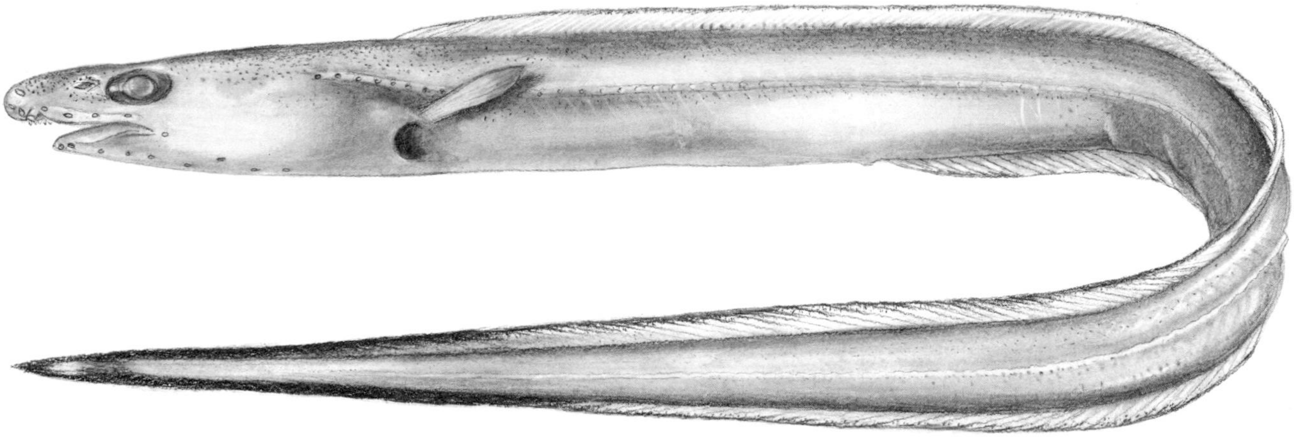

FIGURE 558. *Rhynchoconger flavus*: ANSP 127018, 189 mm TL (tip of tail regenerated).

Rhynchoconger flavus (Goode and Bean, 1896)
Figures 555B, 556B, 558–560, Table 52

Congermuraena flava Goode and Bean, 1896: 138, fig. 159 (original description, 10°37′40″N, 61°42′40″W, lectotype USNM 44612).
Congrellus flavus. Jordan and Evermann, 1900: pl. 58, fig. 100.
Leptocephalus enchodon Lea, 1913: 29, fig. 20, pl. 4, fig. 3 (original description, 36°52′N, 39°55′W, holotype ZMUB). Leptocephalus.
Hildebrandia flava. Jordan and Evermann, 1927: 502.
Congrina flava. Reid, 1934: 7.
Congrina macrosoma Ginsburg, 1951: 443, fig. 1 (original description, 28°19′N, 90°59′W, holotype YPM 3939). Based on a specimen with a regenerated tail.

Distinctive Characters. Fourth supraorbital pore absent; intermaxillary tooth patch as long as or slightly longer than broad; vomerine tooth patch as broad as long; stomach pale; vertebrae 159–172.

Description. Total vertebrae 159–172 (n = 6), predorsal vertebrae 6–9 (17), preanal vertebrae 31–37 (17), precaudal vertebrae 43 (1). Branchiostegal rays 8–9 (7). Pectoral-fin rays 12 (1). Pores: 26–34 (56), POM 10 (10), IO 4 (10) + 1 (10) + 0 (10), SO 1 + 2 (62), STC 1 (9).

Proportions as % of TL: preanal ca 30 (estimated). Of preanal length: predorsal 44–56 (49), head 42–54 (49), depth at anus 11–18 (45), pectoral fin 11–17 (11). Of HL: snout 25–29 (14), eye 19–21 (14), snout-rictus 34–40 (12), gill opening 9–15 (12), interbranchial 16–29 (10), pectoral fin 27–34 (11).

Intermaxillary tooth patch round, about as long as it is broad; vomerine tooth patch short, not elongate.

Color. In preservative brown or yellowish-brown. Vertical fins edged in black posteriorly. Stomach and intestine pale.

Size. The largest specimen examined was 505 mm TL, 173 mm preanal (USNM 268810).

Development and Growth. The leptocephalus has been identified (see p. 747).

Variation. Because of the frequency of regenerated tails among adults of *Rhynchoconger flavus*, variation in vertebral number must be based on myomere counts of leptocephali, which are more readily available. Leptocephali collected south and east of Trinidad have significantly more myomeres than those from the Caribbean and Gulf of Mexico. Forty-eight specimens from the Gulf of Mexico had 157–165 myomeres, 25 spec-

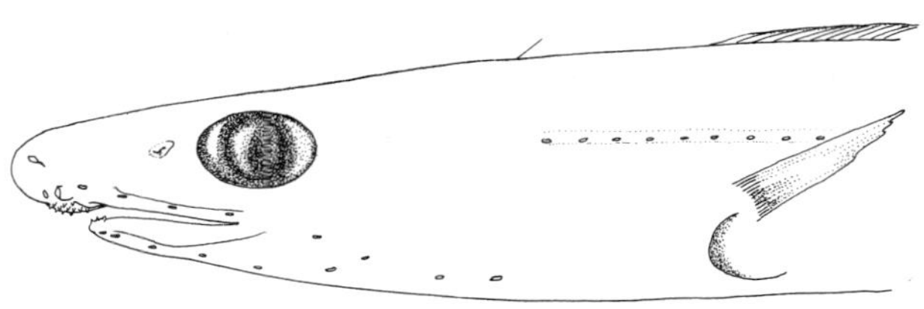

FIGURE 559. *Rhynchoconger flavus*: ANSP 114882, 305 mm TL; head.

TABLE 52. Number of preanal lateral-line pores in specimens of *Rhynchoconger flavus* from different areas.

	LL pores										
Area	26	27	28	29	30	31	32	33	34	n	x̄
Gulf of Mexico		1	2	3	–	1				7	28.7
Caribbean	5	6	10	4	1	1				27	27.7
Guianas			1	4	5	1	3	–	1	15	30.3

imens from the Caribbean and Straits of Florida had 153–163, and eight specimens from the Guianas had ca 165–170. With one exception the few apparently complete vertebral counts from adults follow the same pattern. Two specimens from Colombia and western Venezuela (UMML 28787 and USNM 198681) had 159–161 vertebrae; three specimens from eastern Venezuela, Suriname, and Brazil (USNM 185640, ANSP 127018, and ANSP 138438) had 165–172. The exception is a specimen from the northern Gulf of Mexico off Louisiana (TU 57977) that had 169 vertebrae, a value higher than the myomere counts of any of the northern leptocephali. The anomaly cannot be explained.

A similar pattern is seen in LL pores (Table 52), although the overlap is considerable.

Distribution (Fig. 560). From the Gulf of Mexico through the Caribbean and along the coast of South America at least to the mouth of the Amazon. Depth 26–183 m (mean 73 m).

In the northern Gulf of Mexico, *Rhynchoconger flavus* is rarely taken east of Mobile Bay, and it appears absent from the east coast of Florida. In these areas it is replaced by *R. gracilior.* Although *R. flavus* has been recorded from the West Indies, at least the large islands, it seems to be predominantly a continental species.

Rhynchoconger flavus is one of the most abundant of the western Atlantic congrid eels, perhaps being exceeded in numbers only by *Ariosoma balearicum.* The main centers of abundance are the western Gulf of Mexico, the western part of the Caribbean coast of Colombia, and the area between the mouths of the Orinoco and Amazon Rivers. These areas are all characterized by heavy runoff and silt or mud bottoms.

Remarks. Ginsburg (1951: 445) listed MCZ 28080 among the specimens of *Rhynchoconger flavus* he examined. This specimen, however, is not *R. flavus* but *R. gracilior.*

Etymology. Latin *flavus* (yellow). An adjective.

Study Material. 204 specimens, 70–505 mm TL.

Lectotype: USNM 44612 (239 mm TL, 82 mm preanal), Albatross 2121, off north coast of Trinidad, 10°37′40″N, 61°42′40″W, 62 m, 3 Feb. 1884.

Other material: FLORIDA GULF: FDNR 10265 (1, 265), 10893 (1, 403). ALABAMA: ANSP 114145 (1, 402). LOUISIANA: ANSP 114939 (1, 362). TU 57977 (2, 320–486). UMML 11048 (1, 155 mm preanal). USNM 155000 (1, 286), 155004 (1, 460), 157680 (2, 325–432), 163484 (1, 388), 268809 (1, 395). TEXAS: TCWC 5983.2 (2, 193–218). UMML 11503 (1, 48 mm preanal), 29614 (1, 105 preanal). USNM 154999 (1, 70), 155001 (1, 182), 157963 (2, 300–355), 163483 (3, 137–370). MEXICO GULF: TCWC 6092.2 (1, 300), 4468.10 (1, 240), 3980.1 (1, 157), 3979.1 (2, 253–386). PUERTO RICO: USNM 190537 (1, 294). VENEZUELA: ANSP 114156 (2, 205–213). UMML 29635 (1, 93 mm preanal), 29638 (1, 119 mm preanal), 29640 (2, 41–87 mm preanal), 29643 (1, 132 mm preanal), 29644 (2, 32–128 mm preanal). USNM 185640 (8, 320–434), 198052 (3, 234–485), 198054 (3, 336–360), 198681 (1, 276), 198684 (1, 485), 268816 (2, 445–464), 268810 (1, 505), 268815 (1). COLOMBIA: ANSP 126450 (2, 417–442). UMML 16666 (1, 165 mm preanal), 28786 (1, 86 mm preanal), 28787 (7, 24–49 mm preanal), 28788 (33, 25–130 mm preanal), 28789 (1, 80 mm preanal), 28790 (6, 75–112 mm preanal), 28791 (2, 83–143 mm preanal), 28792 (5, 107–116 mm preanal), 28793 (1, 128 mm preanal), 29527 (1, 85 mm preanal). USNM 198735 (1, 310), 198738 (1, 355), 198739 (1, 266), 268812 (1, 234), 268814 (3, 305–330), 268817 (3, 159–258), 268818 (2, 290–296), 268819 (1, 222). PANAMA: UMML 29602 (21, 75–306), 29604 (8, 85–308). USNM 193597 (4, 433–494). NICARAGUA: UMML 29476 (1, 135 mm preanal). HONDURAS: UMML 29477 (1, 122 mm preanal), 29478 (4, 59–105 mm preanal), 29636 (1, 101 mm preanal). GUYANA: UMML 29637 (2, 99–130 mm preanal). USNM 185630 (1, 416). SURINAME: ANSP 114882 (1, 305), 114923 (1, 293), 114927 (1, 198), 117015 (1, 315), 127018 (1, 189). UMML 29639 (2, 102–110 mm preanal), 29642 (13, 191–270). USNM 196002 (1, 302). FRENCH

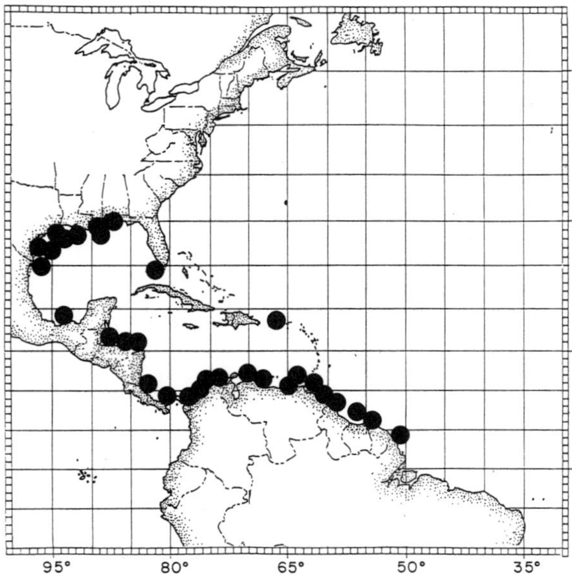

FIGURE 560. Distribution of *Rhynchoconger flavus.*

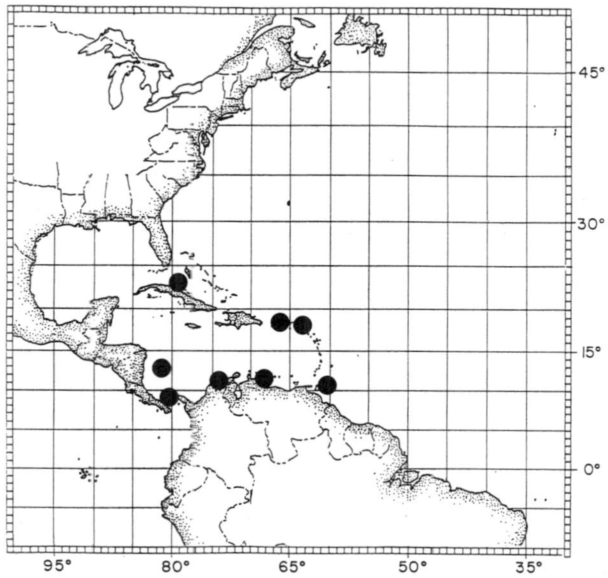

FIGURE 561. Distribution of *Rhynchoconger guppyi.*

GUIANA: UMML 29641 (1, 138 mm preanal). BRAZIL: ANSP 136632 (1, 281), 136638 (4, 265–411), 138438 (1, 292). UMML 13293 (1, 179 mm preanal).

Rhynchoconger guppyi (Norman, 1925)
Figures 555C, 556C, 561

Congromuraena guppyi Norman, 1925: 313 (original description, Tobago, West Indies, holotype BMNH 1924.7.22.21).
Hildebrandia sp. Smith, 1971: 64.
Hildebrandia guppyi. Smith and Kanazawa, 1977: 536, fig. 8.

Distinctive Characters. Fourth supraorbital pore absent. Intermaxillary tooth patch broader than long; vomerine tooth patch slightly elongate. Stomach black. Vertebrae 171–178.

Description. Total vertebrae 171–178 (n = 8), predorsal vertebrae 6–8 (4), preanal vertebrae 30–37 (5). Branchiostegal rays 7–8 (4). Pectoral-fin rays 11–13 (16). Pores: LL to anus 24–33 (21), POM 10 (11), IO 4 (12) + 1 (12) + 0 (12), SO 1 + 2 (19), STC 1 (11).

Proportions as % of TL: preanal 25–32 (7). Of preanal length: predorsal 38–49 (23), head 40–48 (23), depth at anus 9–14 (16), pectoral fin 11–15 (18). Of HL: snout 20–27 (18), eye 11–18 (18), snout-rictus 33–38 (17), gill opening 7–15 (17), interbranchial 19–27 (15), pectoral fin 25–37 (18).

Intermaxillary tooth patch broader than long; vomerine tooth patch longer than broad.

Color. In preservative brown or grayish- brown. Vertical fins edged in black posteriorly. Stomach black, intestine pale.

Size. Rhynchconger guppyi is the largest species of the genus. The largest specimen examined was the holotype, 945 mm TL, 350 mm preanal.

Development and Growth. The leptocephalus has not been conclusively identified. Because the vertebral count of *Rhynchoconger guppyi* overlaps that of *R. gracilior,* however, it is possible that some of the leptocephali identified as *R. gracilior* could be *R. guppyi.* See p. 749.

Variation. One specimen (USNM 213844) had four supraorbital pores on the left side and three on the right. All other specimens examined had three supraorbital pores on each side.

Distribution (Fig. 561). Known only from the Caribbean from Honduras to Tobago and north through the Antilles to the Old Bahama Channel between Cuba and the Great Bahama Bank. Depth 137–458 m (mean 275 m).

Etymology. Named for P. Lechmere Guppy, who collected the holotype and sent it to the British Museum. A noun in the genitive case.

Study Material. 23 specimens, 182–945 mm TL.
Holotype: BMNH 1924.7.22.21 (female, 945), Tobago, West Indies.

Other material: BAHAMAS: USNM 187628 (2, 364–470). PUERTO RICO: ANSP 128764 (1, 537). USNM 190599 (6, 312–535), 196167 (1, 472). LESSER ANTILLES: USNM 213812 (2, 403–543). VENEZUELA: USNM 213844 (3, 347–454). COLOMBIA: ANSP 114152 (1, 790), 126451 (1, 783). PANAMA: USNM 179228 (1, 182), 193599 (2, 510–557). NICARAGUA: USNM 213845 (1, 450). HONDURAS: USNM 185656 (1, 438).

Genus *Rhechias* Jordan, 1921

Rhechias Jordan, 1921: 644 (type species *Rhechias armiger* Jordan, 1921, = *Congermuraena aequoria* Gilbert and Cramer, 1897, by original designation).

Congrina Jordan and Hubbs, 1925: 196 (type species *Congermuraena aequoria* Gilbert and Cramer, 1897, by original designation).

Pseudoxenomystax Breder, 1927: 6 (type species *Pseudoxenomystax dubius* Breder, 1927, by original designation).

Characters. Body moderately elongate, anus near two-fifths TL; tail slender, moderately to markedly attenuate at tip. Dorsal fin begins between base and tip of appressed pectoral fin; pectoral fin well developed.

Snout projects beyond lower jaw. Fleshy part of snout extends anteriorly beyond intermaxillary teeth. Flange present on lower lip; flange on upper lip rudimentary or absent. Anterior nostril tubular, near tip of snout, directed ventrolaterally. Posterior nostril elliptical, in front of middle or upper part of eye. Adnasal pore absent. Pores along upper jaw, second and third supraorbital pores, and third preoperculomandibular pore enlarged.

Teeth moderately large. Intermaxillary teeth in two to several rows, separated from maxillary and vomerine teeth, mostly excluded from closed mouth. Vomerine teeth in a round patch on head of vomer with a median row of enlarged teeth extending posteriorly along shaft, the number in this row varying from about two to seven. Maxillary and mandibular teeth in bands, wider anteriorly, outermost teeth enlarged in some species.

Lateral ethmoid process absent. Supraoccipital present. Preorbital bone with projections from lower edge; four suborbitals. Posterior edge of opercle serrate. Myorhabdoi absent. No epicentral processes on caudal vertebrae. Lateral-line ossicles like those of *Rhynchoconger*. Dorsal- and anal-fin rays segmented.

Digestive tract black. Length of stomach variable. Gas bladder terminates behind anus.

Size. To about 550–600 mm TL.

Distribution. Tropical Atlantic and Indo-West Pacific.

Etymology. From the Greek *rhachos* (thorn bush, brier), referring to the ventral projections of the preorbital bone, from which, in the holotype of *Rhechias armiger*, the skin had dried and pulled away, leaving them exposed. Feminine.

Species. The following ten species are provisionally recognized: *Rhechias dubia* (Breder, 1927), *R. thysanochila* (Reid, 1934), *R. vicinalis* (Garman, 1899), *R. bullisi* Smith and Kanazawa, 1977, *R. polypora* Smith and Kanazawa, 1977, all western North Atlantic; *R. bertini* (Poll, 1953), eastern tropical Atlantic; *R. guttulata* (Günther, 1887), Indo-West Pacific; *R. wallacei* (Castle, 1968), Indian Ocean; *R. retrotincta* (Jordan and Snyder, 1901), Japan; and *R. aequoria* (Gilbert and Cramer, 1897), Hawaii.

These species seem to fall into three more or less distinct groups. The most distinct is that represented by *Rhechias dubia.* It has generally weaker teeth, more or less uniform in size; a less slender tail; a shorter stomach; and a fold of skin on the upper lip that is a remnant of the labial flange.

The Atlantic species *R. thysanochila, R. vicinalis, R. bullisi, R. polypora,* and *R. bertini* compose the second species group. They differ from *R. dubia* by having larger teeth, with the outermost maxillary and mandibular teeth enlarged anteriorly; a more slender tail; a longer stomach; the upper labial flange nearly or entirely absent; and the posterior nostril above mid-eye level. They differ from the Indo-Pacific species by the elevated posterior nostril and by having a smaller leptocephalus with a single row of lateral melanophores.

The Indo-West Pacific species *R. guttulata, R. retrotincta, R. aequoria,* and *R. wallacei* form the third group. They resemble the *thysanochila* group except that the posterior nostril is at mid-eye level and the leptocephalus is larger with three rows of lateral melanophores.

The differences among the species within each

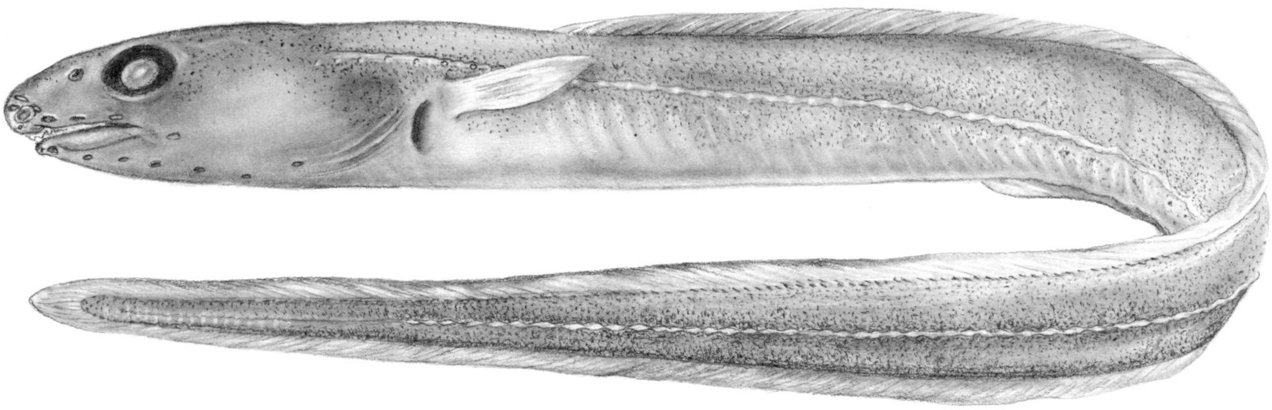

FIGURE 562. *Rhechias dubia*: ANSP 127019, 147 mm TL.

group are slight; about the only useful characters are the number of vertebrae and sensory pores. The Indo-Pacific species especially need critical reevaluation to be properly characterized.

The species treated as *Pseudoxenomystax* by Castle (1960b, 1968c)—*Congermuraena albescens* Barnard, 1923, *Bassanago bulbiceps* Whitley, 1948, and *Pseudoxenomystax hirsutus* Castle, 1960—seem quite different from *P. dubius* Breder and probably should be placed in *Bassanago* Whitley.

KEY TO THE SPECIES OF *RHECHIAS* IN THE WESTERN NORTH ATLANTIC

1a. Posterior nostril at mid-eye level . . . . . *dubia*, p. 533
1b. Posterior nostril opposite upper margin of eye 2
2a. Preanal lateral-line pores fewer than 40 3
2b. Preanal lateral-line pores more than 40 4
3a. Preanal lateral-line pores ca 26–29 . . . . . *thysanochila*, p. 536
3b. Preanal lateral-line pores 33–38 . . . . . *vicinalis*, p. 539
4a. Postorbital pores absent; one supratemporal pore . . . . . *bullisi*, p. 538
4b. Postorbital pores present; three supratemporal pores . . . . . *polypora*, p. 540

Rhechias dubia (Breder, 1927)
Figures 562–564, Tables 53–56

Pseudoxenomystax dubius Breder, 1927: 6 (original description, Belize, holotype YPM 25).
Congrina dubia. Reid, 1934: 7.
Rhechias dubius. Smith, 1971: 69.

Distinctive Characters. Posterior nostril at mid-eye level. Teeth more or less uniform in size. Tail moderately attenuate. Stomach short.

Description (some of these data provided by R. H. Kanazawa). Total vertebrae 120–145 (n = 64), predorsal vertebrae 9–13 (33), preanal vertebrae 29–36 (40), precaudal vertebrae 37–40 (5). Branchiostegal rays 10–11 (9). Pectoral-fin rays ca 11–14 (9). Pores: LL to anus 25–36 (55), POM 9–10 (18), IO 4 (19) + 1 (19) + 0 (19), SO 1 + 2 (19), STC 1 (19).

Proportions as % of TL: preanal 34–40 (47), depth at anus 4–7 (23). Of preanal length: predorsal 43–65 (67), head 37–49 (67), depth at anus 11–18 (23), pectoral fin 8–18 (24). Of HL: snout 21–27 (25), eye 14–20 (24), interorbital 13–20 (11), snout-rictus 29–36 (18), gill opening 9–16 (25), interbranchial 18–30 (24), pectoral fin 16–41 (24).

Tail not greatly attenuate. Posterior nostril at mid-eye level. Weak upper labial flange present.

Teeth moderately small; intermaxillary, maxillary, and mandibular teeth more or less uniform in size. Intermaxillary teeth in 3–4 series, forming a roughly triangular patch, slightly broader than long. Vomerine teeth confluent with

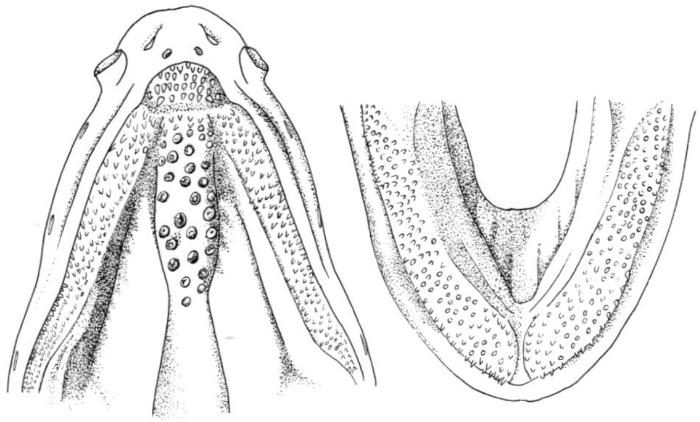

FIGURE 563. *Rhechias dubia*: ANSP 133007, 364 mm TL; dentition.

maxillary teeth; a rounded patch of moderately small teeth on head of vomer; a somewhat irregular median series of larger teeth extending posteriorly along shaft of vomer, terminating at about midlength of maxillary tooth patch. Maxillary and mandibular teeth in narrow bands, wider anteriorly, outer teeth slightly larger than inner but not fang-like.

Stomach short, reaching about halfway to anus.

Color. In preservative brown or yellowish-brown to gray. Vertical fins not conspicuously edged in black.

Size. The largest specimen examined was 443 mm TL, 158 mm preanal (USNM 163528).

Development and Growth. Leptocephali have been identified from the Gulf of Mexico (see p. 752).

Variation. This aptly named species shows a confusing pattern of variation over different parts of its range. Despite the magnitude of this variation and the evidence of sharp genetic discontinuities, I am unable to separate all of the populations sufficiently to be able to name them as species or subspecies. As many as five geographic populations may be present, based on number of vertebrae and lateral-line pores, and predorsal and head length (Tables 53–56). For convenience I will refer to these as Populations I, II, III, and IV.

Population I occurs on the continental slope from Georgia around the Gulf of Mexico to Campeche Bay. It is characterized by high numbers of vertebrae (138–145) and preanal lateral-line pores (30–36), and low predorsal (43–50% preanal) and head lengths (37–44% preanal).

Population II has been found only in the northwestern Caribbean from Belize to Jamaica. It is completely separable from Population I by vertebral count (129–136) and predorsal length (52–57% preanal), almost completely separable by lateral-line pore count (27–30), and significantly different in head length (41–49% preanal). The holotype presumably belongs to this population, although it has a regenerated tail (125 vertebrae present and an undetermined number missing).

Population III is represented by two collections from the southern Caribbean and one from Jamaica. The 13 radiographed specimens have fewer vertebrae (120–126) than the two preceding populations, but in lateral-line pores (26–28) and predorsal (53–60% preanal) and head length (42–46% preanal), they agree with Population II. In dealing with vertebral counts there is always a problem with regenerated tails, but these 13 specimens are so consistent in their low counts that they must be regarded as intact and representing a distinct population.

TABLE 53. Number of vertebrae in various populations of *Rhechias dubia*. 1, Population I; 2, Population II; 3, Population III; 4, Santaren Channel; 5, Population IV.

Population	Vertebrae																	
	120	121	122	123	124	125	126	127	128	129	130	131	132	133	134	135	136	137	138
1																			1
2										1	1	2		4			1		
3	2	1	1	3	2	4	1												
4								2											
5									1								1	2	

TABLE 54. Preanal lateral-line pores in various populations of *Rhechias dubia.* 1, Population I; 2, Population II; 3, Population III; 4, Santaren Channel; 5, Population IV.

Population	LL pores															
	24	25	26	27	28	29	30	31	32	33	34	35	36	37	n	x̄
1							1	5	4	10	8	2	3		33	33.1
2				3	4	1	1								9	28.0
3			2	3	3										8	27.1
4		1													1	25.0
5						1	1	1	3	3					9	31.7

Two specimens from the Santaren Channel have vertebral counts (127) slightly above those of Population III and below those of Population II. It is difficult to interpret this situation with only two specimens available. They may belong to either of the two preceding populations or they may represent a distinct population of their own.

Population IV occurs from western Venezuela to the Guianas. It is intermediate between Populations I and II in all the four characters.

Despite the apparent sympatry of two forms at Jamaica and off western Venezuela, the population structure is not entirely clear. Only Population I is adequately documented. Although it seems completely separable from Populations II and III, Population IV largely bridges this gap. Population III is represented by only three collections, and the only character that separates it from Population II is the number of vertebrae, which can be affected by damage to the tails. The two specimens from the Santaren Channel cannot be confidently assigned to any population. For these reasons, I am provisionally treating this

TABLE 53. Continued.

Vertebrae								
139	140	141	142	143	144	145	n	x̄
	1	3	6	12	10	6	39	143.1
							9	132.1
							14	123.3
							2	127.0
1	1						5	137.8

complex as a single species, *Rhechias dubia,* until it can clearly be demonstrated otherwise.

Distribution (Fig. 564). From Georgia, the east coast of Florida and the Gulf of Mexico south through the Caribbean to the Guianas. Depth 128–886 m (mean 400 m).

Etymology. Latin *dubius* (doubtful, uncertain), "in reference to its doubtful relationships" (Breder, 1925: 7). An adjective.

Study Material. 156 specimens, 113–443 mm TL.

Population I: GEORGIA: GMBL 76-236 (1, 333). FLORIDA ATLANTIC: UMML 29480 (1, 223). FLORIDA GULF: ANSP 112037 (1, 324), 112053 (1, 180), 114847 (1, 434), 126455 (22, 285–381), 133007 (1, 364). USA 03276 (4), 03315 (1). USNM 158560 (1, 265), 163528 (6, 363–443), 164126 (1, 249), 164242 (7, 367–412), 178996 (1, 404), 190542 (15, 240–357), 198047 (3, 288–366), 198048 (1, 377), 198673 (1, 191), 219540 (1, 338). MISSISSIPPI: ANSP 109604 (2, 367–381), 109619 (1, 360), 112056 (1, 229). UMML 11500 (1, 225). USNM 157956 (1, 291), 164240 (1, 330), 164241 (1, 405), 179066 (2, 382–461), 179067 (1, 344), 179068 (7, 336–374), 179227 (1, 368), 187969 (5, 354–390), 187970 (1, 271), 187971 (5, 350–410), 187972 (1, 379), 187973 (1, 352). LOUISIANA: USNM 157890 (2, 377–411). TEXAS: FMNH 45051 (1, 321), 46298 (1, 293). USNM 157959 (1, 331), 198682 (1, 170). MEXICO GULF: USNM 158171 (1, 304).

Population II: *Holotype:* YPM 25 (215 mm TL), north of Glovers Reef, Belize, 886 m, 20 Apr. 1925.

Other material: JAMAICA: UMML 29474 (2, 169–181), 17°55.7′N, 78°00′W, 265 m, 7 July 1970. UMML 29475 (1, 274), 17°27′N, 78°10′W, 603–655 m, 14 July 1970. HONDURAS: USNM 158886 (1, 325), 16°39′N, 81°01′W, 458 m, 24 Aug. 1957. USNM 158887 (1, 224), 16°41′N, 81°02′W, 458 m, 23 Aug. 1957. USNM 158922 (1, 261), 16°38′N, 82°34′W, 375–384 m, 21 Aug. 1957. USNM 193569 (6, 193–242), 17°07′N, 81°38′W, 732 m, 7 June 1962.

Population III: JAMAICA: UMML 29473 (3, 177–197), 17°42.5′N, 77°58′W, 530–558 m, 6 July 1970. VENEZUE-

TABLE 55. Predorsal length in percent of preanal in various populations of *Rhechias dubia*. 1, Population I; 2, Population II; 3, Population III; 4, Santaren Channel; 5, Population IV.

Population	Predorsal length																		n	x̄
	43	44	45	46	47	48	49	50	51	52	53	54	55	56	57	58	59	60		
1	2	6	4	5	5	4	2	3											31	46.3
2										3	1	3	3	2	1				13	54.2
3											1		2	3	2	1		1	10	56.3
4													1						1	55.0
5					2	1	3	2	2	1									11	49.4

LA: UMML 29469 (6, 139–157), 10°54.7′N, 66°17.8′W, 234–280 m, 28 July 1968. COLOMBIA: ANSP 127019 (1, 147) and UMML 28780 (11, 113–165), 11°20′N, 73°48.5′W, 143–174 m, 31 July 1968.

Santaren Channel: UMML 29479 (1, 188), 24°04′N, 79°42′W, 540–558 m, 22 June 1966. UMML uncat. (1, 153, cleared and stained), 23°34′N, 79°12′W, 514–544 m, 15 June 1968.

Population IV: VENEZUELA: UMML 29599 (2, 211–229), 11°18.8′N, 68°22′W, 384–608 m, 26 July 1968. USNM 195903 (7, 262–339), 11°30′N, 62°29′W, 329 m, 18 Apr. 1960. GUYANA: USNM 158915 (1, 265), 09°41′N, 59°47′W, 275 m, 3 Nov. 1957. SURINAME: USNM 158921 (1, 268), 07°34′N, 54°50′W, 366 m, 6 Nov. 1957.

Rhechias thysanochila (Reid, 1934)
Figures 565, 566

Congrina thysanochila Reid, 1934: 7, fig. 2 (original description, 18°32′15″N, 66°17′45″W, holotype USNM 93434).
Rhechias thysanochila. Smith, 1971: 70. Smith and Kanazawa, 1977: 538.

Distinctive Characters. Preanal lateral-line pores ca 26–29; postorbital and interorbital pores absent; one supratemporal pore. Vertebrae 142–149. Predorsal length 45–49% preanal; head length 40–45% preanal. Posterior nostril near level of upper margin of eye. Stomach long.

Description. Total vertebrae 145–149 (n = 7), predorsal vertebrae 8–9 (4), preanal vertebrae 34–35 (4), precaudal vertebrae 44–47 (2). Branchiostegal rays 8 (1). Pectoral-fin rays ca 10–14 (16). Pores: LL to anus ca 26–33 (17), POM 10 (18), IO 4 (19) + 1 (18) + 0 (19), SO 1 + 2 (19), STC 1 (19).

Proportions as % of TL: preanal 32–36 (5). Of preanal length: predorsal 45–49 (22), head 40–45 (22), depth at anus 12–17 (18), pectoral fin 10–14 (18). Of HL: snout 25–29 (18), eye 14–19 (18), snout-rictus 33–41 (17), tip of snout to tip of lower jaw 7–11 (12), gill opening 9–13 (18), interbranchial 18–28 (16), pectoral fin 23–33 (18).

Dorsal fin begins slightly behind base of pectoral fin. Tail attenuate, frequently damaged. Posterior nostril near upper margin of eye.

Teeth moderately large. Intermaxillary teeth in three transverse rows, separated from maxillary and vomerine teeth and exposed when mouth closed. Vomerine teeth in a round patch

TABLE 56. Head length in percent of preanal in various populations of *Rhechias dubia*. 1, Population I; 2, Population II; 3, Population III; 4, Santaren Channel; 5, Population IV.

Population	Head length														n	x̄
	37	38	39	40	41	42	43	44	45	46	47	48	49	50		
1	4	7	6	6	5	1	1	1							31	39.4
2					2	2	5		2		1		1		13	43.6
3						1	2	2	2	3					10	44.4
4									1	1					2	45.5
5			1	2	2	3	1	1		1					11	41.8

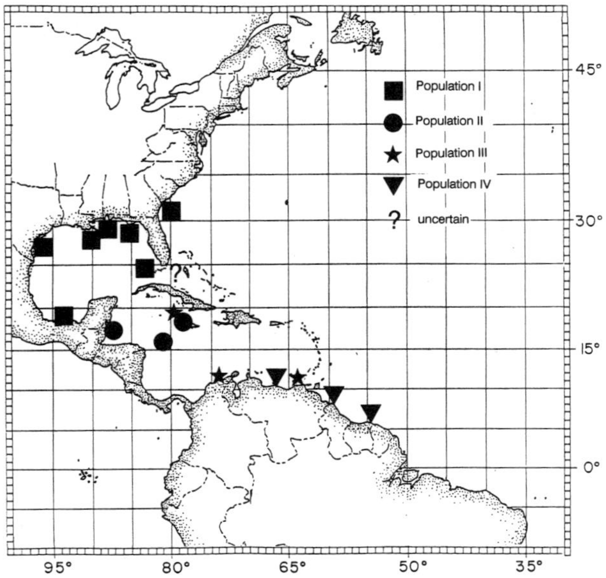

FIGURE 564. Distribution of *Rhechias dubia.*

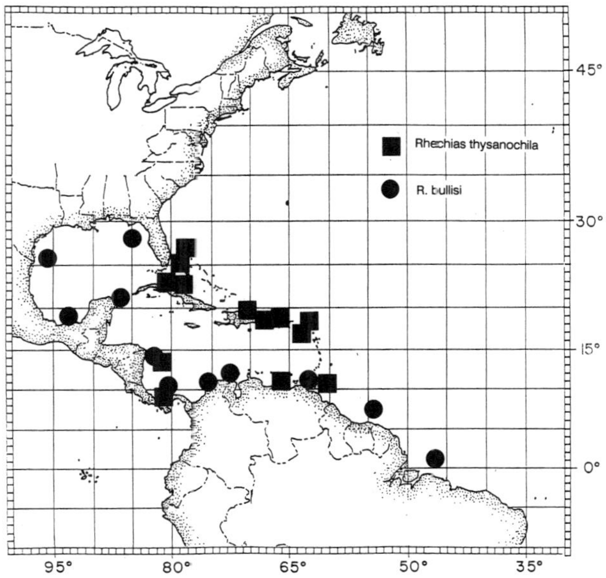

FIGURE 566. Distribution of *Rhechias thysanochila* and *R. bullisi.*

on head of vomer, one or two median teeth distinctly enlarged but not forming a distinct row along shaft of vomer; vomerine teeth more or less confluent with maxillary teeth. Maxillary teeth in narrow bands, wider anteriorly and narrowing to one or two series at posterior end; outer teeth larger than inner, especially anteriorly. Mandibular teeth in bands, wider anteriorly, outer teeth larger than inner.

Stomach long, reaching about three-quarters or four-fifths of way to anus.

Color. In preservative brown or grayish-brown, lighter ventrally. Vertical fins edged in black posteriorly. Stomach and intestine black.

Size. The largest specimen examined was 288 mm TL (MCZ 59212).

Development and Growth. The leptocephalus has been identified (see p. 753).

Distribution (Fig. 566). Bahamas and Caribbean. Depth 104–659 m (mean 374 m).

Rhechias thysanochila has a predominantly insular distribution. It occurs widely in the West Indies and the Bahamas, but it has not been collected on the mainland side of the Florida Straits and seems absent from the Gulf of Mexico. On the continental side of the Caribbean it occurs on the slope off Honduras and Nicaragua, western Panama and eastern Venezuela, all areas with insular-like conditions of low runoff and clear water. It has not been collected from the coast of Colombia, which is characterized by heavy

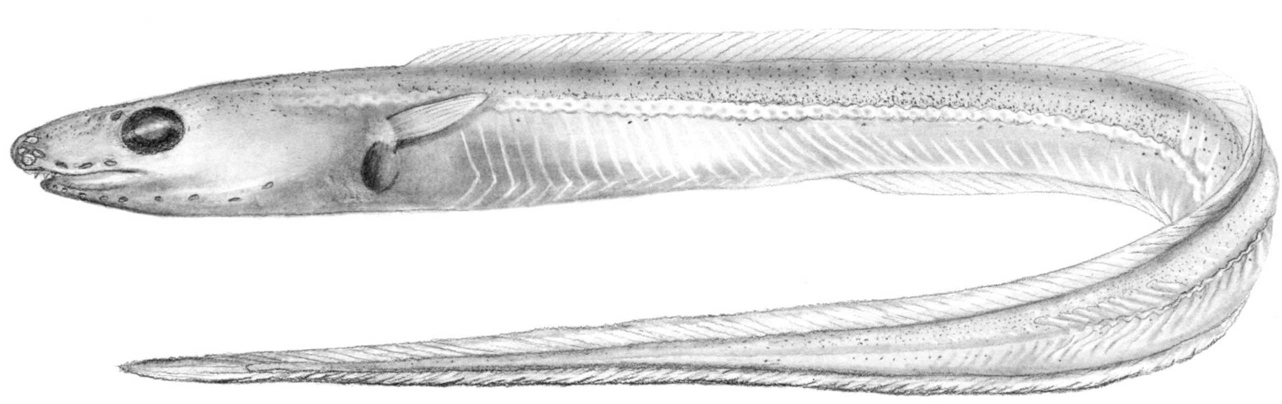

FIGURE 565. *Rhechias thysanochila*: ANSP 127015, 193 mm TL.

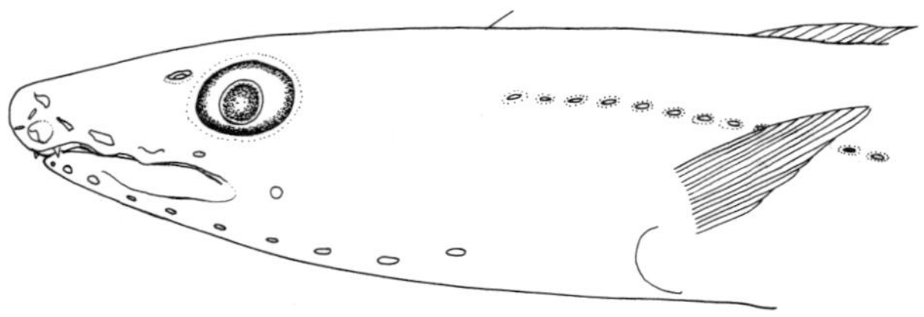

FIGURE 567. *Rhechias bullisi*: ANSP 122923, paratype, 395 mm TL; head (from Smith and Kanazawa, 1977: fig. 9).

runoff and large accumulation of detritus. Under the latter conditions, *R. thysanochila* is replaced by *R. bullisi*.

Etymology. From the Greek *thysanos* (fringe) and *cheilos* (lip), referring to the fringed inner lip. Used as an adjective.

Study Material. 41 specimens, 125–288 mm TL.

Holotype: USNM 93434 (female, 257 mm TL, 95 mm preanal), Caroline 23, north of Puerto Rico, 18°32′15″N, 66°17′45″W, 476–659 m, 4 Feb. 1933.

Other material: BAHAMAS: ANSP 127015 (1, 193). UMML 17721 (1, 185), 20916 (1, 160), 30241 (4, 176–200), 33398 (1, 159). USNM 158923 (3, 190–250), 185637 (1, 195), 187631 (1, 187), 197122 (2, 146–148). CUBA: MCZ 59212 (1, 288), 59213 (1, 270), 64511 (1, 197). USNM 157962 (5, 206–280). JAMAICA: UMML 29597 (1, 249). HISPANIOLA: UMML 29693 (2, 205–210), 29695 (1, 125), 33399 (1, 277), 198670 (1, 257). PUERTO RICO: ANSP 112057 (1, 183). VIRGIN ISLANDS: USNM 268813 (1, 279). LESSER ANTILLES: UMML 29608 (2, 171–228). USNM 190532 (1, 242), 190534 (1, 154), 190535 (1, 214). VENEZUELA: UMML 29606 (1, 156), 33397 (1, 214). PANAMA: USNM 179238 (1, 218). NICARAGUA: USNM 193542 (1, 269).

Rhechias bullisi Smith and Kanazawa, 1977
Figures 566, 567

Rhechias sp. A. Smith, 1971: 71.

Rhechias bullisi Smith and Kanazawa, 1977: 538, fig. 9 (original description, 12°16′N, 72°40′W, holotype USNM 198768).

Distinctive Characters. Preanal lateral-line pores 39–45; postorbital and interorbital pores absent; one supratemporal pore. Vertebrae 179–186. Predorsal length 39–49% preanal, head length 35–40% preanal. Posterior nostril near level of upper margin of eye. Stomach long.

Description. Total vertebrae 179–186 (n = 17), predorsal vertebrae 10–13 (6), preanal vertebrae 45–49 (8). Pectoral-fin rays ca 12–15 (6). Pores: LL to anus 39–45 (30), POM 10 (30), IO 4 (30) + 1 (30) + 0 (30), SO 1 + 2 (30), STC 1 (30).

Proportions as % of TL: preanal 37–41 (15). Of preanal length: predorsal 39–49 (30), head 35–40 (30), depth at anus 12–16 (12), pectoral fin 10–13 (12). Of HL: snout 26–30 (12), eye 15–19 (12), snout-rictus 36–42 (10), gill opening 8–13 (12), interbranchial 21–29 (11), pectoral fin 26–33 (12).

Dorsal fin begins slightly behind base of pectoral fin. Tail attenuate. Posterior nostril near level of upper margin of eye. Teeth as in *R. thysanochila.* Stomach long, reaching three-quarters or four-fifths of way to anus.

Color. In preservative brown or grayish-brown. Vertical fins edged in black posteriorly. Stomach and intestine black.

Size. The largest specimen examined was 606 mm TL, 238 mm preanal (ANSP 109616).

Development and Growth. The leptocephalus has not been identified.

Distribution. From the northern Gulf of Mexico southward along the coast of Central and South America at least to the mouth of the Amazon. Absent from the Bahamas and the West Indies. Depth 55–549 m (mean 326 m).

Rhechias bullisi and *R. thysanochila* have largely complementary distributions. The two species inhabit similar depth ranges, but *R. thysanochila* is primarily an insular species, whereas *R. bullisi* is strictly continental.

Etymology. Named for Harvey R. Bullis of the National Marine Fisheries Service. A noun in the genitive case.

Study Material. 30 specimens, 148–606 mm TL.

Holotype: USNM 198768 (male, 493 mm TL, 180 mm preanal), Oregon 4922, Colombia, 12°16′N, 72°40′W, 366 m, 2 June 1964.

Paratypes: FLORIDA GULF: USNM 178994 (1, 522). TEXAS: USNM 198679 (1, 510). MEXICO GULF: USNM 158172 (1, 452). VENEZUELA: USNM 185653 (1, 303). COLOMBIA: ANSP 114947 (1, 450), 122923 (1, 395), 126526 (5, 395–465). USNM 198674 (1, 510), 198767 (1, 438), 198769 (1, 580), 198770 (1, 511). PANAMA: USNM 193594 (1, 495), 193596 (2, 148–453), 193612 (1, 446). HONDURAS: ANSP 109616 (2, 518–606). MEXICO QUINTANA ROO: UMML 11063 (1, 363). SURINAME: USNM 158911 (1, 418), 179071 (2, 391–395), 185654 (2,

464–507). FRENCH GUIANA: USNM 158908 (1, 346). BRAZIL: USNM 158905 (2, 302–376).

Rhechias vicinalis (Garman, 1899)
Figures 568, 569

Uroconger vicinus (not of Vaillant, 1888). Goode and Bean, 1896: 138, fig. 160.
Uroconger vicinalis Garman, 1899: 304 (original description, 23°10′36″N, 82°20′28″W, holotype USNM 84568). Name applied to the specimen described and figured by Goode and Bean as *Uroconger vicinus.*
Rhechias sp. B. Smith, 1971: 71.
Rhechias bertini (not of Poll, 1953). Blache and Bauchot, 1976a: 397.
Rhechias vicinalis. Smith and Kanazawa, 1977: 539.

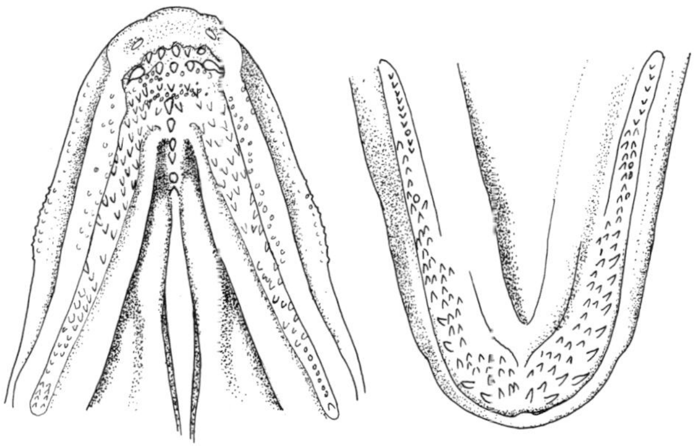

FIGURE 568. *Rhechias vicinalis*: ANSP 112038, 194 mm TL; dentition.

Distinctive Characters. Preanal lateral-line pores 33–38; postorbital and interorbital pores sometimes present; one to three supratemporal pores. Vertebrae 168–176. Predorsal length 42–48% preanal, head length 38–46% preanal. Posterior nostril near level of upper margin of eye. Stomach long.

Description. Total vertebrae 168–176 (n = 8), predorsal vertebrae 10–12 (3), preanal vertebrae 40–47 (3), precaudal vertebrae ca 47–51 (2). Branchiostegal rays 9 (1). Pectoral-fin rays ca 15–18 (14). Pores: LL to anus 33–38 (17), POM 10–13 (17), IO 4 (17) + 1 (17) + 0–3 (17), SO 1 + 2–5 (18), STC 1–3 (18).

Proportions in % TL: preanal 32–38 (8). Of preanal length: predorsal 42–48 (18), head 38–46 (18), depth at anus 12–19 (13), pectoral fin 11–14 (12). Of HL: snout 23–30 (14), eye 14–20 (14), snout-rictus 33–43 (12), gill opening 10–17 (11), interbranchial 19–30 (10), pectoral fin 25–37 (12).

Dorsal fin begins slightly behind base of pectoral fin. Tail attenuate. Posterior nostril near level of upper margin of eye.

Teeth similar to those of *R. thysanochila,* but intermaxillary teeth in two transverse series rather than three. Vomer with 1–5 enlarged teeth forming a median row extending posteriorly to a point slightly before midpoint of maxillary tooth patch; a few smaller teeth around and just before anteriormost enlarged tooth.

Stomach long, reaching about three-quarters or four-fifths of way to anus.

Color. In preservative brown or grayish-brown. Vertical fins edged in black posteriorly. Stomach and intestine black.

Size. The largest specimen examined was 462 mm TL, 176 mm preanal (ANSP 122924).

Development and Growth. The leptocephalus has not been identified.

Variation. This species varies widely in the number and arrangement of head pores, a character that is usually rather stable in congrid eels. Some specimens lack interorbital and postorbital pores, others have various numbers of pores in these positions. Double pores, pores in anomalous positions, and bilaterally asymmetrical arrangements of pores are commonly seen.

Remarks. Blache and Bauchot (1976a: 397) suggested that *Rhechias vicinalis,* cited as *Rhechias* sp. B. Smith, was identical to *R. bertini* from the eastern tropical Atlantic. Although the number of vertebrae and preanal lateral-line pores are similar in the two species, *R. bertini* seems consistent in having the minimum number of head pores (three supraorbital, no postorbital). For this reason I prefer to keep these two species distinct.

Distribution (Fig. 569). Eastern Gulf of Mexico, Bahamas, West Indies, coast of Central and South America at least to mouth of Amazon. Depth 101–503 m (mean 324 m).

Rhechias vicinalis seems to be less selective in its choice of habitat than *R. thysanochila* and *R. bullisi.* It occurs in the Bahamas and West Indies along with *R. thysanochila,* but it has also been

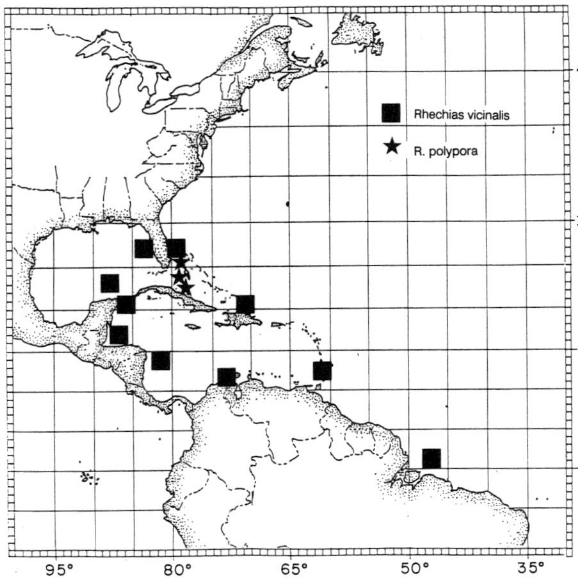

FIGURE 569. Distribution of *Rhechias vicinalis* and *R. polypora.*

collected in the Gulf of Mexico and off the mouth of the Amazon, where *R. bullisi* is found.

Etymology. From the Latin *vicinus* (near) and *-alis* (having the nature of); named to distinguish it from *Uroconger vicinus* Vaillant, as which the type had been misidentified. An adjective.

Study Material. 18 specimens, 98-462 mm TL.

Holotype: USNM 84568 (105 mm TL, 36 mm preanal), Albatross 2161, off northwestern Cuba, 23°10′36″N, 82°20′28″W, 267 m, 30 Apr. 1884.

Other material: FLORIDA GULF: USNM 197738 (1, 259). MEXICO GULF: ANSP 112038 (1, 194), 152117 (1, 225). UMML uncat. Silver Bay 1184 (1, 225). BAHAMAS: UMML 2682 (1, 265), 14152 (1, 242). HISPANIOLA: UMML 29692 (3, 160-225). LESSER ANTILLES: USNM 202951 (1, 112). COLOMBIA: ANSP 122924 (1, 462). NICARAGUA: USNM 158885 (1, 302), 193565 (1, 290). BELIZE: USNM 202952 (1, 263). MEXICO QUINTANA ROO: UMML 29591 (1, 98), 29592 (1, 148). BRAZIL: USNM 158907 (1, 335).

Rhechias polypora Smith and Kanazawa, 1977
Figures 569, 570

Rhechias sp. C. Smith, 1971: 72.

Rhechias polypora Smith and Kanazawa, 1977: 540, fig. 10 (original description, 26°25′N, 79°01′W, holotype USNM 197143).

Distinctive Characters. Preanal lateral-line pores 46–49; postorbital and interorbital pores present;

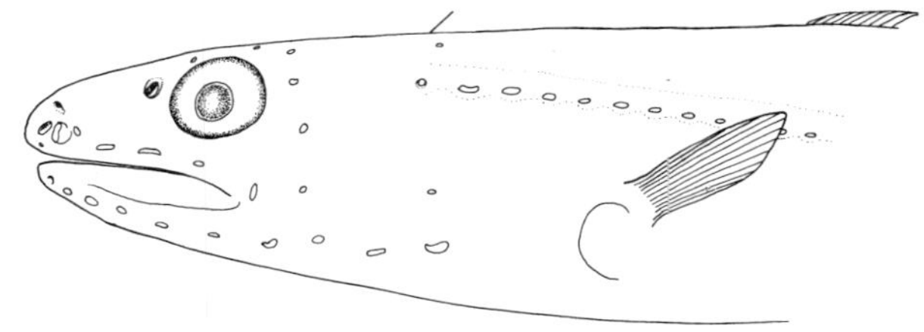

FIGURE 570. *Rhechias polypora*: USNM 157889, paratype, 402 mm TL; head (from Smith and Kanazawa, 1977: fig. 10).

three supratemporal pores. Predorsal length 40–41% preanal, head length 30–32% preanal. Posterior nostril near level of upper margin of eye. Stomach long.

Description. Total vertebrae unknown, predorsal vertebrae 14 (n = 1), preanal vertebrae 52 (1). Pectoral-fin rays 14–15 (3). Pores: LL to anus 46–49 (3), POM 11–12 (3), IO 4 (3) + 1–3 (3) + 2–4 (3), SO 1 + 4 (3), STC 3 (3).

Proportions as % of preanal length: predorsal 40–41 (3), head 30–32 (3), depth at anus 12–15 (3), pectoral fin 7–8 (3). Of HL: snout 24–27 (3), eye 14–16 (3), snout-rictus 39–41 (3), gill opening 11–13 (3), interbranchial 25–29 (2), pectoral fin 24–27 (3).

Dorsal fin begins near or slightly behind tip of appressed pectoral fin. Tail attenuate. Posterior nostril near level of upper margin of eye. Teeth as in *R. thysanochila.* Stomach long, reaching about three-quarters or four-fifths of way to anus.

The missing pore in the supraorbital series is the fourth, i.e., the one between the anterior and posterior nostrils; the fifth and sixth, in the interorbital area, are present. All three specimens have regenerated tails, so total vertebral counts for this species are unavailable.

Color. In preservative grayish-brown. Vertical fins edged in black posteriorly. Stomach and intestine black.

Size. The largest specimen was 420 mm TL, 185 mm preanal (UMML 29672); the smallest was 430 mm TL, 162 mm preanal (holotype). Because the tails are incomplete, total length is not a true indication of size.

Development and Growth. The leptocephalus has not been identified.

Variation. Like *Rhechias vicinalis, R. polypora*

shows considerable variation in the number and arrangement of head pores. The material is too limited, however, to determine the full extent of this variation.

Distribution (Fig. 569). The three specimens were collected on the Bahamian side of the Straits of Florida and off the north coast of Cuba. Depth 439–539 m (mean 502 m).

Etymology. From the Greek *polys* (many) and *poros* (hole), referring to the many pores on the head and the lateral line. Used as an adjective.

Study Material. Three specimens, 402–430 mm TL.

Holotype: USNM 197143 (male, 430 mm TL, 162 mm preanal), Silver Bay 2183, Bahamas, 26°25′N, 79°01′W, 549 m, 9 Nov. 1960.

Paratypes: BAHAMAS: UMML 29672 (1, 420), 25°35′N, 79°24′W, 516–521 m, 27 June 1970. CUBA: USNM 157889 (1, 402), 22°55′N, 79°27′W, 439 m, 15 July 1955.

Genus *Bathyuroconger* Fowler, 1934

Bathyuroconger Fowler, 1934: 273 (type species *Uroconger braueri* Weber and de Beaufort, 1916, by original designation).

Characters. Body moderately elongate, preanal length less than 40% TL; deepest point shortly behind head; tail attenuate. Skin delicate, slimy, rather transparent and loosely attached. Head large in comparison to body. Body remains quite limp in preservative. Dorsal fin begins over or slightly behind base of pectoral fin. Pectoral fin present.

Jaws nearly equal; fleshy part of snout not projecting beyond intermaxillary teeth. Anterior nostril near tip of snout, tubular, directed laterally. Posterior nostril small, in front of mid-eye. Upper and lower labial flange absent. Adnasal pore absent; head pores somewhat enlarged.

Teeth strong, fang-like. Intermaxillary teeth enlarged, in two transverse rows, separated from maxillary and vomerine teeth. Vomerine teeth few, confined to head of vomer; usually a greatly enlarged central tooth, followed by a smaller one in line behind it, and a pair of smaller teeth on either side of midline between the two median teeth. Maxillary teeth in narrow bands, narrower posteriorly, outer teeth largest. Mandibular teeth in two or three irregular rows, the anterior teeth in outer series greatly enlarged.

Lateral ethmoid process absent. Supraoccipital present. Preorbital bone with projections from lower edge; four suborbitals. Posterior edge of opercle serrate. Myorhabdoi absent. Ribs absent. Vertebral centra thin-walled with large central cavity; no epicentral processes on caudal vertebrae. Lateral-line ossicles as in *Rhynchoconger.* Dorsal- and anal-fin rays segmented.

Digestive tract black. Stomach long, reaching almost to anus. Gas bladder short, located anteriorly, extending from slightly before anterior end of stomach to about middle of stomach.

Discussion. Silvesterina Fowler (1934: 274) is very similar to *Bathyuroconger,* differing mainly in the remarkably small gill opening.

Size. To at least 700 mm TL.

Distribution. Eastern and western tropical Atlantic and Indo-West Pacific.

Etymology. From the Greek *bathys* (deep) and *Uroconger* (a related genus). Masculine.

Species. Two nominal species have been recognized, *Bathyuroconger braueri* (Weber and de Beaufort, 1916) from the Indo-West Pacific and *B. vicinus* (Vaillant, 1888) from the Atlantic, but the distinction between them now seems doubtful. The problem is discussed below under *B. vicinus.*

Bathyuroconger vicinus (Vaillant, 1888)
Figures 571–574, Tables 57–59

Uroconger vicinus Vaillant, 1888: 86, pl. 1, fig. 1 (original description, off northwest coast of Africa, syntypes MNHN 1884-436, 1884-437). Alcock, 1892b: 363 (Bay of Bengal); 1896: 338 (Laccadive Sea); 1899: 200. Barnard, 1925: 192 (off Cape Point, South Africa). Castle, 1968c: 713 (redescription of Barnard's specimens). Karrer, 1973: 216 (comparison with *Uroconger lepturus*).

Uroconger lepturus (not of Richardson, 1844). Brauer, 1906: 124 (west coast of Sumatra and east coast of Africa). Sewell, 1912: 12 (Indian Ocean).

Uroconger braueri Weber and de Beaufort, 1916: 266 (original description, Indian Ocean, eight syntypes, see discussion of synonymy). Castle 1968c: 713 (description adapted from Weber and de Beaufort, comparison with *Uroconger lepturus* and *U. vicinus*).

Bathyuroconger braueri. Fowler, 1934: 273 (type of a new

FIGURE 571. *Bathyuroconger vicinus*: ANSP 127016.

subgenus, *Bathyuroconger*). Reid, 1934: 3 (raised *Bathyuroconger* to a genus). Norman, 1939: 40 (Maldives). Blache and Bauchot, 1976a: 410 (tentatively synonymized *Bathyuroconger braueri* with *B. vicinus*.). Karrer, 1982: 37 (expressed doubt that *braueri* was a valid species).

Bathyuroconger vicinus. Smith, 1971: 75 (western Atlantic). Blache and Bauchot, 1976a: 405 (Gulf of Guinea). Blache, 1977: 141 (leptocephali from the Gulf of Guinea). Karrer, 1982: 34.

Misidentifications:

Not *Uroconger vicinus* Vaillant. Goode and Bean, 1896: 138, fig. 160 (*Rhechias vicinalis*).

Not *Uroconger braueri* Weber and de Beaufort. Della Croce and Castle, 1966: 154 (leptocephali, probably *Uroconger lepturus*).

Discussion of Synonymy. Weber and de Beaufort (1916) applied the name *Uroconger braueri* to the specimens reported by Alcock (1892b) and Brauer (1906) as *Uroconger vicinus* and *U. lepturus*, respectively. Alcock had one specimen and Brauer seven. According to Karrer (1982: 37), four of Brauer's specimens cannot be found and another (SMF 2128) is *Rhechias guttulatus*. ZMH 12661 is indeed *U. vicinus*. One other specimen of *U. vicinus* from the Valdivia expedition exists, ZMB 17538, but it is stated to have come from station 207, which was not among those listed by Brauer (Karrer, 1973: 216). I have no information on the location of Alcock's specimen.

Distinctive Characters. Those of the genus.

Description (including data from Karrer, 1982). Total vertebrae 179–206 (n = 18), predorsal vertebrae 10–14 (15), preanal vertebrae 47–55 (15), precaudal vertebrae 58–60 (2). Branchiostegal rays 7–8 (8). Pectoral-fin rays 13–16 (5). Pores: LL to anus ca 40–47 (10), POM 9–10 (29), IO 4 (26) + 1 (29) + 0 (29), SO 1 + 2 (29), STC 1 (27).

Proportions as % of TL: preanal 36 (2). Of preanal length: predorsal 34–41 (42), head 29–38 (42), dorso-pectoral (predorsal minus head) 3–10 (42), depth at anus 7–15 (38), pectoral fin 8–13 (30). Of HL: snout 24–31 (39), eye 10–17 (38), snout-rictus 36–45 (36), gill opening 7–16 (35), interbranchial 11–32 (30), pectoral fin 21–39 (30).

Color. In preservative brown or gray. Interior of branchial chamber black. Vertical fins edged in black posteriorly. In smaller specimens an irregular longitudinal series of small melano-

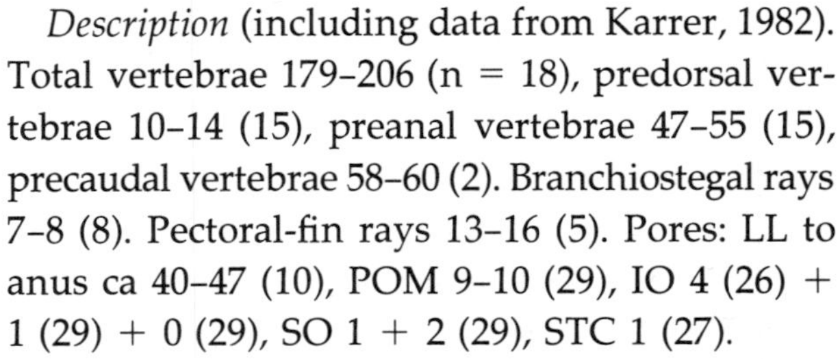

FIGURE 572. *Bathyuroconger vicinus*: ANSP 126130, 430 mm TL; head.

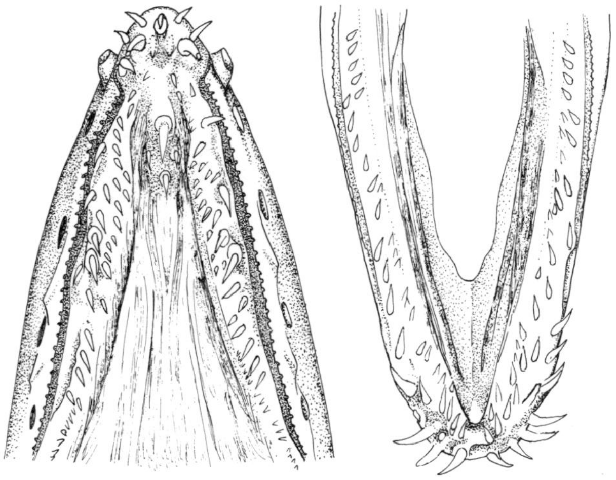

FIGURE 573. *Bathyuroconger vicinus*: ANSP 126130, 430 mm TL; dentition.

phores above and below lateral line and another series of small melanophores on midventral line, which may be remnants of larval pigmentation. Stomach and intestine black.

Size. The largest specimen examined was 880 mm TL (USNM 193611).

Development and Growth. Blache (1977: 141) described leptocephali that he attributed to this species, but the identification should be considered tentative (see p. 756).

Habits and Behavior. Judging by its formidable dentition, *Bathyuroconger vicinus* is an active predator. The attenuate tail, enlarged head pores, and generally frail integument suggest that this is not a burrowing species. The frequency of its collection in bottom trawls is further evidence that it probably spends its time hovering slightly off the bottom in relatively open terrain.

Variation. Bathyuroconger vicinus as recognized here occurs throughout the tropical Atlantic and Indo-West Pacific. The name *B. braueri* (Weber and de Beaufort) has been applied to the Indo-Pacific form, but the distinction does not hold up to close examination. Weber and de Beaufort described *braueri* primarily to distinguish it from *Uroconger lepturus,* to which some of the type specimens had been attributed by Brauer (1906); they did not compare it to the Atlantic *vicinus,* which they may have regarded simply as the Atlantic cognate of *lepturus.* Castle (1968c), working from Weber and de Beaufort's description (he had no specimens of *braueri* available to him), distinguished *braueri* from *vicinus* by the position of the dorsal-fin origin: over the base of the pectoral fin in *braueri* and behind that point in *vicinus.* Blache and Bauchot (1976a) expressed doubt that the two were distinct, and Karrer (1982), after examining some specimens from the Mozambique Channel, concurred. Indeed, the dorsal-origin character cited by Castle does not seem to work. Although in the 14 Indo-Pacific specimens examined by me the dorsal origin was somewhat closer to the pectoral base than in the 28 Atlantic specimens, it was still slightly behind that point. Furthermore, in none of the three characters relating to the relative position of the dorsal origin (predorsal length, head length, and dorso-pectoral distance) can the Indo-Pacific specimens be completely separated from the Atlantic specimens (Tables 57–59). In fact, two of the characters, head length and dorso-pectoral distance (predorsal length minus head length), show a clinal variation from west to east. Vertebral counts are inconclusive. The six Hawaiian

TABLE 57. Predorsal length in *Bathyuroconger vicinus* from various areas (including data from Karrer, 1982).

	Predorsal in % preanal									
Area	34	35	36	37	38	39	40	41	n	x̄
Western Atlantic	1	2	3	3	1	2			12	36.6
Eastern Atlantic			1	2	4	4	3	2	16	38.8
Mozambique Channel						1	2		3	39.7
Philippines						1	–	1	2	40.0
Australia						1	–	1	2	40.0
Hawaii		1	–	2	2	–	2		7	37.9

TABLE 58. HL in *Bathyuroconger vicinus* from various areas (including data from Karrer, 1982).

	HL in % preanal												
Area	28	29	30	31	32	33	34	35	36	37	38	n	x̄
Western Atlantic	1	2	4	4	1							12	30.2
Eastern Atlantic				1	3	5	5	1	–	1		16	33.4
Mozambique Channel									1	1	2	4	37.3
Philippines									1	–	1	2	37.0
Australia									2			2	36.0
Hawaii								4	2	1		7	35.6

specimens have significantly more total vertebrae (201–206) than the 11 Atlantic specimens (179–186) and slightly more preanal vertebrae (52–55 vs 47–53), but only a single count (179—Karrer, 1982: 37) is available for comparison from the Indian Ocean and western Pacific. Because the Atlantic and Indo-Pacific populations cannot be clearly distinguished on the available evidence, they are provisionally treated as a single species, *Bathyuroconger vicinus* (Vaillant).

Distribution (Fig. 574). In the western Atlantic from the eastern Gulf of Mexico south along the coast of Central and South America to the Guianas; one record from the Antilles (USNM 268852). In the eastern Atlantic from about Cape Verde to Angola. In the Indo-West Pacific from South Africa and the Mozambique Channel to Australia, the Philippines, and Hawaii. Depth 229–1318 m (mean 924 m). Relatively common.

Etymology. Latin *vicinus* (near), presumably referring to its perceived relationship to *Uroconger lepturus.* An adjective.

Study Material. 214 specimens, 180–880 mm TL.

FLORIDA GULF: ANSP 109615 (9, 523–814), 117091 (5, 235–417), 126185 (1, 558). UMML 13182 (2, 224–346). USNM 178995 (1, 733), 186048 (2, 230–245), 198059 (1, 349), 198064 (2, 692–795), 268848 (1, 525). ALABAMA: ANSP 109594 (2, 382–450), 109601 (1, 620), 109605 (1, 690), 114161 (1, 305), 126529 (1, 752). USNM 178993 (1, 317), 268849 (1, 245). MISSISSIPPI: UMML 12471 (1, 590). LOUISIANA: USNM 268853 (2, 400–465), 268854 (1, 585). MEXICO GULF: ANSP 126449 (1, 630), 133018 (1, 342). USNM 198740 (1, 704), 268846 (1, 580), 268851 (1, 745). LESSER ANTILLES: USNM 268852 (2, 265–380). VENEZUELA: ANSP 114922 (1, 240), 126130 (1, 430), 126131 (1, 398), 126132 (2, 315–360). USNM 198750 (1, 195 mm preanal), 218914 (1, 435), 268836 (2, 415–615), 268841 (2, 393–430), 268844 (1, 195), 268845 (1, 560), 268855 (1, 588). COLOMBIA: UMML 22582 (1, 235). USNM 268843 (2, 215–580). PANAMA: UMML 22630 (1, 217). USNM 193605 (1, 602), 193611 (7, 620–880). HONDURAS: USNM 193547 (1, 129 mm preanal). SURINAME: ANSP 114865 (7, 306–598), 114867 (1, 555), 114875 (1, 423), 114944 (2, 409–487), 114954 (4, 404–493), 114962 (1, 508), 117025 (2, 433–470), 153562 (1, 402). UMML 4127 (4, 467–695), 29670 (1, 695). USNM 158959 (5, 395–440), 158961 (16, 305–627), 179070 (10, 380–470), 198060 (1, 428), 198062 (6, 380–508). FRENCH GUIANA: ANSP 114843 (1, 410), 114851 (5, 342–545), 114866 (2, 201–500), 114916 (1, 148), 114937 (1, 396), 114948 (3, 396–474), 114964 (1, 517), 117006 (2, 467–515), 153561 (1, 824). USNM 198057 (2, 202–213), 198061 (1, 250), 198063 (9, 318–650), 198065 (13, 515–683), 268837 (1, 245), 268847 (1, 520). GULF OF GUINEA: ANSP 127016 (2, 253–290). UMML 15153 (5, 400–470),

TABLE 59. Dorso-pectoral distance in *Bathyuroconger vicinus* from various areas (including data from Karrer, 1982).

	Dorso-pectoral in % preanal												
Area	0	1	2	3	4	5	6	7	8	9	10	n	x̄
Western Atlantic					3	2	–	3	2	1	1	12	6.5
Eastern Atlantic				2	4	4	3	–	3			16	5.3
Mozambique Channel			1	2								3	2.7
Philippines					2							2	4.0
Australia				1	–	1						2	4.0
Hawaii	1	1	2	2	–	1						7	2.3

15187 (2, 370–449), 16945 (8, 214–295), 21657 (1, 570), 23295 (3, 347–508). USNM 198622 (4, 200–455), 198631 (5, 210–455), 198633 (1, 180), 198636 (4, 504–665). AUSTRALIA: AMS I.20920-011 (2, 600–695). PHILIPPINES: USNM 135270 (2, 447–480). HAWAII: BPBM 21071 (7, 241–315).

Genus *Uroconger* Kaup, 1856

Uroconger Kaup, 1856a: 71 (type species *Congrus lepturus* Richardson, 1845 by monotypy); 1856b: 110.

Congerodon Kaup, 1856a: 73 (type species *Congerodon indicus* Kaup, 1856, by monotypy). Bleeker, 1864b: 28 (synonymized *Congerodon* with *Uroconger*).

Characters. Body moderately elongate, preanal length less than 40% of total; tail greatly attenuate. Dorsal fin begins slightly before to slightly behind base of pectoral fin. Pectoral fin well developed.

Snout projecting slightly beyond lower jaw; fleshy part of snout projecting slightly beyond intermaxillary teeth. Flange present on lower lip but not on upper. Anterior nostril tubular, near tip of snout, directed anterolaterally. Posterior nostril elliptical, at mid-eye level. Adnasal pore absent; three posteriormost pores along upper jaw enlarged, slit-like.

Teeth moderately large, acute. Intermaxillary teeth relatively large, in two transverse rows, separated from maxillary and vomerine teeth. Vomerine teeth in a single median row terminating behind posterior end of maxillary tooth patch; a few anteriormost teeth larger than the others. Maxillary teeth in two rows anteriorly, converging to a single row posteriorly. Mandibular teeth in two rows for most of length, but at anterior end an extra, short inner row; outer rows converge at posterior end.

Lateral ethmoid process absent. Supraoccipital present. Preorbital bone with projections from lower edge; four suborbitals. Posterior edge of opercle serrate. Myorhabdoi absent. Ribs absent; anterior few epineurals fused to neural arch; prominent epicentral processes on caudal vertebrae. Vertebrae numerous, more than 200. Lateral-line ossicles like those of *Rhynchoconger*. Dorsal- and anal-fin rays segmented.

Digestive tract pale. Stomach long, reaching anus. Gas bladder long, relatively thick-walled, beginning well before stomach and extending well past anus, constricted in middle.

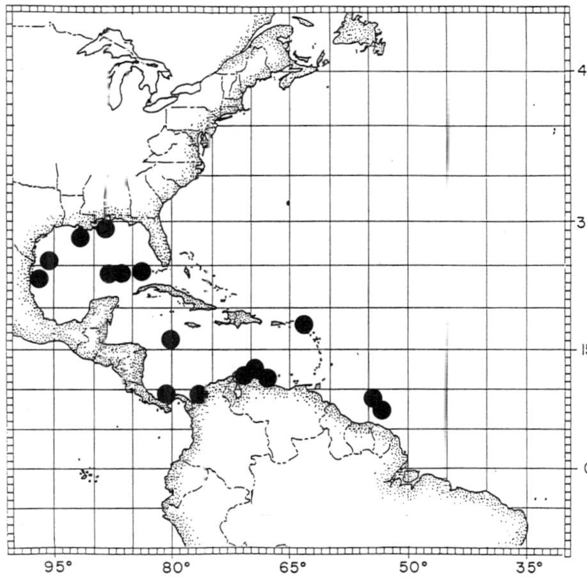

FIGURE 574. Distribution of *Bathyuroconger vicinus.*

Discussion. Ginsburg (1954: 257) expressed doubt that the fish described by Richardson as *Congrus lepturus* (the type species of *Uroconger*) is the same as the species that has been treated under that name subsequently. The original description mentioned only two or three vomerine teeth instead of the long row present on *Uroconger lepturus* of later authors. Castle (1968c: 712) pointed out that the specimen now listed as the type in the British Museum has a long row of vomerine teeth. Because of a difference between the size of the British Museum specimen (260 mm TL) and that reported by Richardson (nine inches or 229 mm), Castle felt that the two might not be the same. There seems no practical reason, however, not to accept the specimen as the type of Richardson's species. It exists, it is clearly labeled as such, it agrees with all later accounts of *Uroconger lepturus,* and the discrepancy in size is relatively minor. Richardson's description agrees in every other way with the *U. lepturus* of later authors, and it seems most reasonable to assume that the description of the vomerine teeth was simply an error.

Paruroconger Blache and Bauchot (1976a: 413, type species *Paruroconger drachi* Blache and Bauchot, 1976) is very close to *Uroconger.* The only differences are stated to be a greater inter-

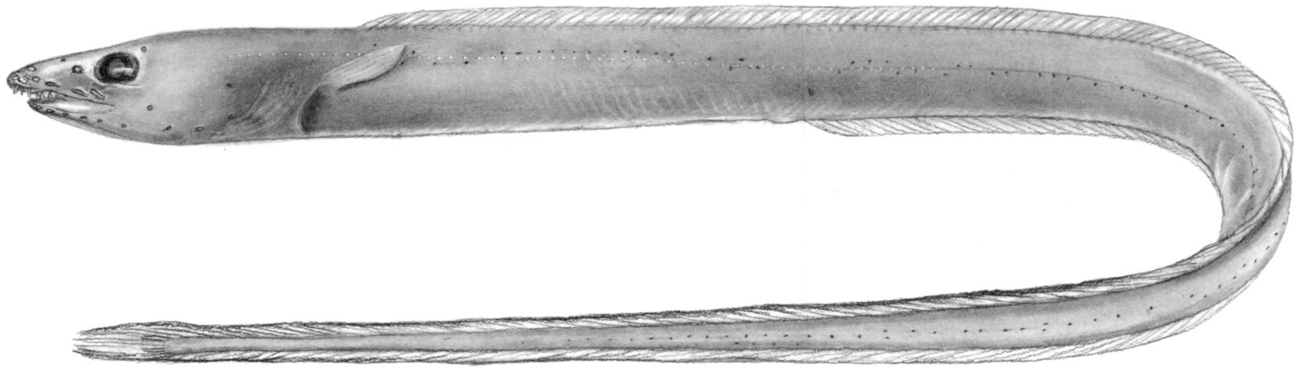

FIGURE 575. *Uroconger syringinus*: ANSP 126971, 163 mm TL.

orbital space and multiserial vomerine dentition in *Paruroconger.*

Size. To about 500 mm TL, usually smaller.

Distribution. Eastern and western tropical Atlantic, Indo-West Pacific from east Africa to Hawaii.

Etymology. From the Greek *oura* (tail) and *Conger* (the conger eel), probably referring to the long, slender tail of this genus. Masculine.

Species. Two species are currently recognized: *Uroconger lepturus* (Richardson, 1845) from the Indo-West Pacific and *U. syringinus* Ginsburg, 1954 from the Atlantic. In reality the species structure of *Uroconger* is not at all clear, and the differences between the two nominal species have yet to be firmly established. The problem is discussed below under *Uroconger syringinus.*

Uroconger varidens Garman, 1899 from the eastern Pacific does not appear to belong in *Uroconger. Uroconger vicinus* Vaillant, 1888 and its synonym *U. braueri* Weber and de Beaufort, 1916 are here assigned to *Bathyuroconger* Fowler.

Uroconger syringinus Ginsburg, 1954
Figures 575–578, Tables 60–63

Uroconger syringinus Ginsburg, 1954: 256, figs. 1–2 (original description, 28°45′N, 89°43′W, holotype USNM 157781).

Uroconger vicinalis (not of Garman, 1899). Castle, 1968c: 712 (stated that *Uroconger syringinus* is "almost certainly a synonym of *U. vicinalis*").

Distinctive Characters. As the only western Atlantic representative of *Uroconger, U. syringinus* can be recognized by the characters of the genus. Differences between this and the other nominal species of the genus are discussed below.

Description. Total vertebrae 220–225 (n = 3), predorsal vertebrae 11 (2), preanal vertebrae 46–49 (3), precaudal vertebrae 86 (1). Branchiostegal rays 9 (5). Pectoral-fin rays 11–12 (12). Pores: LL to anus 40–46 (15), POM 10–11 (12), IO 4 (13) + 1 (13) + 0–2 (13), SO 1 + 2–4 (14), STC 1–3 (13).

Proportions as % of TL: preanal 31–33 (3). Of preanal length: predorsal 39–43 (15), head 36–42 (15), depth at anus 11–14 (15), pectoral fin 8–11 (14). Of HL: snout 25–30 (15), eye 12–16 (15), snout-rictus 36–42 (12), gill opening 13–21 (15), interbranchial 9–15 (13), pectoral fin 20–26 (14).

Color. Brown or grayish-brown. Vertical fins edged in black posteriorly. Stomach and intestine pale. Smaller specimens often retain traces of larval pigmentation in form of small melanophores along midlateral line and ventral midline.

Size. The largest specimen examined was 380 mm TL, 189 mm preanal (UMML 16857).

Development and Growth. The leptocephalus has been identified (see p. 751).

Variation. Uroconger syringinus shows some

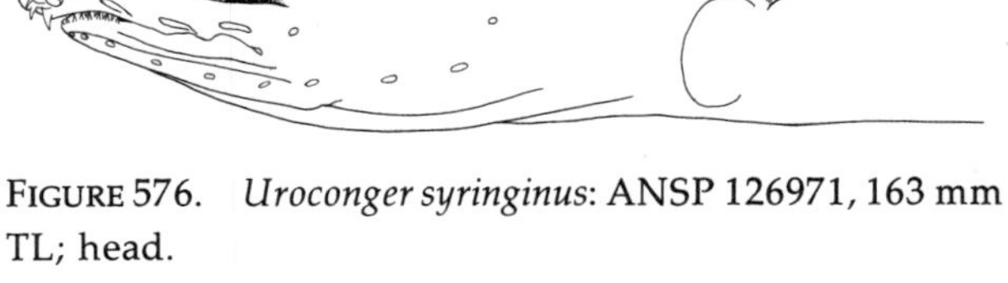

FIGURE 576. *Uroconger syringinus*: ANSP 126971, 163 mm TL; head.

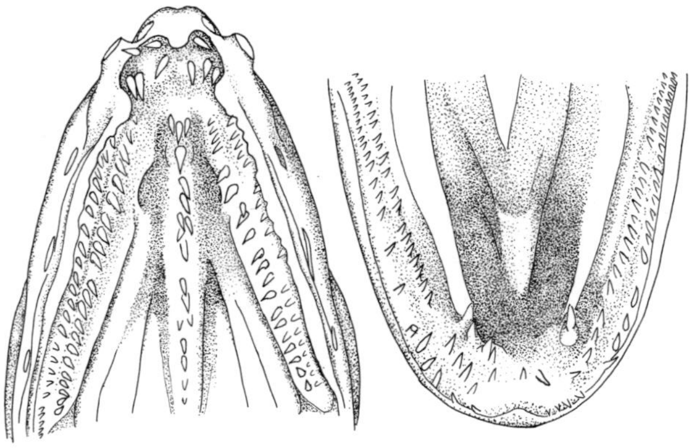

FIGURE 577. *Uroconger syringinus*: ANSP 114159, 385 mm TL; dentition.

variation in the number of head pores, especially in the postorbital and supraorbital series. All but one of the specimens examined had a single supratemporal pore; the exception was the holotype, which had three.

Remarks. Uroconger syringinus is very close to the Indo-Pacific *U. lepturus*, and the differences are not altogether clear. Ginsburg (1954: 257) originally distinguished the two on the basis of one erroneous character (the vomerine teeth; see discussion above under genus *Uroconger*) and several proportional measurements whose value is doubtful due to the frequency of regenerated tails. In one form or another, these species are found from Hawaii to east Africa and on both sides of the Atlantic. Tables 60–63 compare four characters between *Uroconger syringinus* from the eastern and western Atlantic and *U. lepturus* from the Indian Ocean and China. In each case there is a considerable overlap. In predorsal length (Table 60) the eastern Atlantic *syringinus* falls with the western Atlantic *syringinus*, but in the other three characters (Tables 61–63) the geographically intermediate population is also morphologically intermediate.

The only other character that shows any difference is the number of vertebrae, but the scarcity of intact specimens makes it difficult to apply. Three western Atlantic specimens (UMML 28777, USNM 193541, USNM 197118) had 220–225 total vertebrae; the single precaudal count was 86 (UMML 22296). Blache and Bauchot (1976a: 390) recorded 211–218 total and 85–87 precaudal vertebrae in four specimens from the Gulf of Guinea, but they expressed some doubt that the specimens were complete. A specimen of *U. lepturus* from the Indian Ocean (ANSP 112265) had about 216 vertebrae. Asano (1962: 116) gave a range of 203–206 total and 82–84 precaudal vertebrae for Japanese specimens, seemingly a significant difference. However, he also noted three additional specimens that differed in having fewer head pores, fewer precaudal vertebrae (75) and more total vertebrae (214). Head pores are probably too variable to be trusted as a distinguishing character. The total vertebral count of 214 is intermediate between the "standard" Japanese *U. lepturus* and the western Atlantic *syringinus*; the precaudal count is lower than any other on record.

Myomere counts of leptocephali are more readily available and show patterns of variation better. In the western Atlantic, total myomere counts were 216–227 (see page 751). Blache (1977: 124) reported counts of 212–229 for eastern Atlantic leptocephali. Nair and Mohamed (1960c: 184) reported 204–223 myomeres for leptocephali of *U. lepturus* from India. D'Ancona's *Leptocephalus magnaghii* from the Red Sea, which at least in part appears to be a larval *Uroconger*, had 205–

TABLE 60. Predorsal length in *Uroconger syringinus* from the western and eastern Atlantic and *Uroconger lepturus* from the Indo-West Pacific.

Area	Predorsal in % preanal							n	x̄
	39	40	41	42	43	44	45		
Western Atlantic	1	3	5	1	1			11	40.8
Eastern Atlantic	1	1	1	1				4	40.5
Indo-Pacific		3	3	3	2	3	1	15	42.3

TABLE 61. HL in *Uroconger syringinus* from the western and eastern Atlantic and *Uroconger lepturus* from the Indo-West Pacific.

	HL in % preanal													
Area	36	37	38	39	40	41	42	43	44	45	46	47	n	x̄
Western Atlantic	1	2	3	1	2	1	1						11	39.7
Eastern Atlantic					1	2	1						4	41.0
Indo-Pacific				1	1	–	3	1	5	2	–	2	15	43.5

219 myomeres. The counts from the last two sources agree well with the vertebral counts of ANSP 112265 and of Asano's specimens and would seem to indicate a species with 203–223 vertebrae. Castle (1964b: 34, as *Leptocephalus mediopunctatus*), however, reported three specimens from New Caledonia with 235–239 myomeres, Della Croce and Castle (1966: 155) reported five specimens from the Mozambique Channel with 227–236 myomeres, and I have examined two specimens from Hawaii that had ca 240–243 myomeres. This suggests that two species of *Uroconger* may occur in the Indo-Pacific, but further work is needed to clarify the situation. Until the species structure in the Indo-Pacific can be settled, I provisionally retain *Uroconger syringinus* as a distinct species.

Ginsburg (1954: 257) listed the following specimens among the paratypes: "Gulf of Mexico (exact locality uncertain) 3 specimens 255–416 mm (University of Miami)." I found only two specimens of *Uroconger syringinus* at the University of Miami that were collected before 1954. They bear the number UMIM (University of Miami Ichthyological Museum) 897, were collected at 22°14′N, 91°26′W (Oregon station 406), and measure 310–415 mm TL. I cannot determine whether these are Ginsburg's specimens and, if so, what happened to the third specimen.

Distribution (Fig. 578). In the western Atlantic from the east coast of Florida, the western Bahamas, and the Gulf of Mexico south along the coast of Central and South America at least to Suriname; rare in the West Indies. Also the Gulf of Guinea in the eastern Atlantic. Depth 44–384 m (mean 100 m).

Uroconger syringinus is primarily a continental species. There are only three insular records, two

TABLE 62. Interbranchial distance in *Uroconger syringinus* from the western and eastern Atlantic and *Uroconger lepturus* from the Indo-West Pacific.

	Interbranchial in % HL											
Area	6	7	8	9	10	11	12	13	14	15	n	x̄
Western Atlantic					1	1	1	1	4	1	9	13.0
Eastern Atlantic				1	–	1	2				4	11.0
Indo-Pacific	1	5	3	1	4	–	1				15	8.4

TABLE 63. Preanal lateral-line pores in *Uroconger syringinus* from the western and eastern Atlantic and *Uroconger lepturus* from the Indo-West Pacific.

	LL pores										
Area	38	39	40	41	42	43	44	45	46	n	x̄
Western Atlantic					2	–	4	3	2	11	44.3
Eastern Atlantic			1	–	–	2				3	42.0
Indo-Pacific	2	–	–	8	3	–	2			15	41.2

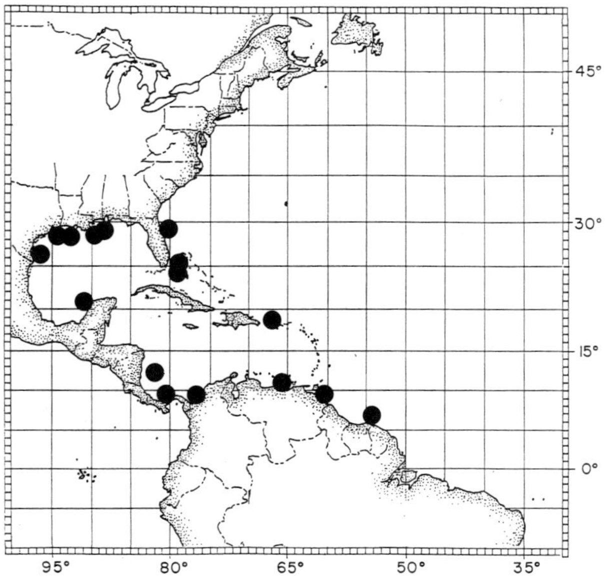

FIGURE 578. Distribution of *Uroconger syringinus.*

from the Bahamas (USNM 197118, USNM 268838) and one from Puerto Rico (ANSP 114159). One of the Bahamian specimens (USNM 268838) is very small and was collected on the surface at night; it probably had just transformed from the leptocephalus.

Etymology. From the Greek *syrinx* (pipe, tube) and the Latin suffix *-inus* (pertaining to, having the nature of). An adjective.

Study Material. 29 specimens, 65–415 mm TL.

Holotype: USNM 157781 (male, 275 mm TL, 103 mm preanal), Oregon 10, Louisiana, 28°45′N, 89°43′W, 92 m, 28 May 1950.

Paratypes: TEXAS: USNM 157789 (1, 333). MEXICO GULF: ? UMIM (UMML) 897 (2, 310–415).

Other material: FLORIDA ATLANTIC: UMML 11228 (1, 234). ALABAMA: USNM 198049 (1, 426). LOUISIANA: USNM 157968 (1, 370), 158254 (1, 340), 196774 (1, 410). TEXAS: TCWC 3982.1 (1, 275). BAHAMAS: USNM 197118 (1, 375), 268838 (1, 65). PUERTO RICO: ANSP 114159 (1, 385). VENEZUELA: UMML 29470 (1, 186), 29471 (1, 170). USNM 185658 (1, 283). COLOMBIA: ANSP 126971 (1, 163), 152116 (1, 217). UMML 22296 (1, 166), 28777 (1, 228). USNM 197138 (2, 353–380), 268831 (1, 255), 268842 (1, 299). PANAMA: USNM 193541 (1, 356). NICARAGUA: USNM 179233 (1, 235). SURINAME: USNM 185643 (1, 397). GULF OF GUINEA: UMML 16762 (1, 236), 16857 (1, 380), 20901 (2, 175–213).

Comparative material (*Uroconger lepturus*): INDIAN OCEAN, ANSP 111996 (4, 129–257), 111999 (3, 215–295), 112265 (7, 165–308). CHINA: BMNH (holotype of *Uroconger lepturus*, 258 mm TL, 85 mm preanal).

TABLE 64. Distinguishing characters in the three species of *Japonoconger* (data on *J. sivicolus* from Asano, 1962: 107; on *J. africanus* from Blache and Bauchot, 1976a: 397).

Character	*sivicolus*	*africanus*	*caribbeus*
PO pores	0	0	0–2
SO pores	4	4	5–8
ST pores	1	1	4–7
LL pores	38–40	34–38	33–39
Total vertebrae	159–164	134–142	160–166
Precaudal vertebrae	52–54	53–57	49

Genus *Japonoconger* Asano, 1958

Japonoconger Asano, 1958: 316 (type species *Arisoma* [sic] *sivicola* Matsubara and Ochiai, 1951, by original designation).

Characters. Body moderately elongate; preanal length less than 40% total; tail not attenuate. Dorsal fin begins slightly before base of pectoral fin. Pectoral fin well developed.

Snout projects beyond lower jaw; fleshy part of snout projects beyond intermaxillary teeth. Anterior nostril tubular, near tip of snout, directed anterolaterally. Posterior nostril elliptical, in front of mid-eye. Flange present on lower lip but not upper. Adnasal pore absent; head pores small.

Teeth small, conical, in granular patches. Intermaxillary tooth patch round, separated from maxillary and vomerine teeth. Vomerine tooth patch elongate, more than half length of maxillary tooth patch. Maxillary and mandibular teeth in bands, wider anteriorly.

Lateral ethmoid process absent. Supraoccipital present. Preorbital bone with projections from lower edge; four suborbitals. Posterior edge of opercle serrate. Upper pharyngeals divided, the anterior element slightly smaller. Lower pharyngeals reduced. Myorhabdoi absent. Ribs present on abdominal vertebrae. Anterior few epineurals fused to neural arch; neural arches of first 20 to 25 vertebrae articulated with centrum rather than fused; no epicentral processes on caudal vertebrae. Lateral-line ossicles as in *Rhynchoconger.* Dorsal- and anal-fin rays segmented.

Stomach reaching halfway to anus. Gas blad-

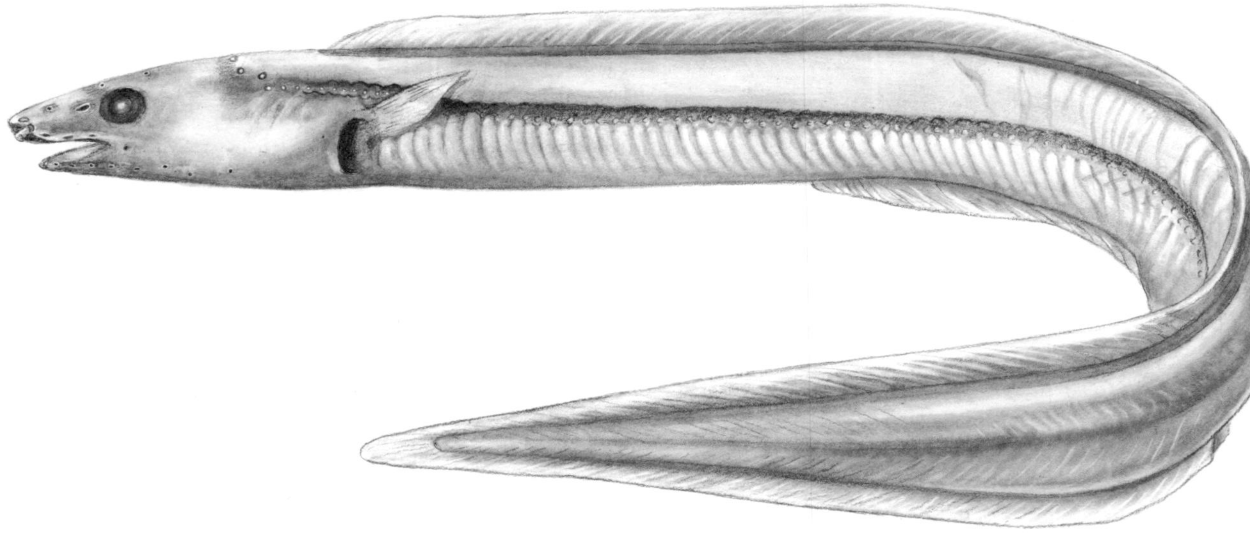

FIGURE 579. *Japonoconger caribbeus*: ANSP 126973, paratype, 356 mm TL.

der short, displaced far posteriorly, beginning about halfway between end of stomach and anus, ending about an equal distance behind anus. Stomach black, intestine pale.

Size. To 500–600 mm TL.

Distribution. Southern Caribbean, Gulf of Guinea, and Japan.

Etymology. Named for Japan, where the type species occurs. Masculine.

Species. Three species are recognized in this genus: *Japonoconger sivicolus* (Matsubara and Ochiai, 1951) from Japan, *J. africanus* (Poll, 1953) from the Gulf of Guinea, and *J. caribbeus* Smith and Kanazawa, 1977 from the Caribbean Sea. Table 64 shows the principal differences between them.

Japonoconger caribbeus
Smith and Kanazawa, 1977
Figures 579–581, Table 64

Japonoconger sp. Smith, 1971: 77.

Japonoconger caribbeus Smith and Kanazawa, 1977: 541, fig. 11 (original description, 11°09′N, 74°26.5′W, holotype USNM 198685).

Distinctive Characters. 0–2 postorbital pores; 5–8 supraorbital pores; 4–7 supratemporal pores, usually five; 33–39 preanal lateral-line pores. Otherwise recognized by the characters of the genus.

Description. Total vertebrae 160–166 (n = 7), predorsal vertebrae 8–9 (5), preanal vertebrae 38–40 (5), precaudal vertebrae 49 (1). Branchiostegal rays 10 (1). Pectoral-fin rays 13–15 (8). Pores: LL to anus 33–39 (14), POM 10 (8), IO 4 (12) + 1 (10) +0–2 (12), SO 1 + 5–8 (11), STC 4–7 (12).

Proportions as % of TL: preanal 33–39 (9). Of preanal length: predorsal 38–45 (14), head 39–45 (14), depth at anus 13–18 (14), pectoral fin 10–12 (14). Of HL: snout 22–29 (14), eye 11–15 (14), snout-rictus 30–33 (12), gill opening 8–13 (14), interbranchial 16–25 (13), pectoral fin 24–32 (14).

Color. In preservative gray or grayish-brown. Vertical fins edged in black posteriorly. Interior of mouth and gill area black. Stomach black, intestine pale.

Size. The largest specimen examined was the holotype, 500 mm TL, 189 mm preanal. Another specimen (USNM 198677) had a slightly greater TL (502 mm) but a lesser preanal length (182 mm).

Variation. Variation occurs in the pores of the postorbital, supraorbital, and supratemporal series. Postorbital pores range from zero to two, the number sometimes differing on the two sides. The supraorbital series contains the full congrid complement of six pores, but variations are common. The supratemporal series usually consists of a single median pore, a pore on each side midway between the median pore and the lateral

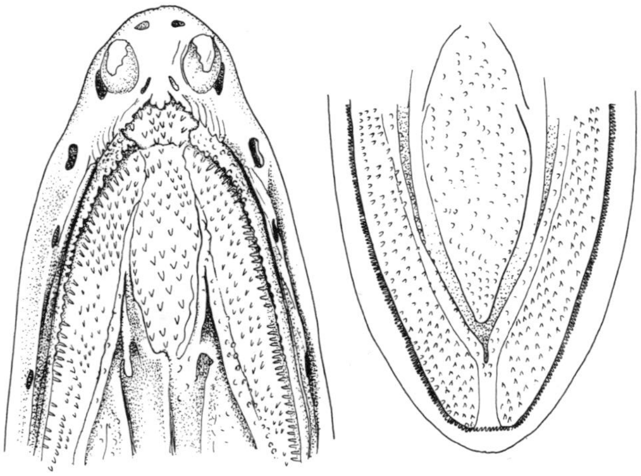

FIGURE 580. *Japonoconger caribbeus*: ANSP 114154, paratype, 395 mm TL; dentition.

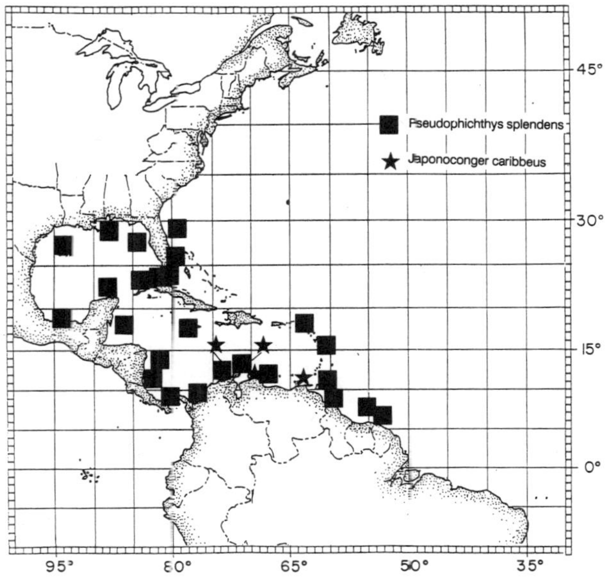

FIGURE 581. Distribution of *Japonoconger caribbeus* and *Pseudophichthys splendens*.

line, and a pore on each side immediately above the first lateral-line pore; variations occur here, too, however. The infraorbital series is fairly constant, but one specimen had two postrictal pores on one side.

Although the tail is not extremely attenuate, several of the specimens examined had regenerated caudal fins. Asano (1962: 107) found a similar situation in *J. sivicolus.*

Development and Growth. The leptocephalus has not been identified.

Distribution (Fig. 581). Known only from the Caribbean coast of Colombia and Venezuela from the mouth of the Magdalena River to Trinidad. Depth 329–576 m (mean 446 m). Uncommon, only 14 specimens known.

Etymology. Named for the Caribbean Sea. An adjective.

Study Material. 14 specimens, 321–502 mm TL.

Holotype: USNM 198685 (female, 500 mm TL, 189 mm preanal), Oregon 4859, Colombia, 11°09'N, 74°26.5'W, 329–348 m, 19 May 1962.

Paratypes: VENEZUELA: ANSP 114154 (1, 395), 11°36'N, 62°46'W, 531 m, 19 April 1960. USNM 198666 (1, 455), 11°10'N, 68°08'W, 403 m, 11 Oct. 1963. USNM 198671 (1, 406), 11°46'N, 69°17'W, 439 m, 5 Oct. 1963. COLOMBIA: ANSP 114153 (1, 383), 11°10'N, 74°27'W, 458–512 m, 17 May 1964. ANSP 126136 (1, 444), 11°27'N, 73°42'W, 403 m, 5 Dec. 1968. ANSP 126973 (1, 356) and UMML 28785 (1, 321), 11°13.3'N, 74°50'W, 408–576 m, 29 July 1968. USNM 198667 (1, 425), 12°28'N, 72°26'W, 549 m, 2 June 1964. USNM 198677 (3, 430–502), 12°16'N, 72°40'W, 366 m, 2 June 1964.

Other material: VENEZUELA: ANSP 130199 (2, 379–444), 12°52'N, 70°43'W, 512 m, 23 Nov. 1970.

Genus *Pseudophichthys* Roule, 1915

Pseudophichthys Roule, 1915: 283 (type species *Pseudophichthys latedorsalis* Roule, 1915, by monotypy).

Characters. Body moderately elongate; preanal length less than 40% total; tail not attenuate. Dorsal fin begins slightly behind tip of appressed pectoral fin; pectoral fin small.

Snout projects well beyond lower jaw, somewhat swollen, fleshy part extends anteriorly beyond intermaxillary teeth. Mouth small, rictus near anterior margin of eye. Anterior nostril tubular, near tip of snout, directed anterolaterally. Upper labial flange absent. Posterior nostril elliptical, in front of mid-eye. Head pores somewhat enlarged; adnasal and supratemporal pores absent.

Teeth moderately small, conical, in broad, cardiform patches. Intermaxillary teeth in several rows, confluent with maxillary teeth. Vomerine teeth in a broad, elongate patch, wider at posterior end, extending posteriorly almost to end of maxillary tooth patch and nearly covering roof of mouth between maxillae; teeth in middle and posterior area larger. Maxillary teeth in moder-

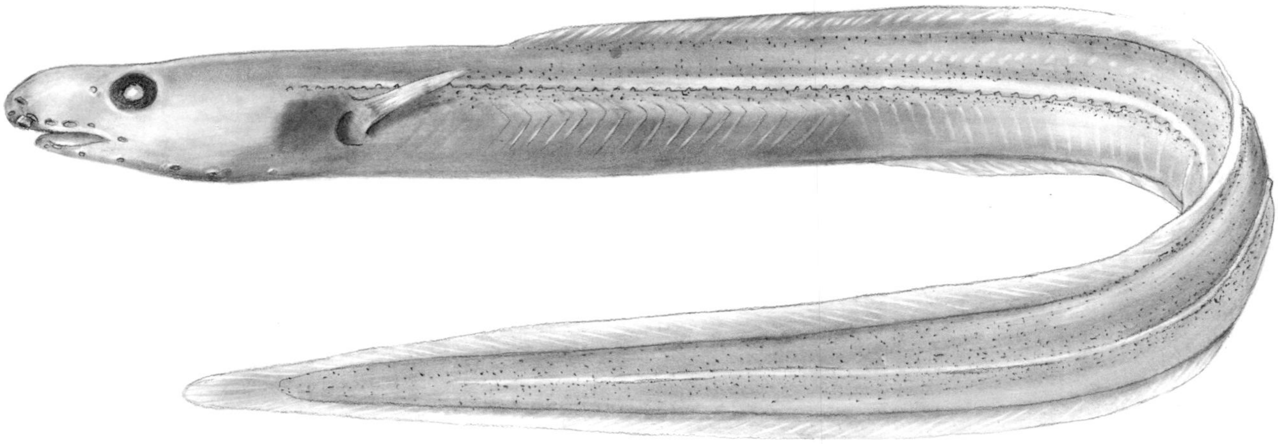

FIGURE 582. *Pseudophichthys splendens*: ANSP 127017, 174 mm TL.

ately broad bands, slightly wider anteriorly. Mandibular teeth in broad bands, widest near middle.

Lateral ethmoid process absent. Supraoccipital absent. Preorbital bone with projections from lower edge; four suborbitals. Posterior edge of opercle serrate. First infrapharyngobranchial with a prominent medially directed process. Myorhabdoi absent. Ribs present on abdominal vertebrae; anterior few epineurals fused to neural arch; neural arches articulated with centrum on about first six vertebrae; no epicentral processes on caudal vertebrae. Lateral-line ossicles as in *Rhynchoconger*. Dorsal- and anal-fin rays segmented.

Stomach short, reaching about a fourth to a third of the way to anus. Gas bladder short, displaced posteriorly, beginning about midway between end of stomach and anus and ending about an equal distance behind anus. Stomach black, intestine pale. Abdomen, inside of mouth and gill area black.

Size. To about 400 mm TL.

Distribution. Eastern and western tropical Atlantic.

Etymology. From the Greek *pseudo* (false) and *Ophichthys* (or *Ophichthus,* a genus of ophichthid eels). Masculine.

Species. A single species, *Pseudophichthys splendens* (Lea, 1913). *Pseudophichthys macroporis* Kotthaus, 1968 from the Indian Ocean does not appear to belong in this genus.

Pseudophichthys splendens (Lea, 1913)
Figures 581–585

Myrus pachyrhynchus (not of Vaillant, 1888). Vaillant, 1888: 81, part.

Leptocephalus splendens Lea, 1913: 29, fig. 19, pl. 4, no. 2 (original description, 28°42′N, 20°00′W, holotype ZMUB). Leptocephalus.

Pseudophichthys latedorsalis Roule, 1915: 283 (original description, Azores, syntypes MOM); 1916: 4 (presented as an original description); 1919: 97, pl. 6, fig. 4a–c (a more complete account than the preceding, including an illustration). Poll, 1953: 143.

Promyllantor latedorsalis. Fowler, 1936a: 260.

Promyllantor schmitti Hildebrand in Longley and Hildebrand, 1940: 226, fig. 1 (original description, Straits of Florida south of Dry Tortugas, holotype USNM 107323).

Pseudophichthys splendens. Smith, 1971: 79. Blache and Bauchot, 1976a: 418.

Pseudophichthys schmitti. Blache and Bauchot, 1976a: 422.

Distinctive Characters. Those of the genus.

Description. Total vertebrae 130–136 (n = 19), predorsal vertebrae 13–17 (9), preanal vertebrae 38–41 (9), precaudal vertebrae 52 (1). Branchiostegal rays 9 (1). Pectoral-fin rays 10–13 (19). Pores: LL to anus ca 33–39 (12), POM 8 (21), IO 4 (23) +1 (23) + 0 (23), SO 1 + 2 (23), STC 0 (23).

Proportions as % of TL: preanal 35–40 (28), predorsal 17–22 (27), head 13–15 (28), depth at anus 4–6 (27), pectoral fin 2–3 (26). Of HL: snout 25–33 (28), eye 13–19 (28), snout-rictus 24–33 (28), gill opening 7–12 (28), interbranchial 18–30 (27), pectoral fin 15–25 (26).

Color. In preservative gray or brown. Abdo-

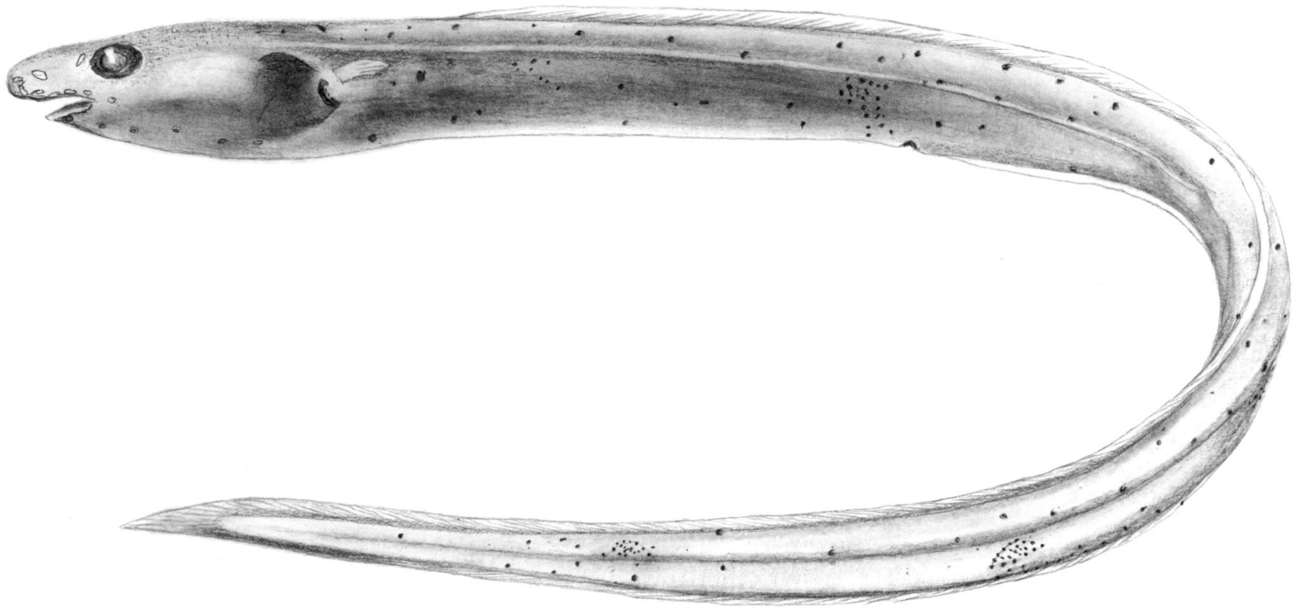

FIGURE 583. *Pseudophichthys splendens*: MCZ 38777, 123 mm TL: showing remnants of larval pigmentation.

men and inside of mouth and gill area black. Vertical fins dark-edged posteriorly. Stomach black, intestine pale. Young specimens retain traces of the larval pigmentation. This consists of melanophores or patches of small melanophores scattered over the surface of the body (Fig. 583).

Size. The largest specimen examined was 387 mm TL (UMML 12394).

Development and Growth. The leptocephalus has been identified (see p. 757).

Remarks. The eastern Atlantic *Pseudophichthys latedorsalis* Roule does not appear to differ in any significant way from the western Atlantic *Promyllantor schmitti* Hildebrand. Measurements of the two syntypes of *P. latedorsalis* taken from Roule (1919: 98) and converted to percentages are as follows: preanal 38–42% TL, predorsal 19% TL, head 13–17% TL, snout 26–31% HL, eye 17–18% HL, depth 6% TL. All these figures are within the range of measurements given above except the one preanal length of 42% TL. In regard to the last, Roule also gave a total vertebral count of 109 for one of the types. This is significantly lower than those of the specimens examined here, but combined with the unusually high preanal length it strongly suggests that the specimen had a regenerated tail. In other characters such as the posterior dorsal-fin origin, small pectoral fin, pore pattern, and dentition, the description of *Pseudophichthys latedorsalis* matches *Promyllantor schmitti* perfectly. Blache and Bauchot (1976a: 422) reached the same conclusion. Both the above names are junior synonyms of *Leptocephalus splendens* Lea.

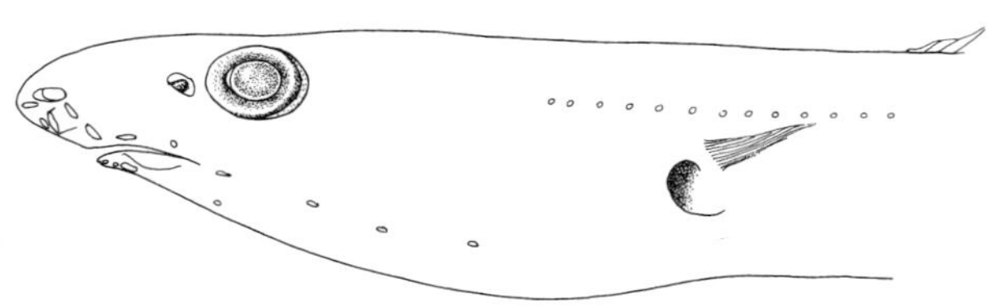

FIGURE 584. *Pseudophichthys splendens*: ANSP 133005, 282 mm TL; head.

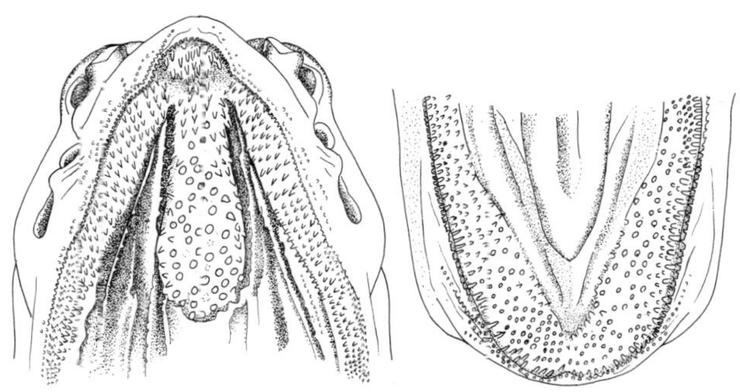

FIGURE 585. *Pseudophichthys splendens*: ANSP 114929, 373 mm TL; dentition.

TABLE 65. Number of vertebrae in the species of *Acromycter*. L and H are the lowest and highest values for *A. nezumi* given by Asano (1962).

	Vertebrae															
Species	159	160	161	162	163	164	165	166	167	168	169	170	171	172	n	x̄
alcocki						1	2	-	1	2	1	1	2	1	11	168.2
atlanticus									1	3	-	2	1		7	168.9
perturbator	2	1	1	1	-	1									6	160.8
nezumi		L	...	...	...	...	...	...	...	...	...	H				

Distribution (Fig. 581). Widely distributed in the western Atlantic from the east coast of Florida and the Gulf of Mexico south through the Caribbean and off the coast of South America at least to the Guianas. In the eastern Atlantic from the Azores to the Gulf of Guinea. Depth 37–1647 m (mean 656 m).

Pseudophichthys splendens is a commonly collected eel in its depth range. This may be partly due to its preference for flat, easily trawled bottoms.

Etymology. Latin *splendens* (splendid), referring to the distinctively pigmented leptocephalus. An adjective.

Study Material. 215 specimens, 121–387 mm TL.

FLORIDA ATLANTIC: ANSP 112039 (1, 323). UMML 16079 (5, 230–246), 18363 (1, 170). USNM 186001 (7). FLORIDA STRAITS: ANSP 127017 (1, 174). UMML 12394 (21, 210–387), 12408 (4, 224–267), 13867 (1, 154), 13927 (8, 158–260), 15507 (8, 172–267), 15649 (7, 161–253), 15729 (4, 218–246), 15733 (7, 142–260), 16618 (1, 121), 16729 (2, 215–251), 29656 (1, 230), 29481 (4, 220–265). USNM 73270 (1), 197120 (1). FLORIDA GULF: ANSP 109611 (7, 245–301), 114158 (1, 268), 114996 (4, 281–305), 114864 (1, 278), 114996 (4). UMML 13181 (1, 285), 18558 (5, 220–282). USNM 107323 (1, 281, holotype of *Promyllantor schmitti*), 164127 (1), 164128 (2), 186049 (2). ALABAMA: ANSP 112042 (2, 315–330), 112045 (1, 305), 114148 (1, 161), 133005 (1, 282), 195904 (25, 195–316). MISSISSIPPI: ANSP 131991 (1, 346), 133001 (1, 304). USNM 179065 (1). TEXAS: USNM 157943 (6). MEXICO GULF: ANSP 117009 (4, 212–253), 126137 (1, 290). USNM 198055 (1). CUBA: MCZ 38777 (1, 123), 38800 (1, 293). JAMAICA: UMML 29595 (2, 145–156), 29650 (2, 250–290). LESSER ANTILLES: ANSP 117014 (1, 224). UMML 29472 (1, 215), 29651 (2, 234–243), 29653 (2, 177–262). VENEZUELA: UMML 22642 (12, 132–340), 22655 (1, 298). USNM 198744 (1). COLOMBIA: UMML 23241 (1, 290), 28782 (2, 274–301), 29652 (1, 282). USNM 198676 (2). PANAMA: ANSP 112051 (1, 169). UMML 22629 (2, 298–322). USNM 198053 (4), 268830 (1, 242). NICARAGUA: USNM 158876 (2), 193604 (1). MEXICO QUINTANA ROO: UMML 29654 (1, 271). SURINAME: USNM 158913 (1). FRENCH GUIANA: ANSP 114879 (1, 317), 114929 (1, 373), 114955 (1, 335), 117002 (2, 330–345). GULF OF GUINEA: UMML 21515 (6, 205–229).

Genus *Acromycter* Smith and Kanazawa, 1977

Acromycter Smith and Kanazawa, 1977: 541 (type species *Ariosoma perturbator* Parr, 1932, by original designation).

Characters. Body moderately elongate; preanal length less than 40% of total; tail not attenuate. Dorsal fin begins near or slightly behind tip of appressed pectoral fin. Pectoral fin small.

Snout somewhat swollen, projects well beyond lower jaw, fleshy part extends anteriorly well beyond intermaxillary teeth. Mouth small, rictus near anterior margin of eye. Eye small. Upper labial flange absent. Anterior nostril tubular, near tip of snout, directed anterolaterally. Posterior nostril elongated anteroposteriorly, located high on head above anterior part of eye. Adnasal pore absent.

Teeth moderately small, conical, in cardiform

TABLE 66. Head length in the species of *Acromycter*. L and H are the low and high values for *A. nezumi* given by Asano (1962).

	HL in % TL					
Species	11	12	13	14	n	x̄
alcocki		1	12	5	18	12.2
atlanticus		1	6		7	11.9
perturbator			4	2	6	12.3
nezumi	L ...	H				

TABLE 67. Snout length in the species of *Acromycter*. L and H are the lowest and highest values for *A. nezumi* given by Asano (1962).

	Snout in % HL												
Species	23	24	25	26	27	28	29	30	31	32	33	34	n	x̄
alcocki				2	2	3	6	2	2	1			18	28.7
atlanticus					1	1	–	1	1	–	2	1	7	30.8
perturbator						1	1	3	–	–	1		6	30.0
nezumi	L	. . .	. . .	. . .	. . .	. . .	H							

patches. Intermaxillary teeth in several rows, confluent with maxillary teeth. Vomerine tooth patch broad, about two-thirds the length of maxillary tooth patch, wider posteriorly. Maxillary teeth in moderately broad bands, wider anteriorly. Mandibular teeth in broad bands, widest at about anterior third of length.

Lateral ethmoid process absent. Supraoccipital present. Preorbital bone with projections from lower edge; three or four suborbitals. Posterior edge of opercle serrate. Myorhabdoi absent. Ribs absent. Anterior few epineurals fused to neural arch; neural arches of about first 25 vertebrae articulated with centrum; no epicentral processes on caudal vertebrae. Lateral-line ossicles as in *Rhynchoconger*. Dorsal- and anal-fin rays segmented.

Stomach short, reaching about a fourth of the way to anus. Gas bladder long, extending from anterior end of stomach to a point slightly behind anus. Stomach black, intestine pale. Abdomen and inside of mouth and gill area black.

Size. To about 400 mm TL.

Distribution. Japan, Hawaii, and the western tropical Atlantic.

Etymology. From the Greek *akron* (summit, top, peak) and *mykter* (nose), referring to the position of the posterior nostril on top of the head. Masculine.

Species. Four species are recognized: *Acromycter nezumi* (Asano, 1958) from Japan, *A. alcocki* (Gilbert and Cramer, 1897) from Hawaii, and *A. perturbator* (Parr, 1932) and *A. atlanticus* n. sp., both from the western North Atlantic. Few characters separate them, and most of these overlap (Tables 65–70).

The two western Atlantic species can be completely separated by the number of vertebrae (Table 65), and they inhabit different depth ranges. They differ slightly in head length and preanal lateral-line pores (Tables 66, 69), but the overlap in these characters reduces their usefulness.

Acromycter alcocki is nearly identical in vertebral count to *A. atlanticus* (Table 65), but it has a distinctly longer pectoral fin (Table 70). The two species also differ slightly in snout length (Table 67), but the overlap is great. *A. alcocki* differs markedly from *A. perturbator* in vertebral count (Table 65). It differs from *A. nezumi* in

TABLE 68. Eye size in the species of *Acromycter*. L and H are the lowest and highest values given for *A. nezumi* by Asano (1962).

	Eye in % HL								
Species	6	7	8	9	10	11	12	n	x̄
alcocki		3	7	7	1			18	7.3
atlanticus		3	1	3				7	7.0
perturbator	1	1	2	1	1			6	7.0
nezumi				L	. . .	. . .	H		

TABLE 69. Number of preanal lateral-line pores in the species of *Acromycter*. L and H are the lowest and highest values for *A. nezumi* given by Asano (1967).

	LL pores								
Species	34	35	36	37	38	39	40	n	x̄
alcocki		1	3	6	1	1		12	36.8
atlanticus		1	–	1	–	1		3	37.0
perturbator	1	2	1	1				5	35.4
nezumi			L	. . .	. . .	. . .	H		

TABLE 70. Pectoral-fin length in species of *Acromycter*. L and H are the lowest and highest values for *A. nezumi* given by Asano (1962). A, *alcocki*. B, *atlanticus*. C, *perturbator*. D, *nezumi*.

	Pectoral fin in % HL																					
Species	11	12	13	14	15	16	17	18	19	20	21	22	23	24	25	26	27	28	29	30	31	n	x̄
A															1	1	3	1	1	3		10	27.9
B	1	-	-	-	1	-	-	1	-	1	-	-	-	-	1							5	17.8
C								2	-	-	1	-	-	1	-	1						5	21.4
D										L	....										H		

several characters, but only the head length (Table 66) seems distinct enough to be useful.

The only character that reliably separates *A. nezumi* from the other species is the head length (Table 66), and even this overlaps slightly. Snout length and eye diameter also differ somewhat (Tables 67, 68), but the overlap is considerable. Based on the specimens listed by Asano (1962: 104), *A. nezumi* grows larger (to 399 mm TL) than any of the other species. The largest specimens of *A. alcocki*, *A. atlanticus*, and *A. perturbator* examined here were 254, 295, and 277 mm TL, respectively.

KEY TO THE SPECIES OF *ACROMYCTER* IN THE WESTERN NORTH ATLANTIC

1a. Total vertebrae 159–164 .... *perturbator*, p. 556
1b. Total vertebrae 167–171 .... *atlanticus*, p. 557

Acromycter perturbator (Parr, 1932)
Figures 586, 587, Tables 65–70

Ariosoma perturbator Parr, 1932: 31 (original description, 23°54′20″N, 77°09′W, 1299–1318 m depth, holotype YPM 2738).
Promyllantor perturbator. Reid, 1934: 3.
New Genus *perturbator*. Smith, 1971: 81.
Acromycter perturbator. Smith and Kanazawa, 1977: 541, fig. 12.

Distinctive Characters. *Acromycter perturbator* is completely separable from the other western Atlantic species, *A. atlanticus*, by its vertebral count (159–164 vs 167–171). Several other characters show significant differences in range and mean value between the two species (Tables 65–70), but the overlap is too great to allow them to be used on individual specimens. Most individuals can probably be identified by the depth of capture; *A. perturbator* has not been recorded in less than 787 m, whereas all the specimens of *A. atlanticus* examined have been collected in less than 724 m.

Description (including data from Parr, 1932). Total vertebrae 159–164 (n = 7), predorsal vertebrae 15 (2), preanal vertebrae 36–38 (4), precaudal vertebrae 48 (1). Dorsal-fin rays ca 230 (1), anal-fin rays ca 190–200 (1), pectoral-fin rays ca 10–11 (3). Branchiostegal rays 8–9 (2). Pores: LL to anus ca 34–37 (6), POM 8 (4), IO 4 (6) + 1 (6) + 0–1 (6), SO 1 + 2 (6), STC 0 (5).

Proportions as % of TL: preanal 34–37 (5), predorsal 16–19 (4), head 13–14 (6), depth at anus 3–5 (5), pectoral fin 2–3 (4). Of HL: snout 27–33 (6), eye 6–10 (6), snout-rictus 26–29 (5), tip of snout to tip of lower jaw 11–13 (5), gill opening 6–11 (6), interbranchial 15–24 (6), pectoral fin 16–26 (5).

Color. In preservative brown or grayish-brown. Inside of mouth and gill area black. Stomach black, intestine pale.

Size. The largest specimen examined was 295 mm TL (MCZ 33212).

Development and Growth. Most of the leptocephali of *Acromycter* examined were *A. atlanticus*, but at least one specimen with 163 myomeres appeared to be *A. perturbator* (see p. 759).

Distribution (Fig. 587). Known from the Atlantic coast off New Jersey, the Straits of Florida, the Bahamas, and Jamaica. Depth 787–1318 m (mean 985 m). *Acromycter perturbator* lives at a greater

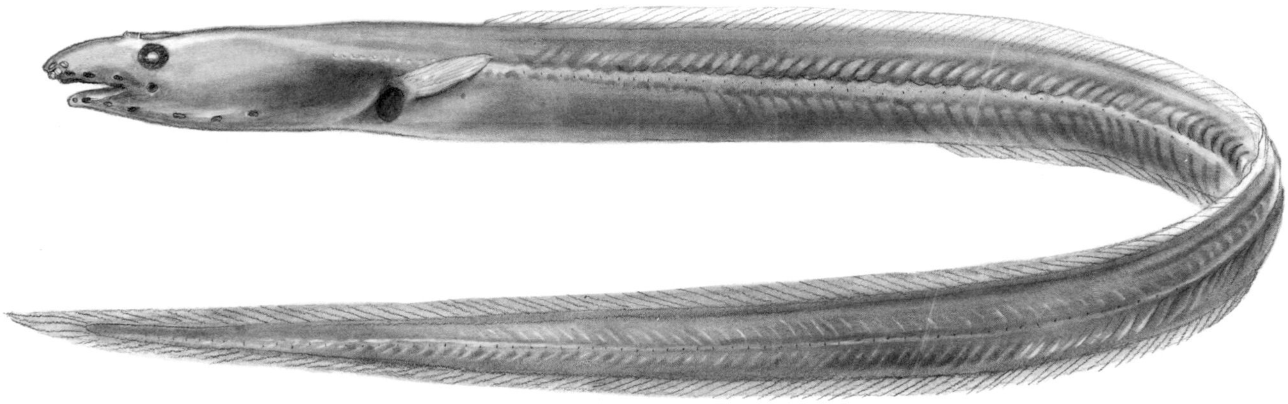

FIGURE 586. *Acromycter perturbator*: ANSP 126976, 277 mm TL.

average depth than any other western Atlantic congrid.

Etymology. From the Latin *perturbo* (confuse, disturb, upset) and *-tor* (action or agency), referring to the trouble it caused Parr in trying to classify it. A noun in apposition.

Study Material. Seven specimens, 207–295 mm TL (holotype not seen).

NEW JERSEY: USNM 206213 (1, 269), 39°37′12″N, 71°54′00″W, 915 m, 15 Aug. 1969. FLORIDA STRAITS: ANSP 126976 (1, 277), 24°23′N, 80°28′W, 787 m, 22 Jan. 1964. UMML 13901 (2, 234–240), 24°28′N, 80°16′W, 805 m, 23 Jan. 1964. UMML 15542 (1, 207), 23°54′N, 81°27′W, 1153–1190 m, 17 Sept. 1964. UMML 24108 (1, 250), 24°28′N, 80°29′W, 920 m, 5 March 1968. JAMAICA: MCZ 33212 (1, 295), 17°28′30″N, 77°30′W, 1116 m, 1880.

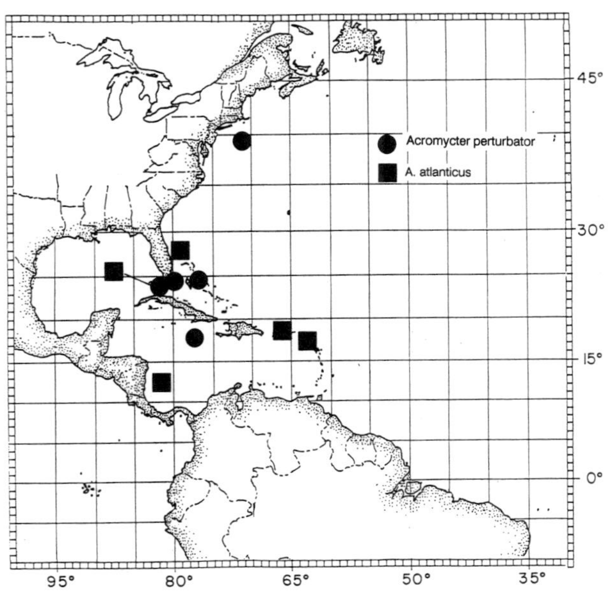

FIGURE 587. Distribution of *Acromycter perturbator* and *A. atlanticus*.

Acromycter atlanticus new species
Figures 587, 588, Tables 65–70

Acromycter alcocki (not of Gilbert and Cramer, 1897). Smith and Leiby, 1980: 392, fig. 2 (leptocephalus).

Distinctive Characters. *Acromycter atlanticus* is distinguished from *A. perturbator* by the number of vertebrae (167–171 vs 159–164). It also differs slightly in head length (12–13% TL vs 13–14% TL) and preanal lateral-line pores (35–39 vs 34–37), but these characters broadly overlap. *Acromycter atlanticus* lives in shallower water (503–724 m) than does *A. perturbator* (787–1318 m).

Acromycter atlanticus has a shorter pectoral fin than *A. alcocki* (11–25% HL vs 25–30% HL), and it has a slightly greater snout length (27–34% HL vs 26–32% HL).

Acromycter atlanticus differs from *A. nezumi* in its slightly greater head length (12–13% TL vs 11–12% TL), greater snout length (27–34% HL vs 23–29% HL), and smaller eye (7–9% HL vs 9–12% HL).

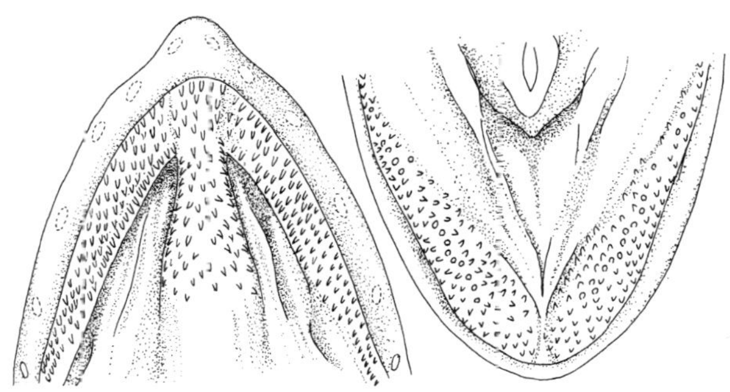

FIGURE 588. *Acromycter atlanticus*: ANSP 157602, paratype, 216 mm TL; dentition.

Description. Total vertebrae 167–171 (n = 7), predorsal vertebrae 12 (1), preanal vertebrae 38 (1). Branchiostegal rays 8–9 (2). Pores: LL to anus 35–39 (3), POM 8 (4), IO 4 (4) + 1 (4) + 0 (4), SO 1 + 2 (4), STC 0 (4).

Proportions as % of TL: preanal 32–35 (7), predorsal 16–17 (7), head 12–13 (7), depth at anus 3–4 (7), pectoral fin 1–3 (5). Of HL: snout 27–34 (7), eye 7–9 (7), snout-rictus 26–32 (3), tip of snout to tip of lower jaw 8–15 (5), gill opening 5–10 (6), interbranchial 15–24 (5), pectoral fin 11–25 (5).

Color. In preservative brown or grayish-brown. Inside of mouth and gill area black. Stomach black, intestine pale.

Size. The largest specimen examined was 295 mm TL (USNM 197131).

Development and Growth. The leptocephalus has been identified (see p. 759).

Distribution (Fig. 587). Known from off the central east coast of Florida, the southern Florida Straits, Puerto Rico, the Lesser Antilles just west of the island of Anguilla, and the western Caribbean off Nicaragua. Depth 503–724 m (mean 616 m).

The northern part of its range overlaps that of *A. perturbator*, but the two species inhabit different depths.

Etymology. Named for the Atlantic Ocean. An adjective.

Study Material. Seven specimens, 163–295 mm TL.

Holotype: USNM 158882 (1, 285), Oregon 1922, Nicaragua, 13°34′N, 81°53′W, 503 m, 13 Sept. 1957.

Paratypes: FLORIDA ATLANTIC: ANSP 157602 (1, 216), 28°03′N, 79°06′W, 715–724 m, 17–18 July 1965. FLORIDA STRAITS: USNM 197131 (1, 295), 24°13′N, 81°24′W, 595 m, 28 Oct. 1960. PUERTO RICO: USNM 93460 (2, 200–227), 18°32′30″N, 65°18′30″W, 641 m, 26 Feb. 1933. LESSER ANTILLES: ANSP 114855 (1, 163), 18°18′N, 63°23′W, 688 m, 6 Dec. 1969. NICARAGUA: USNM 158877 (1, 260), 13°22′N, 82°04′W, 549 m, 13 Sept. 1957.

Comparative material (*Acromycter alcocki*): HAWAII: USNM 47724 (holotype, 254), 21°12′N, 157°49′W, 4 Dec. 1891. USNM 195915 (3 paratypes, too brittle to measure), same data as holotype. ANSP 143063 (13, 151–226) and NMNZ 16527 (1, 225), 21°01.1′N, 156°27.1′W, 187 m, 25 Oct. 1972. USNM 126021 (5, 188–240).

Genus *Xenomystax* Gilbert, 1891

Xenomystax Gilbert, 1891b: 348 (type species *Xenomystax atrarius* Gilbert, 1891, by original designation).

Paraxenomystax Reid, 1940: 1 (type species *Paraxenomystax bidentatus* Reid, 1940, by original designation).

Characters. Body moderately elongate, preanal length about 30–40% TL; tail attenuate. Dorsal fin begins over, slightly before, or slightly behind base of pectoral fin. Pectoral fin well developed.

Mouth large, toothed portion of upper jaw extending beyond posterior margin of eye. Snout projects beyond lower jaw. Fleshy part of snout does not extend anteriorly beyond intermaxillary teeth. Lips reduced, teeth exposed when mouth closed. Upper and lower labial flange absent. Anterior nostril tubular, located just behind tooth-bearing expansion of ethmoid, directed anterolaterally. Posterior nostril elliptical, located at mid-eye level a variable distance in front of eye. Head pores numerous, including five supraorbital, three postorbital, and three supratemporal. Anterior infraorbital and supraorbital pores long and slit-like. Adnasal pore absent.

Teeth slender, conical, recurved. Intermaxillary teeth in several rows, separated from maxillary and vomerine teeth by a broad, transverse groove into which tip of mandible fits when mouth closed; intermaxillary teeth completely exposed when mouth closed. Vomer with a few enlarged median teeth anteriorly flanked by smaller ones, the latter extending posteriorly in a double or single row for a variable distance, sometimes reaching anterior margin of eye. Maxillary and mandibular teeth in bands extending out laterally well beyond confines of closed mouth, exterior teeth directed outward; inner row of teeth on both jaws separated from remainder by a distinct edentulous groove, wider on maxilla than on mandible; inner row of maxillary teeth converging anteriorly with vomerine teeth.

Lateral ethmoid process absent; supraoccipital present or absent. Pterygoid slender. Preorbital bone elongate, poorly ossified, lower edge without projections. Posterior margin of opercle smooth. Lower pharyngeals divided, not fused to ceratobranchial. Myorhabdoi absent. Ribs absent. Caudal vertebrae without epicentral processes. Lateral-line ossicles like those of *Rhynchoconger*. Dorsal- and anal-fin rays segmented.

Stomach variable in length, terminating before, behind, or over anus; stomach and intestine black or pale. Gas bladder long, variable in length, terminating over or behind anus.

Discussion. Peden (1972: 7, fig. 5) described and illustrated a structure in *Xenomystax atrarius* that he felt ". . . must be equivalent to the lateral ethmoid process described by Asano (1962), because a ligament connects this structure to the palatopterygoid." The true lateral ethmoid process, however, is not connected to the pterygoid. The structure illustrated by Peden appears to be the lateral process on the vomer that is typically present in congrids and to which the pterygoid is connected through a ligament.

Reid (1940) created the genus *Paraxenomystax* to contain his new species *P. bidentatus* and *Xenomystax trucidans* Alcock. He distinguished it from *Xenomystax* Gilbert on the basis of two characters that "are considered of generic rank:" the more posterior position of the posterior nostril and the reduced number of branchiostegal rays. Neither character seems to hold up when all the presently known species are compared. The posterior nostril in *X. atrarius* Gilbert is displaced considerably forward, approximately midway between the eye and the anterior nostril or more than one eye diameter from the anterior edge of the eye. In *X. bidentatus* (Reid) and *X. congroides* n. sp. it is about one nostril diameter or less than one-half eye diameter in front of the eye. *Xenomystax austrinus* n. sp. approaches an intermediate condition; its posterior nostril is about two nostril diameters or slightly less than one eye diameter from the eye. Although *X. atrarius* is extreme in its forward displacement of the posterior nostril, the character does vary among the species and hardly seems important enough to base a genus upon.

Xenomystax bidentatus has 7–8 branchiostegal rays, *X. congroides* 7–9, *X. austrinus* 9–10, and *X. atrarius* 11–12 (Peden, 1972). This character obviously forms a graded series among the species and cannot be used to delimit genera.

Xenomystax trucidans from the Indian Ocean is a problem; few data are available for the holotype, and additional specimens have not been reported. Reid included it in his *Paraxenomystax*, presumably because it had the posterior nostril near the eye and a low number of branchiostegal rays. I have examined a single specimen of *Xenomystax* from the Indian Ocean (ANSP 112029) which, as far as can be determined, agrees with *X. trucidans*. This specimen, however, has 12 branchiostegals, the same number as *X. atrarius*. The posterior nostril is located slightly less than one eye diameter in front of the eye.

It is clear that the species treated here do not fall naturally into two groups, and I see no purpose in retaining *Paraxenomystax*. By placing all the species in *Xenomystax*, the natural affinity and distinctiveness of the group is emphasized.

Size. To about 1 m TL.

Distribution. Tropical eastern and western Atlantic; Indian Ocean; eastern Pacific from British Columbia to Chile. Mainly tropical, but extending into temperate waters in the eastern Pacific.

Etymology. From the Greek *xeno-* (strange) and *mystax* (upper lip, mustache), probably referring to the exposed teeth along the upper jaw. Masculine.

Species. Five species of *Xenomystax* are recognized: *X. atrarius* Gilbert, 1891 from the eastern Pacific (synonym *X. rictus* Garman, 1899); *X. trucidans* Alcock, 1894 from the Indian Ocean; *X. bidentatus* (Reid, 1940) and *X. austrinus* n. sp., both from the western Atlantic; and *X. congroides* n. sp., from the western and eastern Atlantic. *Gavialiceps taeniola* Alcock from the Indo-West Pacific resembles the species of *Xenomystax* in many characters, including the separate inner row of maxillary and mandibular teeth. It differs mainly in its more slender head and body and its greatly reduced or absent pectoral fin.

FIGURE 589. *Xenomystax austrinus*: ANSP 109591, paratype, 552 mm TL.

KEY TO THE SPECIES OF *XENOMYSTAX* IN THE WESTERN NORTH ATLANTIC

1a. Posterior nostril between one-half and one eye diameter or two nostril diameters in front of anterior margin of eye . . . . . *austrinus*, p. 560
1b. Posterior nostril less than one-half eye diameter or one nostril diameter in front of anterior margin of eye 2
2a. Intermaxillary tooth patch longer than broad; dorsal fin begins over or slightly behind base of pectoral fin; stomach and intestine pale; vertebrae 175–180 . . . . . *bidentatus*, p. 562
2b. Intermaxillary tooth patch as broad as long; dorsal fin begins before base of pectoral fin; stomach and intestine black; vertebrae 189–219 . . . . . *congroides*, p. 563

Xenomystax austrinus
D. G. Smith and R. H. Kanazawa, new species
Figures 589–592

Distinctive Characters. *Xenomystax austrinus* has the posterior nostril farther from the anterior edge of the eye than do the other western Atlantic species. It is further distinguished from *X. bidentatus* by having more vertebrae (192–196 vs 174–180), more preanal lateral-line pores (37–40 vs 31–34), and more branchiostegal rays (9–10 vs 7–8). It differs from *X. congroides* in having a more acute head and a pale digestive tract. From the eastern Pacific *X. atrarius* it differs in having fewer branchiostegal rays (9–10 vs 11–12), more vertebrae (192–196 vs 171–177; Peden, 1972: 3), and the posterior nostril closer to the eye.

Description. Total vertebrae 192–196 (n = 6), predorsal vertebrae 7–8 (4), preanal vertebrae 40–42 (4). Branchiostegal rays 9–10 (4). Pectoral-fin rays 12–13 (8). Pores: LL to anus 37–40 (14), POM 12–14 (6), IO 5 (7) + 0 (7) + 2–3 (7), SO 1 + 5 (6), STC 3 (12).

Proportions as % TL: preanal 35–39 (6), predorsal 15–18 (6), head 16–18 (6). As % preanal: predorsal 42–46 (13), head 44–48 (15), depth at vent 8–13 (5), pectoral fin 6–10 (9). As % HL: snout 34–39 (15), eye 9–12 (15), posterior nostril to eye 7–10 (14), upper jaw 52–56 (15), gill opening 6–11 (8), interbranchial 8–15 (8), pectoral fin 14–22 (9).

Dorsal fin begins over or slightly before base of pectoral fin. Posterior nostril about two nostril diameters or between one-half and one eye diameter from anterior edge of eye.

Intermaxillary tooth patch triangular to oval, longer than broad. Vomer with a few enlarged median teeth anteriorly flanked by smaller ones continuing posteriorly as two or three irregular rows not quite reaching anterior edge of eye.

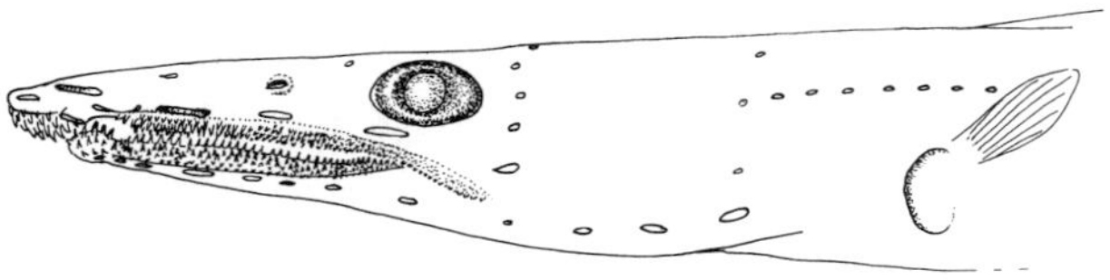

FIGURE 590. *Xenomystax austrinus*: ANSP 109591, paratype, 552 mm TL; head.

Maxillary and mandibular teeth in bands with inner teeth larger; innermost row of teeth separated from outer bands by an edentulous groove.

Color. In preservative brown, somewhat darker dorsally than ventrally. Dorsal and anal fins black-edged posteriorly. Stomach, intestine, and gas bladder pale.

Size. The largest specimen examined was 1000 mm TL, 390 mm preanal (USNM 198771). Based on the material examined, this is the largest of the Atlantic species of *Xenomystax*.

Development and Growth. The leptocephalus has not been identified.

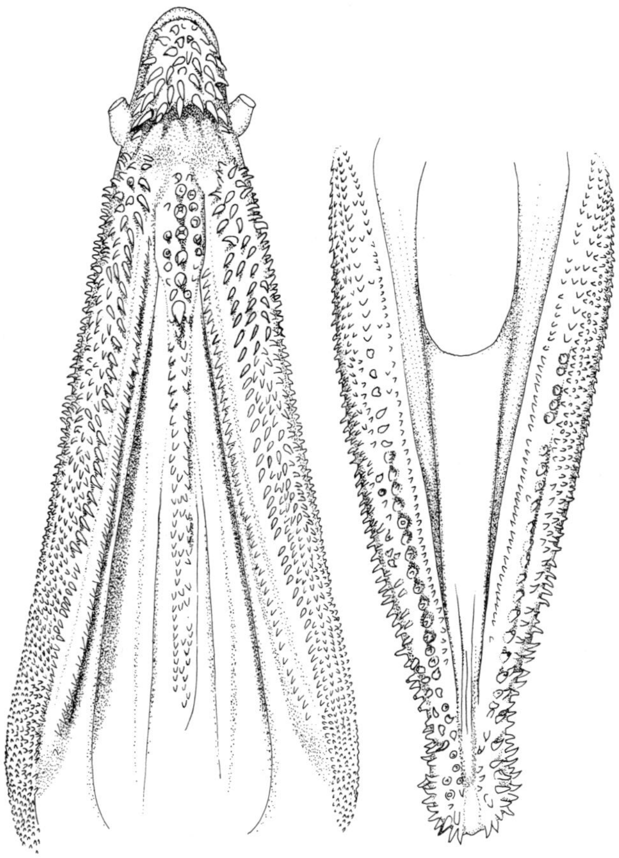

FIGURE 591. *Xenomystax austrinus*: ANSP 126444, paratype, 890 mm TL; dentition.

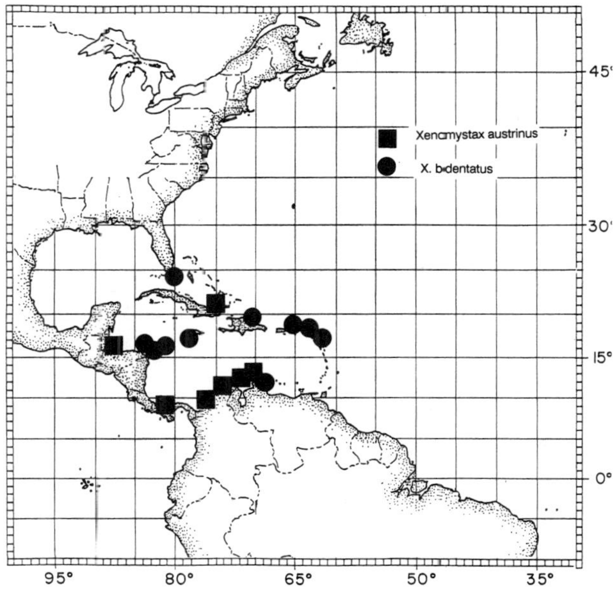

FIGURE 592. Distribution of *Xenomystax austrinus* and *X. bidentatus*.

Variation. The holotype had two postorbital pores on one side and three on the other. All the other specimens had three postorbital pores on both sides.

Distribution (Fig. 592). All but two specimens came from the Caribbean coast of Venezuela, Colombia, and Panama. The other records are from the Gulf of Honduras (USNM 200319) and the northeastern coast of Cuba (MCZ 39419). Depth 458–824 m (mean 602 m).

Etymology. Latin *austrinus* (southern), referring to the predominant distribution in the southern Caribbean Sea. An adjective.

Study Material. 15 specimens, 315–1000 mm TL.

Holotype: USNM 198746 (male, 802 mm TL), Oregon 4902, Colombia, 09°02.4′N, 76°31.5′W, 732 m, 28 May 1964.

Paratypes: CUBA: MCZ 39419 (1, 900), 20°47′N, 75°11′W, 824 m, 20 April 1939. VENEZUELA: ANSP 133011 (1, 723), 12°52′N, 70°43′W, 512 m, 23 Nov. 1970. COLOMBIA: ANSP 109591 (1, 552), 11°10.8′N, 74°28.5′W, 18 May 1964. ANSP 114856 (1, 315), 11°40′N, 73°15′W, 662 m, 4 Dec. 1968. ANSP 114877 (1, 430), 11°26′N, 74°14′W, 576 m, 3 Dec. 1968. ANSP 126444 (1, 890), 08°56′N, 76°53′W, 622 m, 2 Nov. 1970. ANSP 131993 (1, 760), 11°12′N, 74°24′W, 567 m, 16 Nov. 1970. USNM 198680 (1, 742), 10°16.2′N, 75°54.4′W, 549 m, 25 May 1964. USNM 198771 (1, 1000), 12°28′N, 75°26′W, 549 m, 2 June 1964. USNM 201239 (3, 487–765), 12°29′N, 72°19′W, 485 m, 9 Oct. 1965. PANAMA: USNM 193559 (1, 326), 09°03′N, 81°18′W, 549 m, 31

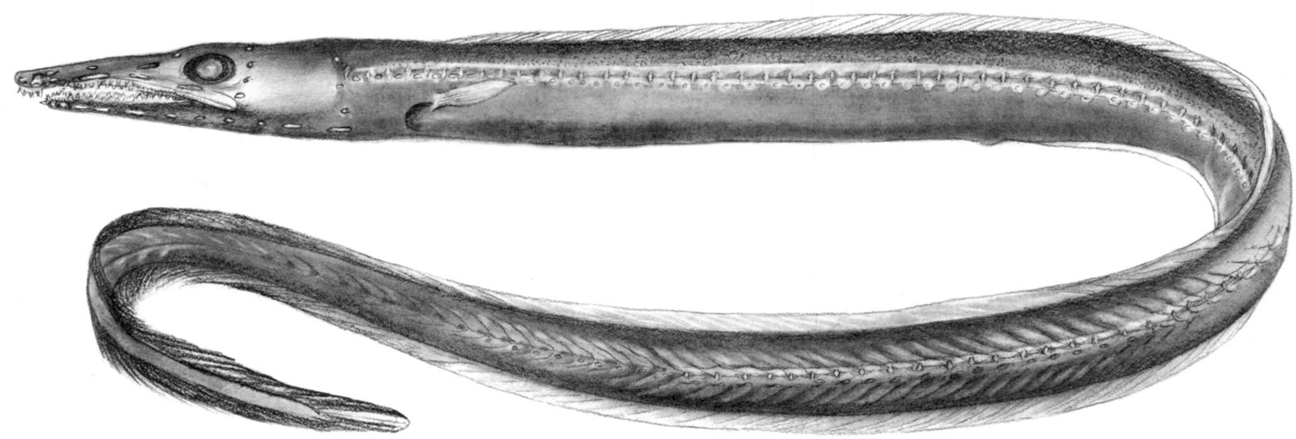

FIGURE 593. *Xenomystax bidentatus*: ANSP 114163, 445 mm TL.

May 1962. BELIZE: USNM 200319 (1, 463), 16°58′N, 87°53′W, 458–732 m, May 1962.

Xenomystax bidentatus (Reid, 1940)
Figures 592–595

Paraxenomystax bidentatus Reid, 1940: 2, fig. 1 (original description, 18°36′N, 65°05.5′W, holotype USNM 108444).

Misidentifications:

Not *Paraxenomystax bidentatus* Reid. Blache, 1968a: 709; 1977: 159. Castle, 1970a: 13. Karmovskaya, 1975: 95. Keller, 1976: 133 (all *Xenomystax congroides* n. sp.).

Distinctive Characters. *Xenomystax bidentatus* has fewer vertebrae (174–180) than the other species found in the western Atlantic (more than 190). It further differs from *X. congroides* in the pale rather than black viscera, the more posterior dorsal-fin origin, the more acute head and snout, and in the dentition. It differs from *X. austrinus* in having the posterior nostril closer to the eye, fewer preanal lateral-line pores (31–34 vs 37–40), and fewer branchiostegal rays (7–8 vs 9–10). It differs from the eastern Pacific *X. atrarius* in having the posterior nostril much closer to the eye and in having fewer branchiostegals (7–8 vs 11–12).

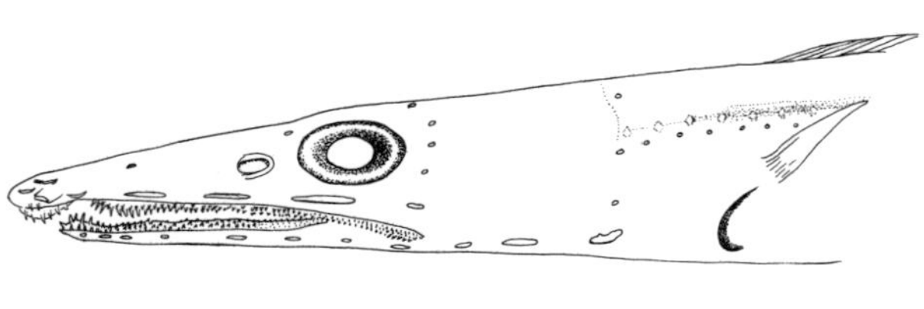

FIGURE 594. *Xenomystax bidentatus*: ANSP 114163, 445 mm TL; head.

Description. Total vertebrae 174–180 (n = 9), predorsal vertebrae 6–7 (5), preanal vertebrae 32–35 (11), precaudal vertebrae 46 (1). Branchiostegal rays 7–8 (6). Pectoral-fin rays 9–11 (17). Pores: LL to anus 31–34 (12), POM 12 (13), IO 5 (8) + 0 (8) +3 (8), SO 1 + 5 (11), STC 3 (18).

Proportions as % of TL: preanal 34–36 (9), predorsal 15 (9), head 14–15 (9). Of preanal length: predorsal 41–45 (24), head 41–44 (24), depth at anus 8–12 (10), pectoral fin 5–8 (19). Of HL: snout 33–40 (24), eye 10–16 (23), posterior nostril to eye 3–6 (15), upper jaw 53–58 (17), gill opening 6–11 (15), interbranchial 8–12 (14), pectoral fin 12–19 (18).

Dorsal fin begins over or slightly behind base of pectoral fin. Posterior nostril about one nostril diameter or less than one-half eye diameter from anterior edge of eye.

Intermaxillary tooth patch triangular, somewhat longer than broad. Vomer with a median row of about 10–14 enlarged teeth anteriorly flanked by smaller teeth which continue posteriorly in an irregular double row to about level of posterior nostril and from there in a single row to level of anterior part of eye. Outer band of maxillary teeth in 3–5 irregular rows, teeth of inner two of these rows enlarged somewhat anteriorly; innermost row of teeth separated from remainder by an edentulous groove, converging with outer rows near posterior end of maxilla. Outer band of mandibular teeth in about five irregular rows, inner teeth largest; innermost row of teeth separated from remainder by a narrow edentulous groove, converging with outer rows near posterior end of mandibular tooth patch.

Supraoccipital present.

Color. In preservative brown, somewhat darker dorsally than ventrally. Margins of dorsal and anal fins black posteriorly. Stomach, intestine, and gas bladder pale.

Size. The largest specimen examined was 525 mm TL, 184 mm preanal (UMML 2769).

Development and Growth. The leptocephalus has been identified (see p. 761).

Distribution (Fig. 592). Known from the Caribbean Sea and the southern Straits of Florida. Depth 384–860 m (mean 617 m).

Etymology. From the Latin *bi* (two) and *dentatus* (toothed), in reference to the arrangement of teeth on the maxilla and mandible, with the inner row separated from the outer rows by an edentulous groove. An adjective.

Study Material. 31 specimens, 244–525 mm TL.

Holotype: USNM 108444 (female, 475 mm TL), Caroline 96, Virgin Islands, 18°36′00″N, 65°05′30″W, 494–604 m, 3 March 1933.

Paratype: PUERTO RICO: USNM 108455 (1, 374).

Other material: FLORIDA STRAITS: USNM 197116 (1, 270). HISPANIOLA: ANSP 114163 (1, 445). JAMAICA: USNM 193603 (1, 488). LESSER ANTILLES: ANSP 114689 (4, 382–464), 117012 (1, 432), 117024 (1, 425), 117096 (3, 244–401), 126519 (1, 272), 138775 (1, 489). UMML 29673 (2, 385–425). USNM 235347 (4, 353–473), 268824 (3, 402–415). VENEZUELA: UMML 29674 (1, 419). HONDURAS: UMML 2769 (1, 525). USNM 158931 (1, 476), 193592 (3, 481–505).

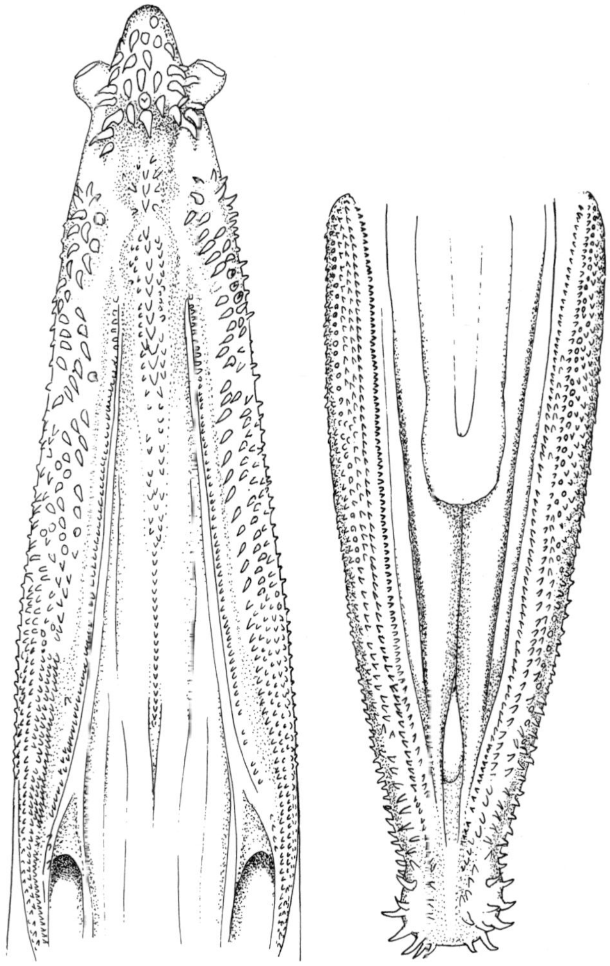

FIGURE 595. *Xenomystax bidentatus*: ANSP 138775, 489 mm TL; dentition.

Xenomystax congroides
D. G. Smith and R. H. Kanazawa, new species
Figures 596–599, Tables 71–74

Paraxenomystax bidentatus (not of Reid, 1934). Blache, 1968a: 709, figs. 14–24 (leptocephali and adults, Gulf of Guinea); 1977: 159, figs. 52b, 53–54 (leptocephali, Gulf of Guinea). Castle, 1970a (leptocephali, equatorial Atlantic). Karmovskaya, 1975: 95 (leptocephali, Gulf of Mexico and Caribbean). Keller, 1976: 133 (leptocephali, Bermuda).

Paraxenomystax sp. Fahay, 1976: 210 (leptocephalus, off New Jersey). Smith, 1979: 31, fig. 47 (leptocephalus).

Distinctive Characters. Xenomystax congroides differs from *X. bidentatus*, *X. austrinus*, and *X. atrarius* by the black digestive tract, the more anterior dorsal-fin origin, and the less acute head and snout. It further differs from *X. bidentatus* and *X. austrinus* in the shorter and broader intermaxillary tooth patch.

Description. Total vertebrae 189–219 (n = 42), predorsal vertebrae 4–6 (20), preanal vertebrae 32–40 (22), precaudal vertebrae 50–60 (10). Branchiostegal rays 7–10 (61). Pectoral-fin rays 10–15 (94). Pores: LL to anus 28–40 (196), POM 13–16

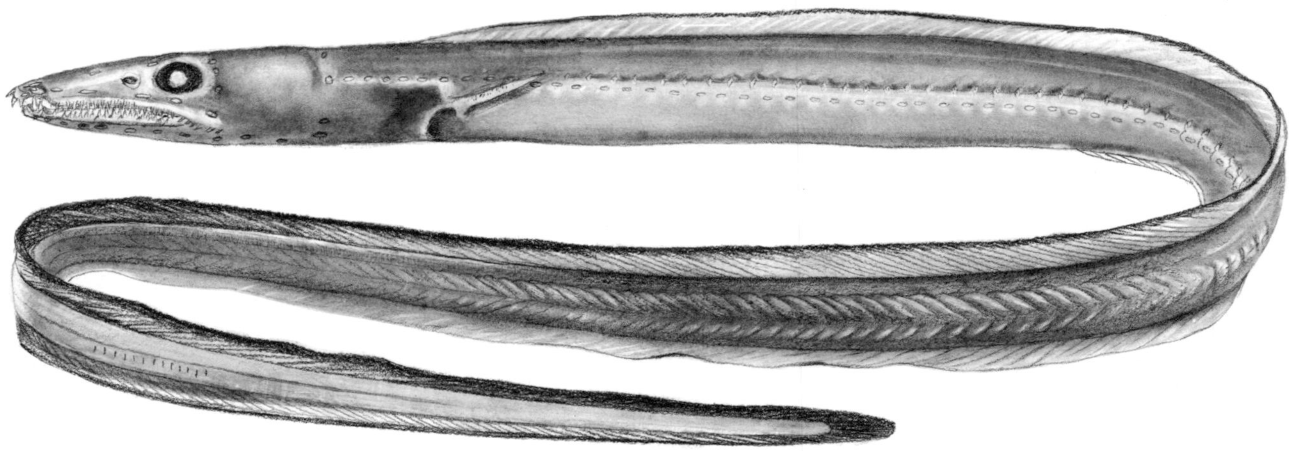

FIGURE 596. *Xenomystax congroides*: ANSP 114951, 475 mm TL.

(41), IO 5 (44) + 0 (44) + 3 (44), SO 1 + 5 (35), STC 3 (44).

Proportions as % of TL: preanal 28–34 (30). Of preanal length: predorsal 30–38 (146), head 36–48 (217), depth at anus 6–13 (45), pectoral fin 8–11 (46). Of HL: snout 30–37 (45), eye 10–14 (45), upper jaw 47–62 (45), gill opening 6–14 (39), interbranchial 5–13 (35), pectoral fin 18–28 (46).

Dorsal fin begins before pectoral-fin base. Head somewhat blunter in profile than in other species of genus. Posterior nostril about one nostril diameter or less than one-half eye diameter from anterior edge of eye.

Intermaxillary tooth patch short and rounded, as broad as long. Vomer with a short row of enlarged median teeth anteriorly, flanked by smaller ones, which continue posteriorly as a distinct double row, converging and terminating at about level of rictus. Outer band of maxillary teeth in three series, teeth of inner two series enlarged; teeth of innermost row small, separated from three outer rows by a wide edentulous groove, converging with outer vomerine teeth anteriorly and with outer band of maxillary teeth posteriorly. Mandible with an outer band of two or three series of teeth and an inner row of small teeth, separated from outer teeth by a narrow groove.

Color. In preservative brown or yellowish. Lin-

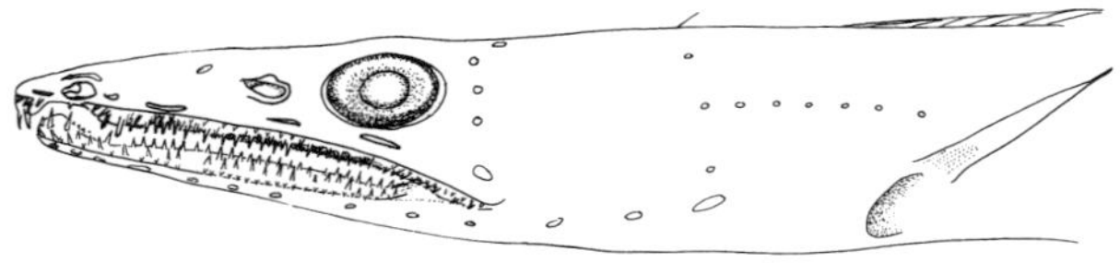

FIGURE 597. *Xenomystax congroides*: ANSP 114951, 475 mm TL; head.

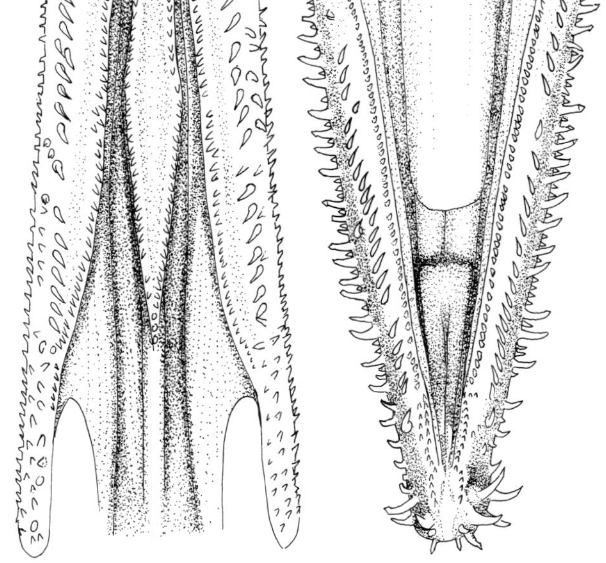

FIGURE 598. *Xenomystax congroides*: ANSP 114951, 475 mm TL; dentition.

TABLE 71. Preanal lateral-line pores in specimens of *Xenomystax congroides* from four areas (each count represents one side of one specimen).

Area	LL pores													Counts	x̄	Speci-mens
	28	29	30	31	32	33	34	35	36	37	38	39	40			
Gulf of Mexico	1	1	3	6	8	2	1	–	–	1				23	31.6	22
Caribbean			1	12	40	55	28	6						142	32.8	137
South America								2	8	13	8	4	2	37	37.3	30
Gulf of Guinea									1	4	4	2		11	37.6	8

ing of mouth and branchial chamber dark. Vertical fins dark-edged posteriorly. Stomach and intestine black, gas bladder silvery.

Size. The largest specimen examined was 876 mm TL, 317 mm preanal (USNM 195902).

Development and Growth. The leptocephalus has been identified (see p. 759).

Variation. Four more or less distinct geographic populations of this species exist. The first is restricted to the Gulf of Mexico; the second inhabits the Caribbean, the Straits of Florida, and the Bahamas; the third is found along the coast of South America from eastern Venezuela to Brazil; and the fourth occurs in the Gulf of Guinea. They are distinguished by four characters, but the characters are distributed in different ways. The number of preanal lateral-line (LL) pores divides the species into two groups: one in the Caribbean and Gulf of Mexico, and the other in South America and West Africa (Table 71). Head length defines the same two groups (Table 72), though with greater overlap. The number of vertebrae, however, shows a different pattern (Table 73). Here the Gulf of Mexico form stands well apart from the others; a gap of six vertebrae exists between the upper limit of the Gulf of Mexico population (195) and the lower limit of the Caribbean population (202). At the upper end of its range, the vertebral count of the Caribbean form overlaps those of the South American and West African forms, which again are nearly identical to each other. The number of branchiostegal rays varies in a stepwise manner among all four populations (Table 74). This is the only character that separates the South American and West African populations.

Except in the case of West Africa, the geographic boundaries between the populations are somewhat uncertain. Five specimens collected about midway between Grenada and Margarita Island (USNM 195902, 196176, 200324) had LL-pore counts typical of the South American-West African form (35–40). Three specimens from the coast of Venezuela just east and west of Caracas (UMML 29675, USNM 217442), however, had LL-pore and vertebral counts typical of the Caribbean form (32 and 202–209, respectively). A specimen from off the northern coast of Cuba (MCZ 38820) had a vertebral count typical of the Caribbean form (ca 207), as did two others from the Bahamian side of the Straits of Florida (UMML 13871 and ANSP 156859, ca 207–208). There are no other records between there and the northern Gulf of Mexico, where the specimens all have

TABLE 72. HL in specimens of *Xenomystax congroides* from four areas.

Area	HL in % preanal												n	x̄
	36	37	38	39	40	41	42	43	44	45	46	47		
Gulf of Mexico		1	2	2	3	7	8	3					25	41.0
Caribbean	3	4	14	24	25	37	24	13	4	2	–	1	151	40.5
South America					2	1	8	2	6	7	2	1	29	43.5
Gulf of Guinea						4	3	1	3	1			12	42.5

TABLE 73. Total vertebrae in specimens of *Xenomystax congroides* from four areas (includes data from Blache, 1968a).

Area	Vertebrae 189	190	191	192	193	194	195	...	202	203	204	205	206	207	208	209	210	211
Gulf of Mexico	2	-	2	1	3	1	2	...										
Caribbean								...	1	-	1	1	2	4	4	2	1	1
South America																	1	-
Gulf of Guinea																		1

low vertebral counts. Because of the inconsistent way in which the distinguishing characters vary, and because of the inadequately defined boundaries, the populations are not named at this time, either as species or subspecies.

Distribution. (Fig. 599). Western Atlantic: Gulf of Mexico from northern Florida and Louisiana to Campeche Bay; eastern and southern Straits of Florida and the Bahamas; West Indies from Cuba to the Leeward Islands; coast of Central and South America from Belize to Brazil. Eastern Atlantic: Gulf of Guinea. Depth 140–825 m (mean 331 m).

This is by far the most common and widespread of the Atlantic species of *Xenomystax.*

Etymology. From the Latinized Greek *Conger* (the conger eel) and *-oides* (like), referring to the overall appearance of this eel, the most congrid-like of the species of *Xenomystax.* An adjective.

Study Material. 376 specimens, 164–817 mm TL.

Holotype: USNM 217386 (male, 492 mm TL), Oregon 2005, Suriname, 07°34′N, 54°50′W, 366 m, 6 Nov. 1957.

Paratypes: ALABAMA: ANSP 109595 (4, 473–515), 29°12.5′N, 88°16′W, 183 m, 23 Oct. 1962. ANSP 109599 (6, 290–415), 29°16.5′N, 88°04.5′W, 146–165 m, 13 July 1960. LOUISIANA: MCZ 49337 (3, 380–415), 28°11′N, 90°08′W, 183 m, 20 Feb. 1964. USNM 217387 (1, 526), 27°58′N, 90°41′W, 275–320 m, 21 Sept. 1955. BAHAMAS: UMML 13871 (1, 200), 25°15′N, 79°15′W, 384 m, 30 Jan. 1964. UMML 17665 (1, 249), 26°13′N, 78°50′W, 403–512 m, 3 March 1965. UMML 33478 (2, 447–540), 23°52′N, 79°11′W, 366–384 m, 6 Nov. 1960. CUBA: MCZ 38783 (2, 244–246), 23°10′N, 79°35′W, 476 m, 1 May 1939. MCZ 38802 (2, 330–415), 22°53′N, 79°10′W, 458 m, 30 April 1939. MCZ 38820 (1, 272), 23°10′N, 81°29′W, 265–421 m, 17 March 1939. PUERTO RICO: ANSP 109622 (8, 290–421), 18°31.5′N, 66°50′W, 329 m, 7 Oct. 1959. USNM 217388 (1, 478), 18°31.5′N, 66°47′W, 293 m, 8 Oct. 1959. VENEZUELA: UMML 29675 (1, 417), 10°54.7′N, 66°17.8′W, 225–234 m, 23 July 1968. USNM 217442 (2, 188–222), 10°42′N, 67°56′W, 210 m, 28 Sept. 1965. BELIZE: USNM 217441 (2, 369–413), 17°28.5′N, 87°57.5′W, 275–329 m, 23 Jan. 1967. SURINAME: ANSP 114952 (2, 346–371), 07°27′N, 54°30′W, 201 m, 16 May 1969. GULF OF GUINEA: UMML

TABLE 74. Branchiostegal rays in specimens of *Xenomystax congroides* from four areas (each count represents one side of one specimen).

Area	Branchiostegal rays 7	8	9	10	Counts	x̄	Specimens
Gulf of Mexico	13	16	1		30	7.6	19
Caribbean	3	21	2		26	8.0	21
South America		3	12		15	8.8	10
Gulf of Guinea		1	4	11	16	9.6	11

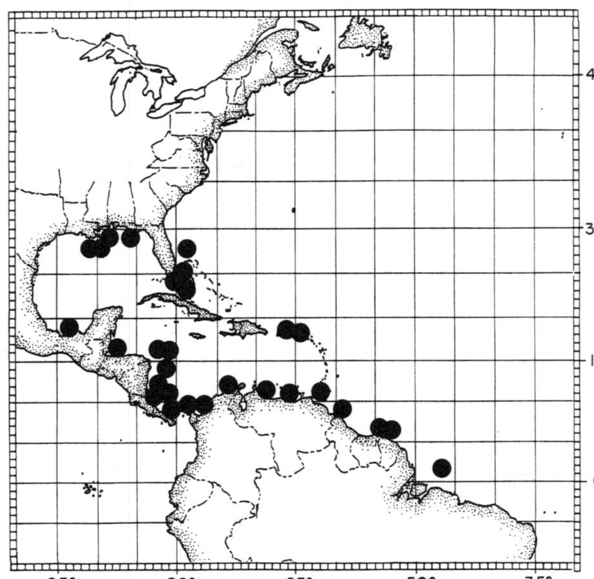

FIGURE 599. Distribution of *Xenomystax congroides.*

TABLE 73. Continued.

Vertebrae								
212	213	214	215	216	217	218	n	$\bar{x}$
							11	192.3
1							18	207.4
1	-	1	2	1			6	213.7
1	2	2	-	-	-	1	7	213.6

16873 (1, 293), 04°59′N, 09°37′W, 156–220 m, 5 June 1964. UMML 20902 (5, 260–417), 03°49′N, 07°38′W, 264–269 m, 13 May 1965.

Other material: FLORIDA STRAITS: USNM 185624 (1, 290), 197133 (1, 405). FLORIDA GULF: USNM 157971 (1, 376). ALABAMA: USNM 198056 (6, 169–490). MISSISSIPPI: ANSP 117026 (3, 410–465). UMML 33481 (1, 551). LOUISIANA: USNM 157891 (8, 368–604), 157973 (1, 515). MEXICO GULF: MBI 60 (1, 264, specimen missing). BAHAMAS: ANSP 156859 (1, 234). UMML 1157 (1, 406), 2685 (1, 195), 14144 (4, 360–450), 18401 (3,484–552). USNM 158642 (1, 420), 158869 (4, 334–405), 185623 (2, 408–457), 185624 (1, 290), 197134 (2, 311–330), 197136 (8, 225–440), 197139 (3, 348–504), 197344 (7, 337–553), 197346 (18, 325–548). CUBA: ANSP 114951 (1, 475). MCZ 38776 (1, 155), 38805 (2, 423–490), 38809 (2, 415–506). USNM 157970 (17, 330–455), 157972 (3, 461–497), 157974 (5, 410–497), 187630 (2, 304–379). PUERTO RICO: ANSP 112036 (1, 320), 114162 (1, 366). USNM 190538 (1, 237), 190539 (6, 360–436), 190540 (10, 229–426). VIRGIN ISLANDS: ANSP 117089 (3, 412–439). USNM 190543 (10, 211–460). LESSER ANTILLES: UMML 29676 (1, 190). USNM 268832 (16, 289–513). VENEZUELA: USNM 158889 (5, 245–500), 158925 (6, 334–437), 195902 (1, 876), 196176 (2, 476–750), 200324 (4, 442–745). COLOMBIA: USNM 198668 (2, 390–438), 198672 (2, 452–490), 198675 (5, 353–538). PANAMA: UMML 33480 (2, 399–401). USNM 193568 (5, 251–410), 193574 (12, 204–461), 193595 (13, 350–420), 193602 (19, 210–445), 217440 (1, 435). NICARAGUA: ANSP 131757 (1, 225). UMML 29677 (6, 372–425). USNM 158888 (7, 211–408), 193544 (1, 164), 193549 (2, 308–316), 193554 (1, 255), 193567 (4, 376–428), 268827 (12, 175–372), 270719 (3, 159–277). HONDURAS: USNM 157879 (5, 329–410), 193545 (1, 397), 193601 (10, 296–405), 268828 (12, 330–430), 268829 (11, 320–409). BELIZE: USNM 193558 (1, 279). SURINAME: USNM 158930 (2, 443–468). FRENCH GUIANA: USNM 158926 (1, 508), 158927 (2, 619–660). BRAZIL: USNM 158924 (3, 212–559). GULF OF GUINEA: USNM 198630 (1, 342), 200322 (1, 476), 200343 (1, 443), 217439 (1, 254).

Family Nettastomatidae

DUCKBILL EELS *or* SORCERERS

DAVID G. SMITH

Acknowledgments. See family Anguillidae.

Introduction. The Nettastomatidae is a family of long-snouted, slender-tailed eels found mainly on the outer continental shelf and continental slope of the world's tropical and warm-temperate seas. Nettastomatids are benthic in habits, but they are not burrowers. Little is known about their biology. Most appear to live in fairly open terrain and feed on a variety of small fishes and invertebrates.

The history of the group is rather convoluted, and only recently has it been recognized in its entirety. Until now, the only comprehensive treatment has been that of Smith and Castle (1982) on the leptocephali. Smith, Böhlke, and Castle (1981) revised the genera *Nettenchelys* and *Nettastoma,* and Lane and Stewart (1968) treated *Hoplunnis.* The family name dates from Jordan and Davis (1891: 649) in the form Nettastomidae.

Methods and Material. Counts and measurements are as in the Congridae except that an upper jaw length is substituted for the snout-rictus distance. The upper jaw is measured from the tip of the snout to the posterior end of the maxilla. In *Venefica,* which has a fleshy proboscis, the bony tip of the ethmovomer is substituted for the external tip of the snout as a point of measurement. Regenerated tails are common, so most proportional measurements are expressed as % of preanal length rather than TL.

Specimens of the following species were cleared and stained for osteological studies: *Hoplunnis macrura* (ANSP 156857), *Hoplunnis megista* (ANSP 138773), *Saurenchelys* sp. (ANSP 156860), *Facciolella* sp. (ANSP 115536, 126520), *Nettenchelys exoria* (ANSP 156856), *Nettastoma melanura* (ANSP 156855), and *Venefica procera* (ANSP 115510, 156853).

Characters. Body elongate, anus before midbody, preanal length less than 40% of TL, tail attenuate. Dorsal and anal fins confluent with caudal fin; dorsal fin begins near gill opening, sometimes slightly ahead or behind. Dorsal- and anal-fin rays segmented. Pectoral fin present or absent. Gill opening moderate in size, lateral in position. Scales absent.

Snout and jaws elongate, upper jaw projects a variable distance beyond lower; no flange on upper or lower lip; teeth exposed in lateral view when mouth closed. Eye well developed. Anterior nostril tubular, near tip of snout, directed laterally; posterior nostril variable in position. Lateralis system complete on head and body; no frontal commissure. Adnasal pore present or absent. Teeth variable in size, from small and al-

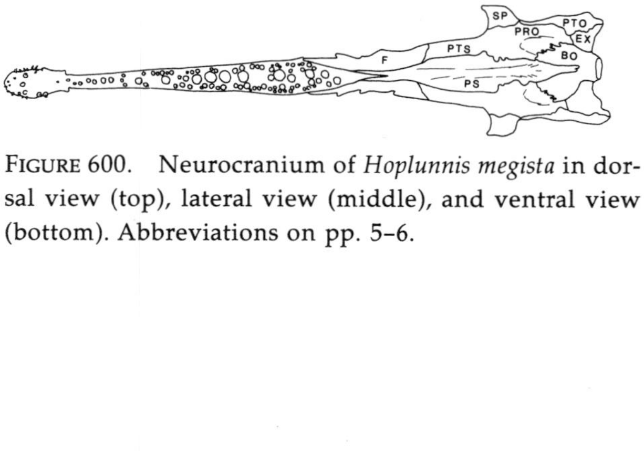

FIGURE 600. Neurocranium of *Hoplunnis megista* in dorsal view (top), lateral view (middle), and ventral view (bottom). Abbreviations on pp. 5–6.

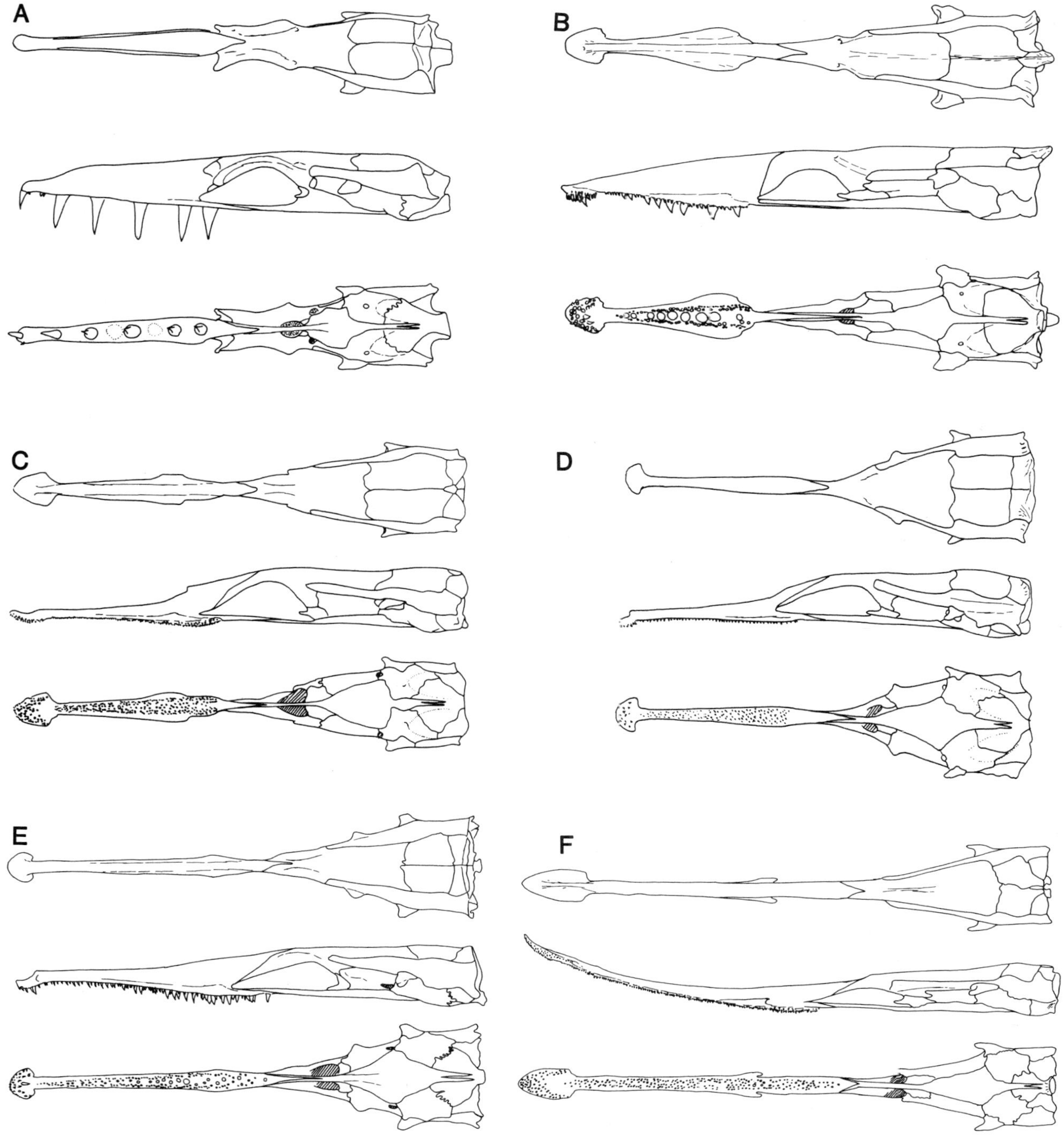

FIGURE 601. Neurocrania. A. *Hoplunnis macrura*. B. *Saurenchelys* sp. C. *Facciolella* sp. D. *Nettenchelys exoria*. E. *Nettastoma melanura*. F. *Venefica procera*.

most granular to large and fanglike; vomerine tooth patch always long.

No apparent sexual dimorphism.

Comparative Osteology. Nettastomatids are characterized by a narrowing and elongation of the skull, particularly the antorbital part. The skeleton on the whole is relatively complete; only the pectoral girdle and, to a slight extent, the gill arches show any reduction or loss of elements.

Neurocranium (Figs. 600, 601): The neurocranium is slender and elongate with a large orbit. The antorbital part is greatly extended by the forward prolongation of the ethmovomerine complex, massive and deep in *Hoplunnis* and

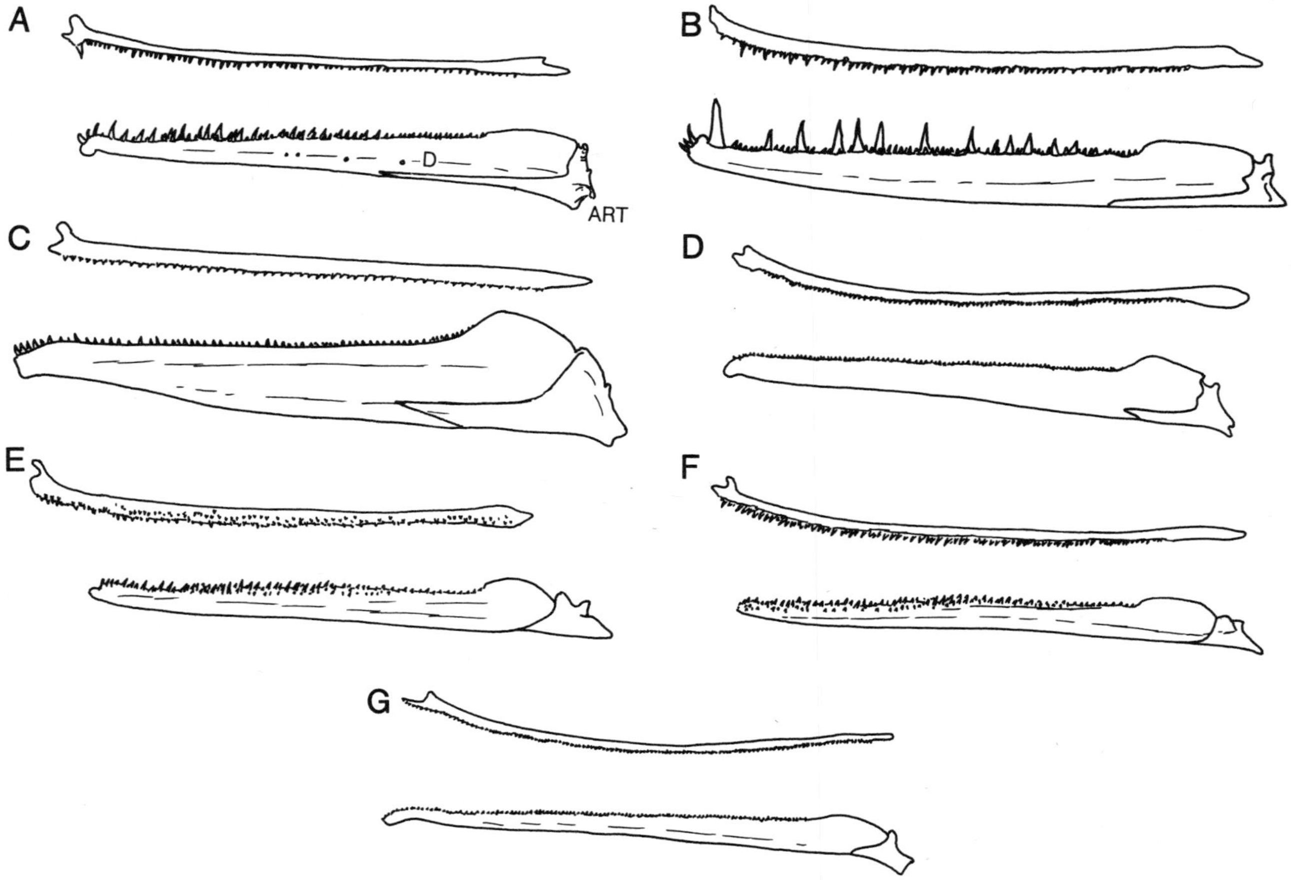

FIGURE 602. Maxilla (above) and mandible (below). A. *Hoplunnis megista*. B. *Hoplunnis macrura*. C. *Saurenchelys* sp. D. *Facciolella* sp. E. *Nettenchelys exoria*. F. *Nettastoma melanura*. G. *Venefica procera*.

Saurenchelys, narrower in the other genera. There is no lateral ethmoid process. The frontals are fused and form most of the dorsal border of the orbit and the anterior roof of the braincase. In *Hoplunnis* the frontals are expanded laterally to form a narrow shelf over the orbits. Behind the frontals are the paired parietals and epiotics. A supraoccipital is present in *Hoplunnis megista*, *Saurenchelys* and *Facciolella*, absent in the others. The posterior wall of the neurocranium is formed by the epiotics dorsally, the pterotics laterally, the exoccipitals medially, and the basioccipital ventrally. The floor is formed by the basioccipital and the parasphenoid, which extends anteriorly over the dorsal side of the vomer to the anterior corner of the orbit. The lateral wall consists of the pterosphenoid, sphenotic, prootic, and pterotic. The pterotic is hollow and houses the temporal canal. A small, median basisphenoid forms the posteroventral boundary of the orbit.

Jaws (Fig. 602): The slender maxilla forms most of the border of the upper jaw. It has a small pedicel at the anterior end and articulates with the ethmovomer just behind the anterior expansion. The mandible is also slender and elongate; almost all the elongation is in the dentary. The

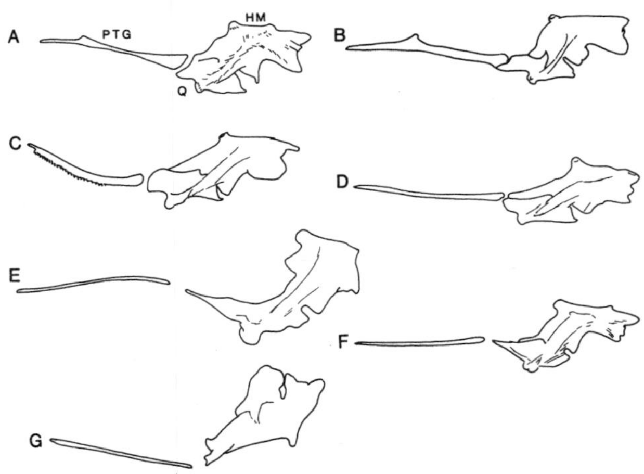

FIGURE 603. Suspensorium. A. *Hoplunnis megista*. B. *Hoplunnis macrura*. C. *Saurenchelys* sp. D. *Facciolella* sp. E. *Nettenchelys exoria*. F. *Nettastoma melanura*. G. *Venefica procera*.

articular is restricted to the posterior extremity of the mandible, and a small coronomeckelian is present on the medial face.

Suspensorium (Fig. 603): The hyomandibula and quadrate are well developed and closely articulated into a somewhat elongate, forwardly inclined, trapezoidal structure. Anteriorly the hyomandibula fits into a socket formed by the sphenotic and prootic, posteriorly into a socket in the pterotic. The opercle articulates with the hyomandibula just below the posterodorsal corner, and the epihyal is connected to the posterior part of the hyomandibula by a ligament. The quadrate articulates ventrally with the mandible via the articular. The pterygoid is broad and stout in *Hoplunnis* and *Saurenchelys*; it is closely articulated with the quadrate and hyomandibula posteriorly and strongly connected by a ligament to the vomer anteriorly. It bears teeth in *Saurenchelys* and *Hoplunnis tenuis*. In the remaining genera, the pterygoid is slender and does not bear teeth, but it still extends from the quadrate-hyomandibula to the vomer, to which it retains a ligamentous attachment.

Hyoid arch (Fig. 604): The hyoid apparatus consists of a median glossohyal and urohyal, paired ceratohyals and epihyals, and about five to nine branchiostegal rays; there are no hypohyals. The ceratohyals and epihyals tend to be short and stout. One or two branchiostegal rays are joined to the ceratohyals, the rest to the epihyals. The urohyal is a simple, slender structure without bifurcations or other modifications.

One of the most characteristic features of nettastomatids is the close connection between the glossohyal and the basibranchials. The pattern varies somewhat between genera, but the result is a compound structure consisting of the glossohyal and one or two closely appressed basibranchials. In *Saurenchelys* the glossohyal is extended both anteriorly and posteriorly from the point of articulation with the ceratohyals. The first basibranchial (B-1) is large and appressed to the posteroventral end of the glossohyal, forming a single unit separated by a suture. The second basibranchial (B-2) is much smaller and abuts the posterior end of B-1. The pattern in *Nettastoma* and *Nettenchelys* is similar, with a long

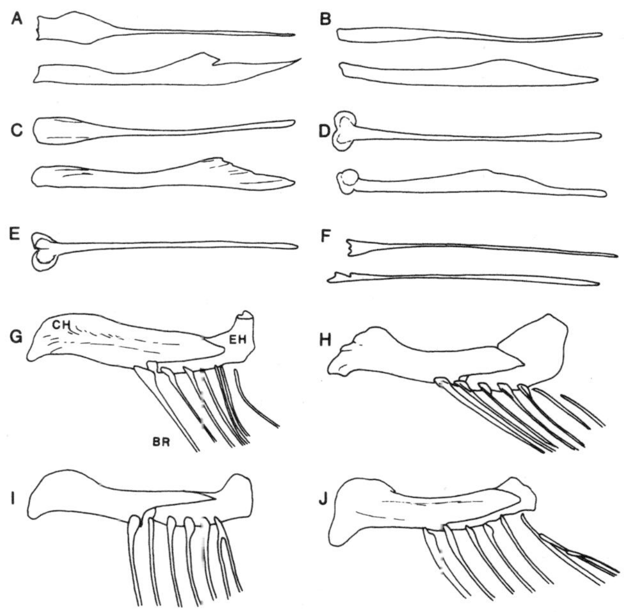

FIGURE 604. Hyoid arch. Urohyal in dorsal (above) and lateral (below) views: A. *Hoplunnis megista*. B. *Hoplunnis macrura*. C. *Saurenchelys* sp. D. *Nettenchelys exoria*. E. *Nettastoma melanura*, dorsal view only. F. *Venefica procera*. Ceratohyal, epihyal, and branchiostegal rays: G. *Hoplunnis megista*. H. *Hoplunnis macrura*. I. *Saurenchelys* sp. J. *Facciolella* sp.

posterior extension of the glossohyal and a close articulation of the relatively large B-1 to its posteroventral surface. The latter two genera, however, lack B-2. In *Venefica* the glossohyal is similar in form to that of the three preceding genera, with a long posterior extension, but B-1 is either absent or fused to the glossohyal. In *Facciolella* the posterior extension of the glossohyal is reduced, but B-1 is enlarged and forms most of the portion of the complex behind the point of articulation of the ceratohyals; B-2 is either absent or rudimentary. In *Hoplunnis*, B-1 is greatly enlarged and seems to have taken over the function of bearing the ceratohyals. The glossohyal extends anteriorly from that point. B-2 is present and articulates with the posteroventral end of B-1.

Gill arches (Fig. 605): The gill arches are relatively complete in nettastomatids, only the basibranchial series showing any reduction. Ventrally there are three hypobranchials and five ceratobranchials. The first two hypobranchials are ossified, the third is cartilaginous. The first four ceratobranchials are large and stout, the fifth small and slender. The lower pharyngeal tooth plates are well developed, nearly as long as the fifth ceratobranchial, and undivided in all but *Hoplunnis megista.* The first basibranchial (B-1) is present and ossified in all genera except *Venefica* and is closely appressed to the posteroventral surface of the glossohyal. In *Venefica* B-1 is either absent or fused to the glossohyal. B-2 is present and ossified in *Hoplunnis* and *Saurenchelys* and abuts the posterior end of B-1. The other genera appear to lack B-2, although *Facciolella* may have a cartilaginous remnant. B-3 and B-4 are present but unossified in all genera. The dorsal elements consist of four epibranchials, two infrapharyngobranchials, and two upper pharyngeals. The posterior upper pharyngeal (UP-4) is broadly oval and relatively large; the anterior one (UP-3) is small and often hidden in dorsal view beneath the infrapharyngobranchial.

Opercular apparatus (Fig. 606): All four elements are present. The opercle is broadly fan- or blade-shaped and articulates with the hyomandibula; the posterior margin is serrate in *Hoplunnis* and *Saurenchelys,* smooth in the others. The subopercle is a narrow, sickle-shaped structure that lies under the posterior margin of the opercle. The preopercle is moderately well developed although distinctly smaller than the opercle; it lies along the posterior margin of the hyomandibula and bears a portion of the preopercular canal. The interopercle is a fairly large triangular structure in *Saurenchelys* and *Facciolella,* almost the size of the preopercle although not fully ossified. It is slightly reduced in *Hoplunnis,* markedly reduced in the remaining genera.

Pectoral girdle (Fig. 607): Only *Hoplunnis* has a pectoral fin. Its pectoral girdle consists of a large cleithrum and a large cartilaginous plate with two centers of ossification representing the scapula and coracoid. *Hoplunnis macrura* has a small, splint-like supracleithrum, but *H. megista* lacks it. Three actinosts were present in the specimen of *H. macrura* examined, four in *H. megista. Saurenchelys* and *Nettenchelys* have a cleithrum and supracleithrum but lack any trace of the scapula

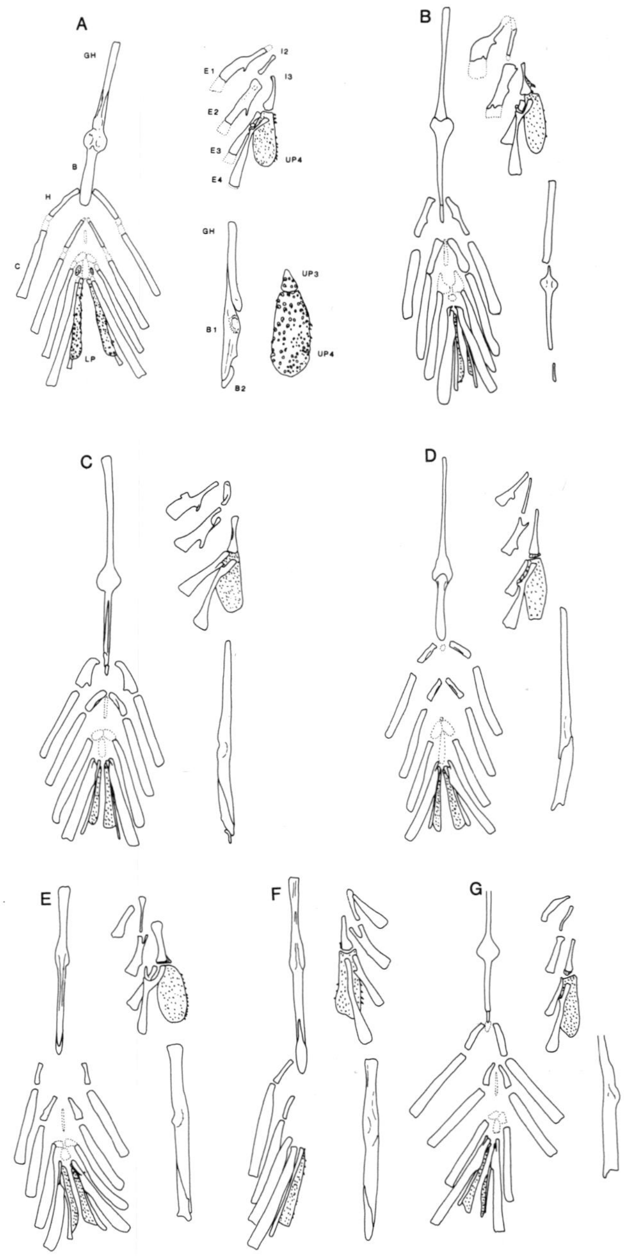

FIGURE 605. Gill arches. Ventral elements in dorsal view, dorsal elements in dorsal view, glossohyal-basibranchials in lateral view, and upper pharyngial in ventral view: A. *Hoplunnis megista.* Ventral elements in ventral view, dorsal elements in dorsal view, glossohyal and basibranchials in dorsal view: B. *Hoplunnis macrura.* As above but with glossohyal and basibranchials in lateral view: C. *Saurenchelys* sp. D. *Facciolella* sp. E. *Nettenchelys exoria.* F. *Nettastoma melanura* (left side of ventral elements not shown). G. *Venefica procera* (basibranchials lacking—anterior end of glossohyal damaged).

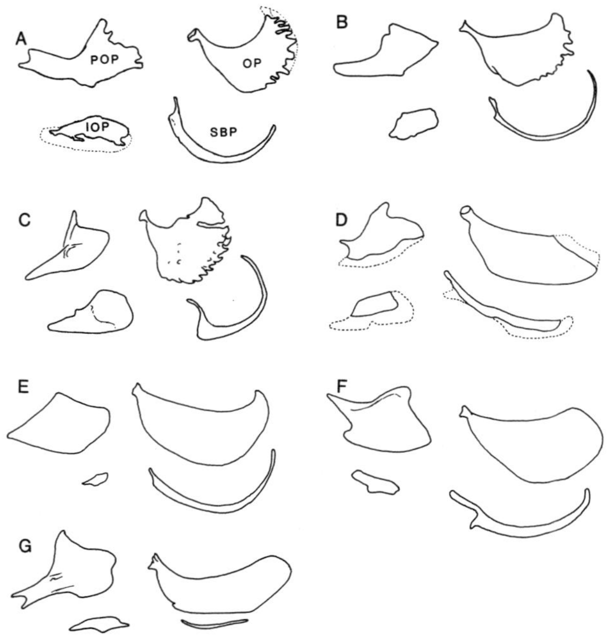

FIGURE 606. Opercular apparatus. A. *Hoplunnis megista.* B. *Hoplunnis macrura.* C. *Saurenchelys* sp. D. *Facciolella* sp. E. *Nettenchelys exoria.* F. *Nettastoma melanura.* G. *Venefica procera.*

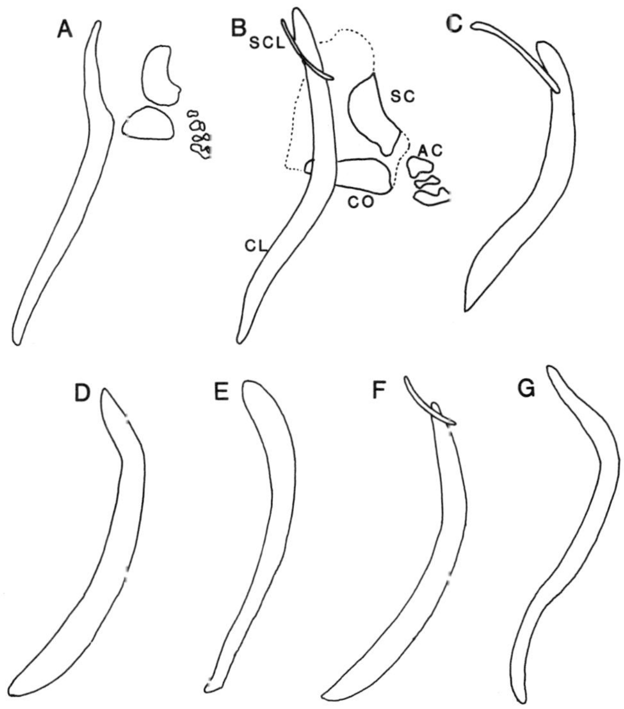

FIGURE 607. Pectoral girdle. A. *Hoplunnis megista.* B. *Hoplunnis macrura.* C. *Saurenchelys* sp. D. *Nettastoma melanura.* E. *Facciolella* sp. F. *Nettenchelys exoria.* G. *Venefica procera.*

and coracoid. The remaining genera have only a cleithrum.

Lateralis system, sensory pores, and associated bones (Figs. 608–610): Nettastomatids have a complete lateralis system with pores in all but the temporal canal (Fig. 608). The right and left sides are connected only by the supratemporal commissure; there is no frontal commissure. There are 4–12 supraorbital pores, including a single ethmoidal pore at the tip of the snout. The infraorbital canal contains a variable number of pores along the upper jaw and three pores behind the eye. *Facciolella, Nettenchelys, Nettastoma,* and *Venefica* have an adnasal pore in the short ascending branch of the infraorbital canal just behind the anterior nostril. Unlike the condition in congrids, the adnasal pore of nettastomatids is located slightly behind the anteriormost maxillary pore. *Hoplunnis* and *Saurenchelys* have the typical congrid pattern of five pores in the infraorbital canal along the upper jaw, although because of the large gape, the last one is still before the rictus. The other genera have more, from eight or nine in *Nettenchelys* to as many as 16 in *Venefica.* This is associated with the elongation of the snout in those genera. There are numerous pores in the preoperculomandibular canal, and from one to three in the supratemporal commissure.

The circumorbital bones (Fig. 609) are delicate and largely cartilaginous. The most prominent elements are the nasal, associated with the anterior part of the supraorbital canal, and the preorbital, which houses the anterior part of the infraorbital canal. The nasal is broad and long

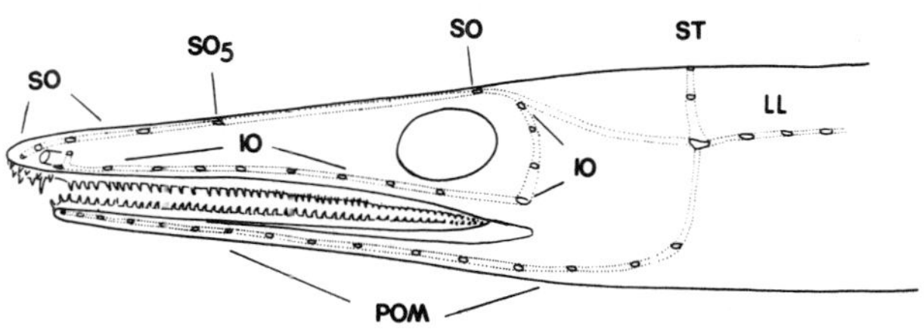

FIGURE 608. Sensory canals and pores on the head of a nettastomatid eel (from Smith, Böhlke, and Castle, 1981: fig. 1).

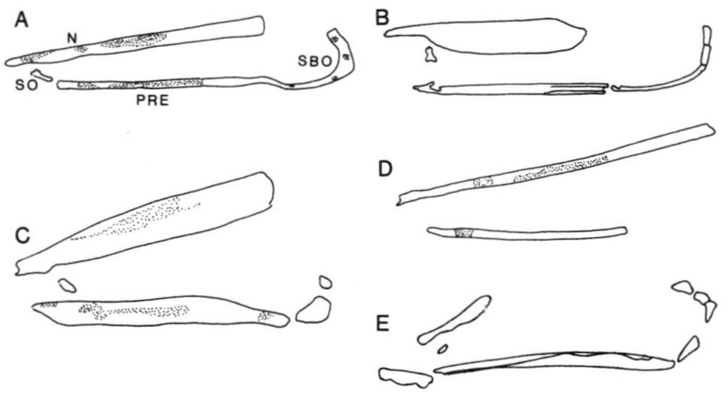

FIGURE 609. Circumorbital bones. A. *Hoplunnis megista.* B. *Hoplunnis macrura.* C. *Saurenchelys* sp. D. *Facciolella* sp. E. *Nettastoma melanura.*

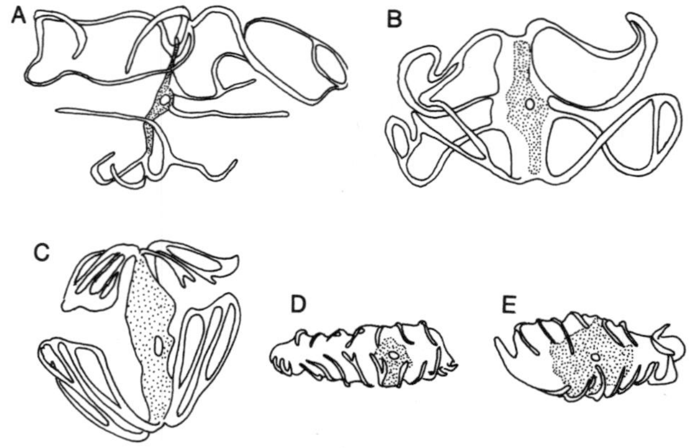

FIGURE 610. Lateral-line ossicles. A. *Hoplunnis megista.* B. *Hoplunnis macrura.* C. *Saurenchelys* sp. D. *Facciolella* sp. E. *Venefica procera.*

in *Hoplunnis* and *Saurenchelys,* long and slender in *Facciolella* and *Venefica,* and short in *Nettastoma.* The preorbital is rather slender in all genera and about as long as the nasal in most. In *Nettastoma* the nasal is much shorter than the preorbital. A small supraorbital is usually present between the anterior ends of the nasal and preorbital. Two or three separate elements, the suborbitals of Asano (1962), are located behind the preorbital and bear the postorbital section of the infraorbital canal.

The lateral-line ossicles of *Hoplunnis* and *Saurenchelys* (Fig. 610A–C) are similar in form to those of the *Rhynchoconger-Uroconger* group of congrids. The body of the ossicle is narrow and vertically elongate with finger-like cartilaginous projections that enclose the canal. The specimen of *Nettastoma melanura* examined (ANSP 156855) was poorly stained, but its lateral-line ossicles appeared oval and vertically elongate; cartilaginous fingers could not be seen. The specimen of *Nettenchelys exoria* (ANSP 156856) was also poorly stained, but its lateral-line ossicles appeared to be of the *Rhynchoconger* type as well. In *Facciolella* and *Venefica* (Fig. 610D, E) the ossicles are horizontally elongate, but only the central part is ossified.

Vertebrae (Figs. 611–613): Similar overall to those of congrids. Ribs are absent. *Nettastoma* has well developed epicentral processes on the caudal vertebrae (Fig. 613F), and *Nettenchelys* has small ones (Fig. 613E); the other genera lack them. Precaudal vertebrae represent less than 35% of the total. There are no myorhabdoi. Dorsal- and anal-fin rays are segmented.

Discussion. The Nettastomatidae appears closely related to the Congridae, in particular the *Uroconger* group. Several trends apparent in *Uroconger* are continued in the nettastomatids. Among these is the attenuation of the tail. At first glance, this may not seem to be a very significant feature in an order of fishes that is characterized by slender, elongate bodies, but in fact these are the only eels (except *Nemichthys,* clearly a special case) in which the posterior part of the tail is drawn out to almost filamentous proportions. Other shared, presumably derived characters include the long and narrow vomerine tooth patch, the high vertebral counts, the form of the lateral-line ossicles, and the serrate opercle. Nettastomatids differ from congrids primarily in the elongation of the snout and jaws, the reduction of the lips, the loss of the pectoral fin in most genera, and

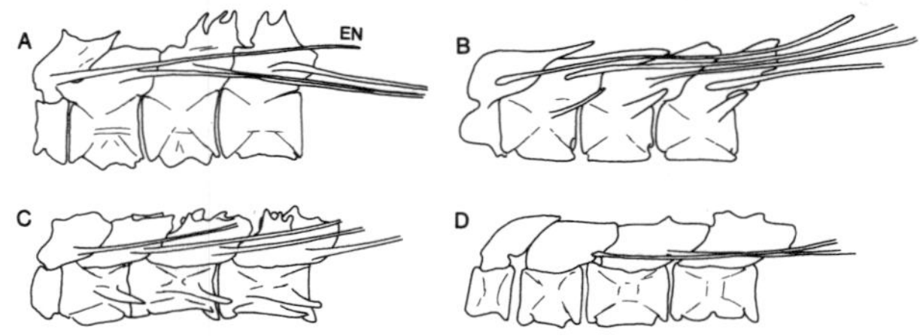

FIGURE 611. Anteriormost few vertebrae. A. *Hoplunnis megista.* B. *Hoplunnis macrura.* C. *Saurenchelys* sp. D. *Facciolella* sp.

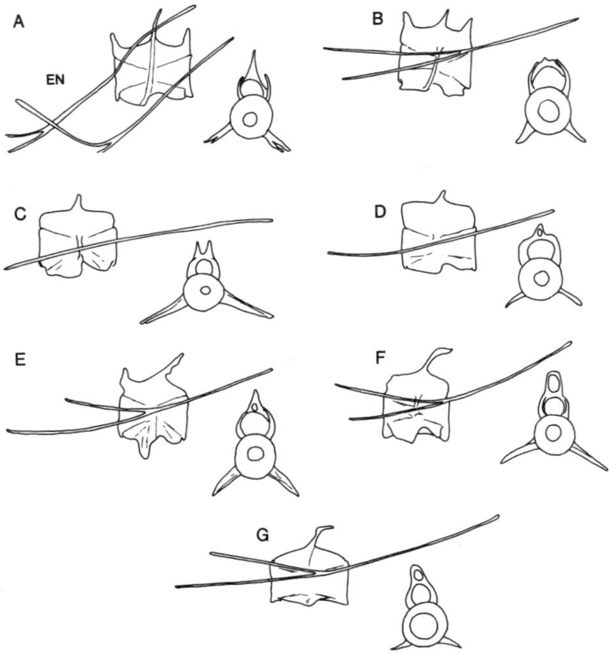

FIGURE 612. Precaudal vertebrae, lateral and front views. A. *Hoplunnis megista.* B. *Hoplunnis macrura.* C. *Saurenchelys* sp. D. *Facciolella* sp. E. *Nettenchelys exoria.* F. *Nettastoma melanura.* G. *Venefica procera.*

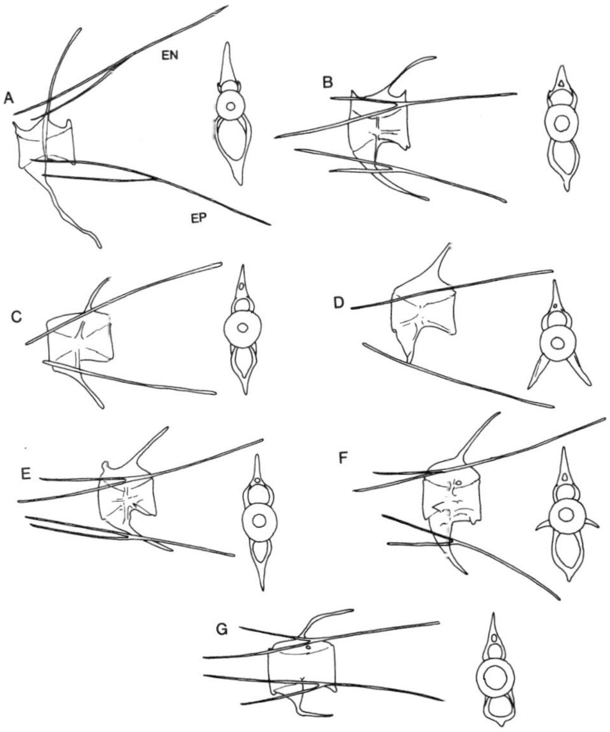

FIGURE 613. Caudal vertebrae, lateral and front views. A. *Hoplunnis megista.* B. *Hoplunnis macrura.* C. *Saurenchelys* sp. D. *Facciolella* sp. E. *Nettenchelys exoria.* F. *Nettastoma melanura.* G. *Venefica procera.*

the peculiar connection between the glossohyal and the basibranchials. The leptocephali are also distinct (see p. 704). The *Xenomystax* group in some ways forms a bridge between the more typical congrids and the Nettastomatidae. Like the nettastomatids, its members have an elongate snout and jaws, and the lips are pulled back from the teeth; one form, *Gavialiceps taeniola,* lacks a pectoral fin. The basibranchials and the glossohyal, however, are separate.

Lane and Stewart (1968: 60) considered the shelf-like expansion of the frontals over the orbit in *Hoplunnis* to be a separate element, which they named the supraorbital and equated with a supposedly similar structure in *Muraenesox.* They used this shared character to indicate a close relationship between the two genera. I found no sutures in stained material of *Hoplunnis* available to me, and there seems no reason to consider the "supraorbital" anything other than an integral part of the frontal. The structure that Lane and Stewart referred to in *Muraenesox* is actually on the ethmovomer and hence cannot be homologous.

Distribution. Worldwide in tropical and warm temperate seas.

Genera. The Nettastomatidae can be readily divided into three subgroups: one containing *Hoplunnis* Kaup, 1860 and *Saurenchelys* Peters, 1864; one containing *Nettenchelys* Alcock, 1898, *Nettastoma* Rafinesque, 1810, and *Venefica* Jordan and Davis, 1891; and one containing *Facciolella* Whitley, 1938.

Hoplunnis and *Saurenchelys* seem, on the whole, the most primitive. Their jaws and snout are only moderately elongate, and they have the typical congrid pattern of five pores along the upper jaw. They have no adnasal pore. The maxillary and mandibular teeth are in two to five rows, and the vomerine teeth consist of a median row of enlarged teeth flanked by a variable number of small teeth. The posterior nostril is at mid-eye level. The pectoral fin is present in *Hoplunnis* but not in *Saurenchelys.* The body is uniformly shallow with no marked deepening at mid-length. The pterygoid is well developed in both genera and bears teeth in *Saurenchelys* and in at least one species of *Hoplunnis*. The second basibranchial is ossified, the posterior margin of the opercle is serrate, and the lateral-line ossicles resemble those of *Uroconger*. The leptocephali are long and slender and have a short gut with two swellings, one associated with the liver and one near the anus.

Nettenchelys, Nettastoma, and *Venefica* have a more elongate snout and jaws, and the number of pores along the upper jaw has increased to 8–15, including an adnasal pore. The teeth are in bands; the vomerine teeth are sometimes enlarged medially but never form a single series running the length of the vomer. The posterior nostril is displaced dorsally to a position near the top of the head or behind the head. The pectoral fin is absent, and the body is deepest in the middle and tapers toward both ends. The pterygoid is slender and never bears teeth. The second basibranchial is cartilaginous or absent, and the posterior margin of the opercle is entire. The leptocephali of this group are deep-bodied but have the same two swellings in the gut that are found in *Hoplunnis* and *Saurenchelys.*

Facciolella differs from all the other genera in having the posterior nostril displaced ventrally and opening into a slit in the upper lip. There are six pores along the upper jaw, including an adnasal pore. The teeth are in bands and none are markedly enlarged. The pectoral fin is absent and the body is deepest in the middle and tapers toward both ends. The pterygoid is slender and toothless, the second basibranchial is reduced or absent, and the third basibranchial is absent. The lateral-line ossicles are trough-like and largely cartilaginous. The leptocephali are slender like those of the *Hoplunnis-Saurenchelys* group, but they lack the two discrete swellings in the gut.

Key to the Genera of Nettastomatidae

1a. Pectoral fin present . . . . *Hoplunnis*, p. 576
1b. Pectoral fin absent 2
2a. Posterior nostril at mid-eye level; pterygoid teeth present . . . . *Saurenchelys*, p. 590
2b. Posterior nostril above or below mid-eye level; pterygoid teeth absent 3
3a. Posterior nostril a slit in upper lip . . . . *Facciolella*, p. 597
3b. Posterior nostril above mid-eye level, located variably from a point in front of anterior margin of eye to dorsal surface of body behind occiput 4
4a. IO pores 11–13 (8–10 along upper jaw); snout 32–37% HL; posterior nostril behind posterior margin of eye . . . . *Nettenchelys*, p. 599
4b. IO pores 14–19 (11–16 along upper jaw); snout 35–55% HL; posterior nostril over or before posterior margin of eye 5
5a. SO pores 4–6; tip of snout without a prominent fleshy proboscis . . . . *Nettastoma*, p. 604
5b. SO pores 8–13; a fleshy proboscis at tip of snout . . . . *Venefica*, p. 609

Genus *Hoplunnis* Kaup, 1860

Hoplunnis Kaup, 1860: 19 (type species *Hoplunnis schmidti* Kaup, 1860, by monotypy).

Characters. Body elongate, round in cross section anteriorly, compressed posteriorly; depth relatively uniform; tail slender, attenuate; anus near anterior quarter or third of TL. Dorsal fin begins over or slightly ahead of gill opening. Pectoral fin present.

Head and snout moderately elongate, conical to somewhat compressed, upper jaw projects slightly beyond lower; fleshy tip of snout does not project beyond bony tip of ethmovomer. Eye moderate. Mouth relatively large, rictus slightly behind posterior margin of eye. Anterior nostril

TABLE 75. Mean vertebral formulae (MVF) and ranges of vertebral counts for western Atlantic species of Nettastomatidae.

Species	MVF	Predorsal Range	Predorsal n	Preanal Range	Preanal n	Total Range	Total n
Hoplunnis diomediana	4-38-228	3–5	20	35–41	25	221–234	17
Hoplunnis macrura	4-36-239	4–5	19	34–38	24	227–245	12
Hoplunnis megista[1]	9-56-274	9–10	4	55–56	4	274	1
Hoplunnis similis[1]	3-40-256	2–4	4	39–42	5	254–260	7
Hoplunnis tenuis	5-37-237	3–6	12	36–38	15	232–244	10
Saurenchelys cognita[1]	9-34-204	8–10	12	32–36	14	199–209	5
Facciolella	11-55-248	10–13	13	53–57	16	243–256	10
Nettastoma melanura	6-47-200	4–8	34	44–50	38	195–207	10
Nettastoma syntresis[1]	5-40-201	5–6	4	40–41	6	201	1
Nettenchelys exoria[1]	7-43-170+	7	3	43–44	3	170+–171+	2
Nettenchelys inion[1]	7-41-210+	7	1	41	1	210–	1
Nettenchelys pygmaea[1]	8-42-235[2]	7–8	3	40–43	3	220–225[2]	—
Venefica procera	7-59-202	6–8	12	57–62	13	200–205	5

[1] Includes holotype.

[2] Estimated from myomere counts of leptocephali.

a short tube just behind tip of snout, directed anterolaterally; posterior nostril a longitudinal slit in front of eye at or slightly below mid-eye level.

Lateral line on body complete to tail. Five supraorbital pores: one (the ethmoidal pore) at anteroventral tip of snout, before anterior nostril; one above this, just in front of anterior nostril; one, slit-like, bordering dorsal side of anterior nostril tube; one between anterior and posterior nostrils; and one near level of posterior margin of eye. Eight infraorbital pores, those along upper jaw slit-like: one just behind anterior nostril tube, two between anterior and posterior nostrils, one between posterior nostril and eye, one below posterior margin of eye, and three behind eye; adnasal pore absent. Three pores in supratemporal commissure. Preoperculomandibular canal with approximately 11–14 pores. Many shallow dermal pits on head, presumably sensory in nature.

Teeth highly variable in size, from small and conical to large and fang-like. Maxillary teeth in two or three rows, small, those of inner row slightly enlarged. Mandibular teeth mostly small, in two to five rows, those of inner row enlarged and widely spaced; one or two enlarged teeth at tip of mandible which fit closely alongside upper jaw between maxillary and intermaxillary teeth when mouth closed. Intermaxillary teeth enlarged and separated from maxillary teeth. Vomer with a median series of conical or lanceolate teeth, some greatly enlarged, flanked on each side by one to three rows of small teeth that are present throughout life in some species but only in very young of others.

Size. Most species do not exceed 500 mm TL, but *H. megista* may reach 1 m.

Discussion. In the past *Hoplunnis* has usually been placed in the Muraenesocidae, primarily because of its large vomerine fangs and its possession of a pectoral fin. Neither character, however, seems very convincing. Loss of the pectoral fin has been a recurring theme in eel evolution. Several families and even some genera have representatives with and without pectoral fins. Enlarged vomerine teeth also occur in several other families. *Hoplunnis* actually has much more in common with the nettastomatids than with *Muraenesox*, and it seems especially close to *Saurenchelys*. Smith and Castle (1982) noted the similarity of the leptocephali. Similarities in adult morphology are described here.

Distribution. Tropical eastern and western Atlantic, including Gulf of Mexico, and eastern tropical Pacific.

Etymology. From the Greek *hoplon* (armor) and *hynis* (vomer), referring presumably to the large vomerine teeth. The name probably should have been spelled *Hoplynnis.* Feminine.

Species. Seven species are now recognized, five in the western Atlantic and one each in the eastern Atlantic and eastern Pacific. The species can be arranged in a sequence based on a progressive reduction in the number of series of teeth combined with an enlargement of individual teeth. At one end of the series is *Hoplunnis megista* n. sp. The lateral vomerine teeth are in as many as three rows posteriorly and the mandibular teeth in five rows. The median vomerine teeth and the inner mandibular teeth are not as large relative to the other teeth as they are in the other species. *Hoplunnis tenuis* Ginsburg, 1951 has three rows of maxillary and mandibular teeth and a single row of close-set, prominent, lateral vomerine teeth. The median vomerine teeth and the inner mandibular teeth are larger relative to the other teeth than in *H. megista* but less so than in the other five species. *Hoplunnis diomediana* Goode and Bean, 1896 and *H. similis* n. sp. still have triserial maxillary and mandibular teeth, but the lateral vomerine teeth are fewer and more widely separated than in *H. tenuis. Hoplunnis macrura* Ginsburg, 1951 has only two rows of maxillary and mandibular teeth. Lateral vomerine teeth are present only in very young individuals. The median vomerine teeth and inner mandibular teeth are very large. *Hoplunnis punctata* Regan, 1915 (eastern Atlantic) and *H. pacifica* Lane and Stewart, 1968 (eastern Pacific) appear most like *H. macrura.*

Hoplunnis diomediana, H. similis, H. macrura, H. punctata, and *H. pacifica* appear to form a closely related group. Here the median vomerine teeth are greatly enlarged while the lateral vomerine teeth are greatly reduced or absent. The snout is relatively deep and not sharply tapered.

Hoplunnis tenuis differs noticeably from the previous five species. The median vomerine teeth are smaller, and the lateral vomerine teeth are much more numerous and better developed. The profile of the head is quite different, the snout being shallower and more sharply tapered. In addition, *H. tenuis* is the only species with pterygoid teeth.

At first glance *Hoplunnis megista* does not appear to belong with the other species. It is a much larger, heavier eel that seems to have little in common with its small, delicate congeners. Yet it shares many critical features with the other species and the differences are more superficial than essential. The dentition is a continuation of the trend that can be followed from *H. macrura* through *H. diomediana* and *H. tenuis.* The shallow, tapering snout strongly resembles that of *H. tenuis.* Several osteological characters resemble the condition in *H. macrura,* particularly in the gill arches, where the first basibranchial is greatly enlarged, closely articulated to the glossohyal, and forms the principal point of articulation for the ceratohyals.

KEY TO THE SPECIES OF *HOPLUNNIS* IN THE WESTERN NORTH ATLANTIC

1a. LL pores to anus more than 50; TL often greater than 600 mm . . . . *megista,* p. 579
1b. LL pores to anus fewer than 45; TL less than 600 mm 2
2a. Pterygoid teeth present except in very young individuals; lateral vomerine teeth close-set and numerous . . . . *tenuis,* p. 581
2b. Pterygoid teeth absent; lateral vomerine teeth few and widely spaced or absent 3
3a. Lateral vomerine teeth present only in very young individuals; stomach pale . . *macrura,* p. 587
3b. Lateral vomerine teeth present at all stages; stomach black 4
4a. Color pale with scattered small, dark spots; intestine pale; vertebrate 222–231 . . . *diomediana,* p. 583
4b. Color darkly mottled with brown spots; intestine black; vertebrae 254–260 . . . . . . *similis,* p. 586

FIGURE 614. *Hoplunnis megista*: USNM 193613, holotype, 695 mm TL.

Hoplunnis megista
D. G. Smith and R. H. Kanazawa,
new species
Figures 614–617, Table 75

Distinctive Characters. Pterygoid teeth absent; lateral vomerine teeth numerous; stomach and intestine black; LL pores 53–57; total vertebrae 274.

Description. Total vertebrae 274 (n = 1), predorsal vertebrae 9–10 (4), preanal vertebrae 55–56 (4). Branchiostegal rays 7 (2). Pectoral-fin rays 11 (1). Pores: LL to anus 53–57 (9), POM 13 (9), IO 8 (8), SO 1 + 4 (9—one specimen has 6 on one side), STC 3 (9).

Proportions as % of TL: preanal 33 (1), predorsal 12 (1), head 12 (1), depth at anus ca 3 (1). Of preanal length: predorsal 34–41 (9), head 34–41 (9), pectoral fin 6–7 (9), depth at anus ca 5–11 (9). Of HL: snout 36–41 (9), eye 7–8 (9), upper jaw 50–59 (9), gill opening 4–8 (9), interbranchial 7–12 (8), pectoral fin 15–19 (9).

Maxillary teeth in three irregular rows, those of inner series slightly enlarged, those of outer row very small. Mandibular teeth in three irregular rows anteriorly, four to five extremely irregular rows posteriorly, those of inner series enlarged anteriorly, becoming progressively smaller posteriorly. Intermaxillary teeth in an oval patch, separated from vomerine teeth. Vomer with a median series of enlarged teeth, becoming somewhat irregular at posterior end; smaller teeth lateral to these, in one row anteriorly, two or three irregular rows posteriorly; in the median series the anteriormost tooth smallest, the next several teeth increase in size posteriorly, then a relatively abrupt increase in size.

Stomach variable in length, somewhat more than half length of intestine. In one specimen stomach doubled over at posterior end, extending almost to anus when straightened out. Gas bladder relatively short, beginning somewhat

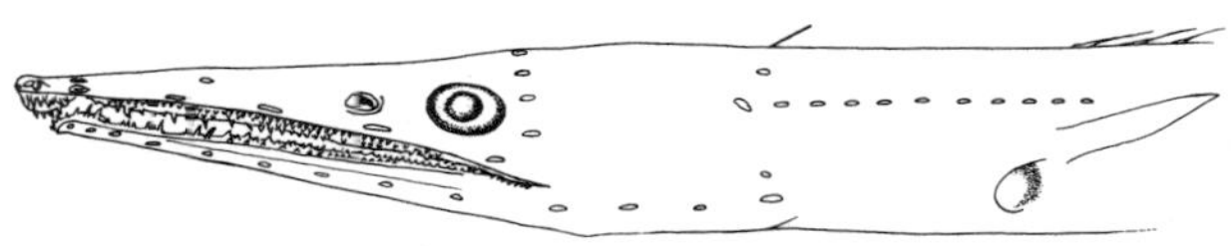

FIGURE 615. *Hoplunnis megista*: USNM 193613, holotype, 695 mm TL; head.

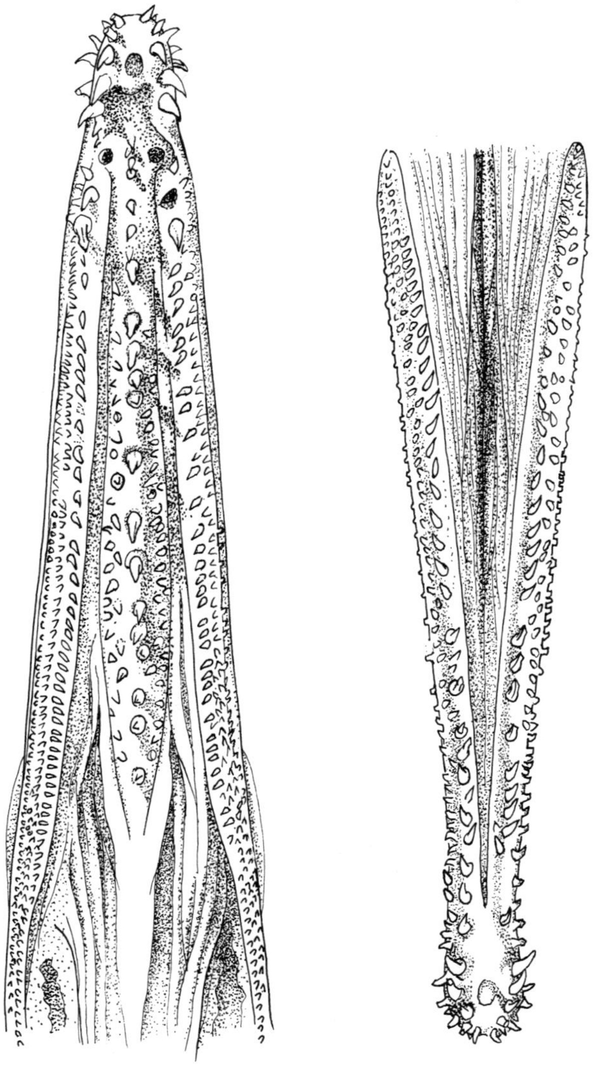

FIGURE 616. *Hoplunnis megista*: ANSP 147544, 770 mm TL; dentition.

behind anterior end of stomach and ending before anus.

Color. In preservative brown or gray, vertical fins edged in black. Stomach and intestine black. Gas bladder pale.

Size. This is by far the largest species of *Hoplunnis.* The largest specimen examined was 1095 mm TL, 405 mm preanal (ANSP 126447).

Development and Growth. The leptocephalus has not been identified.

Remarks. On the whole, this species seems closest to *Hoplunnis tenuis,* although in some ways it stands apart from the other species of the genus. As in *H. tenuis,* the median vomerine teeth and the inner maxillary and mandibular teeth are only slightly enlarged, less so in *H. megista* than in *H. tenuis.* Whereas the other species have no more than three rows of mandibular teeth and one row of lateral vomerine teeth, *H. megista* has up to five and three, respectively. *Hoplunnis tenuis* most closely approaches this condition, but *H. megista* carries the trend to an extreme. *Hoplunnis megista* grows much larger than any other species of *Hoplunnis* and has considerably more vertebrae and LL pores.

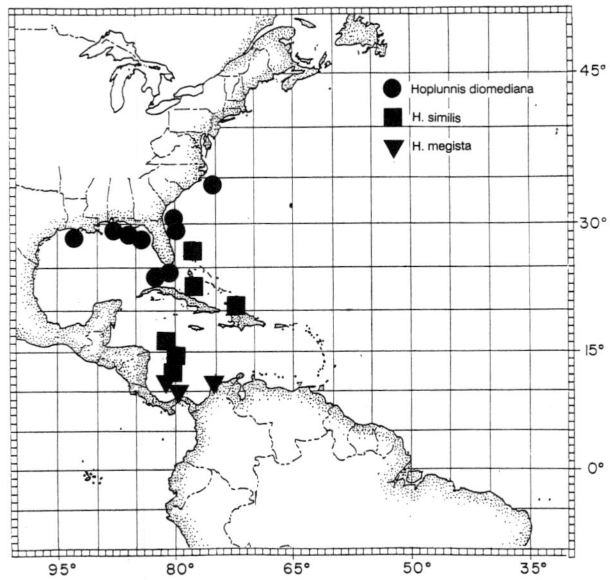

FIGURE 617. Distribution of *Hoplunnis megista, H. diomediana,* and *H. similis.*

Distribution (Fig. 617). Southern Caribbean Sea from Colombia to Nicaragua. Depth 366–421 m. Rare.

Etymology. From the Greek *megistos* (largest), referring to the fact that this is the largest species of *Hoplunnis.* An adjective.

Study Material. Nine specimens, 660–1095 mm TL, 202–405 mm preanal.

Holotype: USNM 193613 (male, 695), Oregon 3598, Panama, 9°03′N, 81°22′W, 366–403 m, 31 May 1962.

Paratypes: COLOMBIA: ANSP 126447 (1, 1095), 11°03′N, 75°18′W, 366 m, 2 Dec. 1968. PANAMA: ANSP 138773 (1, 709) and 147554 (1, 770), same data as holotype. USNM 220039 (2, 910–917), 9°47′N, 79°25′W, 421 m, 19 Oct. 1965. NICARAGUA: USNM 193598 (3, 660–905), 12°23′N, 82°29′W, 366 m, 2 June 1962.

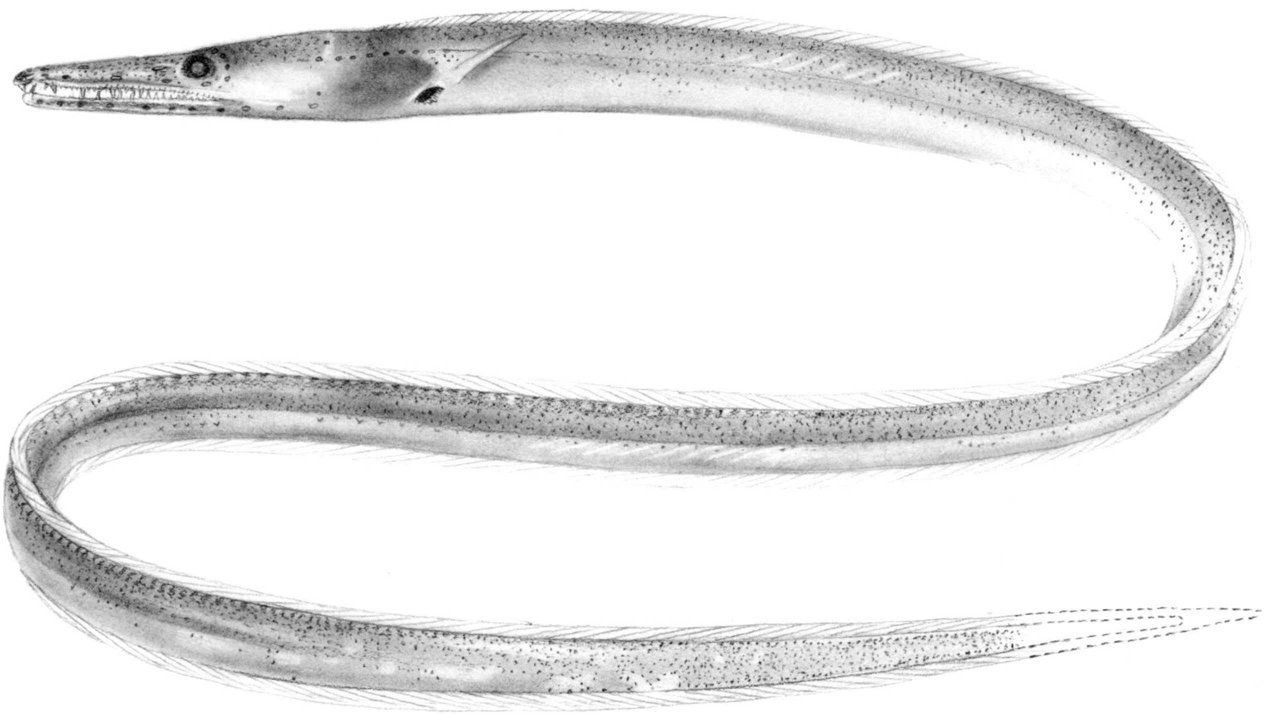

FIGURE 618. *Hoplunnis tenuis*: ANSP 157600, 400+ mm TL.

Hoplunnis tenuis Ginsburg, 1951
Figures 618–621, Table 75

Hoplunnis diomedianus (not of Goode and Bean, 1896). Goode and Bean, 1896: 146, part.
Hoplunnis tenuis Ginsburg, 1951: 448, fig. 4 (original description, 28°36′N, 85°33.5′W, holotype USNM 152574).
Hoplunnis schmidti (not of Kaup, 1860). Miranda Ribeiro, 1961a: 5, fig. 3.
Hoplunnis diomedianus (not of Goode and Bean, 1896). Miranda Ribeiro, 1961b: 6.

Discussion of Synonymy. Ginsburg (1951: 449) pointed out that the original type series of *H. diomediana* Goode and Bean contained two species. He removed one of the two specimens and made it the holotype of *H. tenuis.*

Distinctive Characters. Pterygoid teeth present except in young individuals; lateral vomerine teeth numerous and closely set; maxillary and mandibular teeth in three rows; enlarged mandibular and vomerine teeth relatively smaller than in other species; stomach and intestine black; total vertebrae 232–244.

Hoplunnis tenuis is the only species of the genus that has pterygoid teeth. The lateral vomerine teeth are more numerous than those of *H. diomediana* and *H. similis* and arranged in a more regular order. The head of *H. tenuis* is shaped somewhat differently from that of other species, being more slender and conical in profile.

Description. Total vertebrae 232–244 (n = 10), predorsal vertebrae 3–6 (12), preanal vertebrae 36–38 (15). Branchiostegal rays 7–8 (9). Pectoral-fin rays ca 9–10 (2). Pores: LL to anus 35–38 (24), POM 11–14 (6). IO 8 (12), SO 1 + 4 (11), STC 3 (11).

Proportions as % of TL: preanal ca 25–26 (4), predorsal 8–10 (4), head 10–11 (4), depth at anus 2 (1). Of preanal length: predorsal 33–39 (36), head 39–44 (39), pectoral fin 7–9 (13), depth at anus 6–9 (26). Of HL: snout 35–41 (29), eye 8–11

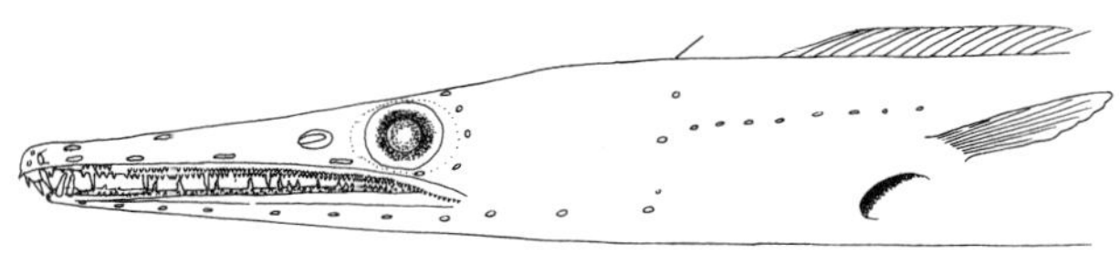

FIGURE 619. *Hoplunnis tenuis*: ANSP 157600, 400+ mm TL; head.

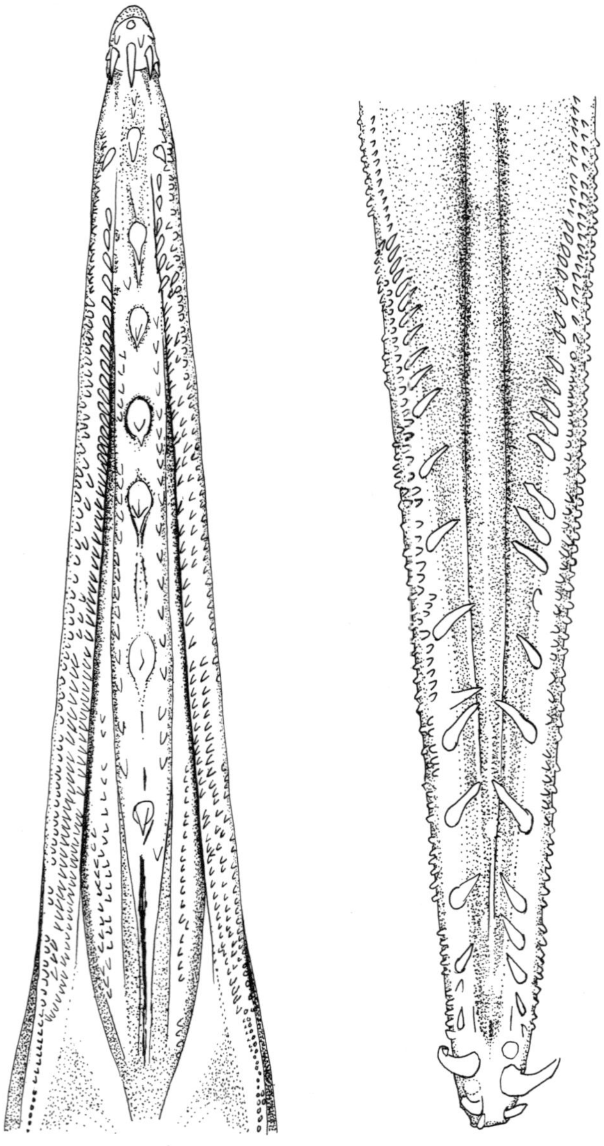

FIGURE 620. *Hoplunnis tenuis*: ANSP 136485, 307 mm TL; dentition.

(29), upper jaw 51–59 (17), gill opening 5–8 (16), interbranchial 5–11 (14), pectoral fin 17–25 (17).

Maxillary teeth in three rows posteriorly, two rows anteriorly, those of inner row slightly enlarged, those of outer row very small and disappearing anteriorly. Mandibular teeth in three irregular rows, reduced to two anteriorly; for most of length teeth of inner series fewer and greatly enlarged; at tip of mandible one or two enlarged teeth on each side with a few very small teeth anterior to them. Intermaxillary teeth in three irregular transverse rows, the posteriormost containing the largest teeth, about five. A single series of teeth on midline of vomer, the anteriormost few small, the remainder enlarged fangs, flanked on each side by a close-set row of small teeth. A slender band of very small teeth on pterygoid, medial to posterior end of maxillary teeth, present only in larger individuals.

Stomach long, ending behind anus. Gas bladder slightly longer than stomach.

Color. In preservative light brown or tan, speckled or mottled with small brown spots dorsally and traces of silvery or golden metallic reflections. Interior of branchial chamber black. In specimens from Caribbean and South America, inside of mouth black. Vertical fins edged in black posteriorly. Stomach and intestine black. Gas bladder silver.

Size. The largest specimens examined were 576 mm TL (MZUSP 11924) and 575 mm TL (USNM 157951).

Development and Growth. The pterygoid teeth are present only in larger specimens, developing at somewhere between 50 and 65 mm preanal length. Three specimens of 28–50 mm preanal length (UMML 28332, USA 03269, VIMS 05941) lacked pterygoid teeth; a specimen of 65 mm preanal length (UMML 1164) had them.

The leptocephalus has been identified (see p. 710).

Variation. Specimens from the Gulf of Mexico, the Caribbean, and the Guianas differ in the total number of vertebrae. Vertebral counts from the Gulf of Mexico were 232–238 (n = 5); from the Caribbean 241–244 (3); from Suriname 233–237 (2). Myomere counts for leptocephali from the same areas are as follows: Gulf of Mexico 234–246 (24), Caribbean 237–250 (29), Guianas and Brazil 232–239 (5) (Smith and Castle, 1982: 22). These counts do not form a simple cline; the highest values are found in the Caribbean population while those of the northern and southern populations are lower. In addition the lining of the mouth is black in the Caribbean and Guianan specimens, whereas in specimens from the Gulf of Mexico it is pale.

Distribution (Fig. 621). *Hoplunnis tenuis* has the widest distribution of any species of the genus. It has been collected most commonly in the Gulf of Mexico and along the coast of Central and South America to the Guianas. It is also known from the Bahamas and the West Indies. The single record from the east coast of the United States is from off New Jersey (VIMS 05941). In addition, the species occurs off southern Brazil (eight specimens, MNRJ and MZUSP, examined and identified by J. E. Böhlke, unpubl.). Smith and Castle (1982: 22) reported a single leptocephalus from the northeastern Atlantic near the Canary Islands (ZMUC Dana 4010 V). Depth 110–468 m.

Etymology. Latin *tenuis* (thin), in reference to the slender body and tail. An adjective.

Study Material. 134 specimens, 147–575 mm TL.

Holotype: USNM 152574 (female, 389), Albatross 2402, Florida Gulf, 28°36′N, 85°33.5′W, 203 m, 14 March 1885.

Other material: NEW JERSEY: VIMS 05941 (1, 188). FLORIDA GULF: ANSP 157600 (1, 400). USA 03269 (6, 168–464), 03278 (10, 131–482). USNM 158152 (1, 204). ALABAMA: ANSP 109596 (9, 374–485), 109597 (5, 308–345), 116640 (1, 310). MISSISSIPPI: ANSP 114963 (1, 575), 117063 (6, 147–297). MEXICO GULF: ANSP 117085 (2, 370–375), 153563 (1, 149). BAHAMAS: UMML 1164 (1, 220). CUBA: USNM 157942 (3, 412–525), 157946 (1, 448), 157951 (13, 400–539). VIRGIN ISLANDS: ANSP 117059 (1, 494). LESSER ANTILLES: USNM 217395 (1, 342). VENEZUELA: ANSP 136754 (1, 438). UMML 28332 (1, 95). USNM 217391 (1, 126), 217394 (2, 162–405), 217397 (8, 176–462). COLOMBIA: ANSP 126135 (1, 317), 136485 (1, 307). USNM 198734 (1, 310–337), 198745 (2, 229–444). PANAMA: USNM 179236 (2, 397–495), 193546 (1, 417), 193553 (2, 286–413), 193562 (3, 359–452), 193566 (5, 334–443), 217396 (3, 289–330). NICARAGUA: ANSP 114958 (1, 448), 133019 (2, 175–340), 138393 (1, 452). UMML 29679 (2, 417–456). USNM 217392 (1, 389), 217393 (3, 424–425). GUYANA: USNM 158895 (1, 306). SURINAME: ANSP 117011 (4, 200–430), 138771 (1, 360). USNM 185636 (11, 323–472). BRAZIL: MNRJ uncat. (1). MZUSP 11923 (1, 468), 11924 (1, 576), 11925 (1, 110), 11926 (4, 115–491).

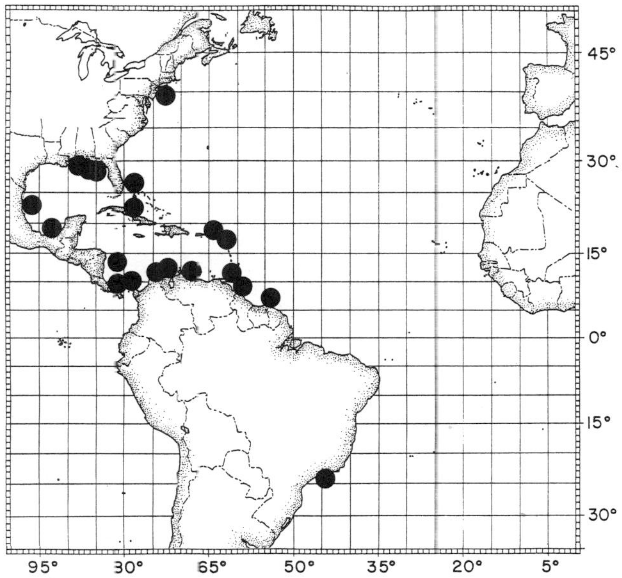

FIGURE 621. Distribution of *Hoplunnis tenuis.*

Hoplunnis diomediana Goode and Bean, 1896
Figures 617, 622–624, Table 75

Hoplunnis diomedianus Goode and Bean, 1896: 146, fig. 163 (original description, 28°16′N, 85°33.5′W, lectotype USNM 44240).

Hoplunnis schmidti (not of Kaup, 1860). Jordan and Evermann, 1900: pl. 58, fig. 151 (figure from Goode and Bean, 1896).

Hoplunnis diomediana. Lane and Stewart, 1968: 54 (emendation).

Misidentifications:

Not *Hoplunnis diomedianus* Goode and Bean. Miranda Ribeiro, 1961b: 6 (*H. tenuis*).

?Not *Hoplunnis diomedianus* Goode and Bean. Uyeno et al., 1983: 118 (probably an undescribed species).

Distinctive Characters. Pterygoid teeth absent; lateral vomerine teeth few and widely spaced; maxillary and mandibular teeth in three rows; body light in color with scattered dark spots; stomach black, intestine pale; total vertebrae 221–234.

In *Hoplunnis diomediana* the lateral vomerine teeth are less conspicuous than in *H. tenuis.* They are few and widely spaced and often have to be felt with a probe. Pterygoid teeth are never present. *Hoplunnis diomediana* differs from *H. macrura* by the presence of lateral vomerine teeth at all stages and by having a black rather than pale stomach. It is closest to *H. similis* among western Atlantic species, differing only in color and number of vertebrae.

Description. Total vertebrae 221–234 (n = 18), predorsal vertebrae 3–5 (20), preanal vertebrae 35–41 (25). Branchiostegal rays 5–7 (9). Pectoral-fin rays 9–10 (3). Pores: LL to anus ca 33–38 (17), POM 12–13 (5), IO 8 (12), SO 1 + 4 (11), STC 3 (12).

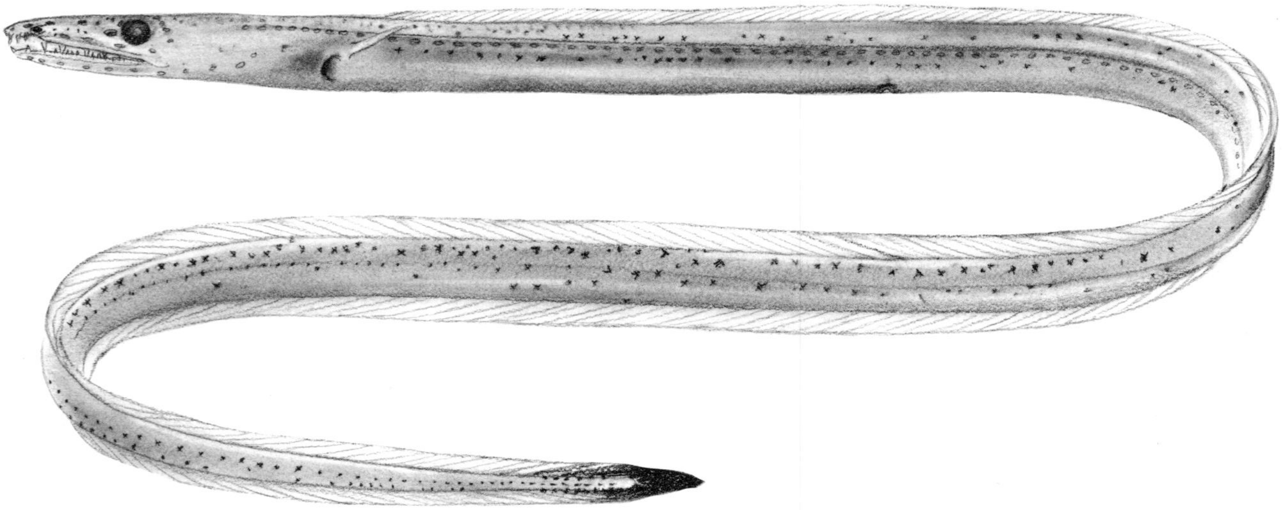

FIGURE 622. *Hoplunnis diomediana*: ANSP 135422, 385 mm TL.

Proportions as % of TL: preanal 23–25 (5), predorsal 8–9 (5), head 9–10 (5). Of preanal length: predorsal 32–38 (16), head 38–43 (19), pectoral fin 6–9 (11), depth at anus 5–8 (11). Of HL: snout 35–38 (13), eye 8–12 (13), upper jaw 52–56 (12), gill opening 4–8 (11), interbranchial 5–8 (10), pectoral fin 15–21 (11).

Maxillary teeth in three rows, those of inner row slightly enlarged, those of outer row very small. Mandibular teeth in three rows, those of inner row markedly enlarged, those of outer row very small; at tip of mandible one or two enlarged teeth with several smaller teeth anterior to them. A semicircular series of six or seven enlarged intermaxillary teeth directed downward and outward, sometimes a single median tooth within the semicircle. Vomer with a single series of teeth on midline, the anteriormost few small, the remainder greatly enlarged; a few small teeth laterally, often concealed by flesh, spaced irregularly, not forming a nearly continuous ridge as in *H. tenuis*.

Length of stomach variable, sometimes extends well behind anus. Gas bladder begins slightly behind anterior end of stomach, ends about one stomach length behind posterior end of stomach.

Color. In preservative light yellowish brown, sparsely sprinkled with small dark spots mainly above lateral line; spots concentrated most densely on dorsal aspect of head. Inside of branchial chamber black. Vertical fins edged in black posteriorly. Stomach black, intestine pale. Gas bladder silver.

Size. The largest specimen examined was 543 mm TL (VIMS 01492).

Development and Growth. The leptocephalus has been identified (see p. 714).

Remarks. Goode and Bean (1896: 146), in the description of *Hoplunnis diomediana*, stated that "a single individual (No. 44240 U.S. Nat. Mus.) was obtained by the Albatross. . . ." Nevertheless, they gave a range of variation for several characters, for example: "Snout three to three and two-thirds times eye" and "a row of six to eight strong, large canines on the vomer." Ginsburg (1951: 449) found that USNM 44240 originally contained three specimens, one of which had been sent to Stanford University. The two specimens remaining in the jar belonged to two species. Ginsburg selected the one that appeared to have been illustrated as the lectotype of *Hop-*

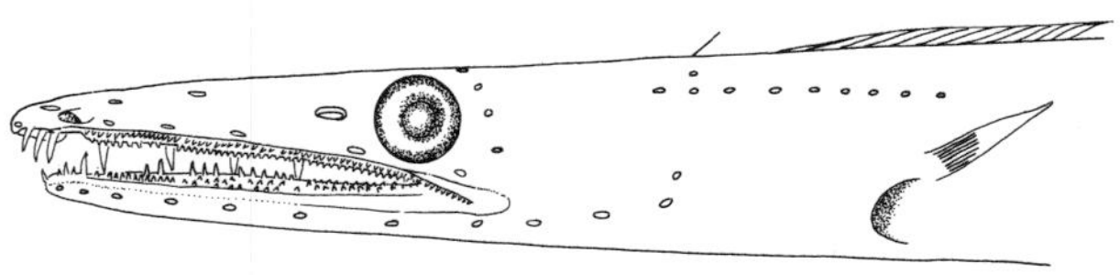

FIGURE 623. *Hoplunnis diomediana*: ANSP 135422, 385 mm TL; head.

lunnis diomediana; the other became the holotype of *H. tenuis.*

Distribution (Fig. 617). Eastern and southern Gulf of Mexico, around the Florida peninsula and north to North Carolina. Depth 33–203 m, mean 102 m.

In the northern Gulf of Mexico, the species occurs mainly in the eastern part. There is only one record west of the Mississippi (ANSP 117070 and USNM 217403, both from the same station). Smith and Castle (1982: 26) reported leptocephali from the southern Gulf just west of Campeche Bank. In the Straits of Florida it occurs only on the continental side.

Uyeno et al. (1983: 118) reported *Hoplunnis diomediana* from 166 m off the Guianas, far outside the known range of the species. The dentition of their specimen is similar to that of *H. diomediana* and *H. similis,* with 7–8 enlarged vomerine teeth flanked by 13 smaller ones. The illustrated specimen resembles *H. diomediana,* but there seems to be some mottling. Vertebral counts were not given. Although these specimens could represent a disjunct population of *H. diomediana,* they are more likely to be a separate, undescribed species. Smith and Castle (1982: 76) found leptocephali of three species of *Hoplunnis* that were identical to those of *H. diomediana* except for myomere counts. One of those (*Hoplunnis* sp. A) turned out to be *H. similis,* but the other two (*Hoplunnis* spp. B and C) remain unidentified. *Hoplunnis diomediana* is thus one of at least four closely related species in the western Atlantic.

Etymology. From *Diomedea* (the albatross) and *-ana* (belonging to). Named for the U.S. Fish Commission steamer ALBATROSS. An adjective.

The original spelling of this name, *diomedianus,* seems based on a misinterpretation of the gender of *Hoplunnis,* perhaps a confusion with *hoplites,* which is indeed masculine. The *-anus* ending is clearly adjectival. Elsewhere in the same work, Goode and Bean used this ending in agreement with the gender of the genus (for example, *Harriotta raleighana*). There seems little basis for interpreting *diomedianus* as a noun, and I follow Lane and Stewart (1968: 54) in emending the name to *diomediana.*

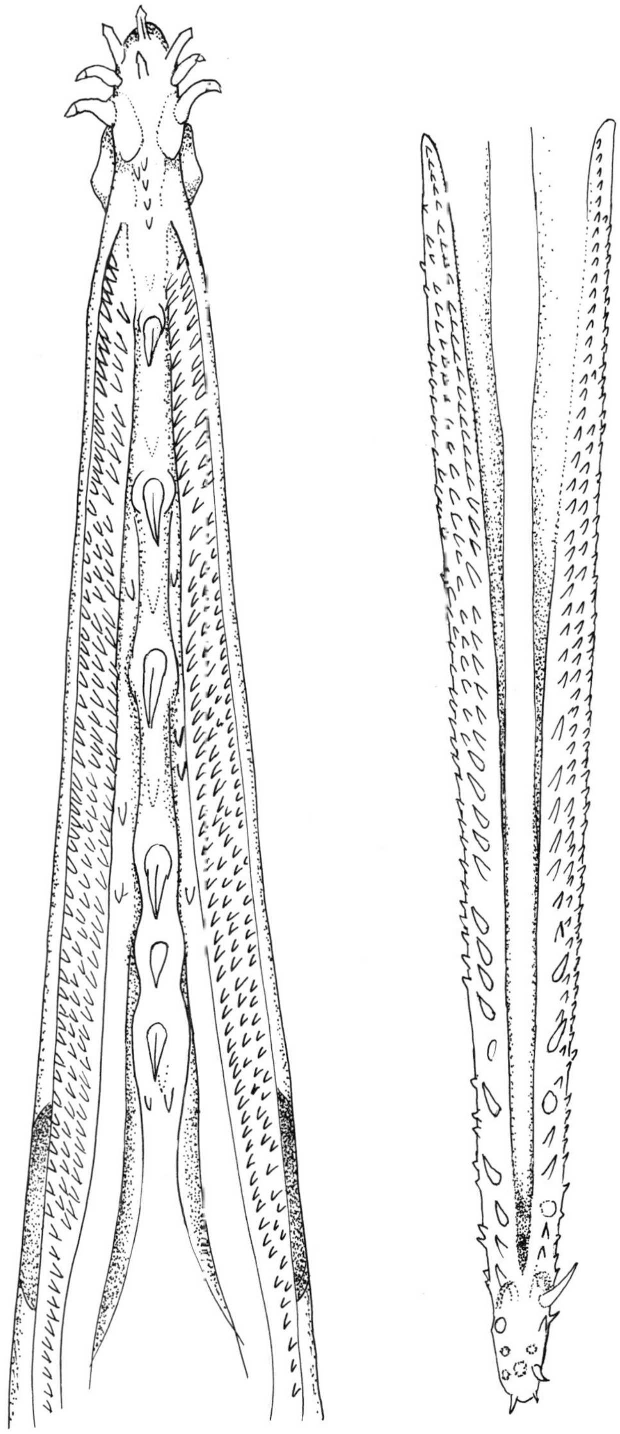

FIGURE 624. *Hoplunnis diomediana*: ANSP 138772, 404 mm TL; dentition.

Study Material. 129 specimens, 203–543 mm TL.

Lectotype: USNM 44240 (male, 418), Albatross 2402, Florida Gulf, 28°36′N, 85°33.5′W, 203 m, 14 March 1885.

Other material: NORTH CAROLINA: VIMS 01492 (2, 460–543). FLORIDA ATLANTIC: ANSP 117075 (1, 267), 128400 (2, 385–496), 128401 (1, 380), 128402 (1, 210),

FIGURE 625. *Hoplunnis similis*: ANSP 130814, holotype, 362 mm TL.

133016 (1, 203), 135422 (1, 385), 138772 (1, 404). GMBL 74-96 (10, 275–371). UMML 7026 (1, 454). USNM 217402 (1, 230). FLORIDA STRAITS: USNM 197123 (1, 356), 197124 (1, 250), 197129 (2, 328–340). FLORIDA GULF: ANSP 138774 (1, 333). USA 03457 (1, 429), 03458 (3), 03459 (3), 03497 (3, 339–438), 03498 (1, 480). USNM 163496 (72, 246–466), 163519 (10, 281–435), 185635 (1, 357). ALABAMA: GCRL 21390 (2, 250–336). USA 03865 (1), 03866 (1, 333). LOUISIANA: ANSP 117070 (2, 292–308). USNM 217403 (1, 217).

Hoplunnis similis new species
Figures 617, 625–627, Table 75

Hoplunnis sp. A. Smith and Castle, 1982: 26 (leptocephalus).

Distinctive Characters. Pterygoid teeth absent; lateral vomerine teeth few and widely spaced; maxillary and mandibular teeth in three rows; body darkly mottled; stomach and intestine black; total vertebrae 254–260.

Hoplunnis similis closely resembles *H. diomediana*. The only external difference is the color. *Hoplunnis diomediana* is basically light with scattered dark spots; *H. similis* has larger spots more densely concentrated, giving a darkly mottled appearance. In *H. diomediana* the spotting on the head is confined to the dorsal surface, whereas in *H. similis* the lateral surface is covered as well. In *H. similis* both the stomach and intestine are black, in *H. diomediana* only the stomach.

Description. Total vertebrae 254–260 (n = 7), predorsal vertebrae 2–4 (4), preanal vertebrae 39–42 (5), precaudal vertebrae 86 (1). Branchiostegal rays 6 (5); one specimen had five on one side. Pectoral-fin rays ca 10 (2). Pores: LL to anus 37–41 (12), POM 11–13 (10), IO 8 (12), SO 1 + 4 (12), STC 3 (8).

Proportions as % of TL: preanal 23–24 (7), predorsal 7–8 (5), head 8–10 (7), depth at anus 1–2 (7). Of preanal length: predorsal 27–32 (18), head 36–41 (20), pectoral fin 6–8 (9), depth at anus 5–8 (11). Of HL: snout 33–38 (15), eye 7–13 (15), upper jaw 48–56 (11), gill opening 3–7 (11), interbranchial 5–8 (7), pectoral fin 16–19 (9).

Dentition as in *H. diomediana*. Stomach and gas bladder end behind anus.

Color. Ground color in preservative light brown or tan, entire body darkly mottled with brown spots. Inside of branchial chamber dark. Vertical fins edged in black posteriorly. Stomach and intestine black, gas bladder silver.

Size. The largest specimen examined was 493 mm TL, 112 mm preanal (USNM 217404).

Development and Growth. The leptocephalus has been identified (see p. 715).

Distribution (Fig. 617). Adults are known from the Bahamas and the coast of Central America. Depth 183–468 m, mean 279 m. Leptocephali have been collected from the Lesser Antilles as well, indicating that the species is widely distributed in the Caribbean.

Hoplunnis similis and *H. diomediana* appear to have disjunct distributions. Although *H. similis* occurs in the northern Bahamas, it has never been collected on the mainland side of the Florida Straits. *Hoplunnis diomediana*, although pres-

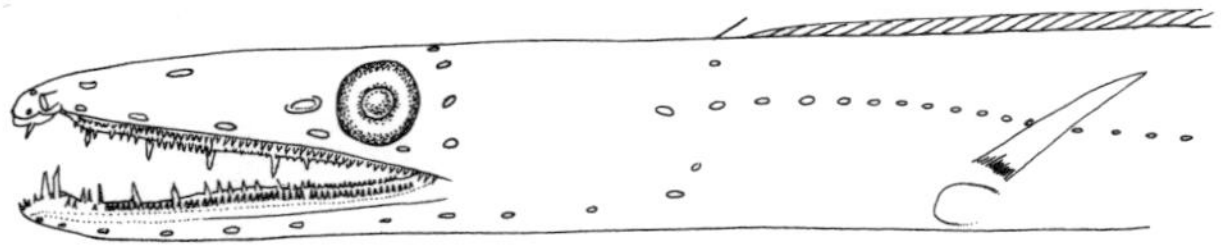

FIGURE 626. *Hoplunnis similis*: ANSP 130814, holotype, 362 mm TL; head.

ent on the Atlantic coast of Florida, has never been recorded from the Bahamas. *Hoplunnis similis* lives at a greater average depth than any of the other species of *Hoplunnis* studied.

Etymology. Latin *similis* (like, resembling), in reference to the resemblance of the species in many characters to *Hoplunnis diomediana*. An adjective.

Study Material. 43 specimens, 229–500 mm TL.

Holotype: ANSP 130814 (male, 362), Oregon 10851, southern Bahamas, 20°53′N, 73°26′W, 287 m, 13 Dec. 1969.

Paratypes: BAHAMAS: ANSP 114933 (1, 450) and 114949 (2, 271–291), 23°12′N, 78°49′W, 275 m, 15 Dec. 1969. UMML 14153 (1, 443), 27°25′N, 78°41′W, 293–311 m, 5 Feb. 1964. USNM 158320 (1, 459), 27°27′N, 78°58′W, 329 m, 2 Feb. 1957. NICARAGUA: ANSP 135423 (1, 345), 14°53.9′N, 81°23.2′W, 296–375 m, 31 Jan. 1971. USNM 217404 (7, 378–493), 14°10′N, 81°57.5′W, 147 m, 12 Feb. 1967. HONDURAS: UMML 29680 (2, 229–444), 15°14′N, 81°26′W, 456–468 m, 31 Jan. 1971. USNM 193548 (4, 321–449), 16°45′N, 81°27′W, 275 m, 7 June 1962.

Other material: NICARAGUA: USNM 179241 (6, 40–483), 193555 (3, 312–501), 217405 (2, 364–479), 217406 (1, 418). HONDURAS: USNM 193556 (1, 375), 193571 (10, 327–500). (Note: because of space limitations in the type section at the USNM, these specimens were not designated as paratypes.)

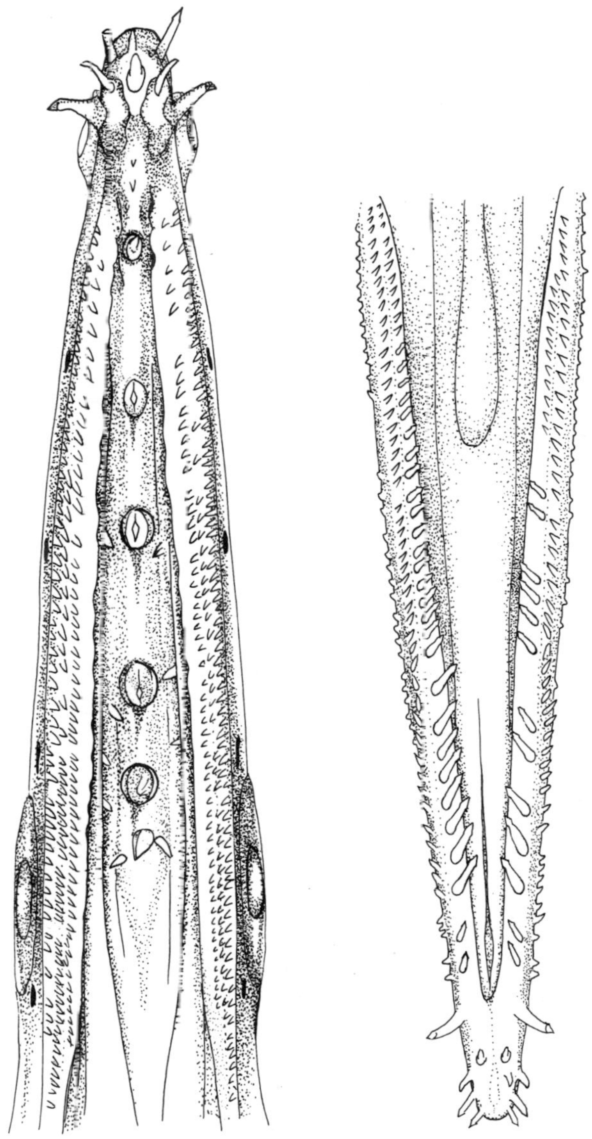

FIGURE 627. *Hoplunnis similis*: ANSP 130814, holotype, 362 mm TL; dentition.

Hoplunnis macrura Ginsburg, 1951
Figures 628–631, Table 75

?*Hoplunnis schmidti* Kaup, 1860: 19 (original description, Puerto Cabello, Venezuela, holotype lost).

Hoplunnis macrurus Ginsburg, 1951: 450, fig. 5 (original description, 29°14′N, 88°35′W, holotype USNM 152565).

Hoplunnis macrura. Lane and Stewart, 1968: 56 (emendation).

Distinctive Characters. Pterygoid teeth absent; lateral vomerine teeth present only in very small individuals; maxillary and mandibular teeth in two rows; stomach and intestine pale; total vertebrae 227–245.

Hoplunnis macrura is the only western Atlantic species of the genus that does not retain the small lateral vomerine teeth as it grows to adult size. It is also the only species with a pale stomach.

Description. Total vertebrae 227–245 (n = 12), predorsal vertebrae 4–5 (19), preanal vertebrae 34–38 (24). Branchiostegal rays 7–8 (11). Pectoral-fin rays 10–11 (5). Pores: LL to anus 32–37 (14), POM 12–14 (7), IO 8 (12), SO 1 + 4 (13), STC 3 (11).

Proportions as % of TL: preanal 22–23 (6), predorsal 9–10 (6), head 10–12 (6), depth at anus 1–2 (5). Of preanal length: predorsal 35–41 (19),

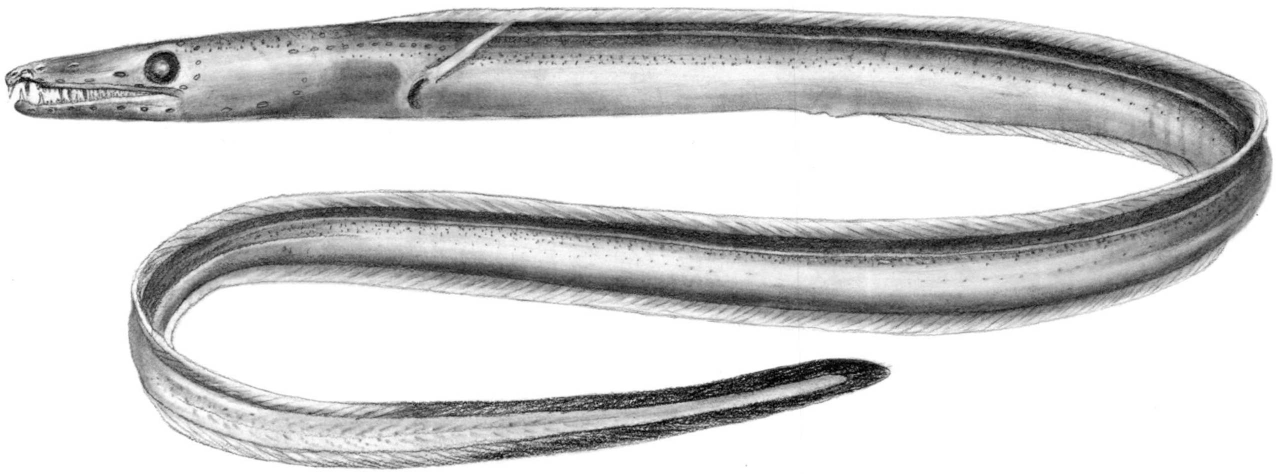

FIGURE 628. *Hoplunnis macrura*: ANSP 130815, 384 mm TL.

head 43–48 (20), pectoral fin 7–10 (8), depth at anus 7–11 (16). Of HL: snout 32–40 (19), eye 9–11 (18), upper jaw 50–57 (18), gill opening 4–7 (10), interbranchial 6–8 (10), pectoral fin 15–23 (10).

Maxillary teeth in two rows, those of inner row slightly enlarged. Mandibular teeth in two rows, those of inner row greatly enlarged; at tip of mandible one or two greatly enlarged teeth on each side with one or two smaller teeth in front or behind. An enlarged, median, downwardly directed intermaxillary tooth flanked by one or two smaller teeth on each side. A single series of greatly enlarged teeth on midline of vomer; small lateral vomerine teeth present only in very small individuals.

In specimen examined (USNM 179247), stomach long, ending behind anus. Gas bladder begins at about midlength of stomach and ends an equivalent distance behind posterior end of stomach.

Color. In preservative light yellowish brown, speckled with small brown spots dorsally. In life a metallic silvery sheen. Interior of branchial chamber black except in very small individuals. Vertical fins edged in black posteriorly. Stomach and intestine pale, gas bladder silver.

Size. The largest specimen examined was 542 mm TL (ANSP 117082). A female of 379 mm TL, 87 mm preanal (USNM 217401) had large eggs.

Development and Growth. The lateral vomerine teeth are present in small individuals but absent in larger ones. The smallest specimen examined, 129 mm TL, 30 mm preanal length (UMML 28795), had fairly well developed lateral vomerine teeth. A specimen of 141+ mm TL, 36 mm preanal (USA 02483) had four lateral vomerine teeth, and another specimen from the same lot, 147 mm TL, 36 mm preanal had four small teeth around the posteriormost enlarged tooth. None of the larger specimens examined had lateral vomerine teeth.

The three small specimens mentioned above also lacked the black pigment in the lining of the branchial chamber.

The leptocephalus has been identified (see p. 712).

Variation. Based on the total number of vertebrae, three geographic populations of *Hoplunnis macrura* can be distinguished. In the Gulf of Mexico the single reliable count was 227. For the Caribbean the counts were 234–239 (n = 5), and for the Guianas and Brazil 241–245 (5). A specimen collected just east of Margarita Island, Ven-

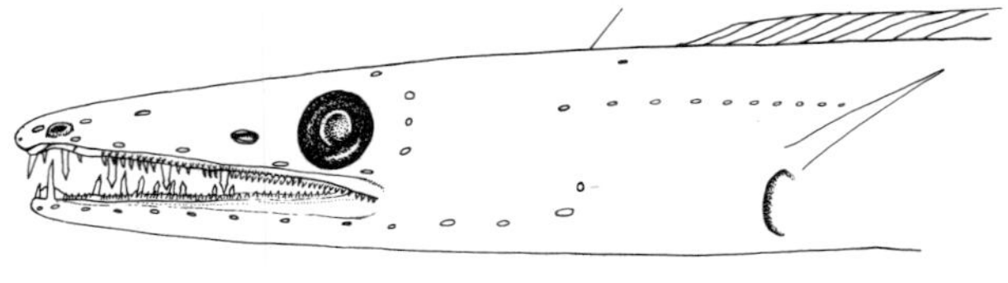

FIGURE 629. *Hoplunnis macrura*: ANSP 130815, 384 mm TL; head.

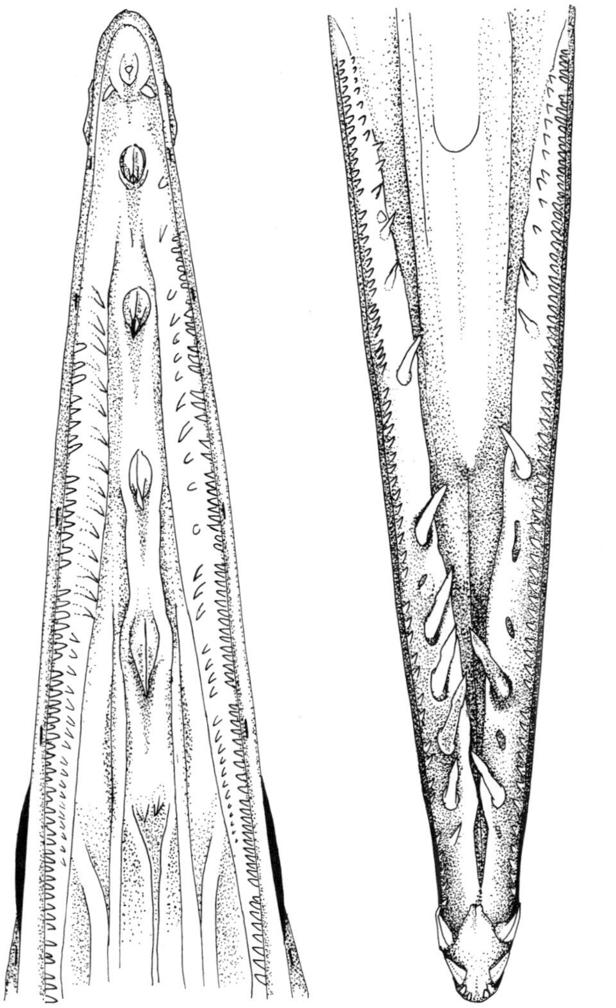

FIGURE 630. *Hoplunnis macrura*: ANSP 130815, 384 mm TL; dentition.

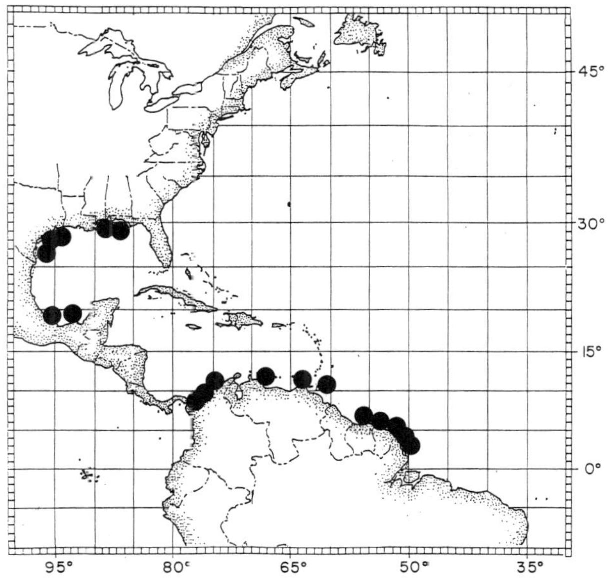

FIGURE 631. Distribution of *Hoplunnis macrura*.

ezuela (UMML 28329) had 245 vertebrae, agreeing with the Guianas-Brazil group. Myomere counts of the leptocephali were as follows: Gulf of Mexico 221–234 (22), Caribbean 229–244 (34), Guianas and Brazil ca 243–252 (8) (Smith and Castle, 1982: 24).

Remarks. Hoplunnis schmidti Kaup, 1860 may be a senior synonym of *H. macrura*, but since the holotype apparently has been lost (Lane and Stewart, 1968: 57), there is no way to be sure. The description of *H. schmidti* (Kaup, 1860: 19) states that there are no lateral vomerine teeth ("... keinen kleineren Zahnreihen ..."), a condition that fits only *H. macrura* among the currently recognized western Atlantic species. In addition, the type locality of *H. schmidti* (Venezuela) is within the range of *H. macrura*. Ginsburg (1951: 447) apparently misread Kaup's statement and attributed "small teeth alongside the palatal fangs" to *H. schmidti*. Being thus unable to match *H. schmidti* with any of the species in front of him, he created a new species, *H. macrura*. Without the holotype, however, some doubt will always remain. Moreover, evidence from leptocephali indicates that additional species of *Hoplunnis* may exist in the western Atlantic besides the five now recognized (Smith and Castle, 1982). In view of the uncertainty, I reject *Hoplunnis schmidti* in favor of the well documented *H. macrura*.

One specimen (UMML 28331) had an isopod in each gill cavity.

Distribution (Fig. 631). Western and southern Gulf of Mexico, Caribbean coast of Colombia, Venezuela, the Guianas, and Brazil at least to the mouth of the Amazon. Depth 20–220 m, mean 67 m.

In the Gulf of Mexico *Hoplunnis macrura* appears restricted to the western part; there are no records east of Pensacola, Florida, nor from anywhere off the Atlantic coast of the United States. Its distribution seems to reflect areas of mud or fine silt. Although there are no records of adults from the West Indies, leptocephali are found there (Smith and Castle, 1982).

Etymology. From the Greek *makros* (long) and *oura* (tail). The correct form of this name, *macrura* or *macrurus,* depends on whether it is treated as a noun or an adjective. If it is a noun, then the original spelling *macrurus* would seem to be required (Article 32 [b] [ii], ICZN), even though it is an alteration of the Greek *oura,* a feminine noun. It seems more likely, however, that Ginsburg intended it as an adjective and held the same misconception as Goode and Bean (1896) that *Hoplunnis* was masculine. The *macrurus* spelling was used six times without variation in the text of the description, so it was certainly intentional. Ginsburg also used the *-us* ending consistently in referring to *H. diomedianus* Goode and Bean. I think it unlikely that Ginsburg would mistakenly or capriciously form a noun by changing the termination of *macrura* to *macrurus.* I therefore follow Lane and Stewart (1968) in emending *macrurus* to *macrura.*

Study Material. 67 specimens, 115–542 mm TL.

Holotype: USNM 152565 (male, 369), Oregon 49, Mississippi, 29°14′N, 88°35′W, 73 m, 13 July 1950.

Paratype: USNM 152566 (1, 405).

Other material: FLORIDA GULF: USA 03460 (2). ALABAMA: ANSP 117051 (1, 435). USNM 186203 (1, 418). MISSISSIPPI: ANSP 117082 (1, 542). USA 02483 (3). USNM 158255 (1, 389). LOUISIANA: USNM 162575 (1, 324), 179247 (1, 279). TEXAS: ANSP 117043 (2, 189–275). UMML 7752 (1, 268). USNM 157945 (2, 200–315), 158891 (2, 200–392). MEXICO GULF: ANSP 126134 (1, 279), 152767 (3, 271–379). VENEZUELA: UMML 28329 (1, 370), 28331 (1, 93 mm preanal), 29678 (1, 442). COLOMBIA: ANSP 112043 (3, 306–365), 114862 (1, 306), 114925 (2, 264–292), 114938 (1, 480), 156857 (1, 287). UMML 28333 (3, 139–278), 28795 (1, 129). USNM 198736 (1, 284), 217398 (2, 420–442), 217399 (2, 331–351), 217400 (2, 294–310), 217401 (1, 379). SURINAME: ANSP 114876 (3, 375–504), 114924 (1, 440), 114941 (1, 384). USNM 158896 (1, 279), 185661 (4, 440–489), 186056 (2, 115–295). FRENCH GUIANA: ANSP 117056 (2, 343–412), 117060 (2, 386–457), 130815 (1, 384), 137638 (1, 398). UMML 11562 (1, 320). BRAZIL: ANSP 138437 (1, 429), 139241 (1, 413).

Genus *Saurenchelys* Peters, 1864

Saurenchelys Peters, 1864: 397 (type species *Saurenchelys cancrivora* Peters, 1864, by monotypy).

Dietrichthys Whitley, 1935: 219 (type species *Chlopsis finitimus* Whitley, 1935, by original designation). Described as a subgenus.

Characters. Body elongate, round in cross section anteriorly, compressed posteriorly, depth relatively uniform, tapering posteriorly to slender, filiform tail; anus near anterior fifth of TL. Dorsal fin begins over or slightly behind level of gill opening. Pectoral fin absent. Gill opening crescentic, on lower half of side.

Head and snout moderately elongate, smooth in profile; upper jaw projects beyond lower. Tip of fleshy snout soft, but not produced, protrudes very slightly beyond intermaxillary teeth. Eye moderate. Mouth relatively large, rictus slightly behind posterior margin of eye. Anterior nostril a low tube just behind tip of snout, directed anteriorly; posterior nostril a longitudinal slit in front of eye at mid-eye level.

Lateral line on body complete to tail. Number of supraorbital pores variable: a small one at anteroventral tip of snout, before anterior nostril (the ethmoidal pore); a small one above the preceding, just before anterior nostril; an enlarged, slit-like pore bordering dorsal side of anterior nostril tube; 0–3 smaller ones behind this on anterior half of snout. Infraorbital canal with eight pores, those along upper jaw elongate and slit-like: one just behind anterior nostril tube, two between anterior and posterior nostrils, one between posterior nostril and eye, one below posterior margin of eye, and three behind eye; adnasal pore absent. Three pores in supratemporal commissure. Preoperculomandibular canal with 13 slit-like pores, nine before rictus and four behind. One pore at junction of supratemporal, preoperculomandibular, and lateral-line canals.

Teeth generally small and conical, in narrow bands. Maxillary teeth in three rows, more or less continuous with intermaxillary teeth, teeth of inner row largest. Intermaxillary tooth patch somewhat spade-shaped, teeth in approximately six transverse rows; a crescent-shaped gap between intermaxillary and vomerine teeth. Vomerine tooth patch long, reaching about opposite posterior nostril, consisting of a median row of enlarged teeth and two to three rows of smaller teeth on each side. Mandibular teeth in about five rows in middle of jaw, about three rows at anterior and posterior ends; a slight gap between left and right tooth patches at tip of jaw; anterior tip of mandibular tooth patches fits into gap between intermaxillary and vomerine tooth patch when mouth closed. Pterygoid with an elongate, multiserial band of small teeth, the band about as long as vomerine patch, beginning lateral to posterior end of vomerine patch, ending about level with posterior end of maxillary tooth patch.

Stomach ends near or slightly behind anus. Gas bladder begins at a point somewhat behind midpoint between gill opening and anus, extends a somewhat greater distance behind anus.

Discussion. Only recently has this genus been unambiguously recognized for what it is. For many years it was confused with the unrelated *Chlopsis* Rafinesque, 1810. This resulted from Jordan and Davis's (1891: 650) unfortunate decision to synonymize the two genera. Their conclusion was accepted without question by American ichthyologists and perpetuated for more than half a century, despite abundant and readily available evidence to the contrary. The matter is discussed in detail elsewhere in this volume (genus *Chlopsis*, p. 92).

As if this were not enough, *Saurenchelys* was also confused with *Facciolella* Whitley, which is at least in the same family. This problem stemmed from uncertainty over the type locality of the type species, *Saurenchelys cancrivora* Peters. In his description of *S. cancrivora*, Peters (1864: 399) stated that the specimen (he had only one) came to him in a collection of Indian fishes but that its true origin must be the Atlantic or Mediterranean because it had a Mediterranean shrimp (*Nika edulis* Risso) in its stomach. This unlikely conclusion was accepted more or less uncritically. Meanwhile, Facciolà had discovered a second nettastomatid species in the Mediterranean in addition to the already known *Nettastoma melanura* Rafinesque, 1810. In a series of four papers (1893, 1899, 1911, and 1914) he diagnosed the two species and named the second one *Nettastomella physonema* (*Nettastomella* was preoccupied and later replaced by *Facciolella* Whitley). In his 1914 paper he specifically and in some detail compared Peters's *Saurenchelys cancrivora* with his own new species and concluded that they were not the same. Others, however, notably Grassi (1913: 164), differed and accepted *Saurenchelys cancrivora*. Grassi's view prevailed, and it was left for Saldanha and Blache (1968) to settle the matter and show that *Saurenchelys* and *Facciolella* are two distinct genera.

It now appears that all records of *Saurenchelys cancrivora* from the Mediterranean are based on *Facciolella*. There is no evidence that *S. cancrivora* occurs in the Mediterranean at all. Referring again to Peters's original description, it seems clear that, shrimp or no shrimp, his specimen probably did come from the Indian Ocean.

Size. These are relatively small eels, probably not exceeding 500–600 mm TL.

Distribution. Worldwide, tropical and subtropical seas.

Etymology. From the Greek *sauros* (lizard) and *enchelys* (eel). Feminine.

Species. The species composition of *Saurenchelys* is uncertain. I have examined specimens of *Saurenchelys* from the Gulf of Mexico, the Caribbean, the Gulf of Guinea, the eastern Pacific, Hawaii, Australia, and the Philippines without finding a single obvious character to divide them into different species. The number of vertebrae, which seems significant in at least some cases, is of limited value because the tails are usually regenerated. The only other character that varies at all between populations is the predorsal length. If no other evidence were available, I would conclude that all the specimens represent a single species and call it *Saurenchelys cancrivora*. Other evidence is available, however, in the form of leptocephali, and it gives a much different picture. Smith and Castle (1982) found nine distinct species of nettastomatid leptocephali that they identified as *Saurenchelys*. Four of these species occur in the Atlantic, one in the eastern Pacific, and five in the Indo-West Pacific; one of them, *Saurenchelys stylura* (Lea), occurs in both the Atlantic and Indo-West Pacific.

This confronts us with another "eel problem." If the identification of the leptocephali as *Saurenchelys* is correct—and the evidence strongly suggests that it is—then there are two possible hypotheses to explain the relationship between the leptocephali and adults. First, all the known adults belong to one of the larval species, and adults of the other leptocephali are unknown. This hypothesis can be rejected almost immediately on the simple grounds of geographical incongruence. No single one of the larval species occurs everywhere adult *Saurenchelys* occur. The second hypothesis is that the species of *Saurenchelys* are simply indistinguishable as adults, at least on the basis of superficial characters. This hypothesis could be tested if it were possible to identify conclusively the adult forms of two distinct larval species. Such identifications can be made with some confidence in two places, Hawaii and the Gulf of Mexico. In both areas only one species of *Saurenchelys* leptocephalus is known; in Hawaii it is *S. stylura* and in the Gulf of Mexico it is *S.* sp. A of Smith and Castle. In each case, moreover, the myomere counts of the leptocephali match the vertebral counts of the adults. I have studied adults from both areas, and only two differences between them are apparent. The Hawaiian specimens have more vertebrae (242) and a lesser predorsal length (38–42% preanal) than those from the Gulf of Mexico (199–209 and 43–48% preanal, respectively). In every other respect they are identical.

Smith and Castle (1982) recorded four species of larval *Saurenchelys* from the western North Atlantic. *Saurenchelys* sp. A is found in the Gulf of Mexico and the Bahamas, *S.* sp. B has been taken only in the Bahamas, *S.* sp. C was collected around the Virgin Islands, and *S. stylura* is found from the eastern Antilles south along the coast of Brazil.

Among the adult *Saurenchelys* examined from the western North Atlantic, there seem to be two or three species. One species has a relatively low vertebral count (ca 199–209) and is found in the Gulf of Mexico and along the Atlantic coast from Florida to North Carolina in depths of less than 200 m. Another species has ca 219 vertebrae and occurs in the Caribbean in depths of 230–460 m. A third species may be present off the Guianas, represented by a single specimen taken in 53 m off Suriname; the specimen is incomplete, and its vertebral count is unknown.

The first of these three species—the one from the U.S. Atlantic and Gulf coasts—is almost certainly the adult of Smith and Castle's *Saurenchelys* sp. A leptocephalus. The leptocephalus and adult are found in the same area, are relatively common, and have identical myomere/vertebral counts. The identity of the other adult species is uncertain. The single vertebral count of 219 from a Caribbean adult could match either *S.* sp. C leptocephali (223–225 myomeres) or *S. stylura* leptocephali (209–227 myomeres). Without a vertebral count, the Suriname specimen cannot be matched with a leptocephalus.

Saurenchelys is difficult to treat taxonomically in the normal way because the species can be recognized, diagnosed, and defined only as leptocephali. The only species that can presently be defined, geographically delimited, and identified with a leptocephalus is the species with the low vertebral count from the Atlantic and Gulf coasts of the United States. It is described here on the basis of leptocephali and adults as *Saurenchelys cognita*. Specimens from the Caribbean and South America are referred to simply as *Saurenchelys* sp.

The five nominal species of *Saurenchelys* were all described from the Indo-West Pacific. They are *Saurenchelys cancrivora* Peters, 1864, *S. petersi* Day, 1878, *S. fierasfer* (Jordan and Snyder, 1901), *S. finitima* (Whitley, 1938), and *S. elongata* (Kotthaus, 1968). None of them was adequately diagnosed or critically compared with the other nominal species, and they appear to have no relation to biological reality.

Note on the Holotype of Saurenchelys cancrivora. Blache et al. (1973: 237) stated that the type specimen of *Saurenchelys cancrivora* was lost. Van Utrecht (1983: 229) repeated this claim, but in fact the specimen (ZMB 5410) still exists. It was loaned to Richard H. Rosenblatt (SIO) in 1959 in answer to his query for information, was later transferred to James E. Böhlke (ANSP) for study

FIGURE 632. *Saurenchelys cognita*: ANSP 158274, 383 mm TL.

for this volume, and has since been returned to the ZMB. The specimen measured approximately 260 mm TL and is glued to a 150 × 265 mm glass plate. Some measurements could be obtained and dentition was readily observable. Radiographs were indistinct, the glass plate obscuring vertebrae in the tail region; dorsal- and anal-fin origins were marked externally by pinning. There were seven predorsal vertebrae and 29 preanal vertebrae; total vertebrae were approximately 209, but the count is probably incomplete.

Saurenchelys cognita new species
Figures 632, 633, Table 75

Saurenchelys sp. A. Smith and Castle, 1982: 27. Leptocephalus.

Distinctive Characters. This species is defined by its low vertebral count (199–209) and the distinctive features of its leptocephalus. Only certain Indo-Pacific leptocephali of the *Leptocephalus lateromaculatus* group regularly have fewer than 210 myomeres. Distinctive features of the leptocephali include the low myomere number and the short, round kidney (see p. 716).

Description. Total vertebrae 199–209 (n = 5), predorsal vertebrae 8–10 (12), preanal vertebrae 32–36 (14), precaudal vertebrae 55 (1). Branchiostegal rays 7 (1). Pores: LL to anus 31–32 (3), POM 13–14 (9), IO 8 (2), SO 1 + 2–5 (4), STC 3 (8).

Proportions as % of TL: preanal 23 (2), predorsal 11(2), head 9–10 (2), depth at anus 2 (2). Of preanal length: predorsal 43–48 (12), head 37–42 (13), depth at anus 6–9 (10). Of HL: snout 35–

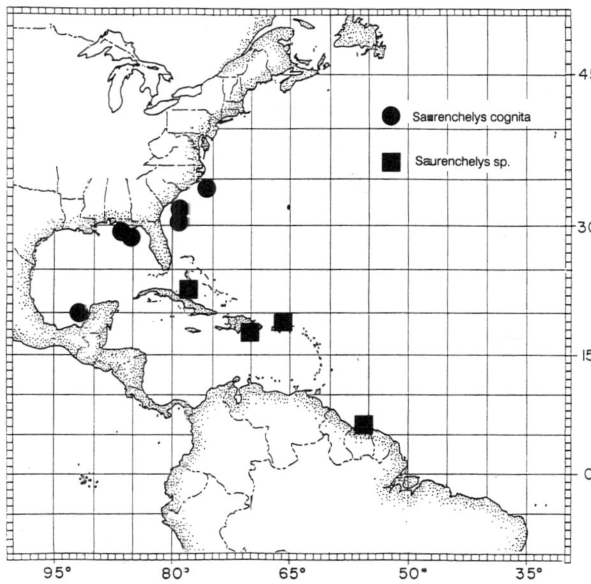

FIGURE 633. Distribution of *Saurenchelys* spp.

FIGURE 634. *Saurenchelys* sp.: ANSP 114934, 497 mm TL.

43 (13), eye 8–11 (13), upper jaw 49–57 (9), gill opening 4–6 (10), interbranchial 8–9 (2).

Dorsal fin begins slightly but distinctly behind level of gill opening.

Color. In preservative light tan to gray, posterior part of vertical fins edged in black. Stomach pale.

Size. The largest specimen examined was 520 mm TL, 122 mm preanal (MCZ 62020).

Development and Growth. For a description of the leptocephali see p. 716.

Distribution (Fig. 633). Adults of this species have been collected in the Gulf of Mexico and off the east coast of Florida, Georgia, and the Carolinas. In the Gulf of Mexico it is found predominantly in the eastern part; most records are from Florida, with one record from the eastern side of Campeche Bay. The leptocephali have essentially the same distribution but with one record from the Bahamas (Smith and Castle, 1982: 27). Depth 59–128 m.

Etymology. Latin *cognitus* (known, proved). This is the only species of *Saurenchelys* that has been identified and defined as a biological entity. An adjective.

Study Material. 28 specimens, 220–520 mm TL.

Holotype: USNM 185621 (1, 360), Oregon 944, Florida Gulf, 29°50′N, 86°30′W, 92 m, 21 March 1954.

Paratypes: NORTH CAROLINA: ANSP 150528 (1, 288), 33°54′N, 76°22′W, 161 m, 10 July 1982. MCZ 62016 (1, 460), 34°10′N, 76°05′W, 184–204 m, 1 Mar. 1985. MCZ 62020 (2, 398–520), 34°33′N, 75°44′W, 178–181 m, Mar. 1985. SOUTH CAROLINA: GMBL 76-244 (1, 272), 32 24.7′N, 78°53.5′W, 92 m, 6 Feb. 1976. GEORGIA: USNM 157940 (1, 310), 31°54′N, 79°28′W, 73 m, 12 Mar. 1956. FLORIDA GULF: ANSP 153303 (1, cleared and stained), 157987 (11, 253–385), and 158274, 29°52′N, 87°06.5′W, 128 m, 28 Aug. 1976. FSU 22118 (2, 254–288), 28°40′N, 85°30′W, 3 Oct. 1970. FSU 22165 (1, 220), 28°56′N, 85°23′W, 1 Dec. 1970. GCRL 21387 (1, 278), 28°26′N, 84°56′W, 92 m, 27 Feb. 1976. USA 03279 (2, 270–290), same data as ANSP 153303. USA 03496, (1), 30°04.5′N, 86°48′W, 92 m, 29 Aug. 1976. MEXICO GULF: UMML uncat. Silver Bay 1158 (1, 498, cleared and stained), 20°03′N, 92°03′W, 59–60 m, 29 Apr. 1959.

Saurenchelys species
Figures 633–636

Distinctive Characters. One specimen has 219 vertebrae.

Description. Total vertebrae 219 (n = 1), predorsal vertebrae 9–10 (2), preanal vertebrae 34–39 (2). Branchiostegal rays 7 (1). Pores: LL to anus ca 30–38 (6), POM 13–14 (4), IO 8 (4), SO 1 + 2–4 (4), STC 3 (6).

Proportions as % of TL: preanal 23 (1), predorsal 11 (1), head 10 (1), depth at anus 2 (1). Of preanal length: predorsal 40–48 (6), head 35–43 (6), depth at anus 5–10 (6). Of HL: snout 37–40 (6), eye 8–10 (6), interorbital 6–7 (2), upper jaw 48–57 (6), gill opening 2–5 (5), interbranchial 7 (3).

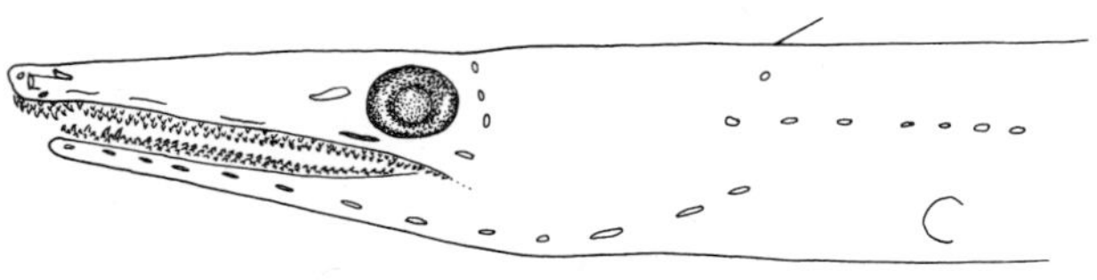

FIGURE 635. *Saurenchelys* sp.: ANSP 114934, 497 mm TL; head.

Dorsal fin begins slightly but distinctly behind level of gill opening.

Color. In preservative light tan or gray, posterior part of vertical fins margined in black. Stomach pale.

Size. The largest specimen examined was 514 mm TL, 133 mm preanal (USNM 187629).

Development and Growth. Leptocephali of two species of *Saurenchelys* are known from the Caribbean and northern South America: *Saurenchelys* sp. C (223–225 myomeres), and *S. stylura* (214–227 myomeres, with one count of 209). It cannot be determined if any of the adults belong with either of these leptocephali.

Remarks. Only six adult specimens of *Saurenchelys* were available from the Caribbean and South America, and five of these had regenerated tails. The one intact specimen came from Puerto Rico and had 219 vertebrae (USNM 190536), significantly more than in *S. cognita.* Furthermore, leptocephali of *S. cognita* are unknown south of the Gulf of Mexico and the northern Bahamas. For these reasons, it seems unlikely that the six specimens treated here are *S. cognita.* It is not even certain that all six belong to the same species. The single specimen from Suriname (USNM 158932) has a slightly lesser predorsal length (40 vs 41–48% preanal), snout length (32 vs 37–40% HL), and upper jaw length (48 vs 52–57% HL) than the five Caribbean specimens. In addition it was collected in shallower water (53 m vs 229–458 m).

Distribution (Fig. 633). Puerto Rico, Hispaniola, off the Great Bahama Bank just south of the Santaren Channel, and Suriname. Depth 229–458 m.

Study Material. Six specimens from the western North Atlantic, 280–514 mm TL, 76–133 mm preanal length. 19 specimens from elsewhere for comparison, 120–530 mm TL.

BAHAMAS: USNM 187629 (2, 490–514), 23°05′N, 78°49′W, 412–458 m, 7 Nov. 1961. HISPANIOLA: UMML 29698 (1, 280), 18°21.4′N, 69°08.7′W, 238–302 m, 9 July 1971. PUERTO RICO: ANSP 114934 (1, 497), 18°26′N, 67°11′W, 229 m, 6 Oct. 1959. USNM 190536 (1, 393), 18°26′N, 67°10.5′W, 275 m, 6 Oct. 1959. SURINAME: USNM 158932 (1, 280+), 6°50′N, 55°22′W, 53 m, 20 July 1957.

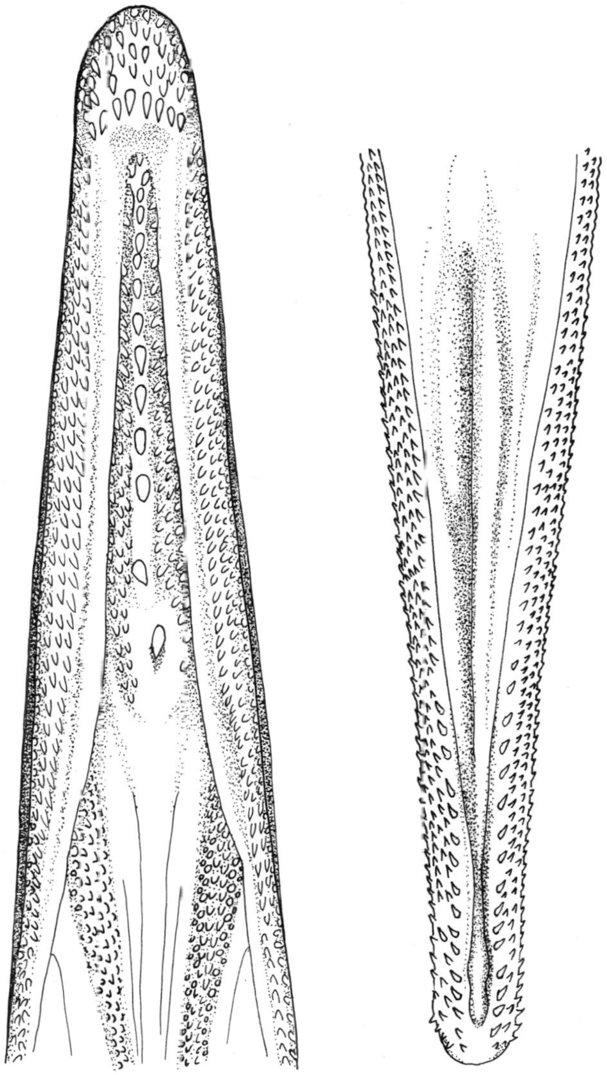

FIGURE 636. *Saurenchelys* sp.: ANSP 114934, 497 mm TL; dentition.

Comparative material: URUGUAY: MZUSP 12299 (1, 392), 34°27′S, 51°50′W, 15 Aug. 1972. GULF OF GUINEA: ANSP 156860 (1, 305, cleared and stained) and UMML 20900 (2, 350–360). EASTERN PACIFIC: CAS 37542 (1, 120), Los Frailes, Baja California Sur, 27–37 m. LACM 32279 (6, 214–329), Costa Rica, Cocos Is., 0.8 km north of Isla Pajara, 5 Apr. 1972. HAWAII: ANSP 156916 (4, 355–530), Pailolo Channel, 220 m, Jan. 1970. AUSTRALIA: AMS IA924 (holotype of *Chlopsis finitimus* Whitley), Queensland, Whitsunday Passage, 21°31′S, 150°16′E. AMS I.15523

(1, 205), 26°31'S, 153°40'E, 137 m, 26 July 1968. PHILIPPINES: MUSORSTOM Philippines sta. 9 (1, 355+). INDIAN OCEAN: ZMB 5410 (1, ca 260, holotype of *Saurenchelys cancrivora* Peters).

Genus *Facciolella* Whitley, 1938

Nettastomella Facciolà, 1911: 277 (type species *Nettastomella physonema* Facciolà, 1914, = *Leptocephalus oxyrhynchus* Bellotti, 1883, by monotypy). Preoccupied by *Nettastomella* Carpenter, 1865, Mollusca.

Facciolella Whitley, 1938: 197 (substitute name for *Nettastomella* Facciolà).

Characters. Body elongate, round in cross section anteriorly, compressed posteriorly, relatively uniform in depth but tapering rather sharply toward tail; tail slender, filiform, frequently damaged; anus near anterior third of TL. Dorsal fin begins slightly but distinctly behind level of gill opening. Pectoral fin absent. Gill opening crescentic, on lower half of side.

Head and snout rather elongate, upper jaw projects beyond lower. Tip of snout fleshy, projecting a short distance beyond bony tip of ethmovomer. Eye moderate. Mouth large, rictus slightly behind posterior margin of eye. Anterior nostril a low tube at level of anterior end of intermaxillary tooth patch, directed anterolaterally; posterior nostril a longitudinal slit in upper lip in front of eye, concealed by a flap that forms outer edge of lip.

Lateral line on body complete to tail. Five supraorbital pores: two on anteroventral surface of snout before anterior nostril, one above and slightly ahead of the other; one above and behind anterior nostril, slightly enlarged and slit-like; a smaller one behind this but before eye; and one at posterior end of canal near level of posterior margin of eye. Usually nine infraorbital pores, including an adnasal pore: the first one fairly large, just below and behind anterior nostril; the second (adnasal) smaller, behind and above first pore, close to larger supraorbital pore; the third on lip just above posterior end of intermaxillary teeth; the fourth just behind this; the fifth on outer surface of lip at anterior end of nasal flap; the sixth below posterior quarter of eye; the last three behind eye. One to three pores in supratemporal commissure. Preoperculomandibular canal with 10–13 pores.

Maxillary teeth small, conical, multiserial, more or less continuous with intermaxillary teeth, inner teeth slightly larger than outer. Intermaxillary teeth slightly larger than maxillary teeth, in a spade-shaped patch, posterior end concave, separated from vomerine teeth. Vomerine teeth in an elongate patch reaching not quite to anterior edge of eye, multiserial, teeth in middle larger than those outside. Mandibular teeth in bands, inner teeth larger; anterior end of mandible with teeth fits into gap between vomerine and intermaxillary teeth.

In specimens examined (ANSP 106197 and 126186) stomach long, ending about a third of its own length behind anus. Gas bladder extremely long, extends from just behind gill opening to a point near end of tail where body cavity ends; large anteriorly, narrows around mid-trunk, abruptly widens again before anus, and then tapers gradually into tail.

Size. Maximum size seems to be about 600–700 mm TL.

Distribution. Worldwide in tropical and warm temperate seas, including the Mediterranean, but it may be absent from the central Pacific.

Etymology. Named for Luigi Facciolà, the Italian ichthyologist who recognized and described the genus. Feminine.

Species. Like those of *Saurenchelys,* the species of *Facciolella* are easily distinguished as leptocephali but not as adults. Smith and Castle (1982) recorded eight species of leptocephali worldwide, of which two inhabit the eastern Atlantic and two the western Atlantic. Adults have been collected from both sides of the Atlantic, the eastern and western Pacific, and the western Indian Ocean, but so far no obvious characters have been found for dividing them into species. One adult specimen from the Mediterranean (MZUF C.4423), two from the Gulf of Guinea (USNM 199580 and 200777) and one from the eastern Pacific (USNM 44391, the holotype of *Chlopsis equatorialis* Gilbert) were available for comparison. Based on the distribution of leptocephali, these and the western Atlantic specimens should

FIGURE 637. *Facciolella* sp.: ANSP 117054, 400 mm TL.

all represent distinct species. The Mediterranean specimen has fewer total vertebrae (237) than the western Atlantic specimens (244–256) but more preanal vertebrae (64 vs 53–57); it also has slightly more preanal lateral-line pores (59 vs 51–55). The eastern Pacific specimen has a greater predorsal length than the western Atlantic specimens (45 vs 34–41% preanal) and a slightly larger eye (11 vs 7–9% head), but the vertebral count is the same (249). The eastern Atlantic specimens show no differences, but vertebral counts were not available.

There are three nominal species of *Facciolella. Facciolella oxyrhynchus* (Bellotti, 1883) is a valid species found in the Mediterranean and adjacent northeastern Atlantic. *Facciolella equatorialis* (Gilbert, 1891) and *F. gilberti* (Garman, 1899) were both described from the eastern Pacific; they have never been critically compared and may not be distinct.

The adult specimens of *Facciolella* in the western Atlantic are treated as a single species here, but this is an assumption based on lack of evidence. Because of regenerated tails, vertebral counts are not available for all specimens, and leptocephali of two species are present in the area. Given the uncertainty, no attempt is made to apply a name.

Facciolella species
Figures 637–640, Tables 75, 76

Distinctive Characters. Those of the genus.

Description (western North Atlantic specimens only). Total vertebrae 243–256 (n = 10), predorsal vertebrae 10–13 (13), preanal vertebrae 53–57 (16). Branchiostegal rays 6–8 (3). Pores: LL to anus 51–55 (25), POM 10–13 (9), IO 9 (11), SO 1 + 4 (12), STC 0–3 (24).

Proportions as % of TL: preanal 30–33 (7), predorsal 11–13 (6), head 11 (6), depth at anus 1–2 (6). Of preanal length: predorsal 34–41 (30), head 31–37 (29), depth at anus 4–7 (30). Of HL: snout 33–39 (28), eye 7–9 (27), upper jaw 48–54 (24), gill opening 2–6 (17), interbranchial 3–11 (12).

Color. In preservative tan to dark brown or gray. Vertical fins edged in black.

Size. The largest specimen examined was 627 mm TL, 231 mm preanal (ANSP 106187).

Variation. Specimens from the Gulf of Mexico

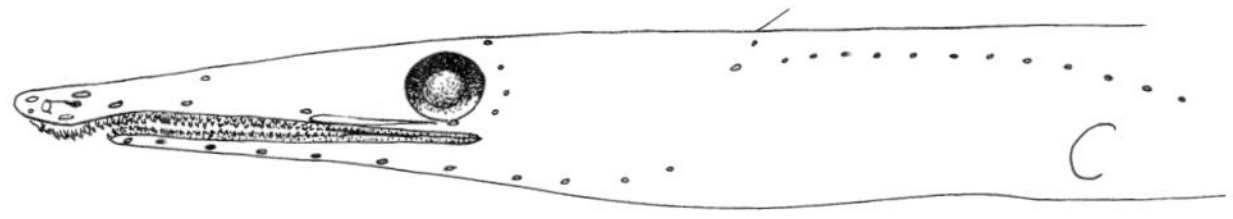

FIGURE 638. *Facciolella* sp.: ANSP 117054, 400 mm TL; head.

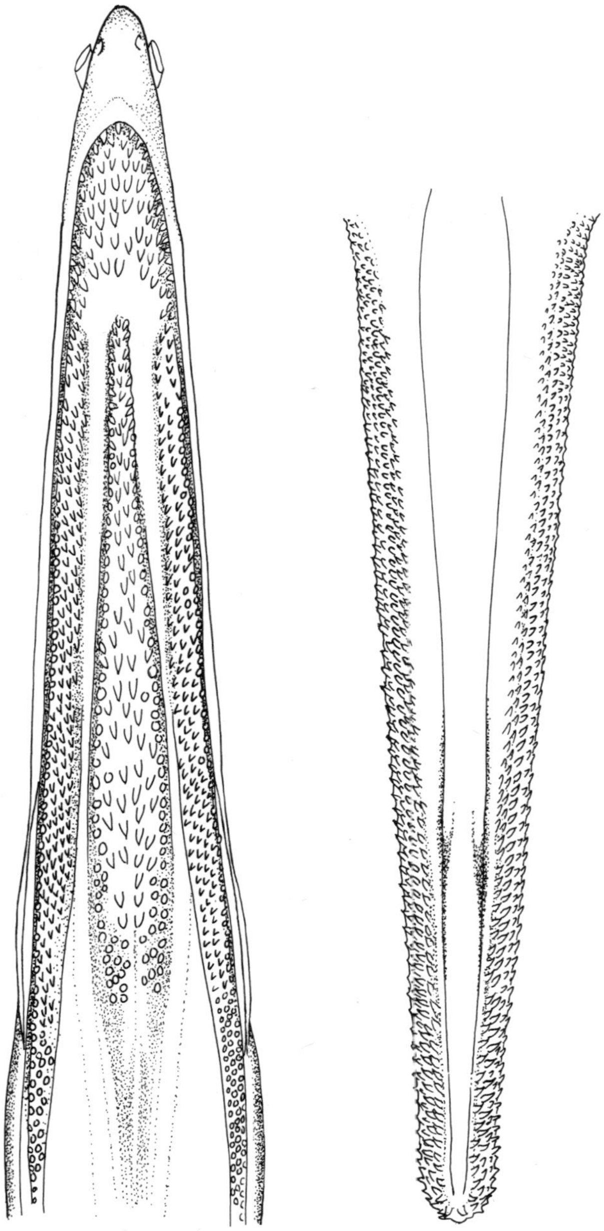

FIGURE 639. *Facciolella* sp.: ANSP 117054, 400 mm TL; dentition.

vary more in the number of STC pores than do specimens from elsewhere in the western Atlantic (Table 76).

Remarks. It is uncertain whether these specimens represent a single species. Smith and Castle (1982) recorded leptocephali of two species of *Facciolella* in the western North Atlantic, designated as *Facciolella* sp. B and sp. C. They were distinguished by differences in pigmentation, gut length, and myomere number. Species B has 240–254 total myomeres, 57–61 preanal myomeres,

TABLE 76. Number of STC pores in *Facciolella* sp. from the Gulf of Mexico, Florida Atlantic, Caribbean, and Bermuda.

	STC pores					
Area	0	1	2	3	n	x̄
Gulf of Mexico	1	3	1	3	7	1.6
Florida Atlantic	-	5	-	1	6	1.3
Caribbean	-	11	-	-	11	1.0
Bermuda	-	1	-	-	1	1.0

and 15–16 predorsal myomeres; for sp. C the values for these characters are 256–264, 43–47, and 18–20, respectively. On the basis of total vertebrae (244–256), the adults would seem to agree with sp. B leptocephali. Preanal vertebrae may be equally significant. In eels the anus sometimes moves forward at metamorphosis and sometimes remains where it is, but there is no known case of it moving backward. The number of preanal vertebrae in the adult, therefore, should be equal to or fewer than the number of preanal myomeres in the leptocephali. The adult *Facciolella* examined here have 53–57 preanal vertebrae, more than the preanal myomeres of sp. C leptocephali but fewer than the preanal myomeres of sp. B. By this criterion also, the adults would belong to sp. B. The problem is that sp. B leptocephali are not fully congruent in distribution with adult *Facciolella* in the western Atlantic and in particular have not been recorded from the Gulf of Mexico, where adult *Facciolella* occur. Species C leptocephali are congruent in distribution and do occur in the Gulf of Mexico, but they have the wrong myomere counts. At this point, the matter must be put aside until more information is available.

Distribution (Fig. 640). In the western North Atlantic *Facciolella* has been collected from the Gulf of Mexico, the Atlantic coast of Florida, Bermuda, Hispaniola, and the southern Caribbean from Nicaragua to Venezuela. There are no records from the Guianas or northern Brazil, but it does occur off southern Brazil (Menezes and Benvegnu, 1976: 272). The depth range is 75–3109 m, but both extremes are based on single records (ANSP 137574 and 106282); if those are excluded, the depth range for the remainder is 329–732 m.

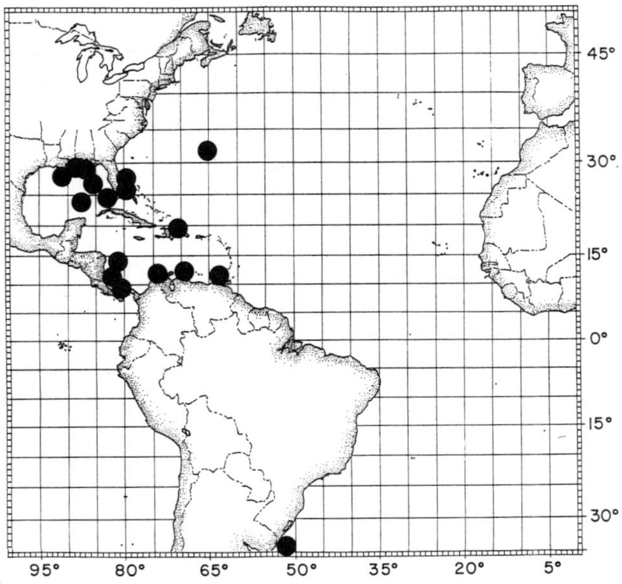

FIGURE 640. Distribution of *Facciolella* sp.

Study Material. 44 specimens, 81–627 mm TL.

BERMUDA: ANSP 137574 (1, 205). FLORIDA ATLANTIC: ANSP 138408 (2, 81–95+). UMML 20590 (1, 285). USNM 185625 (1, 308+), 197125 (1, 563), 197130 (1, 567). FLORIDA GULF: ANSP 106282 (1, 363), 106525 (1, 398), 117042 (1, ca 391), 117054 (1, 400), 133013 (1, 423). UMML 13738 (1, 458). USNM 158323 (1, 390), 163487 (1, 321). ALABAMA: ANSP 114042 (1, ca 391), 117100 (1, 333+). MISSISSIPPI: ANSP 106264 (1, 330). USNM 200789 (1, 329). LOUISIANA: ANSP 106152 (1, 380+). MEXICO GULF: ANSP 106519 (1, 393). HISPANIOLA: ANSP 117053 (1, ca 388). VENEZUELA: ANSP 106151 (1, 400), 106187 (1, 635), 110070 (1, 595), 112044 (1, 384), 115536 (1, cleared and stained), 117050 (1, 320+), 117058 (2, 367–595), 117072 (1, 260+), 117083 (1, 268). MCZ 55541 (1, 418). COLOMBIA: ANSP 126186 (2, 333–550+). PANAMA: USNM 193551 (1, 320). NICARAGUA: ANSP 114861 (1, 265+), 126520 (1, 384). USNM 158937 (1, 388), 193564 (1, 354). LOCALITY UNCERTAIN: UMML 16309 (1, 476).

Comparative material: GULF OF GUINEA: USNM 199580 (1, 334+). USNM 200777 (1, 414). MEDITERRANEAN: MZUF C.4423 (1, 463). EASTERN PACIFIC: USNM 44391 (1, 363, holotype of *Chlopsis equatorialis* Gilbert).

Genus *Nettenchelys* Alcock, 1898
Figure 641

Nettenchelys Alcock, 1898: 149 (type species *Nettenchelys taylori* Alcock, 1898, by monotypy).

Characters. Body elongate, round in cross section anteriorly, compressed posteriorly, deepest in middle and tapering at both ends; tail slender and attenuate, tip delicate and easily damaged; anus near anterior third of TL. Dorsal fin begins above or slightly behind gill opening. Pectoral fin absent. Gill opening crescentic, on lower half of side.

Head moderately elongate, slightly depressed; upper jaw projects beyond lower. Snout length about 33–37% head. Tip of snout fleshy, projects slightly beyond bony tip of ethmovomer. Eye moderate. Mouth large, rictus near posterior margin of eye. Anterior nostril tubular, near tip of snout, directed anterolaterally; posterior nostril located on top of head or body behind posterior margin of eye, degree of posterior displacement varying greatly among the known species.

Lateral line on body complete to tail. Five supraorbital pores: two near tip of snout before anterior nostril, three between anterior nostril and eye, none above or behind eye. Eleven or 12 infraorbital pores: one near ventral base of anterior nostril, one behind this and slightly elevated (the adnasal pore), 6–7 along upper jaw just above edge of lip, and three behind eye. Two

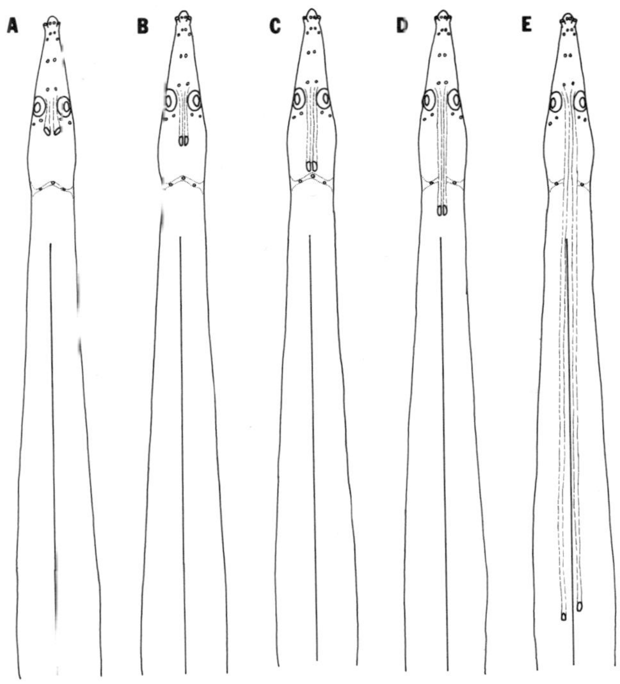

FIGURE 641. Semidiagrammatic dorsal view of the five species of *Nettenchelys* showing position of posterior nostrils (from Smith, Böhlke, and Castle, 1981: fig. 8). A. *N. pygmaea*. B. *N. gephyra*. C. *N. inion* D. *N. exoria*. E. *N. taylori*.

or three pores in supratemporal commissure. Approximately 14–17 preoperculomandibular pores.

Teeth generally small, conical, in narrow bands; enlarged teeth sometimes present on vomer.

Size. Maximum size probably no more than 550 mm TL.

Distribution. Based on the distribution of leptocephali, species of *Nettenchelys* occur widely over the Atlantic and Indo-West Pacific. Most of what we know about the distribution of *Nettenchelys* comes from the leptocephali, which are fairly common. Adults, by contrast, are rarely collected, possibly because of their cryptic habits.

Etymology. From the Greek *netta* (duck) and *enchelys* (eel), referring to the somewhat elongate and depressed snout. The International Commission on Zoological Nomenclature in Opinion 915 (1970) decided that *enchelys* should be feminine. Thus Smith, Böhlke, and Castle (1981: 550) were wrong in considering it masculine.

Species. Five species of *Nettenchelys* are currently recognized: *N. taylori* Alcock, 1898 and *N. gephyra* Castle and Smith, 1981 in the Indo-West Pacific; and *N. pygmaea* Smith and Böhlke, 1981, *N. inion* Smith and Böhlke, 1981, and *N. exoria* Böhlke and Smith, 1981 in the western Atlantic. They can be divided into two groups based on the dentition and the position of the posterior nostril (Figs. 641, 644, 648). *Nettenchelys pygmaea* and *N. gephyra* have a row of enlarged teeth medially on the anterior end of the vomer, slightly separated from the remaining vomerine teeth. The posterior nostril in these two species is located on top of the head shortly behind the eye. The three remaining species lack the enlarged median vomerine teeth, and their posterior nostril is at or behind the occiput.

Uyeno et al. (1983: 117) reported a specimen of *Nettenchelys* (403 mm TL) from off the Guianas that seems to split the diagnostic characters of these groups. Like *N. inion, N. exoria,* and *N. taylori* it lacks the enlarged median vomerine teeth. The posterior nostril, however, is located "between the hinder margins of the eyes." At this point the specimen cannot be identified.

Key to the Species of *Nettenchelys* in the Western North Atlantic

1a. Anterior vomerine teeth forming an enlarged median series (Fig. 644); posterior nostril slightly behind eye (Fig. 641A) . . . . *pygmaea,* p. 600

1b. Anterior vomerine teeth not forming an enlarged median series (Fig. 648); posterior nostril at or behind occiput (Fig. 641C, D) 2

2a. Posterior nostril behind supratemporal canal; median supratemporal pore absent (Fig. 641D) . . . . *exoria,* p. 602

2b. Posterior nostril just before supratemporal canal; median supratemporal pore present (Fig. 641C) . . . . *inion,* p. 603

Nettenchelys pygmaea Smith and Böhlke, 1981
Figures 641A, 642–645, Table 75

Nettenchelys pygmaeus Smith and Böhlke in Smith, Böhlke, and Castle, 1981: 551, figs. 8A, 9, 10 (original description, 29°52′N, 87°06.5′W, holotype ANSP 142277).

Distinctive Characters. Anterior median vomerine teeth enlarged; posterior nostril on top of head just behind posterior margin of eye; mature at 150–200 mm TL.

Description. Total vertebrae ca 220–250 (estimated from myomere counts of leptocephali), predorsal vertebrae 7–8 (n = 3), preanal vertebrae 40–43 (3). Pores: LL to anus ca 40–42 (2), POM uncertain, IO 11 (3), SO 1 + 4 (3), STC 3 (3).

Proportions as % of preanal length: predorsal 33–35 (4), head 30–33 (4), depth at anus 7–8 (4). Of HL: snout 34–37, eye 8–10 (4), upper jaw 53 (2), gill opening 6–7 (3).

Dorsal fin begins slightly behind gill opening. Posterior nostril an ill-defined opening in skin on top of head just behind eye, without a raised rim.

Maxillary and mandibular teeth in bands, teeth

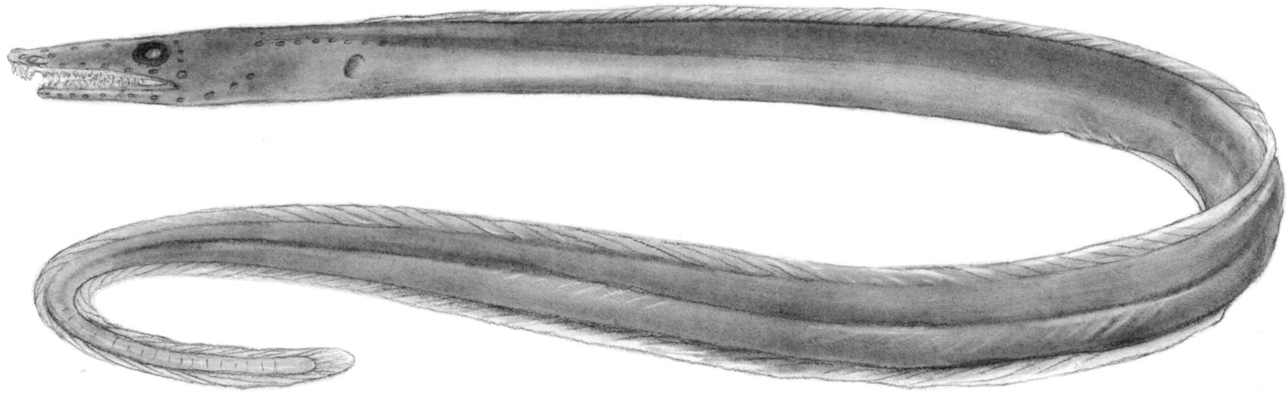

FIGURE 642. *Nettenchelys pygmaea*: ANSP 130812, 167 mm TL.

of innermost series enlarged. Intermaxillary tooth patch short, continuous with maxillary teeth. Vomerine tooth patch long, reaching about to anterior margin of eye; anteriormost few teeth enlarged, forming a median row; remainder of vomerine teeth separated slightly from these, forming a roughly elliptical patch, the inner teeth somewhat larger than the outer.

Color. In preservative brown, darker above than below. One specimen (USA 03460) has a distinct bicolored appearance, dark above and pale below.

Size. This is a dwarf species. The largest specimen (USA 03460) measured only 201 mm TL, 71 mm preanal. The smallest specimen (USNM 200779) was a mature female 151 mm TL, 54 mm preanal, filled with large eggs.

Development and Growth. The leptocephalus has been identified (see p. 708).

Variation. There is evidence from the leptocephali of a slight differentiation between specimens from the Gulf of Mexico and those from the Caribbean and Atlantic. Seven specimens from the Gulf had approximately 220–236 myomeres, whereas 22 specimens from outside the Gulf had 227–251 (Smith and Castle, 1982: 16).

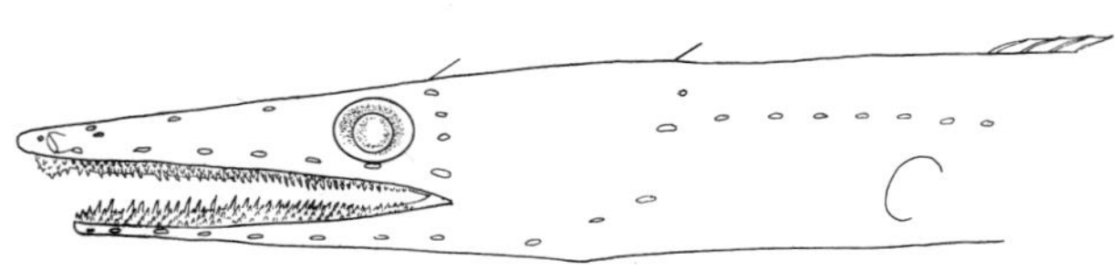

FIGURE 643. *Nettenchelys pygmaea*: ANSP 130812, 167 mm TL; head.

Distribution (Fig. 645). Only four adult specimens are known, two from the northern Gulf of Mexico and two from Venezuela. Leptocephali are relatively common, though, and they are

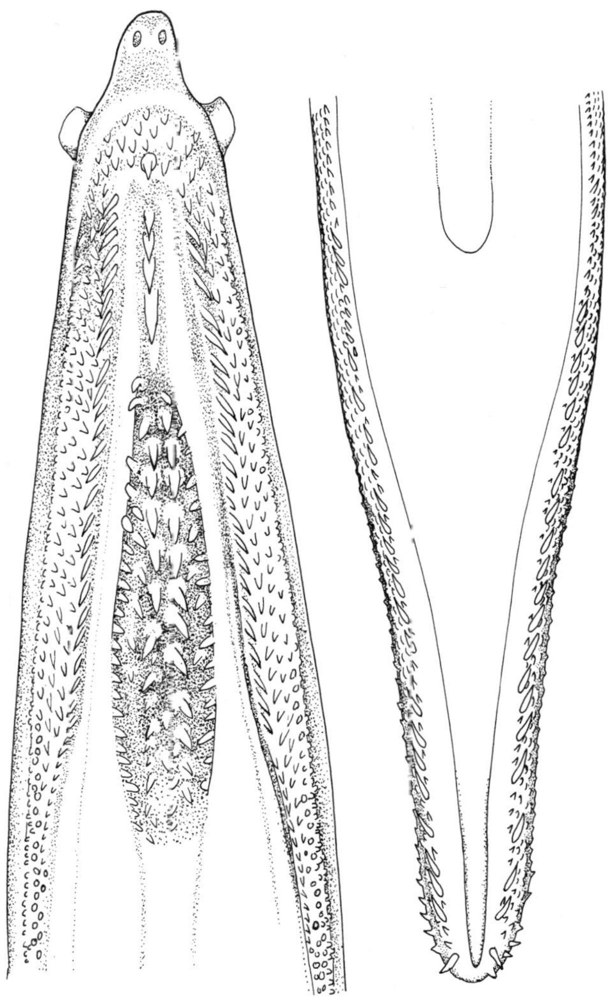

FIGURE 644. *Nettenchelys pygmaea*: ANSP 130812, 167 mm TL; dentition.

found widely in the Gulf of Mexico, the Caribbean, the Gulf Stream, and the Sargasso Sea. The species probably occurs throughout the area, but adults may be difficult to collect because of their small size and possibly cryptic habits. Depth 128–280 m.

Remarks. Because of the small size of this species and the less-than-perfect state of preservation of the specimens, certain counts and measurements were difficult to obtain. Interbranchial distance was particularly hard to measure, and I am still uncertain how many preoperculomandibular pores are present.

Etymology. From the Greek *pygmaios* (small, dwarf), in reference to the small size. An adjective.

Study Material. Four specimens, 151–201 mm TL, 54–71 mm preanal length.

Holotype: ANSP 142277 (male, 175), Florida Gulf, 29°52'N, 87°06.5'W, 128 m, coll. R. L. Shipp et al., 28 Aug. 1976.

Paratypes: FLORIDA GULF: USA 03460 (1, 201), same data as holotype. VENEZUELA: ANSP 130812 (1, 167), 10°54.7'N, 66°17.8'W, 234–280 m, 23 July 1968. USNM 200779 (1, 151), 10°42'N, 67°56'W, 210 m, 28 Sept. 1965.

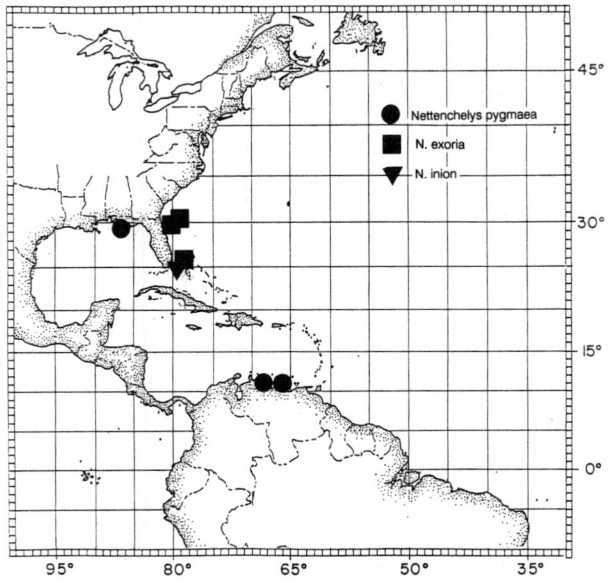

FIGURE 645. Distribution of *Nettenchelys pygmaea, N. exoria,* and *N. inion.*

Nettenchelys exoria Böhlke and Smith, 1981
Figures 641D, 645–648, Table 75

Nettenchelys exorius Böhlke and Smith in Smith, Böhlke, and Castle, 1981: 556, figs. 8D, 9, 12 (original description, 29°41'N, 80°11'W, holotype ANSP 142275).

Distinctive Characters. Posterior nostril behind occiput; median supratemporal pore absent; anterior vomerine teeth not enlarged; predorsal 32–34% preanal, head 31–34% preanal, upper jaw 45–51% HL.

Description. Total vertebrae unknown, predorsal vertebrae 7 (n = 3), preanal vertebrae 43–44 (3). Branchiostegal rays 6 (1). Pores: LL to anus, 40–43 (6), POM 14–16 (3), IO 11–13 (6), SO 1 + 4 (6), STC 2 (6).

Proportions as % of preanal length: predorsal 32–34 (6), head 31–34 (6), depth at anus 8–12 (6). Of HL: snout 34–36 (6), eye 9–11 (6), upper jaw 45–51 (4), gill opening 5–7 (4), interbranchial 15–19 (3).

Posterior nostril behind occiput, between supratemporal canal and dorsal-fin origin. Nasal tube passes over supratemporal canal, apparently obliterating median pore. Maxillary and mandibular teeth in bands, innermost teeth somewhat enlarged. Intermaxillary tooth patch short, continuous with maxillary teeth. Vomerine tooth patch long, reaching about to eye, anterior teeth not enlarged. Stomach moderately short, ends before anus. Gas bladder long, ends behind anus.

Color. In preservative brown, vertical fins dark-edged posteriorly. Stomach and intestine pale.

Size. The largest specimen examined was the holotype, 467 mm TL, 163 mm preanal. It is a female with large eggs. One of the paratypes (ANSP 131739), 329 mm TL, 120 mm preanal, is a male.

Development and Growth. Leptocephali of this species cannot be distinguished from those of *Nettenchelys inion* (see p. 709).

Remarks. The specimens all have regenerated tails, hence the total vertebral count is unknown. Based on myomere counts of leptocephali, total vertebrae should be approximately 209–226.

Distribution (Fig. 645). Five of the six known specimens were collected off the northeast coast of Florida; the sixth came from the junction of the Northeast and Northwest Providence Channels in the Bahamas. Depth 227–494 m.

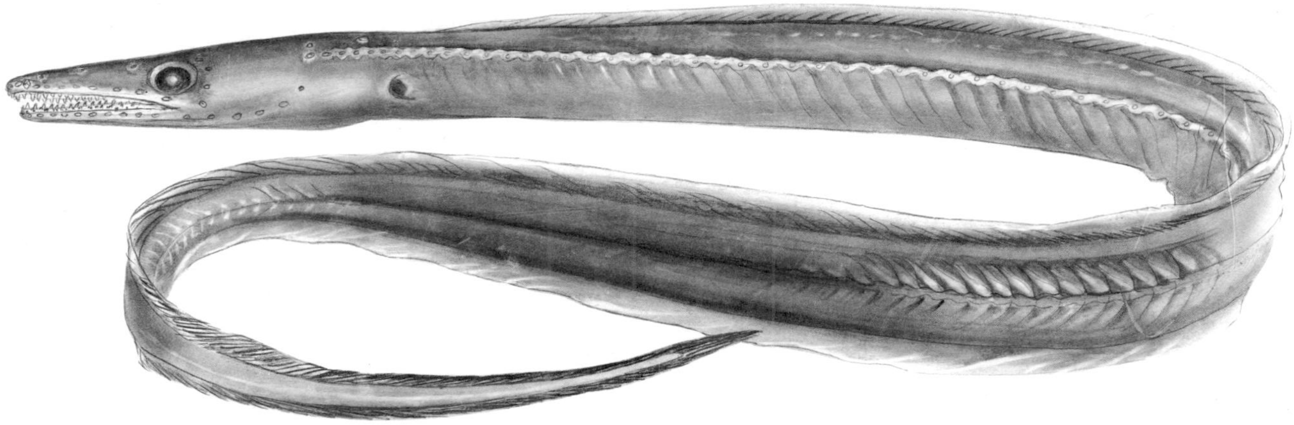

FIGURE 646. *Nettenchelys exoria*: ANSP 130813, 429 mm TL.

Etymology. From the Greek *exorios* (beyond the frontier), in reference to the position of the posterior nostrils, which have migrated beyond the occiput and are located behind the head. An adjective.

Study Material. Six specimens, 182–467 mm TL, 65–163 mm preanal length.

Holotype: ANSP 142275 (female, 467), Silver Bay 3685, Florida Atlantic, 29°41′N, 80°11′W, 347 m, 22 Jan. 1962.

Paratypes: ANSP 130813 (1, 429), 29°40′N, 80°03′W, 458 m, 18 Nov. 1965. 131739 (1, 329), 30°28′N, 79°51′W, 494 m, 19 Jan. 1972.

Other material: ANSP 156856 (1, 253, cleared and stained), 25°52′N, 77°54′W, 277 m, 20 July 1965. FDNR 17770 (1, 182), 30°10.1′N, 80°10.4′W, 227 m, 25 May 1984. 17771 (1, 395+), 30°10.1′N, 80°08.6′W, 271 m, 25 May 1984.

Nettenchelys inion Smith and Böhlke, 1981
Figures 641C, 645, Table 75

Nettenchelys inion Smith and Böhlke in Smith, Böhlke, and Castle, 1981: 555, fig. 8C, 9 (original description, 25°36′N, 79°21′W, holotype ANSP 142274).

Distinctive Characters. Posterior nostril at occiput, just before supratemporal canal; median supratemporal pore present; anterior vomerine teeth not enlarged; predorsal 37% preanal, head 38% preanal, upper jaw 55% HL.

Description (from holotype). Total vertebrae 210

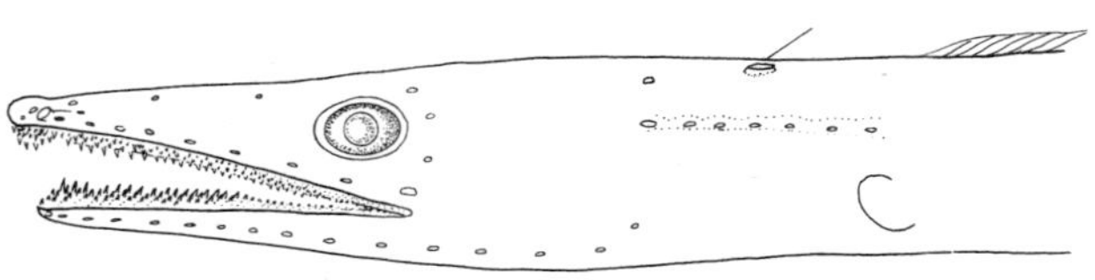

FIGURE 647. *Nettenchelys exoria*: ANSP 130813, 429 mm TL; head (line indicates posterior nostril).

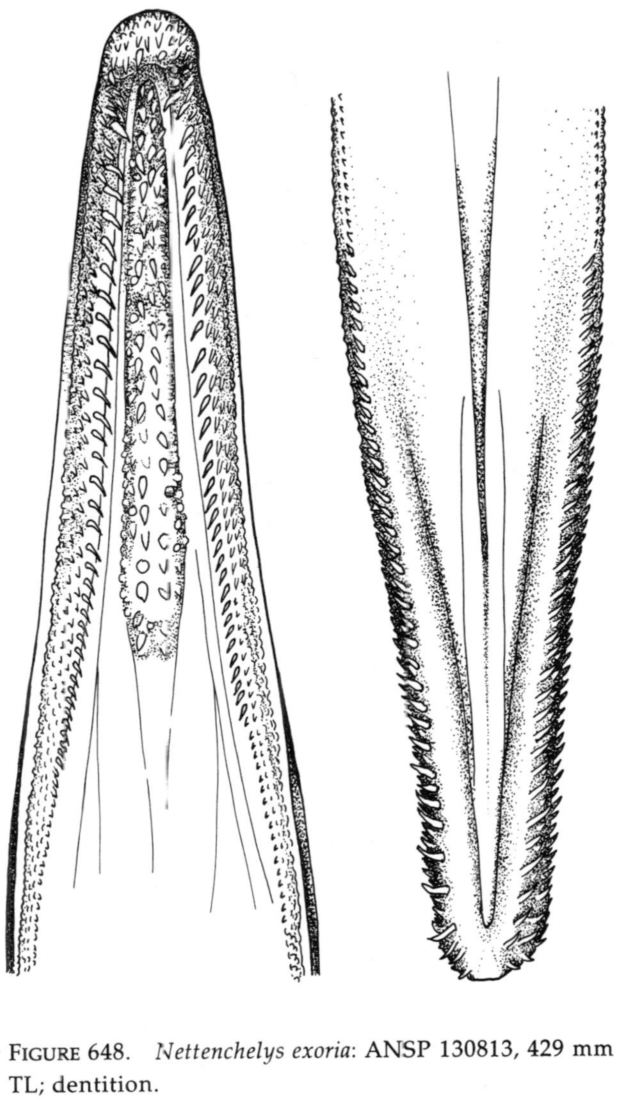

FIGURE 648. *Nettenchelys exoria*: ANSP 130813, 429 mm TL; dentition.

(complete ?), predorsal vertebrae 7, preanal vertebrae 41. Branchiostegal rays 6. Pores: LL to anus 41, POM 14, IO 12, SO 1 + 4, STC 3.

Proportions in % of preanal length: predorsal 37, head 38, depth at anus 10. Of HL: snout 37, eye 10, upper jaw 55.

Posterior nostril located at occiput immediately before supratemporal canal. Supratemporal canal unobstructed, median pore present. Dentition as in *Nettenchelys exoria.*

Color. In preservative brown, vertical fins dark-edged posteriorly.

Size. The only known specimen is 423 mm TL, 127 mm preanal.

Development and Growth. Leptocephali of this species cannot be distinguished from those of *Nettenchelys exoria* (see p. 709).

Remarks. This species is still known only from the holotype. Smith, Böhlke, and Castle (1981) recognized it as a distinct species because of the consistent ways in which it differs from the known specimens of *N. exoria.* The latter all have the posterior nostril behind the supratemporal commissure, and they lack the median supratemporal pore. In addition, the type of *N. inion* has a greater predorsal, head, and upper jaw length than the specimens of *N. exoria.* Nevertheless, the possibility cannot be completely excluded that these differences are within the range of variation of a single species. The species of *Nettenchelys* are known from too few specimens to provide a good estimate of the variability of diagnostic characters.

Distribution (Fig. 645). The specimen came from the eastern edge of the Straits of Florida, just south of Bimini in the Bahamas.

Etymology. Greek *inion* (back of head, occiput), referring to the position of the posterior nostril. A noun in apposition.

Study Material. One specimen, 423 mm TL, 127 mm preanal length.

Holotype: ANSP 142274 (male, 423), Gerda 242, eastern Bahamas, 25°36′N, 79°21′W, 458–531 m, 30 Jan. 1964.

Genus *Nettastoma* Rafinesque, 1810

Nettastoma Rafinesque, 1810a: 66 (type species *Nettastoma melanura* Rafinesque, 1810, by monotypy).

Hyoprorus Kölliker, 1854: 101 (type species *Hyoprorus messinensis* Kölliker, 1854, by monotypy).

Metopomycter Gilbert, 1905: 585 (type species *Metopomycter denticulatus* Gilbert, 1905, by original designation).

Muraenosaurus Osorio, 1909a: 14 (type species *Muraenosaurus Güntheri* Osorio, 1909, by monotypy). Preoccupied by *Muraenosaurus* Seeley, 1874, Reptilia.

Osorina Whitley, 1951: 68 (substitute name for *Muraenosaurus* Osorio).

Characters. Body elongate, round in cross section anteriorly, compressed posteriorly, deepest in middle and tapering at both ends; tail slender and attenuate, tip delicate and easily damaged; anus near anterior one-third to two-fifths of TL. Dorsal fin begins above gill opening. Pectoral fin absent.

Head and snout slender, elongate, slightly depressed; upper jaw projects beyond lower. Snout length about 35–49% HL. Tip of snout fleshy, projects slightly beyond bony tip of ethmovomer. Eye moderate. Mouth large, rictus near posterior margin of eye. Anterior nostril tubular, near tip of snout, directed anterolaterally; posterior nostril elliptical or crescentic, displaced dorsally near anterodorsal or posterodorsal margin of eye.

Lateral line on body complete to tail. Five or six supraorbital pores: two near tip of snout before anterior nostril, two or three along top of snout between anterior nostril and eye, and one at posterior end of canal near posterior margin of eye. Eleven to 14 infraorbital pores: one near ventral base of anterior nostril, one behind this and slightly elevated (the adnasal pore), 6–9 along upper jaw just above edge of lip, and three behind eye. Three pores in supratemporal commissure. Approximately 15–21 preoperculomandibular pores.

Teeth conical, relatively small, in narrow bands.

Size. Maximum size probably about 1 m TL.

Distribution. Western and eastern Atlantic, including Mediterranean, and Indo-West Pacific.

Etymology. From the Greek *netta* (duck) and *stoma* (mouth), referring to the elongate and slightly depressed snout. Neuter.

Species. Four species of *Nettastoma* are currently recognized: *N. melanura* Rafinesque, 1810 and *N. syntresis* Smith and Böhlke, 1981 in the

Atlantic, and *N. solitarium* Castle and Smith, 1981 and *N. parviceps* Günther, 1877 in the Indo-West Pacific. *Nettastoma parviceps* differs from the others in having the posterior nostril near the posterodorsal rather than the anterodorsal margin of the eye.

Key to the Species of *Nettastoma* in the Western North Atlantic

1a. Two pores in supraorbital canal between anterior nostril and eye; LL pores 44–49 .... *melanura*, p. 605

1b. Three pores in supraorbital canal between anterior nostril and eye; LL pores 38–41 .... *syntresis*, p. 608

Nettastoma melanura Rafinesque, 1810
Figures 649–652, Table 75

Nettastoma melanura Rafinesque, 1810a: 66, pl. 16, fig. 1 (original description, Sicily, holotype lost).
Muraenophis saga Risso, 1810: 370, pl. 10, fig. 39 (original description, Nice, holotype MNHN B-849).
Murenophis saga. Risso, 1826: 193.
Muraena saga. Costa, 1844: 10.
Hyoprorus messinensis Kölliker, 1854: 102 (original description, Messina, holotype unknown). Metamorphic leptocephalus.
Leptocephalus longirostris Kaup, 1856b: 150, pl. 18, fig. 14 (original description, Messina, holotype MNHN 2.268). Leptocephalus. According to Bertin (1935a: 102) the holotype is no longer extant.
Nettastoma melanurum. Günther, 1870: 48 (emendation).
Nettastoma mendax Facciolà, 1893: 147 (original description, Mediterranean, holotype not designated).
Nettastoma saga. Facciolà, 1899: 31.
Muraenosaurus Guntheri Osorio, 1909a: 14, figs. 2–4 (original description, Portugal, holotype MB 2066).
Leptocephalus urosema Lea, 1913: 39, fig. 33, pl. 5, no. 5 (original description, North Atlantic near Azores, two syntypes at ZMUB). Leptocephalus.
Osorina guentheri. Whitley, 1951: 68.

Misidentifications:

Not *Nettastoma melanura* Rafinesque. Facciolà, 1893: 146, part (third and fourth developmental stages are *Facciolella oxyrhynchus*); 1899: 30 (*Facciolella oxyrhynchus*). Poll, 1953: 129 (*Facciolella*). Castle, 1964c: 80 (*Nettastoma parviceps*); 1965a: 144 (*Nettastoma parviceps* and *Nettenchelys gephyra*); 1966b: 59, figs. 4, 6 (*Nettenchelys gephyra*); 1970a: 14 (identity uncertain); 1975a: 25 (identity uncertain).

Distinctive Characters. Two supraorbital pores between anterior nostril and eye; LL pores 43–49; median vomerine teeth somewhat enlarged; posterior nostril near anterodorsal margin of eye.

Description. Total vertebrae 195–207 (n = 10), predorsal vertebrae 4–8 (34), preanal vertebrae 44–50 (38). Branchiostegal rays 9 (1). Pores: LL to anus 43–49 (32), POM 18–21 (27), IO 15–17 (29), SO 1 + 4–5 (33), STC 3 (28, one specimen had 5).

Proportions as % of preanal length: predorsal 33–40 (31), head 35–42 (31), depth at anus 5–12 (28). Of HL: snout 35–44 (30), eye 8–11 (30), upper jaw 51–62 (29), gill opening 2–8 (26), interbranchial 13–29 (17).

Maxillary and mandibular teeth in narrow bands, the inner teeth largest. Intermaxillary tooth patch short, continuous with maxillary teeth. Vomerine tooth patch long, reaching to level of eye, median teeth somewhat enlarged but not forming a single row; slight gap between vomerine and intermaxillary teeth.

Stomach ends before anus. Gas bladder large, with moderately thick walls; begins near anterior end of stomach, ends about one snout length behind anus.

Color. In preservative brown or gray, vertical fins edged in black.

Size. The largest specimen examined was 773 mm TL, 303 mm preanal (USNM 179069). Another specimen (USNM 219848) had a slightly greater TL (780 mm) but a lesser preanal length (299 mm). Twelve males had preanal lengths of 130–271 mm. Eight females had preanal lengths of 106–303; one with a preanal length of 268 mm was nearly ripe.

Development and Growth. The leptocephalus has been identified (see p. 705).

Variation. The frequency of regenerated tails makes it difficult to obtain complete vertebral

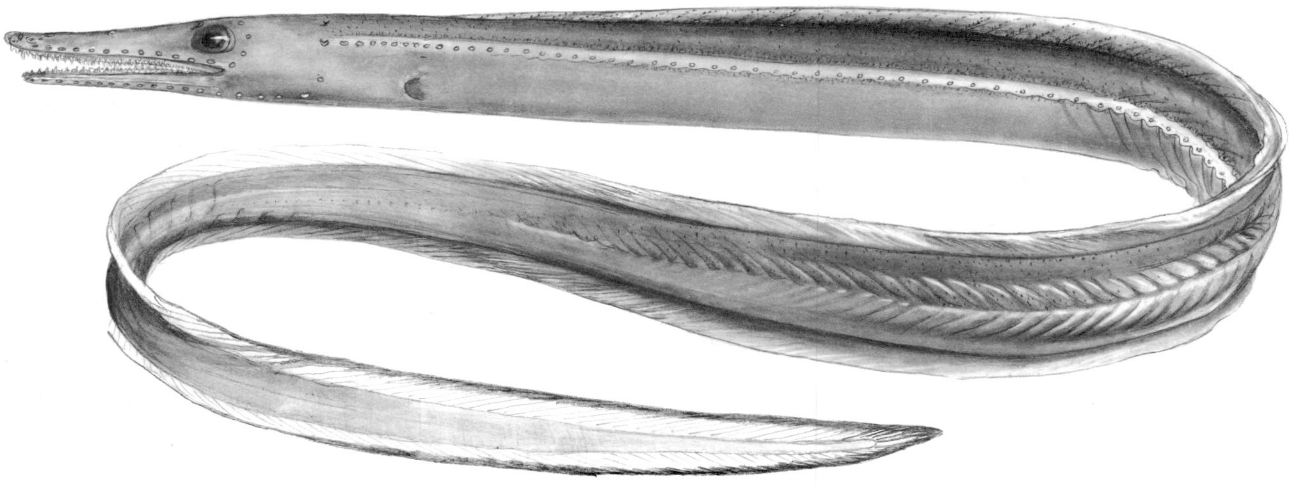

FIGURE 649. *Nettastoma melanura*: ANSP 117074, 581 mm TL.

counts. The ten x-rayed specimens that appeared to be intact yielded counts of 195–207. Myomere counts from 20 leptocephali ranged from 187–211 (Smith and Castle, 1982: 8).

Of 33 specimens examined, 29 had only two pores in the supraorbital canal between the anterior nostril and the eye. Three specimens had a third pore behind these on one side but not the other. One specimen had the extra pore on both sides, but they were not directly opposite each other. A third pore in this location is clearly anomalous in *Nettastoma melanura.*

The scarcity of reliable vertebral counts makes it difficult to assess geographic variation in this species. The variation in myomere counts of leptocephali seems excessive for one species, but no clear pattern is evident. Counts of leptocephali from the western Atlantic were 187–206, from the eastern Atlantic 203–211, and from the Mediterranean 191–204. The single count of 211 among the eastern Atlantic specimens stood somewhat apart from the rest; the next highest count was 206.

Remarks. There is some uncertainty over what species Rafinesque had in front of him when he described *Nettastoma melanura.* Two species of nettastomatid eels occur in the Mediterranean. In one of them the posterior nostril is high on the head near the anterodorsal corner of the eye, and the dorsal fin begins above the gill opening. In the other the posterior nostril opens as a slit in the upper lip, and the dorsal fin begins slightly but distinctly behind the gill opening. Rafinesque did not mention the posterior nostril, but he described and illustrated the dorsal fin as beginning behind the gill opening. Kaup (1856b: 119), Günther (1870: 48), and Moreau (1881: 579), however, applied the name *Nettastoma melanura* to the species with the superior nostril and the anterior dorsal origin.

Facciolà (1893: 147) apparently was the first to realize that two species were involved. He pointed out the differences between them and, on the basis of Rafinesque's description and illustration, concluded that *Nettastoma melanura* was the one with the more posterior dorsal origin and, hence, the inferior nostril. He gave the name *Nettastoma mendax* to the other species (superior nostril, more anterior dorsal origin). Later (Facciolà, 1899) he expanded on this and gave a more complete description of the two species, including vertebral counts. He maintained his position on the identity of *Nettastoma melanura* but decided that his *Nettastoma mendax* was a synonym of Risso's (1810) *Muraenophis saga*, which he transferred to *Nettastoma.*

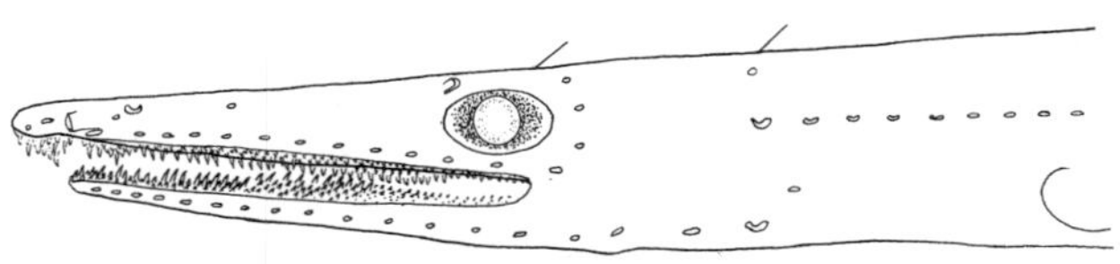

FIGURE 650. *Nettastoma melanura*: ANSP 117074, 581 mm TL; head.

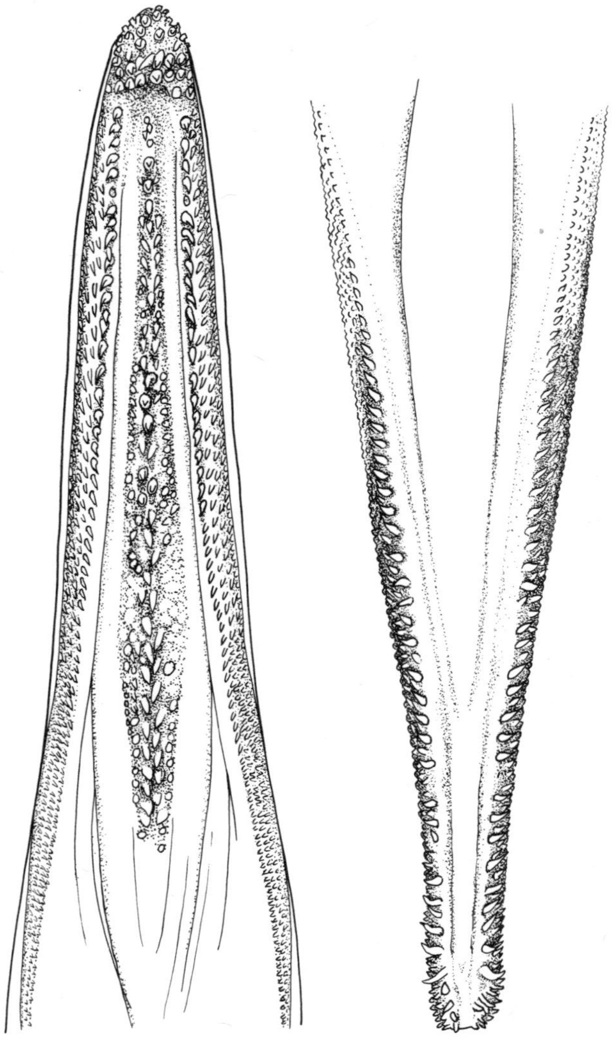

FIGURE 651. *Nettastoma melanura*: ANSP 117074, 581 mm TL; dentition.

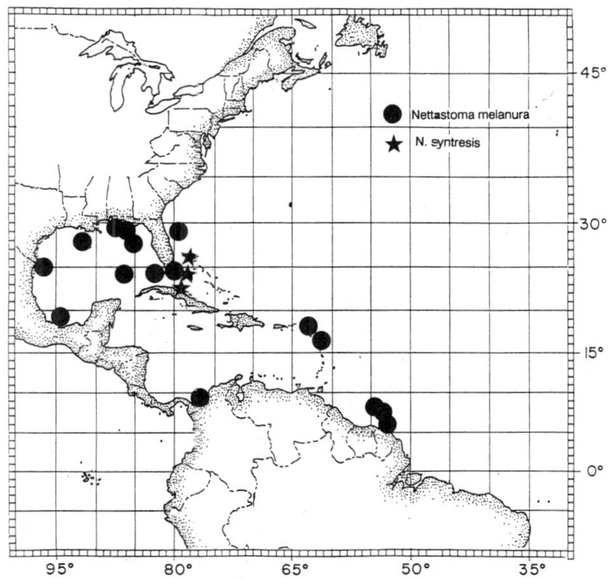

FIGURE 652. Distribution of *Nettastoma melanura* and *N. syntresis*.

Subsequently he changed his mind on two key points. First, he decided that the two species should be placed in different genera. In a footnote (Facciolà, 1911: 277) he created a new genus, *Nettastomella*, which he defined by referring to his 1899 paper and to a study by Supino (1905b). It is evident here that he had also changed his mind about the allocation of *Nettastoma melanura*, although this was not explained until 1914. There (Facciolà, 1914: 44) he referred to a size difference between the two species. Rafinesque gave a length of two to three feet for *Nettastoma melanura*. This approximated the size of the species with the superior nostril, but all the specimens that Facciolà had seen of the species with the inferior nostril were much smaller. He therefore concluded that Rafinesque's species was indeed the one with the superior nostril.

Facciolà seems to have been alone in questioning the traditional interpretation of *Nettastoma melanura*, and eventually he rejoined the majority, yet the suspicion remains that he may have been right. The dorsal-fin origin is the only diagnostic character retrievable from Rafinesque's description, and it is clearly stated and illustrated as being behind the gill opening. Because the type specimen is no longer extant, the question can probably never be answered conclusively. In view of the uncertainty involved, and in the interest of nomenclatural stability, the status quo will be maintained.

Distribution (Fig. 652). Western Atlantic from the east coast of Florida and the northern Gulf of Mexico through the Caribbean to the Guianas, except for a small area around the Bahamas including the Santaren and Northwest Providence Channels, where it is replaced by *Nettastoma syntresis*; eastern Atlantic from the Mediterranean and the coast of Portugal to the Gulf of Guinea. Depth 329–860 m; in addition, there are single records of 37 m (USNM 158933) and 1647 m (USNM 195898). Relatively common.

Etymology. From the Greek *melan* (black, dark)

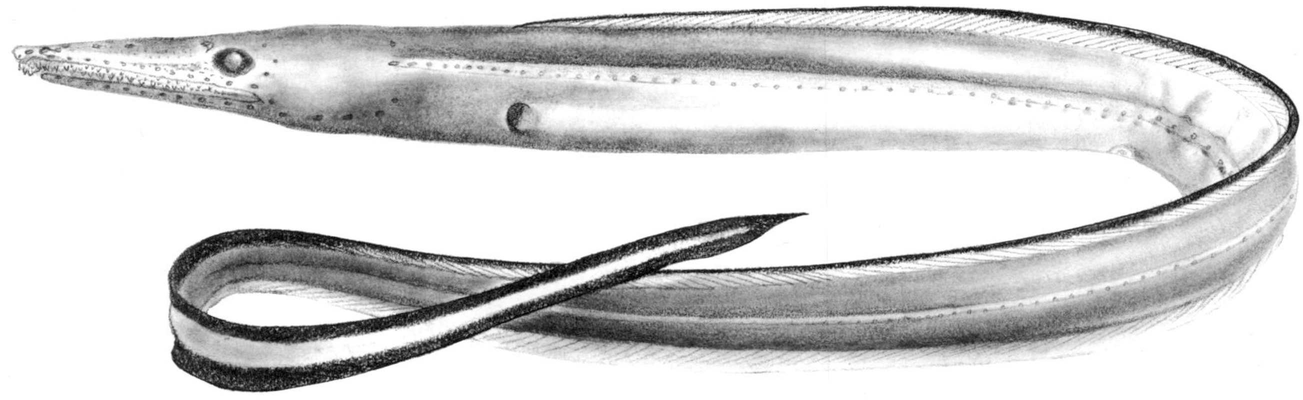

FIGURE 653. *Nettastoma syntresis*: ANSP 158936, 482 mm TL.

and *oura* (tail), presumably referring to the dark-edged posterior portions of the dorsal and anal fins. A noun in apposition.

Study Material. 74 specimens, 200–780 mm TL.

FLORIDA ATLANTIC: UMML 15648 (1, 473). USNM 185651 (1, 294). FLORIDA STRAITS: USNM 197341 (1, 626). FLORIDA GULF: ANSP 114853 (1, 620), 114868 (3, 570–645), 117074 (1, 581), 117090 (2, 375–450), 126187 (1, 606), 126188 (1, 630), 126405 (1, 567), 133014 (1, 655), 156854 (1, 373). UMML 13212 (1, 606). ALABAMA: ANSP 117062 (1, 550), 117065 (1, 610). TU 90367 (1). USNM 178997 (2, 478–615), 179069 (3, 440–773), 195898 (2, 339–544), 198051 (1, 666). LOUISIANA: ANSP 117064 (2, 584–672). MCZ 49338 (1). MEXICO GULF: ANSP 117007 (1, 589), 117066 (2, 567–660), 126174 (1, 468). LESSER ANTILLES: USNM 217390 (1, ca 575), 219848 (3, 598–780). COLOMBIA: ANSP 126445 (1, 635). USNM 198754 (1, 727). SURINAME: ANSP 114869 (2, 529–534). USNM 158934 (1, 557), 185639 (1, 606). FRENCH GUIANA: ANSP 114844 (1, 655). USNM 158933 (1, 483), 158935 (1, 659). LOCALITY UNCERTAIN: UMML 32547 (1).

MEDITERRANEAN: BMNH 1866.9.10.5 (1), 1867.-10.4.12 (1), 1888.6.15.24 (1), 1952.5.26.2 (1). MCZ 40840 (1). MNHN B-2773 (1), 84-427-430 (3). MZUF C.1314 (4). USNM 49345 (1, 602). GULF OF GUINEA: ANSP 156855 (1, 224). UMML 15145 (4, 469–555), 20903 (3, 200–420). USNM 198619 (1, 337), 198629 (1, 406), 199537 (1, 460), 200775 (1, 620+).

Nettastoma syntresis Smith and Böhlke, 1981
Figures 652–655, Table 75

Nettastoma syntresis Smith and Böhlke in Smith, Böhlke, and Castle, 1981: 546, figs. 4, 5B, 6B (original description, 23°40′N, 79°18′W, holotype USNM 219849).

Distinctive Characters. Three supraorbital pores between anterior nostril and eye; LL pores 38–41; median vomerine teeth not enlarged; posterior nostril near anterodorsal margin of eye.

Description. Total vertebrae 201 (n = 1), predorsal vertebrae 5–6 (4), preanal vertebrae 40–41 (6). Branchiostegal rays 7–8 (1). Pores: LL to anus 38–41 (14), POM 18–19 (5), IO 15–16 (11), SO 1 + 4–6 (16), STC 3 (12).

Proportions as % of TL: preanal 35 (1), predorsal 13 (1), head 14 (1), depth at anus 3 (1). Of preanal length: predorsal 37–41 (13), head 39–43 (13), depth at anus 7–13 (11). Of HL: snout 41–44 (13), eye 8–11 (13), upper jaw 57–63 (12), gill opening 4–7 (12), interbranchial 10–22 (9).

Dentition as in *Nettastoma melanura* except that median vomerine teeth are not enlarged.

Color. In preservative brown or gray, vertical fins edged in black.

Size. The largest specimen examined had a TL of 550 mm and a preanal length of 233 mm (USNM 197342); a large part of its tail was missing. Six males had preanal lengths of 95–155 mm. Five females had preanal lengths of 181–233 mm; none seemed fully ripe.

Development and Growth. The leptocephalus has been identified (see p. 707).

Variation. Of 16 specimens examined, 12 had three pores in the supraorbital canal between the

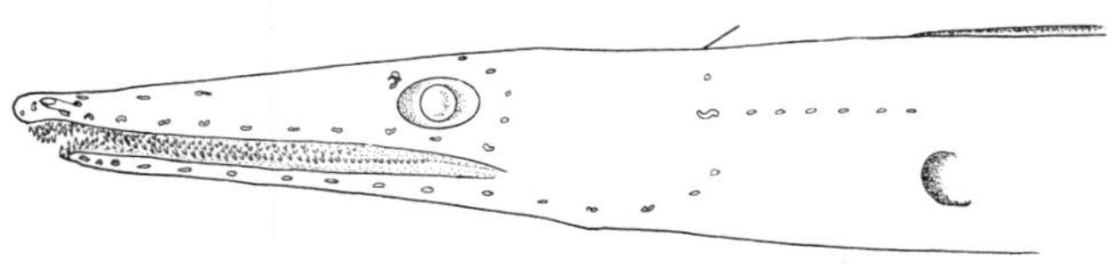

FIGURE 654. *Nettastoma syntresis*: ANSP 114965, 433 mm TL; head.

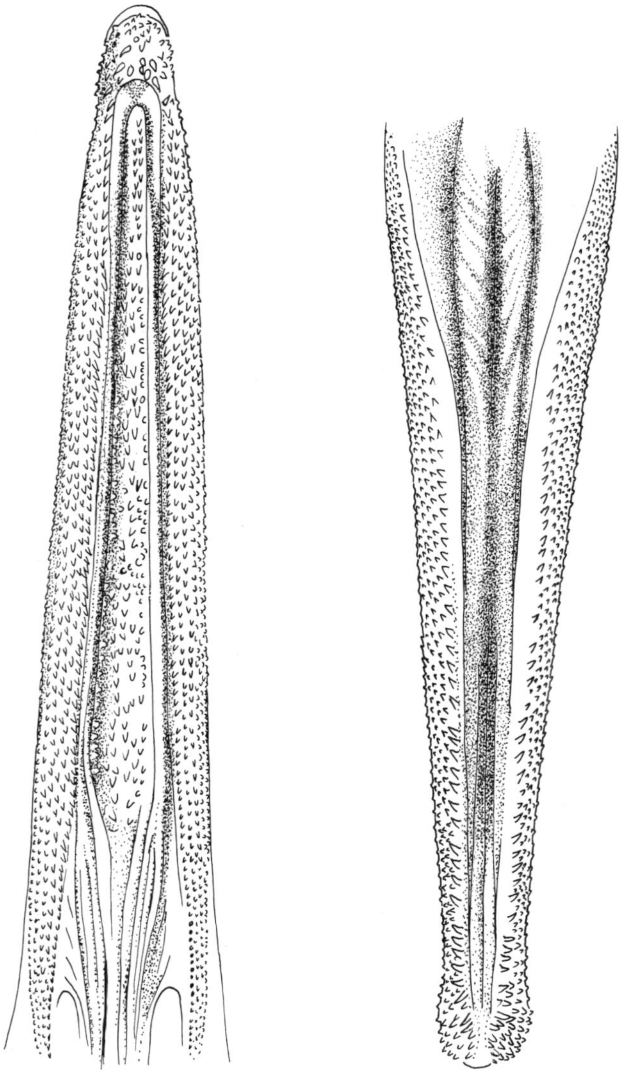

FIGURE 655. *Nettastoma syntresis*: ANSP 158936, 482 mm TL; dentition.

anterior nostril and the eye. One specimen had three pores on one side and two on the other. Three specimens had three pores on one side and four on the other.

Distribution (Fig. 652). This species has been found only in a restricted area off the western edge of the Bahamas, including the Santaren Channel, the eastern part of the Nicholas Channel, the northwestern end of the Old Bahama Channel, and the Northwest Providence Channel. Its distribution seems completely disjunct from that of *N. melanura*.

Etymology. Greek *syntresis* (channel, passage, strait), in reference to the Santaren, Nicholas, and Northwest Providence Channels, where the species occurs. A noun in apposition.

Study Material. 28 specimens, 122–550 mm TL.

Holotype: USNM 219849 (male, 442), Silver Bay 2458, Santaren Channel, 23°40′N, 79°18′W, 531 m, 5 Nov. 1960.

Paratypes: SANTAREN CHANNEL: USNM 158936 (1, 482), 197127 (2, 270–360), 197342 (1, 550), 197345 (6, 251–453), 202953 (1, 260). NICHOLAS CHANNEL: ANSP 114965 (1, 433). USNM 157939 (1, 449). NORTHWEST PROVIDENCE CHANNEL: UMML 31525 (3, 320–409).

Other material: SANTAREN CHANNEL: MCZ 38803 (1, 298). USNM 219869 (6, 174–445). NICHOLAS CHANNEL: MCZ 38781 (1, 122), 38811 (1, 179), 38814 (1, 250). OLD BAHAMA CHANNEL: MCZ 38780 (1, 370).

Genus *Venefica* Jordan and Davis, 1891

Venefica Jordan and Davis, 1891: 651 (type species *Nettastoma procerum* Goode and Bean, 1883, by original designation).

Characters. Body elongate, round in cross section anteriorly, compressed posteriorly, deepest in middle, tapering toward head and tail; tail slender, filiform, frequently damaged; anus near anterior two-fifths of body length. Dorsal and anal fins confluent with caudal fin; dorsal fin begins over, slightly before, or slightly behind gill opening. Pectoral fin absent.

Head and snout slender, elongate; upper jaw projects beyond lower. Tip of snout fleshy, extended into a slender proboscis of variable length; skin of proboscis covered with many pits, presumably sensory in nature, separated from each other by a network of dermal ridges; first two supraorbital pores located on proboscis. Eye moderately small. Mouth large, rictus well behind eye. Anterior nostril a low tube at base of proboscis, directed anterolaterally. Posterior nostril variable in size, elliptical or oblique, located variably before upper part of eye to directly above anterodorsal quarter of eye.

Lateral line on body complete to tail. Pores on head more numerous than in other nettastomatid genera. Twelve to 17 supraorbital pores: the first (ethmoidal pore) on basal portion of proboscis; second also on proboscis, above and slightly before ethmoidal pore, larger and slit-like; approximately 7–12 small supraorbital pores

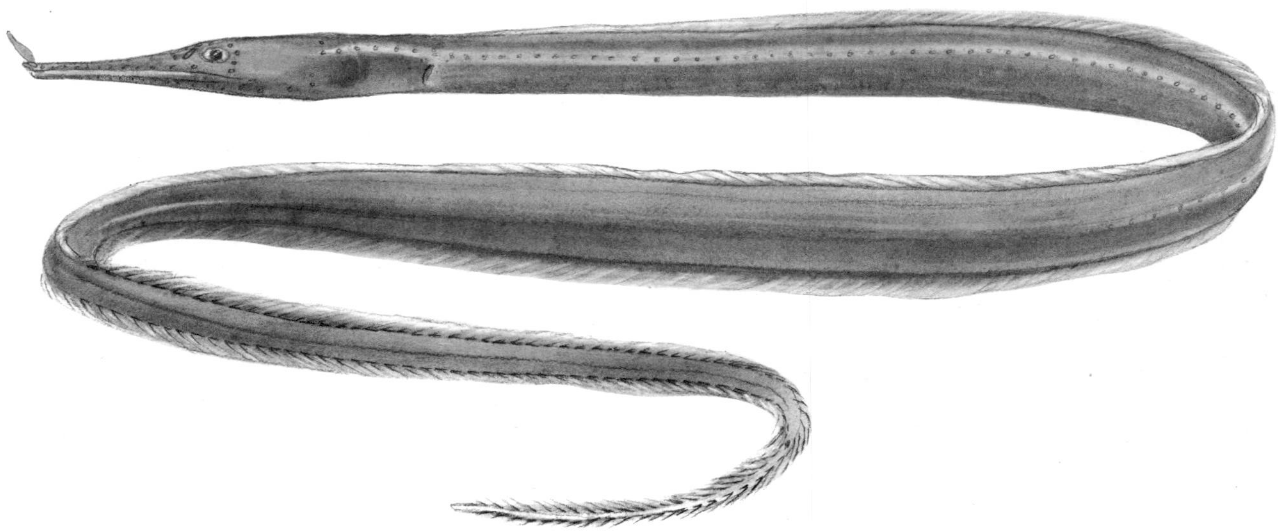

FIGURE 656. *Venefica procera*: composite illustration, body ANSP 126189, 810 mm TL, tail ANSP 106513, 427 or 604 mm TL.

behind anterior nostril. Fourteen to 19 infraorbital pores: one near ventral base of anterior nostril tube; one (the adnasal pore) near dorsal base of anterior nostril tube, above posterior edge of preceding pore; approximately 9–14 behind this along edge of upper lip; three behind eye. Three pores in supratemporal commissure, but occasionally an extra, anomalous pore present. Preoperculomandibular canal with approximately 17–22 pores.

Teeth small, conical, recurved, in narrow bands.

Size. Maximum size probably about 1 m.

Distribution. All oceans, in relatively deep water.

Etymology. Latin *venefica* (sorceress, witch), from the French *sorciere*, a vernacular name for *Nettastoma melanura*. Feminine.

Species. There are five nominal species of *Venefica*: *V. procera* (Goode and Bean, 1883) from the western Atlantic, *V. proboscidea* (Vaillant, 1888) from the eastern Atlantic and Indian Ocean, *V. multiporosa* Karrer, 1982 from the Indian Ocean and the Celebes Sea, and *V. tentaculata* Garman, 1899 and *V. ocella* Garman, 1899, both from the eastern Pacific. The chief distinguishing characters seem to be the length of the proboscis, the form and location of the posterior nostril, the number of head pores, and a few proportional measurements. All of the species need critical reexamination.

Venefica procera (Goode and Bean, 1883)
Figures 656–659, Table 75

Nettastoma procerum Goode and Bean, 1883: 224 (original description, 35°35′N, 76°W, holotype MCZ 25902).
Venefica procera. Jordan and Davis, 1891: 652.

Distinctive Characters. Venefica procera has a shorter proboscis than *V. proboscidea* and *V. tentaculata*. In *V. multiporosa* the posterior nostril is located almost entirely above the eye; in *V. procera* it is before the eye.

Description. Total vertebrae 200–205 (n = 5), predorsal vertebrae 6–8 (12), preanal vertebrae 57–62 (13). Branchiostegal rays 5 (5). Pores: LL to anus ca 60–64 (2), POM 17–22 (16), IO 16–19 (26), SO 1 + 8–12 (25), STC 3–4 (29).

Proportions as % of TL (note: predorsal, head, snout, and upper jaw measured from bony tip of ethmovomer): preanal 38 (1), predorsal 11 (1), head 10 (1), depth at anus 2 (1). Of preanal length: predorsal 22–30 (27), head 24–36 (34), depth at anus 4–7 (34). Of HL: snout 42–55 (34), eye 4–7

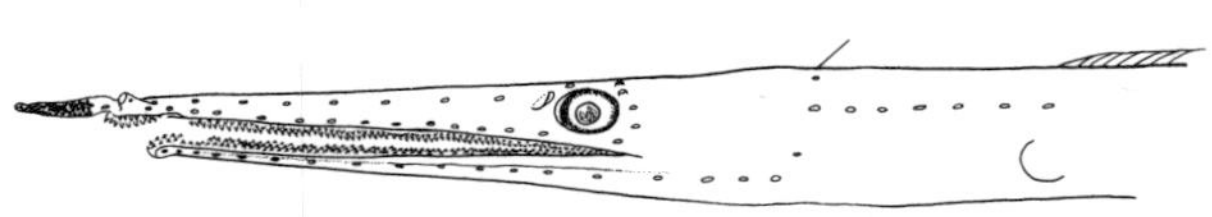

FIGURE 657. *Venefica procera*: ANSP 152112, 762 mm TL; head.

(34), upper jaw 53–67 (33), gill opening 3–5 (18), interbranchial 4–8 (17), proboscis 4–13 (31).

Posterior nostril large, oblique, before anterodorsal quarter of eye, located mainly before anterior margin of eye; covered by a large flap from anterior edge.

Intermaxillary tooth patch somewhat spade-shaped, completely exposed when mouth closed, separated from vomerine teeth by a slight crescent-shaped gap. Vomerine tooth patch long, reaching almost to anterior edge of eye. Maxillary teeth in approximately four rows, separated from intermaxillary teeth. Mandibular tooth patches meet at tip of jaw, where they are expanded slightly and fit into gap between vomerine and intermaxillary teeth when mouth closed.

Stomach moderate in length, ends about ⅔–¾ of distance to anus. Gas bladder slender, reduced, about one-sixth length of stomach; main body located anteriorly, extending posteriorly as a thin ligament which ends before posterior end of stomach.

A very oily fish, oil globules often present in alcohol and in peritoneal cavity.

Color. In preservative brown or gray, vertical fins edged in black. Stomach and intestine covered with black peritoneum. Gas bladder metallic silvery.

Size. The largest specimen examined was 1088 mm TL (MCZ 38818). Six mature females of 308–404 mm preanal length contained eggs 0.4–0.9 mm diameter. The largest male examined was 355 mm preanal (ANSP 117023).

Development and Growth. Despite the apparent abundance of this species, no leptocephali have been identified with it.

Variation. There is some variation in the number of preoperculomandibular, infraorbital, supraorbital, and supratemporal pores. Of these, the supratemporal series is the most constant; 27 of 29 specimens examined had three pores, and two had four pores.

Remarks. Goode and Bean (1883: 224) began their description of *Nettastoma procerum* by saying that two specimens were obtained at Blake station 325 and another, mutilated, specimen was obtained at station 327. They later (Goode and Bean, 1896: 150) said: "The types of *V. procera* were two fishes obtained at station 325. . . ." The description itself, however, was clearly based on a single specimen, and included the following statement: "The total length of the specimen is 727 millimeters. . . ." The two specimens from

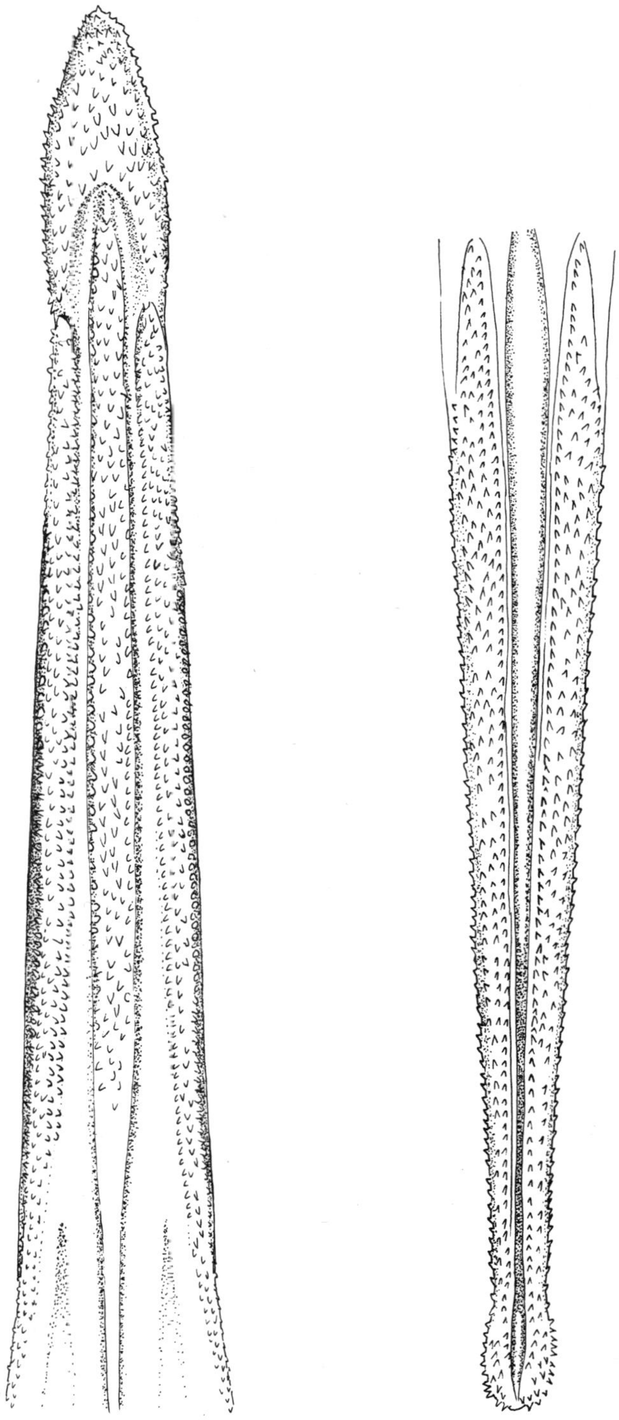

FIGURE 658. *Venefica procera*: ANSP 157598, 650 mm TL; dentition.

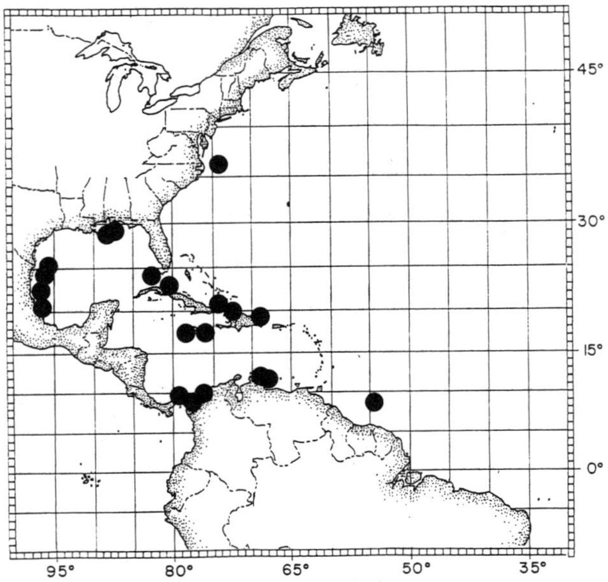

FIGURE 659. Distribution of *Venefica procera.*

Blake 325 are today registered as MCZ 25902 and USNM 153585. The former now measures 625 mm TL, although it is incomplete. The latter is in three pieces, which combine to measure approximately 325 mm TL (K. A. Tighe, pers. comm.). If the specimen described by Goode and Bean measured 727 mm, then it is more likely to be MCZ 25902 than the much smaller USNM 153585. For this reason, I recognize MCZ 25902 as the holotype.

Distribution (Fig. 659). Western Atlantic from North Carolina to Suriname, including the Gulf of Mexico and Caribbean. Depth 832–2304 m. Relatively common.

The record of *Venefica procera* from the Celebes Sea (Weber, 1913: 45) has now been referred to *V. multiporosa* (Karrer, 1982: 66).

Etymology. Latin *procerus* (long), referring to the elongate form. An adjective.

Study Material. 107 specimens, 215–1088 mm TL.

Holotype: MCZ 25902 (female, 630), Blake 325, off North Carolina, 35°35′20″N, 75°14′40″W, 1184 m, 14 July 1880.

Other material: NORTH CAROLINA: ANSP 137879 (1, 712), 137880 (1, 828). FLORIDA STRAITS: ANSP 117038 (1, 215), 117044 (1, 530), 156853 (1, 546). FLORIDA GULF: ANSP 110078 (2, 485–645), 126189 (1, 810). UMML 12476 (1, 610), 16622 (1, 667), 32758 (2, 610–763). ALABAMA: ANSP 106513 (2, 427–604), 106520 (1, 655), 106523 (1, 645), 106524 (4, 460–620), 106526 (1, 635), 108735 (3, 675–795), 108737 (8, 347–815), 110072 (1, 627), 114957 (1, 685), 117023 (3, 400–895), 126452 (2, 650–655), 152112 (1, 762), 157598 (1, 650). UMML 29739 (1, 674). MEXICO GULF: ANSP 106281 (6, 444–530), 106500 (2, 607–750), 106506 (2, 505–535), 106509 (2, 470–590), 106527 (1, 431), 114999 (1, 656), 117029 (3, 485–600), 117049 (2, 580–630), 117052 (1, 491), 117094 (2, 655–810), 117097 (1, 825). CUBA: MCZ 38786 (1, 637), 38790 (2, 508–550), 38818 (1, 1088). JAMAICA: UMML 29681 (4, 295–350 mm preanal), 29682 (3, 603–698). HISPANIOLA: ANSP 117815 (1, 753). UMML 28468 (1, 775). VENEZUELA: ANSP 106521 (1, 920), 107099 (5, 634–798), 110069 (1, 790), 115510 (1, 772), 117816 (1, 970). COLOMBIA: UMML 22808 (1, 563). PANAMA: UMML 22356 (9, 178–306 mm preanal), 22671 (4, 468–605), 23231 (2, 430–735). SURINAME: UMML 24635 (2, 830–835).

Family Serrivomeridae

KENNETH A. TIGHE

Acknowledgments. I thank the following for courtesies extended to me during visits to their home institutions: M. N. Feinberg (AMNH), K. E. Hartel (MCZ), C. R. Robins (UMML), J. E. Craddock and R. H. Backus (WHOI).

Introduction. The family Serrivomeridae contains some of the most common midwater eels of the world's oceans. Juvenile to subadult specimens are commonly taken in midwater trawls in any temperate to tropical region. Their leptocephali are often taken in large numbers and can often constitute a large portion of the leptocephali catch in oceanic waters.

Methods. General methods used for meristic and morphometric data in this study are given on pp. 1–6. The premaxillary projection is measured from the articulation point of the vomer and maxillae to the tip of the vomer. The width of the skull is measured across the sphenotic processes.

Characters. Deep-sea pelagic eels with slender body and moderately attenuate jaws; snout less than half length of head; anterior portion of the vomer projecting well beyond the articulation with the maxillae. Anterior and posterior nostrils set close together just in front of eye; anterior nostril tubular or non-tubular (sexual dimorphism in adults). Anus in advance of middle of body. Body naked, covered with a deciduous silvery or bronze layer and surrounded with a gelatinous sheath. Cephalic lateralis system reduced to three small pores situated between the nostrils supported by two poorly ossified ossicles; several series of infraorbital and supraorbital sensory papillae (may represent reduced cephalic lateralis system); lateral line reduced to a single series of sensory papillae. Anal-fin origin just posterior to anus; dorsal-fin origin near or slightly behind anal origin; dorsal fin relatively low, length of longest ray less than half body depth at anal origin; anal fin higher, length of longest ray approximately equal to body depth at anal origin; dorsal and anal fins confluent with caudal. Caudal filament absent. Maxillary and mandibular teeth small, erect, pointed, set in two or more rows. Vomerine teeth erect, compressed or conical, in two or more rows.

Osteology. Frontals fused along an obvious median line (Fig. 660). Anterior portion of frontal with a bifurcate tubular articulation with the double tubular posterior portion of the ethmovomerine complex. Parietals fused along the median suture. Supraoccipital absent. Exoccipitals and basioccipital forming margins of foramen magnum. Wing-like posteriorly directed processes formed of epiotics, pterotics and hyomandibulars. Anteriorly directed processes on sphenotics; sphenotic process braced by antero-lateral strut from pterotics. Cartilaginous lateral ethmoid process present. Suspensorium (Fig. 661) anteriorly directed (mandibular-quadrate-hyomandibular angle ca 120 degrees). Palato-pterygoid arch well developed with both palatine and pterygoid portions present, although fused. Pterygoid with postero-dorsad directed laminar process interdigitating with a similar antero-dorsad directed process from hyomandibular. Quadrate articulating with pterygoid and hyomandibular with anteriorly and posteriorly directed processes, respectively. Hyomandibular articulates with the neurocranium through condyle and socket joint ventral to the sphenotic process. All elements of opercular series present, although greatly reduced; opercle elongate, blade-like; ventral edge of opercle bordered by subopercle; preopercle filling space between opercle and hyomandibular, with anterior process articulat-

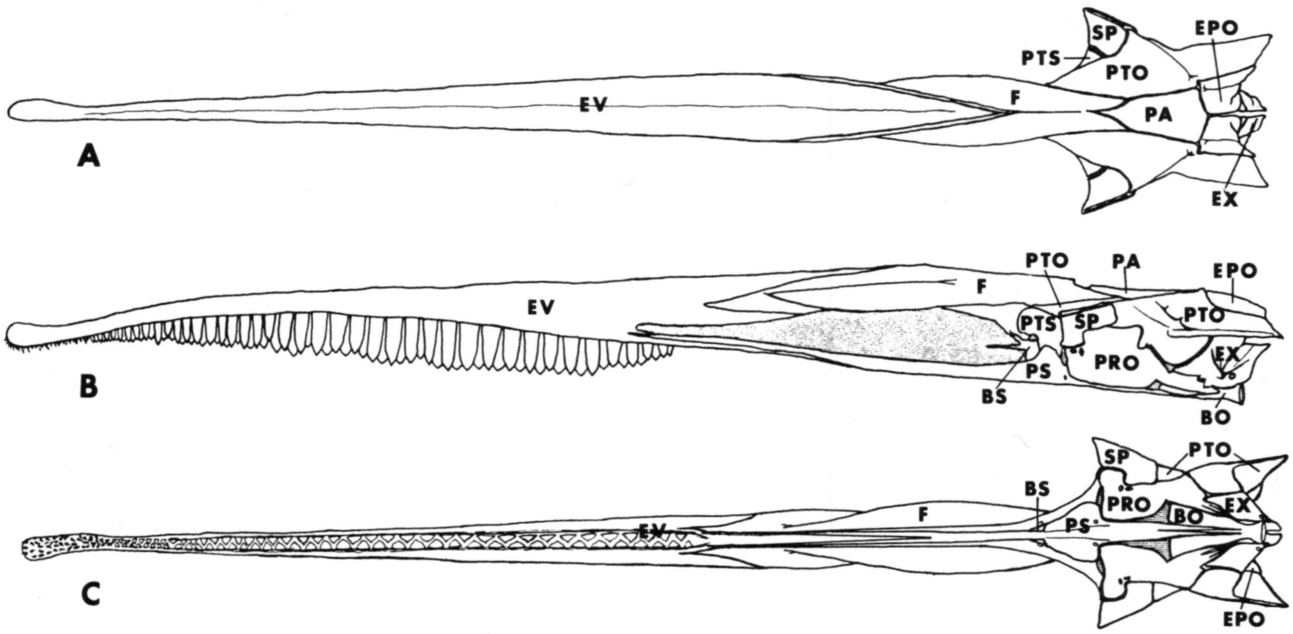

FIGURE 660. *Serrivomer beanii*: neurocranium. A. Dorsal view. B. Lateral view. C. Ventral view. Abbreviations: BO = basioccipital, BS = basisphenoid, EPO = epiotic, EV = ethmo-vomer complex, EX = exoccipital, F = frontal, PA = parietal, PRO = prootic, PS = parasphenoid, PTO = pterotic, PTS = pterosphenoid, SP = sphenotic.

ing with suspensorium along hyomandibular-quadrate joint; interopercle elongate, extending anteriorly along hyoid arch. Hyoid apparatus (Fig. 662) complete with glossohyal, urohyal, and paired hypohyals, ceratohyals and epihyals; hypohyals partially fused ventrally with ceratohyals; 1 pair of branchiostegals attached to ceratohyal; 5–6 pairs attached to epihyal. No osteological connection between hyoid and branchial arches. Branchial arches (Fig. 663) relatively complete and ossified; basibranchial 1 present and well ossified; basibranchials 2 and 4 present as reduced cartilaginous elements; basibranchial 3 absent; hypobranchials 1 and 2 present and well ossified; hypobranchial 3 reduced to cartilaginous element; ceratobranchials and epibranchials of first four arches present and well ossified; ceratobranchial 5 present as a reduced ossified element providing the articulation between the lower pharyngeal tooth plate and ceratobranchial 4; infrapharyngobranchials 2 and 3 present; lower pharyngeal tooth plate with irregular double rows of conical teeth anteriorly with the inner row slightly enlarged, posterior portion of lower tooth plate expanded slightly with an irregular patch of teeth; upper pharyngeal tooth plates irregularly covered with small conical teeth, united by a suture. Cleithrum present, well ossified; coracoid reduced to cartilaginous remnant; scapula, supracleithrum and pectoral radials absent; pectoral rays reduced, poorly ossified. Neurapophyses present on all vertebrae as longitudinal ridges, highly modified on precleithral vertebrae (Fig. 664); anterior portion of neurapophysis of first vertebra modified into pseudoneural arch directed posteriorly under true neural arch (not fused to neural arch in *Serrivomer,* fused in *Stemonidium* producing a double neural spine), numerous crossbraces and foramina producing a lattice-like tube around the spinal cord; neural spine of second or third vertebra vertically or anteriorly directed; neural arches of remaining precleithral vertebrae reduced in complexity, with fewer foramina and crossbraces, and with anteriorly directed spines. Neural arches of preanal vertebrae with crossbrace below apex of arch. Neural spines of remaining predorsal vertebrae elongate, extending posteriorly at least two centra lengths from point of origin. Epineurals present on anterior ninth

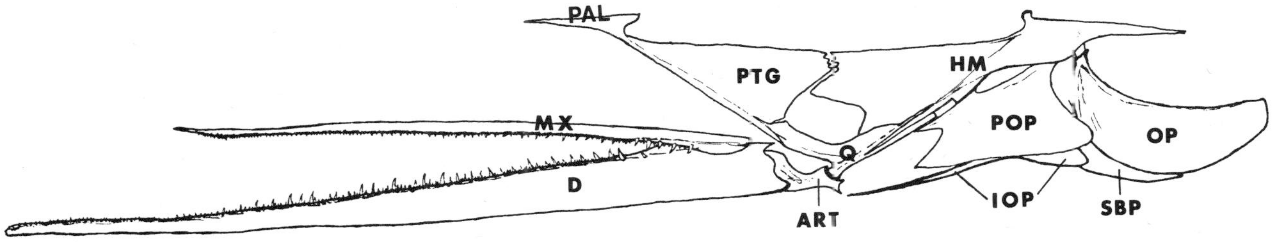

FIGURE 661. *Serrivomer beanii*: suspensorium and jaws (lateral view). Abbreviations: ART = articular, D = dentary, HM = hyomandibula, IOP = interopercle, MX = maxilla, OP = opercle, PAL = palatine portion of palatopterygoid, POP = preopercle, PTG = pterygoid portion of palatopterygoid, Q = quadrate, SBP = subopercle.

to twelfth vertebrae only. Pleural ribs, epipleurals and myorhabdoi absent. Neural arches gradually reduced, becoming open posterior to dorsal origin, again becoming complete arches on posterior caudal vertebrae. Neurapophyses of caudal vertebrae vertically expanded into anterior and posterior wing-like processes. Haemapophyses reduced, represented by low longitudinal ridges on centra. First caudal vertebra distinguished by incomplete haemal arch consisting of a pair of median vertical spines; haemal arches complete on posterior caudal vertebrae. Caudal complex typically anguilliform, with relatively complete but distorted ural centrum and three hypurals; hypural foramen between hypurals 1 and 2.

Distribution. The Serrivomeridae are distributed in temperate, subtropical, and tropical regions of all three major oceans. Serrivomerids may also occur in deep waters associated with the coastal regions of continental and island land masses.

Discussion. When originally described by Gill and Ryder (1883a), serrivomerids were included in the family Nemichthyidae. Roule and Bertin (1929) were the first to recognize the familial status of the serrivomerids, but based this conclusion on only external morphology. Trewavas (1932) provided the first osteological definition of the family based on a cleared and stained specimen of *Serrivomer lanceolatoides* (misidentified as *S. beanii*). Several errors in this osteological definition resulted in the Serrivomeridae being grouped with the Nemichthyidae and Cyematidae within the sub-order Nemichthyiformes (Berg, 1940; Bertin and Arambourg, 1958). Robins and Robins (1971) were the first to point out the fused nature of the frontals (reported by Trewavas to be united by a suture) and suggested

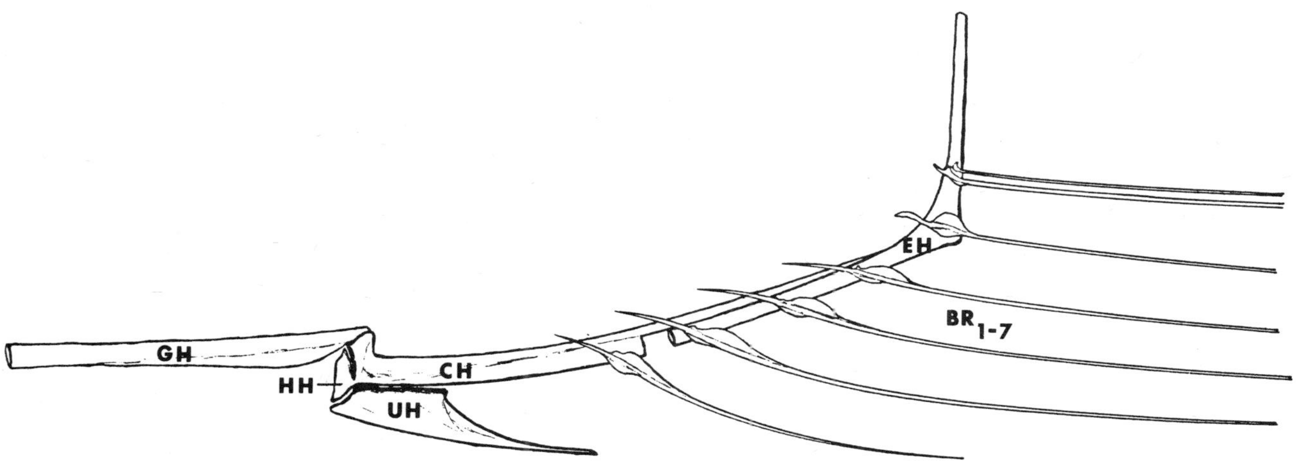

FIGURE 662. *Serrivomer beanii*: hyoid arch and branchiostegal rays (posterior extensions of branchiostegal rays not shown). Abbreviations: BR = branchiostegal rays, CH = ceratohyal, EH = epihyal, GH = glossohyal, HH = hypohyal, UH = urohyal.

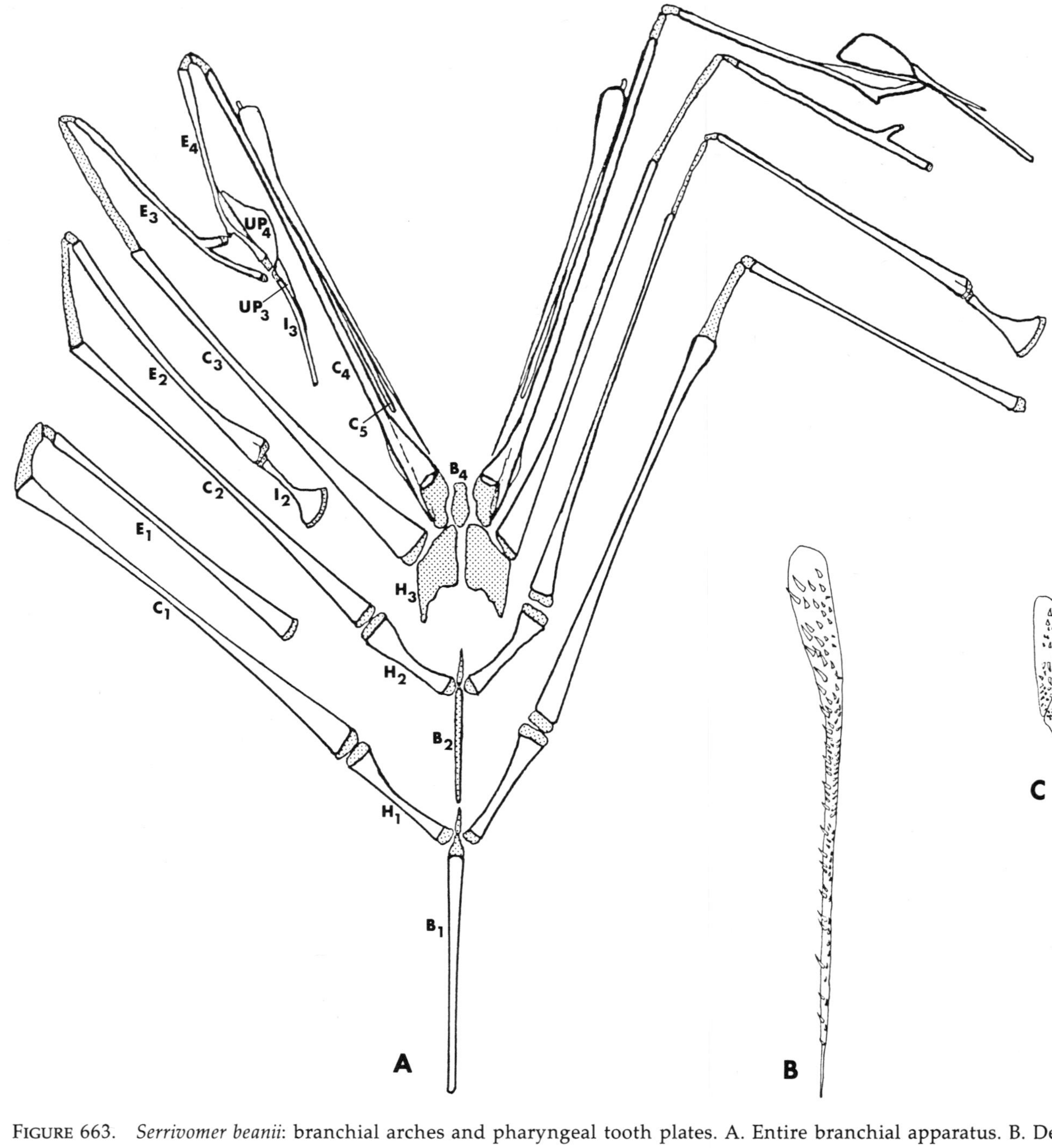

FIGURE 663. *Serrivomer beanii*: branchial arches and pharyngeal tooth plates. A. Entire branchial apparatus. B. Detail of lower pharyngeal tooth plate. C. Detail of upper pharyngeal tooth plate. Abbreviations: B = basibranchial, C = ceratobranchial, E = epibranchial, H = hypobranchial, I = infrapharyngobranchial, LP = lower pharyngeal tooth plate, UP = upper pharyngeal tooth plate.

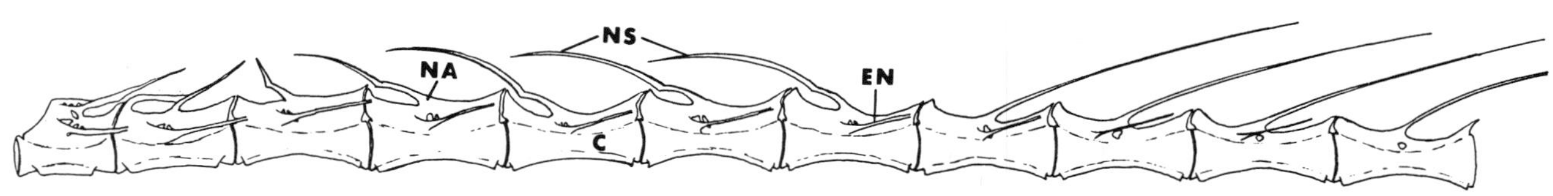

FIGURE 664. *Serrivomer beanii*: anterior vertebral column (posterior extensions of epineurals not shown). Abbreviations: CE = centrum, EN = epineural, NA = neural arch, NS = neural spine.

that the serrivomerids more properly belonged within the Congroidea. They also suggested that the serrivomerids might represent a distinct lineage within the Anguilliformes. The interfamilial relationships of the Serrivomeridae are still unresolved.

The family includes two genera: *Serrivomer*, with about nine species, and *Stemonidium*, with one species. The two genera can be distinguished by proportional measurements and dentition.

Key to the Genera of Serrivomeridae

1a. Dorsal-fin origin posterior to anus; vomerine dentition consisting of two alternating rows of laterally compressed teeth (reduced in adult males of some species) .... *Serrivomer*, p. 617
1b. Dorsal-fin origin over or slightly anterior to anus; vomerine dentition consisting of small, numerous, recurved, conical teeth close set in an irregular pattern .... *Stemonidium*, p. 625

Genus *Serrivomer* Gill and Ryder, 1883

Serrivomer Gill and Ryder, 1883a: 260 (type species *Serrivomer beanii* Gill and Ryder, 1883, by monotypy).
?*Spinivomer* Gill and Ryder, 1883a: 261 (type species *Spinivomer goodei* Gill and Ryder, 1883, by monotypy).
Alcockidia Gilbert, 1905: 586 (type species *Gavialiceps microps* Alcock, 1889, by original designation).
Paraserrivomer Roule and Angel, 1931: 2 (type species *Gavialiceps hasta* Zugmayer, 1911, by original designation).
Platuronides Roule and Bertin, 1924: 64 (type species *Platuronides danae* Roule and Bertin, 1929, by monotypy).

Characters. Body sub-cylindrical or slightly compressed, tapering gradually to tail. Dorsal-fin origin slightly posterior to anus. Vomerine teeth enlarged, erect, laterally compressed, arranged in two alternating rows. All other characters as in family.

Distribution. Species in the genus *Serrivomer* occur in mesopelagic to upper bathypelagic waters of all temperate to tropical oceanic regions of the world.

Etymology. From the Latin *serra* (saw) and *vomer* (plowshare), in reference to the serrate vomerine dentition.

Discussion. Four genera have been synonymized above with *Serrivomer* Gill and Ryder, 1884, while an additional two genera have been considered as synonyms of *Serrrivomer* by previous authors. *Spinivomer* Gill and Ryder, 1884, based on the 147 mm type of *S. goodei*, is probably a juvenile *S. beanii* since the characters separating it from *S. beanii* in the original description are juvenile characters as pointed out by Beebe and Crane (1937a). The type specimen of this species has been lost at the USNM.

Gavialiceps Alcock, 1889, based on the Indian Ocean species *G. microps* and *G. taeniola*, was distinguished from *Serrivomer* by the lack of pectoral fins. As pointed out by Roule and Angel (1931), the pectoral fin is small and delicate, and was either overlooked or lost in the original specimens. *Gavialiceps* has been considered synonymous with *Serrivomer* by past authors (Beebe and Crane, 1937a; Bauchot and Saldahna, 1973) although the name has been used by others (Castle, 1977a; Karrer, 1982) for a valid species of muraenesocid eel, *G. taeniola.* Although Alcock (1892b) subsequently restricted *Gavialiceps* to *G. microps* by the transfer of *G. taeniola* to the genus *Nettastoma*, this does not constitute a designation of type species. The first valid designation of type species was that of Gilbert (1905) in which he selected *G. taeniola* as the type species of *Gavialiceps.* Therefore, *Gavialiceps* does not belong in the synonymy of *Serrivomer.* However, *Alcockidia* Gilbert, 1905, a replacement name based on the serrivomerid *G. microps*, is a junior synonym of *Serrivomer.*

Paraserrivomer Roule and Angel, 1931 was originally distinguished from *Serrivomer* by spatulate enlargements on the tips of the jaws. This character is found in larger specimens of *S. beanii*, but is often lost due to net damage.

Platuronides Roule and Bertin, 1929 was based on an adult male of *S. lanceolatoides* and is therefore synonymized with *Serrivomer*. The status of this generic name is discussed more fully under the section on *S. lanceolatoides*.

Oxystomus Rafinesque, 1810b, believed by Bertin (1935b), Whitley (1948) and Castle (1969c) to be based on a transforming serrivomerid, does not belong in synonymy with *Serrivomer*. The original description leaves little doubt that it is not based on a transforming *Serrivomer* although *O. danae* Bertin, 1935 is described from a transforming *S. beanii*. *Bertinulus* Whitley, 1948, a replacement name for the preoccupied *Oxystomus*, is also not a synonym for *Serrivomer* although Whitley described the genus in the family Serrivomeridae in the mistaken belief that the type species, *O. hyalinus*, was a serrivomerid.

Although the taxonomy of *Serrivomer* has not been critically examined on a worldwide basis, approximately nine nominal species are presently included in the genus. Of these, two species are found in the North Atlantic which can be separated by the mode of branchiostegal attachment and meristics.

Key to the Species of *Serrivomer* in the Western North Atlantic

1a. Anterior tips of at least four of the first five branchiostegals extend beyond their articulation with the hyoid arch; 142 to 173 dorsal rays; 122 to 156 anal rays .... *beanii* Gill and Ryder, 1883, p. 618

1b. Anterior tips of the first five branchiostegals do not extend beyond their articulation with the hyoid arch; 175 to 200 dorsal rays; 165 to 192 anal rays ..... *lanceolatoides* Schmidt, 1916, p. 622

Serrivomer beanii Gill and Ryder, 1883
Figures 665–669

Serrivomer beanii Gill and Ryder, 1883a: 261 (original description, 41°40′N, 65°28′W, holotype USNM 33383). Jordan and Davis, 1891: 653 (compiled). Goode and Bean, 1896: 155 (compiled). Beebe and Crane, 1936: 61 (vertical distribution; life history). Fowler, 1936a: 283 (compiled). Schroeder, 1940: 232 (western North Atlantic). Bertin, 1944: 106 (osteology). Maul, 1948: 145 (Madeira). Bauchot-Boutin, 1953: 366 (generic revision); 1954: 304 (senior synonym of *S. taningi*); 1955: 11. Albuquerque, 1954–1956: 420 (Portugal). Grey, 1955: 291 (Bermuda). Bauchot, 1959: 131 (leptocephali and adults of Dana Expeditions); 1986: 548 (compiled). Quero, 1969: 5 (northeastern Atlantic). Badcock, 1970: 1028 (vertical distribution). Geistdorfer et al., 1971: 1183 (North Atlantic). Bauchot and Saldana, 1973: 229 (checklist). Musick, 1973: 135 (off Gulf of Maine). Smith, 1979: 19 (compiled). Hecht and Appelbaum, 1982: 301 (morphology of otoliths). Post and Tesch, 1982: 348 (Sargasso Sea and North Atlantic). Uyeno and Sasaki, 1983: 119 (off Surinam and French Guiana). McKelvie and Haedrich, 1985: 6 (off Newfoundland and Labrador). van Utrecht and van Utrecht-Cock, 1987: 53 (seasonal variation).

?*Spinivomer goodei* Gill and Ryder, 1884a: 261 (original description, 38°19′N, 68°20′W, holotype USNM 33293, type lost). Jordan and Davis, 1891: 653 (compiled). Goode and Bean, 1896: 155 (compiled). Beebe and Crane, 1937a: 346 (probable synonymy with *S. beanii*).

?*Nemichthys infans* (not of Günther). Vaillant, 1888: 93 (misidentification).

?*Nemichthys richardi* Vaillant, 1888: 385 (original description for specimen identified as *N. infans*, off Azores, holotype MNHN 84-1070).

?*Avocettina richardi*. Jordan and Davis, 1891: 655 (compiled).

?*Serrivomer richardi*. Goode and Bean, 1896: 155 (compiled).

Leptocephalus lanceolatus Strömman, 1896: 37 (original description, 34°N, 65°W, central Atlantic, types ZMUU).

?*Gavialiceps hasta* Zugmayer, 1911a: 7 (original description, 34°14′N, 8°W, off Portugal, type lost). Zugmayer, 1911b: 86 (compiled).

Serrivomer sector type *longidentatus* Roule and Bertin, 1929: 39 (description of Dana Expedition specimens). Parr, 1934: 6 (North Atlantic).

Nemichthys sp. Borodin, 1931: 73 (misidentification).

Gavialiceps microps (not of Alcock). Borodin, 1931: 74.

Paraserrivomer hasta. Roule and Angel, 1931: 3 (neotype; st. 3281, west of Azores, MOM). Roule and Angel, 1933: 70 (compiled). Fowler, 1936b: 1210 (compiled). Belloc, 1949: 4 (type catalogue). Brandes et al., 1953: 48 (occurrence off Iceland). Collins, 1954: 14 (Azores).

Serrivomer sector (not of Garman). Parr, 1932: 2 (in part, off Bahamas and Bermuda).

Oxystomus danae Bertin, 1935b: 1880 (original description,

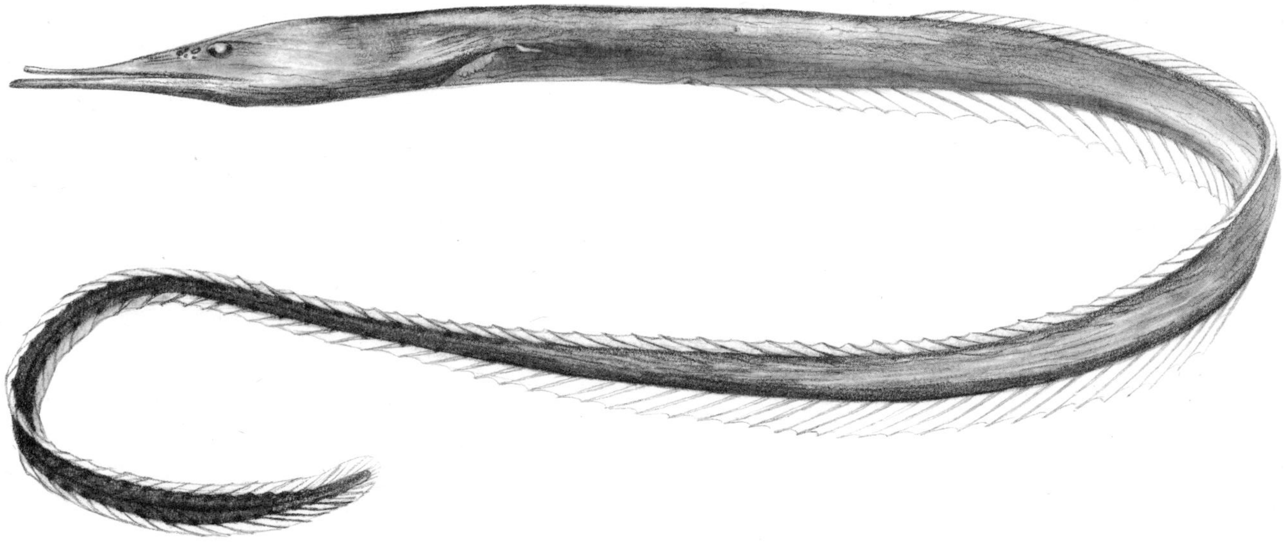

FIGURE 665. *Serrivomer beanii*: MCZ 64719, 495 mm TL.

North Atlantic near Bermuda, type not designated); 1936a: 11 (postlarval stage of *Leptocephalus lanceolatus*).

Nettastoma hasta. Fowler, 1936a: 281 (compiled).

Serrivomer parabeani Bertin, 1940: 78 (original description, types: 36 specimens identified as *Paraserrivomer hasta* by Roule and Angel, 1931 and 1933, MOM). Bertin, 1944: 106 (osteology; key). Bauchot-Boutin, 1953: 366 (generic revision); 1954: 304 (key); 1955: 11 (northeastern Atlantic). Grey, 1956: 88 (part of bathypelagic fauna). Bauchot, 1959: 131 (leptocephali and adults of Dana Expeditions); 1986: 549 (tentatively synonymized with *S. beanii*). Quero, 1969: 5 (northeastern Atlantic). van Utrecht, 1982: 117 (metamorphosis and postlarval development). Uyeno and Sasaki, 1983: 120 (off Surinam and French Guiana).

Bertinulus danae. Whitley, 1948: 74 (transfer from preoccupied genus).

Serrivomer taningi Bauchot-Boutin, 1953: 366 (original description, type not designated). Bauchot-Boutin, 1954: 304 (synonymy with *S. beanii*).

Distinctive Characters. *Serrivomer beanii* can be distinguished from other North Atlantic serrivomerids by the mode of branchiostegal attachment to the hyoid arches, the enlarged vomerine dentition, and the dorsal and anal ray counts.

Description. Dorsal-fin rays 136–175, $\bar{x}$ = 154 (n = 56); anal-fin rays 119–156, $\bar{x}$ = 136 (59); caudal-fin rays 2 + 3 or 3 + 3. Predorsal vertebrae 27–31, $\bar{x}$ = 29 (17); preanal vertebrae 20–22, $\bar{x}$ = 21 (16); precaudal vertebrae 77–103, $\bar{x}$ = 92 (n = 25); total vertebrae 154–162, $\bar{x}$ = 158 (n = 51).

Proportions as % of TL: head 11–17, $\bar{x}$ = 15.9 (n = 26); preanal 22–28, $\bar{x}$ = 24.8 (26); predorsal 27–34, $\bar{x}$ = 31.0 (24); depth at anal origin 0.9–2.5, $\bar{x}$ = 1.4 (18). Of HL: snout 36–40, $\bar{x}$ = 37.7 (16); premaxillary projection 7–13, $\bar{x}$ = 9.4 (16); snout to rictus 37–50, $\bar{x}$ = 42.8 (17), horizontal eye diameter 4–7, $\bar{x}$ = 5.3 (16), width of skull at sphenotic processes 7–11, $\bar{x}$ = 9.1 (16).

Cephalic lateralis pores and nostrils as described for family.

Maxillary dentition (Fig. 667) in 2–4 rows, increasing in size from outer to inner; 20–25 teeth in inner row, approximately 3 times as long as wide. Mandibular teeth in 3–5 rows; inner row consisting of very small teeth, isolated from other rows; next row with 35–40 enlarged teeth, similar to maxillary teeth; outer rows of small replacement teeth, in irregular rows and of irregular size. Vomerine dentition consisting of two alternating rows of laterally compressed teeth, each tooth 3–4 times as long as wide; number of teeth increasing with increasing length, 60–90 in adults.

Color. Dark brown to black with an iridescent silvery epidermal layer. Coloration similar in preserved material.

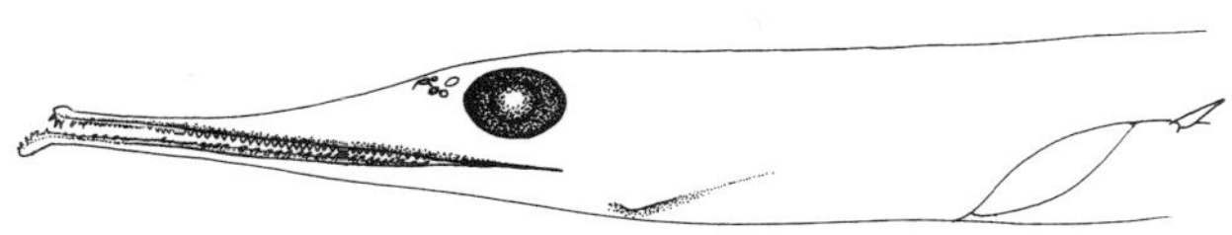

FIGURE 666. *Serrivomer beanii*: ANSP 158179, 549 mm TL; head.

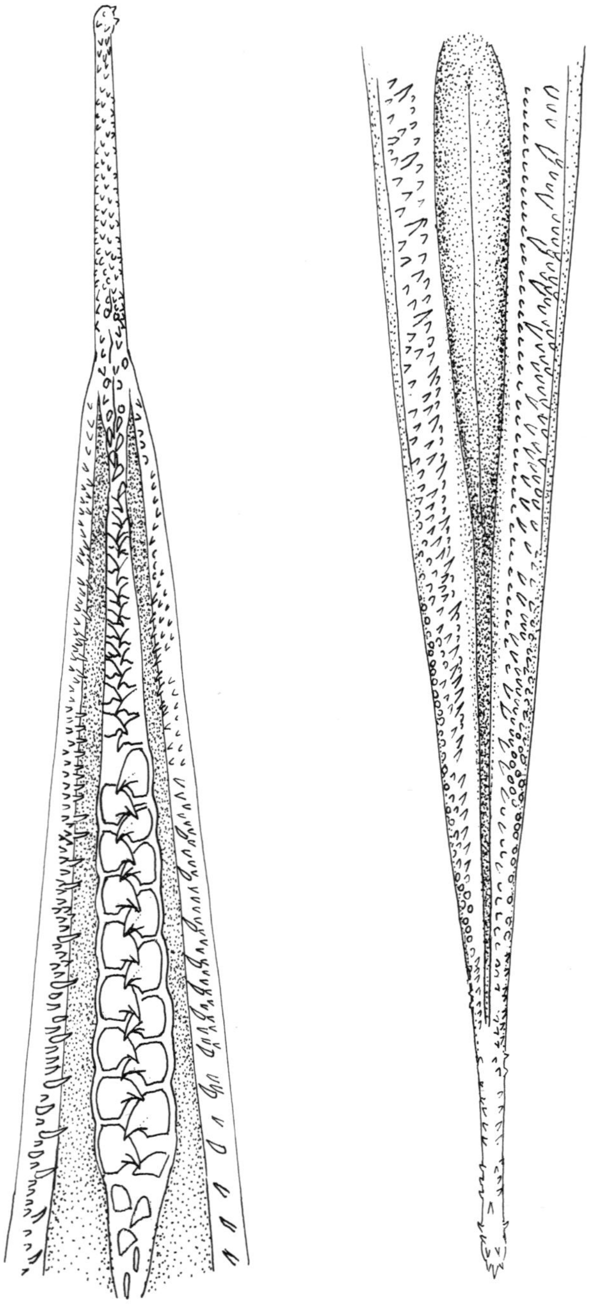

FIGURE 667. *Serrivomer beanii*: ANSP 158179, 325 mm TL; dentition.

Size. The largest specimen examined in this study was 745 mm TL. There is a sexual dimorphism in size, with adult females ranging from 620 to 745 mm TL while adult males were 510 to 640 mm TL.

Reproductive Biology. Adult *Serrivomer beanii* undergo a prereproductive metamorphosis which results in a slightly different morphology from

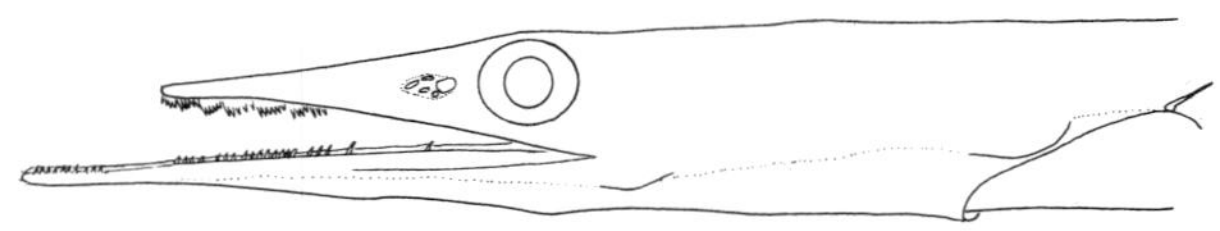

FIGURE 668. *Serrivomer beanii*: MCZ 64720, 439 mm TL, adult male; head.

the juveniles to subadults. The vomerine dentition becomes modified with the vomerine teeth divided into groups of four to six teeth which decrease in length proceeding posteriorly and then abruptly increase in length for the next group of teeth. This results in a double sawtooth effect which can be seen in Figure 668. In addition, the maxillary teeth are reduced to a single, irregular row of small, recurved teeth. The mandibular teeth are also reduced to a single row, but they are enlarged relative to the subadult condition and become long, recurved, and needle-like. The eye becomes enlarged and nearly circular in outline. In males (Fig. 668), the premaxillary projection regresses resulting in a foreshortened snout. Also, the caudal region of males is more laterally compressed and deeper than in females. These modifications in adult morphology are probably adaptations for mate location in the mesopelagic environment, although the function of the modified vomerine dentition and the enlarged mandibular teeth in specimens in breeding condition remains a mystery since there is no evidence to indicate that feeding continues in ripe adults.

Development. Bauchot (1959) identified *Leptocephalus lanceolatus* as the larva of *S. beanii* (see p. 921).

Behavior. Beebe and Crane (1936) reported bathysphere observations of serrivomerid eels that indicate that they are fast swimmers, and are found singly, in pairs, or in groups of four or five. Although the eels could not be identified to species, most of these observations were probably of *S. beanii* as this was the most common species in their study.

Beebe and Crane (1936) have shown that *S. beanii* feeds primarily on large shrimps and euphausiids with occasional small fish and other crustaceans included in the diet.

Distribution (Fig. 669). *Serrivomer beanii* is distributed throughout the North Atlantic although the center of distribution seems to be in the temperate and subtropical regions (Fig. 669). Captures within the Gulf of Mexico and the Caribbean Sea are primarily small juveniles, and are probably strays from the main population.

Detailed studies of the vertical distribution of *S. beanii* have been carried out in the Bermuda Ocean Acre Project (Tighe, 1975). Specimens in late stages of metamorphosis are found rather deep in the water column in the lower ranges of the juvenile depth distribution. Juveniles are found at depths of 551–1000 meters. Subadult to adult specimens were found within the lower depth ranges of the juveniles both day and night. There was no indication of any diel vertical migration of this species.

Etymology. Named for Tarleton H. Bean.

Remarks. In the synonymy above, *Serrivomer parabeani* is considered a junior synonym of *S. beanii*. Bauchot (1959) separated these two species based on the attachment of the first branchiostegal to the hyoid arch. Variation of this character was apparent in a large series of both nominal species from the Bermuda Ocean Acre Project. With the lack of another good differentiating character, these two species are considered conspecific. Bauchot (1986) also tentatively considered *S. parabeani* to be a junior synonym of *S. beanii*.

The South Atlantic population of *Serrivomer* has been described as *S. schmidti* Bauchot-Boutin, 1954. Bauchot (1959) separated these two nominal species based on a higher vertebral number and a caudal ray count of 3 + 3 or 3 + 4 in *S. schmidti*. Post and Tesch (1982) considered *S. beanii* as occurring in the Atlantic from 65 degrees N to the southern subtropical convergence at 42 degrees S. Critical examination of the South Atlantic population is needed to determine the status of *S. schmidti*.

Study Material. A total of 1699 specimens, 73–745 mm TL.

NORTH ATLANTIC: ANSP 114860 (2, 405–430), 114863 (1), 117040 (5, 126–362), 117046 (1, 454), 117069 (1, 395),

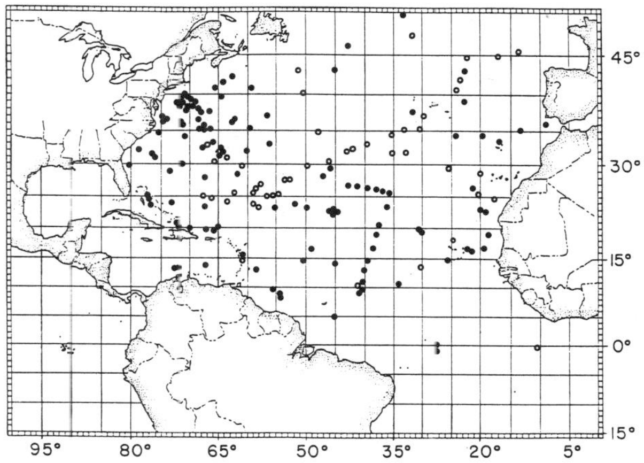

FIGURE 669. *Serrivomer beanii*: distributions based on material examined and literature records from Bauchot-Boutin (1955), Bauchot (1959), and Post and Tesch (1982).

117077 (1, 183), 117084 (8, 131–188), 117087 (6, 97–165), 153494 (1, 510), 158179 (2, 325–549). FMNH 49874 (1), 49877 (1), 63139 (3), 63140 (1), 63142 (1), 63143 (1), 63144 (1), 63146 (1), 63147 (2), 63148 (7, 113–162), 63149 (1), 63150 (1), 65729 (1, 510), 66134 (2, 374–524). MCZ 34651 (1, 604), 34717 (1, 395), 34974 (4, 156–450), 34975 (4, 210–265), 35793 (1, 340), 37526 (1, 510), 39553 (1, 443), 42127 (8, 145–458), 53025 (11, 147–280), 53026 (1, 373), 53027 (2, 240–451), 53028 (6, 155–205), 53029 (2, 247–425), 53030 (5, 272–442), 53031 (5, 250–327), 53033 (6, 104–430), 53036 (6, 295–598), 53037 (2, 173–205), 53039 (1, 302), 53041 (7, 173–286), 53042 (4, 151–210), 53044 (7, 220–560), 53045 (8, 271–435), 64719 (1, 495), 64720 (1, 439), 65257 (1, 725), 65270 (1, 152), 65275 (1, 133), 65276 (1, 100+), 65277 (3, 116–131), 65278 (1, 278), 65279 (3, 89–196), 65280 (1, 140), 65282 (2, 145–175), 65283 (1, 220), 65284 (1, 307), 65285 (4, 153–167), 65286 (2, 228–233), 65288 (8, 160–380), 65546 (1, 120), 65547 (1, 134), 65548 (1, 119), 65549 (1, 160+), 65550 (1, 185+), 65551 (1, 204), 65552 (1, 173), 65553 (1, 175), 65554 (2, 135–169), 65555 (3, 130–160), 65557 (1, 114), 65558 (1, 135), 65559 (1, 120), 65560 (1, 126), 65561 (1, 136), 65562 (1, 128), 65563 (2, 123–129), 65564 (1, 135), 65565 (1, 300), 65566 (2, 132–136), 65567 (1, 143), 65568 (1, 243), 65569 (1, 179+), 65570 (1, 146), 65571 (2, 221–236), 65572 (1, 328), 65573 (1, 120), 65574 (2, 120–198), 65575 (1, 128), 65576 (1, 190), 65577 (17, 128–224), 65578 (1, 295), 65579 (1, 335), 65580 (1, 350), 65581 (1, 355), 65582 (2, 126–171), 65583 (3, 151–245), 65585 (8, 142–175), 65586 (2, 155–158), 65588 (1, 291), 65589 (1, 360), 65590 (1, 405), 65623 (4, 310–535), 65648 (5, 260–640), 65649 (6, 159–194), 65650 (2, 385–453), 65651 (12, 214–672), 65652 (2, 227–241), 65653 (1, 244), 65654 (4, 133–148), 65655 (2, 288–300+), 65656 (2, 207–250+), 65657 (1, 204), 65658 (1, 136), 65659 (1, 140+), 65695 (1, 654), 65828 (5, 375–575), 65829 (56, 128–242), 65831 (1, 440),

65835 (2, 190–215), 65836 (1, 185), 65841 (1, 145), 65846 (5, 118–128), 65851 (1, 310+), 65855 (1, 125), 65857 (1, 123), 65858 (1, 120), 65933 (1, 469), 65952 (3, 116–131), 65986 (1, 633), 65987 (2, 230–305), 65991 (1, 490), 65993 (1, 540), 66125 (1, 139), 66128 (1, 240+). UMML 22689 (1, 220), 22691 (1, 421), 23229 (2, 345–360), 24364 (1, 130), 24441 (1, 300), 24468 (1, 367), 24590 (1, 370), 27528 (1, 250), 27630 (10, 292–578), 27660 (1, 170), 27678 (1, 250), 27704 (1, 165), 32008 (1, 513). Columbus Iselin: CI-49 (2, 153–230), CI-89 (2, 198–332), CI-98 (1, 253), CI-133 (1, 312), CI-149 (1, 250), CI-150 (1, 228), CI-157 (2, 344–456), Pillsbury 1147 (1). USNM 33383 (1, 603, holotype of *Serrivomer beanii*), 33541 (1, 662), 285329 (1, 515), 285332 (2, 425–470). Ocean Acre 1-31N (1, 236), 2-6N (2, 100–226), 3-1N (2, 131–133), 3-2N (2, 226–520), 3-3N (3, 106–203), 3-4N (1, 141), 3-5N (2, 127–139), 3-6N (4, 110–143), 3-7N (2, 116–139), 3-9N (1, 138), 3-10N (9, 130–218), 3-11N (9, 130–745), 3-12N (3, 150–160), 4-31N (1, 311), 5-2N (1, 170), 6-10P (1, 97), 6-11A (1, 85), 6-11P (2, 89–119), 6-12M (1, 160), 6-18A (1, 190), 6-18P (6, 100–180), 6-20N (1, 100), 6-21N (1, 168), 6-24N (2, 84-100), 7-6N (1, 100), 7-8N (1, 140), 7-12N (11, 140–156), 7-13N (8, 121–165), 7-14N (1, 216), 7-15N (1, 111), 7-19N (5, 139–255), 8-3N (3, 195–287), 9-4N (23, 73–180), 9-8N (4, 79–211), 9-9N (4, 78–101), 9-10N (19, 83–153), 9-11N (1, 195), 9-13N (4, 96–175), 9-14N (1, 104), 9-15N (10, 91–210), 9-16N (3, 108–185), 9-21N (4, 75–91), 9-26N (2, 96–106), 9-27N (3, 78–240), 9-30N (1, 201), 9-31N (4, 80–247), 10-1N (2, 124–125), 10-2B (2, 120–135), 10-2C (1, 134), 10-2M (1, 123), 10-4C (1, 130), 10-4M (9, 97–151), 10-5B (4, 140–143), 10-5M (11, 122–146), 10-7A (2, 140–140), 10-7M (3, 120–191), 10-10A (1, 196), 10-10B (1, 120), 10-10M (1, 237), 10-19M (1, 211), 10-20C (1, 120), 10-21N (9, 87–323), 10-23B (1, 123), 10-23M (2, 113–136), 10-24B (2, 130–130), 10-24M (2, 116–205), 10-29M (1, 120), 10-36A (3, 132–198), 10-38N (1, 148), 11-1M (2, 137–151), 11-4C (5, 139–167), 11-4M (1, 175), 11-7M (2, 160–166), 11-9B (1, 324), 11-9M (2, 184–318), 11-13A (1, 190), 11-13B (1, 164), 11-13C (1, 180), 12-2M (2, 145–217), 12-4A (3, 125–140), 12-4B (10, 130–156), 12-4C (9, 128–155), 12-4M (25, 124–160), 12-5A (1, 156), 12-5B (3, 138–153), 12-5C (6, 130–160), 12-5M (2, 133–155), 12-6M (6, 135–153), 12-9A (1, 130), 12-9C (1, 150), 12-12A (1, 180), 12-12B (5, 123–205), 12-12M (4, 140–245), 12-14C (1, 145), 12-16C (1, 160), 12-16M (1, 170), 12-19A (1, 130), 12-21A (4, 133–162), 12-21M (5, 150–155), 12-22B (1, 190), 12-22C (3, 142–198), 12-22M (9, 130–190), 12-23C (1, 150), 12-24A (1, 220), 12-24M (2, 130–230), 12-25M (1, 150), 12-26B (3, 139–234), 12-26C (3, 139–185), 12-26M (4, 142–220), 12-27B (2, 133–200), 12-27C (1, 130), 12-27M (2, 129–140), 12-28B (1, 232), 12-31C (2, 155–177), 12-32B (2, 200–233), 12-33M (5, 130–198), 12-53N (2, 143–155), 12-56N (11, 135–203), 12-57N (242, 130–337), 12-58N (12, 149–250), 12-59N (48, 125–252), 12-62N (35, 134–308), 12-63N (59, 130–307), 12-67N (33, 157–660), 12-71N (1, 125), 12-72N (81, 132–670), 12-74N (89, 131–296), 12-80N (87, 142–640), 12-83N (21, 150–430), 12-84N (53, 129–267), 12-85N (1, 147), 12-88N (18, 146–729), 13-8M (2, 193–266), 13-19P (1, 274), 13-21M (2, 161–268), 13-23P (1, 482), 13-36P (1, 170), 13-39M (4, 124–256), 14-7B (1, 140), 14-7M (1, 133), 14-8B (1, 130), 14-8C (1, 287), 14-10B (1, 118), 14-11B (1, 119), 14-12B (2, 130–138), 14-12C (1, 120), 14-12M (4, 121–143), 14-16P (3, 132–140), 14-17A (1, 142), 14-17M (11, 121–192), 14-19B (1, 143), 14-23B (4, 122–142), 14-23C (4, 132–148), 14-23M (1, 154), 14-24A (1, 181), 14-24B (1, 125), 14-29B (1, 132), 14-29M (1, 126). VIMS 04724 (1, 724), 05918 (1, 494), 05919 (1, 498), 05920 (1, 500), 05922 (3, 230–538), 05923 (9, 325–688), 05925 (3, 392–617), 05926 (12, 327–608), 05928 (1, 680), 05931 (1, 566), 05932 (9, 316–600), 06697 (1, 575), 06792 (1).

Serrivomer lanceolatoides (Schmidt, 1916)
Figures 670–674

Leptocephalus lanceolatoides Schmidt, 1916b: 17 (original description, central portion of the North Atlantic, type not designated).

Serrivomer sector type *brevidentatus* Roule and Bertin, 1929: 39 (original description, North Atlantic, types not designated).

Platuronides danae Roule and Bertin, 1929: 48 (original description, North Atlantic, type ZMUC P 31243). Beebe and Crane, 1937a: 322 (osteology and biology). Bauchot, 1959: 128 (compiled, key).

Serrivomer beanii (not of Gill and Ryder). Trewavas, 1932: 650 (osteology).

Serrivomer sector (not of Garman). Parr, 1932: 2 (in part, off Bahamas and Bermuda).

Platuronides ophiocephalus Parr, 1932: 5 (original description, North Atlantic, holotype YPM 2725). Beebe and Crane, 1937a: 332 (tentatively synonymized with *P. acutus*. Bauchot, 1959: 128 (compiled, key).

Platuronides acutus Parr, 1932: 8 (original description, North Atlantic, holotype YPM 2726). Beebe and Crane, 1937a: 332 (osteology, biology). Bauchot, 1959: 128 (compiled, key).

Serrivomer species. Beebe, 1933: 180 (in part, off Bermuda).

Serrivomer brevidentatus. Beebe and Crane, 1936: 93 (osteology, biology). Bertin, 1944: 106 (osteology, key). Bauchot-Boutin, 1953: 366 (generic revision); 1954: 304 (key). Grey, 1955: 291 (Bermuda). Bauchot, 1959: 131 (leptocephali and adults of Dana Expedition). Post and Tesch, 1982: 349 (Sargasso Sea and North Atlantic).

Distinctive Characters. Serrivomer lanceolatoides can be distinguished from other North Atlantic serrivomerids by the mode of branchiostegal attachment to the hyoid arches, the enlarged vo-

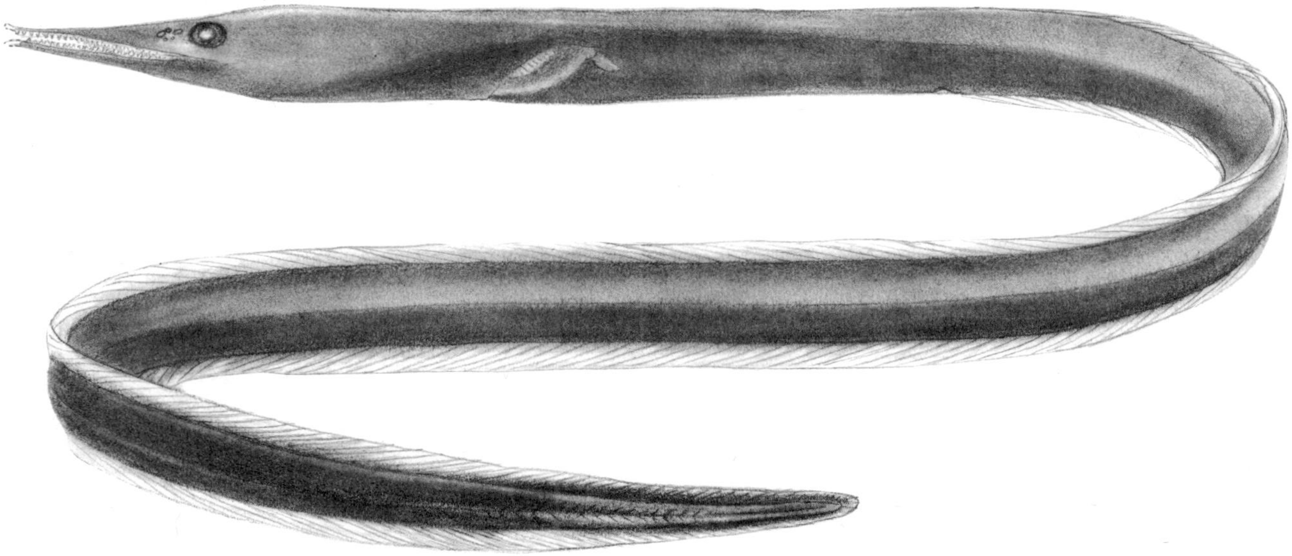

FIGURE 670. *Serrivomer lanceolatoides*: USNM 285185, 520 mm TL.

merine dentition, and the high dorsal and anal ray counts.

Description. Dorsal-fin rays 175–198, x̄ = 184 (n = 23); anal-fin rays 165–192, x̄ = 175 (23); caudal-fin rays 2 + 3 or 3 + 3. Predorsal vertebrae 29–35, x̄ = 32 (15); preanal vertebrae 22–26, x̄ = 24 (15); precaudal vertebrae 89–95, x̄ = 91 (11); total vertebrae 154–163, x̄ = 160 (27).

Proportions as % of TL: head 15–18, x̄ = 16.4 (n = 15); preanal 25–30, x̄ = 26.9 (16); predorsal 30–41, x̄ = 33.2 (14); depth at anal origin 1.4–3.0, x̄ = 2.2 (12). Of HL: snout 30–40, x̄ = 34.0 (10); premaxillary projection 6–10, x̄ = 8.1 (9); snout to rictus 36–43, x̄ = 40.6 (10); horizontal eye diameter 5–10, x̄ = 6.9 (9); width of skull at sphenotic processes 10–12, x̄ = 10.9 (10).

Cephalic lateralis pores and nostrils as described for family.

Maxillary dentition (Fig. 672) in 3–4 rows increasing in size from outer to inner; 30–40 teeth in inner row, only slightly enlarged, approximately twice as long as wide. Mandibular dentition in 4 to 6 rows; inner row consisting of small teeth, slightly isolated from other rows; next row with 30–40 enlarged teeth, similar to maxillary teeth; outer rows of small replacement teeth in irregular rows and irregular in size. Vomerine dentition consisting of two alternating rows of laterally compressed teeth, each tooth approximately twice as long as wide; number of teeth increasing with increasing length, 30–50 in adults.

Color. Dark brown to black with a deciduous, iridescent bronze epidermal layer. Coloration similar in preserved material.

Size. The largest specimen examined in this study was 633 mm TL. Sexual dimorphism is present with adult females ranging from 500 to 633 mm TL while adult males were 471 to 488 mm TL.

Reproductive Biology. Adult male *Serrivomer lanceolatoides* undergo a prereproductive metamorphosis which results in a different morphology from juveniles, subadults and adult females (Fig. 673). The vomerine dentition becomes regressed resulting in a few isolated, conical teeth. The maxillary teeth are reduced to a single, irregular row of small conical teeth while the mandibular teeth are reduced to a single, irregular row of slightly enlarged, recurved teeth. The anterior nostril is produced into an anteriorly di-

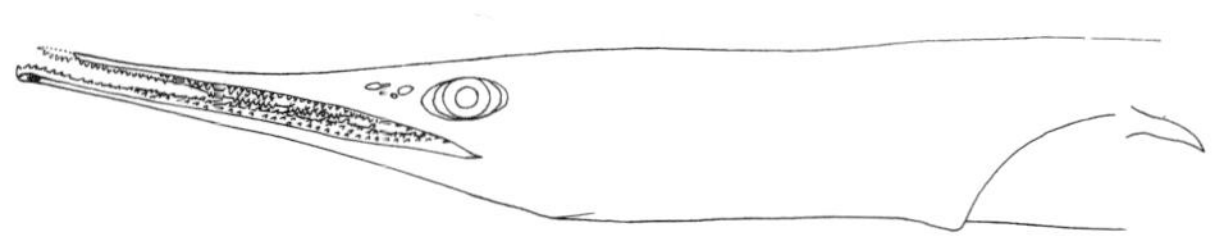

FIGURE 671. *Serrivomer lanceolatoides*: MCZ 64721, 417 mm TL; head.

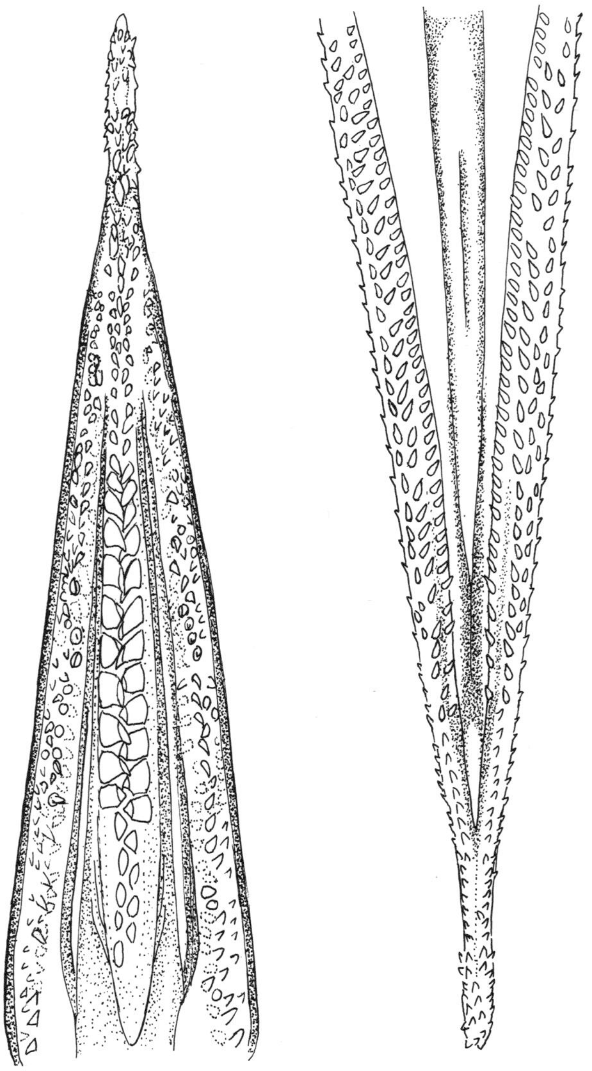

FIGURE 672. *Serrivomer lanceolatoides*: ANSP 158178, 402 mm TL; dentition.

rected tube while the posterior nostril is slightly enlarged. The eye is slightly enlarged relative to that of an adult female, but remains oval in outline. The posterior anal and dorsal rays are stiffened and prolonged into a "pseudocaudal." These modifications are probably adaptations for mate location in the mesopelagic environment. This adult male morphology is so different from the typical *S. lanceolatoides* juveniles and subadults that the first specimen captured was described as *Platuronides danae* by Roule and Bertin (1929).

Development. Bauchot (1959) identified the leptocephalus of this species (see p. 923).

Behavior. Beebe and Crane (1936) reported re-

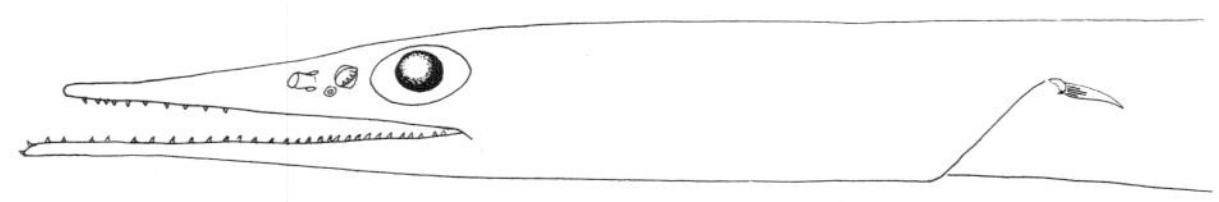

FIGURE 673. *Serrivomer lanceolatoides*: USNM 285186, 487 mm TL; adult male, head.

mains of crustaceans in stomachs of *S. lanceolatoides.* It is probable that *S. lanceolatoides* feeds primarily on shrimp and euphausiids like *S. beanii.*

Distribution (Fig. 674). In the North Atlantic, *S. lanceolatoides* is distributed throughout the tropical and subtropical pelagic regions as well as the Gulf of Mexico. In the Bermuda Ocean Acre Project (Tighe, 1975), specimens in early stages of metamorphosis were taken over a depth range of 500 to 900 m. Day and night captures of juveniles were concentrated around 650 to 850 m in depth while subadult to adults were concentrated deeper in the water column from 800 to 1000 m. There was no indication of any diel vertical migration in the juveniles or adults.

Etymology. From the Latin *lanceolatus* (spear-like) and the Greek *oides* (resembling), due to its resemblance to *Leptocephalus lanceolatus.*

Remarks. This species was known previously as *Serrivomer brevidentatus.* Bauchot (1959) identified *Leptocephalus lanceolatoides* Schmidt, 1916 as the larvae of *S. brevidentatus.* Since the larval

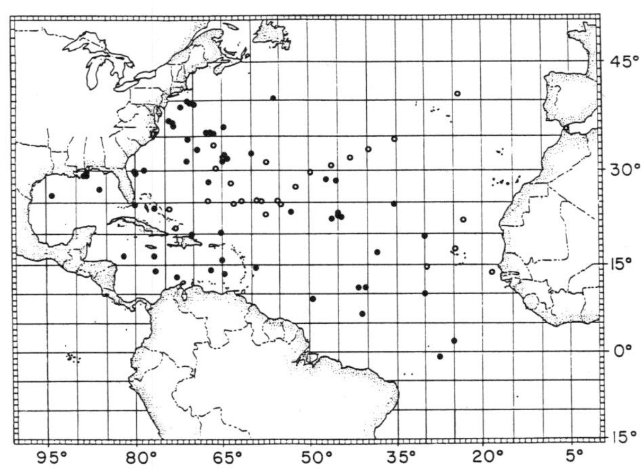

FIGURE 674. *Serrivomer lanceolatoides*: distribution based on material examined and literature records from Bauchot (1959) and Post and Tesch (1982).

name has priority, *S. brevidentatus* is a junior synonym.

Examination of the holotypes of *Platuronides acutus* and *P. ophiocephalus* has shown that these specimens are a juvenile and an adult female of *S. lanceolatoides*, respectively. Examination of the holotype of *P. danae* Roule and Bertin, 1929 has shown that it is an adult male of *S. lanceolatoides*. Therefore, *Platuronides* has been placed in synonymy with *Serrivomer*. However, if certain characters such as the high dorsal and anal ray count, laterally compressed body shape, patterns of dentition, and pattern of adult male metamorphosis are found to be restricted to *S. lanceolatoides* and the closely related Indo-Pacific species *S. neocaledoniensis*, these two species should probably be placed in a separate genus or subgenus, and *Platuronides* would be an available name for that purpose.

Study Material. A total of 127 specimens, 102–633 mm TL.

NORTH ATLANTIC: AMNH 36214 (1, 488). ANSP 117041 (1, 233), 117048 (1, 466), 117055 (1, 370), 117057 (1, 432), 117073 (1, 282), 117078 (1, 302), 117079 (1, 295), 117081 (1, 184), 158178 (1, 405). FDNR B209B 165COD (1, 351). FMNH 49877 (1), 49879 (1, 475), 63145 (1), 66134 (1, 324), 83443 (1, 580+). MCZ 35003 (1, 452), 44208 (1, 382), 53046 (1, 258), 53048 (1, 217), 64721 (1, 417), 65595 (1, 114), 65596 (1, 155), 65597 (1, 190), 65598 (1, 142), 65599 (2, 104–132), 65600 (1, 150), 65601 (1, 123), 65602 (1, 145), 65611 (1, 175), 65612 (2, 135–165), 65995 (1, 343), 65996 (1, 375), 65997 (1, 311), 65998 (1, 275), 65999 (1, 488). UMML 29453 (1, 250). Columbus Iselin: CI-409 (1, 162), CI-413 (1, 402). USNM 285185 (1, 561), 285186 (1, 471), 285330 (1, 330), 285331 (1, 347), 290494 (1, 350). Ocean Acre 3-2N (1, 285), 6-8A (1, 160), 6-11A (1, 120), 6-11B (1, 604), 6-12B (1, 290), 6-18A (2, 125–135), 8-3N (1, 464), 9-10N (1, 121), 9-11N (1, 151), 9-15N (1, 126), 9-16N (2, 130–140), 10-1N (1, 148), 10-4A (1, 135), 10-4M (4, 146–163), 10-21N (1, 469), 11-9A (1, 202), 11-13B (1, 110), 12-14M (1, 118), 12-27A (1, 364), 12-59N (1, 275), 12-62N (5, 171–460), 12-67N (2, 276–500), 12-71N (1, 202), 12-74N (1, 102), 12-80N (1, 110), 12-83N (4, 268–320), 13-8B (1, 172), 13-14B (1, 118), 13-17M (1, 124), 13-21M (2, 139-326), 13-31P (1, 600), 13-33M (1, 126), 13-37M (2, 115–129), 13-39B (1, 116), 13-39M (2, 116–132), 14-7A (1, 353), 14-7B (1, 464). VIMS 03442 (1, 204), 04619 (1, 295), 04628 (1, 633), 04940 (2, 293–536), 05365 (1, 249), 05917 (1, 200), 06790 (1, 293), 06791 (1). YPM 2725 (1, 617, holotype of *Platuronides ophiocephalus*), 2726 (1, 219, holotype of *Platuronides acutus*). ZMUC P31243 (1, 474, holotype of *Platuronides danae*).

Genus *Stemonidium* Gilbert, 1905

Stemonidium Gilbert, 1905: 586 (type species *Stemonidium hypomelas* Gilbert, 1905, by monotypy).

Characters. Body slightly laterally compressed, middle half nearly uniform in depth with deepest portion of body posterior to anus, tapering gradually to the narrower neck and rapidly tapering to the tail. Dorsal origin at or slightly anterior to anal origin. Vomerine teeth small, very numerous, conical with recurved tips directed posteriorly, close set in an irregular banded pattern (similar to Nemichthyidae). All other characters as in the family.

Distribution. Tropical and subtropical waters of all three major oceans.

Etymology. From the Greek *stemon* (thread) and the Latin *idium*, a diminutive suffix, presumably in reference to the long, thin body.

Species. The genus contains a single species: *S. hypomelas*. Examination of Atlantic, Indian Ocean and Pacific specimens as well as the holotype from the Pacific gave no evidence for separation of the three oceanic populations.

Stemonidium hypomelas Gilbert, 1905
Figures 675–678

Stemonidium hypomelas Gilbert, 1905: 586 (original description, Hawaii, vicinity of Niihau Island, holotype USNM 51550). Bertin, 1941: 524 (osteology); 1944: 103 (synopsis of family). Bauchot, 1959: 126 (morphology, distribution and identification of Pacific larvae). Castle, 1965a: 142 (Pacific leptocephali); 1966b: 59 (Southwest Pacific leptocephali); 1969c: 26 (index and bibliography). Parin, 1975: 321 (central Pacific). Parin et al., 1977: 138 (western Pacific). Karmovskaya, 1979: 104 (Pacific leptocephalus); 1983: 136 (Atlantic and Indian Oceans). Clarke, 1984: 249 (ecology and sexual dimorphism).

Distinctive Characters. Stemonidium hypomelas is the only serrivomerid with the dorsal origin at or before the anal origin. Also, the vomerine dentition is distinctive among serrivomerids.

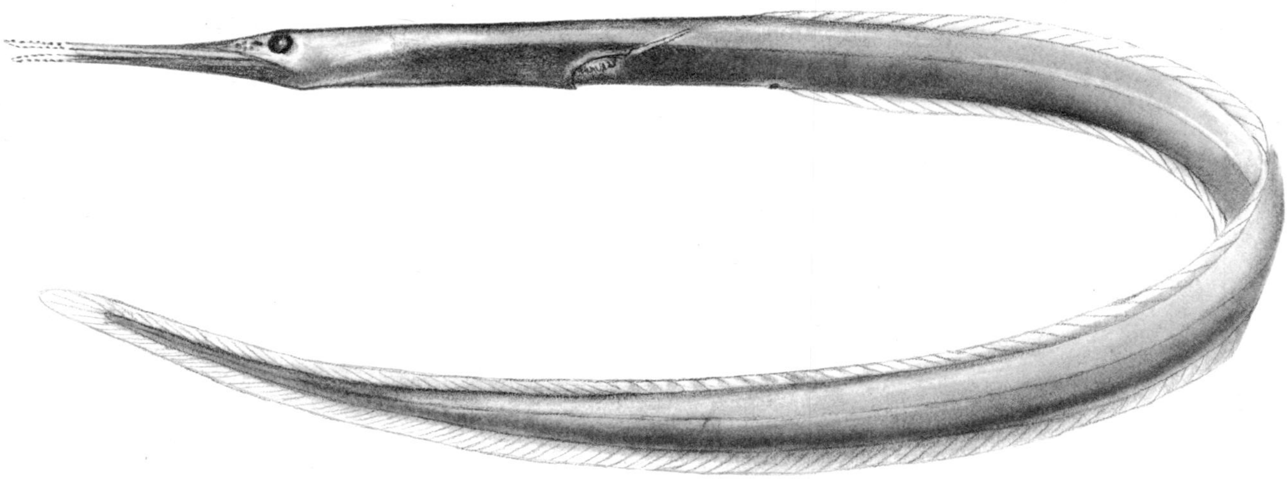

FIGURE 675. *Stemonidium hypomelas*: ANSP 138022, 293 mm TL.

Description (from Atlantic specimens). Dorsal-fin rays 155–184, x̄ = 168 (n = 7); anal-fin rays 139–169, x̄ = 151 (7); caudal-fin rays 3 + 3 or 3 + 3 + 1. Predorsal vertebrae 15–19, x̄ = 17 (12); preanal vertebrae 16–19, x̄ = 17 (13); precaudal vertebrae 67–77, x̄ = 75 (14); total vertebrae 134–141, x̄ = 139 (10).

Proportions as % of TL: head 16–18, x̄ = 16.6 (n = 9); preanal 23–28, x̄ = 25.2 (9); predorsal 23–27, x̄ = 24.7 (9); depth at anal origin 2–3, x̄ = 2.1 (9). Of HL: snout 37–41, x̄ = 39.4 (7); premaxillary projection 10–13, x̄ = 12.7 (7); snout to rictus 42–46, x̄ = 43.8 (7); width of skull at sphenotic processes 9–10, x̄ = 9.7 (7).

Cephalic lateralis pores and nostrils as described for family.

Maxillary dentition (Fig. 677) in 2–4 irregular rows of small, recurved, conical teeth; no rows isolated or enlarged. Mandibular teeth in 3–5 irregular rows of teeth similar to maxillary teeth. Vomerine teeth in 4 or more irregular rows of numerous, small conical teeth with recurved tips.

Color. Light to dark brown with a very light silvery deciduous epidermal layer. Ventral portion of body darker than dorsal. Coloration similar in preserved material.

Size. The largest Atlantic specimen examined in this study was 294 mm TL. Clarke (1984) reported sexual dimorphism in size of specimens taken off Hawaii. Ripe females ranged in size from 282 to 375 mm TL while males were mature at sizes of 175 to 200 mm TL with the largest male (220 mm TL) considerably smaller than the smallest mature female.

Development. Bauchot (1959) identified the leptocephali of *Stemonidium hypomelas* from the Pacific Ocean (see p. 924). However, they have not yet been reported from the Atlantic.

Reproductive Biology. Clarke (1984) discussed sexual dimorphism of this species from off Hawaii. Adult males had relatively larger eyes and nasal rosettes, and relatively deeper caudal regions. He suggested that these sexual differences were adaptations for mate location in the mesopelagic environment.

Behavior. Clarke (1984) reported that *Stemonidium hypomelas* off Hawaii feed primarily on euphausiids with occasional other small crustacea such as sergestid shrimps or amphipods.

Distribution (Fig. 678). In the North Atlantic, *Stemonidium hypomelas* is restricted to equatorial waters. The species also occurs throughout the South Atlantic. It is possible that these few specimens captured in the North Atlantic represent strays from the population centered in the South Atlantic. There are too few captures to determine the vertical distribution of the Atlantic population. However, Clarke (1984) has described the

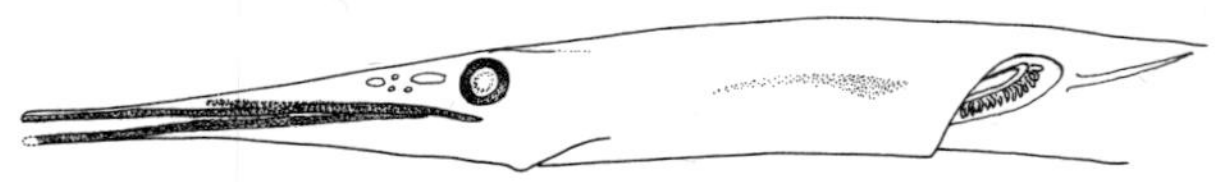

FIGURE 676. *Stemonidium hypomelas*: MCZ 64722, 252 mm TL; head.

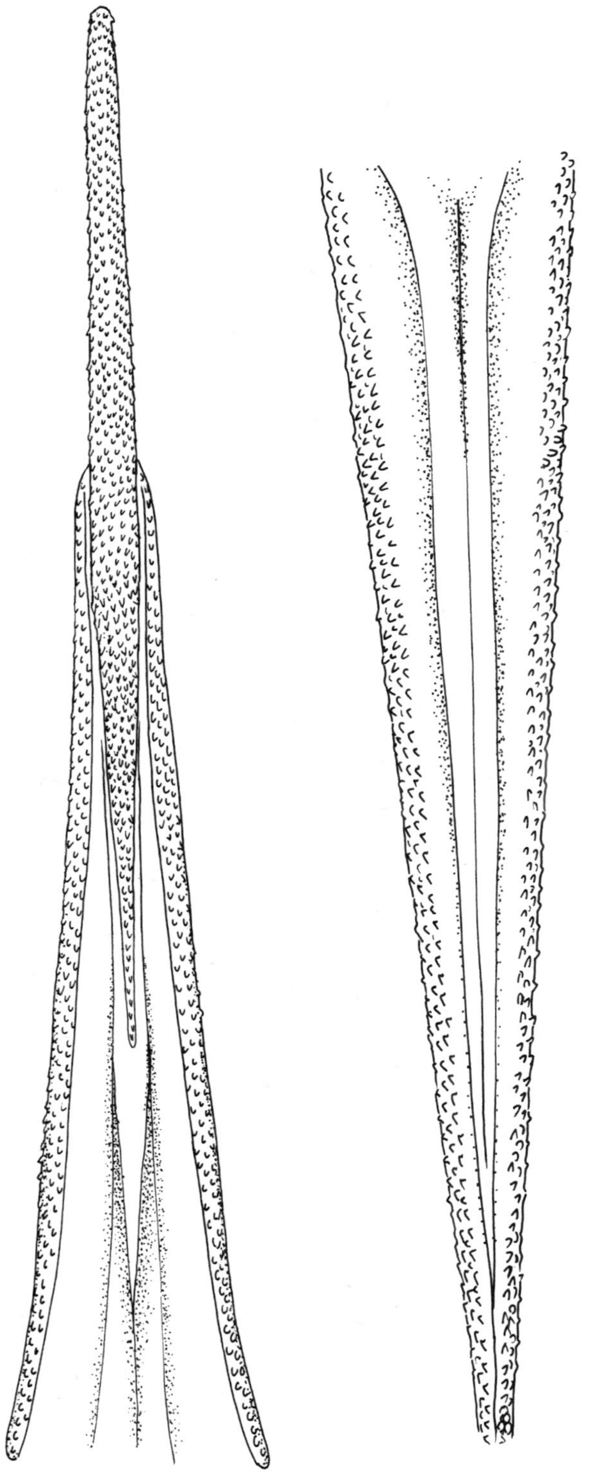

FIGURE 677. *Stemonidium hypomelas*: ANSP 138022, 293 mm TL; dentition.

vertical distribution of this species off Hawaii. Daytime depths for captures of juvenile specimens ranged from 550 to 1200 m with captures concentrated between 700 to 800 m. All specimens over 200 mm TL were taken below 750 m.

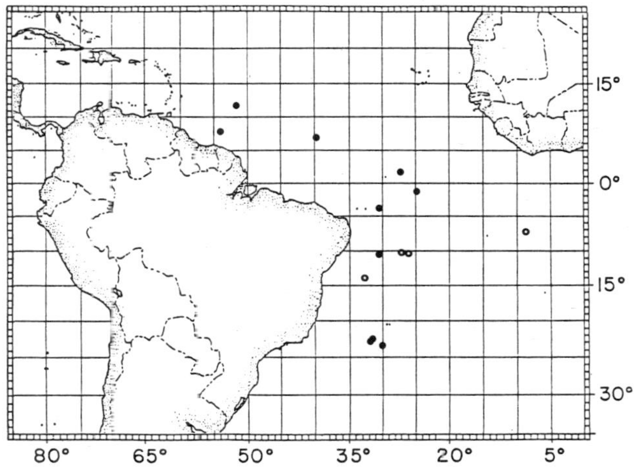

FIGURE 678. *Stemonidium hypomelas*: distribution based on material examined and literature records from Bauchot (1959) and Karmovskaya (1983).

Night captures were primarily within the day depth range with a portion of the population vertically migrating upward to a depth range of 175 to 400 m.

Etymology. From the Greek *hypo* (under) and *melas* (black), in reference to the dark, ventral pigmentation.

Study Material. A total of 15 Atlantic specimens, 74–294 mm TL.

NORTH ATLANTIC: ANSP 138022 (1, 253), 7°50′N, 54°15′W, Oregon 10794, 389 fathoms, 15 Nov. 1969. MCZ 65255 (1, 170), 7°02′N, 39°29′W, R/V Oceanus 22, JEC7735, ca 100 m, 22 Mar. 1977; 66333 (1, 282), 1°35′N, 27°03′W, R/V Chain 35, RHB 975, 0–290 m, 25 Feb. 1963. USNM 290496 (1, 225+), 11°38′N, 51°52′W, R/V Trident 023, MWT-6, 0–1103 m, 31 May 1965.

SOUTH ATLANTIC: MCZ 64722 (1, 294), 1°54′S, 24°55′W, R/V Chain 35, RHB 971, 0–295 m, 22 Feb. 1963; 65253 (1, 154), 35°30′S, 4°43′E, R/V Atlantis II 60, RHB 2235, 80–85 m, 25 Apr. 1971; 65254 (2, 74–159), 23°08′S, 32°22′W, R/V Atlantis II 31, RHB 1432, 100–110 m, 9 Mar. 1967; 65256 (1, 125+), 23°02′S, 32°15′W, R/V Atlantis II 31, RHB 1431, 150–175 m, 9 Mar. 1967; 66334 (1, 275+), 3°55′S, 30°38′W, R/V Chain 35, RHB 980, 0–257 m, 28–29 Feb. 1963; 66335 (1, 182), 10°27′S, 30°35′W, R/V Atlantis II 31, RHB 1421, 0–2030 m, 28–29 Feb. 1967; 66336 (1, 246), 36°03′S, 17°08′E, R/V Atlantis II 60, RHB 2245, 0–410 m, 30 Apr. 1971. USNM 290495 (3, 109–161), 23°30′S, 30°W, R/V Trident 023, MWT-4, 0–1156 m, 12 May 1965.

Comparative material: INDIAN OCEAN: MCZ 52968 (2, 167–300), 52962 (1, 200), 52963 (1, 257).

PACIFIC OCEAN: ANSP 123533 (1, 283), 152332 (2, 185–303), 152339 (4, 189–298), 152581 (1, 312). USNM 51550 (1, 171, holotype of *Stemonidium hypomelas*).

Order Saccopharyngiformes

DAVID G. SMITH

Introduction. Robins (p.13) has presented evidence that the Saccopharyngidae, Eurypharyngidae, and Monognathidae, along with the Cyematidae, constitute an order, the Saccopharyngiformes, that is separate and distinct from the Anguilliformes. He listed 12 specialized characters that distinguish the Saccopharyngiformes from the Anguilliformes: (1) Elongate, posteriorly projecting suspensorium. (2) Deep-bodied leptocephalus with V-shaped rather than W-shaped myomeres. (3) Absence of ventral hyoid elements and branchiostegals. (4) Absence of all opercular elements. (5) Absence of maxilla and premaxilla. (6) A single point of articulation between hyomandibula and neurocranium. (7) Lateral line on body with papillae instead of pores. (8) Gills small, arches rounded, filaments plumelike and set off from pharynx, into which they open through distinct clefts. (9) Efferent branchial arteries united by commisural vessels above and below gill clefts. (10) Median fin rays not segmented or branched. (11) Frontals paired, partly fused or fused. (12) Light organs on tail of saccopharyngids and eurypharyngids.

Character number 5 is interpretational rather than descriptional in nature. All but the monognathids have a tooth-bearing upper jaw; whether it is a maxilla or a pterygoid cannot be determined without further studies, particularly on development. Character 10, unsegmented fin rays, is not unique to this group; congrids of the subfamily Bathymyrinae also have unsegmented fin rays. Character 12 is true only in the two families cited.

The case for linking the Cyematidae with the gulpers, although not beyond argument, seems much stronger than any of the alternatives. Certainly there seems no evidence for a relationship with the Nemichthyidae or any of the other midwater eels. The relationship of that group (cyematids plus gulpers) to the Anguilliformes is less clear. Smith (1984b: 96) considered gulpers to be highly modified midwater eels. Robins (p. 15), however, believes them to be independent derivatives of a separate elopomorph stock. This problem cannot be resolved here. For the purposes of the present study, the group will be treated as an order.

Relationships within the Saccopharyngiformes. For reasons given below, the Saccopharyngidae and Eurypharyngidae seem particularly close to each other, sharing several distinctive, derived characters. The problem then becomes a comparison of three taxa: Saccopharyngidae-Eurypharyngidae, Monognathidae, and Cyematidae. Robins (p. 14) has mentioned the similarity of the gill structure of monognathids to that of the Saccopharyngidae and Eurypharyngidae. The metamorphic monognathid leptocephali described by Raju (1974a: fig. 1A–B) have a single, slightly arched swelling at the posterior end of the gut that is peppered with small melanophores. This condition is strikingly similar to that of *Saccopharynx* and *Eurypharynx* leptocephali but different from the complexly convoluted gut of cyematid leptocephali. In the general form of the head and mouth, despite the loss of the upper jaw, the monognathids resemble the Saccopharyngidae and Eurypharyngidae more than they do the Cyematidae. The cyematids show less modification overall than any of the other families and could be considered the primitive sister group of the other three. If the entire group is recognized as the order Saccopharyngiformes, then two suborders could be recognized: the Cyematoidei, containing only the Cyematidae, and the Saccopharyngoidei, containing the Saccopharyngidae, Eurypharyngidae, and Monognathidae.

Suborder Cyematoidei

Family Cyematidae

DAVID G. SMITH

Acknowledgments. See family Anguillidae.

Introduction. The Cyematidae is a small family of highly modified midwater eels. Only the gulpers (Saccopharyngidae, Eurypharyngidae, and Monognathidae) exceed them in the amount of skeletal reduction. The family name dates from Regan (1912b: 380, 384).

Characters. Body short, stubby, compressed; anus slightly behind midlength; tail blunt. Dorsal and anal fins present, confluent with caudal fin; dorsal fin begins over or slightly before anus. Pectoral fin present. Gill opening crescentic, on lower half of side. Scales absent. No lateral-line pores on body.

Head slender, snout and jaws produced, upper jaw projects slightly beyond lower; no flange on upper or lower lip. Mouth large, rictus far behind eye. Eye small. Nostrils close together in front of eye, distant from tip of snout. Teeth small, granular, in narrow bands.

Osteology (for *Cyema* based on Trewavas, 1933, Bertin, 1937, and Nelson, 1966a; for *Neocyema* based on Castle, 1977b). Neurocranium: *Cyema* (Fig. 679): Slender, antorbital part greatly elongate, rostral part prolonged into a slender beak; intermaxillary tooth patch long; orbit long. Frontals usually described as paired, but in fact they are fused (K. A. Tighe, pers. comm.). Parietals distinctly larger than frontals and form most of roof of skull. No supraoccipital. Ethmovomer slender and greatly prolonged anteriorly; at about its midlength a pair of lateral expansions that mark beginning of intermaxillary portion. Vomer continues back under parasphenoid to about level of eye. Pterotics very long, extending beyond frontals anteriorly and slightly behind level of epiotics and exoccipital posteriorly. No basisphenoid. *Neocyema* (Fig. 680): Similar to *Cyema*, but rostral portion shorter; maxillae join ethmovomer near its tip, intermaxillary tooth patch short; orbit long and narrow. Frontals described as fused anteriorly, divided posteriorly. Parietals divided.

Jaws: *Cyema* (Fig. 679): Maxilla (Robins, p. 14, believes this is the pterygoid) long, slender at anterior end and stouter at posterior end; a spherical ventral prominence present at posterior end, covered with teeth. Mandible long and slender; articular and dentary fused. *Neocyema* (Fig. 680): Maxilla long and slender, without spherical prominence at posterior end. Mandible with dentary and articular fused.

Suspensorium: *Cyema* (Fig. 679): Long and inclined posteriorly, articulates anteriorly with sphenotic, but no articulation posteriorly with pterotic. Continuous with posterior edge of hyomandibula but delimited by a structural line is a triangular lamina which Trewavas (1933: 603) suggested could represent opercle. She also described an interopercle located just behind articulation of lower jaw. Tchernavin (1947a: 345) disagreed with both these conclusions, interpreting

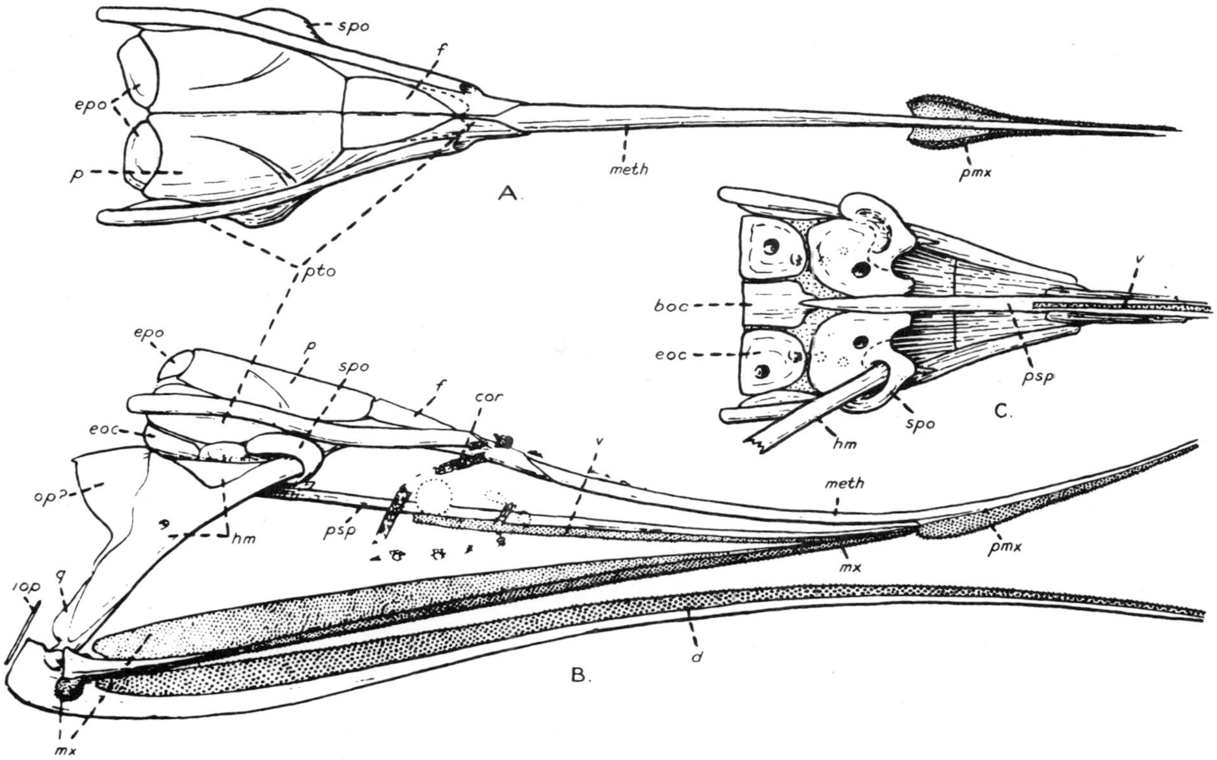

FIGURE 679. Skull of *Cyema atrum* (from Trewavas, 1933: pl. 1). Abbreviations: boc = basioccipital, cor = circumorbital, d = dentary, eoc = exoccipital, epo = epiotic, f = frontal, hm = hyomandibula, iop = interopercle, meth = mesethmoid, mx = maxilla, op = opercle, p = parietal, pmx = premaxilla, psp = parasphenoid, pto = pterotic, q = quadrate, spo = sphenotic, v = vomer. Note: elements labeled opercle and interopercle are probably not those bones.

former as part of hyomandibula and latter as an ossified tendon of abductor mandibulae muscle. No trace of pterygoid, unless Robins (p. 14) is correct about pterygoid being main tooth-bearing bone. *Neocyema* (Fig. 680): Long and slender, inclined posteriorly; articulates anteriorly with sphenotic but not posteriorly with pterotic. No triangular lamina on hyomandibula.

Hyoid arch: In *Cyema* reduced to a single paired rudiment of uncertain homology (Fig. 681). Condition in *Neocyema* unstated.

Gill arches: In *Cyema* (Fig. 681) limited to first four ceratobranchials, epibranchial of fourth arch, a single upper and a single lower pharyngeal tooth plate. In *Neocyema* (Fig. 680) only a single arch, probably fourth, consisting of ceratobranchial, epibranchial, single upper and lower pharyngeal tooth plates.

Opercular apparatus: Apparently absent in both genera.

Pectoral girdle: In *Cyema* (Fig. 681) only a large, crescent-shaped cleithrum and five or six actinosts. *Neocyema* with no pectoral girdle at all, although a rayless fin present.

Circumorbital bones: In *Cyema* an indeterminate number of poorly ossified elements present. Bertin (1937: fig. 13) showed eight elements, including one around adnasal canal (Fig. 682). Absent in *Neocyema*.

Vertebrae: *Cyema* (Figs. 681, 683): Centra thin cylinders of bone, those of anterior part of vertebral column slightly elongate horizontally. Neurapophyses are short laminae produced into short points anteriorly and posteriorly, and a spine with perforated base medially. Spines of first five vertebrae long and directed posteriorly; spines of next few vertebrae variable, those of remaining abdominal vertebrae short. Parapophyses of abdominal vertebrae consist of an anterior and posterior process connected by a low ridge along centrum. Neural and hemal spines present on caudal vertebrae. Caudal skeleton (Fig. 683) somewhat variable; there seem to be two hypural elements, ventral one larger than dorsal. Dorsal

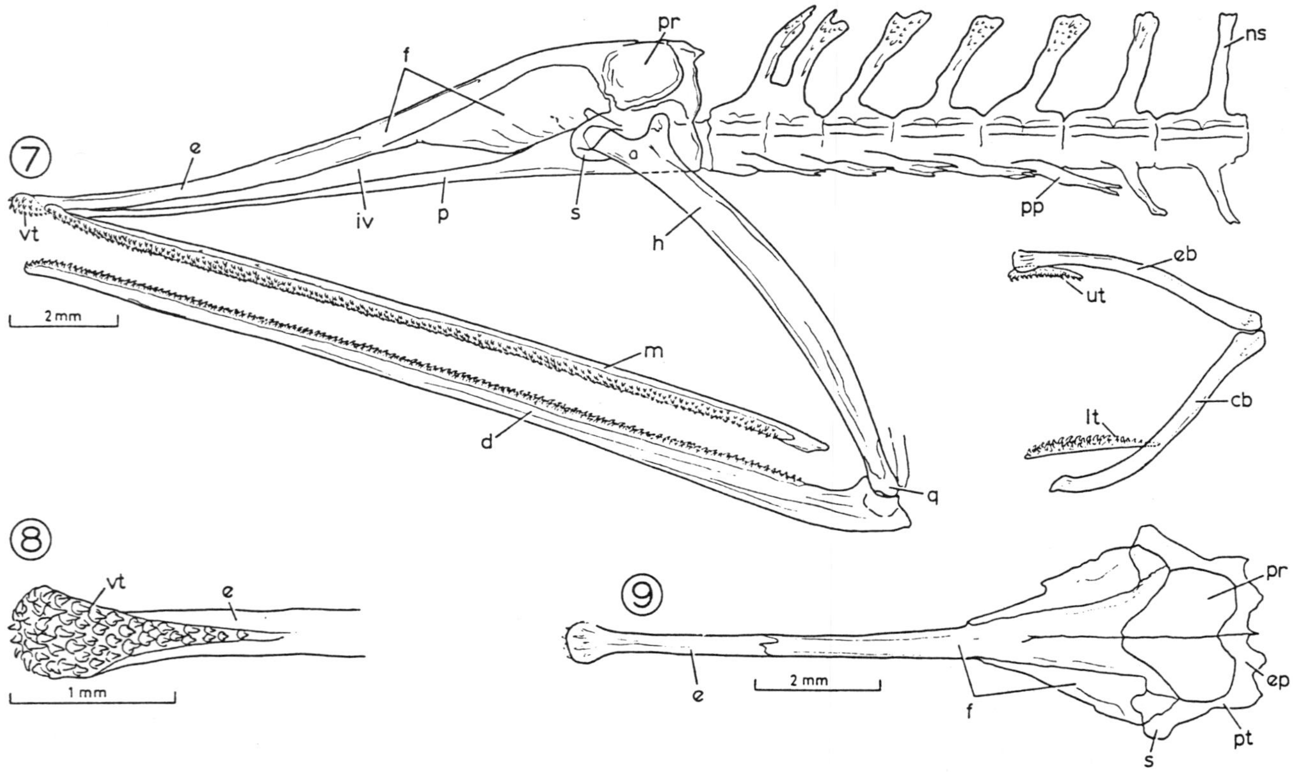

FIGURE 680. Skeleton of *Neocyema erythrosoma* (from Castle, 1977c; figs. 7–9). Abbreviations: cb = ceratobranchial, d = dentary, e = ethmovomer, eb = epibranchial, ep = epiotic, f = frontal, h = hyomandibula, iv = interorbital vacuity, lt = lower pharyngeal tooth plate, m = maxilla, ns = neural spine, p = parasphenoid, pp = parapophysis, pr = parietal, pt = pterotic, q = quadrate, s = sphenotic, ut = upper pharyngeal tooth plate; vt = vomerine teeth.

hypural bears two or three caudal-fin rays, ventral hypural three or four. *Neocyema* (Fig. 680): All vertebrae with strong neural spines, those on anteriormost vertebrae robust and slightly expanded distally, remainder slender. Parapophyses a single process on midcentrum, long and slender, horizontal on anteriormost vertebrae, becoming progressively more ventral posteriorly. Ribs, epineurals, and epipleurals absent.

Discussion. The systematic position of the Cye-

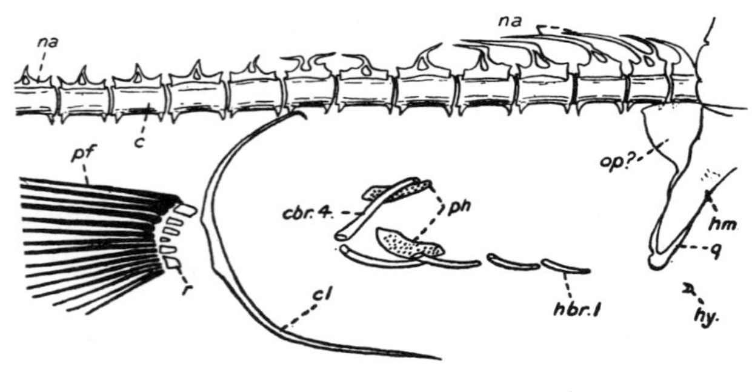

FIGURE 681. Anterior skeletal elements of *Cyema atrum* (from Trewavas, 1933: text-fig. 2). Abbreviations: c = vertebral centrum, cbr = epibranchial (mislabeled by Trewavas as ceratobranchial), cl = cleithrum, hbr = ceratobranchial (mislabeled by Trewavas as hypobranchial), hm = hyomandibula, hy = hyoid element, na = neural arch, op = opercle (probably not really an opercle), pf = pectoral fin, ph = pharyngeal tooth plate, q = quadrate, r = radial (actinost).

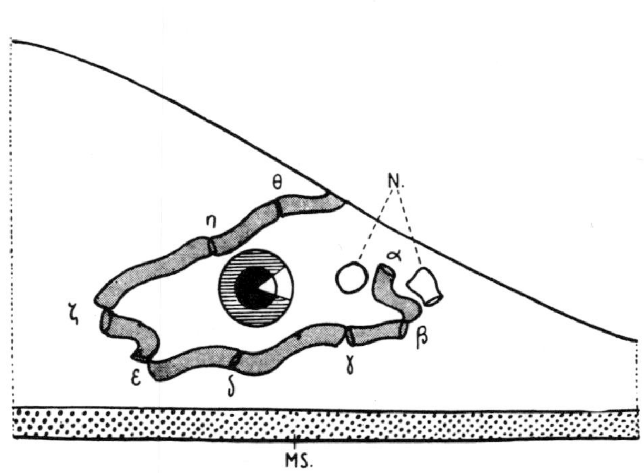

FIGURE 682. Circumorbital bones of *Cyema atrum* (from Bertin, 1937: fig. 13). Abbreviations: ms = upper jaw, n = nostril, Greek letters = pores.

matidae is still somewhat uncertain. Traditionally, it has been allied with the Nemichthyidae, but this is almost certainly wrong. The beak and teeth of *Cyema* resemble those of nemichthyids, but in other characters the two families differ greatly. Nemichthyids are long and slender, cyematids are short and stubby. Nemichthyids have lateral-line pores, cyematids do not. Nemichthyids have numerous vertebrae, cyematids have few. Nemichthyids have a large eye, cyematids a small one. Cyematids show much more skeletal reduction than nemichthyids. In nemichthyids the suspensorium is vertical, and the mouth is enlarged by an anterior prolongation of the ethmovomer and mandibles. In cyematids the mouth is enlarged primarily by a lengthening and posterior inclination of the suspensorium. The file-like tooth bands of cyematids resemble those of nemichthyids, but a nearly identical arrangement is found in the serrivomerid genus *Stemonidium;* this perhaps indicates that the dentition is a convergent feature. Finally, the leptocephali of the two families are about as different as two leptocephali can be: long and slender with a simple gut in nemichthyids, short and deep with a convoluted gut in cyematids (see p. 945).

Bertin (1937: 28) and Raju (1974a: 560) suggested that the affinity of the cyematids lies instead with the gulpers. Robins (p. 13) has summarized the evidence for this and concluded that the Cyematidae should be removed from the Anguilliformes and placed in the Saccopharyngiformes, which he regards as only distantly related to the true eels.

Distribution. All oceans.

Genera. Two genera are currently recognized in the Cyematidae: *Cyema* Günther, 1878 and *Neocyema* Castle, 1977, both monotypic. *Neocyema erythrosoma* Castle is known from two specimens that appear to be either metamorphic leptocephali or neotenic adults. They are reddish in color, compressed and semitransparent. There are no sensory papillae or pores. The skeleton is strongly ossified, however, and developing ova are present. In *Neocyema* there is no beak; the ethmovomer does not project far beyond the

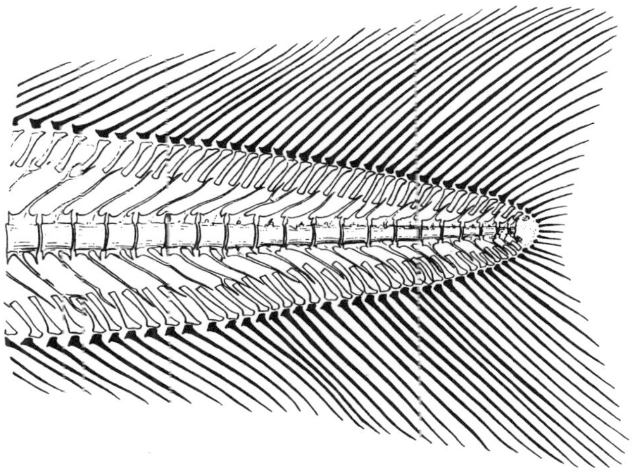

FIGURE 683. Posterior skeleton of *Cyema atrum* (from Trewavas, 1933: text-fig. 3).

junction with the maxillae. The branchial arches are reduced to a single pair, and there is no ossified pectoral skeleton (Castle, 1977b).

Neocyema erythrosoma is known only from the central and eastern South Atlantic. *Cyema atrum* Günther occurs in all oceans. The existence of certain cyematid-like leptocephali—*Leptocephalus holti* Schmidt, 1909, and a similar one described by Raju (1974a: 559, fig. 4B)—suggests that additional species or genera of cyematids await discovery.

Genus *Cyema* Günther, 1878

Cyema Günther, 1878: 251 (type species *Cyema atrum* Günther, 1878, by monotypy).

Characters. Cyema differs from *Neocyema* in the following ways. It is black and opaque rather than red and translucent. The intermaxillary portion of the ethmovomer is prolonged anteriorly and bears an elongate patch of teeth. Teeth on the mandible and maxilla are in multiple oblique rows. Sensory papillae are present along the lateral line and on the head. The tail is notched. The gill arches contain four ceratobranchials and an epibranchial. A pectoral skeleton is present.

Size. Maximum size probably not more than 150 mm TL.

Distribution. All oceans.

Etymology. From the Greek *kyema* (fetus). The

FIGURE 684. *Cyema atrum*: ANSP 152123, 111 mm TL.

fish apparently had a larval or embryonic appearance to Günther. Neuter.

Species. As far as we know, *Cyema* contains a single species, *C. atrum* Günther, 1878.

Cyema atrum Günther, 1878
Figures 684–686

Cyema atrum Günther, 1878: 251 (original description, South Pacific and South Indian Oceans, holotype BMNH 1887.12.7.261).

Distinctive Characters. Those of the genus.

Description (including data from Bertin, 1937). Total vertebrae 70–80 (n = 16), predorsal vertebrae 34–43 (14), preanal vertebrae 35–42 (14), precaudal vertebrae 47 (8). Pectoral-fin rays 12–15 (8), dorsal-fin rays 85–93 (7), anal-fin rays 72–86 (7). Pores: IO 8 (1), SO 1–3 (2).

Proportions as % of distance from posterior margin of eye to tip of hypural: preanal 54–61 (38), predorsal 53–60 (33), eye to base of pectoral fin 8–18 (41), eye to rictus 7–10 (33), pectoral fin 5–7 (33), depth at anus 5–15 (33). Of distance from posterior margin of eye to base of pectoral fin: eye 5–7 (8).

Body short, stubby, compressed; preanal length about two-thirds TL; depth about 6–7% TL. Dorsal and anal fins well developed, confluent with caudal. Dorsal- and anal-fin rays become progressively longer posteriorly until a point just before tip of tail, where they become abruptly shorter; the result, along with short caudal-fin rays, is an emarginate tail fin giving fish an arrow-like appearance. Dorsal fin begins above anus. Pectoral fin present. Fin rays appear to be unsegmented. Gill opening crescentic, on lower half of side, slightly before pectoral fin, dorsal end opposite lower end of pectoral-fin base.

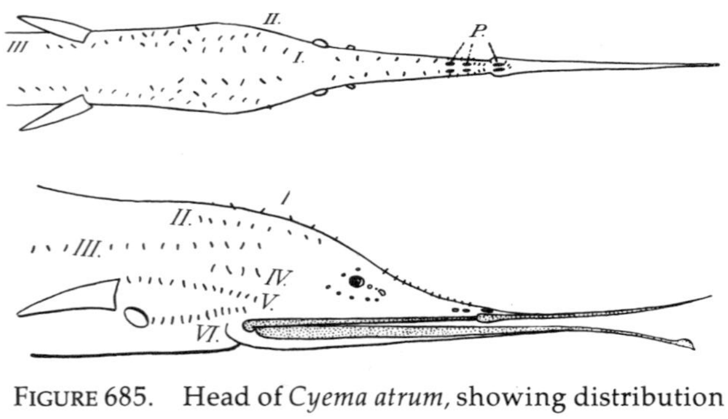

FIGURE 685. Head of *Cyema atrum*, showing distribution of pores and papillae (from Bertin, 1937: figs. 9, 11); dorsal view (top), lateral view (bottom).

Head about 30% TL, dorsal profile convex behind eye, concave before eye. Snout and jaws prolonged into a slender, non-occlusible beak; upper jaw slightly longer than lower; upper jaw ends in a fine point, lower jaw ends in a slight expansion. Eye small. Nostrils close together in front of eye; posterior nostril a simple opening at about mid-eye level, anterior nostril slightly lower and in a low tube.

Teeth small, pointed, recurved, arranged in oblique rows. Intermaxillary tooth patch expanded slightly at point where maxillary joins, then extends anteriorly as a slender band to tip of jaw. Vomerine tooth patch elongate, extends posteriorly to level of eye. Posterior end of maxilla with a ball-like patch of teeth; anterior to this point maxillary teeth in narrow bands. Mandibular teeth in narrow bands, expanded slightly at tip of jaw.

Lateral line on body without pores; instead a series of small, dermal papillae. Several additional series of papillae on head (Fig. 685): two rows above lateral line and behind eye; a short row just below lateral line from about level of rictus to a point midway between rictus and pectoral-fin base; a row between rictus and dorsal end of pectoral-fin base; a row between rictus and gill opening; a row on top of snout between eye and expanded intermaxillary portion of ethmovomer. Pores present around eye (Figs. 682, 685): one in adnasal branch of infraorbital canal

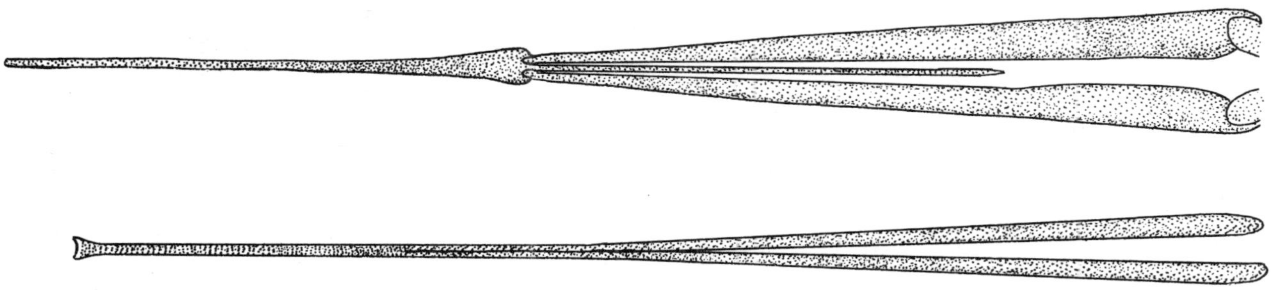

FIGURE 686. *Cyema atrum*: ANSP 152124, 109 mm TL; dentition; upper jaw (top), lower jaw (bottom).

between anterior and posterior nostrils; one below that at anterior extremity of infraorbital canal; one below posterior nostril; one below eye; two below and behind eye where canal curves dorsally; one or two above eye. According to Bertin (1937: figs. 9, 13), three supraorbital pores at about midpoint of snout: one at lateral expansion of ethmovomer, two in line behind it; specimen examined here had only one pore on rostrum, about midway between anterior nostril and lateral expansion of ethmovomer.

Color. Adults are black. Definitive pigmentation begins to appear at metamorphosis, and juvenile specimens darken progressively with development.

Size. The largest specimen recorded by Bertin (1937: 14) from the Dana material was 134 mm TL.

Development and Growth. The transition from leptocephalus to adult has been well documented for *Cyema atrum* (Roule and Bertin, 1929; Bertin, 1937; see also p. 945). There is no indication of sexual dimorphism.

Habits and Behavior. All we know about the habits of *Cyema atrum* is that it is pelagic and it lives in the lower mesopelagic and upper bathypelagic zones of the ocean. The form of the jaws suggests that its feeding habits are similar to those of nemichthyids. Nemichthyids subsist largely on crustaceans. Unlike nemichthyids, however, *Cyema* has very small eyes. It may rely on its lateralis system to locate prey.

Distribution. All oceans. *Cyema atrum* appears to be the deepest-living of all the pelagic eels. Bertin (1937: 26) found that 77% of the specimens available to him came from estimated depths in excess of 1500 m (3000 m of wire). By contrast, only 8% of *Nemichthys*, 33% of *Avocettina*, 36% of *Serrivomer*, and 45% of *Eurypharynx* came from those depths. Eight of Bertin's Dana specimens were captured at estimated depths of 330–3000 m (1000–6000 m of wire). Specimens examined here came from approximately 1800–2000 m.

Etymology. Latin *atrum* (black, dark). An adjective.

Study Material. Eight specimens, 87–114 mm TL.

SARGASSO SEA: ANSP 152123 (1, 111), 26°11'N, 58°26'W, 2000–1800 m, 21 Apr. 1979, 152124 (1, 109), same. ISH 1516/79 (6, 87–114), same.

Suborder Saccopharyngoidei

Families Saccopharyngidae, Eurypharyngidae, *and* Monognathidae

GULPER EELS

E. BERTELSEN, JØRGEN G. NIELSEN, *and* DAVID G. SMITH

Introduction. The deep-sea gulper eels—families Saccopharyngidae, Eurypharyngidae, and Monognathidae—are the most highly modified of all the eels or eel-like fishes. They are characterized most prominently by the enlarged mouth and abdomen, hence the common name. The group was treated in Part Five of this series as the order Lyomeri (Böhlke, 1966b). The following is intended primarily as an update of that account, incorporating new information that has come to light since 1966.

Characters. Gulpers are characterized by the reduction and loss of many structural elements and by an extreme modification of many of the remaining elements. The body is rather typically eel-like, elongate with a long-based dorsal and anal fin reaching to or nearly to the end of the tail. The most obvious specialization is the enormous enlargement of the mouth, brought about by a posterior elongation of the suspensorium and jaws. In saccopharyngids and eurypharyngids the jaws end far behind the cranium, and those of the two sides are connected at their posterior ends by a ligament. At rest, the suspensorium is directed posteroventrally, making an acute angle with the vertebral column. The mouth is opened by swinging the suspensorium forward, downward, and outward; simultaneously, the anterior part of the vertebral column bends upward to a point nearly or fully perpendicular to the remainder of the vertebral column (Fig. 687). In monognathids the suspensorium remains more or less vertical at rest, and the mouth seems to be opened mainly by lowering the mandible.

The neurocranium (Figs. 688–690) is tiny compared to the rest of the head; the orbit and the preorbital portions are especially small. The bones are separated by rather large unossified spaces. The frontals are divided in saccopharyngids and eurypharyngids but fused in monognathids. Only *Saccopharynx* has a separate, ossified ethmoid. Parietals are present in eurypharyngids and monognathids but not in saccopharyngids (Bertin, 1938: figs. 1, 2; Tchernavin, 1947a: 334). The supraoccipital and basisphenoid are missing in all gulpers; saccopharyngids and monognathids lack the vomer and parasphenoid as well.

The suspensorium (Figs. 687, 691, 692) consists of a well-developed and elongate hyomandibula and quadrate. The hyomandibula is articulated to the neurocranium by a single ball-and-socket joint, which permits great freedom of movement. There is some disagreement about whether the upper jaw of saccopharyngids and eurypharyngids is a maxilla or a palatopterygoid element (Gill and Ryder, 1883b: 263; 1884: 61; Regan, 1912c: 348; Tchernavin, 1947a: 313, 331; Robins, p. 14). In saccopharyngids, the mandible contains a distinct dentary and articular, but in eurypharyn-

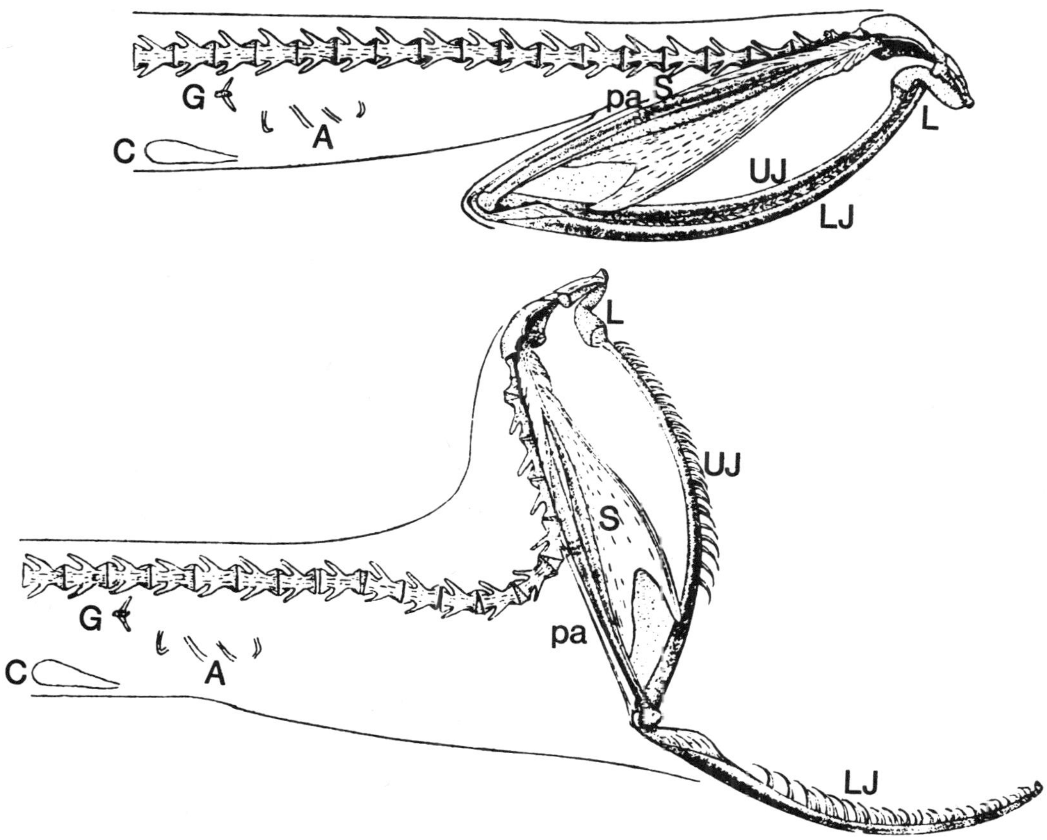

FIGURE 687. Anterior skeletal elements in situ, showing mechanism of jaw opening in *Saccopharynx* (from Tchernavin, 1947a: text-figs. 4–5). Abbreviations: A = gill arches, C = heart, G = pectoral girdle, L = ligament attaching upper jaw to neurocranium, LJ = lower jaw, S = suspensorium, UJ = upper jaw, pa = ossified tendon of adductor mandibulae muscle.

gids they are fused (Tchernavin, 1947a: 311, 331). A separate articular is present in at least one species of monognathid but absent in others (Bertin, 1938: 5). Teeth are variable in size and distribution (Fig. 693).

The ventral hyoid elements are absent, and there are no branchiostegal rays. The gill arches are set a considerable distance behind the cranium and have no connection with the cranium or the jaws (Fig. 687). Saccopharyngids and monognathids have four gill arches (only one is ossified in monognathids); eurypharyngids have five. Basibranchials are absent and the arches of the two sides are not closely connected. There are no opercular elements at all. The elongate gill opening is located on the ventrolateral surface of the body. Unsupported by opercular bones, it cannot be completely closed.

The notochord persists through the entire vertebral column. The vertebral centra are ossified, with neural and hemal arches (Fig. 694). There are no ribs, epineurals, epipleurals, or epicentrals. The fin rays are unsegmented. There is no true caudal fin.

Saccopharyngids and eurypharyngids have a pectoral fin with numerous rays and set on a fleshy pedicel. The pectoral girdle consists of a single, ossified, v-shaped bone, possibly the scapula (Fig. 695). In monognathids the pectoral fin is present as a fleshy flap without skeletal support.

The lateral line is present but without pores in saccopharyngids and eurypharyngids. Instead there is a series of small papillae or tubules, either singly or in groups, running the length of the body and on the head (Figs. 696, 697B). Monognathids lack all trace of a lateral line.

Saccopharyngids and eurypharyngids have a

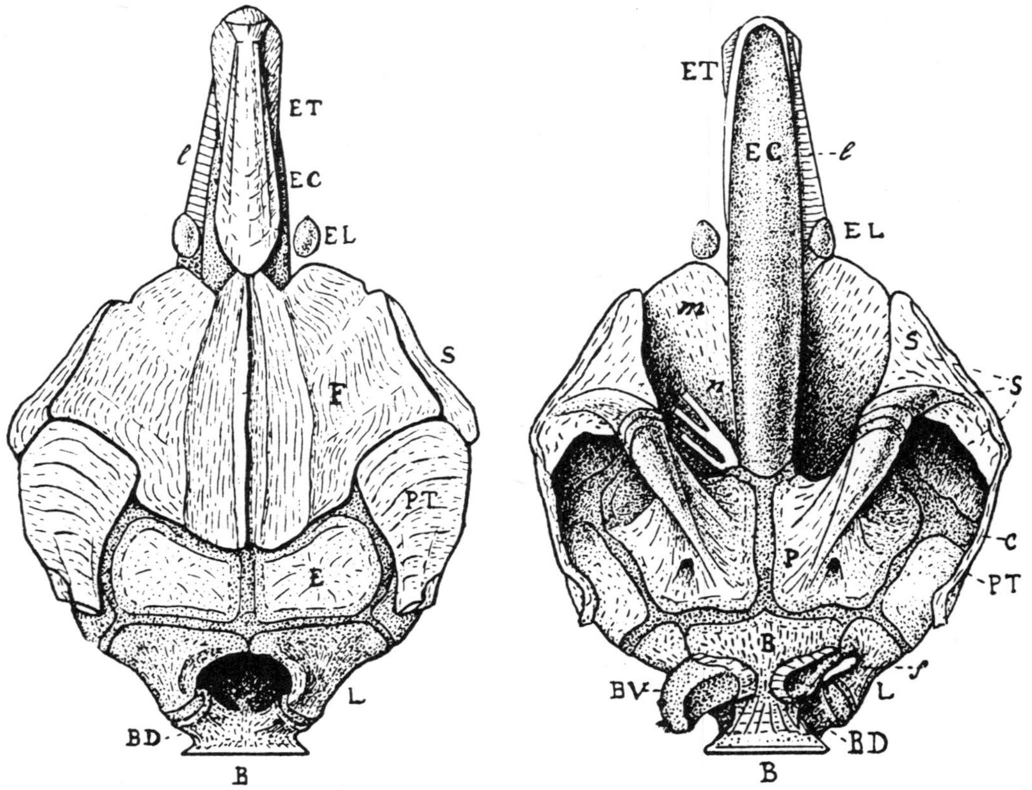

FIGURE 688. Dorsal (left) and ventral (right) view of neurocranium of *Saccopharynx* (from Tchernavin, 1947a: pl. 3). Abbreviations: BD = dorsal process of basioccipital, BV = ventral process of basioccipital, c = cartilage between cartilage pterotic, sphenotic, and prootic, EC = ethmoido-interorbital cartilage, EL = separate cartilage at anterior corner of orbit from which oblique muscles arise, ET = supra-ethmoid, f = foramen for nerves of vagus group, l = ligament uniting cartilage EL with rostrum, m = membrane between interorbital cartilage, sphenotic, and prootic, n = trigeminus nerve; other abbreviations as in Figure 689.

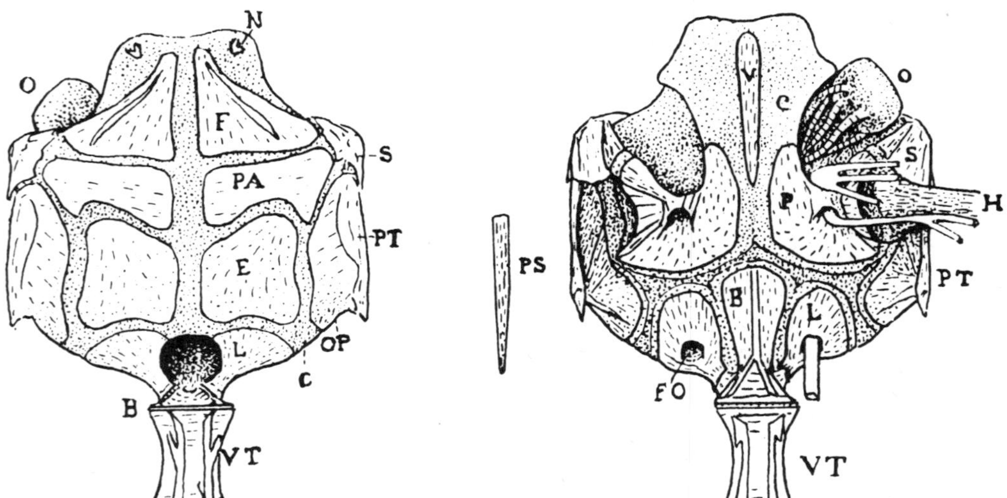

FIGURE 689. Dorsal (left) and ventral (right) views of neurocranium of *Eurypharynx* (from Tchernavin, 1947a: pl. 3). Abbreviations: B = basioccipital, C = ethmoido-interorbital cartilage, E = epiotic, F = frontal, FO = foramen for vagus and glossopharyngeus nerves, H = hyomandibula, L = exoccipital (lateral occipital), N = nasal, O = eye, P = prootic, PA = parietal, PT = pterotic, PS = parasphenoid, S = sphenotic, V = vomer, VT = first vertebra.

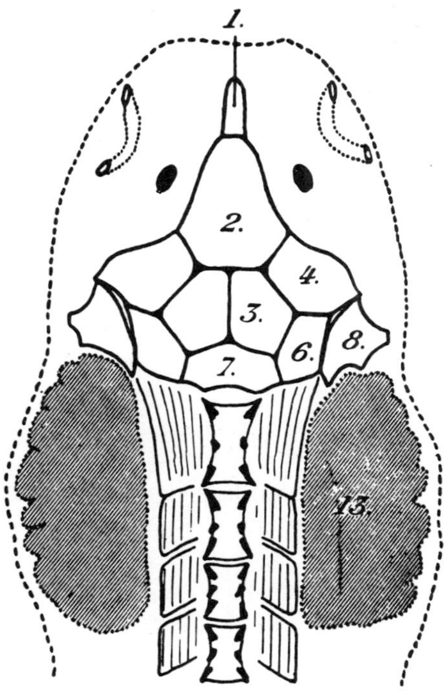

FIGURE 690. Dorsal view of skull of *Monognathus bruuni* (from Bertin, 1938: fig. 1). Abbreviations: 1 = ethmoid, 2 = frontal, 3 = parietal, 4 = sphenotic, 6 = epiotic, 7 = occipital, 8 = hyomandibula, 13 = gill chamber.

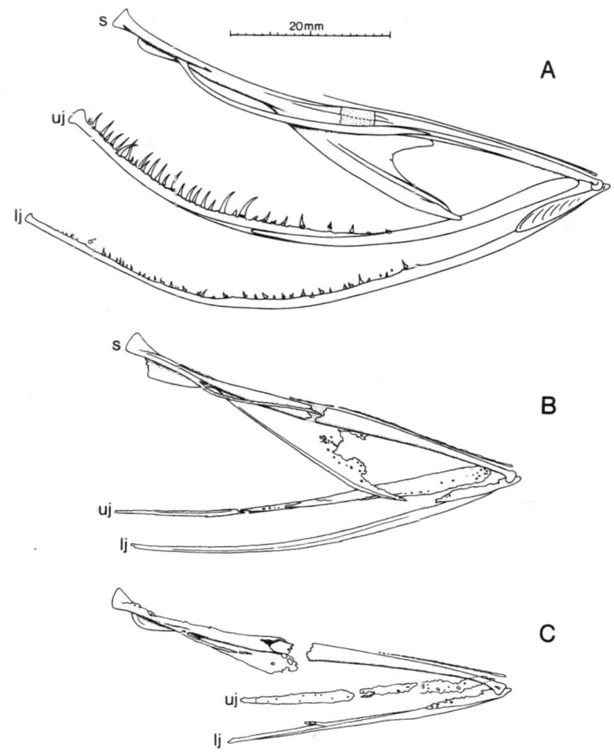

FIGURE 691. Suspensorium and jaws of *Saccopharynx ampullaceus*, left side (from Nielsen and Bertelsen, 1985: fig. 1). A. Mature female. B, C. Mature males. Abbreviations: lj = lower jaw, s = suspensorium, uj = upper jaw.

pair of narrow, white lines or grooves running along each side of the dorsal fin and extending anteriorly some distance in front of the dorsal-fin origin. These lines do not seem to be connected to the lateral line, and their function is uncertain. Saccopharyngids have a variable number of slender, dermal filaments near the dorsal margin of the body. Like the white lines, these appear unconnected to the lateral line.

Saccopharyngids and eurypharyngids have a unique caudal organ, of variable complexity, at or near the tip of the tail (Figs. 697A, 698). It is well supplied with blood vessels and nerves and, in at least some cases, is luminous. In monognathids the tip of the tail ends in a urostyle of variable length (Fig. 699); in some species the posteriormost dorsal- and anal-fin rays are joined around the tip of the tail to form a pseudocaudal fin.

All three families exhibit sexual dimorphism. In eurypharyngids the mature males develop a greatly enlarged olfactory organ that bulges out and changes the anterior profile of the head

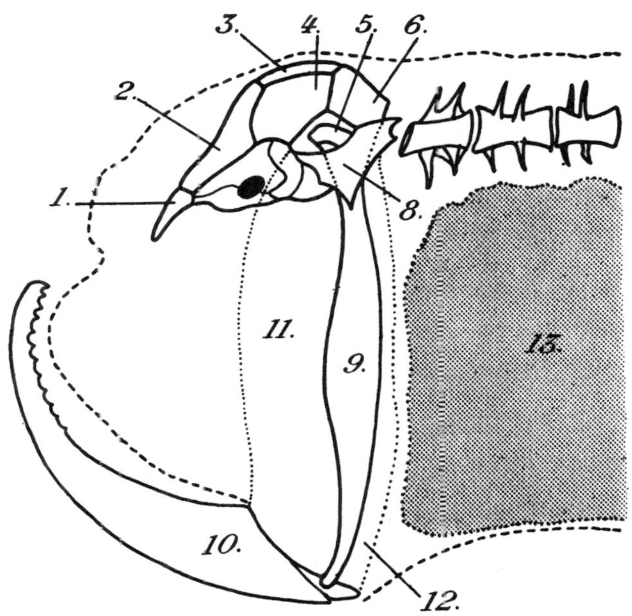

FIGURE 692. Anterior skeletal elements of *Monognathus bruuni* (from Bertin, 1936c: fig. 2; 1938: fig. 2). Abbreviations: 5 = pterotic, 9 = quadrate, 10 = mandible, 11 = elevator muscle, 12 = depressor muscle; other abbreviations as in Figure 690.

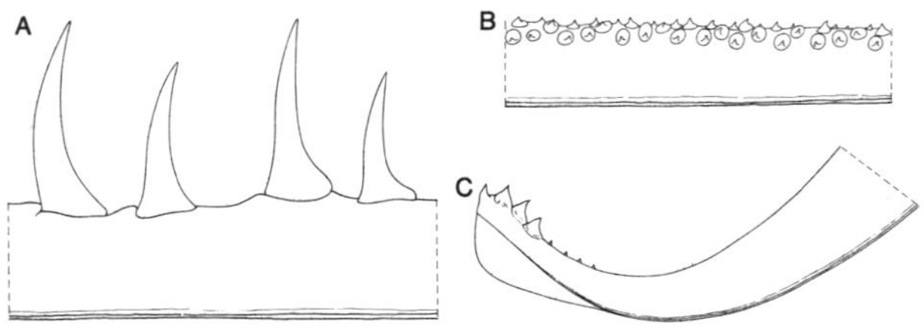

FIGURE 693. Teeth of lower jaw of adult specimens of the three saccopharyngoid families (drawn to the same scale). A. *Saccopharynx ampullaceus*: ZMUC P2340205, 1150 mm TL; section of middle part of left mandible. B. *Eurypharynx pelecanoides*: ZMUC P2340206, 485 mm TL; section of middle part of left mandible. C. *Monognathus jesperseni*: ZMUC P2329851, holotype; left mandible.

(Gartner, 1983) (Fig. 700). In addition, the jaws and teeth are reduced. Mature male saccopharyngids show a reduction of the jaws, with the loss of the teeth and the disintegration or resorption of the bones (Fig. 691B, C). The eye and olfactory organ also enlarge. Ripe female saccopharyngids also show some reduction of the jaws, but to a lesser extent than the males (Nielsen and Bertelsen, 1985: 164). Mature male monognathids have a greatly enlarged olfactory organ and rudimentary lower jaws; the head is covered with a layer of spongy tissue (Fig. 706).

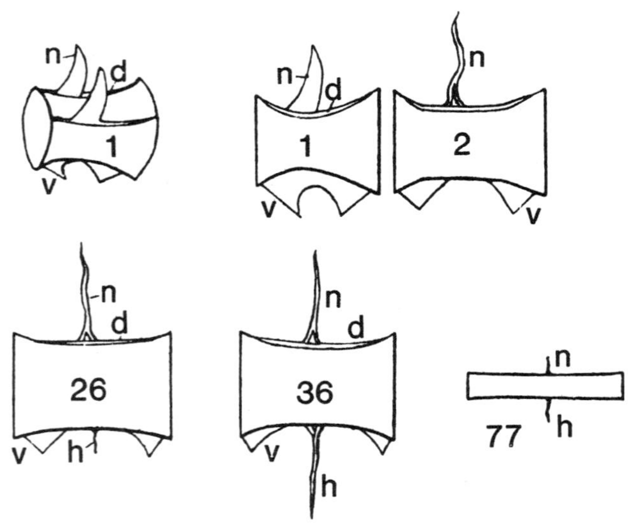

FIGURE 694. Vertebrae of *Eurypharynx pelecanoides*, left side, diagrammatic, showing first, second, 26th, 36th, and 77th vertebra (from Tchernavin, 1947a: text-fig. 10). Abbreviations: d = dorsal ridge of vertebra, n = neural spine, h = hemal spine, v = ventral ridge; numbers are ordinal numbers of vertebrae (anteriormost vertebra = 1).

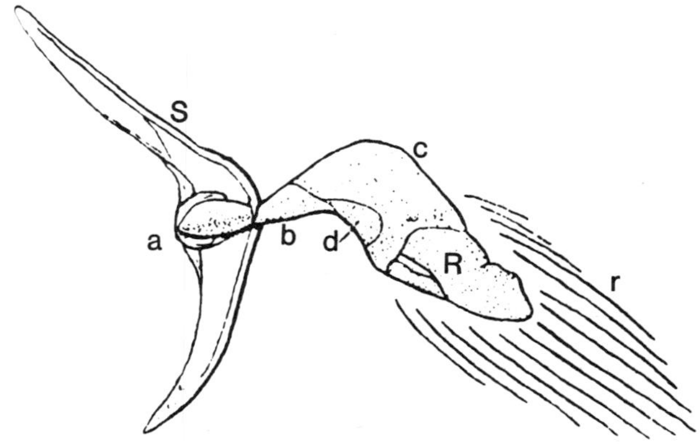

FIGURE 695. Pectoral girdle of *Saccopharynx ampullaceus*, left side (from Tchernavin, 1947a: text-fig. 14). Abbreviations: a = anterior cartilage supporting lobe of pectoral fin, b = part of cartilage a separated from latter by injury (or separate element of lobe?), c = main cartilage of lobe, d = indistinct subdivision of cartilage c, R = radials (actinosts), r = fin rays, S = scapula or scapulacoracoid (ossified).

Discussion. The extreme modification of the gulpers, particularly in the morphology of the head, has long obscured their affinities. At one time or another, they have been allied with such diverse groups as stomiiforms, macrourids, flatfishes, synbranchids, syngnathiforms, lampriforms, myctophiforms, cetomimids, giganturids, and eels. Tchernavin (1947a: 344; 1947b: 392) even expressed doubt that they were bony fishes. Böhlke (1966b: 609) reviewed all the evidence and concluded tentatively ". . . that eels and gulpers extend back in time to a common base and that the characters of general appearance and a very similar leptocephalus larval stage in their early life history are positive factors that indicate such a relationship." Robins (p. 14) argues that

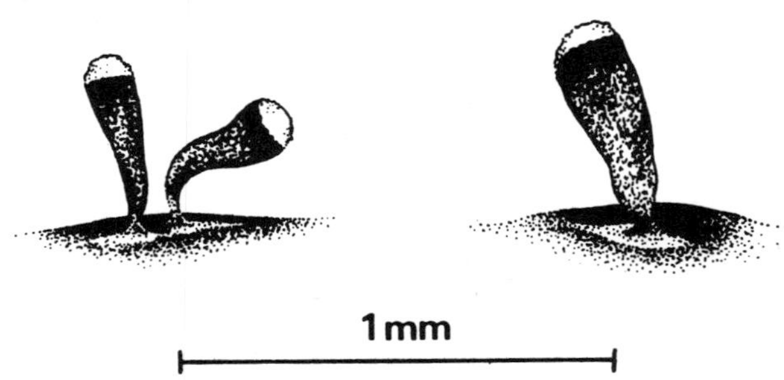

FIGURE 696. Lateral-line papillae from tail of *Saccopharynx ramosus* (from Nielsen and Bertelsen, 1985: fig. 7).

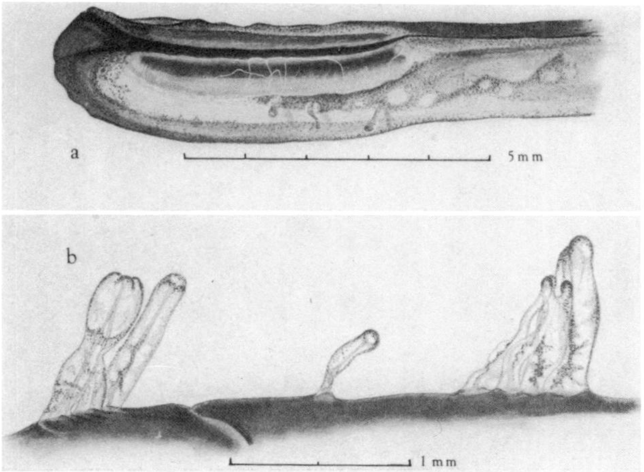

FIGURE 697. Caudal organ and lateral-line tubules of *Eurypharynx pelecanoides* (from Owre and Bayer, 1970: fig. 3). A. Caudal organ, right side in dorsolateral view. B. Tubules of anterior part of lateral line.

the gulpers and cyematids form a separate order remote from the Anguilliformes but with roots within the larger assemblage of elopomorph fishes. Here the matter stands.

Saccopharynx and *Eurypharynx* show many resemblances, and one might question whether they belong in separate families. The main differences are the number of gill arches and clefts and the relative size of the mouth and teeth. The significance of the gill-arch character is uncertain; Böhlke (1966b: 608) suggested that the sixth cleft in *Eurypharynx* is a secondary development. The size of the mouth and teeth seems related to feeding habits; *Saccopharynx* is adapted to taking larger prey than is *Eurypharynx*. Other differences, such as the position of the dorsal-fin origin and the number of vertebrae, seem trivial compared to common characters such as the caudal organ and the dorsal white lines (see above). Nevertheless, convention will be followed here and family status retained.

The monognathids are among the strangest of all fishes. In general appearance, they resemble

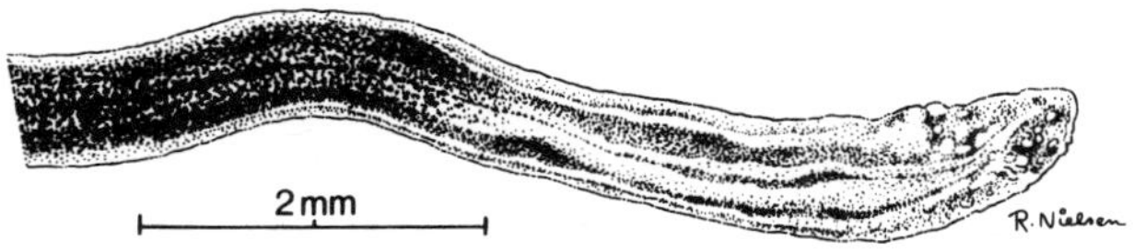

FIGURE 698. Caudal organ of *Saccopharynx ampullaceus* (from Nielsen and Bertelsen, 1985: fig. 11B).

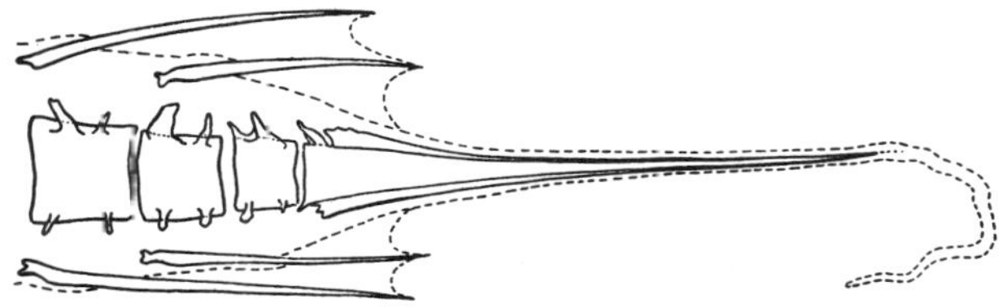

FIGURE 699. Caudal skeleton of *Monognathus taningi* (from Bertin, 1938: fig. 7).

the other gulpers, particularly *Saccopharynx*, but they lack all trace of an upper jaw and they have no caudal organ. They have a fang-like downward projection at the anterior end of the neurocranium, which is hollow and connected to a pair of glands on each side. The frontals are fused (Fig. 690). All the monognathid specimens are small; most hardly seem more than postlarvae. This would ordinarily raise the suspicion that they are merely postlarval saccopharyngids, but individuals have recently been observed with ripe ovaries and testes (Bertelsen and Nielsen, 1987). Their low vertebral counts and their distinctive leptocephali (Raju, 1974a: fig. 1) further indicate that they represent a separate group. The degenerative nature of the monognathids makes it difficult to assess their relationships. The caudal organ and the dorsal white lines could be considered derived characters that unite *Saccopharynx* and *Eurypharynx* against the monognathids. Likewise, the rostral fang and glands and the fused frontals seem to be derived characters peculiar to the monognathids. This would suggest that the Monognathidae is the sister group of the Saccopharyngidae and Eurypharyngidae.

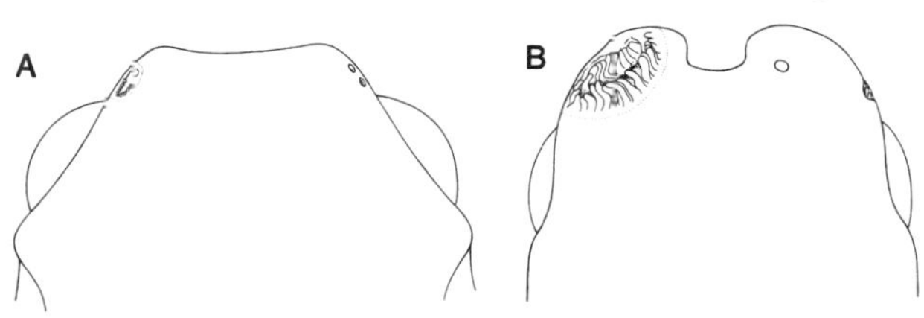

FIGURE 700. Dorsal view of anterior part of head of *Eurypharynx pelecanoides*, showing size of olfactory organ and shape of snout (left side cut away to show lamellae). A. Female, 540 mm TL, ZMUC P2340207. B. Mature male, 420 mm TL, ZMUC P2340209.

Key to the Saccopharyngoid Families

1a. Upper jaw present 2
1b. Upper jaw absent .... Monognathidae, p. 650
2a. Upper jaw less than 40% of preanal length; gill opening closer to tip of snout than to anus; dorsal fin begins behind level of pectoral fin; four gill arches, five gill clefts; more than 150 vertebrae; vomer and parasphenoid absent .... Saccopharyngidae, p. 642
2b. Upper jaw greater than half preanal length; gill opening closer to anus than to tip of snout; dorsal fin begins before pectoral fin; five gill arches, six gill clefts; fewer than 130 vertebrae; vomer and parasphenoid present .... Eurypharyngidae, p. 648

Family SACCOPHARYNGIDAE

Characters (largely from Nielsen and Bertelsen, 1985). Body elongate, preanal portion distinctly deeper than postanal. Anus near anterior one-third to one-fifth of TL. Abdomen distensible, posterior part often protruding ventrally into a pouch that can extend a certain distance behind anus. Tail very slender, terminating in a caudal organ whose structure varies among species. Dorsal fin begins on posterior part of trunk, approximately one-fifth TL behind tip of snout. Anal fin begins shortly behind anus. Dorsal and anal fins not confluent around tail but ending well short of caudal organ, anal fin reaching somewhat farther back than dorsal fin. Caudal fin absent. Pectoral fin present, placed below midside. Scales absent.

Lateral line present but without pores; instead a series of papillae, each narrow at base, expanded distally, tip truncated with opaque, unpigmented tissue (Fig. 696). Papillae usually single, but occasionally in pairs, arising from a more or less distinct white spot on skin; white spots sometimes connected by a narrow, unpigmented groove. Lateral line on body follows level of vertebral column, ends some distance before tip of tail; one papilla or pair of papillae per vertebra. On head (Fig. 701), lateral line curves downward behind eye, then turns posteriorly and splits into two lines, one extending toward pectoral fin and one along suspensorium. An isolated loop above and behind eye. An isolated supraorbital series extends from above eye to above nasal capsule; at its anterior end, it enters a short raised tube with an opening at each end. A short row of papillae extending diagonally from this tube to posterior nostril. An infraorbital series along edge of upper jaw, turning dorsally in front of eye and ending below posterior nostril. A loop beginning just behind anterior nostril, curving ventrally, anteriorly, and dorsally to a point in front of anterior nostril, where it divides into a pair of commissures that cross dorsal surface of snout. No lateralis organs on lower jaw.

A pair of narrow, white lines or grooves immediately adjacent to dorsal fin and extending some distance in front of dorsal-fin origin, separated from dorsal fin and from each other by a narrow strip of pigmented skin. Floor of grooves without any specialized structure.

A series of slender dermal papillae of various lengths, located dorsolaterally, close to ventral edge of white lines, shifting to mid-dorsal line on tail behind dorsal fin. Filaments usually simple, but bead-like distal swellings sometimes present. One species has branched filaments.

Tip of tail partially or completely unpigmented, compressed, vertically expanded to form a lanceolate to spatulate caudal organ (Fig. 698). Upper and lower parts (i.e., above and below vertebral column) generally asymmetrical. Organ contains tissue consisting of assemblages of whitish or yellow tubules, varying in shape from spherical to irregular curved rods. Caudal organ well innervated, with a rich blood supply, appears red in freshly captured specimens. Filaments sometimes present. Caudal organ reported to be luminous.

Head conspicuously large and deep, but most

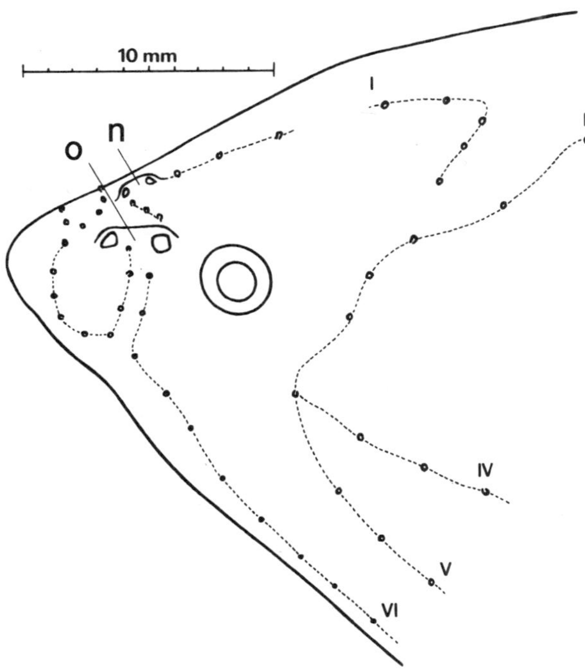

FIGURE 701. Pattern of lateral-line papillae on head of *Saccopharynx ampullaceus* (from Nielsen and Bertelsen, 1985: fig. 4). Abbreviations: n = nasal lateral-line organ, o = olfactory organ.

of its length in postcranial part. Cranium very small, restricted to anteriormost extremity of head, bearing small eyes and olfactory organ. Anterior and posterior nostrils close together, about midway between eye and tip of snout. Snout small, beak-like. Mouth large, length of jaws approximately one-third to one-quarter preanal length. Teeth slender, recurved, depressible, arranged in 3–4 overlapping longitudinal series in both jaws, beginning near tip and extending about two-thirds to three-quarters of the way back, outer rows reaching farther back than inner rows; teeth of inner rows largest (Fig. 693A). Number of teeth approximately equal in upper and lower jaw, maximum number about 50.

Gill opening horizontal, elongate, located near ventral midline, below and slightly before pectoral fin.

Frontals divided; parietals absent; vomer and parasphenoid absent. Four gill arches, five clefts. Vertebrae approximately 150–300.

Maximum size about 2 m TL, but most of this is the slender tail.

Sexual Dimorphism. Mature males show several conspicuous modifications (Fig. 703). The abdominal pouch is reduced, probably reflecting cessation of feeding. Along with this, the entire jaw apparatus shows a marked degeneration, with the loss of the teeth and a resorption or disintegration of the bones (Fig. 691B, C). The eye and the olfactory organ enlarge. The nostrils widen and the olfactory lamellae enlarge, partially protruding through the nostrils. In addition, the tail is more compressed, the white spots around the lateral-line papillae are expanded, and in at least some species the dermal filaments are lost.

Mature females also show some reduction in the jaws and a loss of teeth, although the changes are not as extreme as in the males (Nielsen and Bertelsen, 1985: 164).

Remarks. Because the tail is frequently damaged, proportional measurements are given as a % of SL_{100}. A discussion and definition of this term and of the terms D_{100} and A_{100} is found in the account of the Nemichthyidae (p. 441).

Distribution. All oceans, at depths between approximately 1000 and 3000 m. Absent from the Mediterranean.

Genera. A single genus, *Saccopharynx* Mitchill, 1824.

Genus *Saccopharynx* Mitchill, 1824

Saccopharynx Mitchill, 1824: 86 (type species *Saccopharynx flagellum* Cuvier, 1829, by monotypy).
Ophiognathus Harwood, 1827: 51 (type species *Ophiognathus ampullaceus* Harwood, 1827, by monotypy).
Ophioglossus Duméril, 1856: 218 (lapsus for *Ophiognathus*).

Characters. Those of the family.

Size. Maximum size about 2 m TL, mostly tail.

Distribution. That of the family.

Etymology. From the Greek *sakkos* (bag) and *pharynx* (throat). Masculine.

Species. Nielsen and Bertelsen (1985: 175) recognized nine species. *Saccopharynx ampullaceus* (Harwood, 1827) is the most common and is found across the North Atlantic. *Saccopharynx thalassa* Nielsen and Bertelsen, 1985 is also found on both sides of the Atlantic, though less frequently. The only other species known from the western Atlantic is *S. harrisoni* Beebe, 1932. Four more species occur in the eastern Atlantic: *S. hjorti* Bertin, 1938, *S. paucovertebratis* Nielsen and Bertelsen, 1985, *S. trilobatus* Nielsen and Bertelsen, 1985, and *S. ramosus* Nielsen and Bertelsen, 1985 (also known from the Indian Ocean). *Saccopharynx schmidti* Bertin, 1934 is found in the southwestern Pacific, and *S. lavenbergi* Nielsen and Bertelsen, 1985 in the eastern Pacific.

This is still an imperfectly known group, and the list of species should not be considered definitive. Only the species known from the western North Atlantic are treated here.

KEY TO THE SPECIES OF *SACCOPHARYNX*
(from Nielsen and Bertelsen, 1985)

1a. Origin of anal fin anterior to origin of dorsal fin . . . . . *hjorti* (eastern Atlantic)
1b. Origin of anal fin posterior to origin of dorsal fin 2
2a. Origin of dorsal white lines closer to snout than to pectoral fin; caudal organ slender (length several times depth), continued anteriorly in dorsal and ventral series of isolated spots of luminescent tissue; pectoral-fin rays 28–41 3
2b. Origin of dorsal white lines closer to pectoral fin than to snout; caudal organ short (length less than three times depth), without anterior isolated spots; pectoral-fin rays 40–52 7
3a. Caudal organ distinctly expanded with angular or notched edges; body filaments 1–3 4
3b. Caudal organ indistinctly expanded with smoothly curved edges; many body filaments except in juveniles 5
4a. A single short, papilliform body filament near caudal organ; caudal organ subsymmetrical . . . . . *harrisoni*, p. 647
4b. Two or three slender body filaments on posterior part of tail; caudal organ asymmetrical . . . . . *schmidti* (southwestern Pacific)
5a. Body filaments only on tail; caudal organ with well developed filaments . . . . . *thalassa*, p. 646
5b. Body filaments on both trunk and tail; caudal organ without well developed filaments 6
6a. Skin black; distal third of tail spotted; body filaments 18–90; longest filament located 30–40% TL anterior to caudal tip . . . . . *ampullaceus*, p. 644
6b. Skin brownish-gray; pigmentation of distal part of tail gradually decreasing; body filaments 5–19; longest filament 6–24% TL anterior to caudal tip . . . . . *lavenbergi* (eastern Pacific)
7a. Body filaments on tail; some filaments branched . . . . . *ramosus* (eastern Atlantic and Indian Ocean)
7b. Body filaments absent 8
8a. Caudal organ with three lobes provided with filaments . . . . . *trilobatus* (eastern Atlantic)
8b. Caudal organ undivided, without filaments . . . . . *paucovertebratis* (eastern Atlantic)

Saccopharynx ampullaceus (Harwood, 1827)
Figures 698, 702–704

Saccopharynx Mitchill, 1824: 86 (no specific name given; specimen from North Atlantic, not preserved).

Ophiognathus ampullaceus Harwood, 1827: 49, pl. 7 (original description, 62°N, 57°W, neotype ISH 3288/79).

Saccopharynx ampullaceus. Cuvier, 1829: 355.

Saccopharynx flagellum Cuvier, 1829: 355 (name given to Mitchill's specimen).

FIGURE 702. *Saccopharynx ampullaceus*: ISH 3288/79, neotype, 940 mm TL, immature male (from Nielsen and Bertelsen, 1985: fig. 10).

Saccopharynx chordatus Richardson, 1836: 272 (stated that Mitchill believed his fish to be *Stylephorus chordatus* Shaw and intended to retain the specific name while giving it a new generic name).

Saccopharynx harwoodii South in Smedley, Rose, and Rose, 1845, 7: 388 and Plates to Zoology, Pisces, pl. 9 (new name for *S. ampullaceus*). Stephens and South, *ibid.*, 1845, 24: 250 (spelled *harwoodi*). Plates may have been published earlier (Böhlke, 1966b: 620); if so, name might date from that time.

Saccopharynx mitchilli Stephens and South in Smedley, Rose, and Rose, 1845: 250, pl. 9 (new name for *S. flagellum*).

Saccopharynx johnsoni Bertin, 1934: 45, fig. 41 (original description, 33°N, 17°W, holotype BMNH 1861.6.4.3).

Saccopharynx sp. Jonsson, 1977: 40, fig. 3.

Ophiognathus leei Anonymous, 1983 (a museum name, appeared in print only in a list of type specimens).

Distinctive Characters (from Nielsen and Bertelsen, 1985: 177). Filaments on trunk 8–20 in specimens larger than 500 mm TL, total number of filaments on body 35–90 in specimens larger than 800 mm TL. Distal part of tail spotted in specimens larger than 500 mm TL. Caudal organ simple, slender, with few or no filaments, with anterior series of isolated spots of photogenic tissue, without distinct anterior border. Posteriormost filament to tip of tail 14–40% TL; SL_{100} (tip of snout to anterior edge of 101st vertebra) 49–57% TL; predorsal 16–20% TL (28–42% SL_{100}); preanal 20–24% TL (38–50% SL_{100}), snout to origin of white lines 2–3% TL (4–7% SL_{100}); longest pair of body filaments 114–134% SL_{100} behind tip of snout. Vertebrae more than 200. Pectoral-fin rays 28–41. General color black.

Description (from Nielsen and Bertelsen, 1985: 178). Total vertebrae >200 (n = 1), predorsal vertebrae 23–38 (28), preanal vertebrae 32–45 (29). Dorsal-fin rays 210–247 (18), D_{100} 147–195 (29), anal-fin rays 244–282 (18), A_{100} 139–177 (28), first anal-fin ray below 19th–32nd dorsal-fin ray (25), pectoral-fin rays 28–41 (30).

Proportions as % of TL: SL_{100} 49–57 (23), preanal 20–24 (16), predorsal 16–20 (16). Of SL_{100}: lower jaw in females and immature males 10–13 (34), lower jaw in mature males 6–10 (4), preanal 38–50 (32), predorsal 28–42 (32), snout to origin of white lines 4–7 (23), base of pectoral fin to origin of white lines 9–15 (22).

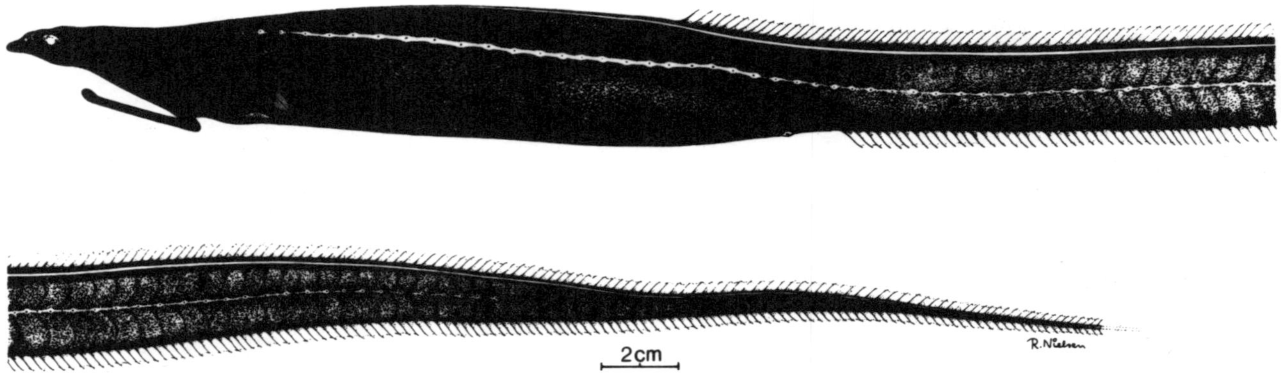

FIGURE 703. *Saccopharynx ampullaceus*: ISH 914/73, 620+ mm TL, mature male (from Nielsen and Bertelsen, 1985: fig. 12).

Caudal organ (Fig. 698) of neotype long and slender with distal 3 mm pointed; greatest depth 2.5 mm; ventral edge of tip with 4–5 rudimentary filaments; dorsal and ventral series of ca 10 isolated, elongate spots of luminescent tissue in front of main concentration of this tissue, anteriormost placed 40 mm (4.3% TL) anterior to caudal tip.

Color. In preservative blackish brown; posterior part of tail irregularly spotted in specimens larger than 500 mm TL.

Size. The 48 specimens examined were 167–1610 mm TL. Four mature males were 620–1100 mm TL, but all may have been incomplete. Two females with large eggs were 800+ and 990+ mm TL; both lacked caudal organs.

Variation. The number of teeth and body filaments increases with TL. Filaments are absent in specimens smaller than 250 mm TL and absent on the trunk in specimens smaller than 500 mm TL. In larger specimens the number of filaments varies from 18–ca 90.

Distribution (Fig. 704). Eastern and western North Atlantic between 10° and 65°N. In the western Atlantic known from the Gulf Stream off North Carolina to Massachusetts, also Bermuda. Depth probably 2000–3000 m. Apparently the most common species in the Atlantic; Nielsen and Bertelsen (1985: 176) recorded 48 specimens.

Etymology. Latin *ampullaceus* (of a flask). An adjective.

Study Material. 48 specimens, 167–1610 mm TL.
Neotype: ISH 3288/79 (immature male, 940), south of Azores, 35°20′N, 30°16′W, 0–1900 m, R/V Anton Dohrn 348/79, 30 April 1979.
Other material: See Nielsen and Bertelsen (1985: 176).

Saccopharynx thalassa Nielsen and Bertelsen, 1985
Figure 704

Saccopharynx flagellum?. Saldanha, 1982: 6.
Saccopharynx sp. Post and Tesch, 1982: 351.
Saccopharynx thalassa Nielsen and Bertelsen, 1985: 185, figs. 15–16 (original description, 26°05′N, 15°54′W, holotype MNHN 1983-526).

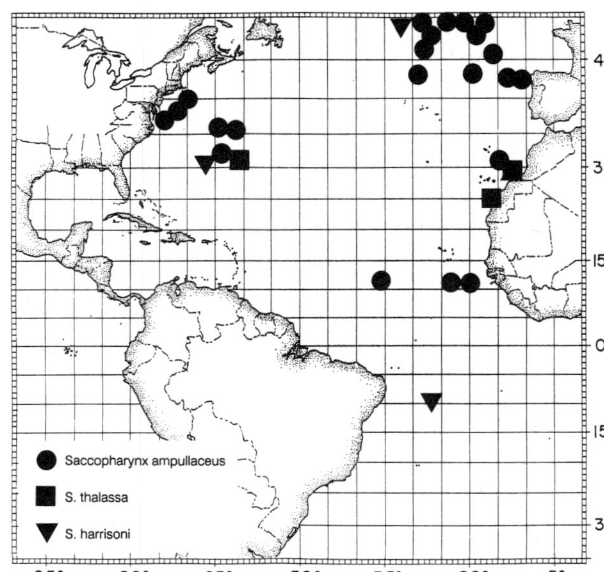

FIGURE 704. Distribution of *Saccopharynx ampullaceus*, *Saccopharynx thalassa*, and *Saccopharynx harrisoni*.

Distinctive Characters (from Nielsen and Bertelsen, 1985: 185). Body filaments 5–19, on tail only, the two longest located 135–145% SL_{100} behind tip of snout. Caudal organ indistinctly expanded, with 20–30 slender filaments. Vertebrae ca 210–225. Anal-fin rays 275–312. Pectoral-fin rays 37–39.

Description (from Nielsen and Bertelsen, 1985: 186). Total vertebrae ca 210–225 (n = 2), predorsal vertebrae 30–35 (3), preanal vertebrae 35–44 (3). Dorsal-fin rays 256–287 (3), D_{100} 171–195 (3), anal-fin rays 275–312 (3), A_{100} 152–192 (3), first anal-fin ray below 23rd–34th dorsal-fin ray (3), pectoral-fin rays 37–39 (3).

Proportions as % TL: SL_{100} 54–59 (2), preanal 23–28 (2), predorsal 21–23 (2), prepectoral 11 (2), lower jaw 7 (2), base of pectoral fin to origin of white lines 9–10 (2). Of SL_{100}: preanal 40–47 (3), predorsal 34–38 (3), prepectoral 17–20 (3), lower jaw 12–13 (3), snout to origin of white lines 4–5 (3), base of pectoral fin to base of white lines 12–17 (3).

Caudal organ of holotype (Fig. 698) pointed and moderately expanded, greatest depth 1.5 mm, ca 5 mm before tip, asymmetrical with nearly straight ventral edge and convex upper edge. Ca 30 thin, simple filaments most densely grouped dorsolaterally and behind deepest part, a few filaments ventrally also. Short series of isolated spots of luminescent tissue dorsally at and before anterior border of organ.

Color. In preservative dark brown or black.

Size. The two intact specimens are 755–1070 mm TL, 167–632 mm SL_{100}. The largest and smallest specimens are immature males, the other a female with small eggs.

Distribution (Fig. 704). Only three specimens are known, one from Bermuda and two from the Canary Islands.

Etymology. Named for the French research vessel THALASSA. A noun in apposition.

Study Material. Three specimens, 260–1070 mm TL.

Holotype: MNHN 1983-526 (female, 755), Canary Islands, 26°05′N, 15°54′W, 0–1000 m, R/V Thalassa X075, 7 Feb. 1971.

Paratypes: ISH 3287/79 (immature male, 1070), 31°11′N, 63°27′W, 0–1700 m, 19 April 1979. ZMUC P2339900 (1, 260+), 30°45′N, 14°30′W, 0–1000 m, 10 Feb. 1971.

Saccopharynx harrisoni Beebe, 1932
Figure 704

Saccopharynx harrisoni Beebe, 1932: 63, figs. 12–13 (original description, off Bermuda, holotype USNM 170939).

Distinctive Characters (from Nielsen and Bertelsen, 1985: 190). Only one short filament, on posterior sixth of body, its length decreasing with increasing TL, from 13 to 1 mm (in the known specimens 1250–1660 mm TL). Caudal organ lanceolate, distally pointed, with dorsal and ventral angular expansions almost symmetrical. SL_{100} less than half TL. Vertebrae 220–310. Anal-fin rays 303–336.

Description (from Nielsen and Bertelsen, 1985: 190). Total vertebrae 220–310 (n = 3), predorsal vertebrae 31–39 (3), preanal vertebrae 40–43 (3), first anal-fin ray below 21st–30th dorsal-fin ray.

Proportions as % of TL: SL_{100} 43–49 (3), preanal 18–24 (3), predorsal 15–22 (3), prepectoral 6–9 (3), lower jaw 5–6 (3), snout to origin of white lines 2–3 (3), base of pectoral fin to origin of white lines 5–6 (3). Of SL_{100}: preanal 42–50 (3), predorsal 34–43 (3), prepectoral 15–18 (3), lower jaw 11–13 (3), snout to origin of white lines 4–6 (3), base of pectoral fin to origin of white lines 12–13 (3).

Caudal organ of holotype with a distal, slender tapering part, greatest depth between two semi-symmetrical, digitiform outgrowths. Two similar but smaller outgrowths present on ventral, unpigmented edge of distal tip. Before this the deepest part of organ unpigmented except for median stripe and filled with luminescent tissue and blood vessels. Farther in front, anterior tapering part of organ pigmented and bears a series of white luminescent papillae along dorsal and ventral edge. Caudal organ of larger female similar to that of holotype, but less extensively pigmented and with a single side branch on distal tip. Caudal organ of male differs from those of females as follows: largely unpigmented; distal part shorter without side branches; anterior series of papillae absent.

Size. The three specimens are 1250–1660 mm TL, 535–800 SL_{100}. The smallest is an immature male, the others are females.

Distribution (Fig. 704). Three specimens known, from Bermuda, the central North Atlantic, and the South Atlantic off Brazil.

Etymology. "I have named this fish in honor of Harrison Williams, Esq., through whose continued interest and support these Bermuda Oceanographic Expeditions have been made possible" (Beebe, 1932: 67).

Study Material. Three specimens, 1250–1660 mm TL.

Holotype: USNM 170939 (female, 1270), Bermuda, 16 km southeast of Nonsuch, 0–1646 m, Bermuda Oceanographic Expedition, 11 June 1931.

Other material: ISH 652/82 (female, 1660), 49°47.9′N, 28°46.8′W, 0–3200 m, 15 June 1982. ISH 1096/68 (immature male, 1250), 9°54′S, 27°11′W, 0–560 m, 5 Feb. 1968.

Family EURYPHARYNGIDAE

Characters (largely from Böhlke, 1966b). Body elongate, caudal part compressed and less deep than trunk. Preanal length about one-third of TL. Abdomen distensible. Tail ending in a rather simple caudal organ, divided longitudinally into a black upper and a white lower portion (Fig. 697A). Dorsal-fin origin just before midpoint of trunk. Anal-fin origin shortly behind anus. Dorsal and anal fins not confluent, ending some distance from caudal organ. Caudal fin absent. Pectoral fin rudimentary with about 11 rays, placed just behind gill opening, low on side. Scales absent. Skin black.

Lateral line well developed on head and body, but without pores; its course marked externally by evenly spaced groups of elevated tubules, each group situated on a white patch. Groups of four tubules usually alternating with a single tubule. Each tubule provided with a narrow, almost translucent stalk up to 2–3 mm long (Fig. 697B). A narrow white line or groove on each side of dorsal fin in its entire length.

Head large and deep with small but distinct eyes. Cranium very small, about 10 times in head. Snout short and broad. Anterior and posterior nostrils close together. Mouth extremely large; length of jaws equal to length of head and more than half preanal length. Membrane between lower jaws greatly distensible. Teeth small, close-set, broad-based with eccentrically placed, recurved, pointed tips (Fig. 693B). Both jaws toothed their entire length. Teeth numerous, maximum number in upper jaw more than 3000. Dentigerous band in upper jaw more than double width of that in lower jaw. Gill opening small, placed near ventral midline, slightly before pectoral fin.

Frontals divided. Parietals, vomer and parasphenoid present. Five gill arches, six clefts. Vertebrae 100–125.

Maximum size 750 mm (Murray and Hjort, 1912a: 699).

Sexual Dimorphism. Mature males show certain morphological changes. The olfactory organ is much enlarged and the snout appears emarginate (Fig. 700B). This was first shown by Gartner (1983: 560), and mature males from the present material support his observation. Furthermore, the lower jaw becomes thin and edentate, and the upper jaw has reduced dentition. Quite a few females from our material have distended ovaries holding eggs with a diameter of 0.8–1.0 mm; this agrees with Raju (1974a: 558), who found the average diameter to be 0.9 mm. None of these females show any of the modifications found among mature males.

Distribution. Common in tropical and temperate areas of all oceans. Most frequently caught in the Atlantic, where it is reported from Iceland in the north to 48°S in the western South Atlantic (Walther Herwig sta. 101/76, 47°45′S, 40°05′W) and to 34°S in the eastern South Atlantic (Walther Herwig sta. 417/71, 34°12′S, 16°35′E). Its upper limit is 0–345 m (Anton Dohrn sta. 331/79, 33°04′N, 39°29′W), caught in a non-closing net during a night station (Post and Tesch, 1982: 351). The deepest catch with a closing net is 1532 m (Owre and Bayer, 1970: 186; Pillsbury sta. 18-1969, 20°37′N, 73°22′W). However, judging from

FIGURE 705. *Eurypharynx pelecanoides*, 393 mm TL (from Böhlke, 1966b: fig. 219).

the numerous specimens caught with a large pelagic trawl during the German Atlantic cruises, we find it reasonable to indicate the lower limit at about 3000 m.

Genera. A single genus, *Eurypharynx* Vaillant, 1882.

Genus *Eurypharynx* Vaillant, 1882

Eurypharynx (Vaillant, 1882: 1226 (type species *Eurypharynx pelecanoides* Vaillant, 1882, by monotypy).
Gastrostomus Gill and Ryder, 1883b: 271 (type species *Gastrostomus bairdii* Gill and Ryder, 1883, by monotypy).
Megalopharynx Brauer in Chun, 1900: 521 (no type species).
Macropharynx Brauer, 1902: 290 (type species *Macropharynx longicaudatus* Brauer, 1902, by monotypy).
Rouleina Fowler, 1925a: 2 (type species *Eurypharynx richardi* Roule, 1914, by original designation). Preoccupied by *Rouleina* Jordan, 1923 (Pisces).
Jordanites Fowler, 1925b: 75. Substitute name for *Rouleina* Fowler, 1925, preoccupied.

Characters. Those of the family.

Size. Maximum size about 750 mm TL.

Distribution. That of the family.

Etymology. From the Greek *eurys* (broad, wide) and *pharynx* (throat). Masculine.

Species. A single species, *Eurypharynx pelecanoides* Vaillant, 1882.

Eurypharynx pelecanoides Vaillant, 1882
Figures 697, 705

Eurypharynx pelecanoides Vaillant, 1882: 1226 (original description, off Morocco, holotype MNHN 83-121).
Saccopharynx flagellum (not of Cuvier, 1829). Goode and Bean, 1883: 136.
Gastrostomus bairdii Gill and Ryder, 1883b: 271 (original description, ca 40°N, 68°W, syntypes USNM 33294, 33295, 33386).
Saccopharynx bairdi. Günther, 1887: 262.
Saccopharynx pelecanoides. Günther, 1887: 262.
Macropharynx longicaudatus Brauer, 1902: 291 (original description, Gulf of Guinea, holotype ZMB 17547).
Gastrostomus pacificus Bean, 1904: 254 (original description, Pacific Ocean halfway between Midway and Guam, holotype USNM 50724).
Eurypharynx (Gastrostomus) pacificus. Roule, 1914c: 1822.
Eurypharynx richardi Roule, 1914c: 1823 (original description, southwest of Madeira, holotype MOM).
Eurypharynx (Rouleina) richardi. Fowler, 1925a: 2.
Jordanites richardi. Fowler, 1925b: 75.

Discussion of Synonymy. For additional references see Böhlke (1966b: 613–617). The following papers published after 1966 and dealing with *Eurypharynx pelecanoides* should be mentioned: Owre and Bayer (1970), Bauchot (1973), Raju (1974a), Gartner (1983).

Distinctive Characters. Those of the family.

Description (from Böhlke, 1966b and from radiographs of 27 specimens of the present material). Total vertebrae 101–113 (n = 16) (97–125, according to Orton, 1963: 14), predorsal vertebrae 11–13 (20), preanal vertebrae 30–34 (22), precaudal vertebrae 29–32 (15), caudal vertebrae 74–81 (11). Dorsal-fin rays 155–196 (37), anal-fin rays 118–147 (37), anteriormost anal-fin ray below dorsal-fin ray 32–38 (9).

Preanal length 26–37% TL (87) with slight but distinct positive allometric growth. Jaw length (tip of snout to angle of jaw) 18–24% TL (87), 54–74% preanal (240) with marked negative allometric growth.

Remarks. Except for a few specimens, the material of *Eurypharynx pelecanoides* is in rather poor condition. The reason is that the long jaws and elongate tail of these fragile fishes easily become entangled in the meshes of the trawl. When removed, the jaws and the tail tend to break and also the delicate skin is often badly torn. Consequently, most measurements are burdened with much inaccuracy and are therefore left out of the description.

Distribution. That of the family.

Etymology. From the Greek *pelekan* (a bird: the pelican) and *-oides* (like), in reference to the large mouth and throat. An adjective.

Study Material. 288 specimens, 160–610 mm TL. NORTH ATLANTIC: ZMUC P2340206–P2340462, P2340566–P2340597.

Family MONOGNATHIDAE

Introduction. The family Monognathidae was unknown in the western North Atlantic until recently, when a single specimen was collected by the British research vessel DISCOVERY southeast of Bermuda. The specimen represented a new species and was described by Bertelsen and Nielsen (1987) in their revision of the Monognathidae. This account is based largely on that work, to which the reader is referred for more detailed information.

Characters. Body elongate, preanal portion distinctly deeper than postanal; postanal portion strongly compressed and tapering gradually to a point. Preanal length one-third to one-quarter of TL. Abdomen distensible, posterior part often protruding ventrally into a pouch that can extend a certain distance behind anus. Tail ending in a urostyle, which is prolonged into a filament in two species (*Monognathus taningi* and *M. ozawai*) (Fig. 699). Scales absent. Lateral line, "white lines," and skin filaments absent. Skin unpigmented or uniformly light to dark brown; larval pigmentation retained in several specimens as midlateral subdermal spots (Raju, 1974a: fig. 1a–c).

Dorsal fin begins at or before middle of trunk, 7.7–20% TL behind tip of snout; in two species (*M. bruuni* and *M. jesperseni*) a section of dorsal fin near middle of body lower than remainder and enclosed in glandular tissue. Anal fin begins shortly behind anus in most species. Dorsal and anal fins reach tip of tail. Pterygiophores absent. Caudal fin absent or in one species represented by a paddle-like expansion without rays; in some species confluent dorsal and anal fins form a pseudocaudal fin around tip of tail. No caudal organ. Pectoral fin absent or reduced to flap without rays.

Head of moderate size bearing rudimentary eyes and tiny olfactory organs, the latter consisting of a short tube open at both ends (larger in mature male, see below). Snout short to moderately long. Upper jaw bones absent. Anterior part of neurocranium tapering into a ventrally curved spine with a bifid tip, more or less freely protruding as a rostral fang anteriorly in roof of mouth (Fig. 692; see also Raju, 1974a: 554; Bertelsen and Nielsen, 1987: figs. 4–7); a gland associated with this spine; spine hollow with paired ducts distal to gland and opening just before tip. Gill openings short, vertical slits located ventrolaterally close before pectoral fins when these are present, unsupported by opercular bones.

Suspensorium and lower jaw greatly prolonged: length of lower jaw 8–14% TL, 25–45% length of trunk, 75–125% HL. Mandible with a single series of 4–16 short teeth (Fig. 693C), compressed at base, triangular with sharp edges and slightly recurved tips; anteriormost 2–6 teeth form a close-set, concave series at tip of jaw, remaining teeth spaced more or less evenly on distal third to two-thirds of jaw.

Four pairs of gills, lobed, with short gill filaments. Stomach a large sac, extending posterior to anus into abdominal pouch; intestine nearly straight; esophagus with a ring of glandular tissue around its middle. Liver highly variable, sometimes relatively large with well developed gall bladder, sometimes small, sometimes absent. Well developed thymus in dorsomedial wall of each gill chamber. Well developed thyroid below anterior part of heart. A small gland on dorsal midline of snout, consisting of a pair of tubes which have closet-set anterior openings and are posteriorly fused, forming an elongate u-shaped sling; one or two layers of glandular cells in posterior part.

Frontals fused, continuous with rostral fang, no sign of a separate ethmoid (Bertin, 1938: 4 to the contrary); parietals present, vomer and parasphenoid absent (Fig. 690). Some species with a rostral cartilage anterior to fang, supporting a modestly elongate snout; other species with a short snout and no cartilage. Maxillae absent. Suspensorium long and slender, vertical or inclined slightly posteriorly, largely cartilaginous with a thin layer of bone; Bertin (1938: 5) described separate hyomandibula and quadrate in *M. bruuni* and *M. jesperseni,* a single element in *M. taningi;* pterygoid absent. A single ossified pair of gill arches, consisting of a single dorsal and ventral element, joined posteriorly in an acute angle; an uncertain number of rudimentary pharyngeal tooth plates present. Opercular bones absent. Pectoral girdle absent. Vertebrae 88–115, cylindrical or rarely hour-glass shaped, variable in relative height; terminal vertebra ends in a urostyle of variable length (Fig. 699); caudal rays absent (the single median caudal ray illustrated by Bertin, 1938: fig. 8 and by Raju, 1974a: fig. 2o appears to be the result of regeneration).

The 70 known specimens are 40–159 mm TL, all postmetamorphic except one or two (Raju, 1974a: fig. 1) that appear to be metamorphic larvae. Most of the specimens examined were immature females with undeveloped ovaries and eggs less than 0.1 mm diameter. Nine females of 55–159 mm TL had developing eggs 0.2–0.8 mm diameter; in seven of these (48–159 mm TL) the walls of the ovaries were broken and the eggs were scattered freely in the body cavity. Three males were found, two immature (61–67 mm TL) and one mature (64 mm TL).

Sexual Dimorphism. The single mature male (tentatively identified as *Monognathus herringi,* ZMUC P23405389) differs markedly from the females and immature males (Fig. 706). The abdomen is not expanded and there is no abdominal pouch. The body reaches a maximum depth at about the two-thirds mark of the TL. At this point the dorsal- and anal-fin rays become markedly longer than they are anteriorly; posteriorly they become shorter and then longer just before the end of the tail. The result is a sort of pseudocaudal fin notched posteriorly. The olfactory organs are greatly hypertrophied, nearly covering the side of the head from the base of the snout to the anterior part of the gills. The posterior nostril is extremely wide; the anterior end of each capsule is prolonged into a large triangular flap shaped like a pointed ear with its grooved inner side leading down to the forwardly directed, relatively narrow anterior nostril. The mandible is reduced to a rudiment. The head is almost completely covered with a layer of spongy tissue of a stiff and fragile structure. A radiograph shows the rostral fang short and blunt, as if it has been partially resorbed. The stomach and intestine appear degenerate. Females show less modification at maturity, the most conspicuous change being a resorption of the anterior end of the mandible and loss of the teeth.

The great enlargement of the olfactory organs in the mature male monognathid indicates that it locates its mate by smell, probably through a pheromone released by the female. The nature of the changes undergone at maturity further suggests that monognathids spawn once and die.

Food and Feeding. The majority of the specimens examined had empty stomachs. Three specimens contained the recognizable remnants of a single large shrimp. In each case the shrimp had been swallowed tail-first and was nearly as long as the preanal length of the fish that had swallowed it. A fourth specimen contained fragments of chitin and muscle that indicated a shrimp of similar relative size. The method of feeding is unknown, but it seems logical to assume that the rostral fang is involved somehow.

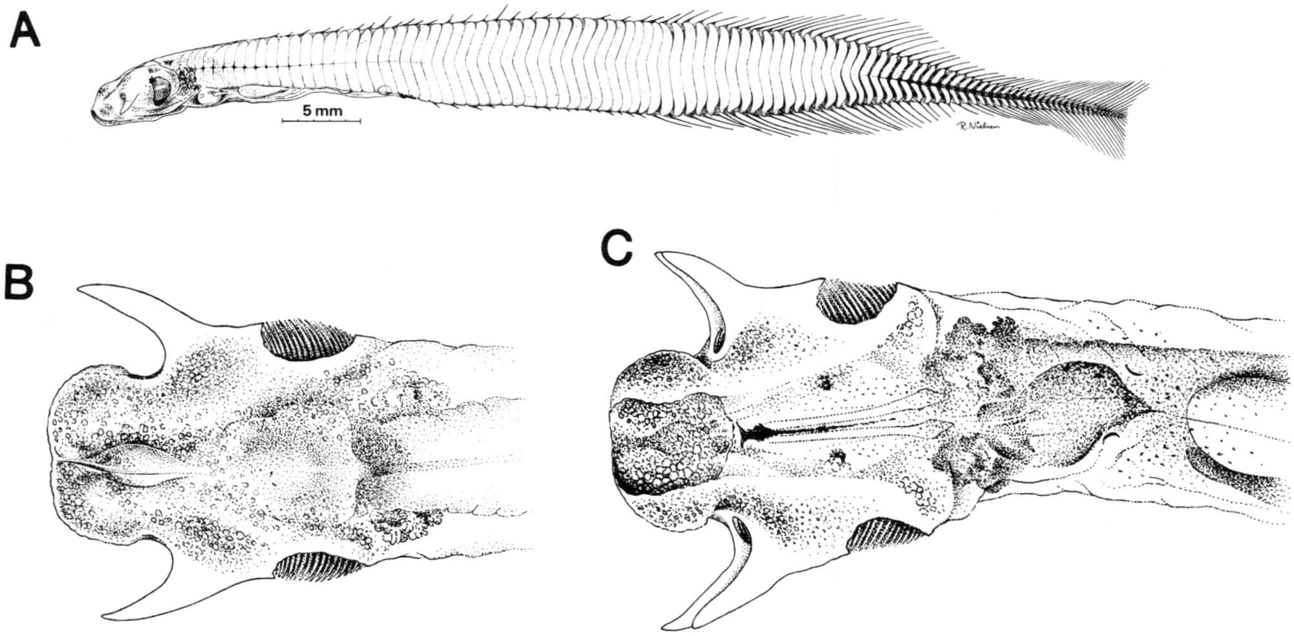

FIGURE 706. Mature male of *Monognathus herringi* (from Bertelsen and Nielsen, 1987: fig. 16). A. Whole view. B. Head in dorsal view. C. Head in ventral view.

An injection of venom could immobilize a shrimp that might otherwise easily escape the grasp of a seemingly feeble predator. How the monognathid finds and captures its prey in the first place is not clear. It is virtually blind, has no lateral line, and only a weakly developed olfactory organ. Bertelsen and Nielsen (1987) suggested that some sort of olfactory attractant (a chemical lure) may be present, but this is speculation. We obviously have much to learn about how this animal functions in its environment.

Distribution. All oceans. Records in the Atlantic from 50°N to 30°S; in the Pacific from 35°N to 23°S and from approximately 130°E to 120°W and from the South China Sea. Forty-seven of the 70 known specimens were caught in closing nets in depths of 2000–5400 m; most of these were taken well off the bottom, although six specimens were taken 10–15 m above the bottom at 2950–5400 m. Only six specimens, which all came from open pelagic nets, were caught in less than 1000 m.

Genera. A single genus, *Monognathus* Bertin, 1936. Myers (1940: 141) created a new genus, *Phasmatostoma,* to contain two of Bertin's species. We agree with Raju (1974a: 555), however, that the present information on the family is insufficient to warrant this or any other separation of the Monognathidae into more than one genus.

Genus *Monognathus* Bertin, 1936

Monognathus Bertin, 1936d: 533 (type species *Monognathus taningi* Bertin, 1936, by subsequent designation of Myers, 1940); 1938: 3 (extended description). Myers, 1940: 141 (designation of *taningi* as type species). Raju, 1974a: 548 (review of genus with description of three new species). Bertelsen and Nielsen, 1987 (revision of genus).

Phasmatostoma Myers, 1940: 141 (type species *Monognathus jesperseni* Bertin, 1936, by original designation).

Characters. Those of the family.

Size. All known postmetamorphic specimens 40–159 mm TL.

Distribution. That of the family.

Etymology. From the Greek *monos* (one) and *gnathos* (jaw), referring to the lack of an upper jaw. Masculine.

Species. Until recently only 11 postmetamorphic specimens had been recorded in the literature, and they had been placed in six species. Bertin (1936d) described three species: *Monognathus taningi* and *M. jesperseni* from the eastern

North Atlantic, and *M. bruuni* from the South China Sea. Raju (1974a) added three more: *M. isaacsi* from the western North Pacific, *M. ahlstromi* from the eastern North Pacific, and *M. jesse* from the central North Pacific. Bertelsen and Nielsen (1987) have added eight more: *M. herringi* and *M. nigeli* from the eastern North Atlantic, *M. boehlkei* from the eastern and western North Atlantic, *M. bertini* from the eastern North and South Atlantic and the western South Pacific, *M. rajui* from the western Indian Ocean, *M. rosenblatti* from the eastern and central North Pacific, *M. smithi* from the central North Pacific, and *M. ozawai* from the western North Pacific. In addition, they reported a 159-mm specimen from the eastern North Atlantic, the largest monognathid known, which they did not assign to a species.

Bertelsen and Nielsen (1987) found that the species of *Monognathus* could be divided into two groups based on the length of the snout and the dorsal profile of the skull. In one group the head is long and relatively low with a long, compressed, pointed snout anterior to the rostral fang supported internally by a cartilage. The rostral fang forms a right angle in lateral view with the dorsal profile of the neurocranium, and the cleft of the mouth is almost horizontal. This group includes *M. rosenblatti, M. nigeli, M. herringi, M. isaacsi,* and *M. smithi.* In the remaining species the head is short and relatively high with a blunt, short snout anterior to the rostral fang. The rostral fang forms nearly a straight line or at most an obtuse angle with the frontal part of the neurocranium. The cleft of the mouth is more or less oblique. *Monognathus taningi* and *M. bertini* have a somewhat longer snout than the other species of this group, and *M. boehlkei* has a sharper angle between the rostral fang and the skull. The second group is thus somewhat more heterogeneous than the first.

Key to the Species of *Monognathus*
(from Bertelsen and Nielsen, 1987)

1a. Snout in front of rostral fang short and blunt (except in *taningi* and *bertini*); dorsal profile of rostral fang forming an almost straight line with skull 2

1b. Snout in front of rostral fang long and pointed; rostral fang forming a right angle with dorsal profile of skull 10

2a. Dorsal fin with a glandular middle part 3

2b. No glandular part in dorsal fin 4

3a. Dorsal-fin rays 72, anal-fin rays 42; 35 rays anterior to glandular part of dorsal fin . . . . *bruuni* (South China Sea)

3b. Dorsal-fin rays 93–97, anal-fin rays 54–55; 45–47 rays anterior to glandular part of dorsal fin . . . . *jesperseni* (northeastern Atlantic)

4a. Dorsal-fin rays 125–150 5

4b. Dorsal-fin rays 72–116 6

5a. Anal-fin rays 59–64, dorsal-fin origin above myomere 3 . . . . *taningi* (northeastern Atlantic)

5b. Anal-fin rays 110–130, dorsal-fin origin above myomeres 13–16 . . . . *boehlkei,* p. 654

6a. Dorsal-fin origin above myomeres 5–7, anal-fin origin below dorsal-fin rays 22–43 7

6b. Dorsal-fin origin above myomeres 11–13, anal-fin origin below dorsal-fin rays 18–29 9

7a. Anal-fin rays 43–58; pectoral fin 5–10% TL; snout present in front of rostral fang . . . . . . *bertini* (eastern Atlantic and southwestern Pacific)

7b. Anal-fin rays 66–68; pectoral fin 0–1% TL; snout absent in front of rostral fang 8

8a. Anal-fin origin below dorsal-fin ray 40; 7 myomeres between anus and anterior anal-fin ray . . . . *ozawai* (northwestern Pacific)

8b. Anal-fin origin below dorsal-fin ray 22; 2 myomeres between anus and anterior anal-fin ray . . . . *rajui* (western Indian Ocean)

9a. Anterior anal-fin ray below dorsal-fin ray 29; lower jaw with 16 teeth . . . . *jesse* (central North Pacific)

9b. Anterior anal-fin ray below dorsal-fin ray 18; lower jaw with 9 teeth . . . . *ahlstromi* (eastern North Pacific)

10a. Skull length 7.0–7.6% TL (TL 68–72 mm) . . . . *smithi* (central North Pacific)

10b. Skull length 5.3–5.6% TL in specimens 64–70 mm TL (5.9–7.9% TL in specimens 45–64 mm TL) 11

11a. Teeth in lower jaw 4; maximum depth of body 9.7% TL . . . . *isaacsi* (western North Pacific)

11b. Teeth in lower jaw 6–8; maximum depth of body 5.4–7.8% TL 12

12a. Skin transparent and no lateral subdermal spots of pigment . . . . *herringi* (eastern North Atlantic)

12b. Skin densely dark-pigmented and/or presence of lateral subdermal pigment 13

13a. 4–5 lateral subdermal, diffuse spots of pigment . . . . *rosenblatti* (central and eastern North Pacific)

13b. 5 lateral subdermal, distinct spots of pigment . . . . *nigeli* (eastern North Atlantic)

Monognathus boehlkei Bertelsen and Nielsen, 1987
Figure 707

Monognathus boehlkei Bertelsen and Nielsen, 1987: 184 (original description, 31°47.2′N, 63°30.9′W, holotype BMNH 1984.12.3.1).

Distinctive Characters. Monognathus boehlkei differs from the other species of the genus by the high number of anal-fin rays (110–130 vs 42–88) and by an obtuse angle between the rostral fang and the dorsal profile of the skull. Other distinguishing characters: head short, snout blunt; skull length 4.2–5.2% TL; dorsal profile of rostral fang slightly concave; dorsal-fin rays 130–150, anal-fin origin below dorsal-fin rays 7–13; pectoral fin 2–3% TL; lower jaw nearly straight; six lateral subdermal spots of pigmentation.

Description (characters of holotype followed by those of paratypes in parentheses). Myomeres 103 (105–109), dorsal-fin rays 138 (130–150), anal-fin rays 110 (111–130), dorsal-fin origin above myomere 13 (15–16), anal-fin origin below dorsal-fin ray 13 (7–13), anal-fin origin below myomere 25 (25–28). Proportions as % of TL: preanal 29.5 (28.5–29.0), predorsal 19.5 (17.5–19.5), depth at anterior anal-fin ray 6.6 (4.7–6.2), maximum depth 7.3 (6.0–7.7), skull length 5.2 (4.2–5.1), suspensorium 6.2 (5.9–7.1), lower jaw 13.0 (11.0–14.0), pectoral fin 2.8.

Holotype: Body elongate, compressed, rather deep. Dorsal fin begins well behind head; anal fin begins far before midpoint of body. Head short with blunt, short snout anterior to rostral fang. Cleft of mouth almost horizontal. Jaws nearly straight with torn tips ending anterior to snout, each with 5–6 evenly spaced teeth. Rostral fang slightly concave dorsally, forming an obtuse angle with skull. Eye black, 0.3 mm diameter. Nostrils small, situated between rostral fang and eye; olfactory lamellae distinct. Abdominal pouch very short. Anus placed on left side of abdominal cavity. One myomere between anus and anal fin. Stomach much distended by a 19-mm shrimp. Testes unripe. Vertebral centra poorly ossified and rectangular in lateral view, much higher than

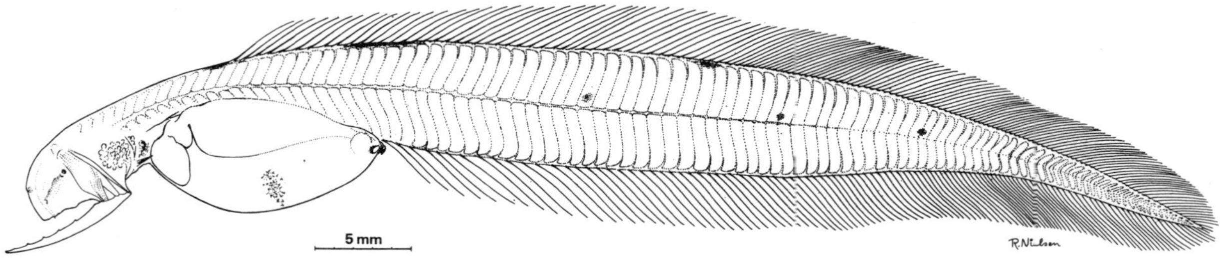

FIGURE 707. *Monognathus boehlkei*: BMNH 1984.12.3.1, holotype, 60 mm TL (from Bertelsen and Nielsen, 1987: fig. 27).

long in precaudal area, where they are provided with small, weak neural processes; caudal vertebrae with almost square centra provided with 3–4 neural and 2 hemal processes, all long.

Paratypes: Similar to holotype in most characters. Skull length slightly less. Olfactory lamellae hardly developed. Smallest paratype (55 mm TL, BMNH 1984.12.3.2) with anus on right side of abdominal cavity and seven myomeres between anus and anal fin.

Size. 55–70 mm TL. The holotype (61 mm TL) is an immature male; the smallest paratype is an immature female with eggs less than 0.1 mm diameter.

Color. Skin unpigmented and transparent; five or six midlateral, subdermal pigment spots located between myomeres 10–72; additional pigment on jaw, base of pectoral fins, and around anus.

Distribution. Western and eastern North Atlantic. All were taken in closing nets, one specimen at 2500–3500 m over a bottom depth of 4563 m, and the others between 10–30 m off the bottom with soundings from 2960–5430 m.

Etymology. Named for the late James E Böhlke in recognition of his most valuable contribution to the study of apodal fishes.

Study Material. Four specimens, 55–70 mm TL.

Holotype: BMNH 1984.12.3.1 (male, 60 mm TL), Discovery sta. 8281-48, southeast of Bermuda, 31°47.2'N, 63°30.9'W, RMT 8, at 2500–3500 m, bottom depth 4563 m, 21 March 1973.

Paratypes: BMNH 1984.12.3.2 (female, 55 mm TL), Discovery sta. 10379-37, southwest of the Azores, 34°58.1'N, 32°56.5'W, RMT 8, at 2960–2985 m, 23 m off bottom, 14 June 1981. ZMUC P2340538 (1, 70), Discovery sta. 11261-65, south of the Azores, 31°28.9'N, 25°21.9'W, RMT 8, at 5410–5430 m, 11–31 m off bottom, 6 July 1985. P2340559 (1, 65), Discovery sta. 11261-46, south of the Azores, 31°22.3'N, 25°24.5'W, RMT 8, at 5325–5427 m, bottom depth ca 5400 m, 2 July 1985.

Fishes of the

Western North Atlantic

Part 9

COMPOSED AND PRINTED BY ALLEN PRESS, INC.

LAWRENCE, KANSAS, U.S.A.

IN AUTOLOGIC PALATINO

IN AN EDITION OF 1000 COPIES

ON P. H. GLATFELTER B16 TEXT PAPER

BOUND IN BOARDS AND FULL CLOTH

BY NICHOLSTONE BOOK BINDERY, NASHVILLE, TENNESSEE

U.S.A.